Physics of Continuous Matter

Exotic and Everyday Phenomena in the Macroscopic Worl

Physics of Continuous Matter
Exotic and Everyday Phenomena in the Macroscopic World

B Lautrup

The Niels Bohr Institute
University of Copenhagen
Denmark

IoP

Institute of Physics Publishing
Bristol and Philadelphia

British Library Cataloguing-in-Publication Data

A catalogue record for this book is available from the British Library.

ISBN 0 7503 0752 8

Library of Congress Cataloging-in-Publication Data are available

Commissioning Editor: Tom Spicer
Editorial Assistant: Leah Fielding
Production Editor: Simon Laurenson
Production Control: Sarah Plenty
Cover Design: Frédérique Swist
Marketing: Louise Higham and Ben Thomas

Published by Institute of Physics Publishing, wholly owned by The Institute of Physics, London

Institute of Physics Publishing, Dirac House, Temple Back, Bristol BS1 6BE, UK

US Office: Institute of Physics Publishing, The Public Ledger Building, Suite 929, 150 South Independence Mall West, Philadelphia, PA 19106, USA

Typeset in LaTeX 2_ε by Text 2 Text Limited, Torquay, Devon
Printed in the UK by J W Arrowsmith Ltd, Bristol

Contents

Preface

Continuum physics is as old as science itself. The decision to write a textbook about this subject has not been easy, because of the feeling that everything has been said and written before. What prompted me to write this book was a one-semester course in which I had to teach the basic principles of continuum physics to students of physics, geophysics and astrophysics. The students had previously been taught mechanics and thermodynamics, and along with this course on continuum physics they were also learning electromagnetism. There is a certain parallelism in the use of partial differential equations in both of these subjects, but basically I could neither assume they knew much about the mathematical methods nor the physics in advance. In the end the book came to contain much more material than could be readily covered in one semester, but its modular layout still makes it fairly easy to select a subset of topics that fits the needs of a particular course.

The book is first and foremost an introduction to the basic ideas and the derivation of the equations of continuum mechanics from Newtonian particle mechanics. The field concept is introduced from the very outset and tensors are used wherever they are necessary. There is no call for shyness in this respect, here nearly 200 years after Cauchy. Although many examples—in particular in the first few chapters—are taken from geophysics and astrophysics, this does not mean that the book is designed only for students of these subjects. All physics students ought to be familiar with the description of the macroscopic world of apparently continuous matter.

Secondly, the necessary mathematical tools are developed along with the physics on a 'need-to-know' basis in order to avoid lengthy and boring mathematical preliminaries seemingly without purpose. The disadvantage of this pedagogical line is of course that the general analytic methods and physical principles, so important later in a physics student's life, become scattered throughout the book. I have attempted to counteract this tendency by structuring the text in various ways and clearly marking out important results, sometimes repeating central material. The three short appendices also help in this respect.

The important thing to learn from this book is how to reason about physics, both qualitatively and—especially—quantitatively. Numerical simulations may be fine for obtaining solutions to practical problems, but are of very little aid in obtaining real understanding. Physicists must learn to think in terms of fundamental principles and generic methods. Solving one problem after another of similar kind seems unnecessary and wasteful. This does not mean that the physicist should not be able to reach a practical result through calculation, but the physical principles behind equations and the conditions underlying approximations must never be lost sight of. Nevertheless, numerical methods are used and explained in some detail whenever it seems natural, including two whole chapters on numeric simulation in elastostatics and fluid mechanics.

The book is divided into five parts; introduction, hydrostatics, deformable solids, basic fluid mechanics and special topics in fluid mechanics. I have made no attempt to balance the space allotted to fluid and solid mechanics; fluid mechanics clearly dominates the book because it is after all conceptually simpler than even the linear mechanics of isotropic solids. Although there may be a certain built-in logic in the way the fundamental equations of continuum physics are derived and presented, the subject of applications is so rich that there can be no canonical order of presentation. Apart from the bare-bone fundamentals, continuum physics appears as a huge collection of interconnected topics, or 'stories'. Any textbook of sufficiently broad scope is therefore necessarily coloured by the interests, predilections and idiosyncracies of its author.

The book should be useful at several levels of teaching. In an introductory second year course one would perhaps choose as curriculum chapters 1–5, 9–12 and 15–20. Later and more advanced courses might wish to include most of the book. The level of difficulty is as much as possible sought to rise steadily within each chapter and in the book as a whole. The chapters are of fairly uniform length, and each chapter has a 'soft' introduction making contact with everyday experience. Historical comments and microbiographies of the major players are sprinkled throughout the text without any attempt at systematics or completeness. Whenever feasible, the mutual interdependence of chapters has—at the cost of some repetition—been reduced in order to facilitate the exclusion of whole chapters in a curriculum. Certain sections and subsections have been marked with a star to indicate that they fall outside the main line of presentation either in subject or in level of difficulty, and may require more teacher support or simply be skipped in a first reading.

As an aid to the text the book has been provided with a large number of marginal vignettes, outlining the salient features of a physical system or a choice of coordinates. They are nothing but the simpleminded sketches that we all draw—or should draw—when trying to learn new physics. Larger figures are of course used whenever the margin turns out to be too narrow. At the end of each chapter there is a collection of problems with answers outlined in the back of the book. Some problems test the use of the formalism in practical settings, others the understanding of the theoretical concepts, and still others develop the theory presented in the chapter further. The system of units is as in any other modern text the international one (SI), although commonly used units strictly speaking outside of this system, for example bar or atm for pressure, are sometimes also employed, though never without a proper definition.

Writing this book over the course of several years I have benefitted from advice and support of many people of whom I can mention only a few. First among them are those that have dared to use the earlier versions of the manuscript in their teaching; Tomas Bohr of the Technical University of Denmark, Predrag Cvitanovic of Georgia Tech and Niels Kjær Nielsen of University of Southern Denmark. For physics input, discussions, and criticism I also thank Anders Andersen, Luiza Angheluta, Andy Jackson, Alex Lande, Morten Olesen, Laurette Tuckerman and all of the students that have followed my course over the years. Special thanks go to Mogens Høgh Jensen without whose generosity I would not have been able to sustain my expensive habit of buying books. I am grateful to my institute and my colleagues for being patient with me, and finally I could not have finished this book without the daily support of my wife Birthe Østerlund.

The book is written for adults with a serious intention to learn physics. I have selected for the readers what I think are the interesting topics in continuum physics, and presented these as pedagogically as I can without trying to cover everything encyclopedically. The list of literature contains pointers to texts that deal with specialized subtopics; mostly the books that I have found useful. I sincerely hope that my own joy in understanding and explaining the physics shines through everywhere.

Benny Lautrup
Copenhagen, October 2004

Part I

Introduction

1

Continuous matter

The everyday experience of the smoothness of matter is an illusion. Since the beginning of the twentieth century it has been known with certainty that the material world is composed of microscopic atoms and molecules, responsible for the macroscopic properties of ordinary matter. Long before the actual discovery of molecules, chemists had inferred that something like molecules had to exist, even if they did not know how big they were. Molecules *are* small—so small that their existence may be safely disregarded in all our daily doings. Although everybody possessing a powerful microscope will note the irregular Brownian motion of small particles in a liquid, it took quite some mental effort and a big conceptual jump from the everyday manipulation of objects to recognize that this is a sign that molecules are really there.

Continuum physics deals with the systematic description of matter at length scales that are large compared to the molecular scale. Most macroscopic length scales occurring in practice are actually huge in molecular units, typically in the hundreds of millions. This enormous ratio of scales isolates continuum theories of macroscopic phenomena from the details of the microscopic molecular world. There might, in principle, be many different microscopic models leading to the same macroscopic physics.

This chapter paints in broad outline the transition from molecules to continuous matter, or mathematically speaking from particles to fields. It is emphasized that the macroscopic continuum description must necessarily be of statistical nature, but that random statistical fluctuations are strongly suppressed by the enormity of the number of molecules in any macroscopic material object. The central theme of this book is the recasting of Newton's laws for point particles into a systematic theory of continuous matter, and the application of this theory to the wealth of exotic and everyday phenomena of the macroscopic material world.

1.1 Molecules

The microscopic world impinges upon the macroscopic almost only through material constants characterizing the interactions between macroscopic amounts of matter, such as coefficients of elasticity and viscosity. It is, of course, an important task for the physics of materials to derive the values of these constants, but this task lies outside the realm of continuum physics. In continuum physics it is nevertheless sometimes instructive to consider the underlying atomic or molecular structure in order to obtain an understanding of the origin of macroscopic phenomena and of the limits to the macroscopic continuum description.

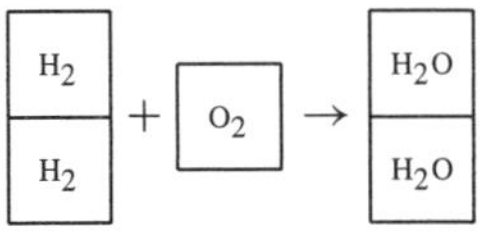

The meaning of a chemical formula.

Molecular weight

Chemical reactions such as $2H_2 + O_2 \rightarrow 2H_2O$ are characterized by simple integer coefficients. Two measures of hydrogen plus one measure of oxygen yield two measures of water without anything left over of the original ingredients. What are these measures? For gases at the same temperature and pressure, it is

Figure 1.1. How continuous matter really looks at the atomic scale. Image of Mica obtained by atomic force microscopy (AFM), approximately 225 Angstrom on a side. Image Courtesy Mark Waner (Mark J. Waner, PhD dissertation, Michigan State University 1998).

simply the volume, so that for example two litres of hydrogen plus one litre of oxygen yield two litres of water vapour, assuming that the water vapour without condensing can be brought to the same temperature and pressure as the gases had before the reaction. In 1811, Count Avogadro of Italy proposed that the simple integer coefficients in chemical reactions between gases could be explained by the rule that equal volumes of gases contain equal numbers of molecules (at the same temperature and pressure).

Lorenzo Romano Amadeo Carlo Avogadro (1776–1856). *Italian philosopher, lawyer, chemist and physicist. Count of Quaregna and Cerratto. Formulated that equal volumes of gas contain equal numbers of molecules. Also argued that simple gases consist of diatomic molecules.*

The various measures do not weigh the same. A litre of oxygen is roughly 16 times heavier than a litre of hydrogen at the same temperature and pressure. The weight of a litre of water vapour must—of course—be the sum of the weights of its ingredients, hydrogen and oxygen, and from the reaction formula it now follows that this becomes roughly $((2 \times 1) + (1 \times 16))/2 = 9$ times the weight of a litre of hydrogen. Such considerations lead to the introduction of the concept of relative molecular weight or mass in the ratio 1:16:9 (or equivalently 2:32:18) for hydrogen, oxygen and water.

In the beginning there was no way of fixing an absolute scale for molecular mass, because that would require knowledge of the number of molecules in a macroscopic amount of a substance. Instead, a unit, called a *mole*, was quite arbitrarily fixed to be one gram of atomic hydrogen (H). This scale is practical for the chemist at work in his laboratory, and the ratios of molecular masses obtained from chemical reactions would then determine the mass of a mole of any other substance. Thus the molar mass of molecular hydrogen (H_2) is 2 grams and that of molecular oxygen (O_2) 32 grams, whereas water has a molar mass of $((2 \times 2) + (1 \times 32))/2 = 18$ grams. This system could be extended to all chemical reactions allowing the determination of molar mass for any substance participating in such processes.

Avogadro's number

We now know that chemical reactions actually describe microscopic interactions between individual molecules built from atoms and that molecular mass is simply proportional to the mass of a molecule. The constant of proportionality was called Avogadro's number by Perrin, who in 1908 carried out the first modern determination of its value from Brownian motion experiments. Perrin's experiments relying on Einstein's recent (1905) theory of Brownian motion were not only seen as a confirmation of this theory

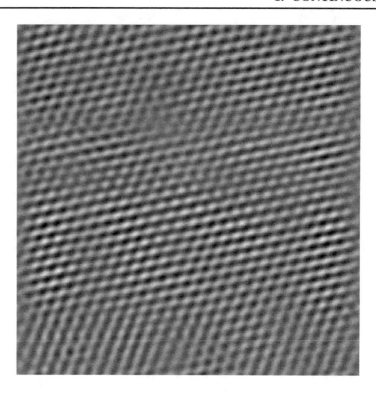

but also as the most direct evidence for the reality of atoms and molecules. Today, Avogadro's number is *defined* to be the number of atoms in exactly 12 grams of the fundamental carbon isotope (^{12}C), and its value is $N_A = 6.022137(3) \times 10^{23}$ molecules per mole[1].

Jean-Baptiste Perrin (1870–1942). *French physicist. Received the Nobel Prize for his work on Brownian motion in 1926. He founded several French scientific institutions, among them the now famous 'Centre National de la Rechèrche Scientifique (CNRS)'.*

Molecular separation

Consider a substance with mass density ρ and molar mass $M_{\rm mol}$. A mole of the substance occupies a volume $M_{\rm mol}/\rho$, and the volume per molecule becomes $M_{\rm mol}/\rho N_A$. A cube with this volume would have sides of length

$$L_{\rm mol} = \left(\frac{M_{\rm mol}}{\rho N_A}\right)^{1/3}, \tag{1.1}$$

which may be called the scale of *molecular separation*. For iron we get $L_{\rm mol} \approx 0.24$ nm, for water $L_{\rm mol} \approx 0.31$ nm, and for air at normal temperature and pressure $L_{\rm mol} \approx 3.4$ nm. For liquids and solids, where the molecules touch each other, this length is roughly the size of a molecule, whereas in gases it may be much larger. There is a lot of vacuum in a gas, in fact about 1000 times the volume of matter at normal temperature and pressure.

Johannes Diederik van der Waals (1837–1923). *Dutch physicist. Developed an equation of state for gases, now carrying his name. Received the Nobel Prize in 1910 for his work on fluids and gases.*

Molecular forces

Apart from the omnipresent gravitational interaction between all bodies, molecular interactions are entirely electromagnetic in nature, from the fury of a tornado to the gentlest kiss. A deeper understanding of the so-called van der Waals forces acting between neutral atoms and molecules requires quantum theory and falls outside the scope of this book.

Generally, however, the forces between neutral atoms and molecules are short-ranged and barely reach beyond nearest molecular neighbours. They are strongly repulsive if the atoms are forced closer than their natural sizes allow and moderately attractive when they are moved apart a little distance, but farther away they quickly die out. When two molecules are near each other, this tug of war between repulsion and attraction leads to a minimum in the potential energy between the molecules. The state of matter depends, broadly speaking, on the relation between the depth of this minimum, called the binding energy, and the average kinetic energy due to the thermal motion of the molecules.

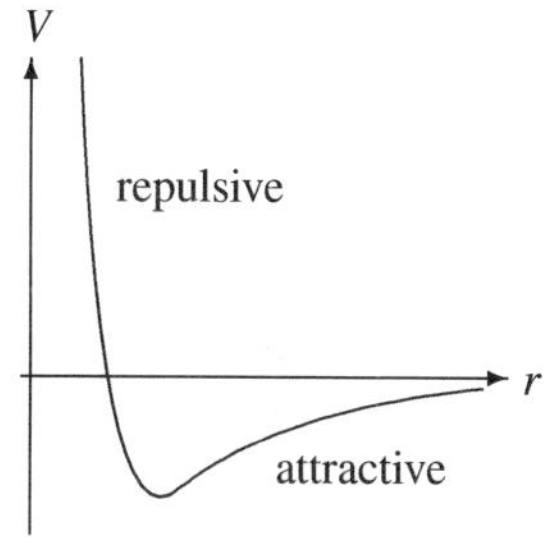

Sketch of the intermolecular potential energy $V(r)$ as a function of intermolecular distance r. It is attractive at moderate range and strongly repulsive at close distance.

Solids, liquids and gases

In *solid matter* the minimum lies so deep that thermal motion cannot overcome the attraction. Each individual atom or molecule is tied to its neighbours by largely elastic forces. The atoms constantly undergo small-amplitude thermal motion around their equilibrium positions, but as long as the temperature is not so high that the solid melts, they are bound to each other. If external forces are applied, solids may *deform* elastically with increasing force, until they eventually become plastic or even fracture. Most of the work done by external forces in deforming elastic solids can be recovered as work when the forces disappear.

In *fluid matter*, liquids and gases, the minimum is so shallow that the thermal motion of the molecules is capable of overcoming the attractive forces between them. The molecules effectively move freely around between collisions, more so in gases than in liquids where molecular conglomerates may form. External forces make fluids *flow*—in liquids a kind of continual fracturing—and a part of the work done by such forces is dissipated into random molecular motion, heat which cannot directly be recovered as work when the forces cease to act.

1.2 The continuum approximation

Whether a given number of molecules is large enough to warrant the use of a smooth continuum description of matter depends on the precision desired. Since matter is never continuous at sufficiently high precision, continuum physics is always an approximation. But as long as the fluctuations in physical quantities caused by the discreteness of matter are smaller than the desired precision, matter may be taken to be

[1] In this book the absolute error on the last digit of a quantity is indicated by means of a parenthesis following the mantissa.

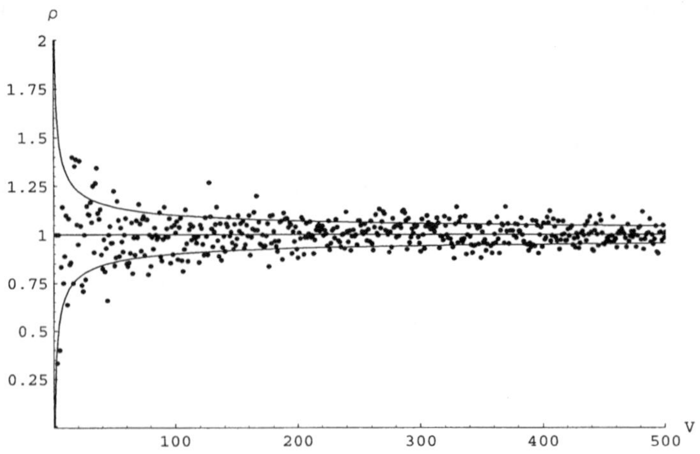

Figure 1.2. Measured density as a function of volume size. A three-dimensional 'universe' consisting of $20 \times 20 \times 20 = 8000$ cells is randomly filled with as many 'molecules'. On average each of the 8000 cells should contain a single molecule, corresponding to a density of $\rho = 1$. A 'material particle' consisting of V cells will in general not contain precisely V molecules, and thus has an actual density that deviates from unity. The plot shows the actual density of a random collection of V cells as a function of V. The drawn curves, $\rho = 1 \pm 1/\sqrt{V}$, indicate the expected fluctuations.

continuous. Continuum physics is, like thermodynamics, a limit of statistical physics where all macroscopic quantities such as mass density and pressure are understood as averages over essentially infinite numbers of microscopic molecular variables.

Luckily, it is only rarely necessary to exploit this connection. In a few cases, such as in the analysis below, it is useful to look at the molecular underpinnings of continuum physics. In doing so, we shall use the simplest 'molecular' description possible. A quite general meta-law of physics says that the physical laws valid at one length scale are not very sensitive to the details of what happens at much smaller scales. Without this law, physics would in fact be impossible, because we never know what lies below our currently deepest level of understanding.

Precision and continuity

Suppose that we want to determine the mass density $\rho = mN/V$ of a gas to a certain relative precision ϵ, say $\epsilon = 1\%$, by counting the number of identical molecules N of mass m in a small volume V. Due to random motion of the gas molecules, the number N will fluctuate and yield a different value if measured again. For a typical fluctuation ΔN in N, the relative fluctuation in density will be the same as in N, or $\Delta\rho/\rho = \Delta N/N$. If the relative density fluctuation should be at most ϵ we must require that $\Delta N \lesssim \epsilon N$. Provided the time between measurements is large compared to the time between molecular collisions, the molecules in the volume V will all be replaced by other molecules, and become an essentially random collection of molecules from the gas at large. In such a random sample the fluctuation in the number is of order $\Delta N \approx \sqrt{N}$, and the condition becomes $\sqrt{N} \lesssim \epsilon N$ or $N \gtrsim \epsilon^{-2}$ (see figure 1.2 and problem 1.1). The smallest allowable number of molecules, $N_{\text{micro}} \approx \epsilon^{-2}$, occupies a cubic volume with side length

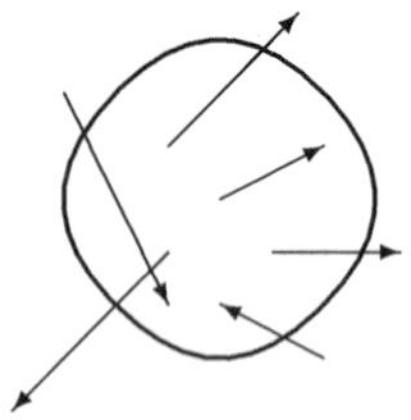

In a gas the molecules move rapidly in and out of a small volume with typical velocities of the order of the speed of sound.

$$L_{\text{micro}} = N_{\text{micro}}^{1/3} L_{\text{mol}} \approx \epsilon^{-2/3} L_{\text{mol}}. \tag{1.2}$$

At a precision level of $\epsilon = 1\%$, the smallest volume under consideration should contain at least $N_{\text{micro}} \approx 10^4$ molecules, and the linear dimension of such a volume will be greater than $L_{\text{micro}} \approx 22 L_{\text{mol}}$. For air under normal conditions this comes to about 80 nm, while for liquids and solids it is an order of magnitude smaller because of the smallness of L_{mol}.

In liquids and especially in solids the molecules do not move around much but oscillate instead randomly around more or less fixed positions, and the density fluctuations in a volume are mainly due to molecules passing in and out of the surface. In problems 1.2 and 1.3 the fluctuations are estimated for a cube and a sphere, resulting in a microscopic length scale of roughly the same form as above, although with an exponent of $-1/2$ instead of $-2/3$.

Mean free path

Another condition for obtaining a smooth continuum description, is that molecules should interact with each other to 'iron out' strong differences in velocities. If there were no interactions, a molecule with a given velocity would keep on moving with that velocity forever. In solids and liquids where the molecules are closely packed, these interactions take place over a couple of molecular separation lengths and put no further restriction on the microscopic length scale.

In gases there is a lot of vacuum and molecules move freely over long distances. The mean free path between collisions may be estimated by considering a spherical molecule or atom of diameter d with its centre tracing out a straight path through the gas. It will hit any other sphere of the same diameter within a 'striking' distance d from the path, i.e. inside a cylinder of diameter $2d$. Since there is on average one molecule in each volume $L_{\rm mol}^3$, the distance the original sphere has to move before being sure of hitting another is on average, $\lambda = L_{\rm mol}^3/\pi d^2$. A more careful analysis leads to the following expression for the *mean free path*,

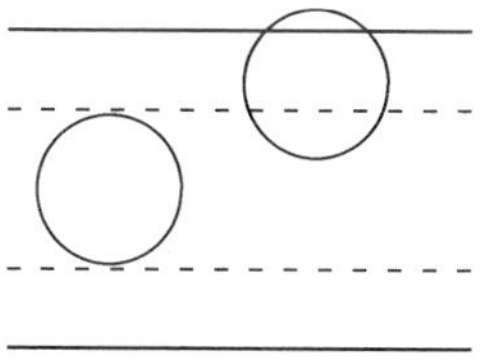

A sphere of diameter d will collide with any other sphere of the same diameter with its centre inside a cylinder of diameter $2d$.

$$\lambda = \frac{L_{\rm mol}^3}{\sqrt{2}\pi d^2} = \frac{M_{\rm mol}}{\sqrt{2}\pi d^2 \rho N_A}, \tag{1.3}$$

with an extra factor $\sqrt{2}$ in the denominator.

For air at normal temperatures we find $\lambda \approx 94$ nm which is not much larger than the microscopic length scale, $L_{\rm micro} \approx 80$ nm (for $\epsilon = 1\%$). For dilute gases the mean free path is much larger than the microscopic scale and sets the length scale for the smallest continuum volumes rather than $L_{\rm micro}$, unless the desired relative precision is very small (see problem 1.5).

Macroscopic smoothness

A smooth continuum description demands that there should be no measurable variation in density between neighbouring microscopic volumes. If L denotes the typical macroscopic length scale for significant variations in density, the relative change in density over the distance $L_{\rm micro}$ will be of magnitude $\Delta\rho/\rho \approx L_{\rm micro}/L$. Demanding that the relative density change is smaller than the measurement precision $\Delta\rho/\rho \lesssim \epsilon$, we conclude that we must have $L \gtrsim L_{\rm macro}$ where

$$L_{\rm macro} \approx \frac{1}{\epsilon} L_{\rm micro}. \tag{1.4}$$

Any region in which the density varies by a significant amount must be larger in size than $L_{\rm macro}$; otherwise the smooth continuum description breaks down. With $\epsilon = 1\%$ we find $L_{\rm macro} \approx 100 L_{\rm micro}$. For air under normal conditions this is about 10 μm, while for solids it is an order of magnitude smaller.

Since interfaces between macroscopic bodies are usually much thinner than $L_{\rm macro}$, these regions of space fall outside the smooth continuum description. In continuum physics interfaces appear instead as surface discontinuities in the otherwise smooth macroscopic description of matter.

Both the micro and macro scales diverge for $\epsilon \to 0$, substantiating the claim that it is impossible to maintain a continuum description to arbitrarily small relative precision. The smallness of both length scales for ordinary matter and for reasonable relative precision shows that there is ample room for a smooth continuum description of everyday phenomena. Nanophysics, however, straddles the border between the continuum and particle descriptions of matter, resulting in a wealth of new phenomena outside the scope of classical continuum physics.

Material particles

In continuum physics we shall generally permit ourselves to speak about *material particles* as the smallest objects that may consistently be considered part of the continuum description within the required precision. A material particle will always contain a large number of molecules but may in the continuum description be thought of as infinitesimal or point-like.

Although we usually shall think of material particles as being similar in different types of matter, they are in fact quite different. In solids, we may with some reservation think of *solid particles* as containing a fixed collection of molecules, whereas in liquids and especially in gases we should not forget that the molecules making up a *fluid particle* at a given instant will shortly be replaced by other molecules. If the molecular composition of the material in the environment of a material particle has a slow spatial variation, this incessant molecular game of 'musical chairs' may slowly change the composition of the material inside the particle. Such *diffusion* processes driven by spatial variations in material properties lie at the very root of fluid mechanics. Even spatial variations in the average flow velocity will drive momentum diffusion, causing internal (viscous) friction in the fluid.

In modern textbooks on continuum physics there has been a tendency to avoid introducing the concept of a material particle. Instead these presentations rely on global conservation laws to deduce the local continuum description—in the form of partial differential equations—by purely mathematical means. Although quite elegant and apparently free of physical interpretation problems, such an approach unfortunately obscures the conditions under which the local laws may be assumed to be valid.

In this book the concept of a material particle has been carefully introduced in the proper physical context of measurement precision. This approach allows us, for example, to conclude that a differential equation involving the density (or any other quantity) cannot be assumed to be physically meaningful at distance scales smaller than the microscopic length scale L_{micro}, and that a spatial derivative of the density must be limited by the macroscopic length scale, $|\partial\rho/\partial x| \lesssim \rho/L_{\text{macro}}$. A further advantage is that the local description may be interpreted as representing the laws valid for the motion of individual material particles. To satisfy both points of view, the interpretation and equivalence of the local and global laws will be carefully elucidated everywhere in this book.

1.3 Newtonian mechanics

In Newtonian mechanics (see appendix A) the basic material object is a point particle with a fixed mass m. Newton's second law is *the* fundamental equation of motion, and states that *mass times acceleration equals force*. Mathematically, it is expressed as a second-order differential equation in time t,

Sir Isaac Newton (1642–1727). *English physicist and mathematician. Founded classical mechanics on three famous laws in his books 'Philosophiae Naturaliz Principia Mathematica' (1687). Newton developed calculus to solve the equations of motion, and formulated theories of optics and of chemistry. He still stands as perhaps the greatest scientific genius of all time.*

$$m\,\frac{d^2\boldsymbol{x}}{dt^2} = \boldsymbol{f}, \tag{1.5}$$

where $\boldsymbol{x}$ the instantaneous position of the particle, and $\boldsymbol{f}$ the force acting on it. In chapter 2 we shall introduce vector calculus to handle quantities like $\boldsymbol{x}$ and $\boldsymbol{f}$ in a systematic way, but for now any understanding of the meaning of a vector will work fine.

Since the force on any given particle can depend on the positions and velocities of the particle itself and of other particles, as well as on external parameters, the dynamics of a collection of particles becomes a web of coupled ordinary second-order differential equations in time. Even if macroscopic bodies *are* huge collections of atoms and molecules, it is completely out of the question to solve the resulting web of differential equations. In addition, there is the problem that molecular interactions are quantum mechanical in nature, and Newtonian mechanics, strictly speaking, does not apply at the atomic level. This knowledge is, however, relatively new and has as mentioned earlier some difficulty in making itself apparent at the macroscopic level. So even if quantum mechanics rules the world of atoms, its special character is rarely amplified to macroscopic proportions.

Global mechanical quantities

In Newtonian particle mechanics, a 'body' is taken to be a fixed collection of point particles, each obeying the second law (1.5). For any body one may define various *global mechanical quantities* which like the total mass are calculated as sums over contributions from each and every particle in the body. Some of the global quantities are *kinematic*: momentum, angular momentum, and kinetic energy. Others are *dynamic*: force, moment of force, and power (rate of work of the forces).

Newton's second law for particles leads to three simple *laws of balance* between the kinematic and dynamic quantities;

- *the rate of change of momentum equals force,*
- *the rate of change of angular momentum equals moment of force,*

- *the rate of change of kinetic energy equals power.*

Even if these laws are insufficient to determine the dynamics of a multiparticle body, they represent seven individual constraints on the motion of any system of point particles, regardless of how complex it is. In particular, they can be taken over to continuum mechanics when exchange of matter between a body and its environment is properly taken into account.

1.4 Continuum physics

In *continuum physics* a macroscopic body is seen as a huge collection of tiny material particles, each of which contains a sufficiently large number of molecules to justify a continuum description. Continuum physics does not 'on its own' go below the level of the material particles. Although the mass density may be calculated by adding together the masses of all the molecules in a material particle and dividing with the volume occupied by it, this procedure falls, strictly speaking, outside continuum physics.

The field concept

In the extreme mathematical limit, material particles are taken to be truly infinitesimal and all physical properties of the particles as well as the forces acting on them are described by smooth functions of space and time. Continuum physics is therefore a theory of *fields*.

Mathematically, a field f is simply a real-valued function $f(x, y, z, t)$ of spatial coordinates x, y, z and time t, representing the value of a physical quantity in this point of space at the given time, for example the mass density $\rho = \rho(x, y, z, t)$. Sometimes a collection of such functions is also called a field and the individual real-valued members are called its components. Thus, the most fundamental field of fluid mechanics, the velocity field $\boldsymbol{v} = (v_x, v_y, v_z)$, has three components, one for each of the coordinate directions.

Besides fields characterizing the state of the material, such as mass density and velocity, it is convenient to employ fields that characterize the forces acting on and within the material. The gravitational acceleration field $\boldsymbol{g}$ is a *body force* field, which penetrates bodies from afar and acts on their mass. Some force fields are only meaningful for regions of space where matter is actually present, as for example the pressure field p, which acts across the imagined contact surfaces that separate neighbouring volumes of a fluid at rest. Pressure is, however, not the only *contact force*. For fluids in motion, for solids and more general materials, contact forces are described by the nine-component stress field, $\boldsymbol{\sigma} = \{\sigma_{ij}\}$, which is a (3×3) matrix field with rows and columns labelled by coordinates: $i, j = x, y, z$.

Mass density, velocity, gravity, pressure and stress are the usual fields of continuum mechanics and will all be properly introduced in the chapters to come. Some fields are thermodynamic, like the temperature T and the specific internal energy density u. Others describe different states of matter, for example the electric charge density ρ_e and current density $\boldsymbol{j}_e$ together with the electric and magnetic field strengths, $\boldsymbol{E}$ and $\boldsymbol{B}$. Like gravity $\boldsymbol{g}$, these force fields are thought to exist in regions of space completely devoid of matter.

There are also fields that refer to *material properties*, for example the coefficient of shear elasticity μ of a solid and the coefficient of shear viscosity η of a fluid. Such fields are usually constant within homogeneous bodies, i.e. independent of space and time, and are mostly viewed as *material constants* rather than true fields.

Field equations

Like all physical variables, fields evolve with time according to dynamical laws, called *field equations*. In continuum mechanics, the central equation of motion descends directly from Newton's second law applied to every material particle. Mass conservation, which is all but trivial and most often tacitly incorporated in particle mechanics, turns into an equation of motion for the mass density in continuum theory. Still other field equations such as Maxwell's equations for the electromagnetic fields have completely different and non-mechanical origins, although they do couple to the mechanical equations of motion.

Mathematically, field equations are *partial differential equations* in both space and time. This makes continuum mechanics considerably more difficult than particle mechanics where the equations of motion are ordinary differential equations in time. On the other hand, this greater degree of mathematical complexity also leads to a plethora of new and sometimes quite unexpected phenomena.

In some field theories, for example Maxwell's electromagnetism, the field equations are *linear* in the fields, but that is not the case in continuum mechanics. The *nonlinearity* of the field equations of fluid mechanics adds a further layer of mathematical difficulty to this subject, making it very different from linear theories. The nonlinearity leads to dynamic instabilities and gives rise to the chaotic and as yet not fully understood phenomenon of *turbulence*, well known from our daily dealings with water and air.

Physical reality of force fields

Whereas the mass density only has meaning in regions actually containing matter, or may be defined to be zero in a vacuum, the gravitational field is assumed to exist and take non-vanishing values even in the vacuum. It specifies the force that would be exerted on a unit mass particle at a given point, but the field is assumed to be there even if no particles are present.

In non-relativistic Newtonian physics, the gravitational field has no independent physical meaning and may be completely eliminated and replaced by non-local forces acting between material bodies. The true physical objects appear to be the material bodies, and the gravitational field just a mathematical convenience for calculating the gravitational force exerted by these bodies. There are no independent dynamical equations that tell us how the Newtonian field of gravity changes with time. When material bodies move around or change their mass distributions, their fields of gravity change instantaneously everywhere as they move around.

In relativistic mechanics, on the other hand, fields take on a completely different meaning. The reason is that instantaneous action-at-a-distance cannot take place. If matter is moved, the current view is that it will take some time before the field of gravity adjusts to the new positions, because no signal can travel faster than light. Due to relativity, fields must travel independently, obey their own equations of motion, and carry physical properties such as energy and momentum. Electromagnetic waves bringing radio and TV signals to us are examples of force fields thus liberated from their origin. Gravitational waves have not yet been observed directly, but indirectly they have been observed in binary neutron star systems which can only be fully understood if gravitational radiation is taken into account.

Even if we shall not deal with relativistic theories of the continuum, and therefore may consider the gravitational field to be merely a mathematical convenience, it may nevertheless be wise, at least in the back of our minds, to think of the field of gravity as having an independent physical existence. Then we shall have no philosophical problem endowing it with physical properties, even in matter-free regions of space.

Is matter *really* discrete or continuous?

Although continuum physics is always an approximation to the underlying discrete atomic level, this is not the end of the story. At a deeper level it turns out that matter is best described by another continuum formalism, relativistic quantum field theory, in which the discrete particles—electrons, protons, neutrons, nuclei, atoms and everything else—arise as quantum excitations in the fields. Relativistic quantum field theory without gravitation emerged in the first half of the twentieth century as *the* basic description of the subatomic world, but in spite of its enormous success it is still not clear how to include gravity.

Just as the continuity of macroscopic matter is an illusion, the quantum field continuum may itself one day become replaced by even more fundamental discrete or continuous descriptions of space, time and matter. It is by no means evident that there could not be a fundamental length in nature setting an ultimate lower limit to distance and time, and theories of this kind have in fact been proposed[2]. It appears that we do not know, and perhaps will never know, whether matter at its deepest level is truly continuous or truly discrete.

Problems

1.1 Consider a small volume of a gas which is a fraction p of a larger volume containing M molecules. The probability for any molecule to find itself in the small volume may be taken to be p.

(a) Calculate the probability that the small volume contains n molecules.

[2]See, for example, J. A. Wheeler, *It from bit, Proceedings of the 3rd International Symposium on Foundations of Quantum Mechanics, Tokyo (1989).*

(b) Show that the average of the number of molecules in the small volume is $N \equiv \langle n \rangle = pM$.

(c) Show that the variance is $\Delta N^2 \equiv \langle (n - \langle n \rangle)^2 \rangle = p(1-p)M \approx N$ for $p \ll 1$.

1.2 Show that a cube containing $N = M^3$ smaller cubes of equal size will have $K = 6M^2 - 12M + 8$ smaller cubes lying on the surface. Estimate the fluctuation ΔN when N molecules in a cube oscillate with amplitude equal to the molecular size.

1.3 A spherical volume contains a large number N of molecules. Estimate the number of molecules N_S situated at the surface and show that the fluctuation in this number is $\Delta N \approx 6^{1/3}\pi^{1/6}N^{1/3} \approx 2.2N^{1/3}$ when they randomly oscillate with amplitude equal to the molecular size.

1.4 Consider a material gas particle containing N identical molecules. Write the velocity of the nth molecule as $\boldsymbol{v}_n = \boldsymbol{v} + \boldsymbol{u}_n$ where $\boldsymbol{v}$ is the centre of mass velocity and $\boldsymbol{u}_n$ is a random contribution from thermal motion. It may be assumed that the average of the random component of velocity vanishes $\langle \boldsymbol{u}_n \rangle = \boldsymbol{0}$, that all random velocities are uncorrelated, and that their fluctuations are the same for all particles $\langle \boldsymbol{u}_n^2 \rangle = v_0^2$. Show that the average of the centre of mass velocity for the fluid particle is $\langle \boldsymbol{v}_c \rangle = \boldsymbol{v}$ and that its fluctuation due to thermal motion is $\Delta v_c = v_0/\sqrt{N}$.

1.5 At what precision is the microscopic scale L_{micro} equal to the mean free path, when the air density is 100 times smaller than normal?

2

Space and time

In classical Newtonian physics space is absolute and eternal, obeying the rules of Euclidean geometry everywhere. It is the perfect stage on which all physical phenomena play out. Time is equally uniform and absolute throughout space from beginning to end, and matter has no influence on the properties of space and time. Rulers to measure length never stretch, clocks to measure time never lose a tick, and both can be put to work anywhere from the deepest levels of matter to the farthest reaches of outer space. It is a clockwork universe, orderly, rigorous and deterministic.

This semblance of perfection was shattered at the beginning of the twentieth century by the theories of relativity and quantum mechanics. Space and time became totally intertwined with each other and with matter, and the determinism of classical physics was replaced by the still disturbing quantum indeterminism. Relativity and quantum mechanics are both theories of extremes. Although in principle they apply to the bulk of all physical phenomena, their special features become dominant only at velocities approaching the velocity of light in the case of relativity, or length scales approaching the size of atoms in the case of quantum mechanics. Newtonian space and time remain a valid, if not 'true', conceptual framework over the vast ranges of length and velocity scales covered by classical continuum physics.

In this chapter the basic ideas behind space and time are introduced in a way which emphasizes the operational aspects of physical concepts. Care is exerted to ensure that the concepts defined here should remain valid in more advanced theories. A certain familiarity with Euclidean geometry in Cartesian coordinates is assumed, and the chapter serves in most respects to define the mathematical notation to be used in the remainder of the book. It may be sampled at leisure, as the need arises.

2.1 Reference frames

Physics is a quantitative discipline using mathematics to relate measurable quantities expressed in terms of real numbers. In formulating the laws of nature, undefined mathematical primitives—for example the points, lines and circles of Euclidean geometry—are not particularly useful, and such concepts have for this reason been eliminated and replaced by numerical representations everywhere in physics. This necessitates a specification of the practical procedures by which these numbers are obtained in an experiment; for example, which units are being used.

Behind every law of nature and every formula in physics, there is a framework of procedural descriptions, a *reference frame*, supplying an operational meaning to all physical quantities. Part of the art of doing physics lies in comprehending this—often tacitly understood—infrastructure to the mathematical formalism. The reference frame always involves physical objects—balances to measure mass, clocks to measure time and rulers to measure length—that are not directly a part of the mathematical formalism. Precisely because they *are* physical objects, they can at least in principle be handed over or copied, and thereby shared among experimenters. This is what is really meant by the objectivity of physics.

The system of units, the *Système Internationale (SI)*, is today fixed by international agreement. But even if our common frame of reference is thus defined by social convention, physics is nevertheless objective. In principle our frames of reference could be shared with any other being in the universe.

The unit of mass, the kilogram, is (still) defined by a prototype stored by the *International Bureau of Weights and Measures* near Paris, France. Copies of this prototype and balances for weighing them can be made to a precision of one part in 10^9 [34].

2.2 Time

Time is the number you read on your *clock*. There is no better definition. Clocks are physical objects which may be shared, compared, copied and synchronized to create an objective meaning of *time*. Most clocks, whether they are grandfather clocks with a swinging pendulum or oscillating quartz crystals, are based on periodic physical systems that return to the same state again and again. Time intervals are simply measured by counting periods. There are also aperiodic clocks, for example hour glasses, and clocks based on radioactive elements. It is especially the latter that allow time to be measured on geological time scales. Beyond such scales the concept of time becomes increasingly more theory-laden.

Like all macroscopic physical systems, clocks are subject to small fluctuations in the way they run. The most stable clocks are those that keep time best with respect to copies of themselves as well as with clocks built on other principles. Grandfather clocks are much less stable than maritime chronometers that in turn are less stable than quartz clocks. The international frame of reference for time is always based on the most stable clocks currently available.

Formerly the unit of time, the *second*, was defined as 1/86 400 of a mean solar day, but the Earth's rotation is not that stable, and since 1966 the second has been defined by international agreement as the duration of 9 192 631 770 oscillations of the microwave radiation absorbed in a certain hyperfine transition in the cesium-133 atom. A beam of cesium-133 atoms is used to stabilize a quartz oscillator at the right frequency by a resonance method, so what we call an atomic clock is really an atomically stabilized quartz clock. The intrinsic relative precision in this time standard is about 4×10^{-14}, or about one second in a million years [34].

In the extreme mathematical limit, time may be assumed to be a real number t, and in Newtonian physics its value is assumed to be universally known.

2.3 Space

It is a mysterious and so far unexplained fact that physical *space* has three dimensions, which means that it takes exactly three real numbers—say x_1, x_2 and x_3—to locate a point in space. These numbers are called the *coordinates* of the point, and the reference frame for coordinates is called the *coordinate system*. It must contain all the operational specifications for locating a point given the coordinates, and conversely obtaining the coordinates given the location. In this way we have relegated all philosophical questions regarding the *real* nature of points and of space to the operational procedures contained in the reference frame.

On Earth everybody navigates by means of a geographical system, in which a point is characterized by latitude, longitude and height. The geographical coordinate system is based on agreed upon fixed points on Earth: the north pole, Greenwich in London, and the average sea level. The modern Global Positioning System uses 'fixed' points in the sky in the form of satellites, and the coordinates of any point on Earth is determined from differences in the time-of-flight of radio signals.

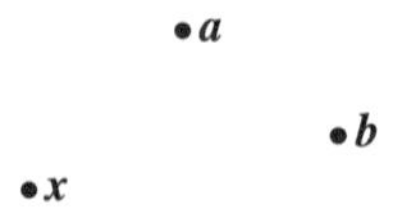

Points may be visualized as dots on a piece of paper.

It is convenient to collect the coordinates x_1, x_2, and x_3 of a point in a single object, a triplet of real numbers

$$\boldsymbol{x} = (x_1, x_2, x_3), \tag{2.1}$$

called the *position* of the point in the coordinate system[1]. The triplet notation is just a notational convenience, so a function of the position $f(\boldsymbol{x})$ is completely equivalent to a function of the three coordinates $f(x_1, x_2, x_3)$.

[1] In almost all modern textbooks it is customary to use boldface notation for triplets of real numbers ('vectors') and we shall also do so here. In calculations with pencil on paper many different notations are used to distinguish such a symbol from other uses, for example a bar ($\overline{x}$), an arrow ($\vec{x}$) or underlining ($\underline{x}$).

There is nothing sacred about the names of the coordinates. In physics and especially in practical calculations, the coordinate variables are often renamed $x_1 \to x$, $x_2 \to y$ and $x_3 \to z$, so that the general point becomes $\boldsymbol{x} \to (x, y, z)$. It is also customary to write $\boldsymbol{a} = (a_x, a_y, a_z)$ for a general triplet, with the coordinate labels used as indices instead of 1, 2 and 3. It is of course of no importance whether the range of indices is labelled x, y, z or 1, 2, or 3 or something else, as long as there are three of them.

Coordinate transformations

Having located a point by a set of coordinates $\boldsymbol{x} = (x_1, x_2, x_3)$ in one coordinate system, the coordinates $\boldsymbol{x}' = (x'_1, x'_2, x'_3)$ of the exact same point in another coordinate system must be calculable from the first

$$
\begin{aligned}
x'_1 &= f_1(x_1, x_2, x_3), \\
x'_2 &= f_2(x_1, x_2, x_3), \\
x'_3 &= f_3(x_1, x_2, x_3).
\end{aligned}
$$

In triplet notation this is written

$$\boldsymbol{x}' = \boldsymbol{f}(\boldsymbol{x}). \tag{2.2}$$

This postulate reflects that physical reality is unique and that different coordinate systems are just different ways of representing the same physical space in terms of real numbers. Conversely, each one-to-one mapping of the coordinates defines another coordinate system. The study of *coordinate transformations* is central to analytic geometry and permits characterization of geometric quantities by the way they transform rather than in abstract terms (see section 2.7).

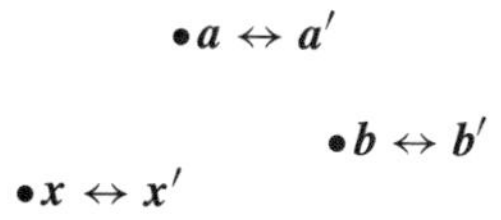

In different coordinate systems the same points have different coordinates.

Length

From the earliest times humans have measured the *length* of a road between two points, say $\boldsymbol{a}$ and $\boldsymbol{b}$, by counting the number of steps it takes to walk along this road. In order to communicate to others the length of a road, the count of steps must be accompanied by a clear definition of a step, for example in terms of an agreed-upon unit of length.

Originally the units of length—inch, foot, span and fathom—were directly related to the human body, but increasing precision in technology demanded better-defined units. In 1793 the metre was introduced as a ten millionth of the distance from equator to pole on Earth, and until far into the twentieth century a unique 'normal metre' was stored in Paris, France. Later the metre became defined as a certain number of wavelengths of a certain spectral line in krypton-86, an isotope of a noble gas which can be found anywhere on Earth (thereby eliminating the need for a trip to Paris!). Since 1983 the metre has been defined by international convention to be the distance travelled by light in exactly 1/299 792 458 of a second [34]. The problem of measuring lengths has thus been transferred to the problem of measuring time, which makes sense because the precision of the time standard is at least a thousand times better than any proper length standard.

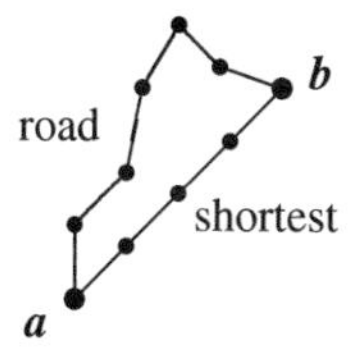

The length of the road between $\boldsymbol{a}$ and $\boldsymbol{b}$ is measured by counting steps along the road. The distance is the length of the shortest road.

This method for determining length may be refined to any desired practical precision by using very short steps. In the extreme mathematical limit, the steps become infinitesimally small, and the road becomes a continuous path. The shortest such path is called a *geodesic* and represents the 'straightest line' between the points. Airplanes and ships travel along geodesics, i.e. great circles on the spherical surface of the Earth.

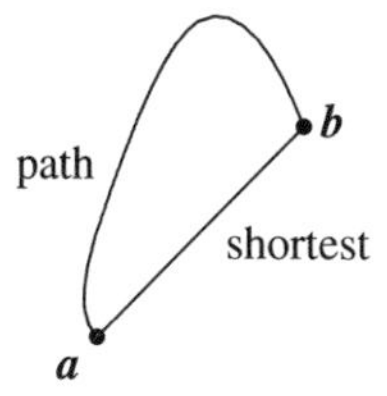

In the mathematical limit the shortest continuous path connecting $\boldsymbol{a}$ and $\boldsymbol{b}$ is called the geodesic ('straight line').

Distance

The *distance* between two points is defined to be the length of the shortest path between them. Since the points are completely defined by their coordinates, $\boldsymbol{a}$ and $\boldsymbol{b}$ relative to the chosen coordinate system, the distance must be a real function $d(\boldsymbol{a}, \boldsymbol{b})$ of the two sets of coordinates. This function is not completely general; certain axioms have to be fulfilled by any distance function (see problem 2.1).

From the definition it is clear that the distance between two points must be the same in all coordinate systems, because it can, in principle, be determined by laying out rulers between points without any reference to coordinate systems. The actual distance function $d'(\boldsymbol{a}', \boldsymbol{b}')$ in a new coordinate system may be different from the old, $d(\boldsymbol{a}, \boldsymbol{b})$, but the numerical values have to be the same,

$$d'(\boldsymbol{a}', \boldsymbol{b}') = d(\boldsymbol{a}, \boldsymbol{b}), \tag{2.3}$$

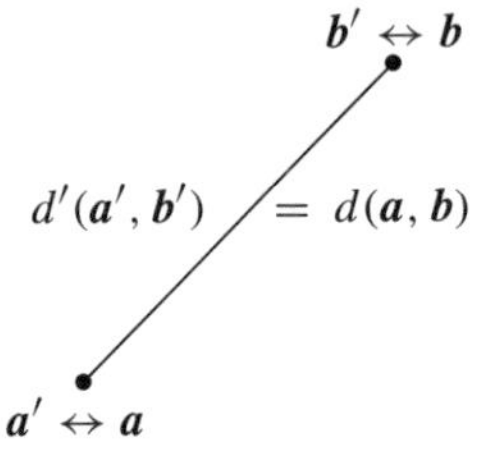

Invariance of the distance.

where $\boldsymbol{a}' = \boldsymbol{f}(\boldsymbol{a})$ and $\boldsymbol{b}' = \boldsymbol{f}(\boldsymbol{b})$ are calculated by the coordinate transformation (2.2). Knowing the distance function $d(\boldsymbol{a}, \boldsymbol{b})$ in one coordinate system, it may be calculated in any other coordinate system by means of the appropriate coordinate transformation.

This expresses the *invariance of the distance* under all coordinate transformations. In the same way as (2.2) may be viewed mathematically as a definition of what is meant by 'another coordinate system', equation (2.3) may be viewed mathematically as a definition of what is meant by distance in the other coordinate system.

René Descartes (1596–1650). *French scientist and philosopher, father of analytic geometry. Developed a theory of mechanical philosophy, later to be superseded by Newton's work. Confronted with doubts about reality, he saw thought as the only argument for existence: 'I think, therefore I am'.*

Cartesian coordinate systems

It is another fundamental physical fact that it is possible (within limited regions of space and time) to construct coordinate systems, in which the distance between any two points, $\boldsymbol{a}$ and $\boldsymbol{b}$, is given by the expression

$$d(\boldsymbol{a}, \boldsymbol{b}) = \sqrt{(a_1 - b_1)^2 + (a_2 - b_2)^2 + (a_3 - b_3)^2}. \tag{2.4}$$

Such coordinate systems were first analysed by Descartes and are called *Cartesian.* The distance function implies that space is *Euclidean* and therefore has all the properties one learns about in elementary geometry.

2.4 Vector algebra

Triplets of real numbers play a central role in everything that follows, and it is now convenient to introduce a set of algebraic rules for these objects. We shall see below (section 2.8) that vectors in Cartesian coordinate systems are triplets that transform in a special way under coordinate transformations.

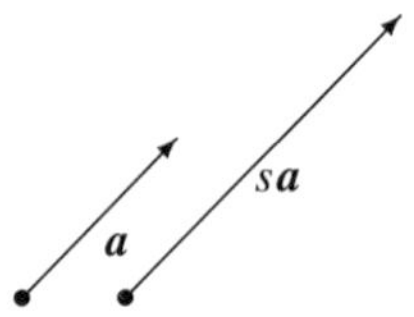

Geometric scaling of a vector.

Basic algebraic rules

The following operations endow triplets with the properties of the familiar geometric vectors. Visualization on paper is of course as useful as ever, so we shall also draw triplets and illustrate their properties by means of arrows.

Linear operations: Linear operations lie at the core of triplet algebra,

$$s\,\boldsymbol{a} = (sa_1, sa_2, sa_3) \qquad \text{(scaling)}, \tag{2.5}$$

$$\boldsymbol{a} + \boldsymbol{b} = (a_1 + b_1, a_2 + b_2, a_3 + b_3) \qquad \text{(addition)}, \tag{2.6}$$

$$\boldsymbol{a} - \boldsymbol{b} = (a_1 - b_1, a_2 - b_2, a_3 - b_3) \qquad \text{(subtraction)}. \tag{2.7}$$

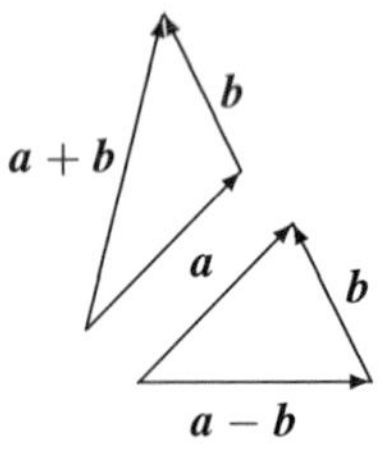

Geometric addition and subtraction of a vector.

These rules tell us that the set of all triplets, also called $\mathbb{R}^3$, mathematically is a three-dimensional *vector space.* A straight line with origin $\boldsymbol{a}$ and direction $\boldsymbol{b}$ is described by the linear function $\boldsymbol{a} + \boldsymbol{b}\,s$ with $-\infty < s < \infty$.

Bilinear products: There are three different bilinear products of triplets, of which the two first are well known from ordinary vector calculus,

$$\boldsymbol{a} \cdot \boldsymbol{b} = a_1b_1 + a_2b_2 + a_3b_3 \qquad \text{(dot product)}, \tag{2.8}$$

$$\boldsymbol{a} \times \boldsymbol{b} = (a_2b_3 - a_3b_2, a_3b_1 - a_1b_3, a_1b_2 - a_2b_1) \qquad \text{(cross product)}. \tag{2.9}$$

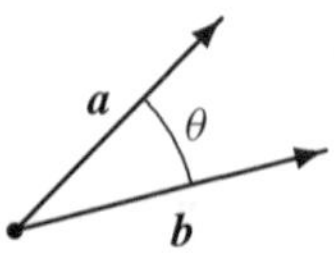

The dot product of two vectors is $|\boldsymbol{a}|\,|\boldsymbol{b}|\cos\theta$ where θ is the angle between them.

Two triplets are said to be *orthogonal* when their dot product vanishes. Note that the cross product is defined entirely in terms of the coordinates and that we do not in the rule itself distinguish between left-handed and right-handed coordinate systems. Whether you use your right or left hand when you draw a cross product on paper does not matter for the triplet product rule, as long as you consistently use the same hand for all such drawings.

The last one,

$$\boldsymbol{ab} = \begin{pmatrix} a_1 \\ a_2 \\ a_3 \end{pmatrix} (b_1, b_2, b_3) = \begin{pmatrix} a_1 b_1 & a_1 b_2 & a_1 b_3 \\ a_2 b_1 & a_2 b_2 & a_2 b_3 \\ a_3 b_1 & a_3 b_2 & a_3 b_3 \end{pmatrix} \quad \text{(tensor product)}, \tag{2.10}$$

called the *tensor product*, is unusual in that it produces a (3×3) matrix from two triplets, but otherwise it is perfectly well defined and useful to have around. It is nothing but an ordinary matrix product of a column-matrix and a row-matrix, also called the *direct product* and sometimes in the older literature the *dyadic product*. In section 2.8 we shall introduce more general geometric objects, called tensors, of which the simplest are matrices of this kind. The tensor product, and tensors in general, cannot be given a simple visualization on paper.

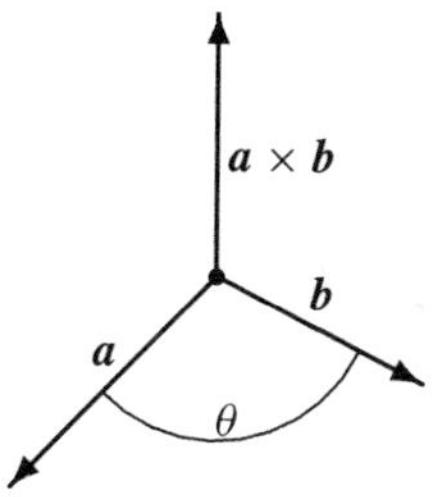

The cross product of two vectors is a vector orthogonal to both, of length $|\boldsymbol{a}|\,|\boldsymbol{b}| \sin\theta$, here drawn using a right-hand rule.

Further definitions

Besides the basic algebraic rules for triplets, a number of other definitions are useful in practical calculations.

Volume product: The trilinear product of three triplets obtained by combining the cross product and the dot product is called the *volume product*,

$$\boldsymbol{a} \times \boldsymbol{b} \cdot \boldsymbol{c} = a_1 b_2 c_3 + a_2 b_3 c_1 + a_3 b_1 c_2 - a_1 b_3 c_2 - a_2 b_1 c_3 - a_3 b_2 c_1. \tag{2.11}$$

The right-hand side shows that the volume product equals the determinant of the matrix constructed from the three vectors,

$$\boldsymbol{a} \times \boldsymbol{b} \cdot \boldsymbol{c} = \begin{vmatrix} a_1 & b_1 & c_1 \\ a_2 & b_2 & c_2 \\ a_3 & b_3 & c_3 \end{vmatrix} \quad \text{(volume product)}. \tag{2.12}$$

The volume product is antisymmetric under exchange of any pair of vectors. Its value is the (signed) volume of the parallelepiped spanned by the vectors.

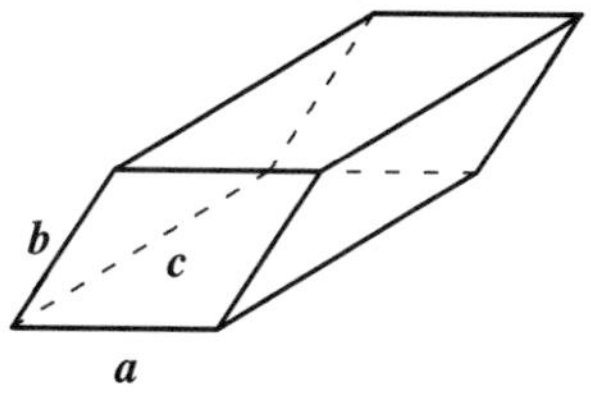

Three vectors spanning a parallelepiped.

Square and norm: The square and the norm are standard definitions

$$\boldsymbol{a}^2 = \boldsymbol{a} \cdot \boldsymbol{a} = a_1^2 + a_2^2 + a_3^2 \quad \text{(square)}, \tag{2.13}$$

$$|\boldsymbol{a}| = \sqrt{\boldsymbol{a}^2} = \sqrt{a_1^2 + a_2^2 + a_3^2} \quad \text{(norm or length)}. \tag{2.14}$$

This definition of the norm is closely related to the form of the Cartesian distance (2.4) which may now be written $d(\boldsymbol{a}, \boldsymbol{b}) = |\boldsymbol{a} - \boldsymbol{b}|$.

Vector derivatives: Various types of derivatives involving triplets may also be defined,

$$\frac{\partial \boldsymbol{a}}{\partial s} = \left(\frac{\partial a_1}{\partial s}, \frac{\partial a_2}{\partial s}, \frac{\partial a_3}{\partial s} \right) \quad \text{(scalar derivative)}, \tag{2.15}$$

$$\frac{\partial}{\partial \boldsymbol{a}} = \left(\frac{\partial}{\partial a_1}, \frac{\partial}{\partial a_2}, \frac{\partial}{\partial a_3} \right) \quad \text{(vector derivative)}. \tag{2.16}$$

In the first line, the derivative of a triplet after a parameter is defined. In the second line, a symbolic notation is introduced for the three derivatives after a triplet's coordinates (see problem 2.9 for simple uses of this notation).

Norm of a matrix: The norm of an arbitrary (3×3) matrix $\mathbf{A} = \{a_{ij}\}$,

$$|\mathbf{A}| = \sqrt{\sum_{ij} a_{ij}^2} \quad \text{(matrix norm)}, \tag{2.17}$$

can sometimes be useful in inequalities. This definition makes sense because the norm of a tensor product is then the product of the norms of the factors, $|\boldsymbol{ab}| = |\boldsymbol{a}|\,|\boldsymbol{b}|$.

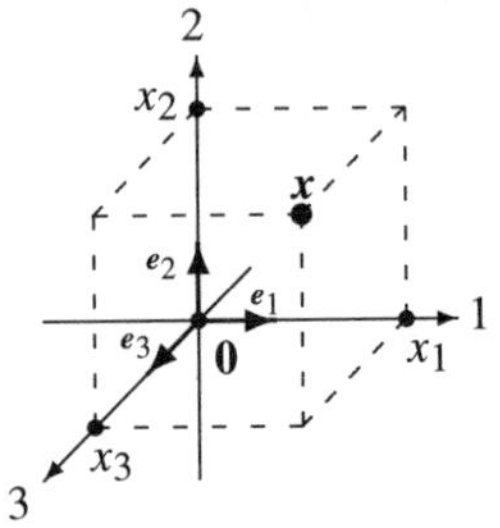

Visualization of the Cartesian coordinate system.

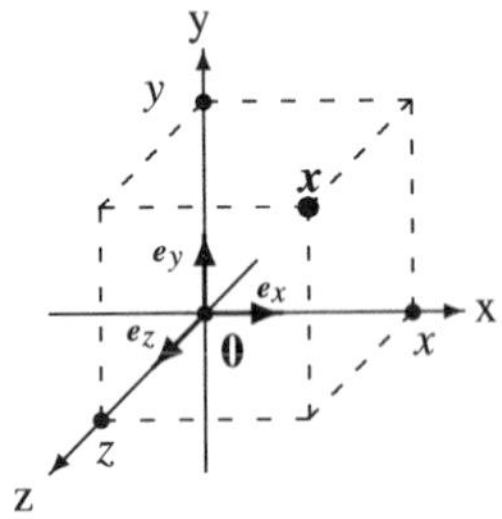

A Cartesian coordinate system with axes labelled x, y and z.

2.5 Basis vectors

The *coordinate axes* of a Cartesian coordinate system are straight lines with a common origin $\mathbf{0} = (0, 0, 0)$ and directions,

$$e_1 = (1, 0, 0), \tag{2.18a}$$
$$e_2 = (0, 1, 0), \tag{2.18b}$$
$$e_3 = (0, 0, 1). \tag{2.18c}$$

These triplets are called the *basis* vectors of the coordinate system[2], or just the basis, and every position x may trivially be written as a linear combination of the basis vectors with the coordinates as coefficients,

$$x = x_1 e_1 + x_2 e_2 + x_3 e_3. \tag{2.19}$$

The basis vectors are *normalized* and mutually *orthogonal*,

$$|e_1| = |e_2| = |e_3| = 1, \tag{2.20}$$
$$e_1 \cdot e_2 = e_2 \cdot e_3 = e_3 \cdot e_1 = 0. \tag{2.21}$$

Using these relations and (2.19) we find

$$x_1 = e_1 \cdot x, \tag{2.22a}$$
$$x_2 = e_2 \cdot x, \tag{2.22b}$$
$$x_3 = e_3 \cdot x, \tag{2.22c}$$

showing that the coordinates of a point may be understood as the normal projections of the point on the axes of the coordinate system.

Completeness of basis

Combining (2.19) with (2.22) we obtain the identity

$$e_1(e_1 \cdot x) + e_2(e_2 \cdot x) + e_3(e_3 \cdot x) = x,$$

valid for all x. Since this is a linear identity, we may remove x and express this *completeness* relation in a compact form by means of the tensor product (2.10),

$$e_1 e_1 + e_2 e_2 + e_3 e_3 = \mathbf{1}, \tag{2.23}$$

where on the right-hand side the symbol $\mathbf{1}$ stands for the (3×3) unit matrix[3].

Handedness

It must be emphasized that the handedness of the coordinate system has not entered the formalism. Correspondingly, the volume of the unit cube,

$$e_1 \cdot e_2 \times e_3 = +1, \tag{2.24}$$

is always $+1$, independent of whether you call the hand you write with the left or the right.

[2] In some texts the basis vectors are symbolized by the coordinate label with a hat above: $\hat{1}$, $\hat{2}$, and $\hat{3}$.

[3] To distinguish a matrix from a triplet, the matrix symbol will be written in heavy unslanted san serif boldface. The distinction is unfortunately not particularly visible in print. With pencil on paper, (3×3) matrices are sometimes marked with a double bar ($\overline{\overline{1}}$) or a double arrow ($\overleftrightarrow{1}$).

2.6 Index notation

Triplet notation for vectors is sufficient in most areas of physics, because physical quantities are mostly scalars (i.e. single numbers like mass) or vectors such as velocity, but sometimes it is necessary to use a more powerful and transparent notation which generalizes better to more complex expressions and quantities. It is called *index notation* or *tensor notation*, and consists in all simplicity of writing out the coordinate indices explicitly wherever they occur. Instead of thinking of a position as a triplet $\boldsymbol{x}$, we think of it as the set of coordinates x_i with the index i running implicitly over the coordinate labels, $i = 1, 2, 3$ or $i = x, y, z$, without having to state it every time.

Algebraic operations

Triplet and index notations coexist quite peacefully as witnessed by the linear operations

$$(s\boldsymbol{a})_i = s a_i, \tag{2.25}$$

$$(\boldsymbol{a} + \boldsymbol{b})_i = a_i + b_i, \tag{2.26}$$

$$(\boldsymbol{a} - \boldsymbol{b})_i = a_i - b_i. \tag{2.27}$$

For the scalar product we let the sum range implicitly over the coordinate labels,

$$\boldsymbol{a} \cdot \boldsymbol{b} = \sum_i a_i b_i. \tag{2.28}$$

In full-fledged tensor calculus even the summation symbol is left out and understood as implicitly present for all indices that occur precisely twice in a term. We shall, however, refrain from doing so here.

The Kronecker delta

The collection of nine scalar products of basis vectors has two indices that each run implicitly over the three coordinate labels, and is written

$$\boldsymbol{e}_i \cdot \boldsymbol{e}_j = \delta_{ij}. \tag{2.29}$$

The expression δ_{ij} is the unit matrix in index notation,

$$\delta_{ij} = \begin{cases} 1 & \text{for } i = j \\ 0 & \text{otherwise} \end{cases}, \tag{2.30}$$

also called the *Kronecker delta*. This is the first example of a true *tensor* of rank 2. Another is the tensor product (2.10) of two vectors, which takes the form

$$(\boldsymbol{a}\boldsymbol{b})_{ij} = a_i b_j. \tag{2.31}$$

It is, in fact, not enough for a tensor just to have two (or more) indices, but it suffices for now. In section 2.8 we shall see what really characterizes tensors.

Leopold Kronecker (1823–1891). *German mathematician, contributed to the theory of elliptic functions, algebraic equations, and algebraic numbers.*

The Levi-Civita symbol

The trilinear volume product (2.11) becomes a triple sum over indices,

$$\boldsymbol{a} \times \boldsymbol{b} \cdot \boldsymbol{c} = \sum_{ijk} \epsilon_{ijk}\, a_i b_j c_k, \tag{2.32}$$

with 27 coefficients,

$$\epsilon_{ijk} = \begin{cases} +1 & ijk = 123\,231\,312, \\ -1 & ijk = 132\,213\,321, \\ 0 & \text{otherwise.} \end{cases} \tag{2.33}$$

The symbol ϵ_{ijk} is, in fact, a tensor of third rank, called the *Levi-Civita symbol*.

Finally, the cross product (2.9) may be written as a double sum over two indices of the form,

$$(\boldsymbol{a} \times \boldsymbol{b})_i = \sum_{jk} \epsilon_{ijk} a_j b_k. \tag{2.34}$$

Mostly we shall avoid this complicated notation, although it does come in handy in general discussions.

Tullio Levi-Civita (1873–1941). *Italian mathematician, contributed to differential calculus, relavitivy, and founded (with Ricci) tensor analysis in curved space.*

2.7 Cartesian coordinate transformations

The same Euclidean world may be described geometrically by different observers with different reference frames. Each observer constructs his own preferred Cartesian coordinate system and determines all positions relative to that. Every observer thinks that his basis vectors have the simple form (2.18) and satisfy the same orthogonality and completeness relations. Every observer believes he is right-handed. How can they ever agree on anything with such a self-centred view of the world?

The answer is—as indicated in section 2.3—that the two descriptions are related by a coordinate transformation (2.2). Since the distance between any two points is independent of the coordinate system, the shortest paths must coincide, and straight lines must be mapped onto straight lines by any Cartesian coordinate transformation. Seen from one Cartesian coordinate system, which we shall call the 'old', the axes of another Cartesian coordinate system, called the 'new', will therefore also appear to be straight lines with a common origin. Furthermore, since the scalar product of two vectors can be expressed in terms of the norm (problem 2.5), it must—like distance—be independent of the specific coordinate system, such that the new axes will also appear to be orthogonal in the geometry of the old coordinate system. Different observers will thus agree that their respective coordinate systems are indeed Cartesian.

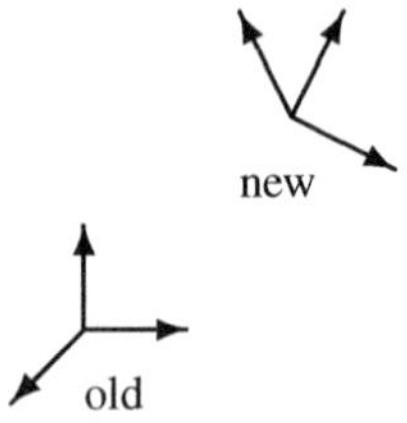

The old and the new Cartesian systems.

Simple transformations

We begin the analysis of coordinate transformations with the familiar elementary transformations: translation, rotation and reflection. These transformations are of the general form (2.2), expressing the coordinates of a geometrical point in the new system as a function of the coordinates of the same point in the old. The simple transformations are related to a special choice of coordinate axes, and similar simple transformations may be defined for other choices.

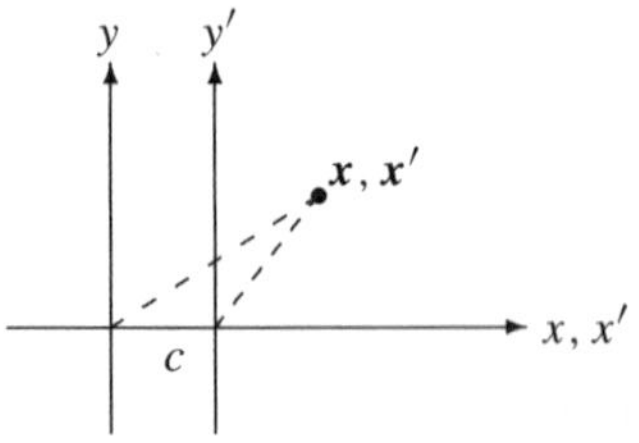

Simple translation of the coordinate system by c along the x-axis.

Simple translation: A simple *translation* of the origin of coordinates along the x-axis by a constant amount c is given by

$$x' = x - c, \tag{2.35a}$$

$$y' = y, \tag{2.35b}$$

$$z' = z. \tag{2.35c}$$

The axes of the new coordinate system are in this case parallel with the axes of the old.

Simple rotation: A simple *rotation* of the coordinate system through an angle ϕ around the z-axis is described by the transformation

$$x' = x\cos\phi + y\sin\phi, \tag{2.36a}$$

$$y' = -x\sin\phi + y\cos\phi, \tag{2.36b}$$

$$z' = z. \tag{2.36c}$$

In this case the z-axes are parallel in the old and the new systems.

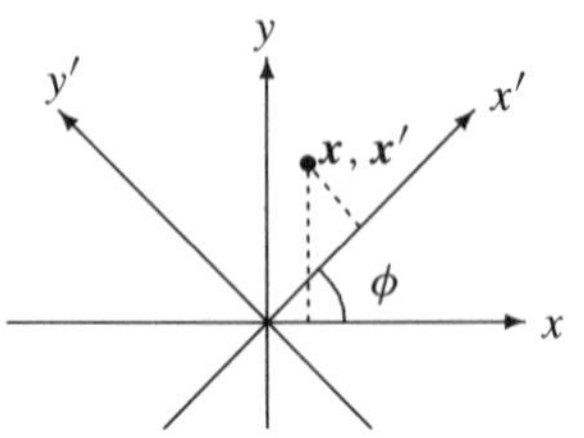

Simple rotation of the coordinate system around the z-axis (pointing out of the paper) through an angle ϕ.

Simple reflection: A simple *reflection* in the yz-plane is described by

$$x' = -x, \quad y' = y, \quad z' = z. \tag{2.37}$$

A simple reflection always transforms a right-handed coordinate system into a left-handed one, and vice-versa, independent of which hand you may claim to be the right one.

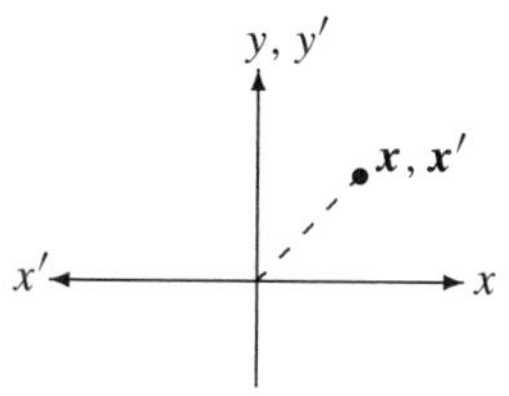

A simple reflection in the yz-plane.

General transformations

Let the new Cartesian coordinate system be characterized (in the old) by its origin $\boldsymbol{c}$ and its three orthogonal and normalized basis vectors $\boldsymbol{a}_1$, $\boldsymbol{a}_2$, and $\boldsymbol{a}_3$, satisfying the usual relations,

$$\left.\begin{array}{c} |\boldsymbol{a}_1| = |\boldsymbol{a}_2| = |\boldsymbol{a}_3| = 1 \\ \boldsymbol{a}_1 \cdot \boldsymbol{a}_2 = \boldsymbol{a}_2 \cdot \boldsymbol{a}_3 = \boldsymbol{a}_3 \cdot \boldsymbol{a}_1 = 0 \end{array}\right\}. \tag{2.38}$$

A position $\boldsymbol{x}' = (x'_1, x'_2, x'_3)$ in the new coordinate system must then correspond to the (old) position,

$$\boldsymbol{x} = \boldsymbol{c} + x'_1\boldsymbol{a}_1 + x'_2\boldsymbol{a}_2 + x'_3\boldsymbol{a}_3. \tag{2.39}$$

The new coordinates are obtained by multiplying from the left with the new basis vectors and using orthonormality (2.38)

$$\begin{aligned} x'_1 &= \boldsymbol{a}_1 \cdot (\boldsymbol{x} - \boldsymbol{c}) = a_{11}(x_1 - c_1) + a_{12}(x_2 - c_2) + a_{13}(x_3 - c_3), \\ x'_2 &= \boldsymbol{a}_2 \cdot (\boldsymbol{x} - \boldsymbol{c}) = a_{21}(x_1 - c_1) + a_{22}(x_2 - c_2) + a_{23}(x_3 - c_3), \\ x'_3 &= \boldsymbol{a}_3 \cdot (\boldsymbol{x} - \boldsymbol{c}) = a_{31}(x_1 - c_1) + a_{32}(x_2 - c_2) + a_{33}(x_3 - c_3), \end{aligned}$$

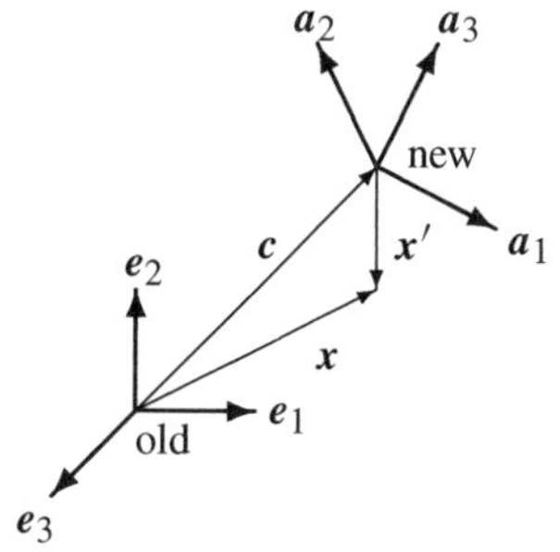

Arrangement of the old and new coordinate systems.

where $a_{ij} = (\boldsymbol{a}_i)_j$ are the coordinates of the new basis vectors. This is the most general coordinate transformation between any two Cartesian coordinate systems. It is not very difficult to show that the most general transformation may be composed from a sequence of simple transformations (problem 2.22).

Using index notation, the general coordinate transformation may be written,

$$\boxed{x'_i = \sum_j a_{ij}(x_j - c_j).} \tag{2.40}$$

It is characterized by the *translation vector* $\boldsymbol{c} = \{c_i\}$ and the *transformation matrix* $\mathbf{A} = \{a_{ij}\}$ with $a_{ij} = (\boldsymbol{a}_i)_j$ having the new basis vectors as rows. In matrix notation the transformation becomes even more compact[4],

$$\boldsymbol{x}' = \mathbf{A} \cdot (\boldsymbol{x} - \boldsymbol{c}). \tag{2.41}$$

The transformation matrix for a simple translation along the x-axis (2.35) is just the unit matrix, $\mathbf{A} = \mathbf{1}$, whereas for a simple rotation around the z-axis (2.36) we obtain the non-trivial matrix,

$$\mathbf{A} = \begin{pmatrix} \cos\phi & \sin\phi & 0 \\ -\sin\phi & \cos\phi & 0 \\ 0 & 0 & 1 \end{pmatrix}. \tag{2.42}$$

A simple reflection in the yz-plane (2.37) is characterized by a diagonal transformation matrix with $(-1, 1, 1)$ along the diagonal.

Orthogonality and completeness of the new basis

The orthogonality and completeness of the new basis vectors imply that

$$\boldsymbol{a}_i \cdot \boldsymbol{a}_j = \delta_{ij}, \tag{2.43}$$

$$\boldsymbol{a}_1\boldsymbol{a}_1 + \boldsymbol{a}_2\boldsymbol{a}_2 + \boldsymbol{a}_3\boldsymbol{a}_3 = \mathbf{1}. \tag{2.44}$$

[4] In ordinary mathematical matrix calculus one would not use the dot to indicate multiplication (nor to indicate a scalar product), but this notation is quite natural for the three-dimensional vectors and matrices that we encounter so often in physics.

In index notation these two relations take the form,

$$\sum_k a_{ik} a_{jk} = \sum_k a_{ki} a_{kj} = \delta_{ij}, \tag{2.45}$$

which in matrix notation become the usual conditions for the matrix to be orthogonal,

$$\mathbf{A} \cdot \mathbf{A}^\top = \mathbf{A}^\top \cdot \mathbf{A} = \mathbf{1}. \tag{2.46}$$

Here $(\mathbf{A}^\top)_{ij} = (\boldsymbol{a}_j)_i = a_{ji}$ is the transposed matrix having the new basis vectors as columns. Since $\mathbf{A}^{-1} = \mathbf{A}^\top$, the second of the above relations actually follows from the first. Orthogonality of the basis implies completeness.

The transposed matrix has the same determinant as the original matrix and the determinant of a product of matrices is the product of the determinants. Calculating the determinant of (2.46) we obtain $(\det \mathbf{A})^2 = 1$, or

$$\det \mathbf{A} = \pm 1. \tag{2.47}$$

The transformation matrices are thus divided into two completely separate classes, those with determinant $+1$, called *rotations* or sometimes *proper rotations*, and those with determinant -1, generically called reflections. Since the simple reflection (2.37) has determinant -1, all reflections may be composed of a simple reflection followed by a rotation.

2.8 Scalars, vectors and tensors

Geometric quantities may be classified according to their behaviour under pure rotations. When you rotate the coordinate system the world stays the same; it is only the way you describe it that changes. Some geometrical quantities, for example the distance between two points, are unaffected by a rotation; others, like the coordinates of your current position, will change.

Scalar quantities

A single quantity S is called a *scalar*, if it is invariant under rotation

$$\boxed{S' = S.} \tag{2.48}$$

Thus the distance, the norm and the dot product are scalars. In physics the natural constants, material constants, as well as mass and charge are scalars.

Vector quantities

Any triplet of quantities, $\boldsymbol{V}$, is called a *vector*, if it transforms under rotation according to

$$\boxed{V_i' = \sum_j a_{ij} V_j,} \tag{2.49}$$

or equivalently in matrix form,

$$\boldsymbol{V}' = \mathbf{A} \cdot \boldsymbol{V}. \tag{2.50}$$

In physics, velocity, acceleration, momentum, force, and many other quantities are vectors in this sense. The coordinates $\boldsymbol{x}$ of a point may also be called a vector according to this definition, but this is only correct in Cartesian coordinate systems, and would be very wrong in curvilinear coordinates or curved spaces.

The above definition of a vector demonstrates that triplets must have special transformation properties to qualify as vectors. A triplet containing your weight, your height, and your age, is not a vector but a collection of three scalars.

Tensor quantities

Using the vector transformation (2.49) the tensor product of two vectors $\boldsymbol{V}\,\boldsymbol{W}$ is found to transform according to the rule,

$$(\boldsymbol{V}'\boldsymbol{W}')_{ij} = V_i' W_j' = \left(\sum_k a_{ik} V_k\right)\left(\sum_l a_{jl} W_l\right) = \sum_{kl} a_{ik} a_{jl} V_k W_l.$$

In the last step we have reordered the sums into a convenient form.

More generally, any set of nine quantities arranged in a matrix

$$\mathbf{T} = \{T_{ij}\} = \begin{pmatrix} T_{11} & T_{12} & T_{13} \\ T_{21} & T_{22} & T_{23} \\ T_{31} & T_{32} & T_{33} \end{pmatrix} \tag{2.51}$$

is called a *tensor of rank 2*, provided it obeys the transformation law,

$$\boxed{T_{ij}' = \sum_{kl} a_{ik} a_{jl} T_{kl},} \tag{2.52}$$

which in matrix form may be written,

$$\mathbf{T}' = \mathbf{A} \cdot \mathbf{T} \cdot \mathbf{A}^\top. \tag{2.53}$$

In physics, the moment of inertia of an extended body and the quadrupole moment of a charge distribution are well-known tensors of second rank.

Tensors of higher rank may be constructed in a similar way. A tensor of rank r has r indices and is a collection of 3^r quantities that transform as the direct product of r vectors. We have so far only met one third rank tensor, the Levi-Civita symbol (2.33) (see problem 2.24). When nothing else is said, a tensor is always assumed to be of rank 2.

2.9 Scalar, vector and tensor fields

In continuum physics the basic quantities are functions of the spatial coordinates, called *fields*, which may also be classified according to their behaviour under rotation. The transformation laws are quite similar to the ones above, the only difference being that the coordinates of the spatial position must also transform.

For scalar, vector and tensor fields we thus have (with $\boldsymbol{x}' = \mathbf{A} \cdot \boldsymbol{x}$),

$$S'(\boldsymbol{x}') = S(\boldsymbol{x}), \tag{2.54a}$$

$$\boldsymbol{V}'(\boldsymbol{x}') = \mathbf{A} \cdot \boldsymbol{V}(\boldsymbol{x}), \tag{2.54b}$$

$$\mathbf{T}'(\boldsymbol{x}') = \mathbf{A} \cdot \mathbf{T}(\boldsymbol{x}) \cdot \mathbf{A}^\top. \tag{2.54c}$$

These definitions express that the new fields in the new position are obtained from the old fields in the old position by transforming them according to their type. In the following chapters we shall meet a number of such fields, such as the scalar mass density field, the vector velocity field and the tensor stress field.

Gradient, divergence and curl

In Cartesian coordinates a special symbol is introduced for the triplet of spatial derivatives, called the *gradient* operator or *nabla*,

$$\nabla = \frac{\partial}{\partial \boldsymbol{x}} = \left(\frac{\partial}{\partial x_1}, \frac{\partial}{\partial x_2}, \frac{\partial}{\partial x_3}\right). \tag{2.55}$$

Apart from being a differential operator, ∇ acts in all respects as a vector (see problem 2.16).

When this operator acts on a scalar field $S(\boldsymbol{x})$ it creates a vector field, called the *gradient* of S,

$$\nabla S = (\nabla_1 S, \nabla_2 S, \nabla_3 S) = \left(\frac{\partial S}{\partial x_1}, \frac{\partial S}{\partial x_2}, \frac{\partial S}{\partial x_3}\right), \tag{2.56}$$

where we for clarity have suppressed the explicit dependence on the spatial coordinates $\boldsymbol{x}$. Similarly, by dotting $\boldsymbol{\nabla}$ with a vector field $\boldsymbol{V}(\boldsymbol{x})$ we obtain a scalar field, called the *divergence* of $\boldsymbol{V}$,

$$\boldsymbol{\nabla}\cdot\boldsymbol{V} = \nabla_1 V_1 + \nabla_2 V_2 + \nabla_3 V_3 = \frac{\partial V_1}{\partial x_1} + \frac{\partial V_2}{\partial x_2} + \frac{\partial V_3}{\partial x_3}. \tag{2.57}$$

Finally, if we use the gradient operator as the left-hand component in the cross product (2.9) with a vector field $\boldsymbol{V}(\boldsymbol{x})$ we obtain another vector field, called the *curl* of $\boldsymbol{V}$,

$$\boldsymbol{\nabla}\times\boldsymbol{V} = (\nabla_2 V_3 - \nabla_3 V_2, \nabla_3 V_1 - \nabla_1 V_3, \nabla_1 V_2 - \nabla_2 V_1). \tag{2.58}$$

Various combinations of these three operations obey important identities (see problem 2.15)[5].

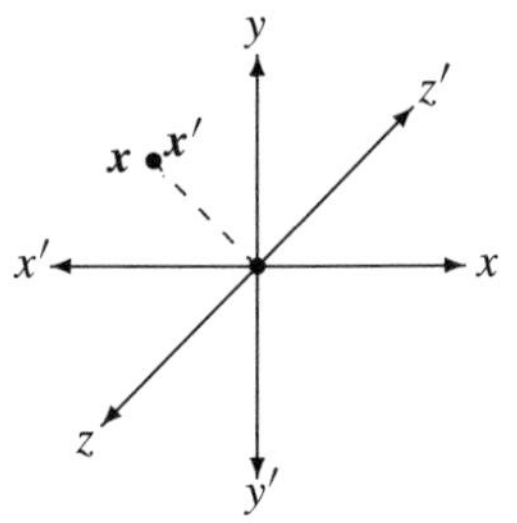

A complete reflection of the coordinate system in the origin. A rotation through π around the x-axis converts this to a simple reflection in the yz-plane.

2.10 Pseudo- and improper quantities *

Geometric quantities may be further subclassified according to their behaviour under reflection and translation. Quantities that transform non-trivially under reflection are called *pseudo-quantities* whereas quantities that transform non-trivially under translation are said to be *improper*.

Classification under reflections

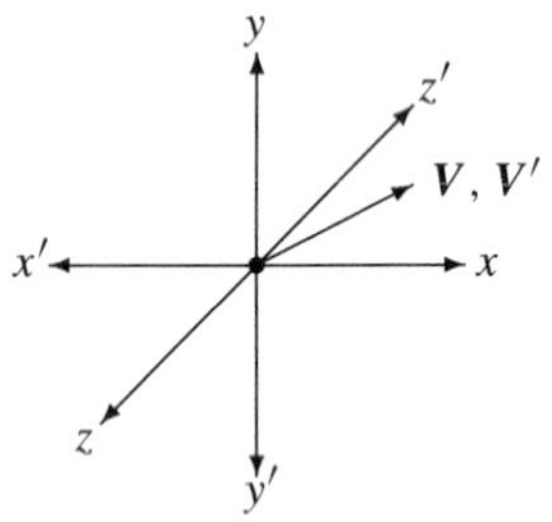

A polar vector retains its geometric placement under a reflection of the coordinate system in the origin.

Instead of a simple reflection in the yz-plane, we shall use a complete reflection of the coordinate system through its origin,

$$\boldsymbol{x}' = -\boldsymbol{x}. \tag{2.59}$$

Geometrically, the reflection in the origin may be viewed as a composite of three simple reflections along the three coordinate axes, or as a simple reflection of a coordinate axis followed by a simple rotation through π around the same axis.

Polar vectors: A vector which obeys the usual transformation equation (2.49) under rotation as well as under reflection is called a *polar* vector. Under a reflection in the origin, the coordinates of a polar vector change sign just like the coordinates of a point, i.e. $\boldsymbol{V}' = -\boldsymbol{V}$. Since the coordinate axes all reverse direction, the geometrical position in space of a polar vector is unchanged by a reflection of the coordinate system and the vector may faithfully be represented by an arrow, also under reflection. In physics, acceleration, force, velocity and momentum are all polar vectors.

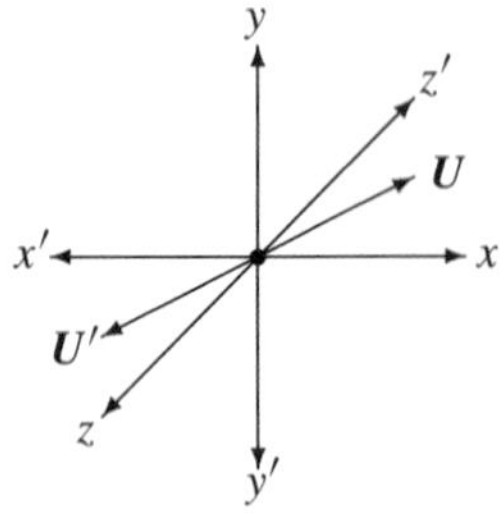

Geometrically, an axial vector has its geometric direction reversed under a reflection of the coordinate system in the origin because it has the same coordinates in the reflected system as in the original.

Axial vectors: There is, however, another possibility. The cross product of two polar vectors, $\boldsymbol{U} = \boldsymbol{V}\times\boldsymbol{W}$, behaves differently than a polar vector under a reflection. According to our rules for calculating the cross product, which are the same in all coordinate systems, we find

$$\boldsymbol{U}' = \boldsymbol{V}'\times\boldsymbol{W}' = (-\boldsymbol{V})\times(-\boldsymbol{W}) = \boldsymbol{V}\times\boldsymbol{W} = \boldsymbol{U}, \tag{2.60}$$

without the expected change of sign. Since $\boldsymbol{U}$ behaves normally under rotation with determinant $+1$, we conclude that the missing minus sign is associated with any transformation with determinant -1, in other words with any reflection. Generalizing, we define an *axial* vector $\boldsymbol{U}$ as a set of three quantities, transforming according to the rule

$$\boxed{U'_i = \det\mathbf{A}\sum_j a_{ij}U_j,} \tag{2.61}$$

under a Cartesian coordinate transformation. The extra determinant eliminates the overall sign change otherwise associated with reflections in the origin. In physics, angular momentum, moment of force and magnetic dipole moments are all axial vectors.

The basis vectors of the old coordinate system have the coordinates, $\boldsymbol{e}'_i = -\boldsymbol{e}_i$, in the new (reflected) coordinate system. Basis vectors are proper geometric quantities and always transform as polar vectors

[5]In the older literature the gradient, divergence and curl are often denoted by $\boldsymbol{\nabla}S = \mathsf{grad}\,S$, $\boldsymbol{\nabla}\cdot\boldsymbol{V} = \mathsf{div}\,\boldsymbol{V}$ and $\boldsymbol{\nabla}\times\boldsymbol{V} = \mathsf{curl}\,\boldsymbol{V}$ or $\boldsymbol{\nabla}\times\boldsymbol{V} = \mathsf{rot}\boldsymbol{V}$. This notation is now all but obsolete.

under reflection. One may easily get confused by the fact that the new basis vectors in the new system have (by definition) the same coordinates $\boldsymbol{e}_i$ as the old basis vectors in the old system, but it would be a mistake to take this to mean that the basis vectors are axial.

The geometric direction of an axial vector depends on what we choose to be right and left. It is for this reason wrong to think of an axial vector as an arrow in space. Geometrically, it has magnitude and direction, but not *sense*, meaning that the positive direction of an axial vector is not a geometric property, but a property fixed by convention which changes under a reflection of the coordinate system. For consistency all humans (even the British) have agreed that one particular coordinate system and all coordinate systems that are obtained from it by proper rotation are right-handed, whereas coordinate systems that are related to this class by reflection are left-handed. We do not know whether non-human aliens would have adopted the same convention, but should we ever meet such beings we would be able to find the correct transformation between our reference frames and theirs.

Pseudo-scalars: The volume product of three polar vectors $P = \boldsymbol{a} \cdot \boldsymbol{b} \times \boldsymbol{c}$ is a scalar quantity which changes sign under a reflection of the coordinate system because the cross product is an axial vector which does not change sign. More generally a *pseudo-scalar* transforms like

$$\boxed{P' = \det \mathbf{A}\ P,} \tag{2.62}$$

under an arbitrary rotation or reflection.

The sign of a pseudo-scalar is not absolute, but depends on the handedness of the coordinate system, and thus on convention. One might think that physics had no use for such quantities, because after all physics itself does not depend on coordinate systems, only its mathematical description does. Nevertheless, magnetic charge, if it is ever found, would be pseudo-scalar, and more importantly some of the familiar elementary particles, for example the pi-mesons, are described by pseudo-scalar fields.

Pseudo-tensors: Axial vectors are also called pseudo-vectors, and one may similarly define pseudo-tensors of higher rank. The Levi-Civita symbol is a pseudo-tensor of third rank (problem 2.24).

Classification under translation

A true vector may always be viewed as the difference between two positions $\boldsymbol{v} = \boldsymbol{b} - \boldsymbol{a}$, and is thus invariant under a pure translation $\boldsymbol{x} \to \boldsymbol{x}' = \boldsymbol{x} - \boldsymbol{c}$. Such vectors are called *proper*. Triplets that transform as vectors under rotations but change under translations, like the position $\boldsymbol{x}$ itself, are called *improper*. In physics electric dipole moments are improper polar vectors, whereas angular momentum, moment of force, and magnetic dipole moments are improper axial vectors.

Problems

2.1 Any distance function must satisfy the axioms

$$d(\boldsymbol{a}, \boldsymbol{a}) = 0, \tag{2.63a}$$

$$d(\boldsymbol{a}, \boldsymbol{b}) = d(\boldsymbol{b}, \boldsymbol{a}), \qquad \text{(symmetry)} \tag{2.63b}$$

$$d(\boldsymbol{a}, \boldsymbol{b}) \leq d(\boldsymbol{a}, \boldsymbol{c}) + d(\boldsymbol{c}, \boldsymbol{b}). \qquad \text{(triangle inequality)} \tag{2.63c}$$

Show that a distance function defined by step counting satisfies these axioms.

2.2 Let $\boldsymbol{a} = (2, 3, -6)$ and $\boldsymbol{b} = (3, -4, 0)$. Calculate

(a) the lengths of the vectors,

(b) the dot product,

(c) the cross product,

(d) and the tensor product.

2.3 Are the vectors $\boldsymbol{a} = (3, 1, -2)$, $\boldsymbol{b} = (4, -1, -1)$ and $\boldsymbol{c} = (1, -2, 1)$ linearly dependent (meaning that there exists a non-trivial set of coefficients such that $\alpha\boldsymbol{a} + \beta\boldsymbol{b} + \gamma\boldsymbol{c} = \boldsymbol{0}$)?

2.4 Calculate the distance between two points on Earth in terms of longitude α, latitude δ and height h over the average sea level.

2.5 Show that

$$|\boldsymbol{a} \cdot \boldsymbol{b}| \leq |\boldsymbol{a}|\,|\boldsymbol{b}|\,, \tag{2.64a}$$

$$|\boldsymbol{a} + \boldsymbol{b}| \leq |\boldsymbol{a}| + |\boldsymbol{b}|\,. \tag{2.64b}$$

2.6 Show that

$$(\boldsymbol{a} \times \boldsymbol{b} \cdot \boldsymbol{c})\,(\boldsymbol{d} \times \boldsymbol{e} \cdot \boldsymbol{f}) = \begin{vmatrix} \boldsymbol{a}\cdot\boldsymbol{d} & \boldsymbol{a}\cdot\boldsymbol{e} & \boldsymbol{a}\cdot\boldsymbol{f} \\ \boldsymbol{b}\cdot\boldsymbol{d} & \boldsymbol{b}\cdot\boldsymbol{e} & \boldsymbol{b}\cdot\boldsymbol{f} \\ \boldsymbol{c}\cdot\boldsymbol{d} & \boldsymbol{c}\cdot\boldsymbol{e} & \boldsymbol{c}\cdot\boldsymbol{f} \end{vmatrix}. \tag{2.65}$$

2.7 Show that

$$\boldsymbol{a} \times \boldsymbol{b} \cdot \boldsymbol{c} = \boldsymbol{b} \times \boldsymbol{c} \cdot \boldsymbol{a} = \boldsymbol{c} \times \boldsymbol{a} \cdot \boldsymbol{b}, \tag{2.66}$$

$$(\boldsymbol{a} \times \boldsymbol{b}) \times \boldsymbol{c} = (\boldsymbol{a} \cdot \boldsymbol{c})\,\boldsymbol{b} - (\boldsymbol{b} \cdot \boldsymbol{c})\,\boldsymbol{a}, \tag{2.67}$$

$$(\boldsymbol{a} \times \boldsymbol{b}) \cdot (\boldsymbol{c} \times \boldsymbol{d}) = (\boldsymbol{a} \cdot \boldsymbol{c})(\boldsymbol{b} \cdot \boldsymbol{d}) - (\boldsymbol{a} \cdot \boldsymbol{d})(\boldsymbol{b} \cdot \boldsymbol{c}), \tag{2.68}$$

$$|\boldsymbol{a} \times \boldsymbol{b}|^2 = |\boldsymbol{a}|^2\,|\boldsymbol{b}|^2 - (\boldsymbol{a} \cdot \boldsymbol{b})^2. \tag{2.69}$$

2.8 Show that (with the normal definition of the matrix product) the following relations make sense for the tensor product

$$(\boldsymbol{ab}) \cdot \boldsymbol{c} = \boldsymbol{a}(\boldsymbol{b} \cdot \boldsymbol{c}) \tag{2.70}$$

$$\boldsymbol{a} \cdot (\boldsymbol{bc}) = (\boldsymbol{a} \cdot \boldsymbol{b})\boldsymbol{c} \tag{2.71}$$

This is sometimes quite useful.

2.9 Show that

$$\frac{\partial(\boldsymbol{a} \cdot \boldsymbol{b})}{\partial \boldsymbol{a}} = \boldsymbol{b} \tag{2.72}$$

and that

$$\frac{\partial\,|\boldsymbol{a}|}{\partial \boldsymbol{a}} = \frac{\boldsymbol{a}}{|\boldsymbol{a}|}. \tag{2.73}$$

2.10 Show that

$$(\boldsymbol{a} \cdot \boldsymbol{b} \times \boldsymbol{c})\boldsymbol{d} = (\boldsymbol{a} \cdot \boldsymbol{d})\boldsymbol{b} \times \boldsymbol{c} + (\boldsymbol{b} \cdot \boldsymbol{d})\boldsymbol{c} \times \boldsymbol{a} + (\boldsymbol{c} \cdot \boldsymbol{d})\boldsymbol{a} \times \boldsymbol{b} \tag{2.74}$$

for arbitrary vectors $\boldsymbol{a}$, $\boldsymbol{b}$, $\boldsymbol{c}$, and $\boldsymbol{d}$.

2.11 Show that

$$\sum_i \delta_{ii} = 3 \tag{2.75}$$

$$\sum_j \delta_{ij}\delta_{jk} = \delta_{ik} \tag{2.76}$$

2.12 Show that

$$\sum_i \nabla_i x_i = 3, \tag{2.77}$$

$$\nabla_i x_j = \delta_{ij}, \tag{2.78}$$

$$\nabla_i \nabla_j (x_k x_l) = \delta_{ik}\delta_{jl} + \delta_{il}\delta_{jk}, \tag{2.79}$$

2.13 Show that the Levi-Civita symbol is completely antisymmetric in all three indices,

$$\epsilon_{ijk} = -\epsilon_{ikj} = -\epsilon_{jik} = -\epsilon_{kji}. \tag{2.80}$$

2.14 Show that the product of two Levi-Civita symbols is (see problem 2.6)

$$\begin{aligned}\epsilon_{ijk}\,\epsilon_{lmn} &= \begin{vmatrix}\delta_{il} & \delta_{im} & \delta_{in}\\ \delta_{jl} & \delta_{jm} & \delta_{jn}\\ \delta_{kl} & \delta_{km} & \delta_{kn}\end{vmatrix}\\ &= \delta_{il}\delta_{jm}\delta_{kn} + \delta_{im}\delta_{jn}\delta_{kl} + \delta_{in}\delta_{jl}\delta_{km}\\ &\quad - \delta_{in}\delta_{jm}\delta_{kl} - \delta_{il}\delta_{jn}\delta_{km} - \delta_{im}\delta_{jl}\delta_{kn}\end{aligned} \tag{2.81}$$

and derive from this

$$\sum_k \epsilon_{ijk}\,\epsilon_{lmk} = \begin{vmatrix}\delta_{il} & \delta_{im}\\ \delta_{jl} & \delta_{jm}\end{vmatrix} = \delta_{il}\delta_{jm} - \delta_{im}\delta_{jl}, \tag{2.82}$$

$$\sum_{jk} \epsilon_{ijk}\,\epsilon_{ljk} = 2\delta_{il}, \tag{2.83}$$

$$\sum_{ijk} \epsilon_{ijk}\,\epsilon_{ijk} = 6. \tag{2.84}$$

2.15 Prove the following relations involving the nabla operator (here Φ is a scalar field and $\boldsymbol{v}$ a vector field),

$$\nabla \cdot (\nabla \times \boldsymbol{v}) = \mathbf{0}, \tag{2.85}$$

$$\nabla \times (\nabla \Phi) = 0, \tag{2.86}$$

$$\nabla \times (\nabla \times \boldsymbol{v}) = \nabla(\nabla \cdot \boldsymbol{v}) - (\nabla \cdot \nabla)\boldsymbol{v}. \tag{2.87}$$

Where in these relations does it make sense to remove the parentheses?

2.16 Show that the nabla operator (2.55) transforms as a vector, $\nabla'_i = \sum_j a_{ij} \nabla_j$ under an arbitrary rotation.

2.17 Show that the trace $\sum_i T_{ii}$ of a tensor is invariant under a rotation.

2.18 Show that the Kronecker delta transforms as a tensor.

2.19 Show that the distance $|\boldsymbol{x} - \boldsymbol{y}|$ is invariant under any transformation between Cartesian coordinate systems.

* **2.20** Show that if $W_i = \sum_j T_{ij} V_j$ and if it is known that $\boldsymbol{W}$ is a vector for all vectors $\boldsymbol{V}$, then T_{ij} must be a tensor. This is called the *quotient rule*.

* **2.21** Consider two Cartesian coordinate systems and make no assumptions about the transformation $\boldsymbol{x}' = \boldsymbol{f}(\boldsymbol{x})$ between them. Show that the invariance of the distance,

$$\left|\boldsymbol{x}' - \boldsymbol{y}'\right| = |\boldsymbol{x} - \boldsymbol{y}|, \tag{2.88}$$

implies that the transformation is of the form $\boldsymbol{x}' = \mathbf{A} \cdot \boldsymbol{x} + \boldsymbol{b}$ where $\mathbf{A}$ is an orthogonal matrix.

$*$ **2.22** Show that the general Cartesian coordinate transformation may be built up from a combination of simple translations, rotations and reflections.

$*$ **2.23** Show that under a simple rotation, a tensor T_{ij} transforms into

$$T'_{xx} = \cos\phi(T_{xx}\cos\phi + T_{xy}\sin\phi) + \sin\phi(T_{yx}\cos\phi + T_{yy}\sin\phi), \tag{2.89a}$$

$$T'_{xy} = \cos\phi(-T_{xx}\sin\phi + T_{xy}\cos\phi) + \sin\phi(-T_{yx}\sin\phi + T_{yy}\cos\phi), \tag{2.89b}$$

$$T'_{xz} = \cos\phi T_{xz} + \sin\phi T_{yz}, \tag{2.89c}$$

$$T'_{yx} = -\sin\phi(T_{xx}\cos\phi + T_{xy}\sin\phi) + \cos\phi(T_{yx}\cos\phi + T_{yy}\sin\phi), \tag{2.89d}$$

$$T'_{yy} = -\sin\phi(-T_{xx}\sin\phi + T_{xy}\cos\phi) + \cos\phi(-T_{yx}\sin\phi + T_{yy}\cos\phi), \tag{2.89e}$$

$$T'_{yz} = -\sin\phi T_{xz} + \cos\phi T_{yz}, \tag{2.89f}$$

$$T'_{zx} = T_{zx}\cos\phi + T_{zy}\sin\phi, \tag{2.89g}$$

$$T'_{zy} = -T_{zx}\sin\phi + T_{zy}\cos\phi, \tag{2.89h}$$

$$T'_{zz} = T_{zz}. \tag{2.89i}$$

$*$ **2.24** Show that

(a) the Levi-Civita symbol satisfies

$$\sum_{lmn} a_{il}a_{jm}a_{kn}\epsilon_{lmn} = \det\mathbf{A}\,\epsilon_{ijk} \tag{2.90}$$

where **A** is an arbitrary matrix.

(b) the Levi-Civita symbol (which by the definition of the cross product must be invariant, $\epsilon'_{ijk} = \epsilon_{ijk}$) obeys the rule

$$\epsilon'_{ijk} = \epsilon_{ijk} = \det\mathbf{A}\sum_{lmn} a_{il}a_{jm}a_{kn}\epsilon_{lmn} \tag{2.91}$$

for an arbitrary coordinate transformation (which has $\det\mathbf{A} = \pm 1$).

(c) the cross product of two vectors $\boldsymbol{W} = \boldsymbol{U} \times \boldsymbol{V}$ must transform like

$$W'_i = \det\mathbf{A}\sum_j a_{ij}W_j \tag{2.92}$$

3 Gravity

The force of gravity determines to a large extent the way we live. It is certainly the force about which we have the best intuitive understanding. We learn the hard way to rise against it as small children, to keep it at bay as adults, only to be brought down by it in the end. A few people have experienced true absence of gravity for longer periods of time in satellites orbiting the Earth or rockets coasting towards the Moon.

Newton gave us the theory of gravity and the mathematics to deal with it. In a world where things only seem to get done by push and pull, man suddenly had to accept that the Earth could act on the distant Moon—and the Moon back on Earth. After Newton everybody had to suppress a feeling of horror for action at a distance and accept that gravity instantaneously could jump across the emptiness of space and tug at distant bodies. It took more than two centuries and the genius of Einstein to undo this learning. There is *no* action at a distance. As we understand it today, gravity is mediated by a field which emerges from massive bodies and in the manner of light takes time to travel through a distance. If the Sun were suddenly to blink out of existence, it would take eight long minutes before daylight was switched off and the Earth set free in space.

In this chapter we shall study the interplay between mass and the instantaneous Newtonian field of gravity, and derive the equations governing this field and its interactions with matter. Some basic knowledge of gravity is assumed in advance, and the presentation in this chapter aims mainly to develop elementary aspects of field theory in the comfortable environment of Newtonian gravity. More advanced concepts will be developed in chapter 6.

3.1 Mass density

In the continuum approximation the mass density is a field, $\rho(\boldsymbol{x}, t)$, assumed to exist everywhere in space and at all times. If there is no mass in a region, the mass density simply vanishes. Knowing this field, we may calculate the mass of a material particle occupying a small volume dV around the point $\boldsymbol{x}$ at time t,

$$\boxed{dM = \rho(\boldsymbol{x}, t)\, dV.} \tag{3.1}$$

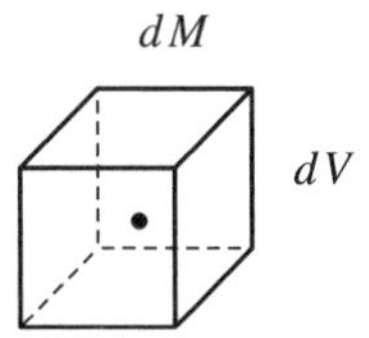

The volume dV occupied by a material particle may take any shape, here cubical. The nominal position of the particle $\boldsymbol{x}$ may be chosen anywhere within the volume.

We shall permit ourselves to suppress the space and time variables and just write $dM = \rho\, dV$, whenever such notation is unambiguous.

Although we usually think of the mass density field as varying smoothly throughout space, it is sometimes convenient to allow for discontinuous boundaries in material bodies (an example is shown in figure 3.1 on page 31). Often these discontinuities are 'real' in the sense that the transition between different materials may happen on the molecular scale, as for example at the interface between two solid bodies that touch each other.

In section 1.2 we discussed the continuum approximation and concluded that there are two length scales, L_{micro} and L_{macro}, that depend on the measurement precision. If $dV \gtrsim L^3_{\text{micro}}$, the

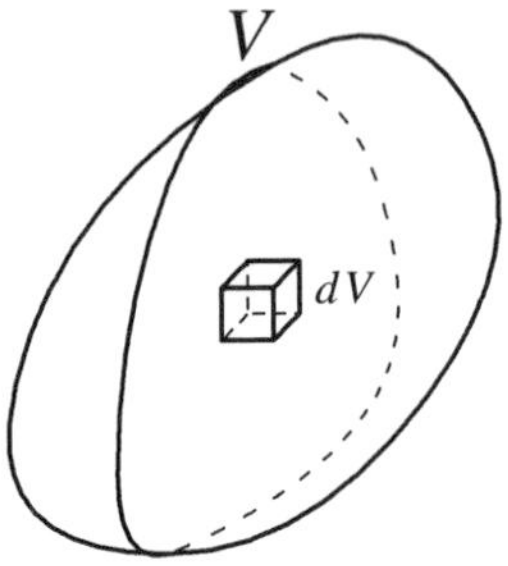

The total mass in a volume is obtained by integrating ('summing') over all material particles in the volume. The integral sign $\int$ is in fact a stylized version of the letter S (for 'sum').

molecular structure may be disregarded, and if the length scale for major density changes is larger than $L_{\rm macro}$, the density is effectively continuous and the shape of the minimal volume dV does not matter.

Total mass and centre of mass

Mass density is a *local* quantity, defined in every point of space. The total mass in a volume V is a *global* quantity obtained mathematically by integrating the mass density over V,

$$M = \int_V dM = \int_V \rho(\boldsymbol{x}, t)\, dV. \tag{3.2}$$

Physically the integral should be understood as an approximation to a huge sum over the tiny, though not truly infinitesimal, material particles contained in the volume. If the density depends on time or if the volume changes shape and size with time, the total mass may depend on time.

In continuum physics the material contained in *any* volume V may be viewed as a 'body', and the *centre of mass* of such a body is naturally defined from the average of the position $\boldsymbol{x}$ over all material particles,

$$\boldsymbol{x}_{\rm M} = \frac{1}{M}\int_V \boldsymbol{x}\, dM = \frac{1}{M}\int_V \boldsymbol{x}\, \rho(\boldsymbol{x}, t)\, dV. \tag{3.3}$$

In the Newtonian mechanics of particles and stiff bodies, the centre of mass of a body plays an important role, because it moves like a point particle under the influence of the total force acting on the body (see appendix A). Although this is also true in continuum mechanics, it is not nearly as useful because the shape of a body may change drastically over longer time-spans. Think for example of a bucket of oil thrown into a waterfall. It may not always be physically meaningful to speak about a well-defined 'body of oil' at later times.

Spherical systems

Planets like Earth or stars like the Sun are in the first approximation spherically symmetric. Spherical symmetry implies that the mass density ρ at any point $\boldsymbol{x}$ only depends on the distance $r = |\boldsymbol{x} - \boldsymbol{c}|$ to the centre $\boldsymbol{c}$, which for symmetry reasons must also be the centre of mass. Usually the origin of the coordinate system is chosen to be at the centre such that $\boldsymbol{c} = \boldsymbol{0}$. For a simple spherical planet with radius a and constant density ρ_0 we have

$$\rho(r) = \begin{cases} \rho_0 & r < a \\ 0 & r > a \end{cases}. \tag{3.4}$$

Such a distribution might be used for analytic calculations for a small rock planet like the Moon, but definitely not for the Earth.

In figure 3.1 the mass density of the Earth is plotted (fully drawn) as a function of the central distance. It cannot be measured directly, but is inferred from a combination of surface observations and modelling. For analytic calculations, its mass distribution may be approximated by two layers of constant density, $\rho_1 = 10.9$ g cm^{-3} in the core and $\rho_2 = 4.4$ g cm^{-3} in the mantle (see problem 3.7).

Galileo Galilei (1564–1642). *Italian natural philosopher, astronomer, mathematician, and craftsman. Carried out gravity experiments with falling objects and inclined planes. Built better telescopes than any before him. The first to see the mountains of the Moon, the large moons of Jupiter, and that the Milky Way is made from stars. Considered the father of the modern scientific method.*

3.2 Gravitational acceleration

Galileo found empirically that all bodies fall in the same way, independent of their mass. In Newton's language this shows that the force of gravity on a body is proportional to its mass. For a material particle of mass $dM = \rho dV$ the force of gravity may be written

$$\boxed{d\boldsymbol{\mathcal{F}} = \boldsymbol{g}(\boldsymbol{x}, t)\, dM = \rho \boldsymbol{g}\, dV,} \tag{3.5}$$

where $\boldsymbol{g}(\boldsymbol{x}, t)$ is called the *gravitational acceleration* field, or just *gravity*. The last form shows that gravity is a *body force* (also called a *volume force*) which acts everywhere in a body with a *density of force* $\boldsymbol{f} = d\boldsymbol{\mathcal{F}}/dV = \rho\boldsymbol{g}$. In figure 3.3 on page 35 the magnitude of Earth's gravity is plotted as a function of the distance from the centre of the Earth.

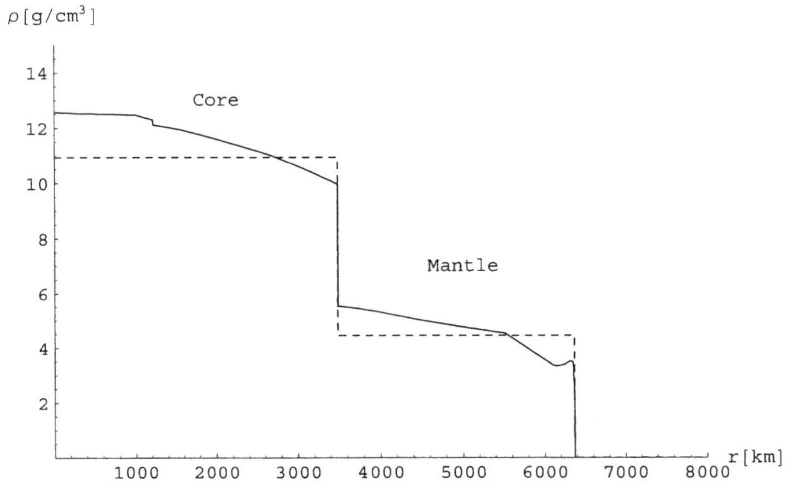

Figure 3.1. The mass density of the Earth as a function of distance r from the centre with the surface at $r = 6371$ km (from the standard Earth model [41]). There is a sharp break in the density at the transition between the iron core and the stone mantle at $r = 3485$ km, and a smaller break at $r = 1216$ km between the outer liquid iron core and the inner solid iron core. The broken lines indicate the average densities in mantle and core.

In Newtonian physics the gravitational field imparts a common acceleration to all bodies. Given the same initial conditions all material bodies will follow the same orbits in a gravitational field. As a consequence there is no way we can distinguish between gravitational forces and the inertial (so-called fictitious) forces experienced in accelerated motion. The identical behaviour of all bodies in a field of gravity allows one to look upon the gravitational field as a property of space and time, rather than simply a vehicle for gravitational interaction. The indistinguishability of gravity and acceleration was raised to a fundamental law, the *Principle of Equivalence*, by Einstein in his *General Theory of Relativity* from 1915, in which gravity reflects the geometric curvature of space and time [78].

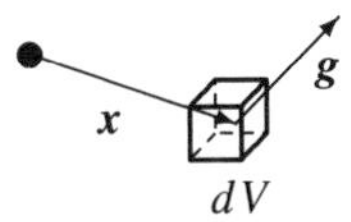

The weight of the matter in a volume element dV is $d\mathcal{F} = \rho \boldsymbol{g}\, dV$.

Total gravitational force and total moment of gravity

The field of gravity specifies the gravitational acceleration at every point of space and every instant of time. The total gravitational force on a body of volume V, the *weight* of the body, is

$$\mathcal{F} = \int_V \rho \boldsymbol{g}\, dV. \tag{3.6}$$

The total force determines how the body as a whole moves. It is independent of the choice of origin of the coordinate system, but depends like any other vector on its orientation. If ρ, $\boldsymbol{g}$, or V change in the course of time, the total force may also change.

The total *moment of force* of gravity relative to the coordinate origin is

$$\mathcal{M} = \int_V \boldsymbol{x} \times \rho \boldsymbol{g}\, dV. \tag{3.7}$$

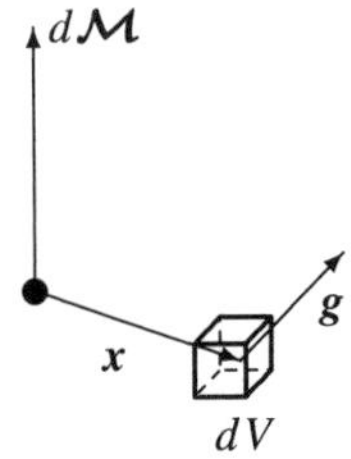

The moment of gravity for a volume element dV is $d\mathcal{M} = \boldsymbol{x} \times \rho \boldsymbol{g}\, dV$.

The total moment determines how a body as a whole rotates around the origin. The moment depends not only on the orientation of the coordinate system, but also on its origin. It is an improper axial vector. In a translated coordinate system with origin in $\boldsymbol{x} = \boldsymbol{c}$, the coordinates are $\boldsymbol{x}' = \boldsymbol{x} - \boldsymbol{c}$, and the moment becomes

$$\mathcal{M}' = \int_{V'} \boldsymbol{x}' \times \rho' \boldsymbol{g}'\, dV' = \int_V (\boldsymbol{x} - \boldsymbol{c}) \times \rho \boldsymbol{g}\, dV = \mathcal{M} - \boldsymbol{c} \times \mathcal{F}, \tag{3.8}$$

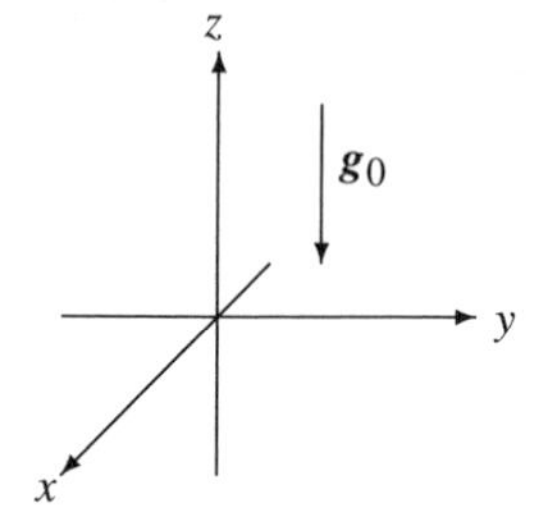

A flat-earth coordinate system.

where the primed quantities all refer to the translated system. This shows that *the total moment is independent of the choice of origin of the coordinate system if (and only if) the total force vanishes.*

Constant gravity

In a constant gravitational field $\boldsymbol{g}(\boldsymbol{x}) = \boldsymbol{g}_0$, the weight (3.6) becomes the familiar

$$\mathcal{F} = M\boldsymbol{g}_0, \tag{3.9}$$

where M is the total mass (3.2). In constant gravity, it is customary to choose a 'flat-earth' coordinate system with vertical z-axis, such that $\boldsymbol{g}_0 = (0, 0, -g_0)$ where g_0 is a positive constant.

At the surface of Earth, gravity is very close to being constant with magnitude equal to the standard gravity, defined by convention to be exactly $g_0 = 9.80665$ m s^{-2} with no uncertainty. The actual gravitational acceleration at the surface of the Earth depends on many factors, for example local mass concentrations and the positions of the Moon and Sun (see figure 7.1 on page 90). The gravitational acceleration has been determined with a relative precision of 3×10^{-9} in an experiment using atom interferometry [56]. Galileo's law was verified in the same experiment to within 7×10^{-9} by comparing the measured values of the gravitational acceleration for a macroscopic body and for a cesium atom, in effect a modern version of his famous 'leaning tower of Pisa' experiment.

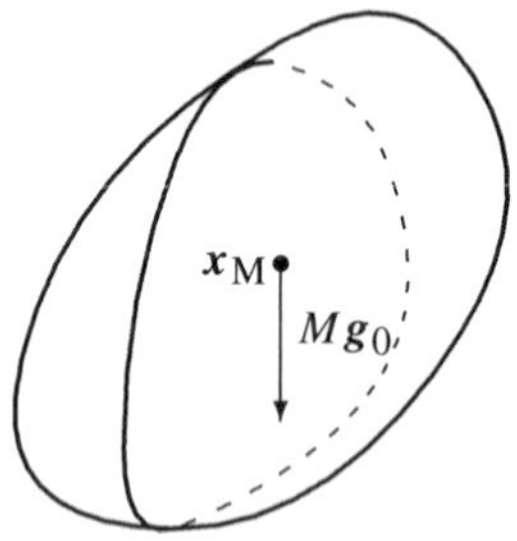

In a constant gravitational field $\boldsymbol{g}_0$, the weight of a body may be viewed as concentrated at the position of the centre of mass, $\boldsymbol{x}_M$.

The moment of force may be expressed in terms of the centre of mass (3.3),

$$\mathcal{M} = \int_V \boldsymbol{x} \times \rho \boldsymbol{g}_0 \, dV = \left(\int_V \rho \boldsymbol{x} \, dV \right) \times \boldsymbol{g}_0 = \boldsymbol{x}_M \times M\boldsymbol{g}_0. \tag{3.10}$$

This shows that in constant gravity, the total moment is the same as that of a point particle with mass equal to the total mass of the body, situated at the centre of mass. In a constant gravitational field, the moment of gravity calculated in a coordinate system with origin at the centre of mass must vanish because $\boldsymbol{x}_M = \boldsymbol{0}$ in these coordinates. The moment of force in constant gravity is important for understanding the stability of floating bodies (chapter 5).

Visualizing the gravitational field

A visual impression of the gravitational field may be given by a picture of the *field lines*, defined to be families of curves that at a given instant t_0 have the gravitational field as tangent (see figure 3.2). This means that the curves are solutions to the first-order differential equation

$$\frac{d\boldsymbol{x}}{ds} = \boldsymbol{g}(\boldsymbol{x}, t_0), \tag{3.11}$$

where s is a running parameter along the curve. This parameter is not the time, but has dimension of time squared because $\boldsymbol{g}$ has dimension of length per unit of time squared. The solutions are of the form $\boldsymbol{x} = \boldsymbol{x}(s, \boldsymbol{x}_0, t_0)$ with $\boldsymbol{x}_0$ being the starting point at $s = 0$. The field lines form an instantaneous picture of the field at time t_0, and cannot be directly related to particle orbits, as illustrated by the nearly circular orbit of a planet which is everywhere orthogonal to the field lines.

Field lines have the very important property that they can never cross. For if two field lines crossed in a point $\boldsymbol{x}$, then by (3.11) there would have to be two different values of the gravitational field at the same point, and that is impossible (except when the field vanishes, as it does in one point of figure 3.2). As will be shown in the following section, all gravitational field lines have to come in from infinity and end on masses, and we shall also see that field lines do not form closed loops.

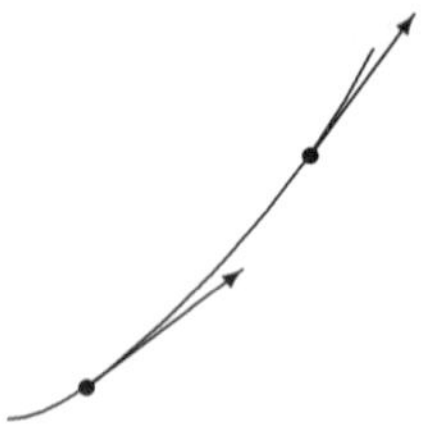

Field lines follow the instantaneous field everywhere. They are very different from the orbits particles would follow through the field. A thrown stone follows a parabolic orbit as it falls to the ground, whereas the field lines on the surface of the Earth are vertical.

3.3 Sources of gravity

The gravitational field tells us how gravity acts on material bodies. But what generates the gravitational field? What is its source? The answer is—as most people are aware—that the field is generated by mass.

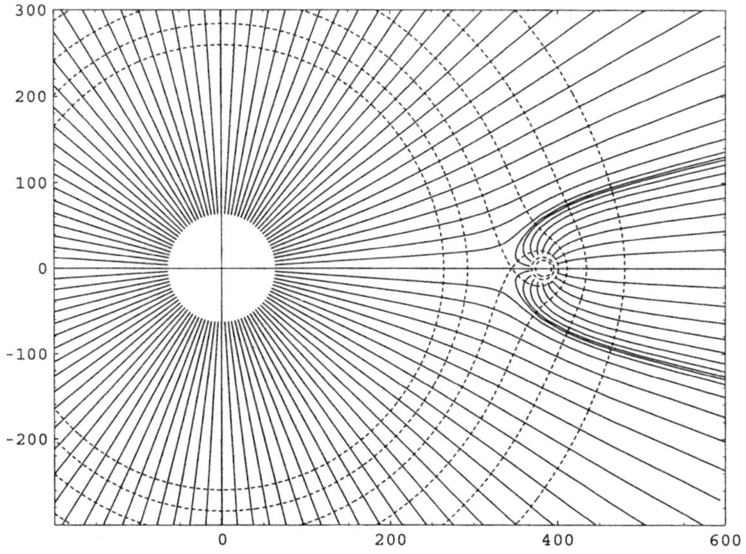

Figure 3.2. The gravitational field (and some nearly circular equipotential surfaces) between Earth and Moon. You should imagine rotating this picture around the Earth–Moon axis. The drawing is to scale, except for two regions of 10 times the sizes of the Earth and the Moon that have been cut out for technical reasons. The field lines are plotted everywhere with a density proportional to the field strength. The numbers on the frame are coordinates centred on Earth in units of 1000 km. The Moon appears to have a streaming 'mane of hair' because all the field lines ending on its surface have to come in from spatial infinity and cannot cross the lines of Earth's field.

Quantitatively this is expressed by *Newton's law of gravity* which says that the field from a point particle of mass M at the origin of the coordinate system is

$$g(x) = -GM\frac{x}{|x|^3}, \tag{3.12}$$

where G is the universal gravitational constant. The negative sign asserts that gravitation is always attractive, or equivalently that field lines always run towards masses. The last factor shows that the strength of gravity decreases with the inverse square of the distance.

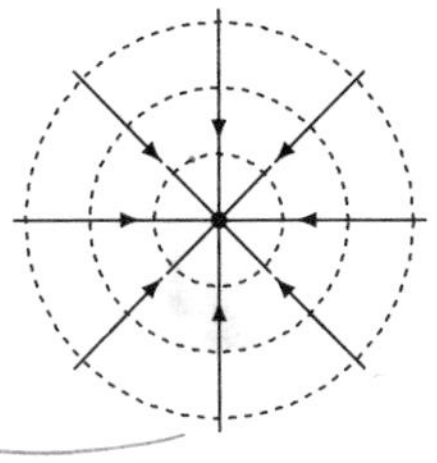

Field lines around a point particle all come in from infinity and converge upon the particle.

PDG 2004 $6.6742(10)\times 10^{-11}$

The gravitational constant is hard to determine to high precision. The recommended [52] 1998 value, $G = 6.673(10) \times 10^{-11}$ N m^2 kg^{-2}, has an embarrassingly large uncertainty of more than one part in 10^3. The inverse square law has been well tested at planetary distances during the last centuries, but only recently at the submillimetre scale (see, for example Long *et al* , *Nature* **421**, (2003) 922). The force of gravity is terribly weak compared to other forces. In the hydrogen atom the ratio of the force of gravity to the electrostatic force (between electron and proton) is 4.4×10^{-40}. The only reason gravity can be observed at all is the nearly complete electric neutrality of macroscopic bodies. Electrostatic and 'gravistatic' forces seem superficially alike in that they are both inversely proportional to the square of the distance (which gives them infinite range; they are, in fact, the only fundamental forces in nature with infinite range). But where electric charge can be both positive and negative, mass is always positive, implying that there are no 'neutral' bodies unaffected by gravity, nor bodies that are repelled by the gravity of other bodies (antigravity). Gravity and electromagnetism are in fact very different at a deeper level, only completely revealed in General Relativity.

Another basic property of gravity is that it is *additive*, so that matter—also from the point of view of its gravity—may be viewed as a collection of material particles of mass $dM = \rho\, dV$, each contributing its own little point-like field to the total. Using (3.12) but shifting the position of the source particle from the

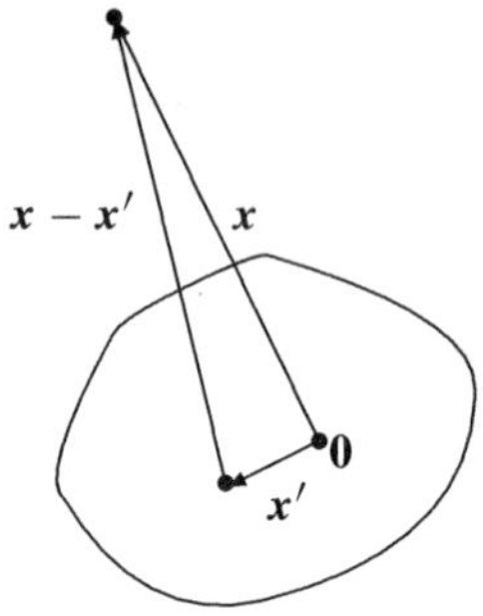

The vectors involved in calculating the field in the position $\boldsymbol{x}$.

origin to an arbitrary point $\boldsymbol{x}'$, the collective gravitational field due to all the material particles in a volume V becomes an integral,

$$\boldsymbol{g}(\boldsymbol{x}) = -G \int_V \frac{\boldsymbol{x} - \boldsymbol{x}'}{|\boldsymbol{x} - \boldsymbol{x}'|^3} \rho(\boldsymbol{x}')\, dV'. \tag{3.13}$$

Note that the integrand has a singularity at $\boldsymbol{x}' = \boldsymbol{x}$ (when $\rho(\boldsymbol{x}) \neq 0$). This singularity is, however, integrable and creates no problem (except problem 3.8).

Forces between extended bodies

The expression for the field (3.13) brings us full circle. We are now able to calculate the gravitational field from a mass distribution, as well as the force that this field exerts on another mass distribution, even on itself. Substituting the field (3.13) into the force (3.6), and renaming the integration variables, the total force by which a mass distribution ρ_2 in the volume V_2 acts on a mass distribution ρ_1 in V_1 becomes,

$$\mathcal{F}_{12} = -G \int_{V_1} \int_{V_2} \frac{\boldsymbol{x}_1 - \boldsymbol{x}_2}{|\boldsymbol{x}_1 - \boldsymbol{x}_2|^3} \rho_1(\boldsymbol{x}_1)\rho_2(\boldsymbol{x}_2)\, dV_1\, dV_2. \tag{3.14}$$

Newton's third law is fulfilled, $\mathcal{F}_{12} = -\mathcal{F}_{21}$, because the integrand is antisymmetric under exchange of $1 \leftrightarrow 2$ due to the first factor. Consequently, the force from a mass distribution acting on itself vanishes, as it ought to. For if this self-force did not vanish a body could, so to speak, lift itself by its bootstraps.

Asymptotic behaviour

The gravitational field from the matter contained in a volume of finite extent has a particularly simple form at large distances. Since $\boldsymbol{x}'$ then only ranges over a finite region in (3.13), it follows for $|\boldsymbol{x}| \to \infty$ that

$$\frac{\boldsymbol{x} - \boldsymbol{x}'}{|\boldsymbol{x} - \boldsymbol{x}'|^3} \to \frac{\boldsymbol{x}}{|\boldsymbol{x}|^3}.$$

Taking this expression outside the integral we obtain

$$\boldsymbol{g}(\boldsymbol{x}) \to -GM \frac{\boldsymbol{x}}{|\boldsymbol{x}|^3}, \tag{3.15}$$

where M is the total mass. At sufficiently large distances the field of an arbitrary spatially bounded mass distribution always approaches that of a point particle containing the total mass of the body.

Field of a spherical body

The mass distribution $\rho(r)$ of a spherically symmetric body is only a function of the distance $r = |\boldsymbol{x}|$ from its centre, which here is taken to be at the origin of the coordinate system. Since gravity according to (3.13) is caused by the mass distribution, it should also be spherically symmetric. For a vector field this means that it is everywhere directed radially away from the centre,

$$\boldsymbol{g}(\boldsymbol{x}) = g(r)\, \boldsymbol{e}_r, \tag{3.16}$$

where $\boldsymbol{e}_r = \boldsymbol{x}/r$ is the radial unit vector and $g(r)$ is a function of r alone. In figure 3.3 the value of $-g(r)$ is plotted for the Earth as a function of the radial distance. One notes the surprising fact that the strength of gravity is actually larger (by about 10%) inside Earth's mantle than on the surface.

In chapter 6 we shall see that spherical gravity takes an extremely simple form. We shall prove that the field strength $g(r)$ is everywhere—inside and outside the distribution—equal to the field of a point particle situated at the centre of the distribution,

$$g(r) = -G\frac{M(r)}{r^2}, \tag{3.17}$$

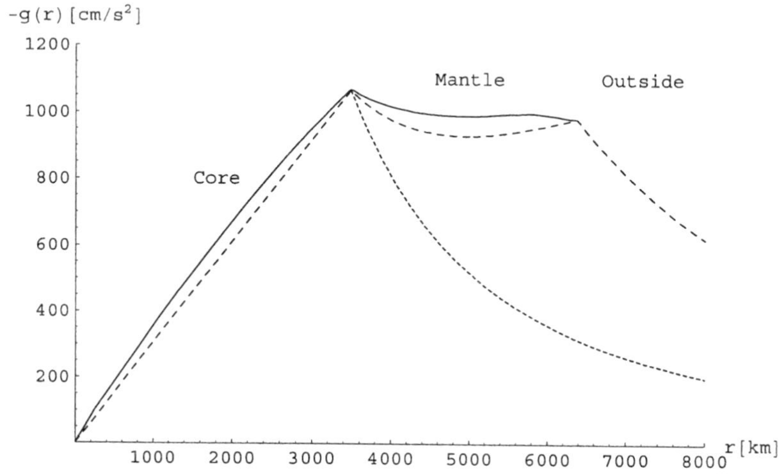

Figure 3.3. The strength of gravity $-g(r)$ inside and outside the Earth as a function of distance from the centre. The solid curve is obtained from the standard Earth data [41]. The strength of gravity grows roughly linearly from the centre to the core/mantle boundary at $r = 3485$ km, and decreases slightly in the mantle due to the sharp drop in mass density at the boundary. The dotted dropping line is the core field itself. The dashed lines are obtained from the two-layer model of the Earth (problem 3.7).

with a mass equaling the total mass inside the radius r,

$$M(r) = \int_0^r \rho(r')4\pi r'^2 \, dr'. \tag{3.18}$$

It is also possible to prove this by direct integration of (3.13) (see problem 3.11).

For a planet with constant density we find from (3.4) that

$$g(r) = -\frac{4}{3}\pi G\rho_0 \begin{cases} r & r < a \\ \dfrac{a^3}{r^2} & r > a. \end{cases} \tag{3.19}$$

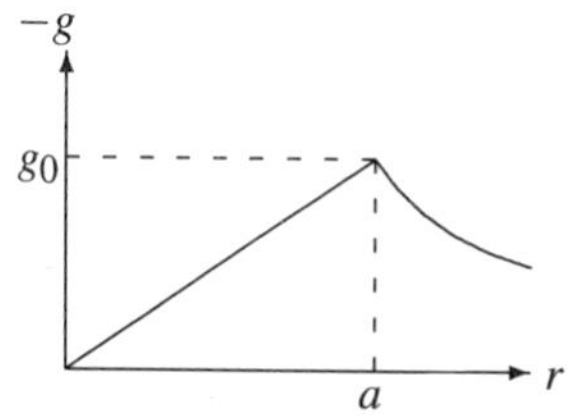

The strength of gravity for a planet with constant density.

The strength of gravity grows linearly with r inside the planet because the total mass grows with the third power of r whereas the field strength decreases with the second power.

It follows from (3.18) that for every point in the vacuum *outside* a spherical mass distribution where $M(r)$ equals the total mass, the field (3.17) is *exactly* the same as that of a point particle at the centre with mass equal to the total mass of the body. We have seen above that the field at great distances from an arbitrary body is always approximately that of a point particle, but now we learn that the field is of this form everywhere around a perfectly spherical body. There are *no near-field corrections to the gravitational field of a spherical body*. Without this wonderful property, Newton could never have related the strength of gravity at the surface of the Earth—iconized by the fall of an apple—to the strength of gravity in the Moon's orbit.

3.4 Gravitational potential

Although the field of gravity is a vector field with three components, there is really only one functional degree of freedom underlying the field, namely the mass distribution giving rise to it. The relationship between these two fields expressed by (3.13) is, however, *non-local*, meaning that $\boldsymbol{g}(\boldsymbol{x})$ in a point $\boldsymbol{x}$ depends on a physical quantity $\rho(\boldsymbol{x}')$ in points $\boldsymbol{x}'$ that may be arbitrarily far away.

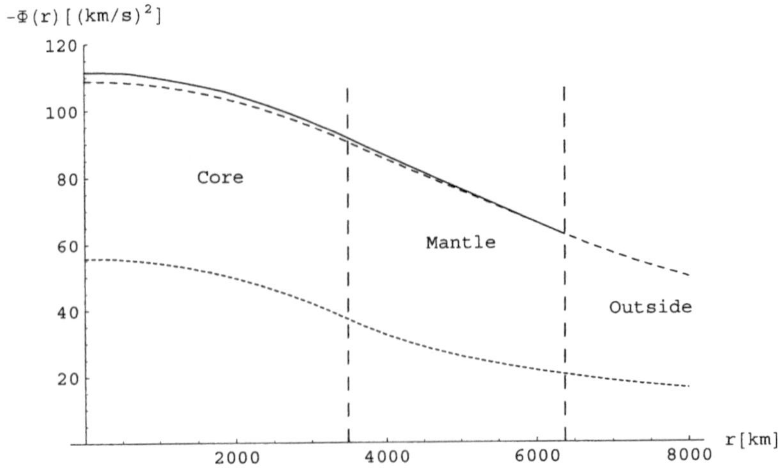

Figure 3.4. The gravitational potential $-\Phi(r)$ of the Earth as a function of distance from the centre. The fully drawn curve is obtained by numerically integrating the standard Earth data [41]. The dotted curve is the potential of the core alone, and the dashed curve is obtained from the two-layer model (problem 3.7). The vertical dashed lines indicate the positions of the sharp transitions in the mass density (see figure 3.1), which have completely disappeared from view in this plot.

Gravity as a gradient field

We shall now prove that the acceleration field can be derived *locally* from a single scalar field $\Phi(\boldsymbol{x})$, called the gravitational *potential*. The relation between the acceleration field and the potential is,

$$\boxed{\boldsymbol{g} = -\nabla\Phi,} \tag{3.20}$$

with a conventional minus sign in front. The gradient operator (nabla) has been defined in equation (2.55). Because of the gradient, the potential is defined only up to an undetermined additive constant.

In order to demonstrate (3.20) we first calculate the gradient of the central distance (see also problem 2.9)

$$\nabla|\boldsymbol{x}| = \nabla\sqrt{\boldsymbol{x}^2} = \frac{1}{2|\boldsymbol{x}|}\nabla\boldsymbol{x}^2 = \frac{1}{2|\boldsymbol{x}|}\nabla(x_1^2 + x_2^2 + x_3^2) = \frac{\boldsymbol{x}}{|\boldsymbol{x}|}, \tag{3.21}$$

and from this

$$\nabla\frac{1}{|\boldsymbol{x}|} = -\frac{1}{|\boldsymbol{x}|^2}\nabla|\boldsymbol{x}| = -\frac{\boldsymbol{x}}{|\boldsymbol{x}|^3}. \tag{3.22}$$

Comparing with (3.12) we conclude that the potential of a point particle of mass M is

$$\Phi(\boldsymbol{x}) = -\frac{GM}{|\boldsymbol{x}|}, \tag{3.23}$$

apart from the arbitrary additive constant which we here choose so that the potential vanishes at spatial infinity. In the same way we derived (3.13), we find the potential from a mass distribution in V

$$\boxed{\Phi(\boldsymbol{x}) = -G\int_V \frac{\rho(\boldsymbol{x}')}{|\boldsymbol{x}-\boldsymbol{x}'|}dV'.} \tag{3.24}$$

Since the mass density is always positive, the gravitational potential is always negative, a direct consequence of the attractive nature of gravity and the normalization to zero potential at infinity.

Even if the mass distribution may jump discontinuously, the gravitational field will always be continuous (as is evident from figure 3.3), because of the integration over the mass distribution in (3.13) or (3.18). The potential is still smoother, because its definition (3.20) requires it to have a continuous derivative. This is also borne out by figure 3.4 which shows the potential of the Earth as a function of central distance. Almost all traces of the original discontinuities have vanished from sight.

The gravitational potential may be visualized by means of surfaces of constant potential, also called *equipotential surfaces*. The field lines are always orthogonal to the equipotential surfaces, and if they are plotted with constant potential difference, the strength of the field will be inversely proportional to the distances between them. A few equipotential surfaces have been shown in the Earth–Moon plot in figure 3.2 on page 33.

Asymptotic behaviour

For a mass distribution of finite extent, the denominator will for $|\boldsymbol{x}| \to \infty$ become independent of $\boldsymbol{x}'$, so that

$$\Phi(\boldsymbol{x}) \to -G\frac{M}{|\boldsymbol{x}|} \tag{3.25}$$

where M is the total mass, in complete accordance with (3.15). At large distances the potential of a finite body thus approaches that of a point mass carrying the total mass of the body (see, however, problems 3.14 and 3.15).

The flat-Earth limit

For a constant gravitational field $\boldsymbol{g}(\boldsymbol{x}) = \boldsymbol{g}_0$ we may take

$$\Phi(\boldsymbol{x}) = -\boldsymbol{x}\cdot\boldsymbol{g}_0. \tag{3.26}$$

This seems to be at variance with expression (3.24) and certainly does not vanish at infinity. Constant gravitational fields should, however, always be understood as an approximation valid within limited regions of space and time, and then the difficulty disappears.

For length scales much smaller than the radius of the Earth, the surface of the sea may be considered to be flat and the gravitational field constant. In a flat-Earth coordinate system with the z-axis vertical and the sea level at $z = 0$, we may take $\boldsymbol{g}_0 = (0, 0, -g_0)$ and find

$$\Phi = g_0 z. \tag{3.27}$$

The gravitational acceleration in the z-direction becomes $g_z = -\partial\Phi/\partial z = -g_0$ and points downwards everywhere, as it should be.

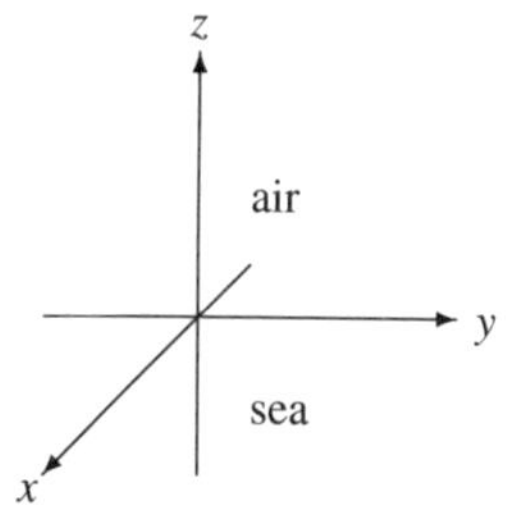

The flat-earth coordinate system with the sea level at $z = 0$.

The spherical case

The potential of a spherical mass distribution can only depend on $r = |\boldsymbol{x}|$. Using $\nabla\Phi(r) = (d\Phi(r)/dr)\,\nabla r$ and $\nabla r = \boldsymbol{x}/|\boldsymbol{x}| = \boldsymbol{e}_r$, we find by comparison with (3.16)

$$\boxed{g(r) = -\frac{d\Phi(r)}{dr}.} \tag{3.28}$$

Conversely, integrating this from r to ∞ and using that the potential vanishes at infinity, we obtain

$$\Phi(r) = \int_r^\infty g(s)\,ds = -G\int_r^\infty \frac{M(s)}{s^2}\,ds, \tag{3.29}$$

where we have also made use of (3.17). Performing a partial integration we obtain

$$\Phi(r) = G\int_r^\infty M(s)\,d\left(\frac{1}{s}\right) = -G\frac{M(r)}{r} - 4\pi G\int_r^\infty s\rho(s)\,ds. \tag{3.30}$$

The final expression is not quite as pretty as (3.17) because of the second term, which is necessary to secure the continuity of the derivative of $\Phi(r)$. Outside the mass distribution the second term vanishes, and the potential becomes, as expected, that of a point particle carrying the total mass of the body.

For a planet with constant mass density we obtain from the above expression and (3.19),

$$\Phi(r) = -\frac{2}{3}\pi G\rho_0 \begin{cases} 3a^2 - r^2 & r < a \\ 2\dfrac{a^3}{r} & r > a. \end{cases} \tag{3.31}$$

One may avoid integrating and instead verify that the potential is continuous at $r = a$ and that the derivative $-d\Phi/dr$ is indeed identical to (3.19).

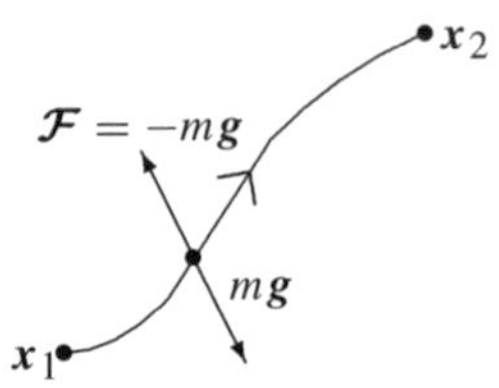

The gravitational force mg must be cancelled by an external force $\mathcal{F} = -mg$ in order to move the particle slowly along any desired path between the end points x_1 and x_2.

3.5 Potential energy

According to the laws of elementary particle mechanics, work is calculated from the product of a force and the distance it moves a particle. It is important to make completely clear who does work on whom. When a particle falls freely in a gravitational field, it is the force of gravity which performs work on the particle while the particle follows the path of its natural motion and gains kinetic energy. If we want the particle to follow any other path, we must 'by hand' cancel the force of gravity with an equal and opposite force, and slowly move the particle along the desired path.

For so-called *conservative* forces, the work we perform on the particle while moving it along the path depends only on its end points and not on where the path goes between the end points. The work we must perform in moving the particle from a fixed position to any desired point x in space is only a function of the end point x of the path. This function, $V(x)$, is called the *potential energy* of the particle at the point x because it represents the work that a particle at x would potentially perform on us, should we move it back to the fixed position.

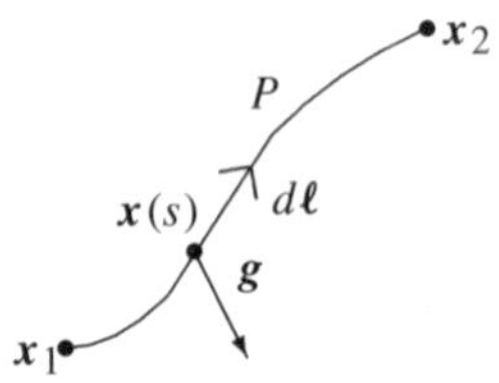

A path $x(s)$ with $s_1 \leq s \leq s_2$ running from x_1 to x_2.

Working against gravity

Suppose we move a point particle of mass m from position x_1 to x_2 along a path P in a static field of gravity, $g(x)$. To keep the particle on this path, we must provide an external force $-mg$ to counter gravity. The work performed by us on the particle while moving it along the curved path is the sum of all the tiny contributions $-mg \cdot d\ell$ from each little path element $d\ell$, and the total work we perform becomes a line integral,

$$W = -\int_P mg \cdot d\ell = -m \int_{s_1}^{s_2} g(x(s)) \cdot \frac{dx(s)}{ds}\, ds. \tag{3.32}$$

In the last expression the line integral along the path $x(s)$ from $x_1 = x(s_1)$ to $x_2 = x(s_2)$ has been made explicit by means of a running parameter s varying in the interval $s_1 \leq s \leq s_2$ along the path.

Because the field of gravity is the gradient of the gravitational potential, the line integral may be carried out. Inserting (3.20) we find

$$W = m\int_{x_1}^{x_2} \nabla\Phi(x) \cdot d\ell = m\int_{s_1}^{s_2} \frac{d\Phi(x(s))}{ds}\, ds \tag{3.33}$$

$$= m\,\Phi(x_2) - m\,\Phi(x_1). \tag{3.34}$$

Place	km s^{-1}
Earth surface	11.2
Mars surface	5.0
Moon surface	2.4
Sun surface	617.6
Earth orbit	42.1
Moon orbit	1.4
Neutron star	1×10^5
Black hole	3×10^5

Escape velocities from some places in the solar system, and a couple of exotic ones. Note that escaping from the orbit of Earth means escaping from the solar system whereas escaping from the orbit of the Moon only gets you free of Earth's gravity. The neutron star is assumed to have solar mass.

Since this result is independent of the actual path of the particle, it follows that the gravitational potential energy of the particle is $m\Phi(x)$, and that the gravitational potential $\Phi(x)$ is the potential energy per unit of mass.

Escape velocity

The total energy of a point particle at x with velocity v is the sum of its kinetic energy and its potential energy, $(1/2)mv^2 + m\Phi(x)$. From elementary mechanics we know that the total energy is conserved, i.e. constant in time. In other words, a particle starting in the point x_0 with velocity v_0 must at all times obey the equation

$$\tfrac{1}{2}v^2 + \Phi(x) = \tfrac{1}{2}v_0^2 + \Phi(x_0). \tag{3.35}$$

Taking $\boldsymbol{x}_0$ at infinity, where the potential vanishes, it follows immediately that in order to escape the grip of gravity with $|\boldsymbol{v}_0| > 0$ from a point $\boldsymbol{x}$ with potential Φ, an object at $\boldsymbol{x}$ must be given a velocity that is larger than

$$v_{\text{esc}} = \sqrt{-2\Phi}. \tag{3.36}$$

Knowing the potential at a point is simply equivalent to knowing the escape velocity from that point.

A particularly interesting case occurs when the potential becomes so deep that the escape velocity equals or surpasses the velocity of light c. In that case the body has turned into a black hole. Using the potential of a point mass (3.23) we find that this happens when the radius a of a spherical mass distribution satisfies

$$a < \frac{2GM}{c^2}. \tag{3.37}$$

Being a non-relativistic calculation this is of course highly suspect, but accidentally it is exactly the same as the correct condition obtained in general relativity [49], where the right-hand side is called the Schwarzchild radius. For the Earth the Schwarzchild radius is about a centimetre, and for the Sun three kilometres.

No closed loops of gravity

For a closed path, C, which begins at the same point as it ends, the line integral must vanish

$$\oint_C \boldsymbol{g} \cdot d\boldsymbol{\ell} = 0 \tag{3.38}$$

because the potential is the same at the end points of the path. This implies that there can be no closed loops of field lines. For if there were, the product $\boldsymbol{g} \cdot d\boldsymbol{\ell}$ would have the same sign all around the loop, because the tangent of a field line is everywhere proportional to the field, i.e. $d\boldsymbol{\ell} \sim \boldsymbol{g}$, and such an integral cannot vanish.

A gravitational field line cannot form a closed path C.

Conversely, if the line integral of a field $\boldsymbol{g}$ around any closed curve vanishes, the field must be a gradient field. To prove this consider the line integral along some path $P(\boldsymbol{x})$ connecting a fixed point, say the origin $\boldsymbol{0}$, with an arbitrary point $\boldsymbol{x}$,

$$\Phi(\boldsymbol{x}) = -\int_{P(\boldsymbol{x})} \boldsymbol{g}(\boldsymbol{x}') \cdot d\boldsymbol{\ell}'. \tag{3.39}$$

This function only depends on the end point $\boldsymbol{x}$ and not on the path because two different paths, P and P', connecting the same points form a closed curve $C = P - P'$, and by (3.38)

$$\oint_C \boldsymbol{g} \cdot d\boldsymbol{\ell} = \int_P \boldsymbol{g} \cdot d\boldsymbol{\ell} - \int_{P'} \boldsymbol{g} \cdot d\boldsymbol{\ell} = 0. \tag{3.40}$$

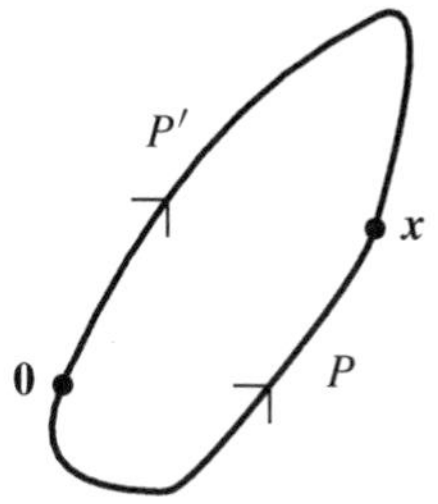

A closed curve $C = P - P'$ can be viewed as the difference between two paths connecting the same points.

Finally, we must show that (3.39) has the gradient $\boldsymbol{g}$. Using the additivity of line integrals we get,

$$\Phi(\boldsymbol{x} + \delta\boldsymbol{x}) - \Phi(\boldsymbol{x}) = -\int_{\boldsymbol{x}}^{\boldsymbol{x}+\delta\boldsymbol{x}} \boldsymbol{g}(\boldsymbol{x}') \cdot d\boldsymbol{\ell}' \approx -\boldsymbol{g}(\boldsymbol{x}) \cdot \delta\boldsymbol{x}. \tag{3.41}$$

Since the left-hand side is $\Phi(\boldsymbol{x}+\delta\boldsymbol{x}) - \Phi(\boldsymbol{x}) \approx \boldsymbol{\nabla}\Phi \cdot \delta\boldsymbol{x}$, and since $\delta\boldsymbol{x}$ is arbitrary, it follows that $\boldsymbol{\nabla}\Phi = -\boldsymbol{g}$.

Problems

3.1 Show that the gravitational field outside a spherical planet may be written

$$g(r) = -g_0 \frac{a^2}{r^2}, \tag{3.42}$$

where a is the radius of the planet and g_0 the magnitude of the surface gravity.

3.2 Show that a satellite moving in a circular orbit around a spherical planet has velocity $v = \sqrt{-\Phi}$, where Φ is the gravitational potential in the satellite's orbit. Calculate the velocity of a satellite moving at ground level.

3.3 Arthur C. Clarke proposed (*Wireless World*, October 1945, pp 305–308) that communication satellites should be put into a circular equatorial orbit with revolution time equal to Earth's rotation period, so that the satellites would stay fixed over a point at the equator. Calculate the height of the orbit above the ground, also taking into account that the Earth circles the Sun in one year.

3.4 A space elevator (fictionalized by Arthur C. Clarke in *Fountains of Paradise* (1978)) can be created if it becomes technically feasible to lower a line down to Earth from a geostationary satellite at height h (see problem 7.5). Assume that the line is unstretchable and has constant cross section A and constant density ρ. Calculate the maximal tension $\sigma = -\mathcal{F}/A$ (force per unit of area) in the line. Determine the numerical value of the ratio of tension to density, σ/ρ, and compare with the tensile strength (breaking tension) of a known material.

3.5 A comet consisting mainly of ice falls to Earth. **(a)** Estimate the minimum energy released in the fall per unit of mass. **(b)** Compare with the estimate of the energy needed to evaporate the comet.

3.6 A stone is set in free fall from rest through a mine shaft going right through the centre of a non-rotating planet with constant density. **(a)** Calculate the speed with which the stone passes the centre. **(b)** Calculate the time it takes to fall to the centre.

3.7 A planet consists of two layers with constant mass density,

$$\rho(r) = \begin{cases} \rho_1 & r \le a_1 \\ \rho_2 & a_1 < r \le a \\ 0 & r > a \end{cases} . \qquad (3.43)$$

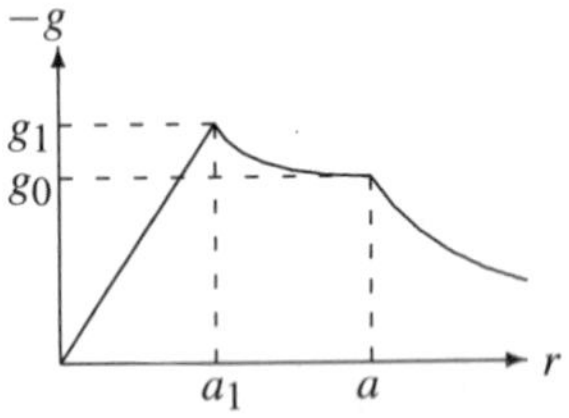

The strength of gravity for a two-layer planet with a dense core.

(a) Calculate the strength of gravity and the potential. **(b)** Show that the strength of gravity at the boundary between the layers is greater than at the surface when

$$\frac{\rho_1 - \rho_2}{\rho_2} > \frac{a^2}{a_1(a + a_1)} . \qquad (3.44)$$

Verify that this is fulfilled for the Earth.

∗ **3.8** Show by direct integration in a small spherical volume around the singularity in (3.13) that it gives a finite contribution to the integral.

∗ **3.9** Show that the mass density is a scalar field.

∗ **3.10** Show that the gravitational field is a vector field.

∗ **3.11** Show that gravitational field of a spherical body (3.17) may be derived by integration of (3.13).

∗ **3.12** A spherical planet has mass distribution of the form $\rho(r) = Ar^\alpha$ for $r \le a$. **(a)** Calculate the gravitational field strength and the potential inside the planet for this distribution. **(b)** For what values of α is the problem solvable with finite planet mass? **(c)** For what value of α does gravity grow stronger towards the centre?

∗ **3.13** An 'exponential star' has a mass density $\rho = \rho_0 e^{-r/a}$, where ρ_0 is the central mass density and a is the 'radius'. Calculate the gravitational field and potential.

∗ **3.14** **(a)** Calculate the gravitational potential and field from a mass distribution shaped like a very thin line (a model of a cosmic string) of length $2a$ with uniform mass λ per unit of length. **(b)** Calculate the behaviour of the potential at infinity orthogonally to the line. **(c)** Discuss what happens in the limit of $a \to \infty$.

∗ **3.15** **(a)** Write an expression for the gravitational potential from a mass distribution shaped like a very thin circular plate of radius a with uniform mass σ per unit of area (a model of the luminous matter of a galaxy). **(b)** Calculate the value of the potential along the central normal of the plate. **(c)** Calculate its form far from the disk. **(d)** What happens for $a \to \infty$?

Part II

Fluids at rest

4

Pressure

If the Sun did not shine, if no heat were generated inside the Earth and no energy radiated into space, all the winds in the air and the currents in the sea would die away, and the air and water on the planet would in the end come to rest in equilibrium with gravity. In the absence of external driving forces or time-dependent boundary conditions, and in the presence of dissipative contact forces, any fluid must eventually reach a state of *hydrostatic equilibrium*, where nothing moves anymore and all fields become constant in time. This state must be the first approximation to the sea, the atmosphere, the interior of a planet or a star.

In a continuous system in mechanical equilibrium there is everywhere a balance between contact forces having zero range and body forces with infinite range. Contact interactions between material bodies or even between parts of the same body take place across *contact surfaces*. A contact force acting on a tiny piece of a surface may in principle take any direction relative to the surface, but can of course be resolved into its normal and tangential components. The normal component is called a *pressure force* and the tangential a *shear force*. Solids and fluids in motion can sustain shear forces, whereas fluids at rest can not.

In this chapter we shall establish the basic concepts and formalism for pressure in hydrostatic equilibrium and apply it to the sea and the atmosphere. Along the way we shall recapitulate some basic rules of thermodynamics. In the following three chapters we shall continue the study of hydrostatic equilibrium for fish, icebergs and ships, the interior of planets and stars, and the shapes of large and small fluid bodies.

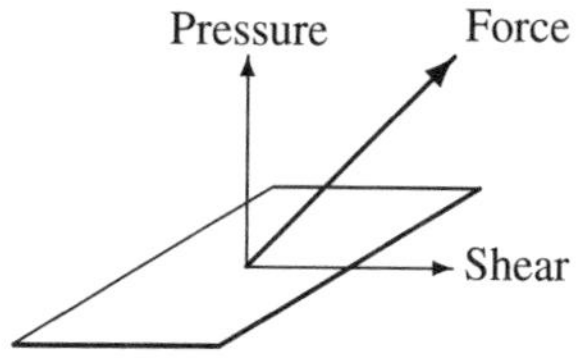

The force on a small piece of a surface can always be resolved as a normal pressure force and a tangential shear force.

4.1 The concept of pressure

A fluid at rest can as mentioned only sustain pressure forces. If shear forces arise, the fluid will flow towards a new equilibrium without shear. This expresses the most basic property of fluids and may be taken as a definition of what constitutes a fluid at the macroscopic level. In this section we shall take a first look at pressure defined as force per unit of area, discuss its microscopic origins, and analyse a couple of elementary cases.

> The SI unit of pressure is pascal (Pa $=$ N m^{-2} $=$ kg m^{-1} s^{-2}), but often pressure is quoted in units of *bars* where 1 bar $= 10^5$ Pa or in *atmospheres* where 1 atm $=$ 1.01325 bar is close to the average air pressure at sea level. Modern meteorologists are now abandoning these units and tend to quote air pressure in hectopascals (hPa) rather than in millibars.

Microscopic origin of pressure

In a liquid the molecules touch each other and the containing solid walls. The number of molecules that are in direct contact with a small area A of a wall is proportional to the size of this area, and so is the total normal force $\mathcal{F}$ that the molecules collectively exert on the area. The pressure defined as the normal force per unit of area, $p = \mathcal{F}/A$, is thus independent of the size of the small area.

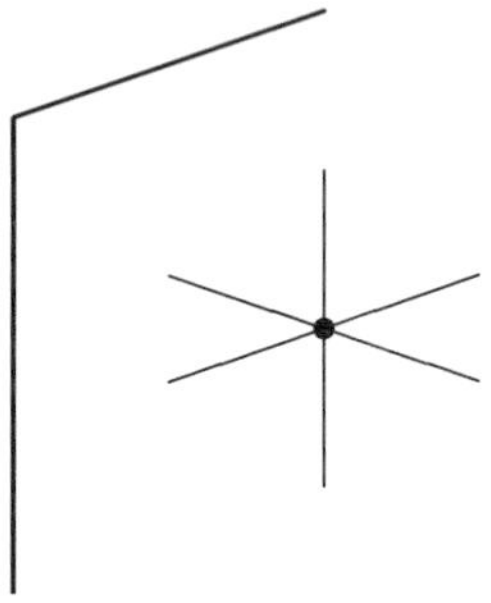

In Bernoulli's model a molecule moves in any of the six possible directions along three orthogonal directions with equal probability.

A gas consists mostly of vacuum with the molecules moving freely around between collisions, and the pressure on a solid wall arises in this case from the incessant molecular bombardment. We shall estimate the pressure using a simple 'molecular' model of a gas going back to Daniel Bernoulli, in which molecules of mass m are only allowed to move back and forth along three orthogonal directions with a fixed velocity v. When a molecule hits a wall orthogonal to one of these directions, it is reflected directly back again and transfers a momentum $2mv$ to the wall. Assuming that all six directions of motion are equally probable, the number of molecules hitting an area A of the wall in a small time interval dt will be $dN = 1/6\rho Avdt/m$, such that the total momentum transfer to the area in the time dt becomes $d\mathcal{P} = 2mvdN = 1/3\rho Av^2dt$. The normal force exerted on the area equals the rate of momentum transfer, $\mathcal{F} = d\mathcal{P}/dt = 1/3\rho Av^2$, and dividing by the area we obtain the pressure,

$$p = \frac{\mathcal{F}}{A} = \frac{1}{3}\rho v^2. \tag{4.1}$$

In the kinetic theory of gases one finds the same result, except that v is replaced by the root-mean-square average of the molecular velocities, $v = \sqrt{\langle \boldsymbol{v}^2 \rangle}$.

Example 4.1.1 (Molecular velocity): At normal pressure, $p = 1$ atm, and temperature, $T = 18\,°$ C $= 291$ K, the density of air is $\rho \approx 1.2$ kg m^{-3}. The molecular velocity calculated from the above expression becomes $v \approx 500$ m s^{-1}. We shall see later that this velocity is related to the velocity of sound.

External and internal pressure

So far we have only defined the external pressure acting on the walls of a container. Is it meaningful to speak about pressure in the middle of a fluid away from containing walls? We could of course insert a tiny manometer to measure the pressure, but then we would just obtain the external pressure acting on another wall, the surface of the manometer. There seems to be no simple way to determine directly what one would call the true internal pressure in the fluid.

Cutting through these 'philosophical' difficulties we shall simply postulate that there is indeed a well-defined pressure field $p(\boldsymbol{x})$ everywhere in the fluid and that it acts along the normal to any surface in the fluid, whether it be a real interface or an imagined cut through the fluid. This postulate is supported by the microscopic view of continuous matter. In a liquid we may define the internal pressure as the total force per unit of area exerted by the molecules at one side of the cut on the molecules at the other side. Similarly, in a gas the internal pressure may be defined by the rate of molecular momentum transfer per unit of area across the cut. In solids internal pressure is more subtle (see chapter 9).

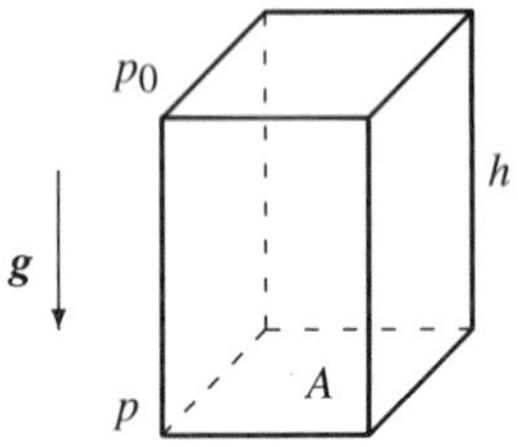

A column of sea water. The pressure difference between bottom and top must carry the weight of the water in the box.

Incompressible sea

Consider now a vertical box with cross-sectional area A and height h in a sea with constant density $\rho(\boldsymbol{x}) = \rho_0$. In hydrostatic equilibrium the difference between the pressure forces pA at the bottom and p_0A at the top must balance the total weight of the water in the box, or

$$pA - p_0A = \rho_0 Ahg_0, \tag{4.2}$$

where g_0 is the constant gravitational acceleration. If this equation were not fulfilled, the total force on the column of water would not vanish, and it would have to move. Dividing by the area A, the pressure difference between bottom and top of the box becomes,

$$p - p_0 = \rho_0 g_0 h. \tag{4.3}$$

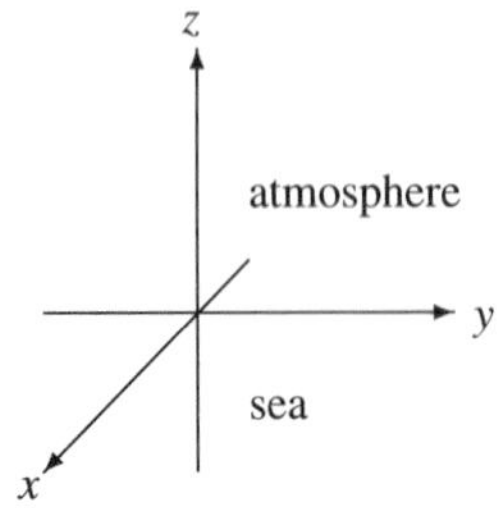

The flat-earth coordinate system.

In a *flat-earth coordinate system* with vertical z-axis and the surface of the sea at $z = 0$, we find by setting $h = -z$,

$$\boxed{p = p_0 - \rho_0 g_0 z,} \tag{4.4}$$

where p_0 is the surface pressure. The linear rise of the pressure with depth $-z$ allows us to immediately calculate the total pressure force on the horizontal and vertical sides of a container. Skew or curved container walls require a more powerful formalism which we shall soon set up.

Using $\rho_0 \approx 1000$ kg m^{-3} and $g_0 \approx 10$ m s^{-2}, the scale of the pressure increase per unit of depth in the sea becomes $\rho_0 g_0 \approx 10^4$ Pa m^{-1}, or about 1 atm/10 m. At the deepest point in the sea, $z \approx -11$ km, the pressure is a little more than 1000 atm. The assumption of constant gravity is well justified even to this depth, because it changes only by about 0.35%, whereas the density of water changes by about 4.5% (see page 54).

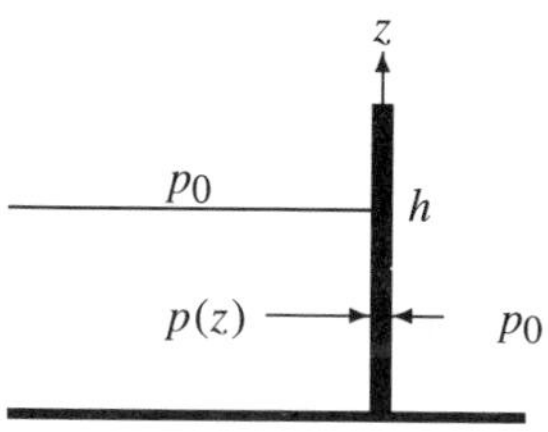

Water stemmed up behind a sluice gate. The pressure varies linearly with height z from the bottom.

Example 4.1.2 (Sluice gate force): Water is stemmed up to height h behind a sluice gate of width L. On the water surface and on the outer side of the gate there is atmospheric pressure p_0. On the inside of the gate the pressure is $p(z) = p_0 + \rho_0 g_0 (h - z)$, so that the total force on the gate becomes

$$\mathcal{F} = \int_0^h (p(z) - p_0)\, L dz = \frac{1}{2}\rho_0 g_0 L h^2. \tag{4.5}$$

Because the pressure rises linearly with depth, this result could have been calculated without an explicit integral. The total force is simply the product of the area of the sluice gate Lh with the average pressure excess $\langle p - p_0 \rangle = 1/2 \rho_0 g_0 h$ acting on the gate.

Incompressible atmosphere?

Since air is compressible, it makes little sense to use expression (4.4) for the pressure in a fluid with constant density (except for very small values of z). If we do so anyway, we find a pressure which falls linearly with height and reaches zero at a height,

$$z = h_0 = \frac{p_0}{\rho_0 g_0}. \tag{4.6}$$

Using $p_0 = 1$ atm and air density $\rho_0 = 1.2$ kg/m^3 we get $h_0 = 8.6$ km, which is a tiny bit lower than the height of Mount Everest. This is, of course, meaningless, since climbers have reached the summit of that mountain without oxygen masks. But as we shall see, this height sets nevertheless the correct scale for major changes in the atmospheric properties.

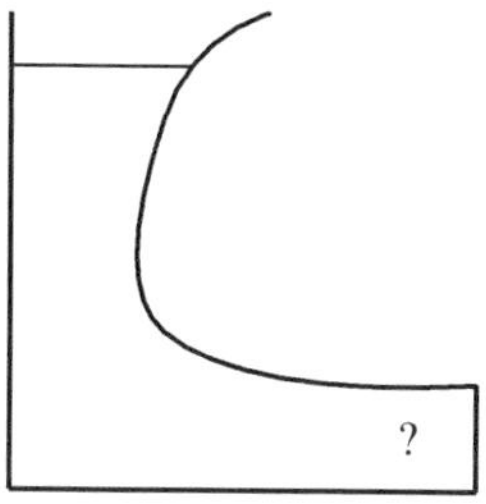

Paradox: the pressure along the vertical wall of the 'boot' rises linearly because it has to carry the weight of the water above, but what about the pressure in the tip of the 'toe'?

Paradox of hydrostatics

The linear rise of water pressure with depth may, as we have seen, be used to calculate the total pressure force on any vertical container wall. For a curved container wall, like that of a vase or a boot, there seems to be no problem, except handling the necessary mathematics. But if the water column does not reach all the way to the surface, as when you fill a boot with water, what is then the pressure at the flat horizontal bottom? Will it be constant along the bottom as in the open sea, or will it vary? And if it is constant, what is it 'up against', since there is only a short column of water above?

The quick answer to this paradox is that the pressure is indeed constant along the horizontal bottom. For if the pressure were lower in the 'toe' than in the 'heel', there would be unbalanced horizontal pressure forces directed towards the toe acting on a horizontal box of water. But that is not allowed in complete mechanical equilibrium. The only possible conclusion is that the material of the boot must supply the necessary forces to compensate for the missing weight of the water column.

4.2 Formal definition of pressure

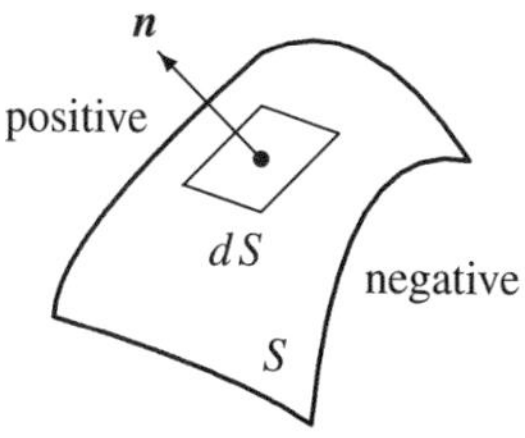

All normals to an oriented open surface have a consistent orientation with common positive and negative sides.

To establish a concise mathematical formalism for pressure we consider a surface S that divides a body into two parts. This surface need not be a real surface where material properties change dramatically but may just be an imaginary surface separating two parts of the same body from each other. A tiny surface element is characterized by its area dS and the direction of its normal $\boldsymbol{n}$. It is convenient to combine these in the vector surface element

$$d\boldsymbol{S} = (dS_x, dS_y, dS_z) = \boldsymbol{n}\, dS. \tag{4.7}$$

There is nothing intrinsic in a surface which defines the orientation of the normal, i.e. whether the normal is really $\boldsymbol{n}$ and not $-\boldsymbol{n}$. A choice must, however, be made, and having done that, one may call the side of the surface element into which the normal points, positive (and the other, of course, negative). Usually neighbouring surface elements are required to be oriented consistently, i.e. with the same positive sides. By universal convention the normal of a closed surface is chosen to be directed out of the enclosed volume, so that the enclosed volume always lies at the negative side of its surface.

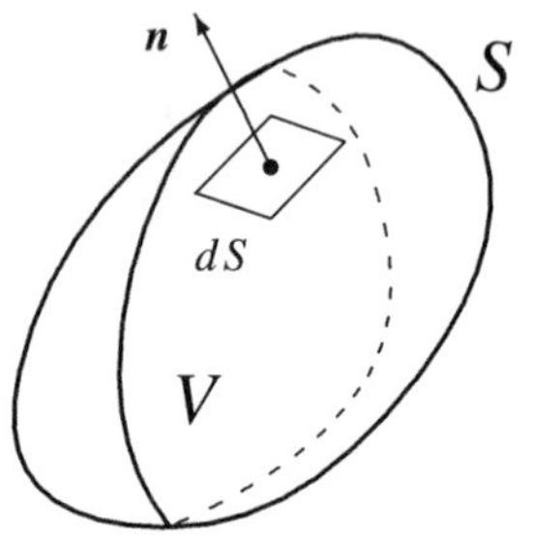

A volume V defined by the closed surface S has all normals oriented towards the outside.

Local and global pressure force

The contact forces due to microscopic molecular interactions have a finite range at the molecular scale, but zero range at macroscopic distances. Across a tiny but still macroscopic piece of surface, the number of neighbouring molecules participating in the interaction as well as the force they exert may for this reason be expected to be proportional to the area of the surface. In a fluid at rest the only contact force is the pressure force acting along the normal to the surface, and the force exerted by the material situated at the positive side of a surface element on the material at the negative side must be of the form

$$\boxed{d\boldsymbol{\mathcal{F}} = -p\, d\boldsymbol{S},} \tag{4.8}$$

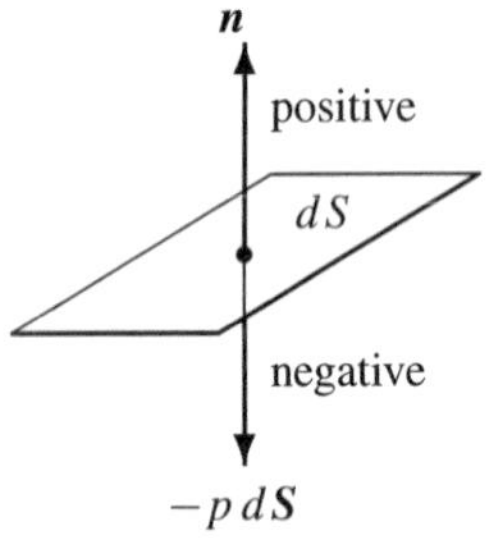

The force on a vector surface element under positive pressure is directed against the normal.

with a coefficient of proportionality p called the *pressure*. Convention dictates that a positive pressure exerts a force directed *towards* the material on the negative side, and this explains the minus sign.

The total pressure force acting on a surface S is obtained by summing up all the little vector contributions from each surface element,

$$\boldsymbol{\mathcal{F}} = -\int_S p\, d\boldsymbol{S}. \tag{4.9}$$

This is the force which acts on the cork in the champagne bottle, moves the pistons in the cylinders of your car engine, breaks a dam, and fires a bullet from a gun. It is also this force that lifts fish, ships, and balloons and thereby cancels their weight so that they are able to float (see chapter 5).

Same pressure in all directions?

Newton's third law guarantees that the material on the negative side of a surface element reacts with an equal and opposite force, $-d\boldsymbol{\mathcal{F}} = -p\,(-d\boldsymbol{S})$, on the material on the positive side (provided there is no surface tension). Since the surface vector seen from the negative side is $-d\boldsymbol{S}$, the above relation shows that the pressure also has the value p on the negative side of the surface. This is part of a much stronger result, called *Pascal's law*, which we shall prove below: *the pressure in a fluid at rest is independent of the direction of the surface element on which it acts.* It implies that pressure $p(\boldsymbol{x})$ cannot depend on the normal $\boldsymbol{n}$, but only on the location $\boldsymbol{x}$ of a surface element, and is therefore a true scalar field.

Blaise Pascal (1623–1662). *French mathematician and physicist. Founded probability theory. Constructed what may be viewed as the first digital calculator. He spent his later years with religious thinking in the Cistercian abbey of Port-Royal. More than one property of pressure goes under the name of Pascal's law.*

The simple reason for pressure being the same in all directions in hydrostatic equilibrium is that the pressure acts on the surface of a body whereas a body force by definition acts on the volume. If we let the body shrink, the contribution from the body force will vanish faster than the contribution from the surface force because the volume vanishes faster than the surface area. In the limit of vanishing body size only the surface force is left, but it must then itself vanish in hydrostatic equilibrium where the total force on all parts of a body has to vanish. This argument will now be fleshed out in mathematical detail.

* **Proof of Pascal's law:** Assume first that the pressure is actually different in different directions. We shall then show that for physical reasons this assumption cannot be maintained. Consider a tiny body in the shape of a tetrahedron with three sides parallel to the coordinate planes. The total pressure force acting on the body is

$$d\boldsymbol{\mathcal{F}} = -p d\boldsymbol{S} - p_x d\boldsymbol{S}_x - p_y d\boldsymbol{S}_y - p_z d\boldsymbol{S}_z, \tag{4.10}$$

where we have denoted the pressures acting on the different faces of the tetrahedron by p, p_x, p_y, and p_z and the outwards pointing normals by $d\boldsymbol{S}$, $d\boldsymbol{S}_x$, $d\boldsymbol{S}_y$, and $d\boldsymbol{S}_z$. It is sufficient to consider infinitesimal bodies of this kind, because an arbitrary body shape can be put together from these. Each of the three triangles making up the sides of the tetrahedron is in fact the projection of the front face onto that plane. By elementary geometry the areas of the three projected triangles are dS_x, dS_y and dS_z, so that their vector surface elements become $d\boldsymbol{S}_x = (-dS_x, 0, 0)$, $d\boldsymbol{S}_y = (0, -dS_y, 0)$, and $d\boldsymbol{S}_z = (0, 0, -dS_z)$.

Inserting this in the above equation we find the total force

$$d\mathcal{F} = ((p_x - p)dS_x, (p_y - p)dS_y, (p_z - p)dS_z). \tag{4.11}$$

In hydrostatic equilibrium, which is all that we are concerned with here, the contact forces must balance body forces,

$$d\mathcal{F} + \boldsymbol{f}\, dV = \boldsymbol{0}, \tag{4.12}$$

where $\boldsymbol{f}$ is the density of body forces.

The idea is now to show that for sufficiently small tetrahedrons the body forces can be neglected and the surface forces $d\mathcal{F}$ must consequently vanish. Consider a geometrically congruent tetrahedron with all lengths scaled by a factor λ. Since the volume scales as the third power of λ whereas the surface areas only scale as the second power, the hydrostatic equation for the scaled tetrahedron becomes $\lambda^2 d\mathcal{F} + \lambda^3 \boldsymbol{f}\, dV = \boldsymbol{0}$ or $d\mathcal{F} + \lambda \boldsymbol{f}\, dV = \boldsymbol{0}$. In the limit of $\lambda \to 0$ it follows that the total contact force must vanish, i.e. $d\mathcal{F} = 0$, and using (4.11) we find,

$$p_x = p_y = p_z = p. \tag{4.13}$$

As promised, the pressure must indeed be the same in all directions.

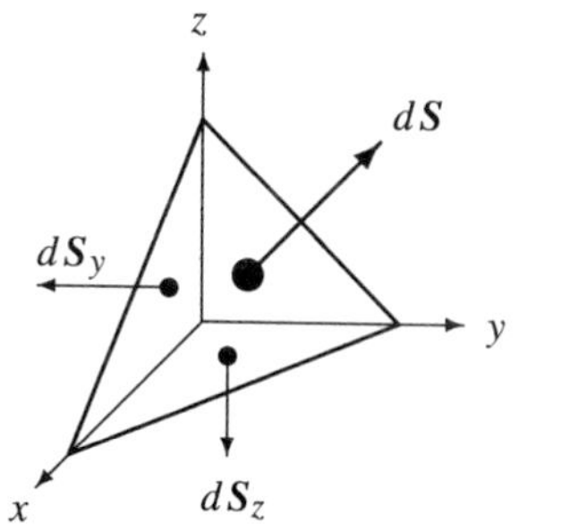

A body in the shape of a tetrahedron. The vector normals to the sides are all pointing out of the body ($d\boldsymbol{S}_x$ is hidden from view). Any body shape can be built up from sufficiently many sufficiently small tetrahedrons.

4.3 Hydrostatic equilibrium

In section 4.1 we intuitively used that in a fluid at rest the weight of a vertical column of fluid should equal the difference in pressure forces between the bottom and the top of the column. We shall now generalize this to an arbitrary macroscopic volume of fluid, often called a *control volume*. The material in a control volume, fluid or solid or whatever, represents the most general 'body' that can be constructed in continuum physics.

Up to this point we have studied only two kinds of forces that may act on the material in a control volume. One is a *body force* described by a force density field $\boldsymbol{f}$ caused by long-range interactions, for example gravity $\boldsymbol{f} = \rho \boldsymbol{g}$. The other is a *contact force*, here the pressure field p, which has zero range and only acts on the surface of the control volume. The total force on the control volume V with surface S is the sum of two contributions

$$\mathcal{F} = \int_V \boldsymbol{f}\, dV - \oint_S p\, d\boldsymbol{S}. \tag{4.14}$$

The first term is for the case of gravity just the weight of the fluid in the volume and the second is the so-called *buoyancy force*. The circle in the symbol for the surface integral is only there to remind us that the surface is closed.

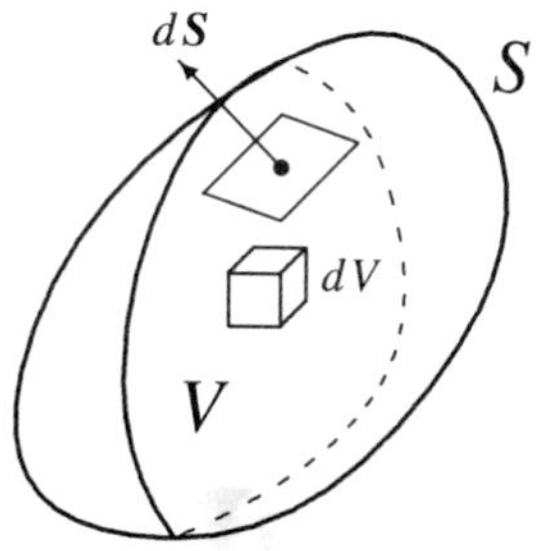

A control volume V with its enclosing surface S, a volume element dV and a surface element $d\boldsymbol{S}$.

Global hydrostatic equilibrium equation

In hydrostatic equilibrium, the total force must vanish for any volume of fluid, $\mathcal{F} = \boldsymbol{0}$, or

$$\boxed{\int_V \boldsymbol{f}\, dV - \oint_S p\, d\boldsymbol{S} = \boldsymbol{0}.} \tag{4.15}$$

This is the *equation of global hydrostatic equilibrium*, which states that buoyancy must exactly balance the total volume force, i.e. the weight. If the cancellation is not exact, as for example when a small volume of water is heated or cooled relative to its surroundings, the fluid *must* start to move, either upwards if the buoyancy force is larger than the weight or downwards if it is smaller.

The problem with the global equilibrium equation is that we have to know the fields $\boldsymbol{f}(\boldsymbol{x})$ and $p(\boldsymbol{x})$ in advance to calculate the integrals. Sometimes symmetry considerations can get us a long way. In constant gravity, the sea on the flat Earth ought to have the same properties for all x and y, suggesting that the pressure $p = p(z)$ can only depend on the depth z. This was, in fact, a tacit assumption used in calculating the pressure in the incompressible sea (4.4), and it is not difficult to formally derive the same result from the equation of global equilibrium (4.15). But, in general, we need to establish a local form of the equations of hydrostatic equilibrium, valid at each point $\boldsymbol{x}$.

Effective force on a material particle

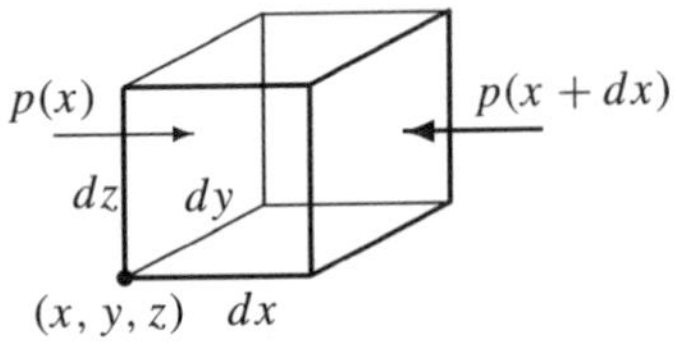

Pressure difference over a small rectangular box.

A material particle is like any other body subject to pressure from all sides, but being infinitesimal it is possible to derive a general expression for the resultant force. Let us choose a material particle in the shape of a small rectangular box with sides dx, dy, and dz, and thus a volume $dV = dxdydz$. Since the pressure is slightly different on opposite sides of the box the resultant pressure force is to leading approximation (in the x-direction)

$$d\mathcal{F}_x \approx (p(x, y, z) - p(x + dx, y, z))dydz \approx -\frac{\partial p}{\partial x}dxdydz.$$

Including the other coordinate directions we obtain

$$d\boldsymbol{\mathcal{F}} = -\boldsymbol{\nabla} p\, dV. \tag{4.16}$$

The result of all pressure forces acting on a tiny material particle is apparently equivalent to a volume force with a density equal to the negative gradient of the pressure. We shall see below that this result does not depend on the shape of the material particle.

If there is also a true volume force, $\boldsymbol{f}$, for example gravity ($\boldsymbol{f} = \rho\boldsymbol{g}$), acting on the material, the total force on a material particle may be written,

$$\boxed{d\boldsymbol{\mathcal{F}} = \boldsymbol{f}^*\, dV,} \tag{4.17}$$

where

$$\boxed{\boldsymbol{f}^* = \boldsymbol{f} - \boldsymbol{\nabla} p,} \tag{4.18}$$

is called the *effective force density*. It must be emphasized that the effective force density is not a true body force, but an expression which for a tiny material particle equals the sum of the true body force and all pressure forces acting on its surface.

Local hydrostatic equilibrium

In hydrostatic equilibrium, the total force on an arbitrary body has to vanish. Applying this to all of the material particles in the body, it follows that the effective density of force must vanish everywhere,

$$\boxed{\boldsymbol{f}^* = \boldsymbol{f} - \boldsymbol{\nabla} p = \boldsymbol{0}.} \tag{4.19}$$

This is the *local equation of hydrostatic equilibrium*. It is a differential equation valid everywhere in a fluid at rest, and it encapsulates in an elegant way all the physics of hydrostatics.

The flat-Earth case

Returning to the case of constant gravity in a flat-earth coordinate system we have $\boldsymbol{f} = \rho\boldsymbol{g}_0 = \rho(0, 0, -g_0)$ and the local equilibrium equation takes the following form when written out explicitly in coordinates,

$$\frac{\partial p}{\partial x} = 0, \tag{4.20a}$$

$$\frac{\partial p}{\partial y} = 0, \tag{4.20b}$$

$$\frac{\partial p}{\partial z} = -\rho g_0. \tag{4.20c}$$

The two first equations show that the pressure does not depend on x and y but only on z, which confirms the previous argument based on symmetry. It also resolves the hydrostatic paradox because we now know that independently of the shape of the container the pressure will always be the same at a given depth in constant gravity. For the special case of constant density, $\rho(z) = \rho_0$, the last equation may immediately be integrated to yield the previous result (4.4) for the pressure in the incompressible sea.

Constant density

More generally, if the density of the fluid is constant, $\rho = \rho_0$, and the body force is due to gravity, $\boldsymbol{f} = \rho_0 \boldsymbol{g} = -\rho_0 \nabla \Phi$, the equation of hydrostatic equilibrium (4.19) takes the form $-\nabla(\rho_0 \Phi + p) = \boldsymbol{0}$, which implies that the quantity,

$$H = \Phi + \frac{p}{\rho_0}, \tag{4.21}$$

is a constant, independent of $\boldsymbol{x}$. The first term is the gravitational potential and the second term is naturally called the *pressure potential*. There is no agreement in the literature about a name for H, but one might call it the *effective potential*, because the effective density of force becomes $\boldsymbol{f}^* = \rho_0 \boldsymbol{g} - \nabla p = -\rho_0 \nabla H$. It is also related to the thermodynamic *enthalpy*.

The constancy of H contains the complete solution of the hydrostatic equation. In constant gravity we thus have $\Phi = g_0 z$ and recover immediately the pressure in the sea (4.4) with $H = p_0/\rho_0$.

Gauss' theorem

The equation of local equilibrium (4.19) has been obtained by applying the global equilibrium equation (4.15) to a tiny material particle. Is it also possible to go the other way and derive the global equation from the local?

The answer to the question is affirmative, because of a purely mathematical theorem due to Gauss (to be proved below), which in its simplest form states that

$$\boxed{\oint_S p\, d\boldsymbol{S} = \int_V \nabla p\, dV,} \tag{4.22}$$

Johann Carl Friedrich Gauss (1777–1855). *German mathematician of great genius. Contributed to number theory, algebra, non-Euclidean geometry, and complex analysis. In physics he developed the magnetometer. The older (cgs) unit of magnetic strength is named after him.*

for an arbitrary function $p(\boldsymbol{x})$. Using Gauss' theorem it follows immediately that the total force (4.14) equals the integral of the effective density of force,

$$\mathcal{F} = \int_V \boldsymbol{f}\, dV - \oint_S p\, d\boldsymbol{S} = \int_V \boldsymbol{f}^*\, dV. \tag{4.23}$$

This equivalence allows us to think of a macroscopic volume of fluid as composed of microscopic material particles acted upon by the effective force density $\boldsymbol{f}^*$.

Gauss' theorem is *a fortiori* also valid for a tiny material particle, where the total pressure force becomes,

$$\oint_S p\, d\boldsymbol{S} = \int_V \nabla p\, dV \approx \nabla p\, V.$$

This confirms that the force on a material particle is indeed independent of its shape as long as the pressure gradient is essentially constant across the particle.

* **Proof of Gauss' theorem:** To prove Gauss' theorem, consider first a volume V described by the inequalities $a(x, y) \le z \le b(x, y)$ where $a(x, y)$ and $b(x, y)$ are two functions defined in some area A of the xy-plane. We then find,

$$\begin{aligned}\int_V \nabla_z p\, dV &= \int_A dxdy \int_{a(x,y)}^{b(x,y)} \frac{\partial p(x, y, z)}{\partial z}\, dz \\ &= \int_A dxdy\big(p(x, y, b(x, y)) - p(x, y, a(x, y))\big) \\ &= \oint_S p(x, y, z)\, dS_z.\end{aligned}$$

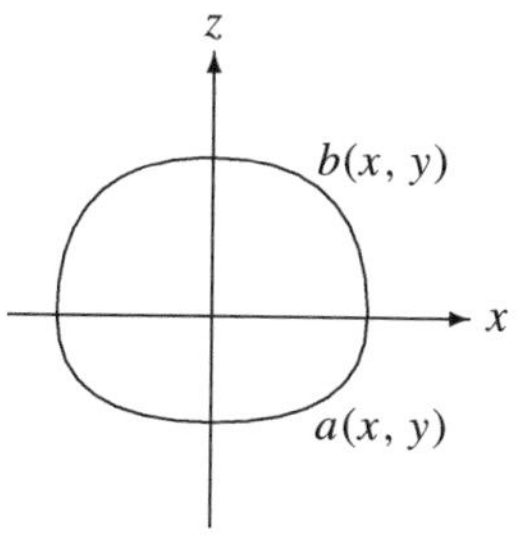

A volume described by $a(x, y) \le z \le b(x, y)$.

It is intuitively rather clear that a general volume may be cut up into pieces of this kind. Adding the surface integrals the contributions from the mutual interfaces between neighbouring pieces will cancel each other in the sum, leaving only the integral over the outermost surface of the total volume on the left-hand side. Similarly the volume integrals add up to a volume integral over the total volume of all the pieces. Gauss' theorem thus holds in full generality.

∗ What about non-gradient forces?

The local equation of hydrostatic equilibrium, $\boldsymbol{f} = \nabla p$, demands that the force density must equal a gradient field, and thus have a vanishing curl $\nabla \times \boldsymbol{f} = \mathbf{0}$ (see problem 2.15). A material with constant density, $\rho = \rho_0$, has a gravitational force density $\boldsymbol{f} = \rho_0 \boldsymbol{g} = -\rho_0 \nabla \Phi$ which is evidently a gradient field. But what happens if a force has a manifestly non-vanishing curl? As an example one can take $\boldsymbol{f} = \boldsymbol{a} \times \boldsymbol{x}$, which has $\nabla \times \boldsymbol{f} = 2\boldsymbol{a}$. Then the only possible conclusion is that hydrostatic equilibrium cannot be established, and the fluid must start to move. A physical example of this phenomenon is an electrically charged fluid with a magnetic field that everywhere increases linearly with time. Such a magnetic field induces a static electric field with non-vanishing curl, which indeed accelerates the charged molecules, and thereby the fluid.

4.4 Equation of state

The local equation of hydrostatic equilibrium is not enough in itself, but needs a relation between density and pressure. In the examples of the preceding section we assumed that the fluid was incompressible with constant density and could then integrate the hydrostatic equation and determine the pressure.

Ordinary thermodynamics [11, 35] provides us with a relationship between density ρ, pressure p, and absolute temperature T, called the *equation of state*, which may be written in many equivalent ways, for example

$$f(\rho, p, T) = 0. \tag{4.24}$$

In continuum physics the equation of state should be understood as a *local* relation, valid at every point $\boldsymbol{x}$,

$$f(\rho(\boldsymbol{x}), p(\boldsymbol{x}), T(\boldsymbol{x})) = 0. \tag{4.25}$$

As usual we shall suppress the explicit dependence on $\boldsymbol{x}$ when it does not lead to ambiguity. The actual form of the equation of state for a particular substance is derived from the properties of molecular interactions, a subject that falls outside the scope of this book.

Benoit Paul Émile Clapeyron (1799–1864). *French engineer and physicist. Formulated the ideal gas law from previous work by Boyle, Mariotte, Charles, Gay-Lussac, and others. Contributed to early thermodynamics by building on Carnot's work. Defined the concept of reversible transformations and formulated the first version of the Second Law of Thermodynamics. Established what is now called Clapeyron's formula for the latent heat in the change of state of a pure substance.*

The ideal gas law

The oldest and most famous equation of state is the *ideal gas law*, credited to Clapeyron (1834) and usually presented in the form,

$$pV = nRT. \tag{4.26}$$

Here n is the number of moles of gas in a small volume V, and $R = 8.31451(6)\ \mathrm{J\,K^{-1}\,mol^{-1}}$ is the *molar gas constant*. This equation of state has played an enormous role in the development of thermodynamics, and an ideal gas is still the best 'laboratory' for understanding materials with a non-trivial thermodynamics. In appendix C the thermodynamics of ideal gases is recapitulated in some detail.

Using that $\rho = M/V = nM_{\mathrm{mol}}/V$, where M is the mass of the gas in V and M_{mol} its molar mass, we obtain the ideal gas law in a form more suited for continuum physics,

$$\boxed{\frac{p}{\rho} = \frac{RT}{M_{\mathrm{mol}}}.} \tag{4.27}$$

The ideal gas law is not only valid for pure gases but also for mixtures of pure gases provided one uses the molar average of the molar mass of the mixture (see problem C.1). Real air with $M_{\mathrm{mol}} = 28.9635\ \mathrm{g/cm^3}$ is quite well described by the ideal gas law, although in precise calculations it may be necessary to include nonlinear corrections as well as corrections due to humidity [34].

4.5 Barotropic fluid states

The problem with the equation of state (4.24) is, however, that it is not a simple relation between density and pressure which may be plugged into the equation of local hydrostatic equilibrium, but also involves temperature. To solve the general problem we need a further equation connecting temperature, pressure and density. Such a *heat equation* is also provided by thermodynamics, and we shall derive it in chapter 30.

Barotropic relationship

At this stage it is, however, best to avoid these complications and for simplicity assume that there exists a so-called *barotropic* relationship between pressure and density,

$$p = p(\rho) \qquad \text{or} \qquad \rho = \rho(p). \tag{4.28}$$

The assumption of a barotropic relationship is not as far-fetched as it might seem at first. The condition of constant density $\rho(\boldsymbol{x}) = \rho_0$ which we used in the preceding section to calculate the pressure in the sea is a trivial example of such a relationship in which the density is independent of the pressure.

A less trivial example is obtained if the walls containing a fluid at rest are held at a fixed temperature T_0. The omnipresent heat conduction will eventually cause all of the fluid to attain this temperature, $T(\boldsymbol{x}) = T_0$, and in this state of *isothermal equilibrium* the equation of state (4.25) simplifies to,

$$f(\rho(\boldsymbol{x}), p(\boldsymbol{x}), T_0) = 0 \tag{4.29}$$

which is indeed a barotropic relationship.

Isothermal atmosphere

Everybody knows that the atmosphere is not at constant temperature, but if we nevertheless assume it to be, we obtain by combining the equation of hydrostatic equilibrium (4.20) with the ideal gas law (4.27),

$$\frac{dp}{dz} = -\rho g_0 = -\frac{M_{\text{mol}}\, g_0}{RT_0}\, p. \tag{4.30}$$

With the initial condition $p = p_0$ for $z = 0$, this ordinary differential equation has the solution

$$p = p_0 e^{-z/h_0}, \tag{4.31}$$

where

$$h_0 = \frac{RT_0}{M_{\text{mol}}\, g_0} = \frac{p_0}{\rho_0 g_0}. \tag{4.32}$$

In the last step we have again used the ideal gas law at $z = 0$ to show that the expression for h_0 is identical to the incompressible scale height (4.6).

In the isothermal atmosphere the pressure thus decreases exponentially with height on a characteristic length scale again set by h_0. Now the pressure at $h = h_0 = 8.6$ km (roughly the top of Mount Everest) is finite and predicted to be $e^{-1} = 37\%$ of the pressure at sea level, or 373 hPa.

Bulk modulus

The archetypal thermodynamics experiment is carried out on a fixed amount $M = \rho V$ of a fluid placed in a cylindrical container with a moveable piston. When you slightly increase the force on the piston, the volume of the chamber decreases, $dV < 0$. The pressure in the fluid must necessarily increase, $dp > 0$. If this were not the case, an arbitrarily small extra force would send the piston to the bottom of the chamber. Since a larger volume diminishes proportionally more for a given pressure increase, we define the *bulk modulus* as the pressure increase dp per *fractional* decrease in volume $-dV/V$, or

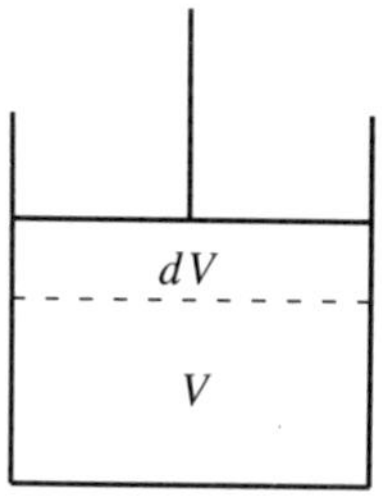

The archetypal thought-experiment in thermodynamics: A cylindrical chamber with a movable piston.

$$K = \frac{dp}{-dV/V} = \frac{dp}{d\rho/\rho} = \rho\frac{dp}{d\rho}. \tag{4.33}$$

In the second step we have used the constancy of the mass $dM = \rho dV + V d\rho = 0$ to eliminate the volume, $dV/V = -d\rho/\rho$. The bulk modulus is a measure of *incompressibility*, and the larger it is, the greater is the pressure increase that is needed to obtain a given fractional increase in density. As a measure of *compressibility* one usually takes $\beta = 1/K$.

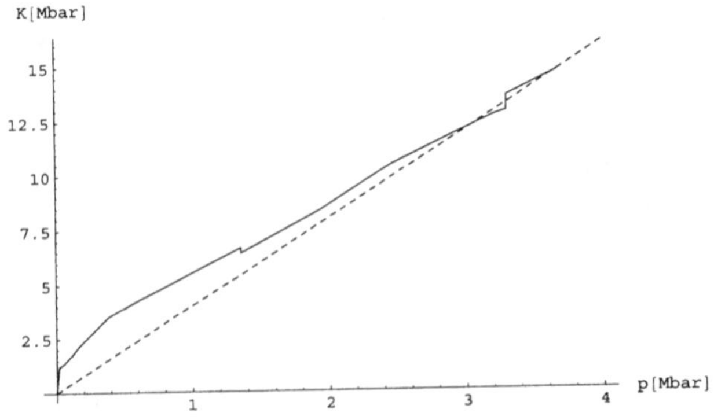

Figure 4.1. Bulk modulus as a function of pressure in the Earth (data from [34]). The surface of the Earth is to the left and the centre to the right in this figure. The bulk modulus varies approximately linearly with pressure, $K \approx 4p$ (the dashed line). The dramatic change in density at the core/mantle boundary (see figure 3.1) is barely visible in the bulk modulus.

	K[GPa]
Mercury	25.0
Glycerol	3.94
Water	2.21
Benzene	1.04
Ethanol	0.89
Methanol	0.82
Hexane	0.60

Bulk modulus for common liquids at normal temperature and pressure.

The definition of the bulk modulus shows that it is measured in the same units as pressure. In water under normal conditions it is fairly constant, $K \approx 22\,000$ atm, doubling only in value between 1 and 3000 atm [34]. As long as the pressure change is much smaller than the bulk modulus, $dp \ll K$, we may estimate the relative change in density from (4.33) to be $d\rho/\rho \approx dp/K$. In the deepest abyss of the sea the pressure is a bit more than 1000 atm, implying that the relative density change is $d\rho/\rho \approx 1/22 \approx 4.5\%$ (see also problem 4.7).

For an isothermal ideal gas it follows from the equation of state (4.27) that the bulk modulus is $K_T = p$ where the index T reminds us that the temperature of the gas must be kept constant. The bulk modulus of the nearly fluid material of the Earth is plotted in figure 4.1 and varies roughly as $K \approx 4p$.

Pressure potential

For a barotropic fluid we may integrate the local equation of hydrostatic equilibrium in much the same way as we did for constant density in (4.21), by defining

$$H = \Phi + w(p), \tag{4.34}$$

where now the *pressure potential* is the integral,

$$w(p) = \int \frac{dp}{\rho(p)}. \tag{4.35}$$

It follows from the chain rule that $\nabla w(p) = (dw/dp)\nabla p = (1/\rho)\nabla p$ and using local hydrostatic equilibrium (4.19) we obtain $\nabla H = \mathbf{0}$. In hydrostatic equilibrium H is always a constant, even when the barotropic fluid is compressible.

For an ideal gas under isothermal conditions, the pressure potential is calculated by means of the ideal gas law (4.27),

$$w = \int \frac{RT_0}{M_{\text{mol}}\, p} dp = \frac{RT_0}{M_{\text{mol}}} \log p. \tag{4.36}$$

In flat-Earth gravity $\Phi = g_0 z$, the constancy of H immediately leads back to the exponentially decreasing pressure in the isothermal atmosphere (4.31).

4.6 The homentropic atmosphere

The assumption that the temperature is the same everywhere in the atmosphere is certainly wrong, as anyone who has ever flown in a modern passenger jet and listened to the pilot can testify. Temperature falls with height instead of staying constant. So the atmosphere is *not* in isothermal equilibrium, and this is perhaps not so surprising, since the 'container walls' of the atmosphere, the ground and outer space, have different temperatures. There must be a heat flow through the atmosphere between the ground and outer space, maintained by the inflow of solar radiation and the outflow of geothermal energy. But air is a bad conductor of heat, so although heat conduction does play a role, it is not directly the cause of the temperature drop in the atmosphere.

Of much greater importance are the indirect effects of solar heating, the *convection* which creates air currents, winds, and local turbulence, continually mixing different layers of the atmosphere. The lower part of the atmosphere, the *troposphere*, is quite unruly and vertical mixing happens at time scales that are much shorter than those required to reach thermal equilibrium. There is in fact no true hydrostatic equilibrium state for the real atmosphere. Even if we disregard large-scale winds and weather systems, horizontal and vertical mixing always takes place at small scales, and a realistic model of the atmosphere must take this into account.

Blob-swapping

Let us imagine that we take a small blob of air and exchange it with another blob of air of the same mass, but taken from a different height with a different volume and pressure. In order to fill out the correct volume, one air mass would have to be compressed and the other expanded. If this is done quickly, there will be no time for heat exchange with the surrounding air, and one air mass will consequently be heated up by compression and the other cooled down by expansion. If the atmosphere initially were in isothermal equilibrium, the temperature of the swapped air would not be the same as the temperature of the surrounding air, and the atmosphere would be brought out of equilibrium.

If, however, the surrounding air initially had a temperature distribution, such that the swapped air after the expansion and compression would arrive at precisely the correct temperatures of their new surroundings, a kind of thermodynamic 'equilibrium' could be established, in which the omnipresent vertical mixing had essentially no effect. Intuitively, it is reasonable to expect that the end result of fast vertical mixing and slow heat conduction might be precisely such a state. It should, however, not be forgotten that this state is not a true equilibrium state but rather a dynamically balanced state depending on the incessant small-scale motion in the atmosphere.

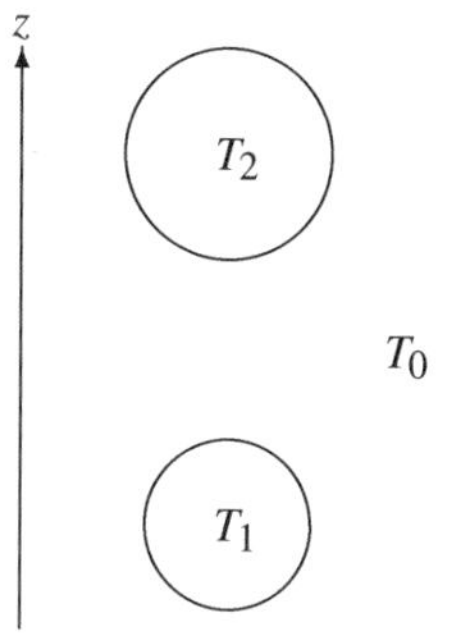

Swapping air masses from different heights. If the air has temperature T_0 before the swap, the compressed air would be warmer $T_1 > T_0$ and the expanded colder $T_2 < T_0$.

Isentropic processes in ideal gases

A process that takes place without exchange of heat between the system and its environment is said to be *adiabatic*. If furthermore the process is *reversible*, it will conserve the entropy and is called *isentropic*. From the thermodynamics of ideal gases (see appendix C) it follows that an isentropic process in a fixed amount M of an ideal gas will leave the expression pV^γ unchanged. Here γ is the so-called *adiabatic index* which for a gas like air with diatomic molecules is approximately $\gamma \approx 7/5 = 1.4$. In terms of the density $\rho = M/V$ an isentropic process thus obeys,

$$p\,\rho^{-\gamma} = C. \tag{4.37}$$

Whereas the 'constants' γ and C keep their values during a local isentropic process, they could in principle vary with the position $\boldsymbol{x}$.

The bulk modulus (4.33) of an isentropic gas is immediately found to be,

$$K_S = \gamma p, \tag{4.38}$$

where the S indicates that the entropy must be kept constant during the compression. The pressure potential (4.35) is similarly obtained,

$$w = \int \frac{dp}{\rho} = \int C\gamma\rho^{\gamma-2}\,d\rho = C\frac{\gamma}{\gamma-1}\rho^{\gamma-1}.$$

By means of (4.37) and the ideal gas law (4.27) this may also be written,

$$w = \frac{\gamma}{\gamma-1}\frac{p}{\rho} = \frac{\gamma}{\gamma-1}\frac{RT}{M_{\text{mol}}} = c_p T, \tag{4.39}$$

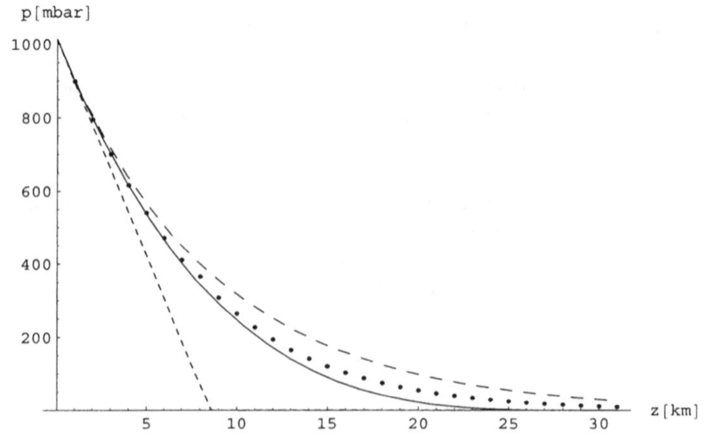

Figure 4.2. Three different models for the atmospheric pressure: constant density (dashed), homentropic (fully drawn) and isothermal (large dashes), plotted together with the standard atmosphere data (dots) [41]. The parameters are $h_0 = 8.6$ km and $\gamma = 7/5$.

where

$$c_p = \frac{dw}{dT} = \frac{\gamma}{\gamma - 1} \frac{R}{M_{\text{mol}}}, \tag{4.40}$$

is the *specific heat at constant pressure* of the gas (see appendix C). For air with $M_{\text{mol}} \approx 29$ g mol^{-1} and $\gamma \approx 7/5$ its value is $c_p \approx 1000$ J K^{-1} kg^{-1}.

The atmospheric temperature lapse rate

The atmosphere of the flat Earth is translationally invariant in the horizontal directions, implying that γ and C can only depend on z, but the blob-swapping argument indicates that both should also be independent of z. The lower atmosphere, the troposphere, is at least approximatively in a so-called *homentropic state* in which (4.37) is valid everywhere with the same values of γ and C.

In constant gravity we find from (4.34) and (4.39) that

$$H = g_0 z + c_p T, \tag{4.41}$$

is independent of z, implying that the temperature drops linearly with height,

$$T = T_0 - \frac{g_0}{c_p} z, \tag{4.42}$$

where $T_0 = H/g_0$ is the temperature at the surface $z = 0$. The magnitude of the vertical temperature gradient $-dT/dz = g_0/c_p \approx 0.01$ K m^{-1} = 10 K km^{-1} is called the *atmospheric temperature lapse rate*.

The above equation may also be written

$$T = T_0 \left(1 - \frac{z}{h_2}\right), \tag{4.43}$$

with the scale height,

$$h_2 = \frac{c_p T_0}{g_0} = \frac{\gamma}{\gamma - 1} h_0. \tag{4.44}$$

For $\gamma = 7/5$ we find $h_2 \approx 30$ km. At this altitude the temperature has dropped to absolute zero, which is, of course, unphysical. It is nevertheless a reasonable scale for the true height of the atmosphere.

The equation of state (4.27) combined with the adiabatic law (4.37) implies that $T\rho^{1-\gamma}$ and $p^{1-\gamma}T^{-\gamma}$ are also constant, and the density and pressure become,

$$\rho = \rho_0 \left(1 - \frac{z}{h_2}\right)^{1/(\gamma-1)}, \qquad p = p_0 \left(1 - \frac{z}{h_2}\right)^{\gamma/(\gamma-1)}. \tag{4.45}$$

Both of these quantities vanish like the temperature for $z = h_2$. At the top of Mount Everest ($z = h_0$) the pressure is now predicted to be 312 hPa whereas the standard atmosphere data says it is 330 hPa.

The real atmosphere of Earth

In figure 4.2 the various atmospheric models for the pressure have been plotted together with data for the standard atmosphere [41]. Even if the homentropic model gives the best fit, it fails at higher altitudes. The real atmosphere is in fact much more complicated than any of these models indicate.

Water vapour is always present in the atmosphere and will condense to clouds in rising currents of air. The latent heat of condensation heats up the air, so that the temperature lapse rate becomes smaller than 10 K km^{-1}, perhaps more like 6–7 K km^{-1}, leading to a somewhat higher estimate for the temperature at the top of Mount Everest. The clouds may eventually precipitate out as rain, and when the dried air afterwards descends again, for example on the lee side of a mountain, the air will heat up at a higher rate than it cooled during its ascent on the windward side and become quite hot, a phenomenon known as *föhn* in the Alps.

The fact that the temperature lapse rate is smaller in the real atmosphere than in the isentropic model has a bearing on the stability of the atmosphere. If a certain amount of air is transported to higher altitude without heat exchange and condensation of water vapour, it will behave like in the isentropic model and become cooler than the surrounding air. Consequently it will also be heavier than the surrounding air and tend to sink back to where it came from, implying that the atmosphere is stable. Conversely, if the real temperature lapse becomes larger than in the isentropic model, the atmosphere becomes unstable and strong vertical currents may arise. This can, for example, happen in thunderstorms.

Problems

4.1 Consider a canal with a dock gate which is 12 m wide and has water depth 9 m on one side and 6 m on the other side.

(a) Calculate the pressures in the water on both sides of the gate at a height z over the bottom of the canal.

(b) Calculate the total force on the gate.

(c) Calculate the total moment of force around the bottom of the gate.

(d) Calculate the height over the bottom at which the total force acts.

4.2 An underwater lamp is covered by a hemispherical glass with a diameter of 30 cm and is placed with its centre at a depth of 3 m on the side of the pool. Calculate the total horizontal force from the water on the lamp, when there is air at normal pressure inside.

4.3 Using a manometer, the pressure in an open container filled with liquid is found to be 1.6 atm at a height of 6 m over the bottom, and 2.8 atm at a height of 3 m. Determine the density of the liquid and the height of the liquid surface.

4.4 An open jar contains two non-mixable liquids with densities $\rho_1 > \rho_2$. The heavy layer has thickness h_1 and the light layer on top of it has thickness h_2. **(a)** An open glass tube is lowered vertically into the liquids towards the bottom of the jar. Describe how high the liquids rise in the tube (disregarding capillary effects). **(b)** The open tube is already placed in the container with its opening close to the bottom when the heavy fluid is poured in, followed by the light. How high will the heavy fluid rise in the tube?

4.5 The equation of state due to van der Waals is

$$\left(P + \frac{n^2 a}{V^2}\right)(V - nb) = nRT \tag{4.46}$$

where a and b are constants. It describes gases and their condensation into liquids. **(a)** Calculate the isothermal bulk modulus. **(b)** Under which conditions can it become negative, and what does it mean?

4.6 The equation of state for water is to a good approximation (for pressures up to 100 000 bar) given by

$$\frac{p + B}{p_0 + B} = \left(\frac{\rho}{\rho_0}\right)^n \tag{4.47}$$

with $B = 3000$ atm, $n = 7$, $p_0 = 1$ atm and $\rho_0 = 1$ g cm^{-3}. **(a)** Calculate the bulk modulus K for water. **(b)** Calculate the density and pressure distribution in the sea. **(c)** What is the pressure and the relative compression of the water at the deepest point in the sea ($z = -10.924$ km)?

4.7 Calculate the pressure and density in the sea, assuming constant bulk modulus. Show that both quantities are singular at a certain depth and calculate this depth.

4.8 A vertical plate is inserted into a liquid at rest with constant density ρ_0 in constant gravity g_0. Introduce a coordinate z going vertically down with the pressure defined to vanish for $z = 0$. In the following we denote the vertical area moments by,

$$I_n = \frac{1}{A}\int_A z^n \, dS, \qquad n = 1, 2, \ldots \tag{4.48}$$

where dS is the surface element. The point $z_M = I_1$ is called the area centre.

(a) Calculate the pressure in the liquid.

(b) Show that $I_2 \geq I_1^2$.

(c) Calculate the total pressure force on the plate.

(d) Calculate the total moment of force of the pressure forces around $z = 0$.

(e) Show that the point of attack of the pressure forces is found below the area centre $z_P \geq z_M$.

(f) A thin isosceles triangle with height h and bottom length b is lowered into the liquid such that its top point is at $z = 0$. Calculate the area centre and the point of attack of the pressure forces.

4.9 Determine the form of the pressure across the core/mantle boundary when the bulk modulus is $K \approx \gamma p$ with $\gamma \approx 4$ throughout the Earth (see figure 4.1).

5

Buoyancy

Fishes, whales, submarines, balloons and airships all owe their ability to float to *buoyancy*, the lifting power of water and air. The understanding of the physics of buoyancy goes back as far as antiquity and probably sprung from the interest in ships and shipbuilding in classic Greece. The basic principle is due to Archimedes. His famous Law states that the buoyancy force on a body is equal and oppositely directed to the weight of the fluid that the body replaces. Actually the Law was not just one law, but a set of four propositions dealing with different configurations of body and liquid [70]. Before his time it was wrongly thought that the shape of a body determined whether it would sink or float.

'Buoy' mostly pronounced 'booe', probably of Germanic origin. A tethered floating object used to mark a location in the sea.

The shape of a floating body and its mass distribution does, however, determine whether it will float stably or capsize. Stability of floating bodies is of importance to shipbuilding, and to anyone who has ever tried to stand up in a small rowboat. Newtonian mechanics not only allows us to derive Archimedes' Principle for the equilibrium of floating bodies, but also to characterize the deviations from equilibrium and calculate the restoring forces. Even if a body floating in or on water is in hydrostatic equilibrium, it will not be in complete mechanical balance in every orientation, because the centre of mass of the body and the centre of mass of the displaced water, also called the centre of buoyancy, do not in general coincide.

Archimedes of Syracuse (287–212 BC). *Greek mathematician. Discovered the formulae for area and volume of cylinders and spheres. Considered the father of fluid mechanics.*

The mismatch between the centers of mass and buoyancy for a floating body creates a moment of force, which tends to rotate the body towards a stable equilibrium. For submerged bodies, submarines, fishes and balloons, the stable equilibrium will always be with the centre of gravity situated directly below the centre of buoyancy. For bodies floating stably on the surface, ducks, ships and dumplings, the centre of gravity is mostly found directly above the centre of buoyancy.

5.1 Archimedes' principle

Mechanical equilibrium takes a slightly different form than global hydrostatic equilibrium (4.15) when a body of another material is immersed in a fluid. If its material is incompressible, the body retains its shape and displaces an amount of fluid with exactly the same volume. If the body is compressible, like a rubber ball, the volume of displaced fluid will be smaller. The body may even take in fluid, like the piece of bread you dunk into your coffee, but then the physics becomes more complicated, and we shall disregard this possibility in the following. A body which is partially immersed may formally be viewed as a body that is fully immersed in a fluid for which the mass density and the equation of state vary from place to place. This also covers the case where part of the body is in vacuum which may be thought of as a fluid with the extreme properties, $\rho = p = 0$.

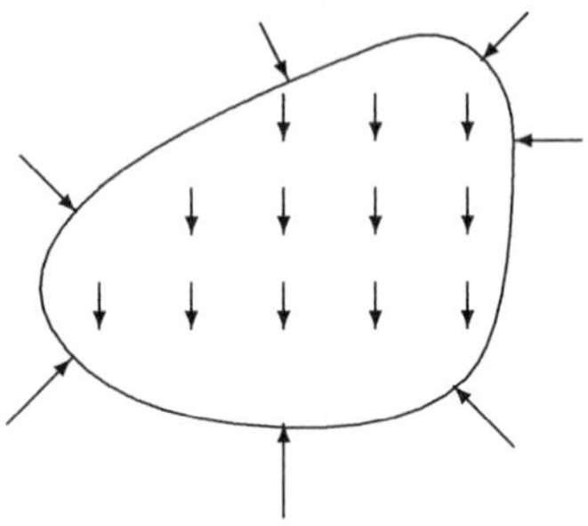

Gravity pulls on a body over its entire volume, while pressure only acts on the surface.

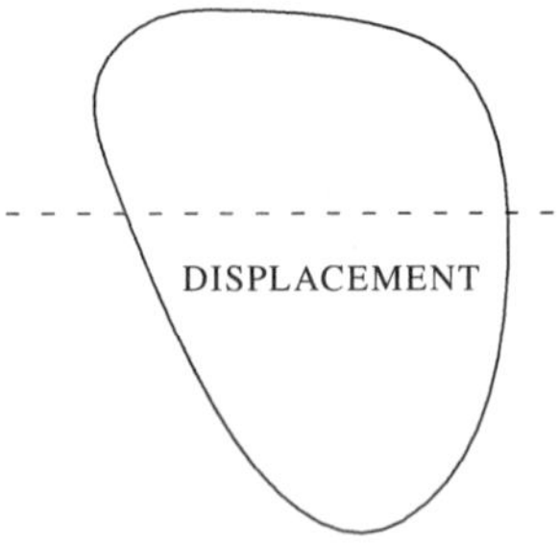

For a body partially submerged in water the displacement is the amount of water that has been displaced by the volume of the body below the waterline.

Karl Friedrich Hieronymus Freiherr von Münchhausen (1720–1797). *German (Hanoveran) soldier, hunter, nobleman, and delightful story-teller. In one of his stories, he lifts himself out of a deep lake by pulling at his bootstraps.*

Weight and buoyancy

Let the actual, perhaps compressed, volume of the immersed body be V with surface S. In the field of gravity an unrestrained body is subject to two forces: its weight

$$\mathcal{F}_G = \int_V \rho_{\text{body}}\, \boldsymbol{g}\, dV, \tag{5.1}$$

and the buoyancy due to pressure acting on its surface,

$$\mathcal{F}_B = -\oint_S p\, d\boldsymbol{S}. \tag{5.2}$$

In general these two forces do not have to be in balance. The resultant $\mathcal{F} = \mathcal{F}_G + \mathcal{F}_B$ determines the direction that the unrestrained body will begin to move. In mechanical equilibrium the two forces must exactly cancel each other so that the body can remain in place.

Assuming that the body does not itself significantly contribute to the field of gravity, the local balance of forces in the fluid (4.19) will be the same as before the body was placed in the fluid. In particular the pressure in the fluid cannot depend on whether the volume V contains material that is different from the fluid itself. The pressure on the surface of the immersed body must for this reason be identical to the pressure on a body of fluid of the same shape. But then the global equilibrium condition (4.15) tells us that the buoyancy force will exactly balance the weight of the displaced fluid, so that

$$\mathcal{F}_B = -\oint_S p\, d\boldsymbol{S} = -\int_V \rho_{\text{fluid}}\, \boldsymbol{g}\, dV. \tag{5.3}$$

This theorem is indeed Archimedes' principle: *the force of buoyancy equals (minus) the weight of the displaced fluid.*

The total force on the body may now be written

$$\mathcal{F} = \mathcal{F}_G + \mathcal{F}_B = \int_V (\rho_{\text{body}} - \rho_{\text{fluid}})\boldsymbol{g}\, dV, \tag{5.4}$$

explicitly confirming that when the body is made from the same fluid as its surroundings, so that $\rho_{\text{body}} = \rho_{\text{fluid}}$, the resultant force vanishes automatically. In general, however, the distributions of mass in the body and in the displaced fluid will be different.

> Note that Archimedes' principle is valid even if the gravitational field varies appreciably across the body. Archimedes principle fails, however, if the body is so large that its own gravitational field cannot be neglected, such as would be the case if an Earth-sized body fell into Jupiter's atmosphere. The extra compression of the fluid near the surface of the body generally increases the buoyancy force. In semblance with Baron von Münchhausen's adventure, the body in effect lifts itself by its bootstraps (see problems 5.6 and 5.7).

Constant field of gravity

If the gravitational field is constant, $\boldsymbol{g}(\boldsymbol{x}) = \boldsymbol{g}_0$, the weight of the body is,

$$\mathcal{F}_G = M_{\text{body}}\, \boldsymbol{g}_0, \tag{5.5}$$

and the buoyancy force becomes

$$\mathcal{F}_B = -M_{\text{fluid}}\, \boldsymbol{g}_0. \tag{5.6}$$

Since the total force is the sum of these contributions, one might say that buoyancy acts as if the displacement were filled with fluid of negative mass $-M_{\text{fluid}}$. In effect the buoyancy force acts as a kind of antigravity.

The total force on an unrestrained object is now,

$$\mathcal{F} = \mathcal{F}_G + \mathcal{F}_B = (M_{\text{body}} - M_{\text{fluid}})\boldsymbol{g}_0. \tag{5.7}$$

If the body mass is smaller than the mass of the displaced fluid, the total force is directed upwards, and the unrestrained body will begin to move upwards. Alternatively, if the body is kept in place, the restraints must deliver a force $-\mathcal{F}$ to prevent the object from moving.

A body can only hover motionlessly in a fluid if its mass equals the mass of the displaced fluid,

$$M_{\text{body}} = M_{\text{fluid}}. \tag{5.8}$$

Fish achieve this balance by adjusting the amount of water they displace through contraction and expansion of an internal air-filled bladder. Submarines, in contrast, change their mass by pumping water in and out of ballast tanks. Curiously, no animals seem to have developed balloons for floating in the atmosphere, although both the physics and chemistry of ballooning appears to be within reach of biological evolution.

5.2 The gentle art of ballooning

The first balloon flights are credited to the Montgolfier brothers who in November 1783 flew a manned hot-air balloon, and to Jacques Charles who on December 1 that same year flew a manned hydrogen gas balloon. In the beginning there was an intense rivalry between the advocates of Montgolfier and Charles type balloons which presented different advantages and dangers to the courageous fliers. Hot air balloons were easier to make but prone to catch fire, while hydrogen balloons had greater lifting power but could suddenly explode. By 1800 the hydrogen balloon had won the day, culminating in the huge (and dangerous) hydrogen airships of the 1930s. Helium balloons are much safer, but also much more expensive to fill. In the last half of the twentieth century hot-air balloons again came into vogue, especially for sports, because of the availability of modern strong lightweight materials (nylon) and fuel (propane).

Joseph Michel Montgolfier (1740–1810). *Experimented (together with his younger brother Jacques Étienne (1745–1799)) with hot-air balloons. On November 21, 1783, the first human flew in such a balloon for a distance of 9 km at a height of 100 m above Paris. Only one of the brothers ever flew, and then only once!*

Gas balloons

A large hydrogen or helium balloon typically begins its ascent being only partially filled, assuming an inverted tear-drop shape. During the ascent the gas expands because of the fall in ambient air pressure, and eventually the balloon becomes nearly spherical and stops expanding (or bursts) because the 'skin' of the balloon cannot stretch further. Since the density of the displaced air falls with height, the balloon will eventually reach a maximum height, a ceiling where it could hover permanently if it did not lose gas. In the end no balloon stays aloft forever.

Jacques Alexandre César Charles (1746–1823). *French physicist. The first to use hydrogen balloons for manned flight, and made in December 1783 an ascent to about 3 km. Discovered Charles' law, a forerunner of the ideal gas law, stating that the ratio of volume to absolute temperature (V/T) is constant for a given pressure.*

Let the total mass of the balloon be M_0, including the mass of the gas, the balloon skin, the gondola, people etc. The condition for upwards flight is then that $M_0 \leq V\rho$ where V is the total volume of air that the balloon displaces and ρ the air density at its actual position. In the homentropic atmospheric model the air density is given by (4.45), and the condition for upwards flight at height z becomes,

$$M_0 < \rho_0 V \left(1 - \frac{z}{h_2}\right)^{1/(\gamma-1)} \tag{5.9}$$

where $\gamma \approx 7/5$ is the adiabatic index of air, $\rho_0 \approx 1.2$ kg m^{-3} its density at sea level, and $h_2 \approx 30$ km the isentropic scale height (4.44). If this inequality is fulfilled on the ground, the balloon will start to rise. During the rise the volume may expand towards a maximal value while the air density falls, and the balloon will keep rising until the inequality is no longer fulfilled, and the balloon has reached its ceiling.

Example 5.2.1: A gas balloon has a maximal spherical diameter of 10 m yielding a volume $V \approx 524$ m^3. For the balloon to lift-off at all, its mass must be smaller than $\rho_0 V = 628$ kg. Taking $M_0 = 400$ kg the ceiling becomes $z \approx 5$ km. At this height the air pressure is $p = 0.53$ atm and the temperature $T = 245$ K $= -29\,^\circ$ C (ignoring humidity). Assuming that the balloon contains hydrogen H_2 (with $M_{H_2} = 2$ g mol^{-1} and $\gamma = 7/5$) at this temperature and pressure, the total mass of the hydrogen is merely 28 kg. The surface area of the balloon is 314 m^2, so if the skin has thickness 2 mm and density 300 kg m^{-3}, its mass becomes 188 kg, which leaves about $400 - 28 - 188 = 184$ kg for the proper payload. Filled with helium He (with $M_{\text{mol}} = 4$ g mol^{-1} and $\gamma = 5/3$), the proper payload would be reduced to 156 kg.

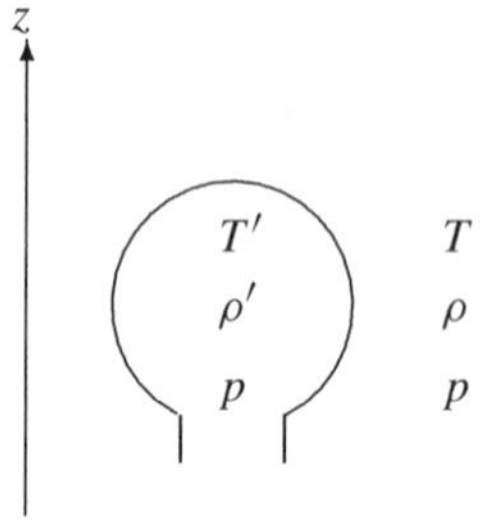

A hot-air balloon has higher temperature $T' > T$ and lower density $\rho' < \rho$ but the same pressure as the surrounding atmosphere because it is open below.

Hot-air balloons

A hot-air balloon is open at the bottom so that the inside pressure is always the same as the atmospheric pressure outside. The air in the balloon is warmer ($T' > T$) than the outside temperature and the density is correspondingly lower ($\rho' < \rho$). If M_0 denotes the total mass of the balloon, including the gondola and the passengers but not the hot air, the condition for flight is now $M_0 < (\rho - \rho')V$. From the ideal gas law (4.27) and the equality of the inside and outside pressures it follows that $\rho' T' = \rho T$, so that the inside density is $\rho' = \rho T/T'$. The condition for flight at height z becomes,

$$M_0 < \left(1 - \frac{T}{T'}\right)\rho V = \left(1 - \frac{T_0}{T'}\left(1 - \frac{z}{h_2}\right)\right)\left(1 - \frac{z}{h_2}\right)^{1/(\gamma-1)} \rho_0 V. \tag{5.10}$$

On the right-hand side we have inserted the expressions (4.43) and (4.45) for the homentropic atmospheric temperature and density.

Example 5.2.2: A spherical hot-air balloon with diameter $d = 15$ m is desired to reach a ceiling of $z = 1000$ m with air temperature $T' = 120°$ C $= 393$ K. When the ground temperature is $T_0 = 20°$ C $= 293$ K and the density $\rho_0 = 1.2$ kg m^{-3}, it follows that this balloon would be capable of lifting $M_0 \approx 547$ kg to the ceiling. Using the same parameters as for the gas balloon, the payload becomes 123 kg.

5.3 Stability of floating bodies

Although a body may be in buoyant equilibrium, such that the total force composed of gravity and buoyancy vanishes, $\boldsymbol{\mathcal{F}} = \boldsymbol{\mathcal{F}}_G + \boldsymbol{\mathcal{F}}_B = \mathbf{0}$, it may not be in complete mechanical equilibrium. The total moment of all the forces acting on the body must also vanish; otherwise an unrestrained body will start to rotate.

Moments of weight and buoyancy

The total moment is like the total force a sum of two contributions,

$$\boldsymbol{\mathcal{M}} = \boldsymbol{\mathcal{M}}_G + \boldsymbol{\mathcal{M}}_B, \tag{5.11}$$

with one contribution from gravity,

$$\boldsymbol{\mathcal{M}}_G = \int_V \boldsymbol{x} \times \rho_{\text{body}} \boldsymbol{g}\, dV, \tag{5.12}$$

and the other from pressure, i.e. buoyancy,

$$\boldsymbol{\mathcal{M}}_B = \oint_S \boldsymbol{x} \times (-p\, d\boldsymbol{S}). \tag{5.13}$$

If the total force vanishes, $\boldsymbol{\mathcal{F}} = \mathbf{0}$, the total moment will be independent of the origin of the coordinate system (page 32).

Assuming again that the presence of the body does not change the pressure distribution in the fluid, the moment of buoyancy is independent of the nature of the material inside V. In hydrostatic equilibrium the total moment on the same volume of fluid must vanish, $\boldsymbol{\mathcal{M}}_G^{\text{fluid}} + \boldsymbol{\mathcal{M}}_B = \mathbf{0}$, such that we get

$$\boldsymbol{\mathcal{M}}_B = -\int_V \boldsymbol{x} \times \rho_{\text{fluid}} \boldsymbol{g}\, dV. \tag{5.14}$$

The moment of buoyancy equals the (minus) moment of gravity of the displaced fluid. This result is a natural corollary to Archimedes' principle, and of immense help in calculating the buoyancy moment. A formal proof of the theorem is found in problem 5.8.

Constant gravity and buoyant equilibrium

In the remainder of this chapter we assume that gravity is constant, $\boldsymbol{g}(\boldsymbol{x}) = \boldsymbol{g}_0$, and that the body is in buoyant equilibrium so that it displaces exactly its own mass of fluid, $M_{\text{fluid}} = M_{\text{body}} = M$. The densities of body and displaced fluid will, however, in general be different, $\rho_{\text{body}} \neq \rho_{\text{fluid}}$.

The moment of gravity (5.12) may as before (page 32) be expressed in terms of the centre $\boldsymbol{x}_G$ of the body mass distribution (here called the *centre of gravity*),

$$\boldsymbol{\mathcal{M}}_G = \boldsymbol{x}_G \times M\boldsymbol{g}_0, \qquad \boldsymbol{x}_G = \frac{1}{M}\int \boldsymbol{x}\,\rho_{\text{body}}\,dV. \tag{5.15}$$

Similarly the moment of the mass distribution of the displaced fluid (5.14) is,

$$\boldsymbol{\mathcal{M}}_B = -\boldsymbol{x}_B \times M\boldsymbol{g}_0, \qquad \boldsymbol{x}_B = \frac{1}{M}\int \boldsymbol{x}\,\rho_{\text{fluid}}\,dV, \tag{5.16}$$

where $\boldsymbol{x}_B$ is the *centre of buoyancy*. Although each of these moments depends on the choice of origin of the coordinate system, the total moment,

$$\boxed{\boldsymbol{\mathcal{M}} = (\boldsymbol{x}_G - \boldsymbol{x}_B) \times M\boldsymbol{g}_0,} \tag{5.17}$$

will be independent, as witnessed by the appearance of the difference of the two centre positions.

As long as the total moment is non-vanishing, the body is not in mechanical equilibrium, but will start to rotate towards an orientation with vanishing moment. Except for the trivial case where the centers of gravity and buoyancy coincide, the above equation tells us that the total moment can only vanish if the centers lie on the same vertical line,

$$\boldsymbol{x}_G - \boldsymbol{x}_B \propto \boldsymbol{g}_0. \tag{5.18}$$

For $\boldsymbol{x}_G \neq \boldsymbol{x}_B$, there are two possible orientations satisfying this condition: one where the centre of gravity lies above the centre of buoyancy, and another where the centre of gravity is lowest. At least one of these will be stable.

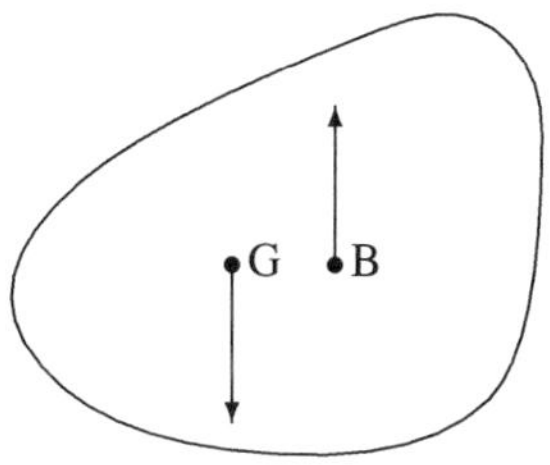

Body in buoyant equilibrium but with non-vanishing total moment which here sticks out of the paper. The moment will for a submerged body tend to rotate it in the anticlockwise direction and thus bring the centre of gravity below the centre of buoyancy.

Submerged body

For a fully submerged rigid body, for example a submarine, both centres are always in the same place relative to the body. If the centre of gravity does not lie directly below the centre of buoyancy, but is displaced horizontally, the direction of the moment will always tend to turn the body so that the centre of gravity is lowered with respect to the centre of buoyancy. The only stable orientation of the body is where the centre of gravity lies vertically below the centre of buoyancy. Any small perturbation away from this orientation will soon be corrected and the body brought back to the equilibrium orientation. A similar argument shows that the other equilibrium orientation with the centre of gravity above the centre of buoyancy is unstable and will flip the body over, if perturbed the tiniest amount.

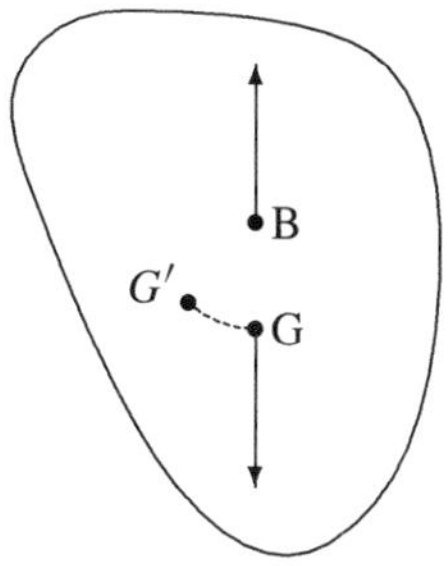

A fully submerged body in stable equilibrium must have the centre of gravity directly below the centre of buoyancy. If G is moved to G' a restoring moment is created which sticks out of the plane of the paper.

> This is why the gondola hangs below an airship or balloon, and why a fish goes belly-up when it dies, because it loses control of the swim bladder which enlarges into the belly and reverses the positions of the centers of gravity and buoyancy. It generally also loses buoyant equilibrium and floats to the surface.

Body floating on the surface

At the surface of a liquid, a body such as a ship or an iceberg will according to Archimedes' principle always arrange itself so that the mass of displaced liquid exactly equals the mass of the body. Here we assume that there is vacuum or a very light fluid such as air above the liquid. The centre of gravity is always in the same place relative to the body, but the centre of buoyancy depends now on the orientation of the body, because the volume of displaced fluid changes place and shape (while keeping its mass constant) when the body orientation changes.

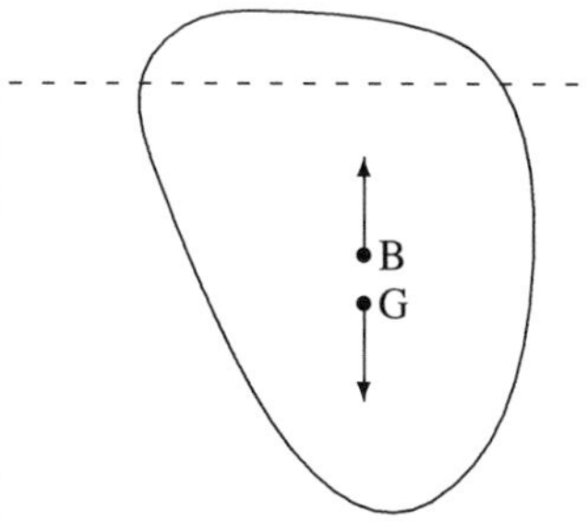

A floating body may have a stable equilibrium with the centre of gravity directly *below* the centre of buoyancy.

Stability can again only occur when the two centres lie on the same vertical line, but there may be more than one stable orientation. A sphere made of homogeneous wood floating on water is stable in all

Figure 5.1. The Flying Enterprise (1952). A body can float stably in many orientations, depending on the position of its centre of gravity. In this case the list to port was caused by a shift in the load which moved the centre of gravity to the port side. The ship and its lonely captain Carlsen became famous because he stayed on board during the storm that eventually sent it to the bottom. Photograph courtesy *Politiken*, Denmark, reproduced with permission.

orientations. None of them are in fact truly stable, because it takes no force to move from one to the other. This is however a marginal case.

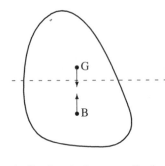

A floating body generally has a stable equilibrium with the centre of gravity directly *above* the centre of buoyancy.

A floating body may, like a submerged body, possess a stable orientation with the centre of gravity directly *below* the centre of buoyancy. A heavy keel is, for example, used to lower the centre of gravity of a sailing ship so much that this orientation becomes the only stable equilibrium. In that case it becomes virtually impossible to capsize the ship, even in a very strong wind.

The stable orientation for most floating objects, such as ships, will in general have the centre of gravity situated directly *above* the centre of buoyancy. This happens always when an object of constant mass density floats on top of a liquid of constant mass density, for example an iceberg on water. The part of the iceberg that lies below the waterline must have its centre of buoyancy in the same place as its centre of gravity. The part of the iceberg lying above the water cannot influence the centre of buoyancy whereas it always will shift the centre of gravity upwards.

How can that situation ever be stable? Will the moment of force not be of the wrong sign? Why don't ducks and tall ships capsize spontaneously? The qualitative answer is that when the body is rotated away from such an equilibrium orientation, the volume of displaced water will change position and shift the centre of buoyancy back to the other side of the centre of gravity, reversing the direction of the moment of force to restore the equilibrium.

5.4 Ship stability

Sitting comfortably in a small rowboat, it is fairly obvious that the centre of gravity lies above the centre of buoyancy, and that the situation is stable with respect to small movements of the body. But many a fisherman has learnt that suddenly standing up may compromise the stability and throw him out among the fishes. There is, as we shall see, a strict limit to how high the centre of gravity may be above the centre of buoyancy. If this limit is violated, the boat becomes unstable and capsizes.

Figure 5.2. The Queen Mary 2 set sail on its maiden voyage on January 2, 2004. It is the world's largest ocean liner—ever—with a length of 345 m, a height of 72 m from keel to funnel, and a width of 41 m. Having a draft of only 10 m, its superstructure rises an impressive 62 m over the waterline. The low average density of the superstructure, including 2620 passengers and 1253 crew, combined with the high average density of the 117 megawatt engines and other heavy facilities close to the bottom of the ship nevertheless allow the stability condition to be fulfilled. Photograph by Daniel Carneiro.

We shall assume that the ship is initially in complete mechanical equilibrium with vanishing total force and vanishing total moment of force. The aim is now to calculate the moment that arises when the ship is brought slightly out of equilibrium. If the moment tends to turn the ship back into equilibrium, the initial orientation is stable, otherwise it is unstable.

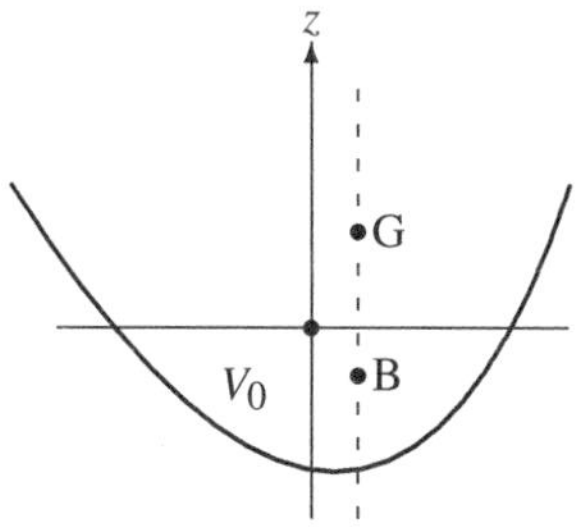

The ship in an equilibrium orientation, stable or unstable, with the aligned centres of gravity and buoyancy. The horizontal line at $z = 0$ indicates the surface of the water.

Displacement geometry

Most ships are mirror symmetric in a plane, but we shall be more general and consider a 'ship' of an arbitrary shape. In a flat earth coordinate system with vertical z-axis the ship displaces in equilibrium a volume V_0 of water below the waterline at $z = 0$. Since water has constant density, the centre of buoyancy is simply the geometric average of the position over the displacement,

$$\boldsymbol{x}_B = \langle \boldsymbol{x} \rangle_0 = \frac{1}{V_0} \int_{V_0} \boldsymbol{x} \, dV. \tag{5.19}$$

In equilibrium the horizontal positions of the centres of buoyancy and gravity must be equal $x_B = x_G$ and $y_B = y_G$, whereas the vertical position z_G of the centre of gravity depends on the actual mass distribution of the ship, determined by its structure and load.

In practical calculations it pays to introduce the ship's equilibrium area function $A(z)$, defined as the horizontal area at depth z below the waterline. The displacement and the height of the centre of buoyancy then become,

$$V_0 = \int_{-d}^{0} A(z)\, dz, \tag{5.20}$$

$$z_B = \frac{1}{V_0} \int_{-d}^{0} z A(z)\, dz, \tag{5.21}$$

where d is the ship's *draft* (maximal depth below the waterline).

Centre of roll

In the waterline the ship covers a horizontal region $A_0 = A(0)$ of arbitrary shape. The geometric centre of this region is defined by the area average of the position,

$$(x_0, y_0) = \frac{1}{A_0} \int_{A_0} (x, y)\, dA, \tag{5.22}$$

where $dA = dx\, dy$ is the area element. Without loss of generality we may always place the coordinate system such that $x_0 = y_0 = 0$. In a ship that is mirror symmetric in a plane, the area centre will also lie in this plane.

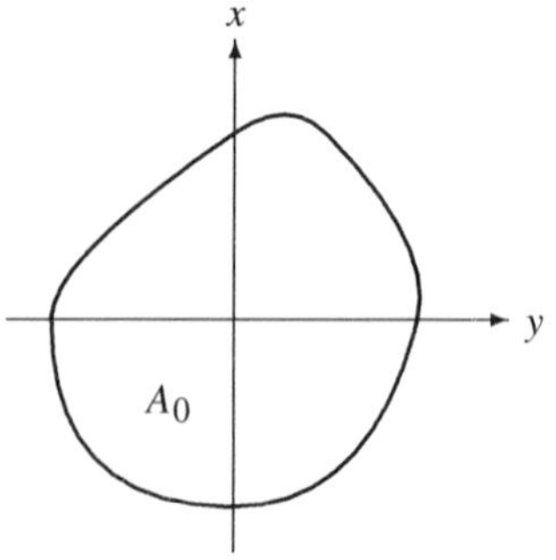

The area A_0 of the ship in the waterline may be of quite arbitrary shape.

To discover the physical significance of the geometric centre of the waterline area, the ship is tilted (or 'heeled' in maritime language) through a tiny angle α around the x-axis, such that the equilibrium waterline area A_0 comes to lie in the plane $z = \alpha y$. The net change in the volume of the displaced water is to lowest order in α given by the difference in volumes of the two wedge-shaped regions between new and the old waterlines. Since the displaced water is removed from the wedge at $y > 0$ and added to the wedge for $y < 0$, the signed volume change becomes

$$\delta V = -\int_{A_0} z\, dA = -\alpha \int_{A_0} y\, dA = 0. \tag{5.23}$$

In the last step we have assumed that the origin of the coordinate system coincides with the geometric centre of the waterline area. There will be corrections to this result of order α^2 due to the actual shape of the hull just above and below the waterline, but they are disregarded here. To leading order the two wedges have the same volume.

Since the direction of the x-axis is quite arbitrary, the conclusion is that the ship may be heeled around any line going through the geometric centre of the waterline area without any first order change in volume of displaced water. This guarantees that the ship will remain in buoyant equilibrium during the tilt. The geometric centre of the waterline area may thus be viewed as the ship's *centre of roll*.

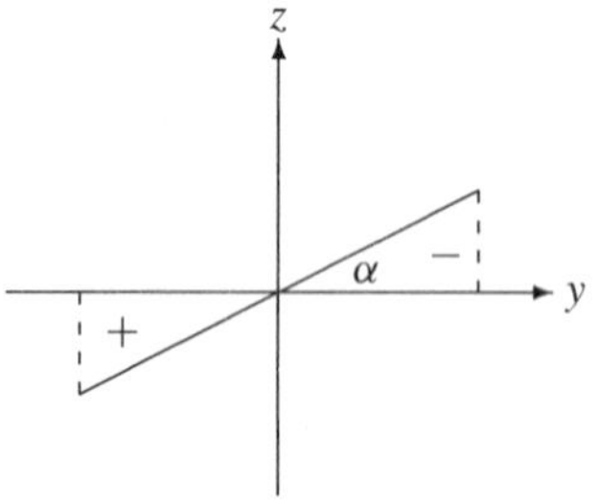

Tilt around the x-axis. The change in displacement consists in moving the water in the wedge to the right into the wedge to the left.

The metacentre

The tilt around the x-axis changes the positions of the centres of gravity and buoyancy. The centre of gravity is (hopefully!) fixed with respect to the ship and is to first order in α shifted horizontally by a simple rotation through the infinitesimal angle α,

$$\delta y_G = -\alpha z_G. \tag{5.24}$$

There will also be a vertical shift, $\delta z_G = \alpha y_G$, but that is of no importance to the stability in the lowest order of approximation.

The centre of buoyancy is at first shifted by the same rule as the centre of gravity by the tilt, but because the displacement also changes there will be another contribution Δy_B, so that we may write

$$\delta y_B = -\alpha z_B + \Delta y_B. \tag{5.25}$$

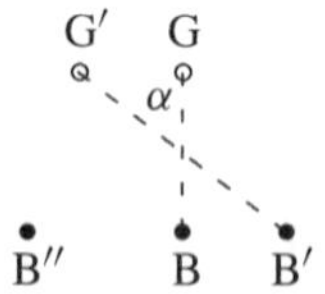

The tilt rotates the centre of gravity from G to G', and the centre of buoyancy from B to B'. In addition, the change in displaced water shifts the centre of buoyancy back to B''. In stable equilibrium this point must for $\alpha > 0$ lie to the left of the new centre of gravity G'.

As discussed above, the change in displacement amounts to moving the water in the wedge at $y > 0$ to the wedge at $y < 0$. The change in displacement is according to (5.19) calculated by averaging y over the volume of the two wedges, taking the sign correctly into account,

$$\Delta y_B = -\langle y \rangle_0 = -\frac{1}{V_0} \int_{A_0} yz\, dA = -\frac{\alpha}{V_0} \int_{A_0} y^2\, dA = -\alpha \frac{I_0}{V_0}$$

where

$$\boxed{I_0 = \int_{A_0} y^2 \, dA\,,} \tag{5.26}$$

is the second "moment of inertia" or *area moment* around the x-axis of the waterline area A.

The total horizontal shift in the centre of buoyancy becomes

$$\delta y_B = -\alpha \left(z_B + \frac{I_0}{V_0} \right). \tag{5.27}$$

This is of exactly the same form as the horizontal shift obtained by rotating the fixed point with vertical coordinate,

$$\boxed{z_M = z_B + \frac{I_0}{V_0}.} \tag{5.28}$$

This point, called the *metacentre*, is usually placed on the straight line (the 'mast') that goes through the centres of gravity and buoyancy. The above calculation shows that when the ship is heeled through a small angle, the centre of buoyancy will move so that it stays vertically below the metacentre.

The metacentre is a purely geometric quantity. The displacement as well as the vertical height of the centre of buoyancy may be calculated from the area function of the ship using (5.20) and (5.21), and the second order moment from the shape of the ship in the waterline. The simplest shapes are,

Rectangular waterline area: For a ship with rectangular waterline area with sides $2a$ and $2b$, the roll centre coincides with the centre of the rectangle, and the second moment around the x-axis becomes,

$$I_0 = \int_{-a}^{a} dx \int_{-b}^{b} dy\, y^2 = \tfrac{4}{3} a b^3. \tag{5.29}$$

If $a > b$ this is the largest moment around any tilt axis.

Elliptic waterline area: If the ship has an elliptical waterline area with axes $2a$ and $2b$, the roll centre coincides with the centre of the ellipse, and the second moment around the x-axis becomes,

$$I_0 = \int_{-a}^{a} dx \int_{-b\sqrt{1-x^2/a^2}}^{b\sqrt{1-x^2/a^2}} y^2 \, dy = \frac{4}{3} a b^3 \int_0^1 (1-t^2)^{3/2} \, dt = \frac{\pi}{4} a b^3. \tag{5.30}$$

Notice that this is only about half of the result for the rectangle.

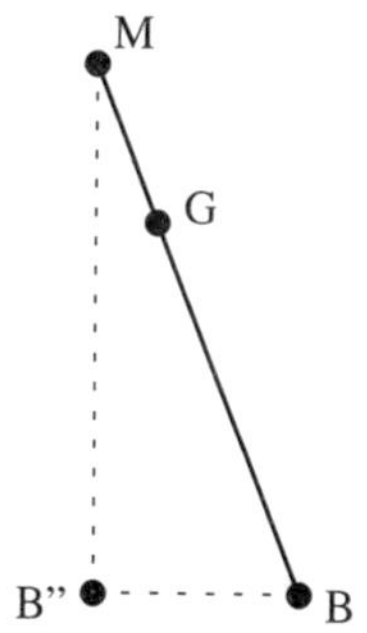

The centre of buoyancy B shifts horizontally by the same amount as the metacentre M. In this case the ship is stable because the metacentre lies above the centre of gravity.

Stability condition

The tilt also generates a *restoring moment* around the x-axis, which may be calculated from (5.17),

$$\mathcal{M}_x = -(y_G - y_B) M g_0. \tag{5.31}$$

Since we have $y_G = y_B$ in the original mechanical equilibrium, the difference in coordinates after the tilt may be written, $y_G - y_B = \delta y_G - \delta y_B$ where δy_G and δy_B are the small horizontal shifts of order α in the centres of gravity and buoyancy, calculated above.

In terms of the metacentric height z_M the restoring moment becomes

$$\boxed{\mathcal{M}_x = -\alpha (z_M - z_G) M g_0.} \tag{5.32}$$

For the ship to be stable, the restoring moment must counteract the tilt and thus have opposite sign of the tilt angle α. Consequently, the stability condition becomes

$$z_M > z_G. \tag{5.33}$$

Evidently, *the ship is stable when the metacentre lies above the centre of gravity.*

The orientation of the coordinate system with respect to the ship's hull was not specified in the analysis which is therefore valid for a tilt around any direction. For a ship to be fully stable, the stability condition

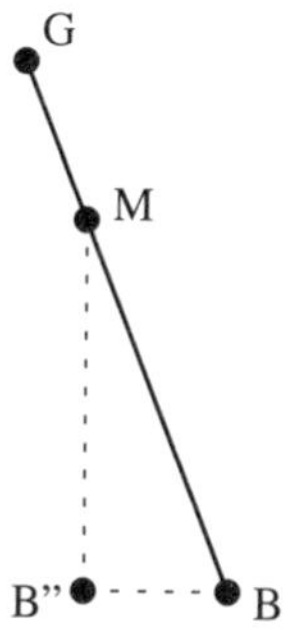

In this case the ship is unstable because the metacentre lies below the centre of gravity.

must be fulfilled for all possible tilt axes. Since the displacement V_0 is the same for all choices of tilt axis, the second moment of the area on the right hand side of (5.28) should be chosen to be the smallest one. Often it is quite obvious which moment is the smallest. Many modern ships are extremely long with the same cross section along most of their length and with mirror symmetry through a vertical plane. For such ships the smallest moment is clearly obtained with the tilt axis parallel to the longitudinal axis of the ship.

The restoring moment (5.32) is proportional to the vertical distance, $z_M - z_G$, between the metacentre and the centre of gravity. The closer the centre of gravity comes to the metacentre, the smaller will the restoring moment be, and the longer will the period of rolling oscillations be. The actual roll period depends also on the true moment of inertia of the ship around the tilt axis (see problem 5.11). Whereas the metacentre is a purely geometric quantity which depends only on the ship's actual draft, the centre of gravity depends on the way the ship is actually loaded. A good captain should always know the positions of the centre of gravity and the metacentre of his ship before he sails, or else he may capsize when casting off.

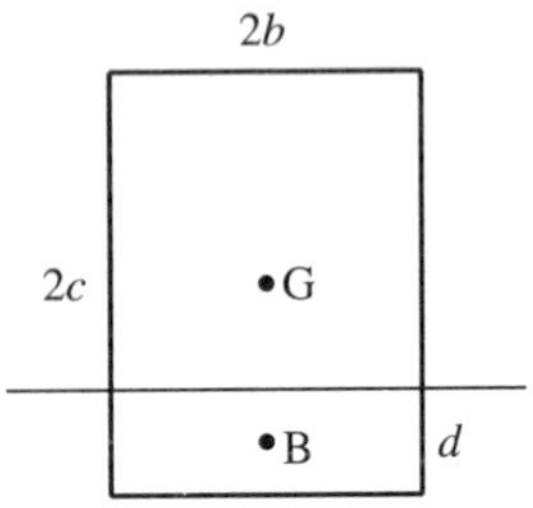

Floating block with height h, draught d, width $2b$, and length $2a$ into the paper.

Example 5.4.1: An elliptical rowboat with vertical sides has major axis $2a = 2$ m and minor axis $2b = 1$ m. The smallest moment of the rectangular area is $I_0 = (\pi/4)ab^3 \approx 0.1$ m^4. If your mass is 75 kg and the boat's is 50 kg, the displacement will be $V_0 = 0.125$ m^3, and the draught $d \approx V_0/4ab = 6.25$ cm, ignoring the usually curved shape of the boat's hull. The coordinate of the centre of buoyancy becomes $z_B = -3.1$ cm and the metacentre $z_M = 75$ cm. Getting up from your seat may indeed raise the centre of gravity so much that it gets close to the metacentre and the boat begins to roll violently. Depending on your weight and mass distribution the boat may even become unstable and turn over.

Floating block

The simplest non-trivial case in which we may apply the stability criterion is that of a rectangular block of dimensions $2a$, $2b$ and $2c$ in the three coordinate directions. Without loss of generality we may assume that $a > b$. The centre of the waterline area coincides with the roll centre and the origin of the coordinate system with the waterline at $z = 0$. The block is assumed to be made from a uniform material with constant density ρ_1 and floats in a liquid of constant density ρ_0.

In hydrostatic equilibrium we must have $M = 4abd\rho_0 = 8abc\rho_1$, or

$$\frac{\rho_1}{\rho_0} = \frac{d}{2c}. \tag{5.34}$$

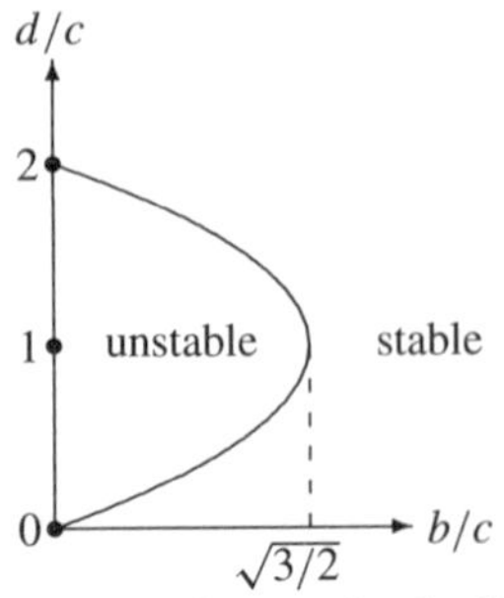

Stability diagram for the floating block.

The position of the centre of gravity is $z_G = c - d$ and the centre of buoyancy $z_B = -d/2$. Using (5.29) and $V_0 = 4abd$, the position of the metacentre is

$$z_M = -\frac{d}{2} + \frac{b^2}{3d}. \tag{5.35}$$

Rearranging the stability condition, $z_M > z_G$, it may be written as

$$\left(\frac{d}{c} - 1\right)^2 > 1 - \frac{2b^2}{3c^2}. \tag{5.36}$$

When the block dimensions obey $a > b$ and $b/c > \sqrt{3/2} = 1.2247\ldots$, the right-hand side becomes negative and the inequality is always fulfilled. On the other hand, if $b/c < \sqrt{3/2}$ there is a range of draught values around $d = c$ (corresponding to $\rho_1/\rho_0 = \frac{1}{2}$),

$$1 - \sqrt{1 - \frac{2}{3}\left(\frac{b}{c}\right)^2} < \frac{d}{c} < 1 + \sqrt{1 - \frac{2}{3}\left(\frac{b}{c}\right)^2}, \tag{5.37}$$

for which the block is unstable. If the draught lies in this interval the block will keel over and come to rest in another orientation (see problem 5.13).

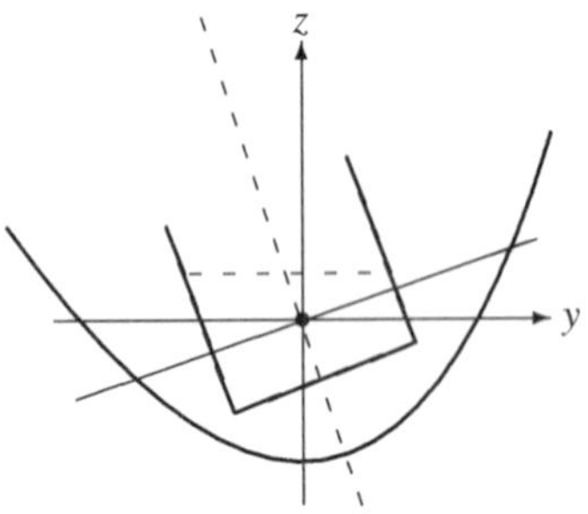

Tilted ship with an open container filled with liquid.

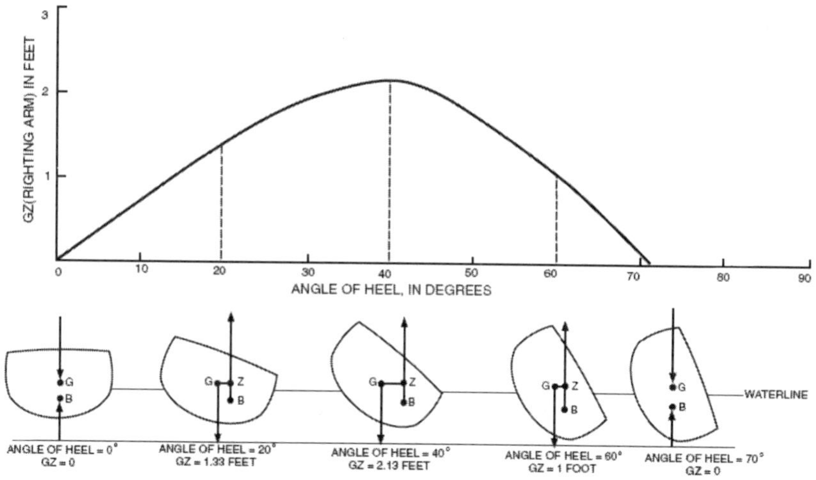

Figure 5.3. Stability curve for large angles of heel. The metacentre is only useful for tiny heel angles where all changes are linear in the angle. For larger angles one uses instead the 'righting arm' which is the distance between the centre of gravity and the vertical line through the actual centre of buoyancy. Instability sets in when the righting arm becomes negative, in the above plot for about 72° heel. Courtesy John Pike, Global Security.

Ship with liquid cargo

Many ships carry liquid cargos, oil, water, etc. When the tanks are not completely filled this kind of cargo may strongly influence the stability of the ship. In heavy weather or due to accidents, car ferries may inadvertently also get a layer of water on the car deck. The main effect of an open liquid surface inside the ship is that the centre of mass is shifted in the same direction by the redistribution of real liquid as the shift in the centre of buoyancy due to the change in displaced water, i.e. towards negative y-values. This disturbs the stability and creates a competition between the liquid carried by the ship and the water displaced by the ship.

For the case of a single open tank the calculation of the restoring moment must now include the liquid cargo. A similar analysis as before shows that there will be a horizontal change in the centre of gravity from the movement of a wedge of real liquid of density ρ_1,

$$\Delta y_G = -\alpha \frac{\rho_1 I_1}{M} = -\alpha \frac{\rho_1}{\rho_0} \frac{I_1}{V_0} \tag{5.38}$$

where I_1 is the second moment of the open liquid surface. The metacentric height now becomes

$$\boxed{z_M = z_B + \frac{I_0}{V_0} - \frac{\rho_1}{\rho_0} \frac{I_1}{V_0}.} \tag{5.39}$$

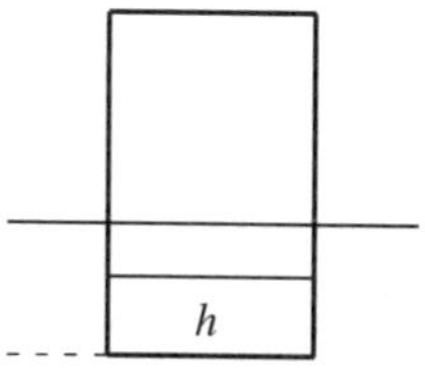

The effect of the moving liquid is to lower the metacentric height with possible destabilization as a result. The unavoidable sloshing of the liquid may further compromise the stability. The destabilizing effect of a liquid cargo is often counteracted by dividing the hold into a number of smaller compartments by means of bulkheads along the ship's principal roll axis.

In car ferries almost any level h of water on the car deck may cause the ferry to capsize because $\rho_1 = \rho_0$ and $I_1 \approx I_0$, making $z_M \approx z_B$ independent of h. As several accidents have shown, car ferries are in fact highly susceptible to the destabilizing effects of water on the car deck. Waterproof longitudinal bulkheads on the car deck of a car ferry are usually avoided because it would hamper efficient loading of the cars.

A 'car ferry' with water on the deck is inherently unstable because the movement of the real water on the deck nearly cancels the stabilizing movement of the displaced water.

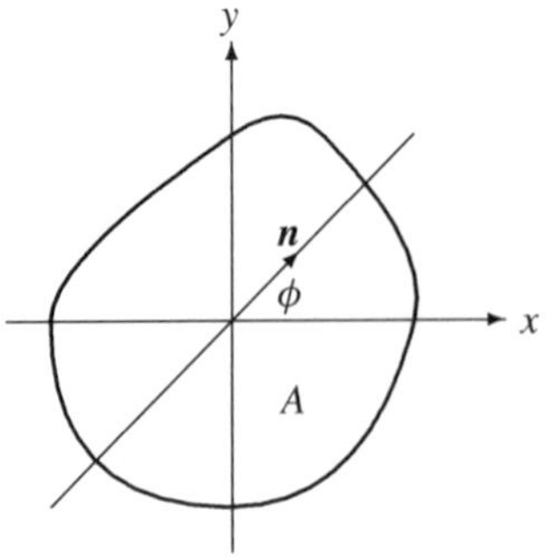

Tilt axis $\boldsymbol{n}$ forming an angle ϕ with the x-axis.

∗ Principal roll axis

It has already been remarked that the metacentre for absolute stability is determined by the smallest second moment of the waterline area. Instead of tilting the ship around the x-axis, it is tilted around an axis $\boldsymbol{n} = (\cos\phi, \sin\phi, 0)$ forming an angle ϕ with the x-axis. Since this configuration is obtained by a simple rotation through ϕ around the z-axis, the transverse coordinate to be used in calculating the second moment becomes $y' = y\cos\phi - x\sin\phi$ (see equation (2.36b)), and we find

$$I_0' = \int_A (y')^2\, dA = I_{xx}\cos^2\phi + I_{yy}\sin^2\phi + 2I_{xy}\sin\phi\cos\phi = \boldsymbol{n}\cdot\mathbf{I}\cdot\boldsymbol{n} \tag{5.40}$$

where I_{xx}, I_{yy} and I_{xy} are the elements of the matrix

$$\mathbf{I} = \begin{pmatrix} I_{xx} & I_{xy} \\ I_{yx} & I_{yy} \end{pmatrix} = \int_A \begin{pmatrix} y^2 & -xy \\ -xy & x^2 \end{pmatrix} dA. \tag{5.41}$$

The extrema of the positive definite quadratic form $\boldsymbol{n}\cdot\mathbf{I}\cdot\boldsymbol{n}$ are found from the eigenvalue equation $\mathbf{I}\cdot\boldsymbol{n} = \lambda\boldsymbol{n}$ (see problem 5.10). The eigenvector corresponding to the smallest eigenvalue is called the *principal roll axis* of the ship and its eigenvalue determines the metacentre for absolute stability.

Problems

5.1 A stone weighs 1000 N in air and 600 N when submerged in water. Calculate the volume and average density of the stone.

5.2 A hydrometer (an instrument used to measure the density of a liquid) with mass $M = 4$ g consists of a roughly spherical glass container and a long thin cylindrical stem of radius $a = 2$ mm. The sphere is weighed down so that the apparatus will float stably with the stem pointing vertically upwards and crossing the fluid surface at at some point. How much deeper will it float in alcohol with mass density $\rho_1 = 0.78$ g cm^{-3} than in oil with mass density $\rho_2 = 0.82$ g cm^{-3}? You may disregard the tiny density of air.

5.3 A cylindrical wooden stick (density $\rho_1 = 0.65$ g cm^{-3}) floats in water (density $\rho_0 = 1$ g cm^{-3}). The stick is loaded down with a lead weight (density $\rho_2 = 11$ g cm^{-3}) at one end such that it floats in a vertical position with a fraction $f = 1/10$ of its length out of the water. **(a)** What is the ratio (M_1/M_2) between the masses of the wooden stick and the lead weight? **(b)** How large a fraction can stick out of the water (disregarding questions of stability)?

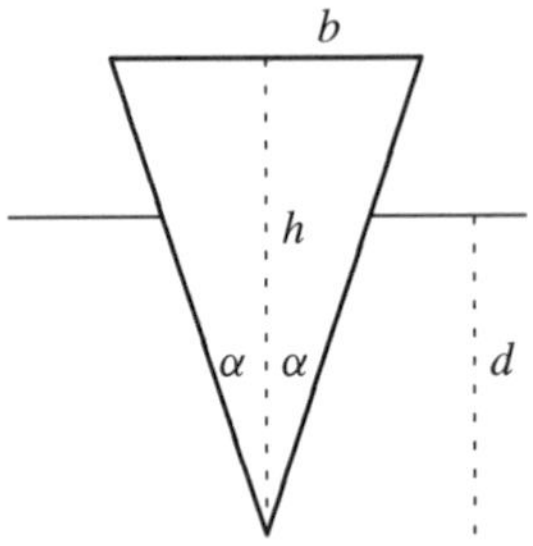

Triangular ship of length L (into the paper) floating with its peak vertically downwards.

5.4 A ship of length L has a longitudinally invariant cross section in the shape of an isosceles triangle with half opening angle α and height h. It is made from homogeneous material of density ρ_1 and floats in a liquid of density $\rho_0 > \rho_1$. **(a)** Determine the stability condition on the mass ratio ρ_1/ρ_0 when the ship floats vertically with the peak downwards. **(b)** Determine the stability condition on the mass ratio when the ship floats vertically with the peak upwards. **(c)** What is the smallest opening angle that permits simultaneous stability in both directions?

5.5 A right rotation cone has half opening angle α and height h. It is made from a homogeneous material of density ρ_1 and floats in a liquid of density $\rho_0 > \rho_1$. **(a)** Determine the stability condition on the mass ratio ρ_1/ρ_0 when the cone floats vertically with the peak downwards. **(b)** Determine the stability condition on the mass ratio when the cone floats vertically with the peak upwards. **(c)** What is the smallest opening angle that permits simultaneous stability in both directions?

5.6 A barotropic compressible fluid is in hydrostatic equilibrium with pressure $p(z)$ and density $\rho(z)$ in a constant external gravitational field with potential $\Phi = g_0 z$. A finite body having a 'small' gravitational field $\Delta\Phi(\boldsymbol{x})$ is submerged in the fluid. **(a)** Show that the change in hydrostatic pressure to lowest order of approximation is

$$\Delta p(\boldsymbol{x}) = -\rho(z)\Delta\Phi(\boldsymbol{x}). \tag{5.42}$$

(b) Show that for a spherically symmetric body of radius a and mass M, the extra surface pressure is $\Delta p = g_1 a \rho(z)$ where $g_1 = GM/a^2$ is the magnitude of surface gravity, and that the buoyancy force is increased.

5.7 Two identical homogenous spheres of mass M and radius a are situated a distance $D \gg a$ apart in a barotropic fluid. Due to their field of gravity, the fluid will be denser near the spheres. There is no other gravitational field present, the fluid density is ρ_0 and the pressure is p_0 in the absence of the spheres. One may assume that the pressure corrections due to the spheres are small everywhere in comparison with p_0. **(a)** Show that the spheres will repel each other and calculate the magnitude of the force to leading order in a/D. **(b)** Compare with the gravitational attraction between the spheres. **(c)** Under which conditions will the total force between the spheres vanish?

* **5.8** Prove without assuming constant gravity that the hydrostatic moment of buoyancy equals (minus) the moment of gravity of the displaced fluid (corollary to Archimedes' law).

* **5.9** Assuming constant gravity, show that for a body not in buoyant equilibrium (i.e. for which the total force $\mathcal{F}$ does not vanish), there is always a well-defined point $\boldsymbol{x}_0$ such that the total moment of gravitational plus buoyant forces is given by $\mathcal{M} = \boldsymbol{x}_0 \times \mathcal{F}$.

* **5.10** Let $\mathbf{I}$ be a symmetric (2×2) matrix. Show that the extrema of the corresponding quadratic form $\boldsymbol{n} \cdot \mathbf{I} \cdot \boldsymbol{n} = I_{xx} n_x^2 + 2 I_{xy} n_x n_y + I_{yy} n_y^2$ where $n_x^2 + n_y^2 = 1$ are determined by the eigenvectors of $\mathbf{I}$ satisfying $\mathbf{I} \cdot \boldsymbol{n} = \lambda \boldsymbol{n}$.

* **5.11** Show that in a stable orientation the angular frequency of small oscillations around around a principal tilt axis of a ship is

$$\omega = \sqrt{\frac{M g_0}{J}(z_M - z_G)}$$

where J is the moment of inertia of the ship around this axis.

* **5.12** A ship has a waterline area which is a regular polygon with $n \geq 3$ edges. Show that the area moment tensor (5.41) has $I_{xx} = I_{yy}$ and $I_{xy} = 0$.

* **5.13** A homogeneous cubic block has density equal to half that of the liquid it floats on. Determine the stability properties of the cube when it floats **(a)** with a horizontal face below the centre, **(b)** with a horizontal edge below the centre, and **(c)** with a corner vertically below the centre. Hint: problem 5.12 is handy for the last case, which you should be warned is quite difficult.

6

Planets and stars

Planets and stars are objects held together by their own gravity, but prevented from collapsing by internal pressure, originating from repulsive atomic or nuclear forces. The more massive a body is, the higher the pressure necessary to prevent collapse. For a sufficiently massive body ultimate gravitational collapse cannot be prevented by any known forces, will eventually occur, and a black hole is born.

So far we have only been able to scratch the surface of our own planet Earth. A little has also been done on the Moon and soon we shall know more about the surface of Mars. Seismic waves created by controlled explosions do allow us to peer deeper into the planet, but mostly we are left with the 'experiments' carried out by nature without any regard to us. Earthquakes generate strong seismic waves, revealing the inner structure of the planet. Continental drift informs us about the mixture of heat and gravity deep inside. Electromagnetic radiation from the surface of a star is almost the only source of information about what goes on below, although neutrino observations have begun to provide a direct window into the deepest core of our Sun, and into the supernovas that explode in our cosmic neighbourhood.

Most of our understanding of the interiors of planets and stars comes from using the laws of physics determined on Earth as an 'analytic drill' allowing us to get insight into processes which cannot be directly observed from the outside. In this chapter, the first turns of this drill consist of applying the equations of hydrostatic equilibrium to these massive self-gravitating bodies. The strongest simplifying assumption we can make about planets and stars, is that they are spherically symmetric, but before we specialize to that case we need to derive a fundamental differential equation connecting a mass distribution and its gravitational field. At the end of the chapter, we apply the formalism to a homentropic star without internal energy production.

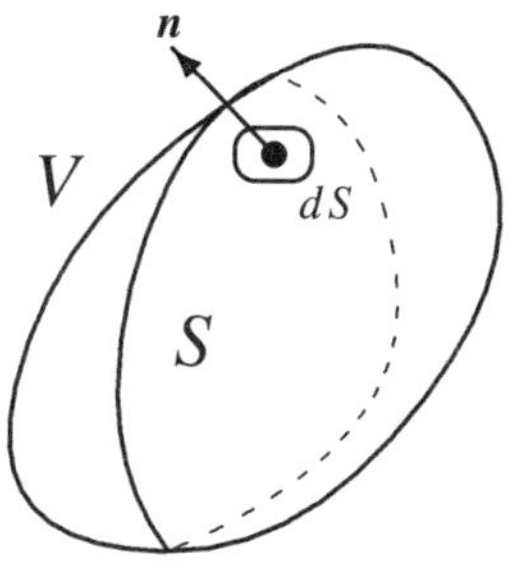

A surface S surrounding a volume V. The direction $\boldsymbol{n}$ of a surface element dS is always oriented outwards from the volume.

6.1 Gravitational flux

Let S be a closed surface surrounding a volume V. We shall as before use the convention that the normal $\boldsymbol{n}$ at a point $\boldsymbol{x}$ of the surface is always a unit vector oriented *outwards* from the surface, and a small surface element of magnitude dS is represented by the vector $d\boldsymbol{S} = \boldsymbol{n}dS$. Seen from the origin of the coordinate system, the solid angle subtended by this surface element is

$$d\Omega = \frac{\boldsymbol{x} \cdot d\boldsymbol{S}}{|\boldsymbol{x}|^3}. \tag{6.1}$$

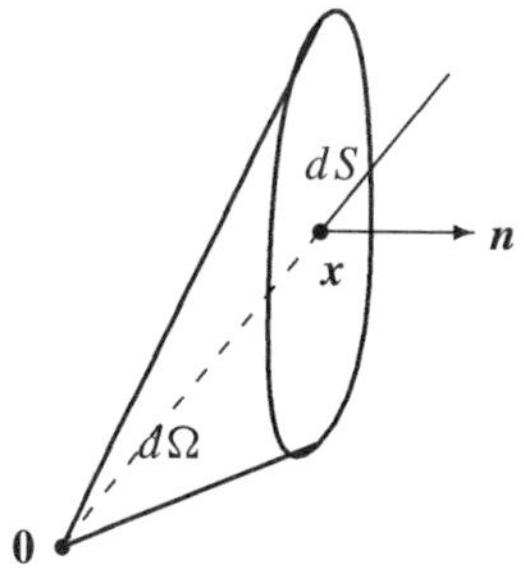

The solid angle subtended by a surface element

Projecting the gravitational field (3.12) from a point mass on to the surface element, one obtains

$$\boldsymbol{g} \cdot d\boldsymbol{S} = -GMd\Omega.$$

This quantity is called the *flux of gravity* through the surface element.

Consider now the total flux through a closed convex surface containing the point mass at the origin. All the little solid angles add up to 4π because the line-of-sight from the particle in any direction crosses the convex surface exactly once. If, on the other hand, the surface does not contain the point mass, the line-of-sight from the particle will always cross the surface twice, and the two contributions to the solid angle will have the same magnitude but opposite sign and thus cancel. In other words,

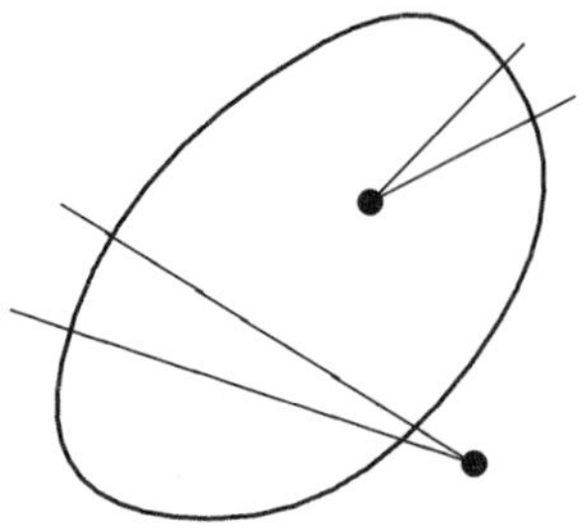

The lines-of-sight from a point inside a convex volume crosses the surface once, whereas they cross twice if the mass is outside.

$$\oint_S \boldsymbol{g} \cdot d\boldsymbol{S} = \begin{cases} -4\pi G M & \text{for } \boldsymbol{0} \in V \\ 0 & \text{otherwise.} \end{cases} \tag{6.2}$$

This result is, in fact, valid for all surfaces, convex or not. For a convoluted surface, the line-of-sight from the inside will instead cross the surface an odd number of times, and since the solid angles are evaluated with sign, all the contributions along the line-of-sight cancel each other except for one. If the particle is outside the volume the line-of-sight will cross an even number of times and all contributions cancel. The conclusion is that the above equation holds in full generality.

Furthermore, this result cannot depend on the particle being at the origin, but must be generally valid for any point particle inside or outside the volume. Adding together the contributions from all the material particles in the volume V, we finally get

$$\boxed{\oint_S \boldsymbol{g} \cdot d\boldsymbol{S} = -4\pi G \int_V \rho \, dV.} \tag{6.3}$$

The integral at the right is simply the total mass (3.2) within the volume, so we may conclude that the gravitational flux through any closed surface is proportional to the total mass contained within the surface, whereas the mass outside the surface does not contribute to the flux.

Gauss' theorem and divergence

We have previously derived a vector relation (4.22) between the surface integral of a scalar field and a volume integral over its gradient. Applying it componentwise to the left-hand side of (6.3) we obtain

$$\oint_S \boldsymbol{g} \cdot d\boldsymbol{S} = \oint_S (g_x dS_x + g_y dS_y + g_z dS_z) = \int_V (\nabla_x g_x + \nabla_y g_y + \nabla_z g_z) \, dV.$$

This is the usual form of Gauss' theorem

$$\boxed{\oint_S \boldsymbol{g} \cdot d\boldsymbol{S} = \int_V \nabla \cdot \boldsymbol{g} \, dV,} \tag{6.4}$$

where the field on the right-hand side

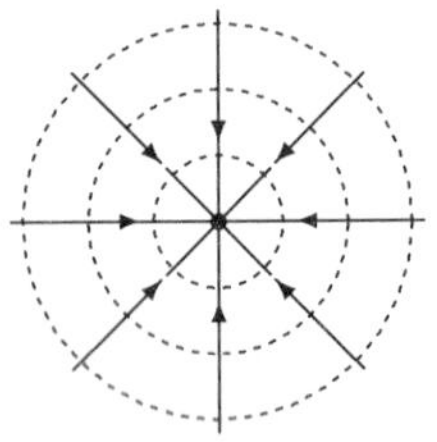

The minus sign in the source equation (6.6) expresses that gravitational field lines always converge upon masses.

$$\nabla \cdot \boldsymbol{g} = \nabla_x g_x + \nabla_y g_y + \nabla_z g_z = \frac{\partial g_x}{\partial x} + \frac{\partial g_y}{\partial y} + \frac{\partial g_z}{\partial z}, \tag{6.5}$$

is the divergence (2.57) of the gravitational field. Its value at a point is a measure of how much field lines diverge away from each other if it is positive, or converge if it is negative.

Gauss' theorem is in this form a general relation between any vector field $\boldsymbol{g}$, not necessarily the gravitation field, and its divergence $\nabla \cdot \boldsymbol{g}$. The two forms, (4.22) and (6.4), are completely equivalent (see problem 6.1).

Poisson's equation

The global equation (6.3) relating the gravitational field to its sources may now, like the global hydrostatic equation (4.15), be converted to a local differential equation. Using Gauss' theorem (6.4) we find from (6.3)

$$\int_V \nabla \cdot \boldsymbol{g} \, dV = -4\pi G \int_V \rho \, dV,$$

which must be valid for all volumes V. That is, however, only possible, if integrands are equal, or

$$\boxed{\nabla \cdot \boldsymbol{g} = -4\pi G \rho.} \tag{6.6}$$

This is one of the fundamental field equations of gravity, expressing that the mass density is the local *source* of the gravitational field.

It is convenient to define the *Laplace operator*

$$\nabla^2 = \nabla \cdot \nabla = \nabla_x^2 + \nabla_y^2 + \nabla_z^2 = \frac{\partial^2}{\partial x^2} + \frac{\partial^2}{\partial y^2} + \frac{\partial^2}{\partial z^2}. \tag{6.7}$$

This operator plays a major role in all field theories.

Using that $\boldsymbol{g} = -\nabla\Phi$ (see section 3.4), the source equation (6.6) may be rewritten in terms of the gravitational potential, and we obtain *Poisson's equation*,

$$\boxed{\nabla^2\Phi = 4\pi G\rho.} \tag{6.8}$$

The linearity of this equation guarantees that if Φ_1 is a particular solution then the most general solution is of the form $\Phi = \Phi_0 + \Phi_1$ where Φ_0 is an arbitrary solution to *Laplace's equation*,

$$\nabla^2\Phi_0 = 0. \tag{6.9}$$

The actual solution selected in a particular problem depends on the boundary conditions.

Pierre Simon marquis de Laplace (1749–1827). *French mathematician, astronomer and physicist. Developed gravitational theory and applied it to perturbations in the planetary orbits and the conditions for stability of the solar system.*

Simeon Denis Poisson (1781–1840). *French mathematician. Contributed to electromagnetism, celestial mechanics, and probability theory.*

Constant mass density: If the universe were uniformly filled with matter at constant density, $\rho(\boldsymbol{x}) = \rho_0$, we would have

$$\nabla^2\Phi = 4\pi G\rho_0. \tag{6.10}$$

It is easy to verify explicitly that a particular solution to this equation is

$$\Phi = \frac{2}{3}\pi G\rho_0 \,|\boldsymbol{x}|^2, \tag{6.11}$$

corresponding to a gravitational acceleration

$$\boldsymbol{g} = -\frac{4}{3}\pi G\rho_0\boldsymbol{x}. \tag{6.12}$$

The gravitational field always points towards the origin of the coordinate system which is thus imbued with an apparently unphysical preferred status. In section 15.7 we shall see that this field appears naturally in Newtonian cosmology.

Hydrostatic equilibrium

One may rightly ask why we need Poisson's equation when the complete connection between a mass distribution and its gravitational potential is already given by the integral (3.24). For compressible matter, however, the mass density depends on the pressure, which in turn depends on gravity through the equation of hydrostatic balance (4.19), and gravity depends in turn on the mass density. Such physical circularity is best handled by means of differential equations.

To see how this works out, we use (4.19) and (6.6) to calculate the divergence of $\boldsymbol{g} = \nabla p/\rho$, and obtain

$$\boxed{\nabla \cdot \left(\frac{1}{\rho}\nabla p\right) = -4\pi G\rho.} \tag{6.13}$$

Together with a barotropic relation of the form $p = p(\rho)$, this becomes a nonlinear, second-order partial differential equation for the density field.

6.2 Spherical bodies

The mass distribution $\rho(r)$ for a spherically symmetric body such as a planet or a star is, as discussed in section 3.3, only a function of the distance $r = |\boldsymbol{x}|$ from its centre, which is taken to be at the origin of the coordinate system. The field of gravity must correspondingly be radial, $\boldsymbol{g}(\boldsymbol{x}) = g(r)\,\boldsymbol{e}_r$, with $\boldsymbol{e}_r = \boldsymbol{x}/r$. Applying the global source equation (6.3) to a spherical surface $S(r)$ of radius r, the surface integral on the left-hand side becomes

$$\oint_{S(r)} \boldsymbol{g} \cdot d\boldsymbol{S} = 4\pi r^2 g(r). \tag{6.14}$$

The volume integral on the right-hand side of (6.3) is simply the integrated mass $M(r)$ given in (3.18), so that we obtain

$$g(r) = -\frac{GM(r)}{r^2}. \tag{6.15}$$

Finally, we have fulfilled the promise of deriving equation (3.17).

The general equation of hydrostatic equilibrium (6.13) simplifies considerably for a spherical system, and becomes an ordinary second-order differential equation for the pressure $p(r)$ or the density $\rho(r)$. Instead of deriving this differential equation from (6.13), it is easier to go back to the original equation of local hydrostatic equilibrium (4.19). Using that

$$\boldsymbol{\nabla} p(r) = \frac{dp(r)}{dr}\boldsymbol{\nabla} r = \frac{dp(r)}{dr}\boldsymbol{e}_r,$$

we get from (4.19)

$$\boxed{\frac{dp(r)}{dr} = g(r)\rho(r) = -G\frac{M(r)}{r^2}\rho(r).} \tag{6.16}$$

Multiplying with r^2/ρ and differentiating after r, we find

$$\frac{d}{dr}\left(\frac{r^2}{\rho(r)}\frac{dp(r)}{dr}\right) = -G\frac{dM(r)}{dr} = -G4\pi r^2 \rho(r),$$

and rearranging, this becomes

$$\boxed{\frac{1}{r^2}\frac{d}{dr}\left(\frac{r^2}{\rho}\frac{dp}{dr}\right) = -4\pi G\rho.} \tag{6.17}$$

Combined with a barotropic equation of state of the form $p = p(\rho)$, this is an ordinary second-order differential equation for the density. In figure 6.1 the Earth's pressure distribution is plotted and compared with a simple model.

Boundary conditions

In principle, a second-order differential equation requires two boundary values (or integration constants), for example the central pressure $p_c = p(0)$ and its first derivative dp/dr for $r = 0$. We shall make the reasonable assumption that the density ρ_c at the centre of the body is finite. Then for 'small' r we have $M(r) \approx \frac{4}{3}\pi r^3 \rho_c$ and equation (6.16) becomes for $r \to 0$,

$$\frac{dp}{dr} \approx -\frac{4}{3}\pi G\rho_c^2\, r,$$

which integrates to

$$p(r) \approx p_c - \frac{2}{3}\pi G\rho_c^2 r^2. \tag{6.18}$$

Thus, under the assumption of finite central density, the pressure is parabolic near the centre with $dp/dr = 0$ for $r = 0$. This shows that under reasonable physical assumptions the hydrostatic equation (6.17) requires in

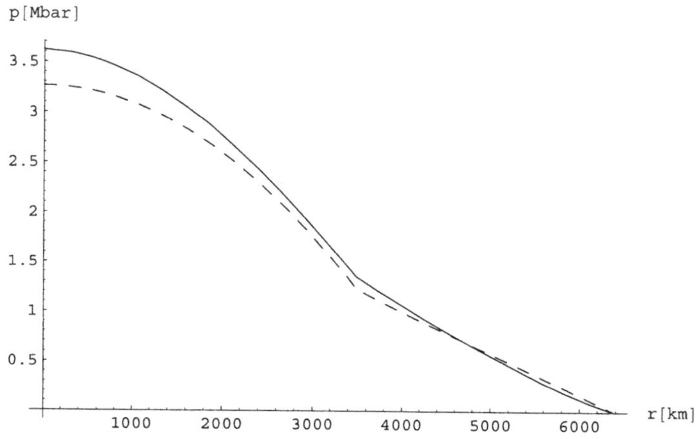

Figure 6.1. Pressure distribution in the Earth. Solid line: data from [41] and dashed: the two-layer model (problem 6.3). The agreement between the model and data is impressive in view of the coarseness of the model.

fact only one boundary condition, for example the central pressure. Knowing p_c together with the equation of state (which also determines ρ_c), the pressure may be calculated throughout the body.

The central pressure and density are, of course, not known for planets and stars, objects that are only accessible from the outside. Most such bodies have a well-defined surface radius, $r = a$, at which the pressure vanishes. We shall arbitrarily call a body a planet, if the density jumps abruptly to zero at the surface, and a star if the density vanishes along with the pressure at the surface. Such a convention makes the gaseous giant planets, Jupiter and Saturn, count as stars even though they probably do not burn much hydrogen.

The requirement of zero pressure at $r = a$ will determine the central pressure. The solutions to the hydrostatic equation can be expressed entirely in terms of the radius of the body and the parameters in the equation of state. In particular the mass M_0 of the body is—as we shall see below—calculable in terms of a (and the state parameters). Conversely, if the mass and radius are known, one of the other unknown parameters may be determined.

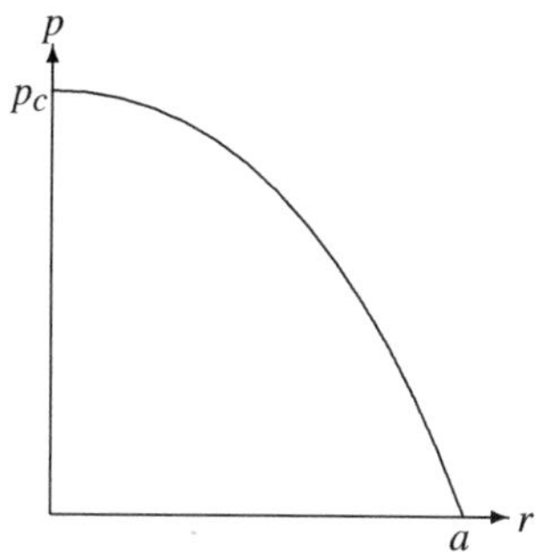

The pressure varies as a parabola in the central region of a spherically invariant body with a finite central density.

Planet with constant density

For a planet with constant density, ρ_0, the assumption of finite central density is exactly valid throughout the planet,

$$p = p_c - \frac{2}{3}\pi G\rho_0^2 r^2. \tag{6.19}$$

At the surface of the planet where the pressure has to vanish this leads to

$$p_c = \frac{2}{3}\pi G\rho_0^2 a^2. \tag{6.20}$$

If the mass and radius are known, the density is obtained from $M_0 = \frac{4}{3}\pi a^3 \rho_0$.

Example 6.2.1: The Moon's mass is 7.3×10^{22} kg and its radius is 1738 km, making the average density 3.34 g cm^{-3} The central pressure is predicted to be 46 500 atm.

6.3 The homentropic star

Stars like the Sun are self-gravitating, gaseous and almost perfectly spherical bodies that generate heat by thermonuclear processes in a fairly small region close to the centre. The heat is transferred to the surface by radiation, conduction and convection and eventually released into space as radiation. Like planets, stars

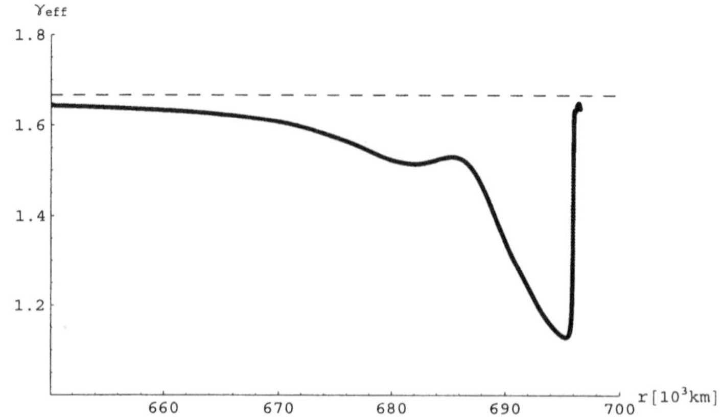

Figure 6.2. The effective adiabatic index γ_{eff} near the surface of the Sun. The fully drawn curve is taken from the 'standard' Sun model [16] and the dashed line is the constant monatomic value $\gamma = 5/3$.

also have a fairly complex structure with several layers differing in chemical composition and other physical properties.

> **Example 6.3.1:** Our Sun consists of a mixture of about 71% hydrogen, 27% helium and 2% other elements. It has a central core of radius 150 000 km, a radiative layer of thickness 350 000 km, and a convection layer of thickness 200 000 km. The 'standard' values [16] for the central parameters are $T_c = 15.7 \times 10^6$ K, $\rho_c = 154$ g cm^{-3}, and $p_c = 2.34 \times 10^{11}$ bar.

The stellar temperature lapse rate

Here we shall completely ignore the layering, heat production and chemical composition, and concentrate solely on hydrostatic equilibrium in a homogeneous star. We shall assume that the whole star consists of an ideal gas with adiabatic index $\gamma = 5/3$, and molar mass $M_{\text{mol}} = 0.5$ g mol^{-1}. This corresponds to fully ionized hydrogen, which consists af 50% hydrogen ions (protons) and 50% essentially massless electrons. Apart from a layer near the surface, the effective adiabatic index, defined by $1 - 1/\gamma_{\text{eff}} = d \log T / d \log p$ is in fact very close to this value throughout the Sun (see figure 6.2 and problem 6.5).

In section 4.6 we argued (for the case of Earth's atmosphere) that—provided the time scale for local mixing is fast compared to heat conduction—a *homentropic* dynamical 'equilibrium' will be established in which $p\rho^{-\gamma}$ takes the same value everywhere (see appendix C). Assuming that the whole star is homentropic and using the ideal gas law $\rho \sim p/T$, we conclude that $p\rho^{-\gamma} \sim p^{1-\gamma}T^{\gamma}$ is also constant. Differentiating $\log(p^{1-\gamma}T^{\gamma})$ after r we obtain,

$$\gamma \frac{1}{T}\frac{dT}{dr} + (1-\gamma)\frac{1}{p}\frac{dp}{dr} = 0,$$

and making use of the hydrostatic equation (6.16) we find the *stellar temperature lapse rate*,

$$\boxed{\frac{dT(r)}{dr} = \frac{g(r)}{c_p},} \tag{6.21}$$

where $g(r) = -GM(r)/r^2$ is the acceleration field, and $c_p = \gamma/(\gamma - 1)\ R/M_{\text{mol}}$ is the specific heat (4.40) of the ideal gas at constant pressure. The only difference is that in the atmosphere the acceleration is constant, whereas in the star it depends on r.

The above equation may be converted to a second-order differential equation,

$$\boxed{\frac{c_p}{r^2}\frac{d}{dr}\left(r^2\frac{dT}{dr}\right) = -4\pi G\rho.} \tag{6.22}$$

On the right-hand side we must use the constancy of $p\rho^{-\gamma} \sim T\rho^{1-\gamma}$ to eliminate the density and make it a differential equation for T only.

Approximative solutions near the centre and the surface

There are many types of solutions to the stellar equations (6.21) or (6.22). Some have infinite central pressure, others have non-vanishing density all the way to infinity (see problem 6.8). We shall limit ourselves to solutions with finite central density and a well-defined radius where the density and pressure vanish.

If the central density ρ_c is finite, the integrated mass becomes $M(r) \approx \frac{4}{3}\pi r^3 \rho_c$ near the centre, and thus $g(r) \approx -\frac{4}{3}\pi G\rho_c r$. From (6.21) we then obtain,

$$T \approx T_c - \frac{2\pi}{3}\frac{G\rho_c}{c_p} r^2, \tag{6.23}$$

where T_c is the central temperature. Evidently, the temperature drops parabolically when one moves away from the centre of the star, and in the leading approximation this is also true for the pressure and the density.

If the density vanishes at the surface, $r = a$, the temperature and pressure must also vanish. From (6.21) it follows that the temperature derivative is finite close to the surface at $r = a$, so that we may make a linear approximation

$$T(r) \approx T_0\left(1 - \frac{r}{a}\right), \tag{6.24}$$

near the surface. Inserting this into (6.21) and taking $r = a$ on the right-hand side, we find

$$T_0 = \frac{g_0 a}{c_p}, \tag{6.25}$$

where $g_0 = GM_0/a^2$ is the magnitude of the star's surface gravity. Note that this temperature, which sets the scale of the temperature gradient at the surface, is calculable in terms of the star's known parameters.

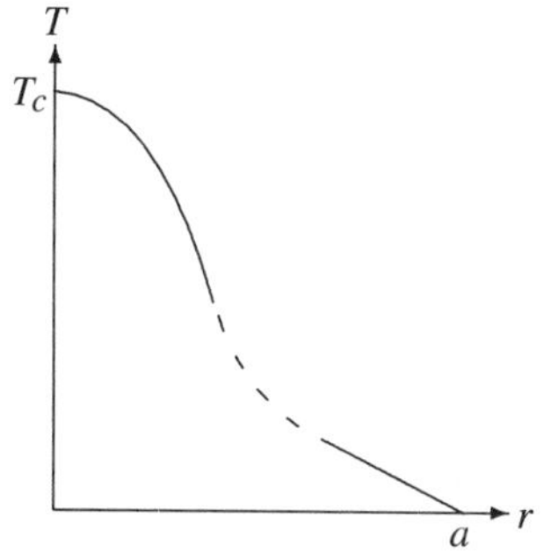

The temperature follows a parabola in the central region and approaches zero linearly near the surface. The dashed curve interpolates between these two extremes.

Example 6.3.2: Putting in the Sun's parameters, $M_0 \approx 2 \times 10^{30}$ kg, $a \approx 7 \times 10^8$ m, and $c_p \approx 4.2 \times 10^4$ J K^{-1} kg^{-1}, we find $g_0 \approx 274$ m s^{-2} and $T_0 \approx 4.6 \times 10^6$ K. Even if the surface approximation is not valid near the centre, T_0 is nevertheless of the same magnitude as the Sun's central temperature.

The Lane–Emden solutions

Having determined the behaviour of the temperature near the centre as well as near the surface, we need to interpolate between these regions. From the general discussion of boundary conditions in section 6.2, we expect that the stellar equation (6.22) will create a connection between the central temperature T_c and the calculable temperature parameter T_0.

Let us introduce the dimensionless variable $\xi = r/\lambda$, where λ is a suitable constant with the dimension of length, and the dimensionless temperature function

$$\theta(\xi) = \frac{T(r)}{T_c}. \tag{6.26}$$

The density is calculated from the homentropic condition $T\rho^{1-\gamma} = T_c\rho_c^{1-\gamma}$,

$$\rho = \rho_c\,\theta^{\frac{1}{\gamma-1}}. \tag{6.27}$$

Choosing the length parameter to be

$$\lambda = \sqrt{\frac{c_p T_c}{4\pi G\rho_c}}, \tag{6.28}$$

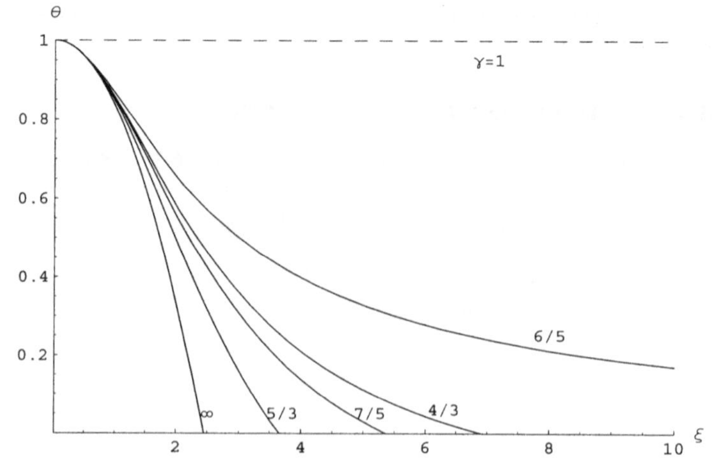

Figure 6.3. The family of Lane–Emden functions for selected values of γ.

γ	ξ_0	T_c/T_0
∞	2.449	0.500
5/3	3.654	1.346
7/5	5.355	2.449
4/3	6.897	3.417

Table of the crossing points ξ_0 and scaled central temperature T_c/T_0 for the Lane–Emden functions at selected values of γ.

the homentropic equation (6.22) becomes the *Lane–Emden equation*,

$$\boxed{\frac{1}{\xi^2}\frac{d}{d\xi}\left(\xi^2\frac{d\theta}{d\xi}\right)+\theta^{\frac{1}{\gamma-1}}=0.} \qquad (6.29)$$

From the solution near the centre of the star we conclude that the boundary conditions for $\theta(\xi)$ are $\theta(0)=1$ and $\theta'(0)=0$. The solutions form a family of functions parametrized by the adiabatic index γ.

Apart from special cases (see below and problem 6.8) this differential equation cannot be solved analytically. In figure 6.3 the Lane–Emden functions have been evaluated numerically for a few relevant values of γ. For $\gamma > 6/5$ it may be shown that the solutions cross the ξ-axis. This means that θ vanishes at this point, which is identified with the boundary of the star and denoted $\xi_0 = \xi_0(\gamma)$. Its precise value may be calculated numerically for all $\gamma > 6/5$. A few relevant ones are given in the table in the margin.

The limiting cases of the Lane–Emden functions are easily determined analytically. For $\gamma \to 1$, corresponding to an isothermal star, the solution is $\theta(\xi) \to 1$ so that $T(r) = T_c$ for all r (with a jump at the surface that makes the star into a planet, according to our definition). For $\gamma \to \infty$, we get from (6.19) and the ideal gas law, $\theta(\xi) \to 1 - \xi^2/6$, which also follows from (6.29). This particular curve crosses the axis at $\xi_0(\infty) = \sqrt{6} \approx 2.45$.

Central values

Knowing $\xi_0 = \xi_0(\gamma)$ the value of the scaling parameter $\lambda = a/\xi_0$ can be calculated from the known radius a of the star. Then from (6.21) at $r = a$ we get

$$\frac{T_c}{\lambda}\theta'(\xi_0) = -\frac{g_0}{c_p},$$

where $\theta'(\xi_0)$ is the slope of the solution at ξ_0. Introducing the temperature scale T_0 from (6.25) this becomes,

$$\frac{T_c}{T_0} = \frac{1}{(-\theta'(\xi_0))\xi_0}. \qquad (6.30)$$

A few selected values are shown in the margin table.

Similarly from (6.28) we find the central density

$$\frac{\rho_c}{\rho_0} = \frac{\xi_0^2}{3}\frac{T_c}{T_0}, \qquad (6.31)$$

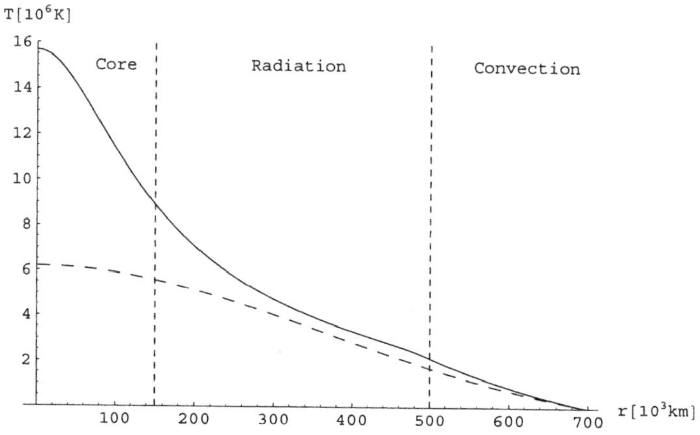

Figure 6.4. The temperature distribution in the Sun as a function of the distance from the centre. The fully drawn curve is from the 'standard' Sun model [16] and the dashed curve is the Lane–Emden solution for $\gamma = 5/3$. The vertical lines are boundaries between various layers of the Sun. The discrepancy between the curves represents the thermonuclear heat production in the centre.

where $\rho_0 = M_0/\frac{4}{3}\pi a^3$ is the average density of the star. Knowing both ρ_c and T_c allows us to determine the central pressure, found from the ideal gas law $p_c = \rho_c R T_c / M_{\text{mol}}$.

Example 6.3.3: For $\gamma = 5/3$ we obtain $T_c/T_0 = 1.35$ and $\rho_c/\rho_0 = 6.0$. For the Sun this leads to a central temperature of $T_c = 6.2 \times 10^6$ K, a central density of $\rho_c = 8.4$ g cm^{-3}, and a central pressure of $p_c = 8.7 \times 10^9$ bar. The temperature distribution is shown in figure 6.4 together with the data from the 'standard' Sun model [16]. The agreement is reasonable, except in the deeper radiative layers and the core where it fails because we have disregarded thermonuclear heat production.

6.4 Gravitational energy

What is the gravitational energy of a planet or a star? Since the gravitational potential of a finite body is always negative and grows more negative the closer one gets to the body, one does not have to perform any work to make such a body grow. It is sufficient to throw material into the general vicinity of the body, and let gravity do the rest. Consequently, the gravitational energy of a body is expected to be negative.

Gravity is in this respect different from most of the other forces we meet in daily life, for example friction, where we have to perform work to get anything done. It does not cost us anything to make matter collapse gravitationally, quite the contrary, we get paid for it (in heat). Matter is inherently unstable because of gravity, and this instability [13] lies at the root of galaxy and star formation, and thus of everything that is.

Assembly work

In chapter 3 it was shown that the work required to move a small particle of mass m from spatial infinity, where the gravitational potential vanishes, to a point $\boldsymbol{x}$, where the potential takes the value $\Phi(\boldsymbol{x})$, is $m\Phi(\boldsymbol{x})$. After you have done this (negative) work, it is conserved as (potential) energy of the particle.

Imagine now that we wish to increase the mass density inside a volume V by an amount $\delta\rho(\boldsymbol{x})$. The total work required to assemble this extra mass by bringing each material particle in from spatial infinity is,

$$\boxed{\delta W = \int_V \Phi\, \delta\rho\, dV.} \tag{6.32}$$

The added mass density $\delta\rho$ will change the potential both inside and outside V by an equally small amount $\delta\Phi$, but its contribution to the work will be of higher order in $\delta\rho$ and can be disregarded. If no other energy is added to or removed from the body, the above work will be stored in the body as gravitational energy.

Gravitational energy in an external field

In an *external* potential Φ, not originating from or influenced by the mass distribution itself, the total work of assembly becomes,

$$W = \int_V \Phi\, \rho\, dV. \tag{6.33}$$

For a constant gravitational field $\boldsymbol{g}_0$ where $\Phi = -\boldsymbol{x} \cdot \boldsymbol{g}_0$ we get,

$$W = -\boldsymbol{x}_M \cdot M \boldsymbol{g}_0, \tag{6.34}$$

where as before $\boldsymbol{x}_M$ is the centre of mass (3.3). With respect to potential energy, a body in a constant gravitational field is also equivalent to a point particle with the total mass situated at the centre of mass.

Gravitational self-energy

For a mass distribution assembled in its own field, the situation is slightly more complicated. Intuitively it is perhaps clear that each particle used to assemble the body on average meets only half the field of the final body. Hence the energy is expected to be only half of (6.33).

To show that there is indeed such a factor $1/2$ we shall employ a frequently used trick. Let us imagine that we build up the mass distribution in such a way that it is everywhere proportional to the final distribution. At any given moment, a certain fraction $\lambda\rho$ of the final distribution is already in place, where $0 < \lambda < 1$. Since the potential is linear in the mass distribution, the current potential will also be the same fraction $\lambda\Phi$ of the final potential. Increasing the fraction of the mass distribution by $\delta\lambda$ will then according to (6.32) cost an amount of work,

$$\delta W = \int (\lambda\Phi)(\delta\lambda\, \rho)\, dV = \lambda\delta\lambda \int \Phi\rho\, dV, \tag{6.35}$$

where the integrals now run over all space. Integrating over λ from 0 to 1, we get the total amount of work we have to perform in building up the complete mass distribution from scratch,

$$\boxed{W = \frac{1}{2}\int \Phi\rho\, dV.} \tag{6.36}$$

This work also equals the total gravitational *self-energy* $\mathcal{E} = W$ stored in the mass distribution. Since the potential of a finite mass distribution normalized to vanish at infinity is always negative, the gravitational energy will also be negative, as we foresaw at the beginning of this section.

Planet with constant density

A planet of radius a with constant density ρ_0 and mass $M_0 = \frac{4}{3}\pi a^3 \rho_0$ has the simple potential (3.19). Carrying out the integral we find the total gravitational self-energy of the planet,

$$\mathcal{E} = -\frac{2}{5}\frac{G M_0^2}{a} = -\frac{2}{5} M_0 g_0 a. \tag{6.37}$$

In the last step we have introduced the surface gravity g_0 of the planet. In spite of the primitive nature of the model, this expression may be used as an order of magnitude estimate of the gravitational energy of a planet, or even a star.

Example 6.4.1: For the Moon we get $\mathcal{E} = -8.3 \times 10^{28}$ J, for Earth $\mathcal{E} = -1.5 \times 10^{32}$ J and for the Sun $\mathcal{E} = -1.5 \times 10^{41}$ J. Since the Sun's energy output is 3.85×10^{26} W, it could only last for 3.9×10^{14} s or about 12.5 million years before the gravitational energy that was converted into heat during its assembly would have been used up. This paradox was resolved in the 1930s with the understanding of the thermonuclear processes responsible for the Sun's energy production.

Field energy density

The total gravitational self-energy of a body (6.36) only receives contributions from regions where the mass density is non-zero. It is possible to transform it into a relation involving only the field strength $\boldsymbol{g}$ by making use of the relationship,

$$\nabla \cdot (\Phi \boldsymbol{g}) = \Phi \nabla \cdot \boldsymbol{g} + (\boldsymbol{g} \cdot \nabla)\Phi, \tag{6.38}$$

which is most easily proven by writing it explicitly out in coordinates. Integrating over a volume V and using Gauss' theorem (6.4) on the left-hand side we obtain

$$\oint_S \Phi \boldsymbol{g} \cdot d\boldsymbol{S} = \int_V \nabla \cdot (\Phi \boldsymbol{g})\, dV = -4\pi G \int_V \Phi\rho\, dV - \int_V \boldsymbol{g}^2,$$

where on the right-hand side the definition of the potential (3.20) and the gravitational source equation (6.6) have also been used.

If we now let the volume V expand to include all of space, the left-hand side will tend towards zero, because at large distance r we have $\Phi \sim 1/r$ and $\boldsymbol{g} \sim 1/r^2$, whereas the surface area expands only as r^2. In the limit we may thus rewrite the gravitational energy (6.36) in the form

$$\boxed{\mathcal{E} = -\frac{1}{8\pi G} \int \boldsymbol{g}^2\, dV.} \tag{6.39}$$

This form also explicitly demonstrates that the gravitational self-energy of a body is always negative.

In the spherical case we use (3.17) and obtain

$$\mathcal{E} = -\frac{1}{2} G \int_0^\infty \frac{M(r)^2}{r^2}\, dr. \tag{6.40}$$

This integral always converges for a body of finite mass, i.e. provided $M(r) \to M_0$ for $r \to \infty$, even if it has no boundary. Inserting $M(r) = \frac{4}{3}\pi r^3 \rho_0$ one immediately recovers (6.37).

Where *is* the energy?

Until now we have calculated the total gravitational energy from the non-local interaction of the mass density with itself through the potential (3.24). It now seems that equation (6.39) tells us that it may also be viewed as arising from a local distribution of energy over all of space. The gravitational *energy density* is $-\boldsymbol{g}(\boldsymbol{x})^2/8\pi G$ which is non-vanishing even in regions of space completely devoid of matter. As we discussed in section 1.4, the question of whether there is *really* energy out there in space depends largely on your theoretical frame of mind. In classical Newtonian physics, rewriting the self-energy as an integral over an energy density is just another mathematical trick.

Problems

6.1 Show that Gauss' theorem in the form (4.22) is equivalent to the usual form (6.4).

6.2 Show that for a barotropic fluid the equation of hydrostatic equilibrium (6.13) may be rewritten

$$\nabla^2 w = -4\pi G\rho \tag{6.41}$$

where $w = \int dp/\rho$ is the pressure potential (4.35). Could you derive this equation without any calculation?

6.3 Calculate the hydrostatic pressure in a two-layer planet (see problem 3.7) and determine the value of the central pressure for Earth.

6.4 Show that for a planet with constant density and fixed mass, the central pressure falls as a^{-4}.

6.5 Show that the adiabatic index for an ideal gas in isentropic equilibrium is given by

$$1 - \frac{1}{\gamma} = \frac{d \log T}{d \log p}. \tag{6.42}$$

6.6 **(a)** Find the power law solutions to the stellar equation (6.22) of the form $T \sim r^{\alpha}$ with $\alpha < 0$. **(b)** Determine the condition for finite mass for $r \to 0$.

6.7 Show that the short distance behaviour of the Lane–Emden functions is $\theta(s) = 1 - \xi^2/6$, independent of γ.

6.8 **(a)** Show that for $\gamma = 6/5$ the solution to the Lane–Emden equation is $\theta(s) = (1 + \xi^2/3)^{-1/2}$. **(b)** Calculate the pressure and density. **(c)** Show that although the star has no boundary, it nevertheless has finite mass.

6.9 Compare the gravitational energy of the Earth to an estimate of how much energy would be needed to melt the Earth. Do you think the Earth melted when its material was accumulated from an early cold cloud around the Sun?

6.10 Compare for a spherical planet with constant mass density the total field energy inside the planet with the field energy outside.

* **6.11** Show that

$$\nabla^2 \frac{1}{|\boldsymbol{x}|} = -4\pi\delta(\boldsymbol{x}), \tag{6.43}$$

where $\delta(\boldsymbol{x})$ is the three-dimensional δ-function, i.e. the mass distribution of a unit mass point particle at the origin.

7 Hydrostatic shapes

It is primarily the interplay between gravity and contact forces that shapes the macroscopic world around us. The seas, the atmosphere, planets and stars all owe their shape to gravity, and even our own bodies bear witness to the strength of gravity at the surface of our massive planet. What physics principles determine the shape of the surface of the sea? The sea is obviously horizontal at short distances, but bends below the horizon at larger distances following the planet's curvature. The Earth as a whole is spherical and so is the sea, but that is only the first approximation. The Moon's gravity tugs at the water in the seas and raises tides, and even the massive Earth itself is flattened by the centrifugal forces of its own rotation.

Disregarding surface tension, the simple answer is that in hydrostatic equilibrium with gravity, an interface between two fluids of different densities, for example the sea and the atmosphere, must coincide with a surface of constant potential, an equipotential surface. Otherwise, if an interface crosses an equipotential surface, there will arise a tangential component of gravity which can only be balanced by shear contact forces which a fluid at rest is unable to supply. An iceberg rising out of the sea does not obey this principle because it is solid, not fluid. Neither is the principle valid for fluids in motion. Waves in the sea are in fact 'waterbergs' that normally move along the surface, but under special circumstances are able to stay in one place, as for example in a river flowing past a big stone.

In this chapter the influence of gravity on the shape of large bodies of fluid is analysed, the primary goal being the calculation of the size and shape of the tides. Centrifugal forces give rise to a gravity-like field, which shapes all rotating fluid bodies with open surfaces, for example a bucket of water or a planet. Surface tension only plays a role for small bodies of fluid and will be discussed in chapter 8.

7.1 Fluid interfaces in hydrostatic equilibrium

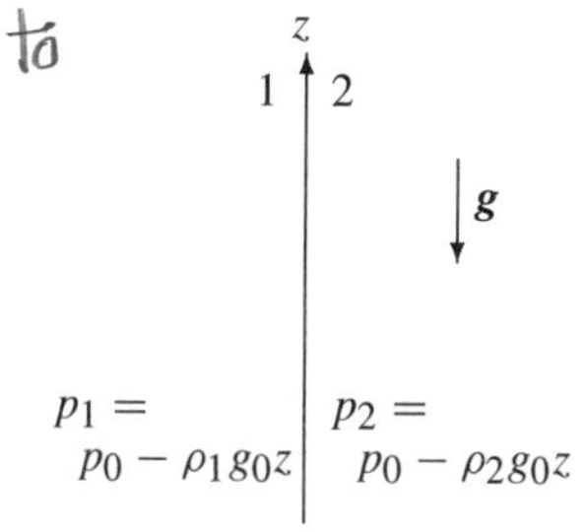

An impossible vertical interface between two fluids at rest with different densities. Even if the hydrostatic pressures on the two sides are the same for $z = 0$ they will be different everywhere else.

The intuitive argument about the impossibility of creating a hydrostatic 'waterberg' must in fact follow from the equations of hydrostatic equilibrium. We shall now show that hydrostatic equilibrium implies that the interface between two fluids with different densities ρ_1 and ρ_2 must be an equipotential surface.

Since the gravitational field is the same on both sides of the interface, hydrostatic balance $\nabla p = \rho \boldsymbol{g}$ implies that there is a jump in the pressure gradient across the interface, because on one side $(\nabla p)_1 = \rho_1 \boldsymbol{g}$ and on the other $(\nabla p)_2 = \rho_2 \boldsymbol{g}$. If the field of gravity has a component tangential to the interface, there will consequently be a jump in the tangential pressure gradient. If the pressures are equal at one point, they must therefore be different a little distance away along the surface. Newton's third law, however, requires pressure to be continuous everywhere, even across an interface (as long as there is no surface tension), so this problem can only be avoided if the tangential component of gravity vanishes everywhere at the interface, implying that it is an equipotential surface.

If, on the other hand, the fluid densities are exactly the same on both sides of the interface but the fluids themselves are different, the interface is not forced to follow an equipotential surface. This is, however, an unusual and highly unstable situation. The smallest deviation from equality in density on the two sides will call gravity in to make the interface horizontal. Stable vertical interfaces between fluids are simply not seen.

Isobars and equipotential surfaces

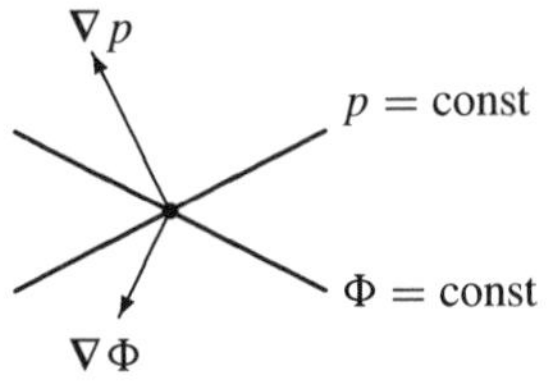

If isobars and equipotential surfaces cross, hydrostatic balance $\nabla p + \rho \nabla \Phi = \mathbf{0}$ becomes impossible.

Surfaces of constant pressure, satisfying $p(\boldsymbol{x}) = p_0$, are called *isobars*. Through every point of space runs one and only one isobar, namely the one corresponding to the pressure at that point. The gradient of the pressure is everywhere normal to the local isobar surface, like gravity, $\boldsymbol{g} = -\nabla \Phi$, is everywhere normal to the local equipotential surface, defined by $\Phi(\boldsymbol{x}) = \Phi_0$. Local hydrostatic equilibrium, $\nabla p = \rho \boldsymbol{g} = -\rho \nabla \Phi$, tells us that the normal to the isobar is everywhere parallel with the normal to the equipotential surface. This can only be the case if *isobars coincide with equipotential surfaces in hydrostatic equilibrium.* If an isobar crossed an equipotential surface anywhere at a finite angle the two normals could not be parallel.

Since the curl of a gradient trivially vanishes, $\nabla \times \nabla f = \mathbf{0}$, it follows from hydrostatic equilibrium that

$$\mathbf{0} = \nabla \times (\rho \boldsymbol{g}) = \nabla \rho \times \boldsymbol{g} + \rho \nabla \times \boldsymbol{g} = -\nabla \rho \times \nabla \Phi. \tag{7.1}$$

This implies that $\nabla \rho \sim \nabla \Phi$, so the surfaces of constant density must also coincide with the equipotential surfaces in hydrostatic equilibrium.

7.2 Shape of rotating fluids

Newton's second law of motion is only valid in *inertial* coordinate systems, where free particles move on straight lines with constant velocity. In rotating, or otherwise accelerated, non-inertial coordinate systems, one may formally write the equation of motion in their usual form, but the price to be paid is the inclusion of certain force-like terms that do not have any obvious connection with material bodies, but derive from the overall motion of the coordinate system (see chapter 20 for a more detailed analysis). Such terms are called *fictitious forces*, although they are by no means pure fiction, as one becomes painfully aware when standing in a bus that suddenly stops. A more reasonable name might be *inertial* forces, since they arise as a consequence of the inertia of material bodies.

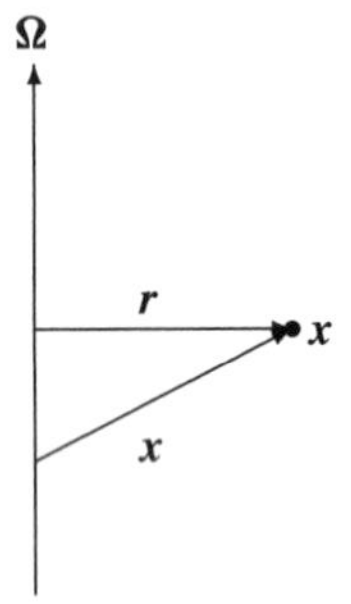

The geometry of a rotating system is characterized by a rotation vector $\boldsymbol{\Omega}$ directed along the axis of rotation, with magnitude equal to the angular velocity. The vector $\boldsymbol{r}$ is directed orthogonally out from the axis to a point $\boldsymbol{x}$.

Antigravity of rotation

A material particle at rest in a coordinate system rotating with constant angular velocity Ω in relation to an inertial system will experience only one fictitious force, the *centrifugal force*. We all know about this from carousels. It is directed perpendicularly outwards from the axis of rotation and of magnitude $M\,r\,\Omega^2$, where r is the shortest distance to the axis.

In a rotating coordinate system placed with its origin on the rotation axis, and z-axis coincident with it, the shortest vector to a point $\boldsymbol{x} = (x, y, z)$ is $\boldsymbol{r} = (x, y, 0)$. The centrifugal force is proportional to the mass of the particle and thus mimics a gravitational field

$$\boldsymbol{g}_{\text{centrifugal}}(\boldsymbol{r}) = \boldsymbol{r}\Omega^2 = (x, y, 0)\Omega^2. \tag{7.2}$$

This fictitious gravitational field may be derived from a (fictitious) potential

$$\Phi_{\text{centrifugal}}(\boldsymbol{r}) = -\frac{1}{2}r^2\Omega^2 = -\frac{1}{2}\Omega^2(x^2 + y^2). \tag{7.3}$$

Since the centrifugal field is directed away from the axis of rotation the centrifugal field is a kind of *antigravity* field, which will try to split things apart and lift objects off a rotating planet. The antigravity field of rotation is, however, cylindrical in shape rather than spherical and has consequently the greatest influence at the equator of Earth. If our planet rotated once in a little less than 1.5 hours, people at the equator could (and would) actually levitate!

Newton's bucket

A bucket of water on a rotating plate is an example going right back to Newton himself. Internal friction (viscosity) in the water will after some time bring it to rest relative to the bucket and plate, and the whole thing will end up rotating as a solid body. In a rotating coordinate system with z-axis along the axis of rotation, the total gravitational field becomes $\boldsymbol{g} = (\Omega^2 x, \Omega^2 y, -g_0)$, including both 'real' gravity and the 'fictitious' centrifugal force. Correspondingly, the total gravitational potential is,

$$\Phi = -\boldsymbol{g} \cdot \boldsymbol{x} = g_0 z - \frac{1}{2}\Omega^2(x^2 + y^2), \tag{7.4}$$

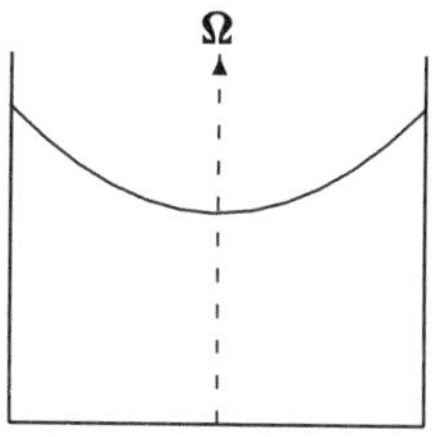

The water surface in rotating bucket forms a parabolic shape because of centrifugal forces.

and the pressure

$$p = p_0 - \rho_0 \Phi = p_0 - \rho_0 g_0 z + \frac{1}{2}\rho_0 \Omega^2 (x^2 + y^2) \tag{7.5}$$

where p_0 is the pressure at the origin of the coordinate system. It grows towards the rim, reflecting everywhere the change in height of the water column.

The isobars and equipotential surfaces are in this case rotation paraboloids,

$$z = z_0 + \frac{\Omega^2}{2g_0}(x^2 + y^2), \tag{7.6}$$

where z_0 is a constant. In a bucket of diameter 20 cm rotating once per second the water stands 2 cm higher at the rim than in the centre.

Example 7.2.1: An ultracentrifuge of radius 10 cm contains water and rotates at $\Omega = 60\,000$ rpm $\approx$ 6300 s^{-1}. The centrifugal acceleration becomes 400 000 times standard gravity and the maximal pressure close to 2000 atm, which is double the pressure at the bottom of the deepest abyss in the sea. At such pressures, the change in water density is about 10%.

* Stability of rotating bodies

Including the centrifugal field (7.2) in the fundamental field equation (6.6), the divergence of the total acceleration field $\boldsymbol{g} = \boldsymbol{g}_{\text{gravity}} + \boldsymbol{g}_{\text{centrifugal}}$ becomes,

$$\nabla \cdot \boldsymbol{g} = -4\pi G\rho + 2\Omega^2. \tag{7.7}$$

Effectively, centrifugal forces create a negative mass density $-\Omega^2/2\pi G$. This is, of course, a purely formal result, but it nevertheless confirms the 'antigravity' aspect of centrifugal forces, which makes gravity effectively repulsive wherever $\Omega^2/2\pi G\rho > 1$.

For a spherical planet stability against levitation at the equator requires the centrifugal force at the equator $\Omega^2 a$ to be smaller than surface gravity, $g_0 = GM/a^2 = \frac{4}{3}\pi\rho_0 Ga$, which leads to the stronger condition,

$$q = \frac{\Omega^2 a}{g_0} = \frac{3}{2}\frac{\Omega^2}{2\pi G\rho_0} < 1. \tag{7.8}$$

Inserting the parameters of the Earth we find $q \approx 1/291$. At the end of section 7.4 the influence of the deformation caused by rotation is also taken into account, leading to an even stricter stability condition.

7.3 The Earth, the Moon and the tides

Kepler thought that the Moon would influence the waters of Earth and raise tides, but Galilei found this notion of Kepler's completely crazy and compared it to common superstition. After Newton we know that the Moon's gravity acts on everything on Earth, also on the water in the sea, and attempts to pull it out of shape, thereby creating the tides. But since high tides occur roughly at the same time at antipodal points of the Earth, and twice a day, the explanation is not simply that the Moon lifts the sea towards itself.

Johannes Kepler (1580–1635). *German mathematician and astronomer. Discovered that planets move in elliptical orbits and that their motion obeys mathematical laws.*

Galileo wrote about Kepler: 'But among all the great men who have philosophized about this remarkable effect, I am more astonished at Kepler than at any other. Despite his open and acute mind, and though he has at his fingertips the motions attributed to the earth, he has nevertheless lent his ear and his assent to the moon's dominion over the waters, and to occult properties, and to such puerilities.' (see [15, p. 145]).

The best natural scientists and mathematicians of the eighteenth and nineteenth centuries worked on the dynamics of the tides, but here we shall only consider the simplest possible case of a quasi-static Moon. For a more complete discussion, including the dynamics of tidal waves, see, for example, Sir Horace Lamb's classic book [36] or [48] for a modern account.

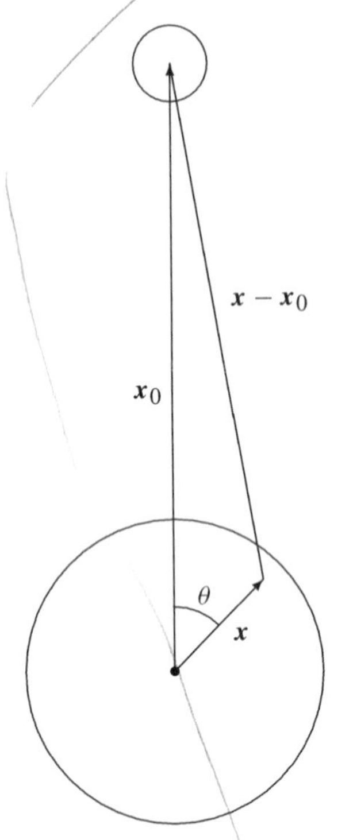

Geometry of the Earth and the Moon (not to scale).

The Earth

We shall limit ourselves to studying the Moon's influence on a liquid surface layer of the Earth. The solid parts of the Earth will, of course, also react to the Moon's field, but the effects are somewhat smaller and are due to elastic deformation rather than flow. This deformation has been indirectly measured to a precision of a few per cent in the daily 0.1 ppm variations in the strength of gravity (see figure 7.1 on page 90). There are also tidal effects in the atmosphere, but they are dominated by other atmospheric motions.

We shall furthermore disregard the changes to the Earth's own gravitational potential due to the shifting waters of the tides themselves, as well as the centrifugal antigravity of Earth's rotation causing it to deviate from a perfect sphere (which increases the tidal range by slightly more than 10%, see section 7.4). Under all these assumptions the gravitational potential at a height h over the surface of the Earth is to first order in h given by

$$\boxed{\Phi_{\text{Earth}} = g_0 h,} \tag{7.9}$$

where g_0 is the magnitude of the surface gravity.

The Moon

The Moon is not quite spherical, but nevertheless so small and far away that we may approximate its potential across the Earth with that of a point particle $-Gm/|\boldsymbol{x}-\boldsymbol{x}_0|$ situated at the Moon's position $\boldsymbol{x}_0$ with the Moon's mass m. Choosing a coordinate system with the origin at the centre of the Earth and the z-axis in the direction of the Moon, we have $\boldsymbol{x}_0 = (0, 0, D)$ where $D = |\boldsymbol{x}_0|$ is the Moon's distance. Since the Moon is approximately 60 Earth radii (a) away, i.e. $D \approx 60a$, the Moon's potential across the Earth (for $r = |\boldsymbol{x}| \le a$) may conveniently be expanded in powers of $\boldsymbol{x}/D$, and we find to second order

$$\begin{aligned}\frac{1}{|\boldsymbol{x}-\boldsymbol{x}_0|} &= \frac{1}{\sqrt{x^2+y^2+(z-D)^2}} = \frac{1}{\sqrt{D^2-2zD+r^2}} \\ &= \frac{1}{D}\frac{1}{\sqrt{1-\dfrac{2z}{D}+\dfrac{r^2}{D^2}}} \\ &\approx \frac{1}{D}\left(1-\frac{1}{2}\left(-\frac{2z}{D}+\frac{r^2}{D^2}\right)+\frac{3}{8}\left(-\frac{2z}{D}\right)^2\right) \\ &= \frac{1}{D}\left(1+\frac{z}{D}+\frac{3z^2-r^2}{2D^2}\right).\end{aligned}$$

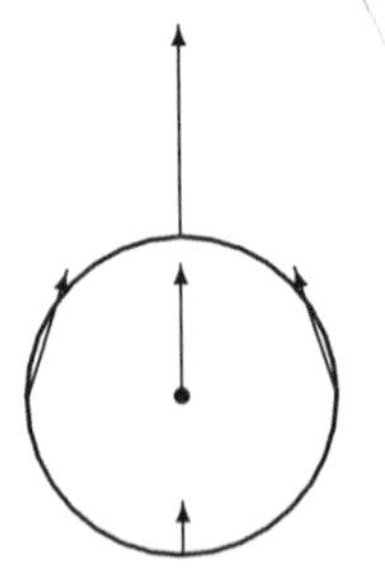

How the Moon's gravity varies over the Earth (exaggerated).

The first term in this expression leads to a constant potential $-Gm/D$, which may of course be ignored. The second term corresponds to a constant gravitational field in the direction towards the moon $g_z = Gm/D^2 \approx 30\ \mu\text{m s}^{-2}$, which is precisely cancelled by the centrifugal force due to the Earth's motion around the common centre-of-mass of the Earth–Moon system (an effect we shall return to below). Spaceship Earth is therefore completely unaware of the two leading terms in the Moon's potential, and these terms cannot raise the tides. Galilei was right to leading non-trivial order, and that's actually not so bad.

Tidal effects come from the *variation* in the gravitational field across the Earth, to leading order given by the third term in the expansion of the potential. Introducing the angle θ between the direction to the

Moon and the observation point on Earth, we have $z = r\cos\theta$, and the Moon's potential becomes (after dropping the two first terms)

$$\Phi_{\text{Moon}} = -\frac{1}{2}(3\cos^2\theta - 1)\left(\frac{r}{D}\right)^2 \frac{Gm}{D}. \tag{7.10}$$

This expansion may of course be continued indefinitely to higher powers of r/D. The coefficients $P_n(\cos\theta)$ are called Legendre polynomials (here $P_2(\cos\theta) = \frac{1}{2}(3\cos^2\theta - 1)$).

The gravitational field of the Moon is found from the gradient of the potential. It is simplest to convert back to Cartesian coordinates, writing $(3\cos^2\theta - 1)r^2 = 2z^2 - x^2 - y^2$, before calculating the gradient. In the xz-plane we get at the surface $r = a$

$$g_{\text{Moon}} = (-\sin\theta, 0, 2\cos\theta)\frac{aGm}{D^3}.$$

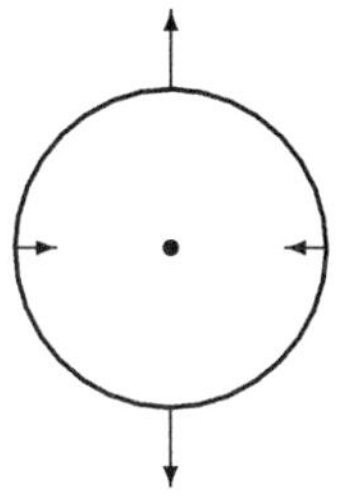

The Moon's gravity with common acceleration cancelled. This also explains why the variations in tidal height have a semi-diurnal period.

Projecting on the local normal $e_r = (\sin\theta, 0, \cos\theta)$ and tangent $e_\theta = (\cos\theta, 0, -\sin\theta)$ to the Earth's surface, we finally obtain the vertical and horizontal components of the gravitational field of the Moon at any point of the Earth's surface

$$g^{\perp}_{\text{Moon}} = g_{\text{Moon}} \cdot e_r = (3\cos^2\theta - 1)\frac{aGm}{D^3}, \tag{7.11}$$

$$g^{\parallel}_{\text{Moon}} = g_{\text{Moon}} \cdot e_\theta = -\sin 2\theta\,\frac{3aGm}{2D^3}. \tag{7.12}$$

The magnitude of the horizontal component is maximal for $\theta = 45^o$ (and, of course, also 135° because of symmetry).

Concluding, we repeat that tide-generating forces arise from variations in the Moon's gravity across the Earth. As we have just seen, the force is generally not vertical, but has a horizontal component of the same magnitude. From the sign and shape of the potential as a function of angle, we see that effectively the Moon lowers the gravitational potential just below its position, and at the antipodal point on the opposite side of the Earth, exactly as if there were shallow 'valleys' at these places. Sometimes these places are called the Moon and anti-Moon positions.

The tides

If the Earth did not rotate and the Moon stood still above a particular spot, water would rush in to fill up these 'valleys', and the sea would come to equilibrium with its open surface at constant total gravitational potential. The total potential near the surface of the Earth is

$$\Phi = \Phi_{\text{Earth}} + \Phi_{\text{Moon}} = g_0 h - \frac{1}{2}(3\cos^2\theta - 1)\left(\frac{a}{D}\right)^2 \frac{Gm}{D}, \tag{7.13}$$

Requiring this potential to be constant we find the tidal height

$$h = h_0 + \frac{1}{2}(3\cos^2\theta - 1)\left(\frac{a}{D}\right)^2 \frac{Gm}{g_0 D}, \tag{7.14}$$

where h_0 is a constant. Since the average over the sphere of the second term is,

$$\frac{1}{4\pi}\int_0^\pi d\theta \int_0^{2\pi} \sin\theta\, d\phi\,(3\cos^2\theta - 1) = \frac{1}{2}\int_{-1}^{+1}(3z^2 - 1)dz = 0,$$

we conclude that h_0 is the average water depth.

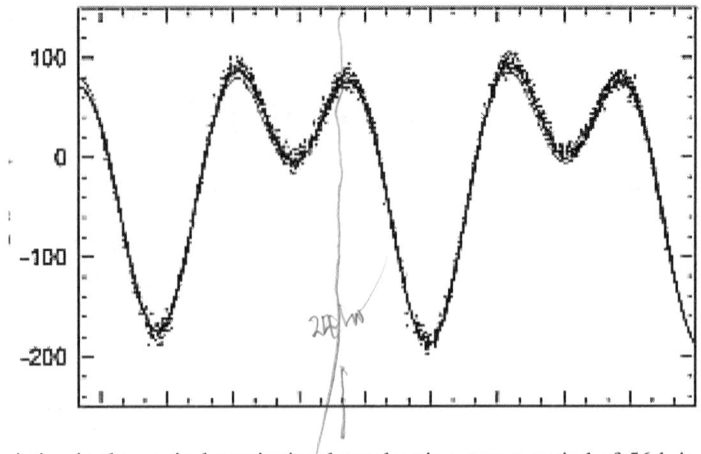

Figure 7.1. Variation in the vertical gravitational acceleration over a period of 56 h in units of $10^{-9} g_0$ (measured in Stanford, California, on December 8–9, 1996 [56]; reproduced here with permission). The semidiurnal as well as diurnal tidal variations are prominently visible as dips in the curves. Modelling the Earth as a solid elastic object and taking into account the effects of ocean loading, the measured data is reproduced to within a few times $10^{-9} g_0$.

Tidal range

The maximal difference between high and low tides, called the tidal range, occurs between the extreme positions at $\theta = 0$ and $\theta = 90°$,

$$\boxed{H_0 = \frac{3}{2}\left(\frac{a}{D}\right)^2 \frac{Gm}{g_0 D} = \frac{3}{2} a \frac{m}{M}\left(\frac{a}{D}\right)^3,} \tag{7.15}$$

where the last equation is obtained using $g_0 = GM/a^2$ with M being the Earth's mass. Inserting the values for the Moon we get $H_0 \approx 54$ cm. Interestingly, the range of the tides due to the Sun turns out to be half as large, about 25 cm. This makes spring tides when the Sun and the Moon cooperate almost three times higher than neap tides when they do not.

For the tides to reach full height, water must move in from huge areas of the Earth as is evident from the shallow shape of the potential. Where this is not possible, for example in lakes and enclosed seas, the tidal range becomes much smaller than in the open oceans. Local geography may also influence tides. In bays and river mouths funnelling can cause tides to build up to huge values. Spring tides in the range of 15 m have been measured in the Bay of Fundy in Canada.

* Quasi-static tidal cycles

The rotation of the Earth cannot be neglected. If the Earth did not rotate, or if the Moon were in a geostationary orbit, it would be much harder to observe the tides, although they would of course be there (problem 7.5). It is, after all, the cyclic variation in the water level observed at the coasts of seas and large lakes which makes the tides observable. Since the axis of rotation of the Earth is neither aligned with the direction to the Moon nor orthogonal to it, the tidal forces acquire a diurnal cycle superimposed on the 'natural' semidiurnal one. This is clearly seen in the data plotted in figure 7.1.

For a fixed position on the surface of the Earth, the dominant variation in the lunar zenith angle θ is due to Earth's diurnal rotation with angular rate $\Omega = 2\pi/24\text{ h} \approx 7 \times 10^{-5}$ radians per second. In addition, there are many other sources of periodic variations in the lunar angle [48], which we shall ignore here.

The dominant such source is the lunar orbital period of a little less than a month. Furthermore, the orbital plane of the Moon inclines about 5° with respect to the ecliptic (the orbital plane of the Earth around the Sun), and precesses with this inclination around the ecliptic in a little less than 19 years.

The Earth's equator is itself inclined about 23° to the ecliptic and precesses around it in about 25 000 years. Due to lunar orbit precession, the angle between the equatorial plane of the Earth and the plane of the lunar orbit will range over $23 \pm 5°$, i.e. between 18° and 28°, in about 9 years.

Let the fixed observer position at the surface of the Earth have (easterly) longitude ϕ and (northerly) latitude δ. The lunar angle θ is then calculated from the spherical triangle formed by the north pole, the lunar position and the observer's position,

$$\cos\theta = \sin\delta \sin\delta_0 + \cos\delta\cos\delta_0\cos(\Omega t + \phi). \tag{7.16}$$

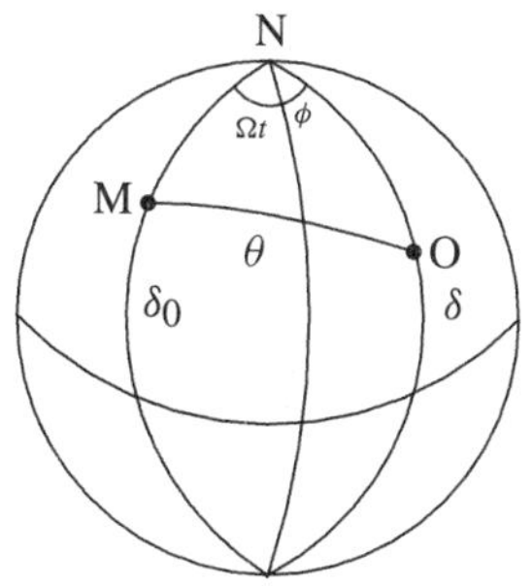

Spherical triangle formed by Moon (M), observer (O) and north pole (N).

Here δ_0 is the latitude of the lunar position and the origin of time has been chosen such that the Moon at $t = 0$ is directly above the meridian $\phi = 0$.

Inserting this into the static expression for the tidal height (7.14), we obtain the quasi-static height variation with time at the observer's place, which becomes the sum of a diurnal and a semidiurnal cycle

$$\boxed{h = \langle h \rangle + h_1 \cos(\Omega t + \phi) + h_2 \cos 2(\Omega t + \phi).} \tag{7.17}$$

Here $\langle h \rangle$ is the time-averaged height, and $h_1 = \frac{1}{2} H_0 \sin 2\delta \sin 2\delta_0$ and $h_2 = \frac{1}{2} H_0 \cos^2\delta \cos^2\delta_0$ are the diurnal and semidiurnal tidal amplitudes. The full tidal range is not quite $2h_1 + 2h_2$, because the two cosines cannot simultaneously take the value -1 (see problem 7.6).

To go beyond the quasi-static approximation, the full theory of fluid dynamics on a rotating planet becomes necessary. The tides will then be controlled not only by the tide-generating forces, but also by the interplay between the inertia of the moving water and friction forces opposing the motion. High tides will no more be tied to the Moon's instantaneous position, but may be both delayed and advanced relative to it.

* Influence of the Earth–Moon orbital motion

A question is sometimes raised concerning the role of centrifugal forces from the Earth's motion around the centre-of-mass of the Earth–Moon system. This point lies at a distance $d = Dm/(m + M)$ from the centre of the Earth, which is actually about 1700 km below the surface, and during a lunar cycle the centre of the Earth and the centre of the Moon move in circular orbits around it. Were the Earth (like the Moon) in bound rotation so that it always turned the same side towards the Moon, one would (in the corotating coordinate system, where the Moon and the Earth have fixed positions) have to add a centrifugal potential to the previously calculated potential (7.13), and the tidal range (see problem 7.7) would become about 14 m!

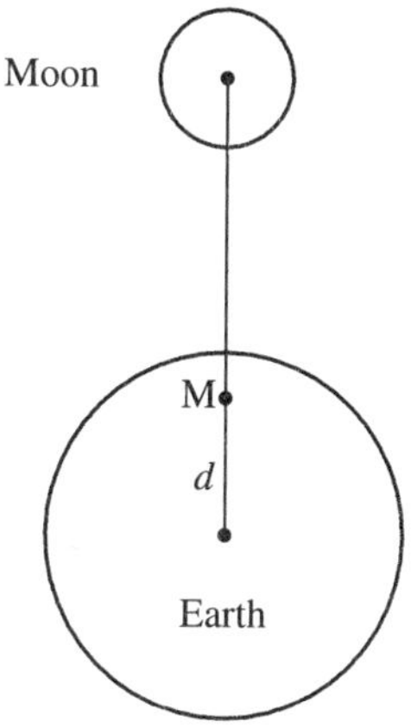

The centre-of-mass of the Earth–Moon system lies below the surface of the Earth.

Luckily, this is not the case. The Earth's own rotation is fixed with respect to the inertial system of the fixed stars (disregarding the precession of its rotation axis). A truly non-rotating Earth would, in the corotating system, rotate backwards in synchrony with the lunar cycle, cancelling the centrifugal potential. Seen from the inertial system, the circular orbital motion imparts the same centripetal acceleration $\Omega^2 d$ (along the Earth–Moon line) to all parts of the Earth. This centripetal acceleration must equal the constant gravitational attraction, Gm/D^2, coming from the linear term in the Moon's potential, and equating the two, one obtains $\Omega^2 = Gm/D^2 d = GM/D^3$, which is the well-known Kepler equation relating the Moon's period of revolution to its mass and distance.

The Moon always turns the same side towards Earth and the bound rotation adds in fact a centrifugal component on top of the tidal field from Earth. Over time these effects have together deformed the Moon into its present egg-like shape.

* 7.4 Shape of a rotating fluid planet

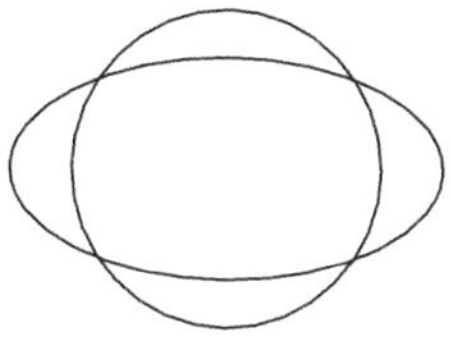

Exaggerated sketch of the change in shape of the Earth due to rotation.

On a rotating planet, centrifugal forces will add a component of 'antigravity' to the gravitational acceleration field, making the road from the pole to equator slightly downhill. At Earth's equator the centrifugal acceleration amounts to only $q \approx 1/291$ of the surface gravity, so a first guess would be that there is a centrifugal 'valley' at the equator with a depth of 1/291 of the Earths radius, which is about 22 km. If such a difference suddenly came to exist on a spherical Earth, all the water would like huge tides run towards the equator. Since there *is* land at the equator, we may conclude that even the massive Earth must over time have flowed into the centrifugal valley. The difference between the equatorial and polar radii is in fact 21.4 km [34], and coincidentally, this is roughly the same as the difference between the highest mountain top and the deepest ocean trench on Earth.

The flattening of the Earth due to rotation is, like the tides, a problem which has attracted the best minds of past centuries [36]. We shall here consider the simplest possible model, which nevertheless captures all the features relevant to slowly rotating planets.

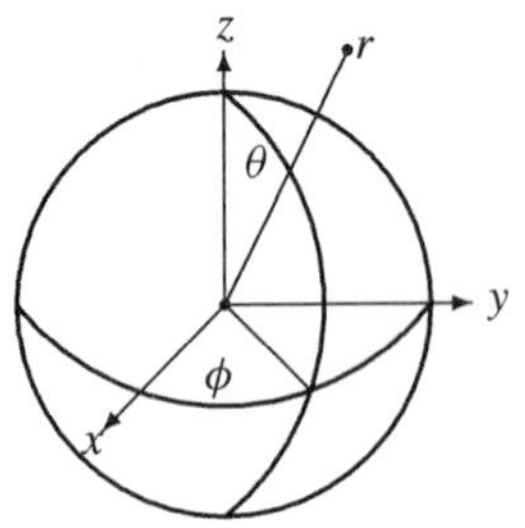

Polar and azimuthal angles.

Rigid spherical planet

If a spherical planet rotates like a stiff body, the gravitational potential above the surface will be composed of the gravitational potential of planet and the centrifugal potential. In spherical coordinates we have,

$$\Phi_0 = -g_0 \frac{a^2}{r} - \frac{1}{2}\Omega^2 r^2 \sin^2\theta \tag{7.18}$$

from which we get the vertical and horizontal components of surface gravity,

$$g_r = -\left.\frac{\partial \Phi_0}{\partial r}\right|_{r=a} = -g_0(1 - q\sin^2\theta), \tag{7.19a}$$

$$g_\theta = -\frac{1}{r}\left.\frac{\partial \Phi_0}{\partial \theta}\right|_{r=a} = g_0\, q \sin\theta\cos\theta, \tag{7.19b}$$

where $q = \Omega^2 a/g_0$ is the 'levitation parameter' defined in (7.8). This confirms that the magnitude of vertical gravity is reduced, and that horizontal gravity points towards the equator. For Earth the changes are all of relative magnitude $q \approx 1/291$.

Fluid planet

Suppose now the planet is made from a heavy fluid which given time will adapt its shape to an equipotential surface of the form,

$$r = a + h(\theta), \tag{7.20}$$

with a small radial displacement, $|h(\theta)| \ll a$. Assuming that the displaced material is incompressible we require,

$$\int_0^\pi h(\theta)\sin\theta\, d\theta = 0. \tag{7.21}$$

If we naively disregard the extra gravitational field created by the displacement of material, the potential is given by (7.18). On the displaced surface this becomes to first order in the small quantities h and q,

$$\Phi_0 \approx g_0 h - g_0 a(1 + \frac{q}{2}\sin^2\theta). \tag{7.22}$$

Demanding that it be constant, it follows that

$$h = h_0\left(\sin^2\theta - \frac{2}{3}\right), \qquad h_0 = \frac{1}{2}aq. \tag{7.23}$$

The $-(2/3)$ in the parenthesis has been chosen such that (7.21) is fulfilled. For Earth we find $h_0 = 11$ km, which is only half the expected result.

Including the self-potential

This result shows that the gravitational potential of the shifted material must play an important role. Assuming that the shifted material has constant density ρ_1, the extra gravitational potential due to the shifted material is calculated from (3.24) by integrating over the (signed) volume ΔV occupied by the shifted material,

$$\Phi_1 = -G\rho_1 \int_{\Delta V} \frac{dV'}{|\boldsymbol{x} - \boldsymbol{x}'|}. \tag{7.24}$$

Since the shifted material is a thin layer of thickness h, the volume element becomes $dV' \approx h(\theta')dS'$ where dS' is the surface element of the original sphere, $|\boldsymbol{x}'| = a$. There are, of course, corrections but they will be of higher order in h. The square of the denominator may be written as $|\boldsymbol{x} - \boldsymbol{x}'|^2 = r^2 + a^2 - 2ra\cos\psi$ where ψ is the angle between $\boldsymbol{x}$ and $\boldsymbol{x}'$. Consequently we have to linear order in h

$$\Phi_1 = -G\rho_1 \oint_S \frac{h(\theta')}{\sqrt{r^2 + a^2 - 2ra\cos\psi}} dS', \tag{7.25}$$

where $\cos\psi = \cos\theta\cos\theta' + \sin\theta\sin\theta'\cos\phi$ and $dS' = a^2\sin\theta' d\theta' d\phi'$. Note that this is exactly the same potential as would have been obtained from a surface distribution of mass with surface density $\rho_1 h(\theta)$.

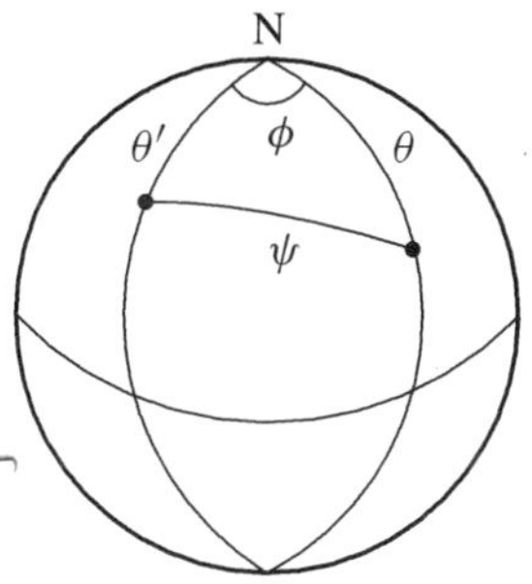

Spherical triangle formed by various angles.

There are various ways to do this integral. We shall use a wonderful theorem about Legendre polynomials, which says that a mass distribution with an angular dependence given by a Legendre polynomial creates a potential with exactly the same angular dependence. So, if we assume that the surface shape is of the form (7.23) with angular dependence proportional to $P_2(\cos\theta) = \frac{1}{2}(3\cos^2\theta - 1) = \frac{1}{2}(2 - 3\sin^2\theta)$, the effective surface mass distribution will be proportional to $P_2(\cos\theta)$, implying that the self-potential will be of precisely the same shape (see problem 7.8),

$$\Phi_1(r, \theta) = F(r)\left(\sin^2\theta - \frac{2}{3}\right). \tag{7.26}$$

The radial function may now be determined from the integral (7.26) by taking $\theta = 0$. Since now $\psi = \theta'$, all the difficult integrals disappear and we obtain

$$\begin{aligned} F(r) = -\frac{3}{2}\Phi_1(r, 0) &= \frac{3}{2}G\rho_1 h_0 a^2 2\pi \int_0^\pi \frac{\sin\theta'\left(\sin^2\theta' - \frac{2}{3}\right)}{\sqrt{r^2 + a^2 - 2ra\cos\theta'}} d\theta' \\ &= \frac{3}{2}G\rho_1 a^2 h_0 2\pi \int_{-1}^1 \frac{\frac{1}{3} - u^2}{\sqrt{r^2 + a^2 - 2rau}} du. \end{aligned}$$

The integral is now standard, and we find

$$F(r) = -\frac{4\pi}{5}\rho_1 G a h_0 \frac{a^3}{r^3} = -\frac{3}{5}g_0 h_0 \frac{\rho_1}{\rho_0}\frac{a^3}{r^3}. \tag{7.27}$$

In the last step we have used $g_0 = GM_0/a^2 = \frac{4}{3}\pi G a \rho_0$ where ρ_0 is the average density of the planet.

Total potential and strength of gravity

The total potential now becomes

$$\Phi = \Phi_0 + \Phi_1 = -g_0\frac{a^2}{r} - \frac{1}{2}\Omega^2 r^2\sin^2\theta - \frac{3}{5}g_0 h(\theta)\frac{\rho_1}{\rho_0}\frac{a^3}{r^3}. \tag{7.28}$$

Inserting $r = a + h$ and expanding to lowest order in h and q, we finally obtain an expression of the form (7.23) with,

$$\boxed{h_0 = \frac{\frac{1}{2}qa}{1 - \frac{3}{5}\frac{\rho_1}{\rho_0}}.} \tag{7.29}$$

For Earth, the average density of the mantle material is $\rho_1 \approx 4.5$ g cm^{-3} whereas the average density is $\rho_0 \approx 5.5$ g cm^{-3}. With these densities and $q = 1/291$ one gets $h_0 = 21.5$ km in close agreement with the quoted value [34]. In the same vein, we may also calculate the influence of the self-potential on the tidal range. Since the density of water is $\rho_1 \approx 1.0$ g cm^{-3}, the denominator increases the tidal range (7.15) by merely a factor of 1.12.

From the total potential we calculate gravity at the displaced surface,

$$g_r = -\left.\frac{\partial\Phi}{\partial r}\right|_{r=a+h} = -g_0\left(1 - q\sin^2\theta - 2\frac{h(\theta)}{a} + \frac{9}{5}\frac{\rho_1}{\rho_0}\frac{h(\theta)}{a}\right), \tag{7.30a}$$

$$g_\theta = -\frac{1}{r}\left.\frac{\partial\Phi}{\partial\theta}\right|_{r=a+h} = g_0\left(q + \frac{6}{5}\frac{h_0}{a}\frac{\rho_1}{\rho_2}\right)\sin\theta\cos\theta. \tag{7.30b}$$

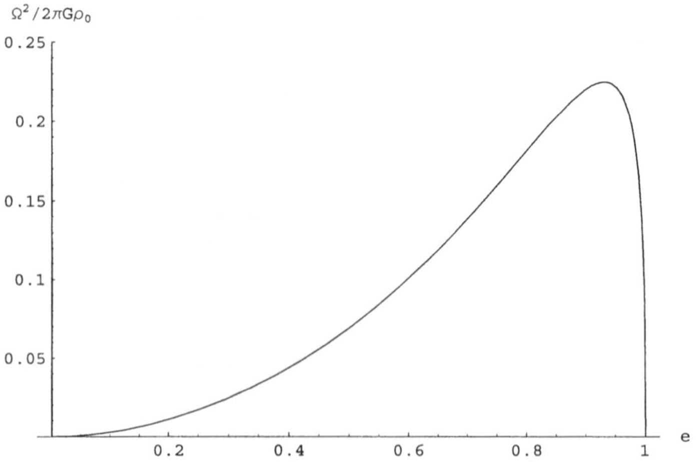

Figure 7.2. The MacLaurin function (right-hand side of equation (7.32)). The maximum 0.225 is reached for $e = 0.93$.

Finally, projecting on the local normal and tangent we find to first order in h and q,

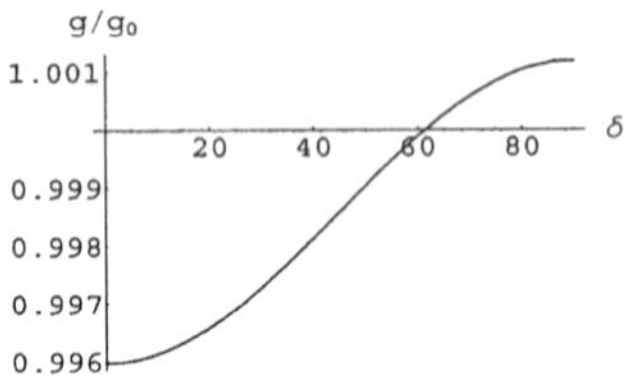

The variation of g/g_0 with latitude δ.

$$g_\perp \approx g_r, \qquad g_\| = g_\theta + g_r \frac{1}{a} \frac{dh(\theta)}{d\theta} \approx 0. \tag{7.31}$$

The field of gravity is orthogonal to the equipotential surface, as it should be.

Note that the three correction terms to the vertical field (7.30a) are due to the centrifugal force, to the change in gravity due to the change in height, and to the displacement of material. All three contributions are of the same order of magnitude, q, because they all ultimately derive from the centrifugal force.

Example 7.4.1 (Olympic games): The dependence of gravity on polar angle (or latitude) given in (7.30a) has practical consequences. In 1968 the Olympic games were held in Mexico City at latitude $\delta = 19°$ N whereas in 1980 they were held in Moscow at latitude $\delta' = 55°$ N. To compare record heights in jumps (or throws), it is necessary to correct for the variation in gravity due to the centrifugal force, the geographical difference in height and air resistance. Assuming that the initial velocity is the same, the height h attained in Mexico city would correspond to a height h' in Moscow, related to h by $v^2 = 2gh = 2g'h'$. Using (7.30a) we find $h/h' = g'/g = 1.00296$. This shows that a correction of -0.3% due to variation in gravity (among other corrections) would have to be applied to the Mexico City heights before they were compared with the Moscow heights.

Fast rotating planet

One of the main assumptions behind the calculations in this section was that the planet should be slowly rotating, meaning that the deformation of the planet due to rotation is small, or $|h(\theta)| \ll a$. Intuitively, it is fairly obvious, that if the rate of rotation of the planet is increased, the flattening increases until it reaches a point, where the 'antigravity' of rotation overcomes the 'true' gravity of planetary matter as well as cohesive forces. Then the planet becomes unstable with dramatic change of shape or even breakup as a consequence.

Colin MacLaurin (1698–1746). *Scottish mathematician who developed and extended Newton's work on calculus and gravitation.*

The study of the possible forms of rotating planets was initiated at a very early stage by Newton and in particular by MacLaurin. It was found that oblate ellipsoids of rotation are possible allowed shapes for rotating planets with constant matter density, ρ_0. An oblate ellipsoid of rotation is characterized by equal-size major axes, $a = b$ and a smaller minor axis $c < a$, about which it rotates.

MacLaurin found that the angular rotation rate is related to the eccentricity $e = \sqrt{1 - c^2/a^2}$ through the formula

$$\frac{\Omega^2}{2\pi G\rho_0} = \frac{1}{e^3}\left(\sqrt{1-e^2}(3-2e^2)\arcsin e - 3e(1-e^2)\right). \tag{7.32}$$

The right-hand side is shown in figure 7.2 and has a maximum 0.225 for $e = 0.93$, implying that stability can only be maintained for $\Omega^2/2\pi G\rho_0 < 0.225$. Actually, various other shape instabilities set in at even lower values of the eccentricity (see [63] for a thorough discussion of these instabilities and their astrophysical consequences). For small e, the MacLaurin(!) expansion of the right-hand side of (7.32) becomes $4e^2/15$. Since $e^2 \approx 2h_0/a$, we obtain $h_0 = 15\Omega^2 a/16\pi G\rho_0 = 5\Omega^2 a^2/4g_0$, in complete agreement with (7.29) for $\rho_1 = \rho_0$.

Problems

7.1 The spaceship Rama (from the novel by Arthur C. Clarke) is a hollow cylinder hundreds of kilometres long and tens of kilometres in diameter. The ship rotates so as to create a standard pseudo-gravitational field g_0 on the inner side of the cylinder. Calculate the escape velocity to the centre of the cylinder for radius $a = 10$ km.

7.2 Calculate the change in sea level if the air pressure locally rises by $\Delta p = 20$ hPa.

7.3 Calculate the changes in air pressure due to tidal motion of the atmosphere **(a)** over sea, and **(b)** over land?

7.4 How much water is found in the tidal bulge (above average height)?

7.5 How heavy must a satellite in geostationary orbit be (problem 3.3) for the tides to be of the same size as the Moon's?

7.6 Calculate the mean value $\langle h \rangle$ and the tidal range in the quasi-static approximation (7.17).

7.7 Estimate the tidal range that would result if the Earth were in bound rotation around the centre-of-mass of the Earth–Moon system.

* **7.8** Show that $\nabla^2[f(r)(3\cos^2\theta - 1)] = g(r)(3\cos^2\theta - 1)$ and determine $g(r)$ as a function of $f(r)$.

7.9 Show that the integral (7.24) may be written explicitly as

$$\Phi_1(\theta) = -\frac{G\rho_1 a}{\sqrt{2}} \int_0^{2\pi} d\phi' \int_0^{\pi} \sin\theta' d\theta' \frac{h(\theta')}{\sqrt{1 - \cos\theta\cos\theta' - \sin\theta\sin\theta'\cos\phi'}}. \tag{7.33}$$

8 Surface tension

At the interface between two materials physical properties change rapidly over distances comparable to the molecular separation scale. The transition layer is, from a macroscopic point of view, an infinitely thin sheet coinciding with the interface. Although the transition layer in the continuum limit thus appears to be a mathematical surface, it may nevertheless possess macroscopic physical properties, such as energy. And where energy is found, forces are not far away. Surface energy is necessarily accompanied by surface forces, because work has to be performed if the area of an interface and thus its surface energy is increased. The surface energy per unit of area or equivalently the force per unit of length is called *surface tension*.

Surface tension depends on the physical properties of both of the interfacing materials, and so is quite different from other material constants, for example the bulk modulus, that normally depend only on the physical properties of just one material. Surface tension creates a finite jump in pressure across the interface, but the typical magnitude of surface tension limits its influence to fluid bodies much smaller than the huge planets and stars discussed in the preceding chapters. When surface tension does come into play, as it does for a drop of water hanging at the tip of an icicle, the shape of the fluid body bears little relation to the gravitational equipotential surfaces that dominate large-scale systems. The characteristic length scale at which surface tension matches standard gravity in strength, the *capillary length*, is merely three mm for the water–air interface. This is the length scale of champagne bubbles, droplets of rain, insects walking on water, and many other phenomena.

In this chapter surface tension is introduced along with the accompanying concept of *contact angle*, and applied to the capillary effect, and to bubble and droplet shapes. In chapter 24 we shall study its influence on surface waves.

8.1 The Young–Laplace law for surface tension

The apparent paradox that a mathematical surface with no volume can possess energy may be resolved by considering a primitive three-dimensional model of a material in which the molecules are placed in a cubic grid with grid length L_{mol}. Each molecule in the interior has six bonds to its neighbours with a total binding energy of $-\epsilon$, but a surface molecule will only have five bonds when the material interfaces with a vacuum. The (negative) binding energy of the missing bond is equivalent to an extra positive energy $\epsilon/6$ for a surface molecule relative to an interior molecule, and thus an extra surface energy density,

$$\alpha \approx \frac{1}{6}\frac{\epsilon}{L_{\text{mol}}^2}. \tag{8.1}$$

The binding energy may be estimated from the specific enthalpy of evaporation H of the material as $\epsilon \approx H M_{\text{mol}}/N_A$. Note that the unit for surface tension is $\text{J m}^{-2} = \text{kg s}^{-2}$.

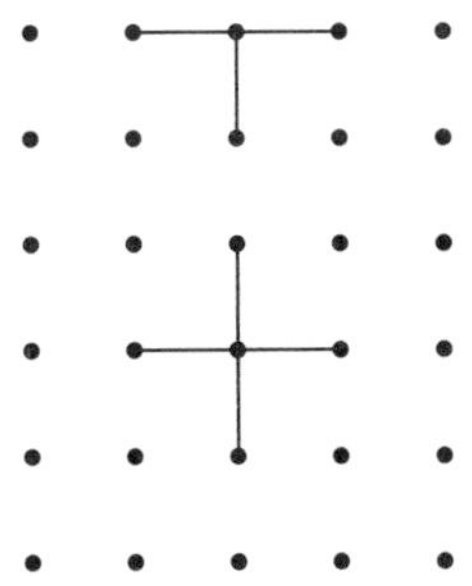

Two-dimensional cross section of a primitive three-dimensional model of a material interfacing to vacuum. A molecule at the surface has only five bonds compared to the six that link a molecule in the interior.

Example 8.1.1: For water the specific evaporation enthalpy is $H \approx 2.2 \times 10^6 \text{ J kg}^{-1}$, leading to the

	α[mN m^{-1}]
Water	72
Methanol	22
Ethanol	22
Bromine	41
Mercury	485

Surface tension of some liquids against air at 1 atm and 25° C in units of mN per metre (from [41]).

estimate $\alpha \approx 0.12$ J m^{-2}. The measured value of the surface energy for water/air interface is in fact $\alpha \approx 0.073$ J m^{-2} at room temperature. Less than a factor of two wrong is not a bad estimate at all!

Surface energy and surface tension

Increasing the area of the interface by a tiny amount dA, takes an amount of work equal to the surface energy contained in the extra piece of surface,

$$\boxed{dW = \alpha\, dA.} \tag{8.2}$$

This is quite analogous to the mechanical work $dW = -p\, dV$ performed against pressure when the volume of the system is expanded by dV. But where a volume expansion under positive pressure takes negative work, increasing the surface area takes positive work. This resistance against extension of the surface shows that the interface has a permanent internal tension, called *surface tension*[1] which we shall now see equals the energy density α.

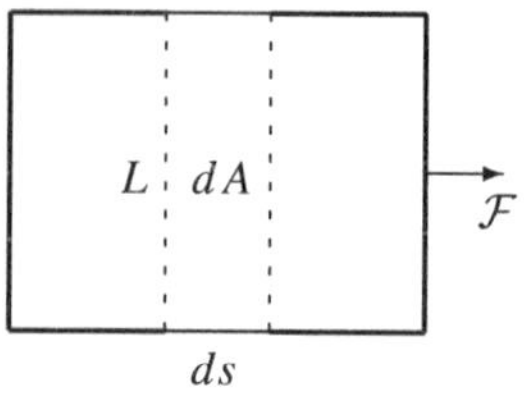

An external force $\mathcal{F}$ performs the work $dW = \mathcal{F}\, ds$ to stretch the surface by ds. Since the area increase is $dA = L ds$, the force is $\mathcal{F} = \alpha\, L$. The force per unit of length, $\alpha = \mathcal{F}/L$, is the surface tension.

Formally, surface tension is defined as the force per unit of length that acts orthogonally to an imaginary line drawn on the interface. Suppose we wish to stretch the interface along a straight line of length L by a uniform amount ds. Since the area is increased by $dA = L ds$, it takes an amount of work $dW = \alpha L ds$, implying that the force acting orthogonally to the line is $\mathcal{F} = \alpha L$, or $\mathcal{F}/L = \alpha$. Surface tension is thus identical to the surface energy density. This is also reflected in the equality of the natural units for the two quantities, N m^{-1} = J m^{-2}.

Since the interface has no macroscopic thickness, it may be viewed as being locally flat everywhere, implying that the energy density cannot depend on the macroscopic curvature, but only on the microscopic properties of the interface. If the interfacing fluids are homogeneous and isotropic—as they normally are—the value of the surface energy density will be the same everywhere on the surface, although it may vary with the local temperature. Surface tension depends, as mentioned, on the physical properties of both of the interfacing materials, which is quite different from other material constants that normally depend only on the physical properties of just one material.

Fluid interfaces in equilibrium are usually quite smooth, implying that α must always be positive. If α were negative, the system could produce an infinite amount of work by increasing the interface area without limit. The interface would fold up like crumbled paper and mix the two fluids thoroughly, instead of separating them. Formally, one may in fact view the rapid dissolution of ethanol in water as due to negative interfacial surface tension between the two liquids. The general positivity of α guarantees that fluid interfaces seek towards the minimal area consistent with the other forces that may be at play, for example pressure forces and gravity. Small raindrops and champagne bubbles are for this reason nearly spherical. Larger raindrops are also shaped by viscous friction, internal flow and gravity, giving them a much more complicated shape.

Pressure excess in a sphere

Consider a spherical ball of liquid of radius a, for example hovering weightlessly in a spacecraft. Surface tension will attempt to contract the ball but is stopped by the build-up of an extra pressure Δp inside the liquid. If we increase the radius by an amount da we must perform the work $dW_1 = \alpha\, dA = \alpha d(4\pi a^2) = \alpha 8\pi a\, da$ against surface tension. This work is compensated for by the thermodynamic work against the pressure excess $dW_2 = -\Delta p\, dV = -\Delta p\, 4\pi a^2\, da$. In equilibrium there should be nothing to gain, $dW_1 + dW_2 = 0$, leading to,

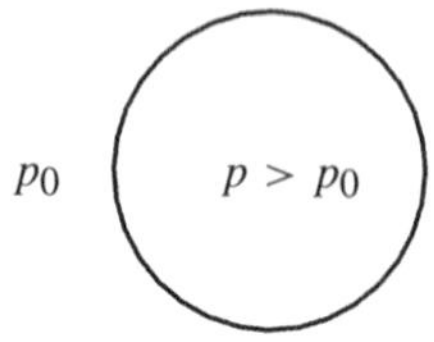

Surface tension increases the pressure inside a spherical droplet or bubble.

$$\Delta p = \frac{2\alpha}{a}. \tag{8.3}$$

The pressure excess is inversely proportional to the radius of the sphere.

It should be emphasized that the pressure excess is equally valid for a spherical raindrop in air and a spherical air bubble in water. A spherical soap bubble of radius a has two spherical surfaces, one from air

[1]There is no universally agreed-upon symbol for surface tension which is variously denoted α, γ, σ, S, Υ and even T. We shall use α, even if it collides with other uses, for example the thermal expansion coefficient.

to soapy water and one from soapy water to air. Each gives rise to a pressure excess of $2\alpha/a$, such that the total pressure inside a soap bubble is $4\alpha/a$ larger than outside.

Example 8.1.2: A spherical raindrop of diameter 1 mm has an excess pressure of only about 300 Pa, which is tiny compared to atmospheric pressure (10^5 Pa). A spherical air bubble the size of a small bacterium with diameter 1 μm acquires a pressure excess due to surface tension a thousand times larger, about 3 atm.

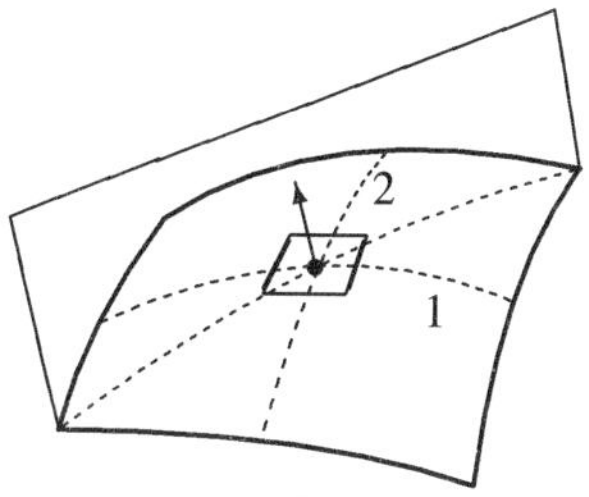

A plane containing the normal at a point intersects the surface in a planar curve with a signed radius of curvature at the point. The extreme values of the signed radii of curvature define the principal directions. The small rectangle has sides parallel with the principal directions.

When can we disregard the influence of gravity on the shape of a raindrop? For a spherical air bubble or raindrop of radius a, the condition is that the change in hydrostatic pressure across the drop should be negligible compared to the pressure excess due to surface tension, i.e. $\rho_0 g_0 2a \ll 2\alpha/a$, where ρ_0 is the density of water (minus the negligible density of air). Consequently, we require

$$a \ll R_c = \sqrt{\frac{\alpha}{\rho_0 g_0}}, \tag{8.4}$$

where the critical radius R_c is called the *capillary length.* It equals 2.7 mm for an air–water interface at 25 ° C and 1.9 mm for mercury.

Derivation of the Young–Laplace law

A smooth surface may, at a given point, be intersected with an infinite number of planes containing the normal to the surface. In each normal plane the intersection is a smooth planar curve which at the given point may be approximated by a circle centred on the normal. The centre of this circle is called the *centre of curvature* and its radius the *radius of curvature* of the intersection. Usually the radius of curvature is given a sign depending on which side of the surface the centre of curvature is situated. As the intersection plane is rotated, the centre of curvature moves up and down the normal between extreme values R_1 and R_2 of the signed radius of curvature, called the *principal radii of curvature.* It may be shown (problem 8.3) that the corresponding principal intersection planes are orthogonal, and that the radius of curvature along any other normal intersection may be calculated from the principal radii.

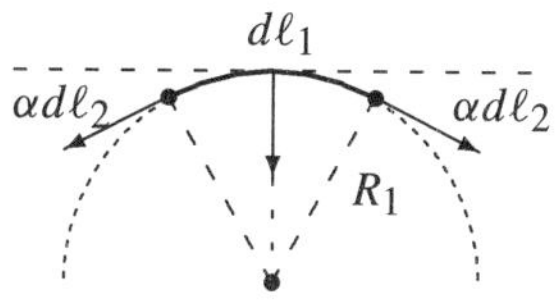

The rectangular piece of the surface of size $d\ell_1 \times d\ell_2$ is exposed to two tension forces along the 1-direction resulting in a normal force pointing towards the centre of the circle of curvature. The tension forces in the 2-direction also contribute to the normal force.

Consider now a small rectangle $d\ell_1 \times d\ell_2$ with its sides aligned with the principal directions, and let us begin with the assumption that R_1 and R_2 are positive. In the 1-direction surface tension acts with two nearly opposite forces of magnitude $\alpha d\ell_2$, but because of the curvature of the surface there will be a resultant force in the direction of the centre of the principal circle of curvature. Each of the tension forces forms an angle $d\ell_1/2R_1$ with the tangent, and projecting both on the normal we obtain the total inwards force $2\alpha d\ell_2 \times d\ell_1/2R_1$. Since the force is proportional to the area $d\ell_1 d\ell_2$ of the rectangle, it represents an excess in pressure $\Delta p = \alpha/R_1$ on the side of the surface containing the centre of curvature. Finally, adding the contribution from the 2-direction we obtain the *Young–Laplace* law for the pressure discontinuity due to surface tension,

$$\boxed{\Delta p = \alpha\left(\frac{1}{R_1} + \frac{1}{R_2}\right).} \tag{8.5}$$

For the sphere we have $R_1 = R_2 = a$ and recover the preceding result (8.3). The Young–Laplace law may be extended to signed radii of curvature, provided it is remembered that *a contribution to the pressure discontinuity is always positive on the side of the surface containing the centre of curvature, otherwise negative.*

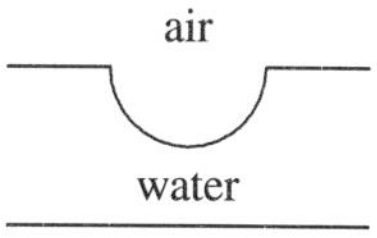

Sketch of the meniscus formed by evaporation of water from the surface of a plant leaf, resulting in a high negative pressure in the water, capable of lifting the sap to great heights.

Example 8.1.3 (How sap rises in plants): Plants evaporate water through tiny pores on the surface of their leaves. This creates a hollow air-to-water surface in the shape of a half-sphere of the same diameter as the pore. Both radii of curvature are negative $R_1 = R_2 = -a$ because the centre of curvature lies outside the water, leading to a negative pressure excess in the water. For a pore of diameter $2a \approx 1$ μm the excess pressure inside the water will be about $\Delta p \approx -3$ atm, capable of lifting sap through a height of 30 m. In practice, the lifting height is considerably smaller because of resistance in the xylem conduits of the plant through which the sap moves. Taller plants and trees need correspondingly smaller pore sizes to generate sufficient negative pressures, even down to -100 atm! Recent research has confirmed this astonishing picture (see M. T. Tyree, *Nature* **423**, (2003) 923).

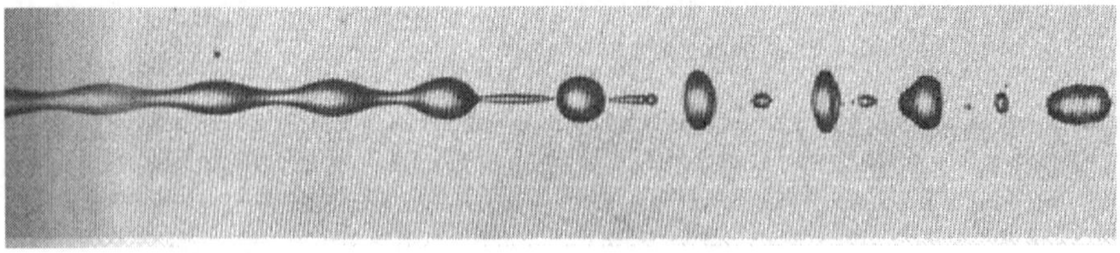

Figure 8.1. Breakup of a water jet emerging from a tube with 4 mm diameter. The wavelength is roughly equal to the Rayleigh–Plateau maximal value of π diameters just before breakup. Reproduced from D. F. Rutland and G. J. Jameson *J. Fluid Mech.* **46** (1971) 267–71.

Joseph Antoine Ferdinand Plateau (1801–1883). *Belgian experimental physicist. Invented the 'phenakistiscope', a stroboscopic device to study motion of vibrating bodies. Stared into the Sun for 20 s in 1829 to study the after effects, an experiment that made him lose his vision permanently in 1843.*

The tangent circle with radius R coincides with the curve up to and including second order in z. It satisfies the equation $z^2 + (r - R)^2 = R^2$, and for small r and z we have $r = z^2/2R$ and thus $r'' = 1/R$.

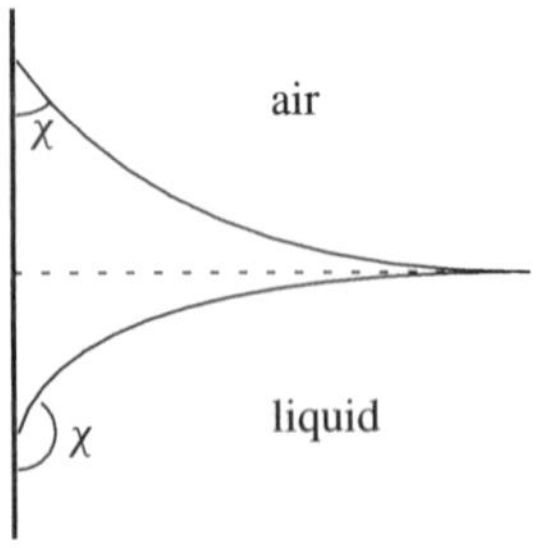

An air/liquid interface meeting a wall. The upper curve makes an acute contact angle, like water, whereas the lower curve makes an obtuse contact angle, like mercury.

The Rayleigh–Plateau instability

The spontaneous breakup of the jet of water emerging from a thin pipe is a well-known phenomenon (first studied by Savart in 1833), although the physical explanation is not. Plateau found experimentally in 1873 that the breakup begins when the length of the water column is longer than its circumference, and it was explained theoretically by Lord Rayleigh in 1879 as due to the interplay between pressure and surface tension. Here we shall verify Plateau's conclusion in a simple calculation without making a full stability analysis involving the equations of fluid dynamics.

Consider an infinitely long cylindrical column of incompressible fluid at rest in the absence of gravity with its axis along the z-axis. If we normalize the pressure outside the cylinder to zero, surface tension will make it larger inside, $p = \alpha/a$ where a is its radius. This uniform pressure excess will attempt to squeeze the column uniformly towards a smaller radius, but if the column is infinitely long, it will be impossible to move the fluid out of the way, which allows us to ignore this problem. What cannot be ignored are local radial variations $r = r(z)$ that make the column thicker in some places and thinner in others, because then the amount of fluid to move out of the way will be finite.

The simplest such variation is a radial wave of wavelength λ and tiny amplitude $b \ll a$, described by $r(z) = a - b\cos(2\pi z/\lambda)$. At any of its narrowest points, for example $z = 0$, one principal radius of curvature is the radius of the cylinder, $R_1 = a - b$, whereas the other radius of curvature is $1/R_2 = r''(0) = b4\pi^2/\lambda^2$, as a geometrical construction shows. Taking into account that its centre of curvature is outside the fluid, the local pressure excess due to the variation in radius at the constriction thus becomes,

$$\delta p = \alpha\left(\frac{1}{a-b} - \frac{1}{a} - b\frac{4\pi^2}{\lambda^2}\right) \approx \alpha\left(\frac{1}{a^2} - \frac{4\pi^2}{\lambda^2}\right) b, \tag{8.6}$$

when expanded to first order in b. For $\lambda > 2\pi a$ the pressure excess is positive and will tend to squeeze fluid away from the constriction, but that will only diminish the radius $a - b$ further and again increase the pressure excess, until the column is cut right through. A column of incompressible fluid is thus unstable against radial disturbances with wavelength larger than its circumference (see problem 8.2 for an alternative derivation of the stability condition).

8.2 Contact angle

An interface between two fluids is a two-dimensional surface which makes contact with a solid wall along a one-dimensional line. Locally the plane of the fluid interface forms a certain *contact angle* χ with the wall. For the typical case of a liquid/air interface, χ is normally defined as the angle inside the liquid. Water and air against glass meet in a small acute contact angle, $\chi \approx 0$, whereas mercury and air meets glass at an obtuse contact angle of $\chi \approx 140\,°$. Due to its small contact angle, water is very efficient in *wetting* many surfaces, whereas mercury has a tendency to make pearls. It should be emphasized that the contact angle is extremely sensitive to surface properties, fluid composition, and additives.

In the household we regularly use surfactants that enable dishwater to wet greasy surfaces on which it otherwise would tend to pearl. After washing our cars we apply a wax which makes rainwater pearl and prevents it from wetting the surface, thereby diminishing rust and corrosion.

Figure 8.2. Spider walking on water. Image courtesy David L. Hu, MIT (Hu, Chan and Bush 2003).

The contact angle is a material constant which depends on the properties of all three materials coming together. Whereas material adhesion can sustain a tension normal to the wall, the tangential tension has to vanish. This yields an equilibrium relation between the three surface tensions,

$$\alpha_{13} = \alpha_{23} + \alpha_{12} \cos \chi . \tag{8.7}$$

This relation is, however, not particularly useful because of the sensitivity of χ to surface properties, and it is better to view χ as an independent material constant.

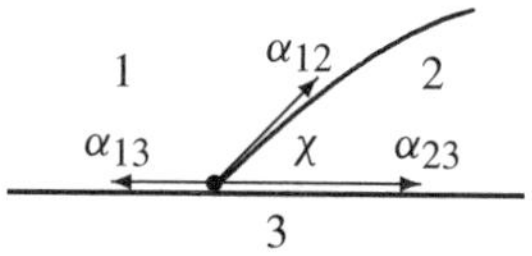

Two fluids meeting at a solid wall in a line orthogonal to the paper. The tangential component of surface tension must vanish.

Example 8.2.1 (Walking on water): Insects capable of walking on the surface of water must 'wax' their feet to obtain an obtuse contact angle and avoid getting wet. They are carried by surface tension (and buoyancy which we disregard here). Denoting by L the length of the total contact perimeter of its feet with water a crude condition for walking on water is $\alpha L > M g_0$ where M is its mass. For an insect with mass $M = 10$ mg the contact perimeter must be larger than 1.3 mm. The ratio $\text{Je} = \alpha L / M g_0$ which must be larger than unity for anyone who wants to walk on water has been called the *Jesus number* [74].

Capillary effect

Water has a well-known tendency to rise above the ambient level in a narrow vertical glass tube which is lowered into the liquid. Closer inspection reveals that the surface inside the tube is concave. This is called the *capillary effect* and is caused by the acute contact angle of water in conjunction with its surface tension which creates a negative pressure just below the liquid surface, balancing the weight of the raised water column. Mercury with its obtuse contact angle displays instead a convex surface shape, creating a positive pressure just below the surface which forces the liquid down to a level where the pressure equals that at the ambient level.

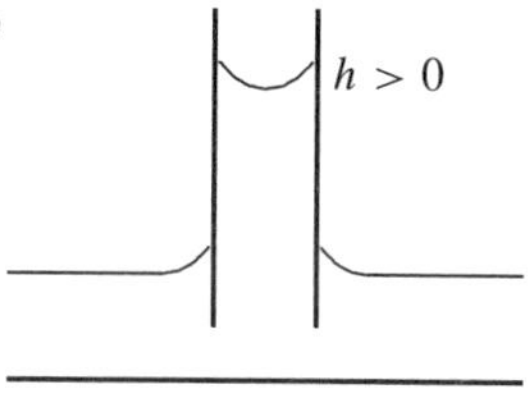

Water rises above the ambient level in a glass tube and displays a concave surface inside the tube.

Let us first calculate the effect for an acute angle of contact. At the centre of the tube the radii of curvature are equal, and since the centre of curvature lies outside the liquid, they are also negative, $R_1 = R_2 = -R_0$ where R_0 is positive. Hydrostatic balance at the centre of the tube then takes the form $\rho_0 g_0 h = 2\alpha / R_0$ where h is the central height, such that

$$h = \frac{2\alpha}{\rho_0 g_0 R_0} = 2 \frac{R_c^2}{R_0} . \tag{8.8}$$

It should be noted that this is an exact relation which does not depend on the surface being spherical. It also covers the case of an obtuse contact angle by taking R_0 to be negative.

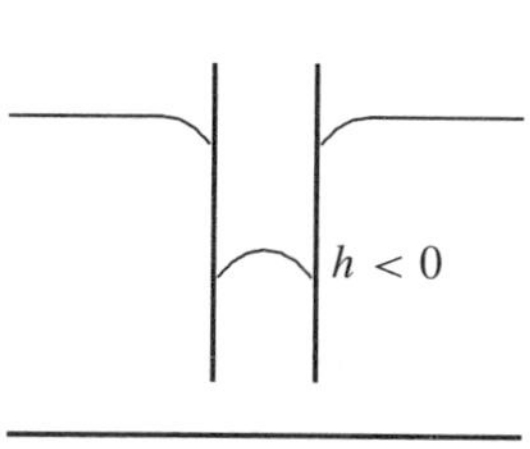

Mercury sinks below the general level in a capillary glass tube.

When the tube radius a is small compared to R_c, gravity has no effect on the shape, and the surface may be assumed to be a sphere of radius R_0. A simple geometric construction shows that $a = R_0 \cos \chi$,

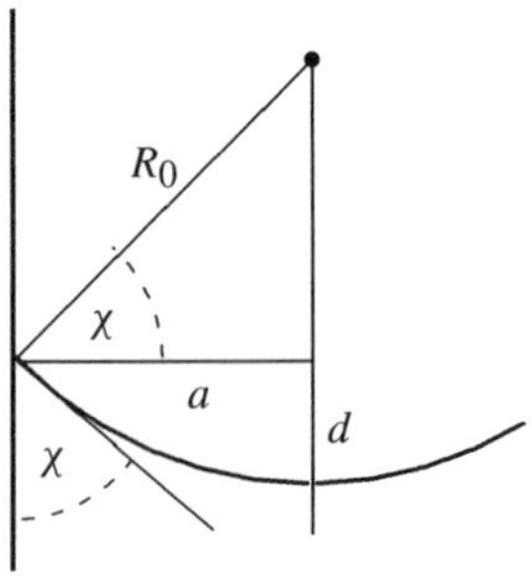

Approximately spherical surface with acute contact angle in a circular tube.

and thus,

$$h = 2\frac{R_c^2}{a}\cos\chi. \tag{8.9}$$

It is as expected positive for acute and negative for obtuse contact angles. From the same geometry it also follows that the depth of the central point of the surface is $d = R_0(1 - \sin\chi)$, or

$$d = a\frac{1 - \sin\chi}{\cos\chi}. \tag{8.10}$$

Both of these expressions are modified for larger radius, $a \gtrsim R_c$ where the surface flattens.

Example 8.2.2 (Capillary tube): A capillary tube has diameter $2a = 1$ mm. Water with $\chi \approx 0$ rises $h = +30$ mm with a surface depth $d = +0.5$ mm. Mercury with contact angle $\chi \approx 140°$ sinks on the other hand to $h = -11$ mm and $d = -0.2$ mm under the same conditions.

8.3 Capillary effect at a vertical wall

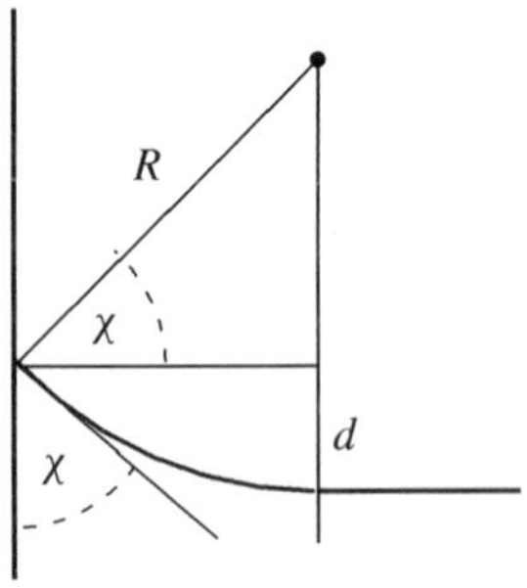

Approximately circular geometry of a liquid surface with acute contact angle near a vertical wall.

In the limit of infinite tube radius the capillary effect only deforms the liquid surface close to the nearly flat vertical wall to accommodate the finite contact angle. Far from the wall the surface is perfectly flat, and there will be no pressure jump due to surface tension, and consequently no general capillary rise of the surface above the ambient level. This is an exactly solvable case which nicely illustrates the mathematics of planar curves.

The rise or drop at the wall may be estimated by a geometric argument of the same kind as at the end of the preceding section. Assuming that the shape is a circle of radius R, the pressure change due to surface tension inside the liquid is $\Delta p = -\alpha/R$ roughly in the middle at $z = d/2$. Hydrostatic balance thus requires $\rho_0 g_0 d/2 \approx \alpha/R$, and since the radius R as before is related by geometry to the depth by $d = R(1 - \sin\chi)$, we find for an acute angle of contact,

$$d \approx R_c\sqrt{2(1 - \sin\chi)} = 2R_c \sin\frac{90° - \chi}{2}. \tag{8.11}$$

The last expression is also valid for an obtuse angle of contact. We shall see below that this expression is, in fact, identical to the exact result.

For water with nearly vanishing angle of contact, we find $d \approx \sqrt{2}R_c \approx 3.9$ mm whereas for mercury with $\chi = 140°$ we get $d \approx -1.6$ mm.

Geometry of planar curves

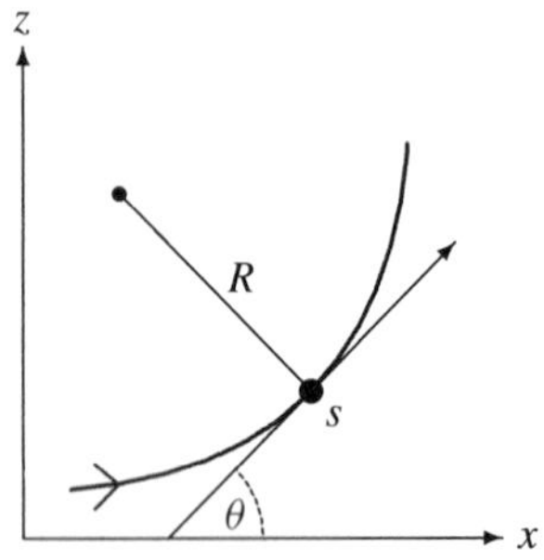

The geometry of a planar curve. The curve is parametrized by the arc length s along the curve. A small change in s generates a change in the elevation angle θ determined by the local radius of curvature. Here the radius of curvature is positive.

Taking the x-axis orthogonal to the wall, and the z-axis vertical, the surface shape may be assumed to be independent of y and described by a simple curve in the xz-plane. The best way to handle the geometry of a planar curve is to use two auxiliary parameters: the arc length s along the interface curve, and the elevation angle θ between the x-axis and the oriented tangent to the curve. From this definition of θ we obtain immediately,

$$\frac{dx}{ds} = \cos\theta, \qquad \frac{dz}{ds} = \sin\theta. \tag{8.12}$$

The radius of curvature may conveniently be defined as,

$$R = \frac{ds}{d\theta}. \tag{8.13}$$

Evidently this *geometric radius of curvature* is positive if s is an increasing function of θ, otherwise it is negative. One should be aware that this sign convention may not agree with the physical sign convention for the Young–Laplace law (8.5). Depending on the arrangement of liquid and air, it may be necessary to introduce an explicit sign to get the physics right.

Hydrostatic balance

Assuming that the air pressure is constant, $p = p_0$, the pressure in the liquid just below the surface is $p = p_0 + \Delta p$ where Δp is given by the Young–Laplace law (8.5). Denoting the local geometric radius of curvature by R we have for an acute angle of contact $R_1 = -R$ and $R_2 = \infty$, because the centre of curvature lies outside the liquid. The pressure is thus negative $\Delta p = -\alpha/R$ just below the surface, and the hydrostatic pressure of the raised surface must balance the drop in pressure, $\rho_0 g_0 z = \alpha/R$, everywhere on the surface. Introducing the capillary radius (8.4), this may be written as $1/R = z/R_c^2$, and we find from (8.13)

$$\frac{d\theta}{ds} = \frac{z}{R_c^2}. \tag{8.14}$$

This equation together with the two definitions (8.12) determine x, z and θ as functions of s.

There are several different types of solution, depending on the boundary conditions. For the surface near the wall the boundary conditions are $x = 0$ and $\theta = \chi - 90°$ for $s = 0$, and $z \to 0$ for $s \to \infty$. Having obtained the solution we may then determine the depth $z = d$ for $x = 0$.

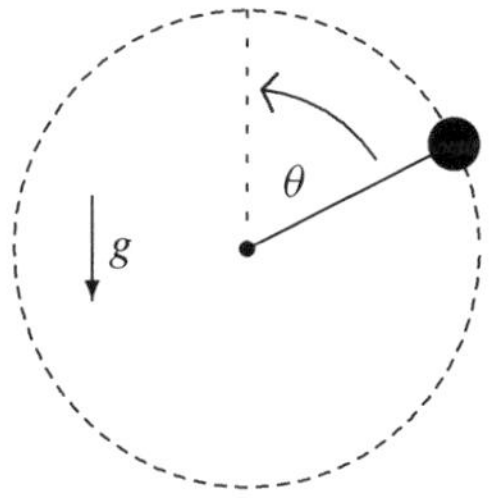

Inverted mathematical pendulum with angle θ moving towards unstable equilibrium at $\theta = 0$. This sketch corresponds to a liquid with acute contact angle. If the contact angle is obtuse, the pendulum has to start from the other side of the unstable equilibrium.

The pendulum connection

Since R_c is a constant for the liquid, we may without loss of generality choose the unit of length such that $R_c = 1$. Differentiating (8.14) once more with respect to s, we obtain the equation of motion for an *inverted mathematical pendulum*,

$$\boxed{\frac{d^2\theta}{ds^2} = \sin\theta.} \tag{8.15}$$

The boundary conditions correspond to the pendulum starting at an angle $\theta = \theta_0 = \chi - 90°$ with velocity $d\theta/ds = d$, according to (8.14) (for $R_c = 1$). The depth d must be chosen precisely such that the pendulum eventually comes to rest in unstable equilibrium at $\theta = 0$.

> If the velocity chosen is larger than the depth, the pendulum will continue through the unstable equilibrium, and the liquid surface will start to rise again. When the pendulum reaches the angle $\theta = -\theta_0$, another vertical wall may be placed there, forming the same angle of contact with the liquid surface. This is the planar analogue of the capillary effect in a circular tube, but this problem is not solvable in terms of elementary functions. The periodic pendulum solutions obtained by letting the pendulum move through stable equilibrium at $\theta = \pi$ correspond to a strip of liquid hanging at the lower edge of the vertical plate.

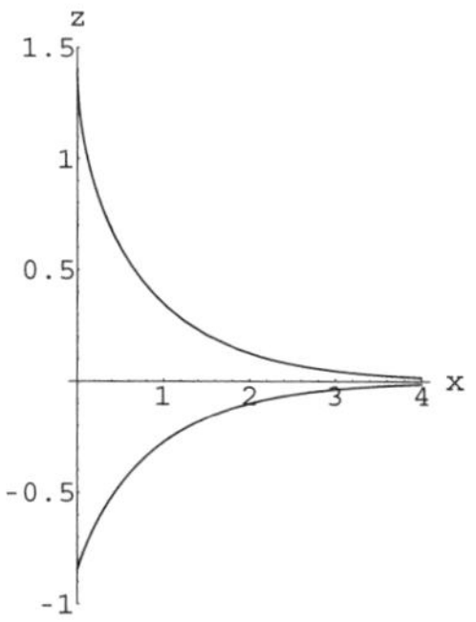

Capillary surface shape of water and mercury in units where $R_c = 1$. Note the exaggerated vertical scale.

To find the solution for the problem at hand, the surface shape near a single wall, we multiply the pendulum equation of motion by $d\theta/ds$ and integrate, to get

$$\frac{1}{2}\left(\frac{d\theta}{ds}\right)^2 = 1 - \cos\theta,$$

where the constant 1 has been determined from the condition that $d\theta/ds = 0$ and $\theta = 0$ for $s \to \infty$. From this equation and (8.14) we derive that,

$$\boxed{z = -2\sin\frac{\theta}{2},} \tag{8.16}$$

independent of whether the contact angle is acute or obtuse. Taking $\theta = \chi - 90°$ we indeed recover (8.11).

The dependence of x on θ is calculated from,

$$\frac{dx}{d\theta} = \frac{dx/ds}{d\theta/ds} = \frac{\cos\theta}{-2\sin\frac{\theta}{2}} = \sin\frac{\theta}{2} - \frac{1}{2\sin\frac{\theta}{2}}.$$

This integrates to

$$\boxed{x = x_0 - 2\cos\frac{\theta}{2} - \log\left|\tan\frac{\theta}{4}\right|,} \tag{8.17}$$

where $x_0 = 2\cos(\theta_0/2) + \log|\tan(\theta_0/4)|$ and $\theta_0 = \chi - 90°$. Together with (8.16) this constitutes a parametric form for the surface shape.

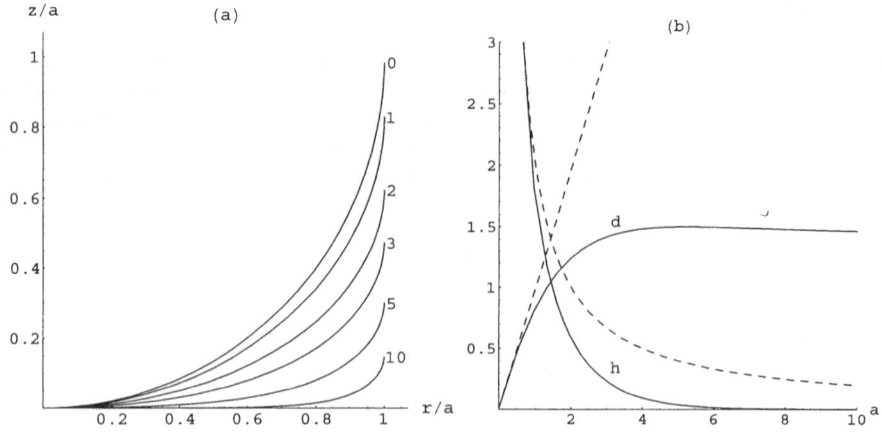

Figure 8.3. Capillary effect in water for a circular tube of radius a in units where $R_c = 1$. **(a)** Surface shape $z(r)/a$ plotted as a function of r/a for $\chi_c = 1°$ and $a = 0, 1, 2, 3, 5, 10$. Note how the shape becomes gradually spherical as the tube radius a approaches 1. For $a \lesssim 1$ the shape is essentially independent of a. **(b)** Computed capillary rise h and depth d as functions of a (solid lines). For $a \gtrsim 1$ the computed values deviate from the spherical surface estimates (8.9) and (8.10) (dashed).

* 8.4 Axially invariant surface shapes

Many static interfaces—capillary surfaces in circular tubes, droplets and bubbles—are invariant under rotation around an axis, allowing us to establish a fairly simple formalism for the shape of the equilibrium surface.

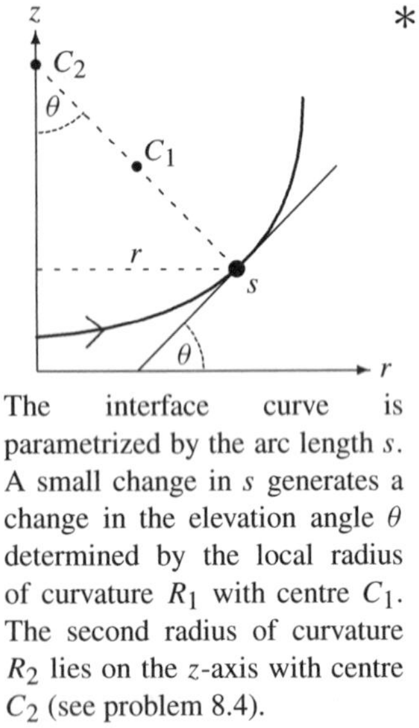

The interface curve is parametrized by the arc length s. A small change in s generates a change in the elevation angle θ determined by the local radius of curvature R_1 with centre C_1. The second radius of curvature R_2 lies on the z-axis with centre C_2 (see problem 8.4).

Geometry of axially invariant interfaces

In cylindrical coordinates an axially invariant interface is a planar curve in the rz-plane. Using again the arc length s along the curve and the angle of elevation θ for its slope, we find as in the planar case,

$$\frac{dr}{ds} = \cos\theta, \qquad \frac{dz}{ds} = \sin\theta. \tag{8.18}$$

The first principal radius of curvature may be directly taken from the planar case, whereas it takes some work to show that the second centre of curvature lies on the z-axis (see problem 8.4), such that

$$R_1 = \frac{ds}{d\theta}, \qquad R_2 = \frac{r}{\sin\theta}. \tag{8.19}$$

One should be aware that the sign convention for these *geometric radii of curvature* may not agree with the physical sign convention for the Young–Laplace law (8.5), and that it may be necessary to introduce explicit signs to get the physics right. This will become clear in the calculations that follow.

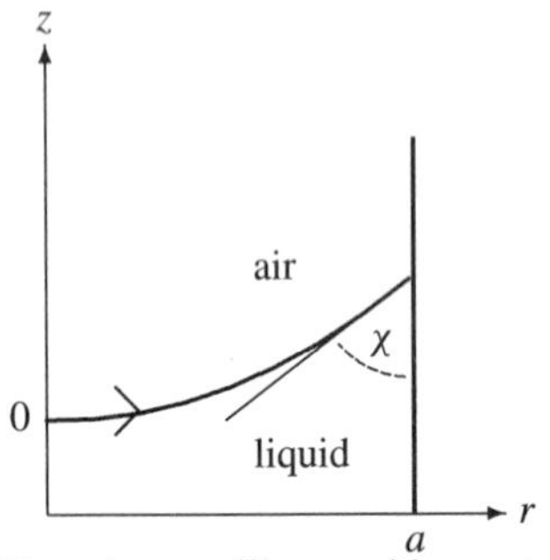

For the capillary with acute angle of contact both centres of curvature lie outside the liquid.

The capillary surface

For the rising liquid/air capillary surface with acute contact angle, both geometric radii of curvature, R_1 and R_2, are positive. Since both centers of curvature lie outside the liquid, the physical radii will be $-R_1$ and $-R_2$ in the Young–Laplace law (8.5). Assuming that the air pressure is constant, hydrostatic balance demands that $H = g_0 z + \Delta p/\rho_0$ is constant, or

$$g_0 z - \frac{\alpha}{\rho_0}\left(\frac{d\theta}{ds} + \frac{\sin\theta}{r}\right) = -\frac{2\alpha}{\rho_0 R_0}.$$

The value of the constant has been determined from the initial condition at the centre, where $r = z = \theta = 0$ and the geometric radii of curvature are equal to the central radius of curvature, $R_1 = R_2 = R_0$. Solving

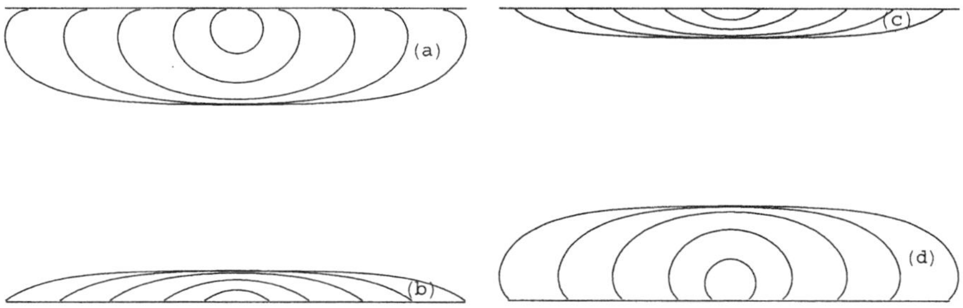

Figure 8.4. Shapes of sessile air bubbles and droplets of water ($R_c = 2.7$ mm, $\chi_c = 1°$) and mercury ($R_c = 1.9$ mm, $\chi_c = 140°$). **(a)** Air bubbles in water under a lid (to scale). **(b)** Water droplet on table plotted with vertical scale enlarged 40 times. **(c)** Air bubbles in mercury (to scale). **(d)** Mercury droplets on a table (to scale).

for $d\theta/ds$ we find,

$$\frac{d\theta}{ds} = \frac{2}{R_0} - \frac{\sin\theta}{r} + \frac{z}{R_c^2}, \tag{8.20}$$

where R_c is the capillary constant (8.4). In the second term one must remember that $r/\theta \to R_0$ for $\theta \to 0$.

Together with the two equations (8.18) we have obtained three first-order differential equations for r, z and θ. Since s does not occur explicitly, and since θ grows monotonically with s, one may eliminate s and instead use θ as the running parameter. Unfortunately these equations cannot be solved analytically, but given R_0 they may be solved numerically with the boundary conditions $r = z = 0$ for $\theta = 0$. The solutions are quasi-periodic curves that spiral upwards forever. The physical solution must however stop at the wall $r = a$ for $\theta = \theta_0 = 90° - \chi$, and that fixes R_0. The numeric solutions are displayed in figure 8.3.

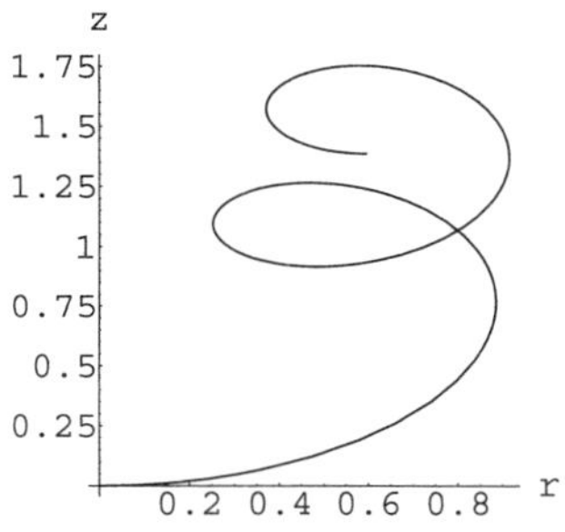

The numeric solution for $R_0 = R_c = 1$ in the interval $0 < \theta < 4\pi$. The spiral is unphysical.

Sessile bubbles and droplets

If a horizontal plate (a 'lid') is inserted into water, air bubbles may come up against its underside, and remain stably sitting (sessile) there. The bubbles are pressed against the plate by buoyancy forces that in addition tend to flatten bubbles larger than the capillary radius. The shape may be obtained from the above solution to the capillary effect by continuing it to $\theta_0 = 180° - \chi$.

Mercury sitting on the upper side of a horizontal plate likewise forms small nearly spherical droplets which may be brought to merge and form large flat puddles of 'quick silver'. In this case the geometric radii of curvature will both be negative while the physical radii of curvature are both positive because the centers of curvature lie inside the liquid. The formalism is consequently exactly the same as before, except that the central radius of curvature R_0 is now negative. The shapes are nearly the same as for air bubbles, except for the different angle of contact.

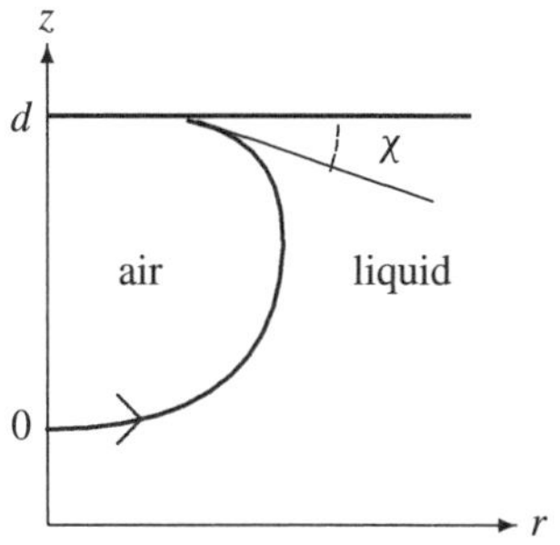

An air bubble in under a horizontal lid with acute angle of contact.

In figure 8.4 the four sessile configurations of bubbles and droplets are displayed. The depth approaches in all cases a constant value for large central radius of curvature R_0, which may be estimated by the same methods as before to be $d = R_c\sqrt{2(1 \pm \cos\chi)}$ for bubbles(+) and droplets(−). Note that the depth of the water droplet (frame 8.4b) is enlarged by a factor 40. If water really has contact angle $\chi = 1°$, the maximal depth of a water droplet on a flat surface is only $d = 0.018 R_c \approx 50\ \mu$m. This demonstrates how efficient water is in wetting a surface, because of its small contact angle.

Pending droplets

Whereas sessile droplets in principle can have unlimited size, hanging liquid droplets will fall if they become too large. Here we shall investigate the shape of a droplet hanging at the end of a thin glass tube, for example a pipette fitted with a rubber bulb which allows us to vary the pressure. When the bulb is slowly squeezed, a droplet emerges at the end of the tube and eventually falls when it becomes large enough.

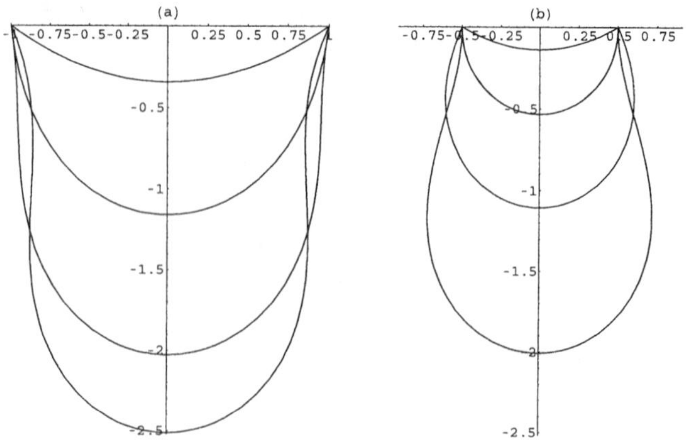

Figure 8.5. Static droplet shapes at the mouth of a circular tube in units of $R_c = 1$. There are no solutions with larger volumes than the largest ones shown here. All contact angles are assumed to be possible at the mouth of the tube. Notice that the origin of z has been redefined. **(a)** Tube with radius $a = 1$ which is larger than the critical radius ($a_1 = 0.918$). **(b)** Tube with radius $a = 0.5$ which is smaller than the critical radius.

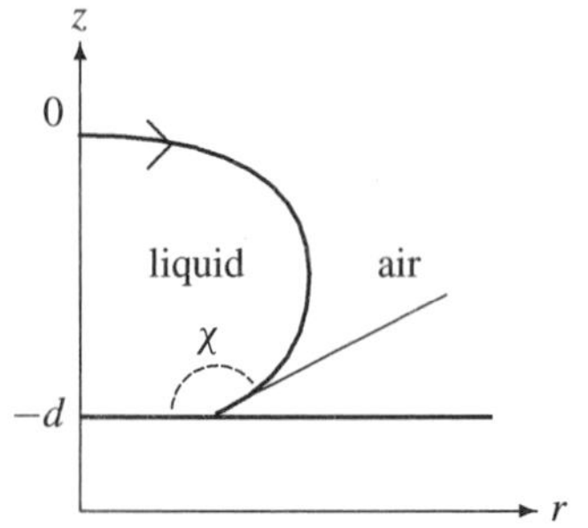

A droplet on a horizontal plate with an obtuse angle of contact.

Both the geometric and physical radii of curvature are positive in this system, such that we get (for $R_c = 1$),

$$\frac{d\theta}{ds} = \frac{2}{R_0} - \frac{\sin\theta}{r} - z, \tag{8.21}$$

with a negative sign of z. In this case θ is not a monotonic function of s, and we may not eliminate s. Assuming that the tube material is very thin, the boundary conditions may be taken to be $r = z = \theta = 0$ for $s = 0$, because the rounded end of the tube material is able to accommodate any angle of contact. The condition that the liquid surface must always make contact with the end of the glass tube at $r = a$ then determines the total curve length s_0 as a function of the central curvature R_0, and thereby the height $d = z(s_0)$.

The volume of a droplet (see problem 8.5),

$$V = \int_0^d \pi r^2 \, dz, \tag{8.22}$$

is a function of R_0. As the bulb is squeezed slowly, the volume of the droplet must grow continuously through a sequence of hydrostatic shapes. At some point instability sets in and the drop falls. It is not possible to determine the point of instability in a purely hydrostatic calculation, but the Rayleigh–Plateau result would indicate that it happens when it develops a 'waist', even if there are static solutions beyond this point.

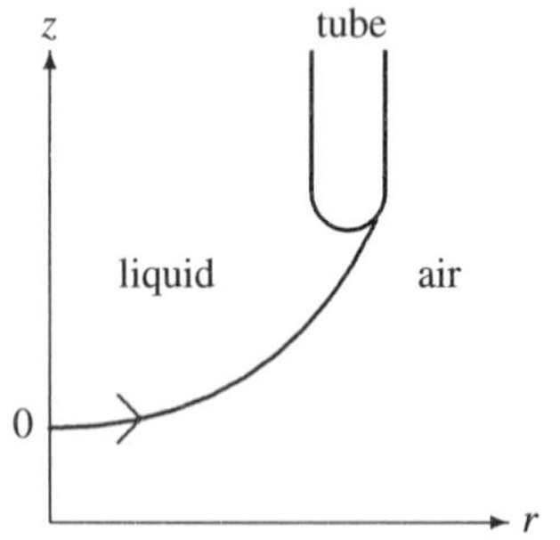

A liquid drop hanging from a tube. Any contact angle can be accommodated by the 180° turn at the end of the tube's material.

The solutions fall into two classes. If R_0 is larger than a certain critical value, $R_0 > R_{01} = 0.778876\ldots$, the radial distance $r(s)$ will grow monotonically with s, but if $R_0 < R_{01}$ the surface will be shaped like an old-fashioned bottle with one or more waists. The critical solution at $R_0 = R_{01}$ has a turning point with vertical tangent, allowing us to locate the critical point by solving $r'(s) = r''(s) = 0$. The result is that at the critical point the curve length is $s_1 = 1.95863\ldots$, the radius $a_1 = 0.917622\ldots$ and the depth $d_1 = 1.47804\ldots$.

In figure 8.5*a* is shown the family of shapes for a droplet with tube radius $a = 1 > a_1$. For large central radius of curvature R_0 the shape is flat, but as R_0 diminishes the droplet grows in volume and develops a 'waist'. It finally reaches a maximum volume of $5.26R_c^3$, beyond which it cannot pass continuously and the drop must surely have fallen. In figure 8.5*b* is shown a family of shapes for a droplet with tube radius $a = 0.5 < a_1$. In this case the droplet may expand beyond the radius of the tube until it reaches a maximal volume of $2.32R_c^3$.

Problems

8.1 A soap bubble of diameter 6 cm floats in air. What is the pressure excess inside the bubble when the surface tension between soapy water and air is taken to be $\alpha = 0.15\ \mathrm{N\,m^{-1}}$? How would you define the capillary length in this case, and how big is it? Do you expect the bubble to keep its spherical shape?

8.2 **(a)** Calculate the leading order change in area and volume of a column of incompressible fluid subject to an infinitesimal radial periodic perturbation with wavelength λ. **(b)** Show that the Rayleigh–Plateau stability condition is equivalent to the requirement that the area should grow for fixed volume. Instability occurs when the area diminishes.

* **8.3** Consider a quadratic surface $z = ax^2 + by^2 + 2cxy$ with a unique extremum at $x = y = 0$. For a suitable choice of coordinates, a smooth function may always be approximated with such a function at any given point. **(a)** Determine the radius of curvature of the surface along a line in the xy-plane forming an angle ϕ with the x-axis. **(b)** Determine the extrema of the radius of curvature as a function of ϕ and show that they correspond to orthogonal directions.

* **8.4** Determine the radii of curvature in section 8.4 by expanding the shape $z = f(r)$ with $r = \sqrt{x^2 + y^2}$ to second order around $x = x_0$, $y = 0$, and $z = z_0$.

* **8.5** Show that the volume of a pending droplet of arc length s is

$$V = \pi r^2 \left(\frac{1}{R_2} - \frac{1}{R_1} \right) = \pi r^2 \left(2\frac{\sin\theta}{r} + z - \frac{2}{R_0} \right), \tag{8.23}$$

where $r = r(s)$ is the radius and $z = z(s)$ the height.

Part III

Deformable solids

9

Stress

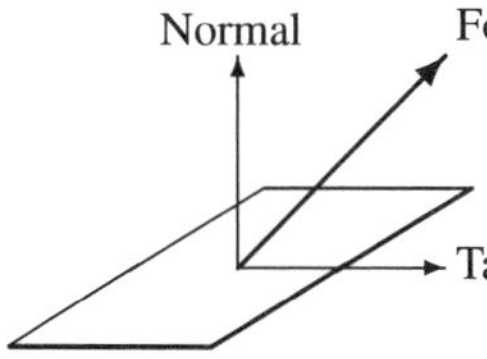

The force on a small piece of a surface can be resolved by a normal pressure-like force and a tangential shear force.

In fluids at rest pressure is the only contact force. For solids at rest or in motion, and for viscous fluids in motion, this simple picture is no longer valid. Besides pressure-like forces acting along the normal to a contact surface, there may also be shear forces acting tangentially to it. In complete analogy with pressure, the relevant quantity turns out to be the *shear stress*, defined to be the shear force per unit of area. Friction forces are always caused by shear stresses.

The two major classes of materials, fluids and solids, react differently to stress. Whereas fluids respond by *flowing*, solids respond by *deforming*. Although the equations of motion in both cases are derived from Newton's second law, fluids and solids are in fact so different that they are usually covered in separate textbooks. In this book, we shall as far as possible maintain a general view of the physics of continuous systems, applicable to all types of materials.

The integrity of a solid body is largely due to internal elastic stresses, both normal and tangential to the surfaces they act on. Together they resist deformation of the material and prevent the body from being pulled apart. Unlike friction, elastic forces do not dissipate energy, and ideally the work done against elastic forces during deformation may be fully recovered. In reality, some elastic energy will always be lost because of the emission of sound waves that eventually decay and turn into heat.

In this chapter the emphasis is on the theoretical formalism for contact forces, independent of whether they occur in solids, fluids, or intermediate forms such as clay or dough. The vector notation used up to this point is not adequate to the task, because contact forces not only depend on the spatial position but also on the orientation of the surface on which they act. A collection of nine stress components, called the *stress tensor*, was introduced by Cauchy in 1822 to describe the full range of contact forces that may come into play.

Baron Augustin-Louis Cauchy (1789–1857). *French mathematician who produced an astounding 789 papers. Contributed to the foundations of elasticity, hydrodynamics, partial differential equations, number theory and complex functions.*

9.1 Friction

The concept of shear stress is best understood through *friction*, a shear force known to us all. We hardly think of friction forces, even though all day long we are served by them and do service to them. Friction is the reason that the objects we hold are not slippery as a piece of soap in the bathtub, but instead allow us to grab and drag, heave and lift, rub and scrub. Most of the work we do is in fact done against friction, from stirring the coffee to making fire by rubbing two sticks against each other.

	μ_0	μ
Glass/glass	0.9	0.4
Rubber/asphalt	0.9	0.7
Steel/steel	0.7	0.6
Metal/metal	0.6	0.4
Wood/wood	0.4	0.3
Steel/ice	0.1	0.05
Steel/teflon	0.05	0.05

Approximate friction coefficients for various combinations of materials.

Static and sliding friction

Consider a heavy crate standing on a horizontal floor. Its weight mg_0 acts vertically downwards on the floor, which in turn reacts back on the crate with an equal and opposite normal force of magnitude $N = mg_0$. If you try to drag the crate along the floor by applying a horizontal force F, called *traction*, you may discover that the crate is so heavy that you are not able to budge it, implying that the force you apply must be fully

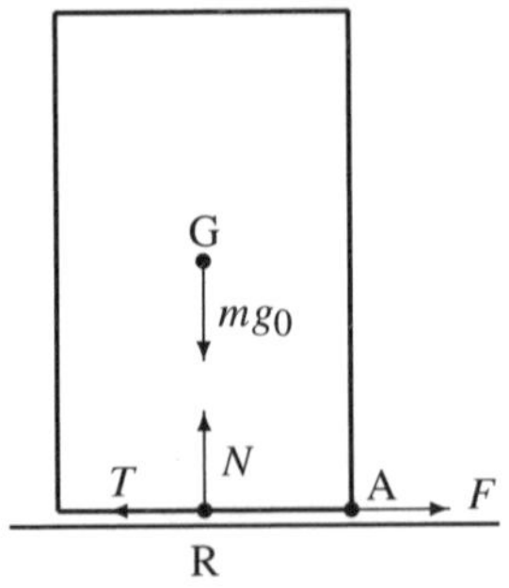

Balance of forces on a crate standing still on a horizontal floor. The point of attack A is here chosen at floor level to avoid creating a moment of force which could turn over the crate.

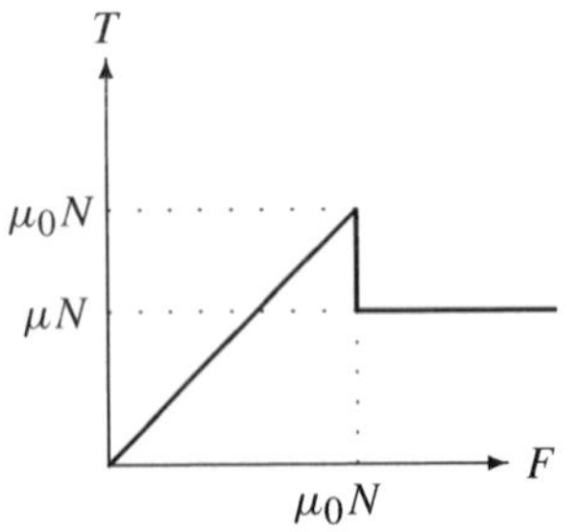

Sketch of tangential reaction T as a function of applied traction F. Up to $F = \mu_0 N$, the tangential reaction adjusts itself to the traction, $T = F$. At $F = \mu_0 N$, the tangential reaction drops abruptly to a lower value, and stays there regardless of the applied traction.

Charles-Augustin de Coulomb (1736–1806). *French physicist, best known from the law of electrostatics and the unit of electric charge that carries his name.*

balanced by a tangential friction force between the floor and the crate of the same magnitude, $T = F$, but of opposite direction.

Empirically, such *static friction* can take any magnitude up to a certain maximum, which is proportional to the normal load,

$$T < \mu_0 N. \tag{9.1}$$

The dimensionless constant of proportionality μ_0 is called the coefficient of static friction which in our daily doings may take a quite sizable value, say 0.5 or greater. Its value depends on what materials are in contact and on the roughness of the contact surfaces.

If you are able to pull with a sufficient strength, the crate suddenly starts to move, but friction will still be present and you will have to do real work to move the crate any distance. Empirically, the *dynamic* (kinetic or sliding) friction is proportional to the normal load,

$$T = \mu N, \tag{9.2}$$

with a coefficient of dynamic friction, μ, that is always smaller than the corresponding coefficient of static friction, $\mu < \mu_0$. This is why you have to heave strongly to get the crate set into motion, whereas afterwards a smaller force suffices to keep it going at constant speed.

> It is at first sight rather surprising that friction is independent of the size of the contact area. A crate on legs is as hard to drag as a box without, provided they weigh the same. Since larger weight generates larger friction, a car's braking distance will be independent of how heavily it is loaded. In braking a car it is also best to avoid skidding because the static (rolling) friction is larger than sliding friction (see problem 9.2). Anti-skid brake systems automatically adjust braking pressure to avoid skidding and thus minimize braking distance.

The law of sliding friction goes back to Coulomb (1779) (and also Amontons (1699)). The full story of friction is complicated, and in spite of our everyday familiarity with friction, there is still no universally accepted microscopic explanation of the phenomenon[1].

9.2 The concept of stress

Shear stress is, just like pressure, defined as force per unit of area, and the standard unit of stress is the same as the unit for pressure, namely the pascal (Pa = N m^{-2}). If a crate on the floor has a contact area A, we may speak both about the average normal stress $\sigma_n = N/A$ and the average tangential (or shear) stress $\sigma_t = T/A$ that the crate exerts on the floor. Depending on the mass distribution of the contents of the crate and the stiffness of its bottom, the local stresses may vary across the contact area A.

External and internal stress

The stresses acting between the crate and the floor are *external* and are found in the true interface between a body and its environment. In analogy with pressure, we shall also speak about *internal* stresses, even if we may be unable to define a practical way to measure them. Internal stresses abound in the macroscopic world around us. Whenever we come into contact with the environment (and when do we not?) stresses are set up in the materials we touch, and in our own bodies. The precise distribution of stress in a body depends not only on the external forces applied to the body, but also on the type of material the body is made from and on other macroscopic quantities such as temperature. In the absence of external forces there is usually no stress in a material, although fast cooling may freeze stresses permanently into certain materials, for example glass, and provoke an almost explosive release of stored energy when triggered by a sudden impact.

[1] D. A. Kessler, Surface physics: A new crack at friction, *Nature* **413** (2001) 260.

Estimating internal stress

In many situations it is quite straightforward to estimate average stresses in a body. Consider, for example, a slab of homogeneous solid material bounded by two stiff flat clamps of area A, firmly glued to it. A tangential force of magnitude F applied to one clamp with the other held fixed will deform the slab a bit in the direction of the applied force. Here we shall not worry about how to calculate the deformation of the slab, but will just assume that the response of the slab is the same throughout, so that there is a uniform shear stress $\sigma = F/A$ acting on the surface of the slab.

The fixed clamp will, of course, act back on the slab with a force of the same magnitude but in an opposite direction. If we make an imaginary cut through the slab parallel with the clamps, then the upper part of the slab must likewise act on the lower with the shear force F, so that the internal shear stress acting on the cut must again be $\sigma = F/A$. If pressure had also been applied to the clamps, we would have gone through the same type of argument to convince ourselves that the normal stress would be the same everywhere in the cut.

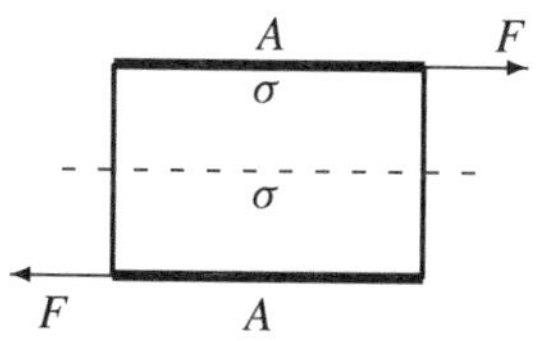

Clamped slab of homogeneous material. The shear force F at the upper clamp is balanced by an oppositely directed fixation force F on the lower clamp. The shear stress $\sigma = F/A$ is the same on all inner surfaces parallel with the clamps.

For bodies with a more complicated geometry and non-uniform external load, internal stresses are not so easily calculated, although their average magnitudes may be estimated. In analogy with friction one may assume that variations in shear and normal stresses are roughly of the same order of magnitude, provided the material and the body geometry are not exceptional.

Example 9.2.1 (Classic gallows): The classic gallows is constructed from a vertical pole, a horizontal beam and sometimes a diagonal strut. A body of mass $M = 70$ kg hangs at the extreme end of the horizontal beam of cross section $A = 100\ \text{cm}^2$. The body's weight must be balanced by a shear stress in the beam of magnitude $\sigma \approx Mg_0/A \approx 70\,000$ Pa ≈ 0.7 bar. The actual distribution of shear stress will vary over the cross section of the beam and the position of the chosen cross section, but its average magnitude should be of the estimated value.

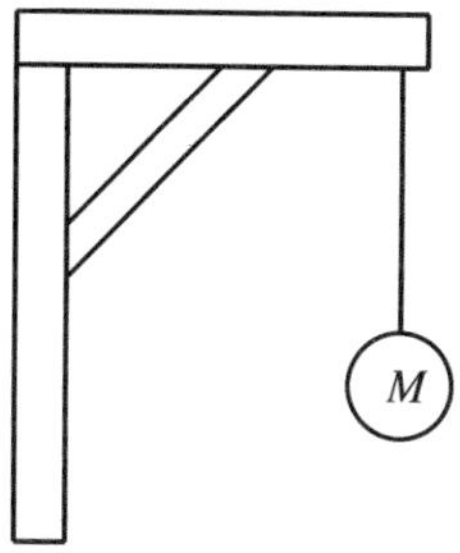

The classic gallows.

Example 9.2.2 (Water pipe): The half-inch water mains in your house have an inner pipe radius $a \approx 0.6$ cm. Tapping water at a high rate, internal friction in the water (viscosity) creates shear stresses opposing the flow, and the pressure drops perhaps by $\Delta p \approx 0.1$ bar $= 10^4$ Pa over a length of $L \approx 10$ m of the pipe. In this case, we may actually calculate the shear stress on the water from the inner surface of the pipe without estimation errors, because the pressure difference between the ends of the pipe is the only other force acting on the water. Setting the force due to the pressure difference equal to the total shear force on the inner surface, we get $\pi a^2 \Delta p = 2\pi a L \sigma$, from which it follows that $\sigma = \Delta p\, a/2L \approx 3$ Pa. This stress is indeed of the same size as we would have estimated from the corresponding pressure drop $\Delta p \cdot a/L$ over a stretch of pipe of the same length as the radius.

Tensile strength

When external forces grow large, a solid body may fracture and break apart. The maximal tension, i.e. negative pressure or pull, a material can sustain without fracturing is called the *tensile strength* of the material. Similarly, the *yield stress* is defined as the stress beyond which otherwise elastic solids begin to undergo permanent deformation. For metals the tensile strength lies typically in the region of several hundred megapascals (i.e. several kilobars) as shown in the margin table. Modern composite carbon fibers can have tensile strengths up to several gigapascals (i.e. several megabars), whereas 'ropes' made from single-wall carbon nanotubes are reported to have tensile strengths of up to 50 gigapascals[2].

	Tensile [MPa]	Yield [MPa]
Titanium	500	450
Nickel	460	
Steel	450	250
Copper	300	
Cast iron	180	
Zink	130	
Lead	17	

Typical tensile strength for common metals and some yield stresses. The values may vary widely for different specimens, depending on heat treatment and other factors.

Example 9.2.3: Plain carbon steel has a tensile strength of 450 MPa. A quick estimate shows that a steel rod with a diameter of 2 cm breaks if loaded with more than 14 000 kg. Adopting a safety factor of 10, one should not load it with more than 1400 kg.

[2] Min-Feng Yu, Bradley S. Files, Sivaram Arepalli and Rodney S. Ruoff, Tensile loading of ropes of single wall carbon nanotubes and their mechanical properties *Phys. Rev. Lett.* **84** (2000) 5552.

9.3 Nine components of stress

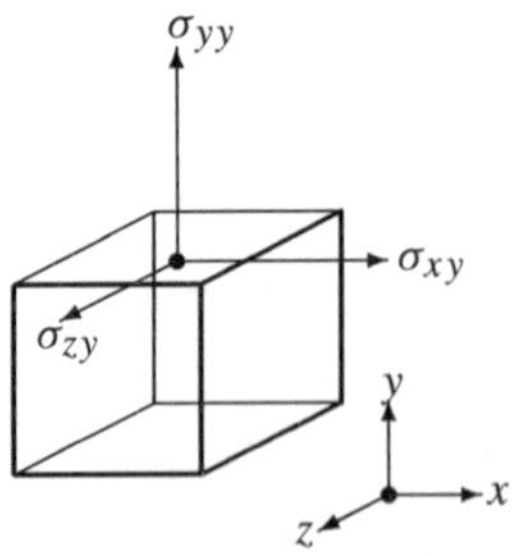

Components of stress acting on a surface element in the xz-plane.

Shear stress is more complicated than normal stress, because there is more than one tangential direction on a surface. In a coordinate system where a force $d\mathcal{F}_x$ is applied along the x-direction to a material surface dS_y with its normal in the y-direction, the shear stress will be denoted $\sigma_{xy} = d\mathcal{F}_x/dS_y$, instead of just σ. Similarly, if the shear force is applied in the z-direction, the stress would be denoted $\sigma_{zy} = d\mathcal{F}_z/dS_y$, and if a normal force had been applied along the y-direction, it would be consistent to denote the normal stress $\sigma_{yy} = d\mathcal{F}_y/dS_y$. By convention, the sign is chosen such that a positive value of σ_{yy} corresponds to a pull or *tension*.

Cauchy's stress hypothesis

Altogether, it therefore appears to be necessary to use at least nine numbers to indicate the state of stress in a given point of a material in a Cartesian coordinate system. *Cauchy's stress hypothesis* (to be proved below) asserts that the force $d\boldsymbol{\mathcal{F}} = (d\mathcal{F}_x, d\mathcal{F}_y, d\mathcal{F}_z)$ on an arbitrary surface element, $d\boldsymbol{S} = (dS_x, dS_y, dS_z)$, is of the form

$$\begin{aligned} d\mathcal{F}_x &= \sigma_{xx} dS_x + \sigma_{xy} dS_y + \sigma_{xz} dS_z, \\ d\mathcal{F}_y &= \sigma_{yx} dS_x + \sigma_{yy} dS_y + \sigma_{yz} dS_z, \\ d\mathcal{F}_z &= \sigma_{zx} dS_x + \sigma_{zy} dS_y + \sigma_{zz} dS_z, \end{aligned} \tag{9.3}$$

where each coefficient $\sigma_{ij} = \sigma_{ij}(\boldsymbol{x}, t)$ depends on the position and time, and thus is a field in the normal sense of the word. Collecting them in a matrix

$$\boldsymbol{\sigma} = \{\sigma_{ij}\} = \begin{pmatrix} \sigma_{xx} & \sigma_{xy} & \sigma_{xz} \\ \sigma_{yx} & \sigma_{yy} & \sigma_{yz} \\ \sigma_{zx} & \sigma_{zy} & \sigma_{zz} \end{pmatrix}. \tag{9.4}$$

The force may be written compactly as a matrix equation,

$$\boxed{d\boldsymbol{\mathcal{F}} = \boldsymbol{\sigma} \cdot d\boldsymbol{S}.} \tag{9.5}$$

The force per unit of area is, $d\boldsymbol{\mathcal{F}}/dS = \boldsymbol{\sigma} \cdot \boldsymbol{n}$, where $\boldsymbol{n}$ is the normal to the surface. It is sometimes called the *stress vector*, although it is *not a vector field* in the usual sense of the word because it depends on the normal.

The stress tensor

Together the nine fields, $\{\sigma_{ij}\}$, make up a single geometric object, called the *stress tensor*, first introduced by Cauchy in 1822. Using index notation, we may write

$$\boxed{d\mathcal{F}_i = \sum_j \sigma_{ij} dS_j.} \tag{9.6}$$

Since the force $d\mathcal{F}_i$ as well as the surface element dS_i are vectors, it follows that σ_{ij} is indeed a tensor in the sense of section 2.8 (see problem 2.20). This collection of nine fields $\{\sigma_{ij}\}$ cannot be viewed geometrically as consisting of nine scalar or three vector fields, but must be considered together as one geometrical object, a *tensor field* $\sigma_{ij}(\boldsymbol{x}, t)$ which is neither scalar nor vector. As for ordinary tensors (see section 2.8), there is unfortunately no simple, intuitive way of visualizing the stress tensor field by geometric means.

Example 9.3.1: A stress tensor field of the form,

$$\{\sigma_{ij}\} = \{x_i x_j\} = \begin{pmatrix} x^2 & xy & xz \\ yx & y^2 & yz \\ zx & zy & z^2 \end{pmatrix}, \tag{9.7}$$

is a tensor product and thus by construction a true tensor. The stress 'vector' acting on a surface with normal in the direction of the x-axis is

$$\boldsymbol{\sigma}_x = \boldsymbol{\sigma} \cdot \boldsymbol{e}_x = \begin{pmatrix} x \\ y \\ z \end{pmatrix} x. \tag{9.8}$$

It does not transform under rotation as a true vector because of the factor x on the right-hand side.

Hydrostatic pressure

For the special case of hydrostatic equilibrium, where the only contact force is pressure, comparison of (9.5) with (4.8) shows that the stress tensor must be

$$\boldsymbol{\sigma} = -p\,\mathbf{1}, \tag{9.9}$$

where $\mathbf{1}$ is the (3×3) unit matrix. In tensor notation this becomes

$$\sigma_{ij} = -p\,\delta_{ij}, \tag{9.10}$$

where δ_{ij} is the index representation of the unit matrix (2.30).

Mechanical pressure

Generally, however, the stress tensor will have both diagonal and off-diagonal non-vanishing components. A diagonal component behaves like a (negative) pressure, and one often defines the pressures along different coordinate axes to be

$$p_x = -\sigma_{xx}, \quad p_y = -\sigma_{yy}, \quad p_z = -\sigma_{zz}. \tag{9.11}$$

Since they may be different, it is not clear what the meaning of *the* pressure in a point should be. Furthermore, it should be remembered that the diagonal elements of a tensor $(\sigma_{xx}, \sigma_{yy}, \sigma_{zz})$ do not have a well-defined geometric meaning, and that the quantity (p_x, p_y, p_z) does not behave like a vector under Cartesian coordinate transformations (see section 2.8 and problem 2.23).

The *mechanical* pressure or simply *the* pressure is defined to be the average of the three pressures along the axes,

$$p = \tfrac{1}{3}(p_x + p_y + p_z) = -\tfrac{1}{3}(\sigma_{xx} + \sigma_{yy} + \sigma_{zz}). \tag{9.12}$$

This makes sense because the sum over the diagonal elements of a matrix, the trace $\mathrm{Tr}\,\boldsymbol{\sigma} = \sum_i \sigma_{ii} = \sigma_{xx} + \sigma_{yy} + \sigma_{zz}$, is invariant under Cartesian coordinate transformations (problem 2.17). Defining pressure in this way ensures that it is a scalar field, taking the same value in all coordinate systems. There is in fact no other scalar linear combination of the stress components.

Example 9.3.2: For the stress tensor given in example 9.3.1 the pressures along the coordinate axes become $p_x = -x^2$, $p_y = -y^2$ and $p_z = -z^2$. Evidently, they do not form a vector, but the average pressure,

$$p = -\tfrac{1}{3}(x^2 + y^2 + z^2), \tag{9.13}$$

is clearly a scalar, invariant under rotations of the Cartesian coordinate system.

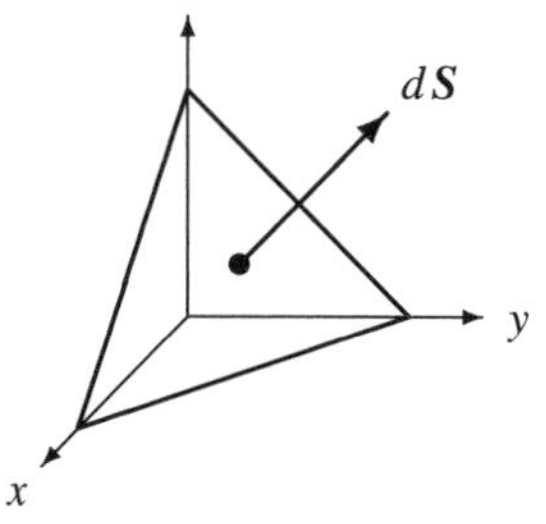

The tiny triangle and its projections form a tetrahedron.

Proof of Cauchy's stress hypothesis

As in the proof of Pascal's law (page 48) we take again a surface element in the shape of a tiny triangle with area vector $d\boldsymbol{S} = (dS_x, dS_y, dS_z)$. This triangle and its projections on the coordinate planes form together a little body in the shape of a tetrahedron. Since we aim to prove the existence of the stress tensor, we cannot assume that it exists. What we know is that the forces acting from the inside of the tetrahedron on the three triangular faces in the coordinate planes are vectors of the form $d\boldsymbol{\mathcal{F}}_x = \boldsymbol{\sigma}_x dS_x$, $d\boldsymbol{\mathcal{F}}_y = \boldsymbol{\sigma}_y dS_y$

and $d\boldsymbol{\mathcal{F}}_z = \boldsymbol{\sigma}_z dS_z$. Denoting the force acting from the outside on the skew face by $d\boldsymbol{\mathcal{F}}$, and adding a possible volume force $\boldsymbol{f} dV$, the equation of motion for the small tetrahedron becomes

$$dM\, \boldsymbol{w} = \boldsymbol{f} dV + d\boldsymbol{\mathcal{F}} - \boldsymbol{\sigma}_x\, dS_x - \boldsymbol{\sigma}_y\, dS_y - \boldsymbol{\sigma}_z\, dS_z, \tag{9.14}$$

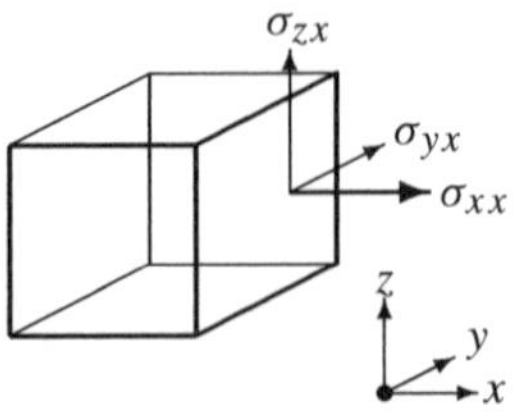

Components of the stress vector $\boldsymbol{\sigma}_x$ acting on a surface element in the yz-plane.

where $\boldsymbol{w}$ is the acceleration of the tetrahedron, and $dM = \rho dV$ its mass, which is assumed to be constant. The signs have been chosen in accordance with the inward direction of the area projections dS_x, dS_y and dS_z.

The volume of the tetrahedron scales like the third power of its linear size, whereas the surface areas only scale like the second power (see section 4.2). Making the tetrahedron progressively smaller, the body force term and the acceleration term will vanish faster than the surface terms. In the limit of a truly infinitesimal tetrahedron, only the surface terms survive, so that we must have

$$d\boldsymbol{\mathcal{F}} = \boldsymbol{\sigma}_x dS_x + \boldsymbol{\sigma}_y dS_y + \boldsymbol{\sigma}_z dS_z. \tag{9.15}$$

This shows that the force on an arbitrary surface element may be written as a linear combination of three basic stress vectors, one for each coordinate axis. Introducing the nine coordinates of the three triplets, $\boldsymbol{\sigma}_x = (\sigma_{xx}, \sigma_{yx}, \sigma_{zx})$, $\boldsymbol{\sigma}_y = (\sigma_{xy}, \sigma_{yy}, \sigma_{zy})$ and $\boldsymbol{\sigma}_z = (\sigma_{xz}, \sigma_{yz}, \sigma_{zz})$, we arrive at (9.3).

9.4 Mechanical equilibrium

Including a volume force density f_i, the total force on a body of volume V with surface S becomes according to (9.6)

$$\boxed{\mathcal{F}_i = \int_V f_i\, dV + \oint_S \sum_j \sigma_{ij}\, dS_j.} \tag{9.16}$$

Using Gauss' theorem (4.22) this may be written as a single volume integral

$$\mathcal{F}_i = \int_V f_i^*\, dV, \tag{9.17}$$

where

$$\boxed{f_i^* = f_i + \sum_j \nabla_j \sigma_{ij},} \tag{9.18}$$

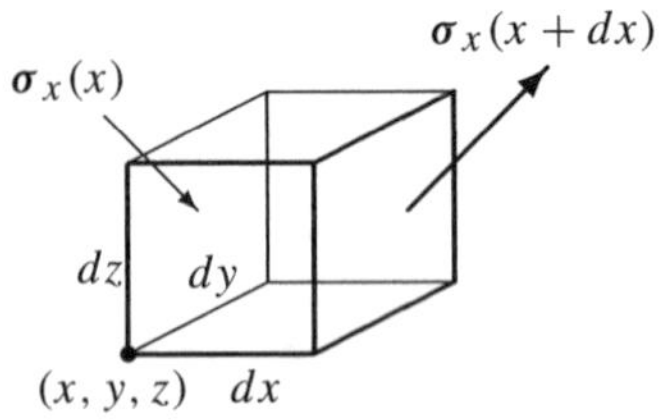

The total contact force on a small box-shaped material particle is calculated from the variations in stress on the sides. Thus $d\boldsymbol{\mathcal{F}} = (\boldsymbol{\sigma}_x(x + dx, y, z) - \boldsymbol{\sigma}_x(x, y, z))dS_x \approx \nabla_x \boldsymbol{\sigma}_x\, dV$ for the stress on dS_x, plus the similar contributions from dS_y and dS_z.

is the *effective force density*. The effective force is not just a formal quantity, because the total force on a material particle is $d\boldsymbol{\mathcal{F}} = \boldsymbol{f}^*\, dV$. As in hydrostatics (page 50) this may be demonstrated by considering a small box-shaped particle.

Cauchy's equation of equilibrium

In mechanical equilibrium, the total force on any body must vanish, for if it does not the body will begin to move. So the general condition is that $\boldsymbol{\mathcal{F}} = \boldsymbol{0}$ for all volumes V. In particular, requiring that the force on each and every material particle must vanish, we arrive at *Cauchy's equilibrium equation(s)*,

$$\boxed{f_i + \sum_j \nabla_j \sigma_{ij} = 0.} \tag{9.19}$$

In spite of their simplicity these partial differential equations govern mechanical equilibrium in all kinds of continuous matter, be it solid, fluid, or whatever. For a fluid at rest where pressure is the only stress component, we have $\sigma_{ij} = -p\,\delta_{ij}$, and recover the equation of hydrostatic equilibrium, $f_i - \nabla_i p = 0$.

It is instructive to explicitly write out the three individual equations contained in Cauchy's equilibrium equation,

$$\begin{aligned} f_x + \nabla_x \sigma_{xx} + \nabla_y \sigma_{xy} + \nabla_z \sigma_{xz} &= 0, \\ f_y + \nabla_x \sigma_{yx} + \nabla_y \sigma_{yy} + \nabla_z \sigma_{yz} &= 0, \\ f_z + \nabla_x \sigma_{zx} + \nabla_y \sigma_{zy} + \nabla_z \sigma_{zz} &= 0. \end{aligned} \tag{9.20}$$

These equations are in themselves not sufficient to determine the state of continuous matter, but must be supplemented by suitable *constitutive* equations connecting stress and state. For fluids at rest, the equation of state serves this purpose by relating hydrostatic pressure to mass density and temperature (chapter 4). In elastic solids, the constitutive equations are more complicated and relate stress to displacement (chapter 11).

Symmetry

There is one very general condition (also going back to Cauchy) which may normally be imposed, namely the symmetry of the stress tensor

$$\boxed{\sigma_{ij} = \sigma_{ji}.} \tag{9.21}$$

Symmetry only affects the shear stress components, requiring

$$\sigma_{xy} = \sigma_{yx}, \qquad \sigma_{yz} = \sigma_{zy}, \qquad \sigma_{zx} = \sigma_{xz}, \tag{9.22}$$

and thus reduces the number of independent stress components from nine to six.

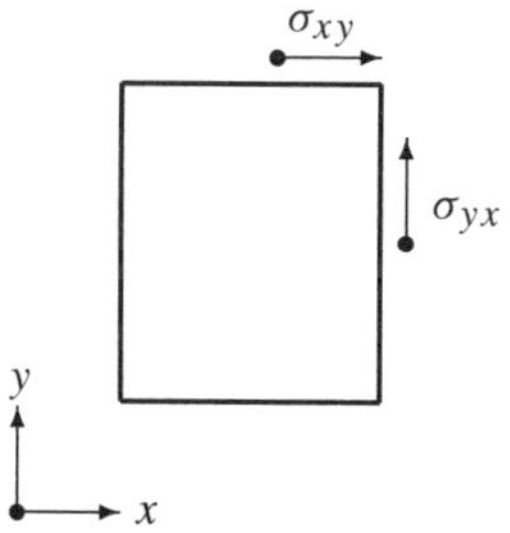

A symmetric stress tensor acts with equal strength on orthogonal faces of a cubic body.

Being thus a symmetric matrix, the stress tensor may be diagonalized. The eigenvectors define the principal directions of stress and the eigenvalues the principal tensions or stresses. In the principal basis, there are no off-diagonal elements, i.e. shear stresses, only pressures. The principal basis is generally different from point to point in space.

Proof of symmetry: There is in fact no general proof of symmetry of the stress tensor, but only some theoretical arguments that allow us to choose the stress tensor to be symmetric in all normal materials. These arguments are presented in section 9.5. Here we shall present a simple argument, only valid in complete mechanical equilibrium.

Consider a material particle in the shape of a tiny rectangular box with sides a, b and c. The force acting in the y-direction on a face in the x-plane is $\sigma_{yx}bc$ whereas the force acting in the x-plane on a face in the y-plane is $\sigma_{xy}ac$. On the opposite faces the contact forces have opposite sign in mechanical equilibrium (their difference is as we have seen of order abc). Since the total force vanishes, the total moment of force on the box may be calculated around any point we wish. Using the lower left corner, we get

$$\mathcal{M}_z = a\,\sigma_{yx}bc - b\,\sigma_{xy}ac = (\sigma_{yx} - \sigma_{xy})abc.$$

This shows that if the stress tensor is asymmetric, $\sigma_{xy} \neq \sigma_{yx}$, there will be a resultant moment on the box. In mechanical equilibrium this cannot be allowed, since such a moment would begin to rotate the box, and consequently the stress tensor must be symmetric. Conversely, when the stress tensor is symmetric, mechanical equilibrium of the forces alone guarantees that all local moments of force will vanish.

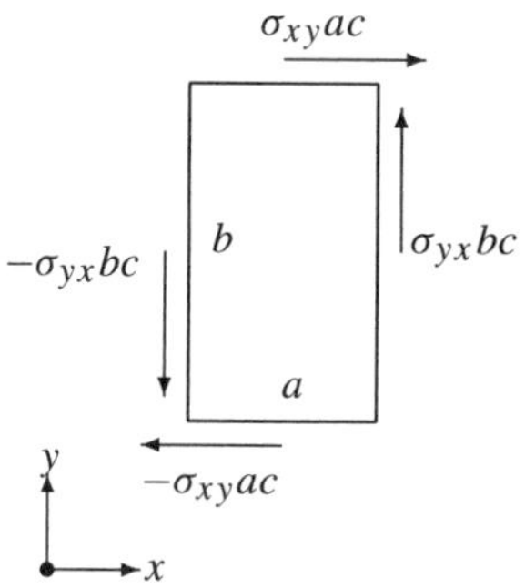

An asymmetric stress tensor will produce a non-vanishing moment of force on a small box (the z-direction not shown).

Boundary conditions

Cauchy's equation of equilibrium constitutes a set of coupled partial differential equations, and such equations require boundary conditions. The stress tensor is a local physical quantity, or rather collection of quantities, and may, like pressure in hydrostatics, be assumed to be continuous in regions where material properties change continuously. Across real boundaries, interfaces, where material properties may change abruptly, Newton's third law only demands that the stress vector, $\boldsymbol{\sigma} \cdot \boldsymbol{n} = \{\sum_j \sigma_{ij} n_j\}$, be continuous across a surface with normal $\boldsymbol{n}$ (in the absence of surface tension). This does *not* mean that all the components of the stress tensor should be continuous. Since Newton's third law is a vector condition, it imposes continuity on three linear combinations of stress components, but leaves for the symmetric stress tensor three other combinations free to jump discontinuously. Surprisingly, it does not follow that the pressure (9.12) is continuous. In full-fledged continuum theory, the pressure in fact loses the appealing intuitive meaning it acquired in hydrostatics.

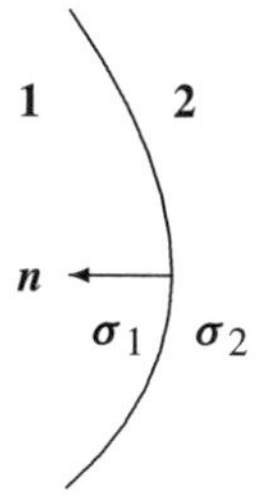

Contact surface separating body 1 from body 2. Newton's third law requires continuity of the stress vector $\boldsymbol{\sigma} \cdot \boldsymbol{n}$ across the boundary, i.e. $\boldsymbol{\sigma}_1 \cdot \boldsymbol{n} = \boldsymbol{\sigma}_2 \cdot \boldsymbol{n}$.

Example 9.4.1: Consider a plane interface in the yz-plane. The stress components σ_{xx}, σ_{yx} and σ_{zx} must then be continuous, because they specify the stress vector on such a surface. Symmetry implies that σ_{xy} and σ_{xz} are likewise continuous. The remaining three independent components σ_{yy}, σ_{zz} and $\sigma_{yz} = \sigma_{zy}$ are allowed to jump at the interface. In particular the average pressure, $p = -(\sigma_{xx} + \sigma_{yy} + \sigma_{zz})/3$, may be discontinuous.

∗ 9.5 'Proof' of symmetry of the stress tensor

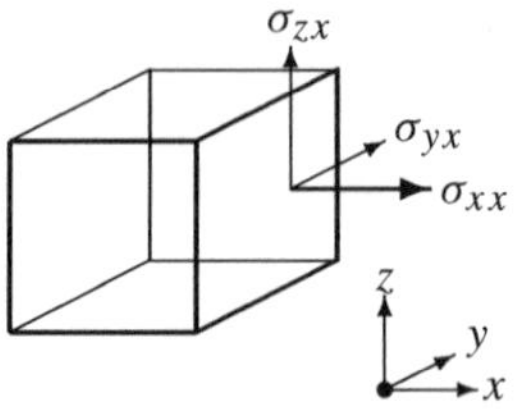

Only the three components of the stress vector need to be continuous at the interface.

If the stress tensor is manifestly *asymmetric*, we shall now show that it is always possible to make it symmetric by exploiting an ambiguity in its definition. The argument which will now be presented is adapted from P. C. Martin, O. Parodi and P. S. Pershan, *Phys. Rev.* **A6**, (1972) 2401 (also used in [37, p. 7]).

The ambiguous stress tensor

The stress tensor was introduced at the beginning of this chapter as a quantity which furnished a complete description of the contact forces that may act on any surface element. But surface elements are not in themselves physical bodies. We can only determine the magnitude and direction of a force by observing its influence on the motion of a real physical body having a volume V and a closed surface S. The resultant of all contact forces acting on the surface of a body is given by (9.17), showing that the relevant quantity for the dynamics of continuous matter is the contribution $\sum_j \nabla_j \sigma_{ij}$ to the effective density of force rather than the stress tensor σ_{ij} itself.

Two stress tensors, σ_{ij} and $\widetilde{\sigma}_{ij}$, are therefore physically indistinguishable, if they give rise to the same effective density of force everywhere. This is, for example, the case if we write

$$\widetilde{\sigma}_{ij} = \sigma_{ij} + \sum_k \nabla_k \chi_{ijk} \tag{9.23}$$

where χ_{ijk} is antisymmetric in j and k,

$$\chi_{ijk} = -\chi_{ikj}. \tag{9.24}$$

For then

$$\sum_j \nabla_j \widetilde{\sigma}_{ij} = \sum_j \nabla_j \sigma_{ij} + \sum_{jk} \nabla_j \nabla_k \chi_{ijk} = \sum_j \nabla_j \sigma_{ij},$$

where the last term in the middle vanishes because of the symmetry of the double derivatives and the assumed antisymmetry of χ_{ijk}.

It remains to show that there exists a tensor χ_{ijk} such that $\widetilde{\sigma}_{ij}$ becomes symmetric. Let us put

$$\chi_{ijk} = \nabla_i \phi_{jk} + \nabla_j \phi_{ik} - \nabla_k \phi_{ij} \tag{9.25}$$

where ϕ_{ij} is an antisymmetric tensor, $\phi_{ij} = -\phi_{ji}$. Evidently this choice of χ_{ijk} has the right symmetry properties (9.24). Inserting it into (9.23) we obtain

$$\widetilde{\sigma}_{ij} = \sigma_{ij} + \sum_k \nabla_k (\nabla_i \phi_{jk} + \nabla_j \phi_{ik}) - \nabla^2 \phi_{ij} .$$

The tensor ϕ_{ij} is now chosen to be a solution to Poisson's equation with the antisymmetric part of the original stress tensor as source,

$$\nabla^2 \phi_{ij} = \tfrac{1}{2}(\sigma_{ij} - \sigma_{ji}). \tag{9.26}$$

Such a solution can in principle always be found, and then we obtain

$$\widetilde{\sigma}_{ij} = \tfrac{1}{2}(\sigma_{ij} + \sigma_{ji}) + \sum_k \nabla_k (\nabla_i \phi_{jk} + \nabla_j \phi_{ik}) \tag{9.27}$$

which is manifestly symmetric. Notice, however, that the new symmetric stress tensor is not just the symmetric part of the old, but contains extra terms.

Non-classical continuum theories

The conclusion is, that if somebody presents you with a stress tensor which is asymmetric, you may always replace it by a suitable symmetric stress tensor, having exactly the same physical consequences.

But even if it is formally possible to choose a symmetric stress tensor, it may not always be convenient, because of the non-locality inherent in the solution to Poisson's equation in (9.26). Asymmetric stress tensors have been used in various generalizations of classical continuum theory, containing elementary volume and surface densities of moments (body couples and couple stresses) and sometimes also intrinsic angular momentum (spin). We shall not go further into these non-classical extensions of continuum theory here; so-called *micropolar* materials are, for example, discussed in [51, p. 493].

Problems

9.1 A crate is dragged over a horizontal floor with sliding friction coefficient μ. Determine the angle α with the vertical of the total reaction force.

9.2 A car with mass m moves with a speed v. Estimate the minimal braking distance without skidding and the corresponding braking time. Do the same if it skids from the beginning to the end. For numerics use $m = 1000$ kg and $v = 100$ km h^{-1}. The static coefficient of friction between rubber and the surface of a road may be taken to be $\mu_0 = 0.9$, whereas the sliding friction is $\mu = 0.7$.

9.3 A strong man pulls a jumbo airplane slowly but steadily exerting a force of $\mathcal{F} = 2000$ N on a rope. The plane has $N = 32$ wheels, each touching the ground in a square area $A = 40 \times 40$ cm^2. **(a)** Estimate the shear stress between the rubber and the tarmac. **(b)** Estimate the shear stress between the tarmac and his feet, each with area $A = 5 \times 25$ cm^2.

9.4 Estimate the maximal height h of a mountain made from rock with a density of $\rho = 3,000$ kg m^{-3} when the maximal stress the material can tolerate before it deforms permanently is $\sigma = 300$ MPa. How high could it be on Mars where the surface gravity is 3.7 m s^{-2}?

9.5 A stress tensor has all components equal, i.e. $\sigma_{ij} = \tau$ for all i, j. Find its eigenvalues and eigenvectors.

9.6 Show that if the stress tensor is diagonal in all coordinate systems, then it can only contain pressure.

9.7 A stress tensor and a rotation matrix are given by,

$$\boldsymbol{\sigma} = \begin{pmatrix} 15 & -10 & 0 \\ -10 & 5 & 0 \\ 0 & 0 & 20 \end{pmatrix}, \qquad \mathbf{A} = \begin{pmatrix} 3/5 & 0 & -4/5 \\ 0 & 1 & 0 \\ 4/5 & 0 & 3/5 \end{pmatrix}.$$

Calculate the stress tensor in the rotated coordinate system $\boldsymbol{x}' = \mathbf{A} \cdot \boldsymbol{x}$.

9.8 **(a)** Show that the average of a unit vector $\boldsymbol{n}$ over all directions obeys

$$\langle n_i n_j \rangle = \frac{1}{3}\delta_{ij}. \tag{9.28}$$

(b) Use this to show that the average of the normal stress acting on an arbitrary surface element in a fluid equals (minus) the mechanical pressure (9.12).

$*$ **9.9** A body of mass m stands still on a horizontal floor. The coefficients of static and kinetic friction between body and floor are μ_0 and μ. An elastic string with string constant k is attached to the body in a point close to the floor. The string can only exert a force on the body when it is stretched beyond its relaxed length. When the free end of the string is pulled horizontally with constant velocity v, intuition tells us that the body will have a tendency to move in fits and starts.

(a) Calculate the amount s that the string is stretched, just before the body begins to move.

(b) Write down the equation of motion for the body when it is just set into motion, for example in terms of the distance x that the point of attachment of the string has moved and the time t elapsed since the motion began.

(c) Show that the solution to this equation is

$$x = \frac{v}{\omega}(\omega t - \sin \omega t) + (1 - r)s(1 - \cos \omega t)$$

where $\omega = \sqrt{k/m}$, $r = \mu/\mu_0$.

(d) Assuming that the string stays stretched, calculate at what time $t = t_0$ the body stops again?

(e) Find the condition for the string to be stretched during the whole motion.

(f) How long time will the body stay in rest, before moving again?

$*$ **9.10** One may define three *invariants*, i.e. scalar functions, of the stress tensor in any point. The first is the trace $I_1 = \sum_i \sigma_{ii}$, the second $I_2 = (1/2)\sum_{ij}(\sigma_{ii}\sigma_{jj} - \sigma_{ij}\sigma_{ji})$ which has no special name, and the determinant $I_3 = \det \boldsymbol{\sigma}$. Show that the characteristic equation for the matrix $\boldsymbol{\sigma}$ can be expressed in terms of the invariants. Can you find an invariant for an asymmetric stress tensor which vanishes if symmetry is imposed?

10 Strain

All materials deform when subjected to external forces, but different materials react in different ways. Elastic materials bounce back again to the original configuration when the forces cease to act. Others are plastic and retain their shape after deformation. Viscoelastic materials behave like elastic solids under fast deformation, but creep like viscous liquid over longer periods of time. Elasticity is itself an idealization, limited to a certain range of forces. If the external forces become excessive, all materials become plastic and undergo permanent deformation, even fracture.

When a body is deformed, the material is displaced away from its original position. Small deformations are mathematically much easier to handle than large deformations, where parts of a body become greatly and non-uniformly displaced relative to other parts, as for example when you crumple a piece of paper. A rectilinear coordinate system embedded in the original body and deformed along with the material of the body becomes a curvilinear coordinate system after the deformation. It can therefore come as no surprise that the general theory of finite deformation is mathematically at the same level of difficulty as general curvilinear coordinate systems. Luckily, our buildings and machines are rarely subjected to such violent treatment, and in most practical cases the deformation may be assumed to be tiny.

The description of continuous deformation inevitably leads to the introduction of a new tensor quantity, the *strain tensor*, which characterizes the state of *local deformation* or *strain* in a material. Strained relations between neighbouring material particles cause tension or stress—as they do among people. In this chapter we shall focus exclusively on the description of strain, and postpone the discussion of the stress-strain relationship for elastic materials to chapter 11.

10.1 Displacement

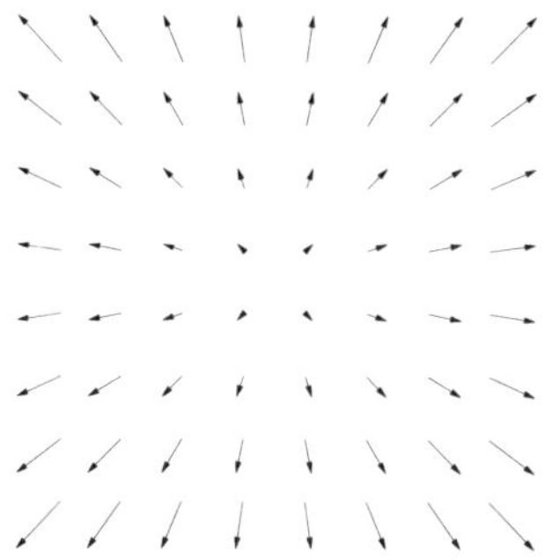

Uniform dilatation. The arrows indicate how material particles are displaced.

The prime example of deformation is a *uniform dilatation*, in which the coordinates of every material particle in a body are multiplied by a constant factor, $\kappa > 1$. Under a dilatation a material particle originally situated in the point $\boldsymbol{x}$ is displaced to the point,

$$\boldsymbol{x}' = \kappa\,\boldsymbol{x}. \tag{10.1}$$

Note that both $\boldsymbol{x}$ and $\boldsymbol{x}'$ refer to the same coordinate system. Contraction is also included for κ-values in the interval $0 < \kappa < 1$. Negative values of κ are physically impossible for macroscopic bodies, because they contain a reflection ($\boldsymbol{x}' = -\boldsymbol{x}$) in the origin of the coordinate system. Although you can see your own reflection in a mirror, the reflection cannot be realized physically.

The only point which does not change place during a uniform dilatation is the origin of the coordinate system. Although it superficially looks as if the origin of the coordinate system plays a special role, this is not really the case. All relative positions of material particles scale in the same way, $\boldsymbol{x}' - \boldsymbol{y}' = \kappa(\boldsymbol{x} - \boldsymbol{y})$, independent of the origin of the coordinate system. There is no special centre for a uniform dilatation, either geometrically or physically. The origin of the coordinate system is simply an *anchor point* for the mathematical description of dilatation.

The displacement field

Under a deformation the centre of mass of a material particle is displaced from its original position $\boldsymbol{x}$ to its actual position $\boldsymbol{x}'$. The *displacement* $\boldsymbol{u} = \boldsymbol{x}' - \boldsymbol{x}$ depends on the particle's position and is thus a field $\boldsymbol{u} = \boldsymbol{u}(\boldsymbol{x})$ in its own right, called the *displacement field*. Given the displacement field, we may calculate the new position,

$$\boxed{\boldsymbol{x}' = \boldsymbol{x} + \boldsymbol{u}(\boldsymbol{x}).} \tag{10.2}$$

For the dilatation (10.1) the displacement field becomes $\boldsymbol{u}(\boldsymbol{x}) = (\kappa - 1)\,\boldsymbol{x}$.

There are actually two equivalent ways of choosing a field to represent the displacement as a function of the particle positions. In the *Lagrange representation* used here the displacement $\boldsymbol{x}' - \boldsymbol{x} = \boldsymbol{u}(\boldsymbol{x})$ is understood as a function of the original position $\boldsymbol{x}$. In the *Euler representation* the displacement $\boldsymbol{x}' - \boldsymbol{x} = \boldsymbol{u}'(\boldsymbol{x}')$ is understood as a function of the actual position $\boldsymbol{x}'$. The two representations represent of course the same displacement, so that $\boldsymbol{u}'(\boldsymbol{x}') = \boldsymbol{u}(\boldsymbol{x})$. Conceptually and mathematically, the Lagrange representation is more convenient for the theory of small elastic deformations. The Euler representation is on the other hand more physical, because the displacement field at a given point represents the state of the material actually found in that position. For small and smooth displacement fields there is very little difference between the Euler and the Lagrange representations.

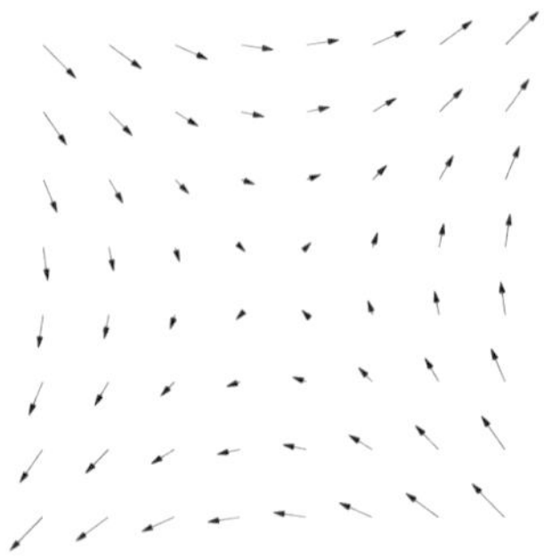

Plot of the two-dimensional linear displacement field $\boldsymbol{u} = (y, x, 0)$ for $-1 < x < 1$ and $-1 < y < 1$. The material is dilated along one diagonal and contracted along the other. These are the principal directions of strain (see problem 10.4). The apparently skew sides are an artefact of the 'small arrows' plot.

Small displacements

The transformation (10.2) is for general $\boldsymbol{u}(\boldsymbol{x})$ a completely arbitrary vector mapping, $\boldsymbol{x} \to \boldsymbol{x}' = \boldsymbol{f}(\boldsymbol{x}) = \boldsymbol{x} + \boldsymbol{u}(\boldsymbol{x})$. The only reason to split off the displacement field, and not work with the mapping function $\boldsymbol{f}(\boldsymbol{x})$ itself, is that the displacement always will be assumed to be 'small'. Since $\boldsymbol{u}(\boldsymbol{x})$ is a quantity with dimension of length, there is no absolute meaning to 'small'. But in a body of size L, the first term in (10.2) will change by L as the position $\boldsymbol{x}$ ranges over the body, and this allows us to define a displacement to be small, if for all $\boldsymbol{x}$,

$$|\boldsymbol{u}(\boldsymbol{x})| \ll L. \tag{10.3}$$

In the following we shall always assume that the displacement is small, except for section 10.5 where the generalization to finite displacements that do not fulfill this condition is briefly outlined.

Linear displacements

A general displacement includes all kinds of ordinary rigid body translations and rotations, and it would be wrong to classify such displacements as deformations. Sailing a submarine at the surface of the water will only displace it, not deform it, whereas taking it to the bottom of the sea will deform it (slightly). A real deformation must involve changes in geometric relationships, i.e. lengths and angles, in the body.

Although the displacement in general will be a nonlinear function of the coordinates, we shall begin by analysing linear displacements akin to the uniform dilatation (10.1). In the most general case, a linear displacement field takes the form

$$\boldsymbol{u}(\boldsymbol{x}) = \mathbf{A} \cdot \boldsymbol{x} + \boldsymbol{b}. \tag{10.4}$$

where $\mathbf{A}$ is a constant matrix and $\boldsymbol{b}$ is a constant vector. There is strong similarity between the class of linear displacements and the transformations of Cartesian coordinates discussed in section 2.7, but the class of linear displacements is larger, because the underlying coordinate transformation is not restricted to be orthogonal. The conceptual difference lies in the interpretation of the displacement field, which represents a physical shift of the material, as opposed to a change in the way the coordinates are calculated.

The general linear displacement may also be resolved into simpler types, namely translation along a coordinate axis, rotation by a fixed angle around a coordinate axis, and scaling by a fixed factor along a coordinate axis. The physically impossible reflections are excluded. We shall not prove here that the general linear displacement may be resolved in this way, but instead rely on geometric intuition.

Simple translation: A rigid body translation of the material through a distance b along the x-axis is described by the displacement field

$$u_x = b,\quad u_y = 0,\quad u_z = 0. \tag{10.5}$$

Since the geometric relationships in a body are unchanged under translation, it is not a deformation.

The displacement is small when $|b| \ll L$.

Simple translation field.

Simple rotation: The geometry of a rigid body rotation through the angle ϕ around the z-axis leads to,

$$x' = x\cos\phi - y\sin\phi,\quad y' = x\sin\phi + y\cos\phi,\quad z' = z. \tag{10.6}$$

The corresponding displacement field is

$$u_x = -x(1-\cos\phi) - y\sin\phi,\quad u_y = x\sin\phi - y(1-\cos\phi),\quad u_z = 0. \tag{10.7}$$

Since all distances in the body are unchanged, this is not a deformation.

The displacement is only small when the angle is small, $|\phi| \ll 1$. Expanding to first order in the angle we find

$$u_x = -y\phi, \qquad u_y = x\phi, \qquad u_z = 0. \tag{10.8}$$

In vector form this may be written $\boldsymbol{u} = (-y, x, 0)\phi$.

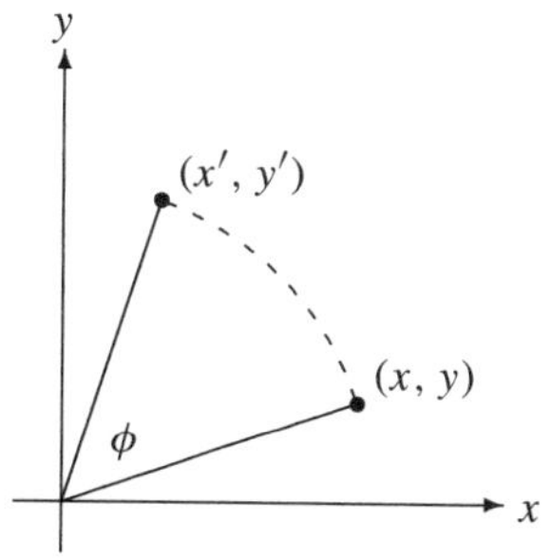

A rigid body rotation through an angle ϕ moves the particle at (x, y) to (x', y').

Simple scaling: Finally, multiplying all distances along the x-axis by the factor κ, the displacement field becomes

$$u_x = kx,\quad u_y = 0,\quad u_z = 0, \tag{10.9}$$

where $k = \kappa - 1$. Uniform dilation (10.1) is a combination of three such scalings by the same factor along the three coordinate axes. Scaling is a true deformation.

The displacement is small when $|k| \ll 1$.

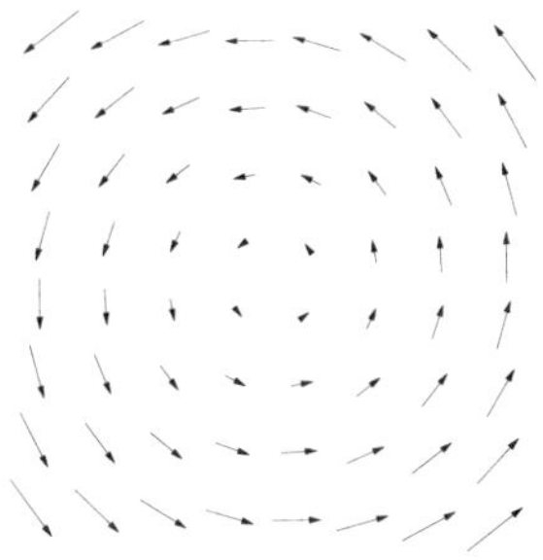

Simple rotation field.

10.2 Local deformation

Displacement is, as demonstrated by the linear examples, not the same as deformation. All the parts of a body could be simultaneously displaced by the same amount, or bodily rotated, without altering the geometric relations between them. What is needed is a measure of the actual change of spatial relations between different parts of the material, also called *strain*.

Large scale deformation can be very complex. Think of all the loops and knots that weavers make from a roll of yarn. We should for this reason not expect to find a simple formalism for global deformation. Weaving, folding, winding, writhing, wringing and squashing may bring particles that were originally far apart into close proximity. Even the wildest weave consists, however, locally of small pieces of straight yarn that have only been translated, rotated, stretched or contracted, but not folded, spindled or mutilated. We may therefore expect to find a much simpler description of deformation for very small pieces of matter.

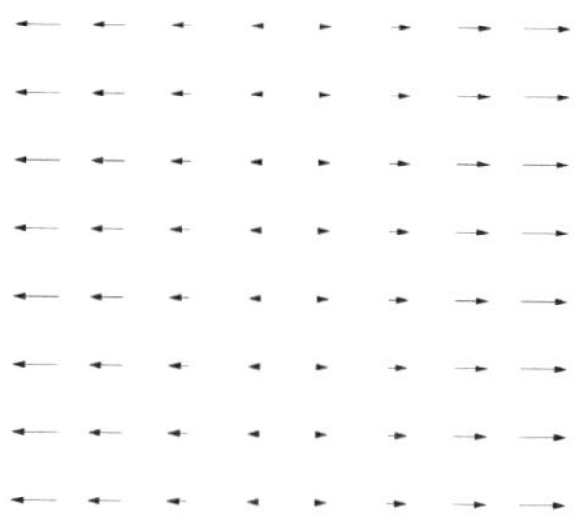

Simple scaling field.

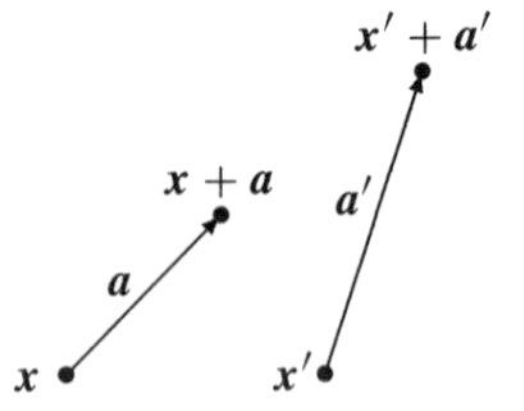

Displacement of a tiny material needle from $\boldsymbol{a}$ to $\boldsymbol{a}'$. It may be translated, rotated, and dilated. Only the latter corresponds to a true deformation.

Displacement of a 'needle'

Consider a tiny elongated piece of matter, a material vector or 'needle', connecting a point $\boldsymbol{x}$ with the point $\boldsymbol{x}+\boldsymbol{a}$. After displacement this needle connects the points $\boldsymbol{x}' = \boldsymbol{x} + \boldsymbol{u}(\boldsymbol{x})$ and $\boldsymbol{x}' + \boldsymbol{a}' = \boldsymbol{x} + \boldsymbol{a} + \boldsymbol{u}(\boldsymbol{x}+\boldsymbol{a})$, and solving for $\boldsymbol{a}'$ we find

$$\boldsymbol{a}' = \boldsymbol{a} + \boldsymbol{u}(\boldsymbol{x}+\boldsymbol{a}) - \boldsymbol{u}(\boldsymbol{x}). \tag{10.10}$$

The needle is now assumed to be so small that we may expand the displacement field $\boldsymbol{u}(\boldsymbol{x}+\boldsymbol{a})$ to first order in $\boldsymbol{a}$. For the x-component of the field, we find

$$\begin{aligned} u_x(\boldsymbol{x}+\boldsymbol{a}) &\approx u_x(\boldsymbol{x}) + a_x \frac{\partial u_x(\boldsymbol{x})}{\partial x} + a_y \frac{\partial u_x(\boldsymbol{x})}{\partial y} + a_z \frac{\partial u_x(\boldsymbol{x})}{\partial z} \\ &= u_x(\boldsymbol{x}) + (\boldsymbol{a}\cdot\boldsymbol{\nabla})u_x(\boldsymbol{x}). \end{aligned}$$

Since $\boldsymbol{a}\cdot\boldsymbol{\nabla}$ is a scalar operator acting in the same way on each component of a vector, we may collect the other components and write the original vector for an infinitesimal needle in the form

$$\boxed{\boldsymbol{a}' = \boldsymbol{a} + (\boldsymbol{a}\cdot\boldsymbol{\nabla})\,\boldsymbol{u}(\boldsymbol{x}).} \tag{10.11}$$

Not surprisingly, since it is a relation between infinitesimal quantities, this transformation is *linear* in $\boldsymbol{a}$. In index notation, it may be written as,

$$a_i' = a_i + \sum_j (\nabla_j u_i) a_j. \tag{10.12}$$

This shows that the coefficients of the linear transformation of $\boldsymbol{a}$ are computed from the derivatives of the displacement field, $\nabla_j u_i$, also called the *displacement gradients*. For a general linear displacement (10.4) we have $\nabla_j u_i = A_{ij}$.

Example 10.2.1: The displacement gradient matrix of a simple rotation (10.7) is,

$$\{\nabla_j u_i\} = \begin{pmatrix} -1+\cos\phi & -\sin\phi & 0 \\ \sin\phi & -1+\cos\phi & 0 \\ 0 & 0 & 0 \end{pmatrix} \tag{10.13}$$

where the index i enumerates the rows and j the columns. If the angle of rotation is small, $|\phi| \ll 1$, the displacement gradients are all small, and the matrix becomes,

$$\{\nabla_j u_i\} = \begin{pmatrix} 0 & -\phi & 0 \\ \phi & 0 & 0 \\ 0 & 0 & 0 \end{pmatrix} \tag{10.14}$$

to lowest order in ϕ.

Small displacement gradients

Since both displacements and coordinates have dimension of length, the displacement gradients are all *dimensionless* quantities, i.e. pure numbers, and this makes it meaningful to speak of small displacement gradients in an absolute way. We shall say that a displacement field is *slowly varying* or *smooth*, if all the displacement gradients are small everywhere, i.e. $|\nabla_j u_i(\boldsymbol{x})| \ll 1$ for all i, j and $\boldsymbol{x}$. In terms of the matrix of displacement gradients, $\boldsymbol{\nabla u} = \{\nabla_j u_i\}$, this may equivalently be written as a condition on the matrix norm (2.17),

$$|\boldsymbol{\nabla u}(\boldsymbol{x})| \ll 1, \tag{10.15}$$

for all $\boldsymbol{x}$ in a body.

A small displacement gradient does not automatically imply that the displacement itself is small compared to the size of the body, because the displacement could include a rigid body translation to the other end of the universe which would not affect its gradient. But relative to a fixed anchor point in the body which is not displaced, a slowly varying field must always be small compared to the size of the body and thus fulfill (10.3). A displacement field satisfying this condition everywhere will in general also be slowly varying, though there are notable exceptions. If you, for example, make a crease in your shirt when you iron it, the displacement gradients will be almost infinitely large in the crease although none of the shirt's material is greatly displaced compared to the size of the shirt.

Except for section 10.5, where a few aspects of the theory of finite deformations are presented, we shall from now on assume that the displacement field is both small and smooth. In that case the change in a needle vector,

$$\boxed{\delta \boldsymbol{a} \equiv \boldsymbol{a}' - \boldsymbol{a} = (\boldsymbol{a} \cdot \nabla)\boldsymbol{u},} \tag{10.16}$$

is also small compared to the length of $\boldsymbol{a}$, i.e. $|\delta \boldsymbol{a}| \ll |\boldsymbol{a}|$.

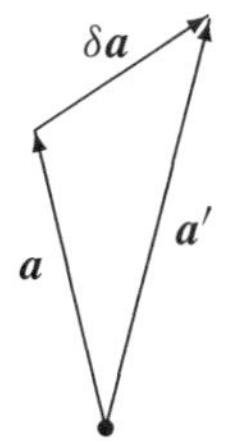

The change in a needle vector is small when the displacement gradients are small.

Cauchy's strain tensor

Since the scalar product of two needles $\boldsymbol{a} \cdot \boldsymbol{b}$ is unchanged by translation and rotation, it ought to be a useful indicator for a change in geometry. Using (10.16), we calculate the change in the scalar product $\delta(\boldsymbol{a} \cdot \boldsymbol{b}) \equiv \boldsymbol{a}' \cdot \boldsymbol{b}' - \boldsymbol{a} \cdot \boldsymbol{b}$ to first order in the small displacement gradients

$$\begin{aligned} \delta(\boldsymbol{a} \cdot \boldsymbol{b}) &= \delta \boldsymbol{a} \cdot \boldsymbol{b} + \boldsymbol{a} \cdot \delta \boldsymbol{b} \\ &= (\boldsymbol{a} \cdot \nabla)\boldsymbol{u} \cdot \boldsymbol{b} + (\boldsymbol{b} \cdot \nabla)\boldsymbol{u} \cdot \boldsymbol{a} \\ &= \sum_{ij} (\nabla_i u_j + \nabla_j u_i)\, a_i b_j. \end{aligned}$$

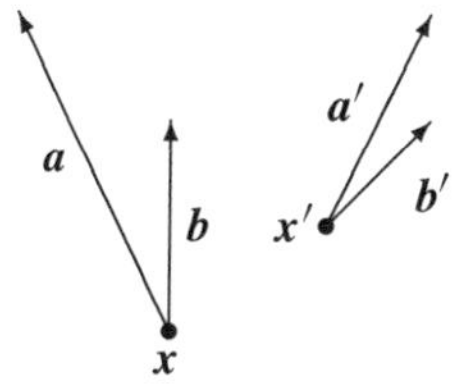

Displacement of a pair of infinitesimal material needles may affect their lengths as well as the angle between them.

In the last line we have cast the rather ugly vector expression in the much more elegant index notation, replacing all dot products by explicit sums. We may now (with a conventional factor 2) write,

$$\delta(\boldsymbol{a} \cdot \boldsymbol{b}) = 2 \sum_{ij} u_{ij} a_i b_j, \tag{10.17}$$

where

$$\boxed{u_{ij} = \frac{1}{2}(\nabla_i u_j + \nabla_j u_i),} \tag{10.18}$$

is the symmetric combination of displacement gradients, called *Cauchy's strain tensor* (or just the *strain tensor* when that is unambiguous). The *strain tensor contains all the information about geometric changes in the displacement.* It must, however, be emphasized that the above expression *is only valid for small displacement gradients.* For finite displacement gradients, a more complicated expression must be used (see section 10.5).

It pays to write out all the components of the strain tensor explicitly, once and for all. On the diagonal, they are

$$u_{xx} = \nabla_x u_x, \quad u_{yy} = \nabla_y u_y, \quad u_{zz} = \nabla_z u_z, \tag{10.19}$$

whereas off the diagonal,

$$\begin{aligned} u_{xy} = u_{yx} &= \tfrac{1}{2}(\nabla_x u_y + \nabla_y u_x) \\ u_{yz} = u_{zy} &= \tfrac{1}{2}(\nabla_y u_z + \nabla_z u_y) \\ u_{zx} = u_{xz} &= \tfrac{1}{2}(\nabla_z u_x + \nabla_x u_z) \end{aligned} \tag{10.20}$$

Had we not assumed that the displacement was slowly varying, there would also have been quadratic terms in the displacement gradients, and the strain tensor might take large values. But with our assumption of small displacement gradients (10.15), the strain tensor field is likewise small, $|u_{ij}(\boldsymbol{x})| \ll 1$ for all i, j and $\boldsymbol{x}$.

Example 10.2.2 (Simple translation): The displacement gradients of a bodily translation, $\boldsymbol{u}(\boldsymbol{x}) = \boldsymbol{b}$, vanishes trivially, and so does the strain tensor. This confirms that a translation is not a deformation.

Example 10.2.3 (Simple rotation): For small angles of rotation, $|\phi| \ll 1$, the displacement field of a simple rotation is $\boldsymbol{u} = (-y, x, 0)\phi$. The corresponding matrix of displacement gradients (10.14) is antisymmetric. Cauchy's strain tensor, which is the symmetric part of this matrix, vanishes, confirming that a rotation is not a deformation.

Example 10.2.4: The linear displacement $\boldsymbol{u} = (2\alpha\, y, \alpha\, x, 0)$ with $|\alpha| \ll 1$ has a matrix of displacement gradients

$$\{\nabla_j u_i\} = \begin{pmatrix} 0 & 2\alpha & 0 \\ \alpha & 0 & 0 \\ 0 & 0 & 0 \end{pmatrix}. \tag{10.21}$$

Cauchy's strain tensor becomes

$$\{u_{ij}\} = \begin{pmatrix} 0 & \frac{3}{2}\alpha & 0 \\ \frac{3}{2}\alpha & 0 & 0 \\ 0 & 0 & 0 \end{pmatrix}. \tag{10.22}$$

Since the strain tensor does not vanish, this is a true deformation.

Symmetry of the strain tensor

According to its definition (10.18) the strain tensor is *symmetric* in its indices

$$\boxed{u_{ij} = u_{ji}.} \tag{10.23}$$

It differs in this respect from the stress tensor, which for symmetry required further assumptions (see section 9.5). The symmetry implies that the strain tensor may be *diagonalized* at every point. The eigenvectors of the strain tensor at a given point are called the *principal axes* of strain, and form an orthonormal basis at every point. Whereas the angles between the principal axes are unchanged under the displacement, the signs and magnitudes of the eigenvalues determine how much the material is being stretched or contracted along the principal axes. It should, however, be remembered that the principal basis varies from point to point in space.

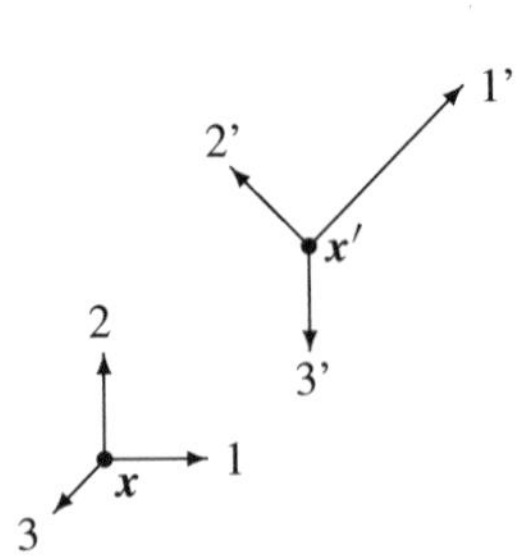

Principal basis of deformation of a point $\boldsymbol{x}$. Apart from translation and rotation, this basis is only subject to scale changes along the principal axes.

10.3 Geometrical meaning of the strain tensor

The strain tensor contains all the relevant information about changes in geometric relationships, such as lengths of material needles and the angles between them. Other geometric quantities, for example volume, are also changed under a deformation.

Changes in lengths and angles

It is useful for the following discussion to define the *projection* u_{ab} of a tensor u_{ij} on the directions of two arbitrary vectors $\boldsymbol{a}$ and $\boldsymbol{b}$,

$$u_{ab} = \frac{\sum_{ij} u_{ij} a_i b_j}{|\boldsymbol{a}|\,|\boldsymbol{b}|}. \tag{10.24}$$

Then we may simply write,

$$\delta(\boldsymbol{a}\cdot\boldsymbol{b}) = 2\,|\boldsymbol{a}|\,|\boldsymbol{b}|\,u_{ab}, \tag{10.25}$$

for the change in a scalar product (10.17).

Change in length: The change in length of a needle is obtained by setting $\boldsymbol{b} = \boldsymbol{a}$ in (10.25), and using that $2\,|\boldsymbol{a}|\,\delta\,|\boldsymbol{a}| = \delta(\boldsymbol{a}^2) = 2\,|\boldsymbol{a}|^2\,u_{aa}$ we get,

$$\boxed{\frac{\delta\,|\boldsymbol{a}|}{|\boldsymbol{a}|} \equiv \frac{|\boldsymbol{a}'| - |\boldsymbol{a}|}{|\boldsymbol{a}|} = u_{aa}.} \tag{10.26}$$

The diagonal strain projection u_{aa} is thus the *fractional change of lengths* in the direction $\boldsymbol{a}$. Obtaining this relation is the reason behind the conventional factor 2 in the definition (10.18) of the strain tensor.

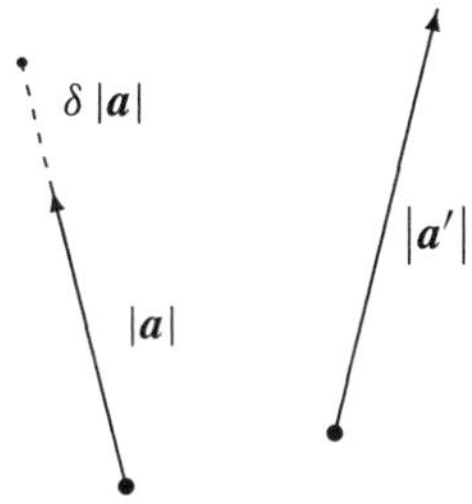

The change in length $\delta\,|\boldsymbol{a}|$ is determined by the diagonal projection of the strain tensor.

Change in angle: Introducing the angle ϕ_{ab} between two needles we have $\boldsymbol{a}\cdot\boldsymbol{b} = |\boldsymbol{a}|\,|\boldsymbol{b}|\cos\phi_{ab}$, and thus

$$\delta(\boldsymbol{a}\cdot\boldsymbol{b}) = \delta\,|\boldsymbol{a}|\,|\boldsymbol{b}|\cos\phi_{ab} + |\boldsymbol{a}|\,\delta\,|\boldsymbol{b}|\cos\phi_{ab} - |\boldsymbol{a}|\,|\boldsymbol{b}|\sin\phi_{ab}\delta\phi_{ab}.$$

Solving for $\delta\phi_{ab}$, we obtain by means of (10.25) and (10.26)

$$\delta\phi_{ab} \equiv \phi'_{ab} - \phi_{ab} = -\frac{2u_{ab}}{\sin\phi_{ab}} + (u_{aa} + u_{bb})\cot\phi_{ab}. \tag{10.27}$$

For $\phi_{ab} \to 0$ the vectors become parallel and the expression diverges, but the divergence is only apparent because $u_{aa} = u_{bb} = u_{ab}$ for parallel vectors. For originally orthogonal vectors, such as the coordinate axes, we have $\phi_{ab} = \pi/2$, and the change in angle simplifies to

$$\boxed{\delta\phi_{ab} = -2u_{ab}.} \tag{10.28}$$

The off-diagonal projections of the strain tensor thus determine the change in angle between originally orthogonal needles.

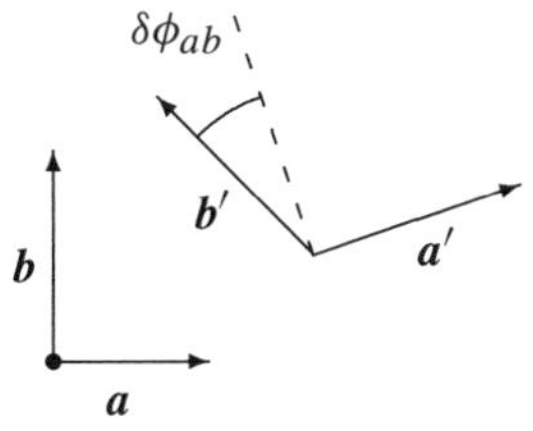

The off-diagonal projections of the strain tensor determine the change in angle for originally orthogonal needles.

Change in volume and density

Under a displacement all the material particles in a volume V are simultaneously moved to fill out another volume V' which for small displacements lies in the vicinity of the original volume. The smallness of the displacement permits us to calculate the change in volume from the small displacement of its surface S. Since a surface element $d\boldsymbol{S}$ at $\boldsymbol{x}$ is displaced through the vector distance $\boldsymbol{u}(\boldsymbol{x})$, it scoops up a tiny volume $\boldsymbol{u}(\boldsymbol{x})\cdot d\boldsymbol{S}$ (counted with sign), so that the total change in the volume V becomes,

$$\delta V = V' - V = \oint_S \boldsymbol{u}\cdot d\boldsymbol{S}. \tag{10.29}$$

Converting the surface integral into a volume integral by means of Gauss' theorem (6.4), we obtain,

$$\delta V = \int_V \boldsymbol{\nabla}\cdot\boldsymbol{u}\,dV. \tag{10.30}$$

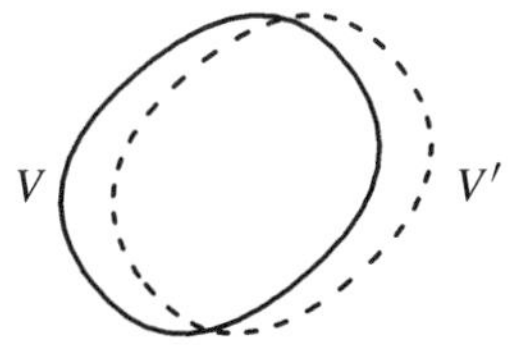

Displacement of a volume of material. The change in volume is given by the thin surface layer between the dashed and solid curves.

This shows that we may equivalently think of the change in volume V as the sum of the local volume changes, $\delta(dV) = \boldsymbol{\nabla}\cdot\boldsymbol{u}\,dV$, of each and every material particle contained in V. Dividing this equation by dV we get,

$$\boxed{\frac{\delta(dV)}{dV} = \boldsymbol{\nabla}\cdot\boldsymbol{u},} \tag{10.31}$$

saying that the local fractional volume change equals the divergence of the smooth displacement field.

The change in volume of a material particle induces a change in the local density of the displaced matter. Since the displacement does not change the mass of a material particle, we find,

$$\delta(dM) = \delta(\rho dV) = \rho\,\delta(dV) + \delta\rho\,dV = 0,$$

and making use of (10.31) the change in density becomes,

$$\boxed{\delta\rho \equiv \rho' - \rho = -\rho\boldsymbol{\nabla}\cdot\boldsymbol{u}.} \tag{10.32}$$

The fractional change in density $\delta\rho/\rho$ is equal to minus the divergence of the smooth displacement field. If the divergence vanishes, there is no change of volume or density associated with the deformation.

Example 10.3.1: A translation $\boldsymbol{u} = \boldsymbol{b}$ does not change the density. Likewise for an infinitesimal rotation around the z-axis through a small angle ϕ where $\boldsymbol{u} = (-y, x, 0)\phi$, has vanishing divergence. A dilatation $\boldsymbol{u} = k\boldsymbol{x}$ has $\delta\rho/\rho = -3k$. If $k > 0$ the volume increases and the density diminishes as it should with a contribution k from each dimension.

* 10.4 Work and energy

Deforming a body takes work, a fact known to everyone who has ever kneaded clay or dough. In these cases, the work you perform seems to be lost inside the material, but in other cases, as for example when you squeeze an elastic rubber ball, the material appears to store the work and release it again when you relinquish your grip. Many ball games like Ping-Pong or tennis rely entirely on the elasticity of the ball. No material is, however, perfectly elastic. Some energy is always lost to sound waves that are radiated away and eventually degenerate to heat. A hard steel ball may jump many times on a hard floor, but eventually it loses all its energy and comes to rest, partly due to air resistance, partly due to losses in the ball and, perhaps more importantly, in the floor. But even when your work seems to get lost in the dough, this is not really the case. The energy you have put into the clay has in the end been converted into heat which, however, cannot easily be recovered.

In continuum physics it can be quite subtle to derive the correct energy relations. The simplest way to proceed is to *follow the work*, as we did in the qualitative argument above. This is quite analogous to the admonition, 'follow the money', often used with success to uncover economic or political fraud.

Virtual displacement

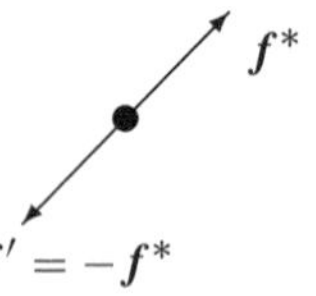

Every material particle can be kept in place by acting on it with an additional external force that balances the already existing effective body force on the particle.

Suppose we act on the external surface S of a body with some distribution of contact forces under our control, as we for example do when we knead dough. Any mechanical system that is not in equilibrium, is (literally) forced to move. If we nevertheless wish to keep it in place in one particular non-equilibrium state, we must act on its material with an extra volume distribution of external forces, $\boldsymbol{f}' = -\boldsymbol{f}^*$, to compensate for the effective internal forces $\boldsymbol{f}^*$ acting on the material particles. This procedure freezes the system in any desired state for any length of time, and is quite analogous to what we did to determine the potential energy in a gravitational field in section 3.5 on page 38.

Imagine now that the material of the body is displaced by an infinitesimal displacement field, $\delta\boldsymbol{u} = \delta\boldsymbol{u}(\boldsymbol{x})$, perhaps on top of an already existing displacement. The work of all the external forces under this so-called *virtual* displacement is then,

$$\boxed{\delta W = -\int_V \delta\boldsymbol{u}\cdot\boldsymbol{f}^*\,dV + \oint_S \delta\boldsymbol{u}\cdot\boldsymbol{\sigma}\cdot d\boldsymbol{S}.} \tag{10.33}$$

Applying Gauss' theorem to the work of the last term we find,

$$\oint_S \delta\boldsymbol{u}\cdot\boldsymbol{\sigma}\cdot d\boldsymbol{S} = \oint_S \sum_{ij}\delta u_i\sigma_{ij}dS_j = \int_V \sum_{ij}\nabla_j(\delta u_i\sigma_{ij})dV$$
$$= \int_V \sum_{ij}\delta u_i\,\nabla_j\sigma_{ij}\,dV + \int_V \sum_{ij}\sigma_{ij}\nabla_j\delta u_i dV,$$

and inserting the expression for the effective force (9.18) the virtual work may be written

$$\boxed{\delta W = -\int_V \delta\boldsymbol{u}\cdot\boldsymbol{f}\,dV + \int_V \sum_{ij}\sigma_{ij}\nabla_j\delta u_i dV.} \tag{10.34}$$

Since the system is at rest during the virtual displacement, both of these terms can only contribute to its *internal energy*.

The first term represents the part of the virtual displacement work that is spent *against* the true body forces,

$$\boxed{\delta W_{\text{body}} = -\int_V \delta\boldsymbol{u} \cdot \boldsymbol{f}\, dV.} \tag{10.35}$$

In a field of gravity this work contributes to the *gravitational energy* of the body, because the potential energy of material particle with mass dM is indeed changed by,

$$\big(\Phi(\boldsymbol{x} + \delta\boldsymbol{u}) - \Phi(\boldsymbol{x})\big) dM \approx -\boldsymbol{g} \cdot \delta\boldsymbol{u}\, dM = -\delta\boldsymbol{u} \cdot \boldsymbol{f}\, dV$$

under an infinitesimal displacement $\delta\boldsymbol{u}$.

The last term in (10.34) represents the part of the virtual displacement work that is spent against the *internal stresses* in the body,

$$\boxed{\delta W_{\text{deform}} = \int_V \sum_{ij} \sigma_{ij} \nabla_j \delta u_i dV = \int_V \sum_{ij} \sigma_{ij} \delta u_{ij}\, dV.} \tag{10.36}$$

Here we have in the second step used the (assumed) symmetry of the stress tensor to express the integrand in terms of the infinitesimal change in the strain tensor $\delta u_{ij} = \frac{1}{2}(\nabla_i \delta u_j + \nabla_j \delta u_i)$. Evidently this work is associated with deformation of the material and contributes to the *deformation energy* of the body.

In the special case that the stresses are only due to pressure, $\sigma_{ij} = -p\delta_{ij}$, the deformation work becomes

$$\delta W_{\text{deform}} = -\int_V p\, \nabla \cdot \delta\boldsymbol{u}\, dV. \tag{10.37}$$

Since $\delta(dV) = \nabla \cdot \delta\boldsymbol{u}\, dV$ is the change in volume of a material particle, we see that the deformation work is identical to the local thermodynamic work $-p\,\delta(dV)$ integrated over all material particles. This confirms the interpretation of δW_{deform} as the part of the virtual work spent on deforming the body.

In chapter 11 we shall find explicit expressions for the gravitational and deformation energies. In chapter 22 energy balance is discussed without using the concept of virtual work.

10.5 Finite deformations

When the condition (10.15) for slowly varying displacement is not fulfilled, we can no longer use the simple Cauchy strain tensor (10.18). The local description of finite deformation (see for example [27]) is essentially equivalent to the formalism of general curvilinear coordinate systems, but because space is Euclidean the description is not quite as complicated as that of truly non-Euclidean spaces.

Although many aspects of finite deformation theory were developed in the nineteenth century, the subject was not fully established until the mid twentieth century through Rivlin's work on nonlinear materials. Here we shall only touch briefly on the most general aspects of finite deformation theory which is a mathematically rather difficult subject [17, 26].

Ronald Samuel Rivlin (1915–). *British born mathematician and physicist. Contributed to the understanding of nonlinear materials during the 1940s and 1950s. Discovered exact nonlinear solutions for isotropic materials.*

The Lagrange representation

In the Lagrange representation a finite displacement is simply an arbitrary mapping of the original position of the material before the displacement to the actual position after the displacement,

$$\boldsymbol{x} \to \boldsymbol{x}' = \boldsymbol{f}(\boldsymbol{x}) \equiv \boldsymbol{x} + \boldsymbol{u}(\boldsymbol{x}). \tag{10.38}$$

There is in fact no reason to split off a special displacement field $\boldsymbol{u}(\boldsymbol{x})$, and we shall only do so to keep contact with the preceding analysis.

The global mapping of coordinates is considerably simplified in infinitesimal regions of space, where the local mapping is described by the differential transformation,

$$dx'_i = \sum_j \frac{\partial x'_i}{\partial x_j} dx_j, \tag{10.39}$$

which is what we called the transformation of a 'needle' in section 10.2. The matrix of derivatives $\partial x'_i/\partial x_j$ is the *Jacobian* of the transformation.

The length of the deformed (Cartesian) differential $d\boldsymbol{x}'$ may then be written as a double sum,

$$d\boldsymbol{x}'^2 = \sum_{jk} g_{jk} dx_j dx_k, \tag{10.40}$$

where the coefficients,

$$g_{jk} = \sum_i \frac{\partial x'_i}{\partial x_j} \frac{\partial x'_i}{\partial x_k} \tag{10.41}$$

constitute the *deformation tensor*. This tensor field, $g_{ij} = g_{ij}(\boldsymbol{x})$, contains all the information about local deformation of the material expressed in the coordinates of the undeformed body.

Inserting the actual displacement field (10.38) the Jacobian becomes

$$\frac{\partial x'_i}{\partial x_j} = \delta_{ij} + \nabla_j u_i. \tag{10.42}$$

Splitting off the strain tensor by writing,

$$g_{ij} = \delta_{ij} + 2u_{ij}, \tag{10.43}$$

we find

$$\boxed{u_{ij} = \frac{1}{2}\left(\nabla_i u_j + \nabla_j u_i + \sum_k \nabla_i u_k \nabla_j u_k\right).} \tag{10.44}$$

This is the generalization of Cauchy's strain tensor (10.18) to finite displacements, also called *Green's strain tensor*.

George Green (1793–1841). *Largely self-taught English mathematician and mathematical physicist. Contributed to hydrodynamics, electricity and magnetism and partial differential equations.*

Example 10.5.1 (Uniform dilatation): For a uniform dilatation $\boldsymbol{x}' = \kappa\, \boldsymbol{x}$ we have

$$\nabla_j u_i = (\kappa - 1)\delta_{ij}, \tag{10.45}$$

and the strain tensor becomes,

$$u_{ij} = \tfrac{1}{2}\left[(\kappa - 1)\delta_{ij} + (\kappa - 1)\delta_{ij} + (\kappa - 1)^2\delta_{ij}\right] = \tfrac{1}{2}(\kappa^2 - 1)\delta_{ij}, \tag{10.46}$$

for any value of κ. It vanishes for $\kappa = \pm 1$, i.e. for no displacement and for a pure reflection in the origin.

The scalar product of two needles then becomes,

$$\boldsymbol{a}' \cdot \boldsymbol{b}' = \kappa^2 \boldsymbol{a} \cdot \boldsymbol{b}, \tag{10.47}$$

and just reflects that all vectors are scaled by the same amount κ.

The Euler representation

In the Euler representation, the displacement field is defined from the mapping of the actual position $\boldsymbol{x}'$ to the original coordinates $\boldsymbol{x}$,

$$\boldsymbol{x}' \to \boldsymbol{x} = \boldsymbol{f}'(\boldsymbol{x}') \equiv \boldsymbol{x}' - \boldsymbol{u}'(\boldsymbol{x}'). \tag{10.48}$$

By comparison with (10.38) it follows that $\boldsymbol{u}'(\boldsymbol{x}') = \boldsymbol{u}(\boldsymbol{x})$, and that the mapping $\boldsymbol{f}'$ is the inverse of the original mapping, $\boldsymbol{f}' = \boldsymbol{f}^{(-1)}$.

The analysis now proceeds exactly in the same way as for the Euler representation. The infinitesimal transformation is,

$$dx_i = \sum_j \frac{\partial x_i}{\partial x'_j} dx'_j, \tag{10.49}$$

with Jacobian

$$\frac{\partial x_i}{\partial x'_j} = \delta_{ij} - \nabla'_j u'_i. \tag{10.50}$$

The Cartesian length of a needle dx becomes,

$$dx^2 = \sum_{jk} g'_{jk} dx'_j dx'_k \tag{10.51}$$

with deformation tensor,

$$g'_{jk} = \sum_i \frac{\partial x_i}{\partial x'_j}\frac{\partial x_i}{\partial x'_k}. \tag{10.52}$$

Inserting (10.48) we obtain

$$g'_{ij} = \delta_{ij} - 2u'_{ij}, \tag{10.53}$$

where the Euler strain tensor becomes,

$$u'_{ij} = \frac{1}{2}\left(\nabla'_i u'_j + \nabla'_j u'_i - \sum_k \nabla'_i u'_k \nabla'_j u'_k\right). \tag{10.54}$$

It was introduced by Almansi in 1911 and Hamel in 1912 [14].

Emilio Almansi (1869–1948). *Italian mathematical physicist. Worked on nonlinear elasticity theory, electrostatics and celestial mechanics.*

Georg Hamel (1877–1954). *German mathematician. Solved one of the famous Hilbert problems in his doctoral thesis under Hilbert (1901).*

Example 10.5.2 (Uniform dilatation): For a uniform dilatation $\boldsymbol{x}' = \kappa\, \boldsymbol{x}$ we have

$$\nabla'_j u'_i = \left(1 - \frac{1}{\kappa}\right)\delta_{ij}, \tag{10.55}$$

and the Euler strain tensor becomes,

$$u'_{ij} = \frac{1}{2}\left[\left(1-\frac{1}{\kappa}\right)\delta_{ij} + \left(1-\frac{1}{\kappa}\right)\delta_{ij} - \left(1-\frac{1}{\kappa}\right)^2\delta_{ij}\right] = \frac{1}{2}\left(1-\frac{1}{\kappa^2}\right)\delta_{ij}, \tag{10.56}$$

for any value of κ.

Euler versus Lagrange

The difference between the Euler and Lagrange representations of displacement lies only in the nonlinear terms of the strain tensor. Since the mappings are each other's inverses they are related by relatively simple expressions.

The two mutually inverse mappings satisfy the following elementary relations between the Jacobians,

$$\sum_k \frac{\partial x'_i}{\partial x_k}\frac{\partial x_k}{\partial x'_j} = \sum_k \frac{\partial x_i}{\partial x'_k}\frac{\partial x'_k}{\partial x_j} = \delta_{ij}. \tag{10.57}$$

Using this and (10.52) we derive

$$\sum_{ij} g'_{ij}\frac{\partial x'_i}{\partial x_k}\frac{\partial x'_j}{\partial x_l} = \delta_{kl} \tag{10.58}$$

and inserting (10.53) we obtain,

$$u_{kl} = \sum_{ij} u'_{ij}\frac{\partial x'_i}{\partial x_k}\frac{\partial x'_j}{\partial x_l}. \tag{10.59}$$

The inverse relation is,

$$u'_{ij} = \sum_{kl} u_{kl} \frac{\partial x_i}{\partial x'_k} \frac{\partial x_j}{\partial x'_l}. \tag{10.60}$$

The Euler and Lagrange representations are thus completely equivalent, though in the general case connected by fairly complicated expressions.

Example 10.5.3 (Uniform dilatation): For a uniform dilatation $\boldsymbol{x}' = \kappa\, \boldsymbol{x}$ we have

$$\frac{\partial x'_i}{\partial x_j} = \kappa \delta_{ij}, \tag{10.61}$$

and the relation between the Euler and Lagrange strains becomes,

$$u_{ij} = \kappa^2 u'_{ij}, \tag{10.62}$$

which is trivially fulfilled by the strain tensors of the preceding examples.

Problems

10.1 Calculate displacement gradients and the strain tensor for the transformation,

$$u_x = \alpha(5x - y + 3z),$$
$$u_y = \alpha(x + 8y),$$
$$u_z = \alpha(-3x + 4y + 5z),$$

where α is small.

10.2 A displacement field is given by

$$u_x = \alpha(x + 2y) + \beta x^2,$$
$$u_y = \alpha(y + 2z) + \beta y^2,$$
$$u_z = \alpha(z + 2x) + \beta z^2,$$

where α and β are 'small'. Calculate the divergence and curl of this field. Calculate Cauchy's strain tensor.

10.3 Calculate the strain tensor for the displacement field $\boldsymbol{u} = (Ax + Cy, Cx - By, 0)$ where A, B, C are small constants. Under what condition will the volume be unchanged?

10.4 Calculate the strain tensor for $\boldsymbol{u} = \alpha(y, x, 0)$ where $0 < \alpha \ll 1$. Determine the principal directions of strain and the change in length scales along these.

10.5 **(a)** Calculate the displacement gradients and the strain tensor for the displacement field $\boldsymbol{u} = \alpha(y^2, xy, 0)$ with $|\alpha| \ll 1/L$, where L is the size of the body. **(b)** Calculate the principal directions of strain and the dilatation factors.

10.6 Show that the change in a scalar product under a deformation is derivable from changes in length, i.e. from the diagonal projections u_{aa} of the strain tensor.

10.7 Show that the general displacement rule for a an infinitesimal needle (10.11) may be written

$$\boldsymbol{a}' = \boldsymbol{a} + \boldsymbol{\phi} \times \boldsymbol{a} + \mathbf{U} \cdot \boldsymbol{a} \tag{10.63}$$

where $\boldsymbol{\phi} = (1/2)\nabla \times \boldsymbol{u}$ and $\mathbf{U} = \{u_{ij}\}$ is Cauchy's strain tensor (10.18). What does the second term mean?

10.8 Show that the most general solution, for which Cauchy's strain tensor (10.18) vanishes, is

$$u_x = A + Dy + Ez$$
$$u_y = B - Dx + Fz$$
$$u_z = C - Ex - Fy$$

where A, B, C are arbitrary constants and D, E, F are small.

10.9 A deformable material undergoes two successive displacements, $\boldsymbol{x}' = \boldsymbol{x} + \boldsymbol{u}(\boldsymbol{x})$ and $\boldsymbol{x}'' = \boldsymbol{x}' + \boldsymbol{u}'(\boldsymbol{x}')$, both having small strain. Calculate the final strain tensor for the total deformation u''_{ij} relative to the original reference state.

* **10.10** Show that Cauchy's strain tensor satisfies the relation

$$\nabla_i \nabla_j u_{kl} + \nabla_k \nabla_l u_{ij} = \nabla_i \nabla_l u_{kj} + \nabla_k \nabla_j u_{il}. \tag{10.64}$$

[Conversely, if this relation is fulfilled for a symmetric tensor field u_{ij} then there is a displacement field such that the strain tensor is given by (10.18).]

* **10.11** Show that for finite deformations

$$\delta_{ij} + 2u_{ij} = \sum_k (\delta_{ik} + \nabla_i u_k)(\delta_{jk} + \nabla_j u_k), \tag{10.65}$$

and use this to prove that the matrix $\{\delta_{ij} + 2u_{ij}\}$ is positive definite. Show that

$$\det\{\delta_{ij} + \nabla_i u_j\} = \sqrt{\det\{\delta_{ij} + 2u_{ij}\}}. \tag{10.66}$$

* **10.12** Show that the only finite displacements with vanishing strain tensor are the rigid body translations and rotations.

11 Linear elasticity

When you bend a stick the reaction grows notably stronger the further you go—until it perhaps breaks with a snap. If you release the bending force before it breaks, the stick straightens out again and you can bend it again and again without it changing its reaction or its shape. That is elasticity.

In elementary mechanics the elasticity of a spring is expressed by Hooke's law which says that the force necessary to stretch or compress a spring is proportional to how much it is stretched or compressed. In continuous elastic materials Hooke's law implies that stress is a linear function of strain. Some materials that we usually think of as highly elastic, for example rubber, do not obey Hooke's law except under very small deformations. When stresses grow large, most materials deform more than predicted by Hooke's law. The proper treatment of nonlinear elasticity goes beyond the simple linear elasticity which we shall discuss in this book.

Robert Hooke (1635–1703). *English physicist. Worked on elasticity, built telescopes, and the discovered diffraction of light. The famous law which bears his name is from 1660. He already stated in 1678 the inverse square law for gravity, over which he got involved in a bitter controversy with Newton.*

The elastic properties of continuous materials are determined by their underlying molecular structure, but the relation between material properties and the molecular structure and arrangement in solids is complicated, to say the least. Luckily, there are broad classes of materials that may be described by a few material parameters which can be determined by macroscopic experiments. The number of such parameters depends on the complexity of the crystalline structure of the material, but we shall almost exclusively concentrate on the properties of structureless, isotropic elastic materials, described by just two material constants, Young's modulus and Poisson's ratio.

In this chapter, the emphasis will be on matters of principle. We shall derive the basic equations of linear elasticity, but only solve them in the simplest possible cases. In chapter 12 we shall solve these equations in generic situations of more practical interest.

Thomas Young (1773–1829). *English physician, physicist and egyptologist. He observed the interference of light and was the first to propose that light waves are transverse vibrations, explaining thereby the origin of polarization. He contributed much to the translation of the Rosetta stone.*

11.1 Hooke's law

Ideal massless elastic springs obeying Hooke's law are a mainstay of elementary mechanics. If a relaxed spring of length L is anchored at one end and pulled in the other with a force $\mathcal{F}$, its length is increased to $L + x$. Hooke's law states that there is proportionality between force and change in length,

$$\mathcal{F} = kx, \tag{11.1}$$

with a constant of proportionality, k, called the *spring constant*.

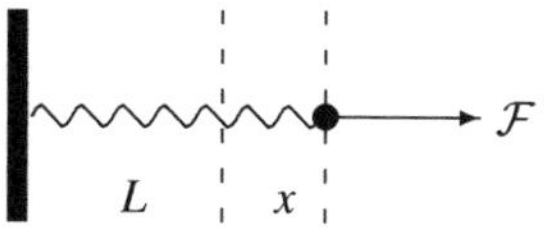

A spring anchored at the left and pulled towards the right by a force F will be stretched by the amount $x = F/k$.

Young's modulus

Real springs are physical bodies with mass, shape and internal molecular structure. Almost any solid body, anchored at one end and pulled at the other, will react like a spring, when the pull is not too strong. Basically, this reflects that interatomic forces are approximately elastic, when the atoms are only displaced slightly away from their positions (problem 11.1). Many elastic bodies that we handle daily, for example rubber

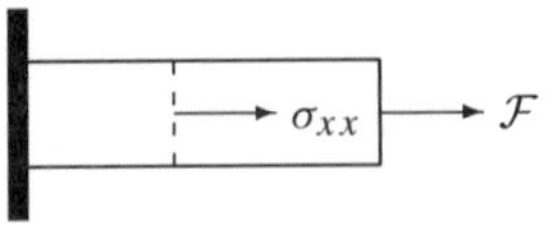

The same tension must act on any cross section of the string.

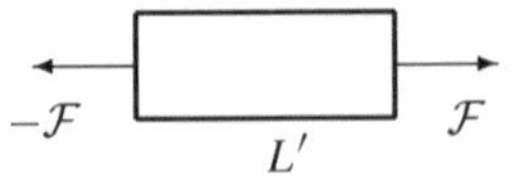

The force acts in opposite directions at the terminal cross sections of a smaller slice of the string. The extension is proportionally smaller.

Material	E [GPa]	ν [%]
Wolfram	411	28
Nickel (hard)	219	31
Iron (soft)	211	29
Plain steel	205	29
Cast iron	152	27
Copper	130	34
Titanium	116	32
Brass	100	35
Silver	83	37
Glass (flint)	80	27
Gold	78	44
Quartz	73	17
Aluminium	70	35
Magnesium	45	29
Lead	16	44

Young's modulus and Poisson's ratio for various isotropic materials (from [34]). These values are typically a factor 1000 larger than the tensile strength. Single wall carbon nanotubes have been reported with a Young's modulus of up to 1500 GPa (see footnote 2 on page 113).

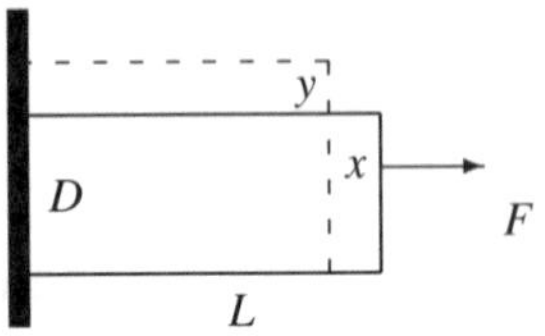

A string normally contracts in transverse directions when pulled at the ends.

bands, piano wire, sticks or water hoses, are long string-like objects with constant cross section, typically made from homogeneous and *isotropic* material without any particular internal structure. Their uniform composition and simple form make material strings convenient models for real springs.

The force $\mathcal{F}$ necessary to extend the length of a real string by a small amount x must be proportional to the area, A, of the string cross section, because if we bundle N such such strings loosely together to make a thicker string of area NA, the total force will have to be $N\mathcal{F}$ in order to get the same change of length. This shows that the relevant quantity to speak about is not the force $\mathcal{F}$ itself, but rather the (average) normal stress or tension, $\sigma_{xx} = N\mathcal{F}/NA = \mathcal{F}/A$, which is independent of the number of substrings N and consequently of the size A of the cross section. Since the same force $\mathcal{F}$ acts on any cross section of the string, the *string tension* σ_{xx} must be the same at each point along the string. For a smaller slice of the string of length $L' < L$, the uniformity implies that it will be stretched proportionally less, i.e. $x'/L' = x/L$. This indicates that the relevant parameter is not the absolute change of length x but rather the relative longitudinal extension or strain $u_{xx} = x/L$, which is independent of the length L of the string.

Putting everything together, we conclude that the quantity,

$$E = \frac{\sigma_{xx}}{u_{xx}} = \frac{\mathcal{F}/A}{x/L} = k\frac{L}{A} \tag{11.2}$$

must be independent of both the length L of the spring, the area A of its cross section, and the extension x. It is a material parameter, called the *modulus of extension* or *Young's modulus* (1807). Given Young's modulus we may calculate the actual spring constant,

$$k = E\frac{A}{L}, \tag{11.3}$$

for any string of length L and cross section A made from this particular material.

Young's modulus characterizes the behaviour of the material of the spring, when stretched in one direction. The relation (11.2) also tells us that a unidirectional tension σ_{xx} creates a relative extension,

$$u_{xx} = \frac{\sigma_{xx}}{E}, \tag{11.4}$$

in the material. Evidently, Hooke's law leads to a *linear* relation between stress and strain, and materials with this property are generally called linear.

Young's modulus is by way of its definition (11.2) measured in units of pressure, and typical values for metals are, like the bulk modulus (4.33), of the order of 10^{11} Pa $= 1$ Mbar. In the same way as the bulk modulus is a measure of the incompressibility of a material, Young's modulus is a measure of the *instretchability*. The larger it is, the harder it is to stretch the material. In order to obtain a large strain $u_{xx} \approx 100\%$, one would have to apply stresses of magnitude $\sigma_{xx} \approx E$, as shown by (11.4). Such strains are, of course, not permitted in the theory of small deformations, but Young's modulus nevertheless sets the scale.

Example 11.1.1 (Rope pulling contest): At company outings, employees often play the game of pulling in teams at each end of a rope. Before the inevitable terminal instability sets in, there is often a prolonged period where the two teams pull with almost equal force $\mathcal{F}$. If the teams each consist of 10 persons, all pulling with about their average weight of 70 kg, the total force becomes $\mathcal{F} = 7000$ N. For a rope diameter of 5 cm, the stress becomes quite considerable, $\sigma_{xx} \approx 3.6$ MPa. For a reasonable value of Young's modulus, say $E = 36$ MPa, the rope will stretch by $u_{xx} \approx 10\%$.

Poisson's ratio

Normal materials will contract in directions transverse to the direction of extension. If the transverse size, the 'diameter' D of a string changes by y, the transverse strain becomes of the order of $u_{yy} = y/D$, and will in general be negative for a positive stretching force $\mathcal{F}$. In linear materials, the transverse strain is also proportional to $\mathcal{F}$, so that the ratio u_{yy}/u_{xx} will be independent of $\mathcal{F}$. The negative of this ratio is called *Poisson's ratio* (1829)[1],

$$\nu = -\frac{u_{yy}}{u_{xx}}, \tag{11.5}$$

[1] Poisson's ratio is also sometimes denoted σ, but that clashes too much with the symbol for the stress tensor. Later we shall in the context of fluid mechanics also use ν for the kinematic viscosity, a choice which does not clash seriously with the use here.

and is another parameter characterizing isotropic materials. It is dimensionless, and typical values lie around 0.30 in metals. We shall see below that it cannot exceed 0.5 in isotropic materials.

Whereas longitudinal extension can be understood as a consequence of elastic atomic bonds being stretched, it is harder to understand why materials should contract transversally. The reason is, however, that in an isotropic material there are atomic bonds in all directions, and when bonds that are not purely longitudinal are stretched, they create a transverse tension which can only be relieved by contracting the material.

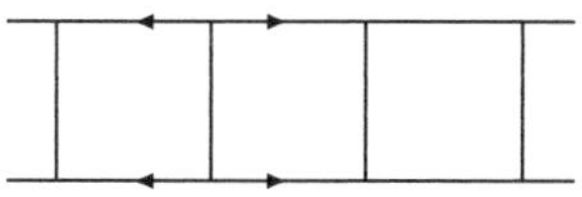

Stretching a ladder with purely transverse rungs will not create transverse forces.

A ladder constructed from ideal springs, with rungs orthogonal to the sides, will not experience a transverse contraction when stretched. If, on the other hand, some of the rungs are skew (making the ladder unusable), they will be stretched along with the ladder. But that will necessarily generate forces that tend to contract the ladder, i.e. a negative transverse tension, which either has to be balanced by external forces or relieved by contraction of the ladder.

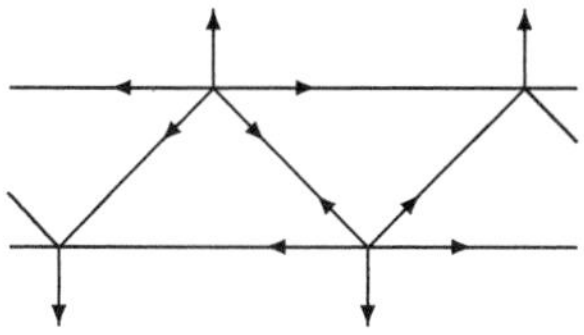

Stretching a ladder with skew rungs creates transverse forces which must be balanced by external forces at the boundary (as here) or relieved by transverse contraction.

11.2 Hooke's law in isotropic materials

For a stretched string-like object laid out along the x-direction of the coordinate system, the only non-vanishing stress component is a tension P along x,

$$\sigma_{xx} = P. \tag{11.6}$$

From (11.4) and (11.5) we obtain the corresponding diagonal strain components,

$$u_{xx} = \frac{P}{E}, \qquad u_{yy} = u_{zz} = -\nu\frac{P}{E}. \tag{11.7}$$

All the shear strains vanish, $u_{xy} = u_{yz} = u_{zx} = 0$, in this coordinate system.

In an arbitrary coordinate system, the components of the strain tensor will be completely mixed up with each other, as illustrated by the simple rotation (2.89), and there will also arise shear stresses and strains, but the relation between the stress tensor and the strain tensor will still be linear. To determine the form of the most general linear relationship between stress and strain in an isotropic material, we note that in such a material there are no internal directions which can be used to construct a linear relation between the tensors σ_{ij} and u_{ij}. In that case only two tensors can appear in the relation: one is the strain tensor u_{ij} itself, another is the Kronecker delta δ_{ij} multiplied with the trace of the strain tensor $\sum_k u_{kk}$. This is the only possible factor, because the trace is the only scalar quantity that can be formed from a linear combination of the strain tensor components.

Gabriel Lamé (1795–1870). *French mathematician. Worked on curvilinear coordinates, number theory and mathematical physics.*

Thus, we conclude that the most general strictly linear tensor relation between stress and strain tensors in an isotropic material is of the form (Cauchy 1822; Lamé 1852),

$$\boxed{\sigma_{ij} = 2\mu\, u_{ij} + \lambda\, \delta_{ij} \sum_k u_{kk},} \tag{11.8}$$

which is the generalization of Hooke's law to arbitrary isotropic materials. The coefficients λ and μ are material constants, called *elastic moduli* or *Lamé coefficients*. The coefficient λ has no special name, whereas μ is called the *shear modulus* or the *modulus of rigidity*, because it controls the magnitude of shear (off-diagonal) stresses. Since the strain tensor is dimensionless, the Lamé coefficients are, like the stress tensor itself, measured in units of pressure, and we shall soon see that the Lamé coefficients are directly related to Young's modulus and Poisson's ratio.

In practical calculations it always pays to write out Hooke's law explicitly. Its diagonal elements are,

$$\sigma_{xx} = (2\mu + \lambda)u_{xx} + \lambda(u_{yy} + u_{zz}), \tag{11.9a}$$
$$\sigma_{yy} = (2\mu + \lambda)u_{yy} + \lambda(u_{zz} + u_{xx}), \tag{11.9b}$$
$$\sigma_{zz} = (2\mu + \lambda)u_{zz} + \lambda(u_{xx} + u_{yy}), \tag{11.9c}$$

and its off-diagonal elements,

$$\sigma_{xy} = \sigma_{yx} = 2\mu\, u_{xy}, \tag{11.10a}$$
$$\sigma_{yz} = \sigma_{zy} = 2\mu\, u_{yz}, \tag{11.10b}$$
$$\sigma_{zx} = \sigma_{xz} = 2\mu\, u_{zx}. \tag{11.10c}$$

For $\mu = 0$ there are no shear stresses and all pressures become equal $p_x = p_y = p_z = p$, just as in a fluid.

The arguments leading to the general form of Hooke's law depend strongly on our understanding of tensors as geometric objects in their own right with well-defined transformation properties under rotations. As soon as we have cast a law of nature in the form of a scalar, vector or tensor relation, its validity in all Cartesian coordinate systems is immediately guaranteed. The form invariance of the natural laws under transformations that relate different observers has since Einstein been an important guiding principle in the development of modern theoretical physics.

Young's modulus and Poisson's ratio

Young's modulus and Poisson's ratio must be functions of the two Lamé coefficients. To derive the relations between these material parameters we insert the stresses (11.6) and strains (11.7) for the simple stretching of a material spring into the general relations (11.9a,b), and get,

$$P = (2\mu + \lambda)\frac{P}{E} - 2\lambda\nu\frac{P}{E}, \qquad 0 = -(2\mu + \lambda)\nu\frac{P}{E} + \lambda\left(-\nu\frac{P}{E} + \frac{P}{E}\right).$$

These equations are solved for E and ν,

$$\boxed{E = \frac{\mu(3\lambda + 2\mu)}{\lambda + \mu}, \qquad \nu = \frac{\lambda}{2(\lambda + \mu)}.} \tag{11.11}$$

Conversely, we may also express the Lamé coefficients in terms of Young's modulus and Poisson's ratio,

$$\boxed{\lambda = \frac{E\nu}{(1 - 2\nu)(1 + \nu)}, \qquad \mu = \frac{E}{2(1 + \nu)}.} \tag{11.12}$$

The 'engineering' constants, Young's modulus and Poisson's ratio, are in practice what is found in tables, and the above relations allow us to immediately calculate the Lamé coefficients.

Average pressure and bulk modulus

The trace of the stress tensor (11.8) becomes

$$\sum_i \sigma_{ii} = (2\mu + 3\lambda)\sum_i u_{ii}, \tag{11.13}$$

because the trace of the Kronecker delta is $\sum_i \delta_{ii} = 3$. Since the stress tensor in Hooke's law represents the change in stress due to the deformation we find the change in average pressure (9.12) caused by the deformation,

$$\Delta p = -\frac{1}{3}\sum_i \sigma_{ii} = -\left(\lambda + \frac{2}{3}\mu\right)\sum_i u_{ii}. \tag{11.14}$$

The trace of the strain tensor was previously shown in (10.32) to be proportional to the relative local change in density, $\sum_i u_{ii} = \nabla \cdot \boldsymbol{u} = -\Delta\rho/\rho$, and using the definition of the bulk modulus (4.33) $K = \rho dp/d\rho \approx \Delta p/(\Delta\rho/\rho)$, one finds,

$$\boxed{K = \lambda + \frac{2}{3}\mu = \frac{E}{3(1 - 2\nu)}.} \tag{11.15}$$

The bulk modulus equals Young's modulus for $\nu = 1/3$, which is in fact a typical value for ν in many materials. The Lamé coefficients and the bulk modulus are all proportional to Young's modulus and thus of the same order of magnitude. The values of the elastic moduli in metals are huge on ordinary scales, of the order of 10^{11} Pa $= 100$ GPa $= 10^6$ bar.

Inverting Hooke's law

Hooke's law (11.8) may be inverted so that strain is instead expressed as a linear function of stress. Solving (11.8) for u_{ij} and inserting $\sum_k u_{kk} = \sum_k \sigma_{kk}/(3\lambda + 2\mu)$ from (11.13), we get

$$u_{ij} = \frac{1}{2\mu}\sigma_{ij} - \frac{\lambda}{2\mu(3\lambda + 2\mu)}\delta_{ij}\sum_k \sigma_{kk}. \tag{11.16}$$

Introducing Young's modulus and Poisson's ratio (11.11), this takes the simpler form

$$\boxed{u_{ij} = \frac{1+\nu}{E}\sigma_{ij} - \frac{\nu}{E}\delta_{ij}\sum_k \sigma_{kk}.} \tag{11.17}$$

Explicitly, we find for the diagonal components

$$u_{xx} = \frac{1}{E}\sigma_{xx} - \frac{\nu}{E}(\sigma_{yy} + \sigma_{zz}), \tag{11.18a}$$

$$u_{yy} = \frac{1}{E}\sigma_{yy} - \frac{\nu}{E}(\sigma_{zz} + \sigma_{xx}), \tag{11.18b}$$

$$u_{zz} = \frac{1}{E}\sigma_{zz} - \frac{\nu}{E}(\sigma_{xx} + \sigma_{yy}), \tag{11.18c}$$

and for the off-diagonal ones,

$$u_{xy} = \frac{1+\nu}{E}\sigma_{xy}, \tag{11.19a}$$

$$u_{yz} = \frac{1+\nu}{E}\sigma_{yz}, \tag{11.19b}$$

$$u_{zx} = \frac{1+\nu}{E}\sigma_{zx}. \tag{11.19c}$$

Evidently, if the only stress is $\sigma_{xx} = P$, we obtain immediately from (11.18) the correct relations for simple stretching, (11.6) and (11.7).

Positivity constraints

The bulk modulus $K = \lambda + 2\mu/3$ cannot be negative, because a material with negative K would expand when put under pressure. Imagine what would happen to such a strange material if placed in a closed vessel also containing normal material like air or water. The hydrostatic pressure would make the strange material expand, causing a further pressure increase followed by expansion until the whole thing blew up. Likewise, materials with negative shear modulus, μ, would mimosa-like pull away from a shearing force instead of yielding to it. Formally, it may be shown (see section 11.4) that the conditions $3\lambda + 2\mu > 0$ and $\mu > 0$ follow from the elastic energy density being bounded from below. Although λ, in principle, may assume negative values, natural materials always have $\lambda \geq 0$.

Young's modulus cannot be negative because of these constraints, and this confirms that strings always stretch when pulled at the ends. If there were materials with the ability to contract when pulled, they would also behave magically. As you begin climbing up a rope made from such material, it pulls you further up. Presumably, if such materials were ever created, they would spontaneously contract into nothingness at the first possible occasion or at least into a state with a normal relationship between stress and strain.

Poisson's ratio $\nu = \lambda/2(\lambda + \mu)$ depends only on the ratio λ/μ and reaches its maximum $\nu \to 1/2$ for $\lambda/\mu \to \infty$. In the extreme limit $\nu = 1/2$ corresponding to $\mu = 0$ there are no shear stresses in the material which therefore behaves like a fluid at rest. Since the bulk modulus $K = 2\mu + 3\lambda$ is positive we have $\lambda/\mu > -2/3$, and the minimal value of Poisson's ratio $\nu = -1$ is obtained for $\lambda/\mu = -2/3$. Although most materials shrink in the transverse directions when stretched, and thus have $\nu > 0$, there might actually exist materials which expand transversally without violating the laws of physics. In practice all natural materials have $\nu > 0$, but it is fairly easy to construct artificial models of materials that expand when pulled, for example a grid of connected umbrellas.

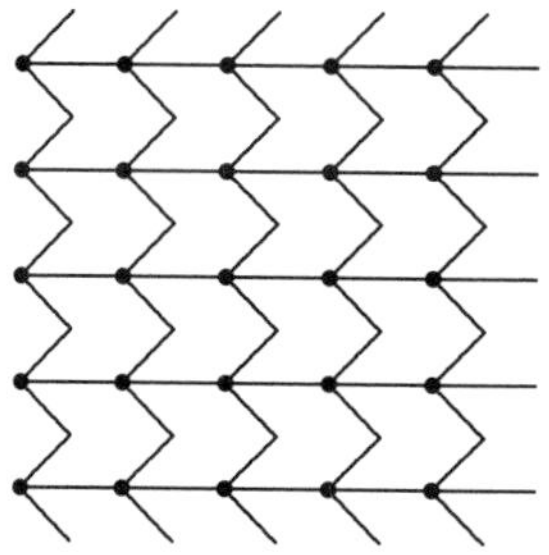

Grid of connected 'umbrellas'. If anchored to the left and pulled to the right, the umbrellas will open and expand the grid in the transverse direction.

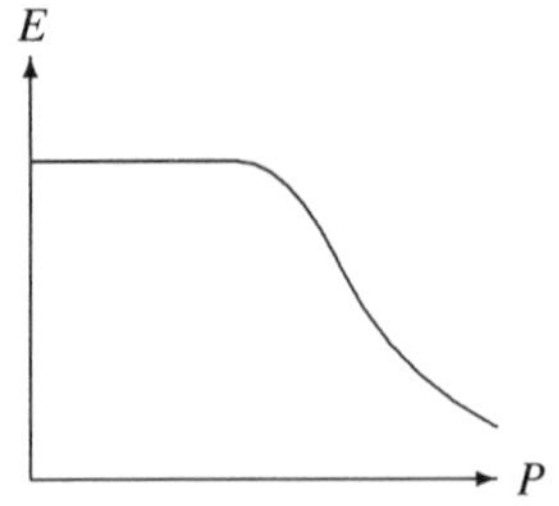

Sketch of how Young's modulus might vary as a function of increased tension. Beyond the proportionality limit, its effective value becomes generally smaller.

Limits to Hooke's law

Hooke's law in isotropic materials, expressed by the linear relationships, (11.8) or (11.17), between stress and strain, is only valid for stresses up to a certain value, called the *proportionality limit*. Beyond the proportionality limit, nonlinearities set in, and the present formalism becomes invalid. Eventually, one reaches a point, called the *elasticity limit*, where the material ceases to be elastic and undergoes permanent deformation, or even fracture, without much further increase of stress. Hooke's law is, however, a very good approximation for most metals under normal conditions where stresses are tiny compared to the elastic moduli.

* Anisotropic materials

In this book we shall limit the discussion to isotropic materials, for which Hooke's law takes the simple form (11.8). *Anisotropic* (also called *aeolotropic*) materials having different properties in different directions are of great technical importance.

For generally anisotropic materials, the linear relation between the stress and strain tensors, the generalized Hooke's law, is of the form,

$$\sigma_{ij} = \sum_{kl} \lambda_{ijkl} u_{kl}, \tag{11.20}$$

where the coefficients λ_{ijkl} form a tensor of rank 4, called the *elasticity tensor*. For isotropic materials the elasticity tensor is of the form,

$$\lambda_{ijkl} = \lambda \delta_{ij} \delta_{kl} + \mu(\delta_{ik}\delta_{jl} + \delta_{jk}\delta_{il}). \tag{11.21}$$

In general there are more parameters depending on the intrinsic complexity of the material structure.

Requiring only that the stress tensor is symmetric, the elasticity tensor connects the six independent components of stress with the six independent components of strain and can in principle contain $(6\times6) = 36$ independent parameters. If one further demands that an elastic energy function should exist (see problem 11.9) this (6×6) array of coefficients must itself be symmetric under the exchange $ij \leftrightarrow kl$, i.e. $\lambda_{ijkl} = \lambda_{klij}$, and can thus only contain $6 + (6 \times 5)/2 = 21$ independent parameters. The orientation of an anisotropic material relative to the coordinate system takes three parameters (Euler angles), so altogether there may be up to 18 independent constants characterizing the intrinsic elastic properties of a general anisotropic material, a number actually realized by triclinic crystals [37, 26].

11.3 Static uniform deformation

To see how Hooke's law works for continuous systems, we now turn to the extremely simple case of a *static uniform deformation* in which the strain tensor u_{ij} takes the same value everywhere in a body at all times. Hooke's law (11.8) then ensures that the stress tensor is likewise constant throughout the body, so that all its derivatives vanish, $\nabla_k \sigma_{ij} = 0$. From the condition for mechanical equilibrium, $f_i + \sum_j \nabla_j \sigma_{ij} = 0$, it follows that $f_i = 0$ so that uniform deformation necessarily excludes body forces. Conversely, in the presence of body forces, there must always be non-uniform deformation of an isotropic material, a quite reasonable conclusion.

Furthermore, at the boundary of a uniformly deformed body, the stress vector $\boldsymbol{\sigma} \cdot \boldsymbol{n}$ is as always required to be continuous, and this puts strong restrictions on the form of the external forces that may act on the surface of the body. Uniform deformation is for this reason only possible under very special circumstances, but when it applies the analytic solution for the displacement field is nearly trivial.

Uniform compression

In a fluid at rest with a constant pressure P, the stress tensor is $\sigma_{ij} = -P\delta_{ij}$ everywhere in the fluid. If a solid body made from isotropic material is immersed into this fluid, the natural guess is that the pressure will also be P inside the body. Inserting $\sigma_{ij} = -P\delta_{ij}$ into (11.17) and using that $\sum_k \sigma_{kk} = -3P$, the strain may be written,

$$u_{ij} = -\frac{P}{3K}\delta_{ij}. \tag{11.22}$$

Since

$$u_{xx} = \nabla_x u_x = -\frac{P}{3K}, \tag{11.23}$$

we may immediately integrate this equation (and the similar ones for u_{yy} and u_{zz}) and obtain a particular solution to the displacement field,

$$u_x = -\frac{P}{3K}x, \qquad u_y = -\frac{P}{3K}y, \qquad u_z = -\frac{P}{3K}z. \tag{11.24}$$

The most general solution is obtained by adding an arbitrary small rigid body displacement to this solution.

We arrived at this result by making an educated *guess* of the form of the stress tensor inside the body. It could in principle be wrong, but is in fact correct due to a uniqueness theorem to be derived in section 11.4. The theorem guarantees, in analogy with the uniqueness theorems of electrostatics, that provided the equations of mechanical equilibrium and the boundary conditions are fulfilled by the guess (which they are here), there is essentially only one solution to any *elastostatic* problem. The only liberty left is an arbitrary rigid body displacement which may always be added to the solution.

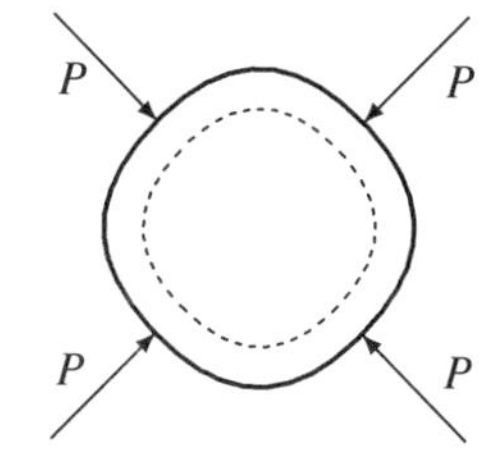

A body made from isotropic, homogenous material subject to a uniform external pressure will be uniformly compressed.

Uniform stretching

At the beginning of this chapter we investigated the reaction of a string-like material body stretched along its main axis, say the x-direction, by means of a tension $\sigma_{xx} = P$ acting uniformly over its cross section. If there are no other external forces acting on the body, the natural guess is that the only non-vanishing component of the stress tensor is $\sigma_{xx} = P$ throughout the body. Inserting that into (11.16), we obtain as before the strains (11.7).

The corresponding displacement field is again found by integrating $\nabla_x u_x = u_{xx}$ etc, and we find the particular solution

$$u_x = \frac{P}{E}x, \qquad u_y = -\nu\frac{P}{E}y, \qquad u_z = -\nu\frac{P}{E}z. \tag{11.25}$$

The solution describes as expected a simple dilatation along the x-axis and a contraction along the other axes.

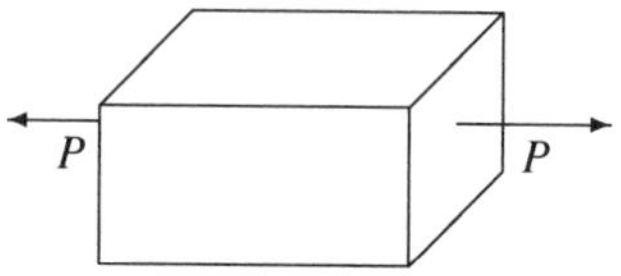

Uniformly stretched body with a constant tension P.

Uniform shear

Finally we return to the example from section 9.2 of a clamped slab of homogeneous, isotropic material in the xz-plane, subjected to a shear force in the x-direction. As we argued, the shear stress $\sigma_{xy} = P$ must be constant throughout the material. Assuming that there are no other stresses, the only strain component becomes $u_{xy} = P/2\mu$, and using that $2u_{xy} = \nabla_x u_y + \nabla_y u_x$, we find a particular solution

$$u_x = \frac{P}{\mu}y, \qquad u_y = u_z = 0. \tag{11.26}$$

As expected, the displacement in the x-direction vanishes for $y = 0$ and grows linearly with y. Evidently, this is why μ is called the *shear modulus*. In problem 11.8 the displacement field is calculated without making the assumption of small strains.

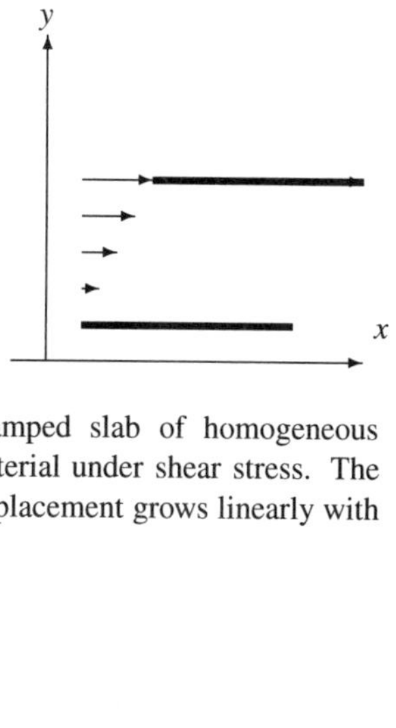

Clamped slab of homogeneous material under shear stress. The displacement grows linearly with y.

11.4 Energy of deformation

The work performed by the external force in extending a spring further by the amount δx is $dW = \mathcal{F}\delta x = kx\delta x$. Integrating this expression, we obtain the total work $W = (1/2)kx^2$, which is identified with the well-known expression for the elastic energy, $E = (1/2)kx^2$, stored in a stretched spring. Calculated per unit of volume $V = AL$ for a material string, we find the density of elastic energy in the material

$$\epsilon = \frac{\mathcal{E}}{V} = \frac{kx^2}{2V} = \frac{1}{2}Eu_{xx}^2 = \frac{P^2}{2E}. \tag{11.27}$$

Note that Poisson's ratio ν does not appear. The transverse contraction controlled by Poisson's ratio can play no role in building up the elastic energy of a stretched string, when there are no forces acting on the sides of the string.

Elastic energy

In the general case, strains and stresses vary over the body, and the calculation becomes more complicated. To determine the general expression for the elastic energy density we use the expression derived in section 10.4,

$$\delta W_{\text{deform}} = \int_V \sum_{ij} \sigma_{ij} \delta u_{ij} \, dV. \tag{11.28}$$

This is the work that must be performed in order to change the internal strain field by an infinitesimal amount δu_{ij}.

If the stress tensor is a linear function of the strain tensor one must build up the deformation in infinitesimal steps, as we did for the gravitational self-energy in section 6.4. The elastic self-energy, i.e. the energy of a deformation in its own stress field, becomes in this way quadratic in the strain tensor with a factor 1/2 to represent the average value of the final stress field,

$$\boxed{\mathcal{E} = \frac{1}{2} \int_V \sum_{ij} \sigma_{ij} u_{ij} \, dV,} \tag{11.29}$$

where the integral runs over the volume V of the undeformed material. The corrections due to the change of volume are negligible for small and smooth displacements.

In the general linear case the *elastic energy density* becomes,

$$\epsilon = \frac{1}{2} \sum_{ij} \sigma_{ij} u_{ij} = \frac{1}{2} \sum_{ijkl} \lambda_{ijkl} u_{ij} u_{kl}. \tag{11.30}$$

For anisotropic materials the mere existence of such an energy function will, as mentioned, impose the further condition that λ_{ijkl} must be symmetric under exchange $ij \leftrightarrow kl$ (see problem 11.9).

For isotropic materials σ_{ij} is given by Hooke's law (11.8) and the energy density simplifies to,

$$\boxed{\epsilon = \mu \sum_{ij} u_{ij}^2 + \frac{1}{2} \lambda \left(\sum_i u_{ii} \right)^2 = \frac{1+\nu}{2E} \sum_{ij} \sigma_{ij}^2 - \frac{\nu}{2E} \left(\sum_i \sigma_{ii} \right)^2.} \tag{11.31}$$

Inserting the stresses for uniform stretching where the only non-vanishing stress is $\sigma_{xx} = P$, all the dependence on Poisson's ratio ν cancels out, bringing us back to the energy density in a material string (11.27).

The energy density must be bounded from below, for if it were not, elastic materials would be unstable, and an unlimited amount of work could be obtained by increasing the state of deformation. Imagine for a moment how magically a body made from such material would behave when squeezed. From the boundedness, it follows immediately that the shear modulus must be positive, $\mu > 0$, because otherwise we might make one of the off-diagonal components of the strain tensor, say u_{xy}, grow without limit without performing any work. The condition on λ is more subtle because the diagonal components of the strain tensor are involved in both terms. For a uniform deformation with $u_{ij} = k\delta_{ij}$ we get the energy density $\epsilon = (3/2)(3\lambda + 2\mu)k^2$, implying that $3\lambda + 2\mu > 0$, i.e. that the bulk modulus (11.15) is positive. In problem 11.7 it is shown that there are no stronger conditions.

Total energy in an external gravitational field

In an external gravitational potential, $\Phi(\boldsymbol{x})$, not depending on the displacement field, the change in gravitational energy due to the displacement of a material particle of fixed mass dM from $\boldsymbol{x}' = \boldsymbol{x} + \boldsymbol{u}(\boldsymbol{x})$ to $\boldsymbol{x}$ is to leading order in the displacement,

$$\big(\Phi(\boldsymbol{x}') - \Phi(\boldsymbol{x})\big) dM \approx -\boldsymbol{g}(\boldsymbol{x}) \cdot \boldsymbol{u}(\boldsymbol{x}) \, dM.$$

The last expression is obtained under the assumption that the displacement is small compared to the length scale for variations in the gravitational potential. The total (potential) energy of a deformed body in a gravitational field is therefore the sum of the extra gravitational energy of the deformation and the internal energy of deformation,

$$\boxed{\mathcal{E} = -\int_V \rho \boldsymbol{g} \cdot \boldsymbol{u} \, dV + \frac{1}{2} \int_V \sum_{ij} \sigma_{ij} u_{ij} \, dV,} \tag{11.32}$$

where $\rho(\boldsymbol{x})$ is the mass density of the body. This expression for the total potential energy of an elastic body in a gravitational field will be useful when we develop the theory behind numerical computation of the displacement field in mechanical equilibrium in chapter 13.

Suppose we vary the displacement field by an infinitesimal amount, $\delta u_i(\boldsymbol{x})$. The associated change in energy becomes,

$$\begin{aligned} \delta\mathcal{E} &= -\int_V \rho \boldsymbol{g} \cdot \delta \boldsymbol{u} \, dV + \frac{1}{2} \int_V \sum_{ij} \sigma_{ij} \delta u_{ij} \, dV + \frac{1}{2} \int_V \sum_{ij} \delta\sigma_{ij} u_{ij} \, dV \\ &= -\int_V \rho \boldsymbol{g} \cdot \delta \boldsymbol{u} \, dV + \int_V \sum_{ij} \sigma_{ij} \delta u_{ij} \, dV. \end{aligned}$$

In the second step we used the assumed symmetry of the coefficients in the generalized Hooke's law (11.20). This is exactly the same as the work in a virtual displacement work (10.34), such that we have now explicitly proven that the virtual work contributes to the internal energy, $\delta\mathcal{E} = \delta W$.

Absolute minimum of energy in mechanical equilibrium

In mechanical equilibrium, $\boldsymbol{f}^* = \boldsymbol{0}$, it follows from (10.33) that the virtual work is given entirely by the external work of the contact forces on the surface of the body,

$$\delta\mathcal{E} = \delta W = \oint_S \delta \boldsymbol{u} \cdot \boldsymbol{\sigma} \cdot d\boldsymbol{S}. \tag{11.33}$$

If the external work also vanishes the total deformation energy will be unchanged or *stationary*, $\delta\mathcal{E} = 0$, under the variation in the displacement. In practice it often happens that some parts of the external boundary of a body are held fixed with vanishing displacement, $\boldsymbol{u} = 0$, while other parts are left free with vanishing stress vector, $\boldsymbol{\sigma} \cdot \boldsymbol{n} = \boldsymbol{0}$. In that case the integrand of (11.33) will vanish everywhere on the surface of the body, and the energy will be stationary under any variations respecting these boundary conditions.

Assuming that the elastic energy density (11.30) is a positive definite quadratic polynomial in the strains, it follows that *when no external work is performed on a body, the total potential energy must have a unique minimum corresponding to mechanical equilibrium.* Below we shall prove that this result also guarantees the uniqueness of solutions in elastostatics, and we shall see in chapter 13 that it constitutes the foundation for numerical calculations.

Proof of uniqueness of elastostatics solutions: To prove the uniqueness of the solutions to the mechanical equilibrium equations (9.19) for linearly elastic materials we assume that we have two solutions $\boldsymbol{u}^{(1)}$ and $\boldsymbol{u}^{(2)}$ that both satisfy the equilibrium equations and the boundary conditions for a specific problem. Due to the linearity, the difference between the solutions $\boldsymbol{u} = \boldsymbol{u}^{(1)} - \boldsymbol{u}^{(2)}$ generates a difference in strain $u_{ij} = (1/2)(\nabla_i u_j + \nabla_j u_i)$ and a difference in stress $\sigma_{ij} = \sum_{kl} \lambda_{ijkl} u_{kl}$. Under the assumption that the body forces are identical for the two fields the change in stress satisfies the equation, $\sum_j \nabla_j \sigma_{ij} = 0$, and by means of Gauss' theorem (4.22) we obtain as before,

$$0 = \int_V \sum_{ij} u_i \nabla_j \sigma_{ij} \, dV = \oint_S \sum_{ij} u_i \sigma_{ij} \, dS_j - \int_V \sum_{ij} \nabla_j u_i \sigma_{ij} \, dV.$$

Here the surface integral vanishes because of the boundary conditions, which either specify the same displacements for the two solutions at the surface, i.e. $u_i = 0$, or the same stress vectors, i.e. $\sum_j \sigma_{ij} n_j = 0$. Using the symmetry of the stress tensor we get,

$$0 = \int_V \sum_{ij} \nabla_j u_i \sigma_{ij} \, dV = \int_V \sum_{ij} u_{ij} \sigma_{ij} \, dV = \int_V \sum_{ijkl} \lambda_{ijkl} u_{ij} u_{kl} \, dV.$$

The integrand is of the same form as the energy density (11.30), which is always assumed to be positive definite, and consequently, the integral can only vanish if the strain tensor for the difference field vanishes everywhere in the body, i.e. $u_{ij} = 0$.

Given the boundary conditions, there is essentially only one solution to the equations of mechanical equilibrium in linear elastic materials. Although the two displacement fields may, in principle, differ by a rigid body displacement, they must give rise to identical deformations everywhere in the body. If we have somehow guessed a solution satisfying the equations of mechanical equilibrium and the boundary conditions, it will necessarily be the right one.

Problems

11.1 Two particles interact with a smooth distance dependent force $\mathcal{F}(r)$. Show that the force obeys Hooke's law in the neighbourhood of an equilibrium configuration.

11.2 **(a)** Show that we may write (11.8) in the form

$$\sigma_{ij} = 2\mu\left(u_{ij} - \frac{1}{3}\delta_{ij}\sum_k u_{kk}\right) + K\delta_{ij}\sum_k u_{kk}. \tag{11.34}$$

(b) Show that first term gives no contribution to the average pressure.

11.3 A displacement field is given by

$$\begin{aligned} u_x &= \alpha(x+2y) + \beta x^2, \\ u_y &= \alpha(y+2z) + \beta y^2, \\ u_z &= \alpha(z+2x) + \beta z^2, \end{aligned}$$

where α and β are 'small'.

(a) Calculate the divergence and curl.
(b) Calculate Cauchy's strain tensor.
(c) Calculate the stress tensor in a linear elastic medium with Lamé coefficients λ and μ for the special case $\beta = 0$.

11.4 A beam with constant cross section is fixed such that its sides cannot move. One end of the beam is also held in place while the other end is pulled with a uniform tension P. Determine the strains, stresses and the displacement field in the beam.

11.5 A rectangular beam with its axis along the x-axis is fixed on the two sides orthogonal to the y-axis but left free on the two sides orthogonal to the z-axis. The beam is held fixed at one end and pulled with a uniform tension P the other. Determine the strains, stresses and the displacement field in the beam.

11.6

(a) Show that Cauchy's strain tensor always fulfills the condition,

$$\nabla_y^2 u_{xx} + \nabla_x^2 u_{yy} = 2\nabla_x\nabla_y u_{xy}. \tag{11.35}$$

(b) Assume now that the only non-vanishing components of the stress tensor are σ_{xx}, σ_{yy} and $\sigma_{xy} = \sigma_{yx}$. Formulate Cauchy's equilibrium equations for this stress tensor in the absence of volume forces.

(c) Show that the following stress tensor is a solution to the equilibrium equations,

$$\sigma_{xx} = \nabla_y^2\phi, \qquad \sigma_{yy} = \nabla_x^2\phi, \qquad \sigma_{xy} = -\nabla_x\nabla_y\phi, \tag{11.36}$$

where ϕ is an arbitrary function of x and y.

(d) Calculate the strain tensor in terms of ϕ in an isotropic elastic medium, and show that the condition (11.35) implies that ϕ must satisfy the biharmonic equation,

$$\nabla_x^4 \phi + \nabla_y^4 \phi + 2\nabla_x^2 \nabla_y^2 \phi = 0. \tag{11.37}$$

(e) Determine a solution of the displacement field (u_x, u_y, u_z), when $\phi = xy^2$ and Young's modulus is set to $E = 1$. Hint: begin by integrating the diagonal elements of the strain tensor, and afterwards add to extra terms to u_x to get the correct off-diagonal elements.

$*$ **11.7** Show that one may write the energy density (11.30) in the following form

$$\epsilon = \tfrac{1}{2}[\lambda - 2\mu(3\alpha^2 - 2\alpha)]\left(\sum_i u_{ii}\right)^2 + \mu \sum_{ij}\left(u_{ij} - \alpha \sum_k u_{kk}\delta_{ij}\right)^2 \tag{11.38}$$

where α is arbitrary. Use this to argue that $3\lambda + 2\mu > 0$ and that this is the strictest condition on λ.

$*$ **11.8** Consider a shear deformation of a slab of elastic material in the xz-plane by a force in the x-direction. Assume that the sides of the slab are kept free to move, so that the only non-vanishing components of the strain tensor are $u_{xy} = u_{yx} = \alpha$. Show that the displacement becomes

$$u_x = \alpha y, \tag{11.39}$$

$$u_y = -\left(1 - \sqrt{1 - \alpha^2}\right) y. \tag{11.40}$$

for a deformation which is not assumed to be small. Describe what happens for $\alpha \to 1$.

$*$ **11.9** The most general linear relation between stress and strain is of the form (11.20). **(a)** Show that the elasticity tensor is symmetric in the first two and last two indices. **(b)** Show that for the elastic energy to take the form (11.29), the elasticity tensor must obey the further symmetry relation $\lambda_{ijkl} = \lambda_{klij}$.

12
Solids at rest

Flagpoles, bridges, houses and towers are built from elastic materials, and are designed to stay in one place with at most small excursions away from equilibrium due to wind and water currents. Ships, airplanes and space shuttles are designed to move around, and their structural integrity depends crucially on the elastic properties of the materials from which they are made. Almost all human constructions and natural structures depend on elasticity for stability and ability to withstand external stresses [74, 76].

It can come as no surprise that the theory of static elastic deformation, *elastostatics*, is a huge engineering subject. Engineers must know the deformation and internal stresses in their constructions in order to predict risk of failure and set safety limits, and that is only possible if the elastic properties of the building materials are known, and if they are able to solve the equations of elastostatics, or at least get decent approximations to them. Today computers aid engineers in getting precise numeric solutions to these equations and allow them to build critical structures, such as submarines, supertankers, airplanes and space vehicles, in which over-dimensioning of safety limits is deleterious to fuel consumption as well as to construction costs. In chapter 13 the basic technique behind the numeric solution of elastostatics problems is presented and applied to the relatively simple two-dimensional problem of how a long block of elastic material settles under gravity.

In this chapter we shall develop the theory of elastostatics for bodies made from isotropic materials and apply it to generic cases with simple body geometries for which analytic solutions can be obtained. The field equations of elastostatics and their solutions are in many respects similar to the field equations and solutions of electrostatics and magnetostatics. They also bear comparison to the equations of stationary fluid flow that will be taken up in later chapters.

12.1 Equations of elastostatics

Combining the results of the preceding three chapters we arrive at the fundamental equations for elastostatics,

$$u_{ij} = \tfrac{1}{2}(\nabla_i u_j + \nabla_j u_i), \qquad \text{Cauchy's strain tensor (10.18)} \qquad (12.1a)$$

$$\sigma_{ij} = 2\mu\, u_{ij} + \lambda\delta_{ij} \sum_k u_{kk}, \qquad \text{Hooke's law (11.8)} \qquad (12.1b)$$

$$f_i^* = f_i + \sum_j \nabla_j \sigma_{ij} = 0, \qquad \text{mechanical equilibrium (9.19).} \qquad (12.1c)$$

Inserting Hooke's law and Cauchy's strain tensor into the effective force we get,

$$f_i^* = f_i + 2\mu \sum_j \nabla_j u_{ij} + \lambda \nabla_i \sum_j u_{jj}$$
$$= f_i + \mu \sum_j \nabla_j^2 u_i + (\lambda + \mu) \nabla_i \sum_j \nabla_j u_j.$$

Rewriting this in vector notation, we finally arrive at *Navier's equation of equilibrium*, also called the *Navier–Cauchy equilibrium equation*,

Claude Louis Marie Henri Navier (1785–1836). *French engineer, worked on applied mechanics, elasticity, fluid mechanics and suspension bridges. Formulated the first version of the elastic equilibrium equation in 1821, a year before Cauchy gave it its final form.*

$$\boxed{\boldsymbol{f} + \mu \nabla^2 \boldsymbol{u} + (\lambda + \mu) \boldsymbol{\nabla}\, \boldsymbol{\nabla} \cdot \boldsymbol{u} = \boldsymbol{0}.} \tag{12.2}$$

After all the troubles with tensor notation, we end up with a relatively simple linear field equation for the vector displacement field! One should, however, not be taken in by its apparent simplicity. There is a surprising richness hidden in its compact form.

It should be noted that the linearity of the equilibrium equation permits us to *superpose* solutions to it. For example, if you both compress and stretch a body uniformly, the displacement field for the combined operation will be the sum of the respective displacement fields, (11.25) and (11.24).

There is a subtlety in the derivation of the Navier–Cauchy equation which we have quietly ignored. The problem arises from the use of the Lagrange representation where the displacement field is viewed as a function of the original coordinates $\boldsymbol{x}$ whereas the derivatives in the equation of mechanical equilibrium (12.1c) refer to the actual coordinates $\boldsymbol{x}'$. Using the definition of the displacement field (10.2) we find the following relation between the two kinds of derivatives,

$$\nabla_i = \frac{\partial}{\partial x_i} = \sum_j \frac{\partial x'_j}{\partial x_i} \frac{\partial}{\partial x'_j} = \nabla'_i + \sum_j \nabla_i u_j \nabla'_j. \tag{12.3}$$

Under the assumption of small displacement derivatives (10.15) the second term may be consistently ignored in the linear approximation, such that $\nabla_i \approx \nabla'_i$.

Estimates

Confronted with partial differential equations, it is always useful to get a rough idea of the order of magnitude of a particular solution. It should be emphasized that such estimates just aim to find the right orders of magnitude of the fields, and that there may be special circumstances in a particular problem which invalidate them. If that is the case, or if precision is needed, there is no way around analytic or numeric calculation.

Imagine, for example, that a body made from elastic material is subjected to surface stresses of a typical magnitude P and no body forces. A rough guess on order of magnitude of the stresses in the body is then also $\sigma_{ij} \sim P$. The elastic moduli λ, μ, E and K are all of the same magnitude, so that the deformation is of the order of $u_{ij} \sim P/E$. Since the deformation is calculated from gradients of the displacement field, the variation in displacement across a body of typical size L may be estimated to be of the order of $u_i \sim L u_{ij} \sim LP/E$.

Example 12.1.1 (Deformation of a chair): Standing with your full weight of 70 kg on the seat of a chair supported by four wooden legs, each 7 cm^2 in cross section and 50 cm long, you exert a stress $P \approx 250$ kPa $= 2.5$ bar on the legs. Taking $E \approx 10^9$ Pa, the deformation will be about 2.5×10^{-4} and the maximal displacement about 0.12 mm. The squashing of the legs of the chair due to your weight is barely visible.

In mechanical equilibrium (12.1c), there is balance between local changes in stress and body forces. This allows us to estimate the change in stress over a distance L due to, for example gravity $\boldsymbol{f} = \rho \boldsymbol{g}$ to be $\sigma_{ij} \approx \rho g L$. The corresponding variation in strain becomes $u_{ij} \sim L\rho g/E$ for non-exceptional materials. Since u_{ij} is dimensionless, it is convenient to define a deformation scale,

$$D \sim \frac{E}{\rho g}, \tag{12.4}$$

so that $u_{ij} \sim L/D$. The quantity D has dimension of length and sets the scale for major changes in deformation (of order unity). Small deformations require $L \ll D$. Finally, we estimate the variation in the displacement over a distance L to be of magnitude $u_i \sim Lu_{ij} \sim L^2/D$, which depends quadratically on L.

Example 12.1.2 (Settling of a tall building): How much does a tall building settle under its own weight when it is built? Let the height of the building be $L = 413$ m, its ground area $A = 63 \times 63$ m^2, and the average mass density $\rho = 100$ kg m^{-3}, including walls, columns, floors, office equipment and people. The weight of it all is carried by steel columns taking up about $f = 1\%$ of its ground area.

The total mass of the building is $M = \rho A L = 1.6 \times 10^8$ kg and the stress in the supports at ground level is $P = Mg_0/fA \approx 400$ bar. Taking Young's modulus to be that of steel, $E = 2 \times 10^{11}$ Pa, the deformation scale becomes huge, $D \sim fE/\rho g_0 \approx 2000$ km so that the strain is about 2×10^{-4}. The top of the building settles thus by merely $L^2/D \approx 8$ cm.

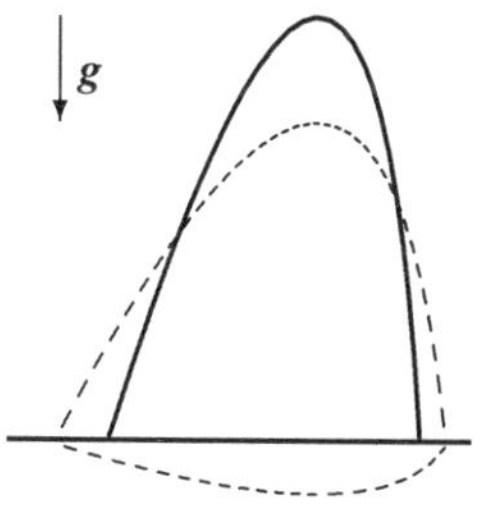

Settling of a body under the influence of gravity.

12.2 Standing up to gravity

Solid objects, be they mountains, bridges, houses or coffee cups, standing on a surface are deformed by gravity, and deform in turn, by their weight, the supporting surface. Intuition tells us that gravity makes such objects settle towards the ground and squashes their material so that it bulges out horizontally, unless prevented by constraining walls. In a fluid at rest, each horizontal surface element has to carry the weight of the column of fluid above it, and this determines the pressure in the fluid. In a solid at rest, this is more or less also the case, except that shear elastic stresses in the material are able to distribute the vertical load in the horizontal directions.

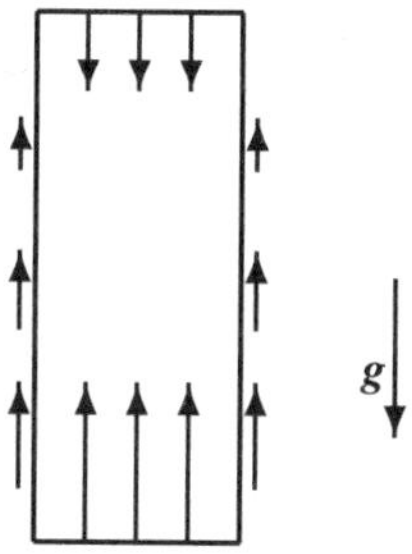

Shear stresses may aid in carrying the weight of a vertical column of elastic material.

Uniform settling

An infinitely extended slab of homogeneous and isotropic elastic material placed on a horizontal surface is a kind of 'elastic sea', which like the fluid sea may be assumed to have the same properties everywhere in a horizontal plane. In a flat-Earth coordinate system, where gravity is given by $\boldsymbol{g} = (0, 0, -g_0)$, we expect a uniformly vertical displacement, which only depends on the z-coordinate,

$$\boldsymbol{u} = (0, 0, u_z(z)) = u_z(z)\, \boldsymbol{e}_z. \tag{12.5}$$

In order to realize this 'elastic sea' in a finite system, it must be surrounded by fixed, vertical and slippery walls. The vertical walls forbid horizontal but allow vertical displacement, and at the bottom, $z = 0$, we place a horizontal supporting surface which forbids vertical but allows horizontal displacement. At the top, $z = h$, the elastic material is left free to move without any external forces acting on it.

The only non-vanishing strain is $u_{zz} = \nabla_z u_z$. From Hooke's law (11.9), we obtain the non-vanishing stresses

$$\sigma_{xx} = \sigma_{yy} = \lambda u_{zz}, \qquad \sigma_{zz} = (\lambda + 2\mu) u_{zz}, \tag{12.6}$$

and Cauchy's equilibrium equation (12.1c) simplifies in this case to

$$\nabla_z \sigma_{zz} = \rho_0 g_0. \tag{12.7}$$

Using the boundary condition $\sigma_{zz} = 0$ at $z = h$, this equation may immediately be integrated to

$$\sigma_{zz} = -\rho_0 g_0 (h - z). \tag{12.8}$$

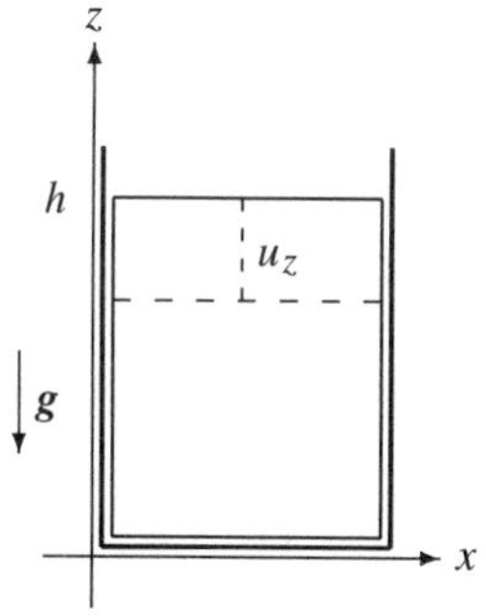

Elastic 'sea' of material undergoing a uniform downwards displacement because of gravity. The container has fixed, slippery walls.

The vertical pressure $p_z = -\sigma_{zz} = \rho_0 g_0 (h - z)$ is positive and rises linearly with depth $h - z$, just as in the fluid sea. It balances everywhere the full weight of the material above, but this was expected since there are no shear stresses to distribute the vertical load. The horizontal pressures $p_x = p_y = p_z \lambda/(\lambda + 2\mu)$ are also positive but smaller than the vertical, because both λ and μ are positive in normal materials. The horizontal pressures are eventually balanced by the fixed vertical walls.

The strain

$$u_{zz} = \nabla_z u_z = \frac{\sigma_{zz}}{\lambda + 2\mu} = -\frac{\rho_0 g_0}{\lambda + 2\mu}(h - z) \tag{12.9}$$

is negative, corresponding to a compression. The characteristic length scale for major deformation is in this case

$$D = \frac{\lambda + 2\mu}{\rho_0 g_0} = \frac{1-\nu}{(1+\nu)(1-2\nu)} \cdot \frac{E}{\rho_0 g_0}. \tag{12.10}$$

Integrating the strain with the boundary condition $u_z = 0$ for $z = 0$, we finally obtain

$$\boxed{u_z = -\frac{h^2 - (h-z)^2}{2D}}. \tag{12.11}$$

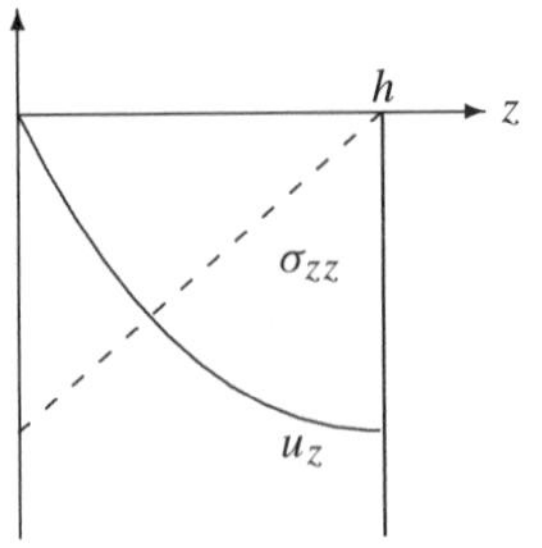

Sketch of the displacement (solid curve) and stress (dashed) for the elastic 'sea in a box'.

The displacement is always negative, largest in magnitude at the top, $z = h$, and varies quadratically with height h at the top, as expected from the estimate in the preceding section.

Shear-free settling

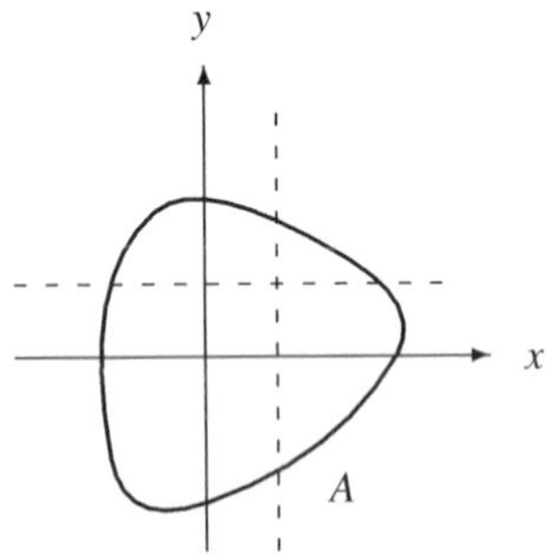

Horizontal cross section of elastic block of 'jelly'. Straight lines running parallel with the axes of the coordinate system must cross the outer perimeter in at least two places.

What happens if we remove the container walls? A fluid will of course spill out all over the place, whereas an elastic material is only expected to settle a bit more while bulging horizontally out where the walls were before. Jelly on a flat plate is perhaps the best image to have in mind. In spite of the basic simplicity of the problem, there seems to be no simple analytic solution. But if one cannot find the right solution to a problem, it is common practice in physics to redefine the problem to fit a solution which one *can* get! What we can obtain is a solution with no shear stresses (in the chosen coordinates), but the price we pay is that the vertical displacement will not vanish across the bottom of the container, as it ought to.

The equilibrium equations (12.1c) with all shear stresses set to zero, i.e. $\sigma_{xy} = \sigma_{yz} = \sigma_{zx} = 0$, simplify now to

$$\nabla_x \sigma_{xx} = 0, \qquad \nabla_y \sigma_{yy} = 0, \qquad \nabla_z \sigma_{zz} = \rho_0 g_0. \tag{12.12}$$

The first equation says that σ_{xx} does not depend on x, or in other words that σ_{xx} is constant on straight lines parallel with the x-axis. But such lines must always cross the vertical sides, where the x-component of the stress vector, $\sigma_{xx} n_x$, has to vanish, and consequently we must have $\sigma_{xx} = 0$ everywhere. In the same way it follows that $\sigma_{yy} = 0$ everywhere. Finally, the third equation tells us that σ_{zz} is linear in z, and using the condition that $\sigma_{zz} = 0$ for $z = h$ we find

$$\sigma_{zz} = -\rho_0 g_0 (h - z), \tag{12.13}$$

implying that every column carries the weight of the material above it. This result was again to be expected because there are no shear stresses to redistribute the vertical load.

From the inverse Hooke's law (11.16), the non-vanishing strain components are found to be

$$u_{xx} = u_{yy} = \nu \frac{\rho_0 g_0}{E}(h - z), \qquad u_{zz} = -\frac{\rho_0 g_0}{E}(h - z), \tag{12.14}$$

where E is Young's modulus and ν Poisson's ratio. The typical scale of major deformation is in this case

$$D = \frac{E}{\rho_0 g_0}. \tag{12.15}$$

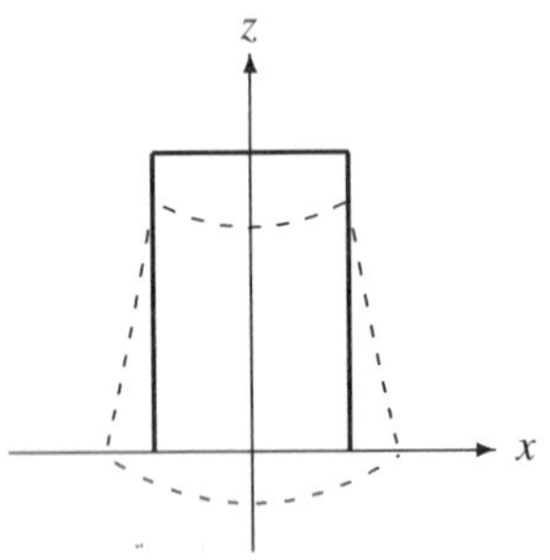

Simple model for the gravitational settling of a block of elastic material ('jelly on a plate'). The model is not capable of fulfilling the boundary condition $u_z = 0$ at $z = 0$ and describes a cylindrical block which partly settles into the supporting surface.

Using that $u_{xx} = \nabla_x u_x$, etc, the strains may be readily integrated, but in doing so, one must remember that all shear strains have to vanish. One may verify that the following displacement field leads to the strains above,

$$\boxed{\begin{aligned} u_x &= \frac{\nu}{D}(h - z)x, \\ u_y &= \frac{\nu}{D}(h - z)y, \\ u_z &= -\frac{1}{2D}\left(h^2 - (h - z)^2 + \nu(a^2 - x^2 - y^2)\right), \end{aligned}} \tag{12.16}$$

where a is a constant. The peculiar last term in u_z is forced upon us by the requirement of vanishing shear strains, u_{xz} and u_{yz}. The most general solution is obtained by adding a small translation of the body in the xy-plane and a small rotation of the body around the z-axis.

The trouble with this solution is that we cannot impose flatness at the bottom, i.e. $u_z = 0$ for $z = 0$, as we would like to do. The vertical displacement is negative everywhere at the bottom except at the circle $x^2 + y^2 = a^2$, where it vanishes. In the xy-plane, however, the horizontal displacement represents a uniform expansion in all horizontal directions with a z-dependent scale factor, which vanishes on top and is maximal at the bottom. Instead of describing the deformation of a block of material sitting on a hard and flat horizontal surface, we have obtained a solution which seems to describe a cylindrical block sinking into the supporting surface!

Adhémar Jean Claude Barré de Saint-Venant (1797–1886). *French engineer. Worked on mechanics, elasticity, hydrostatics and hydrodynamics. Rederived the Navier–Stokes equations in 1843, avoiding Navier's molecular approach, but did not get credited for these equations with his name.*

There can be only one explanation, namely that the initial assumption about the shear-free stress tensor is wrong. What seems to be needed to find a solution including a hard, flat supporting surface is an extra vertical pressure distribution from the supporting surface which is able to 'shore up' the sagging underside of the shear-free solution and make it flat. We expect that this extra pressure will be accompanied by shear stresses, enabling the inner part of the block close to $x = 0$ to carry more than its share of the weight of the material above it, and the outer part at $x = a$ to carry less. This intuition is in fact verified by numerical solution (see section 13.3).

The extra stress distribution can presumably only exert influence on the deformation up to a height of the same order of magnitude as the horizontal dimensions of the block. For a tall block, the shear-free solution may thus be expected to be practically valid everywhere except in a region near the bottom of the same height as the horizontal dimensions of the block. This argument is an application of a rule known as *Saint-Venant's principle*, which states that the effects of a particular application of external forces is only notable near the regions where these forces are applied.

12.3 Bending a beam

Bending a beam by wrenching it at the ends.

Sticks, rods, girders, struts, masts, towers, planks, poles and pipes are all examples of a generic object, which we shall call a *beam*. Geometrically, a beam consists of a bundle of straight parallel lines or *rays*, covering the same cross section in any plane orthogonal to the lines. Physically, the beam is assumed to be made from homogeneous and isotropic elastic material.

Uniform pure bending

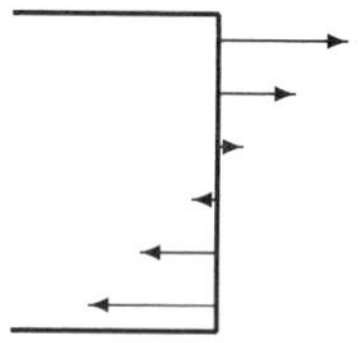

A bending couple may be created by varying normal stresses applied to a terminal.

There are many ways to bend a beam. A cantilever is a beam that is fixed at one end and bends like a horizontal flagpole or a fishing rod. A beam may also be weighed down in the middle like a bridge, but the cleanest way to bend it is probably to grab it close to the ends and wrench it like a pencil so that it adopts a uniformly curved shape. Ideally, in *pure bending*, external stresses should not be applied to the sides of the beam, but only to the terminal cross sections, and on average these stresses should neither stretch nor compress the beam, but only provide external couples (moments) at the terminals. It should be noted that such couples do not require shear stresses, but may be created by normal stresses alone which vary in strength over the terminal cross sections. If you try, you will realize that it is in fact rather hard to bend a pencil in this way. Bending a rubber eraser by pressing it between two fingers is somewhat easier, but tends to add longitudinal compression as well.

The bending of the beam is also assumed to be *uniform*, such that the physical conditions, stresses and strains, will be the same everywhere along the beam. This is only possible if the originally straight beam of length L is deformed to become a section of a (huge) circular ring of radius R with every ray becoming part of a perfect circle. In that case, it is sufficient to consider just a tiny slice of the beam in order to understand uniform bending for a beam of any length. We shall see below that non-uniform bending can also be handled by piecing together little slices with varying radius of curvature. Furthermore, by appealing to Saint-Venant's principle and linearity, we may even calculate the properties of a beam subject to different types of terminal loads by judicious superposition of displacement fields.

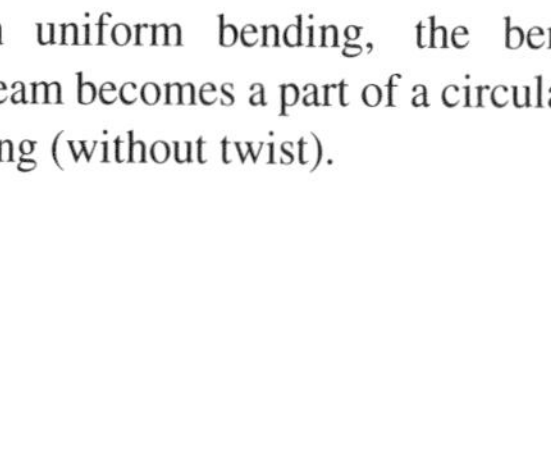

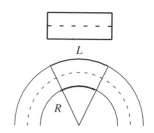

In uniform bending, the bent beam becomes a part of a circular ring (without twist).

Centring the beam

In a Cartesian coordinate system, we align the undeformed beam with the z-axis, and put the terminal cross sections at $z = 0$ and $z = L$. The length L of the beam may be chosen as small as we please. The cross section A in the xy-plane may be of arbitrary shape, but we may, without loss of generality, position the

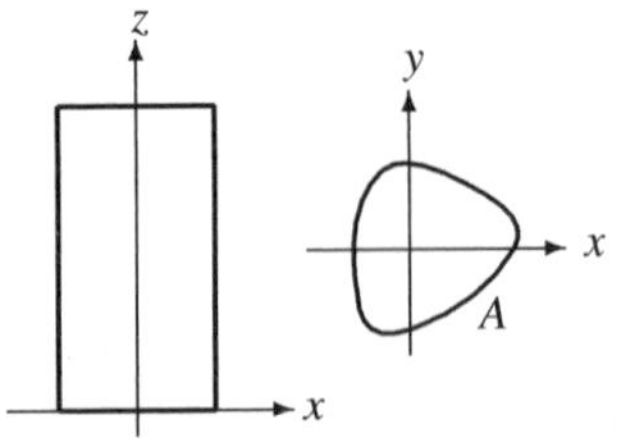

The unbent beam is aligned with the z-axis, and its cross section, A, in the xy-plane is the same for all z. The z-axis goes through the centre of the cross section.

coordinate system in the xy-plane such that its origin coincides with the *centre* of the area, defined by,

$$\int_A x\,dxdy = \int_A y\,dxdy = 0. \tag{12.17}$$

We require in other words that the z-axis, $x = y = 0$, coincides with the *central ray* of the unbent beam. Finally, we fix the remaining degree of freedom by requiring the central ray after bending to be part of a circle in the xz-plane with radius R and its centre on the x-axis at $x = R$. The radius R is obviously the length scale for major deformation.

Shear-free solution

What precisely happens in the beam when it is bent depends on the way the actual stresses are distributed on its terminals, although by Saint-Venant's principle, the details should only matter near the terminals. In the simplest case we may view the beam as a loose bundle of thin elastic strings that do not interact with each other, but are stretched or compressed individually according to their position in the beam without generating shear stresses. Let us fix the central string so that it does not change its length L, when bent into a circle. A simple geometric construction then shows that a nearby ray in position x will change its length to $L' = L(1 - x/R)$, and consequently experience a longitudinal strain,

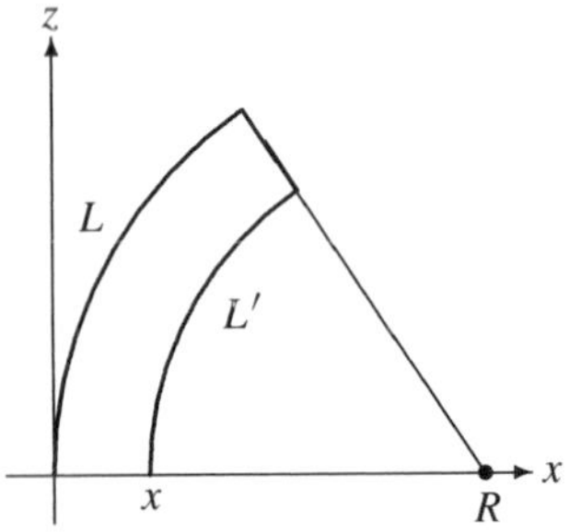

The length of the arc at x must satisfy $L'/(R - x) = L/R$.

$$u_{zz} = \frac{L' - L}{L} = -\frac{x}{R}. \tag{12.18}$$

For negative x the material of the beam is being stretched, while for positive x it is being compressed.

Under the assumption that the bending is done without shear and that there are no forces acting on the sides of the beam, it follows as in the preceding section that $\sigma_{xx} = \sigma_{yy} = 0$. The only non-vanishing stress is $\sigma_{zz} = E u_{zz}$, and the non-vanishing strains are as before found from the inverted Hooke's law (11.16),

$$u_{xx} = u_{yy} = -\nu u_{zz} = \nu\frac{x}{R}, \tag{12.19}$$

where ν is Poisson's ratio. This shows that the material is being stretched horizontally for $x < 0$ and compressed for $x > 0$.

Using $u_{xx} = \nabla_x u_x$, etc, a particular solution is found to be

$$\boxed{\begin{aligned} u_x &= \frac{1}{2R}\left(z^2 + \nu\left(x^2 - y^2\right)\right), \\ u_y &= \frac{\nu}{R}xy, \\ u_z &= -\frac{1}{R}xz. \end{aligned}} \tag{12.20}$$

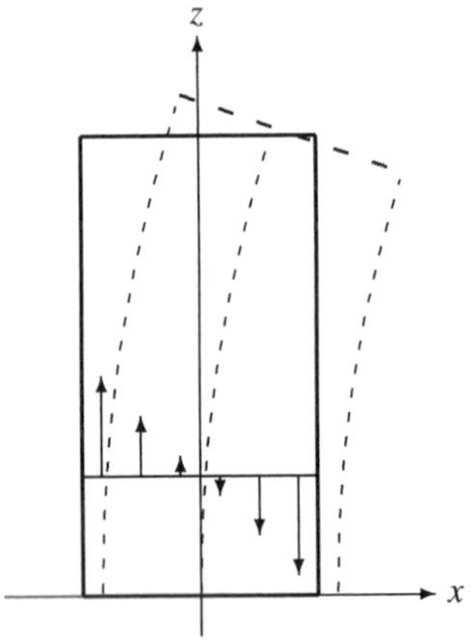

Sketch of the bending of a beam towards positive x-values. The arrows indicate the strain u_{zz}.

The quadratic terms dependent on y and z in u_x are as in the preceding section forced upon us by the requirement of no shear stresses (and strains). In order for displacement gradients to be small, all dimensions of the beam have to be small compared to R. Note that the beam's actual dimensions do not appear in the displacement field, so that this may be considered to be an entirely local solution to the bending problem.

Total force

The only non-vanishing stress component is as mentioned before

$$\sigma_{zz} = E u_{zz} = -\frac{E}{R}x. \tag{12.21}$$

It is a tension for negative x, and we consequently expect the material of the beam to first break down at the extreme point of the cross section opposite to the direction of bending, as common experience also tells us.

The total force acting on a cross section vanishes

$$\mathcal{F}_z = \int_A \sigma_{zz}\,dS_z = -\frac{E}{R}\int_A x\,dxdy = 0, \tag{12.22}$$

because of the conditions (12.17). The total force would of course vanish even if the beam was not centred, because the stress would then be $\sigma_{zz} = -E(x - x_0)/R$ where x_0 is the x-coordinate of the centre.

If the terminal load is not a pure moment, the total force on a cross section will not vanish, but the linearity of the Navier–Cauchy equilibrium equation (12.2) allows us as usual to superpose solutions. A uniform normal force $\mathcal{F}_z$ may, for example, be included by adding the stretching displacement field (11.25) which in the present coordinates becomes,

$$u_x = -\nu\frac{P}{E}\,x, \qquad u_y = -\nu\frac{P}{E}\,y, \qquad u_z = \frac{P}{E}\,z, \tag{12.23}$$

where $P = \mathcal{F}_z/A$. Similarly, uniform shear forces $\mathcal{F}_x$ and $\mathcal{F}_y$ may be included by adding the shear displacement field (11.26), suitably expressed in the present coordinates.

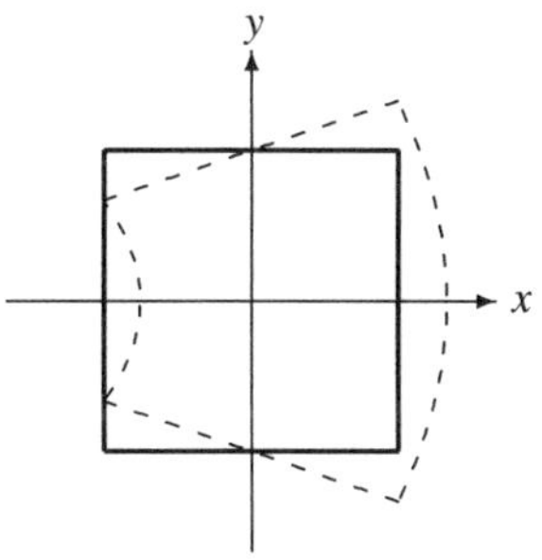

Sketch of the deformation in the xy-plane of a beam with rectangular cross section (exaggerated). This peculiar deformation may easily be observed by bending a rubber eraser.

Total moment

The non-vanishing moments of the longitudinal stress in a cross section are

$$\mathcal{M}_x = \int_A y\sigma_{zz}\,dS_z = -\frac{E}{R}\int_A xy\,dxdy, \tag{12.24}$$

$$\mathcal{M}_y = -\int_A x\sigma_{zz}\,dS_z = \frac{E}{R}\int_A x^2\,dxdy. \tag{12.25}$$

It is often the case that the undeformed beam is mirror symmetric under reflection in the xz-plane or in the yz-plane, so that $\int_A xy\,dxdy = 0$, and this implies that $\mathcal{M}_x = 0$. The important moment is the one in the bending plane, i.e. $\mathcal{M}_y$.

The expression for $\mathcal{M}_y$ is the same as the *Bernoulli–Euler* law,

$$\boxed{\frac{1}{R} = \frac{\mathcal{M}}{EI},} \tag{12.26}$$

where

$$I = \int_A x^2\,dA \tag{12.27}$$

is the 'moment of inertia' of the beam cross section around the direction of $\mathcal{M}$ (analogous to the area moment of the ship's waterline in equation (5.26) on page 67). Note that constant shear or normal stresses, if present, do not contribute to the cross-section moment, and in many engineering applications the Bernoulli–Euler law combined with linearity and Saint-Venant's principle is enough to give a reasonable idea of how much a beam is deformed by external loads.

Rectangular beam: For a rectangular beam with $A = 2a \times 2b$, we get

$$I = \int_{-b}^{b} dy \int_{-a}^{a} x^2 dx = \frac{4}{3}a^3 b. \tag{12.28}$$

It grows more rapidly with the width of the beam in the direction of bending (x) than orthogonally to it (y). This agrees with the common experience that to obtain a given bending radius R it is much harder to bend a thick beam than a thin. In lorries and train wagons flat steel springs are often used to soften the impact forces due to irregularities in the road or rails.

Elliptic beam: For an elliptical beam with axes $2a$ and $2b$ along x and y, the moment of inertia becomes,

$$I = \int_{-b}^{b} dy \int_{-a\sqrt{1-y^2/b^2}}^{a\sqrt{1-y^2/b^2}} x^2\,dx = \frac{4}{3}a^3 b \int_0^1 (1-t^2)^{3/2}\,dt = \frac{\pi}{4}ab^3, \tag{12.29}$$

which is only about half of the rectangular result. An elliptical spring would thus bend about twice the amount of a flat spring.

Threshold for buckling

A walking stick must be chosen with care. Too sturdy, and it will be heavy and unyielding; too slender, and it may buckle or even collapse under the weight you put on it. More generally, all kinds of columns and struts are used to support buildings and other structures, and buckling can lead to dangerous collapse. Luckily, however, buckling does not happen until the forces on the beam terminals exceed a certain threshold, first determined by Euler.

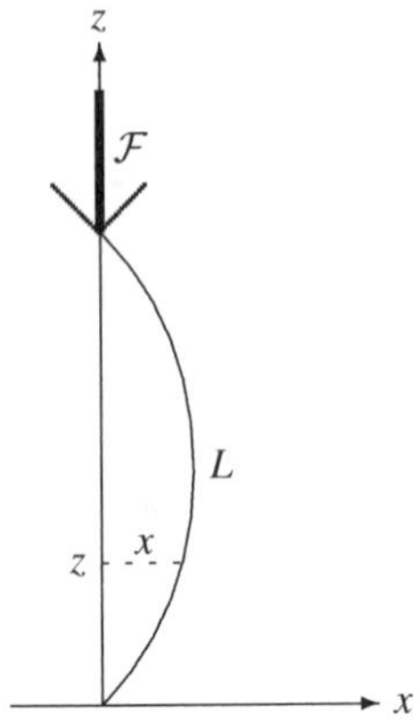

A strut with a longitudinal terminal load may buckle when the force $\mathcal{F}$ exceeds a certain threshold. There is of course an equal and opposite force from the 'ground' at $z = 0$. The local moment exerted in a strut cross section is $\mathcal{M} = x\mathcal{F}$.

Let an undeformed slender beam of length L be placed 'vertically' along the z-axis. Applying a constant terminal force $\mathcal{F}$ along the negative z-direction, it follows from geometry that the moment exerted by the upper part of the deformed beam on a cross section is $\mathcal{M} = x\mathcal{F}$, where x is the 'horizontal' displacement of the cross section at 'height' z. There will also be non-vanishing normal and shear forces at play in the cross section, but as discussed above they do not contribute to the moment. Evidently, the moment is not constant throughout the deformed beam, so this cannot be a uniform pure bending of the beam. We may nevertheless exploit the Bernouilli–Euler law (12.26) and relate the local moment to the local curvature,

$$\frac{1}{R} = \frac{\mathcal{F}}{EI} x. \tag{12.30}$$

This result is valid for any deformation of the terminally loaded beam and shows that the curvature is largest in the middle of the beam and smallest at the ends where x vanishes.

To determine the threshold for buckling, we assume that the beam is only slightly displaced, such that the curvature is given by the double derivative, $1/R \approx -d^2x/dz^2$. The above expression then becomes an equation for the shape,

$$\frac{d^2x}{dz^2} = -k^2 x, \tag{12.31}$$

where $k = \sqrt{\mathcal{F}/EI}$. This is nothing but the standard harmonic equation with 'wavenumber' k, and its general solution is $x = A \sin kz + B \cos kz$ where A and B are constants. Applying the boundary conditions that $x = 0$ for $z = 0$ and $z = L$, the solution becomes,

$$x = A \sin \frac{n\pi z}{L}, \tag{12.32}$$

where n is an arbitrary integer. This shows that $k = n\pi/L$ and since $\mathcal{F} = k^2 EI$ we arrive at Euler's result for the force,

$$\mathcal{F} = n^2\pi^2 \frac{EI}{L^2}. \tag{12.33}$$

This is quite surprising: the slightly displaced shape of a terminally loaded beam is only mechanically stable for certain values of the applied force, corresponding to integer values of n.

Actually, this conclusion cannot be right because everyday experience tells us that we can load a walking stick with a small terminal force and still keep it in stable equilibrium. This is in fact what we do when we lean on it. What must happen is that a small force cannot bend the beam but only compress it longitudinally, an effect we have not taken into account in the above calculation. As the applied force increases, it will eventually reach the threshold value corresponding to $n = 1$ above,

$$\boxed{\mathcal{F}_E = \pi^2 \frac{EI}{L^2},} \tag{12.34}$$

and at this point the longitudinal compression mode becomes unstable and the first buckling solution takes over at the slightest provocation. To prove that this is indeed what takes place requires a stability analysis, which we shall not go into here. In practice only the lowest mode is seen, unless a strong force is rapidly applied, in which case the beam may crumble and collapse completely. That is why you wouldn't choose a straw as your walking stick!

Example 12.3.1 (Wooden stick): A wooden walking stick has length $L = 1$ m and circular cross section of radius $a = 1$ cm. Taking Young's modulus $E = 10^{10}$ Pa and using (12.29), the buckling threshold becomes $\mathcal{F}_E = 775$ N, corresponding to the weight of 79 kg. If you weigh more, it would be prudent to choose a slightly thicker stick. Since the threshold grows as the fourth power of the radius, increasing the radius by just one millimetre raises the threshold mass to 116 kg which is sufficient for most people.

∗ Shape of a relaxed stringed bow

If the applied force continues to increase beyond the threshold value, the displacement grows rapidly and the approximation used in Euler's calculation ceases to be valid. A relaxed stringed bow may, for example, be viewed as a beam that has been brought beyond the buckling threshold and is kept in equilibrium by the tension in the bowstring. In this case, the overall displacement of the beam is not small, even if the local strains are small everywhere.

To determine the equilibrium shape we may directly use the description of planar curve geometry introduced on page 102,

$$\frac{dx}{ds} = \cos\theta, \qquad \frac{dz}{ds} = \sin\theta, \qquad \frac{d\theta}{ds} = \frac{1}{R}. \tag{12.35}$$

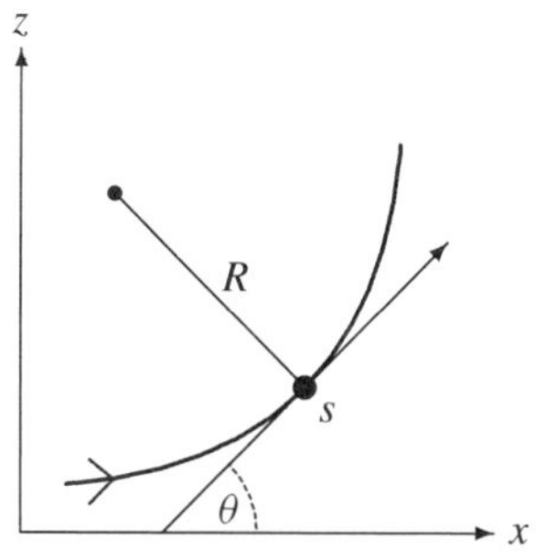

The geometry of a planar curve. The curve is parametrized by the arc length s along the curve. A small change in s generates a change in the elevation angle θ determined by the local radius of curvature. Here the radius of curvature is positive.

Differentiating the last equation once more and using (12.30), we obtain the equilibrium equation,

$$\boxed{\frac{d^2\theta}{ds^2} = k^2\cos\theta} \tag{12.36}$$

where as before $k = \sqrt{\mathcal{F}/EI}$. This is the equation for a mathematical pendulum in a slight disguise.

Integrating the pendulum equation with the boundary condition that $d\theta/ds = 1/R = 0$ for $x = 0$, we find

$$\left(\frac{d\theta}{ds}\right)^2 = 2k^2(\sin\theta - \sin\alpha), \tag{12.37}$$

where α is the elevation angle at the ends of the beam. The inverse derivative $ds/d\theta$ may now be integrated to yield the length L of the beam,

$$L = \frac{1}{k}\int_{\alpha}^{\pi-\alpha} \frac{d\theta}{\sqrt{2(\sin\theta - \sin\alpha)}}. \tag{12.38}$$

This is an elliptic integral which is easy to evaluate numerically. It provides a relation between the ratio $\mathcal{F}/F_E = (Lk/\pi)^2$ and α (see figure 12.1).

Similarly, we may calculate $dx/d\theta$ and $dz/d\theta$ and integrate to find the coordinates of the displaced beam,

$$x = \frac{1}{k}\sqrt{2(\sin\theta - \sin\alpha)}, \qquad z = \frac{1}{k}\int_{\alpha}^{\theta} \frac{\sin\theta'}{\sqrt{2(\sin\theta' - \sin\alpha)}}\,d\theta'. \tag{12.39}$$

Together these expressions define the shape of the beam parametrized by θ. In the margin three different bow shapes are plotted.

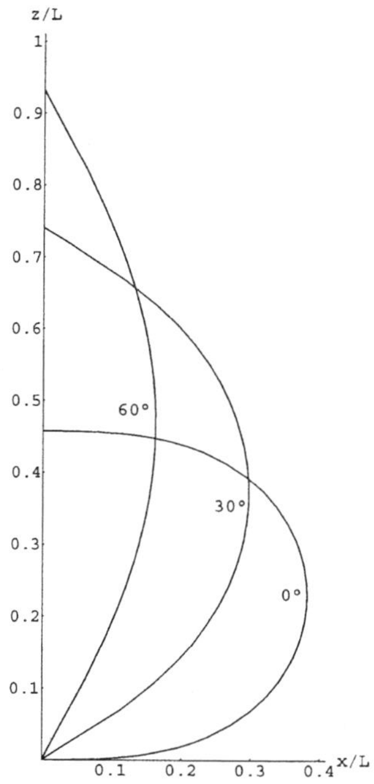

Plots of three different relaxed bow shapes. The numbers indicate the bow elevation angle α in degrees.

Example 12.3.2 (Steel bow): A longbow is constructed from a rectangular steel beam of length $L = 200$ cm with dimensions $2a = 10$ mm and $2b = 20$ mm. The bow is stringed with an opening angle of 30 ° corresponding to $\alpha = 60$ °. The maximal string distance from the bow becomes $d \approx 32$ cm and the stringed height $h \approx 186$ cm. The moment of inertia becomes $I \approx 1.67 \times 10^{-9}$ m^4, and taking $E = 2 \times 10^{11}$ Pa the Euler threshold becomes $\mathcal{F}_E \approx 822$ N. Since $\mathcal{F}/F_E \approx 1.035$ we find $\mathcal{F} = 851$ N, corresponding to a weight of 87 kg. It would take 'Little John's' strength to fit the string to this bow and shoot an arrow. If the thickness were reduced by 20%, it would only take half the strength to string it.

12.4 Twisting a shaft

The drive shaft in older cars connects the gear box to the differential and transmits engine power to the rear wheels. In characterizing engine performance, maximum torque is often quoted, because it creates the largest shear force between wheels and road and therefore maximal acceleration, barring wheel-spin. Although the shaft is made from steel, it will nevertheless undergo a tiny deformation, a *torsion* or *twist*.

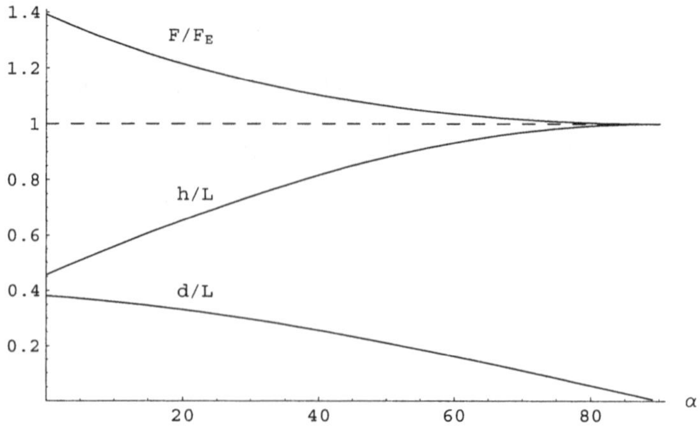

Figure 12.1. Plots of the force ratio $\mathcal{F}/\mathcal{F}_E$, the stringed height h/L and the maximal string displacement d/L, as functions of the bow elevation angle α (in degrees).

Pure torsion

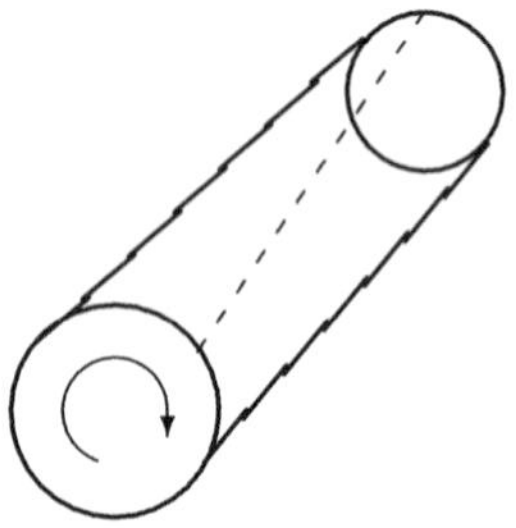

Pure torsion consists of rotating every cross section by a fixed amount per unit of length.

The shaft is assumed to be a circular beam with radius a and axis coinciding with the z-axis. The deformation is said to be a *pure torsion* if the shaft's material is rotated by a constant amount τ per unit of length, such that a given cross section at the position z is rotated by an angle τz relative to the cross section at $z = 0$. The constant τ which measures the rotation angle per unit of length is sometimes called the *torsion angle*.

The uniform nature of pure torsion allows us to consider just a small slice of the shaft of length L which is only twisted through a tiny angle, $\tau L \ll 1$. Since the physical conditions are the same in all such slices, we can later put them together to make a shaft of any length. To lowest order in the angle τz, the displacement field in the slice becomes

$$\boldsymbol{u} = \tau z \boldsymbol{e}_z \times \boldsymbol{x} = \tau z(-y, x, 0). \tag{12.40}$$

Not surprisingly, it is purely tangential and is always much smaller than the radius, a, of the shaft because $\tau L \ll 1$.

Strain and stress

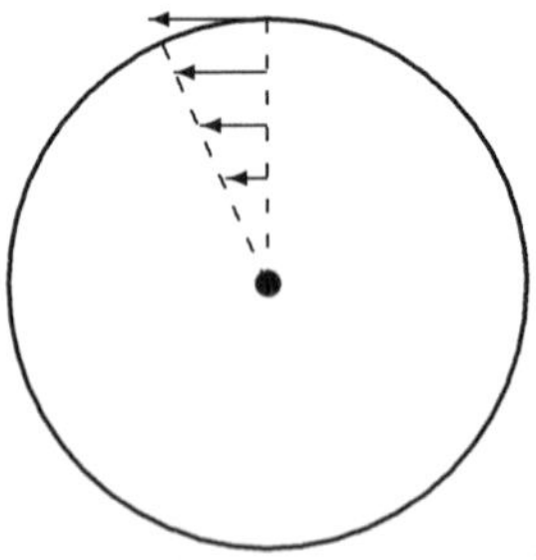

The displacement field for a rotation through a tiny angle τz (exaggerated here) is purely tangential and grows linearly with the radial distance.

From the displacement field we calculate displacement gradient tensor

$$\{\nabla_j u_i\} = \begin{pmatrix} 0 & -\tau z & -\tau y \\ \tau z & 0 & \tau x \\ 0 & 0 & 0 \end{pmatrix}. \tag{12.41}$$

For this matrix to be small, we must also require $\tau a \ll 1$, or in other words that the twist must be small over a length of the shaft comparable to its radius.

The only non-vanishing strains are,

$$u_{xz} = -\tfrac{1}{2}\tau y, \qquad u_{yz} = \tfrac{1}{2}\tau x. \tag{12.42}$$

The corresponding stresses are obtained from Hooke's law (11.8),

$$\sigma_{xz} = \sigma_{zx} = -\mu\tau y, \qquad \sigma_{yz} = \sigma_{zy} = \mu\tau x. \tag{12.43}$$

Inserting these stresses into the equilibrium equations (12.1c), it is seen that it is trivially fulfilled. At the cylindrical surface of the shaft, the normal is $(x, y, 0)/a$, and the stress vector vanishes, i.e. $\sigma_{zx}x/a + \sigma_{zy}y/a = 0$, as it should when there are no external forces acting there.

This solution was first obtained by Coulomb in 1787, whereas the corresponding solution for rods with non-circular cross section (see [37, p. 59] or [65, p. 109]) was obtained by Saint-Venant in 1855.

In order to realize a pure torsion, the correct stress distribution (12.43) must be applied to the ends of the shaft. Applying a different stress distribution by, for example, grabbing one end of the shaft with a monkey-wrench, leads to a different solution near the end, but the pure torsion solution should according to Saint-Venant's principle still be valid far away from the ends.

Torque

In any cross section we may calculate the total moment of force around the shaft axis, in this context called the *torque*. On a surface element $d\boldsymbol{S}$, the moment is $d\boldsymbol{\mathcal{M}} = \boldsymbol{x} \times d\boldsymbol{\mathcal{F}} = \boldsymbol{x} \times \boldsymbol{\sigma} \cdot d\boldsymbol{S}$. Since the cross section lies in the xy-plane, we have $d\boldsymbol{S} = \boldsymbol{e}_z dxdy$, and the z-component of the moment becomes,

$$\begin{aligned}\mathcal{M}_z &= \int_A (x\sigma_{yz} - y\sigma_{xz})\,dxdy = \int_A \mu\tau(x^2+y^2)dxdy \\ &= \int_0^a \mu\tau r^2 \cdot 2\pi r dr = \frac{\pi}{2}\mu\tau a^4. \end{aligned} \tag{12.44}$$

The quantity

$$\boxed{C = \frac{\mathcal{M}_z}{\tau} = \frac{\pi}{2}\mu a^4,} \tag{12.45}$$

is called the *torsional rigidity* of the shaft. The torsional rigidity depends only the radius of the shaft and the shear modulus but not on the applied torque. Knowing the torsional rigidity and the torque $\mathcal{M}$, one may calculate the torsion angle $\tau = \mathcal{M}/C$, and conversely.

Transmitted power

If the shaft rotates with constant angular velocity Ω, the material at the point (x, y, z) will have velocity $\boldsymbol{v}(x, y) = \Omega\boldsymbol{e}_z \times \boldsymbol{x} = \Omega(-y, x, 0)$. The shear stresses acting on an element of the cross section, $d\boldsymbol{S} = \boldsymbol{e}_z dxdy$, will transmit a *power* (i.e. work per unit of time) of $dP = \boldsymbol{v} \cdot d\boldsymbol{\mathcal{F}} = \boldsymbol{v} \cdot \boldsymbol{\sigma} \cdot d\boldsymbol{S}$. Integrating over the cross section the total power becomes,

$$P = \int_A \boldsymbol{v} \cdot \boldsymbol{\sigma} \cdot d\boldsymbol{S} = \int_A \Omega(x\sigma_{yz} - y\sigma_{xz})dxdy = \Omega\mathcal{M}_z. \tag{12.46}$$

As the derivation shows, this relation $P = \boldsymbol{\Omega} \cdot \boldsymbol{\mathcal{M}}$ is generally valid.

Example 12.4.1 (Car engine): The typical torque delivered by a family car engine can be of the order of 100 Nm. If the shaft rotates with 3000 rpm, corresponding to an angular velocity of $\Omega \approx 314\ \text{s}^{-1}$, the transmitted power is about 31.4 kW, or 42 horsepower. For a drive shaft made of steel with radius $a = 2$ cm, the torsional rigidity is $C \approx 2 \times 10^4\ \text{Nm}^2$. In direct drive without gearing, the torsion angle becomes $\tau \approx 0.005\ \text{m}^{-1} = 0.3\,°\ \text{m}^{-1}$. For a car with rear-wheel drive, the length of the drive shaft may be about 2 m, and the total twist amounts to about 0.6 °. The maximal shear stress in the material is $\mu\tau a \approx 8 \times 10^6$ Pa = 80 bar at the rim of the shaft.

12.5 Tube under pressure

Elastic tubes carrying fluids under pressure are found everywhere, in living organisms and in machines, not forgetting the short moments of intense pressure in the barrel of a gun or canon. How much does a tube expand because of the pressure, and how is the deformation distributed? What are the stresses in the material and where will it tend to break down?

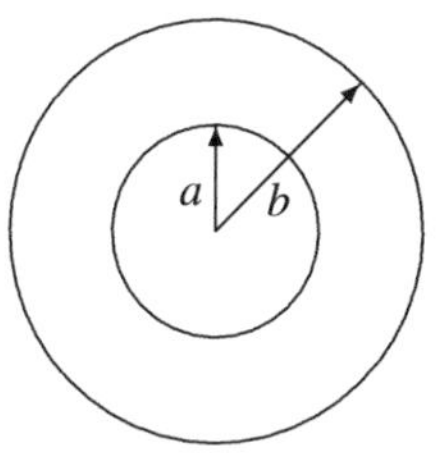

Tube cross section.

Uniform radial displacement

The ideal tube is a beam in the shape of a circular cylinder with inner radius a, outer radius b and length L, made from homogeneous and isotropic elastic material. When subjected to a uniform internal pressure, the tube is expected to expand radially and perhaps also contract longitudinally. The latter may be prevented by clamping the ends of the tube, so to simplify matters we shall assume that the displacement field is uniformly radial, of the form[1]

$$\boldsymbol{u} = u_r(r)\,\boldsymbol{e}_r, \tag{12.47}$$

where $u_r(r)$ is only a function of the radial distance $r = \sqrt{x^2 + y^2}$, and $\boldsymbol{e}_r$ is the radial unit vector at the point (x, y, z),

$$\boldsymbol{e}_r = \frac{(x, y, 0)}{r}. \tag{12.48}$$

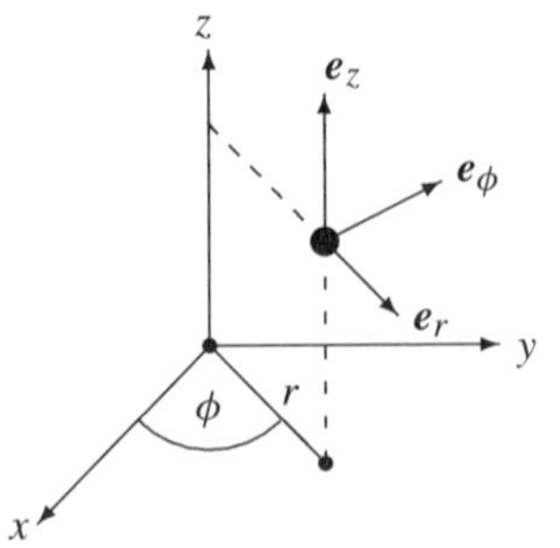

Cylindrical coordinates and basis vectors.

It is tempting here to introduce true cylindrical coordinates, (r, ϕ, z), instead of the Cartesian coordinates, (x, y, z), but although more systematic it would in fact make the following analysis harder. The only other element we need here is the angular unit vector

$$\boldsymbol{e}_\phi = \frac{(-y, x, 0)}{r}, \tag{12.49}$$

which is orthogonal to both $\boldsymbol{e}_r$ and $\boldsymbol{e}_z = (0, 0, 1)$. The three unit vectors form together a local orthogonal basis for cylindrical geometry (see appendix B for details).

Displacement gradients

In Cartesian coordinates, the displacement field takes the form

$$u_x = \frac{x}{r} u_r(r) \tag{12.50}$$

$$u_y = \frac{y}{r} u_r(r). \tag{12.51}$$

It is then straightforward to calculate the non-vanishing displacement gradients

$$\nabla_x u_x = \frac{x^2}{r}\frac{d}{dr}\left(\frac{u_r}{r}\right) + \frac{u_r}{r} = \frac{x^2}{r^2}\frac{du_r}{dr} + \frac{y^2}{r^2}\frac{u_r}{r}, \tag{12.52a}$$

$$\nabla_y u_y = \frac{y^2}{r}\frac{d}{dr}\left(\frac{u_r}{r}\right) + \frac{u_r}{r} = \frac{y^2}{r^2}\frac{du_r}{dr} + \frac{x^2}{r^2}\frac{u_r}{r}, \tag{12.52b}$$

$$\nabla_x u_y = \nabla_y u_x = \frac{xy}{r}\frac{d}{dr}\left(\frac{u_r}{r}\right) = \frac{xy}{r^2}\frac{du_r}{dr} - \frac{xy}{r^2}\frac{u_r}{r}, \tag{12.52c}$$

where we have also used $\partial r/\partial x = x/r$, etc. Adding the two first equations we obtain the divergence of the displacement field,

$$\nabla \cdot \boldsymbol{u} = \frac{du_r}{dr} + \frac{u_r}{r} = \frac{1}{r}\frac{d(ru_r)}{dr}, \tag{12.53}$$

where the last expression has been rewritten for later convenience.

Equilibrium equation

Since $\nabla r = \boldsymbol{e}_r$, it follows from the radial assumption (12.47) that the displacement field may be written as the gradient of another field

$$\boldsymbol{u} = \nabla\psi(r), \qquad \psi(r) = \int u_r(r)\,dr. \tag{12.54}$$

Now the gradient of its divergence becomes,

$$\nabla_i \nabla \cdot \boldsymbol{u} = \nabla_i \nabla^2 \psi = \nabla^2 \nabla_i \psi = \nabla^2 u_i, \tag{12.55}$$

[1]The index on u_r is redundant and could be dropped, but we keep it systematically so as to remind ourselves that it is the radial component of the displacement field.

and the Navier–Cauchy equation (12.2) takes the much simpler form

$$\boldsymbol{f} + (2\mu + \lambda)\nabla\nabla \cdot \boldsymbol{u} = \boldsymbol{0}. \tag{12.56}$$

Using the expression (12.53) for the divergence, the Navier–Cauchy equation becomes,

$$\boldsymbol{f} + (2\mu + \lambda)\boldsymbol{e}_r \frac{d}{dr}\left(\frac{1}{r}\frac{d(ru_r)}{dr}\right) = \boldsymbol{0}. \tag{12.57}$$

This shows that the body force density, if present, must also be radial,

$$\boldsymbol{f} = f_r(r)\boldsymbol{e}_r, \tag{12.58}$$

and we finally arrive at the ordinary second-order differential equation in r,

$$\boxed{f_r + (\lambda + 2\mu)\frac{d}{dr}\left(\frac{1}{r}\frac{d(ru_r)}{dr}\right) = 0.} \tag{12.59}$$

Given a radial body force, this equation may be integrated to yield the radial displacement.

General solution without body forces

In the simplest case, $f_r = 0$, we find immediately

$$\frac{1}{r}\frac{d(ru_r)}{dr} = 2A, \tag{12.60}$$

where A is a constant. Integrating once more we obtain

$$\boxed{u_r(r) = Ar + \frac{B}{r},} \tag{12.61}$$

where B is another constant. These constants will be determined by the boundary conditions imposed on the tube.

Strain and stress

Expressed in the tensor product notation (2.10), the displacement gradients (12.52) may be compactly written

$$\nabla\boldsymbol{u} = \frac{du_r}{dr}\boldsymbol{e}_r\boldsymbol{e}_r + \frac{u_r}{r}\boldsymbol{e}_\phi\boldsymbol{e}_\phi.$$

Since the right-hand side is a symmetric matrix, it is identical to Cauchy's strain tensor, which accordingly has only two non-vanishing projections on the basis vectors,

$$u_{rr} = \frac{du_r}{dr} = A - \frac{B}{r^2}, \tag{12.62}$$

$$u_{\phi\phi} = \frac{u_r}{r} = A + \frac{B}{r^2}. \tag{12.63}$$

Finally, the non-vanishing stress tensor components are found from Hooke's law (11.8) by projecting on the basis vectors

$$\sigma_{rr} = 2\mu u_{rr} + \lambda(u_{rr} + u_{\phi\phi}) = 2A(\lambda + \mu) - \frac{2\mu B}{r^2}, \tag{12.64a}$$

$$\sigma_{\phi\phi} = 2\mu u_{\phi\phi} + \lambda(u_{rr} + u_{\phi\phi}) = 2A(\lambda + \mu) + \frac{2\mu B}{r^2}, \tag{12.64b}$$

$$\sigma_{zz} = \lambda(u_{rr} + u_{\phi\phi}) = 2A\lambda. \tag{12.64c}$$

Here we have used that the trace of the strain tensor is independent of the basis, so that $\sum_k u_{kk} = u_{xx} + u_{yy} + u_{zz} = u_{rr} + u_{\phi\phi} + u_{zz}$. One should note that a longitudinal stress, σ_{zz}, appears as a consequence of the fixed clamps on the ends of the cylinder.

Solution for the pressurized clamped tube

The boundary conditions are $\sigma_{rr} = -P$ at the inside surface $r = a$ and $\sigma_{rr} = 0$ at the outside surface $r = b$. The minus sign may be a bit surprising, but remember that the normal to the inside surface of the tube is in the direction of $-\boldsymbol{e}_r$, so that the stress vector $\sigma_{rr}(-\boldsymbol{e}_r) = P\boldsymbol{e}_r$ points in the positive radial direction, as it should. Solving the boundary conditions for A and B,

$$2A(\lambda + \mu) - \frac{2\mu B}{a^2} = -P \tag{12.65}$$

$$2A(\lambda + \mu) - \frac{2\mu B}{b^2} = 0 \tag{12.66}$$

we find the integration constants,

$$A = \frac{1}{2(\lambda + \mu)} \frac{a^2}{b^2 - a^2} P = (1 + \nu)(1 - 2\nu) \frac{a^2}{b^2 - a^2} \frac{P}{E}, \tag{12.67a}$$

$$B = \frac{1}{2\mu} \frac{a^2 b^2}{b^2 - a^2} P = (1 + \nu) \frac{a^2 b^2}{b^2 - a^2} \frac{P}{E}, \tag{12.67b}$$

where E is Young's modulus and ν Poisson's ratio.

Displacement field: The displacement field becomes

$$u_r(r) = (1 + \nu) \frac{a^2}{b^2 - a^2} \left((1 - 2\nu) r + \frac{b^2}{r} \right) \frac{P}{E}. \tag{12.68}$$

Since $\nu \leq 1/2$, it is always positive and monotonically decreasing. It reaches its maximum at the inner surface, $r = a$, confirming the intuition that the pressure in the tube should push the innermost material farthest away from its original position.

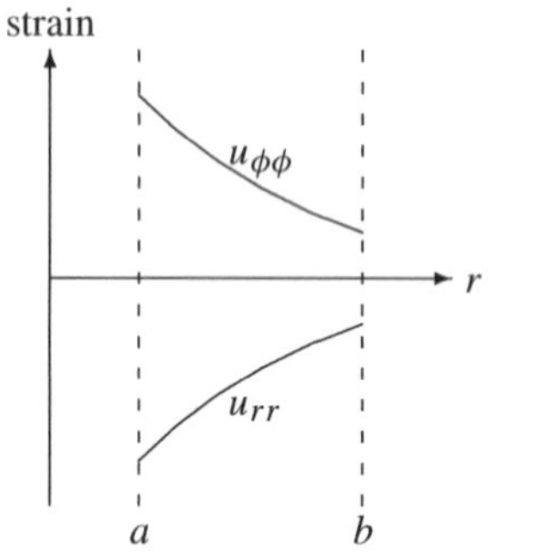

Sketch of strain components in the tube.

Strain tensor: The non-vanishing strain tensor components become

$$u_{rr} = (1 + \nu) \frac{a^2}{b^2 - a^2} \left(1 - 2\nu - \frac{b^2}{r^2} \right) \frac{P}{E}, \tag{12.69a}$$

$$u_{\phi\phi} = (1 + \nu) \frac{a^2}{b^2 - a^2} \left(1 - 2\nu + \frac{b^2}{r^2} \right) \frac{P}{E}. \tag{12.69b}$$

For normal materials with $0 < \nu \leq 1/2$, the radial strain u_{rr} is negative, corresponding to a compression of the material, whereas the tangential strain $u_{\phi\phi}$ is always positive, corresponding to an extension. There is no longitudinal strain because of the clamping of the ends of the tube.

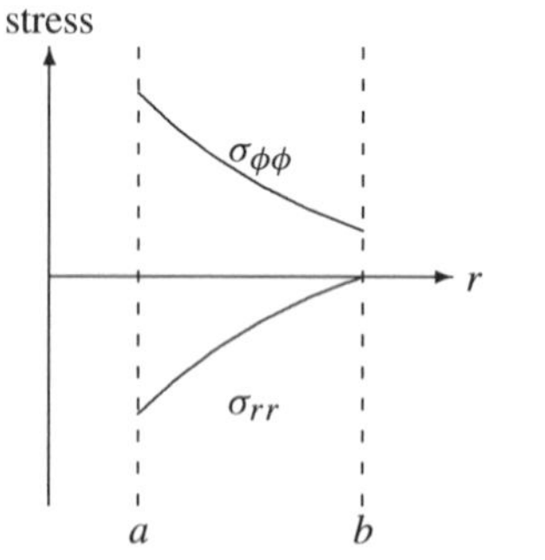

Sketch of stress components in the tube.

The scale of the strain is again set by the ratio P/E. For normal materials under normal pressures, for example an iron pipe with $E \approx 1$ Mbar subject to a water pressure of a few bars, the strain is only of the order of parts per million, whereas the strains in the walls of your garden hose or the arteries in your body are much larger. When the walls become thin, i.e. for $d = b - a \ll a$, the strains grow stronger because of the denominator $b^2 - a^2 \approx 2da$, and actually diverge towards infinity in the limit. This is in complete agreement with agreement with our understanding that the walls of a tube need to be of a certain thickness to withstand the internal pressure.

Stress tensor: The non-vanishing stress tensor components become

$$\sigma_{rr} = -\frac{a^2}{b^2 - a^2} \left(\frac{b^2}{r^2} - 1 \right) P, \tag{12.70a}$$

$$\sigma_{\phi\phi} = \frac{a^2}{b^2 - a^2} \left(\frac{b^2}{r^2} + 1 \right) P, \tag{12.70b}$$

$$\sigma_{zz} = 2\nu \frac{a^2}{b^2 - a^2} P. \tag{12.70c}$$

Note that the transversal stresses, σ_{rr} and $\sigma_{\phi\phi}$, are independent of the material properties of the tube, E and ν, and that the longitudinal stress, σ_{zz}, only depends on Poisson's ratio, ν.

Pressure: The radial pressure, $p_r = -\sigma_{rr}$ can never become larger than P, because we may write

$$\frac{p_r}{P} = \frac{b^2 - r^2}{b^2 - a^2}\frac{a^2}{r^2}, \tag{12.71}$$

which is the product of two factors, both smaller than unity for $a < r < b$. The tangential pressure $p_\phi = -\sigma_{\phi\phi}$ and the longitudinal pressure $p_z = -\sigma_{zz}$ are both negative (tensions), and can become large for thin-walled tubes. The average pressure

$$p = \frac{1}{3}(p_r + p_\phi + p_z) = -\frac{2}{3}(1+\nu)\frac{a^2}{b^2 - a^2}P \tag{12.72}$$

is also negative and like the longitudinal pressure is constant throughout the material. Note that the average pressure is not continuous with the pressure outside the tube. This confirms the suspicion voiced on page 115 that the pressure may behave differently in a solid with shear stresses than the pressure in a fluid at rest, which has to be continuous across boundaries in the absence of surface tension.

Blowup: A tube under pressure develops cracks and eventually blows up if the material is extended beyond a certain limit. Compression does not matter, except for enormous pressures. The point where the tube breaks is primarily determined by the point of maximal local tension and extension. As we have seen, this occurs at the inside of the tube for $r = a$ in the tangential direction. A crack will develop where the material has a small weakness, and the tube then blows up from the inside!

Unclamped tube

The constancy of the longitudinal tension (12.70c) permits us to solve the case of an unclamped tube by superposing the above solution with the displacement field for uniform stretching (11.25). In the cylindrical basis the field of uniform stretching becomes (after interchanging x and z)

$$u_r = -\nu r\frac{Q}{E}, \qquad u_z = z\frac{Q}{E}, \tag{12.73}$$

where Q is the tension applied to the ends. Choosing Q equal to the longitudinal tension (12.70c) in the clamped tube,

$$Q = 2\nu\frac{a^2}{b^2 - a^2}P, \tag{12.74}$$

and subtracting the stretching field from the clamped tube field (12.68), we find for the unclamped tube

$$u_r = \frac{a^2}{b^2 - a^2}\left((1-\nu)r + (1+\nu)\frac{b^2}{r}\right)\frac{P}{E}, \tag{12.75a}$$

$$u_z = -2\nu\frac{a^2}{b^2 - a^2}z\frac{P}{E}. \tag{12.75b}$$

The strains are likewise obtained from the clamped strains (12.69a) by subtracting the strains for uniform stretching, and we get

$$u_{rr} = \frac{a^2}{b^2 - a^2}\left(1 - \nu - (1+\nu)\frac{b^2}{r^2}\right)\frac{P}{E}, \tag{12.76a}$$

$$u_{\phi\phi} = \frac{a^2}{b^2 - a^2}\left(1 - \nu + (1+\nu)\frac{b^2}{r^2}\right)\frac{P}{E}, \tag{12.76b}$$

$$u_{zz} = -2\nu\frac{a^2}{b^2 - a^2}\frac{P}{E}. \tag{12.76c}$$

The superposition principle guarantees that the radial and tangential stresses are the same as before and given by (12.70), while the longitudinal stress now vanishes, $\sigma_{zz} = 0$.

Thin wall approximation

Most tubes have thin walls relative to their radius. Let us introduce the wall thickness, $d = b - a$, and the radial distance, $s = r - a$, from the inner wall. In the thin-wall approximation, these quantities are small compared to a, and we obtain the following expressions to leading order for the unclamped tube.

The radial displacement field is constant in the material whereas the longitudinal one is linear in z,

$$u_r \approx a\frac{a}{d}\frac{P}{E}, \qquad u_z \approx -z\nu\frac{a}{d}\frac{P}{E}. \tag{12.77}$$

The corresponding strains become

$$u_{rr} \approx -\nu\frac{a}{d}\frac{P}{E}, \qquad u_{\phi\phi} \approx \frac{a}{d}\frac{P}{E}, \qquad u_{zz} \approx -2\nu\frac{a}{d}\frac{P}{E}. \tag{12.78a}$$

The strains all diverge for $d \to 0$, and the condition for small strains is now $P/E \ll d/a$. The ratio a/d amplifies the strains beyond naive estimates. Finally, we get the non-vanishing stresses

$$\sigma_{rr} \approx -\left(1 - \frac{s}{d}\right)P, \tag{12.79a}$$

$$\sigma_{\phi\phi} \approx \frac{a}{d}P. \tag{12.79b}$$

The radial pressure $p_r = -\sigma_{rr}$ varies between 0 and P as it should when s ranges from 0 to d. It is always positive and of order P, whereas the tangential tension $p_\phi = -\sigma_{\phi\phi}$ diverges for $d \to 0$. Blowups always happen because the tangential tension becomes excessive.

Problems

12.1 Show that Navier's equation of equilibrium may be written as

$$\nabla^2 \boldsymbol{u} + \frac{1}{1-2\nu}\nabla\nabla\cdot\boldsymbol{u} = -\frac{1}{\mu}\boldsymbol{f}, \tag{12.80}$$

where ν is Poisson's ratio.

12.2 A body made from isotropic elastic material is subjected to a body force in the z-direction, $f_z = kxy$. Show that the displacement field

$$u_x = Ax^2yz, \qquad u_y = Bxy^2z, \qquad u_z = Cxyz^2, \tag{12.81}$$

satisfies the equations of mechanical equilibrium for suitable values of A, B and C.

12.3 A certain gun has a steel barrel of length of $L = 1$ m, a bore diameter of $2a = 1$ cm. The charge of gunpowder has length $x_0 = 1$ cm and density $\rho_0 = 1\ \mathrm{g\,cm^{-3}}$. The bullet in front of the charge has mass $m = 5$ g. The expansion of the ideal gases left by the explosion of the charge at $t = 0$ is assumed to be isentropic with index $\gamma = 7/5$. **(a)** Determine the velocity $\dot{x}$ as a function of x for a bullet starting at rest from $x = x_0$. **(b)** Calculate the pressure just after the explosion and when the bullet leaves the muzzle with a velocity of $U = 800\ \mathrm{m\,s^{-1}}$. **(c)** Calculate the initial and final temperatures when the average molar mass of the gases is $M_{mol} = 30\ \mathrm{g\,mol^{-1}}$. **(d)** Calculate the maximal strains in the steel on the inside of the barrel when it has thickness $d = b - a = 5$ mm and compare with the tensile strength of the steel. Will the barrel blow up?

12.4 Show that the most general solution to the uniform shear-free bending of a beam is

$$u_x = a_x - \phi_z y + \phi_y z - \alpha\nu x + \tfrac{1}{2}\beta_x\left(z^2 - \nu(x^2 - y^2)\right) - \beta_y \nu xy, \tag{12.82a}$$

$$u_y = a_y + \phi_z x - \phi_x z - \alpha\nu y + \tfrac{1}{2}\beta_y\left(z^2 - \nu(y^2 - x^2)\right) - \beta_x \nu xy, \tag{12.82b}$$

$$u_z = a_z - \phi_y x + \phi_x y + \alpha z - \beta_x xz - \beta_y yz, \tag{12.82c}$$

and interpret the coefficients.

12.5 Calculate the displacement, strain and stress for an evacuated tube with fixed ends subject to an external pressure P.

12.6 A massive cylindrical body with radius a and constant density ρ_0 rotates around its axis with constant angular frequency Ω. **(a)** Find the centrifugal force density in cylindrical coordinates rotating with the cylinder. **(b)** Calculate the displacement for the case where the ends of the cylinder are clamped to prevent change in length and the sides of the cylinder are free. **(c)** Show that the tangential strain always corresponds to an expansion, whereas the radial strain corresponds to an expansion close to the centre and a compression close to the rim. Find the point, where the radial strain vanishes. **(d)** Where will the breakdown happen?

12.7 Show that a shift in the x-coordinate, $x \to x - \alpha$, in the shear-free bending field (12.20) corresponds to adding in a uniform stretching deformation (plus a simple translation).

13 Computational elastostatics

Historically, almost all of the insights into elasticity were obtained by means of analytic calculations, carried out by some of the best scientists of the time using the most advanced methods available to them, sometimes even inventing new mathematical concepts and methods along the way. Textbooks on the theory of elasticity are often hard to read because of their demands on the reader for command of mathematics [44, 27, 50, 17, 66].

In the last half of the twentieth century, the development of the digital computer has changed the character of this field completely. Faced with a problem in elastostatics, modern engineers quickly turn to numerical computation. The demand for prompt solutions to design problems has over the years evolved these numerical methods into a fine art, and numerous commercial and public domain programs are now available to assist engineers in understanding the elastic properties of their constructions.

In this chapter we shall illustrate how it is possible to solve a concrete problem numerically, providing sufficient detail that a computer program can be implemented. It is not the intention here to expose the wealth of tricks of the trade, but just present the basic reasoning behind the numerical approach and the various steps that must be carried out in order to make a successful numerical simulation. First, one must decide on the field equations and boundary conditions that should be implemented, and what simplifications can be made to these from the outset. Second, the infinity of points in continuous space must be replaced by a finite set, often a regular grid or lattice, and the fundamental equations must be approximated on this set. Third, a method must be adopted for an iterative approach towards the desired solution, and finally one needs to choose convergence criteria that enable one to monitor the progress of the computation and calculate error estimates that give confidence in the solution.

13.1 Relaxing towards equilibrium

As we do not, from the outset, know the solution to the problem we wish to solve by numerical means, we must begin by making an educated guess about the initial displacement field. This guess should preferably satisfy the boundary conditions, but unless we are incredibly lucky it will fail to satisfy the mechanical equilibrium equation, resulting in a non-vanishing effective body force $f_i^* = f_i + \sum_j \nabla_j \sigma_{ij}$. The idea is now to create an iterative procedure which through a sequence of tiny displacements $\delta \boldsymbol{u}$ will proceed from an arbitrary initial state towards the desired equilibrium state, satisfying $\boldsymbol{f}^* = \boldsymbol{0}$ with the right boundary conditions.

Work and relaxation

It was shown in section 11.4 that the (potential) energy of a displaced elastic body in an external gravitational field $\boldsymbol{f} = \rho \boldsymbol{g}$ is the sum of gravitational and elastic contributions,

$$\mathcal{E} = -\int_V \boldsymbol{f} \cdot \boldsymbol{u} \, dV + \frac{1}{2} \int_V \sum_{ij} \sigma_{ij} u_{ij} \, dV. \tag{13.1}$$

It was also shown that the change in energy under an arbitrary variation in displacement is given by,

$$\boxed{\delta\mathcal{E} = -\int_V \delta\boldsymbol{u} \cdot \boldsymbol{f}^* \, dV + \oint_S \delta\boldsymbol{u} \cdot \boldsymbol{\sigma} \cdot d\boldsymbol{S}.} \tag{13.2}$$

Here the first term represents the work done *against* the non-equilibrium effective body forces under the change in displacement, and the second term represents the work of the external stresses acting on the surface of the body. The second term evidently vanishes if every point of the surface of the body is either held *fixed*, $\delta\boldsymbol{u} = \boldsymbol{0}$, or left *free* to vary with no stresses, $\boldsymbol{\sigma} \cdot \boldsymbol{n} = \boldsymbol{0}$. We shall assume this to be the case in the following, such that the change in energy under any variation respecting these boundary conditions is given by the first term, which vanishes in equilibrium.

The iterative procedure now consists of designing each infinitesimal displacement to drain energy away from the body, such that it continually moves towards states of lower energy while respecting the boundary conditions. Since we perform negative work against the non-equilibrium forces, i.e. they perform work on us, such a procedure is said to *relax* the body. When $\delta\mathcal{E} < 0$ in each iteration the energy of the body will decrease until it reaches a minimum, which in section 11.4 for linear elastic materials was shown to be unique. Having arrived at the minimum, the internal energy must be stationary $\delta\mathcal{E} = 0$ for all variations in displacement $\delta\boldsymbol{u}(\boldsymbol{x})$, and that is only possible if $\boldsymbol{f}^*(\boldsymbol{x}) = \boldsymbol{0}$ for all $\boldsymbol{x}$. Thus, in the end the relaxation procedure will arrive at an equilibrium state.

Gradient descent

A common relaxation procedure is to select the change in displacement to always be proportional to the effective force,

$$\delta\boldsymbol{u} = \epsilon \boldsymbol{f}^*, \tag{13.3}$$

where ϵ is a positive quantity, called the *step-size*. Relaxing the displacement in this way by 'running along' with the effective force guarantees that the density of work, $-\boldsymbol{f}^* \cdot \delta\boldsymbol{u} = -\epsilon(\boldsymbol{f}^*)^2$, is negative everywhere and thus drains energy away from every material particle in the body that is not already in equilibrium. Since the displacement 'walks downhill' *against* the gradient of the total energy (in the space of all allowed displacement fields), it is naturally called *gradient descent*.

Gradient descent is not a foolproof method, even when the energy (as in linear elastic media) is a quadratic function of the displacement field with a unique minimum. In particular the step-size ϵ must be chosen judiciously. Too small, and the procedure may never seem to converge; too large, it may overshoot the minimum and go into oscillations or even diverge. Many fine tricks have been invented to get around these problems and speed up convergence [59, 10], for example conjugate gradient descent in which the optimal step-size is calculated in advance by searching for a minimum along the chosen direction of descent. Here, however, we shall just use the straightforward technique of the dedicated downhill skier, always looking for the steepest gradient.

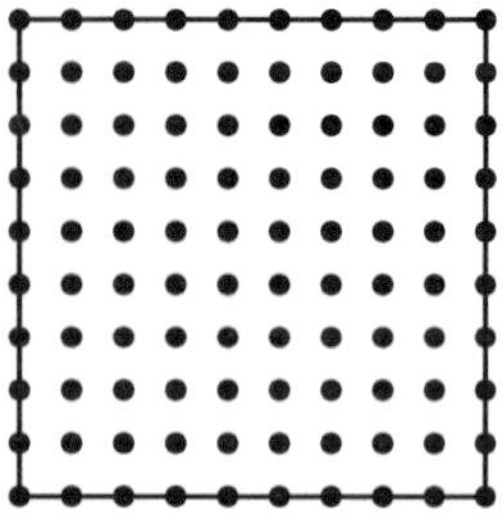

A two-dimensional (10 × 10) square grid. There are 36 points at the boundary and 64 inside. Small grids have a lot of boundary.

13.2 Discretization of space

The infinity of points in space cannot be represented in a finite computer. In numerical simulations of the partial differential equations of continuum physics, smooth space is nearly always replaced by a finite collection of points, a grid or lattice, on which the various fields 'live'. In Cartesian coordinates the most convenient grid for a rectangular volume $a \times b \times c$ is a rectangular lattice with $(N_x+1) \times (N_y+1) \times (N_z+1)$

points that are equally spaced at coordinate intervals $\Delta x = a/N_x$, $\Delta y = b/N_y$, and $\Delta z = c/N_z$. The grid coordinates are numbered by $n_x = 0, 1, \ldots, N_x$, $n_y = 0, 1, \ldots, N_y$ and $n_z = 0, 1, \ldots, N_z$, and the various fields can only exist at the positions $(x, y, z) = (n_x \Delta x, n_y \Delta y, n_z \Delta z)$.

There are many other ways of discretizing space besides using rectangular lattices, for example triangular, hexagonal or even random lattices. The choice of grid depends on the problem itself, as well as on the field equations and the boundary conditions. The coordinates in which the system is most conveniently described may not be Cartesian but curvilinear, and that leads to quite a different discretization. The surface of the body may or may not fit well with the chosen grid, but that problem may be alleviated by making the grid very dense at the cost of computer time and memory. When boundaries are irregular, as they usually are for real bodies, an adaptive grid that can fit itself to the shape of the body may be the best choice. Such a grid may also adapt to put more points where they are needed in regions of rapid variation of the displacement field.

Finite difference operators with first-order errors

In a discrete space, coordinate derivatives of fields such as $\nabla_x f(x, y, z)$ must be approximated by finite differences between the field values at the allowed points. Using only the nearest neighbours on the grid there are two basic ways of forming such differences at a given *internal* point of the lattice, namely forwards and backwards

$$\widehat{\nabla}_x^+ f(x) = \frac{f(x + \Delta x) - f(x)}{\Delta x}, \tag{13.4a}$$

$$\widehat{\nabla}_x^- f(x) = \frac{f(x) - f(x - \Delta x)}{\Delta x}. \tag{13.4b}$$

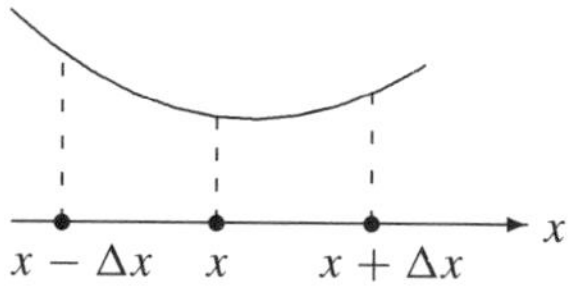

Forward and backward finite differences can be very different, and may as here even have opposite signs.

Here and in the following we suppress for clarity the 'sleeping' coordinates y and z and furthermore assume that finite differences in these coordinates are defined analogously.

According to the rules of differential calculus, both of these expressions will in the limit of $\Delta x \to 0$ converge towards $\nabla_x f(x)$. Inserting the Taylor expansion

$$f(x + \Delta x) = f(x) + \Delta x \nabla_x f(x) + \frac{1}{2} \Delta x^2 \nabla_x^2 f(x) + \frac{1}{6} \Delta x^3 \nabla_x^3 f(x) + \frac{1}{24} \Delta x^4 \nabla_x^4 f(x) + \cdots,$$

we find indeed

$$\widehat{\nabla}_x^\pm f(x) = \nabla_x f(x) \pm \frac{1}{2} \Delta x \nabla_x^2 f(x) + \cdots,$$

with an error that is of first order in the interval Δx.

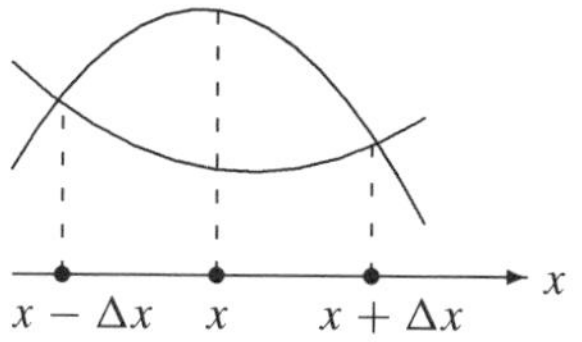

The central difference is insensitive to the value at the centre. The two curves shown here have the same symmetric difference but behave quite differently.

Finite difference operators with second-order errors

It is clear from the above expression that the first-order error may be suppressed by forming the average of right- and left-difference operators, called the *central difference*,

$$\widehat{\nabla}_x f(x) = \frac{f(x + \Delta x) - f(x - \Delta x)}{2\Delta x}. \tag{13.5}$$

Expanding the function values to third order we obtain

$$\widehat{\nabla}_x f(x) = \nabla_x f(x) + \frac{1}{6} \Delta x^2 \nabla_x^3 f(x) + \cdots,$$

with errors of second order only. The central difference does not involve the field value at the central point x, so one should be wary of possible 'leapfrog' or 'flipflop' numeric instabilities in which half the points of the lattice behave differently than the other half.

On a boundary, the central difference cannot be calculated, and one is forced to use one-sided differences. If the grid is rectangular with a rectangular border, one must use the right-hand difference on the boundary to the left and the left-hand difference on the boundary to the right. In order to consistently avoid $\mathcal{O}(\Delta x)$ errors one may instead of (13.4) use the one-sided two-step difference operators (see problem 13.1),

$$\widehat{\nabla}_x^+ f(x) = \frac{-f(x+2\Delta x) + 4f(x+\Delta x) - 3f(x)}{2\Delta x}, \tag{13.6a}$$

$$\widehat{\nabla}_x^- f(x) = \frac{f(x-2\Delta x) - 4f(x-\Delta x) + 3f(x)}{2\Delta x}. \tag{13.6b}$$

The coefficients are chosen here such that the leading order corrections vanish. Expanding to third order we find

$$\widehat{\nabla}_x^{\pm} f(x) = \nabla_x f(x) \mp \frac{1}{3}\Delta x^2 \nabla_x^3 f(x) + \cdots,$$

which shows that both one-sided differences represent the derivative at the point x with leading errors of $\mathcal{O}\left(\Delta x^2\right)$ only.

Other schemes involving more distant neighbours to suppress even higher order errors are of course also possible.

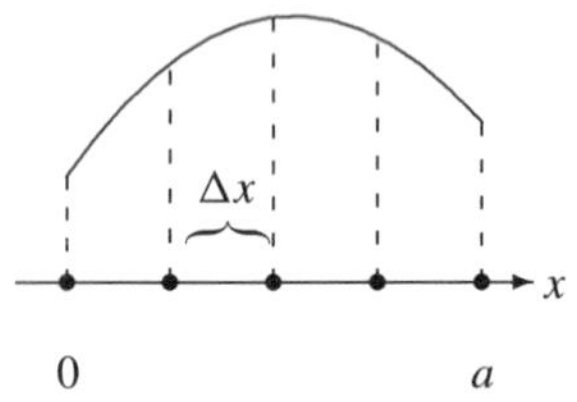

The interval $0 \le x \le a$ has four subintervals of size Δx numbered $n = 0, 1, 2, 3$.

Numeric integration

In simulations it will also be necessary to calculate various line, surface and volume integrals over discretized space. Since the fields are only known at the points of the discrete lattice, the integrals must be replaced by suitably weighted sums over the lattice points.

Let us, for example, consider a one-dimensional integral over an interval, say $\int_0^a f(x)\,dx$, on a regular grid with coordinates $x_n = n\Delta x$ where $n = 0, 1, \ldots, N$. The contribution to the integral from the nth subinterval $x_n \le x \le x_{n+1}$ is approximated by the trapezoidal rule [59, p. 131]

$$\int_{x_n}^{x_{n+1}} f(x)\,dx = \frac{1}{2}\left(f(x_n) + f(x_{n+1})\right)\Delta x + \mathcal{O}\left(\Delta x^3\right),$$

which is analogous to the central difference in suppressing the leading order error. Adding the contributions from the N subintervals together, we obtain the well-known extended trapezoidal rule for numerical integration

$$\int_0^a f(x)\,dx \approx \frac{1}{2} f(0)\,\Delta x + \sum_{n=1}^{N-1} f(n\Delta x)\,\Delta x + \frac{1}{2} f(a)\,\Delta x + \mathcal{O}\left(\Delta x^2\right). \tag{13.7}$$

In higher dimensions one may integrate each dimension according to this formula.

Again there exist schemes for numerical integration on a regular grid with more complicated weights and correspondingly smaller errors, for example Simpson's famous formula [59, p. 134] which is correct to $\mathcal{O}\left(\Delta x^4\right)$.

13.3 Gravitational settling in two dimensions

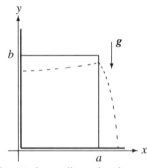

Expected two-dimensional gravitational settling. If the wall at $x = a$ is removed, the elastic material will bulge out, because of its own weight.

One of the simplest non-trivial problems that does not seem to admit an exact analytic solution is the gravitational settling of a rectangular block of elastic material in a long open box of dimensions $a \times b \times c$ with one of the sides removed (see section 12.2 for the case where all sides are removed).

In this case we follow the conventions normally used in two dimensions and take the y-axis to be vertical. The wall that is removed is situated at $x = a$ whereas the wall at $x = 0$ remains in place. It is reasonable to assume that the clamping in the z-direction at $z = 0$ and $z = c$ prevents any displacement in that direction, i.e. $u_z = 0$ everywhere. Since the block is assumed to be very long in this direction, it is also reasonable to assume that the displacements u_x and u_y only depend on x and y, but not on z. The problem has become effectively two-dimensional, although there are vestiges of the three-dimensional problem, for example the non-vanishing stress along the z-direction which is taken up by the walls at the ends.

Equations

The components of the two-dimensional strain tensor are

$$u_{xx} = \nabla_x u_x, \tag{13.8a}$$

$$u_{yy} = \nabla_y u_y, \tag{13.8b}$$

$$u_{xy} = \tfrac{1}{2}(\nabla_x u_y + \nabla_y u_x). \tag{13.8c}$$

The corresponding stresses are found from Hooke's law (11.9) and (11.10),

$$\sigma_{xx} = (2\mu + \lambda)u_{xx} + \lambda u_{yy}, \tag{13.9a}$$

$$\sigma_{yy} = (2\mu + \lambda)u_{yy} + \lambda u_{xx}, \tag{13.9b}$$

$$\sigma_{xy} = \sigma_{yx} = 2\mu\, u_{xy}. \tag{13.9c}$$

Finally, the components of the effective force are

$$f_x^* = \nabla_x \sigma_{xx} + \nabla_y \sigma_{xy}, \tag{13.10a}$$

$$f_y^* = \nabla_x \sigma_{xy} + \nabla_y \sigma_{yy} - \rho_0 g_0. \tag{13.10b}$$

Note that only first-order partial derivatives are used in these equations.

We could of course substitute the equations into each other to express the effective force in terms of second-order derivatives of the displacement fields

$$f_x^* = (\lambda + 2\mu)\nabla_x^2 u_x + \mu \nabla_y^2 u_x + (\lambda + \mu)\nabla_x \nabla_y u_y, \tag{13.11a}$$

$$f_y^* = (\lambda + 2\mu)\nabla_y^2 u_y + \mu \nabla_x^2 u_y + (\lambda + \mu)\nabla_x \nabla_y u_x - \rho_0 g_0. \tag{13.11b}$$

Although there are excellent numerical methods to solve such (elliptic) differential equations, the boundary conditions that involve stresses (see below) are not so easy to implement.

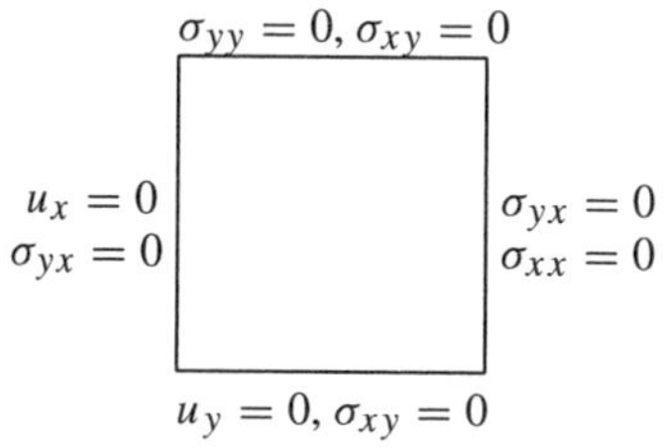

Boundary conditions for the rectangular block.

Boundary conditions

The boundary consists of the two fixed surfaces at $x = 0$ and $y = 0$ and the free surfaces at $x = a$ and $y = b$. We shall adopt the following boundary conditions,

$$\sigma_{xx} = 0, \quad \sigma_{yx} = 0 \qquad \text{free surface at } x = a, \tag{13.12a}$$

$$\sigma_{yy} = 0, \quad \sigma_{xy} = 0 \qquad \text{free surface at } y = b, \tag{13.12b}$$

$$u_x = 0, \quad \sigma_{yx} = 0 \qquad \text{fixed wall at } x = 0, \tag{13.12c}$$

$$u_y = 0, \quad \sigma_{xy} = 0 \qquad \text{fixed wall at } y = 0. \tag{13.12d}$$

Here we have assumed that the fixed surfaces are slippery, so that the shear stress must vanish. That is however not the only choice.

Had we instead chosen the fixed walls to be sticky so that the elastic material were unable to slip along the sides, the tangential displacements at these boundaries would have to vanish, i.e. $u_y = 0$ at $x = 0$ and $u_x = 0$ at $y = 0$. The tangential stress $\sigma_{xy} = \sigma_{yx}$ would, on the other hand, be left free to take any value determined by the field equations. Whereas freedom appears to be unique, there is always more than one way to constrain it.

Shear-free solution

Since the shear stress vanishes at all boundaries, it is tempting to solve the equations by requiring the shear stress also to vanish throughout the block, $\sigma_{xy} = \sigma_{yx} = 0$, as we did for the three-dimensional settling in section 12.2. One may verify that the following field solves the field equations

$$u_x = \frac{\nu}{(1-\nu)D}(b - y)x \tag{13.13}$$

$$u_y = -\frac{1}{2D}\left(b^2 - (b-y)^2 + \frac{\nu}{1-\nu}(a^2 - x^2)\right), \tag{13.14}$$

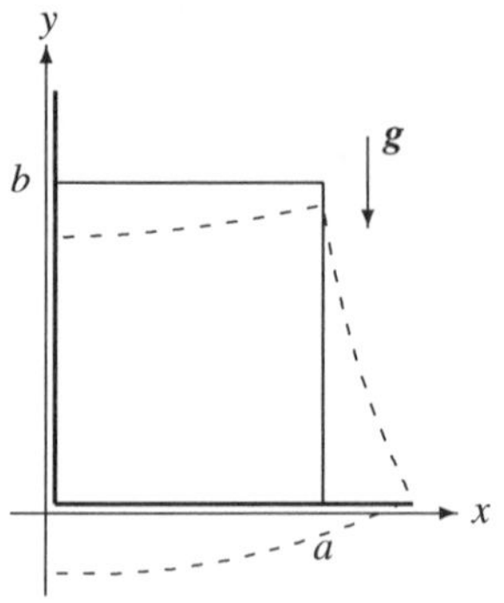

The shear-free solution sinks into the bottom of the box. An extra vertical stress distribution is needed from below in order to fulfill the boundary conditions.

where

$$D = \frac{4\mu(\lambda+\mu)}{(2\mu+\lambda)\rho_0 g_0} = \frac{E}{(1-\nu^2)\rho_0 g_0} \tag{13.15}$$

is the characteristic deformation scale. The solution is of the same general form as in the three-dimensional case (12.16), but the dependence on Poisson's ratio ν is different because of the two-dimensionality. As before, this solution also fails to meet the boundary conditions at the bottom, here $y = 0$.

Convergence measures

The approach towards equilibrium may, for example, be monitored by means of the integral over the square of the effective force field which must converge towards zero, if the algorithm works. We shall choose the monitoring parameter to be

$$\chi = \frac{1}{\rho_0 g_0}\sqrt{\frac{1}{ab}\int_0^a dx \int_0^b dy \left(f_x^{*2} + f_y^{*2}\right)}. \tag{13.16}$$

It is normalized such that $\chi = 1$ in the undeformed state where $u_x = u_y = 0$ and thus $f_x^* = 0$ and $f_y^* = -\rho_0 g_0$. The integral is calculated as a sum over the two-dimensional lattice (with appropriate weighting of the boundaries). The iterative process can then be stopped when the value falls below any desired accuracy, say $\chi \lesssim 0.01$.

Another possibility is to calculate the total energy (11.32),

$$\mathcal{E} = \int_0^a dx \int_0^b dy \left[\frac{1}{2}(u_{xx}\sigma_{xx} + u_{yy}\sigma_{yy} + 2u_{xy}\sigma_{xy}) + \rho_0 g_0 u_y\right], \tag{13.17}$$

with the integral replaced by a double sum over the lattice points. This quantity should decrease monotonically towards its minimum and, since it, like χ, also has a well-defined continuum limit, its value should be relatively independent of how fine-grained the discretization is, as long as the lattice is large enough. It is, however, harder to determine the accuracy attained.

Iteration cycle

Assuming that the discretized displacement field on the lattice (u_x, u_y) satisfies the boundary conditions, we may calculate the strains (u_{xx}, u_{yy}, u_{zz}) from (13.8) by means of the discrete derivatives, and the stresses $(\sigma_{xx}, \sigma_{yy}, \sigma_{xy})$ from Hooke's law (13.9). Stress boundary conditions are then imposed and the effective force field (f_x^*, f_y^*) is calculated from (13.10). At this point the monitoring parameter χ may be checked and if below the desired accuracy, the iteration process is terminated. If not, the corrections

$$\delta u_x = \epsilon f_x^* \tag{13.18}$$

$$\delta u_y = \epsilon f_y^* \tag{13.19}$$

are added into the displacement field, boundary conditions are imposed on the displacement field, and the cycle repeats.

The iteration process may be viewed as a dynamical process which in the course of (computer) time makes the displacement field converge towards its equilibrium configuration. The true dynamics of deformation (see chapter 14) go on in real time and are quite different. Since dissipation in solids is not included here, the true dynamics are unable to eat away energy and make the system relax towards equilibrium. Releasing the block from the undeformed state, as we do here, would instead create vibrations and sound waves that reverberate forever throughout the system.

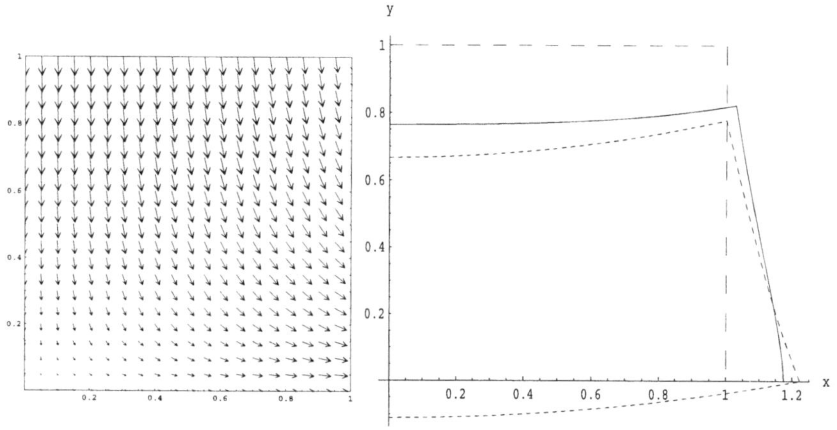

Figure 13.1. Computed deformation of a square two-dimensional block. On the left the equilibrium displacement field is plotted by means of little arrows (not to scale). On the right is plotted the outline of the deformed block. The displacement vanishes as it must at the fixed walls. The protruding material (solid line) has a slightly convex shape rather than the concave shape in the shear-free approximation (small dashes).

Choice of parameters

Since we are mostly interested in the shape of the deformation, we may choose convenient values for the input parameters. They are the box sides $a = b = 1$, the lattice sizes $N_x = N_y = 20$, Young's modulus $E = 2$, Poisson's ratio $\nu = 1/3$ and the force of gravity $\rho_0 g_0 = 1$. The step-size is chosen of the form

$$\epsilon = \frac{\omega}{E} \frac{\Delta x^2 \Delta y^2}{\Delta x^2 + \Delta y^2} \tag{13.20}$$

where ω is called the *convergence parameter*. The reason for this choice is that the effective force is proportional to Young's modulus E and (due to the second-order spatial derivatives) to the inverse squares of the grid spacings, say $1/\Delta x^2 + 1/\Delta y^2 = (\Delta x^2 + \Delta y^2)/\Delta x^2 \Delta y^2$. The convergence parameter ω is consequently dimensionless and may be chosen to be of order unity to get fastest convergence. In the present computer simulation, the largest value that could be used before numeric instabilities set in was $\omega = 1$.

Programming hints

The fields are represented by real arrays, containing the field values at the grid points, for example

$$UX[i, j] \Leftrightarrow u_x(i\Delta x, j\Delta y), \tag{13.21}$$

$$UY[i, j] \Leftrightarrow u_y(i\Delta x, j\Delta y), \tag{13.22}$$

and similarly for the strain and stress fields. Allocating separate arrays for strains and stresses may seem excessive and can be avoided, but when lattices are as small as here, it does not matter. Anyway, the days of limited memory are over.

The iteration cycle is implemented as a loop, containing a sequence of calls to subroutines that evaluate strains, stresses, effective forces and impose boundary conditions, followed by a step that evaluates the monitoring parameters and finally updates the displacement arrays before the cycle repeats. The iteration loop is terminated when the accuracy has reached the desired level, or the number of iterations has exceeded a chosen maximum.

Results

After about 2000 iteration cycles the monitoring parameter χ has fallen from 1 to about 0.01 where it seems to remain without further change. This is most probably due to the brute enforcing of boundary values. The

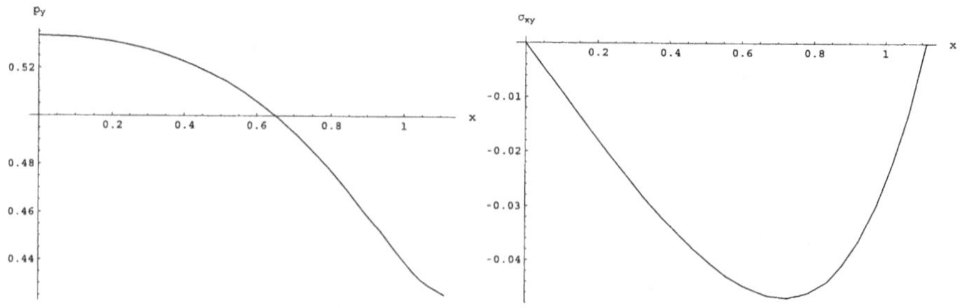

Figure 13.2. The computed vertical pressure, $p_y = -\sigma_{yy}$, is plotted on the left for $y = 0.5$ as a function of the true x. On the right, the corresponding shear stress is plotted at the same height. The pressure is higher in the central region than the shear-free estimate ($p_y = 0.5$) and the shear stress is negative (but small) and thus adds to the force exerted by gravity. The curves have been linearly interpolated between the data points. If the grid is made denser, there will be more detail in the region of the protrusion ($x \gtrsim 1$).

limiting value of χ diminishes with increasing lattice volume $N = N_x N_y$, in accordance with the lessened importance of the boundary which decreases like $1/\sqrt{N}$ relative to the volume.

The final displacement field and its influence on the outline of the original box is shown in figure 13.1. One notes how the displacement does not penetrate into the fixed bottom wall as it did in the shear-free approximation. In figure 13.2 the vertical pressure $p_y = -\sigma_{yy}$ is plotted as a function of x in the middle of the block ($y = 0.5$). Earlier we argued that there would have to be an extra normal reaction from the bottom in order to push up the sagging solution to the shear-free equations. This is also borne out by the plot of p_y which has roughly the same shape throughout the block. Since the vertical pressure is now larger than the weight of the column of material above, we expect that there must be a negative shear stress on the sides of the column to balance the extra vertical pressure, as is also evident from figure 13.2.

Problems

13.1 Show that the coefficients in the one-sided two-step differences (13.6) are uniquely determined.

14

Elastic vibrations

Sound is the generic term for harmonic pressure waves in matter, be it solid, liquid or gaseous. Our daily existence as humans, communicating in and out of sight, is strongly dependent on sound transmission in air, and only rarely—as for example in the dentists chair—do we note the primary effects of sound in solids. What we do experience in our daily lives are mostly secondary effects of vibrations in solids transferred to air as sound waves, for example a mouse scratching on the other side of a wooden wall, or more insidiously the neighbour's drilling into concrete. There are also wave motions in elastic solids, for example caused by earthquakes that we would hardly call sound, except sometimes one speaks about infrasound. We do not hear these phenomena directly but rather experience an earthquake as a motion of the ground, though usually accompanied by audible sound.

There are actually two kinds of vibrations in isotropic elastic solids: *longitudinal* pressure waves and *transverse* shear waves. The two kinds of waves are transmitted with different phase velocities because elastic materials respond differently to pressure and shear stress. Vibrations in ideal elastic materials do not dissipate energy, but energy can be lost to spatial infinity through radiation of sound. A church bell or tuning fork may ring for a long time but eventually stops because of radiative and dissipative losses.

In this chapter we shall use Newton's second law to derive the basic equations for small-amplitude vibrations in isotropic elastic materials and then apply them to a few generic situations. Elastic vibrations constitute a huge subfield of continuum physics which cannot be given just treatment in a single chapter. The chapter is, however, important because it is the first time we encounter continuous matter in motion, the main theme for the remainder of this book.

14.1 Elastodynamics

The instantaneous state of a deformable material is described by a time-dependent displacement field $\boldsymbol{u}(\boldsymbol{x}, t)$ which indicates how much a material particle at time t is displaced from its original position $\boldsymbol{x}$. The field $\boldsymbol{u}(\boldsymbol{x}, t)$ should as before be understood as the displacement from a chosen reference state which may itself already be highly stressed and deformed. There are, for example, huge static stresses in balance with gravity in the pylons and girders of a bridge, but when the wind acts on the bridge, small-amplitude vibrations may arise around the static state.

In this section we shall first establish the fundamental equation of motion for isotropic elastic matter and then draw some general conclusions about the nature of its solutions. Some of the results to be derived will have much wider application than just for elastic waves.

Navier's equation of motion

The actual position of a displaced particle is $\boldsymbol{x}' = \boldsymbol{x} + \boldsymbol{u}(\boldsymbol{x}, t)$, and since its original position $\boldsymbol{x}$ is time-independent, its actual velocity is $\boldsymbol{v}(\boldsymbol{x}, t) = \partial \boldsymbol{u}(\boldsymbol{x}, t)/\partial t$ and its acceleration $\boldsymbol{w}(\boldsymbol{x}, t) = \partial^2 \boldsymbol{u}(\boldsymbol{x}, t)/\partial t^2$.

Newton's second law—mass times acceleration equals force—applied to every material particle in the body takes the form, $dM\boldsymbol{w} = \boldsymbol{f}^* dV$. Dividing by dV and reusing the effective force density for an isotropic homogeneous elastic material from the left-hand side of the equation of equilibrium (12.2), we arrive at *Navier's equation of motion* (1821),

$$\rho \frac{\partial^2 \boldsymbol{u}}{\partial t^2} = \boldsymbol{f} + \mu \nabla^2 \boldsymbol{u} + (\lambda + \mu) \nabla \nabla \cdot \boldsymbol{u}. \tag{14.1}$$

Here λ, μ, and ρ are, as before, assumed to be material parameters that do not depend on space and time. In the case that they depend on the spatial position $\boldsymbol{x}$, as they do in Earth's solid mantle, Navier's equation of motion takes a somewhat different form (see problem 14.1). The above equation of motion reduces by construction to Navier's equilibrium equation for a time-independent displacement. As in elastostatics, the displacement field and the stress vector must be continuous across material interfaces.

It must be emphasized that Navier's equation of motion is only valid in the limit of small and smooth displacement fields. If the displacement gradients are large, nonlinear terms will first of all appear in the strain tensor (10.44), but there will also arise nonlinear terms from the derivatives of the stress tensor in the effective force, as demonstrated by equation (12.3). In chapter 15 we shall derive the correct equations of motion for continuous matter (in the Euler representation) with all such terms included.

Driving forces, dissipation and free waves

Time-dependent displacement is often caused by contact forces that—like the wind on the bridge—impose time-dependent stresses on the surface of a body. If you hit a nail with a hammer or stroke the strings of a violin, time-varying displacement fields are also set up in the material. Body forces may likewise drive time-dependent displacements. The Moon's tidal deformation of the rotating Earth is caused by time-dependent gravitational body forces, acting on top of the static gravitational force of Earth itself. Magnetostrictive, electrostrictive and piezoelectric materials deform under the influence of electromagnetic fields, and are, for example, used in loudspeakers to set up vibrations that can be transmitted to air as sound.

The omnipresent forces of dissipation—not included in Navier's equation of motion—will in the end make all vibrations die out and turn their energy into heat. Sustained vibrations in any body can strictly speaking only be maintained by time-dependent external forces continually performing work by interacting with the body. Dissipation is nevertheless so small in most elastic materials that it, to a very good approximation, can be omitted, as it is in Navier's equation of motion. This argument justifies the study of *free elastic waves* in a body subject only to time-independent external forces. Due to the linearity of Navier's equation of motion, a time-independent body force may be removed by means of a suitable time-independent displacement, such that the general equation of motion for free elastic waves becomes,

$$\rho \frac{\partial^2 \boldsymbol{u}}{\partial t^2} = \mu \nabla^2 \boldsymbol{u} + (\lambda + \mu) \nabla \nabla \cdot \boldsymbol{u}. \tag{14.2}$$

Although free waves must at some point in time have been created by time-dependent driving forces, they will in a finite, isolated, perfectly elastic body continue indefinitely after the driving forces cease to act.

The violin paradox: How can stroking a violin string with a horsehair bow at constant speed make the string vibrate at a nearly constant frequency, when we claim that sustained vibration demands time-dependent driving forces?

Although the external force delivered by your arm to the bow is nearly constant for the length of the stroke, the interaction between the bow and the string develops time-dependence because of the finite difference between static and dynamic friction forces (section 9.1). The string sticks to the bow when it starts to move until the restoring elastic force in the string surpasses the static friction force, whereupon the string slips and begins to move with much smaller or even no friction (if it lifts off the bow). Swinging once back and forth the string eventually again matches the speed of the bow and sticks. Since it only sticks for a very short time, the frequency generated in this way is very nearly equal to the natural oscillation frequency of a taught but otherwise free string. This *stick-slip* mechanism underlies many oscillatory phenomena apparently generated by steady driving agents (see for example problem 9.9 on page 120).

Longitudinal and transverse waves

An arbitrary vector field may always be resolved into *longitudinal* and *transverse* components (see problem 14.5),

$$\boldsymbol{u} = \boldsymbol{u}_L + \boldsymbol{u}_T, \tag{14.3}$$

where the longitudinal component $\boldsymbol{u}_L$ has no curl, and the transverse component $\boldsymbol{u}_T$ has no divergence,

$$\nabla \times \boldsymbol{u}_L = \boldsymbol{0} \qquad \nabla \cdot \boldsymbol{u}_T = 0. \tag{14.4}$$

By the 'double-cross' rule (2.67) on page 26 it follows that $\nabla \times (\nabla \times \boldsymbol{u}_L) = \nabla(\nabla \cdot \boldsymbol{u}_L) - \nabla^2 \boldsymbol{u}_L = \boldsymbol{0}$, or $\nabla\nabla \cdot \boldsymbol{u}_L = \nabla^2 \boldsymbol{u}_L$, so that the wave equation (14.2) specialized to purely longitudinal and transverse free waves becomes,

$$\rho \frac{\partial^2 \boldsymbol{u}_L}{\partial t^2} = (\lambda + 2\mu)\nabla^2 \boldsymbol{u}_L, \qquad \rho \frac{\partial^2 \boldsymbol{u}_T}{\partial t^2} = \mu \nabla^2 \boldsymbol{u}_T. \tag{14.5}$$

Conversely it may be shown that the longitudinal and transverse components of any mixed field (14.3) must also satisfy these equations (see problem 14.5).

Both of these equations are in the form of the standard wave equation,

$$\frac{\partial^2 \boldsymbol{u}}{\partial t^2} = c^2 \nabla^2 \boldsymbol{u}, \tag{14.6}$$

for non-dispersive waves with phase velocity c. For longitudinal and transverse waves the phase velocities are,

$$\boxed{c_L = \sqrt{\frac{\lambda + 2\mu}{\rho}}, \qquad c_T = \sqrt{\frac{\mu}{\rho}}.} \tag{14.7}$$

In typical elastic materials the phase velocities are a few kilometres per second which is an order of magnitude greater than the velocity of sound in air, but roughly of the same magnitude as the sound velocity in liquids, such as water.

The ratio between the transversal and longitudinal velocities is a useful dimensionless parameter,

$$q = \frac{c_T}{c_L} = \sqrt{\frac{\mu}{\lambda + 2\mu}} = \sqrt{\frac{1 - 2\nu}{2(1 - \nu)}}. \tag{14.8}$$

It depends only on Poisson's ratio ν, and is a monotonically decreasing function of ν. Its maximal value $(1/2)\sqrt{3} \approx 0.87$ is obtained for $\nu = -1$, implying that the transverse velocity is always smaller than the longitudinal one. In practice there are no materials with $\nu < 0$, so the realizable upper limit to the ratio is instead $(1/2)\sqrt{2} \approx 0.71$. For a typical value $\nu = 1/3$ we get $q = 1/2$ and hence longitudinal waves typically propagate with double the speed of transverse waves.

The tiny pressure change (11.14) generated by the displacement field is $\Delta p = -K \nabla \cdot \boldsymbol{u}$, where $K = \lambda + 2/3\mu$ is the bulk modulus. Since $\nabla \cdot \boldsymbol{u} = 0$ for transverse waves, only the longitudinal waves are accompanied by an oscillating pressure. They are for this reason also called *pressure* waves or *compressional* waves. Transverse waves generate no pressure changes in the material, only shear, and are therefore called *shear* waves.

Finally, it must be emphasized that although the longitudinal and transverse displacement fields individually satisfy the standard wave equation, the boundary conditions on the surface of a body must be applied to the complete displacement field (14.3). The boundary conditions will thus in general couple the longitudinal and transverse components, the only exception being plane waves in an infinitely extended medium.

Material	c_L [km s^{-1}]	q [%]
Aluminium	6.4	48
Titanium	6.1	51
Iron	5.9	54
Nickel	5.8	52
Magnesium	5.8	54
Quartz	5.5	63
Wolfram	5.2	55
Copper	4.7	49
Silver	3.7	45
Gold	3.6	33
Lead	2.1	33

Longitudinal sound speed and the ratio of transverse to longitudinal speed, $q = c_T/c_L$, for various isotropic materials. The lightest and hardest materials generally have the largest longitudinal sound speed.

Earthquake wave types: In earthquakes (see figure 14.1) pressure waves are denoted P (for *primary*), because they arrive first due to the higher longitudinal phase velocity in any material. Typically they move at speeds of 4–7 km s^{-1} in the Earth's crust. Shear waves move at roughly half the speed and thus arrive later at the seismometer. They are for this reason denoted by S (for *secondary*). In fluid material, such as the Earth's liquid core, shear waves cannot propagate. Besides these *body waves*, earthquakes are also accompanied by *surface waves* to be discussed in section 14.3.

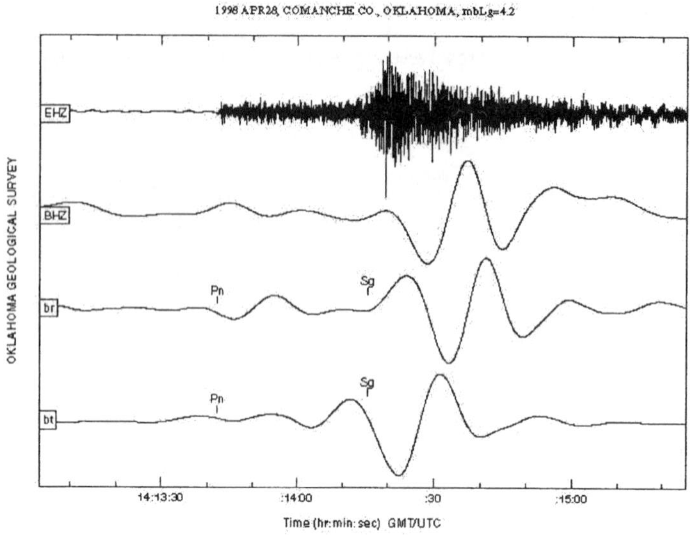

Figure 14.1. Seismogram of an earthquake of strength 4.2 that took place in Comanche county, Oklahoma on April 28, 1998. [Reproduced here with the permission of the Oklahoma Geological Survey]. The four traces are: **EHZ** vertical earth velocity at all frequencies, **BHZ** the low-frequency vertical component, **br** the low-frequency horizontal compressional component (Rayleigh waves) and **bt** the low-frequency horizontal shear component (Love waves). The times labelled Pn and Sg represent the onset of the primary and secondary disturbances.

Harmonic analysis

A general mathematical theorem due to Fourier tells us that any time-dependent function may be resolved as a superposition of *harmonic* or *monochromatic* components, each oscillating with a single frequency. For linear differential equations—ordinary or partial—with time-independent coefficients this is particularly advantageous because it reduces the time-dependent problem to a time-independent one (for each frequency).

A real harmonic displacement field with *circular frequency* ω and *period* $2\pi/\omega$ satisfies the equation,

$$\frac{\partial^2 \boldsymbol{u}}{\partial t^2} = -\omega^2 \boldsymbol{u}. \tag{14.9}$$

The most general solution is a linear superposition of two time-independent *standing wave fields* $\boldsymbol{u}_1(\boldsymbol{x})$ and $\boldsymbol{u}_2(\boldsymbol{x})$,

$$\boldsymbol{u}(\boldsymbol{x}, t) = \boldsymbol{u}_1(\boldsymbol{x}) \cos \omega t + \boldsymbol{u}_2(\boldsymbol{x}) \sin \omega t. \tag{14.10}$$

Instead of working with two real fields it is often most convenient to collect them in a single *complex* time-independent standing-wave field,

$$\boldsymbol{u}(\boldsymbol{x}) = \boldsymbol{u}_1(\boldsymbol{x}) + i\boldsymbol{u}_2(\boldsymbol{x}). \tag{14.11}$$

The harmonic displacement field then becomes the real part of a complex field,

$$\boldsymbol{u}(\boldsymbol{x}, t) = \mathcal{R}e\left[\boldsymbol{u}(\boldsymbol{x})\, e^{-i\omega t}\right]. \tag{14.12}$$

The displacement velocity is correspondingly given by the imaginary part,

$$\frac{\partial \boldsymbol{u}(\boldsymbol{x}, t)}{\partial t} = \omega\, \mathcal{I}m\left[\boldsymbol{u}(\boldsymbol{x})\, e^{-i\omega t}\right], \tag{14.13}$$

as may easily be verified.

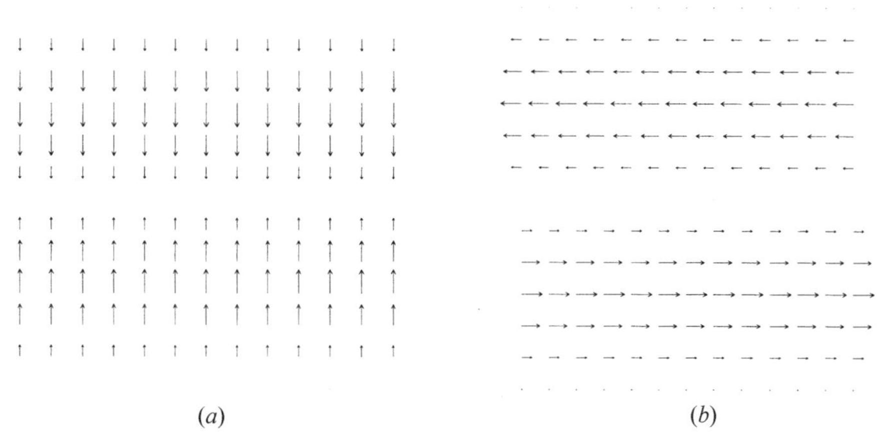

Figure 14.2. Displacement fields for plane waves moving towards the top of the page. **(a)** Longitudinal wave; the displacement oscillates in the direction of motion. **(b)** Transversal wave; the displacement oscillates orthogonally to the direction of motion.

Since the wave equation (14.2) is linear in $\boldsymbol{u}$, it is also satisfied by the velocity field $\partial \boldsymbol{u}/\partial t$ and thus by both the real and imaginary part of the complex field $\boldsymbol{u}(\boldsymbol{x})e^{-i\omega t}$, i.e. by the whole complex field itself. Inserting this field into the wave equation we obtain a single time-independent equation for the complex standing-wave field $\boldsymbol{u}(\boldsymbol{x})$,

$$\boxed{-\rho\omega^2 \boldsymbol{u} = \mu \nabla^2 \boldsymbol{u} + (\lambda + \mu)\nabla\nabla \cdot \boldsymbol{u}.} \tag{14.14}$$

It may be viewed as an *eigenvalue equation* for the operator $\mu\delta_{ij}\nabla^2 + (\lambda + \mu)\nabla_i\nabla_j$ with eigenfunction $\boldsymbol{u}(\boldsymbol{x})$ and $-\rho\omega^2$ as the eigenvalue. It may be shown that ω^2 is always real and positive (problem 14.4). In a finite body, the boundary conditions only allow solutions for a discrete set of eigenfrequencies, whereas in an infinite medium the eigenfrequencies normally form a continuum.

The harmonic analysis may immediately be extended to Navier's equation of motion with a time-dependent body force field $\boldsymbol{f}(\boldsymbol{x}, t)$. This will only add the complex harmonic amplitude $\boldsymbol{f}(\boldsymbol{x})$ of the force field to the right-hand side of (14.14).

Plane waves

Plane waves have infinite extension, and infinitely extended material bodies do not exist. Nevertheless, deeply inside a finite body, far from the boundaries, conditions are almost as if the body were infinite, and the displacement field may be resolved into a superposition of independent longitudinal or transverse plane waves. The condition for this to be possible is that the typical wavelengths contained in the wave should be much smaller than the dimensions of the body or the distance to boundaries.

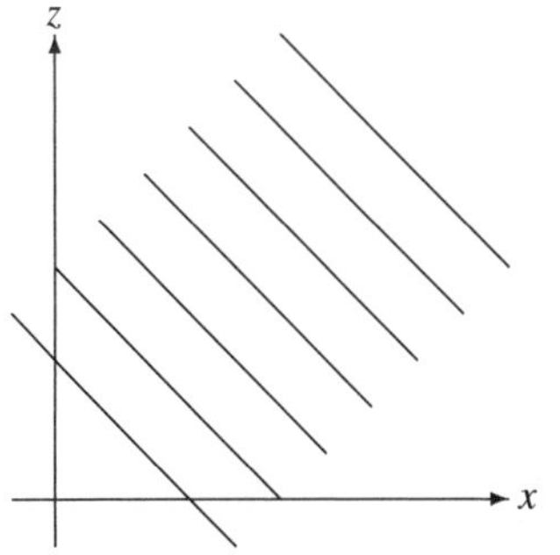

A plane wave has constant phase on planes orthogonal to the wave vector $\boldsymbol{k}$ (here with $k_y = 0$). The phase is spatially periodic with wavelength $\lambda = 2\pi/|\boldsymbol{k}|$.

It is instructive to carry through the harmonic analysis for a plane harmonic wave, described by (the real part of) a complex harmonic field of the form,

$$\boldsymbol{u} = \boldsymbol{a}\, e^{i(\boldsymbol{k}\cdot\boldsymbol{x} - \omega t)}. \tag{14.15}$$

Here $\boldsymbol{a}$ is the generally complex *amplitude* or *polarization vector*, $\boldsymbol{k}$ the *wave vector* and ω the *circular frequency*. The wave's *direction of propagation* is $\boldsymbol{k}/|\boldsymbol{k}|$, its *wavelength* $2\pi/|\boldsymbol{k}|$ and its *period* $2\pi/\omega$. The *phase* of the wave is $\boldsymbol{k}\cdot\boldsymbol{x} - \omega t$ and its *phase velocity* $\omega/|\boldsymbol{k}|$. Inserting this field into (14.2) (or just $\boldsymbol{a}e^{i\boldsymbol{k}\cdot\boldsymbol{x}}$ into (14.14)), we obtain,

$$\rho\omega^2 \boldsymbol{a} = \mu \boldsymbol{k}^2 \boldsymbol{a} + (\lambda + \mu)\boldsymbol{k}\boldsymbol{k}\cdot\boldsymbol{a}. \tag{14.16}$$

This is a simple eigenvalue equation for the real symmetric (3×3) matrix $\mu\boldsymbol{k}^2\delta_{ij} + (\lambda + \mu)k_i k_j$, with eigenvector $\boldsymbol{a}$ and eigenvalue $\rho\omega^2$.

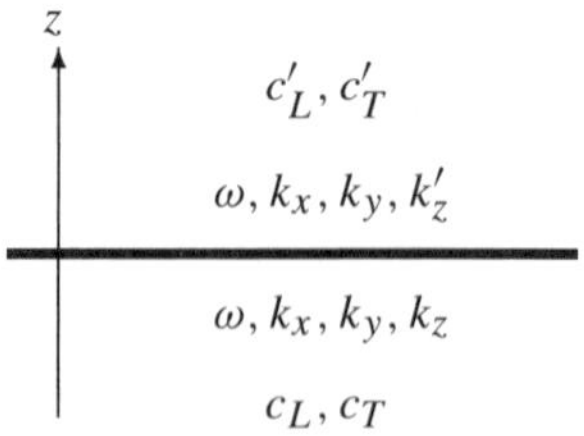

A plane interface between two media. The material properties are different on the two sides of the interface, but the frequency and the wavenumbers components along the interface are the same.

The eigenvectors are easily found. One is *longitudinal* with amplitude proportional to the wave vector itself, $\boldsymbol{a} \sim \boldsymbol{k}$. Inserting this into (14.16) we obtain $\rho\omega^2 = (\lambda + 2\mu)\boldsymbol{k}^2$, showing that a general longitudinal harmonic plane wave is of the form (see figure 14.2(a)),

$$\boldsymbol{u}_L = A\,\boldsymbol{k}\,e^{i(\boldsymbol{k}\cdot\boldsymbol{x}-\omega t)}, \qquad |\boldsymbol{k}| = \frac{\omega}{c_L}, \tag{14.17}$$

where c_L is given in (14.7) and A is an arbitrary complex number representing the longitudinal amplitude. The two other eigenvectors are *transverse* with amplitudes orthogonal to the wave vector, i.e. $\boldsymbol{k}\cdot\boldsymbol{a} = 0$, and it follows from (14.16) that $\rho\omega^2 = \mu\boldsymbol{k}^2$. The transverse harmonic plane wave is therefore of the form (see figure 14.2(b)),

$$\boldsymbol{u}_T = \boldsymbol{a}_T\,e^{i(\boldsymbol{k}\cdot\boldsymbol{x}-\omega t)}, \qquad |\boldsymbol{k}| = \frac{\omega}{c_T}, \tag{14.18}$$

where $\boldsymbol{a}_T$ is an arbitrary vector orthogonal to $\boldsymbol{k}$. All of the transverse directions orthogonal to $\boldsymbol{k}$ propagate with the same phase velocity, and are thus *degenerate* eigenvectors. We may write $\boldsymbol{a}_T = A_1\boldsymbol{n}_1 + A_2\boldsymbol{n}_2$ where $\boldsymbol{n}_1$ and $\boldsymbol{n}_2$ are mutually orthogonal transverse vectors (both orthogonal to $\boldsymbol{k}$), and A_1 and A_2 are arbitrary complex numbers representing the transverse amplitudes.

Fourier's theorem applied to both space and time variables tells us that the most general solution to the wave equation (14.2) is (the real part of) a superposition of longitudinal and transverse plane waves with different frequencies, directions of propagation and amplitudes.

14.2 Refraction and reflection

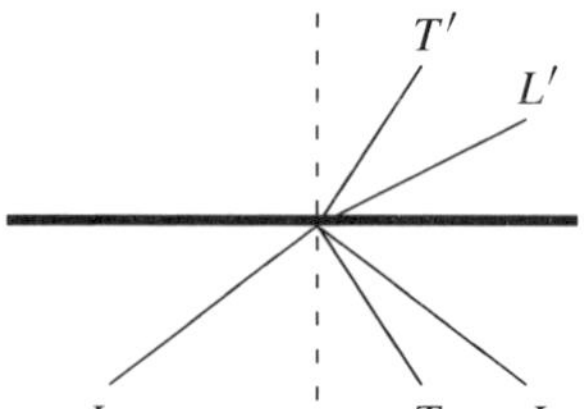

An incident longitudinal wave L_{in} is refracted into longitudinal and transverse components L', T' and also reflected into L and T.

The simplest system which differs from an infinitely extended medium consists of two semi-infinite media interfacing along a plane. The materials on both sides of the interface are homogeneous and isotropic, but have different longitudinal and transverse phase velocities, c_L, c_T and c'_L, c'_T. A plane wave incident on one side of the interface will give rise to both a *refracted* wave on the other side and a *reflected* wave on the same side. Even if the incident wave is purely longitudinal or purely transverse, the refracted and reflected waves will in general be superpositions of longitudinal and transverse waves propagating in different directions. In this section we shall investigate some aspects of these waves which even in this simplest non-trivial case are rather complicated.

Snell's law

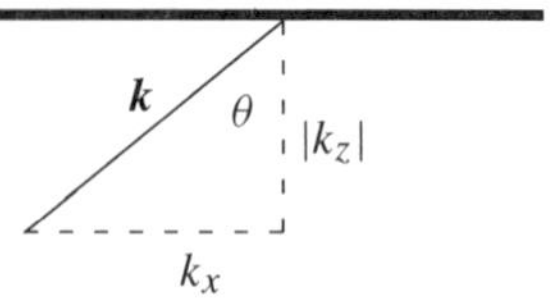

Geometry for determining the angle between the direction of propagation and the normal to the interface.

Willebrord van Roijen Snell (1580–1626). *Dutch mathematician. Contributed to geodesy (triangulation), and discovered the law of refraction.*

Taking the interface to be the xy-plane, $z = 0$ of the coordinate system, the planar geometry is translationally invariant in all directions along x and y. That permits us to resolve the displacement field on either side into a superposition of plane waves of the form (14.15) where all the components have the same fixed values of ω, k_x and k_y on both sides of the interface, whereas in the z-direction the waves may have different values of k_z and k'_z. From this we conclude that the refracted and reflected waves propagate in the same plane as the incident wave. In the following we shall, without loss of generality, choose the waves to propagate in the xz-plane with $k_y = 0$ and $k_x \geq 0$.

A simple geometric construction shows that the angle between the normal to the interface and the direction of propagation of any plane wave with phase velocity $c = \omega/|\boldsymbol{k}|$ is given by

$$\sin\theta = \frac{k_x}{|\boldsymbol{k}|} = \frac{k_x c}{\omega}. \tag{14.19}$$

From the geometry it also follows that

$$|k_z| = k_x \cot\theta = \sqrt{\frac{\omega^2}{c^2} - k_x^2}. \tag{14.20}$$

For $k_x < \omega/c$ the last expression is real and $\theta < 90\,°$. We shall later discuss what happens for $k_x > \omega/c$ where the square root becomes imaginary.

Since k_x and ω are the same for any plane wave component, the angles of incidence of two different wave components with phase velocities c_1 and c_2 must be related by *Snell's law*,

$$\boxed{\frac{\sin\theta_2}{\sin\theta_1} = \frac{c_2}{c_1}.} \tag{14.21}$$

This relation applies to any combination of plane wave components whether they are longitudinal or transverse, on the same side (ipsilateral) as for reflection or on opposite sides (contralateral) as for refraction. Since $c_L > c_T$ we always have $\theta_L > \theta_T$ for the ipsilateral longitudinal and transverse components of a refracted or reflected wave. Reflected and incident waves of the same type will have the same angles with the normal. The angles of contralateral components are determined by the different material properties of the interfacing media, and cannot be generally characterized.

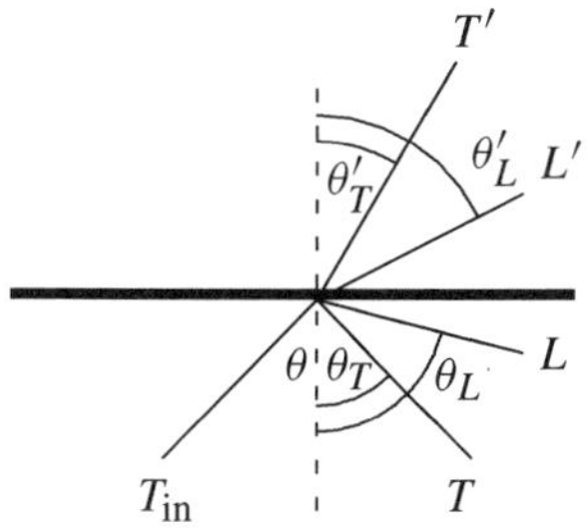

An incident transverse wave produces a reflected transverse wave having the same angle with the normal, $\Theta_T = \theta$, but may also produce a longitudinal wave with larger angle. Snell's law connects all these angles with the phase velocities.

Snell's law takes the same form for elastic, acoustic and electromagnetic waves. The fact that light moves with smaller phase velocity in water than in air immediately tells us that a light ray passing the plane water surface has a smaller angle with the normal in water than in air, thereby explaining the familiar observation that a straight rod apparently breaks when it is partially immersed into water. If the interface is curved we expect that Snell's law will be valid for wavelengths much smaller than the radii of curvature of the interface.

A peculiar thing happens when a refracted wave passes from lower to higher phase velocity, $c' > c$ (which it will always do from one side of the interface). By increasing the angle of incidence there will be a maximal incidence angle $\theta_{\rm max}$ satisfying $\sin\theta_{\rm max} = c/c'$ where the refraction angle becomes $\theta' = 90\,°$, and the wave appears to crawl along the interface. For $\theta > \theta_{\rm max}$ the incident wave is completely unable to penetrate the interface and is totally reflected. Comparing with (14.20) *total reflection* is seen to correspond to imaginary values of the refracted wave vector component k'_z. A similar phenomenon takes place for reflection in isotropic elastic media when the incident wave is transverse. The reflected longitudinal wave always has larger velocity than the reflected transverse wave, and there will be a maximal incident angle for longitudinal reflection. Beyond that angle, the reflected wave will be purely transverse.

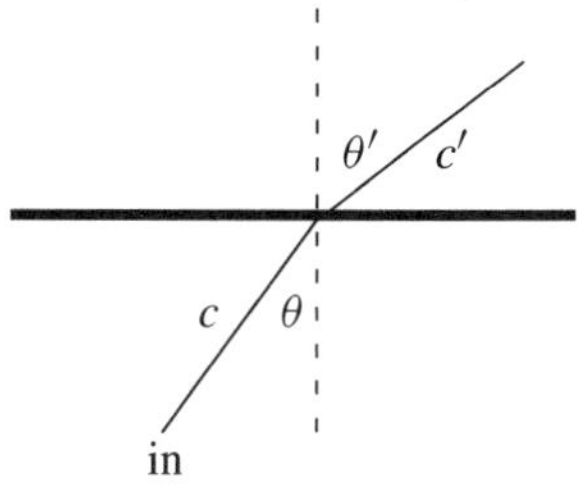

Refraction into a medium with higher phase velocity $c' > c$.

Total reflection is well known to divers looking at the water surface from below, or to fish looking at you from inside their aquarium. It is also of great importance for the functioning of optical fibers where total reflection guarantees that light sent down the fiber stays inside it even if it bends and winds.

Boundary conditions

At the interface $z = 0$, the boundary conditions demand continuity of the displacement fields and the stress vectors on the two sides of the interface,

$$u'_x = u_x, \qquad u'_y = u_y, \qquad u'_z = u_z, \tag{14.22a}$$

$$\sigma'_{xz} = \sigma_{xz}, \qquad \sigma'_{yz} = \sigma_{yz}, \qquad \sigma'_{zz} = \sigma_{zz}. \tag{14.22b}$$

A single incident longitudinal or transverse wave can, in principle, generate one longitudinal and two transverse waves on either side of the interface. The amplitudes of the six waves are determined by the six boundary conditions. Intuitively it is fairly clear that 'supertransverse' waves polarized orthogonally to the plane of incidence (i.e. along the y-direction) must decouple from the others which only involve the x- and z-directions. We therefore only face four equations with four unknowns for the waves with polarization in the plane of incidence, or two equations with two unknowns for the 'super-transverse' waves. It is still an unpleasant task to solve four equations with four unknowns, so in the remainder of this section we shall limit the analysis to a couple of cases resulting in only two equations with two unknowns.

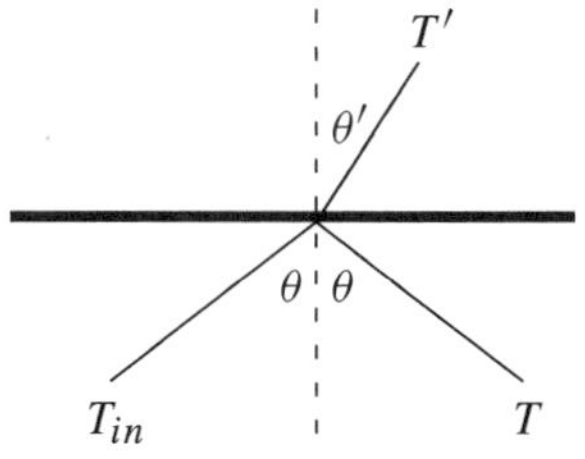

An incident supertransverse wave $T_{\rm in}$ is refracted and reflected into supertransverse waves, T' and T.

'Supertransverse' waves at an interface

In a 'supertransverse' wave the incident as well as the reflected and refracted components are polarized along $\boldsymbol{e}_y = (0, 1, 0)$. It is convenient to set $k_x = k$ and define $k_T = \sqrt{(\omega/c_T)^2 - k^2}$ and $k'_T = \sqrt{(\omega/c'_T)^2 - k^2}$. Leaving out the common factor $e^{i(kx-\omega t)}$, the only non-vanishing displacement components are,

$$u_y = e^{ik_T z} + Ae^{-ik_T z}, \qquad u'_y = A' e^{ik'_T z}. \tag{14.23}$$

The first term in u_y represents the incident field, normalized to unity, while the second term represents the reflected wave with amplitude A. The field u'_y consists entirely of the refracted wave with amplitude A'.

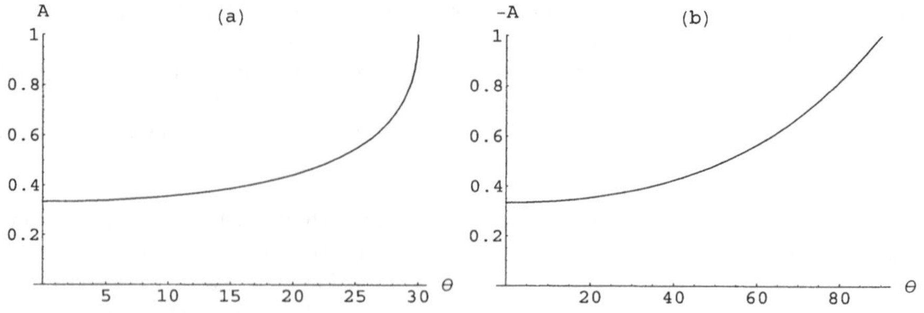

Figure 14.3. 'Supertransverse' waves at an interface with $\mu' = \mu$. The refracted amplitudes are obtained by adding 1. **(a)** Reflected amplitude for $c'_T = 2c_T$. The maximum angle is 30° before total reflection sets in. **(b)** Reflected amplitude for $c'_T = \frac{1}{2}c_T$. The maximum angle is 90°. Note that the amplitude is plotted as $-A$.

Since all the diagonal strains vanish, the boundary conditions are $u'_y = u_y$ and $\sigma'_{yz} = \sigma_{yz}$ at $z = 0$. Using that $\sigma_{yz} = \mu \nabla_z u_y$ and $\sigma'_{yz} = \mu' \nabla_z u'_y$ we are led to the equations,

$$A' = 1 + A, \qquad \mu' k'_T A' = \mu k_T (1 - A). \tag{14.24}$$

The solution is,

$$A = \frac{\mu k_T - \mu' k'_T}{\mu k_T + \mu' k'_T}, \qquad A' = \frac{2\mu k_T}{\mu k_T + \mu' k'_T}, \tag{14.25}$$

and using that $k_T = k \cot\theta$ and $k'_T = k \cot\theta'$, the reflected and refracted amplitudes may be expressed in terms of the angles and the ratio μ'/μ (called the *Fresnel equations*),

Jean Augustin Fresnel (1788–1827). *French physicist. Derived the equations for the amplitudes of reflected and transmitted light. Rejected Newton's corpuscular theory of light in favour of an ether theory.*

$$\boxed{A = \frac{\cot\theta - \frac{\mu'}{\mu}\cot\theta'}{\cot\theta + \frac{\mu'}{\mu}\cot\theta'}, \qquad A' = \frac{2\cot\theta}{\cot\theta + \frac{\mu'}{\mu}\cot\theta'}.} \tag{14.26}$$

Snell's law, $\sin\theta'/\sin\theta = c'_T/c_T$, connects the two angles, such that

$$\cot\theta' = \sqrt{\frac{1}{\sin^2\theta'} - 1} = \sqrt{\left(\frac{c_T}{c'_T \sin\theta}\right)^2 - 1}. \tag{14.27}$$

In figure 14.3 the reflected amplitude A is plotted as a function of the incident angle θ for two choices of material parameters.

If $c'_T > c_T$ and $k > \omega/c'_T$, the refracted wavenumber becomes imaginary, $k'_T = i\kappa_T$ with $\kappa'_T = \sqrt{k^2 - (\omega/c'_T)^2}$. The refracted wave now decays with increasing z as $\exp(-\kappa'_T z)$ and only penetrates a finite distance into the upper half space. It has become a surface wave. The amplitudes are in this case,

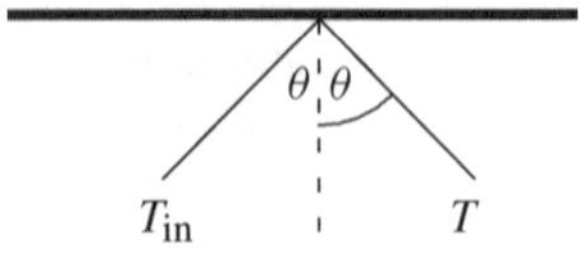

Reflection of a super-transverse wave at a free boundary.

$$A = \frac{\mu k_T - i\mu' \kappa'_T}{\mu k_T + i\mu' \kappa'_T} = e^{-i\phi}, \qquad A' = 1 + e^{-i\phi}, \tag{14.28}$$

where $\tan 1/2\phi = \mu' \kappa'_T / \mu k_T$. Since the complex modulus is unity, $|A| = 1$, the totally reflected wave has the same intensity as the incident wave, although *phase shifted* by ϕ relative to the incident wave.

Reflection from a free surface

A longitudinal or transverse wave incident on a free surface can only be reflected, and the boundary conditions reduce in this case to the vanishing of the stress vector on the boundary $z = 0$,

$$\sigma_{xz} = \sigma_{yz} = \sigma_{zz} = 0. \tag{14.29}$$

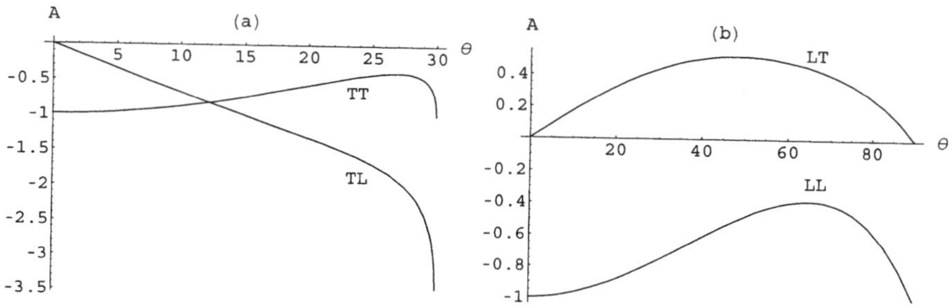

Figure 14.4. Reflected amplitudes as functions of the incident angle θ for $q = \frac{1}{2}$. **(a)** Transverse incident wave. The maximum angle with both transverse and longitudinal reflected waves is $30°$. **(b)** Longitudinal incident wave. The maximum angle is in this case $90°$.

For the 'supertransverse' wave (14.23) the solution is trivial. We simply take $A' = 0$ and find $A = 1$, such that $u_y \sim \cos(k_T z)$.

For a 'normal' transverse wave the field is more complicated, because it can contain both longitudinal and transverse reflected components. Apart from an overall oscillating factor $e^{i(kx-\omega t)}$ the field is of the form,

$$\boldsymbol{u} = (-k_T, 0, k)e^{ik_T z} + A_T(k_T, 0, k)e^{-ik_T z} + A_L(k, 0, -k_L)e^{-ik_L z}, \tag{14.30}$$

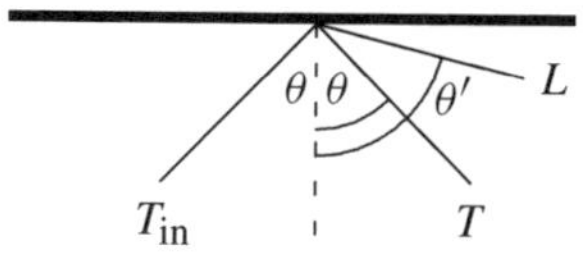

Reflection of a transverse wave at a free boundary. The transverse reflection angle is always the same as the angle of incidence, whereas the longitudinal reflection angle is larger. There is a maximal value $\theta_{max} = \arcsin q$ for which a longitudinal reflection is possible.

where A_T and A_L are the amplitudes of the reflected longitudinal and transverse fields, and where as before $k_T = \sqrt{(\omega/c_T)^2 - k^2}$ and $k_L = \sqrt{(\omega/c_L)^2 - k^2}$. It may readily be verified that the longitudinal field is a gradient, and the transverse fields have no divergence (remembering the oscillating factor $e^{i(kx-\omega t)}$).

From the above field we obtain the surface stresses (apart from the oscillating factor),

$$\sigma_{xz} = \frac{1}{2} i\mu \left((k^2 - k_T^2)(1 + A_T) - 2kk_L A_L\right), \tag{14.31a}$$

$$\sigma_{zz} = i\left(2\mu k k_T(1 - A_T) + ((\lambda + 2\mu)k_L^2 + \lambda k^2)A_L)\right). \tag{14.31b}$$

Using the relation $\rho\omega^2 = (\lambda + 2\mu)(k_L^2 + k^2) = \mu(k_T^2 + k^2)$, and requiring these stresses to vanish, we obtain the equations,

$$2kk_L A_L - (k^2 - k_T^2)A_T = k^2 - k_T^2, \tag{14.32a}$$

$$(k^2 - k_T^2)A_L + 2kk_T A_T = 2kk_T, \tag{14.32b}$$

with the straightforward solution

$$A_T = \frac{4k^2 k_L k_T - (k^2 - k_T^2)^2}{4k^2 k_L k_T + (k^2 - k_T^2)^2}, \qquad A_L = \frac{4kk_T(k^2 - k_T^2)}{4k^2 k_L k_T + (k^2 - k_T^2)^2}. \tag{14.33}$$

Setting $k_T = k \cot\theta$ and $k_L = k\cot\theta'$, the solution may be cast into a convenient form depending only on the two angles,

$$\boxed{A_T = \frac{4\cot\theta\cot\theta' - (1 - \cot^2\theta)^2}{4\cot\theta\cot\theta' + (1 - \cot^2\theta)^2}, \qquad A_L = \frac{4\cot\theta(1 - \cot^2\theta)}{4\cot\theta\cot\theta' + (1 - \cot^2\theta)^2}.} \tag{14.34}$$

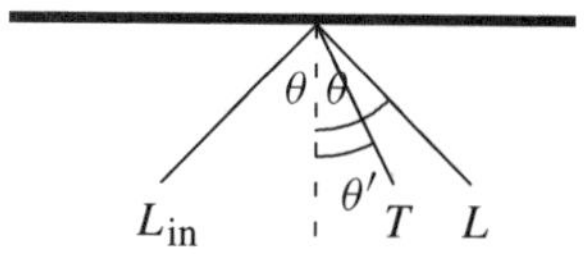

Reflection of a longitudinal wave at a free boundary. The longitudinal reflection angle is the same as the angle of incidence, whereas the transverse reflection angle is smaller.

Snell's law $\sin\theta/\sin\theta' = c_T/c_L = q$, connects as before the two angles,

$$\cot\theta' = \sqrt{\frac{1}{\sin^2\theta'} - 1} = \sqrt{\frac{q^2}{\sin^2\theta} - 1}. \tag{14.35}$$

This clearly shows that a transverse incident wave produces a longitudinal reflected wave for $\sin\theta < q$, whereas for $\sin\theta > q$ only a reflected transverse wave is obtained.

The case of a longitudinal incident wave is very similar and is analysed in problem 14.2. In figure 14.4 the intensities of the reflected wave amplitudes are shown for both a transverse and longitudinal incident wave as a function of the angle of incidence.

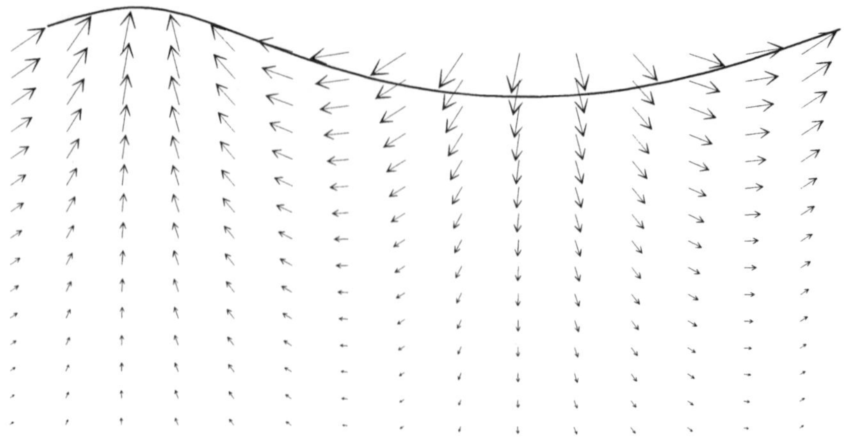

Figure 14.5. One period of a Rayleigh wave moving to the right. The shear component creates a wave-like motion of the surface. Note the exponential decay of the wave under the surface.

* 14.3 Surface waves

At a material interface there are special types of waves which do not penetrate into the bulk of the materials, but decay exponentially with the distance from the interface. We have already seen in the preceding section, how such wave components can arise in the refraction of an incident wave into a material with larger phase velocity when the angle of incidence becomes large enough. In this section we shall consider two kinds of free surface waves, *Rayleigh waves* and *Love waves*. Both have geophysical significance, in particular in relation to earthquakes where they arise when seismic bulk waves encounter the surface of the Earth.

Rayleigh waves

The most general exponentially decaying superposition of normal transverse and a longitudinal surface wave in the lower half space $z < 0$ is (apart from the oscillating factor $\exp(i(kx - \omega t))$ which is common to both terms),

$$\boldsymbol{u} = A_T(i\kappa_T, 0, k)e^{\kappa_T z} + A_L(k, 0, -i\kappa_L)e^{\kappa_L z}, \tag{14.36}$$

where A_T and A_L are generally complex constants. It is obtained from the general expression (14.30) by leaving out the incident wave (which diverges exponentially) and setting $k_T = i\kappa_T$ with $\kappa_T = \sqrt{k^2 - (\omega/c_T)^2}$ and $k_L = i\kappa_L$ with $\kappa_L = \sqrt{k^2 - (\omega/c_L)^2}$. Both κ_T and κ_L are real for $k > \omega/c_T$, because $c_T < c_L$. One may verify directly that the longitudinal wave is indeed a gradient field, and that the transverse wave is free of divergence.

John William Strutt, 3rd Baron Rayleigh (1842–1919). *Discovered and isolated the rare gas Argon for which he got the Nobel Prize (1904). Published the influential book 'The Theory of Sound' on vibrations in solids and fluids in 1877–78.*

The free surface boundary conditions are the same as in the preceding section $\sigma_{xz} = \sigma_{zz} = 0$ for $z = 0$, and leaving out the terms due to the incident wave on the right-hand side of (14.32), we find from the left-hand side,

$$2ik\kappa_L A_L - (k^2 + \kappa_T^2)A_T = 0, \tag{14.37}$$

$$(k^2 + \kappa_T^2)A_L + 2ik\kappa_T A_T = 0. \tag{14.38}$$

Since $iA_L/A_T = (k^2+\kappa_T^2)/2k\kappa_L = 2k\kappa_T/(k^2+\kappa_T^2)$, these equations only have a non-vanishing solution for,

$$(k^2 + \kappa_T^2)^2 = 4k^2\kappa_T\kappa_L. \tag{14.39}$$

Defining the phase velocity along the surface $c = \omega/k$, this condition turns into an equation for c,

$$\boxed{\left(2 - \frac{c^2}{c_T^2}\right)^2 = 4\sqrt{\left(1 - \frac{c^2}{c_T^2}\right)\left(1 - \frac{c^2}{c_L^2}\right)}.} \tag{14.40}$$

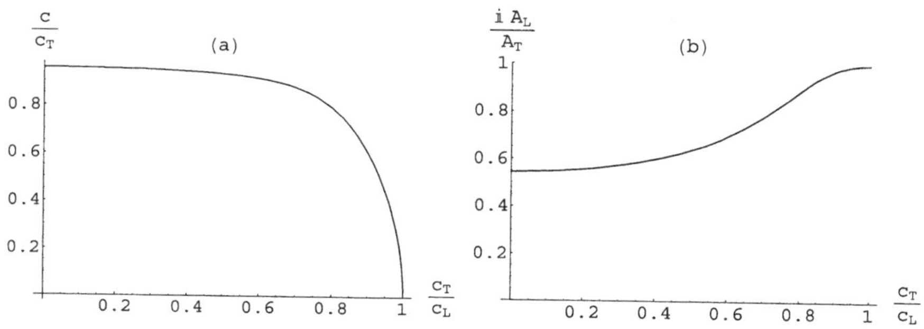

Figure 14.6. Rayleigh waves. **(a)** Phase velocity as a function of $q = c_T/c_L$. For all possible physical values, $q < 1/\sqrt{2} \approx 0.7$, the phase velocity is nearly equal to the phase velocity of transverse waves. **(b)** Ratio of longitudinal and transverse amplitudes as a function of q.

The simplest way to solve this equation is to square it and isolate the ratio $q = c_T/c_L$ in terms of the ratio $\xi = c/c_T$,

$$q = \sqrt{\frac{16 - 24\xi^2 + 8\xi^4 - \xi^6}{16(1-\xi^2)}}. \qquad (14.41)$$

In figure 14.6(a) the phase velocity ξ of Rayleigh waves has been plotted as a function of q. The maximal value $\xi_0 = 0.955313\ldots$ is the real root of the polynomial in the numerator under the square root.

Typical values of q are around 0.5, showing that the value of ξ is close to unity in all practical cases. Expanding (14.41) to lowest order near $\xi = 1$ we find the approximation

$$\xi = 1 - \frac{1}{2(11 - 16q^2)}$$

which for $q = 0.5$ is better than 1%. The phase velocity of Rayleigh waves is thus normally just a little below the phase velocity of free transverse waves.

Seismic waves created deep inside the Earth's crust are reflected from the surface. If the angle of incidence is large enough, the transverse components will excite Rayleigh waves running along the surface. Since their speed is slightly lower than the transverse waves, they arrive even later than S-waves at a seismometer (if they originate in the same point). During the passing of a Rayleigh wave, the surface suffers a combination of compressional and vertical shear displacements, much like a wave rolling across the sea (see figures 14.5 and 14.1). Horizontal shear displacements are absent in a Rayleigh wave.

Augustus Edward Hough Love (1863–1940). *British scholarly physicist. Contributed to the mathematical theory of elasticity, and to the understanding and analysis of the waves created by earthquakes.*

Love waves

One could think that there might be 'supertransverse' free surface waves, either at a free surface or at a material interface, but neither of these types are in fact possible (see problem 14.3). Love found, however, in 1911 that supertransverse free waves may be created if the surface material is heterogeneous with elastic properties that change with height z.

The simplest geometry is obtained by placing a layer of material of thickness h situated on top of a material filling the half-space $z < 0$. Under the conditions that $c_T > c'_T$ and $\omega/c_T < k < \omega/c'_T$ there will be a solution which is exponentially damped in the lower half-space and has running waves in the upper layer. Anticipating that the stress, $\sigma'_{yz} = \mu'\nabla_z u'_y$, must vanish at the top of the layer, $z = h$, a supertransverse field takes the following form in the two media,

$$u_y = Ae^{\kappa_T z}, \qquad u'_y = A'\cos k'_T(h - z), \qquad (14.42)$$

where $\kappa_T = \sqrt{k^2 - (\omega/c_T)^2} = k\sqrt{1 - (c/c_T)^2}$ and $k'_T = \sqrt{(\omega/c'_T)^2 - k^2} = k\sqrt{(c/c'_T)^2 - 1}$. The remaining boundary conditions are as before $u'_y = u_y$ and $\sigma'_{yz} = \sigma_{yz}$ at $z = 0$, leading to

$$A = A'\cos(k'_T h), \qquad \mu\kappa_T A = \mu' k'_T A' \sin(k'_T h). \qquad (14.43)$$

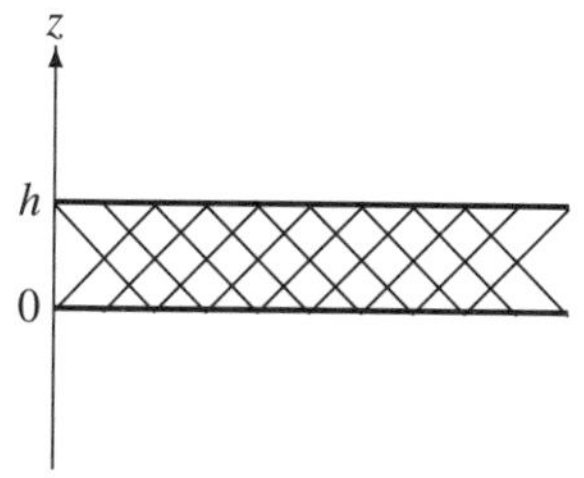

Love waves may arise in a layer of thickness h on top of a half-space $z < 0$ of other material.

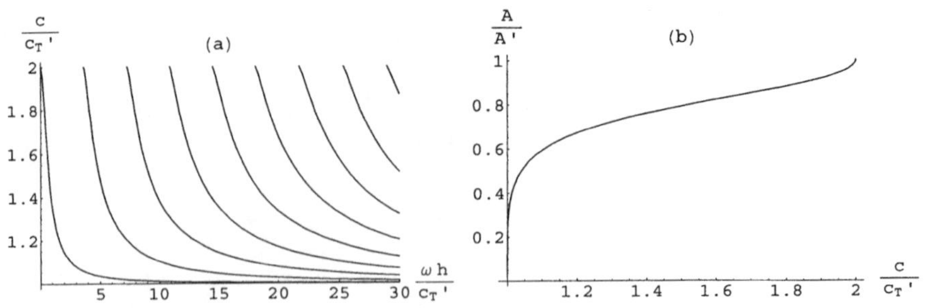

Figure 14.7. Love waves for $c'_T = 0.5c_T$ and $\mu' = \mu$. **(a)** Phase velocity c/c'_T as a function of $\omega h/c'_T$. For a given frequency there is a finite number of possible phase velocities. **(b)** Amplitude ratio A/A' as a function of phase velocity c/c'_T.

A non-trivial solution can only exist for,

$$\mu\kappa_T = \mu' k'_T \tan(k'_T h). \tag{14.44}$$

Introducing $c = \omega/k$ we have $\kappa_T = k\sqrt{1-(c/c_T)^2}$ and $k'_T = k\sqrt{(c/c'_T)^2 - 1}$. Solving the above equation for kh, we obtain

$$kh = \frac{1}{\sqrt{\frac{c^2}{c'^2_T} - 1}} \arctan \frac{\mu\sqrt{1 - \frac{c^2}{c^2_T}}}{\mu'\sqrt{\frac{c^2}{c'^2_T} - 1}}. \tag{14.45}$$

For any value of the phase velocity in the interval $c'_T < c < c_T$ this permits us to calculate the value of kh and—since h is assumed known—of $\omega = ck$. Conversely, for a given frequency ω one may solve this equation for the allowed values of c. The infinity of branches of inverse tangent yields an infinite number of possible frequencies for a given phase velocity. Conversely, there are only a finite number of allowed phase velocities for a given frequency. The solutions c/c'_T are plotted in figure 14.7(*a*) as functions of the dimensionless frequency parameter $\omega h/c'_T$ for a choice of material parameters. In figure 14.7(*b*) the amplitude ratio A/A' is plotted as a function of the phase velocity parameter c/c'_T.

A surface layer thus acts like a wave guide for Love waves. Contrary to Rayleigh waves, Love waves are *dispersive*, with phase velocity that depends on the wavelength (or frequency). Since $c'_T < c < c_T$ Love waves move faster than Rayleigh waves in the surface layer, but slower than shear waves in the bulk. Love waves thus arrive before the Rayleigh waves originating in the same point (see figure 14.1). In earthquakes Love waves are the most destructive because of the shearing motion of the surface layer which is not well tolerated by buildings.

Problems

14.1 Derive the form of Navier's equation of motion when the Lamé coefficients depend on position.

14.2 Show that the reflection amplitudes for a longitudinal incident wave on a free surface are,

$$A_L = \frac{4\cot\theta\cot\theta' - (1-\cot^2\theta')^2}{4\cot\theta\cot\theta' + (1-\cot^2\theta')^2}, \qquad A_T = -\frac{4\cot\theta(1-\cot^2\theta')}{4\cot\theta\cot\theta' + (1-\cot^2\theta')^2} \tag{14.46}$$

with $\cot\theta' = \sqrt{-1 + (q\sin\theta)^{-2}}$.

14.3 Show that supertransverse Love waves cannot exist at an interface (or at a free surface).

* **14.4** Show that the eigenvalues ω^2 of the amplitude equation for free waves (14.14) are real and positive when the boundary conditions specify the vanishing of the displacement field or of the stress vector.

* **14.5** **(a)** Show that an arbitrary vector field may be resolved into (not necessarily unique) longitudinal and transverse components, and that the longitudinal component may be chosen to be a gradient. **(b)** Show that the individual components of a mixed field (14.3) may always be chosen to satisfy Navier's equation of motion individually.

Part IV

Basic hydrodynamics

15 Fluids in motion

The running water in a brook, the streaming wind and the rolling sea are all examples of fluids in motion. A waterfall is a vivid illustration of the richness of fluid motion. It is one of the wonders of nature that all this richness is 'just' a consequence of Newton's equations of motion applied to continuous matter. Easy to write down, these equations only have analytic solutions in a number of idealized and highly constrained situations. Nevertheless, such solutions offer valuable insights into fluid dynamics which is otherwise accessible only by experiments and computer calculations.

The motion of solids is generally less rich than fluid motion, and it is precisely for this reason that solids are used to build structures like houses, bridges and machines. It is also the reason that the study of continuous systems in this book started with elasticity. Fluids and solids are extremes in the world of continuous matter, and there are many transition materials with properties in between. In this chapter we shall analyse matter in motion without distinguishing between particular types of matter, although in the back of our minds we shall mostly think of fluids.

Two basic mechanical equations govern the motion of continuous matter. One concerns conservation of mass and states that the only way the mass of a volume of matter can change is through flow of material across its surface. The other is Newton's second law applied to continuous systems. Together with suitable expressions for the forces at play in the material, we arrive at a complete set of equations of motion for the mass density and velocity fields. In this chapter we shall only apply them to the whole universe, and show that they lead to a surprisingly sensible cosmology. Later chapters will deal with much more earthly aspects of fluid dynamics.

15.1 The velocity field

In trying to define a velocity field $\boldsymbol{v}(\boldsymbol{x}, t)$ we are faced with the problem that the molecules, especially in the gaseous state, move rapidly around among each other, even in matter 'at rest', and this motion must somehow be averaged out. Newtonian particle mechanics tells us that *the centre of mass of a collection of particles moves as a single particle with mass equal to the total mass of all the particles and acted upon by the total force* (see appendix A). The only meaningful definition of the velocity $\boldsymbol{v}$ of a material particle is, therefore, that it should represent the centre of mass velocity of the collection of molecules contained in it, for then the total momentum of a material particle of mass dM occupying a volume dV near the point $\boldsymbol{x}$ at time t will be,

$$\boxed{d\boldsymbol{\mathcal{P}} = \boldsymbol{v}(\boldsymbol{x}, t)\, dM = \boldsymbol{v}\, \rho\, dV.} \tag{15.1}$$

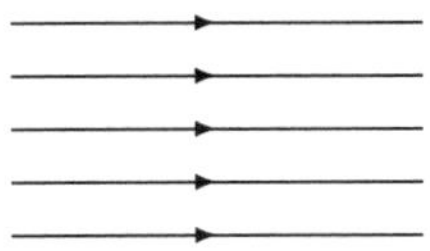

Streamlines of the velocity field for rigid body translation.

Identifying the velocity field with the centre of mass velocity of material particles permits us to apply Newtonian particle mechanics to the material particles themselves without thinking too much about their internal structure.

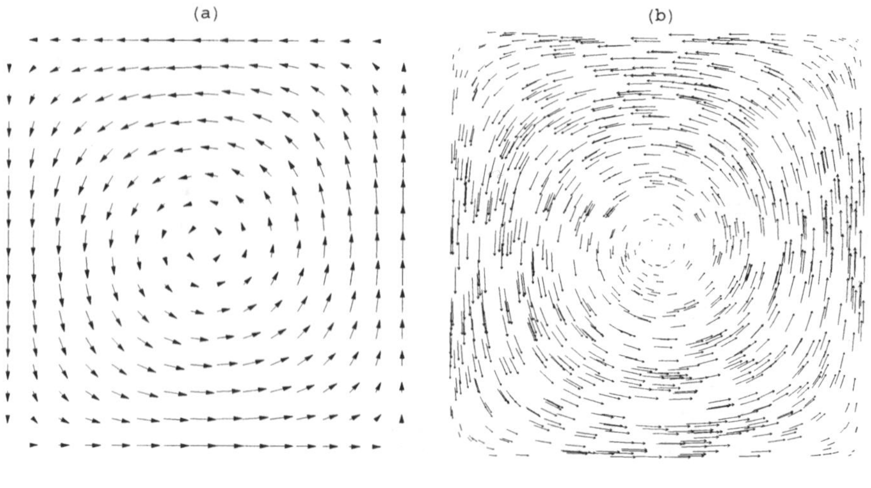

Figure 15.1. Plots of little arrows for the incompressible flow $\boldsymbol{v} = (\sin x \cos y, -\cos x \sin y, 0)$ in the region $0 \leq x \leq \pi$ and $0 \leq y \leq \pi$ (see also example 15.2.2). The whirl (and its mirror images) are repeated periodically throughout the xy-plane. **(a)** Regular plot on a Cartesian (16×16) grid. The apparent skewness is a graphical artifact. **(b)** Random plot of the field in 1000 points.

The velocity field is like the mass density a fluctuating quantity obtained from an average over nearly random individual molecular contributions. In problem 1.4 on page 11 it was shown that, provided the linear dimension of the material particle is larger than the micro scale (1.2), the fluctuation in centre of mass velocity relative to the typical molecular velocity will be smaller than the measurement precision.

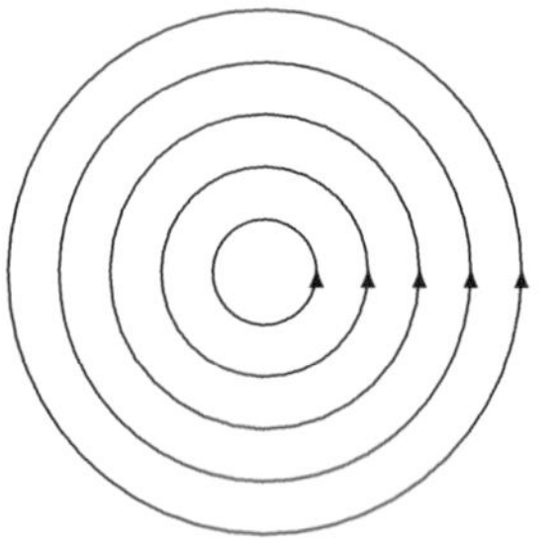

Streamlines for the velocity field of rigid body rotation.

Example 15.1.1: The velocity field of a non-rotating rigid body moving with constant velocity $\boldsymbol{U}$ is $\boldsymbol{v}(\boldsymbol{x}, t) = \boldsymbol{U}$. If instead the body is rotating around the origin of the coordinate system with constant angular velocity vector $\boldsymbol{\Omega}$, the velocity field is $\boldsymbol{v}(\boldsymbol{x}, t) = \boldsymbol{\Omega} \times \boldsymbol{x}$.

Methods of flow visualization

Wind and water currents are normally invisible unless polluted by foreign matter. A gentle breeze in the air can be observed from the motion of dust particles dancing in the sunshine or the undulations of smoke from a cigarette. Even a tornado first becomes visible when water vapour condenses near its centre or debris is picked up and thrown around. Modern technology does, on the other hand, permit us indirectly to 'see' velocity fields. Doppler radar is used for tracking and visualizing damaging winds in violent storms, and likewise, Doppler acoustics is used to visualize blood flow in the heart.

Little arrows: The instantaneous velocity field is often visualized by means of little arrows attached to a regular grid of points, each of a length and direction proportional to the velocity field in the point (see figure 15.1). Sometimes it is more illustrative and permits finer flow details to be seen if the arrows are drawn from a random selection of points, because the density of arrows can be higher.

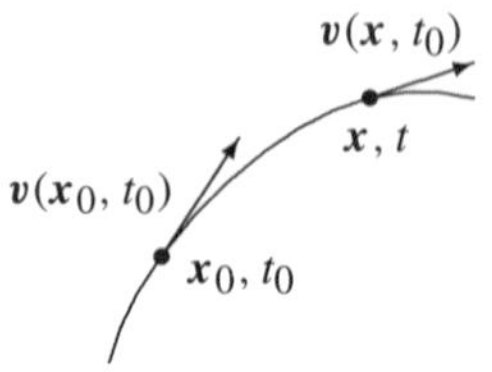

A streamline is everywhere tangent to the velocity field at a given time t_0.

Streamlines: In section 3.2 we discussed how the gravitational field could be visualized by means of field lines, defined to be curves that everywhere had the gravitational field at a fixed time as tangent. Similar field lines, called *streamlines*, can be defined for the velocity field as curves that are everywhere tangent to the velocity field at a fixed time. Such curves are solutions to the ordinary differential equation

$$\frac{d\boldsymbol{x}}{dt} = \boldsymbol{v}(\boldsymbol{x}, t_0), \tag{15.2}$$

where the velocity field is calculated for a fixed value of time t_0. Starting at any point $\boldsymbol{x}_0$ at $t = t_0$ we may use this equation to determine the path $\boldsymbol{x} = \boldsymbol{x}(t, \boldsymbol{x}_0, t_0)$ of a streamline. Because the velocity field is evaluated at a fixed moment in time t_0 there will be only one tangent and thus only one streamline through every point of space. Streamlines depict the velocity field at a single instant in time and can never intersect.

Particle trajectories: Imagine you drop a tiny particle—a speck of dust—into a fluid, and watch how it is carried along with the fluid in its motion. The speck of dust should be neutrally buoyant in the fluid, and so small that its inertia and mass play no role but on the other hand so large that it is not buffeted around much by collisions with individual molecules (i.e. by Brownian motion). The path $\boldsymbol{x} = \boldsymbol{x}(t)$ which it follows is called a *particle trajectory* or orbit and is determined by the differential equation,

$$\frac{d\boldsymbol{x}}{dt} = \boldsymbol{v}(\boldsymbol{x}, t). \tag{15.3}$$

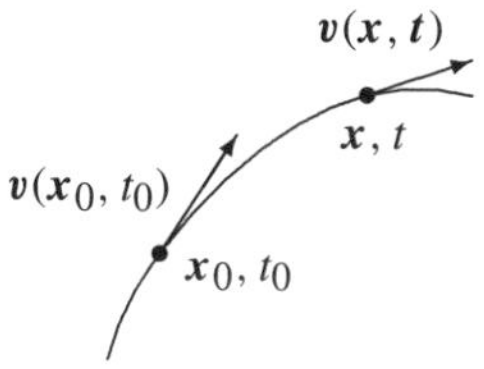

A particle orbit is everywhere parallel with the instantaneous velocity field.

Given a starting point $\boldsymbol{x}_0$ and a starting time t_0, the path $\boldsymbol{x}(t, \boldsymbol{x}_0, t_0)$ may be calculated from this equation for all times t. There will be only one particle orbit going through each point in space at a fixed instant, but different orbits may cross each other and even themselves as long as this occurs at different times.

Streaklines: A standard method for visualizing fluid flow in, for example, wind tunnels, is to inject smoke (or dye) into the fluid at a constant rate. This leads to *streaklines* of smoke weaving through the fluid. Since smoke particles are tiny and light they must follow particle orbits, so that a streakline is obtained from the particle orbit $\boldsymbol{x}(t, \boldsymbol{x}_0, t_0)$ by varying the start time t_0 while keeping fixed the observation time t and the point $\boldsymbol{x}_0$ from which the smoke emanates.

Relating the various flow lines

For a time-dependent velocity field the relationship between the three types of lines can be hard to visualize. For *steady flow*, where the velocity field is independent of time, $\boldsymbol{v}(\boldsymbol{x}, t) = \boldsymbol{v}(\boldsymbol{x})$, the particle orbits evidently coincide with the streamlines, and since the streamlines in this case can only depend on the time difference $t - t_0$, the streaklines will also coincide with them. One should always remember that streamlines are quite misleading for unsteady flow.

Example 15.1.2 (Streamlines, streaklines, particle orbits): To illustrate the difference between the three types of field lines consider a spatially uniform two-dimensional velocity field of the form

$$\boldsymbol{v}(\boldsymbol{x}, t) = (a, bt, 0) \tag{15.4}$$

where the y-component everywhere grows linearly with t. The stream lines are in this case determined by

$$\frac{d\boldsymbol{x}}{dt} = (a, bt_0, 0), \tag{15.5}$$

and are straight lines

$$x = x_0 + a(t - t_0), \qquad y = y_0 + bt_0(t - t_0). \tag{15.6}$$

The particle orbits are determined by

$$\frac{d\boldsymbol{x}}{dt} = (a, bt, 0), \tag{15.7}$$

and are parabolas

$$x = x_0 + a(t - t_0), \qquad y = y_0 + \tfrac{1}{2}b(t^2 - t_0^2). \tag{15.8}$$

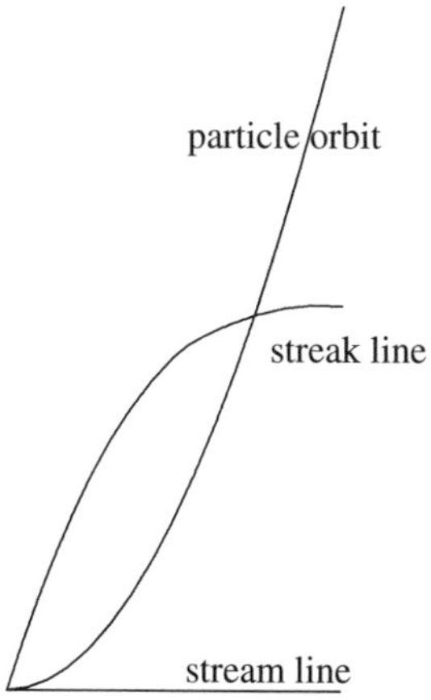

The three types of flow lines from example 15.1.2.

Varying t_0 in the interval $-\infty < t_0 < t$ the streak lines are seen also to be parabolas curving the opposite way to the particle orbits.

Taking $x_0 = y_0 = 0$ and $a = b = 1$, the streamline at $t_0 = 0$ runs along the x-axis, $y = 0$, and the particle orbit becomes $y = x^2/2$. At a given moment of time t, the corresponding streakline, obtained by varying the start time in the interval $0 < t_0 < t$ in (15.8), is described by $y = (1/2)(t^2 - t_0^2) = (1/2)(t^2 - (t - x)^2) = xt - (1/2)x^2$, a curve with maximum at $x = t$.

15.2 Incompressible flow

Most solids and liquids are fairly incompressible under ordinary circumstances and, as we shall see later, gases will often be effectively incompressible when flow speeds are much smaller than the velocity of sound. One should, however, not forget that all materials are in principle compressible. Incompressibility is always an approximation, and should be viewed as a condition on the flow rather than an absolute material property. It is nevertheless such an important condition that we shall devote large parts of the remainder of this book to the study of incompressible flows.

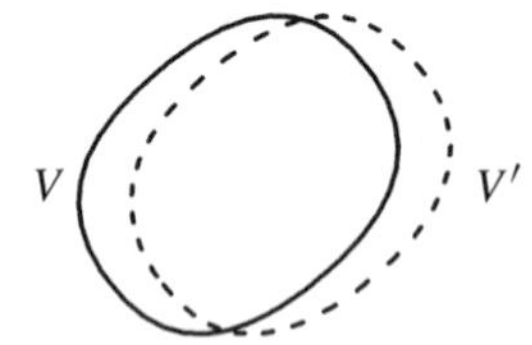

Displacement of a volume of material in a small time interval. The change in volume is given by the thin surface layer between the dashed and solid curves.

Material rate of change of volume

Let V be a fixed volume of matter with surface S. In a small time interval δt, all the material particles in V are simultaneously displaced by $\boldsymbol{v}(\boldsymbol{x}, t)\,\delta t$ to fill out another volume V' in the vicinity of the original volume. Since a surface element $d\boldsymbol{S}$ is displaced through the vector distance $\boldsymbol{v}\delta t$, it scoops up an infinitesimal volume $\boldsymbol{v}\delta t \cdot d\boldsymbol{S}$ (counted with sign), so that the total change in volume becomes

$$\delta V = V' - V = \oint_S \boldsymbol{v}\delta t \cdot d\boldsymbol{S}.$$

Dividing by δt we get the so-called *material rate of change of volume*,

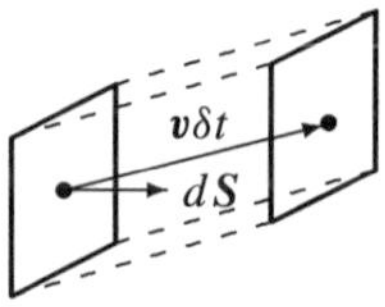

A surface element $d\boldsymbol{S}$ scoops up the volume $\boldsymbol{v}\delta t \cdot d\boldsymbol{S}$ in a small time interval δt.

$$\boxed{\frac{DV}{Dt} = \oint_S \boldsymbol{v} \cdot d\boldsymbol{S}.} \tag{15.9}$$

The reason for the special notation for the material time derivative, D/Dt, is that this is not an ordinary time derivative[1]. It represents the would-be instantaneous rate of change of a volume following along with the flow of matter, a so-called *comoving volume*. Here it is calculated using a fixed volume, but since the integral only depends on the instantaneous shape of the volume, it does not matter whether afterwards it will move along with the movement of matter or not. In chapter 22 we shall generalize the formalism to volumes that move in any way we desire.

By means of Gauss' theorem (6.4) the surface integral may be converted to a volume integral,

$$\frac{DV}{Dt} = \int_V \nabla \cdot \boldsymbol{v}\, dV \tag{15.10}$$

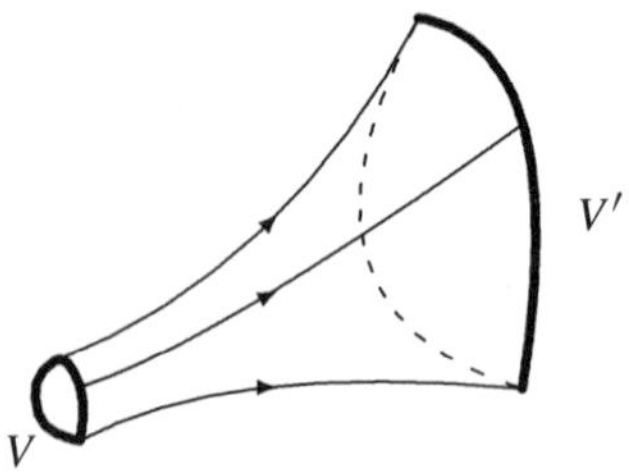

A comoving volume expands in a diverging velocity field, and contracts in a converging field.

showing that a volume expands in a diverging velocity field with $\nabla \cdot \boldsymbol{v} > 0$, and contracts in a converging field with $\nabla \cdot \boldsymbol{v} < 0$. Applied to the infinitesimal volume dV of a material particle, it follows that

$$\boxed{\frac{D(dV)}{Dt} = \nabla \cdot \boldsymbol{v}\, dV.} \tag{15.11}$$

The volume of a comoving material particle thus swells and shrinks according to the divergence or convergence of the velocity field.

Example 15.2.1 (Radial expansion): Let $\boldsymbol{v} = \kappa \boldsymbol{x}$ be a radially expanding velocity field. The comoving rate of change of the volume of a sphere of radius r centred at the origin becomes,

$$\frac{DV}{Dt} = \oint_{|\boldsymbol{x}|=r} \kappa \boldsymbol{x} \cdot d\boldsymbol{S} = 4\pi\kappa r^3, \tag{15.12}$$

because the surface area of the sphere is $4\pi r^2$ and its normal points along $\boldsymbol{x}$. This could also have been obtained by multiplying the volume $(4/3)\pi r^3$ with the constant divergence $\nabla \cdot \boldsymbol{v} = 3\kappa$.

[1] There is no general agreement in the literature on how to denote the material time derivative. Some texts use the ordinary differential operator d/dt and others use a notation like $(d/dt)_{\text{system}}$, but it seems as if the notation D/Dt used in this book is the most common.

Incompressibility condition

Incompressibility means that any comoving volume of matter must be constant in time, $DV/Dt = 0$, or

$$\boxed{\oint_S \boldsymbol{v} \cdot d\boldsymbol{S} = 0,} \tag{15.13}$$

for all closed surfaces. Incompressible matter cannot accumulate anywhere, and equal volumes of incompressible matter must move in and out of any closed surface per unit of time.

Using Gauss' theorem on (15.13) or equivalently (15.11) we find that *the divergence of the velocity field must vanish for incompressible flow*,

$$\boxed{\nabla \cdot \boldsymbol{v} = 0.} \tag{15.14}$$

A divergence-free field is sometimes called *solenoidal*. Note that in this local formulation the condition of incompressibility does not refer to any volume of matter, but only to the field itself.

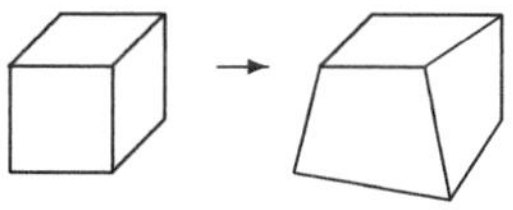

A material particle is also deformed by the flow because the velocity field varies from place to place within it.

Example 15.2.2: The flow described by the time-independent velocity field

$$\boldsymbol{v} = (\sin x \cos y, -\cos x \sin y, 0) \tag{15.15}$$

is incompressible, because

$$\frac{\partial v_x}{\partial x} + \frac{\partial v_y}{\partial y} + \frac{\partial v_z}{\partial z} = \cos x \cos y - \cos x \cos y = 0. \tag{15.16}$$

Due to the periodicity in both x and y, the flow forms a regular array of stationary whirls, one of which is shown in figure 15.1 on page 190. There is probably no practical way of generating this flow pattern.

Leonardo's law

Leonardo da Vinci knew—and used—that the water speed decreases when a canal or river becomes wider and increases when it becomes narrower [70]. He discovered the simple law that the product of the cross-sectional area of a canal and the flow velocity in the canal is always the same.

Leonardo da Vinci (1452–1519). *Italian renaissance artist, architect, scientist and engineer. A universal genius who made fundamental contributions to almost every field. Also a highly practical man who concerned himself with the basic mechanical principles behind everyday machines, and sometimes also future machines, such as the helicopter.*

Consider, for example, an aqueduct or canal and mark two fixed planar cross sections A_1 and A_2, both orthogonal to the general direction of flow. Leonardo's law then says that the water velocities v_1 and v_2 through these cross sections must obey the relation

$$A_1 v_1 = A_2 v_2. \tag{15.17}$$

The law expresses the rather self-evident fact that the same volume of incompressible water has to pass any cross section of the canal per unit of time. We use it all the time when dealing with water.

Leonardo's law follows from the global condition of incompressibility (15.13). Together with the sides of the canal, the two cross sections define a volume to which the condition can be applied. Since no water can flow through the sides of the canal, the surface integral only receives contributions from the two cross sections, and consequently, taking both normals along the general direction of the canal, we get

$$\oint \boldsymbol{v} \cdot d\boldsymbol{S} = \int_{A_2} \boldsymbol{v} \cdot d\boldsymbol{S} - \int_{A_1} \boldsymbol{v} \cdot d\boldsymbol{S} = \boldsymbol{0}. \tag{15.18}$$

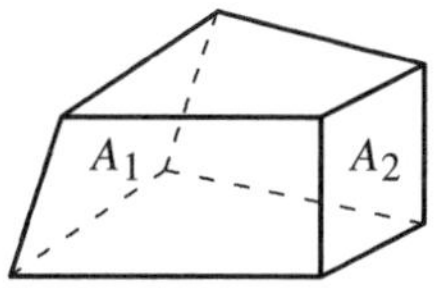

Aqueduct with varying cross section. The volume of water passing through the cross section A_1 is the same as the amount of water passing through A_2, or $A_1 v_1 = A_2 v_2$ where v_1 and v_2 are average flow velocities.

The average flow velocity through a cross section A of the canal is

$$v_A = \frac{1}{A} \int_A \boldsymbol{v} \cdot d\boldsymbol{S}. \tag{15.19}$$

It then follows from (15.18) that the product $A\, v_A$ is the same everywhere along the canal. This modern formulation of Leonardo's law is valid independent of whether the flow is orderly or turbulent.

Example 15.2.3 (Hypodermic syringe): A hypodermic syringe contains a few cm^3 of liquid in a small chamber about 1 cm in diameter and a few centimetres long. The liquid is injected through a hollow needle with an inner diameter of about 1 mm in the course of a few seconds. Since the ratio of cross sections is 100, Leonardo's law tells us that the speed of the liquid in the needle is about 100 times larger than the speed with which the piston of the syringe is pushed, i.e. of the order of metres per second. No surprise that it sometimes hurts to have an injection.

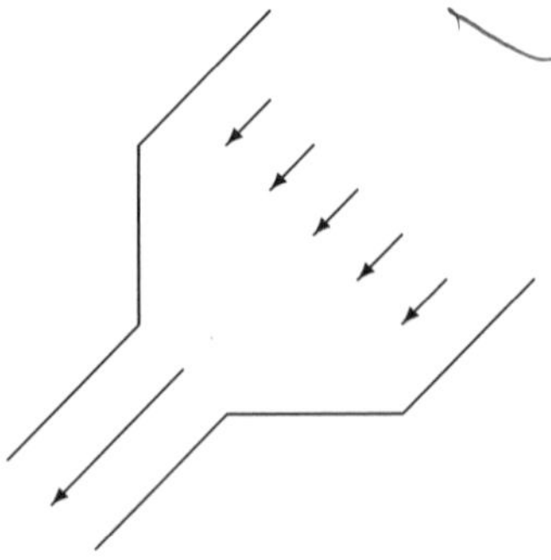

The flow velocity in the thin needle of a syringe is much higher than in the liquid chamber.

Leondardo's law is definitely *not* valid for compressible fluids. When you pump your bicycle by pushing a piston into a cylindric chamber filled with air, the average flow velocity will decrease towards the end of the chamber where it has to vanish, because no air can pass through there. But the cross section of the chamber is constant, so the product of cross section and average velocity *cannot* be constant throughout the chamber.

15.3 Mass conservation

In Newtonian mechanics, mass is *conserved.* The mass of a collection of point particles ('molecules') can only change by addition or removal of particles. Since all matter is made from molecules, this must mean that the only way the mass in a given volume of continuous matter can change is by mass flowing in or out of the volume through its surface. This almost trivial remark leads to the first of the two central equations of continuum dynamics, the *equation of continuity*.

Mass flux

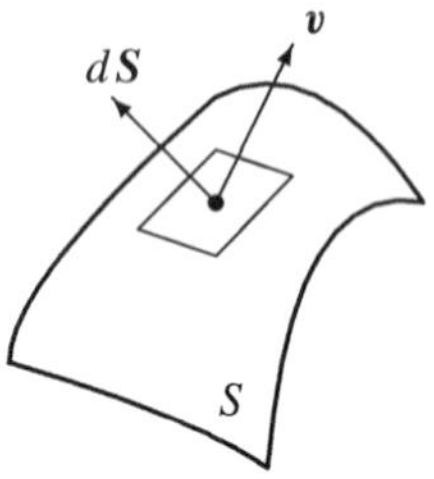

Matter moves with velocity $\boldsymbol{v}(\boldsymbol{x}, t)$ through every surface element $d\boldsymbol{S}$ near $\boldsymbol{x}$.

Consider a piece of a fixed open or closed surface, S. The (signed) amount of mass passing through a tiny surface element $d\boldsymbol{S}$ in the time interval dt will be $\rho \boldsymbol{v} dt \cdot d\boldsymbol{S}$. Integrating over the whole surface and dividing by dt, we find the total amount of mass transported through S per unit of time,

$$Q = \int_S \rho\, \boldsymbol{v} \cdot d\boldsymbol{S}, \tag{15.20}$$

also called the *flux of mass* through the surface S. According to the definition of the velocity field (15.1) the quantity $\rho\, \boldsymbol{v}$ is the *density of momentum*, but now we learn that it also specifies how much mass that flows through a unit area in a unit of time, a quantity called the *current density of mass*.

Example 15.3.1 (Water hose): Water is discharged from a fixed hose at a rate of $Q = 1\ \mathrm{kg\ s^{-1}}$. The hose has cross section $A = 1\ \mathrm{cm^2}$, so that the average current density of mass at the exit of the tube is $\langle \rho v \rangle = Q/A = 10\,000\ \mathrm{kg\ m^{-2}\ s^{-1}}$. Since the water density is constant, $\rho = 1000\ \mathrm{kg\ m^{-2}}$, the average flow speed is $\langle v \rangle = 10\ \mathrm{m\ s^{-1}}$. Ignoring air resistance, the water will reach a height of $h \approx \langle v \rangle^2 / 2g_0 = 5$ m, if directed vertically upwards.

Global mass conservation

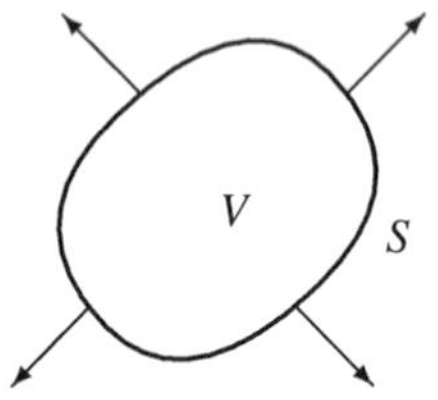

The only way the amount of mass can diminish in a fixed volume V is by a net outflow through its closed surface S.

Since mass can neither be created nor destroyed, the rate of gain of mass in a fixed volume, V, must equal the flux of mass *into* the volume through its closed surface S (as usual oriented outwards),

$$\boxed{\frac{d}{dt} \int_V \rho\, dV = -\oint_S \rho\, \boldsymbol{v} \cdot d\boldsymbol{S}.} \tag{15.21}$$

This is the *global equation of mass conservation* for an arbitrary *fixed* control volume. It expresses the obvious fact that the mass you gain when you eat equals the mass of the food you pass into your mouth.

The material rate of change of mass is the sum of the rate of gain of mass in the fixed volume and the rate at which mass would be incorporated through its surface, were it comoving,

$$\frac{DM}{Dt} = \frac{d}{dt} \int_V \rho\, dV + \oint_S \rho\, \boldsymbol{v} \cdot d\boldsymbol{S}. \tag{15.22}$$

Comparing with mass conservation (15.21) we find,

$$\boxed{\frac{DM}{Dt} = 0.} \tag{15.23}$$

The material rate of change of mass vanishes, expressing simply that the amount of mass in a comoving volume does not change with time, as one would expect.

Local mass conservation

Since the control volume is fixed, we may move the time derivative inside the volume integral and use Gauss' theorem (6.4) on the surface integral, to obtain

$$\frac{DM}{Dt} = \int_V \left(\frac{\partial \rho}{\partial t} + \nabla \cdot (\rho \boldsymbol{v}) \right) dV = 0.$$

As this has to be true for any volume V, we conclude that mass conservation requires,

$$\boxed{\frac{\partial \rho}{\partial t} + \nabla \cdot (\rho \boldsymbol{v}) = 0,} \tag{15.24}$$

for all points $\boldsymbol{x}$ in space and all times t. This is the *equation of continuity*, first obtained by Euler (1753). Although derived from global mass conservation applied to a fixed volume, it is itself a local relation completely without reference to macroscopic volumes.

The continuity equation applies to all locally conserved quantities, for example electric charge. In general it relates the local rate of change of the density of a quantity, here $\partial \rho / \partial t$, to the divergence of its current density, here $\nabla \cdot (\rho \boldsymbol{v})$.

Example 15.3.2 (Bicycle pump): A piston is pushed into an air-filled cylindric chamber with constant cross section A. Let its distance from the end wall be $x = a(t)$ at time t. If the piston moves slowly enough (compared to the velocity of sound), the density is the same throughout the chamber, $\rho(t) = M/Aa(t)$, where M is the (constant) mass of the air in the chamber. It is reasonable to assume that only the x-component of the velocity field is non-vanishing, and the equation of continuity becomes (with a dot denoting the time derivative),

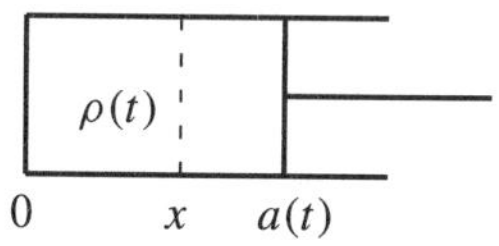

Bicycle pump. The velocity field varies linearly with x.

$$\dot{\rho}(t) + \rho(t) \frac{\partial v_x(x, t)}{\partial x} = 0.$$

The solution to this differential equation, which vanishes for $x = 0$, is

$$v_x(x, t) = -x \frac{\dot{\rho}(t)}{\rho(t)} = x \frac{\dot{a}(t)}{a(t)}. \tag{15.25}$$

When you pull at the piston, $\dot{a} > 0$, the velocity grows linearly with the position x along the chamber, and conversely.

15.4 Moving along with the flow

How does the environment look from the point of view of a small object riding along with the motion of the material? A speck of dust being sucked into a vacuum cleaner will find itself in a region with higher air velocity and lower pressure and density than outside, even if the flow of air is completely steady with air velocity, pressure and density being constant in time everywhere (because you have stopped moving the head of the cleaner). The ambient flow of matter may thus contribute to changes in physical quantities in the neighbourhood of *comoving* particles.

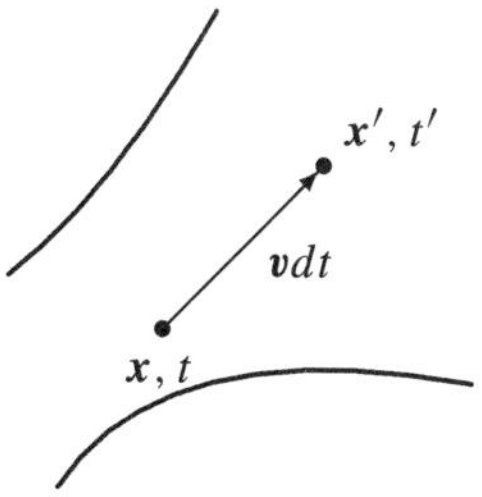

A particle may be swept along with the flow into a region of different density and velocity.

Local material rate of change of density

A particle near the point $\boldsymbol{x}$ at time t riding along with the flow will at time $t' = t + \delta t$ have been displaced to the point $\boldsymbol{x}' = \boldsymbol{x} + \boldsymbol{v}(\boldsymbol{x}, t)\delta t$. Expanding to first order in δt we find the change in mass density

$$\begin{aligned}
\delta \rho &= \rho(\boldsymbol{x} + \boldsymbol{v}\delta t, t + \delta t) - \rho(\boldsymbol{x}, t) \\
&= v_x \delta t \frac{\partial \rho(\boldsymbol{x}, t)}{\partial x} + v_y \delta t \frac{\partial \rho(\boldsymbol{x}, t)}{\partial y} + v_z \delta t \frac{\partial \rho(\boldsymbol{x}, t)}{\partial z} + \delta t \frac{\partial \rho(\boldsymbol{x}, t)}{\partial t} \\
&= \left(\frac{\partial \rho}{\partial t} + (\boldsymbol{v} \cdot \nabla)\rho \right) \delta t.
\end{aligned}$$

As before we shall introduce a special notation for this *local material rate of change* of the density,

$$\boxed{\frac{D\rho}{Dt} = \frac{\partial \rho}{\partial t} + (\boldsymbol{v} \cdot \nabla)\rho.} \tag{15.26}$$

Using the relation

$$\nabla \cdot (\rho \boldsymbol{v}) = (\boldsymbol{v} \cdot \nabla)\rho + \rho \nabla \cdot \boldsymbol{v}, \tag{15.27}$$

we may write the equation of continuity (15.24) as

$$\frac{D\rho}{Dt} = -\rho\, \nabla \cdot \boldsymbol{v}. \tag{15.28}$$

Returning to the question of incompressibility, we see that setting $\nabla \cdot \boldsymbol{v} = 0$ forces the comoving rate of change to vanish. The density must thus be constant in the neighbourhood of a particle that moves along with the incompressible flow, even if the density should vary from place to place.

Rolling boulder: A rigid body, for example a boulder rolling down a mountainside, is by all counts incompressible. Although its mass density may vary according to the mineral composition of the rock, it stays—of course—the same in the neighbourhood of any particular mineral grain, independent of how the boulder rolls.

Constant mass of material particle

The material rate of change of the mass dM of a material particle can be calculated from the material rate of change (15.26) of the density and the material volume rate of change (15.11),

$$\frac{D(dM)}{Dt} = \frac{D(\rho\, dV)}{Dt} = \frac{D\rho}{Dt} dV + \rho \frac{D(dV)}{Dt} = -\rho \nabla \cdot \boldsymbol{v}\, dV + \rho \nabla \cdot \boldsymbol{v}\, dV = 0.$$

In the last step we used the equation of continuity in the form (15.28). Conversely, this calculation shows that we could have arrived at the continuity equation by postulating the constancy of the mass for every comoving material particle.

Material time derivative operator

The material rate of change of the density (15.26) is obtained by applying the differential operator,

$$\boxed{\frac{D}{Dt} = \frac{\partial}{\partial t} + \boldsymbol{v}(\boldsymbol{x}, t) \cdot \nabla,} \tag{15.29}$$

to the density field. This operator, called the *local material time derivative*, is a mixed differential operator in time and space which can be applied to any field. The first term $\partial/\partial t$ represents the *local rate of change* of the field in a fixed point $\boldsymbol{x}$ whereas the second part $\boldsymbol{v} \cdot \nabla$, called the *advective* part, represents the effect of following the motion of the material in the environment of the point[2].

For the trivial vector field $\boldsymbol{x}$ we find, for example,

$$\frac{D\boldsymbol{x}}{Dt} = (\boldsymbol{v}(\boldsymbol{x}, t) \cdot \nabla)\boldsymbol{x} = \boldsymbol{v}(\boldsymbol{x}, t). \tag{15.30}$$

The material rate of change of position thus equals the velocity field.

[2]There appears to be no universally accepted name for the advective term which in some other texts is called the *convective* or *inertia* term.

Eulerian displacement field

In principle we may trace back any material particle to its origin at some instant in time and speak about its (Eulerian) displacement $\boldsymbol{u}(\boldsymbol{x}, t)$ as a function of its actual position $\boldsymbol{x}$ at time t. Since the point of origin, $\boldsymbol{x}_0 = \boldsymbol{x} - \boldsymbol{u}(\boldsymbol{x}, t)$, is fixed for any particle trajectory, we have $D\boldsymbol{x}_0/Dt = \boldsymbol{0}$, and thus from (15.30)

$$\boldsymbol{v} = \frac{D\boldsymbol{u}}{Dt} = \frac{\partial \boldsymbol{u}}{\partial t} + (\boldsymbol{v} \cdot \nabla)\boldsymbol{u}. \tag{15.31}$$

The velocity field is thus the material derivative of the Eulerian displacement field, but note that this is an implicit equation for the velocity field. The displacement field is important for the mechanics of large elastic deformations and for general *viscoelastic* materials having both elastic and fluid properties. In this book we shall not discuss such materials.

15.5 Continuum dynamics

Newton's Second Law states that 'mass times acceleration equals force' for a point particle of fixed mass. Continuum physics is not concerned with point particles, but instead with volumes of matter of finite extent. The smallest such volumes are the material particles, and since a comoving material particle has constant mass it comes closest to the concept of a fixed-mass point particle.

Material acceleration field

For a comoving particle, the material derivative of the velocity field defines the *material acceleration field*,

$$\boldsymbol{w} = \frac{D\boldsymbol{v}}{Dt} = \frac{\partial \boldsymbol{v}}{\partial t} + (\boldsymbol{v} \cdot \nabla)\boldsymbol{v}. \tag{15.32}$$

The first term, the *local acceleration* $\partial\boldsymbol{v}/\partial t$, is most important for rapidly varying small-amplitude velocity fields, such as sound waves in solids or fluids (see section 16.2). The second term, the *advective acceleration* $(\boldsymbol{v} \cdot \nabla)\boldsymbol{v}$, is most important for flows dominated by strong spatial variations in the velocity field.

If the flow is steady, such that the local acceleration vanishes, $\partial\boldsymbol{v}/\partial t = \boldsymbol{0}$, the advective term will be the only cause of acceleration. We become acutely aware of the advective acceleration in a little boat approaching the rapids in a narrowing river.

Newton's Second Law for continuous matter

Applying Newton's Second Law to an arbitrary comoving material particle with mass $dM = \rho\, dV$, we find

$$dM\boldsymbol{w} = d\boldsymbol{\mathcal{F}}, \tag{15.33}$$

where $\boldsymbol{w}$ is the material acceleration (15.32) and $d\boldsymbol{\mathcal{F}}$ is the total force acting on the particle. It was shown in section 9.4 on page 116, that the total force on a material particle can be written as $d\boldsymbol{\mathcal{F}} = \boldsymbol{f}^*\, dV$ where

$$f_i^* = f_i + \sum_j \nabla_j \sigma_{ij}, \tag{15.34}$$

is the *effective force density*. Dividing (15.33) by dV and inserting the acceleration field (15.32) we arrive at *Cauchy's equation* (1827),

$$\boxed{\rho\frac{D\boldsymbol{v}}{Dt} = \rho\left(\frac{\partial \boldsymbol{v}}{\partial t} + (\boldsymbol{v} \cdot \nabla)\boldsymbol{v}\right) = \boldsymbol{f}^*.} \tag{15.35}$$

In conjunction with the continuity equation (15.24), *this equation governs the dynamics of all continuous matter.*

Different types of materials, gases, liquids, solids etc, are characterized by different expressions for the effective force density (15.34), in particular what concerns the part due to contact forces. The last two and a half centuries of continuum physics have essentially 'only' been an exploration of the rich ramifications of this dynamical equation.

Field equations of motion

The equation of continuity (15.24) and Cauchy's dynamic equation may be written in the form of *equations of motion* for the four field components, ρ and the three components of $\boldsymbol{v}$,

$$\frac{\partial \rho}{\partial t} = -\nabla \cdot (\rho \boldsymbol{v}), \tag{15.36a}$$

$$\frac{\partial \boldsymbol{v}}{\partial t} = -(\boldsymbol{v} \cdot \nabla)\boldsymbol{v} + \frac{1}{\rho}\boldsymbol{f}^*. \tag{15.36b}$$

Knowing the density and velocity fields at a given time together with the effective body force density (which usually also depends on these fields), the above equations allow us to calculate the rate of change of the fields. If the forces depend on non-mechanical fields, for example the temperature, special equations of motion are also needed for those fields to make the system complete (see chapter 30).

* Global dynamics

Cauchy's equation (15.35) has been derived by applying Newton's second law to material particles. It might as well have been obtained by postulating the global law that the material rate of change of momentum of any volume of matter must equal the total force acting on that volume,

$$\boxed{\frac{D\mathcal{P}}{Dt} = \mathcal{F}.} \tag{15.37}$$

For a fixed volume, the material rate of change is as before defined as the normal rate of change in the volume plus the loss of momentum through the surface. For the ith component of momentum we have

$$\frac{D\mathcal{P}_i}{Dt} = \frac{d}{dt}\int_V \rho v_i \, dV + \oint_S \rho v_i \, \boldsymbol{v} \cdot d\boldsymbol{S}. \tag{15.38}$$

Passing the time derivative into the integral and using Gauss' theorem, we obtain a volume integral with the integrand resembling the continuity equation,

$$\frac{\partial(\rho v_i)}{\partial t} + \nabla \cdot (\rho v_i \boldsymbol{v}) = v_i \left(\frac{\partial \rho}{\partial t} + \nabla \cdot (\rho \boldsymbol{v})\right) + \rho \left(\frac{\partial v_i}{\partial t} + (\boldsymbol{v} \cdot \nabla) v_i\right). \tag{15.39}$$

Here the first term vanishes because of the continuity equation for mass, and the second term becomes f_i^* because of Cauchy's equation, thereby proving (15.37). In chapter 22 global laws of this kind will be derived systematically and used to obtain approximate solutions to fluid mechanics problems.

15.6 Big Bang

A cloud of non-interacting particles, grains or fragments, is perhaps a poor model for continuous matter, but it is nevertheless of interest to study the equations of motion for the velocity field in this most simple case where all volume and contact forces are absent. It may even be used as a primitive model for the expanding universe with galaxies playing the role of grains. But the lack of interaction violates the continuum conditions discussed in chapter 1, and no dynamic smoothing of the fields by collisions will occur. Given a certain initial velocity any grain will like a ghost continue unhindered with the same velocity through the cloud for all time, and thus have infinite mean free path. The model should definitely be taken with a grain of salt.

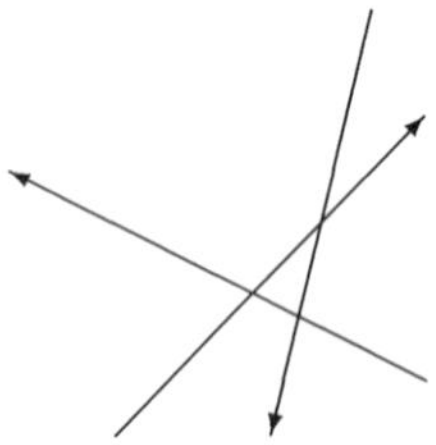

In a cloud of free particles all particles move in straight lines with constant velocity forever.

Explosion field

Suppose that the cloud is created in an explosion where the fragments have stopped interacting immediately after the event and now move freely away from each other. Since there are neither contact forces nor body forces, we have $\boldsymbol{f}^* = \boldsymbol{0}$, and every material particle is unaccelerated. Consequently, the comoving acceleration (15.32) must vanish everywhere,

$$\boxed{\frac{\partial \boldsymbol{v}}{\partial t} + (\boldsymbol{v} \cdot \nabla)\boldsymbol{v} = 0.} \tag{15.40}$$

In spite of this being the simplest possible dynamical equation, it looks complicated enough, and if presented without any other explanation, we would have some difficulty solving it because of its nonlinearity.

Underneath, however, we know that it only implements the law of inertia, with all particles moving at constant velocity along straight lines. If the explosion happened at the point $\boldsymbol{x} = \mathbf{0}$ at time $t = 0$, the fragments were almost instantly given random velocities in all directions. After the explosion the fragments will be separated according to their velocities with the fastest fragments being farthest away. At time t, the velocity of any fragment found at $\boldsymbol{x}$ must be

$$\boxed{\boldsymbol{v}(\boldsymbol{x}, t) = \frac{\boldsymbol{x}}{t},} \tag{15.41}$$

independent of how the explosion started out. To see that this indeed satisfies (15.40), we calculate the x-component

$$\begin{aligned}\frac{\partial v_x}{\partial t} + (\boldsymbol{v} \cdot \nabla) v_x &= \frac{\partial (x/t)}{\partial t} + \frac{x}{t}\frac{\partial (x/t)}{\partial x} + \frac{y}{t}\frac{\partial (x/t)}{\partial y} + \frac{z}{t}\frac{\partial (x/t)}{\partial z} \\ &= -\frac{x}{t^2} + \frac{x}{t^2} \\ &= 0,\end{aligned}$$

and similarly for the y and z components.

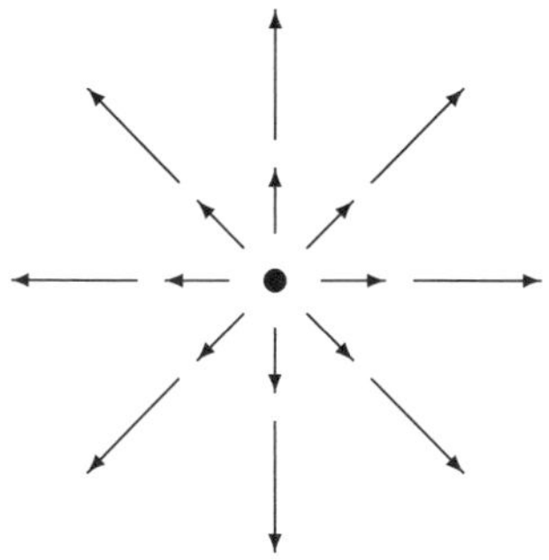

Explosion fragments become separated according to initial velocity, because those fragments that accidentally have the largest initial velocity will ever after be farthest away from the centre.

Hubble's law

The explosion field (15.41) is of the same form as Hubble's law, which states that all galaxies move away from us (and each other) with speeds that are proportional to their distances, in our notation written as,

$$\boldsymbol{v} = H_0 \boldsymbol{x}, \tag{15.42}$$

where H_0 is called the *Hubble constant*.

In general relativity, this law is understood as a consequence of a uniform expansion of space itself since the initial Big Bang. In the primitive model used here, the Hubble constant equals the inverse age of the universe $H_0 = 1/t$. Although first determined in 1929, it has been very difficult for astronomy to settle on a reliable experimental value for the Hubble constant. The latest value[3] is 75 kilometres per second per megaparsec or $H_0 \approx 2.4 \times 10^{-18}\ \text{s}^{-1}$ with an uncertainty of about 10%. The inverse comes to $1/H_0 \approx 13$ billion years which agrees comfortably with the age of 12.5 billion years for the oldest stars[4]. The problem is, however, that the age of the universe determined from cosmology is model-dependent (see section 15.7), and more conservative estimates places the age of the universe somewhere between 12 and 20 billion years.

Edwin Powell Hubble (1889–1953). *American astronomer. Although originally obtaining a BS-degree in astronomy and mathematics, he continued in law but only practiced it for a year before turning back to astronomy. Demonstrated in 1924 that the observed spiral clouds were galaxies, far outside the Milky Way. Discovered the distance-velocity relationship that now carries his name in 1929.*

* Evolution of the mass density

The shape of the localized mass distribution just before the explosion at $t = 0$ does not matter much for what happens later. An explosion is a cataclysmic event where large unknown forces distribute essentially random velocities to all the fragments of the 'body' of mass M_0 that existed before the explosion.

After the explosion these fragments become separated according to their velocities, as described by the explosion field (15.41). Let us, for example, assume that the probability that a fragment has a velocity $\boldsymbol{v}$ in a small neighbourhood $d^3\boldsymbol{v} = dv_x dv_y dv_x$ around $\boldsymbol{v}$ is $f(\boldsymbol{v})\, d^3\boldsymbol{v}$, where the probability density is normalized,

$$\int f(\boldsymbol{v})\, d^3\boldsymbol{v} = 1. \tag{15.43}$$

[3] L. V. E. Koopmans, T. Trey, C. D. Fassnacht, R. D. Blandford and G. Surpi, The Hubble constant from the gravitational lens $B1608 + 656$, *The Astrophysical Journal* **599**, (2003) 70.

[4] R. Cayrel, V. Hill, T. C. Beers, B. Barbuy, M. Spite, F. Spite, B. Plez, J. Andersen, P. Bonifacio, P. François, P. Molaro and B. Nordström, Measurement of stellar age from uranium decay, *Nature* **409**, (2001) 691.

The fragments given the velocity $\boldsymbol{v}$ in $d^3\boldsymbol{v}$ will have mass $dM = M_0 f(\boldsymbol{v}) d^3\boldsymbol{v}$, and since they are found at the position $\boldsymbol{x} = \boldsymbol{v}\,t$ at time t, the mass is also $dM = \rho\, dV = \rho t^3 d^3\boldsymbol{v}$. Solving for the mass density we find,

$$\rho(\boldsymbol{x}, t) = \frac{M_0}{t^3} f\left(\frac{\boldsymbol{x}}{t}\right). \tag{15.44}$$

It is not hard to show (problem 15.7) that the total mass equals M_0 and that this density satisfies the equation of continuity (15.24) with the velocity field (15.41).

15.7 Newtonian cosmology

Continuing our investigation of the expanding universe, we now wish to include the gravitational field in the dynamical equation (15.35), but still allow no contact forces, so that $\boldsymbol{f}^* = \rho \boldsymbol{g}$. The equation of motion then becomes

$$\frac{\partial \boldsymbol{v}}{\partial t} + (\boldsymbol{v} \cdot \nabla)\, \boldsymbol{v} = \boldsymbol{g}, \tag{15.45}$$

instead of (15.40). Despite being non-relativistic, this model captures essential elements of cosmology, although a proper understanding does require general relativity [78].

Cosmic democracy

In days of old, the Earth was thought to be at the centre of the universe. Since Copernicus, this thinking has been increasingly replaced by the more 'democratic' view that the Earth, the Sun, the Galaxy are common members of the universe of no particular distinction (except that we live here!). The end of this line of thought is the extreme Copernican view that for cosmological considerations every place in the universe is as good as any other. As we shall see, this 'Cosmological Principle' or 'Principle of Cosmic Democracy' is quite useful.

Nicolaus Copernicus (1473–1543)). *Polish astronomer. Studied when young both medicine and astronomy. Revolutionized the understanding of the solar system with his unauthored booklet* Little Commentary *from 1514, and especially with his life's work, the 200 page book* De revolutionibus orbium coelestium, *published in latin in the year of his death (1543). His views of the solar system were widely criticized but supported by both Kepler and Galileo.*

Mass density: It immediately follows from this principle that at a particular instant of time t, the (average) mass density cannot depend on where you are, and must thus be the same everywhere,

$$\boxed{\rho(\boldsymbol{x}, t) = \rho(t).} \tag{15.46}$$

Velocity field: The Hubble expansion of the universe,

$$\boxed{\boldsymbol{v}(\boldsymbol{x}, t) = H(t)\, \boldsymbol{x},} \tag{15.47}$$

with a time-dependent Hubble 'constant', $H(t)$, does not look 'democratic' because it seems to single out the centre of the coordinate system. It is, in fact, completely democratic, because

$$\boldsymbol{v}(\boldsymbol{x}, t) - \boldsymbol{v}(\boldsymbol{y}, t) = H(t)(\boldsymbol{x} - \boldsymbol{y}). \tag{15.48}$$

This means that an observer in a galaxy at point $\boldsymbol{y}$ will also see the other galaxies recede from him according to the same Hubble law as ours.

Field of gravity: Can gravity be democratic? In Newtonian cosmology it is not possible to view the universe as a homogeneous whole when it comes to gravity. What is gravity in an infinite universe? Symmetry would seem to argue that it should vanish, because there is as much matter pulling from one side as from the opposite, but is that right?

To overcome this problem, let us for a while think of the universe as a huge sphere with vacuum outside and centred somewhere, perhaps right here, and let us put the origin of the coordinate system at the centre of this sphere. In that case, we have seen in section 6.2 that the strength of gravity at a given point $\boldsymbol{x}$ depends only on the amount of mass, $M(r) = (4/3)\pi r^3 \rho$, inside the sphere with radius $r = |\boldsymbol{x}|$, whereas one may forget the mass outside this radius. In other words, the field of gravity is

$$\boxed{\boldsymbol{g}(\boldsymbol{x}, t) = -\frac{4\pi}{3} G\rho(t)\,\boldsymbol{x}.} \tag{15.49}$$

Interestingly, by the same argument as for the velocity, an observer in another galaxy at $\boldsymbol{y}$ will see a similar gravitational acceleration field around himself

$$\boldsymbol{g}(\boldsymbol{x}, t) - \boldsymbol{g}(\boldsymbol{y}, t) = -\frac{4}{3}\pi G\rho(t)(\boldsymbol{x} - \boldsymbol{y}), \tag{15.50}$$

as if he/she/it were also at the centre of the universe. This observer may, however, not think of his universe as a huge sphere centred on himself, but must concede that our galaxy is special, at least as long as he subscribes to Newtonian physics. In general relativity, this problem happily goes away.

Cosmological equations

Using $\nabla \cdot \boldsymbol{x} = 3$, we obtain from (15.47) and the equation of continuity (15.24)

$$\boxed{\dot{\rho} = -3H\rho,} \tag{15.51}$$

where as before a dot denotes differentiation with respect to time. Similarly, inserting (15.47) into (15.45) and using $(\boldsymbol{x} \cdot \nabla)\boldsymbol{x} = \boldsymbol{x}$, this equation becomes, after removal of a common factor $\boldsymbol{x}$

$$\boxed{\dot{H} + H^2 = -\frac{4\pi}{3} G\rho.} \tag{15.52}$$

Newtonian cosmology thus reduces to just two coupled ordinary differential equations for the mass density and the Hubble 'constant'. Note that the reference to the centre of the universe has disappeared completely, and we may from now on again think of a truly infinite universe with equal rights for all observers.

The cosmic scale factor

The simplest way to solve these equations is by introducing a new quantity with the dimension of length, $a(t)$, called the *cosmic scale factor*, satisfying

$$\dot{a} = Ha. \tag{15.53}$$

From the equation of continuity (15.51) we get

$$\frac{d}{dt}(\rho a^3) = -3H\rho\, a^3 + \rho\, 3a^2 Ha = 0, \tag{15.54}$$

and this shows that the mass, $M = (4/3)\pi\rho(t)a(t)^3$, in an expanding sphere of radius $a(t)$ is constant in time. Eliminating H from (15.52), we obtain the following differential equation for cosmic scale factor,

$$\ddot{a} = -G\frac{M}{a^2}, \tag{15.55}$$

which is identical to the equation of motion for a particle moving radially in the gravitational field of a point mass M.

Critical density

The 'equation of motion' (15.55) implies that the 'energy'

$$E = \frac{1}{2}\dot{a}^2 - \frac{GM}{a}, \tag{15.56}$$

must be conserved in the time evolution of the scale factor, i.e. $\dot{E} = 0$, as may easily be verified. Eliminating the mass, and using (15.53) to eliminate $\dot{a}$, it may be written in the form

$$E = \frac{4}{3}\pi G a^2 (\rho_c - \rho), \tag{15.57}$$

where the *critical density*,

$$\boxed{\rho_c = \frac{3H^2}{8\pi G},} \tag{15.58}$$

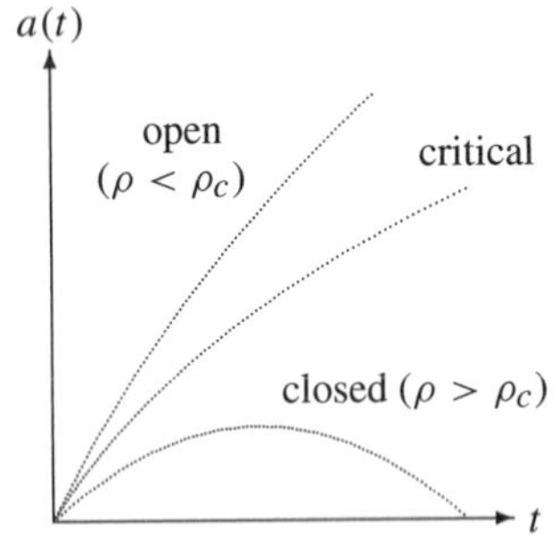

Time evolution of the cosmic scale factor depending on the actual average mass density compared to the critical density.

can be calculated from present-day observation of the Hubble 'constant'. Using the latest value $H_0 = 75\ \mathrm{km\ s^{-1}\ Mpc^{-1}}$ one finds $\rho_c \approx 1.1 \times 10^{-26}\ \mathrm{kg\ m^{-3}}$, corresponding to about six protons per cubic metre.

From the particle analogy, we know that the scale factor will 'escape' to infinity for $E > 0$, but turn around and 'fall back' for $E < 0$. This means that the expansion will continue forever for $\rho < \rho_c$, but eventually must turn around and become a contraction if $\rho > \rho_c$. The two types of universe are called *open* or *closed*, respectively. For the critical case, $\rho = \rho_c$, the expansion also continues forever at the slowest possible pace.

Age of the universe

The energy equation (15.56) may be solved for $\dot{a}$ with the result,

$$\dot{a} = \sqrt{2\left(E + \frac{GM}{a}\right)}.$$

Demanding that $a = 0$ for $t = 0$, the solution is given implicitly by

$$t = \int_0^a \frac{dr}{\sqrt{2\left(E + \frac{GM}{r}\right)}} \tag{15.59}$$

which must be the time elapsed since the scale factor was zero, i.e. since the Big Bang. For the critical case, $E = 0$, the integral is easy to evaluate and we find

$$t = \frac{2}{3}\frac{a^{3/2}}{\sqrt{2GM}} \tag{15.60}$$

from which it follows that $H = \dot{a}/a = 2/3t$ or $t = 2/3H$. With the present day value of H the age of the universe comes to about 8 billion years, disagreeing strongly with the age of the oldest stars.

The cosmological constant

The problem is that the actual age of the universe determined from cosmology is model-dependent. If the universe, for example, besides ordinary matter were filled with a ghostly 'dark matter' (or perhaps 'dark energy') with a positive mass density, ρ_0, constant in both space and time, then the 'energy' (15.57) should be replaced by

$$E = \frac{1}{2}\dot{a}^2 - \frac{G(M + M_0)}{a} = \frac{4\pi}{3}Ga^2(\rho_c - (\rho + \rho_0)), \tag{15.61}$$

where $M_0 = (4/3)\pi\rho_0 a^3$ and ρ as before is the density of ordinary matter still obeying the continuity equation (15.51).

The new dynamic equation is found by differentiating the energy E after time (remembering that ρ_c is *not* constant),

$$\begin{aligned}\dot{E} &= \frac{4\pi}{3}G2a\dot{a}(\rho_c - (\rho + \rho_0)) + \frac{4\pi}{3}Ga^2(\dot{\rho}_c - \dot{\rho}) \\ &= \frac{4\pi}{3}Ga^2\left(2H(\rho_c - (\rho + \rho_0)) + \frac{3H\dot{H}}{4\pi G} + 3H\rho\right) \\ &= a^2H\left(\dot{H} + H^2 + \frac{4\pi}{3}G(\rho - 2\rho_0)\right).\end{aligned}$$

Since the energy must be constant, the new dynamic equation is,

$$\boxed{\dot{H} + H^2 = -\frac{4\pi}{3}G(\rho - 2\rho_0).} \tag{15.62}$$

For sufficiently large 'dark' density, satisfying $2\rho_0 > \rho$, the gravitational attraction apparently turns into a *gravitational repulsion.*

The real understanding of how a positive mass density ρ_0 can give rise to an effective gravitational repulsion can only be obtained from relativistic theory [78, p. 614], which correctly takes into account the huge pressure accompanying ρ_0. Einstein already introduced in 1917 the so-called *cosmological constant* (which is proportional to ρ_0) for the explicit purpose of permitting static solutions to the cosmological equations of general relativity. Clearly, the new cosmological equations admit a non-expanding, static solution having $\rho = 2\rho_0$ and $H = 0$. Since that time we have learned that our universe is not static but instead expands with a finite Hubble constant. For many years the case for the cosmological constant seemed to be closed, but recent discoveries may have changed that.

In an open universe, the following peculiar scenario is possible when $\rho_0 < 1/3\rho_c$. For $2\rho_0 < \rho < \rho_c - \rho_0$, the energy is positive and the effective gravity is attractive, so that the expansion will decelerate with time, i.e. with $\ddot{a} = a(\dot{H} + H^2) < 0$. But sooner or later, the expansion will make the density fall below the critical value $\rho = 2\rho_0$, and the effective gravity becomes repulsive. The expansion will then begin to accelerate with $\ddot{a} > 0$, and continue to do so forever thereafter. Recent observations seem to indicate that the cosmic expansion is in fact accelerating and that it may have been decelerating in the past[5].

Problems

15.1 Draw streamlines, particle orbits and streaklines for a rotating velocity field $\boldsymbol{v} = a(\cos\omega t, \sin\omega t, 0)$. This resembles the velocity field of a pad sander, used to sand wooden surfaces.

15.2 The wind suddenly turns from south to west. Draw streamlines, particle orbits and streaklines before and after the event.

15.3 A water pipe with diameter 1 inch branches into two pipes with diameters 3/4 inch and 1/2 inch. Water is tapped from the largest branch at double the rate as from the other. What is the ratio of velocities in the pipes?

15.4 Calculate the time-derivative of $\rho(\boldsymbol{x}(t), t)$ where $\boldsymbol{x}(t)$ is a particle orbit and show that it is identical to the comoving derivative.

15.5 Consider an incompressible steady flow in a stream with constant depth, $z = d$, bounded on one side by a straight line, $y = 0$, and on the other side by a curve $y = h(x)$, which is slowly varying $|dh/dx| \ll 1$. **(a)** Calculate the average flow velocity in the x-direction (for fixed x). **(b)** Approximately calculate the comoving acceleration in the flow. **(c)** What should the shape of the curve be in order for the comoving acceleration to be independent of x?

[5] See for example M. S. Turner and A. G. Riess, Do Type Ia supernovae provide direct evidence for past deceleration of the universe?, *The Astrophysical Journal* **569**, (2002) 18.

15.6 Consider an incompressible steady flow in a circular tube along the x-axis with a slowly changing radius $r = a(x)$. **(a)** Calculate the average flow velocity in the x-direction. **(b)** Approximately calculate the comoving acceleration in the tube, and **(c)** determine what shape of tube will lead to constant comoving acceleration.

15.7 Show that the explosion density (15.44) satisfies the equation of continuity.

$*$ **15.8** Prove (15.11) directly from the change of the infinitesimal volume of a material particle without making use of Gauss' theorem (hint: use the Jacobi determinant of the infinitesimal displacement).

$*$ **15.9** Consider a universe in which matter is created everywhere at a constant rate, J, per unit of volume and time (Bondi and Gold (1948), Hoyle (1949)). Show that this allows for a steady-state cosmological solution with constant mass density and Hubble constant, and determine the rate of mass creation.

16 Nearly ideal flow

The most important fluids of our daily life, air and water, are lively and easily set into irregular motion. Getting out of a bathtub creates visible turbulence in the soapy water, whereas we have to imagine the unruly air behind us when we jog. Internal friction, or *viscosity*, seems to play only a minor role in these fluids. Other fluids, like honey and grease, are highly viscous, do not easily become turbulent, and would certainly be very hard to swim in.

Being lively or sluggish is, however, not an absolute property of a fluid, but rather a condition of the circumstances under which it flows. Lava may be very sluggish in small amounts, but when it streams down a mountainside it appears to be quite lively. We shall see later that there is a way of characterizing fluid flow by means of a real number, called the *Reynolds number*, which is typically large for lively flow and small for sluggish flow.

The earliest modern quantitative model of fluid behaviour goes back to Euler about 250 years ago and did not include viscosity. Although Newton introduced the concept, viscosity first entered fluid mechanics in its modern formulation almost a century later. Fluids with no viscosity have been called *ideal* or *perfect*, or even dry. An ideal fluid does not hang on to solid surfaces but is able to slip along container walls with finite velocity, whereas a real fluid has to adjust its velocity field so that it matches the surface speed of the solid objects that are in contact with it.

Although truly ideal fluids do not really exist, except for a component of superfluid helium close to zero kelvin, viscous fluids may nevertheless flow with such high Reynolds number that they behave as nearly ideal. Being ideal is thus more a property of the flow conditions than of the fluid itself. Independent of how large the Reynolds number of the flow, there will always be viscous *boundary layers* around solid obstacles and near the walls of fluid conduits. In this chapter we focus on nearly ideal flow, and postpone the introduction of viscosity to chapter 17 and of boundary layers to chapter 28.

16.1 The Euler equation

Leonhard Euler (1707–83). *Swiss mathematician who made fundamental contributions to calculus, geometry, number theory, and to practical ways of solving mathematical problems. His books on differential calculus (1755) and integral calculus (1768–70) have been especially useful for physics.*

In 1755 Euler was the first to write down Newton's second law of motion for fluids without viscosity. In such a fluid the only forces at play are pressure and gravity, but in distinction to hydrostatics, where these two forces are in balance, they now give rise to a non-vanishing effective density of force $\boldsymbol{f}^* = \rho \boldsymbol{g} - \nabla p$. Inserting this into the dynamic equation (15.35) and dividing by the density ρ, we obtain the *Euler equation* for ideal fluids,

$$\boxed{\frac{\partial \boldsymbol{v}}{\partial t} + (\boldsymbol{v} \cdot \nabla)\, \boldsymbol{v} = \boldsymbol{g} - \frac{1}{\rho} \nabla p.} \tag{16.1}$$

Together with the equation of continuity (15.24) which we repeat here,

$$\frac{\partial \rho}{\partial t} + \nabla \cdot (\rho \boldsymbol{v}) = 0, \tag{16.2}$$

and a barotropic equation of state $p = p(\rho)$, we have arrived at a closed set of five equations for the five fields, v_x, v_y, v_z, ρ and p (assuming that $\boldsymbol{g}$ is known). If the equation of state also depends on temperature, $p = p(\rho, T)$, a heat equation for the temperature field must be added to the set (see chapter 30).

Boundary conditions

Partial differential equations need boundary conditions. As in hydrostatics, Newton's third law demands that the pressure must be continuous across any material interface (in the absence of surface tension). Furthermore, since moving fluids are normally contained in tubes, pipes, or other kinds of conduits that are impenetrable to the fluid, it follows that the velocity component normal to a containing surface at rest must vanish,

$$\boldsymbol{v} \cdot \boldsymbol{n} = 0. \tag{16.3}$$

There are, however, no conditions on the tangential velocity in non-viscous flow. Ideal fluids are thus able to slip along container surfaces with finite tangential velocity, whereas the omnipresent viscosity of real fluids also demands that the tangential velocity must vanish at a containing surface at rest.

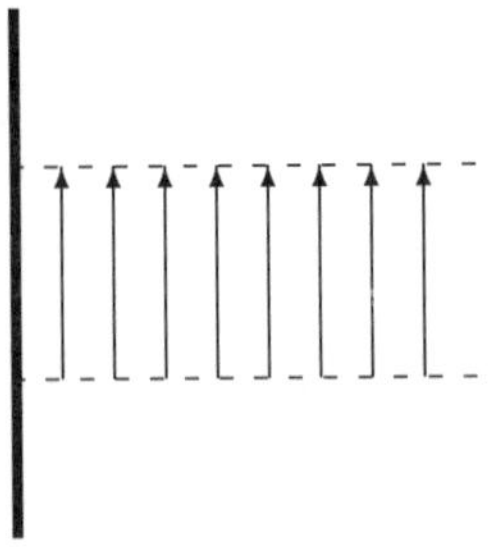

In an ideal fluid the tangential component of the velocity field is non-vanishing all the way to the static boundary.

Exploring the extremes

Although we are now in possession of the fundamental equations for ideal fluids, solving them is another matter. The bad news about nonlinear partial differential equations is that they are very hard to solve and that makes it imperative to explore their usually simpler extreme limits. One such limit is the *linear approximation* in which the nonlinearities are dropped, another is *incompressible flow* in which the density is taken to be a constant, and still another is *steady flow* where all fields are assumed to be time-independent. In the following sections these limits will be discussed in detail.

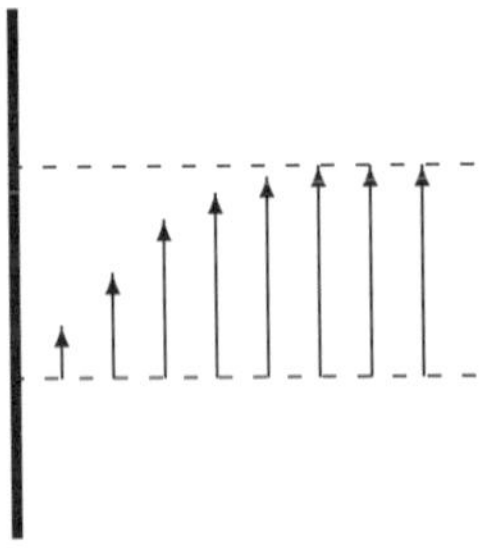

In a real fluid the tangential component of the velocity field rises linearly from a static boundary and joins smoothly with the flow at large.

16.2 Small-amplitude sound waves

When you clap your hands, you create momentarily a small disturbance in the air which propagates to your ear and tells you that something happened. The diaphragm of the loudspeaker in your radio vibrates in tune with the music carried by the radio waves and the electric currents in the radio and transfers its vibrations to the air where they continue as *sound*. No significant bulk movement of air takes place over longer distances, but locally the air oscillates back and forth with small spatial amplitude, and the velocity, density and pressure fields oscillate along with it.

Wave equation

For simplicity we assume that there is no gravity (see however problem 16.17). Before the sound starts, the fluid is assumed to be in hydrostatic equilibrium with constant density ρ_0 and constant pressure p_0. We now disturb the equilibrium by setting the fluid into motion with a tiny velocity field $\boldsymbol{v}(\boldsymbol{x}, t)$. The disturbance generates a small change in the density, $\rho = \rho_0 + \Delta\rho$, and in the pressure $p = p_0 + \Delta p$. Inserting this into the Euler equations we obtain to first order in the small quantities, $\boldsymbol{v}$, Δp, and $\Delta\rho$,

$$\frac{\partial \boldsymbol{v}}{\partial t} = -\frac{1}{\rho_0}\nabla \Delta p, \qquad \frac{\partial \Delta\rho}{\partial t} = -\rho_0 \nabla \cdot \boldsymbol{v}. \tag{16.4}$$

Differentiating the second equation and inserting the first, we obtain

$$\frac{\partial^2 \Delta\rho}{\partial t^2} = \nabla^2 \Delta p. \tag{16.5}$$

Provided the fluid obeys a barotropic equation of state $p = p(\rho)$ we get a relation between the pressure and density corrections. From the definition of the bulk modulus (4.33) on page 53 we get to first order,

$$\Delta p = \left.\frac{dp}{d\rho}\right|_0 \Delta\rho = \frac{K_0}{\rho_0}\Delta\rho, \tag{16.6}$$

where K_0 is the bulk modulus in hydrostatic equilibrium. Multiplying (16.5) with K_0/ρ_0 we get a *standard wave equation* for the pressure correction,

$$\boxed{\frac{\partial^2 \Delta p}{\partial t^2} = c_0^2 \nabla^2 \Delta p,} \tag{16.7}$$

where we, for convenience, have introduced the constant,

$$c_0 = \sqrt{\frac{K_0}{\rho_0}}. \tag{16.8}$$

It has the dimension of a velocity, and may as we shall see below be identified with the *speed of sound*. For water with $K_0 \approx 2.3$ GPa and $\rho_0 \approx 10^3$ kg m^{-3} the sound speed comes to about $c_0 \approx 1500$ m s^{-1} $\approx$ 5500 km h^{-1}.

Fluid	T [° C]	c_0 [m s^{-1}]
Glycerol	25	1920
Sea water	20	1521
Fresh water	20	1482
Lube Oil	25	1461
Mercury	25	1449
Ethanol	25	1145
Helium	0	973
Water vapour	100	478
Neon	30	461
Humid air	20	345
Dry air	20	343
Oxygen	30	332
Argon	0	308
Nitrogen	29	268

Sound speeds in various fluids (from [34]).

Isentropic sound speed in an ideal gas

Sound vibrations in air are normally so rapid that temperature equilibrium is never established, allowing us to assume that the oscillations take place without heat conduction, i.e. adiabatically. The bulk modulus (4.38) of an isentropic ideal gas is $K_0 = \gamma p_0$ where γ is the adiabatic index, and we obtain

$$\boxed{c_0 = \sqrt{\frac{\gamma p_0}{\rho_0}} = \sqrt{\gamma \frac{R T_0}{M_{\rm mol}}}.} \tag{16.9}$$

In the last step we have used the ideal gas law $p_0 = \rho_0 R T_0 / M_{\rm mol}$.

Example 16.2.1 (Sound speed in the atmosphere): For air at 20° C with $\gamma = 7/5$ and $M_{\rm mol} =$ 29 g mol^{-1}, this comes to $c_0 \approx 343$ m s^{-1} ≈ 1235 km h^{-1}. Since the temperature of the homentropic atmosphere falls linearly with height according to (4.43), the speed of sound varies with height z above the ground as

$$c = c_0 \sqrt{1 - \frac{z}{h_2}}, \tag{16.10}$$

where c_0 is the sound speed at sea level and $h_2 \approx 30$ km is the homentropic scale height (4.44). At the flying altitude of modern jet aircraft, $z \approx 10$ km, the sound speed has dropped to $c \approx 280$ m s^{-1} $\approx$ 1000 km h^{-1}. At greater heights this expression begins to fail because the homentropic model of the atmosphere fails.

Plane wave solution

An elementary plane pressure wave moving along the x-axis with wavelength λ, period τ and amplitude $p_1 > 0$ is described by a pressure correction of the form,

$$\Delta p = p_1 \cos(kx - \omega t), \tag{16.11}$$

where $k = 2\pi/\lambda$ is the wavenumber and $\omega = 2\pi/\tau$ is the circular frequency. Inserting the plane wave into the wave equation (16.7), we obtain $\omega^2 = c_0^2 k^2$ or $c_0 = \omega/k = \lambda/\tau$. The surfaces of constant pressure are planes orthogonal to the direction of propagation, satisfying $kx - \omega t =$ const. Differentiating this equation with respect to time, we see that the planes of constant pressure move with velocity $dx/dt = \omega/k = c_0$, also called the *phase velocity* of the wave. This shows that c_0 given by (16.8) may indeed be identified with the speed of sound in the material.

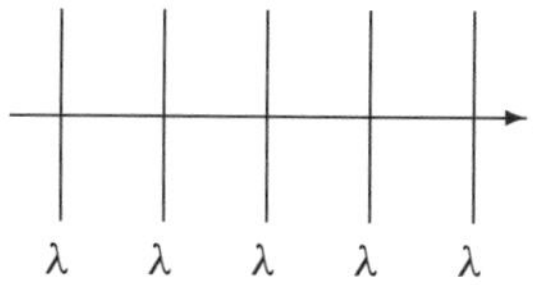

Plane pressure wave propagating along the x-axis with wavelength λ. There is constant pressure in all planes orthogonal to the direction of propagation.

Inserting the plane pressure wave into the x-component of the Euler equation (16.4), we find the velocity field

$$v_x = v_1 \cos(kx - \omega t), \qquad v_1 = \frac{p_1}{\rho_0 c_0} = \frac{p_1}{K_0} c_0. \tag{16.12}$$

Since $v_y = v_z = 0$, the velocity field of a sound wave is always *longitudinal*, i.e. parallel to the direction of wave propagation. The corresponding spatial displacement field u_x satisfying $v_x = \partial u_x/\partial t$ becomes amplitude $u_x = -a_1 \sin(kx - \omega t)$ with amplitude $a_1 = v_1/\omega$.

Validity of the approximation

It only remains to check whether the approximation of disregarding the nonlinear terms is valid. The actual ratio between the magnitudes of the advective and local accelerations is,

$$\frac{|(\boldsymbol{v}\cdot\nabla)\boldsymbol{v}|}{|\partial\boldsymbol{v}/\partial t|} \approx \frac{kv_1^2}{\omega v_1} = \frac{v_1}{c_0}. \tag{16.13}$$

The condition for the approximation is thus that *the amplitude of the velocity oscillations should be much smaller than the speed of sound,* $v_1 \ll c_0$, or equivalently that $p_1 \ll K_0$ or $a_1 \ll \lambda/2\pi$.

Example 16.2.2 (Loudspeaker): A certain loudspeaker transmits sound to air at a frequency $\omega/2\pi = 1000$ Hz with diaphragm displacement amplitude of $a_1 = 1$ mm. The velocity amplitude becomes $v_1 = a_1\omega \approx 6\ \mathrm{m\,s^{-1}}$, and since $v_1/c_0 \approx 1/57$ the approximation of leaving out the advective acceleration is well justified.

16.3 Steady incompressible flow

In many practical uses of fluids, the flow does not change with time, and is said to be *steady* or *stationary*. In this section we shall for simplicity also assume that the fluid is incompressible with constant density. Steady flow in compressible fluids will be analysed in section 16.4 where we shall learn that *a fluid in steady flow is effectively incompressible when the flow velocity is everywhere much smaller than the speed of sound.*

Taking $\partial\boldsymbol{v}/\partial t = \boldsymbol{0}$ and $\rho = \rho_0$ in Eulers equations we now find,

$$\boxed{(\boldsymbol{v}\cdot\nabla)\,\boldsymbol{v} = \boldsymbol{g} - \frac{1}{\rho_0}\nabla p, \qquad \nabla\cdot\boldsymbol{v} = 0.} \tag{16.14}$$

Truly steady flow is like true incompressibility an idealization, only valid to a certain approximation. A river may flow steadily for days and weeks, while over several months seasonal changes in rainfall makes its water level rise and subside again. If one empties a cistern filled with water through a narrow pipe, the flow is almost steady in the pipe for a time, except that the level of water in the cistern slowly goes down and thereby reduces the flow speed in the pipe. In such cases, the flow should rather be called *nearly steady* or *quasi-stationary*.

Daniel Bernoulli (1700–82). *Dutch born mathematician who made major contributions to the theory of elasticity, fluid mechanics and the mechanics of musical instruments. Bernoulli pointed out the relation between pressure and velocity in the world's first book on hydrodynamics,* Hydrodynamica, *which he published in 1738. Actually it was not Bernoulli who formulated the quantitative theorem which now bears his name, but rather Lagrange in his famous book on analytic mechanics from 1788 [70].*

How steady flow is reached: This explanation begs the question of how a steady flow is ever established, for example in a calm river downstream from a waterfall. Since all flows start out being time-dependent, viscosity must be capable of removing the surplus energy from a lively flow to calm it down, but even when we include it (in chapter 17) there is no guarantee that the flow will become steady after a long time. You just have to open the water faucet wide and watch the persistent turbulent splashing in the kitchen sink to realize that steady flow may not always be the result of steady or even static conditions.

Bernoulli's theorem for incompressible fluid

The negative sign of the pressure gradient in Euler's steady-flow equation shows that in the absence of gravity a flow accelerating in a certain direction must be accompanied by a drop in pressure in the same direction. Regions of high flow velocity must generally have a lower pressure than regions of low velocity. Bernoulli's theorem makes this qualitative conclusion quantitative.

For an incompressible fluid with constant density $\rho = \rho_0$, Bernoulli's theorem (to be proved below) states that the field,

$$\boxed{H = \frac{1}{2}v^2 + \Phi + \frac{p}{\rho_0},} \tag{16.15}$$

is *constant along streamlines*. In the absence of gravity Φ, the constancy of the Bernoulli field H implies as we foresaw that an increase in velocity along a streamline must be compensated by a similar decrease in pressure, and conversely. The Bernoulli field may be viewed as an extension of the effective potential

(4.21) to fluid in motion. Note that the first two terms in the Bernoulli field make up the total mechanical (i.e. kinetic plus potential) energy of a unit mass particle, also called the *specific mechanical energy*. We shall later discuss how the Bernoulli field is related to energy (section 22.9).

The proof of the theorem is straightforward. The material rate of change of H along a particle orbit is given by the material derivative (15.29), and since all fields are time-independent in steady flow we get

$$
\begin{aligned}
\frac{DH}{Dt} &= \boldsymbol{v} \cdot \frac{D\boldsymbol{v}}{Dt} + \frac{D\Phi}{Dt} + \frac{1}{\rho_0}\frac{Dp}{Dt} \\
&= \boldsymbol{v} \cdot (\boldsymbol{v} \cdot \nabla)\boldsymbol{v} + (\boldsymbol{v} \cdot \nabla)\Phi + \frac{1}{\rho_0}(\boldsymbol{v} \cdot \nabla)p \\
&= \boldsymbol{v} \cdot \left((\boldsymbol{v} \cdot \nabla)\boldsymbol{v} - \boldsymbol{g} + \frac{1}{\rho_0}\nabla p\right) \\
&= 0,
\end{aligned}
$$

where we used in the last step the steady-flow Euler equation (16.14). This shows that H is constant along any particle orbit, and thus along any streamline since streamlines coincide with particle orbits in steady flow. Bernoulli's theorem for compressible fluids will be discussed in section 16.4.

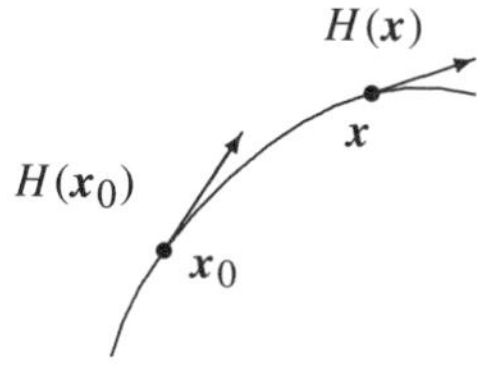

H is constant along a streamline in steady flow, i.e. $H(\boldsymbol{x}) = H(\boldsymbol{x}_0)$.

Terminology

The importance of Bernoulli's theorem for many practical hydrodynamical applications has led to several different terminologies. Since $\rho_0 H$ has the dimension of pressure, the term $(1/2)\rho_0 \boldsymbol{v}^2$ is often called the *dynamic pressure* as opposed to the *static pressure* p. The combination $p + \rho_0\Phi$ is called the *effective pressure*. A point where the fluid has zero velocity, $\boldsymbol{v} = \boldsymbol{0}$, is called a *stagnation point* for the flow. In the absence of gravity, the pressure at a stagnation point is $p_0 = \rho_0 H$, also called the *stagnation pressure*.

An often encountered engineering terminology is used in constant gravity, $\boldsymbol{g} = (0, 0, -g_0)$, where the Bernoulli field in flat-Earth coordinates becomes,

$$
H = \frac{1}{2}\boldsymbol{v}^2 + g_0 z + \frac{p}{\rho_0}. \qquad (16.16)
$$

The vertical height z of a point on the streamline above some fixed reference level $z = 0$ is called the *static head*. Similarly $p/\rho_0 g_0$ is called the *pressure head*, and $v^2/2g_0$ is called the *velocity head*. The total head, H/g_0, is the sum of these three terms, and is by Bernoulli's theorem the same everywhere along a streamline.

Uses of Bernoulli's theorem

Bernoulli's theorem is highly useful because most of the flows that we deal with in our daily lives are nearly ideal and nearly incompressible. Bernoulli's theorem often provides us with a first idea about the behaviour of a flow in a given geometry. The drop in pressure accompanying an increase in flow velocity lies, for example, at the root of lift generation for both animals and machines, whether they swim or fly.

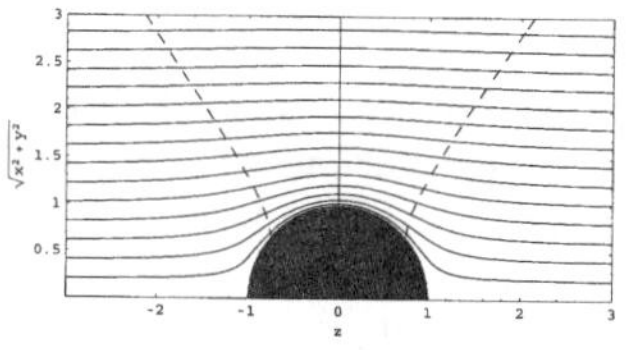

The flow has to quicken around an obstacle (here a half sphere) on the bottom of a stream and by Bernoulli's theorem, there will be a lower pressure above the obstacle, i.e. a lift.

Example 16.3.1 (Life of a flatfish): Lift is usually thought of as beneficial, but that may not always be the case. Some fish hide by burrowing superficially into the sandy bottom of a stream. The fish's curved upper surface forces the passing water to speed up, leading to a pressure drop above the fish that grows with the square of the flow velocity. If the stream velocity increases, the fish may be lifted out of the sand, whether it wants to or not. To avoid that may be why flatfish are indeed—flat. We shall return to this example and calculate the lift on a half-buried spherical fish in example 16.9.1.

In many cases of practical interest (see, for example [24, 80]), fluids stream through a network of ducts with inlets and outlets that are controlled externally. Assuming that there are well-defined streamlines connecting inlets with outlets, Bernoulli's theorem is typically used to relate the velocities and pressures at inlets and outlets, even if nothing is known about details of the the internal flow in the system. Viscosity is, however, never completely absent, but is mostly negligible well away from the boundaries and containers that we use to handle fluids. Exploiting the constancy of H along a streamline is always an approximation, and in any realistic problem there will be what the engineers call 'head loss' due to viscosity, to secondary flow, and to turbulence generated by surface irregularities.

Typically it is also assumed that the flow velocity U is the same all over the duct cross-section A. Such a pattern is called *plug flow* and the volume flux in the duct is simply $Q = AU$. It is by no means guaranteed to be correct, and engineers often put in a *discharge factor* C to take care of the lower velocities at the sides of a duct, writing $Q = CAU$ where U is now the velocity of a central streamline (see problem 16.7).

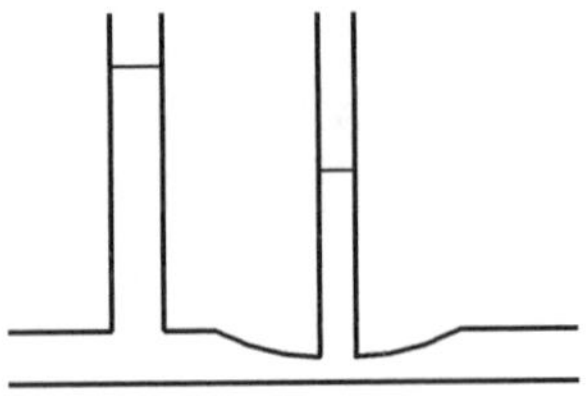

Sketch of a Venturi experiment. Water streams from the left through a constriction in the horizontal tube where the pressure is lower than in the tube outside the constriction because of the Venturi effect, as shown by the lower water level in the second vertical tube.

The Venturi effect

A simple duct with slowly varying cross section carries a constant volume flux Q of incompressible fluid. For simplicity we assume the duct is horizontal, such that gravity can be disregarded. Taking a streamline running horizontally through the duct, we obtain from Bernoulli's theorem that,

$$H = \frac{1}{2}v^2 + \frac{p}{\rho_0}, \tag{16.17}$$

takes the same value everywhere in the duct. Approximating the velocity with its average $U = Q/A$ over the cross-section A, the pressure becomes,

$$p = \rho_0 \left(H - \frac{Q^2}{2A^2} \right). \tag{16.18}$$

This demonstrates the *Venturi effect*: the pressure falls when the duct cross section decreases, and rises when it increases.

Giovanni Batista Venturi (1746–1822). *Italian physicist and engineer. Studied how fluids behave in a duct with a constriction.*

Torricelli's Law

Consider a barrel of wine with a little spout close to the bottom. If the plug in the spout is suddenly removed, the pressure causes the wine to stream out with a considerable speed. Provided the spout is narrow compared to the size of the barrel, a nearly steady flow will soon establish itself. This is a case where Bernoulli's theorem readily yields an expression for the velocity of the outflow.

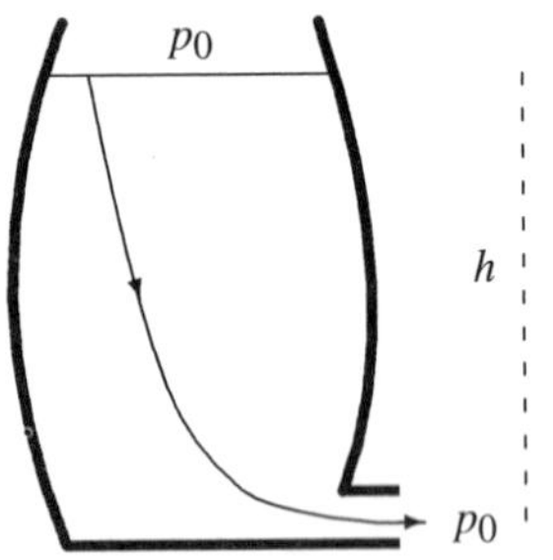

Wine running out of a barrel. The wine emerges with the same speed as it would have obtained by falling freely through the height h of the fluid in the barrel.

Consider a streamline starting near the top of the barrel and running with the flow down through the middle of the spout. Near the top at a height $z = h$ over the position of the spout, the fluid is almost at rest, i.e. $v \approx 0$. The pressure is atmospheric, $p = p_0$, and the gravitational potential may be taken to be $g_0 h$ so that

$$H_{\text{top}} = g_0 h + \frac{p_0}{\rho_0}. \tag{16.19}$$

Just outside the spout the fluid has some horizontal velocity v, and the pressure is also atmospheric, $p = p_0$, with no contribution from gravity, because the potential has been chosen to vanish here. Hence

$$H_{\text{bottom}} = \frac{1}{2}v^2 + \frac{p_0}{\rho_0}. \tag{16.20}$$

Equating the values of H at the top and the bottom we find

$$\frac{1}{2}v^2 + \frac{p_0}{\rho_0} = \frac{p_0}{\rho_0} + g_0 h,$$

which has the solution

$$\boxed{v = \sqrt{2g_0 h}.} \tag{16.21}$$

Evangelista Torricelli (1608–1647). *Italian physicist. Constructed the first mercury barometer and noted that the barometric pressure varied from day to day. Served as companion and secretary for Galileo in the last months of Galileo's life.*

Surprisingly, this is exactly the same velocity as a drop of wine would have obtained by falling freely from the top of the barrel to the spout. This result is called *Torricelli's Law* (1644), and preceding Bernoulli's theorem by more than a century it was in its time a major step forward in the understanding of fluid mechanics. We could, in fact, have come to the same conclusion simply by converting the potential energy of a water particle into kinetic energy.

In a sense the barrel acts as a device for diverting the vertical momentum of the falling liquid into the horizontal direction (see chapter 22). All streamlines yield the same result, as long as we avoid choosing a streamline running very near to the walls of the barrel and spout, where the unavoidable viscosity slows down the flow in the boundary layers. Even if the spout is replaced by a pipe that is not even horizontal but may turn and twist, the exit velocity at the exit of the pipe will equal the free-fall velocity from the fluid surface at the top of the barrel.

Example 16.3.2 (Wine barrel): A large cylindrical wine barrel has diameter 1 m and height 2 m. According to Torricelli's law the wine will emerge from the spout with the free-fall speed of about 6.3 m s^{-1}. If the spout opening has diameter 5 cm, about 12.3 litres of wine will be spilled on the floor per second. At this rate it would take 2 min to empty the barrel, but we shall see below that it actually takes double because the level sinks.

Quasi-stationary emptying of a wine barrel

If the barrel has constant cross-section A_0, Leonardo's law (15.17) tells us that the average vertical flow velocity in the barrel is $v_0 = vA/A_0$ where $A \ll A_0$ is the cross section of the spout and $v = \sqrt{2g_0 z}$ the average horizontal flow velocity through the spout when the water level is z. Since $dz/dt = -v_0$, we obtain the following differential equation for quasi-stationary emptying of the barrel,

$$\frac{dz}{dt} = -\frac{A}{A_0}\sqrt{2g_0 z}. \tag{16.22}$$

Integrating this equation with initial value $z = h$ for $t = 0$, we obtain the time it takes to empty the barrel (see also problem 16.6),

$$T = \int_h^0 \frac{dt}{dz}\,dz = -\int_h^0 \frac{A_0}{A}\frac{dz}{\sqrt{2g_0 z}} = \frac{A_0}{A}\sqrt{\frac{2h}{g_0}}. \tag{16.23}$$

This time equals the free-fall time from height h multiplied by the usually huge ratio of the barrel and spout cross sections. For the cylindrical wine barrel of example 16.3.2 the freefall time is 0.64 s and it takes 400 times longer, about 4 min, to empty the barrel.

The Pitot tube

Henri Pitot (1695–1771). *French mathematician, astronomer and hydraulic engineer. Invented the Pitot tube around 1732 to measure the flow velocity in the river Seine.*

Fast aircraft often have a sharply pointed nose which on closer inspection is seen to end in a little open tube. On other aircraft the tube may stick orthogonally out from the side and is bent forward into the oncoming airstream. This device is called a *Pitot tube*, and is used in many variants to measure flow speeds in gases and liquids. In its simplest and original form, the Pitot tube is just an open glass tube bent through a right angle. The tube is lowered into a river streaming steadily with velocity U, with one end turned horizontally towards the current and the other vertically in the air above. The flow will stem water up into the vertical part of the tube, until the hydrostatic pressure of the water column balances the dynamic pressure from the flow. After the flow has steadied, the water in the tube rises to a height h above the river surface.

In the steady state the water speed must vanish at the entrance to the horizontal part of the tube, and a horizontal streamline arriving here from afar must come to an end in a so-called *stagnation point*. The gravitational potential is constant everywhere along the horizontal streamline, and can be disregarded, so that Bernoulli's theorem for this streamline reads,

$$\frac{p_0}{\rho_0} = \frac{1}{2}U^2 + \frac{p}{\rho_0}, \tag{16.24}$$

where p is the pressure at infinity, and p_0 is the stagnation pressure. The excess stagnation pressure $\Delta p = p_0 - p = (1/2)\rho_0 U^2$ must also equal the extra hydrostatic pressure from the water column above the water surface, $\Delta p = \rho_0 g_0 h$, so that

$$U = \sqrt{2g_0 h}. \tag{16.25}$$

Again we find the simple and surprising result that the speed of the water is exactly the same as it would be after a free-fall from the height h. At a typical flow speed of 1 m s^{-1} the water in the tube is raised 5 cm above the river level.

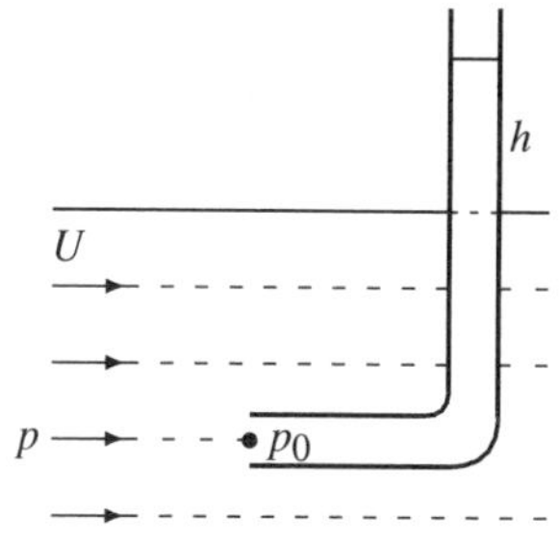

The principle of the Pitot tube. The pressure increase along the stagnating streamline must equal the weight of the raised water column.

Farmer's wisdom: Farmers know better than to leave the barn door open towards the wind in a storm, even if they do not know Bernoulli's theorem. A gust of wind will not only decrease the pressure above the barn roof because the wind has to move faster above the roof than at the ground, but the Pitot effect will also increase the pressure inside, giving the roof a double reason to blow off.

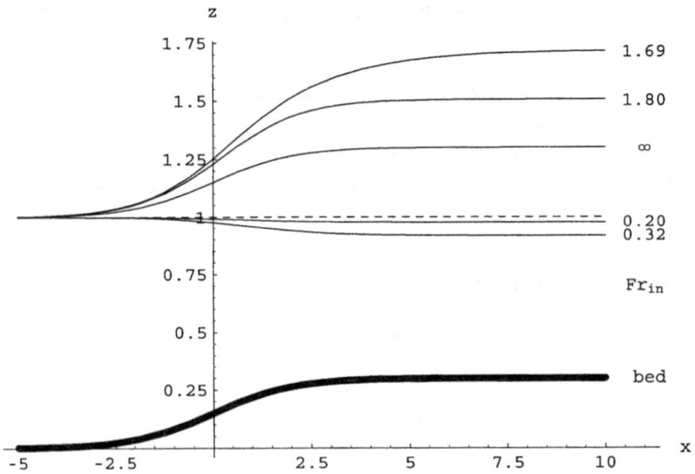

Figure 16.1. Numeric solution of (16.26) for a soft step $b(x) = 0.15(1 + \tanh(0.5x))$ (thick line) in the bed of a stream of initial depth $d_{in} = 1$ (dashed) for various values of the incoming Froude number Fr_{in}, shown on the right. For $\text{Fr}_{in} \geq 1.69\ldots$, the surface rises above the shape of the step, and for $\text{Fr}_{in} \to \infty$ the bump in the bottom is simply copied. For $\text{Fr}_{in} \leq 0.32\ldots$ the surface drops below the incoming level. There are no acceptable real solutions to (16.26) in the interval $0.32 < \text{Fr}_{in} < 1.69$.

Example 16.3.3 (Water scoop): Forest fires are sometimes combated by aircraft dropping large amounts of sea or lake water. To avoid landing and take-off, the aircraft collects water by lowering a scoop into the water while flying slowly at very low altitude. If the scoop turns directly forward and the aircraft velocity is U, it can like the Pitot tube raise the water to a maximal height, $h = U^2/2g_0$. Even for a speed as low as $U = 120\ \text{km h}^{-1} \approx 33\ \text{m s}^{-1}$, this comes to $h = 55$ m. In practice, the height of the water tank over the lake surface is much smaller, so that the water should ideally arrive in the tank with nearly maximal speed U. A scoop with an opening area of just $A \approx 300\ \text{cm}^2$ can deliver water at a rate of $Q = UA \approx 1\ \text{m}^3\ \text{s}^{-1}$. Turbulence lowers this somewhat, but typically such an aircraft can collect 6 m^3 in just 12 s.

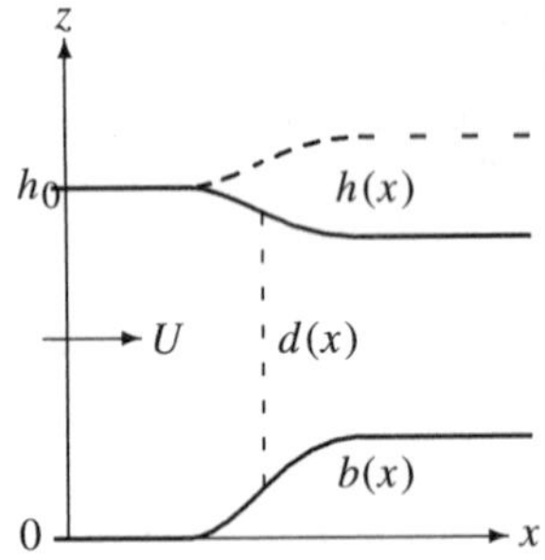

Does a rise in the river bed $z = b(x)$ generate a drop or a rise in the height of the surface?

∗ Shallow river with slowly changing bed

The surface of a shallow river flowing steadily along reflects the underlying structure of the river bed, as everybody has probably noted. It is, however, not so clear how the level of the water surface and the average water speed react to a change in the shape of the river bed. Will these quantities rise or fall when the otherwise horizontal river bed, for example, rises gently?

For simplicity we model the river as a very wide, essentially two-dimensional stream along x in which the bottom height $z = b(x)$ depends only on x. The depth is assumed to vary so slowly that its derivative is small, $|b'(x)| \ll 1$. In that case the flow will be dominated by the x-component of the velocity, $|v_x| \gg |v_z|$, and it may be reasonable to assume that the river carries a plug flow characterized by $v_x = U(x)$, independent of y and z. Denoting the local depth by $d(x)$, the height of the open surface is $h(x) = b(x) + d(x)$. Mass conservation together with Bernoulli's theorem applied to a streamline along the surface (where the pressure is constant) imply that,

$$Q = U(x)d(x), \tag{16.26a}$$

$$H = \tfrac{1}{2}U(x)^2 + g_0(b(x) + d(x)), \tag{16.26b}$$

are both independent of x. In principle these two equations can be solved for $U(x)$ and $d(x)$ in terms of $b(x)$. Unfortunately the resulting third degree polynomial equation has a rather messy solution.

As usual it is easier to understand the local structure. Differentiating both equations with respect to x

we get

$$\frac{U'}{U} + \frac{d'}{d} = 0, \qquad UU' + g_0(b' + d') = 0. \tag{16.27}$$

Solving for d' we obtain,

$$d' = \frac{g_0 d}{U^2 - g_0 d} b', \qquad \frac{U'}{U} = -\frac{d'}{d}. \tag{16.28}$$

Since the denominator is singular for $U = \sqrt{g_0 d}$, it is useful to introduce the dimensionless (local) *Froude number*,

$$\boxed{\mathrm{Fr}(x) = \frac{U(x)}{\sqrt{g_0 d(x)}}.} \tag{16.29}$$

William Froude (1810–79). *English engineer and naval architect. Discovered what are now called scaling laws, allowing predictions of ship performance to be made from studies of much smaller model ships.*

In chapter 24 we shall see that $\sqrt{g_0 d}$ is the velocity of shallow-water surface waves. Intuitively one would expect something dramatic to happen when the flow velocity passes through the shallow wave speed, because this is akin to passing through the speed of sound in the atmosphere. Eliminating the velocity in favour of the Froude number we obtain the differential equations,

$$d'(x) = \frac{b'(x)}{\mathrm{Fr}(x)^2 - 1}, \qquad \frac{\mathrm{Fr}'(x)}{\mathrm{Fr}(x)} = -\frac{3}{2}\frac{d'(x)}{d(x)}. \tag{16.30}$$

The behaviour of the river depends crucially on whether the local Froude number is smaller or greater than the *critical* value $\mathrm{Fr} = 1$.

A soft step in the river bed with $b'(x) > 0$ causes the depth to decrease, $d' < 0$, if the incoming Froude number before the step is smaller than unity, $\mathrm{Fr_{in}} < 1$, and conversely the depth will increase if $\mathrm{Fr_{in}} > 1$. The water surface height, $h = b + d$, follows the same pattern. The local Froude number will itself in both cases come closer to the nonphysical singular value $\mathrm{Fr} = 1$. Whether this value is actually reached depends on the height of the step and the incoming Froude number. There will, for a given step height, always be an interval of incoming Froude numbers, $\mathrm{Fr_1} < \mathrm{Fr_{in}} < \mathrm{Fr_2}$ with $\mathrm{Fr_1} < 1$ and $\mathrm{Fr_2} > 1$ in which the simple model presented here breaks down and no acceptable solution can be found.

Physically, the smooth water surface will break and froth at the step in the forbidden Froude number interval. For allowed values outside the forbidden interval the local Froude number must stay on the same side of the singularity everywhere along the step. A smooth crossing through $\mathrm{Fr} = 1$ can only take place at a point where $b'(x) = 0$, i.e. at a soft bump in the river bed.

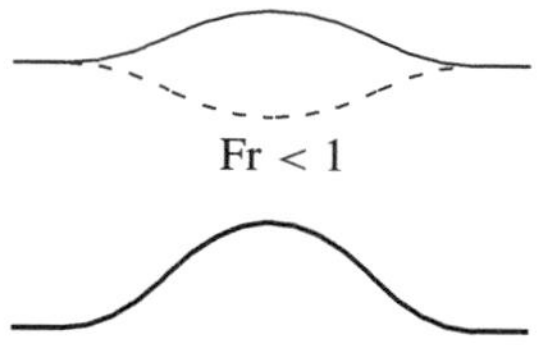

Flow over a bump with Froude number below and above the critical value 1.

Example 16.3.4 (Shallow river): In figure 16.1 a river of initial depth $d_{\mathrm{in}} = 1$ m passes a soft step of height $b_{\mathrm{out}} - b_{\mathrm{in}} = 0.3$ m. The forbidden incoming Froude number interval is $0.32 < \mathrm{Fr_{in}} < 1.69$, corresponding to a forbidden incoming velocity interval $1.0\ \mathrm{m\,s^{-1}} < U_{\mathrm{in}} < 5.3\ \mathrm{m\,s^{-1}}$. The *allowed* changes in the height of the water surface lie in the intervals $-8\ \mathrm{cm} < h_{\mathrm{out}} - h_{\mathrm{in}} < 0$ for $\mathrm{Fr_{in}} < \mathrm{Fr_1}$ and $30\ \mathrm{cm} < h_{\mathrm{out}} - h_{\mathrm{in}} < 72\ \mathrm{cm}$ for $\mathrm{Fr_{in}} > \mathrm{Fr_2}$.

16.4 Steady compressible flow

In steady compressible flow, the velocity, pressure and density are independent of time. The Euler equation and the continuity equation take the form,

$$\boxed{(\boldsymbol{v} \cdot \nabla)\, \boldsymbol{v} = \boldsymbol{g} - \frac{1}{\rho}\nabla p, \qquad \nabla \cdot (\rho \boldsymbol{v}) = 0.} \tag{16.31}$$

Here we shall, for simplicity, assume that the fluid is in a barotropic state, $p = p(\rho)$ or $\rho = \rho(p)$, such that we again have a closed set of five field equations for the five fields, v_x, v_y, v_z, ρ and p. In this section gravity will mostly be ignored.

Effective incompressibility

First we shall demonstrate the claim made in the preceding section that *in steady flow a fluid is effectively incompressible when the flow speed is everywhere much smaller than the local speed of sound.* The ratio of the local flow speed $\boldsymbol{v}$ (relative to a static solid object or boundary wall) and the local sound speed c is called the (local) *Mach number*,

$$\boxed{\mathrm{Ma} = \frac{|\boldsymbol{v}|}{c}.} \tag{16.32}$$

In terms of the Mach number, the claim is that a steady flow is effectively incompressible when $\mathrm{Ma} \ll 1$ everywhere. Conversely, the flow is truly compressible if the local Mach number somewhere is comparable to unity or larger, $\mathrm{Ma} \gtrsim 1$.

Ernst Mach (1838–1916). *Austrian positivist philosopher and physicist. Made early advances in psycho-physics, the physics of sensations. His rejection of Newton's absolute space and time prepared the way for Einstein's theory of relativity. Proposed the principle that inertia results from the interaction between a body and all other matter in the universe.*

The essential step in the proof is to relate the gradient of pressure to the gradient of density, $\nabla p = (dp/d\rho)\nabla\rho = c^2\nabla\rho$, where $c = \sqrt{dp/d\rho}$ is the local speed of sound. Writing the divergence condition in the form $\nabla\cdot(\rho\boldsymbol{v}) = \rho\nabla\cdot\boldsymbol{v} + (\boldsymbol{v}\cdot\nabla)\rho = 0$ and making use of the Euler equation without gravity, we find the exact result,

$$\nabla\cdot\boldsymbol{v} = -\frac{1}{\rho}(\boldsymbol{v}\cdot\nabla)\rho = -\frac{1}{\rho c^2}(\boldsymbol{v}\cdot\nabla)p = \frac{\boldsymbol{v}\cdot(\boldsymbol{v}\cdot\nabla)\boldsymbol{v}}{c^2}. \tag{16.33}$$

Applying the Schwarz inequality to the numerator (see problem 16.16) we get

$$|\nabla\cdot\boldsymbol{v}| \leq \frac{|\boldsymbol{v}|^2}{c^2}|\nabla\boldsymbol{v}| = \mathrm{Ma}^2\,|\nabla\boldsymbol{v}|, \tag{16.34}$$

where $|\nabla\boldsymbol{v}|$ is the norm (2.17) of the velocity gradient matrix. This relation clearly demonstrates that for $\mathrm{Ma}^2 \ll 1$ the divergence is much smaller than the local velocity gradients, making the incompressibility condition, $\nabla\cdot\boldsymbol{v} = 0$, a good approximation. Typically, a flow will be taken to be incompressible when $\mathrm{Ma} \lesssim 0.3$ everywhere so that $\mathrm{Ma}^2 \lesssim 0.1$. In the remainder of this section we assume that this condition is *not* fulfilled (everywhere).

Example 16.4.1 (Mach numbers): Waving your hands in the air, you generate flow velocities at most of the order of metres per second, corresponding to $\mathrm{Ma} \approx 0.01$. Driving a car at 120 km h^{-1} $\approx$ 33 m s^{-1} corresponds to $\mathrm{Ma} \approx 0.12$. A passenger jet flying at a height of 10 km with velocity about 900 km h^{-1} $\approx$ 250 m s^{-1} has $\mathrm{Ma} \approx 0.9$ because the velocity of sound is only about 1000 km h^{-1} at this height (see example 16.2.1). Even if this speed is subsonic, considerable compression of the air must occur especially at the front of the wings and body of the aircraft. The Concorde and modern fighter aircraft operate at supersonic speeds at Mach 2–3, and the Space Shuttle enters the atmosphere at the hypersonic speed of Mach 25. The strong compression of the air at the frontal parts of such aircraft create shock waves that appear to us as a sonic boom (see chapter 25).

Bernoulli's theorem for barotropic fluids

For compressible fluids, Bernoulli's theorem is still valid in a slightly modified form. If the fluid is in a barotropic state with $\rho = \rho(p)$, the Bernoulli field becomes,

$$H = \tfrac{1}{2}v^2 + \Phi + w(p), \tag{16.35}$$

where

$$w(p) = \int \frac{dp}{\rho(p)} \tag{16.36}$$

is the *pressure potential*, previously defined in (4.35). The proof of the modified Bernoulli theorem is elementary and follows the same lines as before, using $Dw/Dt = (dw/dp)Dp/Dt = (1/\rho)(\boldsymbol{v}\cdot\nabla)p$.

The most interesting barotropic fluid is an ideal gas with adiabatic index γ, for which it has been shown on page 55 that

$$w = \frac{\gamma}{\gamma-1}\frac{p}{\rho} = c_p T \qquad c_p = \frac{\gamma}{\gamma-1}\frac{R}{M_{\mathrm{mol}}}. \tag{16.37}$$

Thus, in the absence of gravity, a drop in velocity along a streamline in isentropic flow is accompanied by a rise in temperature (as well as a rise in pressure and density).

Isentropic steady flow: There is a conceptual subtlety in understanding isentropic steady flow because of the unavoidable heat conduction that takes place in all real fluids. Since truly steady flow lasts 'forever', one might think that there would be ample time for a local temperature change to spread throughout the fluid, regardless of how badly it conducts heat. But remember that steady flow is not static, and fresh fluid is being compressed or expanded adiabatically all the time. So, provided the flow is sufficiently fast, heat conduction will have little effect. The physics of heat and flow is discussed in chapter 30.

Stagnation point temperature rise in an ideal gas

An object moving through a fluid has at least one stagnation point at the front where the fluid is at rest relative to the object. There is also at least one stagnation point at the rear of a body, but vortex formation and turbulence will generally disturb the flow so much in this region that the streamlines get tangled and we cannot use Bernoulli's theorem to relate velocity and pressure at the rear of the body.

A static airfoil in an airstream coming in horizontally from the left. The pictured streamline (dashed) ends at the forward stagnation point.

At the forward stagnation point the gas is compressed and the temperature will always be higher than in the fluid at large (and similarly at the rear stagnation point if it exists). In the frame of reference where the object is at rest and the fluid asymptotically moves with constant speed U and temperature T, the flow is steady, and we find from (16.35) and (16.37) in the absence of gravity,

$$\tfrac{1}{2}U^2 + c_p T = c_p T_0,$$

where T_0 is the stagnation point temperature. Accordingly the temperature rise due to adiabatic compression becomes,

$$\boxed{\Delta T = T_0 - T = \frac{U^2}{2c_p} = \frac{1}{2}U^2 \frac{\gamma - 1}{\gamma} \frac{M_{\text{mol}}}{R}.} \tag{16.38}$$

Note that the stagnation temperature rise depends only on the velocity difference between the body and the fluid far from the body, and not on the pressure or density of the gas.

Example 16.4.2: Taking $\gamma = 7/5$ we obtain for a car moving at 100 km h^{-1} a stagnation point temperature rise of merely 0.4 K. At the front of a passenger jet travelling at 900 km h^{-1} the stagnation point rises a moderate 31 K, whereas a supersonic aircraft travelling at 2300 km h^{-1} suffers a stagnation point temperature rise of about 200 K. When a re-entry vehicle hits the dense atmosphere with a speed of 3 km s^{-1} the stagnation point temperature rise becomes nearly 4500 K. At full orbital speed, 8 km s^{-1}, the stagnation point temperature would formally become 32 000 K, but the density of the air in the outer reaches of the atmosphere is so small that the continuum assumption breaks down.

Whether the tip of a moving object actually attains the stagnation point temperature depends primarily on how efficiently heat is conducted away from this region by the material of the object. The moving object is usually solid with a vastly greater heat capacity than the air near the stagnation point. In addition to adiabatic compression viscous friction also produces heat, and for extreme aircraft such as the Space Shuttle it has been necessary to use special ceramic materials to withstand temperatures that are otherwise capable of melting and burning metals. Freely falling meteors appear as shooting stars in the sky, mainly because of viscous friction.

Stagnation point properties

It is often convenient to express the ratio of the stagnation point temperature and the local temperature in terms of the local Mach number $\text{Ma} = |\boldsymbol{v}|/c$ where $c = \sqrt{\gamma RT/M_{\text{mol}}}$ is the local sound velocity. From (16.38) we obtain

$$\frac{T_0}{T} = 1 + \frac{1}{2}(\gamma - 1)\text{Ma}^2. \tag{16.39}$$

Even if the streamline does not actually end in a stagnation point, T_0 will be a constant for the streamline because it represents the value of the Bernoulli function, $H = c_p T_0$, which is constant along the streamline. The stagnation temperature may be thought of as the temperature that would be obtained if a tiny object were inserted into the flow far downstream of the observation point.

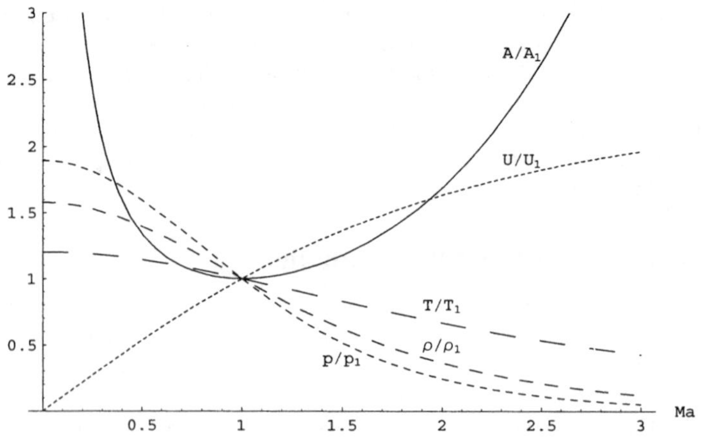

Figure 16.2. Flow in a slowly varying duct. Plot of the ratio of local to sonic values as a function of the Mach number. The ratio A/A_1 is a solid line, T/T_1 has large dashes, ρ/ρ_1 medium dashes, p/p_1 small dashes and U/U_1 dotted.

The stagnation density ρ_0 and pressure p_0 are obtained from the isentropic conditions,

$$T\rho^{1-\gamma} = T_0\rho_0^{1-\gamma}, \qquad T^\gamma p^{1-\gamma} = T_0^\gamma p_0^{1-\gamma},$$

which may be written,

$$\frac{\rho_0}{\rho} = \left(\frac{T_0}{T}\right)^{1/(\gamma-1)}, \qquad \frac{p_0}{p} = \left(\frac{T_0}{T}\right)^{\gamma/(\gamma-1)}. \tag{16.40}$$

Again it follows that ρ_0 and p_0 are constants for any given streamline.

Sonic point properties

A point where the velocity of a steady flow passes through the local velocity of sound is called a *sonic point*. The collection of sonic points typically form a sonic surface, for example in the duct flow to be discussed below. The sonic point temperature may be calculated from the stagnation point temperature (16.39) by setting Ma = 1,

$$\frac{T_1}{T_0} = \frac{2}{\gamma+1}. \tag{16.41}$$

For $\gamma = 7/5$, the ratio is $T_1/T_0 = 5/6$. Multiplying (16.39) by T_1/T_0 and rearranging the expression, we obtain the sonic temperature in terms of the local temperature and Mach number,

$$\frac{T_1}{T} = 1 + \frac{\gamma-1}{\gamma+1}\left(\mathrm{Ma}^2 - 1\right). \tag{16.42}$$

The sonic temperature T_1 is, like the stagnation temperature T_0, a constant for any streamline, independent of whether the flow actually becomes sonic on this streamline. The sonic density ρ_1 and pressure p_1 may similarly be obtained from stagnation values and related to the local Mach number by relations like (16.40).

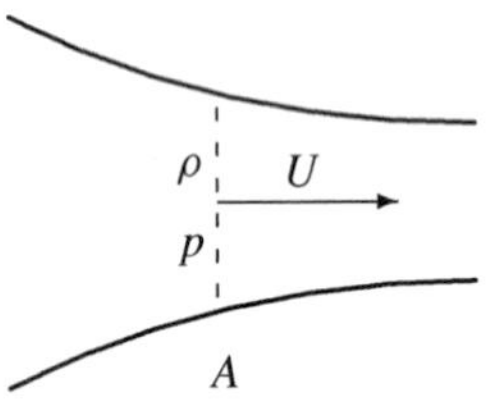

Flow in a slowly converging duct. All parameters are assumed constant on every planar cross section A.

Ideal gas flow in duct with slowly varying cross section

Consider now an ideal gas flowing through a duct with a slowly varying cross section such that the temperature T, density ρ, pressure p and normal velocity U may be assumed to be constant over any given (but otherwise arbitrary) planar duct cross section of area A. We are interested in determining the conditions under which the flow may become sonic in the duct.

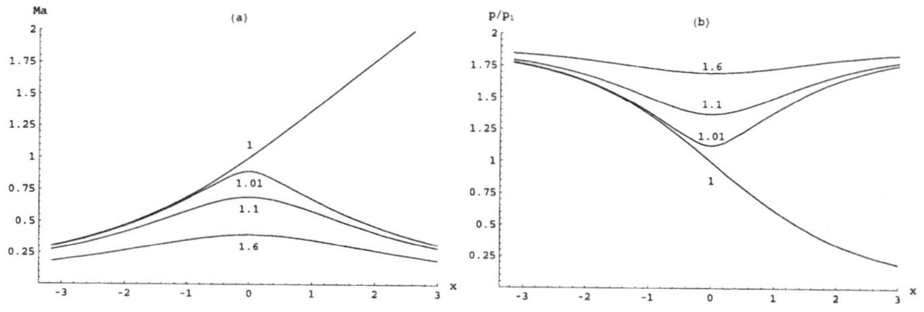

Figure 16.3. Simple model of a constricted duct, $A(x) = A_{\rm throat} + kx^2$, with $A_{\rm throat} = 1$ and $k = 0.1$ (and $\gamma = 7/5$). **(a)** Plot of the Mach number Ma(x) as a function of the duct coordinate x. The different curves are labelled with the ratio $A_{\rm throat}/A_1$. **(b)** The pressure ratio $p(x)/p_1$ under the same conditions. Note that the pressure is lowest in the throat (the Venturi effect) for all $A_1 < A_{\rm throat}$, but drops to much lower values for $A_1 = A_{\rm throat}$ when the flow becomes supersonic.

The constancy of the mass flux along the duct,

$$Q = \rho A U, \tag{16.43}$$

provides us with a relation between the duct area and the local Mach number. At the sonic point we have $\rho AU = \rho_1 A_1 U_1$, and using $U = \text{Ma}\, c$ and $U_1 = c_1$ where c and c_1 are the local and sonic sound velocities, we find

$$\frac{A}{A_1} = \frac{\rho_1 U_1}{\rho U} = \frac{1}{\text{Ma}}\frac{\rho_1}{\rho}\frac{c_1}{c} = \frac{1}{\text{Ma}}\frac{\rho_1}{\rho}\sqrt{\frac{T_1}{T}} = \frac{1}{\text{Ma}}\left(\frac{T_1}{T}\right)^{(1/2)(\gamma+1)/(\gamma-1)}.$$

In the last step we also used $\rho_1/\rho = (T_1/T)^{1/(\gamma-1)}$. Inserting T_1/T from (16.42), we arrive at the required relation,

$$\boxed{\frac{A}{A_1} = \frac{1}{\text{Ma}}\left(1 + \frac{\gamma-1}{\gamma+1}\left(\text{Ma}^2 - 1\right)\right)^{(1/2)(\gamma+1)/(\gamma-1)}.} \tag{16.44}$$

This function is shown (for $\gamma = 7/5$) as the solid curve in fig. 16.2 together with the corresponding temperature, density, pressure and velocity ratios.

The main observation to make from fig. 16.2 is that the local area A has a minimum at Ma $= 1$. This shows that it is only possible to make a *smooth* transition from subsonic to supersonic flow by sending the gas through a constriction in which the duct first converges and then diverges, forming a *throat* at the point where its area is minimal. If the flow parameters are set up such that the sonic area A_1 precisely equals the physical area $A_{\rm throat}$ of the throat, the subsonic flow entering the converging part of the duct will travel down the left-hand branch of the fully drawn curve in fig. 16.2, increasing in velocity until the Mach number reaches unity exactly at the throat. Having passed the throat, the flow is now supersonic and travels up the right-hand branch of the fully drawn curve while increasing in velocity further in the diverging part of the constriction.

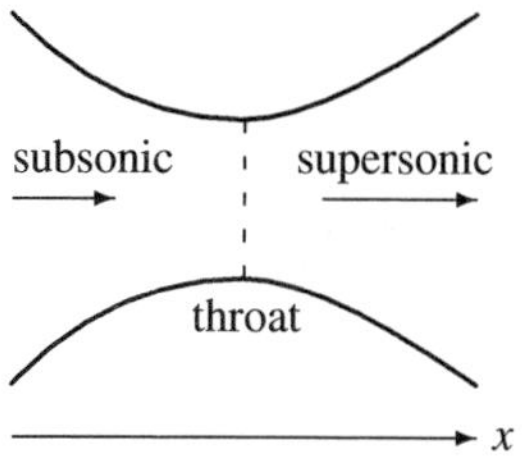

A subsonic flow may become supersonic in a duct with a constriction where it changes from converging to diverging. The transition happens at the narrowest point, the throat, where the cross section $A(x)$ is minimal.

The sonic point is not always reached. After all, flutes and other musical instruments, including our voices, have constrictions in the airflow that do not give rise to supersonic flow (which would surely destroy the music). If the physical throat area is larger than the sonic area, $A_{\rm throat} > A_1$, the Mach number does not reach unity at the throat. The maximal Mach number at the throat, $\text{Ma}_{\rm throat}$, is determined from (16.44) by setting $A = A_{\rm throat}$. Graphically, it may be read off from the left-hand branch of the solid curve in figure 16.2 at the point it crosses through $A_{\rm throat}/A_1$. In figure 16.3 this is illustrated in a simple model.

Suppose now that the duct is carrying a subsonic flow through the throat with $A_{\rm throat} > A_1$, and that we begin to *throttle* the flow by diminishing the throat area $A_{\rm throat}$ without changing the flow parameters at the entry. Since the sonic area A_1 is determined by the entry values, it will not change, so the throttling may continue until $A_{\rm throat} = A_1$. At this point the flow becomes sonic at the throat. If we now continue throttling, the entry parameters are at least forced to change in such a way that the the sonic area follows the diminishing throat area, $A_1 = A_{\rm throat}$, and the flow stays sonic at the throat. A duct with a throat operating at the sonic point is said to be *choked* because there is no way you can increase the mass flow through the throat by varying the entry parameters (for further details see [22, 80]).

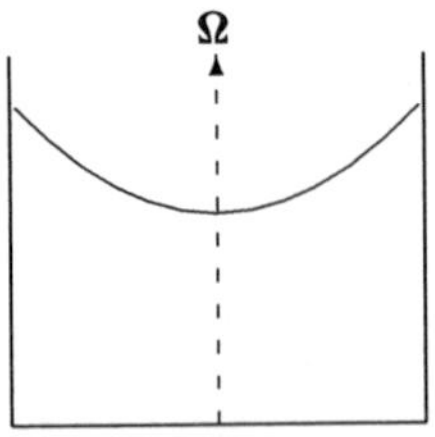

The water surface is parabolic in a bucket rotating with angular velocity Ω.

16.5 Vorticity

The value of the Bernoulli field $H(\boldsymbol{x})$ at a point $\boldsymbol{x}$ is only a function of the streamline going through this point. Different streamlines will in general have different values of H. This can be illustrated by considering the case of Newton's rotating bucket which was discussed before on page 87.

Bernoulli field in Newton's bucket

In the corotating (bucket) coordinate system, the fluid is at rest and subject to both the force of gravity and the centrifugal force. Since the velocity vanishes, the Bernoulli field becomes

$$H_0 = g_0 z - \frac{1}{2}\Omega^2 r^2 + \frac{p}{\rho_0}, \tag{16.45}$$

where the middle term is the centrifugal potential and $r = \sqrt{x^2 + y^2}$ is the distance from the axis of rotation. Hydrostatic balance ensures that $\nabla H_0 = 0$ so that H_0 is a true constant, independent of both r and z. Solving for the pressure we obtain

$$\frac{p}{\rho_0} = H_0 - g_0 z + \frac{1}{2}\Omega^2 r^2. \tag{16.46}$$

The constancy of the pressure on the open surface determines its parabolic shape $z = \Omega^2 r^2/2g_0 + \text{const}$.

In the non-rotating (laboratory) system, there is no centrifugal force, but the fluid moves steadily with velocity $v = r\Omega$, and the streamlines are concentric circles. The Bernoulli field becomes in this system,

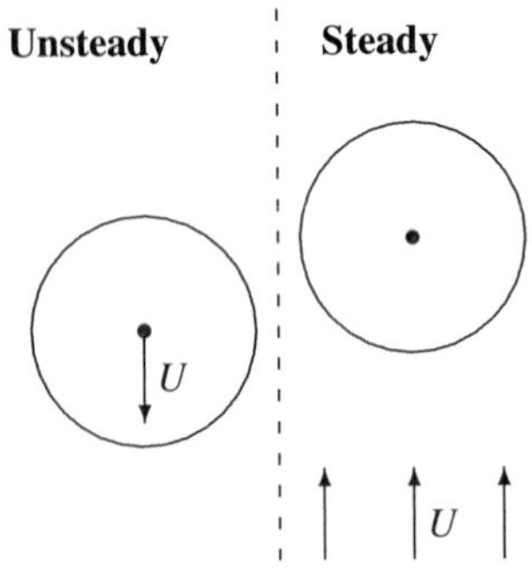

A body moving at constant speed through a fluid is physically equivalent to the same body being at rest in a steady flow which is asymptotically uniform.

$$H(r) = \frac{1}{2}\Omega^2 r^2 + g_0 z + \frac{p}{\rho_0} = \Omega^2 r^2 + H_0 \tag{16.47}$$

where in the last step we have made use of (16.45) to eliminate the pressure. The Bernoulli function depends in this case on the radial distance from the axis, but is of course constant on the circular streamlines as Bernoulli's theorem assures us. H is evidently independent of z, and we shall see below that this also follows from general rules.

Asymptotically uniform flow

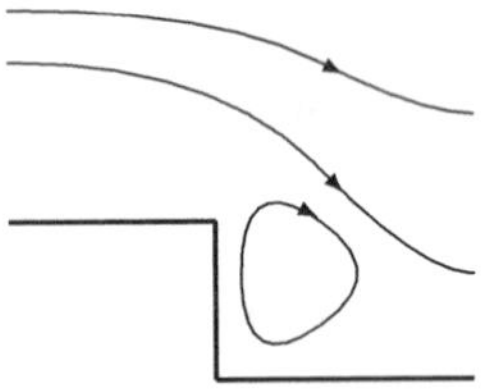

Closed streamlines may appear when a fluid flows past an edge.

It is often possible to relate the values of H for different streamlines. A general and frequently occurring example is a body moving with constant velocity through a fluid otherwise at rest, were it not for the disturbance created by the body. The relativity of motion in Newtonian mechanics tells us that this unsteady flow is physically equivalent to a steady flow around a stationary body in a fluid which, far away from the body, moves with constant velocity under constant pressure. We tacitly used this equivalence before in example 16.3.3 on page 212.

At great distances from the body the flow will have the same velocity $\boldsymbol{U}$, and in hydrostatic balance in the rest system it will also have a constant value of the effective pressure $P^* = p + \rho_0\Phi$. Consequently, the Bernoulli field must take the same value

$$H_0 = \frac{1}{2}U^2 + \frac{P^*}{\rho_0} \tag{16.48}$$

for all streamlines coming in from afar. Thus, if one can be sure that all streamlines around the body originate infinitely far away, then H must take the same value over all space, i.e. $H(\boldsymbol{x}) = H_0$.

The vorticity field

The simple result that H is spatially constant for asymptotically uniform flow is spoilt by the possibility that there may be streamlines forming closed curves unconnected with the flow at infinity. This was the case for Newton's bucket in the non-rotating coordinate system where the streamlines were concentric circles, but common experience indicates that such circulating flow may occur in the wake of the disturbance created by the shape of a container or a moving body. We are thus naturally led to the study of local circulation in a fluid, and we shall see that H *is* in fact globally constant if there is no local circulation anywhere.

Using the Euler equation for steady incompressible non-viscous flow (16.14), the gradient of the Bernoulli field becomes,

$$\nabla_i H = \frac{1}{2}\nabla_i \boldsymbol{v}^2 + \nabla_i \Phi + \frac{1}{\rho_0}\nabla_i p = \boldsymbol{v}\cdot\nabla_i \boldsymbol{v} - \boldsymbol{v}\cdot\nabla v_i = (\boldsymbol{v}\times(\nabla\times\boldsymbol{v}))_i .$$

This result, which is also valid for non-viscous, barotropic, compressible flow, allows us to write

$$\boxed{\nabla H = \boldsymbol{v}\times\boldsymbol{\omega},} \tag{16.49}$$

where we have defined a new field, the *vorticity field*, which is simply the curl of the velocity field

$$\boldsymbol{\omega} = \nabla\times\boldsymbol{v}. \tag{16.50}$$

The vorticity field also goes back to Cauchy (1841) and is a quantitative measure of the local circulation in the fluid. In any region where the vorticity field vanishes completely, we have $\nabla H = 0$, so the Bernoulli field must take one value only in that region, i.e. $H(\boldsymbol{x}) = H_0$. Flow completely free of vorticity is called *irrotational* flow and leads to a particularly simple formalism which we shall present in section 16.7.

Example 16.5.1: The curl of the field $\boldsymbol{v} = (x^2, 2xy, 0)$ is $\boldsymbol{\omega} = (0, 0, 2y)$. The curl of $\boldsymbol{v} = (y^2, 2xy, 0)$ vanishes, so this velocity field is irrotational.

Vorticity and local rotation

A trivial example of a flow with vorticity is a steadily rotating rigid body, for example Newton's bucket in the laboratory system. If the rotation vector of the body is $\boldsymbol{\Omega}$, the velocity field becomes $\boldsymbol{v} = \boldsymbol{\Omega}\times\boldsymbol{x}$, and the vorticity

$$\nabla\times(\boldsymbol{\Omega}\times\boldsymbol{x}) = \boldsymbol{\Omega}(\nabla\cdot\boldsymbol{x}) - (\boldsymbol{\Omega}\cdot\nabla)\boldsymbol{x} = 3\boldsymbol{\Omega} - \boldsymbol{\Omega} = 2\boldsymbol{\Omega}. \tag{16.51}$$

The vorticity is in this case constant and equal to twice the rotation vector.

The factor of 2 seems a bit strange but is a general result which may be verified by calculating the gradient of H for Newton's bucket. We find using (16.49)

$$\nabla H = (\boldsymbol{\Omega}\times\boldsymbol{x})\times 2\boldsymbol{\Omega} = 2\Omega^2(x, y, 0),$$

which agrees with the gradient of the Bernoulli field calculated from (16.47).

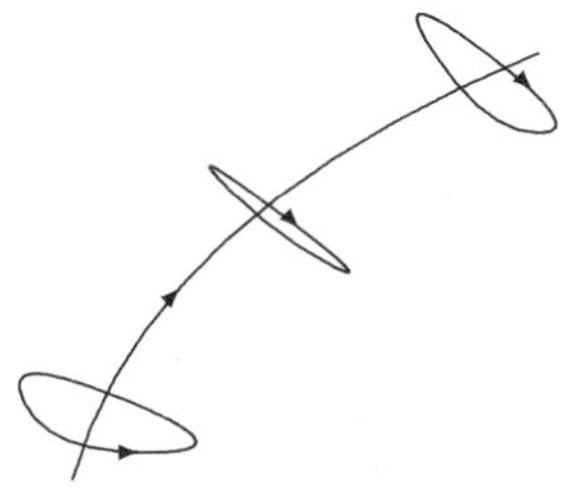

Around a vortex line there is local circulation of fluid.

Vortex lines

The field lines of the vorticity field are called *vortex lines*, and are defined as curves that are always tangent to the vorticity field. They are solutions to the ordinary differential equation

$$\frac{d\boldsymbol{x}}{ds} = \boldsymbol{\omega}(\boldsymbol{x}, t_0), \tag{16.52}$$

where s is a running parameter of the curve. For streamlines we could identify the running parameter with time, but this is not the case here, where s has the dimension of time multiplied by length. In steady flow these lines are fixed curves in space, just like the streamlines.

Bernoulli's theorem, $(\boldsymbol{v}\cdot\nabla)H = 0$, follows immediately from (16.49) by dotting with $\boldsymbol{v}$. Similarly, by dotting with $\boldsymbol{\omega}$ we obtain

$$\boxed{(\boldsymbol{\omega}\cdot\nabla)H = 0,} \tag{16.53}$$

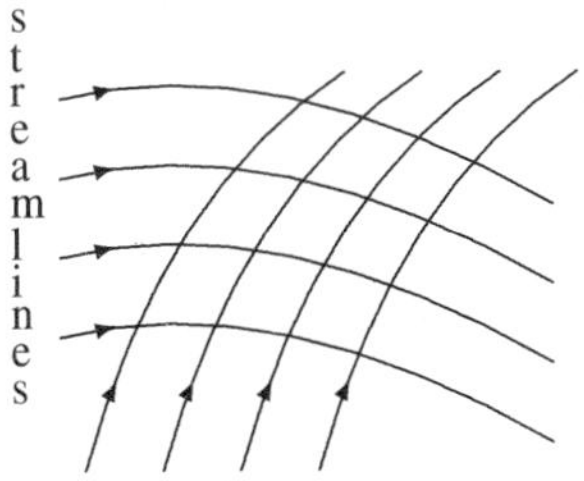

The Bernoulli field is constant on surfaces made from vortex lines and streamlines.

showing that the Bernoulli field is also constant along vortex lines.

Together these results imply that *the Bernoulli field is constant on the two-dimensional surfaces formed by combining vortex lines and streamlines*, sometimes called *Lamb surfaces*, and sometimes *Bernoulli surfaces*. Since streamlines for circulating fluids tend to form closed curves, these surfaces will usually be long tubes, called *vortex tubes*. For Newton's bucket the vortex tubes are cylinders concentric with the axis of rotation, and this explains why the Bernoulli field cannot depend on z, as we noted earlier.

Equation of motion for vorticity

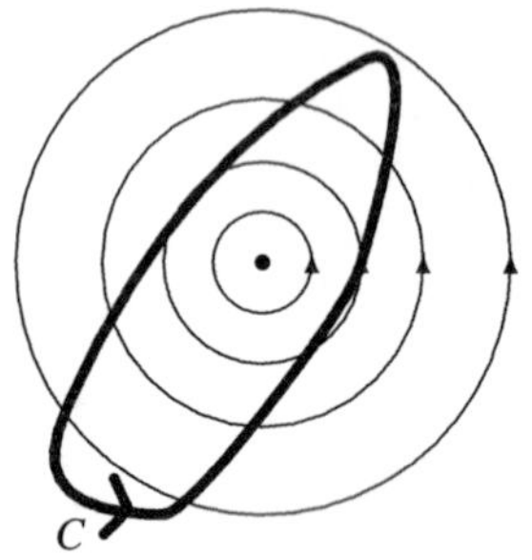

A closed curve C encircling a whirl.

The vorticity field is derived from the velocity field, so the equation of motion for the vorticity field must follow from the equation of motion for the velocity field, (16.1). Eliminating the pressure by means of the Bernoulli field (16.15) and retracing the steps leading to (16.49), the Euler equation for a non-viscous incompressible fluid may be written,

$$\frac{\partial \boldsymbol{v}}{\partial t} = -(\boldsymbol{v}\cdot\nabla)\boldsymbol{v} - \frac{1}{\rho_0}\nabla p + \boldsymbol{g} = \boldsymbol{v}\times\boldsymbol{\omega} - \nabla H. \tag{16.54}$$

The final result is also valid for any non-viscous compressible barotropic fluid. For steady flow where $\partial \boldsymbol{v}/\partial t = \boldsymbol{0}$, we recover of course (16.49). The equation of motion for vorticity is obtained by taking the curl of both sides of this equation, and using that the curl of a gradient vanishes,

$$\boxed{\frac{\partial \boldsymbol{\omega}}{\partial t} = \nabla\times(\boldsymbol{v}\times\boldsymbol{\omega}).} \tag{16.55}$$

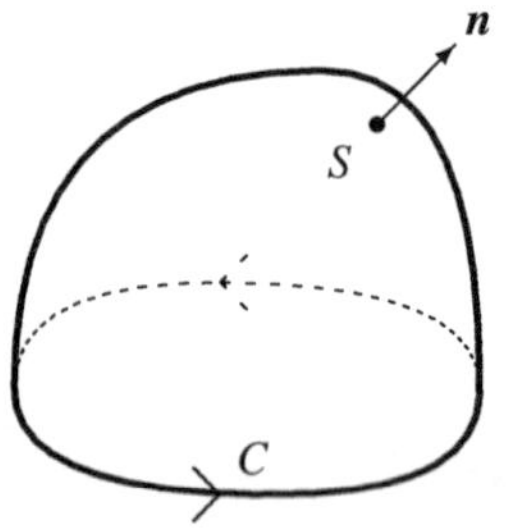

A surface S with perimeter C. The normal to the surface is consistent with the orientation of C (here using a right-hand rule).

There is a major lesson to draw from this equation. If the vorticity vanishes identically, $\boldsymbol{\omega}(\boldsymbol{x},t) = \boldsymbol{0}$, throughout a region V of space at an instant of time t, then we have $\partial\boldsymbol{\omega}/\partial t = \boldsymbol{0}$ for all $\boldsymbol{x}$ in V at t. Thus, the vorticity field will not change in the next instant, and continuing this argument, we conclude that if the vorticity field vanishes in the volume V at time t, it will vanish in this volume forever after.

In other words: in the absence of viscosity, *vorticity cannot be generated by the flow of a non-viscous fluid but must be present from the outset*. Thus, if you accelerate a body from rest in a truly ideal fluid, the flow will remain without vorticity, because $\boldsymbol{\omega} = \boldsymbol{0}$ at the beginning. The whirling air that trails a speeding car or an airplane must for this reason somehow be caused by viscous forces, independent of how tiny the viscosity is.

16.6 Circulation

Vorticity is a local property of a fluid, indicating how much the fluid rotates in the neighbourhood of a point. The corresponding global concept is called *circulation* and is formally defined to be the integrated projection of the velocity field on the line elements of a closed curve C,

$$\boxed{\Gamma(C,t) = \oint_C \boldsymbol{v}(\boldsymbol{x},t)\cdot d\boldsymbol{\ell}.} \tag{16.56}$$

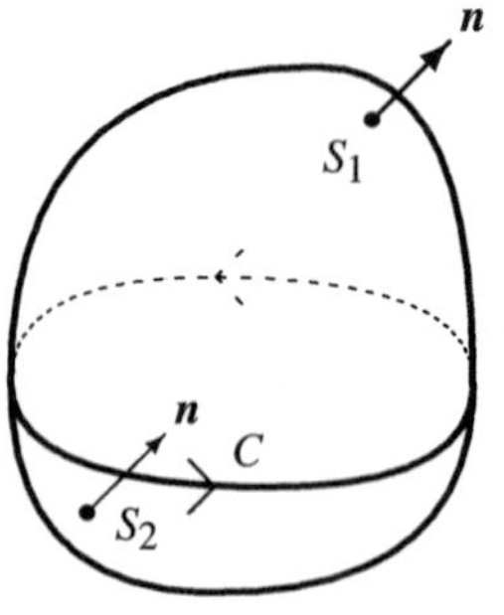

If C encircles a region of whirling fluid, the projection of the velocity field onto the curve will tend to be of the same sign all the way around. Whether it is positive or negative depends on whether the curve runs with the whirling flow or against it. We emphasize that the circulation may be calculated for any curve, not just a streamline encircling a whirl, although that might be the natural thing to do in some situations.

Stokes' theorem

The flux of vorticity is the same for any two surfaces with the same bounding curve. The normals to the surfaces are both consistent with the orientation of the curve.

The most important theorem about circulation is due to Stokes. It is completely general and states that the circulation of the velocity field around a closed curve is equal to the flux of the vorticity through any surface bounded by the curve,

$$\boxed{\oint_C \boldsymbol{v}\cdot d\boldsymbol{\ell} = \int_S \nabla\times\boldsymbol{v}\cdot d\boldsymbol{S}.} \tag{16.57}$$

It does not matter which surface S the flux is calculated for, as long as it has C as the boundary, lies entirely within the fluid, and is oriented consistently with the orientation of C. Stokes' theorem is like Gauss' theorem (6.4) valid for any vector field. The vorticity field $\boldsymbol{\omega} = \nabla \times \boldsymbol{v}$ is thus a measure of *local* circulation in the fluid.

Example 16.6.1 (Solid rotation): A fluid rotates steadily like a solid body with velocity $v = \Omega r$. In the non-rotating 'laboratory' system the circulation around a circle (which is also a streamline) with radius r is obtained by multiplying the constant velocity with the circumference of the circle,

$$\Gamma(r) = 2\pi r\ \Omega r = 2\pi \Omega r^2 = 2\Omega\ \pi r^2. \tag{16.58}$$

The last expression confirms Stokes' theorem because the circulation is the product of the constant vorticity 2Ω of a solid rotation body and the area of the circle.

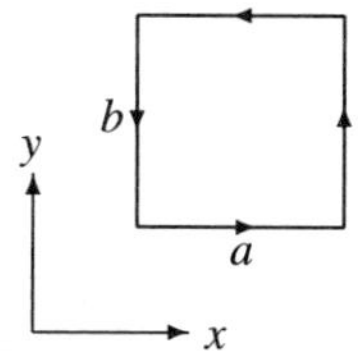

Circulation around a small rectangle of dimensions $a \times b$.

Proof of Stokes' theorem: As before the relation between global and local quantities is established by calculating the global quantity for an infinitesimal geometric figure, in this case a tiny rectangle in the xy-plane with sides a and b. To first order in the sides we find the circulation (suppressing z and t)

$$\begin{aligned}
\oint_{a\times b} \boldsymbol{v}\cdot d\boldsymbol{\ell} &= \int_x^{x+a} v_x(x', y)\, dx' + \int_y^{y+b} v_y(x+a, y')\, dy' \\
&\quad - \int_x^{x+a} v_x(x', y+b)\, dx' - \int_y^{y+b} v_y(x, y')\, dy' \\
&\approx -\int_x^{x+a} b\, \nabla_y v_x(x', y)\, dx' + \int_y^{y+b} a\, \nabla_x v_y(x, y')\, dy' \\
&\approx ab\big(\nabla_x v_y(x,y) - \nabla_y v_x(x,y)\big) = ab\, (\nabla \times \boldsymbol{v})_z.
\end{aligned}$$

The last expression is indeed the projection $(\nabla \times \boldsymbol{v}) \cdot d\boldsymbol{S}$ of the vorticity field on the small vector surface element of the rectangle, $d\boldsymbol{S} = (0, 0, ab)$.

Consider now a not necessarily planar open surface built up from little rectangles of this kind. Adding together the circulation for each rectangle, the contributions from the inner common edges cancel and one is left only with the circulation around the outer perimeter of the surface, which is Stokes theorem. The proof also implies that the shape of the surface S does not matter (see problem 16.9).

William Thomson, alias Lord Kelvin (1824–1907). *Scottish mathematician and physicist. Instrumental in the development of thermodynamics, in particular the relation between the Second Law and irreversibility. Viewed electromagnetic forces as elastic strains in the ether.*

* Kelvin's circulation theorem

Kelvin's famous circulation theorem from 1868 states that in a *non-viscous* fluid the circulation around a closed *comoving* curve is independent of time. In other words, if $C(t)$ is a comoving closed curve, then

$$\boxed{\frac{D\Gamma}{Dt} \equiv \frac{d\Gamma(C(t), t)}{dt} = 0.} \tag{16.59}$$

A comoving closed curve is washed along with the fluid and may thus change shape dramatically without a change in its circulation. In steady flow the circulation around any fixed closed curve is independent of time, but the theorem concerns a curve following the material of the fluid whether the motion is steady or not.

Kelvin's theorem applies only to ideal or nearly ideal flow. In a viscous fluid the circulation will change at a rate proportional to the viscosity, and viscosity will act both as dissipator and generator of vorticity. It is as mentioned before impossible to generate vorticity—or to get rid of it—without the aid of viscosity.

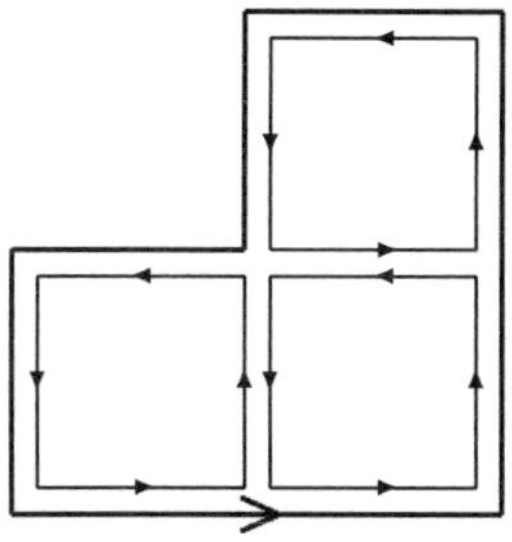
Adding rectangles together, the circulation cancels along the edges where two rectangles meet. This is also valid if the rectangles bend into the other coordinate directions.

Proof of Kelvin's theorem: The proof of the theorem is straightforward. We shall carry it through for an incompressible fluid, but the proof is easily extended to a barotropic compressible fluid (problem 16.11).

Let $C(t)$ be the comoving closed curve. In a small time interval δt the circulation along this curve changes by

$$\begin{aligned}\delta\Gamma(C(t),t) &= \Gamma(C(t+\delta t), t+\delta t) - \Gamma(C(t),t)\\ &= \oint_{C(t+\delta t)} \boldsymbol{v}(\boldsymbol{x}, t+\delta t)\cdot d\boldsymbol{\ell} - \oint_{C(t)} \boldsymbol{v}(\boldsymbol{x},t)\cdot d\boldsymbol{\ell}\\ &= \oint_{C(t)} (\boldsymbol{v}(\boldsymbol{x}+\boldsymbol{v}(\boldsymbol{x},t)\delta t, t+\delta t) - \boldsymbol{v}(\boldsymbol{x},t))\cdot d\boldsymbol{\ell}\\ &= \delta t \oint_{C(t)} \left(\frac{\partial \boldsymbol{v}}{\partial t} + (\boldsymbol{v}\cdot\nabla)\boldsymbol{v}\right)\cdot d\boldsymbol{\ell}\\ &= \delta t \oint_{C(t)} \left(\boldsymbol{g} - \frac{\nabla p}{\rho_0}\right)\cdot d\boldsymbol{\ell}.\end{aligned}$$

In the third step we have used that the point $\boldsymbol{x}$ at time t of a comoving curve is found at $\boldsymbol{x}+\boldsymbol{v}(\boldsymbol{x},t)\delta t$ at time $t+\delta t$, and this generates the comoving derivative in the fourth line. The Euler equation (16.1) is then used in the fifth step. Finally, since gravity and the pressure term are both gradient fields, their integrals around a closed curve vanish, and we arrive at Kelvin's theorem.

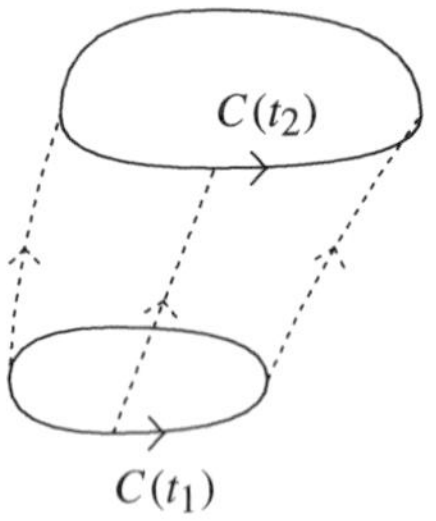

In ideal flow the circulation around the comoving curve $C(t)$ is the same at t_1 as at t_2.

16.7 Potential flow

A flow that is irrotational in some region of space will in the absence of viscosity stay irrotational at all times. From Stokes' theorem it follows that the circulation must vanish around *any* closed curve, and from the general result about vector fields with vanishing circulation (proven on page 39), we conclude that the velocity field is a gradient,

$$\boxed{\boldsymbol{v} = \nabla\Psi.} \tag{16.60}$$

The scalar field Ψ is called the *flow potential* or the *velocity potential*[1]. Potential flow obeys a much simpler formalism than flow with vorticity, in particular when it is also incompressible. The results to be derived below for non-viscous incompressible flow can be generalized to compressible flow, although much of the simplicity is lost (see problem 16.12).

Incompressible potential flow

In an incompressible fluid, the vanishing of the divergence of the velocity field implies that the flow potential must satisfy the Laplace equation,

$$\boxed{\nabla^2\Psi = 0.} \tag{16.61}$$

Typically, the boundary conditions consist of requiring the normal velocity $\boldsymbol{n}\cdot\boldsymbol{v} = \boldsymbol{n}\cdot\nabla\Psi$ to vanish at all impermeable solid walls. If the flow has constant velocity $\boldsymbol{v}(\boldsymbol{x}) = \boldsymbol{U}$, the flow potential becomes $\Psi = \boldsymbol{U}\cdot\boldsymbol{x}$, which evidently satisfies the Laplace equation.

Inserting $\boldsymbol{\omega} = \boldsymbol{0}$ and $\boldsymbol{v} = \nabla\Psi$ into the Euler equation in the form (16.54) we immediately find $\nabla(H + \partial\Psi/\partial t) = \boldsymbol{0}$. This means that $H + \partial\Psi/\partial t$ can only depend on time. Inserting the Bernoulli function (16.15) and solving for the pressure we obtain,

$$\boxed{p = C - \rho_0\left(\frac{1}{2}v^2 + \Phi + \frac{\partial\Psi}{\partial t}\right)} \tag{16.62}$$

where C is a possibly time-dependent 'constant' and $\boldsymbol{v} = \nabla\Psi$. In steady flow the time derivative $\partial\Psi/\partial t$ is absent. Thus, for incompressible potential flow the pressure is simply obtained from the solution of the linear Laplace equation (16.61) with suitable boundary conditions. All the original nonlinearity of the Euler equation has thus been relegated to the expression for the pressure.

[1] There is no general agreement on the symbol for the velocity potential, although the preferred notation appears to be ϕ. In this book we use Ψ to avoid clashes with other uses.

16.8 Potential flow for cylinder in cross-wind

A circular cylinder of radius a is an infinitely extended three-dimensional object which is invariant under translations along as well as rotations around its axis. We choose as usual a coordinate system with the z-axis coinciding with the cylinder (see section B.1 on page 531). An asymptotically uniform 'cross-wind' U along the x-axis does not break the longitudinal symmetry, which makes it natural to look for a velocity potential, $\Psi = \Psi(x, y)$, that is independent of z. Alternatively, the potential may be expressed in plane polar coordinates $\Psi = \Psi(r, \phi)$.

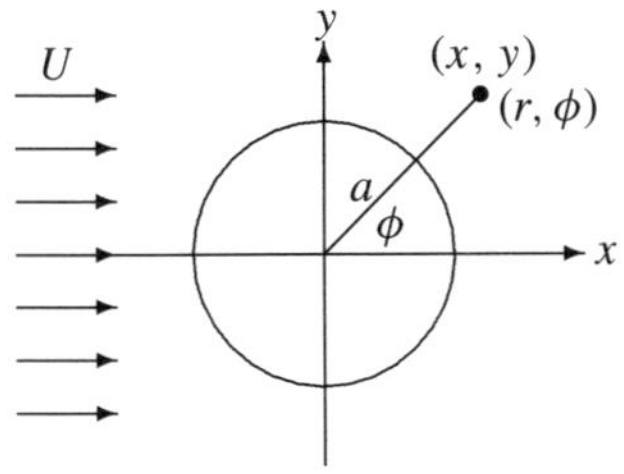

Cylinder of radius a in an asymptotically uniform cross-wind U.

Potential and velocity field

Asymptotically, for $r \to \infty$, the potential must approach the field of a constant uniform cross-wind, $\Psi \to Ux = Ur\cos\phi$. The linearity of the Laplace equation (16.61) demands that the potential everywhere is linear in the asymptotic field,

$$\Psi = U\cos\phi f(r), \tag{16.63}$$

where $f(r)$ is a so far unknown function of r which behaves as $f(r) \to r$ for $r \to \infty$. From the Laplacian in cylindrical coordinates (B.9), we obtain

$$\frac{d^2 f}{dr^2} + \frac{1}{r}\frac{df}{dr} - \frac{f}{r^2} = 0. \tag{16.64}$$

Since all three terms are of order $1/r^2$, we should look for power law solutions of the form, $f \sim r^\alpha$. Inserting this into the equation we find $\alpha = \pm 1$ so that the most general solution is of the form $f = Ar + B/r$, where A and B are arbitrary constants. The asymptotic condition implies $A = 1$, and B is determined by requiring the radial velocity field $v_r = \nabla_r \Psi$ to vanish at the surface of the cylinder, $r = a$. This leads to $B = a^2$, so that the solution is

$$\boxed{\Psi = Ur\cos\phi\left(1 + \frac{a^2}{r^2}\right).} \tag{16.65}$$

Calculating the gradient in cylindrical coordinates by means of (B.5) we finally obtain the velocity field

$$v_r = \nabla_r \Psi = \quad U\cos\phi\left(1 - \frac{a^2}{r^2}\right), \tag{16.66a}$$

$$v_\phi = \nabla_\phi \Psi = -U\sin\phi\left(1 + \frac{a^2}{r^2}\right). \tag{16.66b}$$

This flow is plotted in figure 16.4. The radial flow vanishes at the surface of the cylinder as it should, whereas the tangential flow, $v_\phi|_{r=a} = -2U\sin\phi$, only vanishes at the front and rear stagnation points $\phi = 0, \pi$.

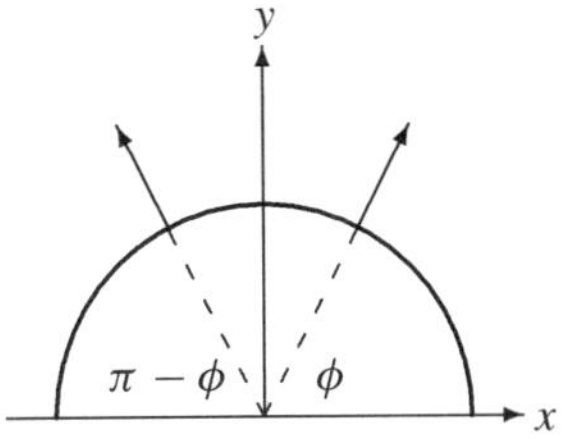

The projection of the pressure force on the x-axis is equal and opposite for ϕ and $\pi - \phi$.

Pressure, lift and drag

The pressure is obtained from (16.62). In the absence of gravity and normalized to vanish at infinity, it becomes

$$p = \frac{1}{2}\rho_0\left(U^2 - \boldsymbol{v}^2\right) = \frac{1}{2}\rho_0 U^2 \frac{a^2}{r^2}\left(4\cos^2\phi - 2 - \frac{a^2}{r^2}\right), \tag{16.67}$$

which on the surface of the cylinder simplifies to,

$$p_a = p|_{r=a} = \frac{1}{2}\rho_0 U^2\left(4\cos^2\phi - 3\right). \tag{16.68}$$

It is negative for $30° < \phi < 150°$. The up/down invariance of the pressure (under $\phi \to 2\pi - \phi$) shows that the total force in the y-direction, called *lift*, must vanish. What is more surprising is that due to the

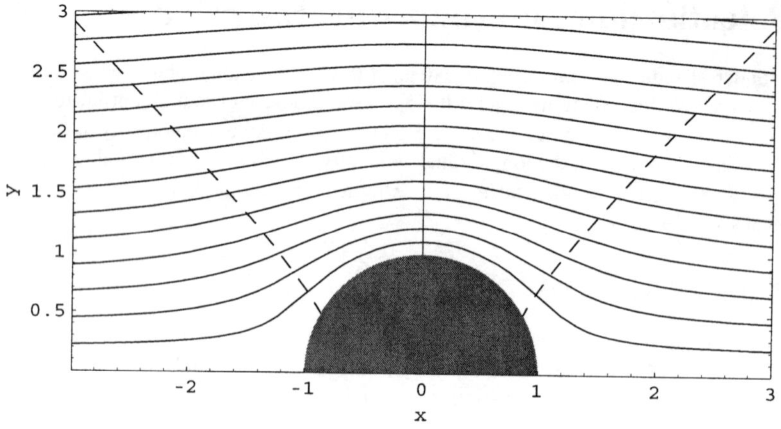

Figure 16.4. Potential flow around a cylinder with $a = 1$ and $U = 1$. Only the upper half is shown here (the lower half is the mirror image). The pressure vanishes on the dashed lines. The streamlines have been obtained by numeric integration of the differential equation for streamlines (15.2) with the velocity field given by the solution (16.66) converted to Cartesian coordinates. The streamlines are initialized to be equidistantly spaced by $\Delta y = 0.1$ for $x = -20$.

forwards/backwards invariance of the pressure (under $\phi \to \pi - \phi$) the total force along the x-direction, called *drag*, must also vanish.

For the upper half of the cylinder the drag also vanishes, but there is a non-vanishing lift (on a stretch of the cylinder of length L),

$$\mathcal{L} = -\int_{\phi=0}^{\pi} p_a dS_y = -\int_{\phi=0}^{\pi} p_a aL \sin\phi \, d\phi = \frac{5}{3}\rho_0 U^2 La. \tag{16.69}$$

If the density of the cylinder is $\rho_1 > \rho_0$, the lift-to-weight ratio becomes

$$\frac{\mathcal{L}}{Mg_0} = \frac{5}{3\pi}\frac{\rho_0}{\rho_1 - \rho_0}\frac{U^2}{ag_0}, \tag{16.70}$$

where we have also taken buoyancy into account. Evidently, there is a critical flow speed beyond which the lift becomes larger than the weight.

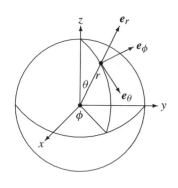

Spherical coordinates and their basis vectors.

16.9 Potential flow around a sphere in a stream

The natural coordinates for a sphere of radius a inserted into an asymptotically uniform flow are of course spherical (section B.2 on page 533) with the z-axis along the asymptotic velocity U. The solution follows along exactly the same lines as for the cylinder above. The symmetry of the problem implies that the velocity potential cannot depend on the azimuthal angle, so that $\Psi = \Psi(r, \theta)$.

Potential and velocity field

Asymptotically, for $r \to \infty$, the velocity potential has to approach the uniform flow $\Psi \to Ur\cos\theta$, and the linearity of the Laplace equation (16.61) requires the velocity potential to be linear in the asymptotic flow,

$$\Psi = U\cos\theta f(r). \tag{16.71}$$

Inserting this into the spherical Laplacian (B.16) one obtains an ordinary differential equation

$$\frac{d^2f}{dr^2} + \frac{2}{r}\frac{df}{dr} - \frac{2}{r^2}f = 0, \tag{16.72}$$

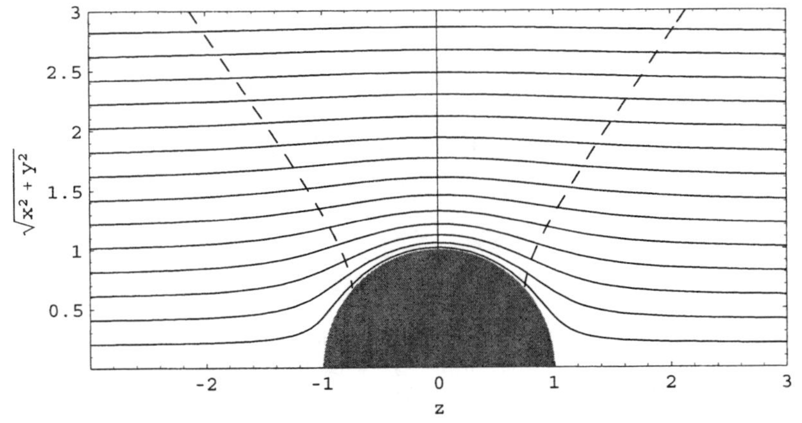

Figure 16.5. Potential flow around a sphere with $a = 1$ and $U = 1$. Only the upper half is shown here (the lower half is the mirror image). The streamlines have been obtained by numeric integration of the differential equation for streamlines (15.2) with the velocity field given by the solution (16.74) converted to Cartesian coordinates ($z, s = \sqrt{x^2 + y^2}$). The streamlines are initialized to be equidistantly spaced by $\Delta s = 0.1$ for $z = -20$. The field appears qualitatively different from the flow around a cylinder in figure 16.4 and hugs much closer to the surface of the sphere. The pressure vanishes on the dashed lines.

for which the most general solution is $f = Ar + B/r^2$. The asymptotic condition implies that $A = 1$, and the vanishing of the radial field $v_r = \nabla_r \Psi$ at the surface of the sphere requires $f'(a) = 0$, leading to $B = (1/2)a^3$. The velocity potential around a sphere is thus

$$\boxed{\Psi = Ur\cos\theta\left(1 + \frac{a^3}{2r^3}\right),} \tag{16.73}$$

and the velocity field is calculated from the spherical representation of the gradient (B.13),

$$v_r = \nabla_r \Psi = \quad U\cos\theta\left(1 - \frac{a^3}{r^3}\right), \tag{16.74a}$$

$$v_\theta = \nabla_\theta \Psi = -U\sin\theta\left(1 + \frac{a^3}{2r^3}\right), \tag{16.74b}$$

$$v_\phi = \nabla_\phi \Psi = 0. \tag{16.74c}$$

The streamlines are shown in figure 16.5.

Pressure, lift and drag

The pressure is obtained from (16.62),

$$p = \frac{1}{2}\rho_0 U^2 \frac{a^3}{r^3}\left(3\cos^2\theta - 1 - \frac{1}{4}(1 + 3\cos^2\theta)\frac{a^3}{r^3}\right). \tag{16.75}$$

On the surface of the sphere the pressure becomes

$$p_a = p|_{r=a} = \frac{1}{2}\rho_0 U^2 \frac{9\cos^2\theta - 5}{4}. \tag{16.76}$$

It is negative in the interval $42° \lesssim \theta \lesssim 138°$. Again the symmetry $\phi \to 2\pi - \phi$ shows that there is no lift on the sphere, and the symmetry $\theta \to \pi - \theta$ that there is no drag on the sphere (and not even on half a sphere).

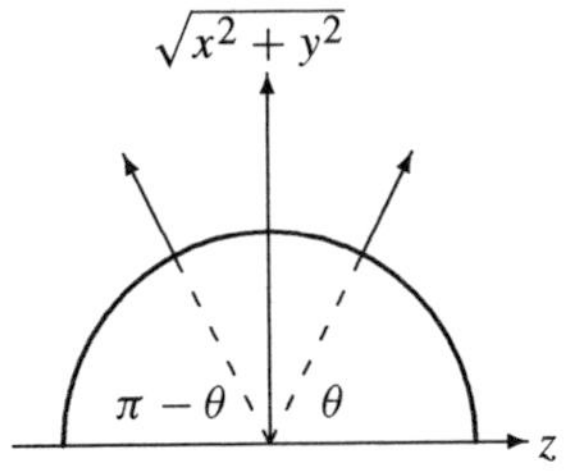

There is no drag because the projection of the pressure force on the z-axis is equal and opposite for θ and $\pi - \theta$.

Using that $dS_y = \sin\theta \sin\phi \, dS = a^2 \sin^2\theta \sin\phi \, d\theta \, d\phi$, the lift on a half sphere becomes,

$$\mathcal{L} = -\int_{\phi=0}^{\pi} p_a \, dS_y = -a^2 \int_0^{\pi} d\theta \int_0^{\pi} d\phi \, p_a \sin^2\theta \sin\phi = \frac{11\pi}{32} \rho_0 U^2 a^2. \tag{16.77}$$

If the density of the sphere is $\rho_1 > \rho_0$, the lift-to-weight ratio becomes,

$$\frac{\mathcal{L}}{M g_0} = \frac{33}{128} \frac{\rho_0}{\rho_1 - \rho_0} \frac{U^2}{a g_0}. \tag{16.78}$$

As for the half cylinder, there is a critical stream velocity beyond which the lift is greater than the weight.

Example 16.9.1 (Half-buried fish): A spherical fish with radius $a = 10$ cm and average density 10% higher than water (with the swim-bladder deflated) lies half buried in the sand at the bottom of a stream. The critical velocity for lift-off is merely $U \approx 0.6 \text{ m s}^{-1}$, so this fish would do much better by flattening its shape or burrowing deeper.

16.10 D'Alembert's paradox

Jean le Rond d'Alembert (1717–83). French mathematician. Introduced the concept of partial differential equations and was the first to solve such an equation.

The absence of drag in steady potential flow which we have explicitly verified for the cylinder and sphere may be formally shown to be true for any body shape (see page 454). But since everyday experience tells us that a steadily moving object is subject to drag from the fluid that surrounds it, even if the viscosity is vanishingly small, we have exposed a problem called *d'Alembert's paradox*.

Drag from the trailing wake

The resolution of the paradox illustrates the risks inherent in the assumption of potential flow. Although a tiny viscosity may not give rise to an appreciable friction force between body and fluid, it will generate vorticity close to the surface of the body. The vorticity will then spread into the fluid and produce a *trailing wake* behind the moving body, carrying a non-vanishing kinetic energy, and this constant loss of kinetic energy produces a drag on the body.

In potential flow around an object, the fluid does not create a wake but returns to its original state with no loss of kinetic energy, implying that there is no resultant drag. Potential flow may be a mathematically correct solution, but it misses in this case important aspects of the physics of real flow. We shall see in chapter 29 that d'Alembert's paradox may in fact be viewed positively as a statement about the smallness of drag compared to lift for streamlined bodies in nearly ideal fluids, with important consequences for the emergence of powered flight.

Effective mass in unsteady potential flow

Even if no kinetic energy is lost from the fluid in steady potential flow, there will be kinetic energy in the flow around the body. In the reference frame where the fluid is asymptotically at rest and the sphere moves with velocity $-\boldsymbol{U}$, the total kinetic energy of a sphere and the fluid around it becomes (problem 16.14)

$$\mathcal{T} = \frac{1}{2} M U^2 + \int_{r \geq a} \frac{1}{2} \rho_0 (\boldsymbol{v} - \boldsymbol{U})^2 \, dV = \frac{1}{2} \left(M + \frac{2\pi}{3} a^3 \rho_0 \right) U^2, \tag{16.79}$$

where M is the mass of the sphere and $\boldsymbol{v}$ is the potential flow velocity field (16.74). This shows that the *effective mass* of the sphere plus fluid is $M + 1/2m$ where $m = (4\pi/3)a^3 \rho_0$ is the mass of the fluid displaced by the sphere. Apart from the factor of 1/2, Archimedes would have loved this result!

If the sphere moves steadily with constant velocity, the total kinetic energy is also constant, and no external forces need to perform any work. If the sphere is accelerated by means of an external force $\mathcal{F}$, the rate of work of this force is $\mathcal{F} U = \dot{\mathcal{T}} = (M + m/2) U \dot{U}$. The change in the flow pattern around an accelerated sphere thus produces an apparent *drag force* of magnitude $\mathcal{F} = (1/2) m \dot{U}$ against the direction of motion.

Problems

16.1 Show that the ratio between the estimate of the molecular velocity obtained from (4.1) and the velocity of sound in an ideal gas is $v_0/c_0 \approx \sqrt{3/\gamma}$. Calculate the ratio for air.

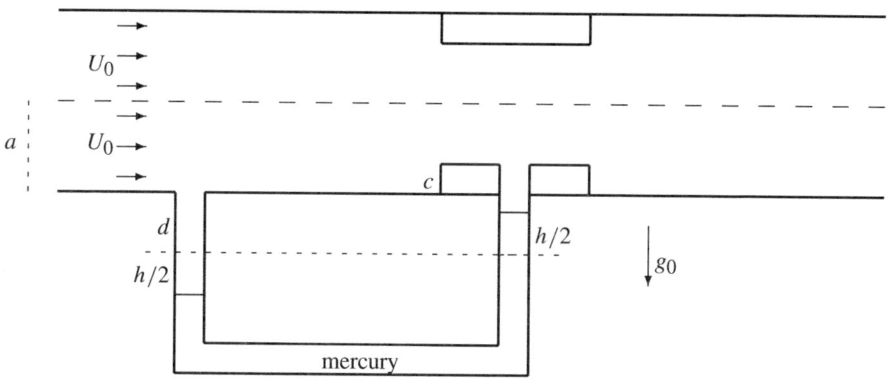

16.2 An ideal incompressible liquid (water) of density ρ_0 streams through a horizontal tube of radius a. To determine the average flow velocity U_0 a small ring is welded into the tube. The ring has outer radius a and inner radius $b = a - c$. As shown in the drawing a manometer is built into the system in the form of a small bypass partially filled with mercury of density ρ_1. The mercury surface lies a distance d below the tube's inner surface before the liquid is set into motion. The pressure in the liquid may everywhere be assumed to be constant across the tube cross section.

The following numbers may be used: $a = 15$ cm, $c = 1$ cm, $d = 5$ cm, $g_0 = 981$ cm^2/s, $\rho_0 = 1$ g/cm^3, $\rho_1 = 13.6$ g/cm^3, $U_0 = 5$ m/s.

(a) Calculate the average velocity of the water when it passes the ring.
(b) Determine the pressure difference between the bypass openings.
(c) Calculate the difference h in the mercury levels in the bypass.
(d) Find the maximal velocity U_0, which can be allowed under the condition that the mercury does not enter the mainstream.

16.3 A common way of stealing gasoline from a car is by means of a siphon. A tube of diameter 1 cm is inserted into the gasoline tank and sucked full of gasoline. The filled tube is quickly lowered into the opening of a 10 litre canister which is about 20 cm below the level of gasoline in the tank. How long does it take to fill the canister?

16.4 There is a small correction to the flow from the wine barrel (page 210) because the velocity of the flow does not vanish exactly on the top of the barrel. Estimate this correction from the ratio of the barrel cross section A_0 and the spout cross section A.

16.5 A wine barrel has two spouts with different cross sections A_1 and A_2 at the same horizontal level. Show that under steady flow conditions the wine emerges with the same speed from the two spouts.

16.6 Consider the quasi-stationary emptying of the wine barrel. Determine how the actual height z varies as a function of time.

16.7 An incompressible non-viscous fluid flows out of a cistern with water level h through a small circular drain with radius a. The flow through the drain is non-uniform with velocity at a radius r from the centre given by

$$v(r) = U\left(1 - \frac{r^2}{a^2}\right)^{\kappa} \tag{16.80}$$

where $U > 0$ is the central velocity and $\kappa > 0$ is a non-uniformity parameter. Calculate the total volume flux through the drain. Assuming that Bernoulli's theorem is valid for the central streamline, determine the reduction factor in the volume flux (in relation to Torricelli's law) due to the non-uniformity.

16.8 Assume (somewhat unrealistically) that air is an ideal gas with constant temperature T_0. **(a)** Calculate the relation between pressure and air velocity for a Pitot tube which is closed at one end. **(b)** Estimate the pressure increase relative to outside pressure in the Pitot tube of a passenger jet with speed $250\ \mathrm{m\ s^{-1}}$ in air of temperature $T_0 = -50\,^\circ$ C.

16.9 Show explicitly that Stokes' theorem is independent of the shape of the surface.

16.10 Show that the vector area of a surface bounded by a closed curve C is given by

$$\int_S d\boldsymbol{S} = \frac{1}{2}\oint_C \boldsymbol{x} \times d\boldsymbol{\ell}. \tag{16.81}$$

16.11 Show that Kelvin's theorem is valid for a barotropic compressible fluid.

16.12 Show that for unsteady, compressible potential flow in a barotropic fluid with $\rho = \rho(p)$, the equations of motion may be chosen to be,

$$\frac{\partial \Psi}{\partial t} + \frac{1}{2}\boldsymbol{v}^2 + \Phi + w(p) = 0 \tag{16.82}$$

$$\frac{\partial \rho}{\partial t} + (\boldsymbol{v}\cdot\nabla)\rho = -\rho\nabla^2\Psi \tag{16.83}$$

$$\tag{16.84}$$

where $\boldsymbol{v} = \nabla\Psi$ and $w(p) = \int dp/\rho(p)$.

16.13 A cylindrical worm with radius 3 mm lies half buried in sand at the bottom of the sea. Its density is 10% higher than the density of water. Calculate the critical speed at which the worm is lifted out of the water.

16.14 Carry out the integral of the effective mass in (16.79).

* **16.15** Incompressible fluid flows along x in an open channel along x. Show that if the horizontal flow $v_x(x)$ is independent of z, the vertical flow will be

$$v_z = v_x \frac{(h-z)b' + (z-b)h'}{h-b} \tag{16.85}$$

where $b(x)$ and $h(x)$ are the bottom surface heights.

* **16.16** Use the Schwarz inequality

$$\left|\sum_n A_n B_n\right|^2 \leq \sum_n A_n^2 \sum_m B_m^2 \tag{16.86}$$

to derive (16.34).

* **16.17** Consider a non-viscous barotropic fluid in an external time-independent gravitational field $\boldsymbol{g}(\boldsymbol{x})$ with $\nabla\cdot\boldsymbol{g} = 0$. Let $\rho_0(\boldsymbol{x})$ and $p_0(\boldsymbol{x})$ be density and pressure in hydrostatic equilibrium. **(a)** Show that the wave equation for small-amplitude pressure oscillations around hydrostatic equilibrium becomes,

$$\frac{\partial^2 \Delta p}{\partial t^2} = c_0^2\nabla^2\Delta p - c_0^2(\boldsymbol{g}\cdot\nabla)\frac{\Delta p}{c_0^2}, \tag{16.87}$$

where c_0^2 is the local sound velocity in hydrostatic equilibrium. **(b)** Estimate under which conditions the extra term can be disregarded in standard gravity for an atmospheric wave of wavelength λ.

17

Viscosity

All fluids are viscous, except for a component of liquid helium close to absolute zero in temperature. Air, water and oil all put up resistance to flow, and a part of the money we spend on transport by plane, ship or car goes to overcome fluid friction, and eventually to heating the atmosphere and the sea.

It is primarily the interplay between the mechanical inertia of a moving fluid and its viscosity which gives rise to all the interesting and beautiful phenomena, the whirling and the swirling that we are so familiar with. If a volume of fluid is set into motion, inertia would dictate that it continue in its original motion, were it not checked by the action of internal shear stresses. Viscosity acts as a brake on the free flow of a fluid and will eventually make it come to rest in mechanical equilibrium, unless external driving forces continually supply energy to keep it moving. In an Aristotelian sense the 'natural' state of a fluid is thus at rest with pressure being the only stress component. Disturbing a fluid at rest slightly, setting it into motion with spatially varying velocity field, will to first order of approximation generate stresses that depend linearly on the spatial derivatives of the velocity field. Fluids with a linear relationship between stress and velocity gradients are called Newtonian, and the coefficients in this linear relationship are material constants that characterize the strength of viscosity.

In this largely theoretical chapter the formalism for Newtonian viscosity will be set up and we shall derive the famous Navier–Stokes equations for fluids, the central theme for the remainder of the book. Superficially simple, these nonlinear differential equations remain a formidable challenge to engineers, physicists and mathematicians.

17.1 Shear viscosity

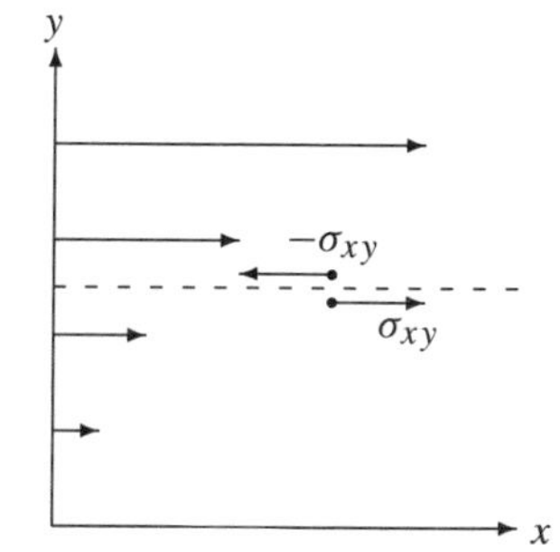

Laminar flow. If the fluid moves faster above the dashed line than below, it will exert a positive shear stress σ_{xy} on the material below in the direction of flow. By Newton's third law the fluid below will exert an opposite stress $-\sigma_{xy}$ on the material above.

Consider a fluid flowing steadily along the x-direction with a velocity field $v_x(y)$ which is independent of x, but may change with y. Such a field could, for example, be created by enclosing a fluid between moving plates, and is an elementary example of *laminar* or layered flow. If the velocity field has no y-dependence there should not be any internal friction stresses, because the fluid is then in uniform motion along the x-axis. If, on the other hand, the velocity grows with y, so that its gradient is positive $dv_x(y)/dy > 0$, we expect that the fluid immediately *above* a plane $y = \text{const.}$ will drag along the fluid immediately *below* because of friction and thus exert a *positive* shear stress, $\sigma_{xy}(y) > 0$, on this plane.

It also seems reasonable to expect that a larger velocity gradient will evoke stronger stress, and in *Newton's law of viscosity* the shear stress is simply made proportional to the gradient,

$$\boxed{\sigma_{xy}(y) = \eta \frac{dv_x(y)}{dy}.} \tag{17.1}$$

The constant of proportionality, η, is called the coefficient of *shear viscosity*, the *dynamic viscosity*, or simply *the* viscosity. It is a measure of how strongly the moving layers of fluid are coupled by friction,

Table 17.1. Table of density and dynamic and kinematic viscosity for common substances (at the indicated temperature and at atmospheric pressure). Some of the values are only estimates. Note that air has greater kinematic viscosity than water and hydrogen greater than olive oil. Glass is usually viewed as a solid, but there are (not very well substantiated) claims that it flows very slowly like a liquid over long periods of time, even at normal temperatures (estimates of its viscosity run even as high as 10^{27} Pa s).

	T [°C]	ρ [kg m^{-3}]	η [Pa s]	ν [m^{2}s^{-1}]
Hydrogen	20	0.084	8.80×10^{-6}	1.05×10^{-4}
Air	20	1.18	1.82×10^{-5}	1.54×10^{-5}
Water	20	1.00×10^3	1.00×10^{-3}	1.00×10^{-6}
Ethanol	25	0.79×10^3	1.08×10^{-3}	1.37×10^{-6}
Mercury	25	13.5×10^3	1.53×10^{-3}	1.13×10^{-7}
Whole blood	37	1.06×10^3	2.7×10^{-3}	2.5×10^{-6}
Olive oil	25	0.9×10^3	6.7×10^{-2}	7.4×10^{-5}
Castor oil	25	0.95×10^3	0.7	7.4×10^{-4}
Glycerol	20	1.26×10^3	1.41	1.12×10^{-3}
Honey(est)	25	1.4×10^3	14	1×10^{-2}
Glass (est)	20	2.5×10^3	$10^{18} - 10^{21}$	$10^{15} - 10^{18}$

and a material constant of the same nature as the shear modulus for elastic materials. We shall see below (section 17.5) that there is also a bulk coefficient of viscosity corresponding to the elastic bulk modulus, but that turns out to be rather unimportant in ordinary applications.

The viscosities of naturally occurring fluids range over many orders of magnitude (see table 17.1). Since dv_x/dy has dimension of inverse time, the unit for viscosity η is Pa s (pascal seconds). Although this unit is sometimes called Poiseuille, there is in fact no special name for it in the standard system of units (SI). In the older cgs-system it used to be called poise = 0.1 Pa s.

Molecular origin of viscosity in gases

In gases where molecules are far apart, internal stresses are caused by the incessant molecular bombardment of a boundary surface, transferring momentum in both directions across it. In liquids where molecules are in closer contact, internal stress is caused partly by molecular motion as in gases, and partly by intermolecular forces. The resultant stress in a liquid is a quite complicated combination of the two effects, and we shall for this reason limit the following discussion to the molecular origin of shear stress in gases.

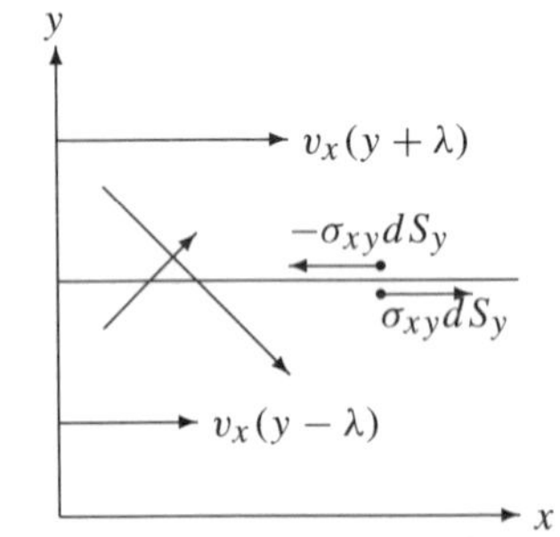

Layers of fluid moving with different velocities give rise to shear forces because they exchange molecules with different average velocities.

Gas molecules move nearly randomly in all directions at speeds much higher than the velocity field $\boldsymbol{v}(\boldsymbol{x}, t)$, which should be understood as the centre-of-mass velocity of a large collection of molecules, and represents the average non-random component of the molecular motion. In steady laminar planar flow with velocity $v_x(y)$ and positive velocity gradient $dv_x(y)/dy > 0$, a molecule of mass m crossing a surface element in the plane $y =$ const from above will carry along an average excess of momentum $mv_x(y)$ in the x-direction and therefore exert a force in the x-direction on the material below. Similarly, the material below will exert an equal and opposite force on the material above.

Let the typical distance between molecular collisions in the gas be λ and the typical time between collisions τ. The excess of momentum in the x-direction above an area element dS_y in a layer of thickness λ is of the order of

$$d\mathcal{P}_x \approx \frac{1}{2}(v_x(y+\lambda) - v_x(y))\rho\lambda dS_y \approx \frac{1}{2}\rho\lambda^2 \frac{dv_x(y)}{dy} dS_y.$$

This excess of momentum will be carried along by the fast molecular motion in all directions and about half of it will cross the surface in time τ. The shear stress may be estimated from the momentum transfer per unit of time and area, $\sigma_{xy} = (1/2)d\mathcal{P}_x/\tau dS_y$, and indeed takes the form of Newton's law of viscosity

(17.1) with a rough estimate of the shear viscosity,

$$\eta \approx \rho \frac{\lambda^2}{4\tau}. \tag{17.2}$$

Using the mean free path of gases (1.3) and identifying $v_{mol} \approx \lambda/\tau$ as the average molecular velocity (4.1), this estimate is of the right order.

Example 17.1.1: Air at normal temperature and pressure has $\rho \approx 1.2$ kg m^{-3}, $\lambda \approx 100$ nm, and $v_{mol} \approx 500$ m s^{-1}, leading to $\eta \approx 1.5 \times 10^{-5}$ Pa s in decent agreement with the tabulated value.

Temperature dependence of viscosity

The viscosity of any material depends on temperature. Common experience from kitchen and industry tells us that most liquids become 'thinner' when heated, indicating that the viscosity falls with temperature. Gases on the other hand become more viscous at higher temperatures, simply because the molecules move faster at random and thus transport momentum across a surface at a higher rate.

For a gas, the mean free path between collisions is given by (1.3). According to that expression $\lambda\rho$ is a combinations of constants, so that the viscosity $\eta \sim \lambda\rho \cdot \lambda/\tau$ will depend on the thermodynamic parameters in the same way as the molecular velocity $v_{mol} = \lambda/\tau$. From the gas pressure estimate (4.1) we have $v_{mol} \approx \sqrt{3p/\rho} = \sqrt{3RT/M_{mol}}$ for an ideal gas. Thus, if the viscosity is η_0 at temperature T_0, it will simply be

$$\eta = \eta_0 \sqrt{\frac{T}{T_0}} \tag{17.3}$$

at temperature T, independent of the pressure. The ideal gas viscosity depends only on the absolute temperature, not on the pressure.

Kinematic viscosity

The viscosity estimate (17.2) seems to point to another measure of viscosity, called the *kinematic viscosity*[1],

$$\boxed{\nu = \frac{\eta}{\rho}.} \tag{17.4}$$

Since the estimate, $\nu \sim \lambda^2/\tau$, does not depend on the unit of mass, this parameter is measured in purely kinematic units of m^2 s^{-1} (in the older cgs-system, the corresponding unit was called stokes = cm^2 s^{-1} = 10^{-4} m^2 s^{-1}). In fluids with constant density, it is a material constant at equal footing with the dynamic viscosity η (see table 17.1). It should be remembered that in an ideal gas we have $\rho \propto p/T$, so that the kinematic viscosity will depend on both temperature and pressure, $\nu \propto T^{3/2}/p$. In isentropic gases it always decreases with temperature (problem 17.1).

It is as we shall see the kinematic viscosity which appears in the dynamic equations for the velocity field, rather than the dynamic viscosity. Normally, we would think of air as less viscous than water and hydrogen as less viscous than olive oil, but under suitable conditions it is really the other way around. If a flow is driven by inflow of fluid with a certain velocity rather than being controlled by external pressure, air behaves in fact as if it were 10–20 times more viscous than water. But subject to a given pressure, air is much easier to set into motion than water because it is a thousand times lighter, and that is what fools our intuition.

17.2 Velocity-driven planar flow

Before turning to the derivation of the Navier–Stokes equations for viscous flow, we shall explore the concept of shear viscosity a bit further for the simple case of planar flow. Let us, as before, assume that the

[1]The use of ν for kinematic viscosity as well as for Poisson's ratio should not cause confusion.

flow is laminar and planar with the only non-vanishing velocity component being $v_x = v_x(y, t)$, now also allowing for time dependence. It is rather clear that there can be no advective acceleration in such a field, and formally we also find $(\boldsymbol{v} \cdot \nabla)v_x = v_x \nabla_x v_x = 0$. In the absence of volume and pressure forces, the Newtonian shear stress (17.1) will be the only non-vanishing component of the stress tensor, and Cauchy's dynamical equation (15.35) reduces to

$$\rho \frac{\partial v_x}{\partial t} = f_x^* = \nabla_y \sigma_{xy} = \eta \frac{\partial^2 v_x}{\partial y^2}.$$

Dividing by the density (which is assumed to be constant) we get

$$\boxed{\frac{\partial v_x}{\partial t} = \nu \frac{\partial^2 v_x}{\partial y^2},} \tag{17.5}$$

where ν is the kinematic viscosity (17.4). This is a simplified version of the Navier–Stokes equations, particularly well suited for the discussion of the basic physics of shear viscosity.

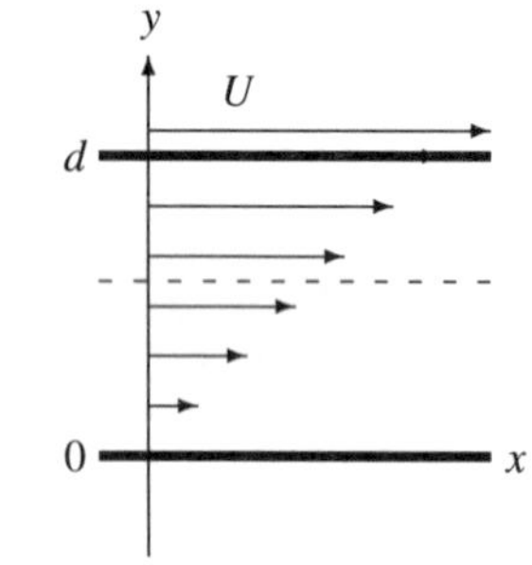

A Newtonian fluid with spatially uniform properties between moving parallel plates. The velocity field varies linearly between the plates and satisfies the no-slip boundary condition that the fluid is at rest relative to both plates. The stress must be the same on any plane in the fluid parallel with the plates (dashed).

Steady planar flow

Let us first return to the case of steady planar laminar flow which this chapter began with. In steady flow the left-hand side of (17.5) vanishes, and from the vanishing of the right-hand side it follows that the general solution must be linear, $v_x = A + By$, with arbitrary integration constants A and B. We shall imagine that the flow is maintained between (in principle infinitely extended) solid plates, one kept at rest at $y = 0$ and one moving with constant velocity U at $y = d$. Where the fluid makes contact with the plates, we require it to assume the same speed as the plates, in other words $v_x(0) = 0$ and $v_x(d) = U$ (this *no-slip* boundary condition will be discussed in more detail later). Solving these conditions we find $A = 0$ and $B = U/d$ such that the field between the plates becomes

$$v_x(y) = \frac{y}{d} U, \tag{17.6}$$

independent of the viscosity. From this expression we obtain the shear stress,

$$\sigma_{xy} = \eta \frac{dv_x}{dy} = \eta \frac{U}{d}, \tag{17.7}$$

which is independent of y, as one might have expected because in stationary flow the balance of forces requires the stress on any plane parallel with the plates to be the same.

Viscous friction

A thin layer of viscous fluid may be used to lubricate the interface between solid objects. From the above solution we may calculate the friction force, or *drag*, exerted on the body by the layer of viscous lubricant (see also chapter 27). Let the would-be contact area between the body and the surface on which it slides be A, and let the thickness of the fluid layer be d everywhere. If the layer is thin, $d \ll \sqrt{A}$, we may disregard edge effects and use the planar stress (17.7) to calculate the drag force,

$$\mathcal{D} \approx -\sigma_{xy} A = -\frac{\eta U A}{d}. \tag{17.8}$$

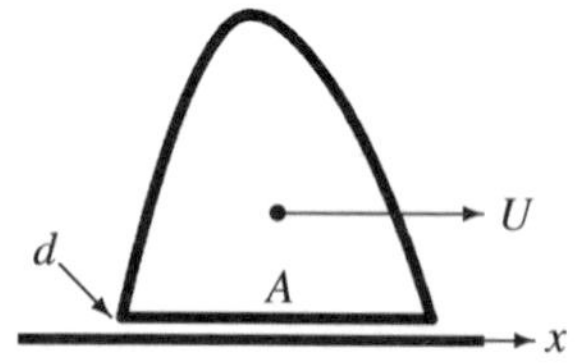

A solid object sliding on a plane lubricated surface.

The velocity-dependent viscous drag is quite different from the constant drag experienced in solid friction (see section 9.1 on page 111). The decrease in drag with falling velocity makes the object seem to want to slide 'forever', and this is presumably what makes ice sports such as skiing, skating, sledging and curling interesting. A thin layer of liquid water acts here as the lubricant. Likewise, it is scary to brake a car on ice or to aquaplane, because the fall in viscous friction as the speed drops makes the car appear to run away from you.

The quasi-steady horizontal equation of motion for an object of mass M, not subject to forces other than viscous drag, is

$$M \frac{dU}{dt} = -\eta \frac{A}{d} U. \tag{17.9}$$

Assuming that the thickness of the lubricant layer stays constant (and that is by no means evident) the solution to (17.9) is

$$U = U_0 e^{-t/t_0}, \qquad t_0 = \frac{Md}{\eta A}, \tag{17.10}$$

where U_0 is the initial velocity and t_0 is the characteristic exponential decay time for the velocity. Integrating this expression we obtain the total stopping distance

$$L = \int_0^\infty U\,dt = U_0 t_0 = \frac{U_0 M d}{\eta A}. \tag{17.11}$$

Although it formally takes infinite time for the sliding object to come to a full stop, it does so in a finite distance! The stopping length grows with the mass of the object which is quite unlike solid friction, where the stopping length is independent of the mass. This effect is partially compensated by the dynamic dependence of the layer thickness $d \sim 1/\sqrt{M}$ on the mass (problem 27.1).

Example 17.2.1 (Curling): In the ice sport of *curling*, a 'stone' with mass $M \approx 20$ kg is set into motion with the aim of bringing it to a full stop at the far end of an ice rink of length $L \approx 40$ m. The area of the highly polished contact surface to the ice is $A \approx 700\ \mathrm{cm}^2$ and the initial velocity about $U_0 \approx 3\ \mathrm{m\,s^{-1}}$. From (17.11) we obtain the thickness of the fluid layer $d \approx 43\ \mu\mathrm{m}$ which does not seem unreasonable, and neither does the decay time $t_0 \approx 13$ s. The players' intense sweeping of the ice in front of the moving stone presumably serves to smooth out tiny irregularities in the surface, which could otherwise slow down the stone.

Momentum diffusion

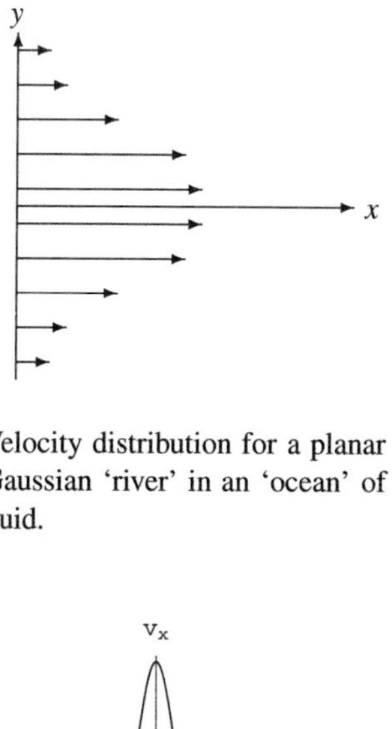

Velocity distribution for a planar Gaussian 'river' in an 'ocean' of fluid.

The dynamic equation (17.5) is a typical *diffusion equation* with diffusion constant equal to the kinematic viscosity, ν, also called *momentum diffusivity*. In general, such an equation leads to a spreading of the distribution of the diffused quantity, which in this case is the velocity field, or perhaps better, the momentum density ρv_x. The generic example of a flow with momentum diffusion is the Gaussian 'river',

$$v_x(y,t) = U\frac{a}{\sqrt{a^2+4\nu t}}\exp\left(-\frac{y^2}{a^2+4\nu t}\right), \tag{17.12}$$

which may be verified to be a solution to (17.5) by direct insertion. This river starts out at $t = 0$ with Gaussian width a and maximum velocity U, and spreads with time so that at time t it has width $\sqrt{a^2+4\nu t}$. Although momentum diffuses away from the centre of the river, the total momentum must remain constant because there are no external forces acting on the fluid. Kinetic energy is on the other hand dissipated and ends up as heat (see problem 17.3).

For sufficiently large times, $t \gg a^2/4\nu$, the shape of the Gaussian becomes independent of the original width a. This is, in fact, a general feature of *any* bounded 'river' flow: it eventually becomes proportional to $\exp(-y^2/4\nu t)$ (see problem 17.6). The Gaussian factor drops sharply to zero for $y \gtrsim \sqrt{4\nu t}$ and it appears as if momentum diffusion has a fairly well-defined front, which for example may be taken to be $y = 2\sqrt{\nu t}$ where the Gaussian is $e^{-1} = 37\%$ of its central value. Depending on the application, it is sometimes convenient to choose a more conservative estimate for the spread of momentum, for example $y = k\sqrt{\nu t}$, where for $k = 3$ the Gaussian factor is only 10% of its central value.

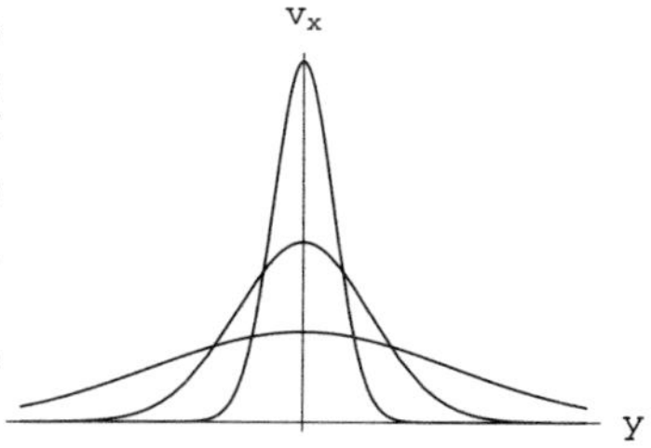

A Gaussian 'river' widens and slows down in the course of time because of viscosity.

Momentum diffusion may equivalently be characterized by the time it takes for a velocity disturbance to spread through a distance L by diffusion,

$$\boxed{t \approx \frac{L^2}{4\nu},} \tag{17.13}$$

or a correspondingly more conservative estimate. It must be emphasized that momentum diffusion (in this case) takes place orthogonally to the general direction of motion of the fluid. In spite of the fact that momentum diffuses away from the centre in the y-direction, there is no mass flow in the y-direction because $v_y = 0$. In less restricted flows there may be more direct competition between mass flow and diffusion. If the velocity scale of a flow is $|\boldsymbol{v}| \sim U$, it would take the time $t_{\rm flow} \sim L/U$ for the fluid to move through the distance L, and the ratio of the the diffusion time scale $t_{\text{diff}} \sim L^2/\nu$ to the mass flow time scale becomes a dimensionless number $\text{Re} \approx t_{\text{diff}}/t_{\text{flow}} \sim UL/\nu$, first used by Reynolds to classify different flows. When this number is large compared to unity, momentum diffusion takes a much longer time than mass flow and plays only a small role, whereas when Re is small momentum diffusion wins over mass flow and dominates the flow pattern. The Reynolds number is a very useful parameter which will be discussed in more detail in section 17.4.

Shear sound waves

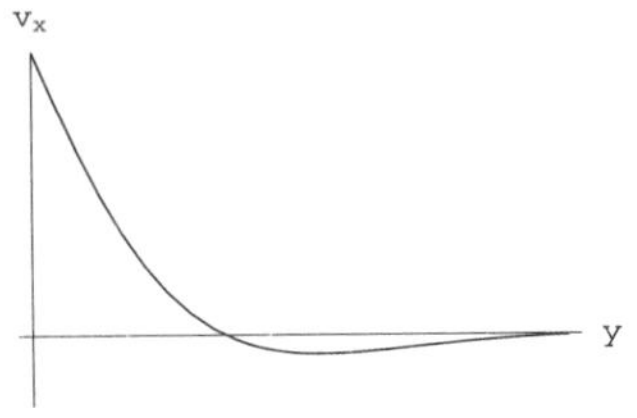

The shape of a transverse wave.

Consider an infinitely extended plate in contact with an infinite sea of fluid. Let the plate oscillate with circular frequency ω, so that its instantaneous velocity in the x-direction is $U(t) = U_0 \cos \omega t$. The motion of the plate is transferred to the neighbouring fluid because of the no-slip condition and then spreads into the fluid at large. How far does it go? By direct insertion into (17.5) it is found that

$$v_x(y,t) = U_0 e^{-ky} \cos(ky - \omega t), \qquad k = \sqrt{\frac{\omega}{2\nu}}, \tag{17.14}$$

satisfies the planar flow equation (17.5) as well as the no-slip boundary condition $v_x = U(t)$ for $y = 0$. Evidently, this is a damped wave spreading from the oscillating plate into the fluid. Since the velocity oscillations take place in the x-direction whereas the wave propagates in the y-direction it is a *transverse* or *shear* wave. The wavenumber k both determines the wavelength $\lambda = 2\pi/k$ and the decay length of the exponential, also called the *penetration depth* $d = 1/k = \lambda/2\pi$. The wave is critically damped and penetrates less than one wavelength into the fluid, so it is really not much of a wave. Although longitudinal (pressure) waves are also attenuated by viscosity, they propagate over much greater distances (see section 17.6).

Example 17.2.2: A shear sound wave in air of frequency 1000 Hz has wavelength 0.4 mm, whereas in water it is 0.1 mm.

17.3 Incompressible Newtonian fluids

Numerous everyday fluids obey Newton's law of viscosity (17.1), for example water, air, oil, alcohol and antifreeze. A number of common fluids are only approximatively Newtonian, for example paint and blood, and others are strongly non-Newtonian, for example tomato ketchup, jelly and putty. There also exist *viscoelastic* materials that are both elastic and viscous, sometimes used in toys that can be deformed like clay but also jump like a rubber ball.

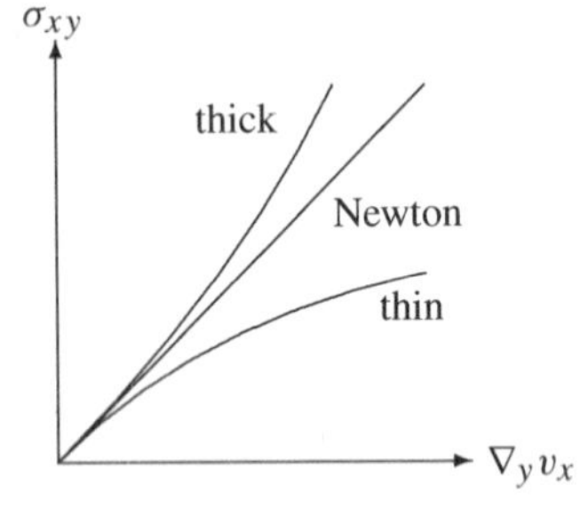

In Newtonian fluids the shear stress σ_{xy} increases linearly with the strain-rate $\nabla_y v_x$, whereas non-Newtonian fluids mostly become thinner and only a few become thicker.

In Newtonian fluids the shear stress σ_{xy} is directly proportional to the velocity gradient $\nabla_y v_x$ (also called the shear strain rate) with proportionality constant equal to the constant shear viscosity η. Most non-Newtonian fluids become thinner as the shear strain rate increases, implying that the shear stress as a function of shear strain rate grows slower than linearly. Even the most Newtonian of fluids, water, becomes thinner at shear strain rates above $10^{12}\ \text{s}^{-1}$. Only few fluids (for example some starches stirred in water) appear to thicken with increasing strain rate. The science of the general flow properties of materials is called *rheology*.

Most everyday liquids are incompressible, and gases are effectively so when the flow velocities are much smaller than the velocity of sound (see section 16.4). We shall in this section only establish the general dynamical equations for the simpler case of incompressible, isotropic Newtonian fluids and postpone the analysis of the slightly more complicated compressible fluids to section 17.5.

Isotropic viscous stress

Newton's law of viscosity (17.1) is a linear relation between the stress and the velocity gradient, only valid in a particular geometry. As for Hooke's law for elasticity (page 137) we want a more general definition of viscous stress which takes the same form for any geometry and in any Cartesian coordinate system, leaving us free to choose our own reference frame.

Most fluids are not only Newtonian, but also *isotropic*. Liquid crystals are anisotropic, but so special that we shall not consider them here. In an isotropic fluid at rest there are no internal directions at all and the stress tensor is determined by the pressure, $\sigma_{ij} = -p\,\delta_{ij}$. When such a fluid is set in motion, the velocity field $v_i(\boldsymbol{x}, t)$ defines a direction in every point of space, but as we have argued before the velocity at a point cannot itself provoke stress in the fluid. It is the variation in velocity from point to point that causes stress. Viscous stress is in other words determined by the velocity gradient tensor $\nabla_i v_j$.

In an incompressible fluid, the trace of the velocity gradient matrix vanishes, $\sum_i \nabla_i v_i = \boldsymbol{\nabla} \cdot \boldsymbol{v} = 0$, so the most general symmetric tensor one can construct from the velocity gradients is of the form,

$$\boxed{\sigma_{ij} = -p\,\delta_{ij} + \eta\,(\nabla_i v_j + \nabla_j v_i).} \tag{17.15}$$

This is the natural generalization of Newton's law of viscosity (17.1) for incompressible flow in arbitrary Cartesian coordinate systems. It may readily be verified that the coefficient η is indeed the shear viscosity introduced in Newton's law by inserting the field of a steady planar flow $\boldsymbol{v} = (v_x(y), 0, 0)$, because for such a field the only shear stress is $\sigma_{xy} = \sigma_{yx} = \eta \nabla_y v_x(y)$.

The Navier–Stokes equations for incompressible fluid

The right-hand side of Cauchy's general equation of motion (15.35) equals the effective density of force $f_i^* = f_i + \sum_j \nabla_j \sigma_{ij}$. Inserting the stress tensor (17.15) and using again $\boldsymbol{\nabla} \cdot \boldsymbol{v} = 0$, we find

$$\sum_j \nabla_j \sigma_{ij} = -\nabla_i p + \eta \left(\sum_j \nabla_i \nabla_j v_j + \sum_j \nabla_j^2 v_i \right) = -\nabla_i p + \eta \nabla^2 v_i.$$

Here we have tacitly assumed that the fluid is homogeneous such that the shear viscosity (like the density ρ) does not depend on $\boldsymbol{x}$. If the temperature or chemical composition of the fluid varies in space, the right-hand side must be modified.

Inserting this expression into Cauchy's equation of motion and converting to ordinary vector notation we finally obtain the *Navier–Stokes equation* for incompressible fluid (Navier (1822), Stokes (1845))

George Gabriel Stokes (1819–1903). *British mathematician and physicist. Contributed to the development of fluid dynamics, optics and heat conduction. In mathematical physics a famous theorem bears his name (page 220).*

$$\boxed{\frac{\partial \boldsymbol{v}}{\partial t} + (\boldsymbol{v} \cdot \boldsymbol{\nabla})\boldsymbol{v} = \boldsymbol{g} - \frac{1}{\rho_0} \boldsymbol{\nabla} p + \nu \nabla^2 \boldsymbol{v},} \tag{17.16}$$

where ρ_0 is the constant density, $\nu = \eta/\rho_0$ is the kinematic viscosity and $\boldsymbol{g} = \boldsymbol{f}/\rho_0$ is the acceleration field of the volume forces (normally due to gravity). The only difference from the Euler equation (16.1) is the last term on the right-hand side. Besides this equation, we have also the divergence condition for incompressibility,

$$\boxed{\boldsymbol{\nabla} \cdot \boldsymbol{v} = 0.} \tag{17.17}$$

Given the acceleration field $\boldsymbol{g}$, we now have four equations for the four fields, v_x, v_y, v_z, and p. Note, however, that whereas the three velocity fields obey truly dynamic equations with each field having its own time derivative, this is not the case for the pressure which is only determined indirectly through the divergence condition.

Relatively simple to look at, the Navier–Stokes equations contain all the complexity of real fluid flow, including that of Niagara Falls! It is therefore clear that one cannot in general expect to find simple solutions. Exact solutions can only be found in strongly restricted geometries and under simplifying assumptions concerning the nature of the flow, as in the planar laminar flow examples in the preceding section, and the examples to be studied in chapter 18.

Millenium Prize Problem: Among the seven Millenium Prize Problems set out by the Clay Mathematics Institute of Cambridge, Massachusetts, one concerns the existence of smooth, non-singular solutions to the Navier–Stokes equations (even for the simpler case of incompressible flow). The prize money of one million dollars illustrates how little we know and how much we would like to know about the general features of these equations which appear to defy the standard analytic methods for solving partial differential equations.

Boundary conditions

Field equations that are first order in time, like the Navier–Stokes equation (17.16), need initial values of the fields (and their spatial derivatives) in order to predict their values at later times. But what about physical boundaries, the containers of fluids, or even internal boundaries between different fluids? How do the fields behave there? Let us discuss the various fields that we have met one by one.

Density: The density is easy to dispose of, since it is allowed to be discontinuous and jump at a boundary between two materials, so this provides us with no condition at all. It is evident from the Navier–Stokes equation that a jump in density across a fluid boundary must somehow be accompanied by a jump in the derivatives of the other fields, but we shall not go into this question here.

Pressure: Newton's third law requires the pressure to be continuous across any boundary. This simple picture is, however, complicated by surface tension, which can give rise to a discontinuous jump in pressure across an interface between two materials.

Velocity: The normal component $v_n = \boldsymbol{v}\cdot\boldsymbol{n}$ of the velocity field must be continuous across any boundary, for the simple reason that what goes in on one side must come out on the other. If this were not the case, material would collect at the boundary or holes would develop in the fluid. The latter kind of breakdown can actually happen in extreme situations (cavitation).

The tangential component of velocity $\boldsymbol{v}_t = \boldsymbol{n}\times(\boldsymbol{v}\times\boldsymbol{n})$ must also be continuous, but for different reasons. The linear relationship (17.1) between stress and velocity gradient implies that a tangential velocity field which changes rapidly along the direction normal to a surface, must create very large and rapidly varying shear stress. In the extreme case of a discontinuous jump in velocity, the shear stress would become infinite. Although large stresses may be created, for example by hitting a fluid container with a hammer, they can however not be maintained for long, but are rapidly smoothed out by viscous momentum diffusion. Only if the continuum approximation breaks down, shear slippage may occur, for example in extremely rarified gases.

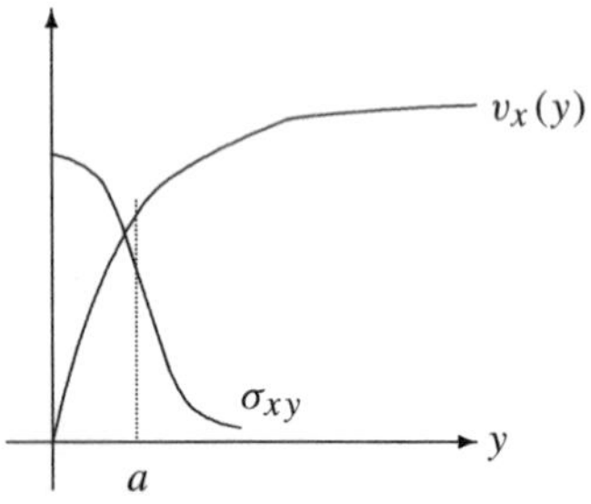

Sketch of rapidly varying velocity and shear stress in a region of size a near a boundary. For $a \to 0$ the velocity develops an abrupt jump and the stress becomes singular. The decrease in shear stress away from the discontinuity leads to spreading of the sharp discontinuity.

Usually the whole velocity field, normal as well as tangential components, will therefore be assumed to be continuous across any boundary between Newtonian fluids. Since a solid wall may be viewed as an extreme Newtonian fluid with infinite viscosity, we recover the previously mentioned *no-slip condition*: a fluid has zero velocity relative to its containing walls. Viscous fluids never slip along the containing boundaries but adhere to them, and this is part of the reason that viscous fluids are *wet*. Another reason for a fluid to be able to wet a solid surface is an acute contact angle (see chapter 8).

* Viscous dissipation

When you gently and steadily stir a pot of soup, the fluid will after some time settle down into a nearly steady flow. The fact that you still have to perform work while you stir steadily, shows that there must be viscous friction forces at play in the soup. The friction forces between the sides of the pot and the soup cannot perform any work because the fluid is at rest there, due to the no-slip condition. All the work you perform must for this reason be spent against the *internal* friction forces, the shear stresses acting between the moving layers of the fluid. If you stop stirring the kinetic energy of the soup is quickly *dissipated* into heat. We shall return to dissipation in chapters 22 and 30.

To make a quick calculation of the dissipative rate of work, we turn back to the discussion of deformation work on page 129. Since a fluid particle is displaced by $\delta\boldsymbol{u} = \boldsymbol{v}\,\delta t$ in a small time interval δt, fluid motion may be seen as a continuous sequence of infinitesimal deformations with strain tensor, $\delta u_{ij} = (1/2)(\nabla_i \delta u_j + \nabla_j \delta u_i) = (1/2)(\nabla_i v_j + \nabla_j v_i)\,\delta t$. The symmetrized velocity gradients $v_{ij} \equiv \delta u_{ij}/\delta t = (1/2)(\nabla_i v_j + \nabla_j v_i)$ may thus be understood as the *rate of deformation* or *rate of strain* of the fluid material. The rate of work $\dot{W} = \delta W/\delta t$ performed against the internal stresses is correspondingly given by (10.36),

$$\dot{W}_{\text{int}} = \int_V \sum_{ij} \sigma_{ij} \nabla_j v_i \, dV = \int_V \sum_{ij} \sigma_{ij} v_{ij} \, dV = \int_V 2\eta \sum_{ij} v_{ij}^2 \, dV, \tag{17.18}$$

where in the last step we have inserted the Newtonian stress tensor (17.15). Evidently, the rate of work against internal shear stresses is always positive. It always costs work to keep things moving against friction forces.

Non-locality of pressure

For incompressible fluids, the pressure is not given by an equation of state, but rather determined by the divergence condition, and that leads to special difficulties. Calculating the divergence of both sides of (17.16), we obtain a Poisson equation for the pressure,

$$\nabla^2 p = \rho_0\big(-\nabla \cdot ((\boldsymbol{v} \cdot \nabla)\boldsymbol{v}) + \nabla \cdot \boldsymbol{g}\big). \tag{17.19}$$

A solution to the Poisson equation is generally of the same form as the gravitational potential from a mass distribution (3.24) and depends non-locally on the field (the source) on the right-hand side. The non-local pressure therefore instantly communicates any change in the velocity field to the rest of the fluid.

Like true rigidity, true incompressibility is an ideal which cannot be reached with real materials, where the velocity of sound sets an upper limit to small-amplitude signal propagation speed. The above result nevertheless means that any local change in the flow will be communicated by the non-local pressure to any other parts of the fluid at the speed of sound. This phenomenon is in fact well-known from everyday experience where the closing of a faucet can result in rather violent so-called 'water-hammer' responses from the house piping. The non-locality of pressure is also a major problem in numerical simulations of the Navier–Stokes equations for incompressible fluids (see chapter 21).

17.4 Classification of flows

The most interesting phenomena in fluid dynamics arise from the competition between inertia and viscosity, represented in the Navier–Stokes equation (17.16) by the advective term $(\boldsymbol{v}\cdot\nabla)\boldsymbol{v}$ and the viscous term $\nu\nabla^2\boldsymbol{v}$. Inertia attempts to continue the motion of a fluid once it is started whereas viscosity acts as a brake. If inertia is dominant we may leave out the viscous term, arriving again at Euler's equation (16.1) describing lively, *non-viscous* or *ideal* flow (see chapter 16). If on the other hand viscosity is dominant, we may drop the advective term, and obtain the basic equations for sluggish *creeping* flow (see chapter 19).

Osborne Reynolds (1842–1912). *British engineer and physicist. Contributed to fluid mechanics in general, and to the understanding of lubrication, turbulence and tidal motion in particular.*

The Reynolds number

As a measure of how much an actual flow is lively or sluggish, one may make a rough estimate, called the *Reynolds number*, for the magnitude of the ratio of the advective to the viscous terms. To get a simple expression we assume that the velocity is of typical size $|\boldsymbol{v}| \approx U$ and that it changes by a similar amount over a region of size L. The order of magnitude of the first-order spatial derivatives of the velocity will then be of magnitude $|\nabla\boldsymbol{v}| \approx U/L$, and the second-order derivatives will be $\left|\nabla^2\boldsymbol{v}\right| \approx U/L^2$. Consequently, the Reynolds number becomes

$$\mathrm{Re} \approx \frac{|(\boldsymbol{v}\cdot\nabla)\boldsymbol{v}|}{|\nu\nabla^2\boldsymbol{v}|} \approx \frac{U^2/L}{\nu U/L^2} = \frac{UL}{\nu}. \tag{17.20}$$

For small values of the Reynolds number, Re $\ll 1$, advection plays no role and the flow creeps along, whereas for large values, Re $\gg 1$, viscosity can be ignored and the flow tends to be lively. The streamline pattern of creeping flow is orderly and layered, also called *laminar*, well known from the kitchen when mixing cocoa into dough to make a chocolate cake (although dough is hardly Newtonian!). The laminar flow pattern continues quite far beyond Re $\simeq 1$, but depending on the flow geometry and other circumstances, there will be a Reynolds number, typically in the region of thousands, where *turbulence* sets in with its characteristic tumbling and chaotic behaviour.

It is often quite easy to estimate the Reynolds number from the geometry and boundary conditions of a flow pattern, as is done in the following examples and in table 17.2.

Example 17.4.1 (Bathtub turbulence): Getting out of a bathtub you create flows with speeds of say $U \approx 1$ m s^{-1} over a distance of $L \approx 1$ m. The Reynolds number becomes Re $\approx 10^6$ and you are definitely creating visible turbulence in the water. Similarly, when jogging you create air flows with

Table 17.2. Table of Reynolds numbers for some moving objects calculated on the basis of typical values of lengths and speeds. Viscosities are taken from table 17.1 on page 230. It is perhaps surprising that a submarine operates at a Reynolds number that is larger than that of a passenger jet at cruising speed, but this is mostly due to the kinematic viscosity of air being larger than that of water.

	Fluid	Size L [m]	Velocity U [ms^{-1}]	Reynolds number
Submarine	water	100	15	1.7×10^9
Airplane	air	50	200	6.3×10^8
Blue whale	water	30	10	3.4×10^8
Car	air	5	30	9.4×10^6
Swimming human	water	2	1	2.3×10^6
Running human	air	2	3	3.8×10^5
Herring	water	0.3	1	3.8×10^5
Golf ball	air	0.043	40	2.2×10^5
Ping-pong ball	air	0.040	10	5×10^4
Fly	air	0.01	1	600
Flea	air	0.001	3	190
Gnat	air	0.001	0.1	6
Bacterium	water	10^{-6}	10^{-5}	10^{-5}
Bacterium	blood	10^{-6}	4×10^{-6}	10^{-9}

$U \approx 3$ m s^{-1} and $L \approx 1$ m, leading to a Reynolds number around 2×10^5, and you know that you must leave all kinds of little invisible turbulent eddies in the air behind you. The fact that the Reynolds number is smaller in air than in water despite the higher velocity is a consequence of the kinematic viscosity being larger for air than for water.

Example 17.4.2 (Curling): For planar flow between two plates (section 17.1), the velocity scale is set by the velocity difference U between the plates whereas the length scale is set by the distance d between the plates. In the curling example 17.2.1 on page 233 we found $U \approx 3$ m s^{-1} and $d \approx 43$ μm, leading to a Reynolds number $\text{Re} = Ud/\nu \approx 140$. Although not truly creeping flow, it is definitely laminar and not turbulent.

Example 17.4.3 (Water pipe): A typical 1/2 inch water pipe has diameter $d \approx 1.25$ cm and that sets the length scale. If the volume flux of water is $Q = 100$ cm^3 s^{-1}, the average water speed becomes $U = Q/\pi a^2 \approx 0.8$ m s^{-1} and we get a Reynolds number $\text{Re} = Ud/\nu \approx 10^4$ which brings the flow well into the turbulent regime. For olive oil under otherwise identical conditions we get $\text{Re} \approx 0.15$, and the flow would be creeping.

Hydrodynamic similarity

What does it mean if two flows have the same Reynolds number? A stone of size $L = 1$ m sitting in a steady water flow with velocity $U = 2$ m s^{-1} has the same Reynolds number as another stone of size $L = 2$ m in a steady water flow with velocity $U = 1$ m s^{-1}. It even has the same Reynolds number as a stone of size $L = 4$ m in a steady airflow with velocity $U = 8$ m s^{-1}, because the kinematic viscosity of air is about 16 times larger than of water (at normal temperature and pressure). We shall now see that provided the stones are geometrically similar, i.e. have congruent geometrical shapes, flows with the same Reynolds numbers are also *hydrodynamically similar* and only differ by their overall length and velocity scales, so that their flow patterns visualized by streamlines will look identical.

In the absence of volume forces, steady incompressible flow is determined by (17.16) with $\boldsymbol{g} = \boldsymbol{0}$ and

$\partial \boldsymbol{v}/\partial t = \mathbf{0}$, or

$$(\boldsymbol{v} \cdot \boldsymbol{\nabla})\boldsymbol{v} = -\frac{1}{\rho_0}\boldsymbol{\nabla} p + \nu \boldsymbol{\nabla}^2 \boldsymbol{v}. \tag{17.21}$$

Let us rescale all the variables by means of the overall scales ρ_0, U, and L, writing

$$\boldsymbol{v} = U\hat{\boldsymbol{v}}, \quad \boldsymbol{x} = L\hat{\boldsymbol{x}}, \quad p = \rho_0 U^2 \hat{p}, \quad \boldsymbol{\nabla} = \frac{1}{L}\hat{\boldsymbol{\nabla}}, \tag{17.22}$$

where the hatted symbols are all dimensionless. Inserting these variables, the steady flow equation takes the form,

$$(\hat{\boldsymbol{v}} \cdot \hat{\boldsymbol{\nabla}})\hat{\boldsymbol{v}} = -\hat{\boldsymbol{\nabla}} \hat{p} + \frac{1}{\mathrm{Re}}\hat{\boldsymbol{\nabla}}^2 \hat{\boldsymbol{v}}. \tag{17.23}$$

The only parameter appearing in this equation is the Reynolds number which may be interpreted as the inverse of the dimensionless kinematic viscosity. The pressure is as mentioned not an independent dynamic variable and its scale is here fixed by the velocity scale, $P = \rho_0 U^2$. If the flow instead is driven by external pressure differences of magnitude P rather than by velocity, the equivalent flow velocity scale is given by $U = \sqrt{P/\rho_0}$.

In congruent flow geometries, the no-slip boundary conditions will also be the same, so that any solution of the dimensionless equation can be scaled back to a solution of the original equation by means of (17.22). The three different flows around stones mentioned at the beginning of this subsection may thus all be obtained from the same dimensionless solution if the stones are geometrically similar and the Reynolds numbers identical.

Even if the flows are similar in air and water, the forces exerted on the stones will, however, not be the same. The shear stress magnitude may be estimated as $\sigma \approx \eta\, |\boldsymbol{\nabla} \boldsymbol{v}| \sim \eta U/L$, times a function of the Reynolds number. The viscous drag on an object of size L will then be of magnitude $\mathcal{D} \approx \sigma L^2 \sim \eta U L = \eta \nu \mathrm{Re}$. Since the Reynolds numbers are assumed to be the same in the two cases, the ratio of the viscous drag on the stone in air to that in water is about $\mathcal{D}_{\text{air}}/\mathcal{D}_{\text{water}} \approx (\eta\nu)_{\text{air}}/(\eta\nu)_{\text{water}} \approx 0.27$.

Example 17.4.4 (Flight of the robofly): The similarity of flows in congruent geometries can be exploited to study the flow around tiny insects by means of enlarged slower moving models, immersed in another fluid. It is, for example, hard to study the air flow around the wing of a hovering fruit fly, when the wing flaps $f = 50$ times per second. For a wing size of $L \approx 4$ mm flapping through $180\,°$ the average velocity becomes $U \approx \pi L f \approx 1.3\ \text{m s}^{-1}$ and the corresponding Reynolds number $\mathrm{Re} \approx UL/\nu \approx 160$. The same Reynolds number can be obtained (see J. M. Birch and M. H. Dickinson, *Nature* **412**, (2001) 729) from a 19 cm plastic wing of the same shape, flapping once every 6 s in mineral oil with kinematic viscosity $\nu = 1.15\ \text{cm}^2\ \text{s}^{-1}$, allowing for easy recording of the flow around the wing.

Example 17.4.5 (High-pressure wind tunnels): In the early days of flight, wind tunnels were extensively used for empirical studies of lift and drag on scaled-down models of wings and aircraft. Unfortunately, the smaller geometrical sizes of the models reduced the attainable Reynolds number below that of real aircraft in flight. A solution to the problem was obtained by operating wind tunnels at much higher than atmospheric pressure. Since the dynamic viscosity η is independent of pressure (page 231), the Reynolds number $\mathrm{Re} = \rho_0 UL/\eta$ scales with the air density and thus with pressure. The famous *Variable Density Tunnel (VDT)* built in 1922 by the US National Advisory Committee for Aeronautics (NACA) operated on a pressure of 20 atm and was capable of attaining full-scale Reynolds numbers for models only 1/20th of the size of real aircraft [3, p. 301]. The results obtained from the VDT had great influence on aircraft design in the following 20 years.

In the presence of external volume forces, for example gravity, or for time-dependent inflow, the flow patterns will depend on further dimensionless quantities besides the Reynolds number. We shall only introduce such quantities when they arise naturally in particular cases. Flows in different geometries can only be compared in a coarse sense, even if they have the same Reynolds number. A running man has the same Reynolds number as a swimming herring, and a flying gnat the same Reynolds number as a man swimming in castor oil (which cannot be recommended). In both cases the flow geometries are quite different, leading to different streamline patterns. Here the Reynolds number can only be used to indicate the character of the flow which tends to be turbulent around the running man and laminar around the flying gnat.

17.5 Compressible Newtonian fluids

When flow velocities become a finite fraction of the velocity of sound, it is no longer possible to maintain the simplifying assumption of effective incompressibility. Whereas submarines and ships never come near such velocities, passenger jets routinely operate at speeds up to 90% of the velocity of sound, and rockets, military aircraft, the Concorde and the Space Shuttle, are all capable of flying at supersonic and even hypersonic speeds. In these cases it is necessary to modify the Navier–Stokes equation.

Shear and bulk viscosity

In compressible fluids the divergence of the velocity field is non-vanishing. This opens up the possibility of adding a term proportional to $(\nabla \cdot \boldsymbol{v})\delta_{ij}$ to the isotropic stress tensor (17.15),

$$\sigma_{ij} = -p\,\delta_{ij} + \eta\left(\nabla_i v_j + \nabla_j v_i\right) + a\nabla \cdot \boldsymbol{v}\,\delta_{ij}. \tag{17.24}$$

Demanding as usual that the pressure is the average of the three normal stresses, $p = -\sum_i \sigma_{ii}/3$, the trace of this expression becomes $-3p = -3p + 2\eta\nabla \cdot \boldsymbol{v} + 3a\nabla \cdot \boldsymbol{v} = 0$, so that we must have $a = -(2/3)\eta$. The complete stress tensor thus takes the form,

$$\boxed{\sigma_{ij} = -p\,\delta_{ij} + \eta\left(\nabla_i v_j + \nabla_j v_i - \frac{2}{3}\nabla \cdot \boldsymbol{v}\,\delta_{ij}\right).} \tag{17.25}$$

This stress tensor may be viewed as a first-order expansion in the velocity gradients. In the same approximation the pressure may also depend linearly on the velocity gradients, but since the pressure is a scalar it can only depend on the scalar divergence $\nabla \cdot \boldsymbol{v}$, so the most general form of the mechanical pressure must be

$$\boxed{p = p_e - \zeta\nabla \cdot \boldsymbol{v},} \tag{17.26}$$

with coefficients p_e and ζ that may depend on the density ρ and temperature T. In mechanical equilibrium, $\boldsymbol{v} = \boldsymbol{0}$, we have $p = p_e = p_e(\rho, T)$, which must be the thermodynamic pressure given by the equation of state.

The new parameter ζ is called the *bulk viscosity* or the *expansion viscosity*. Its presence implies that a viscous fluid in motion exerts an extra *dynamic pressure* of size $-\zeta\nabla \cdot \boldsymbol{v}$. The dynamic pressure is negative in regions where the fluid expands ($\nabla \cdot \boldsymbol{v} > 0$), positive where it contracts ($\nabla \cdot \boldsymbol{v} < 0$), and vanishes for incompressible fluids. Bulk viscosity is hard to measure, because one must set up physical conditions such that expansion and contraction become important, for example by means of high frequency sound waves. In the following section we shall calculate the viscous attenuation of sound in fluids, and see that it depends on the bulk modulus. The measurement of attenuation of sound is quite complicated and yields a rather frequency-dependent bulk viscosity, although it is generally of the same magnitude as the coefficient of shear viscosity.

The Navier–Stokes equation

Inserting the modified stress tensor (17.25) into Cauchy's equation of motion we obtain the field equation,

$$\boxed{\rho\left(\frac{\partial \boldsymbol{v}}{\partial t} + (\boldsymbol{v} \cdot \nabla)\boldsymbol{v}\right) = \boldsymbol{f} - \nabla p + \eta\left(\nabla^2 \boldsymbol{v} + \frac{1}{3}\nabla(\nabla \cdot \boldsymbol{v})\right).} \tag{17.27}$$

This is the most general form of the *Navier–Stokes equation*. Together with the equation of continuity (15.24), which we repeat here for convenience,

$$\boxed{\frac{\partial \rho}{\partial t} + \nabla \cdot (\rho \boldsymbol{v}) = 0,} \tag{17.28}$$

we have obtained four dynamic equations for the four fields v_x, v_y, v_z and ρ, and one constitutive relation (17.26) for the pressure p as a function of the thermodynamic variables and the velocity field.

* Viscous dissipation

For compressible fluids the rate of work against internal stresses is slightly more complicated than for incompressible fluids (17.18). Defining the traceless shear strain rate

$$v_{ij} = \frac{1}{2}\left(\nabla_i v_j + \nabla_j v_i - \frac{2}{3}\nabla\cdot\boldsymbol{v}\,\delta_{ij}\right), \tag{17.29}$$

the stress tensor (17.25) takes the form $\sigma_{ij} = -p\delta_{ij} + 2\eta v_{ij}$. Using the symmetry and tracelessness of v_{ij} we have $\sum_{ij}\sigma_{ij}\nabla_j v_i = -p\nabla\cdot\boldsymbol{v} + 2\eta\sum_{ij} v_{ij}^2$, and inserting the mechanical pressure from (17.26) we obtain the total rate of work against internal stresses,

$$\boxed{\dot{W}_{\text{int}} = \int_V \sum_{ij}\sigma_{ij}\nabla_j v_i\,dV = \int_V \left(-p_e\nabla\cdot\boldsymbol{v} + \zeta(\nabla\cdot\boldsymbol{v})^2 + 2\eta\sum_{ij} v_{ij}^2\right) dV.} \tag{17.30}$$

This expression reduces of course to (17.18) for $\nabla\cdot\boldsymbol{v} = 0$. The first term represents the familiar thermodynamic rate of work on the fluid. It is positive during compression ($\nabla\cdot\boldsymbol{v} < 0$) and negative during expansion, and may in principle be recovered completely under under adiabatic conditions. The last two terms represent the work against internal viscous stresses and are always positive. In particular we note that the second term due to the bulk viscosity is positive whether the flow expands or contracts. Both the bulk and shear terms contribute to the attenuation of sound, as we shall see in the following section.

17.6 Viscous attenuation of sound

It has previously (on page 234) been shown that free shear waves do not propagate through more than about one wavelength from their origin in any type of fluid. In nearly ideal fluids such as air and water, free pressure waves are as everybody knows capable of propagating over many wavelengths. Viscous dissipation (and many other effects) will nevertheless slowly sap their strength, and in the end all of the kinetic energy of the waves will be converted into heat.

In this section we shall calculate the rate of attenuation from damped-wave small-amplitude solutions to the Navier–Stokes equations. The attenuation may as well be calculated from the general expression for the dissipative work (17.30).

The wave equation

As in the discussion of unattenuated pressure waves in section 16.2 on page 206 we assume to begin with that a barotropic fluid is in hydrostatic equilibrium without gravity, $\boldsymbol{v} = \boldsymbol{0}$, so that its density $\rho = \rho_0$ and pressure $p = p(\rho_0)$ are constant throughout. Consider now a disturbance in the form of a small-amplitude motion of the fluid, described by a velocity field $\boldsymbol{v}$ which is so tiny that the nonlinear advective term $(\boldsymbol{v}\cdot\nabla)\boldsymbol{v}$ can be completely disregarded. This disturbance will be accompanied by tiny density corrections, $\Delta\rho = \rho - \rho_0$, and pressure corrections $\Delta p = p - p_0$, which we assume to be of first order in the velocity. Dropping all higher order terms, the linearized Navier–Stokes equations become (in the absence of gravity),

$$\rho_0\frac{\partial\boldsymbol{v}}{\partial t} = -\nabla\Delta p + \eta\left(\nabla^2\boldsymbol{v} + \frac{1}{3}\nabla(\nabla\cdot\boldsymbol{v})\right), \tag{17.31a}$$

$$\frac{\partial\Delta\rho}{\partial t} = -\rho_0\nabla\cdot\boldsymbol{v}, \tag{17.31b}$$

$$\Delta p = c_0^2\Delta\rho - \zeta\nabla\cdot\boldsymbol{v}. \tag{17.31c}$$

The last equation has been obtained from (17.26) by expanding to first order in the small quantities, and using that the velocity of sound is $c_0 = \sqrt{dp_e/d\rho}$.

Differentiating the second equation with respect to time and making use of the first, we obtain

$$\frac{\partial^2 \Delta\rho}{\partial t^2} = \nabla^2 \Delta p - \frac{4}{3}\eta \nabla^2 \nabla \cdot \boldsymbol{v} = \nabla^2 \Delta p + \frac{4}{3}\frac{\eta}{\rho_0}\nabla^2 \frac{\partial \Delta\rho}{\partial t}.$$

Now substituting the pressure correction (17.31c) we arrive at the following equation for the density corrections (for the pressure corrections, see problem 17.5),

$$\frac{\partial^2 \Delta\rho}{\partial t^2} = c_0^2 \nabla^2 \Delta\rho + \frac{\zeta + \frac{4}{3}\eta}{\rho_0}\nabla^2 \frac{\partial \Delta\rho}{\partial t}. \tag{17.32}$$

If the last term on the right-hand side were absent, this would be a standard wave equation of the form (16.7) describing free density (or pressure) waves with phase velocity c_0. It is the last term which causes viscous attenuation.

The ratio of the coefficients of the first to the second terms has dimension of inverse time and defines a circular frequency scale,

$$\omega_0 = \frac{c_0^2 \rho_0}{\zeta + \frac{4}{3}\eta}. \tag{17.33}$$

Taking $\zeta \sim \eta$, the right-hand side is of the order of $\omega_0 \approx 3 \times 10^9\ \mathrm{s}^{-1}$ in air and $\omega_0 \approx 10^{12}\ \mathrm{s}^{-1}$ in water. In terms of c_0 and ω_0, the wave equation may now be written more conveniently,

$$\boxed{\frac{1}{c_0^2}\frac{\partial^2 \Delta\rho}{\partial t^2} = \nabla^2 \Delta\rho + \frac{1}{\omega_0}\nabla^2 \frac{\partial \Delta\rho}{\partial t}.} \tag{17.34}$$

The time derivative in the last term is of order $\omega \Delta\rho$ for a wave with circular frequency ω, and in view of the huge values of the viscous frequency scale ω_0, the ratio of the terms ω/ω_0 will be small, implying that attenuation is weak for normal sound, including ultrasound in the megahertz region.

Damped plane wave

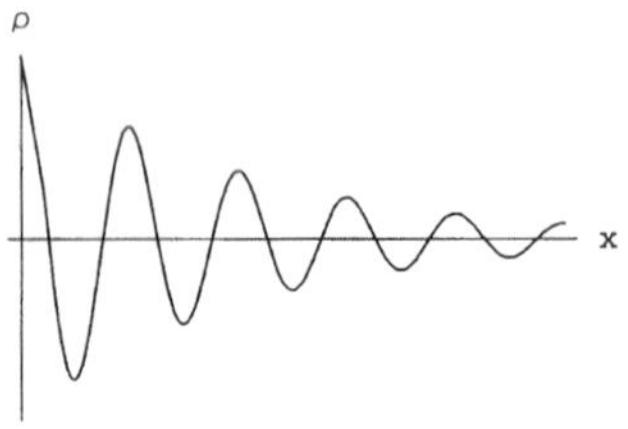

Damped density wave.

Let us assume that a wave is created by an infinitely extended plane, a 'loudspeaker', situated at $x = 0$ and oscillating in the x-direction with a small amplitude at a definite circular frequency ω. The fluid near the plate has to follow the plate and will be alternately compressed and expanded, thereby generating a damped density (or pressure) wave of the form,

$$\Delta\rho = \rho_1 e^{-\kappa x} \cos(kx - \omega t), \tag{17.35}$$

where $\rho_1 \ll \rho_0$ is the small density amplitude, k is the wavenumber, and κ is the *viscous amplitude attenuation coefficient*. In view of the weak attenuation of normal sound, we expect that $\kappa/k \ll 1$ and $\omega/\omega_0 \ll 1$. Inserting this wave into (17.34), we get to lowest order in both κ/k and ω/ω_0,

$$-\frac{\omega^2}{c_0^2}\cos(kx - \omega t) = -k^2 \cos(kx - \omega t) + 2\kappa k \sin(kx - \omega t) - k^2 \frac{\omega}{\omega_0}\sin(kx - \omega t).$$

This can only be fulfilled when the wavenumber has the usual free-wave relation to frequency, $k = \omega/c_0$, and

$$\boxed{\kappa = \frac{k\omega}{2\omega_0} = \frac{\omega^2}{2\omega_0 c_0} = \frac{\omega^2}{2\rho_0 c_0^3}\left(\zeta + \frac{4}{3}\eta\right).} \tag{17.36}$$

The viscous amplitude attenuation coefficient grows quadratically with the frequency, causing high frequency sound to be attenuated much more by viscosity than low frequency sound.

In air the viscous attenuation length determined by this expression is huge, about $1/\kappa \approx 50$ km, at a frequency of 1000 Hz. At 1 MHz it is a million times shorter, about $1/\kappa \approx 5$ cm. Diagnostic imaging typically uses ultrasound between 1 and 15 MHz, but since living tissue is mostly water with higher density and sound velocity, the attenuation length is much longer than in air. The drastic reduction in attenuation length $1/\kappa$ with increased frequency is also what makes measurements of the attenuation coefficient much easier at high frequencies. From the viscous attenuation coefficient one may in principle extract the value of the bulk viscosity, but this is complicated by several other fundamental mechanisms that also attenuate sound, such as thermal conductivity, and excitation of molecular rotations and vibrations.

In the real atmosphere, many other effects contribute to the attenuation of sound. First, sound is mostly emitted from point sources rather than from infinitely extended vibrating planes, and that introduces a quadratic drop in amplitude with distance. Other factors like humidity, dust, impurities and turbulence also contribute, in fact much more than viscosity at the relatively low frequencies that human activities generate (see for example [22, appendix] for a discussion of the basic physics of sound waves in gases).

Problems

17.1 Calculate the temperature dependence of the kinematic viscosity for an isentropic gas. What is the exponent of the temperature for monatomic, diatomic and multiatomic gases?

17.2 A car with $M = 1000$ kg moving at $U_0 = 100$ km h^{-1} suddenly hits a patch of ice and begins to slide. The total contact area between each wheel and the water is 800 cm^2 and it is observed to slide to a full stop in about 300 m. Calculate the thickness of the water layer and discuss whether it is a reasonable value. What is the time scale for stopping the car?

17.3 Consider planar momentum diffusion (page 233). Assume that the flow of the incompressible 'river' vanishes fast at infinity, as in the Gaussian case. **(a)** Show that for any river flowing along x the total volume flux per unit of length in the z-direction is independent of time. **(b)** Show that the total momentum per unit of length in the z-direction is likewise constant. **(c)** For the Gaussian river, calculate the kinetic energy per unit of length in the x and z directions as a function of time. What happens for $t \to \infty$?

17.4 Estimate the Reynolds number for (a) an ocean current, (b) a water fall, (c) a weather cyclone, (d) a hurricane, (e) a tornado, (f) lava running down a mountainside and (g) plate tectonic motion.

17.5 Show that the pressure correction Δp also satisfies the wave equation for attenuated sound (17.34).

17.6 Show that the general solution to the momentum diffusion equation (17.5) is

$$v_x(y, t) = \frac{1}{2\sqrt{\pi \nu t}} \int_{-\infty}^{\infty} \exp\left(-\frac{(y - y')^2}{4\nu t}\right) v_x(y', 0)\, dy'. \tag{17.37}$$

Use this to show that any bounded initial velocity distribution which is non-vanishing only for $|y| < a$ at $t = 0$ is Gaussian for $|y| \to \infty$ at any later time.

18 Plates and pipes

Even the most common of fluid flows, the water coming out of the kitchen tap or the draught of air around you, are so full of little eddies that their description seems totally beyond exact analysis. Fluid mechanics is the story of coarse approximation, gross oversimplification, and if accurate results are required, numeric computation.

Analytic solutions of the Navier–Stokes equations are few and hard to come by, even for steady flow. They are always found in geometries characterized by a high degree of symmetry, for example planar, cylindrical or spherical. Then again, even if the geometry of a problem—the containing boundaries—is perfectly symmetric, there is no guarantee that the analytic solutions obeying these symmetries are the ones that the actual fluid selects as its flow pattern.

Symmetry is only a guideline. In any geometry there may and will be flows that are not maximally symmetric. At low Reynolds number one expects that the maximally symmetric solution might be stable against disturbances, but at large Reynolds number this will not be the case. The simple laminar flow pattern of the maximally symmetric steady flow is then broken spontaneously or by little irregularities in the geometry, and replaced by time-dependent flow with less than maximal symmetry, or no symmetry at all.

In this chapter we shall study steady laminar flow in the simplest of geometries: planar and cylindrical. The exact solutions presented here are all effectively one-dimensional and thus lead to ordinary differential equations that are easy to solve. Although the solutions are all of infinite extent they nevertheless provide valuable insight into the behaviour of viscous fluids for moderate Reynolds number. Many flows of interest are more complicated than these solutions, and we include in this chapter a discussion of the phenomenology of turbulent pipe flow and secondary flow between rotating cylinders.

18.1 Steady, incompressible, viscous flow

Most of the fluids encountered in daily life, water, air, gasoline and oil, are effectively incompressible as long as flow velocities are well below the speed of sound, and often they flow steadily through the channels and pipes that we use to guide them. In looking for exact solutions for viscous flow, we shall therefore make the simplifying assumptions that the flow is incompressible and steady, satisfying the Navier–Stokes equation (17.16) and the divergence condition,

$$(\boldsymbol{v} \cdot \nabla)\boldsymbol{v} = \boldsymbol{g} - \frac{1}{\rho_0}\nabla p + \nu\nabla^2\boldsymbol{v} \tag{18.1a}$$

$$\nabla \cdot \boldsymbol{v} = 0 \tag{18.1b}$$

where ρ_0 is the constant density and ν is the kinematic viscosity (17.4). These four partial differential equations determine in principle the four fields, v_x, v_y, v_z and p, throughout a volume, when provided with suitable boundary conditions that fix the values of the fields or their derivatives on the surface of the volume.

In practice, however, we often solve these equations by making assumptions about the form of the fields, based on the symmetries of the particular problem at hand. Symmetry assumptions limit the possible boundary conditions that may be imposed on the fields. In the absence of gravity, the field equations are, for example, always solved by the extremely 'symmetric' solution, $\boldsymbol{v} = \boldsymbol{0}$ and $p = 0$, but this assumption prevents us from selecting any non-zero values for the fields or their derivatives on the boundaries. Symmetry assumptions also make the solution blind to the symmetry breakdown that may happen in the real world of fluids. So it should never be forgotten that although symmetric solutions can be beautiful, they may also be completely irrelevant!

18.2 Pressure-driven planar flow

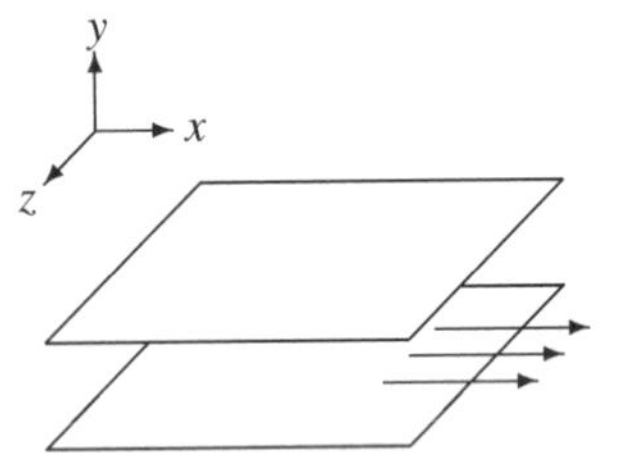

Planar flow between parallel plates.

In section 17.2 we analysed velocity-driven planar flow between infinitely extended parallel plates, where one of the plates was moving at constant velocity with respect to the other. Here we shall solve the case where the plates are fixed and fluid is driven between them by means of a pressure difference. For simplicity we assume that there is no gravity; gravity-driven flow will be analysed in the following section.

The coordinate system is chosen with the x-axis pointing along the direction of flow and the y-axis orthogonal to the plates. A velocity field respecting the planar symmetry is of the form,

$$\boldsymbol{v} = (v_x(y), 0, 0) = v_x(y)\,\boldsymbol{e}_x. \tag{18.2}$$

An infinitely extended flow like this is of course unphysical, but should nevertheless offer an approximation to the real flow between plates of finite extent, provided the dimensions of the plates are sufficiently large compared to their mutual distance.

General solution

With the assumed form of the flow, the incompressibility condition (18.1b) is automatically fulfilled, $\boldsymbol{\nabla} \cdot \boldsymbol{v} = \nabla_x v_x(y) = 0$. The advective acceleration vanishes likewise, $(\boldsymbol{v} \cdot \boldsymbol{\nabla})\boldsymbol{v} = v_x(y)\nabla_x v_x(y)\boldsymbol{e}_x = \boldsymbol{0}$. The Navier–Stokes equation (18.1a) then takes the form,

$$\boldsymbol{\nabla} p = \eta\, \boldsymbol{e}_x \boldsymbol{\nabla}^2 v_x(y).$$

From the y and z components of this equation we get $\nabla_y p = \nabla_z p = 0$, implying that the pressure cannot depend on y and z, or in other words $p = p(x)$. The x-component then takes the form,

$$\frac{dp(x)}{dx} = \eta \frac{d^2 v_x(y)}{dy^2}. \tag{18.3}$$

The left-hand side depends only on x and the right-hand side only on y, and that is only possible, if both sides take the same value independently of both x and y. Denoting the common value $-G$, we may immediately solve each of the equations $dp/dx = -G$ and $\eta d^2 v_x/dy^2 = -G$ with the result,

$$p = p_0 - Gx, \tag{18.4a}$$

$$v_x = -\frac{G}{2\eta} y^2 + Ay + B, \tag{18.4b}$$

where p_0, A, and B are integration constants. The only freedom left in the planar flow problem lies in the integration constants, and they will be fixed by the boundary conditions of the specific flow configuration.

Specific solution

Let the plates be positioned a distance d apart, for example at $y = 0$ and $y = d$. Applying the no-slip boundary conditions, $v_x(0) = v_x(d) = 0$, to the general solution, we obtain $A = Gd/2\eta$ and $B = 0$, so the velocity field becomes,

$$\boxed{v_x = \frac{G}{2\eta} y(d - y).} \tag{18.5}$$

It has a characteristic parabolic shape with the maximal velocity $v_x|_{\max} = Gd^2/8\eta$ in the middle of the gap for $y = d/2$.

The velocity profile depends on the (negative) pressure gradient G in the direction of flow. It may be calculated from the pressure drop between $x = 0$ where the pressure is p_0 and $x = L$ where the pressure is p_1, where $\Delta p = p_0 - p_1$ is the magnitude of the pressure drop over the distance L in the direction of flow. For finite plates of length L and width W with a small mutual distance, $d \ll L, W$, the flow should be reasonably well described by this solution.

$$G = \frac{\Delta p}{L}, \tag{18.6}$$

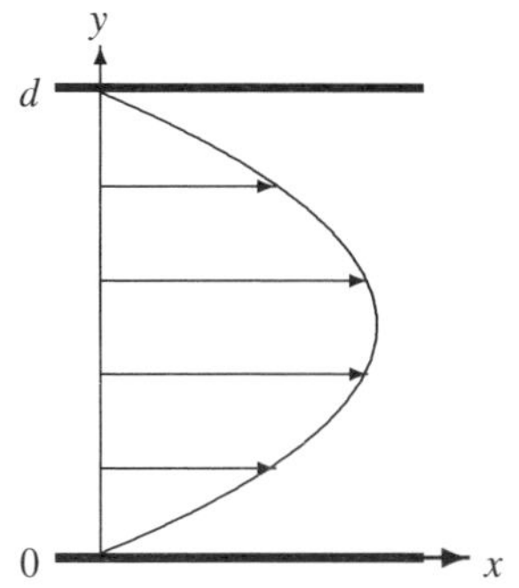

Characteristic parabolic velocity profile between the plates, driven by a pressure which is higher to the left than to the right.

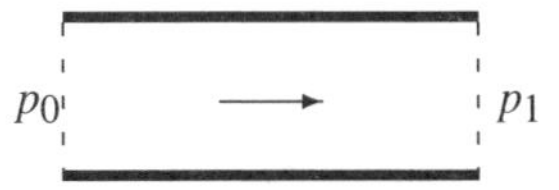

Pressure driven flow between static plates. When $p_0 > p_1$, the flow goes to the right.

Discharge rate and average velocity

The total volume of fluid passing between the plates per unit of time, also called the *volumetric discharge rate* or simply the volume flux, is obtained by integrating the velocity field over the area $d \times W$ orthogonal to the flow,

$$Q = \int_0^d v_x(y)\, W dy = \frac{GWd^3}{12\eta}. \tag{18.7}$$

From the discharge rate we may calculate the average velocity of the flow,

$$U = \frac{1}{d}\int_0^d v_x(y)\, dy = \frac{Q}{Wd} = \frac{Gd^2}{12\eta}. \tag{18.8}$$

It is not surprising that the average velocity grows with d, because with increasing distance the friction from the walls becomes less and less important.

Reynolds number

The Reynolds number (17.20) has been defined in general as the ratio between advective and viscous terms in the Navier–Stokes equation, but since the advective acceleration $(\boldsymbol{v} \cdot \nabla)\boldsymbol{v}$ always vanishes in planar flow, the Reynolds number must strictly speaking be zero. How can that be?

The apparent paradox is resolved when we consider what happens when the driving pressure is increased. Laminar flow is then replaced by irregular, time-dependent, turbulent flow with non-vanishing advective acceleration. The Reynolds number should be understood as a dimensionless characterization of the ratio of advective to viscous forces in terms of the speed and geometry of the general flow. Since the velocity changes by about U across the gap, the Reynolds number for pressure driven flow between parallel plates is conventionally defined as,

$$\mathrm{Re} = \frac{Ud}{\nu}, \tag{18.9}$$

independent of whether the flow is laminar or turbulent. Pressure-driven planar flow is stable[1] towards infinitesimal perturbations in the xy-plane for $\mathrm{Re} < 5772$, but empirically the flow becomes turbulent for $\mathrm{Re} \gtrsim 1000$–2000, accompanied by breakdown of the two-dimensionality of the flow. The phenomenology of turbulent pressure-driven planar flow is quite similar to that of turbulent pressure-driven pipe flow (section 18.5).

Example 18.2.1: Oil with $\eta = 2 \times 10^{-2}$ Pa s and $\rho_0 = 800$ kg m^{-3} is driven between plates that are $d = 1$ cm apart by a pressure drop $\Delta p = 10^3$ Pa over a distance $L = 1$ m. The average velocity becomes in this case $U \approx 0.4$ m s^{-1} corresponding to a Reynolds number of $\mathrm{Re} \approx 167$. The discharge rate per unit of length orthogonal to the flow $Q/W \approx 4 \times 10^{-3}$ m^2/s.

[1] S. A. Orzag, Accurate solution of the Orr–Sommerfeld stability equation, *J. Fluid. Mech.* **50**, (1971) 689. See also [43].

Drag

The fluid exerts a friction force on the bounding walls which is determined by the *wall shear stress*

$$\sigma_{\text{wall}} = \sigma_{xy}\big|_{y=0} = \eta\ \nabla_y v_x\big|_{y=0} = \tfrac{1}{2}Gd. \tag{18.10}$$

The total friction force on the walls, also called the *drag*, is obtained by multiplying the wall shear stress by the total wall area of both plates,

$$\mathcal{D} = \sigma_{\text{wall}}\, 2LW = GLWd = \Delta p\, Wd. \tag{18.11}$$

The last result shows that the drag equals the total pressure force on the fluid calculated from the pressure drop times the inlet area Wd. That this must be so could have been foreseen because the only forces acting on the non-accelerated volume of fluid between the plates are the pressure forces and friction, and they must therefore balance each other.

Since the pressure is constant over the inlet area, the rate of work of the pressure forces on the fluid is similarly obtained by multiplying the pressure force by the average velocity,

$$P = U\Delta p W d = U\mathcal{D}. \tag{18.12}$$

Since no other forces perform work on the fluid, this must equal the rate at which viscous forces produce heat in the fluid, also called *dissipation*.

18.3 Gravity-driven planar flow

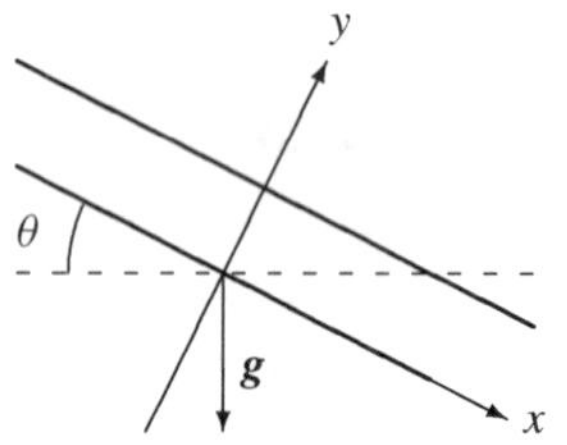

Parallel plates inclined an angle θ to the horizon.

Gravity may also drive the flow between parallel plates if they are inclined an angle θ to the horizon. We choose again a coordinate system with the x-axis in the direction of flow and the y-axis orthogonal to the plates. Assuming constant gravity, the gravitational field is in this coordinate system inclined an angle θ to the negative y-axis, so that $\boldsymbol{g} = g_0(\sin\theta, -\cos\theta, 0)$.

Flow between inclined parallel plates

The y, z-components of the Navier–Stokes equation (18.1a) take the form

$$\nabla_y p = -\rho_0 g_0 \cos\theta, \qquad \nabla_z p = 0. \tag{18.13}$$

The solution is clearly $p = p_0(x) - \rho_0 g_0 y\cos\theta$ where $p_0(x)$ is an arbitrary function of x. Inserting this into the x-component of (18.1a) we get,

$$\frac{1}{\rho_0}\nabla_x p_0(x) = g_0 \sin\theta + \nu\frac{d^2 v_x(y)}{dy^2}. \tag{18.14}$$

As in the preceding section it follows that $\nabla_x p_0(x)$ is a constant, and assuming that there is no pressure difference between the ends of the plates at $x = 0$ and $x = L_x$, the pressure itself must be constant, $p_0(x) = p_0$. Applying the no-slip boundary conditions, $v_x(0) = v_x(d) = 0$, the complete solution becomes,

$$v_x = \frac{g_0 \sin\theta}{2\nu} y(d-y), \tag{18.15a}$$

$$p = p_0 - \rho_0 g_0 y \cos\theta, \tag{18.15b}$$

where p_0 is the pressure on the bottom plate. The pressure is for all θ simply the hydrostatic pressure in a constant field of gravity of strength $g_0\cos\theta$. This solution could have been trivially found from the solution of the preceding section by including the gravitational field in the effective pressure $p^* = p + \rho_0\Phi$ such that $G = -dp^*/dx = \rho_0 g_x = \rho_0 g_0 \sin\theta$, but we have refrained from doing so here because it obscures the physics of the true pressure p (see problem 18.1).

Flow with an open surface

A liquid layer of constant thickness h flowing down an inclined plate with an open surface is another example of purely gravity-driven flow. On the inclined plate, $y = 0$, the no-slip condition again demands $v_x(0) = 0$, whereas on the open surface, $y = h$, the pressure must be constant and the shear stress must vanish, $dv_x/dy = 0$. Everything follows along the same lines as above, and we find

$$v_x = \frac{g_0 \sin\theta}{2\nu} y(2h - y), \tag{18.16a}$$

$$p = p_0 + \rho_0 g_0 (h - y)\cos\theta, \tag{18.16b}$$

where p_0 is the surface pressure. In retrospect we could have derived it from (18.15) by setting $d = 2h$, because the shear stress vanishes for $y = d/2$ between parallel plates.

The profile is parabolic as before, but now has its maximum at the open surface. The average velocity is,

$$U = \frac{1}{h}\int_0^h v_x(y)\,dy = \frac{g_0 h^2 \sin\theta}{3\nu}. \tag{18.17}$$

The Reynolds number is naturally defined in terms of the layer thickness,

$$\mathrm{Re} = \frac{Uh}{\nu}. \tag{18.18}$$

The maximal velocity is $v_x|_{\max} = (3/2)U$, and the total flux in a swath of width W is $Q = UhW$.

Example 18.3.1 (Water film): A water film of thickness $a = 0.1$ mm flows down an inclined plate at $\theta = 30°$. The average velocity becomes $U = 16$ mm s^{-1} and the Reynolds number $\mathrm{Re} = 1.6$.

* Stability of open-surface flow

Most people are familiar with the unstable nature of a layer of liquid flowing down an inclined surface. Here we shall make a simple estimate which nevertheless captures the correct form of the stability criterion. The basic idea is to calculate the effective surface tension α created by the liquid layer and determine the condition under which it is positive. As discussed in section 8.1 a negative real surface tension will make the surface crumble up, and we shall assume that the same will be the case for the sheet of liquid, such that the sheet can only be stable for $\alpha > 0$.

In steady flow the potential energy gained by the downhill flow of the liquid is completely dissipated into heat by viscous friction, and need not be discussed further. Apart from that, the mechanical energy of an area $A = L \times W$ of the liquid layer is the sum of the kinetic and potential energies,

$$\mathcal{E} = LW\rho_0 \int_0^h \left(\frac{1}{2}v_x^2 + g_0 y\cos\theta\right) dy = \left(\frac{3}{5}U^2 + \frac{1}{2}g_0 h\cos\theta\right)\rho_0 LWh, \tag{18.19}$$

where the integral has been carried out by means of (18.15).

Using (18.17) this expression may be rewritten in terms of the volume flux $Q = UhW$ and the Reynolds number $\mathrm{Re} = Q/\nu W$. After some algebra one arrives at the following expression for the energy

$$\mathcal{E} = \tfrac{3}{10} QL(\tfrac{1}{3}\nu g_0 \sin\theta)^{1/3}(5\mathrm{Re}^{-1/3}\cot\theta + 2\mathrm{Re}^{2/3}), \tag{18.20}$$

which is most easily verified by eliminating Q and Re.

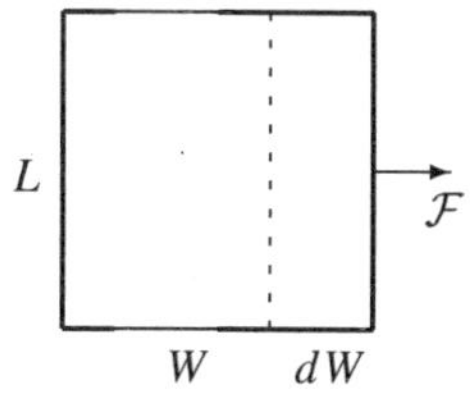

Increasing the flow area by $L\,dW$ changes its energy by $d\mathcal{E} = \alpha\, L\,dW$ where α is the surface tension.

To determine the effective surface tension we calculate the change in energy under a change of width W with the volume flux Q and the length L held constant. In that case only the terms in parenthesis depend on the width W of the liquid layer through the Reynolds number $\mathrm{Re} = Q/\nu W$. In this way we obtain the effective surface tension,

$$\alpha = \frac{1}{L}\frac{\partial \mathcal{E}}{\partial W} = -\frac{Q}{\nu W^2 L}\frac{\partial \mathcal{E}}{\partial \mathrm{Re}} = K\left(\frac{5}{3}\mathrm{Re}^{-4/3}\cot\theta - \frac{4}{3}\mathrm{Re}^{-1/3}\right), \tag{18.21}$$

where K is positive. The condition for the surface tension to be positive is therefore that the expression in parenthesis is positive, or[2]

$$\boxed{\text{Re} \tan\theta < \frac{5}{4}.} \tag{18.22}$$

If the Reynolds number is large, $\text{Re} \gg 1$, the angle of inclination must be tiny, $\theta \lesssim 1/\text{Re}$, for the flow to be stable, and similarly, for a given angle of inclination the Reynolds number must obey $\text{Re} \lesssim \cot\theta$. A vertical flow ($\theta = 90\,°$) with an open surface is never stable. For the thin water film in example 18.3.1 we have $\text{Re}\tan\theta = 0.94$ and the film will be stable.

Jean-Louis-Marie Poiseuille (1799–1869). *French physician who studied blood circulation and performed experiments on flow in tubes. Presumably the first to have measured blood pressure by means of a mercury manometer.*

Gotthilf Heinrich Ludwig Hagen (1797–1884). *German hydraulic engineer, specialized in water-works, harbors and dikes.*

18.4 Pipe flow

Cylindrical pipes carrying effectively incompressible fluids are ubiquitous, in industry, in the home and in our own bodies. Household water and almost all other fluids are transported under pressure in long pipes with circular cross sections. The question of how much fluid a given pressure can drive through a circular tube is one of the most basic problems in fluid mechanics and was first addressed quantitatively by Poiseuille around 1841 (and unbeknown to the physics community at that time, independently by Hagen in 1839)[3]. Today *Poiseuille flow* is often used to denote any pressure-driven laminar flow, for example also the planar flow between fixed plates discussed in section 18.2.

An infinitely long circular cylindrical tube is invariant both under translations along its axis and rotations around it. In a coordinate system with the z-axis running along the cylinder axis, a flow pattern respecting this symmetry is,

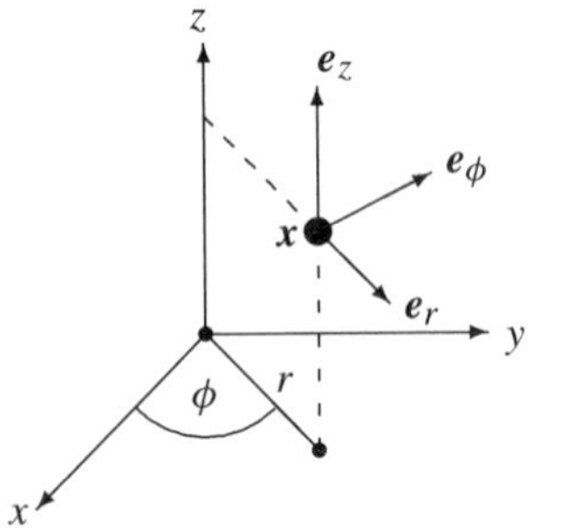

Cylindrical coordinates and basis vectors.

$$\boldsymbol{v} = (0, 0, v_z(r)) = v_z(r)\boldsymbol{e}_z, \tag{18.23}$$

where $r = \sqrt{x^2 + y^2}$ is the radial distance from the cylinder axis (see section B.1 on page 531). In section 18.6, we shall analyse another maximally symmetric flow which instead circles around the cylinder axis.

General solution

The analysis follows along the same lines as for planar flow in section 18.2. The streamlines are all parallel with the cylinder axis, implying that the assumed field (18.23) must be free of divergence, $\nabla \cdot \boldsymbol{v} = \nabla_z(v_z(r)\boldsymbol{e}_z) = 0$. Similarly, the advective acceleration also vanishes for the same reason, $(\boldsymbol{v} \cdot \nabla)\boldsymbol{v} = v_z(r)\,\nabla_z(v_z(r)\boldsymbol{e}_z) = \boldsymbol{0}$. In the absence of gravity, the Navier–Stokes equation (18.1a) now simplifies to,

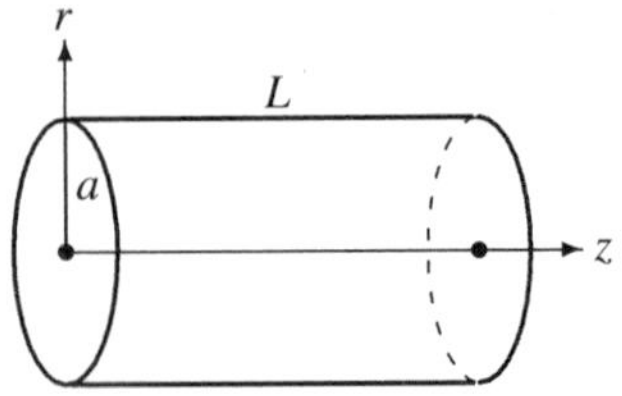

Section of a circular pipe of inner radius a and length L. There is a pressure drop Δp between $z = 0$ and $z = L$ which drives the flow of fluid through the pipe.

$$\nabla p = \eta\, \boldsymbol{e}_z \nabla^2 v_z(r). \tag{18.24}$$

From the x and y-components of this equation we get $\nabla_x p = \nabla_y p = 0$, and consequently the pressure can only depend on z, i.e. $p = p(z)$. The Laplacian on the right-hand side is given by (B.9) and we find

$$\frac{dp(z)}{dz} = \eta\nabla^2 v_z(r) = \eta\left(\frac{d^2 v_z(r)}{dr^2} + \frac{1}{r}\frac{dv_z(r)}{dr}\right) = \eta\frac{1}{r}\frac{d}{dr}\left(r\frac{dv_z(r)}{dr}\right).$$

The left-hand side depends only on z whereas the right-hand side depends only on r, so neither side can depend on r and z. Denoting the common constant value by $-G$, and solving the two resulting ordinary differential equations, we obtain in the same way as in section 18.2,

$$p = p_0 - Gz, \tag{18.25a}$$

$$v_z = -\frac{G}{4\eta}r^2 + A\log r + B \tag{18.25b}$$

where p_0, A, and B are integration constants.

[2] D. J. Benney, Long waves in liquid films, *J. Math. Phys.* **45**, (1966) 150.

[3] There is some confusion in the literature on the precise years. The original references are: G. H. L. Hagen, Über die Bewegung des Wassers in engen cylindrischen Rohren, *Poggendorfs Annalen der Physik und Chemie* **16** (1839) and J. L. Poiseuile, *Recherches experimentales sur le mouvement des liquides dans les tubes de tres petits diametres*, Compte-rendus hebdomadaire des Seances de l'Academie des Sciences (1841).

Poiseuille solution

Consider now a pipe with inner radius a. The fluid velocity cannot be infinite at $r = 0$, and consequently we must have $A = 0$ in the general solution (18.25). The no-slip boundary condition requires that $v_z(a) = 0$, and this fixes the last integration constant to $B = Ga^2/4\eta$, so that the velocity profile becomes,

$$\boxed{v_z = \frac{G}{4\eta}(a^2 - r^2).} \tag{18.26}$$

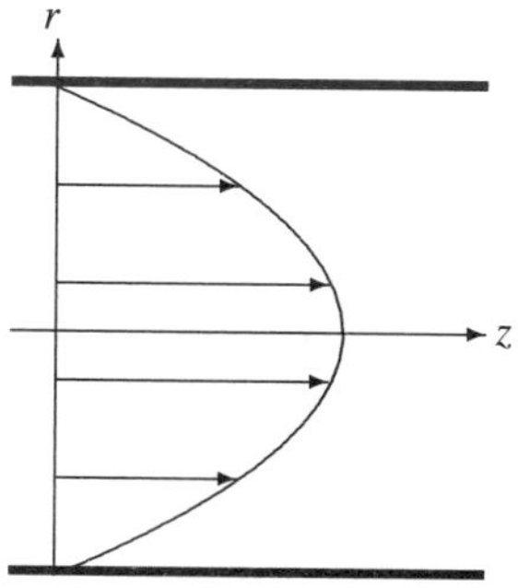

Velocity profile for laminar flow through a circular pipe.

It is parabolic as for planar flow and reaches, as one would expect, its maximal value $U_{\max} = Ga^2/4\eta$ at the centre of the pipe.

The (negative) pressure gradient G may be related to the pressure drop between the ends of a pipe of length $L \gg 2a$. When the fluid flows in the positive z-direction, the pressure p_0 at the entry at $z = 0$ must be higher than the pressure p_1 at the exit, $z = L$, so that

$$G = \frac{\Delta p}{L} \tag{18.27}$$

where $\Delta p = p_0 - p_1$ is the pressure drop.

The Hagen–Poiseuille law

The volumetric discharge rate, i.e. the volume of fluid carried through the pipe per unit of time, may immediately be calculated by integrating the velocity field over the cross section of the pipe,

$$\boxed{Q \equiv \int_0^a v_z(r)\, 2\pi r\, dr = \frac{\pi G a^4}{8\eta}.} \tag{18.28}$$

This is the famous *Hagen–Poiseuille law* (see footnote on page 250). As could have been expected, the throughput grows linearly with the pressure gradient, and inversely with viscosity. The dramatic fourth-power growth with radius could of course have been deduced from dimensional arguments since it is the only missing factor. It expresses that with growing radius the friction from the walls becomes less and less important in holding back the fluid.

Reynolds number

The velocity of the flow averaged over the cross section of the pipe may be calculated from the rate of discharge,

$$U = \frac{Q}{\pi a^2} = \frac{Ga^2}{8\eta}. \tag{18.29}$$

It is exactly half the maximal velocity found at the centre of the pipe.

As for planar flow, there is a formal problem in defining the Reynolds number, because the advective acceleration vanishes, but usually the Reynolds number is defined as,

$$\mathrm{Re} = \frac{2aU}{\nu} = \frac{\rho_0 G a^3}{4\eta^2}. \tag{18.30}$$

The choice of the pipe diameter $d = 2a$ as length scale is purely a matter of convention. Reynolds himself actually used the radius [75, p. 85].

Empirically, pipe flow remains laminar until turbulence sets in (see section 18.5). Typically the transition happens at a Reynolds number between 2000 and 4000, with 2300 as a 'nominal' value for smooth pipes. At that point the otherwise linear relationship between volume discharge Q and pressure gradient G becomes nonlinear.

Example 18.4.1 (Aortic flow): Human blood is not a particularly Newtonian fluid, but its viscosity may approximatively be taken to be $\eta = 2.7 \times 10^{-3}$ Pa s and its density near that of water. The blood flow rate in the aorta (averaged over a heartbeat) is about $Q \approx 70\ \text{cm}^3\ \text{s}^{-1}$. Since the diameter of the aorta is $2a \approx 35$ mm the average velocity becomes $U \approx 7.3\ \text{cm s}^{-1}$ and the Reynolds number Re ≈ 940, well below the turbulent region. The pressure gradient becomes $G \approx 5\ \text{Pa m}^{-1}$, showing that the pressure drop is negligible in the large arteries compared to systolic blood pressure, $p \approx 120$ mmHg $\approx 16\,000$ Pa.

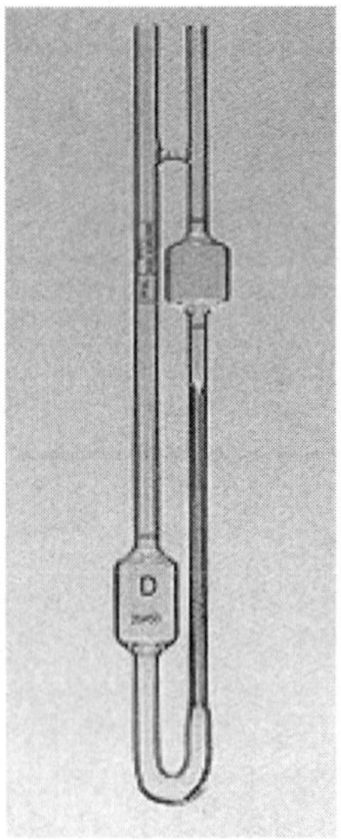

Commercial Ostwald viscometer (produced by Poulten Selfe & Lee, Great Britain). One measures the time it takes the liquid surface to pass between the marks on the container on the right (on its way to the container on the left through the capillary pipe section) and compares it with the corresponding time for a calibration liquid of known viscosity.

Wilhelm Ostwald (1853–1932). *German scientist. Received the Nobel prize in chemistry in 1909. Considered the father of modern physical chemistry. Invented the Ostwald process for synthesizing nitrates still used in the manufacture of explosives.*

Example 18.4.2 (Water pipe): Household water supply has to reach the highest floor in apartment buildings with pressure 'to spare'. Pressures must therefore be of the order of bars when water is not tapped. The typical discharge rate from a kitchen faucet is around $Q \approx 100\ \text{cm}^3\ \text{s}^{-1}$, leading to an average velocity in a half-inch pipe of about $U \approx 0.8\ \text{m s}^{-1}$. The Reynolds number is about Re $\approx 10\,000$ which is well inside the turbulent regime. The pressure gradient calculated from the Hagen–Poiseuille law, $G \approx 160\ \text{Pa m}^{-1}$, is for this reason untrustworthy.

Example 18.4.3 (Hypodermic syringe): A hypodermic syringe has a cylindrical chamber with a diameter of about $2b = 1$ cm and a hollow needle with an internal diameter of about $2a = 0.5$ mm. During an injection, about 5 cm^3 of the liquid (here assumed to be water) is gently pressed through the needle in a time of $\Delta t = 10$ s, such that the volume rate is about $Q = 0.5\ \text{cm}^3\ \text{s}^{-1}$.

The average fluid velocity in the needle becomes $U \approx 2.5\ \text{m s}^{-1}$, corresponding to a Reynolds number Re ≈ 1300, which is in the laminar region somewhat below the onset of turbulence. One is thus justified in using the Poiseuille solution for the flow through the needle. The pressure gradient necessary to drive this flow is found from (18.29) and becomes rather large, $G \approx 3.3\ \text{bar m}^{-1}$. For a needle of length $L \approx 5$ cm the pressure drop is $\Delta p \approx 0.16$ bar $= 16\,000$ Pa. The pressure drop in the fluid chamber can be completely ignored, because the chamber's diameter is 20 times that of the needle.

Ostwald viscometer

The Hagen–Poiseuille law (18.28) may be used experimentally to determine the viscosity of a fluid from a measurement of the pressure drop $\Delta p = GL$ and the total volume $\Delta V = Q\Delta t$ of fluid discharged from the pipe in the time interval Δt. The simplest way to create a precisely defined effective pressure drop is to place the pipe vertically in the gravitational field, such that the effective pressure gradient becomes $G = \rho_0 g_0$. Inserting this into (18.28) we find the kinematic viscosity

$$\boxed{\nu = \frac{\eta}{\rho_0} = \frac{\pi a^4 G}{8\rho_0 Q} = \frac{\pi a^4 g_0 \Delta t}{8\Delta V}.} \tag{18.31}$$

The *Ostwald viscometer* is made entirely from glass with no moving parts, and one obtains the kinematic viscosity of a liquid by measuring the passage time for a known volume of liquid through a narrow section of pipe.

The viscosity determined in this way is only meaningful for laminar flow which requires the Reynolds number to be smaller than about 2000. From (18.30) we obtain the Reynolds number for the viscometer,

$$\text{Re} = \frac{g_0 a^3}{4\nu^2} \tag{18.32}$$

and this shows that for water-like liquids with $\nu \approx 10^{-6}\ \text{m}^2\ \text{s}^{-1}$ the tube radius must be smaller than 1 mm in order to avoid turbulence.

Pipe resistance

The pressure drop, $\Delta p = GL$, necessary to drive a given mass flux $\rho_0 Q$ through a section of a pipe is completely analogous to the voltage drop necessary to drive an electric current through a conducting wire. In analogy with Ohm's law, the *pipe resistance* is defined as,

$$R = \frac{\Delta p}{\rho_0 Q} = \frac{8\nu L}{\pi a^4}. \tag{18.33}$$

It depends only on the kinematic viscosity ν of the fluid and the dimensions of the pipe and is measured in the strange unit $\text{m}^{-1}\ \text{s}^{-1}$. Like electric resistance, pipe resistance is additive for pipes connected in series and reciprocally additive for pipes connected in parallel (see problem 18.10).

Drag

The fluid only interacts with the pipe at its surface $r = a$ and attempts to drag it along with a total force or drag $\mathcal{D}$ along z. Using the Poiseuille velocity profile (18.26) the shear wall stress on the pipe becomes,

$$\sigma_{\text{wall}} = -\ \sigma_{zr}|_{r=a} = -\ \eta\nabla_r v_z(r)|_{r=a} = \frac{1}{2}Ga = 4\eta\frac{U}{a}. \tag{18.34}$$

Multiplying with the area $2\pi a L$ of the inner pipe surface the drag becomes

$$\boxed{\mathcal{D} = 2\pi a L \sigma_{\text{wall}} = 8\pi\eta U L.} \tag{18.35}$$

The linear growth of drag with velocity is characteristic of laminar flow at Reynolds numbers below unity (see chapter 19). At Reynolds numbers above unity the nonlinear effects of the advective acceleration ought to set in, but due to the simple geometry of pipe flow the nonlinearity first shows up when turbulence sets in at Re $\gtrsim$ 2000.

In steady flow the amount of momentum of the fluid in the pipe is constant, and consequently the total force on the volume must vanish. Besides the drag $-\mathcal{D}$ from the inner pipe surface, the only other force acting on the volume of fluid in the pipe is the external pressure force $\mathcal{F} = \pi a^2 \Delta p$ due to the difference in pressure between the ends of the volume, and this force must therefore equal the drag, $\mathcal{F} = \mathcal{D}$. Using the Poiseuille relation (18.29) we may verify that the forces indeed balance each other: $\mathcal{F} = \pi a^2 \Delta p = \pi a^2 L G = 8\pi\eta U L = \mathcal{D}$.

Example 18.4.3 (Hypodermic syringe, continued): To overcome the pressure drop in the hypodermic needle, one must act with a force of about $\mathcal{F} = \pi b^2 \Delta p \approx 1.3$ N on the piston, corresponding to the weight of about 130 g. To this force must be added the force arising from the solid friction between the piston and the wall of the fluid chamber.

Dissipation

The rate of work (or power) of the external pressure forces acting on the fluid is easily calculated because the pressure is constant across the pipe. Multiplying $\mathcal{F} = \pi a^2 \Delta p$ by the average velocity U we obtain,

$$\dot{W} = \pi a^2 \Delta p\, U = Q\Delta p = 8\pi\eta U^2 L. \tag{18.36}$$

The shear stress of the fluid on the pipe wall performs no work because the no-slip condition requires the velocity field to vanish at the wall. In steady flow the kinetic energy of the fluid in the pipe is constant, so the rate of work of the pressure forces must be exactly equal to the rate of kinetic energy loss to internal friction forces in the fluid, which is the same as the total power $P = \dot{W}$ dissipated into heat. A direct calculation of the dissipated power (17.18) confirms this claim (see problem 18.23).

The average rate of dissipation per unit of fluid mass $M = \pi a^2 L \rho_0$ in the pipe is an important parameter, called the *average specific rate of dissipation*, which may be calculated from the total rate of dissipation $P = \dot{W}$,

$$\varepsilon = \frac{P}{M} = 8\nu\frac{U^2}{a^2}. \tag{18.37}$$

Dissipation does not, however, take place uniformly across the pipe volume but is by far the largest at the inner pipe surface where the velocity gradient is largest (see problem 18.23). In chapter 32 we shall see that the specific rate of dissipation plays a major role in the theory of turbulence.

Example 18.4.3 (Hypodermic syringe, continued): In the hypodermic syringe, the piston velocity is $U_0 = Q/\pi b^2 \approx 0.6\ \text{cm s}^{-1}$ such that the rate of work on the piston is $P = \mathcal{F}U_0 = Q\Delta p \approx 8$ mW, corresponding to an average specific rate of dissipation $\varepsilon = 830\ \text{W kg}^{-1}$. Assuming that the fluid is heated uniformly (which it isn't) and taking the specific heat capacity to be that of water $c_p = 4.2 \times 10^3\ \text{J K}^{-1}\ \text{kg}^{-1}$, this dissipation would raise the temperature of the fluid passing through the needle by merely $\Delta T \approx P/(c_p\rho U) \approx 4$ mK.

* Entrance length

The Poiseuille velocity profile is not established immediately at the entrance to a pipe of finite length. The friction exerted on the fluid from the pipe wall will take some time to slow the fluid near the wall, but eventually the influence of the wall reaches all the way to the centre of the pipe. Although it is difficult to make an analytic calculation, we may estimate how far down the pipe the influence of the entrance can be felt. In chapter 21 we shall calculate the entrance length numerically for pressure-driven planar flow.

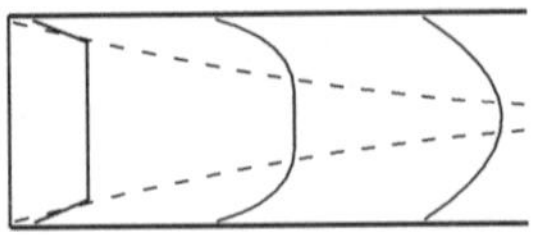

Sketch of the expected shape of the velocity profile at various distances downstream from the entrance. The influence of the wall is indicated by the dashed lines.

At the entrance proper the flow is assumed to be *plug flow* with the same velocity U over the pipe cross section. Close to the pipe wall, where the velocity must vanish, momentum is removed from the fluid by diffusion. As discussed on page 233 the diffusion front moves the distance $2\sqrt{\nu t}$ in the time t, so that in the time $t = a^2/4\nu$ the diffusion front will reach the centre of the pipe, and the average flow will have moved a distance $\ell = Ut = Ua^2/4\nu$ downstream from the entrance. Identifying ℓ with the entrance length we find,

$$\boxed{\frac{\ell}{2a} \approx \frac{\text{Re}}{16} \approx 0.06\text{Re}.} \tag{18.38}$$

At the transition to turbulence, say Re $\approx$ 2300, this formula yields an entrance length of about 140 diameters. In the creeping regime Re $\lesssim$ 1 there are corrections to this estimate [79, p. 293], and in the turbulent regime Re $\gtrsim$ 2300 a different formula is needed.

Laminar drain

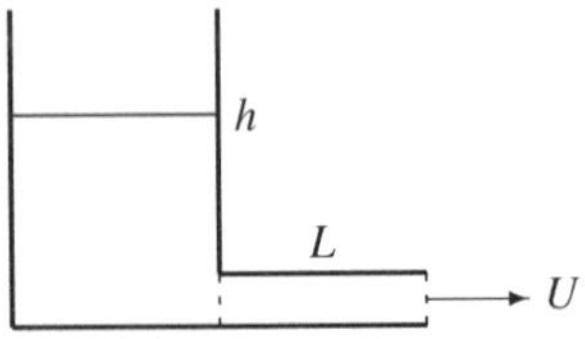

Viscous liquid streaming from a cistern through a long pipe.

Returning to Torricelli's law for the draining of a cistern (page 210), we are now able to take into account the slowing down of the flow due to viscosity. For simplicity we imagine that the liquid level in the cistern is kept constant by replenishing the cistern with the same amount of fluid as exits from the pipe. In this case we use energy conservation to calculate the average velocity.

The rate at which work is performed against gravity in moving the fluid from the exit level to the cistern level is,

$$\dot{W} = \rho_0 g_0 h Q, \tag{18.39}$$

where $Q = U\pi a^2$ is the volume flux. Assuming that the liquid exits with the Poiseuille velocity profile, which we write as $v_z = 2U(1-r^2/a^2)$ where U is the average velocity, the flux of kinetic energy becomes,

$$\dot{\mathcal{T}} = \int_0^a \frac{1}{2}\rho_0 v_z(r)^2 \cdot v_z(r)\, 2\pi r\, dr = \rho_0 U^3 \pi a^2 = \rho_0 U^2 Q. \tag{18.40}$$

This is actually twice the rate $(1/2)\rho_0 U^2 Q$ that one would find in plug flow with the same velocity across the exit, but such a result must be expected when the velocity varies.

Energy conservation implies that the work performed against gravity either goes to provide the fluid with kinetic energy or is dissipated into heat,

$$\dot{W} = \dot{\mathcal{T}} + P, \tag{18.41}$$

where P is the rate of dissipation. We now make the somewhat shaky assumption that the exit pipe is so long that the entrance length can be disregarded, and that the cistern is so large that the dissipation in it can be ignored. Then we may use (18.36) and find from energy conservation a quadratic equation for the average velocity,

$$g_0 h = U^2 + \frac{8\nu L}{a^2} U, \tag{18.42}$$

where ν is the kinematic viscosity. The solution is

$$\boxed{U = \sqrt{g_0 h + \left(\frac{4\nu L}{a^2}\right)^2} - \frac{4\nu L}{a^2}.} \tag{18.43}$$

For $4\nu L/a^2 \ll \sqrt{g_0 h}$ the solution becomes $U \approx \sqrt{g_0 h}$, which is a factor $\sqrt{2}$ smaller than the Torricelli expression for nearly ideal flow (page 210), but that is only an apparent paradox. Rewriting the inequality as $L/2a \ll \text{Re}/16$ this is seen to imply that the length of the tube must be much smaller than the entrance length (18.38), which invalidates the use of the Poiseuille velocity profile. The above formula is only valid when the length of the exit pipe is somewhat greater than the entrance length, $L \gtrsim \ell$ or $4\nu L/a^2 \gtrsim \sqrt{g_0 h}$. For very long pipes, $4\nu L/a^2 \gg \sqrt{g_0 h}$, we have approximately $U \approx g_0 h a^2/8\nu L$.

18.5 Phenomenology of turbulent pipe flow

Osborne Reynolds was in 1883 the first to investigate the nature of turbulent pipe flow. Empirically, laminar flow is only possible up to a certain value of the Reynolds number, beyond which turbulence sets in. Before turbulence is fully developed there is, however, first a transition region characterized by intermittent behaviour. After turbulence has become fully established the flow goes for increasing Reynolds number through at least two distinct stages where different physical processes dominate the character of the flow. Although to this day there is no established theory of turbulent pipe flow, there exist efficient semiempirical expressions for the general behaviour as a function of the Reynolds number.

Average velocity and Reynolds number

In what is customarily called 'fully developed turbulence' the true velocity field varies rapidly in time as well as from place to place throughout the tube. If dye is injected into the fluid, it quickly spreads through the whole volume, quite different from laminar flow where it makes orderly streaks.

The volumetric discharge rate Q is nevertheless expected to be relatively steady in turbulent flow, albeit with some fluctuations. The average velocity calculated from the discharge rate,

$$U = \frac{Q}{\pi a^2}, \tag{18.44}$$

is thus a measure of the rate at which the turbulent fluid progresses down the pipe. As for laminar flow, the Reynolds number is defined by

$$\text{Re} = \frac{2aU}{\nu}, \tag{18.45}$$

with a conventional factor 2. It is important to realize that independent of the character of the flow, the discharge rate Q—and therefore U and Re—are experimentally easy to determine by measuring the volume of fluid discharged from the pipe over a sufficiently long time interval.

Empirically, turbulence sets in around Re $\approx$ 2000–4000, depending on the conditions under which this region is approached, with the 'nominal' value being 2300. For a smooth pipe under *very* carefully controlled conditions, the transition to turbulence can in fact be delayed until a Reynolds number of the order of 100 000. Above that value, the flow is so sensitive to disturbances that it becomes practically impossible to avoid turbulence, even if the Poiseuille solution is believed to be stable towards infinitesimal perturbations (in the rz-plane) for all Reynolds numbers[4].

Friction factor

Intuitively, one expects that turbulence causes increased resistance to the flow and therefore requires a larger pressure drop Δp than laminar flow to obtain the same rate of discharge Q under steady conditions. The drag is as before determined from the pressure drop by the balance of forces, $\mathcal{D} = \pi a^2 \Delta p$, and is accordingly also expected to be larger than the laminar expression (18.35). Since the Reynolds number is the only dimensionless quantity that may be constructed from the average velocity (problem 18.9), we may always write the drag as the laminar expression times a dimensionless *friction factor* $f(\text{Re})$, depending only on the Reynolds number,

$$\boxed{\mathcal{D} = 8\pi\eta U L f(\text{Re}).} \tag{18.46}$$

The friction factor is evidently anchored with the value $f(\text{Re}) = 1$ in laminar flow, and as we shall see below it grows monotonically with the Reynolds number in the turbulent region, although in the transition to turbulence it is not too well defined because of intermittent shifts between laminar and turbulent flow.

[4]See for example B. Hof, A. Juel and T. Mullin, Scaling of the turbulence transition in a pipe, *Phys. Rev. Lett.* **91**, (2003) 244502-1.

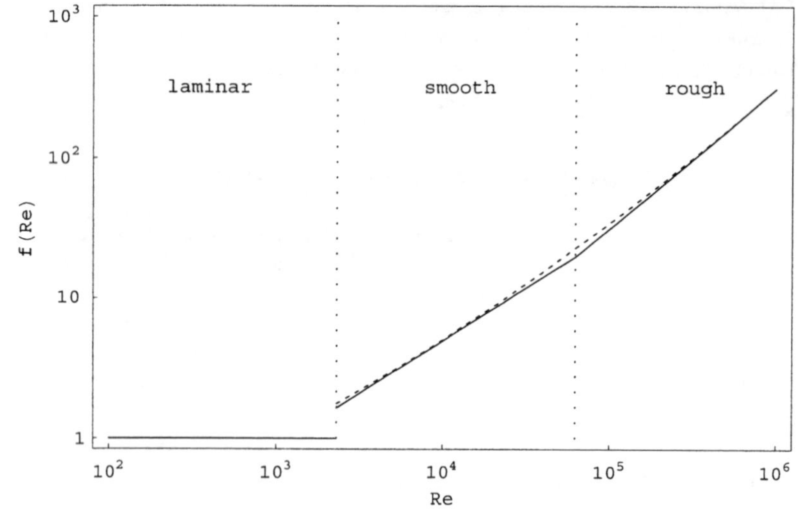

Figure 18.1. Schematic behaviour of the friction factor $f(\text{Re})$ as a function of Reynolds number Re for a pipe with roughness parameter $k/2a = 10^{-3}$ (where k is the average height of the irregularities in the pipe wall). The transition from laminar to smooth-pipe turbulent flow happens around $\text{Re} \approx 2300$ but the friction factor is not too well-defined for $2000 \lesssim \text{Re} \lesssim 4000$. The transition from smooth-pipe turbulence to rough-pipe turbulence happens in this case for $\text{Re} \approx 63\,000$ and is quite soft. The dashed line is the interpolation (18.53).

The friction factor can be directly determined from the measured average velocity U and the measured average pressure gradient $G = \Delta p/L$,

$$f(\text{Re}) = \frac{a^2 G}{8\eta U}. \tag{18.47}$$

In figure 18.1 this function is schematically plotted across the laminar and turbulent regions.

Friction coefficient

John Fanning. *Not much is available about him, except that he apparently introduced the friction coefficient in a treatise on water supply in 1877.*

Henri-Philibert-Gaspard Darcy (1803–1858). *French engineer. Pioneered the understanding of fluid flow through porous media and established Darcy's law which is used in hydrogeology.*

Like laminar drag, turbulent drag also arises from the interaction between the fluid and the pipe wall. In terms of the friction factor the average shear stress on the inner pipe wall becomes,

$$\sigma_{\text{wall}} = \frac{\mathcal{D}}{2\pi a L} = 4\eta \frac{U}{a} f(\text{Re}). \tag{18.48}$$

Dividing by the dynamic pressure, $(1/2)\rho_0 U^2$, we obtain a dimensionless measure of the wall stress, called the *friction coefficient*,

$$C_f(\text{Re}) = \frac{\sigma_{\text{wall}}}{(1/2)\rho_0 U^2} = \frac{16}{\text{Re}} f(\text{Re}). \tag{18.49}$$

In the technical literature C_f is sometimes called the *Fanning friction factor*, and often one also meets the *Darcy friction factor* which equals $4C_f$.

It is a matter of taste whether to use the friction coefficient $C_f(\text{Re})$ or the friction factor $f(\text{Re})$. For fairly low Reynolds numbers the friction factor is the most natural choice, because it is anchored in the laminar region. The friction coefficient on the other hand tends to become constant for infinite Reynolds number, except for perfectly smooth pipes. It is thus anchored in the region of extremely turbulent flow, but its limiting value $C_f(\infty)$ depends on the roughness of the pipe surface.

Smooth pipe case

For a perfectly smooth inner pipe surface a decent semiempirical approximation was given by Blasius in 1911, in which the friction coefficient decreases slowly as the $-1/4$ power of the Reynolds number, while the friction factor increases as the $3/4$ power,

$$C_f \approx 0.079\mathrm{Re}^{-1/4}, \qquad\qquad f \approx 0.0049\mathrm{Re}^{3/4}. \tag{18.50}$$

Turbulence theory (chapter 32) produces a more complicated expression for C_f which agrees numerically with the above expression in the interval $4000 \lesssim \mathrm{Re} \lesssim 100\,000$.

One may wonder whether the smooth pipe expression has any relevance except for glass pipes that are in fact quite smooth. Industrially produced pipes are typically made from rubber, plastic, iron or concrete with different degrees of inner surface roughness. But it turns out that any not too irregular surface will tend to behave approximately like a perfectly smooth surface in some range of Reynolds numbers beyond the transition to turbulence. The reason is that when turbulence sets in, the no-slip condition still requires the fluid velocity to vanish at a solid wall. This leads to the formation of viscous wall layers which screen the bulk of the flow from the influence of the surface roughness, thus making the surface appear to be smooth. The wall layers, however, become thinner and the screening less effective for very high Reynolds numbers. Above a certain roughness-dependent value, typically for $\mathrm{Re} \gtrsim 100\,000$, the smooth pipe approximation generally loses its usefulness.

Example 18.5.1 (Water pipe): A quite normal use of household water was shown in example 18.4.2 to lead to a Reynolds number of about $\mathrm{Re} \approx 10\,000$, which is well inside the turbulent regime, though not in the rough pipe region. From the smooth pipe formula (18.50) we find the friction factor $f(10\,000) = 5$. The pressure gradient must therefore be 5 times larger than the Hagen–Poiseuille value (which is 160 Pa m^{-1}), or $G \approx 800$ Pa m^{-1}. Drag and dissipation are similarly augmented.

Turbulence makes the pipes hiss or 'sing' when you tap water at full speed, though most of the noise probably comes from the narrow passages of the faucet, where the water speed and Reynolds number are highest, and the big drop from main to atmospheric pressure takes place.

Rough pipe limit

At sufficiently high Reynolds number the irregular turbulent flow is caused by the fluid literally slamming into the small protrusions of the rough pipe surface, rather than by the viscous friction. In this limit the drag becomes independent of the actual viscosity, even if it in the end must owe its very existence to viscosity! Correspondingly, the friction coefficient becomes a constant, $C_f(\infty) = \lim_{\mathrm{Re}\to\infty} C_f(\mathrm{Re})$, which depends on the character of the rough surface.

The surface roughness may be characterized by the average height k of the surface irregularities, a parameter that depends on the material of the pipe and how it is produced. A decent semi-empirical expression for the limiting friction coefficient is also given by a power law,

$$C_f(\infty) \approx 0.028 \left(\frac{k}{2a}\right)^{1/4}. \tag{18.51}$$

	$k[m]$
Glass	3×10^{-7}
Steel	4.5×10^{-5}
Rubber	1×10^{-5}
Cast Iron	2.6×10^{-4}
Concrete	1.5×10^{-3}

Characteristic roughness constants for various materials (from [79]). Note that machining of a surface (sawing, planing, milling, grinding and polishing) will have great influence on its roughness.

Again it must be remarked that turbulence theory yields a more complicated expression which however agrees reasonably with the power law in the interval $10^{-4} \lesssim k/2a \lesssim 10^{-2}$. The deviations reach the 30% level about an order of magnitude beyond this interval in both directions.

The smooth pipe expression crosses the rough pipe expression for

$$\mathrm{Re} \approx 63\frac{2a}{k}. \tag{18.52}$$

The transition from smooth to rough pipe turbulence is rather soft, certainly not as sharp as the transition from laminar to turbulent flow (see figure 18.1). A simple interpolation which covers both regimes could be,

$$C_f = 0.028 \left(\frac{63}{\mathrm{Re}} + \frac{k}{2a}\right)^{1/4}, \tag{18.53}$$

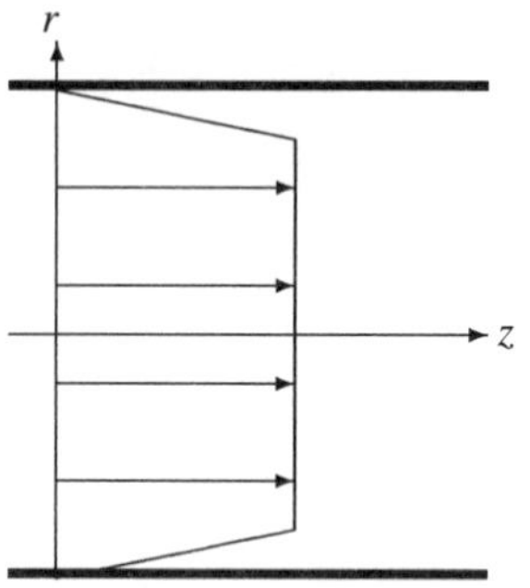

Assumed velocity profile for fully developed turbulent flow through a circular pipe. Apart from thin boundary layers (which we ignore), the velocity field is approximately constant across the pipe.

and yields a decent approximation for $10^{-4} \lesssim k/2a \lesssim 10^{-2}$ and $\text{Re} \gtrsim 4 \times 10^3$. More precise semi-empirical expressions and diagrams useful for engineering design purposes may be found in the technical literature [80, p. 348].

Example 18.5.2 (Rubber roughness): Rubber has $k \approx 0.01$ mm, so that for a 10 mm rubber hose we get $k/2a \approx 10^{-3}$ and $C_f(\infty) \approx 5 \times 10^{-3}$. The smooth and rough pipe expressions cross each other at $\text{Re} \approx 63\,000$ where the friction coefficient is $C_f \approx 6 \times 10^{-3}$.

Turbulent drain

The calculation of the turbulent drain velocity proceeds in much the same way as for laminar viscous flow (page 254), except that we do not know the actual velocity distribution in the pipe. Although there will also be boundary layers in turbulent pipe flow, we shall for simplicity assume that the bulk of the turbulent flow proceeds through the pipe as a *turbulent plug* with roughly the same average speed across the pipe.

The rate of loss of kinetic energy from the pipe exit is then $\dot{\mathcal{T}} = (1/2)\rho_0 U^2 Q$ whereas the work to replenish the cistern is unchanged $\dot{W} = \rho_0 g_0 h Q$. Writing the dissipative loss as $P = \mathcal{D}U$ with $\mathcal{D}$ given by (18.46), energy conservation (18.41) takes the form

$$g_0 h = \frac{1}{2}U^2 + \frac{8\nu U L}{a^2} f\left(\frac{2aU}{\nu}\right). \tag{18.54}$$

Were it not for the second term on the right-hand side, we would indeed recover Torricelli's law, $U = \sqrt{2g_0 h}$. In the general case this equation has to be solved numerically (see problem 18.16).

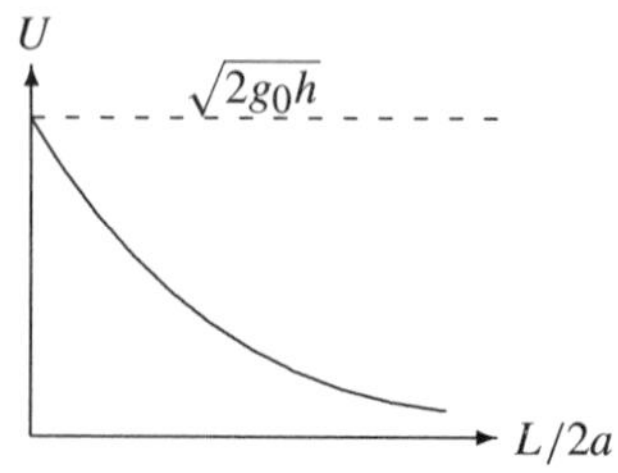

Turbulent drain speed as a function of pipe length. For sufficiently long pipes the flow becomes laminar.

In the limit of $\nu \to 0$ we may use the rough pipe form of the friction factor $f(\text{Re}) \approx C_f(\infty)\text{Re}/16$ to obtain the solution,

$$U = \sqrt{\frac{2g_0 h}{1 + 4C_f(\infty)L/2a}}. \tag{18.55}$$

Clearly, Torricelli's result is only obtained for sufficiently short pipes satisfying $4C_f(\infty)L/2a \ll 1$. In the rubber tube example 18.5.2 we have $4C_f(\infty) \approx 0.02$ and the condition becomes $L/2a \ll 50$. For long pipes with $4C_f(\infty)L/2a \gg 1$, the terminal velocity falls as $1/\sqrt{L}$. If by some method the pipe is elongated this will eventually bring the Reynolds number down into the smooth pipe and laminar regions.

Example 18.5.3 (Barrel of wine, continued): For the barrel of wine (example 16.3.2 on page 211) which empties through a wooden spout of length $L = 20$ cm and diameter $2a = 5$ cm the Reynolds number is about 300 000, well into the rough pipe region. Disregarding entry corrections and assuming a wood roughness of $k \approx 0.5$ mm, corresponding to $k/2a = 0.01$, we find $C_f(\infty) \approx 10^{-2}$ and $4C_f(\infty)L/2a \approx 0.16$. The Toricelli value for the exit velocity is thus only reduced by about 8% by turbulence.

18.6 Circulating cylindrical flow

In cylindrical geometry there is another exact solution with maximal symmetry, in which the fluid circles around the cylinder axis with a velocity field of the form,

$$\boldsymbol{v} = v_\phi(r)\boldsymbol{e}_\phi, \tag{18.56}$$

where $\boldsymbol{e}_\phi$ is the tangential unit vector in cylindrical coordinates (see appendix B). The field lines are concentric circles, and the field is also in this case invariant under rotation around the cylinder axis and translation along it.

General solution

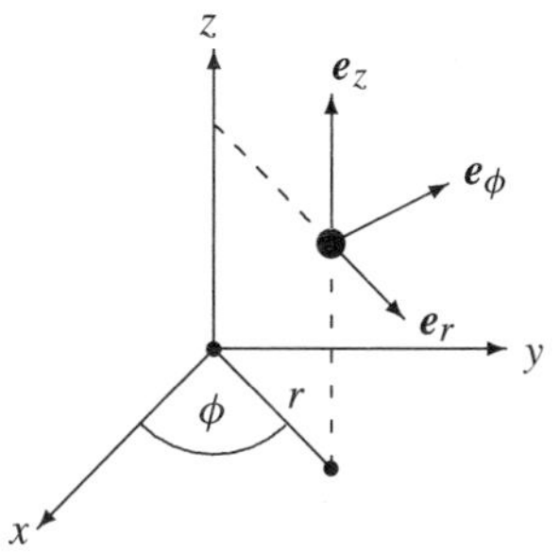

Cylindrical coordinates and basis vectors (see appendix B).

The simplest way to handle this field is to make use of the gradient in cylindrical coordinates (B.6) to calculate its 'tensor product' with the velocity,

$$\nabla v = (e_r \nabla_r + e_\phi \nabla_\phi + e_z \nabla_z) v_\phi(r) e_\phi = e_r e_\phi \frac{dv_\phi}{dr} - e_\phi e_r \frac{v_\phi}{r}. \tag{18.57}$$

Here we have used that the basis vectors only depend on ϕ and that $\nabla_\phi e_\phi = -e_r/r$. The trace of this tensor yields the divergence, $\mathrm{Tr}[\nabla v] = \nabla \cdot v = 0$, which vanishes because ∇v only has off-diagonal components in the cylindrical basis. This is in agreement with the elementary observation that streamlines neither diverge nor converge in this flow. Dotting from the left with v we find the advective acceleration,

$$(v \cdot \nabla) v = v \cdot (\nabla v) = -e_r \frac{v_\phi^2}{r}.$$

One should not be surprised: the centripetal acceleration in a circular motion with velocity v_ϕ is indeed directed radially inwards and of size v_ϕ^2/r. Finally, using the cylindrical Laplacian (B.9) and $\partial^2 e_\phi/\partial\phi^2 = -e_\phi$, we obtain

$$\nabla^2 v = e_\phi \frac{d^2 v_\phi}{dr^2} + e_\phi \frac{1}{r}\frac{dv_\phi}{dr} - e_\phi \frac{v_\phi}{r^2} = e_\phi \frac{d}{dr}\left(\frac{1}{r}\frac{d(r v_\phi)}{dr}\right), \tag{18.58}$$

where the last rewriting is done for later convenience. The Navier–Stokes equation (18.1a) then becomes (in the absence of gravity),

$$-\rho_0 e_r \frac{v_\phi^2}{r} = -\nabla p + \eta\, e_\phi \frac{d}{dr}\left(\frac{1}{r}\frac{d(r v_\phi)}{dr}\right). \tag{18.59}$$

Maurice Frédéric Alfred Couette (1858–1943). *French physicist from the provincial university of Angers. Published only seven papers, all from 1888 to 1900. Invented what is now called the Couette viscometer.*

Projecting it on the three cylindrical basis vectors, e_r, e_ϕ and e_z, we obtain

$$-\rho_0 \frac{v_\phi^2}{r} = -\frac{\partial p}{\partial r}, \tag{18.60a}$$

$$0 = -\frac{1}{r}\frac{\partial p}{\partial \phi} + \eta \frac{d}{dr}\left(\frac{1}{r}\frac{d(r v_\phi)}{dr}\right), \tag{18.60b}$$

$$0 = -\frac{\partial p}{\partial z}. \tag{18.60c}$$

The last equation shows that the pressure is independent of z, and from the second equation we see by differentiating after ϕ and using that v_ϕ only depends on r, that $\partial^2 p/\partial\phi^2 = 0$, which means that p can at most be linear in ϕ, i.e. of the form $p = p_0(r) + p_1(r)\phi$. But here we must require $p_1 = 0$, for otherwise the pressure would have different values for $\phi = 0$ and $\phi = 2\pi$, and that is impossible. For this reason the pressure cannot depend on ϕ but only on r, and thus it disappears completely from (18.60b).

With the pressure out of the way, the integration of (18.60b) has become almost trivial. The general solution is

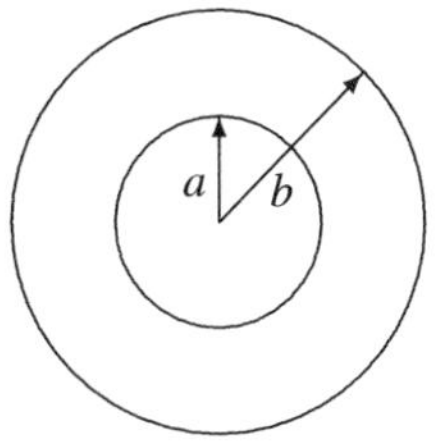

General Couette flow geometry in the xy-plane.

$$v_\phi = Ar + \frac{B}{r}, \tag{18.61}$$

where A and B are integration constants. The first equation (18.60a) expresses that there is a positive radial pressure gradient which supplies the centripetal force necessary to keep the fluid in its circular motion. Inserting the general solution into (18.60a) and integrating over r, we find the pressure

$$\frac{p}{\rho_0} = C + \frac{1}{2}A^2 r^2 - \frac{1}{2}\frac{B^2}{r^2} + 2AB \log r, \tag{18.62}$$

where C is a third integration constant.

Couette flow between rotating cylinders

Suppose the fluid is contained between two long coaxial material cylinders with the innermost having radius a and outermost radius $b > a$. This problem was solved by Couette (1890), but today *Couette flow* is often used to denote any kind of motion-driven laminar flow, for example also laminar channel flow between moving parallel plates (p. 232).

Here we shall for simplicity only study the case where the outer cylinder is held fixed and the inner cylinder rotates like a spindle with constant angular velocity Ω. The boundary conditions, $v_\phi(a) = a\Omega$ and $v_\phi(b) = 0$, then determine A and B, so the velocity profile becomes

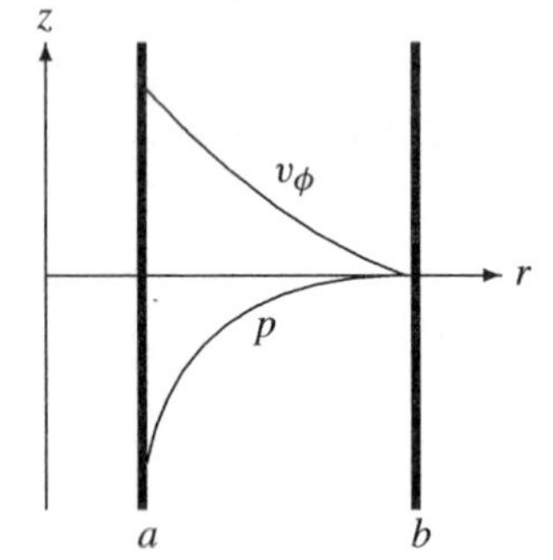

Sketch of the velocity and pressure between cylinders in Couette flow with the outer cylinder held fixed.

$$\boxed{v_\phi = \frac{\Omega\, a^2}{r}\,\frac{b^2 - r^2}{b^2 - a^2}.} \tag{18.63}$$

The velocity field decreases monotonically from its value $a\Omega$ at the inner cylinder to zero at the outer.

The pressure is found from (18.62)

$$p^* = p_0 + \frac{1}{2}\rho_0 \left(\frac{\Omega\, a^2}{b^2 - a^2}\right)^2 \left(r^2 - \frac{b^4}{r^2} + 4b^2 \log\frac{b}{r}\right), \tag{18.64}$$

where p_0 is its value at the outer cylinder. It increases monotonically towards p_0.

Stress, torque and dissipation

From (18.57) it follows that the only non-vanishing shear stress component is

$$\sigma_{\phi r} = \eta(\boldsymbol{e}_\phi \cdot (\boldsymbol{\nabla}\boldsymbol{v}) \cdot \boldsymbol{e}_r + \boldsymbol{e}_r \cdot (\boldsymbol{\nabla}\boldsymbol{v}) \cdot \boldsymbol{e}_\phi) = \eta\left(\frac{dv_\phi}{dr} - \frac{v_\phi}{r}\right), \tag{18.65}$$

which upon insertion of the Couette solution (18.63) becomes,

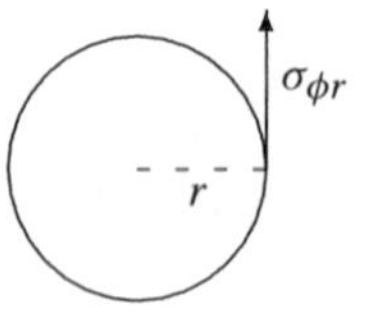

The shear stress acts tangentially on the cylinder surface at r.

$$\sigma_{\phi r} = -2\eta\,\Omega \frac{a^2 b^2}{b^2 - a^2}\frac{1}{r^2}. \tag{18.66}$$

It represents the friction between the layers of circulating fluid, and the sign is negative because the fluid *outside* the radius r acts as a brake on the motion of the fluid *inside* (which has normal $\boldsymbol{e}_r$).

In order to maintain the steady rotation of the inner cylinder it is necessary to act on it with a moment of force or torque $\mathcal{M}_z$ (and with an opposite moment at the outer cylinder). Multiplying the shear stress with the moment arm r and the area $2\pi r L$ of the cylinder at r, we finally obtain the torque with which the fluid *inside* r acts on the fluid *outside*,

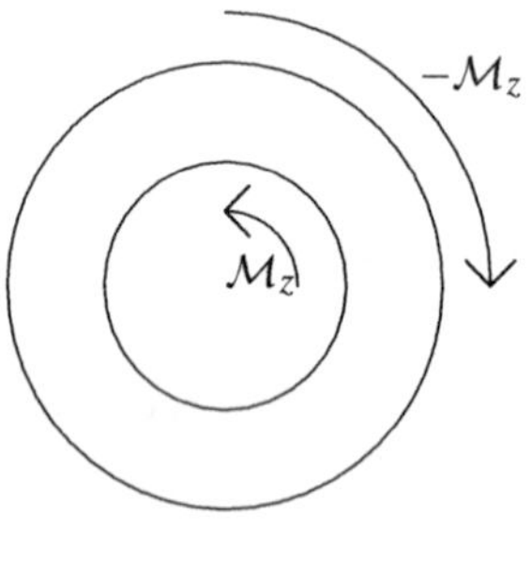

The angular momentum of the fluid between the two cylinder surfaces is constant, which implies that the total moment of force on the fluid must vanish.

$$\mathcal{M}_z = r\,(-\sigma_{\phi r})2\pi r L = 4\pi\eta\Omega L \frac{a^2 b^2}{b^2 - a^2}. \tag{18.67}$$

We could have foreseen that it would be independent of r from angular momentum conservation. In steady flow the total angular momentum of the fluid between any two cylindrical surfaces is constant and there is no transport through the cylindrical surfaces because they are orthogonal to the velocity. Consequently the total moment of force has to vanish, implying that the moments of shear stress acting on the two cylindrical surfaces must be equal.

The rate of work that must be done against the friction from the fluid to keep the inner cylinder rotating is obtained by multiplying the stress $-\sigma_{\phi r}$ with the velocity $r\Omega$ and the area of the cylinder,

$$P = (-\sigma_{\phi r})\, r\Omega\, 2\pi r L = \mathcal{M}_z \Omega = 4\pi\eta\Omega^2 L \frac{a^2 b^2}{b^2 - a^2}. \tag{18.68}$$

Since the kinetic energy of the fluid is constant in steady flow and since no fluid enters or leaves the system, this must equal the rate of energy dissipation in the fluid.

Couette viscometer: In the Couette viscometer the viscosity of the fluid is determined from the extra motor power necessary to drive the inner cylinder at constant angular velocity. A more stable arrangement (see page 264) consists of hanging the inner cylinder by a torsion wire and rotating the outer cylinder at constant angular velocity. When the chamber is filled with viscous fluid there will arise a torque on the inner cylinder, which is counteracted by the torsion in the wire so that at equilibrium a certain deflection angle is obtained. Measuring the deflection angle then leads to a value for the viscosity (problem 18.20). The torsional rigidity (page 157) of the wire may be determined from the period of free oscillations of the inner cylinder when the chamber is empty.

Unloaded journal bearing

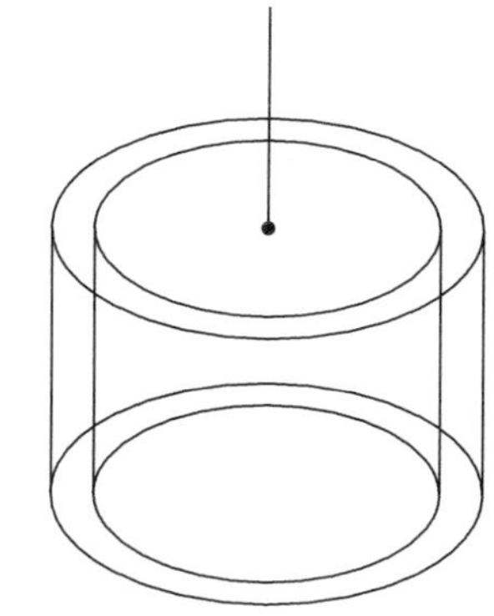

Sketch of torsion wire Couette viscometer.

In a lubricated *journal bearing* a liquid, say oil, is trapped in a tiny gap between a rotating shaft (or journal) and its bearing (or bushing). Normally such systems carry a load which brings the shaft off-centre (see section 27.4) but here we shall assume that the journal and its bushing are concentric cylinders, such that we may apply the Couette formalism in the limit of very small distance $d = b - a \ll a$ between the cylinders.

To lowest order in d and the distance $s = r - a$ from the shaft, we find,

$$v_\phi = \Omega a \left(1 - \frac{s}{d}\right). \tag{18.69}$$

The velocity field is linear in s, just as for planar Couette flow. This is not particularly surprising since the tiny gap between the cylinders looks very much like the gap between parallel plates. The natural definition of the Reynolds number is correspondingly,

$$\mathrm{Re} = \frac{\Omega a d}{\nu}, \tag{18.70}$$

and we expect turbulence to arise for Re $\gtrsim$ 2000, as in the planar case.

The pressure is found from (18.64) by expanding the logarithm to third order in s and d, and it becomes in the leading approximation,

$$p = p_0 - \frac{1}{3} a d \Omega^2 \left(1 - \frac{s}{d}\right)^3. \tag{18.71}$$

The pressure variation across the gap is proportional to the size d of the gap and thus normally tiny, unless the quadratic growth with angular velocity overwhelms it. The dissipation becomes on the other hand (to lowest order in d),

$$P = \frac{2\pi \eta \Omega^2 a^3 L}{d}. \tag{18.72}$$

It is inversely proportional to the thickness of the layer of fluid and grows, like the pressure variation, quadratically with the angular velocity. This is why even a lubricated bearing may burst into flames at sufficiently high rotation speed.

Example 18.6.1 (Burning bush): Consider a vertical wooden shaft in a wooden bushing with diameter $2a = 10$ cm, such as could have been used in an old water mill. Let the gap be $d = 1$ mm and let the lubricant be heavy grease with $\eta \approx 10$ Pa s, corresponding to $\nu \approx 0.01$ m^2 s^{-1}. The power dissipated per unit of lubricant volume is

$$\frac{P}{2\pi a L d} = \frac{\eta \Omega^2 a^2}{d^2}. \tag{18.73}$$

For a modest speed of one rotation per second, $\Omega \approx 2\pi\ s^{-1}$ with a Reynolds number of Re ≈ 0.06, this comes to about 1 J cm^{-3} s^{-1}. If the mill rotates 10 times faster because of torrential rains, this number becomes instead 100 J cm^{-3} s^{-1} with a Reynolds number of 0.6. Since wood is a bad heat conductor, the grease may ignite in a matter of minutes even taking into account that its viscosity decreases with temperature.

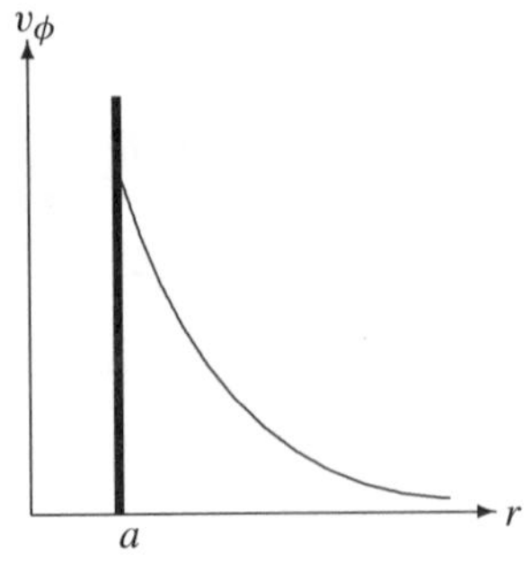

Characteristic $1/r$ velocity profile of spindle-driven vortex.

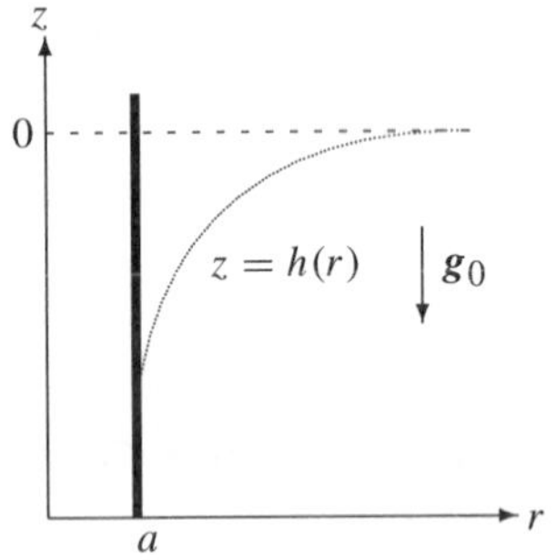

Open liquid surface of a spindle-driven vortex in constant gravity varies like $1/r^2$.

Spindle-driven vortex

Suppose a long cylindrical spindle of radius a is inserted into an infinite sea of fluid, and that the spindle is rotated steadily with constant angular velocity Ω. Provided the spindle does not rotate so fast that it creates turbulence, the flow will eventually become steady with an azimuthal velocity field that may be found from the Couette solution with $b \to \infty$,

$$v_\phi = \frac{\Omega a^2}{r}, \qquad p = -\frac{1}{2}\rho_0 \frac{\Omega^2 a^4}{r^2}, \tag{18.74}$$

where the pressure is normalized to vanish at infinity. A singular vortex with $v_\phi = C/r$ for all r is called a *line vortex*, and in this case we have $C = \Omega a^2$.

In a gravitational field g_0 pointing downwards along the cylinder axis we must add the hydrostatic pressure, to get

$$p = -\frac{1}{2}\rho_0 \frac{\Omega^2 a^4}{r^2} - \rho_0 g_0 z. \tag{18.75}$$

If the fluid is a nearly ideal liquid, the true pressure at an open surface must equal the (constant) air pressure at the interface, just as in hydrostatics. Choosing the asymptotic level of fluid to be $z = 0$, we obtain the following shape of the surface,

$$z = -\frac{\Omega^2 a^4}{2 g_0 r^2}. \tag{18.76}$$

This is quite different from the parabolic shape of the liquid surface in a rotating bucket (page 87).

Example 18.6.2: A spindle with radius $a = 1$ cm making 10 turns per second, corresponding to $\Omega = 63\ \mathrm{s}^{-1}$, would make a depression in the liquid surface, $d = L - h(a) \approx 2$ cm, at the spindle.

The rule about continuity of pressure across any surface is only strictly valid in hydrostatics (in the absence of surface tension), whereas in hydrodynamics it is instead the full stress vector that has to vanish at the open surface. For the spindle-driven vortex the non-vanishing stresses are

$$\sigma_{\phi r} = -2\eta \frac{\Omega a^2}{r^2}, \qquad \sigma_{rr} = \sigma_{zz} = \sigma_{\phi\phi} = -p. \tag{18.77}$$

Whereas the diagonal stresses vanish at the surface (18.76), the shear stress is always non-zero. Strictly speaking the above solution only describes a vortex with an open surface in the limit of vanishing viscosity. Unfortunately we encounter another problem in this limit, because without viscosity it is impossible to crank up the vortex from a fluid at rest. The structure of driven as well as free vortices will be analysed in detail in chapter 26.

* 18.7 Secondary flow and Taylor vortices

Real machinery cannot be infinite in any direction. Suppose the cylinders are capped with plates fixed to the outer non-rotating cylinder so that only the inner cylinder rotates. The no-slip condition forces the rotating fluid to slow down and come to rest not only at the outer cylinder, but also at both the end caps, implying that the assumption about a simple circulating flow (18.61) with its z-independent azimuthal velocity $v_\phi(r)$ cannot be right everywhere. In fact it cannot be right anywhere! Clean Couette flow is like clean Poiseuille flow an idealization that can only be approximately realized far from the ends of the apparatus. *Secondary flow* with non-vanishing radial and longitudinal velocity components, v_r and v_z, will have to arise near the container walls in order to satisfy the no-slip boundary conditions at the end caps.

Direction of secondary flow

We shall begin with assuming that the length of the apparatus is comparable to the gap, say $L \approx 2d$, and that the gap is completely filled with fluid without open surfaces. In the bulk of the fluid, the Couette solution is still a good approximation, and the main job of the pressure here is to provide centripetal force for the circulating fluid as shown by the radial gradient (18.60a).

The longitudinal pressure gradient $\nabla_z p$ determines how rapidly the pressure can vary in the z-direction, and from the z-component of the full Navier–Stokes equation (18.1a) we obtain (still disregarding gravity)

$$\nabla_z p = -\rho_0 (\boldsymbol{v} \cdot \nabla) v_z + \eta \nabla^2 v_z.$$

The right-hand side is expected to be small throughout the fluid because the longitudinal velocity v_z must nearly vanish in the Couette bulk flow in the middle of the apparatus and also has to vanish at the bottom end cap (which the fluid cannot penetrate). The vertical variation in pressure near the bottom is therefore not able to seriously challenge the large values of the centripetally dominated bulk pressure. The pressure is for this reason rather 'stiff' as we approach the bottom and remains more or less equal to the pressure in the bulk with only small corrections.

Near the bottom, the still intact bulk pressure will have an inwards directed radial gradient, $-\nabla_r p$, capable of delivering the centripetal force to keep the rapidly rotating fluid in the bulk moving in a circle. But the no-slip condition requires the azimuthal velocity to vanish at the bottom, $v_\phi \to 0$ for $z \to 0$, so that the actually required centripetal acceleration $-v_\phi^2/r$ is much smaller there than in the bulk. As a result, the combination of the stiff pressure gradient and the smaller centripetal acceleration gives rise to a net force directed towards the axis, and that force will in turn drive a radial flow inwards along the bottom. The same argument shows that there must also be an inwards flow along the top and since the moving fluid has to go somewhere, it tends to form two oppositely rotating ring-shaped (toroidal) vortices that encircle the axis.

If the axis is vertical and the liquid has an open surface, only a single vortex needs to be formed. It is this kind of secondary flow that drives the tea leaves along the bottom towards the centre of a cup of tea after stirring it. As the above argument shows the flow is independent of the direction of stirring. You cannot, for example, make the tea leaves move out again by 'unstirring' your tea.

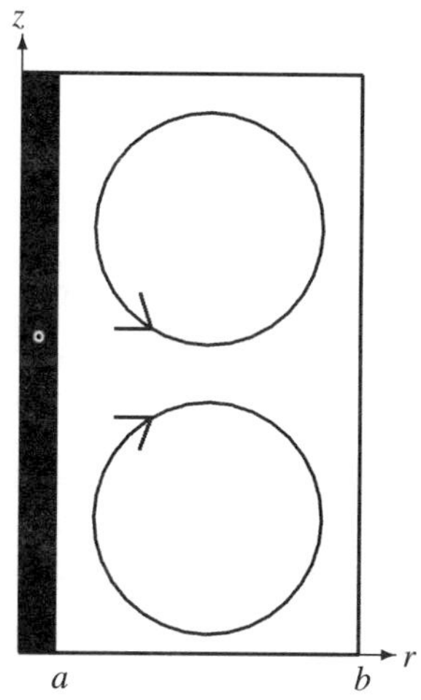

Sketch of the secondary flows that arise in a Couette apparatus as a consequence of its finite longitudinal size. Both at the bottom and the top end caps fluid is driven towards the central region

Rayleigh's stability criterion

In the above discussion, the cylinder length was assumed to be comparable to the gap size. For a long cylinder, the preceding argument seems to indicate that a secondary flow with two highly elongated 'vortices' should form which satisfy the boundary conditions on the end caps, or just one if there is an open surface. Such elongated vortices are, however, unstable and tend to break up in a number of smaller and more circular vortices.

The instability can be understood by an argument analogous to the discussion of atmospheric stability on page 57. For simplicity we shall, as Rayleigh did, assume that there is no viscosity. Suppose now that a unit mass fluid particle in the bulk of the rotating fluid is shifted outwards to a larger value of r by a tiny amount. During the move the fluid particle will conserve its angular momentum and will in general arrive in its new position with a speed which is different from the ambient speed of the fluid there. The angular momentum of a unit mass particle in the ambient fluid is $r v_\phi$, and depending on how the fluid rotates, it may grow or decrease with increasing distance. If it is constant, such that $d(r v_\phi)/dr = 0$, then the moved particle fits snugly into its new home with just the right velocity. This is a marginal case, realized by the spindle-driven vortex (page 262) which has $r v_\phi = \Omega a^2$.

In general, however, the angular momentum $r v_\phi$ of a unit mass particle in the ambient fluid will not be a constant. If it locally grows in magnitude with distance, then a fluid particle displaced to a slightly larger radius will arrive with its original angular momentum and thus have a smaller velocity than its new surroundings. The inward pressure gradient in the ambient fluid will therefore be larger than required to keep the particle in a circular orbit with lower velocity, and it will be forced back to where it came from. This is *Rayleigh's stability criterion*,

$$\frac{d\,|r v_\phi|}{dr} > 0. \tag{18.78}$$

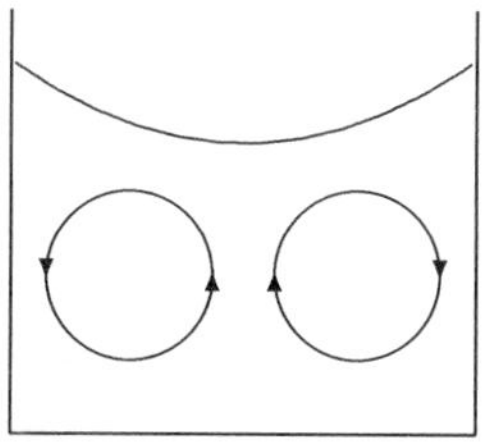

The secondary flow responsible for collecting tea-leaves at the centre of the bottom of a stirred cup. Note the parabolic shape of the surface of the rotating liquid (see page 87).

The stability criterion will trivially be fulfilled if the inner cylinder is held fixed and the outer rotated, because then the azimuthal velocity $|v_\phi|$ and thus $|r v_\phi|$ must necessarily grow with the radial distance. Such a situation is perfectly stable. Any little radial fluctuation in the flow is quickly 'ironed out' and the flow remains well-ordered and laminar until turbulence sets in.

Taylor vortices

Conversely, if the above inequality is not fulfilled everywhere, the flow will be unstable and reconfigure itself at the first opportunity. For the Couette solution (18.63) we have

$$\frac{d\,|rv_\phi|}{dr} = -\frac{2a^2\Omega r}{b^2 - a^2}, \tag{18.79}$$

which is evidently negative, and more negative the faster the inner cylinder rotates and the narrower the gap. Viscosity, which has been ignored so far, will delay the onset of instability, but it requires a more careful analysis to determine the precise criterion (see for example [43, 1, 13, 19]).

For sufficiently large angular driving velocity Ω, the Couette flow in a small gap between the rotating inner and fixed outer cylinder becomes unstable and a string of ring-shaped vortices will arise in the gap. They are called *Taylor vortices* in honour of G. I. Taylor who first analysed their properties, both experimentally and theoretically. The Taylor vortices are all mutually counter-rotating and match up with the sense of flow dictated by the boundary conditions on the ends (for pictures see [5]).

The formation of secondary flow and the instability leading to toroidal (ring-shaped) Taylor vortices represents a breakdown of the longitudinal symmetry (along z) of the cylinder. The azimuthal symmetry (in ϕ) is to begin with maintained but also breaks down with increasing speed of rotation. The regular Taylor vortices are then replaced by wavy vortices that undulate up and down (in z) while they encircle the axis. Increasing the speed further will increase the number of undulations until the flow finally, after an infinity of such transitions, becomes chaotic and turbulent.

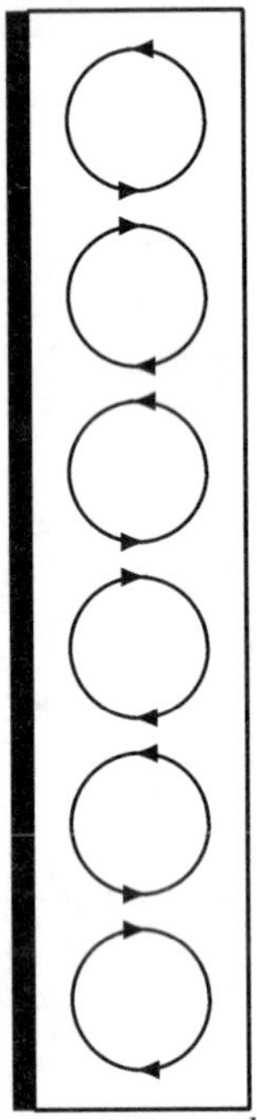

Taylor vortices are toroidal rings of fluid encircling the axis of rotation. In this case both end caps are fixed and the secondary flow must be directed inwards at both top and bottom. That is only possible if an even number of counter-rotating vortices are formed.

Problems

18.1 Define the effective pressure

$$p^* = p + \rho_0\Phi, \tag{18.80}$$

and show that the gravitational field can be eliminated from the Navier–Stokes equation.

18.2 Consider a pressure-driven laminar flow between infinitely extended plates moving with constant velocity with respect to each other in the same direction as the pressure gradient. Determine under which conditions the maximal flow velocity will be found between the plates.

18.3 Determine the planar flow field in the case that pressure gradient is orthogonal to the plate motion.

18.4 Why is it impossible to make a pressure-driven planar fluid sheet with an open surface?

18.5 A thin fluid sheet (water) of density ρ_1, viscosity η_1 and thickness d_1 runs steadily down a plate inclined at an angle θ to the horizon. Another plate with the same inclination is placed at a distance $d_1 + d_2$ from the first, and there is another fluid (air) in the gap with density ρ_2, viscosity η_2 and thickness d_2. Assuming laminar flow, calculate the pressure and velocity fields in the water and in the air between the plates.

18.6 Calculate drag and dissipation for pressure driven planar flow in terms of the average velocity.

18.7 Calculate drag and dissipation for gravity driven planar flow with an open surface on an inclined plate in terms of the average velocity.

18.8 Calculate laminar gravity driven flow through a vertical pipe. Determine the radius of the pipe as a function of the Reynolds number when the fluid is water. Determine the largest radius before onset of turbulence.

18.9 **(a)** Show that in pipe flow the only dimensionless combination of the parameters Q, a, ρ_0, η is the Reynolds number (up to trivial transformations). **(b)** Can you make another dimensionless parameter by using $G = \Delta p/L$ instead of Q?

18.10 Show that pipe resistance is additive for pipes connected in series and reciprocally additive for pipes connected in parallel.

18.11 Show that for the Poiseuille profile,

$$\int_0^a v_z(r)^n \, 2\pi r \, dr = \frac{2^{n-1}}{n+1} \rho_0 U^n \pi a^2 \tag{18.81}$$

where U is the average velocity.

18.12 **(a)** Show that for a cylindrical tube with annular cross section with radii $a < b$, the solution is given by the general Poiseuille solution (18.25) with

$$A = \frac{G}{4\eta} \frac{b^2 - a^2}{\log b - \log a}$$

$$B = \frac{G}{4\eta} \frac{a^2 \log b - b^2 \log a}{\log b - \log a}.$$

(b) Determine the drag on the tube per unit of length when a fluid is pressed through the tube under a pressure gradient G.

18.13 A pipe has elliptic cross section with major and minor semi-axes a and b. **(a)** Show that laminar pressure driven flow of an incompressible fluid with pressure gradient G is given by,

$$v_z = \frac{G}{2\eta} \frac{a^2 b^2}{a^2 + b^2} \left(1 - \frac{x^2}{a^2} - \frac{y^2}{b^2}\right), \tag{18.82a}$$

$$p = p_0 - Gz. \tag{18.82b}$$

(b) Show that the fields of the circular pipe and the parallel plates are contained in this expression. **(c)** Calculate the volume flux (Hint: use elliptic coordinates $x = ar\cos\theta$ and $y = br\sin\theta$.) **(d)** Calculate the average flow velocity. **(e)** Calculate the drag per unit of length.

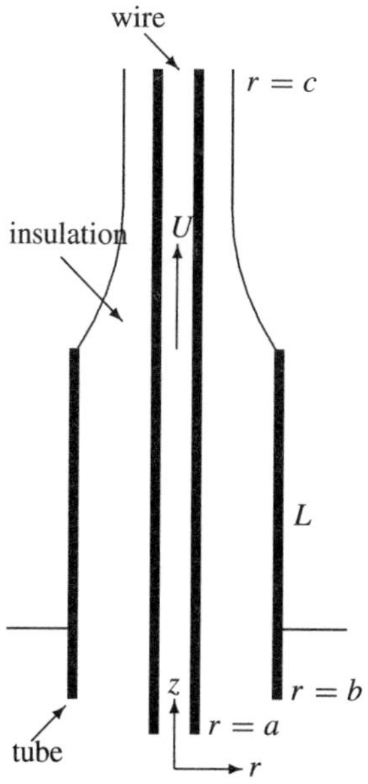

18.14 An electric lead consists of a metal wire covered with an insulating material. The insulation is applied by pulling the wire vertically up through a tube standing in a bath of melted material. The wire drags along some of the melt which quickly solidifies without volume change after the wire has emerged from the tube.

The metal wire has radius $a = 1$ mm, the inner radius of the tube is $b = 5$ mm, and its length $L = 20$ cm. The wire is concentric with the tube at the constant velocity $U = 1$ m/s. The melted insulation is incompressible with mass density $\rho_0 = 1$ g/cm^3 and viscosity $\eta = 0.1$ Pa s.

The flow may be assumed to be cylindrical and laminar with a velocity profile that is independent of z (in cylindrical coordinates). All end effects are disregarded. Gravity is also disregarded and the pressure may be taken as p_0 everywhere inside and outside the system.

(a) Determine the flow field.

(b) Calculate the volume discharge rate.

(c) Determine the radius c of the final insulation layer.

(d) Calculate the shear stress on the surface of the wire and the total friction that the liquid exerts on the wire.

(e) Estimate if it was correct to disregard gravity.

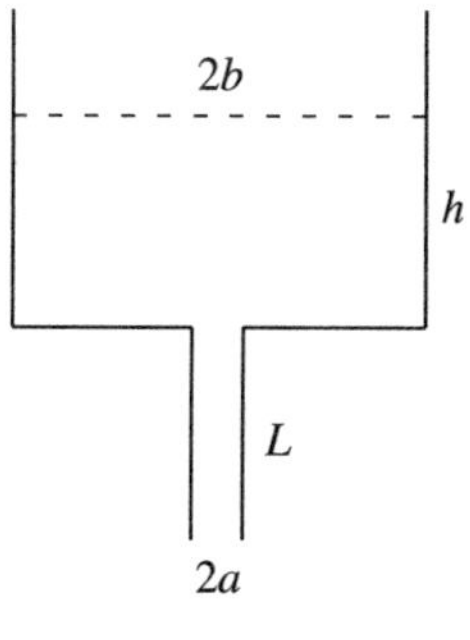

Water tank with a pipe.

18.15 A cylindrical water tank with radius b and initial water height h_0 empties quasi-steadily through a vertical cylindrical pipe of length L with radius $a \ll b, L$. **(a)** Assume Poiseuille flow in the pipe and write a differential equation for the height h as a function of time. **(b)** Calculate the time it takes to empty the tank.

18.16 Show that for turbulent draining of a cistern (page 258), one obtains the following equation for the Reynolds number

$$\mathrm{Re}^2 + 64k\mathrm{Re}\, f(\mathrm{Re}) = \mathrm{Re}_0^2 \tag{18.83}$$

where $k = L/2a$, $\mathrm{Re}_0 = 2aU_0/\nu$ and $U_0 = \sqrt{2g_0 h}$.

18.17 Show that the coefficients of the general solution to circulating flow may be written as

$$A = \frac{\Omega_b b^2 - \Omega_a a^2}{b^2 - a^2} \tag{18.84}$$

$$B = \frac{a^2 b^2 (\Omega_a - \Omega_b)}{b^2 - a^2} \tag{18.85}$$

where Ω_a and Ω_b are the angular velocities of the inner and outer cylinders.

18.18 **(a)** Show that the rotating volume flux in Couette flow with rotating inner cylinder and static outer becomes

$$Q = \frac{1}{2} L a^2 \Omega \left(\frac{b^2}{b^2 - a^2} \log \frac{b^2}{a^2} - 1 \right) \tag{18.86}$$

where L is the length of the cylinders. This is equivalent to the volumetric discharge rate for pipe flow, although no fluid is actually discharged here. **(b)** Show that the kinetic energy of the fluid is

$$\mathcal{T} = \pi \rho_0 L \frac{a^4 b^2 \Omega^2}{b^2 - a^2} \left(\frac{1}{4} \frac{a^2}{b^2} - \frac{3}{4} + \frac{b^2}{b^2 - a^2} \log \frac{b}{a} \right).$$

18.19 A permeable circular cylindrical pipe carrying fluid under pressure leaks fluid through its surface into the surrounding ocean at a constant rate. Determine the steady state, maximally symmetric form of the radial flow pattern outside the pipe.

18.20 **Couette viscometer:** Consider a torsion-wire Couette viscometer with radii a and $b = a + d$ with a narrow gap ($d \ll a$), and axial length $L \gg d$. Let the torque exerted by the torsion wire (from which the inner cylinder hangs) be $-\tau\phi$ where ϕ is the deflection angle. Assume that the inner cylinder has total mass M and is made from a very thin shell of homogeneous material. Let the cylinder perform damped torsion oscillations in the presence of the fluid. **(a)** Show that the equation of motion for the torsion angle becomes

$$M a^2 \frac{d^2\phi}{dt^2} = -\tau\phi - 2\pi\eta a^3 \frac{L}{d} \frac{d\phi}{dt}, \tag{18.87}$$

under the assumption that the fluid flow may be considered quasi-steady at all times. **(b)** Find the solution to the equation of motion and discuss it.

18.21 The drive shaft in a truck has a diameter of 15 cm and a bearing of length 10 cm with 1 mm of lubricant of viscosity $\eta = 0.01$ Pa s. Calculate the rate of heat production when the shaft rotates 10 revolutions per second.

18.22 Show that the vorticity field of the spindle vortex (18.74) vanishes everywhere.

* **18.23** **(a)** Show that the local specific rate of dissipation is

$$\varepsilon(r) = \frac{G^2}{4\eta\rho_0} r^2 \tag{18.88}$$

in laminar pipe flow. **(b)** Show the total dissipated power in a stretch of a circular tube is given by (18.36).

19 Creeping flow

Viscosity may be so large that a fluid only flows with difficulty. Heavy oils, honey, even tight crowds of people, show insignificant effects of inertia, and are instead dominated by internal friction. Such fluids do not make spinning vortices or become turbulent, but rather ooze or creep around obstacles. Fluid flow which is dominated by viscosity is quite appropriately called creeping flow.

Since there is no absolute meaning to 'large' viscosity, creeping flow is more correctly characterized by the Reynolds number being small, $\text{Re} \ll 1$. Creeping flow may occur in any fluid, as long as the typical velocity and geometric extent of the flow combine to make a small Reynolds number. Blood flowing through a microscopic capillary can be as sluggish as heavy oil. Tiny organisms like bacteria live in air and water like ourselves, but theirs is a world of creeping and oozing rather than whirls and turbulence, and movement requires special devices, for example oar-like cilia or whip-like flagella [75]. Some bacteria have even mounted a rotating helical tail in a journal bearing (the only one known to biology), which like a corkscrew allows them to advance through the thick fluid that they experience water to be. A spermatozoan pushes forward like a slithering snake in the grass by undulating its tail.

In this chapter we shall study creeping flow around moving bodies far from containing boundaries. For any creature in creeping flow, the most important quantity is the fluid's resistance against motion, also called *drag*. The drag on a body has two components of comparable magnitude in creeping flow, one being *skin drag* from viscous friction between the fluid and the body, and the other *form drag* from the variations in fluid pressure across the body. The contact forces may also produce *lift* orthogonally to the direction of motion, but in creeping flow far from containing boundaries lift is of the same order of magnitude as drag. A tiny creature that must already overcome a drag many times its weight in order to move freely has no problem with flying. That's why bacteria do not have wings.

19.1 Steady incompressible creeping flow

Leaving out the advective acceleration, we obtain from (18.1) the following equations for steady incompressible creeping flow, also called *Stokes flow*,

$$\nabla p = \rho_0 \boldsymbol{g} + \eta \nabla^2 \boldsymbol{v}, \tag{19.1a}$$

$$\nabla \cdot \boldsymbol{v} = 0, \tag{19.1b}$$

valid for $\text{Re} \ll 1$. The divergence of the first equation implies that in the absence of gravitational sources, $\nabla \cdot \boldsymbol{g} = 0$, the pressure must satisfy the Laplace equation,

$$\nabla^2 p = 0. \tag{19.2}$$

This is sometimes convenient in practical calculations.

Creeping flow is mathematically (and numerically) much easier to handle than general flow because of the absence of nonlinear terms that tend to spontaneously break the natural symmetry of the solutions in time as well as space with turbulence as the extreme result. The linearity of the creeping flow equations allows us to express solutions to more complicated flow problems as linear superpositions of simpler solutions. We may for example eliminate the static field of gravity by subtracting the hydrostatic pressure $-\rho_0\Phi$ from the actual pressure. In the following we shall assume $\boldsymbol{g} = \boldsymbol{0}$, but must at the same time not forget to subtract the buoyancy force from the weight of an immersed body.

Drag and lift on a moving body

Consider a body moving with constant velocity U through a static fluid. Cruising along the body creates a temporary disturbance in the fluid that disappears again some time after the body has passed a fixed observation point. But seen from the body, the fluid appears to move in a steady pattern which at sufficiently large distances has uniform magnitude U. Newtonian relativity guarantees that these situations are physically equivalent, so we may use the creeping flow equations (19.1).

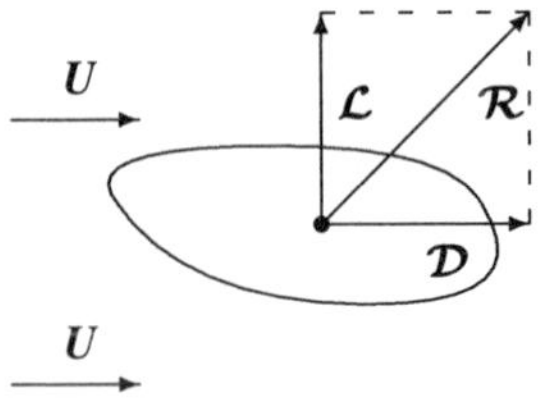

The total reaction force $\mathcal{R}$ from the fluid on the body is composed of lift $\mathcal{L}$ and drag $\mathcal{D}$.

The only way the fluid can influence a solid body is through contact forces acting on its surface. In all such cases it is convenient to resolve the total contact force from the fluid on the body, also called the *reaction force*, into two components,

$$\mathcal{R} \equiv \oint_S \boldsymbol{\sigma} \cdot d\boldsymbol{S} = \mathcal{D} + \mathcal{L}. \tag{19.3}$$

The first component $\mathcal{D}$ is called the *drag* and acts in the direction of the asymptotic flow, whereas the other is called the *lift* and acts orthogonally to the direction of asymptotic flow. Thus we have

$$\mathcal{D} = \mathcal{D}\boldsymbol{e}_U, \qquad \mathcal{L} \cdot \boldsymbol{e}_U = 0, \tag{19.4}$$

where $\boldsymbol{e}_U = \boldsymbol{U}/U$ is a unit vector in the direction of the asymptotic velocity.

Estimates

We shall think of the body as roughly potato-shaped with typical diameter L. Since (for $\boldsymbol{g} = \boldsymbol{0}$) the pressure only appears in (19.1) in the combination p/η, and since the no-slip boundary conditions on the surface of the body do not involve the pressure, the velocity field cannot depend on η, but only on the asymptotic velocity $\boldsymbol{U}$ and on the shape and orientation of the body. The linearity of creeping flow guarantees that the velocity and pressure must both be proportional to the asymptotic velocity, such that $|\boldsymbol{v}| \sim U$ and $p/\eta \sim U$.

Skin drag is produced by the viscous shear stresses acting on the surface (skin) of a body. The velocity gradients near the surface are of magnitude $|\nabla \boldsymbol{v}| \sim U/L$ so the magnitude of the shear stress becomes $|\boldsymbol{\sigma}| \sim \eta U/L$. Multiplying by the surface area of the body which is of magnitude L^2 we obtain an estimate,

$$\mathcal{D}_{\text{skin}} \approx f_{\text{skin}} \eta U L, \tag{19.5}$$

with an unknown dimensionless prefactor f_{skin} of order unity. This factor cannot depend on the Reynolds number but only on the geometry of the body. We may similarly estimate the pressure gradient near the surface from (19.1) to be $|\nabla p| \approx \eta U/L^2$, and multiplying with the length L of the body, the pressure variations over the body become of the same size as the shear stresses, $|\Delta p| \sim \eta U/L$. Multiplying by the surface area L^2, the estimate of the form drag takes the same form as skin drag,

$$\mathcal{D}_{\text{form}} \approx f_{\text{form}} \eta U L. \tag{19.6}$$

In general, it has another geometric prefactor. The total drag is the sum of skin and form drag, and becomes of order of magnitude

$$\mathcal{D} \approx f_{\text{drag}} \eta U L, \tag{19.7}$$

with $f_{\text{drag}} = f_{\text{skin}} + f_{\text{form}}$.

Lift is produced by the same contact forces as drag and is therefore also of this magnitude,

$$|\mathcal{L}| = f_{\text{lift}} \eta U L. \tag{19.8}$$

The lift vanishes for a symmetric body like a sphere or cylinder orthogonal to the asymptotic flow, but for asymmetric bodies the geometric prefactor f_{lift} is normally strongly dependent on the orientation of the body with respect to the direction of motion.

Note that lift is not necessarily related to the direction of gravity, but can point in any direction orthogonal to the direction of motion. Near a solid boundary lift can be much larger than drag and may in fact serve to keep the body away from the boundary. This is the secret behind lubrication, which is taken up in chapter 27. Lift in nearly ideal fluids (aerodynamic lift) may also be much larger than drag and will be discussed at length in chapter 29.

19.2 Creeping flow around a solid ball

A solid spherical ball moving at constant speed through a viscous fluid is the centerpiece of steady creeping flow (Stokes, 1851). The solution may be worked out starting from the field equations (19.1) and the boundary conditions (see problem 19.10), but even if deriving the solution is a bit complicated the result is fairly simple.

The Stokes solution

In spherical coordinates with polar axis in the direction of the asymptotic flow $\boldsymbol{U}$, the solution for a sphere of radius a is (see figure 19.1),

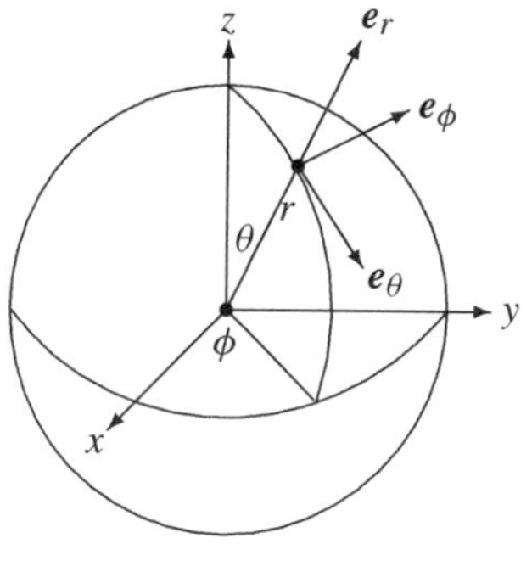

Spherical coordinates and their tangent vectors.

$$v_r = \left(1 - \frac{3}{2}\frac{a}{r} + \frac{1}{2}\frac{a^3}{r^3}\right) U \cos\theta, \tag{19.9a}$$

$$v_\theta = -\left(1 - \frac{3}{4}\frac{a}{r} - \frac{1}{4}\frac{a^3}{r^3}\right) U \sin\theta, \tag{19.9b}$$

$$v_\phi = 0, \tag{19.9c}$$

$$p = -\frac{3}{2}\eta\frac{a}{r^2} U \cos\theta. \tag{19.9d}$$

Note that the fields are independent of the azimuthal angle ϕ and has $v_\phi = 0$ as one would expect from symmetry. The velocity field vanishes for $r = a$ and thus fulfills the no-slip boundary condition on the surface of the sphere. Asymptotically for $r \to \infty$ it becomes $v_r \to U\cos\theta$ and $v_\theta \to -U\sin\theta$, which is the expected homogeneous velocity field $\boldsymbol{U} = \boldsymbol{e}_z U$.

The pressure is forward-backwards asymmetric, highest on the part of the sphere that turns towards the incoming flow (at $\theta = \pi$). This asymmetry is in marked contrast with the potential flow solution (16.74) where the pressure is symmetric. The flow pattern is independent of the viscosity of the fluid, whereas the pressure is proportional to η, as expected for creeping flow problems where the boundary conditions do not involve the pressure.

Stokes law for drag on a sphere

For symmetry reasons the sphere can only be subject to drag. To find it we first calculate the normal and shear stress tensor components. Using the standard Newtonian stress tensor, $\sigma_{ij} = -p\delta_{ij} + \eta(\nabla_i v_j + \nabla_j v_i)$, we find the radial stress

$$\sigma_{rr} = \boldsymbol{e}_r \cdot \boldsymbol{\sigma} \cdot \boldsymbol{e}_r = -p + 2\eta\frac{\partial v_r}{\partial r} = \frac{3}{2}\frac{a}{r^2}\left(3 - 2\frac{a^2}{r^2}\right)\eta U \cos\theta.$$

The only non-vanishing shear stress is,

$$\sigma_{\theta r} = \boldsymbol{e}_\theta \cdot \boldsymbol{\sigma} \cdot \boldsymbol{e}_r = \eta\left((\nabla_\theta \boldsymbol{v}) \cdot \boldsymbol{e}_r + (\nabla_r \boldsymbol{v}) \cdot \boldsymbol{e}_\theta\right)$$

$$= \eta\left(\frac{1}{r}\left(\frac{\partial v_r}{\partial \theta} - v_\theta\right) + \frac{\partial v_\theta}{\partial r}\right) = \eta\frac{1}{r}\left(2 - 3\frac{a}{r} - \frac{1}{2}\frac{a^3}{r^3}\right) U \sin\theta.$$

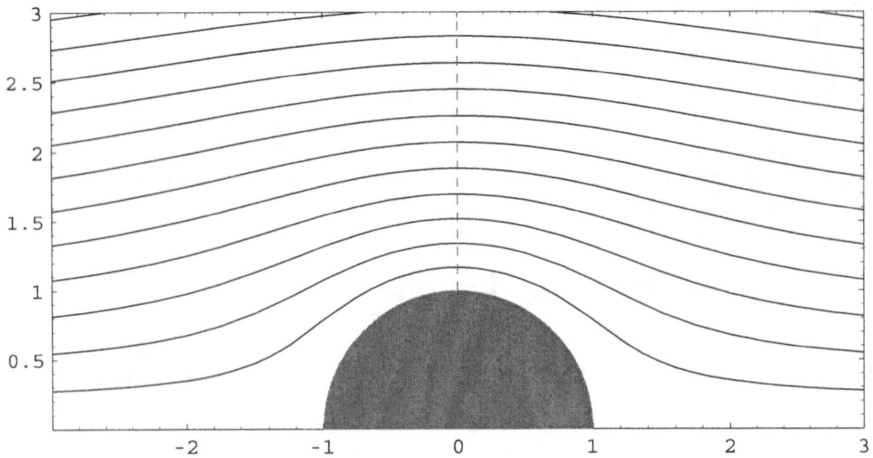

Figure 19.1. Creeping flow around a unit sphere with only the upper half shown. Streamlines have been obtained by numeric integration of equation (15.2) starting equidistantly at $z = -100$ to avoid problems of long-range terms (see problem 19.7 for how to obtain the streamlines in an efficient way). The pressure vanishes at the dashed line. Note how this pattern is qualitatively different from the potential flow pattern shown in figure 16.5 with the influence of the sphere stretching much farther.

At the surface of the sphere, $r = a$, the two stress components become

$$\sigma_{rr}|_{r=a} = \frac{3}{2a}\eta U \cos\theta, \qquad \sigma_{\theta r}|_{r=a} = -\frac{3}{2a}\eta U \sin\theta, \tag{19.10a}$$

from which we get the stress vector

$$\boldsymbol{\sigma}_r|_{r=a} = \boldsymbol{e}_r\sigma_{rr} + \boldsymbol{e}_\theta\sigma_{\theta r} = \frac{3}{2}\eta\frac{\boldsymbol{U}}{a}. \tag{19.11}$$

Surprisingly, it is of constant magnitude and across the entire surface points in the direction of the asymptotic flow.

Due to this result, the total reaction force is simply obtained by multiplying the constant stress vector by the area $4\pi a^2$ of the sphere. Since the force is proportional to $\boldsymbol{U}$, it is a pure drag,

$$\boxed{\mathcal{D} = 6\pi\eta a U.} \tag{19.12}$$

This is the famous *Stokes law* from 1851. The symmetry of the sphere could have told us in advance that there would be no lift, because there is no other global direction than $\boldsymbol{U}$ defined for the problem. Taking $L = 2a$ to represent the size of the sphere, this form of the drag was already predicted in (19.7) with a geometric prefactor $f_{\text{drag}} = 3\pi$. In problem 19.5 it is shown that 2/3 of the total is skin drag and 1/3 is form drag.

Terminal velocity

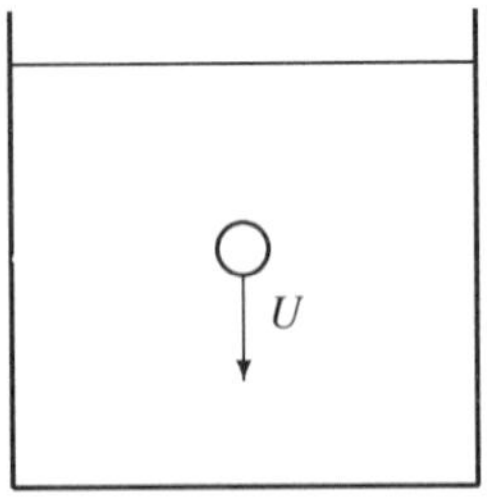

Sphere falling through viscous fluid at constant terminal speed U.

Although Stokes law has been derived in the rest system of the sphere, it is also valid in the rest system of the asymptotic fluid, because forces are invariant under translation with constant velocity. The terminal velocity of a falling solid sphere may be obtained by equating the force of gravity (minus buoyancy) with the Stokes drag,

$$(\rho_1 - \rho_0)\frac{4}{3}\pi a^3 g_0 = 6\pi\eta a U, \tag{19.13}$$

where g_0 is the gravitational acceleration and ρ_1 is the average density of the sphere. Solving for U we find

$$U = \frac{2}{9}\left(\frac{\rho_1}{\rho_0} - 1\right)\frac{a^2 g_0}{\nu} \tag{19.14}$$

where $\nu = \eta/\rho_0$ is the kinematic viscosity.

Example 19.2.1 (Sedimentation): Sand grains of diameter 0.1 mm and density 2.5 times that of water float towards the bottom of the sea with a terminal velocity of 0.8 cm s^{-1}, as calculated from Stokes' law. The corresponding Reynolds number is about 0.8, which is at the limit of the region of validity of Stokes' law.

Robert Andrews Millikan (1868–1953). *American experimental physicist, prolific author and excellent teacher. Awarded the Nobel prize in 1923 for the determination of the charge of the electron. Verified in 1916 Einstein's expressions for the photoelectric effect, and investigated the properties of cosmic radiation.*

Conversely, in the *falling sphere viscometer*, equation (19.14) may be used to determine the viscosity ν from a measurement of the terminal velocity for a sphere of known radius and density. One may also, as Millikan did in his famous electron charge experiments (1909–1913), determine the radius of a falling sphere (an oil drop) from a measurement of the terminal velocity, provided the viscosity of the fluid is known (it was actually necessary to include corrections to Stokes law for the internal motion of the oil in the tiny drops). In Millikan's experiment the oil drops were charged and could be made to hover, rise or fall in the field of gravity as slowly as desired by means of an electric field of suitable strength. Knowing the radius, the electric force on an oil drop could be compared to gravity, allowing the charge to be determined[1].

Stokes law has been derived under the assumption of creeping flow, which can only be valid under the condition that the Reynolds number

$$\mathrm{Re} = \frac{\rho_0\, 2aU}{\eta} \tag{19.15}$$

is small compared to unity, $\mathrm{Re} \ll 1$. For a falling sphere where the terminal velocity is determined from the balance of forces (19.13), the Reynolds number varies as the third power of the radius. As long as the sphere is sufficiently small, the conditions for creeping flow can always be fulfilled.

Limits to Stokes flow

The Reynolds number (19.15) is, however, only a rough estimate for the ratio between advective and viscous terms in the Navier–Stokes equations for a particular geometry. Having obtained an explicit solution we may actually calculate this ratio throughout the fluid to see if it is indeed small.

Close to the sphere, the velocity is very small because of the no-slip condition. Problems are only expected to arise at large distances, where the leading corrections to the uniform flow are provided by the a/r terms in the solution (19.9). Disregarding the angular dependence, the leading advective terms are at large distances,

$$|\rho_0 \boldsymbol{v}\cdot\boldsymbol{\nabla}\boldsymbol{v}| \sim \rho_0 U^2 \frac{a}{r^2}, \tag{19.16}$$

whereas the viscous terms become,

$$|\eta \boldsymbol{\nabla}^2 \boldsymbol{v}| \sim \eta U \frac{a}{r^3}. \tag{19.17}$$

The actual ratio between advective and viscous terms is then,

$$\frac{|\rho_0 \boldsymbol{v}\cdot\boldsymbol{\nabla}\boldsymbol{v}|}{|\eta \boldsymbol{\nabla}^2 \boldsymbol{v}|} \sim \frac{\rho_0 U r}{\eta} \sim \frac{r}{a}\mathrm{Re} \tag{19.18}$$

where Re is the global Reynolds number (19.15). As this expression grows with r, we conclude that however small the Reynolds number may be, there will always be a distance $r \gtrsim a/\mathrm{Re}$ where the advective force begins to dominate the viscous. This clearly illustrates that the global Reynolds number is just a guideline, not a guarantee that creeping flow will occur everywhere in a system. Since any finite potato-like body at sufficiently large distances looks like a sphere, this problem must in fact be present in all creeping flows.

The slowly decreasing a/r terms in the Stokes solution (19.9) also indicate the influence of the containing vessel. The relative magnitude of these terms at the boundary of the vessel is estimated as $2a/D$ where D is the size (diameter) of the vessel. In the falling sphere viscometer a 1% measurement of viscosity thus requires that the sphere diameter must be smaller than 1% of the vessel size. That can be hard to fulfil in highly viscous fluids where a measurable terminal speed demands rather large and heavy spheres.

[1] The viscosity of air used in Millikan's experiment was determined by Couette viscometer measurements (see p. 260). At that time, there was actually a systematic error of about 0.5% in the viscosity of air deriving mainly from the failure to take into account the end caps of the Couette viscometer, an error first corrected more than 20 years later (see, for example J. A. Bearden, *Phys. Rev.* **56**, (1939) 1023).

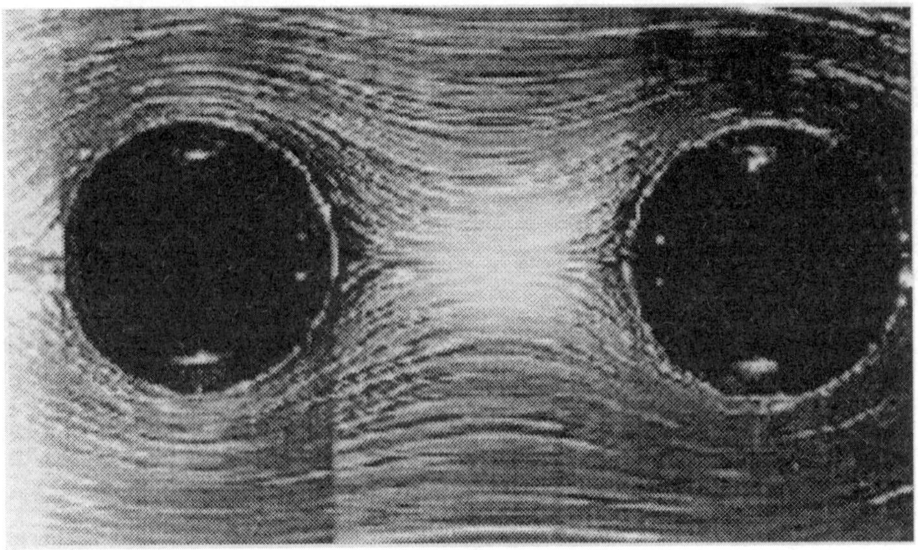

Figure 19.2. Laminar flow around two spheres at Re = 20. Nearly indistinguishable from the flow in figure 19.1, this is not creeping flow and the friction factor is somewhat larger than unity. Reproduced from S. Tanada *J. Phys. Soc. Jpn* **46** (1979) 1935–42.

19.3 Beyond Stokes' law

Stokes law has been derived in the limit of vanishing Reynolds number (19.15), and is empirically valid for Re $\lesssim 1$. For larger values of Re the simplicity of the problem nevertheless allows us to make a general analysis, as we did for turbulent pipe flow (section 18.5 on page 255), even if we cannot solve the Navier–Stokes equations explicitly.

Friction factor

Since the only parameters defining the problem are the radius a, the velocity U, the viscosity η and the density of the fluid ρ, we may for any Reynolds number write the drag on the sphere in the form of the Stokes law multiplied by a dimensionless factor $f(\mathrm{Re})$,

$$\mathcal{D} = 6\pi\eta a U f(\mathrm{Re}). \tag{19.19}$$

This factor accounts for the deviations from Stokes law and is evidently anchored at unity for vanishing Reynolds number, i.e. $f(0) = 1$. As for pipe flow we shall call it the *friction factor*, although as mentioned above only 2/3 of the drag on the sphere is in fact due to friction and 1/3 to the front-to-back pressure asymmetry (problem 19.5).

Drag coefficient

In the opposite limit at large Reynolds number Re $\gg 1$, far beyond the creeping flow region, the sphere literally plows its way through the fluid, leaving a wake of highly disturbed and turbulent fluid. The drag may, for example, be estimated from the rate of loss of momentum from the incoming fluid that is disturbed by the sphere. The incoming fluid carries a momentum density $\rho_0 U$ and since the sphere presents a cross-sectional area πa^2 to the flow, the rate at which momentum impinges on the sphere is $\rho_0 U \cdot \pi a^2 U = \rho_0 \pi a^2 U^2$. Alternatively, the scale of the drag may be obtained from the stagnation pressure increase, $\Delta p = (1/2)\rho_0 U^2$, times the cross-section area πa^2.

The drag at high Reynolds numbers is thus expected to grow with the square of the velocity (at subsonic speeds). But the fluid in the wake trailing the sphere is not completely at rest, such that only a certain

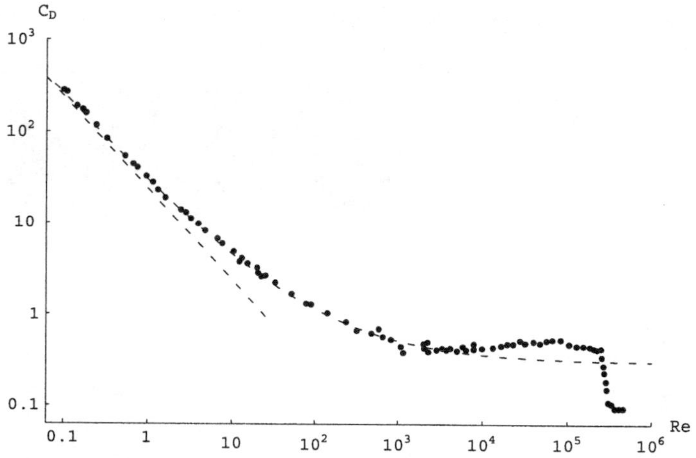

Figure 19.3. Drag coefficient for a smooth ball (data extracted from [61, fig. 1.19]). The dashed line corresponds to Stokes law $C_D = 24/\text{Re}$ and the dashed curve is the interpolation (19.22). The sharp drop (the 'drag crisis') at $\text{Re} = 2.5 \times 10^5$ signals the onset of turbulence in the boundary layer on the front half of the sphere and an accompanying shift in the shape of the trailing wake. After the drop the drag coefficient rises again. The terminal value of the drag coefficient and the Reynolds number for the onset of the drag crisis both increase with increasing Mach number. For Mach numbers above 0.8 there is no drag crisis [61, p. 12].

fraction of the incoming momentum will be lost. It is for this reason customary to define the dimensionless *drag coefficient*

$$C_D = \frac{\mathcal{D}}{(1/2)\rho_0 \pi a^2 U^2}, \tag{19.20}$$

with a conventional factor 1/2 in the denominator, perhaps inspired by the form of the stagnation pressure. Empirically, the drag coefficient for a sphere is $C_D \approx 0.5$ in the interval $10^4 \lesssim \text{Re} \lesssim 2.5 \times 10^5$, implying that about 25% of the incoming fluid momentum is lost to drag.

Inserting the definition of the friction factor for a sphere (19.19) we find

$$C_D = \frac{24}{\text{Re}} f(\text{Re}) \tag{19.21}$$

for all Reynolds numbers. It is a matter of taste whether one prefers to describe the drag on a sphere by means of the drag coefficient or the friction factor. At small Reynolds numbers, it seems a bit pointless to use the drag coefficient, because it introduces a strong artificial variation $C_D \approx 24/\text{Re}$ for $\text{Re} \to 0$ without a corresponding strong variation in the physics (which is simply described by Stokes law). At large Reynolds numbers the drag coefficient becomes a constant, which is more convenient to use than the linearly rising friction factor, although its value depends on the roughness of the ball surface.

Interpolation

One may join the regions of low and high Reynolds numbers by the simple interpolating expression

$$C_D = \frac{24}{\text{Re}} + \frac{5}{\sqrt{\text{Re}}} + 0.3. \tag{19.22}$$

The first term is Stokes' result and the last is a constant terminal form drag. The middle term may be understood as due to friction in a thin laminar boundary layer (see chapter 28) on the forward half of the sphere. There are a number of different formulae in the literature covering the same empirical data. As seen in figure 19.3, this formula agrees decently with data for all Reynolds numbers up to $\text{Re} \approx 10^4$, where the drag coefficient first rises to about 0.5. Above $\text{Re} \approx 2.5 \times 10^5$ the drag coefficient drops sharply by more than a factor 2, after which it begins to rise again, a phenomenon called the 'drag crisis'.

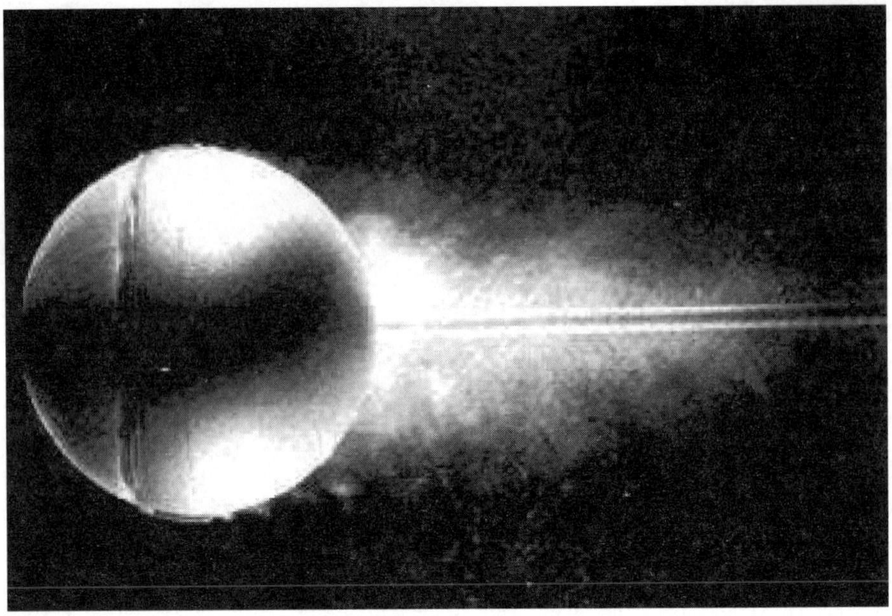

Figure 19.4. Turbulent boundary layer tripped by wire ('seam') on a sphere at Re = 30 000. The turbulent layer separates on the rear of the sphere whereas the laminar boundary layer separates slightly before the crest. ONERA photograph, H. Werle *Rech. Aerospace* **198-5** (1980) 35–49.

The drag crisis

Alexandre-Gustave Eiffel (1832–1923). *French structural engineer. Famous for building the Eiffel tower in Paris. Worked on aerodynamics for the last 21 years of his life. Built wind tunnels and performed numerous studies of drag and lift.*

The dramatic *drag crisis* at Re $\approx 2.5 \times 10^5$ was first carefully observed (and published) by Gustave Eiffel in 1914 [3, p. 282]. The crisis is caused by a transition from laminar to turbulent flow in the boundary layer of the forward-facing half of the sphere, accompanied by a front-to-back shift in the separation point for the turbulent wake that trails the sphere. At a Reynolds number just beyond the drag crisis, the wake is narrower than before, entailing a smaller loss of momentum, i.e. smaller drag. At still higher Reynolds numbers the drag coefficient regains part of its former magnitude, while at supersonic speeds the drag crisis disappears and the terminal drag rises to around unity.

The Reynolds number at which the boundary layer becomes turbulent depends on the surface properties of the sphere. Roughness tends to facilitate the generation of a turbulent boundary layer and moves the onset of turbulence to sometimes much lower Reynolds number. This is the deeper reason for manufacturing golf balls with surface dimples. A golf ball flying at a typical speed of 30 m s^{-1} has Re $\approx 1.6 \times 10^5$ which is below the drag crisis for a smooth ball, but not for a dimpled one. The lower drag on a dimpled ball permits it to fly considerably longer for a given initial thrust. The seams of a tennis ball serve the same function, whereas a ping-pong ball is quite smooth. A ping-pong ball flying at 10 m s^{-1} has Re $\approx 5 \times 10^4$, and even if turbulence could be triggered seams or dimples would probably interfere with the game.

Terminal speed

The variation in drag with Reynolds number makes the calculation of the terminal speed of a spherical ball somewhat more complicated than equation (19.14). Equating the force of gravity with the drag we obtain instead of (19.13) for $\rho_1 > \rho_0$,

$$(\rho_1 - \rho_0)\tfrac{4}{3}\pi a^3 g_0 = \tfrac{1}{2} C_D \rho_0 \pi a^2 U^2,$$

and solving for the terminal speed we get,

$$U = \sqrt{\frac{8}{3}\left(\frac{\rho_1}{\rho_0} - 1\right)\frac{a g_0}{C_D}}. \qquad (19.23)$$

Since C_D depends on the Reynolds number, this is an implicit equation for the terminal velocity except for very large Reynolds numbers where C_D is constant. The terminal velocity grows with the square root of the density, implying that the terminal kinetic energy of a falling sphere grows with the square of the density. This is why projectiles and bombs provided with heavy metal jackets (made, for example, from depleted uranium) are used to penetrate concrete structures and rock.

Example 19.3.1 (Skydiver terminal speed): A skydiver weighing 70 kg curls up like a ball with radius $a = 50$ cm and average density $\rho_1 \approx 133$ kg m^{-3}. Taking $C_D \approx 0.3$ and $\rho_0 \approx 1$ kg m^{-3}, we obtain $U = 76$ m s^{-1} $= 273$ km h^{-1}. The corresponding Reynolds number is $\text{Re} = 4.7 \times 10^6$, confirming the approximation used.

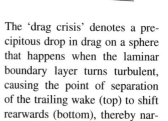

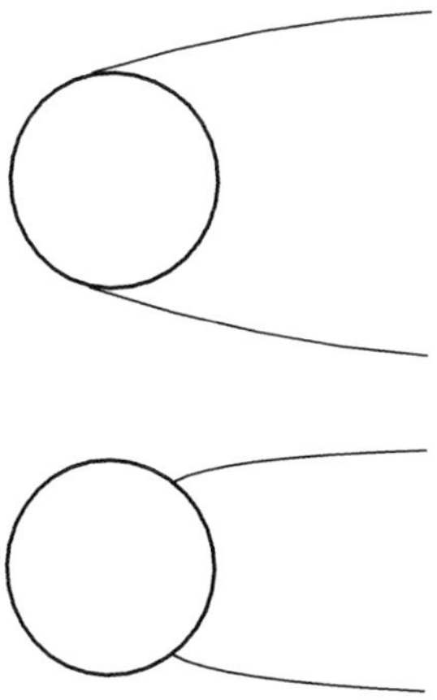

The 'drag crisis' denotes a precipitous drop in drag on a sphere that happens when the laminar boundary layer turns turbulent, causing the point of separation of the trailing wake (top) to shift rearwards (bottom), thereby narrowing the trailing wake.

19.4 Beyond spherical shape

The general arguments given in section 19.1 showed that in creeping flow the drag on an arbitrary body is proportional to viscosity, velocity, size and a shape-dependent dimensionless factor. Using the Stokes' drag (19.12) as a baseline, we may write the drag on an arbitrary body as

$$\mathcal{D} = 6\pi\eta a_S U, \tag{19.24}$$

where a_S is a characteristic length, sometimes called the *Stokes radius*. The Stokes radius may be determined from a measurement of the terminal speed of the falling object,

$$a_S = \frac{m' g_0}{6\pi\eta U}, \tag{19.25}$$

where m' is the mass of the body corrected for buoyancy.

The Stokes radius may be calculated analytically for some simple bodies, for example a circular disk of radius a. In the two major orientations of the disk, we have [79, p. 178],

$$\frac{a_S}{a} = \begin{cases} \dfrac{8}{3\pi} \approx 0.85 & \text{disk orthogonal to flow} \\[2ex] \dfrac{16}{9\pi} \approx 0.57 & \text{disk parallel to flow.} \end{cases} \tag{19.26}$$

In spite of the vast geometric difference between the two cases, the Stokes radii are of the same order of magnitude. For general bodies of non-exceptional geometry, one may estimate the Stokes radius from the surface area or the volume of the body.

At high Reynolds numbers, where viscosity plays a diminishing role, the drag coefficient may be defined as

$$\boxed{C_D = \frac{\mathcal{D}}{(1/2)\rho_0 A U^2},} \tag{19.27}$$

where A is some area that the body presents to the flow. For blunt bodies the area may be taken to be the 'shadow' of the body on a plane orthogonal to the direction of motion. For a given shape, the drag force

$$\mathcal{D} = C_D \tfrac{1}{2}\rho_0 A U^2, \tag{19.28}$$

exposes the dependence on the fluid (ρ_0), the body size (A), and the state of motion (U). Barring the presence of a drag crisis, typical values for C_D are of the order of unity for bodies of non-exceptional geometry. A flat circular disk orthogonal to the flow has $C_D \approx 1.17$, whereas a circular cup with its opening towards the flow has $C_D \approx 1.4$ [80, p. 460]. If on the other hand the disk is infinitely thin and oriented parallel with the flow, there will be no asymptotic form drag, and the drag coefficient will vanish (as $1/\sqrt{\text{Re}}$) in the limit of infinite Reynolds number.

Streamlining

Drag reduction by *streamlining* is important in the construction of all kinds of moving vehicles, such as cars and airplanes. Car manufacturers have over the years reduced the drag coefficient on a car to lower than 0.4, and there is still room for improvement. In modern times drag-reducing helmets have also appeared on the heads of bicycle racers and speed skaters, giving the performers of these sports quite alien looks. Among animals, drag reduction yields evolutionary advantages, which has led to the beautiful outlines of fast flyers and swimmers, like falcons and sharks. The density of water is about a thousand times that of air, implying a thousand times larger form drag (19.28) in water than in air, although this effect is partly offset by the higher velocities necessary for flight. This has forced swimming animals towards extremes of streamlined shapes. The mackerel has thus reduced the drag coefficient of its sleek form to the astonishingly low value of 0.0043 at Re $\approx 10^5$, an order of magnitude lower than for a swimming human [75, p. 143]. Since muscular power is roughly the same, birds should typically be capable of moving about $\sqrt{1000} \approx 30$ times faster than fish of the same size and shape, which seems quite reasonable.

Problems

19.1 Tiny unicellular animals in the sea die and their carcasses settle slowly towards the bottom. Assume that an animal's carcass is spherical with diameter 10 μm and average density 1.2 times that of water. Calculate the time it takes for the carcass to settle to the bottom of the deep sea (depth 4 kilometres).

19.2 A microscopic spherical grain of vulcanic (or cosmic) dust of radius a enters the homentropic atmosphere (section 4.6) at a height z. Determine the time t_0 it takes for the grain to settle to the ground under the assumption that the density of the grain is much larger than the density of air, and that Stokes law is valid at any height with the actual density and viscosity of air at that height. Evaluate the settling time numerically for reasonable values of the parameters.

19.3 A spherical particle begins to fall from rest in constant gravity. Determine how the particle reaches terminal speed (19.14) under the assumption that the Reynolds number is always small.

19.4 An object with constant drag coefficient C_D falls from rest in constant gravity. Determine how the particle reaches terminal speed under the assumption that buoyancy can be disregarded.

19.5 Calculate the pressure contribution to the drag from Stokes solution for a sphere.

19.6 Consider Stokes flow around a sphere of radius a. **(a)** Calculate the volume discharge of fluid passing a concentric annular disk of radius $b > a$ placed orthogonally to the flow. **(b)** Calculate the volume in relation to the volume that would pass through the same disk if the sphere were not present. **(c)** Justify qualitatively why the ratio vanishes for $b \to a$.

* **19.7** Consider spherical Stokes flow and

(a) show that the stream lines are determined by the solutions to

$$\frac{dr}{dt} = v_r = A(r)U\cos\theta$$
$$\frac{d\theta}{dt} = \frac{v_\theta}{r} = -B(r)U\sin\theta$$

with

$$A(r) = 1 - \frac{3}{2}\frac{a}{r} + \frac{1}{2}\frac{a^3}{r^3} = \left(1 - \frac{a}{r}\right)^2\left(1 + \frac{1}{2}\frac{a}{r}\right)$$
$$B(r) = \frac{1}{r}\left(1 - \frac{3}{4}\frac{a}{r} - \frac{1}{4}\frac{a^3}{r^3}\right) = \frac{1}{r}\left(1 - \frac{a}{r}\right)\left(1 + \frac{1}{4}\frac{a}{r} + \frac{1}{4}\frac{a^2}{r^2}\right).$$

(b) Show that this leads to a solution of the form

$$\sin\theta = e^{-\int B/A\,dr} = \frac{d}{(r-a)\sqrt{1+a/2r}} \tag{19.29}$$

where d is an integration constant. This is the equation for the streamlines in polar coordinates.

(c) Show that d is the asymptotic distance of the flow line from the polar axis (also called the impact parameter).

(d) Find the relation between d and the point of closest approach of the flow line to the sphere.

* **19.8** Show that the rate of work of the contact forces exerted by a steadily moving body on the fluid through which it moves is $\mathcal{D}U$, where $\mathcal{D}$ is the total drag.

* **19.9** **(a)** Show that the creeping flow solution around a fixed body in an asymptotically uniform steady velocity field $\boldsymbol{U}$ must be of the form

$$\boldsymbol{v}(\boldsymbol{x}) = \mathbf{A}(\boldsymbol{x}) \cdot \boldsymbol{U} \tag{19.30}$$

$$p(\boldsymbol{x}) = \eta \boldsymbol{Q}(\boldsymbol{x}) \cdot \boldsymbol{U} \tag{19.31}$$

where $\mathbf{A}(\boldsymbol{x})$ is a tensor field and $\boldsymbol{Q}(\boldsymbol{x})$ is a vector field, both independent of $\boldsymbol{U}$.
(b) Determine the field equations and the boundary conditions for $\mathbf{A}$ and $\boldsymbol{Q}$.
(c) Calculate the total reaction force on the body.

* **19.10 Analytic solution of Stokes flow for a sphere**

(a) Use the field equations and symmetry to show that the solution must be of the form (in spherical coordinates)

$$\boldsymbol{v}(\boldsymbol{x}) = a(r)\boldsymbol{U} + b(r)\boldsymbol{x}\boldsymbol{U} \cdot \boldsymbol{x}, \tag{19.32a}$$

$$p(\boldsymbol{x}) = \eta c(r)\boldsymbol{U} \cdot \boldsymbol{x}, \tag{19.32b}$$

where $a(r)$, $b(r)$ and $c(r)$ are functions only of r.

(b) Use the boundary conditions to show that $a(r)$ and $b(r)$ must vanish at $r = a$, and that at $r \to \infty$ they satisfy $a(r) \to 1$ and $r^2 b(r) \to 0$.

(c) Show that the field equations lead to the ordinary differential equations

$$\frac{d^2a}{dr^2} + \frac{2}{r}\frac{da}{dr} + 2b = c, \tag{19.33a}$$

$$\frac{d^2b}{dr^2} + \frac{6}{r}\frac{db}{dr} = \frac{1}{r}\frac{dc}{dr}, \tag{19.33b}$$

$$\frac{1}{r}\frac{da}{dr} + r\frac{db}{dr} + 4b = 0. \tag{19.33c}$$

(d) Show that these equations are homogeneous in r, and justify that one should look for power solutions

$$a = Ar^{\alpha}, \tag{19.34a}$$

$$b = Br^{\beta}, \tag{19.34b}$$

$$c = Cr^{\gamma}. \tag{19.34c}$$

(e) Show that

$$\beta = \gamma = \alpha - 2 \tag{19.35}$$

and

$$\alpha(\alpha + 1)A + 2B = C, \tag{19.36a}$$

$$(\alpha - 2)(\alpha + 3)B = (\alpha - 2)C, \tag{19.36b}$$

$$\alpha A + (\alpha + 2)B = 0. \tag{19.36c}$$

(f) Show that a non-trivial solution requires

$$(\alpha - 2)\alpha(\alpha + 1)(\alpha + 3) = 0, \tag{19.37}$$

such that the allowed powers are $\alpha = 2, 0, -1, -3$.

(g) Show that the most general solution is

$$a(r) = Kr^2 + L + \frac{M}{r} + \frac{N}{r^3}, \tag{19.38a}$$

$$b(r) = -\frac{1}{2}K + \frac{M}{r^3} - 3\frac{N}{r^5}, \tag{19.38b}$$

$$c(r) = 5K + 2\frac{M}{r^3}, \tag{19.38c}$$

where K, L, M, and N are the four integration constants.

(h) Show that the boundary conditions at infinity require $K = 0$ and $L = 1$, and that the boundary conditions at $r = a$ require $M = -3/4a$ and $N = -1/4a^3$, leading to the desired solution (19.9).

20

Rotating fluids

The conductor of a carousel knows about fictitious forces. Moving from horse to horse while collecting tickets, he not only has to fight the centrifugal force trying to kick him off, but also has to deal with the dizzying sideways Coriolis force. On a fast carousel with a five metre radius and turning once every six seconds, the centrifugal force is strongest at the rim where it amounts to about 50% of gravity. Walking across the carousel at a normal speed of one metre per second, the conductor experiences a Coriolis force of about 20% of gravity. Provided the carousel turns anticlockwise seen from above, as most carousels seem to do, the Coriolis force always pulls the conductor off his course to the right. The conductor seems to prefer to move from horse to horse against the rotation, and this is quite understandable, since the Coriolis force then counteracts the centrifugal force.

The Earth's rotation creates a centrifugal 'antigravity' field, reducing effective gravity at the equator by 0.3%. This is hardly a worry, unless you have to adjust Olympic records for geographic latitude. The Coriolis force is even less notable at the Olympics. You have to move as fast as a jet aircraft for it to amount to 0.3% of a per cent of gravity. Weather systems and sea currents are so huge and move so slowly compared to the Earth's local rotation speed that the weak Coriolis force can become a major player in their dynamics. The Coriolis force guarantees that low pressure weather cyclones on the northern half of the globe always turn anticlockwise around low pressure regions, whereas high pressure cyclones turn clockwise.

In this chapter we shall investigate the strange behaviour of fluids in rotating containers, mainly due to the Coriolis force. At the end we shall debunk the persistent 'urban legend' about the sense of rotation of bathtub vortices.

20.1 Fictitious forces

When dealing with a moving object, for example a rotating planet, it is often convenient to give up the inertial coordinate system and instead attach a fixed coordinate system to the moving object. In a non-inertial, accelerated coordinate system, the laws of mechanics take a different form. The acceleration on the left-hand side of Newton's second law must be corrected by terms deriving from the motion of the coordinate system, and when these terms are shifted to the right-hand side they can be interpreted as forces. As these forces apparently have no objective cause, in contrast to forces caused by other bodies, they are called *fictitious*. A better name might be *inertial* forces, since there is nothing fictitious about the jerk you experience when the bus suddenly stops.

Velocity and acceleration in steady rotation

In this chapter we shall only be concerned with steadily rotating coordinate systems, such as those we use on Earth. Consider a Cartesian coordinate system rotating with constant angular velocity Ω relative to an inertial system. Without loss of generality we may take the z-axis to be the axis of rotation. The coordinates

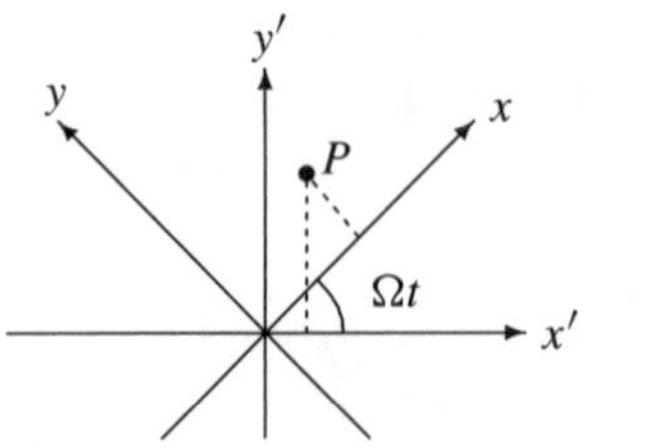

Transformation of coordinates of a point particle P from a rotating to an inertial coordinate system.

(x, y, z) of a point particle in the rotating system are then related to the coordinates (x', y', z') by a simple rotation (2.36) through an angle $\phi = \Omega t$,

$$x = x' \cos \Omega t + y' \sin \Omega t, \tag{20.1a}$$
$$y = -x' \sin \Omega t + y' \cos \Omega t, \tag{20.1b}$$
$$z = z'. \tag{20.1c}$$

Differentiating once with respect to time t, we find the particle velocity $\boldsymbol{v} = d\boldsymbol{x}/dt$ in the rotating system expressed in terms of its velocity $\boldsymbol{v}' = d\boldsymbol{x}'/dt$ in the inertial system, plus extra terms arising from differentiation of the sines and cosines. For example

$$v_x = v'_x \cos \Omega t + v'_y \sin \Omega t - \Omega x' \sin \Omega t + \Omega y' \cos \Omega t.$$

The last two terms become Ωy, and treating the other velocity components in the same way we get,

$$v_x = v'_x \cos \Omega t + v'_y \sin \Omega t + \Omega y, \tag{20.2a}$$
$$v_y = -v'_x \sin \Omega t + v'_y \cos \Omega t - \Omega x, \tag{20.2b}$$
$$v_z = v'_x. \tag{20.2c}$$

Differentiating once more after time, we obtain the particle acceleration $\boldsymbol{w} = d\boldsymbol{v}/dt = d^2\boldsymbol{x}/dt^2$ in the rotating system expressed in terms of its acceleration $\boldsymbol{w}' = d\boldsymbol{v}'/dt = d^2\boldsymbol{x}'/dt^2$ in the inertial system,

$$w_x = w'_x \cos \Omega t + w'_y \sin \Omega t + \Omega^2 x + 2\Omega v_y, \tag{20.3a}$$
$$w_y = -w'_x \sin \Omega t + w'_y \cos \Omega t + \Omega^2 y - 2\Omega v_x, \tag{20.3b}$$
$$w_z = w'_z. \tag{20.3c}$$

The extra terms on the right-hand side arising from differentiation of the sines and cosines are the origin of the 'fictitious' forces in Newton's second law.

These results may be expressed in a more compact and transparent form by means of the rotation matrix (2.42) for a simple rotation with angle $\phi = \Omega t$,

$$\boldsymbol{x} = \mathbf{A} \cdot \boldsymbol{x}', \tag{20.4a}$$
$$\boldsymbol{v} = \mathbf{A} \cdot \boldsymbol{v}' - \boldsymbol{\Omega} \times \boldsymbol{x}, \tag{20.4b}$$
$$\boldsymbol{w} = \mathbf{A} \cdot \boldsymbol{w}' - \boldsymbol{\Omega} \times (\boldsymbol{\Omega} \times \boldsymbol{x}) - 2\boldsymbol{\Omega} \times \boldsymbol{v}. \tag{20.4c}$$

Although the equations have been derived for a special choice of coordinates, they must in this form be valid for arbitrary Cartesian coordinate systems with the common origin on the rotation axis, but where the rotation vector $\boldsymbol{\Omega}$ is not necessarily aligned with the z-axis.

Centrifugal and Coriolis forces

Gaspard-Gustave Coriolis (1792–1843). *French mathematician. Worked on friction and hydraulics. Introduced the terms 'work' and 'kinetic energy' in the form used today. Defined what is now called the Coriolis force in 1835.*

The force $\boldsymbol{f}'$ acting on a point particle in the inertial system depends in general on the position $\boldsymbol{x}'$ and velocity $\boldsymbol{v}' = d\boldsymbol{x}'/dt$ of the particle. Think, for example, of the force of gravity $\boldsymbol{f}' = -GmM\boldsymbol{x}'/|\boldsymbol{x}'|^3$ from a point particle of mass M which depends on $\boldsymbol{x}'$ (but not on $\boldsymbol{v}'$). In the rotating system the transformed force is naturally defined to be

$$\boldsymbol{f} = \mathbf{A} \cdot \boldsymbol{f}'. \tag{20.5}$$

Since $|\boldsymbol{x}| = |\boldsymbol{x}'|$, this definition makes the transformed gravitational force take the same form in the rotating and inertial systems, $\boldsymbol{f} = -GmM\boldsymbol{x}/|\boldsymbol{x}|^3$. If the velocity appears in the force, it must be transformed using (20.4b).

Multiplying (20.4c) by m, Newton's second law in the inertial system, $m\boldsymbol{w}' = \boldsymbol{f}'$, becomes in the rotating system,

$$m\boldsymbol{w} = \boldsymbol{f} - m\boldsymbol{\Omega} \times (\boldsymbol{\Omega} \times \boldsymbol{x}) - 2m\boldsymbol{\Omega} \times \boldsymbol{v}. \tag{20.6}$$

Writing this as an equation of motion for the particle trajectory $\boldsymbol{x}(t)$, it becomes

$$\boxed{m\frac{d^2\boldsymbol{x}}{dt^2} = \boldsymbol{f} - m\boldsymbol{\Omega}\times(\boldsymbol{\Omega}\times\boldsymbol{x}) - 2m\boldsymbol{\Omega}\times\frac{d\boldsymbol{x}}{dt}.} \tag{20.7}$$

The extra terms on the right-hand side are the fictitious forces:

- the centrifugal force $-m\boldsymbol{\Omega}\times(\boldsymbol{\Omega}\times\boldsymbol{x})$,
- the Coriolis force $-2m\boldsymbol{\Omega}\times\frac{d\boldsymbol{x}}{dt}$.

Note that both of these forces are proportional to the mass of the particle and resemble in this respect ordinary gravity. In a completely general moving coordinate system there will appear three further fictitious terms on the right-hand side (see problem 20.3 and equation (20.41)).

Fictitious forces on Earth

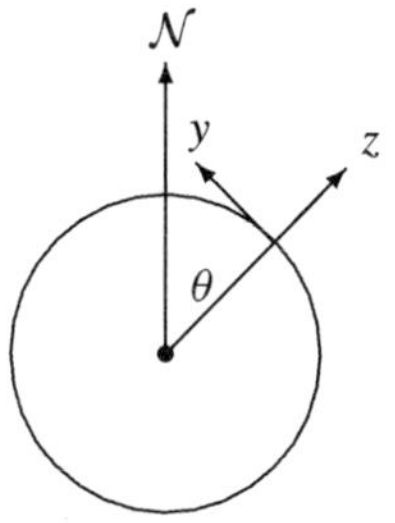

Local flat-Earth coordinate system and its relation to the polar angle.

The effect of the centrifugal force on Earth is primarily to flatten the spherical shape in such a way that it conforms to an equipotential surface. It was shown in section 7.4 on page 91 that the direction of the combined gravitational and centrifugal force is always orthogonal to the equipotential surface everywhere on Earth, and thus by definition vertical. The magnitude of the centrifugal force is only a fraction $q = \Omega^2 a/g_0 \approx 1/291$ of standard gravity, and thus generally negligible. If precision is needed, one must use the expression (7.31) for the effective surface gravity which properly takes into account the centrifugal flattening.

The Coriolis force is different. Let us introduce a local flat-Earth coordinate system tangential to the surface at a given point with the x-axis towards the east and the y-axis towards the north. In this reference frame the Earth's rotation vector is $\boldsymbol{\Omega} = \Omega\,(0, \sin\theta, \cos\theta)$ where θ is the polar angle. The components of the Coriolis acceleration $\boldsymbol{g}^C = -2\boldsymbol{\Omega}\times\boldsymbol{v}$, expressed in terms of the velocity $\boldsymbol{v} = d\boldsymbol{x}/dt$ then become,

$$g_x^C = 2\Omega\cos\theta\, v_y - 2\Omega\sin\theta\, v_z, \tag{20.8a}$$

$$g_y^C = -2\Omega\cos\theta\, v_x, \tag{20.8b}$$

$$g_z^C = 2\Omega\sin\theta\, v_x. \tag{20.8c}$$

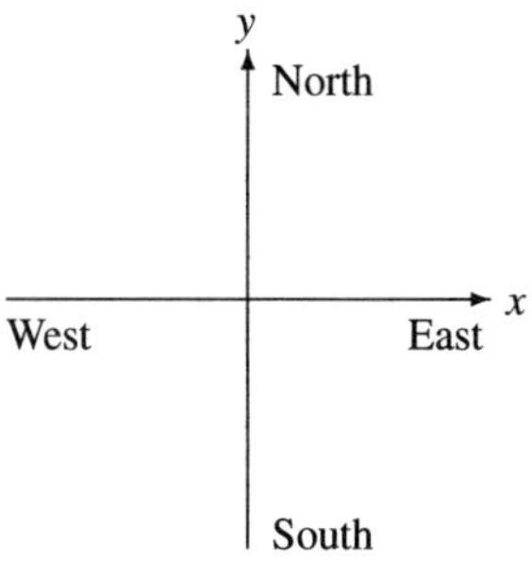

Flat-Earth local coordinate system at a point on the surface.

At the poles $\theta = 0, 180^\circ$ the Coriolis force is always horizontal, and at the equator $\theta = 90^\circ$ it is always vertical for horizontal motion with $v_z = 0$.

The vertical Coriolis force g_z^C is very small compared to gravity. Even for a modern jet aircraft flying on an east-west course at middle latitudes at a speed close to the velocity of sound, it amounts to only about 0.3% of gravity, which accidentally is of the same order of magnitude as the correction due to the centrifugal force. So we may ordinarily ignore the vertical Coriolis force along with the centrifugal force in most Earthly matters. We shall also ignore the Coriolis force due to the vertical component of velocity v_z. A jet plane experiences again only a Coriolis acceleration in a westerly direction of magnitude 0.3% of gravity during upwards vertical flight at the speed of sound.

The conclusion is that under normal circumstances only the first terms in g_x^C and in g_y^C need to be considered, leading to a purely two-dimensional Coriolis acceleration which we write in the form,

$$g_x^C = 2\Omega_\perp v_y, \qquad g_y^C = -2\Omega_\perp v_x, \qquad g_z^C = 0 \tag{20.9}$$

where $\Omega_\perp$ is the *local angular velocity*

$$\Omega_\perp = \Omega\cos\theta. \tag{20.10}$$

For all practical purposes, the Coriolis force in a local flat-Earth coordinate system looks as if the Earth were indeed flat and rotated around the local vertical with the local angular velocity $\Omega_\perp$. A good number to know is that at middle latitudes $\theta \approx 45^\circ$ we have $2\Omega_\perp \approx 10^{-4}\ \mathrm{s}^{-1}$. At this latitude the period of rotation of a Foucault pendulum is $2\pi/\Omega_\perp \approx 34$ h, irrefutably proving that the Earth rotates under your feet.

Jean-Bernard Léon Foucault (1819–1868). *French physicist. Set up an enormous pendulum on the Paris Panthéon in 1851. Invented the gyroscope, the reflecting telescope, and measured the velocity of light in absolute units* (km s^{-1}).

20.2 Flow in a rotating system

In a steadily rotating coordinate system, the Navier–Stokes equation must also include the fictitious forces. For an incompressible fluid we find in a coordinate system with the origin on the rotation axis,

$$\frac{\partial \boldsymbol{v}}{\partial t} + (\boldsymbol{v} \cdot \nabla)\boldsymbol{v} = \boldsymbol{g} - \boldsymbol{\Omega} \times (\boldsymbol{\Omega} \times \boldsymbol{x}) - 2\boldsymbol{\Omega} \times \boldsymbol{v} - \nabla \frac{p}{\rho_0} + \nu \nabla^2 \boldsymbol{v}. \tag{20.11}$$

For compressible flow (17.27) the fictitious terms must similarly be added to the right-hand side.

As discussed in the preceding section, the centrifugal force can generally be ignored at the surface of the Earth, though not in laboratory experiments with rotating containers, such as Newton's bucket (page 87). Formally, the gravitational and centrifugal fields may in an incompressible fluid be included in the *effective pressure*,

$$p^* = p + \rho_0 \Phi - \tfrac{1}{2}\rho_0 (\boldsymbol{\Omega} \times \boldsymbol{x})^2, \tag{20.12}$$

such that the Navier–Stokes equation simplifies to,

$$\boxed{\frac{\partial \boldsymbol{v}}{\partial t} + (\boldsymbol{v} \cdot \nabla)\boldsymbol{v} = -2\boldsymbol{\Omega} \times \boldsymbol{v} - \nabla \frac{p^*}{\rho_0} + \nu \nabla^2 \boldsymbol{v}.} \tag{20.13}$$

At Earth's surface the Coriolis term can to a good approximation be calculated using only the vertical local angular velocity vector $\boldsymbol{\Omega}_\perp$.

The Rossby number

Carl-Gustav Arild Rossby (1898–1957). *Swedish born meteorologist who mostly worked in the US. Contributed to the understanding of large-scale motion and general circulation of the atmosphere.*

Let us again characterize the flow by a length scale L and a velocity scale U. For nearly ideal steady flow with large Reynolds number, $\text{Re} = UL/\nu \gg 1$, the advective term dominates the viscous term (except near boundaries). The interesting quantity is therefore the ratio between the advective acceleration and the Coriolis acceleration, called the *Rossby number*

$$\text{Ro} = \frac{|(\boldsymbol{v} \cdot \nabla)\boldsymbol{v}|}{|2\boldsymbol{\Omega} \times \boldsymbol{v}|} \approx \frac{U^2/L}{2\Omega U} = \frac{U}{2\Omega L}. \tag{20.14}$$

For nearly ideal flow the Coriolis force is significant only if $\text{Ro} \lesssim 1$, in other words for $U \lesssim 2\Omega L$. Thus, the general rule is that the Coriolis force only matters when the flow velocity U is of the same magnitude or smaller than the typical variation ΩL in the local rotation velocity across the system.

Ocean currents and weather cyclones are relatively steady phenomena. The characteristic speeds are metres per second for the ocean currents and tens of metres per second for winds over distances of the order of a thousand kilometres. With a local angular velocity $2\Omega \approx 10^{-4}\ \text{s}^{-1}$ one gets a Rossby number $\text{Ro} \approx 0.01$ for ocean currents and $\text{Ro} \approx 0.1$ for weather cyclones. Both of these phenomena are thus dominated by the Coriolis force, but the ocean currents by far the most.

When you (of size $L = 1$ m) swim with a speed of $U \approx 1\ \text{m s}^{-1}$, the Rossby number becomes $\text{Ro} \approx 10^4$, and the Coriolis force can be completely neglected. The water draining out of your bathtub moves with similar speeds over similar distances, making the Rossby number just as large and the Coriolis force just as insignificant as for swimming (see also section 20.5). In space stations designed for long-term habitation, gravity must for health reasons be simulated by rotation, and the large Coriolis force will play havoc in ballistic games like Ping-Pong, football, and basket ball (see problem 20.1).

The Ekman number

Vagn Walfrid Ekman (1874–1954). *Swedish physical oceanographer. Contributed to the understanding of the dynamics of ocean currents.*

The ratio between viscous and Coriolis forces is called the *Ekman number*,

$$\text{Ek} = \frac{|\nu \nabla^2 \boldsymbol{v}|}{|2\boldsymbol{\Omega} \times \boldsymbol{v}|} \approx \frac{\nu U/L^2}{2\Omega U} = \frac{\nu}{2\Omega L^2} = \frac{\text{Ro}}{\text{Re}}. \tag{20.15}$$

When the Ekman number is small, the viscous force can be neglected relative to the Coriolis force. For large Reynolds number and moderate Rossby number, the Ekman number is automatically small. In a natural system on Earth, such as the sea and the atmosphere, the huge Reynolds number makes the Ekman number extremely small. The Ekman number is normally only of order unity close to boundaries where viscosity always comes to dominate over advection. Here the interplay between viscous and Coriolis forces gives rise to a highly interesting boundary layer, called the *Ekman layer* which we shall analyse in section 20.4.

20.3 Geostrophic flow

In natural large-scale systems, the Rossby number and Ekman numbers are so small, Ro, Ek $\ll 1$, that one can in the first approximation ignore both the viscous and the advective terms in the Navier–Stokes equations. Such a flow completely dominated by the Coriolis force is said to be *geostrophic*.

Dropping both advective and viscous terms in (20.11), we arrive at the remarkably simple Navier–Stokes equation for steady geostrophic flow,

$$\boxed{-2\boldsymbol{\Omega} \times \boldsymbol{v} - \frac{1}{\rho_0}\nabla p^* = \mathbf{0}.} \tag{20.16}$$

Much like in hydrostatics, it states that the effective pressure gradient must always balance the Coriolis force but because of the special form of the Coriolis force, this equation has strong and peculiar consequences. The most spectacular is that geostrophic flow is essentially two-dimensional.

Water level in an open canal

The Coriolis force may tilt an otherwise horizontal surface of a moving fluid. Let, for example, a constant water current flow with velocity U through a canal of width d. If the canal runs along the x-direction and the angular rotation is Ω around the z-direction, the effective pressure will according to (20.16) satisfy $\nabla_x p^* = \nabla_z p^* = 0$ and thus be independent of both x and z. From the y-component of (20.16) we get,

$$\frac{1}{\rho_0}\frac{\partial p^*}{\partial y} = -2\Omega U. \tag{20.17}$$

The solution is, apart from an unimportant constant, $p^* = -2\rho_0\Omega U y$. For positive Ω and U, the Coriolis force wants to turn the water towards the right, creating thereby an increasing effective pressure in the negative y-direction. Intuitively this seems to indicate that the Coriolis force will also raise the water level for negative y and lower it for positive.

In order to check our intuition we use (20.12) with a gravitational potential $\Phi = g_0 z$ and no centrifugal force. Apart from an unimportant constant the true pressure is $p = p^* - \rho_0 g_0 z = -\rho_0(2\Omega U y + g_0 z)$. The water level $z = h(y)$ at the open surface is determined from the requirement that the true pressure p should be constant at the free surface, so with the boundary condition $h(0) = 0$ we get,

$$h(y) = -\frac{2\Omega U}{g_0}y. \tag{20.18}$$

This confirms our intuition; on the northern half of the globe the Coriolis force always makes the water level highest on the right-hand bank of a stream.

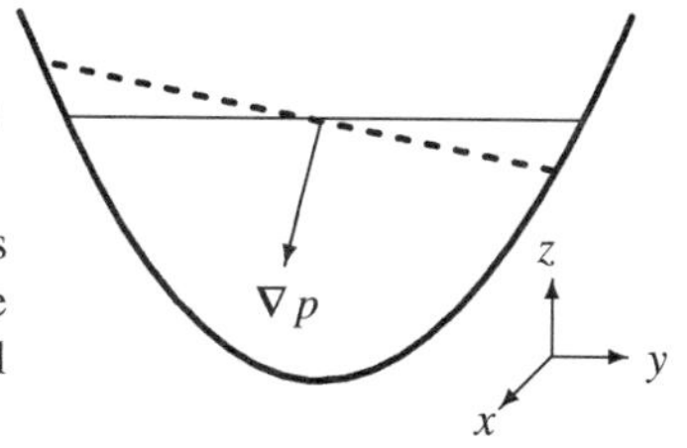

The water level is tilted by the Coriolis force, such that the surface stays orthogonal to the gradient of the true pressure. Here the Earth's rotation is positive around z and water flows out of the picture in the positive x-direction.

> **Example 20.3.1:** For a 10 km wide strait and a current flowing at 1 m s^{-1}, the difference in water level at the two sides of the strait due to the Coriolis force is about 10 cm at middle latitudes. Although Ro $= U/2\Omega d \approx 1$, the use of the geostrophic equation (20.16) is nevertheless justified, because the advective term vanishes for an x-independent flow in the x-direction (like in Poiseuille flow), and the Reynolds number, Re $\approx 10^{10}$ is so large that viscous forces can be ignored.

Isobaric flow and weather maps

An immediate consequence of the geostrophic equation (20.16) is that

$$\boldsymbol{v} \cdot \nabla p^* = 0, \tag{20.19}$$

which means that the effective pressure is constant along streamlines. In horizontal motion the gravitational force plays no role and the effective pressure is the same as the hydrostatic pressure. This means that *streamlines and isobars coincide in geostrophic flow*. This is also well known from weather maps where wind directions can be read off from the isobars.

To read off the correct sign for the wind direction one must also use the fact that the Coriolis force makes winds on the northern hemisphere turn anti-clockwise around low pressure regions (cyclones) and clockwise around regions with high pressure (anti-cyclones). Quantitatively we may invert the geostrophic equation (20.16) to calculate the wind velocities from the pressure gradients. Using $\boldsymbol{\Omega} \cdot \boldsymbol{v} = 0$ we find $\boldsymbol{\Omega} \times (\boldsymbol{\Omega} \times \boldsymbol{v}) = -\Omega^2 \boldsymbol{v}$, and then

$$\boldsymbol{v} = \frac{\boldsymbol{\Omega} \times \nabla p^*}{2\Omega^2 \rho_0}. \tag{20.20}$$

Were it not for the Coriolis force, air masses would stream along the negative pressure gradient, from high pressure towards low pressure regions. On the northern hemisphere, the Coriolis force generates cyclones by making the air that begins to stream towards the low pressure veer to the right until it is aligned with the isobars. Funnily enough, the same mechanism creates anticyclones.

Two-dimensionality of geostrophic flow

The geostrophic equation (20.16) is an extremely serious constraint on the flow. Forming the scalar product with $\boldsymbol{\Omega}$ we get

$$(\boldsymbol{\Omega} \cdot \nabla) p^* = \Omega \frac{\partial p^*}{\partial z} = 0. \tag{20.21}$$

The effective pressure is evidently constant along the axis of rotation. In a constant gravitational field g_0 anti-parallel with the axis of rotation, we may use (20.12) to obtain the hydrostatic pressure (ignoring again the centrifugal contribution)

$$p = p^*(x, y) - \rho_0 g_0 z, \tag{20.22}$$

just as we did in the case of the open canal.

Geoffrey Ingram Taylor (1886–1975). *British physicist, mathematician and engineer. Had great impact on all aspects of 20th century fluid mechanics from aircraft to explosions. Devised a method to determine the bulk viscosity of compressible fluids. Studied the movements of unicellular marine creatures.*

Joseph Proudman (1888–1975). *British mathematician and oceanographer. First proved the Taylor–Proudman theorem in 1915.*

The invariance of the effective pressure under translations along the axis of rotation actually extends to the whole flow. To verify this, we use that the curl of a gradient vanishes, and find from the geostrophic equation (20.16) and the divergence condition $\nabla \cdot \boldsymbol{v} = 0$,

$$0 = \nabla \times (\boldsymbol{\Omega} \times \boldsymbol{v}) = \boldsymbol{\Omega}\, \nabla \cdot \boldsymbol{v} - (\boldsymbol{\Omega} \cdot \nabla)\boldsymbol{v} = -(\boldsymbol{\Omega} \cdot \nabla)\boldsymbol{v} = -\Omega \frac{\partial \boldsymbol{v}}{\partial z}.$$

The flow field $\boldsymbol{v}$ is also a function of x and y only, implying that the flow is essentially two-dimensional. This result is the *Taylor–Proudman theorem.* To the extent that weather cyclones satisfy the conditions for steady geostrophic flow, one may conclude that the same large-scale wind patterns are found all way up through the atmosphere [55].

Taylor columns

The Taylor–Proudman theorem is a strange result, which predicts that if one disturbs the flow of a rotating fluid at, say, $z = 0$, then the pattern of the disturbance will, after all time-dependence has died away, have been copied to all other values of z. This is even true when the disturbance is caused by a three-dimensional object with finite extent in the z-direction. A so-called Taylor column (of disturbed flow) is created in the rotating fluid. Taylor columns are sometimes also called Proudman pillars.

In spite of the strangeness, many experiments beginning with Taylor's own in 1923 have amply verified the existence of Taylor columns. Even a body moving steadily along the axis of rotation, such as a falling sphere, will push a long column of fluid in front of itself, and trail another behind it. The mechanics underlying the formation of Taylor columns derives from a complicated interplay between the advective and viscous terms left out in the geostrophic equation (20.16), and it is not easy to give an explanation in simple physical terms [7].

20.4 The Ekman layer

Boundary layers arise around a body in nearly ideal flow, because viscous forces must necessarily come into play to secure the no-slip boundary condition. Steady boundary layers are normally asymmetric with respect to the direction of the 'slip-flow' outside the layer, being thinnest at the leading edge of a body and thickening towards the rear. This happens even at an otherwise featureless body surface and may be understood as a cumulative effect of the slowing down of the fluid by the contact with the boundary. The further downstream from the leading edge of a body, the longer time will the fluid have been under the influence of shear forces from the boundary, and the thicker the boundary layer will be. In chapter 28 we shall analyse boundary layers in general, but in this section we shall solve the much simpler case of the Ekman boundary layer in a rotating system.

A fluid in geostrophic flow must also form boundary layers around the bodies immersed in it. For small bodies the effect of the rotation is negligible, but for larger bodies with flow velocity of the same scale as the local rotation speed, i.e. for $\text{Ro} = U/2L\Omega \lesssim 1$, the Coriolis force comes to play a major role in the formation of boundary layers. As the flow velocity rises from zero at the boundary to its asymptotic value in the geostrophic slip-flow outside the boundary, the Coriolis force becomes progressively stronger, making the flow veer more and more to the right (for anti-clockwise rotation). This geostrophic cross-wind effectively 'blows away' accumulated fluid and prevents the downstream growth of the boundary layer. Such a boundary layer, confined to a finite thickness by the Coriolis force, is called an *Ekman layer*.

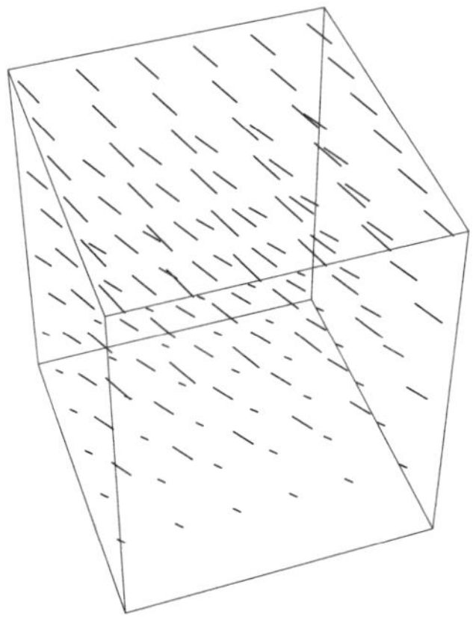

Ekman flow pattern in a vertical box. Note how the fluid close to the ground flows to the left of the geostrophic flow higher up. The angle between the direction of flow close to the ground and the geostrophic flow is in fact 45°.

Ekman layer solution

Outside the Ekman layer we assume that there is a steady geostrophic flow in the x-direction with velocity $v_x = U$, accompanied by an effective pressure $p^* = -2\rho_0\Omega U y$ in the y-direction, as for canal flow (page 283). In looking for a solution which interpolates between the static boundary and the geostrophic flow, we again exploit the symmetry of the problem. As long as the centrifugal acceleration can be ignored (which it can), the equations of motion as well as the boundary conditions are independent of the exact position in x and y, i.e. invariant under arbitrary translations in these coordinates. It is then natural to guess that there may be a maximally symmetric solution $\boldsymbol{v} = \boldsymbol{v}(z)$ which is also independent of x and y and only depends on the height z.

With this assumption, mass conservation $\nabla_z v_z = 0$ implies that the vertical velocity component is a constant. Since it has to vanish on the non-permeable boundary $z = 0$ it vanishes everywhere, $v_z = 0$. The flow in the transition layer is horizontal and independent of x and y, but varies with height z. Since the advective term vanishes, $(\boldsymbol{v} \cdot \boldsymbol{\nabla})\boldsymbol{v} = (v_x\nabla_x + v_y\nabla_y)\boldsymbol{v}(z) = \boldsymbol{0}$, the Navier–Stokes equation including the Coriolis force becomes,

$$0 = 2\Omega v_y - \nabla_x p^*/\rho_0 + \nu\nabla_z^2 v_x, \tag{20.23a}$$

$$0 = -2\Omega v_x - \nabla_y p^*/\rho_0 + \nu\nabla_z^2 v_y, \tag{20.23b}$$

$$0 = -\nabla_z p^*/\rho_0. \tag{20.23c}$$

From the last equation it follows that the effective pressure is independent of height z and consequently for all z equal to its value $p^* = -2\rho_0\Omega U y$ in the geostrophic flow outside the boundary layer. Inserting this result, the equations of motion now simplify to

$$\nu\nabla_z^2 v_x = -2\Omega v_y \qquad \nu\nabla_z^2 v_y = -2\Omega(U - v_x). \tag{20.24}$$

This is a pair of coupled homogenous differential equations for $U - v_x$ and v_y. From the first we get $v_y = -(\nu/2\Omega)\nabla_z^2 v_x$ and inserting this into the second, we get a single fourth-order equation

$$\nabla_z^4(U - v_x) = -\frac{4\Omega^2}{\nu^2}(U - v_x). \tag{20.25}$$

The general solution to such a linear fourth-order differential equation is a linear combination of four terms of the form e^{kz} where $k^4 = -4\Omega^2/\nu^2$. Defining

$$\boxed{\delta = \sqrt{\frac{\nu}{\Omega}},} \tag{20.26}$$

the roots of $k^4 = -4/\delta^4$ are $k = \pm(1 \pm i)/\delta$. The roots with positive real part are of no use, because the solution e^{kz} then grows exponentially for $z \to \infty$. The most general acceptable solution is then of the form,

$$U - v_x = Ae^{-(1+i)z/\delta} + Be^{-(1-i)z/\delta}, \tag{20.27}$$

$$v_y = i\left(Ae^{-(1+i)z/\delta} - Be^{-(1-i)z/\delta}\right), \tag{20.28}$$

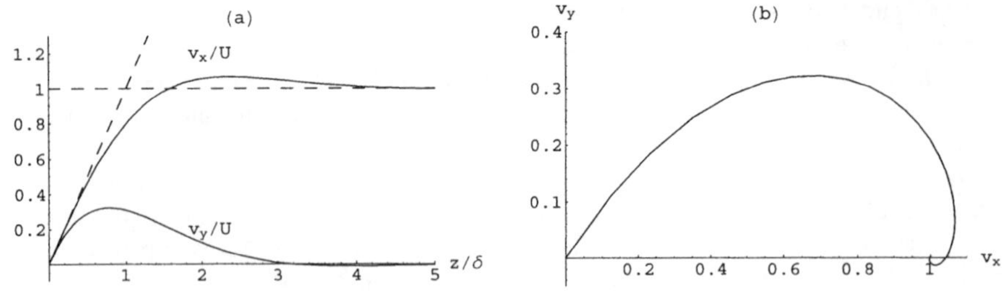

Figure 20.1. Plot of Ekman layer velocity components. **(a)** The velocity components as a function of height z. **(b)** Parametric plot of the velocities as a function of z leads to the characteristic Ekman spiral.

where A and B are integration constants and where the last expression is obtained from the first equation in (20.24).

Applying the no-slip boundary condition, $v_x = v_y = 0$ for $z = 0$, we find $A = B = U/2$, and the final solution becomes,

$$\boxed{\begin{aligned} v_x &= U\left(1 - e^{-z/\delta}\cos z/\delta\right), \\ v_y &= U e^{-z/\delta}\sin z/\delta. \end{aligned}} \tag{20.29}$$

Evidently δ is a measure of the thickness of the Ekman layer. Note that δ is independent of U, so the Ekman layer has the same thickness everywhere even if the velocity of the geostrophic flow should vary from place to place.

In figure 20.1*a* the velocity components are plotted as a function of scaled height z/δ in units of the asymptotic flow U. One notes that v_x first overshoots its asymptotic value, and then quickly returns to it. The y-component also oscillates but is 90° out of phase with the x-component. The direction of the velocity close to $z = 0$ is 45° to the left of the asymptotic geostrophic flow. Plotted parametrically as a function of height, the velocity components create a characteristic spiral, called the *Ekman spiral*, shown in figure 20.1*b*. The damping is however so strong that only the very first turn in this spiral is visible.

The presence of an Ekman layer of the right thickness has been amply confirmed by laboratory experiments. For the atmosphere at middle latitudes the thickness (20.26) becomes $\delta = 55$ cm when the diffusive viscosity $\nu = 1.54 \times 10^{-5}\ \mathrm{m^2\,s^{-1}}$ is used. This disagrees with the measured thickness of the Ekman layer in the atmosphere which is more like a kilometre. The reason is that atmospheric flow tends to be turbulent rather than laminar with an effective viscosity that can be up to a million times larger than the diffusive viscosity [55]. We shall not go further into this question here.

* Ekman upwelling and suction

If the geostrophic flow does not run along the x-direction, but has components U_x and U_y, the Ekman flow (20.29) becomes instead,

$$v_x = U_x\left(1 - e^{-z/\delta}\cos z/\delta\right) - U_y e^{-z/\delta}\sin z/\delta, \tag{20.30a}$$

$$v_y = U_y\left(1 - e^{-z/\delta}\cos z/\delta\right) + U_x e^{-z/\delta}\sin z/\delta. \tag{20.30b}$$

If furthermore the velocity components $U_x(x, y)$ and $U_y(x, y)$ change slowly with x and y on a large scale $L \gg \delta$, this expression should still be valid, because the thickness (20.26) is independent of the velocity of the asymptotic geostrophic flow.

The presence of an Ekman layer underneath a slowly varying geostrophic flow generates in fact a non-vanishing asymptotic flow $U_z(x, y)$ in the z-direction. To find it we calculate the derivative of the vertical flow component v_z inside the Ekman layer using mass conservation $\nabla \cdot \boldsymbol{v} = \nabla_x v_x + \nabla_y v_y + \nabla_z v_z = 0$, and we find from (20.30)

$$\nabla_z v_z = -\nabla_x v_x - \nabla_y v_y = \left(\nabla_x U_y - \nabla_y U_x\right) e^{-z/\delta}\sin z/\delta.$$

Here we have used that geostrophic flow also has to satisfy the divergence condition, $\nabla_x U_x + \nabla_y U_y = 0$. We recognize the factor in parenthesis on the right-hand side as the geostrophic vorticity in the z-direction. Integrating the above equation over z and using that $v_z = 0$ for $z = 0$, we obtain

$$v_z = \tfrac{1}{2}\delta(\nabla_x U_y - \nabla_y U_x)[1 - e^{-z/\delta}(\cos z/\delta + \sin z/\delta)], \tag{20.31}$$

an equation which is most easily verified by differentiation after z. Evidently, the vertical velocity is of size $U\delta/L$, which is always smaller than the geostrophic flow U by a factor $\delta/L \ll 1$.

For $z \gg \delta$ the exponential falls away, and there remains a vertical component in the asymptotic geostrophic flow

$$\boxed{U_z = \tfrac{1}{2}\delta(\nabla_x U_y - \nabla_y U_x).} \tag{20.32}$$

Since it is independent of z, it is not at variance with the geostrophic nature of the exterior flow or the Taylor–Proudman 'vertical copy' theorem.

If the geostrophic vorticity $\omega_z = \nabla_x U_y - \nabla_y U_x$ is positive, i.e. of the same sign as the global rotation, fluid wells up from the Ekman layer (without changing its thickness). This is, for example, the case for a low-pressure cyclone, where the cross-isobaric flow inside the Ekman layer towards the centre of the cyclone is accompanied by upwelling of fluid. Conversely if the geostrophic vorticity is negative, as in high-pressure anticyclones, fluid is sucked down into the Ekman layer from the geostrophic flow. Both of these effects tend to equalize the pressure between the centre and the surroundings of these vast vortices.

20.5 Steady bathtub vortex in rotating container

In the laboratory, gravity-sustained vortices may be created by letting a liquid, typically water, run freely out through a small drain-hole in the centre of a slowly rotating cylindrical container. The liquid lost through the drain is constantly pumped back into the container. In the steady state the pump provides the kinetic energy of the liquid falling out through the drain, and its angular momentum is provided by the motor rotating the container. The container is drained through a very small hole and we shall for simplicity disregard the influence of the outer container wall.

In the following we shall repeatedly refer to the experiment[1] shown in figure 20.2 with the parameters given in the figure caption. In this experiment, a steady flow pattern with a beautiful central vortex is established after about half an hour. The vortex is remarkably stable and its flow can be studied experimentally by modern imaging techniques. The needle-like central depression is accompanied by a very rapid central rotation of more than 150 turns per second, or nearly 10 000 rpm, which is as fast as a formula one car engine rotates at full throttle!

Experimentally, the bulk of the vortex outside the surface depression is found to take the shape of a *line vortex* with azimuthal velocity, $v_\phi = C/r$ in the rotating coordinate system of the container. The azimuthal Reynolds number of the line vortex is independent of r,

$$\mathrm{Re}_\phi = \frac{r v_\phi}{\nu} = \frac{C}{\nu}, \tag{20.33}$$

and in the experiment it has the value $\mathrm{Re}_\phi \approx 1600$, which is well below the onset of turbulence.

Rossby radius

The local Rossby number of the line vortex at a distance r from its axis is,

$$\mathrm{Ro} = \frac{v_\phi}{2\Omega r} = \frac{C}{2\Omega r^2}. \tag{20.34}$$

It decreases rapidly with growing r and drops below unity for $r \gtrsim R$ where

$$R = \sqrt{\frac{C}{2\Omega}}. \tag{20.35}$$

[1]A. Andersen, T. Bohr, B. Stenum, J. J. Rasmussen and B. Lautrup: Anatomy of a bathtub vortex, *Phys. Rev. Lett.* **91**, (2003) 103402-1.

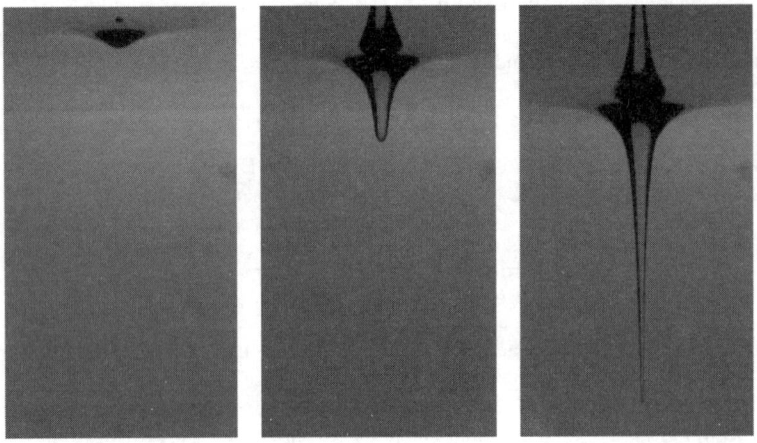

Figure 20.2. Water vortex in a rotating container at 6, 12 and 18 rpm. The upper part is the reflection of the vortex in the water surface. The radius of the drain-hole is 1 mm, the asymptotic water level is $L = 11$ cm and the container radius 20 cm. At 18 rpm the central dip is 6 cm, and the volume discharge through the drain is 3.16 $\text{cm}^3\ \text{s}^{-1}$, corresponding to an average drain velocity of 101 cm s^{-1}. The circulation constant is measured to be $C = 16.0\ \text{cm}^2\ \text{s}^{-1}$ and the inner core rotates about 150 turns per second! A. Anderson *et al*, *Phys. Rev. Lett.* **91** (2003) 103402-1.

We shall call this the *Rossby radius*, and in the experiment of figure 20.2 we find $R = 2.1$ cm. Well inside the Rossby radius for $r \ll R$, the Coriolis force is small compared to the advective force, and the vortex will resemble the bathtub-like vortices to be discussed in chapter 26, except that there is marked upflow from the Ekman layer at the bottom outside the drain. At the other extreme, well beyond the Rossby radius for $r \gg R$, the flow will be purely geostrophic.

Cylindrical geostrophic flow

The Taylor–Proudman 'copycat' theorem guarantees that in the geostrophic regime the flow cannot depend on the vertical height z well away from the boundaries. Assuming further that the flow is rotationally invariant, it follows that the velocity components in cylindrical coordinates, $v_{r,\phi,z}$, are only functions of r. The geostrophic equation (20.16) now decomposes into the three equations,

$$\nabla_r p^* = 2\rho_0 \Omega v_\phi, \qquad \nabla_\phi p^* = -2\rho_0 \Omega v_r, \qquad \nabla_z p^* = 0. \tag{20.36}$$

By the usual argument, the uniqueness of the pressure forbids any ϕ-dependence, so the second equation leads to $v_r = 0$. The effective pressure can then be obtained by integrating the two other equations. Note that whereas the radial flow always has to vanish, we find apparently no restrictions on the azimuthal flow v_ϕ or the upflow v_z. In particular, it does not follow from this argument that the azimuthal flow must take the form of a line vortex, $v_\phi = C/r$.

The main conclusion is that *geostrophic flow can never carry any inflow* towards the drain. This is, in fact, equivalent to the previously derived result that flow lines and isobars coincide in geostrophic flow. The only way inflow can occur is through deviations from clean geostrophic flow, and that happens primarily in the Ekman layer close to the bottom of the container, and inside the Rossby radius.

The Ekman layer valve

The asymptotic upflow from the Ekman layer (20.32) is controlled by the vorticity of the geostrophic flow. Since the Taylor–Proudman theorem guarantees that the upflow is independent of z, and since there can be no geostrophic inflow, this upflow will reach unabated the nearly horizontal open surface at top of the container. But there it has nowhere to go, so the only possibility that remains is for the upflow and thus the vorticity of the geostrophic flow to vanish, and this is only possible for $v_\phi = C/r$ (see problem 18.22). To complete the argument, one also has to verify that the Ekman layer which will also form at the upper open surface is incapable of diverting any large upflow towards the drain. Therefore it is essentially the finite vertical extent of the rotating fluid that forces the flow to be that of a line vortex!

The constant thickness of the Ekman layer $\delta = \sqrt{\nu/\Omega}$ is merely 0.7 mm in the experiment (figure 20.2(*c*)). The Ekman fields may be directly taken over from (20.29) with the asymptotic azimuthal velocity $U = C/r$,

$$v_\phi = \frac{C}{r}\left(1 - e^{-z/\delta}\cos z/\delta\right), \qquad v_r = -\frac{C}{r}e^{-z/\delta}\sin z/\delta. \qquad (20.37)$$

The total radial inflow rate can now be calculated from v_r,

$$Q = \int_0^\infty (-v_r) 2\pi r \, dz = \pi C \delta. \qquad (20.38)$$

This is a fundamental result which connects the primary circulating flow with the secondary inflow towards the drain. The Ekman layer in effect acts as a valve that only allows a certain amount of fluid to flow towards the drain per unit of time. In the experiment of figure 20.2 the circulation constant was measured to be $C \approx 16.0 \text{ cm}^2 \text{ s}^{-1}$, leading to a total predicted inflow of $Q \approx 3.66 \text{ cm}^3 \text{ s}^{-1}$. Since the whole inflow has to go down the drain, this is in fair agreement with the measured drain flow of $Q = 3.16 \text{ cm}^3 \text{ s}^{-1}$, especially in view of the difficulties in measuring the circulation constant. If at all significant, the discrepancy could be caused by the finite radial size of the container which has been ignored here.

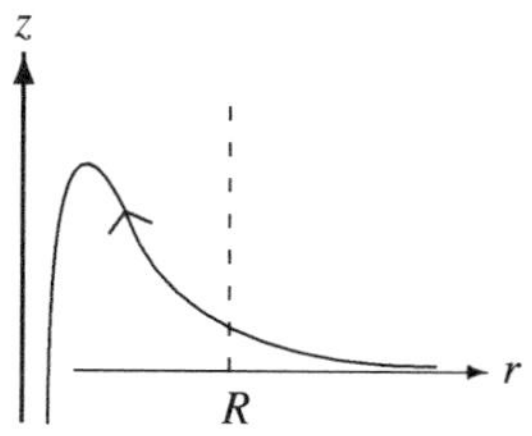

Typical flowline for fluid streaming in through the Ekman layer outside the Rossby radius R, welling up inside R and finally falling through the drain.

Inner vortex

Inside the Rossby radius the nonlinear advective forces take over, but there will still exist a thin—in fact thinner—Ekman-like layer close to the bottom[2]. But even if the circulating primary flow is that of a line vortex, the nonlinearities will now cause an upwelling of fluid from the bottom layer. The bulk flow is no more geostrophic, so there is no injunction against the upflow turning into an inflow directed towards the centre of the vortex. Thus, the general picture is that the secondary flow creeps inwards along the bottom through the Ekman layer outside the Rossby radius, flares up vertically from the bottom inside, and then turns towards the centre. Finally it dives sharply down into the drain.

20.6 Debunking an urban legend

Let us now turn to the 'urban legend' concerning the direction of rotation of real bathtub vortices and their dependence on the Earth's rotation. The legend originates in the correct physical theory of the Coriolis force, amply confirmed by the everyday observation of weather cyclones. So the urban legend can only be debunked by quantitative arguments, usually not given much attention in urban circles.

Suppose to begin with that our bathtub is essentially infinitely large and that the water level is $L = 50$ cm. Bernoulli's theorem tells us that the drain velocity is at most $W = \sqrt{2g_0L} \approx 300 \text{ cm s}^{-1}$. Taking the drain radius to be $a = 2.5$ cm, the maximal drain discharge rate becomes $Q \approx 6$ litres per second. This seems not unreasonable for bathtubs that typically contain hundreds of litres of water. Assuming furthermore that the bulk of the flow is perfectly laminar, we find the Ekman thickness $\delta \approx 14$ cm due to Earth's rotation at middle latitudes. From (20.38) we get the vortex circulation constant, $C \approx 140 \text{ cm}^2 \text{ s}^{-1}$, corresponding to a Rossby radius of $R = 12$ m. Most bathtubs are not that big and this shows that the Earth's rotation can only have a little influence on a real bathtub vortex, in spite of the many claims to the contrary. A swimming pool of Olympic dimensions is on the other hand of the right scale. What matters for the man-sized bathtub is much more the bather's accidental deposition of angular momentum in the water while getting out.

There is, however, the objection that the effect of the Earth's rotation could show up, if the water were left to settle down for some time before the plug is pulled. For that to happen, the Rossby number (20.14) would have to become comparable to unity. Taking the diameter of a real tub to be $A \approx 1$ m, this implies that the water velocity near the rim of the tub should not be larger than $2\Omega A \approx 0.1 \text{ mm s}^{-1}$. This seems terribly small, not much larger than the thickness of a human hair per second!

[2]This layer is analysed in A. Andersen, B. Lautrup and T. Bohr: Averaging method for nonlinear laminar Ekman layers, *J. Fluid. Mech.* **487**, (2003) 81.

The following argument indicates what patience is needed to carry out a successful experiment. After the initial turbulence from filling the tub has died out, the water settles into a laminar flow which is further slowed down under the action of viscous forces (although the Reynolds number for a flow with velocity 0.1 mm s^{-1} is still about 100). Viscosity not only smoothes out local velocity differences but also secures that the fluid eventually comes to rest with respect to the container. The typical viscous diffusion time over a distance L is $t \approx L^2/4\nu$, as we have seen for momentum diffusion on page 233. In a bathtub with a water level of $L = 50$ cm, the time it takes for the bottom of the container to influence the water at the top is about $t \approx 17$ h at middle latitudes! To make the experiment work, you must not only let the water settle for a few times 17 hours, but also secure that no heat is added to the water which may generate convection, and that no air drafts are present in and around the container[3].

Problems

20.1 A space station is built in the form of a wheel with diameter 100 m. **(a)** Calculate the revolution time necessary to obtain standard gravity at the outer rim. **(b)** How and how much does the Coriolis force influence the game of Ping Pong? **(c)** What about basket ball?

20.2 The Danish Great Belt is a strait with a width of 20 km and a typical current velocity of 1 m s^{-1}. There are two layers of water, a lower and slower saline layer with a lighter and faster brackish layer on top. Assuming a density difference of about 4% and a velocity difference of about 25%, calculate the difference in water levels for the separation surface between saline and brackish water.

* **20.3** In an inertial Cartesian system the coordinates of a point are denoted $\boldsymbol{x}'$ whereas in an arbitrarily moving Cartesian system the coordinates of the same point are denoted $\boldsymbol{x}$. The relation between the coordinates may be written

$$\boldsymbol{x} = -\boldsymbol{c} + \sum_i \boldsymbol{a}_i x'_i, \tag{20.39}$$

where $-\boldsymbol{c}(t)$ are the coordinates of the origin of the inertial system and $\boldsymbol{a}_i(t)$ its Cartesian basis (all seen from the moving system). **(a)** Show that there exists a vector $\boldsymbol{\Omega}(t)$ such that the time derivative of the basis vectors takes the form (the choice of sign is for later convenience),

$$\dot{\boldsymbol{a}}_i = -\boldsymbol{\Omega} \times \boldsymbol{a}_i . \tag{20.40}$$

(b) Calculate the velocity and acceleration in the moving coordinate system.
(c) Show that Newton's second law in the moving system becomes

$$m\ddot{\boldsymbol{x}} = \boldsymbol{f} - m\ddot{\boldsymbol{c}} - m\boldsymbol{\Omega} \times \dot{\boldsymbol{c}} - m\dot{\boldsymbol{\Omega}} \times \boldsymbol{x} - 2m\boldsymbol{\Omega} \times \dot{\boldsymbol{x}} - m\boldsymbol{\Omega} \times (\boldsymbol{\Omega} \times \boldsymbol{x}) \tag{20.41}$$

where $\dot{\boldsymbol{c}}$ and $\ddot{\boldsymbol{c}}$ are the velocity and acceleration of the origin of the moving coordinate system (seen from the moving system). The three first fictitious force terms vanish for a coordinate system in steady rotation.

[3]Some experimenters *are* patient and careful enough to observe the effect. See, for example, the very enjoyable paper by L. M. Trefethen, R. W. Bilger, P. T. Fink, R. E. Luxton and R. I. Tanner, The bath-tub vortex in the southern hemisphere, *Nature* **207**, (1965) 1084-5.

21 Computational fluid dynamics

Computational fluid dynamics (CFD) is a whole field in its own right [29, 2, 73]. Swift modern computers have to a large extent replaced wind tunnels and wave tanks for the design of airplanes, ships, cars, bridges and in fact any human construction that is meant to operate in a fluid. The same richness of phenomena which makes analytic solutions to the equations of fluid mechanics difficult to obtain also makes these equations hard to handle by direct numerical methods. Secondary flows, instabilities, vortices of all sizes and turbulence complicate matters and may require numerical precision that can be hard to attain. The infinite speed of sound in incompressible fluids creates its own problems, and on top of that there are intrinsic approximation errors and instabilities.

As in computational elastostatics (chapter 13) there are a number of steps that must be carried out in any simulation. First, it is necessary to clarify which equations one wishes to solve and even there make simplifications to the problem or class of problems at hand. Secondly, continuous space must be discretized, and here there are a variety of methods based on finite differences, finite elements or finite volumes. Thirdly, a discrete dynamic process must be set up which guides the initial field configuration towards the desired solution. Most often this process emulates the time evolution of fluid dynamics itself, as described by the Navier–Stokes equations. Finally, convergence criteria and error estimates are needed to monitor and gain confidence in the numerical solutions.

The methods presented in this chapter are applicable to a variety of steady and unsteady flow problems. Here we shall only compute two-dimensional laminar flow in a channel of finite length between parallel plates and determine how it turns into the well-known parabolic Poiseuille profile downstream from the entrance, and how far the influence of the entrance reaches.

21.1 Unsteady, incompressible flow

In numeric elastostatics (chapter 13) we were able to set up an artificial dissipative dynamic, called gradient descent, that guided the displacement field towards a static solution with minimal elastic energy. This technique cannot be transferred to computational fluid dynamics, because a solution to the steady-state equations does not correspond to an extremum of any bounded quantity (problem 21.1). Instead we shall attempt to copy nature by simulating the complete set of time-dependent Navier–Stokes equations (17.27). Appealing to the behaviour of real fluids, the natural viscous dissipation built into these equations should hopefully guide the velocity field towards a steady-state solution.

There is, however, no guarantee—either from Nature or from the equations—that the flow will always settle down and become steady, even when the boundary conditions are time-independent. We are all too familiar with the unsteady and sometimes turbulent flow that may spontaneously arise under even the steadiest of circumstances, as for example a slow river narrowing down or even worse, coming to a water fall. But this is actually not so bad, because a forced steady-state solution is completely uninteresting when the real fluid refuses to end up up in that state. We do not, for example, care much for the Poiseuille solution (18.26) to pipe flow at a Reynolds number beyond the transition to turbulence, or for that matter the laminar flow around a sphere in an ideal fluid (section 16.7). If steady flow is desired, one must keep the Reynolds number so low that there is a chance for it to become established.

Field equations

Not to complicate matters we shall only consider incompressible fluids with constant density $\rho = \rho_0$, for which the divergence vanishes at all times,

$$\nabla \cdot \boldsymbol{v} = 0. \tag{21.1}$$

The Navier–Stokes equation (17.16) is written as an equation of motion for the velocity field,

$$\frac{\partial \boldsymbol{v}}{\partial t} = \boldsymbol{F} - \nabla \widetilde{p}, \tag{21.2}$$

with

$$\boldsymbol{F} = -(\boldsymbol{v} \cdot \nabla)\boldsymbol{v} + \nu \nabla^2 \boldsymbol{v} + \boldsymbol{g}, \tag{21.3}$$

where $\nu = \eta/\rho_0$ is the kinematic viscosity, and $\widetilde{p} = p/\rho_0$ might be called the 'kinematic pressure'. For convenience we have also introduced the special symbol $\boldsymbol{F}$ to denote the local acceleration arising from inertia, viscosity and gravity, but not pressure. The gravitational field could in principle be included in an effective pressure $p^* = p + \rho_0 \Phi$, although that would obscure the boundary conditions we might want to impose on the real pressure. In the following we shall simply assume that the field of gravity is constant.

Poisson equation for pressure

In incompressible flow the pressure is determined indirectly through the vanishing of the divergence of the velocity field, as discussed on page 237. Calculating the divergence of both sides of (21.2), we find a Poisson equation for $\widetilde{p}$,

$$\nabla^2 \widetilde{p} = \nabla \cdot \boldsymbol{F}. \tag{21.4}$$

This is the condition which must be fulfilled in order that the velocity field remains free of divergence at all times. Knowing the velocity field $\boldsymbol{v}$ at a given time we can calculate the right-hand side and solve this equation with suitable boundary conditions to determine the pressure everywhere in the fluid at that particular instant of time.

Solutions to the Poisson equation are, however, non-local functions of the source, basically of the same form as the gravitational potential (3.24). Any local change in the velocity field at a point $\boldsymbol{x}'$ is via the Poisson equation above *instantaneously* communicated to the pressure at any other point $\boldsymbol{x}$ in the fluid, albeit damped by the $1/\left|\boldsymbol{x} - \boldsymbol{x}'\right|$ dependence on distance. The non-local changes in pressure are then communicated back to the velocity field via the Navier–Stokes equation (21.2). The pressure thus links the velocity field at any instant of time non-locally to its immediately preceding values, even for infinitesimally small time intervals. Pressure instantly informs the world at large about the present state of the incompressible velocity field.

Physically, unpleasant non-local behaviour is caused by the assumption of absolute incompressibility (21.1), which is just as untenable in the real world of local interactions as absolute rigidity. So, we have again come up against a physical limit arising from our simplifying assumptions. As pointed out before, incompressibility should be viewed as a property of the flow rather than of the fluid itself. The unavoidable compressibility of real matter will in fact limit the rate at which pressure changes can propagate through a fluid to the speed of sound. Nevertheless, such a conclusion does not detract from the practical usefulness of the divergence condition (21.1) for 'normal' flow speeds well below the speed of sound.

Boundary conditions

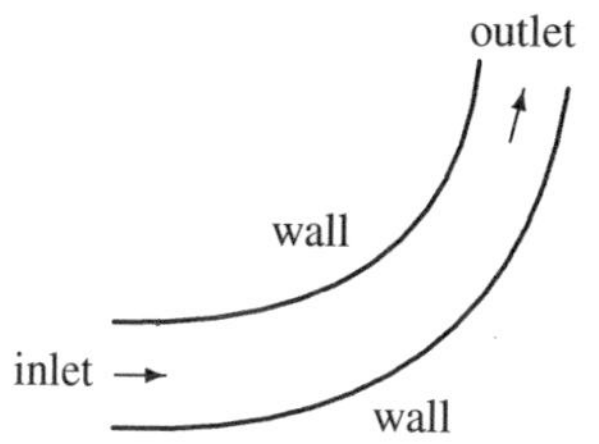

In many fluid dynamics problems, fixed impermeable walls guide the fluid between openings where it enters and leaves the system. At fixed walls the velocity field must vanish at all times, because of the impermeability and no-slip conditions which respectively require the normal and tangential components to vanish. Setting $\boldsymbol{v} = \boldsymbol{0}$ on the left-hand side of the equation of motion (21.2) we obtain a boundary condition for pressure: $\nabla\widetilde{p} = \boldsymbol{F}$ on any fixed boundary. The same is the case at a fluid inlet, where the velocity field is typically fixed to an externally defined constant value, $\boldsymbol{v} = \boldsymbol{U}$. Outlet velocities are usually not controlled externally, and as a boundary condition on the velocity field one may choose the vanishing of the normal derivative, $(\boldsymbol{n}\cdot\nabla)\boldsymbol{v} = \boldsymbol{0}$. Alternatively, the stress vector may be required to vanish, but that is a bit harder to implement.

21.2 Temporal discretization

Suppose the current velocity field is $\boldsymbol{v}(\boldsymbol{x}, t)$ and the current pressure field $\widetilde{p}(\boldsymbol{x}, t)$. From these fields we can calculate the current acceleration field $\boldsymbol{F}(\boldsymbol{x}, t)$ and then use the equation of motion (21.2) to move the velocity field forward in time through a small but finite time step Δt,

$$\boldsymbol{v}(\boldsymbol{x}, t+\Delta t) = \boldsymbol{v}(\boldsymbol{x}, t) + \big(\boldsymbol{F}(\boldsymbol{x}, t) - \nabla\widetilde{p}(\boldsymbol{x}, t)\big)\Delta t. \tag{21.5}$$

Taylor expansion of the left-hand side shows that the error is $\mathcal{O}\left(\Delta t^2\right)$, but less error-prone higher-order schemes are also possible [2, 29]. Provided the velocity field is free of divergence and the pressure satisfies (21.4), the new velocity field obtained from this equation will also be free of divergence.

Divergence suppression

But approximation errors cannot be avoided in any finite step algorithm. Since for this reason the current velocity field may not be perfectly free of divergence, it is more appropriate to demand that the new velocity field is free of divergence, i.e. $\nabla\cdot\boldsymbol{v}(\boldsymbol{x}, t+\Delta t) = \boldsymbol{0}$. Calculating the divergence of both sides of (21.5) we obtain a modified Poisson equation for the pressure

$$\nabla^2\widetilde{p} = \nabla\cdot\boldsymbol{F} + \frac{\nabla\cdot\boldsymbol{v}}{\Delta t}. \tag{21.6}$$

The factor $1/\Delta t$ amplifies the divergence errors, so that this Poisson equation will primarily be concerned with correction of divergence errors and only when they have been suppressed will the acceleration field $\boldsymbol{F}$ gain influence on the pressure. In practice, the stepping algorithm (21.5) can get into trouble if the divergence becomes too large. It is for this reason important to secure that the initial velocity field is reasonably free of divergence. In a complicated flow geometry that can in fact be quite hard to attain.

Stability conditions

There are essentially only two possibilities for what can happen to the approximation errors in the course of many time steps. Either the errors will become systematically larger, in which case the computation goes straight to the land of meaningless results, or the errors will diminish or at least stay constant and 'small', thereby keeping the computation on track. It takes careful mathematical analysis to determine a precise value for the upper limit to the size of the time step. The result depends strongly on both the spatial and temporal discretization (see for example [2, 59]), and may range from zero to infinity depending on the particular algorithm that is implemented.

Here we shall present an intuitive argument for the stability conditions that apply to the straightforward numerical simulation (21.5) on a spatial grid with typical coordinate spacings Δx, Δy and Δz. The conditions to be derived may be understood from the physical processes that compete in displacing fluid particles in, say, the x-direction. One is momentum diffusion due to viscosity which effectively displaces the particle by $\sqrt{\nu\Delta t}$ in a time interval Δt (see page 233). Another is advection with velocity v_x which displaces the particle a distance $|v_x|\,\Delta t$. Finally, there is the gravitational field which typically displaces the particle by $g_x \Delta t^2$. Intuitively, it seems reasonable to demand that all these displacements be smaller than the grid spacing,

$$\sqrt{\nu \Delta t} \lesssim \Delta x, \quad |v_x| \, \Delta t \lesssim \Delta x, \quad g_x \Delta t^2 \lesssim \Delta x. \tag{21.7}$$

Taking into account that the maximal velocity provides the most stringent advective condition, the global condition may be taken to be

$$\Delta t \lesssim \min \left(\frac{\Delta x^2}{\nu}, \frac{\Delta x}{|v_x|_{\max}}, \sqrt{\frac{\Delta x}{g_x}} \right), \tag{21.8}$$

and similarly for the other coordinate directions. In practice trial-and-error may be used to determine where in the neighbourhood of the smallest of these limits instability actually sets in.

The diffusive condition is most restrictive for large viscosity, the advective for high velocities, and the gravitational in strong gravity. The allowed time step is generally largest for the coarsest spatial grid, which on the other hand is blind to finer details of the flow. Thus, there is a payoff between the detail desired in the simulation and its rate of progress. High detail entails slow progress, and thus a high cost in computer time.

21.3 Spatial discretization

In the discussion of numeric elastostatics we described a method based on finite differences with errors of only second order in the grid spacings (section 13.2 on page 166). Although it is possible to solve simple flow problems using this method, most such problems will benefit from a somewhat more sophisticated treatment. The method of *staggered grids* to be presented here comes at essentially no cost in computer memory or time, but does complicate matters a bit. A number of applications of this method are given in [29].

Restriction to two dimensions

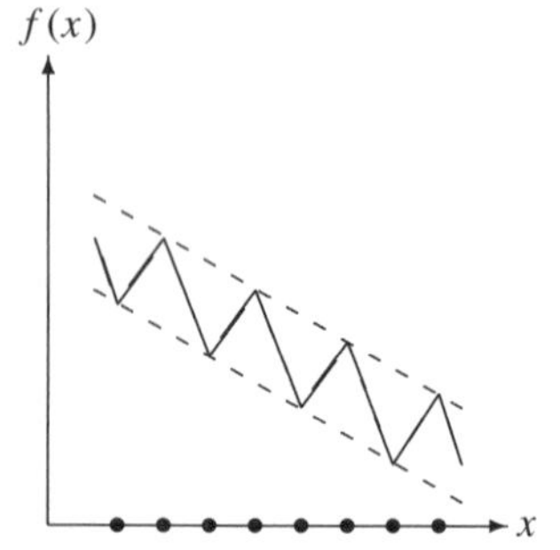

The solid zigzag-curve $f(x)$ between the two straight lines has constant central difference (13.5) on the grid! Such zigzag behaviour is typical of the 'leapfrog' errors that may arise when using the naive central differencing scheme of section 13.2.

For simplicity we shall limit the following discussion to two-dimensional flow in the xy-plane, as exemplified by the flow in a channel between parallel plates (section 18.2 on page 246). The restriction to two dimensions still leaves ample room for interesting applications. Generalization to three dimensions is straightforward.

Two-dimensionality is taken to mean that the fields can only depend on x and y and that $v_z = 0$. The equations of motion now simplify to

$$\frac{\partial v_x}{\partial t} = F_x - \nabla_x \widetilde{p}, \qquad \frac{\partial v_y}{\partial t} = F_y - \nabla_y \widetilde{p}, \tag{21.9}$$

with

$$F_x = -v_x \nabla_x v_x - v_y \nabla_y v_x + \nu \left(\nabla_x^2 + \nabla_y^2 \right) v_x + g_x, \tag{21.10a}$$

$$F_y = -v_x \nabla_x v_y - v_y \nabla_y v_y + \nu \left(\nabla_x^2 + \nabla_y^2 \right) v_y + g_y, \tag{21.10b}$$

and the divergence condition becomes

$$\nabla_x v_x + \nabla_y v_y = 0. \tag{21.11}$$

The stresses may of course be calculated, but are not as important here as in numeric elastostatics, because boundary conditions are most often specified directly in terms of the velocities.

Midpoint differences

The main objection to the central difference $\widehat{\nabla}_x f(x)$ defined in (13.5) is that it spans twice the fixed interval Δx around the central point x to which it 'belongs'. As was remarked there, this opens for 'leap-frog' or 'flip-flop' instabilities in which neighbouring grid points behave quite differently. The problem becomes particularly acute in the interplay between the equations of motion and the instantaneous Poisson equation for pressure.

One way out is to recognize that the difference in field values between two neighbouring grid-points properly 'belongs' to the midpoint of the line that connects them. If we denote the coordinates of the points by $x \pm (1/2)\Delta x$ the central difference around the midpoint x,

$$\widehat{\nabla}_x f(x) = \frac{f(x + (1/2)\Delta x) - f(x - (1/2)\Delta x)}{\Delta x}, \tag{21.12}$$

has errors of only second order in the grid spacing. This is what is really meant by saying that the difference 'belongs' to the midpoint.

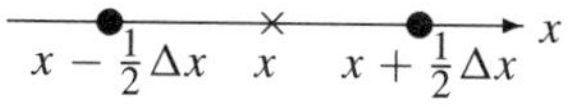

The finite difference between the grid points $x \pm (1/2)\Delta x$ belongs to the point x.

But the midpoints between grid points are not themselves part of the grid. We could of course double the grid and use $(1/2)\Delta x$ as grid-spacing, thereby including the midpoints, but that would just bring us back to the situation we started out to correct. Instead we shall think of the midpoint values of fields and their derivatives as being *virtual*, meaning that they may arise during a calculation but are not retained as part of the information kept about the fields on the grid.

Staggered grids

To see how this works out, let us discretize the divergence condition (21.11) writing it in terms of midpoint differences in the point (x, y)

$$\widehat{\nabla}_x v_x(x, y) + \widehat{\nabla}_y v_y(x, y) = 0. \tag{21.13}$$

Since (x, y) has to be the common midpoint for both coordinate directions, we should have direct access to v_x in $(x \pm (1/2)\Delta x, y)$ and v_y in $(x, y \pm (1/2)\Delta y)$, but not necessarily in (x, y). Repeating this argument throughout space (see figure 21.1), we conclude that the grids for the fields v_x and v_y do not overlap anywhere, but are systematically shifted, or *staggered*, with respect to each other.

In effect we *have* doubled the grid in both spatial directions, but the new grid is viewed as composed of four interlaced grids of the original type, each carrying some of the fields and their derivatives. Systematically, the coordinates of the four grids may be written

$$x = x_0 + i_x \Delta x + j_x \frac{1}{2} \Delta x \tag{21.14a}$$

$$y = y_0 + i_y \Delta y + j_y \frac{1}{2} \Delta y \tag{21.14b}$$

where i_x, i_y are integers and j_x, j_y are binary, taking only the values 0 or 1. The grid of common midpoints used in calculating the divergence is arbitrarily chosen to be $j_x = j_y = 0$, so that we may denote the four grids by 00, 10, 01, and 11 (marked with different symbols in figure 21.1). Thus, v_x is defined on the 10-grid and v_y on the 01-grid. We shall see below that the pressure p naturally belongs to the 00-grid, whereas there is no fundamental field associated with the 11-grid, only derivatives of the fields (see however problem 21.3).

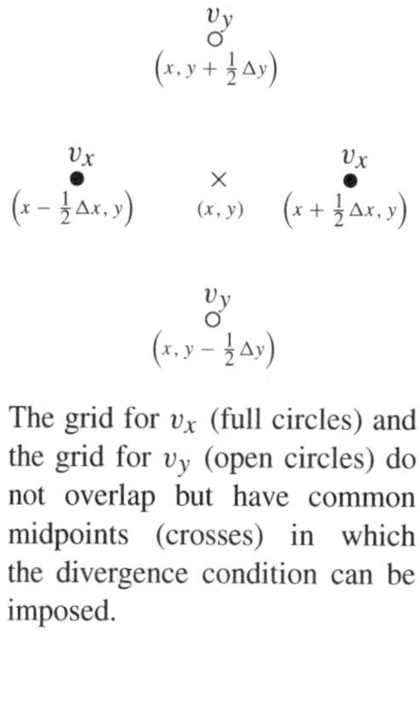

The grid for v_x (full circles) and the grid for v_y (open circles) do not overlap but have common midpoints (crosses) in which the divergence condition can be imposed.

The four staggered grids create a *tiling* of the plane with rectangular cells numbered by i_x and i_y, each cell containing four grid points numbered by j_x and j_y. Generalization of this scheme to three dimensions is straightforward, though harder to visualize. In three dimensions there will be eight staggered grids characterized by three integers and three binary variables.

Double differences

A double derivative, say ∇_x^2, is particularly simple when represented by midpoint differences. Combining the two levels of midpoint differencing we obtain (see figure 21.2)

$$\widehat{\nabla}_x^2 f(x) = \frac{f(x + \Delta x) + f(x - \Delta x) - 2f(x)}{\Delta x^2}, \tag{21.15}$$

and because of the symmetry in Δx the errors are of second order only. Geometrically the two levels of midpoint differences bring the double derivative back to the point it 'originated in', so that both $\widehat{\nabla}_x^2 v_x$ and $\widehat{\nabla}_y^2 v_x$ belong to the same grid as v_x (i.e. the 10-grid).

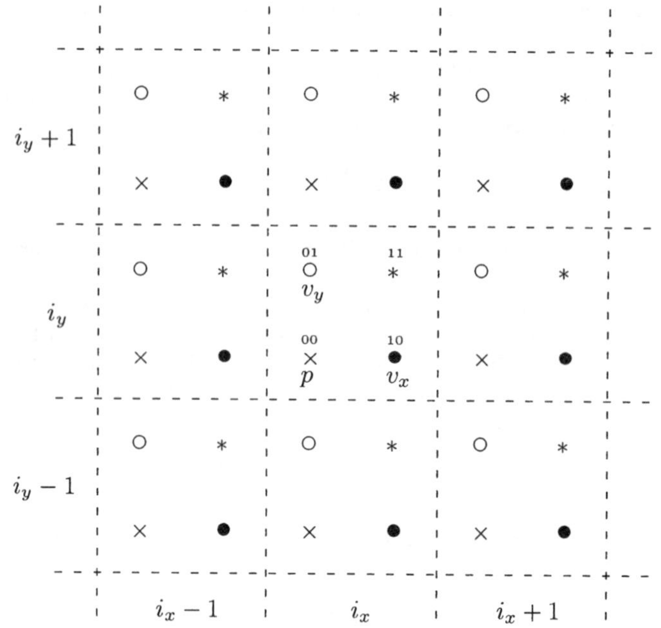

Figure 21.1. Staggered grids. Four rectangular grids with uniform coordinate spacings Δx and Δy are shifted with respect to each other by half intervals. Three of the grids are naturally associated with the fundamental fields: the 00-grid (crosses) carries p, the 01-grid (full circles) carries v_x and the 10-grid (open circles) carries v_y. The 11-grid (asterisks) carries only derivatives of the fundamental fields. The whole grid may be viewed as as a tiling of the plane with congruent cells of size $\Delta x \times \Delta y$, numbered by integers i_x and i_y, as shown.

Discretized equations of motion

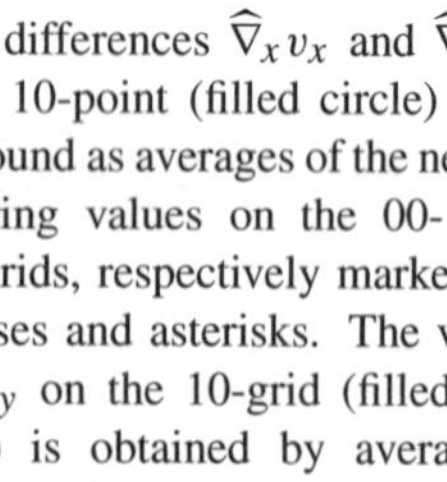

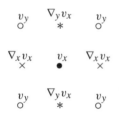

The differences $\widehat{\nabla}_x v_x$ and $\widehat{\nabla}_y v_x$ in a 10-point (filled circle) may be found as averages of the neighbouring values on the 00- and 11-grids, respectively marked by crosses and asterisks. The value of v_y on the 10-grid (filled circles) is obtained by averaging over all four nearest neighbours on the 01-grid (open circle).

The local acceleration fields F_x and F_y should be discretized on the same grids as v_x and v_y (i.e. on 10 and 01) for the equations of motion (21.9) to be fulfilled with fields and derivatives calculated in the same point.

The double derivatives in the viscous terms present in this respect no problems, and neither does the gravitational acceleration nor the pressure gradient which automatically ends up on the right grids. In the advective term, $-v_x \nabla_x v_x$, there is the problem that $\widehat{\nabla}_x v_x$ belongs to the 00-grid and not to the 10-grid as we would like it to. In order to keep errors to second order one must form the average of $\widehat{\nabla}_x v_x$ over the two neighbouring 00-values. Similarly, in the advective term, $-v_y \nabla_y v_x$, the derivative $\widehat{\nabla}_y v_x$ is naturally found on the 11-grid, it must also be averaged over the two nearest neighbours to get its 10-value. The worst case is v_y for which the value on the 10-grid is obtained as the average over the four nearest neighbours on the 01-grid.

Marking the averaged quantities with brackets we may write the discretized acceleration fields in the form (see problem 21.4 for the explicit expressions)

$$F_x = -v_x \left\langle \widehat{\nabla}_x v_x \right\rangle - \left\langle v_y \right\rangle \left\langle \widehat{\nabla}_y v_x \right\rangle + \nu \left(\widehat{\nabla}_x^2 + \widehat{\nabla}_y^2 \right) v_x + g_x, \tag{21.16a}$$

$$F_y = -\left\langle v_x \right\rangle \left\langle \widehat{\nabla}_x v_y \right\rangle - v_y \left\langle \widehat{\nabla}_y v_y \right\rangle + \nu \left(\widehat{\nabla}_x^2 + \widehat{\nabla}_y^2 \right) v_y + g_y. \tag{21.16b}$$

It should be noted that there is more than one way of calculating the advective terms, even if errors are required to be of second order only.

Finally, the discretized equations of motion (21.5) become

$$v_x(x, y, t + \Delta t) = v_x(x, y, t) + \left(F_x(x, y, t) - \widehat{\nabla}_x \widetilde{p}(x, y, t) \right) \Delta t, \tag{21.17a}$$

$$v_y(x, y, t + \Delta t) = v_y(x, y, t) + \left(F_y(x, y, t) - \widehat{\nabla}_y \widetilde{p}(x, y, t) \right) \Delta t, \tag{21.17b}$$

to be evaluated on the 10- and 01-grids, respectively.

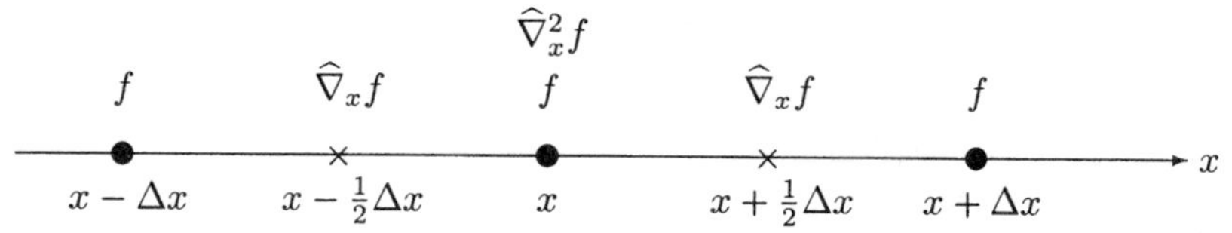

Figure 21.2. A double derivative is represented by the midpoint difference of the two neighbouring single derivatives, themselves represented by midpoint differences. Conveniently, it ends up on the same grid as the field itself.

Solving the discrete Poisson equation

During an iteration cycle the accelerations (21.16) are calculated from the current values of the discrete fields at time t, and afterwards the corresponding pressure at time t is calculated by solving the discretized version of the Poisson equation (21.6),

$$\left(\widehat{\nabla}_x^2 + \widehat{\nabla}_y^2\right) \widetilde{p} = \widehat{\nabla}_x F_x + \widehat{\nabla}_y F_y + \frac{\widehat{\nabla}_x v_x + \widehat{\nabla}_y v_y}{\Delta t}. \tag{21.18}$$

The solution to this equation may be found by means of relaxation methods, for example gradient descent (see section 13.1 on page 165), in which the pressure $\widetilde{p}(x, y, t)$ undergoes successive changes of the form

$$\delta \widetilde{p} = \epsilon \left(\left(\widehat{\nabla}_x^2 + \widehat{\nabla}_y^2\right) \widetilde{p} - s\right), \tag{21.19}$$

where $\epsilon > 0$ is the step size and $s(x, y, t)$ is the source (the right-hand side of (21.18)). The relaxation algorithm converges towards a solution to $\nabla^2 \widetilde{p} = s$ for sufficiently small ϵ because it descends along the steepest downwards gradient towards the unique minimum of a quadratic 'energy'-function (see problem 21.2).

Denoting the nth approximation to the solution by $\widetilde{p}_n(x, y, t)$, the discretized relaxation process may be written explicitly as

$$\widetilde{p}_{n+1}(x, y, t) = \widetilde{p}_n(x, y, t) + \epsilon \left(\left(\widehat{\nabla}_x^2 + \widehat{\nabla}_y^2\right) \widetilde{p}_n(x, y, t) - s(x, y, t)\right). \tag{21.20}$$

Starting with some field configuration $\widetilde{p}_0(x, y, t)$ and imposing boundary conditions after each step, this process will eventually lead to the desired solution, $\widetilde{p}(x, y, t)$. The problem is, however, that simple gradient descent is slow, too slow in fact to be applied inside every time step. *Conjugate gradient descent* [59, p. 420] offers considerable speed-up by calculating the optimal step-size directly, but as it turns out there are still faster methods.

From the double difference operator (21.15) we see that the coefficient of $\widetilde{p}_n$ in (21.20) is $1 - 2\epsilon(1/\Delta x^2 + 1/\Delta y^2)$, and this suggests the following reparametrization of the step-size

$$\epsilon = \frac{\omega}{2}\left(\frac{1}{\Delta x^2} + \frac{1}{\Delta y^2}\right)^{-1}, \tag{21.21}$$

where ω is the dimensionless *convergence parameter*. This choice allows a precise definition of what is meant by *underrelaxation* ($\omega < 1$) and *overrelaxation* ($\omega > 1$). Straightforward gradient descent, in which the new field ($\widetilde{p}_{n+1}$) is calculated all over the grid before replacing the old ($\widetilde{p}_n$), only converges when underrelaxed.

In *successive overrelaxation* or SOR (see [2, p. 231] and [29, p. 35]), the new value $\widetilde{p}_{n+1}(x, y, t)$ at a grid point replaces the old value $\widetilde{p}_n(x, y, t)$ as soon as it is calculated during a sweep of the grid. The method converges for $1 < \omega < 2$ and in practice the best value for ω may be located by trial-and-error, usually not far below the upper limit, say $\omega = 1.7$–1.9. Since this algorithm sweeps sequentially through the grid, one should be aware that it may create small asymmetry errors in an otherwise symmetric situation. But fast it is, on small grids often converging in just a few iterations, after the initial phase has passed.

21.4 Channel entrance flow

A simple and—from the look of it—well-behaved problem concerns the steady entrance flow pattern in a channel between two parallel plates with gap width d. Directly at the entrance, the flow is thought to be uniform with a flat velocity distribution which downstream smoothly turns into the characteristic parabolic Poiseuille shape. The typical distance for this to happen is the so-called *entrance length*, L'.

Momentum diffusion due to viscosity (page 233) tends to iron out all velocity gradients unless they are maintained by external forces. Near the entrance to the channel where the fluid speed is uniformly U, the velocity gradients and stresses on the sides are very large but soften progressively as the fluid moves downstream. The action of viscosity slows the fluid down near the sides of the channel, and since the volume flux must be the same throughout the channel, it has to speed up in the centre.

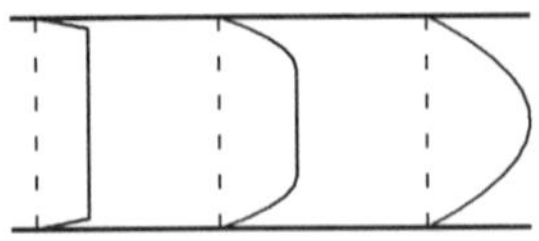

Sketch of the expected shape of the velocity profile at various distances downstream from the entrance.

Estimates

We have previously seen (page 233) that the diffusive momentum spread in a time interval t has a typical 95% range of $\delta \approx 3.5\sqrt{\nu t}$ and reaches the middle of the channel from both sides for $3.5\sqrt{\nu t} \approx d/2$, or $t \approx 0.02\, d^2/\nu$. Multiplying by the velocity U we obtain an estimate for the entrance length, $L' \approx 0.02\, U d^2/\nu$, and introducing the Reynolds number $\text{Re} = Ud/\nu$, we get

$$\frac{L'}{d} \approx k\,\text{Re}, \tag{21.22}$$

with $k \approx 0.02$. For $\text{Re} \approx 100$ which is squarely in the laminar region, the entrance length is thus estimated to be about twice the channel width, and about 40 times the channel width for $\text{Re} \approx 2000$. In the turbulent regime the entrance length on the contrary decreases with growing Reynolds number. The influence of the entrance is always expected to be notable at least for a length of the same size as the channel width d. For smaller values of the Reynolds number, say for $k\text{Re} \lesssim 1$, we consequently expect that L'/d becomes a fixed constant, independent of the Reynolds number.

In the present numerical simulation we shall attempt to verify the linear growth with Reynolds number in the laminar regime and determine the magnitude of the coefficient k as well as the constant value of L'/d for $k\,\text{Re} \lesssim 1$.

Boundary conditions

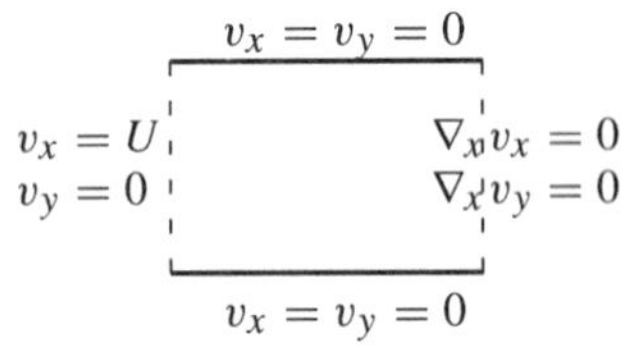

Boundary conditions for channel flow.

The channel is two-dimensional with the fluid coming in from the 'west' and leaving to the 'east' with the plates to the 'south' and 'north'. At the western entrance to the channel, $x = 0$, the velocity field is uniform $v_x = U$ with no cross flow, $v_y = 0$. The exit flow at $x = L$ is determined by the dynamics, and we shall just assume that the flow has stabilized in this region with longitudinally constant velocities, $\nabla_x v_x = \nabla_x v_y = 0$, as in ideal Poiseuille flow. On the impermeable southern and northern walls of the channel we must of course have $v_y = 0$, together with the no-slip condition, $v_x = 0$. In velocity-driven flow, the boundary conditions on pressure follow from the velocity conditions, and will be discussed below.

Initial data

The equations of motion must be supplied with initial data that fulfill the spatial boundary conditions and the condition of vanishing divergence. This is not nearly as simple as it sounds, even if there is great freedom in the choice of initial data and even if the final steady state is supposed to be independent of this choice. The problem becomes particularly acute if the boundaries are of irregular shape which they will always be in any realistic flow problem.

Here we shall choose the initial velocity and pressure fields (at $t = 0$) to be

$$v_x = U, \quad v_y = 0, \quad \widetilde{p} = 0 \tag{21.23}$$

everywhere inside the channel. This certainly fulfills the divergence condition, but has a discontinuous jump on the sides of the channel due to the no-slip boundary conditions. The initial fields also fulfill the Poisson equation for pressure (21.6).

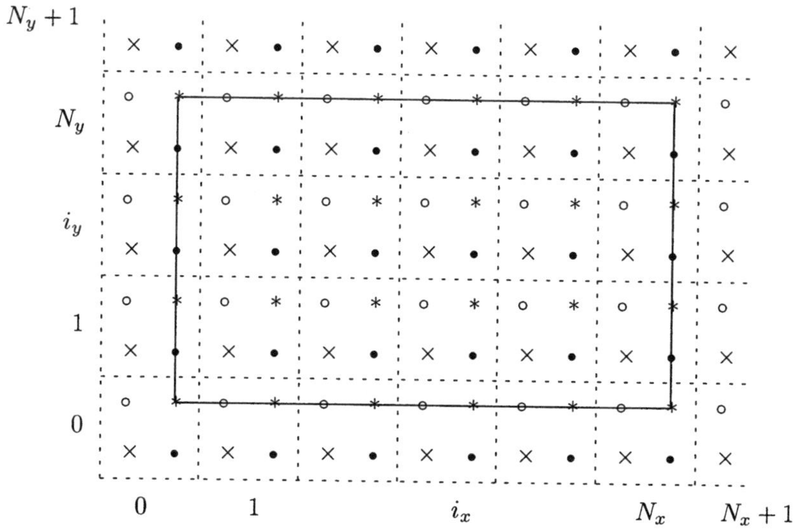

Figure 21.3. A rectangular region (solid line) is discretized using staggered grids, here with $N_x = 5$ and $N_y = 3$. Boundary conditions necessitate one layer of data outside the region on all sides, requiring (7×5) basic cells. The velocity v_x is defined on the 10-points (filled circles), v_y on the 01-points (open circles) and the pressure $\widetilde{p}$ on the 00-points (crosses).

The grid

The rectangular region of size $L \times d$ is discretized using staggered grids with coordinate intervals $\Delta x = L/N_x$ and $\Delta y = d/N_y$, where N_x and N_y are integers. This requires an array of $(N_x+2) \times (N_y+2)$ basic cells to cover the region of interest. In a computer program, any field value belonging to a basic cell is represented by an array, $f(x, y) \leftrightarrow f[i_x, i_y]$, indexed by the cell indices. At the eastern and northern borders only half the cell data is needed, as indicated in figure 21.3.

The discrete boundary conditions have to reflect that not all field values are known precisely on the border. At the entrance, the v_x-condition remains $v_x = U$ because this field is known on the border, but the v_y-condition must be implemented as $\langle v_y \rangle = 0$, where the average is over the nearest points on both sides of the border. On the northern and southern walls, the roles are reversed, and we have $\langle v_x \rangle = 0$ and $v_y = 0$. At the eastern exit, the condition $\nabla_x v_y = 0$ become $\widehat{\nabla}_x v_y = 0$. Since there is no data on v_x east of the border, the condition $\widehat{\nabla}_x v_x = 0$ can only be approximatively implemented as a backwards difference, $\widehat{\nabla}_x^- v_x = 0$.

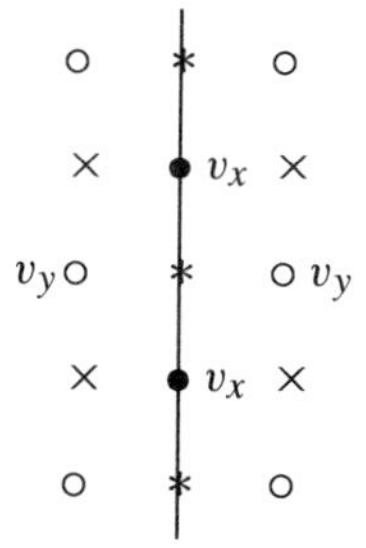

On the western boundary the value of the velocity v_x may be implemented directly, whereas the boundary value of the velocity v_y is calculated as an average over the two nearest neighbours on both sides.

General theory of the Poisson equation tells us that we either need to know the pressure itself or its normal derivative on the boundary. Since at all times $v_x = U$ on the western border, the time-step equation (21.17) implies that $\widehat{\nabla}_x \widetilde{p} = F_x$ there. The value of F_x on the western boundary (obtained from (21.16)) requires knowledge of v_x further to the west, data that is not available on the grid. As it turns out, we do not in fact need to know F_x on the border. To see this, the discretized Poisson equation (21.18) is written as

$$\widehat{\nabla}_x \left(\widehat{\nabla}_x \widetilde{p} - F_x\right) + \widehat{\nabla}_y \left(\widehat{\nabla}_y \widetilde{p} - F_y\right) = \frac{\widehat{\nabla}_x v_x + \widehat{\nabla}_y v_y}{\Delta t}. \tag{21.24}$$

Since $\widehat{\nabla}_x \widetilde{p} = F_x$ on the border, the boundary value of $\widehat{\nabla}_x \widetilde{p} - F_x$ is always zero wherever it appears in the first term, independent of the boundary value of F_x. Thus, the value of F_x at the fluid inlet never appears in the Poisson equation and has therefore no influence on the solution. In practice, it is convenient to choose the boundary value $F_x = 0$, and correspondingly $\widehat{\nabla}_x \widetilde{p} = 0$.

On the solid walls $y = 0, d$, similar arguments lead to $\widehat{\nabla}_y \widetilde{p} = 0$. Finally, at $x = L$ we only know that $\widehat{\nabla}_x v_x = 0$ and consequently $\widehat{\nabla}_x^2 \widetilde{p} = \widehat{\nabla}_x F_x$ from the equation of motion (21.17). Again it follows from analysis of the Poisson equation that the actual boundary value of F_x cannot influence the solution. In this case one may either choose the pressure to be constant, $\langle \widetilde{p} \rangle = 0$, or better require $F_x = 0$ and $\widehat{\nabla}_x \widetilde{p} = 0$ at the exit.

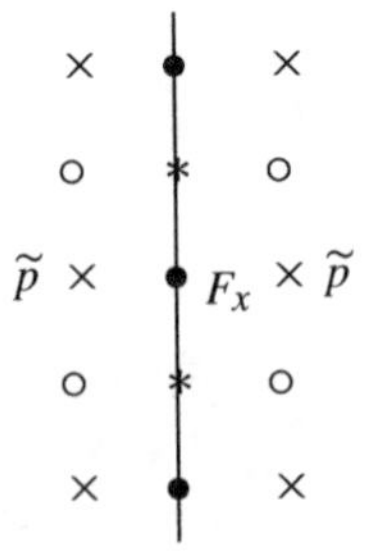

On the western boundary the normal difference of the pressure $\widehat{\nabla}_x \tilde{p}$ is determined by F_x.

Summarizing, the discrete boundary conditions are taken to be

$$v_x = U, \quad \langle v_y \rangle = 0, \quad \widehat{\nabla}_x \tilde{p} = 0, \quad \text{for } x = 0, \tag{21.25a}$$

$$\widehat{\nabla}_x v_x = 0, \quad \widehat{\nabla}_x v_y = 0, \quad \widehat{\nabla}_x \tilde{p} = 0, \quad \text{for } x = L, \tag{21.25b}$$

$$\langle v_x \rangle = 0, \quad v_y = 0, \quad \widehat{\nabla}_y \tilde{p} = 0, \quad \text{for } y = 0, d. \tag{21.25c}$$

In the second line the difference $\widehat{\nabla}_x v_x$ actually belongs to the points inside the channel with $x = L - \Delta x/2$.

Monitoring the process

The most important quantity to monitor is the divergence, which ideally should vanish. A convenient parameter is

$$\chi = \sqrt{\frac{\sum(\widehat{\nabla}_x v_x + \widehat{\nabla}_y v_y)^2}{\sum(\widehat{\nabla}_x v_x)^2 + (\widehat{\nabla}_y v_y)^2}} \tag{21.26}$$

where the sum runs over all internal grid points. It is dimensionless, independent of the grid for $N_x, N_y \to \infty$, and measures how well the two differences cancel each other in the divergence.

The convergence of the Poisson relaxation process (21.20) may be monitored by a similar parameter

$$\chi' = \Delta t \sqrt{\frac{\sum((\widehat{\nabla}_x^2 + \widehat{\nabla}_y^2)\tilde{p} - s)^2}{\sum(\widehat{\nabla}_x v_x)^2 + (\widehat{\nabla}_y v_y)^2}}, \tag{21.27}$$

because this quantity is a dimensionless estimate of the average of the future divergence, $\nabla \cdot \boldsymbol{v}(t + \Delta t) = -(\nabla^2 \tilde{p} - s)\Delta t$.

Iteration cycle

The grid arrays for all the fields, $v_x[i_x, i_y]$, $v_y[i_x, i_y]$, $\tilde{p}[i_x, i_y]$, $F_x[i_x, i_y]$ and $F_y[i_x, i_y]$ are first cleared to zero, and then the velocity is initialized to $v_x[i_x, i_y] = U$ for $i_x = 0, \ldots, N_x$ and $i_y = 1, \ldots, N_y$.

Assuming that we have obtained a current set of discrete fields at time t, the following iteration cycle produces a new set of fields at $t + \Delta t$.

1. Calculate the new velocities v_x and v_y at time $t+\Delta t$ from (21.17) in all internal points (i.e. not on the boundary). Explicitly the internal grid points are given by $i_x = 1, \ldots, N_x - 1$ and $i_y = 1, \ldots, N_y$ for v_x, and $i_x = 1, \ldots, N_x$ and $i_y = 1, \ldots, N_y - 1$ for v_y.

2. Use the boundary conditions (21.25) to determine boundary values of the velocities. Explicitly they become

$$v_x[0, i_y] = U, \quad v_x[N_x, i_y] = v_x[N_x - 1, i_y], \quad i_y = 1, \ldots, N_y,$$

$$v_y[0, i_y] = -v_y[1, i_y], \quad v_y[N_x + 1, i_y] = v_y[N_x, i_y], \quad i_y = 1, \ldots, N_y - 1,$$

$$v_x[i_x, 0] = -v_x[i_x, 1], \quad v_x[i_x, N_y + 1] = -v_x[i_x, N_y], \quad i_x = 0, \ldots, N_x,$$

$$v_y[i_x, 0] = 0, \quad v_y[i_x, N_y] = 0, \quad i_x = 0, \ldots, N_x + 1.$$

   Note the care that is necessary in the specifications of index ranges.

3. Calculate the new acceleration fields, F_x and F_y, at all internal points (with the same index ranges as for the velocities) from (21.16) using the new velocity fields. The boundary values of the accelerations remain zero.

4. Calculate the source of the Poisson equation (21.18) from the new fields at all internal points.

5. Solve the Poisson equation iteratively by means of the following subcycle.

   (a) Apply the boundary conditions to the pressure. Explicitly they are

$$\tilde{p}[0, i_y] = \tilde{p}[1, i_y], \quad \tilde{p}[N_x + 1, i_y] = \tilde{p}[N_x, i_y], \quad i_y = 1, \ldots, N_y,$$

$$\tilde{p}[i_x, 0] = \tilde{p}[i_x, 1], \quad \tilde{p}[i_x, N_y + 1] = \tilde{p}[i_x, N_y], \quad i_x = 0, \ldots, N_x + 1.$$

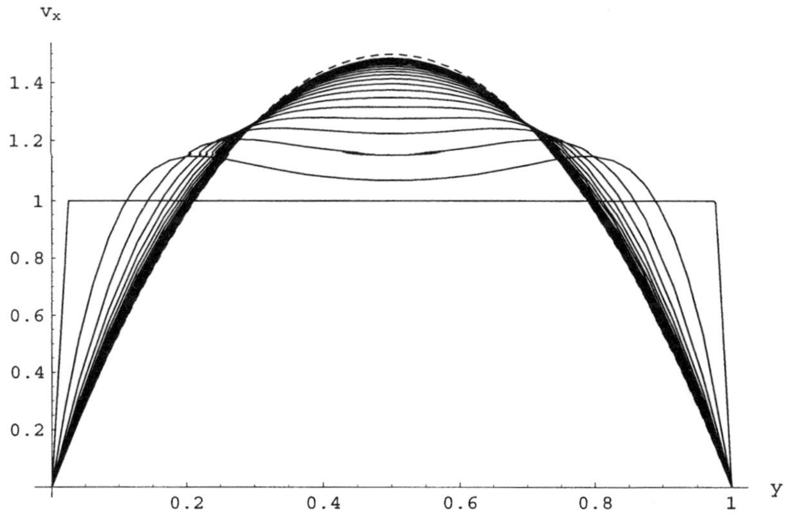

Figure 21.4. Transformation of the steady velocity profile from the initial square shape to the parabolic Poiseuille shape (dashed) downstream from the entrance. The Reynolds number is chosen to be Re $= 100$, the channel length $L = 10$ and width $d = 1$, and the curves are separated by $\Delta x = 1/N_x$ (with $N_x = 40$) in the interval $0 \leq x \leq L/2$.

(b) Calculate the new pressure in all internal points using successive overrelaxation (SOR).

(c) Repeat until the desired precision (χ') or the iteration limit is reached.

6. Repeat until the required time or iteration limits are reached.

At chosen intervals the grid arrays may be displayed graphically or written out to a file for later treatment.

Results

We shall fix the mass scale by choosing unit density $\rho_0 = 1$, the length scale by choosing unit plate distance $d = 1$, and the time scale by setting the entry velocity $U = 1$. With these units the only parameter left in the problem are the (now dimensionless) kinematic viscosity ν and the (also dimensionless) length L of the channel. The Reynolds number is, for example, the reciprocal viscosity, Re $= 1/\nu$. Here we shall mainly present results for Re $= 100$. In view of the estimate (21.22) which predicts the entrance length to be 3 in this case, we choose $L = 10$. For Re $\gtrsim 20$ the length is chosen to be $L =$ Re$/10$ whereas for Re $\lesssim 20$ it is chosen to be $L = 2$, because the entrance length is expected to be constant. The grid dimensions are everywhere chosen to be $N_x = N_y = 40$.

In figure 21.5(a) the time evolution of the exit velocity is plotted together with the rise of the velocity in the channel. Allowing for maximally 100 SOR iterations, the process converges to $\chi' = 1\%$ in about 50 time steps, corresponding to $t = 0.3$. It reaches 95% of the Poiseuille maximal velocity $1.5U$ in 400 time steps corresponding to $t = 2.48$. The same is the case for the downstream rise of the steady flow velocity plotted in figure 21.5(b) which reaches 95% of its maximum at $x = 2.34$. The downstream evolution of the velocity profile towards the parabolic Poiseuille shape is shown in figure 21.4. In figure 21.6(a) the pressure in the middle of the channel is plotted as a function of x, and it also reaches the Poiseuille form with constant gradient for $x \approx 2.5$.

Finally, in figure 21.6(b) the entrance length, defined as the point where v_x has reached 95% of maximum, has been plotted as a function of Reynolds number. It is remarkable that the same program with identical convergence parameters covers a range of Reynolds numbers from nearly 0 to 5000. For Re $\lesssim 10$ the entrance length becomes constant, $L' = 0.43d$, and this makes sense because the influence of the entrance must always be notable at a distance compared to the channel width. At high Reynolds numbers, Re $\gtrsim 200$, the linearity of the estimate (21.22) is confirmed and we obtain $k = 0.0197$ in perfect agreement with the rough estimate. There seems to be no sign of turbulence for Reynolds number between

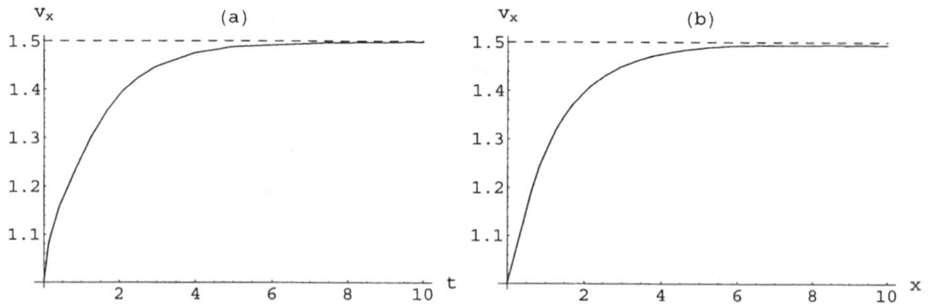

Figure 21.5. Channel entrance flow at Reynolds number 100. Both curves approach the Poiseuille maximal velocity of 1.5 times the average velocity. **(a)** The rise of the exit velocity $v_x(L, d/2, t)$ as a function of time. **(b)** The rise of the steady state velocity along the middle of the channel $v_x(x, d/2)$ as a function of x.

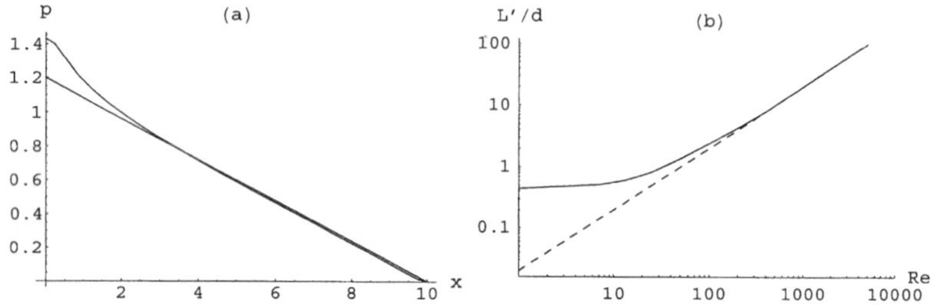

Figure 21.6. **(a)** The pressure as a function of distance x from the entrance. Its gradient becomes constant at about $x = 2.5$. **(b)** The 95% entrance length is plotted as a function of Reynolds number on the right. The dashed line corresponds to $L'/d = k\,\mathrm{Re}$ with $k = 0.0197$. For $\mathrm{Re} \lesssim 10$ the curve approaches the constant value 0.43.

2000 and 5000, but as mentioned on page 247 planar pressure-driven flow is in fact stable towards small perturbations for $\mathrm{Re} < 5772$.

Problems

* **21.1** Show that it is not possible to find an integrated quantity F for which the equation for incompressible steady flow (18.1) corresponds to an extremum. Hint: show that there is no integral F for which the variation is of the form

$$\delta F = \int [(\boldsymbol{v} \cdot \nabla)\boldsymbol{v} - \nu \nabla^2 \boldsymbol{v} + \nabla \widetilde{p}] \cdot \delta \boldsymbol{v}\, dV \tag{21.28}$$

which vanishes for all $\boldsymbol{v}$ satisfying (18.1).

21.2 Show that the Poisson equation

$$\nabla^2 \widetilde{p} = s \tag{21.29}$$

is the minimum of the quadratic 'energy' function

$$\mathcal{E} = \int_V \left(\frac{1}{2} (\nabla \widetilde{p}(\boldsymbol{x}))^2 + \widetilde{p}(\boldsymbol{x}) s(\boldsymbol{x}) \right) dV, \tag{21.30}$$

under suitable boundary conditions. Use this result to devise a gradient descent algorithm towards the minimum.

21.3 Indicate in figure 21.1 which staggered grids naturally carry the various stress components, σ_{xx}, σ_{yy} and σ_{xy}.

21.4 Verify that the various averages in F_x in terms of the grid arrays become

$$\left\langle \widehat{\nabla}_x v_x \right\rangle [i_x, i_y] = \frac{v_x[i_x+1, i_y] - v_x[i_x-1, y]}{2\Delta x}$$

$$\left\langle v_y \right\rangle [i_x, i_y] = \frac{v_y[i_x, i_y] + v_y[i_x+1, i_y] + v_y[i_x, i_y-1] + v_y[i_x+1, i_y-1]}{4}$$

$$\left\langle \widehat{\nabla}_y v_x \right\rangle [i_x, i_y] = \frac{v_x[i_x, i_y+1] - v_x[i_x, y-1]}{2\Delta y}$$

and in F_y

$$\langle v_x \rangle [i_x, i_y] = \frac{v_x[i_x, i_y] + v_x[i_x-1, i_y] + v_x[i_x, i_y+1] + v_x[i_x-1, i_y+1]}{4}$$

$$\left\langle \widehat{\nabla}_x v_y \right\rangle [i_x, i_y] = \frac{v_y[i_x+1, i_y] - v_y[i_x-1, y]}{2\Delta x}$$

$$\left\langle \widehat{\nabla}_y v_y \right\rangle [i_x, i_y] = \frac{v_y[i_x, i_y+1] - v_y[i_x, y-1]}{2\Delta y}.$$

Part V

Special topics

22 Global laws of balance

Momentum, angular momentum, and kinetic energy are purely mechanical quantities that, like mass, are carried along with the movement of material. Besides these mechanical quantities, there are thermodynamic quantities such as internal energy and entropy, also transported along with the material. The mechanical and thermodynamic quantities mentioned here are all *extensive*, meaning that the amount in a composite body is the sum of the amounts in the parts. Other so-called *intensive* quantities, for example pressure and temperature, are not additive like the extensive quantities.

We have already seen that mass is transported without actual loss or gain, as expressed by the local and global equations for mass conservation. Extensive quantities are generally not conserved (except under special conditions to be discussed), but have *sources* that create and destroy them. Forces create momentum, moments of force create angular momentum, work creates kinetic energy. The global laws for extensive quantities express the balance between creation and accumulation of a quantity, and basically state the obvious: *in any volume of matter the net amount of a quantity produced by the sources is either accumulated in the volume or leaves it through the surface.*

In this chapter we shall derive the global laws of mechanical balance in a systematic form which clearly exposes their relation to the corresponding laws in Newtonian particle mechanics, and which is also suitable for practical applications. For each of the laws we shall discuss the circumstances under which the corresponding quantity is conserved. We shall see that mass, momentum and angular momentum are in general conserved for isolated systems, whereas kinetic energy is not. Only when potential and internal energies are included, do we arrive at a total energy which is conserved for isolated systems. This chapter is mainly theoretical with only one key application for each quantity. In chapter 23 we shall apply the laws of balance to a number of generic examples.

22.1 Connected tubes

We shall begin with a simple example of the use of mechanical energy conservation to derive how water moves in one of the most basic experiments in fluid dynamics, analysed by Newton himself in *Principia* [53, proposition 44].

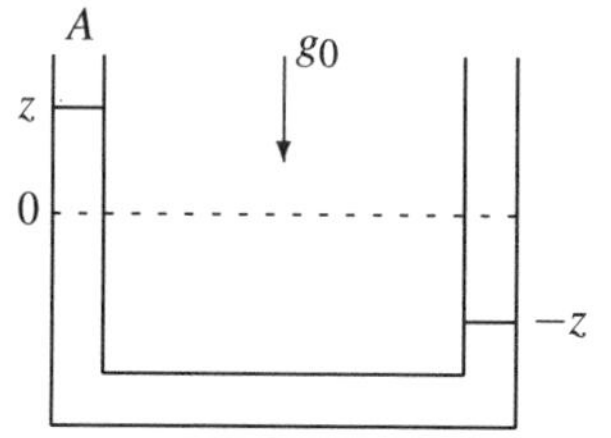

When the water is at rest, the water level will be the same in connected vertical tubes. The tube may have any shape between the two vertical sections.

Consider a long straight tube with cross section A, bent through $180°$ somewhere in the middle and placed with the open sections vertically upwards. Water is poured into the system, and as everybody knows, gravity will eventually make the levels of water equal in the two vertical tubes. Before reaching equilibrium the water sloshes back and forth with diminishing amplitude. Basic physics knowledge tells us that the energy originally given to the water oscillates between being kinetic and potential, while slowly draining away because of internal friction in the water.

Even if we do not know the exact solution to the fluid flow problem, we are nevertheless able to make a reasonable quantitative estimate of the behaviour of the water. When the water level in one vertical tube is raised by z relative to the equilibrium level, mass conservation tells us that it is lowered by the same amount in the other vertical tube. Denoting the instantaneous velocity of the water in the tube by v and the full length of the water column by L, the total mass of the water column will be $\rho_0 A L$, where ρ_0 is the constant density of water, and the kinetic energy becomes $(1/2)\rho_0 A L v^2$. Since a small water column of height z and weight $g_0\rho_0 A z$ has effectively been moved from one vertical tube to the other and thereby raised by z relative to equilibrium, the potential energy becomes $g_0\rho_0 A z^2$. Adding these contributions we get the following estimate for the total energy,

$$\mathcal{E} = \tfrac{1}{2}\rho_0 A L v^2 + g_0\rho_0 A z^2. \tag{22.1}$$

For simplicity we assume that there is no friction to eat away the energy. The total energy must therefore be conserved, so that its time-derivative has to vanish,

$$\frac{d\mathcal{E}}{dt} = \rho_0 A L v \frac{dv}{dt} + 2g_0\rho_0 A z \frac{dz}{dt} = 0. \tag{22.2}$$

Using that $v = dz/dt$, we find the differential equation,

$$\frac{d^2 z}{dt^2} = -\frac{2g_0}{L} z. \tag{22.3}$$

The solution to this harmonic equation is of the form

$$z = a \cos \omega t, \tag{22.4}$$

with amplitude a and circular frequency $\omega = \sqrt{2g_0/L}$. As noted by Newton, this is the frequency of the small-amplitude oscillations of a pendulum with length $L/2$, but in contrast to the pendulum the motion of the water is purely harmonic, also for large amplitudes.

22.2 Overview of the global laws

Before going into detailed discussion of the global laws, we shall here outline the contents of these laws. We shall see that in continuum mechanics, the global laws of balance take nearly the same form as the global laws of Newtonian particle mechanics (see appendix A).

Mechanical quantities

There are four global mechanical quantities that may be calculated for the material contained in any volume V: the total mass M, the total momentum $\mathcal{P}$, the total angular momentum $\mathcal{L}$ and the total kinetic energy $\mathcal{T}$. These quantities are defined as integrals over densities of the form,

$$M = \int_V \rho \, dV, \qquad \mathcal{P} = \int_V \rho \, \boldsymbol{v} \, dV, \tag{22.5a}$$

$$\mathcal{L} = \int_V \rho \, \boldsymbol{x} \times \boldsymbol{v} \, dV, \qquad \mathcal{T} = \int_V \frac{1}{2}\rho \, \boldsymbol{v}^2 \, dV. \tag{22.5b}$$

Each of these definitions expresses that the quantity may be understood as the sum over the material particles making up the body.

Mechanical laws of balance

Each of the mechanical quantities obeys a global law of balance completely analogous to the global laws of Newtonian particle mechanics. In continuum physics the laws of balance of mass, momentum, angular momentum and kinetic energy take the form,

$$\frac{DM}{Dt} = 0, \qquad \frac{D\mathcal{P}}{Dt} = \mathcal{F}, \tag{22.6a}$$

$$\frac{D\mathcal{L}}{Dt} = \mathcal{M}, \qquad \frac{D\mathcal{T}}{Dt} = P, \tag{22.6b}$$

where $\mathcal{F}$ is the total force (9.16), $\mathcal{M}$ the total moment of force and P the total power (rate of work). The first two have already been derived in sections 15.3 and 15.5, whereas angular momentum balance and kinetic energy balance will be derived below[1].

The global laws all follow from Cauchy's equation of motion for continuous matter (15.35), and are for this reason automatically fulfilled for any solution to the field equations. Having found such a solution, one should not worry about the balance of mass, momentum, angular momentum or kinetic energy. But when one is unable to solve the field equations and has to guess a simplified form of the solution, the laws of balance impose two scalar and two vector constraints on the solution, in many cases sufficient to make a decent estimate of the system's behaviour.

22.3 The control volume

In Newtonian particle mechanics, a 'body' is understood as a collection of a fixed number of particles. In continuum mechanics the notion of a body is much more general: any volume—usually called a *control volume*—may be viewed as a 'body' at a given time. Intuitively we think of bodies as made from different materials, but the surface of the control volume does not have to correspond to an interface between materials with different properties, although it often is convenient to choose it to coincide with such an interface. In the course of time the control volume $V(t)$ with surface $S(t)$ may be moved around and deformed any way we desire, which is why it is called a *control* volume.

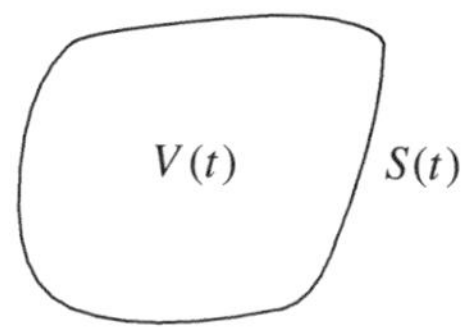

A 'body' consists of all the matter contained in an arbitrary time-dependent control volume $V(t)$ with surface $S(t)$.

> One may wonder whether it is really necessary to consider bodies this generally. Previously, for example in the discussion of mass conservation and momentum balance, we have only considered arbitrary fixed volumes which do not change with time. A quick review of the example of connected tubes in section 22.1 reveals, however, that the control volume we instinctively used there encompasses all the water in the system and moves along with it. Not permitting arbitrarily moving control volumes would in fact put unreasonable constraints on our freedom to analyse the physics of continuous systems.

Specific quantities

Before proceeding to a discussion of the global laws of balance for mass, momentum, angular momentum and kinetic energy, it is necessary to establish some general relations. For an arbitrary time-dependent volume $V(t)$ we let

$$Q(t) = \int_{V(t)} q(\boldsymbol{x}, t)\, dM = \int_{V(t)} \rho(\boldsymbol{x}, t)\, q(\boldsymbol{x}, t)\, dV, \tag{22.7}$$

denote any of the eight components of the global mechanical quantities M, $\mathcal{P}_x$, $\mathcal{P}_y$, $\mathcal{P}_z$, $\mathcal{L}_x$, $\mathcal{L}_y$, $\mathcal{L}_z$ and $\mathcal{T}$. Since the densities of all of the global quantities are proportional to the mass density, it is convenient to factor it out and instead introduce the *specific quantity*, $q = dQ/dM$, as the local amount per unit of mass. Whereas the specific mass is trivially unity, the specific momentum is the velocity $\boldsymbol{v}$, the specific angular momentum $\boldsymbol{x} \times \boldsymbol{v}$, and the specific kinetic energy $(1/2)\boldsymbol{v}^2$. One might even say that the specific volume is $1/\rho$.

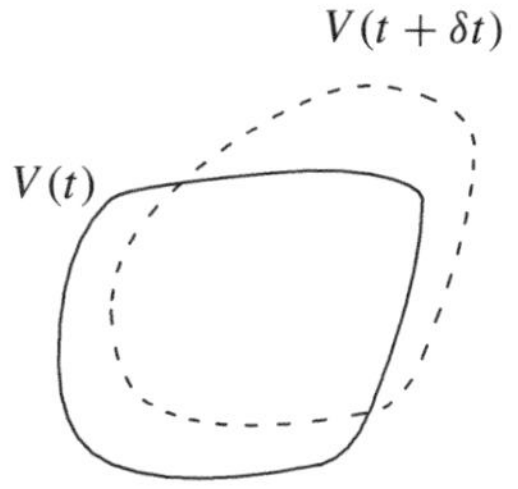

A moving volume changes with time. The (signed) change in volume is the region between the dashed and solid surface outlines.

Rate of change of a quantity

The time dependence of a global quantity Q has two origins: the changing density ρq and the changing control volume V. In a small time interval δt the control volume changes from $V(t)$ to $V(t+\delta t)$. Expanding

[1] There is an unfortunate clash in the use of the same symbol $\mathcal{L}$ for angular momentum and lift (page 268). Total momentum $\mathcal{P}$ and total power P are distinguished by typography.

to first order in the small quantities, we find

$$\begin{aligned} Q(t+\delta t) - Q(t) &= \int_{V(t+\delta t)} \rho q(\boldsymbol{x}, t+\delta t)\, dV - \int_{V(t)} \rho q(\boldsymbol{x}, t)\, dV \\ &\approx \delta t \int_{V(t)} \frac{\partial(\rho q(\boldsymbol{x}, t))}{\partial t}\, dV + \int_{V(t+\delta t)-V(t)} \rho q(\boldsymbol{x}, t)\, dV \\ &\approx \delta t \int_{V(t)} \frac{\partial(\rho q(\boldsymbol{x}, t))}{\partial t}\, dV + \delta t \oint_{S(t)} \rho q(\boldsymbol{x}, t)\, \boldsymbol{v}_S(\boldsymbol{x}, t) \cdot d\boldsymbol{S}, \end{aligned}$$

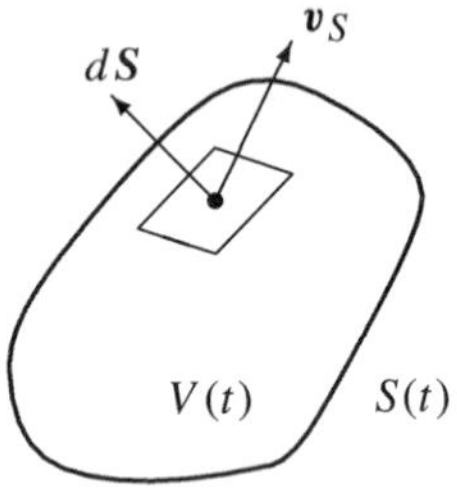

Every surface element $d\boldsymbol{S}$ of the control volume may move with a different velocity $\boldsymbol{v}_S(\boldsymbol{x}, t)$ and scoops up the volume $\boldsymbol{v}_S \delta t \cdot d\boldsymbol{S}$ in a small time interval δt.

where $\boldsymbol{v}_S(\boldsymbol{x}, t)$ is the velocity of a surface element $d\boldsymbol{S}$ near the point $\boldsymbol{x}$ at time t. This velocity needs only be defined on the surface itself and not all over space. It is not a field in the usual sense of the word.

Dividing by δt and suppressing the explicit dependence on space and time, the rate of change of the quantity in the moving control volume becomes,

$$\boxed{\frac{dQ}{dt} = \int_V \frac{\partial(\rho q)}{\partial t}\, dV + \oint_S \rho q\, \boldsymbol{v}_S \cdot d\boldsymbol{S}.} \tag{22.8}$$

The first term is the contribution from the local change in density, and the second is the signed amount of mass which the surface of the moving control volume scoops up per unit of time.

Reynolds' transport theorem

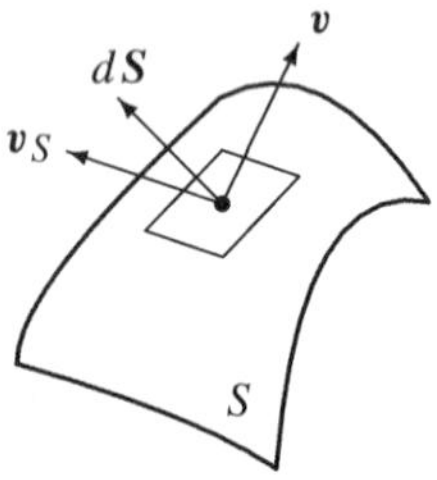

Matter moves with velocity $\boldsymbol{v}$ through the surface element $d\boldsymbol{S}$ which itself moves with velocity $\boldsymbol{v}_S$. The velocity of the matter relative to the moving surface is $\boldsymbol{v} - \boldsymbol{v}_S$.

If the surface of the control volume always follows the material, $\boldsymbol{v}_S = \boldsymbol{v}$, it is said to be *comoving* (in which case it is actually not under our control!). We shall, as before, use the symbol D/Dt for the *material time derivative*, defined as the rate of change of a quantity in a comoving volume, and find from (22.8) with $\boldsymbol{v}_S = \boldsymbol{v}$,

$$\frac{DQ}{Dt} = \int_V \frac{\partial(\rho q)}{\partial t}\, dV + \oint_S \rho q\, \boldsymbol{v} \cdot d\boldsymbol{S}. \tag{22.9}$$

Combining this equation with the general expression (22.8) for the rate of change, we may for an arbitrary control volume write

$$\boxed{\frac{DQ}{Dt} = \frac{dQ}{dt} + \oint_S \rho\, q\, (\boldsymbol{v} - \boldsymbol{v}_S) \cdot d\boldsymbol{S}.} \tag{22.10}$$

This equation goes under the name of *Reynolds' transport theorem* (1903) and is the general basis for global analysis in continuum physics. It has a fairly clear intuitive content: *independent of how the control volume moves, the material rate of change of an extensive quantity equals the actual rate of change of the quantity in the control volume plus its net rate of loss through the surface.*

Local and global rates of change

Using Gauss' theorem on the surface integral in (22.9) we obtain a single volume integral with the integrand,

$$\frac{\partial(\rho q)}{\partial t} + \nabla \cdot (\rho q \boldsymbol{v}) = q\left(\frac{\partial \rho}{\partial t} + \nabla \cdot (\rho \boldsymbol{v})\right) + \rho\left(\frac{\partial q}{\partial t} + (\boldsymbol{v} \cdot \nabla) q\right).$$

Here the first parenthesis vanishes because of the continuity equation (15.24) and the second parenthesis is recognized as the local material time derivative (15.29) of the specific quantity q. Putting it all together we arrive at the important relation,

$$\boxed{\frac{DQ}{Dt} = \int_V \rho \frac{Dq}{Dt}\, dV.} \tag{22.11}$$

The material derivative of a global quantity equals the integral of the density times the local material derivative Dq/Dt of the specific quantity. This result facilitates enormously the derivation of the global laws of balance.

22.4 Mass balance

Since the specific mass corresponds to $q = 1$ we immediately re-derive the *equation of global mass balance*,

$$\boxed{\frac{DM}{Dt} = \frac{dM}{dt} + \oint_S \rho(\boldsymbol{v} - \boldsymbol{v}_S) \cdot d\boldsymbol{S} = 0.} \tag{22.12}$$

Since the source vanishes, mass is always conserved in a comoving volume. Previously this relation was obtained for a fixed volume (page 194), but now we see that it may actually be applied in an arbitrary moving control volume when Reynolds' transport theorem is used to calculate the material rate of change.

Down the drain

Suppose a cistern filled with water is flushed through a short drain pipe in the bottom of the container. The drain pipe has cross section A, and the cistern is vertical with cross section A_0. We choose a control volume encompassing all the water in the cistern but not the pipe. As the cistern drains the horizontal water surface moves whereas the other parts of the control surface are fixed. This is an example of a control volume which is neither fixed nor comoving.

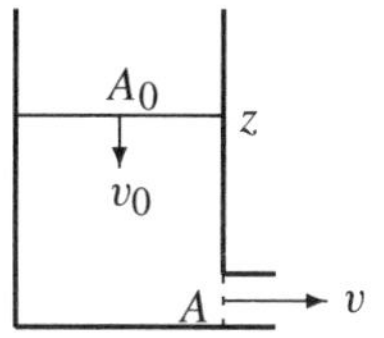

The water leaves the container through the drain. The control volume is bounded by the fixed cistern walls, the moving horizontal water surface, and the fixed entrance surface to the pipe.

When the water level is z, the total mass of the water is $M = \rho_0 A_0 z$, and the rate of mass loss through the drain is $\rho_0 A v$ where v is the average velocity of the drain flow. There is no mass flux through the moving open surface of the cistern. Global mass balance (22.12) and Reynolds' transport theorem (22.10) with $q = 1$ now leads to,

$$\frac{DM}{Dt} = \frac{dM}{dt} + \rho_0 A v = \rho_0 A_0 \frac{dz}{dt} + \rho_0 A v = 0. \tag{22.13}$$

Since $dz/dt = -v_0$ where v_0 is the average downwards velocity of the water in the cistern, this becomes $A_0 v_0 = A v$ which is Leonardo's law (15.18). Although this looks like a bit of an overkill to get a well-known result, it illustrates the way the global law works for a non-trivial control volume.

22.5 Momentum balance

The total momentum in a control volume is obtained by integrating the momentum density $\rho \boldsymbol{v}$ over the volume,

$$\boldsymbol{\mathcal{P}} = \int_V \rho\, \boldsymbol{v}\, dV. \tag{22.14}$$

It represents as mentioned before the sum of the momenta of all the material particles in the control volume.

Total force

Using Cauchy's equation, $\rho D\boldsymbol{v}/Dt = \boldsymbol{f}^*$, it follows from (22.11) with $q \to \boldsymbol{v}$ that the *global equation of momentum balance* is,

$$\boxed{\frac{D\boldsymbol{\mathcal{P}}}{Dt} = \boldsymbol{\mathcal{F}},} \tag{22.15}$$

where

$$\frac{D\boldsymbol{\mathcal{P}}}{Dt} = \frac{d\boldsymbol{\mathcal{P}}}{dt} + \oint_S \rho \boldsymbol{v}(\boldsymbol{v} - \boldsymbol{v}_S) \cdot d\boldsymbol{S}, \qquad \boldsymbol{\mathcal{F}} = \int_V \boldsymbol{f}^*\, dV. \tag{22.16}$$

It has been previously shown (page 116) that

$$\boldsymbol{\mathcal{F}} = \int_V \boldsymbol{f}\, dV + \oint_S \boldsymbol{\sigma} \cdot d\boldsymbol{S}. \tag{22.17}$$

Momentum balance thus tells us that *the total momentum of a comoving volume is conserved if and only if the total force vanishes.*

Internal and external force

Whereas the contact forces acting on the surface of the control volume are always external, the body forces acting in the volume V may be split into external and internal parts

$$\boldsymbol{f} = \boldsymbol{f}_{\text{int}} + \boldsymbol{f}_{\text{ext}}, \tag{22.18}$$

where $\boldsymbol{f}_{\text{int}}$ is produced by the material inside the volume, and $\boldsymbol{f}_{\text{ext}}$ by the material outside. Gravitational and electromagnetic forces are two-particle body forces where a particle of volume dV' situated at $\boldsymbol{x}'$ acts on a particle of volume dV at $\boldsymbol{x}$ with a force $d^2\boldsymbol{\mathcal{F}} = \boldsymbol{f}(\boldsymbol{x}, \boldsymbol{x}')dV dV'$, such that

$$\boldsymbol{f}_{\text{int}} = \int_V \boldsymbol{f}(\boldsymbol{x}, \boldsymbol{x}')\, dV', \qquad \boldsymbol{f}_{\text{ext}} = \int_{V'} \boldsymbol{f}(\boldsymbol{x}, \boldsymbol{x}')\, dV', \tag{22.19}$$

where V' is the volume of the material outside.

Both gravitational and electrostatic forces obey Newton's third law, $\boldsymbol{f}(\boldsymbol{x}, \boldsymbol{x}') = -\boldsymbol{f}(\boldsymbol{x}', \boldsymbol{x})$, and then the total internal force vanishes,

$$\boldsymbol{\mathcal{F}}_{\text{int}} = \int_V \boldsymbol{f}_{\text{int}}\, dV = \int_V \int_V \boldsymbol{f}(\boldsymbol{x}, \boldsymbol{x}')\, dV\, dV' = \boldsymbol{0}, \tag{22.20}$$

because of the antisymmetry of the integrand. Thus, only the external body forces contribute to the total force (22.17), and in particular it follows that *the total momentum is conserved if a system is not subject to any external forces.*

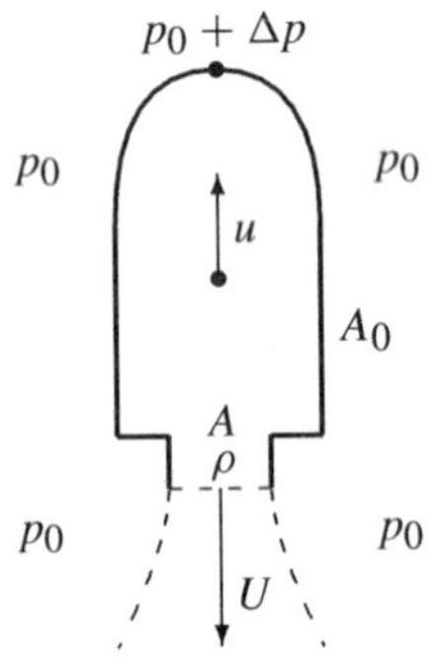

Sketch of a rocket moving upwards with vertical velocity u while emitting material of density ρ and relative velocity U through an opening of cross section A. The control volume moves with the rocket and follows the outside of the hull, cutting across the exhaust opening. There is ambient air pressure p_0 everywhere around the rocket, also at the nozzle outlet. The small rise in pressure Δp at the front of the rocket is the main cause of drag.

Launch of a small rocket

As a demonstration of how to use global momentum balance, we consider the launch of a small fireworks rocket. The rocket accelerates vertically upwards by burning chemical fuel and spewing the hot reaction gases downwards. The gases are emitted from the rocket through an opening (nozzle) with cross section A, and we assume for simplicity that the density of the gas ρ and its velocity U relative to the rocket remain unchanged during the burn. We also assume that the gas velocity is much smaller than the velocity of sound, which it is in toy rockets, so that the pressure at the exit may be taken to be equal to the ambient atmospheric pressure p_0.

In this case it is most convenient to choose a control volume which follows the *outside* of the rocket, cutting across the nozzle outlet. Such a control volume always moves with the instantaneous speed u of the rocket. It contains at any moment all the material of the rocket, including the fuel and the burning gases, but not the gases that have been exhausted through the nozzle. The rate of loss of mass through the nozzle, $Q \approx \rho U A$, is constant by the assumptions we have made. This implies that the rocket mass must decrease linearly from its initial value M_0 at $t = 0$, so that its mass at time t is,

$$M = M_0 - Q\,t. \tag{22.21}$$

At the end of the burn, when all the fuel has been spent, a 'payload' mass $M_1 < M_0$ remains. The time it takes to burn the fuel mass $M_0 - M_1$ is $t_1 = (M_0 - M_1)/Q$. After the burn the rocket flies ballistically like a cannon ball, subject only to the forces of gravity and air resistance. Here we are only interested in establishing the equation of motion valid from lift-off to burn out.

The absolute vertical velocity (relative to the ground) of the exhaust gases is $v = u - U$, allowing us to estimate the loss of vertical momentum (the surface integral in Reynolds' transport theorem (22.10)) as $\rho v(-U)(-A) = (u - U)Q$. Assuming that the centre of gravity remains fixed relative to the rocket (and that is by no means sure), the total momentum is $\mathcal{P} = Mu$, and the material derivative becomes, according to Reynolds theorem,

$$\frac{D(Mu)}{Dt} = \frac{d(Mu)}{dt} + (u - U)Q = M\frac{du}{dt} - UQ. \tag{22.22}$$

In the last step we have used the expression (22.21) for the rocket mass.

The total vertical force on the rocket is the sum of its weight $-Mg_0$ and the resistance or drag caused by the interaction of the rocket's hull with the air. Air drag has two components: *skin drag* from viscous friction between air and hull, and *form drag* from the changes in pressure at the hull caused by the rocket 'punching' through the atmosphere. Form drag can, for example, be estimated from the Bernoulli stagnation pressure

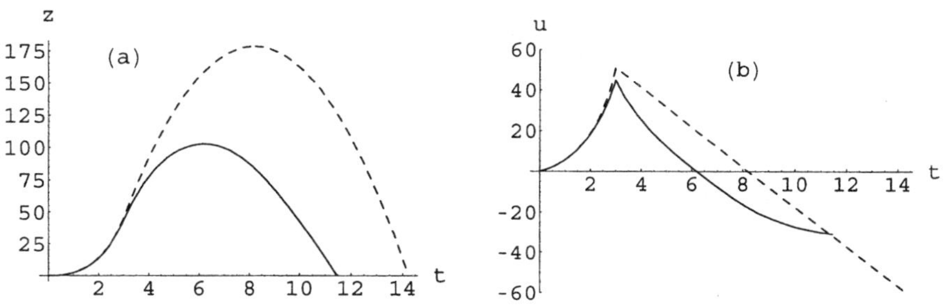

Figure 22.1. Height **(a)** and velocity **(b)** as a function of time during vertical flight of a small fireworks rocket with form drag (solid) and without (dashed). The cusp in the velocity happens at the end of the burn and signals the transition to ballistic flight. The rocket has diameter 6 cm, total mass 1 kg, payload 0.2 kg and form drag coefficient $C_D = 1$. The rocket burns for 3 s, emitting gases at a speed of 50 m s^{-1}. The numeric solution of the rocket equation (22.24) shows that with form drag the rocket reaches a velocity of 45 m s^{-1} at the end of the burn and a height of 43 m. During the subsequent ballistic flight, it reaches a maximum height of 103 m about 6 s after start, and finally it falls back and hits the ground again with a speed of 30 m s^{-1} about 11.5 s after start. Without form drag the rocket would reach a maximum height of 178 m about 8 s after start, and it would hit the ground with a speed of 60 m s^{-1} after 14 s.

increase at the tip of the rocket, $\Delta p \approx (1/2)\rho_0 u^2$, where ρ_0 is the density of air. Multiplying by the cross section of the rocket A_0, the estimate of the form drag becomes $-(1/2)\rho_0 u^2 A_0$ (with opposite sign if u is negative). Form drag thus grows quadratically with rocket speed, and at high velocity it dominates skin drag which only grows linearly with velocity. Leaving out skin drag, the total vertical force on the control volume may be written

$$\mathcal{F} = -Mg_0 - \tfrac{1}{2}\rho_0 u^2 A_0 C_D \tag{22.23}$$

where we have included a dimensionless factor C_D, called the *drag coefficient*, which takes into account the actual shape of the rocket.

Equating the material derivative (22.22) with the total force (22.23), and dividing by the mass (22.21), we find the 'rocket equation' for vertical flight during the burn,

$$\boxed{\frac{du}{dt} = -g_0 + \frac{UQ - 1/2\rho_0 u^2 A_0 C_D}{M_0 - Qt}.} \tag{22.24}$$

This shows that the rocket will take off from rest $u = 0$ at $t = 0$ provided the initial acceleration is positive, i.e. $UQ > M_0 g_0$. The differential equation can only be solved numerically (except in vacuum for $C_D = 0$ where it can be solved analytically; see problem 22.2), and the results are shown for a typical fireworks rocket in figure 22.1 with $C_D = 1$. The figure also includes the period of ballistic flight which follows the burn and brings the rocket to a maximal height before it turns around and falls back to earth. One notes how the form drag reduces the maximum height for this rocket to a little more than half of what it would be in vacuum (dashed curve). If the rocket shape were made highly streamlined, the form drag coefficient could be made considerably smaller than unity, allowing it to attain greater heights for the same amount of fuel (see page 449 for a general discussion of form drag).

22.6 Angular momentum balance

The *angular momentum* (sometimes called the moment of momentum) of the matter contained in a control volume V is,

$$\boldsymbol{\mathcal{L}} = \int_V \rho\, \boldsymbol{x} \times \boldsymbol{v}\, dV. \tag{22.25}$$

Angular momentum depends on the origin of the coordinate system. If we shift the origin by $\boldsymbol{x} \to \boldsymbol{x} + \boldsymbol{a}$, the angular momentum shifts by $\mathcal{L} \to \mathcal{L} + \boldsymbol{a} \times \mathcal{P}$, unless the total momentum vanishes which is generally not the case in fluid systems. This definition is, however, not without subtlety (see problem 22.7).

Total moment of force

The material time derivative of the specific angular momentum $\boldsymbol{x} \times \boldsymbol{v}$ is

$$\rho \frac{D(\boldsymbol{x} \times \boldsymbol{v})}{Dt} = \rho \frac{D\boldsymbol{x}}{Dt} \times \boldsymbol{v} + \rho \boldsymbol{x} \times \frac{D\boldsymbol{v}}{Dt} = \boldsymbol{x} \times \boldsymbol{f}^*,$$

because $D\boldsymbol{x}/Dt = \boldsymbol{v}$ and $\rho D\boldsymbol{v}/Dt = \boldsymbol{f}^*$. Using (22.11) with $q \to \boldsymbol{x} \times \boldsymbol{v}$ we arrive at the equation of *global angular momentum balance*,

$$\boxed{\frac{D\mathcal{L}}{Dt} = \mathcal{M},} \tag{22.26}$$

where

$$\frac{D\mathcal{L}}{Dt} = \frac{d\mathcal{L}}{dt} + \oint_S \rho(\boldsymbol{x} \times \boldsymbol{v}) \cdot d\boldsymbol{S}, \qquad \mathcal{M} = \int_V \boldsymbol{x} \times \boldsymbol{f}^* \, dV. \tag{22.27}$$

The source $\mathcal{M}$ is the total moment of the effective forces acting on the material particles in the control volume. Under a shift $\boldsymbol{x} \to \boldsymbol{x} + \boldsymbol{a}$ the total moment transforms as $\mathcal{M} \to \mathcal{M} + \boldsymbol{a} \times \mathcal{F}$ where $\mathcal{F}$ is the total force. Combined with momentum balance (22.15), both sides of the angular momentum balance (22.26) thus shift in the same way so that the form of this relation is independent of the choice of origin.

From angular momentum balance it follows immediately that *the total angular momentum is conserved if and only if the total moment of force vanishes*. It is, however, a bit more complicated to derive the global form of the total moment than it was for the total force. Inserting the effective force (9.18), we find the identity

$$(\boldsymbol{x} \times \boldsymbol{f}^*)_i - (\boldsymbol{x} \times \boldsymbol{f})_i = \sum_{jkl} \epsilon_{ijk} x_j \nabla_l \sigma_{kl} = \sum_{jkl} \epsilon_{ijk} \nabla_l (x_j \sigma_{kl}) - \sum_{jk} \epsilon_{ijk} \sigma_{kj}.$$

The first term on the right-hand side is a divergence which by Gauss' theorem leads to a surface integral, and the last term vanishes when the stress tensor is symmetric, which we assume it is. After integration the total moment of force may be written,

$$\boxed{\mathcal{M} = \int_V \boldsymbol{x} \times \boldsymbol{f} \, dV + \oint_S \boldsymbol{x} \times \boldsymbol{\sigma} \cdot d\boldsymbol{S}.} \tag{22.28}$$

In complete analogy with the total force, the total moment thus consists of the moment of the body forces plus the moment of the contact forces acting on the surface of the volume.

Internal moment of force

Splitting the body force density into internal and external contributions (22.18), and assuming again that we are dealing with two-particle body forces $\boldsymbol{f}(\boldsymbol{x}, \boldsymbol{x}')$ obeying Newton's third law, $\boldsymbol{f}(\boldsymbol{x}, \boldsymbol{x}') = -\boldsymbol{f}(\boldsymbol{x}', \boldsymbol{x})$, the moment of the internal body forces may be written as

$$\begin{aligned}
\mathcal{M}_{\text{int}} &= \int_V \boldsymbol{x} \times \boldsymbol{f}_{\text{int}} \, dV \\
&= \int_V \int_V \boldsymbol{x} \times \boldsymbol{f}(\boldsymbol{x}, \boldsymbol{x}') \, dV \, dV' = \int_V \int_V \boldsymbol{x}' \times \boldsymbol{f}(\boldsymbol{x}', \boldsymbol{x}) \, dV \, dV' \\
&= \frac{1}{2} \int_V \int_V (\boldsymbol{x} - \boldsymbol{x}') \times \boldsymbol{f}(\boldsymbol{x}, \boldsymbol{x}') \, dV \, dV'.
\end{aligned}$$

In the second line we interchanged $\boldsymbol{x}$ and $\boldsymbol{x}'$ and in the third we used the antisymmetry of the two-particle force. If the two-particle forces are furthermore *central*, $\boldsymbol{f}(\boldsymbol{x}, \boldsymbol{x}') \sim \boldsymbol{x} - \boldsymbol{x}'$, which is the case for gravity and electrostatics, the cross product vanishes, and we conclude that the internal moment of force vanishes, $\mathcal{M}_{\text{int}} = \boldsymbol{0}$. Under these conditions, internal forces can simply be ignored in the calculation of the moment of force, and we have shown that *the angular momentum is conserved for a system not subject to an external moment of force*.

Spinning a rotating lawn sprinkler

As an illustration of the use of angular momentum balance we now calculate how a rotating lawn sprinkler spins after the pressure is turned on. Some rotating lawn sprinklers are constructed with two horizontal arms of length R mounted on a common pivot. Each arm is a tube carrying water from the pivot towards a nozzle that is bent by $90\,°$ with respect to the tube and elevated through an angle α against the azimuthal direction. The nozzle outlet has cross section A and the water emerges with velocity U relative to the nozzle. When the water is turned on, the sprinkler starts to rotate and reaches after a while a steady situation in which it rotates with constant angular velocity, determined by the friction moment from the pivot.

We choose a control volume which follows the outer surface of the sprinkler, cutting across the nozzle outlets and horizontally through the pivot. In cylindrical coordinates with the z-axis along the axis of rotation, the angular momentum of the sprinkler around the z-axis is $\mathcal{L}_z = I\,\Omega$ where I is the moment of inertia of the whole sprinkler plus water and $\Omega > 0$ is the instantaneous angular velocity. The moment of inertia should be calculated in the usual way from the mass distribution in the arms, the pivot and the water contained in the system. For incompressible water the moment of inertia is time-independent.

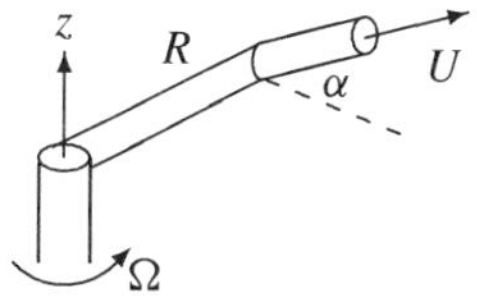

Arm of a lawn sprinkler. The nozzle is at right angles to the arm and has elevation angle α against the azimuthal direction. The control volume envelops the rotating sprinkler and cuts across the nozzle outlet and the pivot inlet.

The water entering vertically through the pivot carries no angular momentum, so the pivot does not contribute to the surface integral in Reynolds' transport theorem (22.10) with $q \to \boldsymbol{x} \times \boldsymbol{v}$. In cylindrical coordinates we have $(\boldsymbol{x} \times \boldsymbol{v})_z = r v_\phi$, and since the azimuthal velocity of the water emerging from a nozzle is $v_\phi \approx R\Omega - U\cos\alpha$, the material derivative of $\mathcal{L}_z$ becomes (taking both arms into account),

$$\frac{D\mathcal{L}_z}{Dt} \approx \frac{d(I\Omega)}{dt} + 2\rho_0 R(R\Omega - U\cos\alpha)UA. \qquad (22.29)$$

Disregarding air resistance, the only moment of force acting on the sprinkler arises from the (deliberately imposed) friction in the pivot. Since the friction moment must be negative, $\mathcal{M}_z = -N$, the angular momentum balance becomes

$$I\frac{d\Omega}{dt} + R(R\Omega - U\cos\alpha)Q = -N \qquad (22.30)$$

where $Q = 2\rho_0 UA$ is the total mass flux through the system. With initial condition $\Omega = 0$ at $t = 0$, and assuming that the friction moment is constant, the solution to this linear differential equation is,

$$\Omega = \frac{RUQ\cos\alpha - N}{R^2 Q}\left(1 - e^{-\lambda t}\right) \qquad (22.31)$$

with time constant

$$\lambda = \frac{R^2 Q}{I}. \qquad (22.32)$$

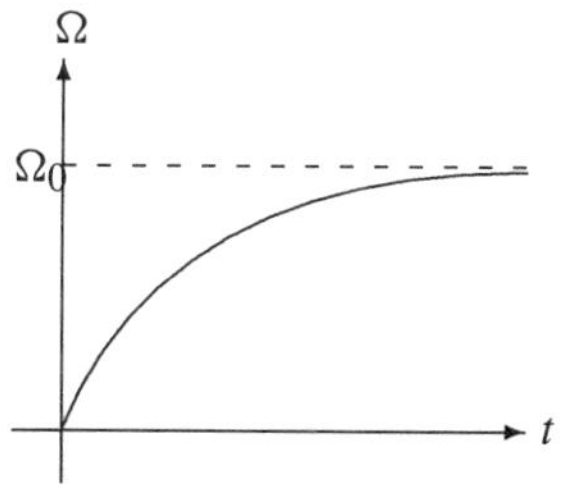

Spin of the lawn sprinkler.

Evidently, the condition for the angular velocity to be positive is that $RUQ\cos\alpha > N$, and the steady state angular velocity is

$$\Omega_0 = \frac{RUQ\cos\alpha - N}{R^2 Q}, \qquad (22.33)$$

independent of the moment of inertia.

Example 22.6.1 (Lawn sprinkler): A lawn sprinkler with radius $R = 20$ cm, nozzle outlet diameter $d = 4$ mm, and nozzle elevation $\alpha = 30\,°$, is designed to pass $Q/\rho_0 = 0.25$ litre water per second through its two arms and rotate once every 2 s in the steady state. The nozzle area becomes $A = \frac{1}{4}\pi d^2 \approx 12.5\ \text{mm}^2$, the outlet velocity $U = Q/2A\rho_0 \approx 10\ \text{m s}^{-1}$, and the steady state angular velocity $\Omega_0 = \pi\ \text{s}^{-1}$. The required friction moment that the pivot must deliver, $N = RUQ\cos\alpha - R^2 Q\Omega_0 = 0.4$ Nm. The horizontal velocity of the water relative to the lawn is $v_\phi = R\Omega_0 - U\cos\alpha \approx -8\ \text{m s}^{-1}$, and the vertical velocity of the water is $v_z = U\sin\alpha \approx 5\ \text{m s}^{-1}$. Assuming a ballistic orbit for the water flying across the lawn (disregarding air resistance) the diameter of the sprinkled region comes to about $D \approx 4\,|v_\phi|\,v_z/g_0 \approx 16$ m. The sprinkled area, $A_0 = (1/4)\pi D^2 \approx 200\ \text{m}^2$, receives about $Q/\rho_0 A_0 \approx 4$ mm rain per hour from the sprinkler.

22.7 Kinetic energy balance

The total kinetic energy of the material in a control volume is defined to be,

$$\mathcal{T} = \int_V \frac{1}{2}\rho \boldsymbol{v}^2 \, dV. \tag{22.34}$$

It must be emphasized that this is only the kinetic energy of the mean flow of matter represented by the velocity field $\boldsymbol{v}$, and that there is a 'hidden' kinetic energy associated with the fast thermal motion of the molecules relative to the mean flow. We shall return to this question in section 22.9 (see also problem 22.7).

Total power

The material derivative of the specific kinetic energy $(1/2)\boldsymbol{v}^2$ is calculated by means of the fundamental dynamical equation of continuous matter (15.35),

$$\rho \frac{D\left((1/2)\boldsymbol{v}^2\right)}{Dt} = \rho \boldsymbol{v} \cdot \frac{D\boldsymbol{v}}{Dt} = \boldsymbol{v} \cdot \boldsymbol{f}^*.$$

Using (22.11) with $q \to (1/2)\boldsymbol{v}^2$ we arrive at the *global equation of kinetic energy balance*

$$\boxed{\frac{D\mathcal{T}}{Dt} = P,} \tag{22.35}$$

where

$$\frac{D\mathcal{T}}{Dt} = \frac{d\mathcal{T}}{dt} + \oint_S \frac{1}{2}\rho \boldsymbol{v}^2 (\boldsymbol{v} - \boldsymbol{v}_S) \cdot d\boldsymbol{S}, \qquad P = \int_V \boldsymbol{v} \cdot \boldsymbol{f}^* \, dV. \tag{22.36}$$

The source of kinetic energy P is the total rate of work, or *power*, of the effective forces acting on the material particles in the control volume. Formally, we may conclude that the kinetic energy is conserved when the total power vanishes, but that is, as we shall see, not very useful, because the power rarely vanishes.

To derive a global expression for the power we use the expression (9.18) for the effective force density to write,

$$\boldsymbol{v} \cdot \boldsymbol{f}^* - \boldsymbol{v} \cdot \boldsymbol{f} = \sum_{ij} v_i \nabla_j \sigma_{ij} = \sum_{ij} \nabla_j (v_i \sigma_{ij}) - \sum_{ij} \sigma_{ij} \nabla_j v_i.$$

Converting the integral of the first term to a surface integral by means of Gauss' theorem (6.4), we find,

$$\boxed{P = \int_V \boldsymbol{v} \cdot \boldsymbol{f} \, dV + \oint_S \boldsymbol{v} \cdot \boldsymbol{\sigma} \cdot d\boldsymbol{S} - \int_V \sum_{ij} \sigma_{ij} \nabla_j v_i \, dV.} \tag{22.37}$$

The total power thus consists of the power of the volume forces (external as well as internal), the power of the external contact forces, plus a third contribution which we (including the sign) interpret as the *power of the internal stresses.*

Internal rate of work

Although Newton's third law guarantees that the contact forces between neighbouring material particles cancel, the power of these forces does not cancel because neighbouring material particles have slightly different velocities, as witnessed by the appearance of the velocity gradients $\nabla_j v_i$ in the last term of (22.37).

Given that the displacement of a comoving material particle in a small time interval δt is $\delta u_i = v_i \delta t$, we may employ (10.36) on page 129 to obtain the rate of work *against* the internal stresses,

$$\boxed{\dot{W}_{\text{int}} = \int_V \sum_{ij} \sigma_{ij} \nabla_j v_i \, dV.} \tag{22.38}$$

Taking the sign into account, this is of course consistent with the interpretation of the last term in the total power (22.37). We previously used this expression to calculate the rates of dissipative loss for incompressible fluids (17.18), as well as for compressible fluids (17.30).

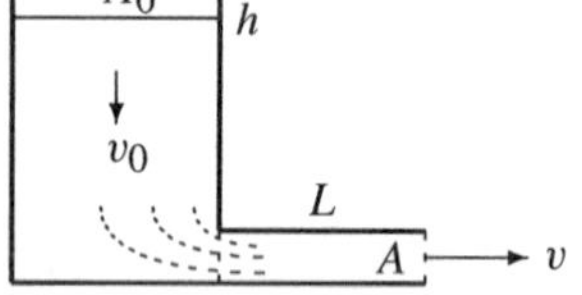

The water first accelerates through the pipe while pouring out through the exit. The kinetic energy of the water in the transition region between cistern and pipe (dashed flowlines) is assumed to be negligible.

Pulling the plug

As an illustration of the use of kinetic energy balance, we return to the familiar example of a cistern being drained through a narrow pipe. When the plug is pulled at the bottom of the cistern, the fluid goes through a short initial acceleration before steady flow is established. We shall now calculate how the water flow becomes steady and how long it takes.

The drain is a horizontal pipe of length L and cross section A, and the cistern is vertical with large cross section $A_0 \gg A$, continuously being refilled to maintain a constant water level h. Initially the pipe is full and all the water is assumed to be at rest. The control volume is fixed and contains all the water in the system between the open surface, the pipe outlet and the walls of the cistern and pipe. Assuming that the water is non-viscous and incompressible, the kinetic energy balance (22.35) in a fixed control volume takes the form

$$\frac{d\mathcal{T}}{dt} + \oint_V \frac{1}{2}\rho_0 \boldsymbol{v}^2 \boldsymbol{v} \cdot d\boldsymbol{S} = \int_V \rho_0 \boldsymbol{v} \cdot \boldsymbol{g}\, dV - \oint_S p\, \boldsymbol{v} \cdot d\boldsymbol{S}. \tag{22.39}$$

On the left-hand side we have the material time derivative $D\mathcal{T}/Dt$ expressed through Reynolds' transport theorem (22.10) with $q \to (1/2)\boldsymbol{v}^2$. On the right-hand side we find the total power, consisting of the rate of work of constant gravity and the rate of work of the contact forces, in this case only the pressure. For non-viscous incompressible fluids, the rate of work of the internal contact forces vanishes according to (17.18).

To estimate the various terms in this equation we shall again use the *plug flow* approximation in which the flow is assumed to have the same vertical velocity throughout the cistern and the same horizontal velocity throughout the pipe. In this approximation the average of a product of velocity fields equals the product of the averages. We have previously seen (page 254) that for laminar viscous flow in pipes much longer than the entrance length, such an approximation is rather poor, but for non-viscous or turbulent flow it is not so bad.

Ignoring the transition region from the cistern to the pipe (see problem 22.5), the total kinetic energy of the water in the system is approximatively given by the sum of the kinetic energies of the water moving vertically down through the cistern with average velocity v_0 and the water moving horizontally through the pipe with average velocity v. Using mass conservation (Leonardo's law), $v_0 A_0 = vA$, to eliminate v_0 we find,

$$\mathcal{T} \approx \frac{1}{2}\rho_0 A_0 h\, v_0^2 + \frac{1}{2}\rho_0 A L v^2 = \frac{1}{2}\rho_0 A v^2 \left(L + h\frac{A}{A_0}\right) = \frac{1}{2}\rho_0 A L_0 v^2, \tag{22.40}$$

where

$$L_0 = L + h\frac{A}{A_0} \tag{22.41}$$

is the effective total pipe length, including the kinetic energy of the descending water in the cistern. Even if $A/A_0 \ll 1$ the water level may be so high that there is a significant contribution from the cistern.

The time derivative of the kinetic energy is thus,

$$\frac{d\mathcal{T}}{dt} = \rho_0 A L_0 v \frac{dv}{dt}. \tag{22.42}$$

Kinetic energy is only lost through the outlet A and gained through the inlet A_0, and in the same approximation as above the net rate of loss of kinetic energy from the system is,

$$\oint_S \frac{1}{2}\rho_0\, \boldsymbol{v}^2\, \boldsymbol{v} \cdot d\boldsymbol{S} \approx \frac{1}{2}\rho_0 A v^3 - \frac{1}{2}\rho_0 A_0 v_0^3 = \frac{1}{2}\rho_0 A v^3 \left(1 - \frac{A^2}{A_0^2}\right) \approx \frac{1}{2}\rho_0 A v^3,$$

where we in the last step have used $A \ll A_0$.

Gravity only performs work on the descending water in the cistern at the rate,

$$\int_V \rho_0 \boldsymbol{v} \cdot \boldsymbol{g}\, dV \approx \rho_0 v_0 g_0 A_0 h = \rho_0 g_0 v A h. \tag{22.43}$$

At both the open surface and at the pipe outlet, there is atmospheric pressure p_0, such that

$$-\oint p\,\boldsymbol{v}\cdot d\boldsymbol{S} = -p_0(-v_0)A_0 - p_0 v A = 0. \tag{22.44}$$

Atmospheric pressure performs no work, because of mass conservation.

Putting all four terms together in (22.39), kinetic energy balance results in the differential equation for the drain pipe velocity,

$$\frac{dv}{dt} = \frac{2g_0 h - v^2}{2L_0}. \tag{22.45}$$

The right-hand side vanishes for $v = \sqrt{2g_0 h}$, which is the well-known terminal velocity obtained from Torricelli's law (16.21). Solving the differential equation with the initial condition $v = 0$ at $t = 0$, we find

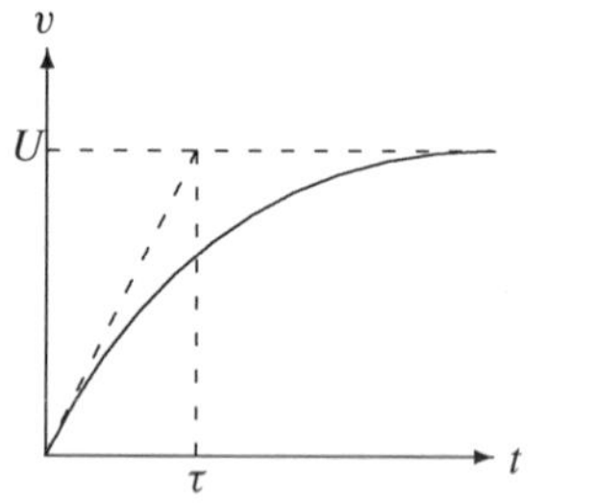

Rise of the velocity in the drain pipe towards terminal velocity $U = \sqrt{2g_0 h}$.

$$\boxed{v(t) = \sqrt{2g_0 h}\,\tanh\frac{t}{\tau},} \tag{22.46}$$

where the characteristic rise time towards terminal velocity is,

$$\tau = \frac{2L_0}{\sqrt{2g_0 h}}. \tag{22.47}$$

Quite reasonably, τ equals the time it takes the water to pass through the pipe at half the terminal speed. At $t = 2\tau$, the water has reached 96% of terminal speed.

Example 22.7.1 (Barrel of wine): For a cylindrical barrel of wine (example 16.3.2) with diameter 1 m and height 2 m, emptied through a spout with diameter 5 cm and length $L = 20$ cm, the effective length becomes $L_0 = 20.5$ cm. The terminal speed is about 6 m s^{-1}, and the water obtains 96% of this speed after just 0.13 s! If you try to put back the plug in the spout after merely half a second, there will nevertheless be about 5.6 litres of wine on the floor (see problem 22.3).

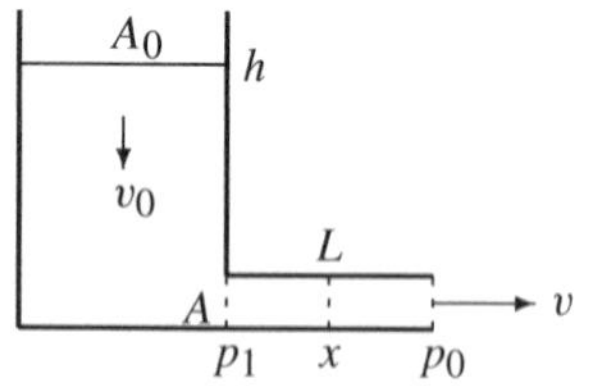

During acceleration there is a pressure drop $\Delta p = p_1 - p_0$ between entry and exit to the pipe.

Until now we have completely escaped the problem of how the pressure behaves inside the cistern and in the pipe. Taking the x-axis along the pipe, and using that v does not depend on x, there will be no comoving acceleration in the pipe, and the x-component of Euler's equation (16.1) simply becomes

$$\frac{dv}{dt} = -\frac{1}{\rho_0}\frac{\partial p}{\partial x}. \tag{22.48}$$

Since dv/dt takes the same value throughout the pipe, the pressure gradient will be independent of x. Expressed in terms of the pressures p_1 and p_0 at the ends of the pipe, the gradient is $\partial p/\partial x = (p_0 - p_1)/L$. Finally, inserting the acceleration (22.45) and the solution (22.46) we obtain the pressure difference between the pipe entry and the exit (or the cistern surface)

$$\Delta p = p_1 - p_0 = \frac{1}{2}\rho_0(2g_0 h - v^2) = \frac{\rho_0 g_0 h}{\cosh^2(t/\tau)}. \tag{22.49}$$

This shows that at $t = 0$ the entry pressure p_1 equals the hydrostatic pressure of the water in the cistern, as one would expect, and that it approaches atmospheric pressure p_0 in the pipe in the characteristic rise time τ. In reality there will remain a residual pressure drop in the pipe due to viscous friction (see section 18.4).

22.8 Mechanical energy balance

Even if kinetic energy is not conserved, kinetic energy balance can as shown in the preceding section be very useful in obtaining an estimate of the solution to a fluid mechanics problem by calculating the rate of work of the forces acting on the system. In numerous practical situations the body force is entirely due to a *static external field of gravity*, $\boldsymbol{f} = \rho\boldsymbol{g} = -\rho\nabla\Phi$, and in that case it is, as in ordinary mechanics, convenient to introduce the potential energy in the gravitational field,

$$\mathcal{V} = \int_V \rho\,\Phi\,dV. \tag{22.50}$$

From Reynolds' transport theorem (22.10) with $q \to \Phi$ we find,

$$\frac{D\mathcal{V}}{Dt} = \int_V \rho \frac{D\Phi}{Dt}\, dV = \int_V \rho(\boldsymbol{v}\cdot\nabla)\Phi\, dV = -\int_V \rho \boldsymbol{v}\cdot\boldsymbol{g}\, dV, \tag{22.51}$$

where we have used the fact that the potential is time-independent $\partial\Phi/\partial t = 0$ and that $\nabla\Phi = -\boldsymbol{g}$.

Total contact power

Since $D\mathcal{V}/Dt$ is the opposite of the body force contribution to the total power (22.37), it is convenient to define the total mechanical energy as the sum of the kinetic and potential energies,

$$\mathcal{E} = \mathcal{T} + \mathcal{V}. \tag{22.52}$$

It satisfies the *global equation of mechanical energy balance*,

$$\boxed{\frac{D\mathcal{E}}{Dt} = \widetilde{P},} \tag{22.53}$$

where the right-hand side is,

$$\widetilde{P} = \oint_S \boldsymbol{v}\cdot\boldsymbol{\sigma}\cdot d\boldsymbol{S} - \int_V \sum_{ij} \sigma_{ij}\nabla_j v_i\, dV. \tag{22.54}$$

The first term is the power of the *external* and the last the power of the *internal* contact forces. Evidently, *the mechanical energy is conserved for a comoving volume when the total power of the contact forces vanishes.*

For a non-viscous, incompressible fluid where $\sigma_{ij} = -p\,\delta_{ij}$, the last term in (22.54) vanishes, such that the mechanical energy will be conserved if the pressure does no net work on the surface. This was the principle used intuitively in the analysis of connected tubes at the beginning of this chapter (section 22.1). For a viscous incompressible fluid, the last term is given by (17.18) and is always negative, showing that viscosity always eats away the mechanical energy of a fluid.

22.9 Energy balance in elastic fluids

The global equation of mechanical energy balance (22.53) arises—so to speak—by 'moving over' the rate of work of gravity from the right-hand side of the kinetic energy balance (22.35) to the left-hand side as a potential energy. In general, it is also possible to 'move over' the power of the internal contact forces, the last term in the contact power (22.54), such that the total energy takes the form

$$\boxed{\mathcal{E} = \mathcal{T} + \mathcal{V} + \mathcal{U},} \tag{22.55}$$

where

$$\mathcal{U} = \int_V \rho\, u\, dV, \tag{22.56}$$

is called the *total internal energy* and u the *specific internal energy*. The specific internal energy contains essentially all the information about the thermodynamics of the material in the control volume and is not known in general. For particular types of matter, the most important being ideal gases, there are simple explicit expressions for the specific internal energy (see below and appendix C).

Specific internal energy for non-viscous barotropic fluids

Here we shall only carry through the analysis for non-viscous barotropic fluids, for which the internal energy is analogous to the potential energy stored in a compressed spring. Such fluids are for this reason often called *elastic*. In the absence of viscosity, $\sigma_{ij} = -p\delta_{ij}$, the contact power (22.54) becomes,

$$\widetilde{P} = -\oint_S p\,\boldsymbol{v}\cdot d\boldsymbol{S} + \int_V p\nabla\cdot\boldsymbol{v}\, dV. \tag{22.57}$$

The task is now to find an expression for the specific internal energy which cancels the last term.

A fluid in a barotropic state has a unique relationship between density and pressure, $\rho = \rho(p)$, and the so-called *pressure potential*,

$$w(p) = \int \frac{dp}{\rho(p)}, \tag{22.58}$$

plays an important role in the dynamics (see page 54). Using the chain rule we obtain,

$$\rho \frac{D}{Dt}\left(w - \frac{p}{\rho}\right) = \rho\left(\frac{1}{\rho}\frac{Dp}{Dt} - \frac{1}{\rho}\frac{Dp}{Dt} + \frac{p}{\rho^2}\frac{D\rho}{Dt}\right) = \frac{p}{\rho}\frac{D\rho}{Dt} = -p\boldsymbol{\nabla}\cdot\boldsymbol{v},$$

where in the last step the continuity equation $D\rho/Dt = -\rho\boldsymbol{\nabla}\cdot\boldsymbol{v}$ in the form (15.28) on page 196 was used.

Since the final result apart from sign is identical to the integrand of the last term in (22.57), we define the *specific internal energy of a non-viscous barotropic fluid* to be (apart from an additive constant),

$$\boxed{u = w - \frac{p}{\rho}.} \tag{22.59}$$

An incompressible non-viscous fluid has constant density and pressure potential $w = p/\rho$, such that its internal energy may formally be chosen to vanish. For such fluids, there is only mechanical energy.

For an isentropic ideal gas with pressure potential (16.37), the specific internal energy becomes (see also appendix C),

$$u = \frac{1}{\gamma - 1}\frac{p}{\rho} = c_v T, \qquad c_v = \frac{1}{\gamma - 1}\frac{RT}{M_{\text{mol}}}. \tag{22.60}$$

Here γ is the adiabatic index and c_v is the specific heat at constant volume (i.e. constant density).

Total energy balance

With the specific internal energy of an elastic fluid given by (22.59) the complete specific energy becomes,

$$\epsilon = \tfrac{1}{2}\boldsymbol{v}^2 + \Phi + u. \tag{22.61}$$

From (22.11) with $q \to \epsilon$ we obtain the equation of *global energy balance*,

$$\boxed{\frac{D\mathcal{E}}{Dt} = \dot{W},} \tag{22.62}$$

where

$$\frac{D\mathcal{E}}{Dt} = \frac{d\mathcal{E}}{dt} + \oint_S \epsilon(\boldsymbol{v} - \boldsymbol{v}_S)\cdot d\boldsymbol{S}, \qquad \dot{W} = -\oint_S p\,\boldsymbol{v}\cdot d\boldsymbol{S}. \tag{22.63}$$

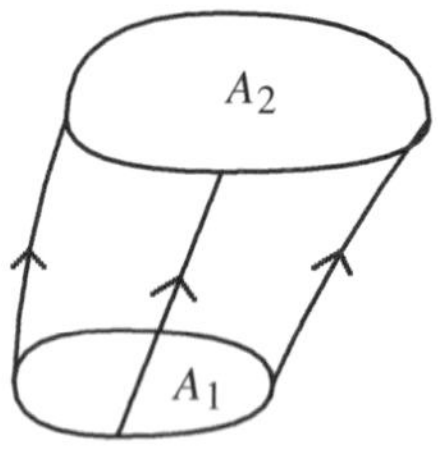

A stream tube consisting of all the streamlines that enter the area A_1 and exit through A_2.

Since $\delta V = \boldsymbol{v}\delta t \cdot d\boldsymbol{S}$ is the (signed) volume swept out by a comoving surface element $d\boldsymbol{S}$ in the time δt, the expression for $\dot{W}$ is identical to the rate of the thermodynamic work $-p\,\delta V$ performed on the surface of the control volume.

Concluding, we have shown that for a barotropic non-viscous fluid *the total energy $\mathcal{E}$ is conserved when the pressure performs no work on the surface of the control volume.* We shall not present any applications of this result, because it is equivalent to Bernoulli's theorem.

Bernoulli's theorem and energy balance

To find the relationship between energy balance and Bernoulli's theorem we assume that the flow is steady and apply energy balance to a fixed control volume in the form of a tiny stream tube consisting of all the streamlines that go into a tiny area A_1 and leave through the equally tiny area A_2. Denoting the average entry and exit velocities by v_1 and v_2, the mass flux through the tube is,

$$Q = \rho_1 A_1 v_1 = \rho_2 A_2 v_2, \tag{22.64}$$

because no fluid passes through the sides of a tube consisting of streamlines.

The total energy of the fluid in the stream tube must be constant in steady flow, i.e. $d\mathcal{E}/dt = 0$, such that the material derivative obtained from Reynolds' transport theorem (22.10) is entirely determined by the rate of loss of energy through the surface of the tube. For a sufficiently narrow stream tube we have,

$$\frac{D\mathcal{E}}{Dt} = \rho_2\epsilon_2 v_2 A_2 - \rho_1\epsilon_1 v_1 A_1, \tag{22.65}$$

and similarly the rate of work (22.63),

$$\dot{W} = -p_2 v_2 A_2 + p_1 v_1 A_1. \tag{22.66}$$

Again there is no contribution to the work integral from the sides of the stream tube, because $\boldsymbol{v}\cdot d\boldsymbol{S}$ vanishes there.

Energy balance (22.62) implies now in conjunction with mass conservation (22.64),

$$\epsilon_1 + \frac{p_1}{\rho_1} = \epsilon_2 + \frac{p_2}{\rho_2}. \tag{22.67}$$

Since the specific energy (22.61) may be written

$$\epsilon = H - \frac{p}{\rho} \tag{22.68}$$

where H is the Bernoulli function (16.35), this relation is Bernoulli's theorem for a compressible fluid, $H_1 = H_2$, applied to any streamline running through the stream tube. Bernoulli's theorem is thus completely equivalent to energy balance for steady flow in a non-viscous barotropic fluid. Under these circumstances, we do not gain further information about the system from energy balance than from applying Bernoulli's theorem.

Problems

22.1 Before the advent of modern high precision positioning systems, standard height levels could be transmitted across a strait by means of a long tube filled with water. **(a)** Calculate the oscillation time of the water in the tube for a strait about 20 km wide when viscosity is disregarded. **(b)** Use energy balance to calculate the influence of viscosity on the oscillations for a pipe of radius 1 cm (hint: use Poiseuille dissipation). How long does it take for the water to come to rest? **(c)** Would it be better to use a wider tube?

22.2 Solve the rocket flight problem analytically in the absence of air resistance.

22.3 Calculate the volume of fluid that has emerged from the draining cistern (page 317) after time t.

22.4 Use mechanical energy balance (22.53) rather than kinetic energy balance to analyse the initial acceleration of the water in a cistern, when the plug is pulled.

22.5 Consider a circular drain of radius a in the bottom of a large circular cistern of radius $b \gg a$. Assume that the average velocity of the water at any half-sphere of radius r centred at the drain is $v(r) = (a/r)^2 v$ where v is the average velocity at the drain, so that the same amount of water passes through the half-sphere for all $r > a$. Calculate the kinetic energy associated with this velocity distribution and compare it with the kinetic energies of the water in the cistern and in the pipe.

22.6 A container with (constantly refilled) water level h and cross section A_0 is drained through a very long pipe of cross section A forming an angle α with the horizontal. A valve is placed a good distance L down the pipe, but the pipe continues downwards with the same slope far beyond the valve. Initially there is water in the pipe up to the valve. Find the equation of motion for the water front after the valve is opened and it has progressed a distance x past the valve.

∗ **22.7 Molecular model for a material particle.** A volume V contains N point-like particles of mass m_n with instantaneous positions $\boldsymbol{x}_n$ and velocities $\boldsymbol{v}_n$. The total mass is $M = \sum_n m_n$ and the centre of mass is $\boldsymbol{x} = \sum_n m_n \boldsymbol{x}_n / M$. Define the relative positions $\boldsymbol{x}'_n = \boldsymbol{x}_n - \boldsymbol{x}$ and the relative velocities $\boldsymbol{v}'_n = \boldsymbol{v}_n - \boldsymbol{v}$ where $\boldsymbol{v} = \sum_n m_n \boldsymbol{v}_n / M$ is the centre-of-mass velocity.

Assume that the relative positions and velocities are random and average out to zero, such that

$$\langle \boldsymbol{x}'_n \rangle = \boldsymbol{0} \tag{22.69}$$

$$\langle \boldsymbol{v}'_n \rangle = \boldsymbol{0}. \tag{22.70}$$

Also assume that they are independent, uncorrelated and that the velocities are uniformly and spherically distributed, such that

$$\langle (\boldsymbol{x}'_n)_i (\boldsymbol{v}'_m)_j \rangle = 0 \tag{22.71}$$

$$\langle (\boldsymbol{v}'_n)_i (\boldsymbol{v}'_m)_j \rangle = U^2 \delta_{ij} \delta_{nm} \tag{22.72}$$

where U is a constant with dimension of velocity.

(a) Show that the total angular momentum of all the particles in the system is

$$\mathcal{L} = M\boldsymbol{x} \times \boldsymbol{v} + \sum_n m_n \boldsymbol{x}'_n \times \boldsymbol{v}'_n \tag{22.73}$$

and calculate its average.

(b) Show that the total kinetic energy of all the particles is

$$\mathcal{T} = \frac{1}{2} M \boldsymbol{v}^2 + \frac{1}{2} \sum_n m_n (\boldsymbol{v}'_n)^2 \tag{22.74}$$

and calculate its average.

23 Reaction forces and moments

A rocket accelerating in empty space while spewing hot gases in the opposite direction is at first sight rather mysterious, because there are no external forces to account for its acceleration. On reflection we understand from momentum conservation that the high-speed gas streaming out of the rocket nozzle carries momentum away from the rocket and thereby adds momentum to the rocket itself. But why are rockets and jet aircraft said to be 'reaction-driven'? What is reacting to what?

Fluids flowing through tubes, pipes, ducts and conduits are important components in many of the machines used in the home, in transport and industry. Everywhere in such a system the conduit walls will act on the fluid with contact forces that confine and guide the fluid. By Newton's third law the fluid must respond with an equal and opposite *reaction force*. This definition is purely a matter of convention; there is no deeper distinction between the agents in an action/reaction pair. With this convention one may say that a rocket is driven by the reaction of the gas against the confining contact forces exerted on it by the inner walls of the rocket engine.

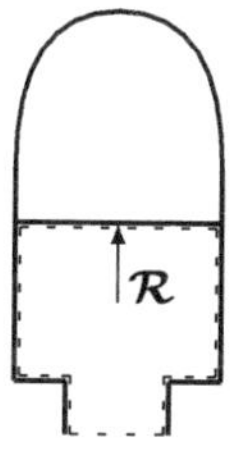

The pressure from the burning gases exerts a force $\mathcal{R}$ on the inside of the rocket engine which propels the rocket upwards in this picture. Conventionally, $\mathcal{R}$ is called a reaction force.

Lawn sprinklers, fans, windmills, hairdryers, propellers, pumps, water turbines, compressors and jet engines are all examples of *turbomachinery* used everywhere today. In such machines the fluid will likewise exert a *reaction moment* of force on the main rotating part, called the *rotor*. Depending on the sign of the reaction moment energy is either taken out of the fluid as in a water turbine or put into it as in an air blower.

In this chapter we shall develop the formalism for reaction forces and moments, and apply it to a number of generic examples. We shall see that it is sometimes quite a subtle task to unravel what is the true reaction force on a given part of a machine, because forces may be transmitted to this part from far away through stiff machine housing and supports.

23.1 Reaction forces

Formally, the cause of a rocket's acceleration in empty space may be exposed by calculating the rate of momentum change from momentum balance (22.15) and (22.16),

$$\frac{d\mathcal{P}}{dt} = \mathcal{F} - \oint_S \rho \boldsymbol{v}\,(\boldsymbol{v} - \boldsymbol{v}_S) \cdot d\boldsymbol{S}. \tag{23.1}$$

The integral on the right-hand side (including the sign) represents the net momentum flux *into* the control volume through its surface. Although this term simulates the action of an external force, it is not a reaction against another force, but more akin to the 'fictitious' centrifugal and Coriolis forces.

Total force in steady flow

In steady flow, it is natural to choose a fixed control volume. Since the flow pattern always remains the same, the total momentum of the material in the fixed control volume will be constant, $d\boldsymbol{\mathcal{P}}/dt = \mathbf{0}$, and from momentum balance (23.1) it follows that the total external force on the control volume must be,

$$\boxed{\boldsymbol{\mathcal{F}} = \oint_S \rho \boldsymbol{v}\, \boldsymbol{v} \cdot d\boldsymbol{S}.} \tag{23.2}$$

The total force is still given by the sum of the volume and contact forces (22.17), so this relation should be seen as a constraint which necessarily must be satisfied by these forces to secure a steady flow. Contrary to hydrostatics ($\boldsymbol{v} = \mathbf{0}$) where the total force on any volume has to vanish, the environment is required to exert a generally non-vanishing total force on the control volume, determined by the net flux of momentum *out* of the control volume. Since the integrand vanishes at solid walls where $\boldsymbol{v} \cdot d\boldsymbol{S} = 0$, it may always be calculated from the flow through the openings where fluid is let in and out of the control volume.

From the total force, which is perfectly unambiguous, it is often fairly easy to determine the actual reaction force according to one's personal preferences concerning what acts and what reacts. We shall now present some examples of how to use this method.

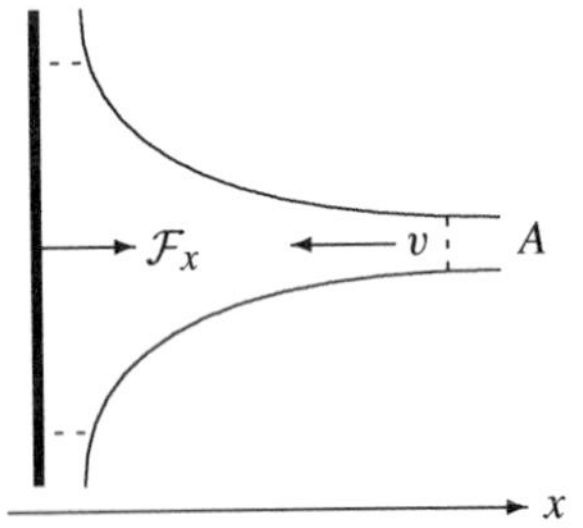

A jet of water hitting a wall. The control volume is limited by the jet's surface and by the dashed lines.

Reaction from a jet hitting a wall

A steady jet of water hits a fixed wall and produces no backsplash in the process, but just streams uniformly away over the wall (when gravity is ignored). A fixed control volume is chosen to contain all of the water in the jet with inlet far from the wall and the outlet far from the centre of the jet. The total force $\mathcal{F}_x$ normal to the wall equals the momentum flux in the x-direction through the inlet because the outflow is approximately parallel with the wall and thus orthogonal to x. Let the inlet area be A, and let the magnitude of the average inlet velocity be v, so that the total volume flux through the inlet in the x-direction becomes $-vA$. The momentum density along x is similarly $-\rho_0 v$, leading to the following estimate of the total force on the water,

$$\mathcal{F}_x \approx \rho_0 v^2 A. \tag{23.3}$$

This force can only be due to the contact force from the wall and may be viewed as the wall's reaction to the impact of the incoming fluid. Alternatively, $-\mathcal{F}_x$ may be viewed as the fluid's reaction to the confining forces from the wall. The argument also shows that $\mathcal{F}_x$ will be even larger, if there is backsplash from the wall.

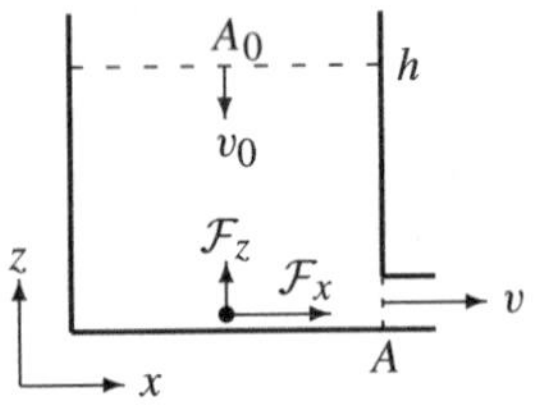

Example 23.1.1 (Water cannon): Police use water cannons to control crowds. With a jet diameter of 1.5 inch and volume flux of 20 litres (5 gallons) per second, the water speed becomes about 18 m s^{-1} and the jet reaction about 350 N, corresponding to the weight of 35 kg. You do not stay on your feet for long after being hit by such a jet, although it will probably not hurt you seriously.

Draining cistern. The loss of momentum from the water leaving the container through the drain requires a horizontal reaction force $\mathcal{F}_x$ from the cistern. There is also a much smaller vertical reaction force F_z.

Reaction from draining cistern

A cistern with cross section A_0 drains steadily with velocity v through a spout of cross section $A \ll A_0$ while being continually refilled to maintain a constant water level h above the spout. Assuming nearly ideal flow, we may use the Toricelli result (16.21) that the outlet speed equals the free-fall speed from height h, i.e. $v \approx \sqrt{2g_0h}$. In the steady state, the required horizontal force on the water is as in the preceding example,

$$\mathcal{F}_x \approx \rho_0 v^2 A \approx 2\rho_0 g_0 A h. \tag{23.4}$$

This force must be ultimately provided by clamps that hold the cistern into place, or by the static friction between the cistern and the floor that supports it.

The water moving slowly down through the cistern with vertical velocity $v_0 = vA/A_0$ requires a much smaller force,

$$\mathcal{F}_z \approx \rho_0 v_0^2 A_0 = \frac{A}{A_0} \mathcal{F}_x. \tag{23.5}$$

The positive sign can be understood from the fact that the water transports a negative volume flux $-v_0 A_0$ along the z-direction with a negative momentum density $-\rho_0 v_0$. The true vertical reaction force from the ground on the bottom of the cistern is in this case $\mathcal{F}_z + M g_0$ where M is the total mass of the cistern plus water.

Example 23.1.2 (Barrel of wine): With the parameters of the wine barrel (example 16.3.2), i.e. $A \approx 20\ \text{cm}^2$, $A_0 \approx 0.8\ \text{m}^2$, and $h \approx 2$ m, one finds $\mathcal{F}_x \approx 77$ N, corresponding to the weight of almost 8 litres of wine. The vertical force becomes $\mathcal{F}_z = 0.2$ N which only amounts to the weight of a small thimbleful of wine.

Around the bend

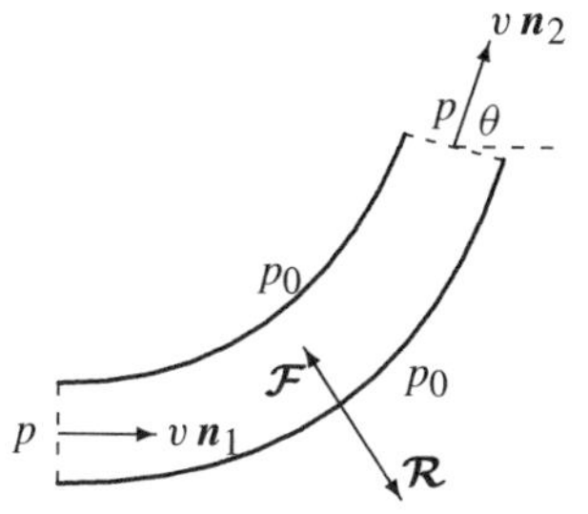

A pipe bending through an angle θ. There is ambient pressure p_0 outside the system and pressure p in the fluid.

A pipe with cross section A bends horizontally through an angle θ, and incompressible fluid flows steadily through the pipe with velocity v. The control volume is chosen to contain all the fluid between inlet and outlet, leading to a total force

$$\mathcal{F} \approx (\boldsymbol{n}_2 - \boldsymbol{n}_1)\rho_0 v^2 A, \tag{23.6}$$

where $\boldsymbol{n}_1$ and $\boldsymbol{n}_2$ are the inlet and outlet directions of the flow. In this case there is also an external force from the pressure p on the inlet and outlet, such that the true reaction force from the fluid on the inner pipe surface is,

$$\mathcal{R} = -\mathcal{F} + (\boldsymbol{n}_1 - \boldsymbol{n}_2)pA = (\boldsymbol{n}_1 - \boldsymbol{n}_2)(p + \rho_0 v^2)A. \tag{23.7}$$

The precise reasons for including the pressure are given in section 23.2.

Example 23.1.3: A 90° bend in a 1 inch pipe with water flowing at 3 m s^{-1} requires a total force of magnitude $|\mathcal{F}| = 6.5$ N.

Nozzle puzzle

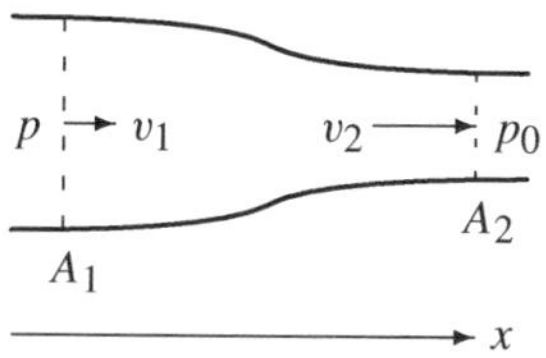

A nozzle formed by narrowing down a pipe from the inlet A_1 to the open outlet A_2.

Although it is always unambiguous what the required total force should be for a fixed control volume in steady flow, it can sometimes be hard to decide how a piece of equipment will act on the external supports that keep it in place.

As an illustration we consider a nozzle—the metal device sitting at the end of a firehose—formed by narrowing down the cross section of a straight pipe from A_1 to A_2 over a fairly short distance. Mass conservation guarantees that the same volume flux of liquid goes in and out,

$$Q = v_1 A_1 = v_2 A_2, \tag{23.8}$$

so that the total force in the downstream direction becomes

$$\mathcal{F}_x = \rho_0 v_2^2 A_2 - \rho_0 v_1^2 A_1 = \rho_0 Q^2 \left(\frac{1}{A_2} - \frac{1}{A_1}\right). \tag{23.9}$$

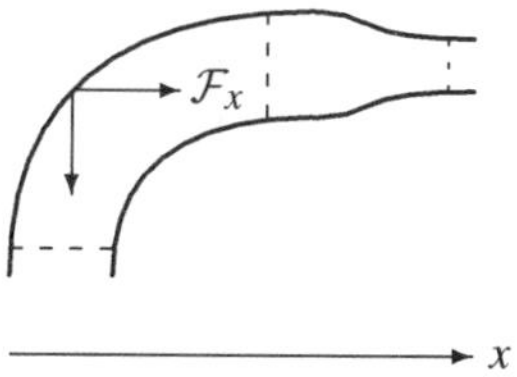

A nozzled firehose with an unsupported 90° bend. The extra force required by the bend is transmitted down from the nozzle through the soft but unstretchable material of the hose.

The required external force is positive for $A_2 < A_1$ and it follows that the fireman, as expected, must push forward on the nozzle.

For $A_1 = A_2$ the total force vanishes, and that seems to contradict the common gardening experience that it is always necessary to hold firmly on to an open hose (even without nozzle) because it otherwise will flail back and drench you. So what is the actual extra force that must be applied by the hands that hold a nozzle when water is pouring out?

The puzzle is resolved when one realizes that a soft but unstretchable hose always meanders through bends and turns upstream of the nozzle. If, for example, the hose has an unsupported 90° bend, we find by including the bend in the control volume that the required total force along the nozzle equals the rate of loss of momentum through the outlet without any contribution from the inlet,

$$\mathcal{F}_x = \rho_0 v_2^2 A_2 = \frac{\rho_0 Q^2}{A_2}. \tag{23.10}$$

The required total force is always positive in agreement with common experience.

The explanation is apparently that the moving fluid in the unsupported bend demands an external force for the bend to remain in place, a force which in this case is transmitted down to the bend from the nozzle through the unstretchable material of the hose. It is definitely better to use this value for the purpose of calculating the required strength of the supports of the nozzle. In the following section we will determine the true reaction force on the nozzle itself.

Example 23.1.4 (Firehose): Older firehoses with 2.5 inch diameter equipped with a 1.5 inch nozzle can typically deliver 20–40 litres of water per second. At a rate of 20 litres per second, the required nozzle force (with a bend) is 350 N which is just about manageable for a single firefighter. A modern 5 inch firehose can deliver up to 100 litres of water per second through a 3 inch nozzle, leading to a required force of about 2200 N. This nozzle cannot even be handled by three firefighters, but should be firmly anchored in a boat or truck.

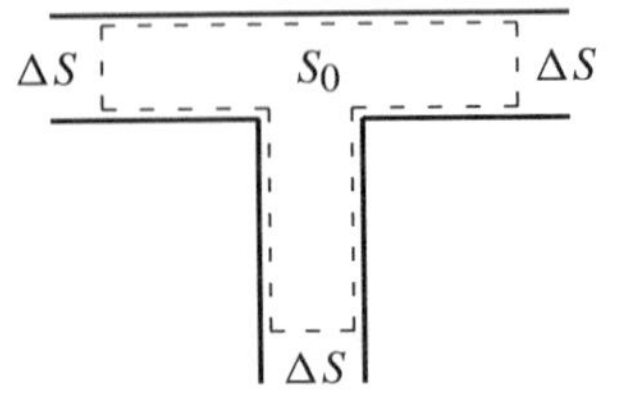

A control volume for a T-junction containing all the fluid between the inner conduit walls S_0 and the inlets and outlets ΔS.

23.2 Formal definition of reaction force

In most cases fluid flows in and out of the control volume through well-defined openings which together constitute a piece ΔS of the control volume surface S. The closed surface is thus composed of two open parts, $S = S_0 + \Delta S$, where S_0 is chosen to follow the inside of the fluid conduits.

Absolute reaction force

The *absolute reaction force* is defined to be the resultant of all the contact forces that the contents of the control volume exert on S_0,

$$\mathcal{R} = -\int_{S_0} \boldsymbol{\sigma} \cdot d\boldsymbol{S}. \tag{23.11}$$

The surface contribution to the total force (22.17) may now be split into the reaction force plus a part from the openings ΔS. Using the momentum balance (23.1) we obtain an exact expression for the reaction force,

$$\boxed{\mathcal{R} = -\frac{d\mathcal{P}}{dt} + \int_V \boldsymbol{f}\, dV + \int_{\Delta S} \boldsymbol{\sigma} \cdot d\boldsymbol{S} - \int_{\Delta S} \rho \boldsymbol{v}\, (\boldsymbol{v} - \boldsymbol{v}_S) \cdot d\boldsymbol{S}.} \tag{23.12}$$

The last term (including the sign) represents as before the net momentum flux *into* the control volume. Although it in itself is not a true force, momentum balance makes it appear as a contribution to the reaction force.

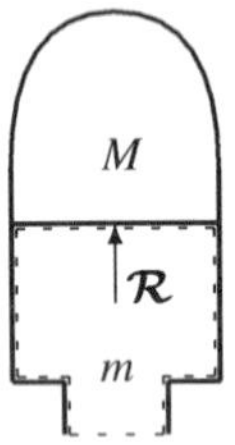

The control volume encircles the chamber in the rocket where the fuel burns. The reaction force $\mathcal{R}$ is the total force exerted on the walls by the gas in the chamber. It is this reaction force which lifts the solid structure of the rocket.

Finally, we are able quantitatively to clear up the question of why rockets are said to be *reaction-driven*. Choosing the control volume to encompass only the chamber in which the gases burn, the reaction force that these gases exert on the walls of the chamber is given by the above expression. In this case, the first two terms are tiny because they are proportional to the mass m of the burning gasses, which is normally much smaller than the mass M of the rocket. The third term can also be disregarded because there is essentially no stress in the exhaust (at least at subsonic speeds). Thus, for the rocket in vacuum the true reaction force may to a very good approximation be calculated from the last term in (23.12), representing the net gain of momentum from the gases expelled through the exhaust nozzle.

Reaction force in steady flow

Several simplifications are possible in the above expression for the reaction force. In steady flow with a fixed control volume, the first term vanishes, $d\mathcal{P}/dt = \boldsymbol{0}$. Usually the only volume force is constant gravity $\boldsymbol{g}_0 = (0, 0, -g_0)$, making the volume integral equal to the weight $M\boldsymbol{g}_0$ where M is the total mass of the control volume. Although there may be shear stresses at play in the openings ΔS, they tend to be small, so the stress tensor may be replaced by the pressure $-p$ in the third term (at least for high Reynolds number). With these simplifications the absolute steady-flow reaction force on a fixed control volume may be written,

$$\boxed{\mathcal{R} \approx M\boldsymbol{g}_0 - \int_{\Delta S} p\, d\boldsymbol{S} - \int_{\Delta S} \rho \boldsymbol{v}\, \boldsymbol{v} \cdot d\boldsymbol{S}.} \tag{23.13}$$

The first term is the well-known weight reaction which for incompressible flow is independent of the velocity and thus equal to the static weight of the fluid. For compressible flow the mass of the fluid could depend on the velocity of the flow; if, for example, you increase the speed of the gas in a natural gas pipeline by increasing the inlet pressure, the gas in the pipeline becomes denser and thus weighs more.

Role of the ambient atmosphere

Most machines are immersed in an ambient 'atmosphere' of air (or water), which for simplicity is assumed to be at rest with a hydrostatic pressure distribution p_0. If the fixed control volume were filled with ambient fluid at rest, the absolute (and static) reaction force would be,

$$\mathcal{R}_0 = M_0 g_0 - \int_{\Delta S} p_0 \, d\boldsymbol{S}, \tag{23.14}$$

where M_0 is the mass of the ambient fluid in the control volume.

Usually, we are interested in the change $\boldsymbol{\Delta} R = \mathcal{R} - \mathcal{R}_0$ in the reaction force that occurs when the ambient fluid is replaced with the intended moving fluid, for example when water enters the firehose. This *relative reaction force* becomes,

$$\boxed{\Delta\mathcal{R} \approx (M - M_0) g_0 - \int_{\Delta S} (p - p_0) \, d\boldsymbol{S} - \int_{\Delta S} \rho \boldsymbol{v}\, \boldsymbol{v} \cdot d\boldsymbol{S}.} \tag{23.15}$$

Evidently, the relative reaction force is obtained by reducing the mass of the fluid by the buoyancy of the displaced ambient fluid and calculating all pressures relative to the ambient pressure. The difference $p - p_0$ between absolute pressure and the ambient pressure is called the *gauge* (or *gage*) pressure, because that is what you would read on a manometer gauge.

Example 23.2.1 (Gas pipeline): A natural gas pipeline at the bottom of the sea transports gas between two islands. During construction the engineers allowed sea water to fill the pipeline which rested comfortably at the bottom under its own weight. But when the system was filled with gas, the buoyancy of the displaced water made the pipeline float to the surface. The engineers got fired for failing to correctly calculate the relative reaction force.

Nozzle puzzle, continued

We are now in position to calculate the relative reaction force on the firehose nozzle discussed on page 325. Disregarding gravity and taking the outlet pressure equal to the ambient pressure p_0, the relative reaction force becomes

$$\Delta\mathcal{R}_x = (p - p_0) A_1 - \rho_0 v_2^2 A_2 + \rho_0 v_1^2 A_1. \tag{23.16}$$

For nearly ideal flow, Bernoulli's theorem allows us to find the pressure difference between inlet and outlet,

$$\frac{1}{2} v_1^2 + \frac{p}{\rho_0} = \frac{1}{2} v_2^2 + \frac{p_0}{\rho_0}. \tag{23.17}$$

Together with mass conservation (23.8), everything can be expressed in terms of the inlet and outlet areas, with the result,

$$\Delta\mathcal{R}_x = \frac{(A_1 - A_2)^2}{2 A_1 A_2} \frac{\rho_0 Q^2}{A_2}. \tag{23.18}$$

Somewhat surprisingly the fluid appears to exert a positive reaction force on the nozzle along the direction of flow.

On reflection this is in fact in agreement with our expectations because the pressure on the inside of the nozzle constriction is larger than the ambient pressure surrounding the nozzle. The reaction force $\Delta\mathcal{R}_x$ is the force that the fireman would have to oppose, should the hose suddenly break.

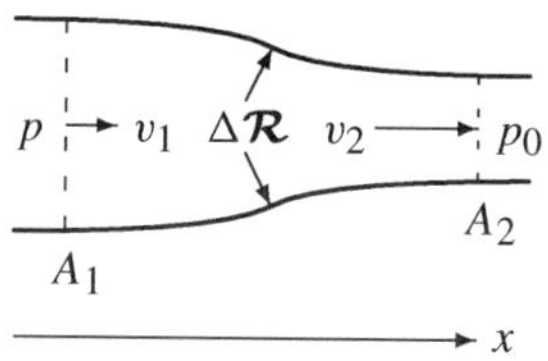

The true reaction force on a nozzle points along the direction of flow because there is a higher pressure before and in the constriction than at the outlet.

Example 23.1.4 (Firehose, continued): For the older firehose the required handle force will change instantaneously from +350 N to −200 N if the hose breaks. Since these forces are rather large, there is a real risk that the fireman will fly off together with the nozzle.

* 23.3 Reaction moments

A rotating lawn sprinkler spins and gains angular momentum by expelling water from its nozzles. As for reaction forces, this is clearly exposed by using angular momentum balance (22.26) and (22.27),

$$\frac{d\mathcal{L}}{dt} = \mathcal{M} - \oint_S \rho\, \boldsymbol{x} \times \boldsymbol{v}\, (\boldsymbol{v} - \boldsymbol{v}_S) \cdot d\boldsymbol{S}. \tag{23.19}$$

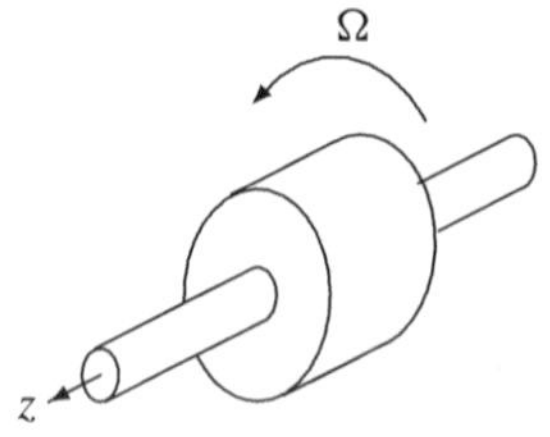

Sketch of a rotor on its shaft, rotating steadily with angular velocity Ω around the z-axis.

The last term (including the sign) represents the net flux of angular momentum *into* the control volume through its surface. It must be emphasized that this term is *not* a true moment of force acting on the system, nor the reaction to another moment, but rather a 'fictitious' moment akin to the moment created by the Coriolis force in a rotating frame of reference. As for reaction forces, this term may be shown to contribute to the true reaction moment although we shall not carry out the analysis here.

The rotor

In turbomachines there is always a solid rotating part, generically called the *rotor*, which diverts the flow and either puts energy into the fluid or takes it away. The rotor is usually mounted on a rigid shaft, and the energy is transmitted to or from the rotor through an external moment of force, or *torque*, applied to the shaft. The rotor is carefully designed with a number of channels that guide the flow in and out, and mostly the channels are constructed from thin blades that obstruct the flow as little as possible. Rotors essentially only differ in the way their channels are designed. In an *axial flow* rotor, the channels run along the axis, and in a *radial flow* rotor the channels are radial. Rotor design is a highly evolved engineering discipline and most real turbomachines employ a mixture of axial and radial flow.

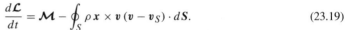

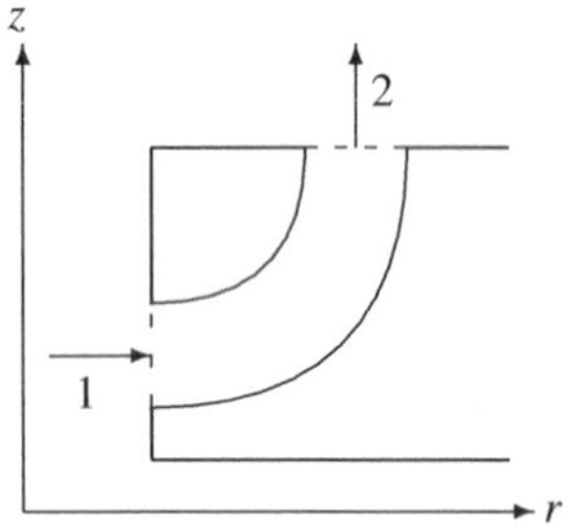

A rotor channel with radial inflow at 1 and axial outflow at 2.

Steady flow rotor physics

Even when the rotor spins steadily, the flow in the rotor will strictly speaking be unsteady seen from the inertial 'laboratory' frame because of the moving blades. In the rotating frame of the rotor, the flow will however be steady, apart from the turbulence that may arise in rapid flow. It is for this reason natural to choose a rotating control volume which is fixed with respect to the rotor and contains all the fluid in the rotor.

When the rotor spins with angular velocity Ω around the z-axis, the control volume rotates with velocity $\boldsymbol{v}_S = \Omega\, r\, \boldsymbol{e}_\phi$ in the laboratory frame (using cylindrical coordinates with z-axis along the rotor axis). The flow velocity relative to the rotor is denoted

$$\boldsymbol{u} = \boldsymbol{v} - \boldsymbol{v}_S = \boldsymbol{v} - \Omega\, r\, \boldsymbol{e}_\phi \tag{23.20}$$

where $\boldsymbol{v}$ is the 'absolute' flow velocity in the non-rotating laboratory frame.

In steady operation the total angular momentum contained in the control volume must be constant, $d\mathcal{L}/dt = \boldsymbol{0}$, in spite of the fact that it rotates. Using that the specific angular momentum along the cylinder axis is $(\boldsymbol{x} \times \boldsymbol{v})_z = r v_\phi$, the total axial moment of force which has to be exerted by the environment on the fluid in the rotor is obtained from angular momentum balance (23.19),

$$\mathcal{M}_z = \oint_S r v_\phi\, \rho\, \boldsymbol{u} \cdot d\boldsymbol{S}. \tag{23.21}$$

The environment may contribute in several ways to this moment, for example through friction as we saw for the lawn sprinkler, but if friction can be disregarded, $\mathcal{M}_z$ will be the moment that must be applied to the shaft to keep the rotor spinning steadily (and $-\mathcal{M}_z$ is the reaction moment from the fluid). In the following we shall for simplicity assume this is the case.

The total rate of work, or *power*, that the environment must supply to or take away from the solid shaft is obtained by multiplying the moment $\mathcal{M}_z$ with the angular velocity Ω (see problem 23.3),

$$P = \Omega \mathcal{M}_z. \tag{23.22}$$

If the shaft power is positive, the rotor takes energy from the shaft which is put into the flow, as for example in a hair dryer. If on the other hand the shaft power is negative, the rotor produces energy and acts like a turbine by taking energy out of the flow.

Turbomachines running full of fluid, such as a lawn sprinklers, are driven by pressure differences between inlets and outlets. Taking only into account the pseudo-gravitational potential $-(1/2)\Omega^2 r^2$ of the centrifugal force (see equation (7.3)), the general Bernoulli function (16.35) becomes in the rotating frame,

$$H = \tfrac{1}{2}\boldsymbol{u}^2 - \tfrac{1}{2}\Omega^2 r^2 + w(p). \qquad (23.23)$$

Here $w(p)$ is the pressure function (16.36), which for incompressible fluids is simply $w = p/\rho_0$. The constancy of H along a streamline in nearly ideal steady flow creates a relation between inlet and outlet pressures. Bernoulli's theorem should, however, be used with some caution and will not be fulfilled if a significant amount of energy is dissipated into heat through turbulence.

Example 23.3.1 (Lawn sprinkler): Returning to the lawn sprinkler in example 22.6.1 on page 315 we now assume that the water feeds into the sprinkler pivot through a 1/2 inch tube. Ignoring the square of the small inlet velocity (about 2 m s^{-1}), Bernoulli's theorem implies that in the steady state the inlet pressure excess is

$$\Delta p = p - p_0 \approx \tfrac{1}{2}\rho_0(U^2 - \Omega_0^2 R^2) \approx 0.5 \text{ bar}. \qquad (23.24)$$

The total power consumed by the sprinkler is determined by the pressure excess (and thus paid for by the waterworks). It is $P_0 = \Delta p\, Q/\rho_0 \approx 12$ W and mostly goes to supply kinetic energy to the sprinkled water. The total shaft power (lost to friction) becomes $P = -N\Omega_0 \approx -1.3$ W which is only a tenth of the total power consumption.

Radial flow rotor

Radial flow rotors are mostly used in high-pressure turbomachinery, such as pumps, compressors and hydraulic turbines. Although the following analysis concerns a turbine, it is also valid for a pump. We assume that the flow is effectively incompressible and nearly ideal.

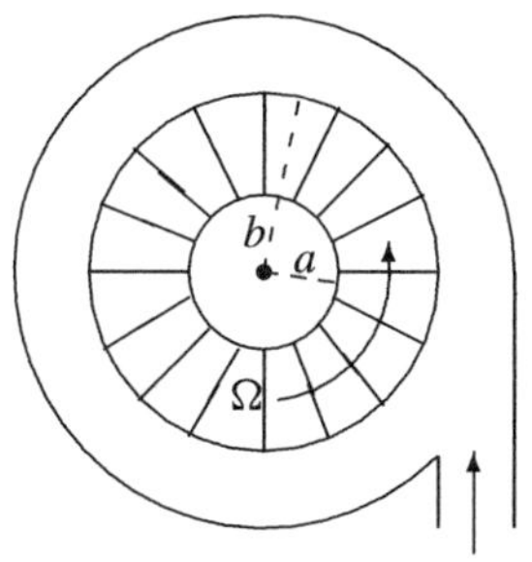

Sketch of a radial flow turbine rotating steadily with angular velocity Ω. The z-axis comes out of the paper. Fluid enters the rotor at $r = b$, moves radially inwards, and leaves at $r = a$ where it is diverted away along the axis.

Let the rotor have inner radius a, outer radius b and axial length L. We shall for simplicity assume that it is perfectly cylindrical with the same properties along the whole of its length $0 < z < L$, such that the channels have the same shape for all z. In a turbine, high-speed water is fed in at the outer perimeter $r = b$ with (negative) radial velocity $u_r(b) = v_r(b)$ and removed at the inner perimeter $r = a$ with (negative) radial velocity $u_r(a) = v_r(a)$. In a pump the radial velocities are positive. Disregarding the blocking from the thin blades separating the otherwise identical channels, the total volume flux is approximately,

$$Q = 2\pi a L u_r(a) = 2\pi b L u_r(b). \qquad (23.25)$$

The integral in the shaft moment (23.21) only receives contributions from the radial surfaces at $r = a, b$. Since v_ϕ is approximatively independent of ϕ and z, we find from (23.22) the shaft power (*Euler's turbine equation*),

$$P = (b\, v_\phi(b) - a\, v_\phi(a))\rho_0 \Omega Q. \qquad (23.26)$$

For a turbine where Q is negative, the shaft power will be negative as long as the angular momentum density is larger at the entry than at the exit, $b\, v_\phi(b) > a\, v_\phi(a)$. The fluid understandably has to lose angular momentum for the rotor to produce the work that may drive an electric generator. In a pump, it is the other way around.

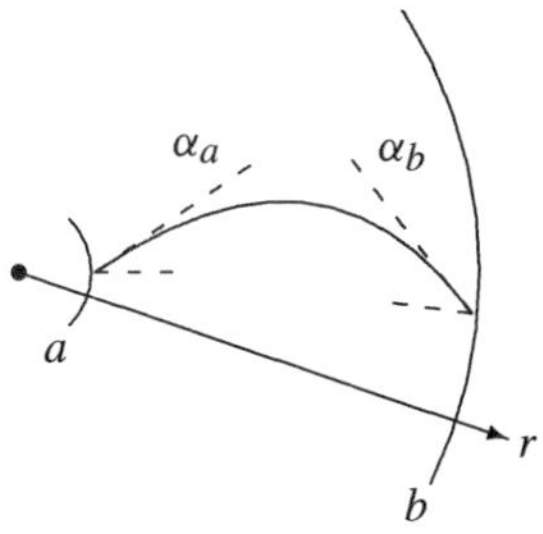

Sketch of a curved blade with slope α_a at $r = a$ and α_b at $r = b$ relative to the radial direction. Here α_a is positive and α_b negative.

The rotor consists of a number of thin blades mounted radially on a hub like the spokes of a wheel. Let the rotor blades be designed such that the geometric slopes with respect to the radial direction are α_a at $r = a$ and α_b at $r = b$. In smooth operation we assume that the steady flow enters and leaves tangentially along the blades. Although a rotor can still operate if this condition is not fulfilled, it is much more liable to generate turbulence and even cavitation accompanied by loss of power, because of the sudden change in flow direction, especially at the entry. The smoothness condition provides the following relations between the relative radial and azimuthal velocities,

$$u_\phi(a) = \alpha_a u_r(a), \qquad u_\phi(b) = \alpha_b u_r(b). \qquad (23.27)$$

Using that the absolute azimuthal velocity is $v_\phi(r) = u_\phi(r) + \Omega r$, and eliminating the azimuthal velocities by means of these equations and the radial velocities by means of mass conservation (23.25), we obtain

$$\boxed{P = \left((b^2 - a^2)\Omega + \frac{Q}{2\pi L}(\alpha_b - \alpha_a)\right)\rho_0 \Omega Q.} \qquad (23.28)$$

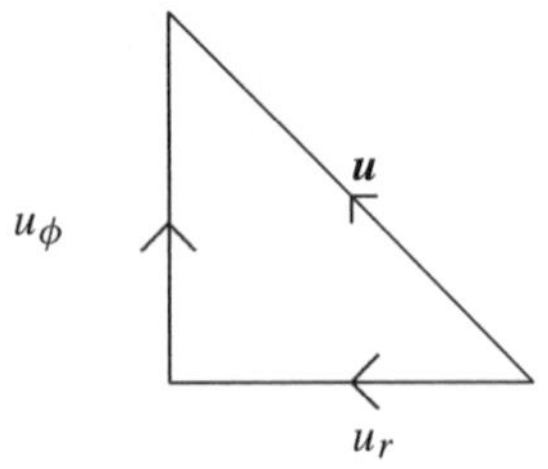

The relation between radial velocity $u_r = v_r$ and azimuthal velocity $u_\phi = v_\phi - \Omega r$ in the rest system of the rotor. The slope is defined to be $\alpha = u_\phi/u_r$, and is negative in this picture.

Even if the blades have no curvature, $\alpha_a = \alpha_b$, the shaft power will be non-vanishing. If $\alpha_b < \alpha_a$, both terms in parenthesis are positive for a turbine where $Q < 0$.

Applying Bernoulli's theorem in the rotating frame to a streamline passing through a channel, it follows from (23.23) that

$$\frac{1}{2}\boldsymbol{u}(a)^2 - \frac{1}{2}\Omega^2 a^2 + \frac{p(a)}{\rho_0} = \frac{1}{2}\boldsymbol{u}(b)^2 - \frac{1}{2}\Omega^2 b^2 + \frac{p(b)}{\rho_0}.$$

Using $\boldsymbol{u}^2 = u_r^2 + u_\phi^2$ we obtain after elimination of the velocities

$$p(b) - p(a) = \frac{1}{2}\rho_0\left(\Omega^2\left(b^2 - a^2\right) + \frac{Q^2}{(2\pi L)^2}\left(\frac{1+\alpha_a^2}{a^2} - \frac{1+\alpha_b^2}{b^2}\right)\right). \tag{23.29}$$

The first term is due to the centrifugal force and the second to the change in velocity through the rotor channel. If the slopes are of nearly equal magnitudes, $|\alpha_a| \approx |\alpha_b|$ both terms will be positive, so the pressure will always be largest at the outer rim of the rotor. This is the pressure head which drives a hydraulic turbine powered by high-pressure water coming down from a reservoir. Conversely, this must also be the pressure head created by the pump used to lift water up into a reservoir.

Example 23.3.2 (Hydraulic turbine): A large hydraulic turbine is driven by water piped down from a reservoir $h = 110$ m above the turbine. The rotor axis is vertical and the inner and outer radii are $a = 2$ m and $b = 3$ m. The axial length is $L = 0.5$ m, and the rotor rotates at 60 rpm, i.e. once per second or $\Omega = 2\pi$ s^{-1}. The blades are constructed with vanishing entry slope $\alpha_b = 0$, so that the azimuthal inlet velocity is $v_\phi(b) = \Omega b \approx 19$ m s^{-1}. At the exit the slope is chosen to be $\alpha_a = 1/3$, and the flow is required to be purely radial, $v_\phi(a) = 0$, leading to $u_r(a) = -\Omega a/\alpha_a \approx -38$ m s^{-1} and $u_r(b) = a u_r(a)/b \approx -25$ m s^{-1}. The absolute inlet velocity of the water becomes $v(b) \approx 31$ m s^{-1}, corresponding to a free-fall from 50 m height. The flux of water may now be calculated from (23.25) and becomes $Q \approx -237$ m^3 s^{-1}. The total shaft power (23.28) is $P \approx -84$ MW. The pressure drop through the turbine is about $\Delta p = p(b) - p(a) \approx 5.7$ bar which is a bit more than half the static pressure head from the reservoir. The total rate of work of the excess pressure is $P_0 = \Delta p\ Q \approx 136$ MW, so the shaft work of the turbine is about $P/P_0 \approx 62\%$ of the total work of the water. The remainder is found in the kinetic energy of the water coming out of the turbine at higher speed than it entered. In this calculation we have disregarded losses due to friction and turbulence which can be quite significant.

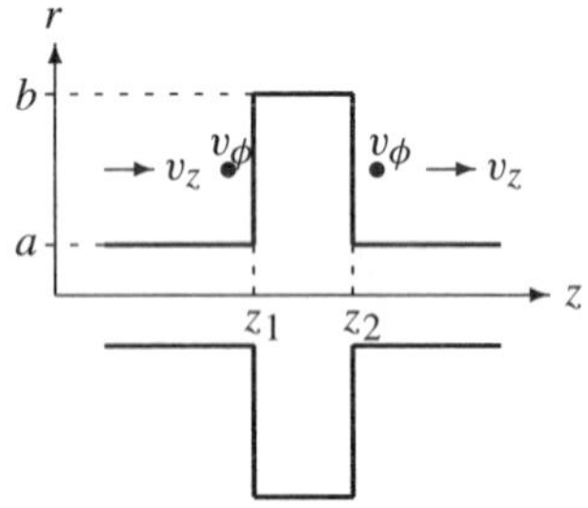

Sketch of axial flow rotor. The control volume is defined by $z_1 < z < z_2$ and $a < r < b$. The axial flow $v_z(r)$ is independent of z, whereas the azimuthal flow $v_\phi(r,z)$ may depend on both r and z.

Axial flow rotor

Axial flow rotors are typically used in high-volume turbomachinery, such as fans and low-pressure turbines. The fluid is also in this case taken to be nearly ideal and effectively incompressible with constant density, $\rho = \rho_0$. Since there is no radial flow in this design, mass conservation guarantees that the axial velocity $v_z = v_z(r)$ is independent of z. The rotor is assumed to be cylindrical with inner and outer radii a and b, and thin blades that do not significantly obstruct the flow. The fluid enters the rotor at $z = z_1$ and leaves at $z = z_2$. We shall for simplicity assume that the axial flow is uniform $v_z = U$, so that the total volume flux through the rotor becomes

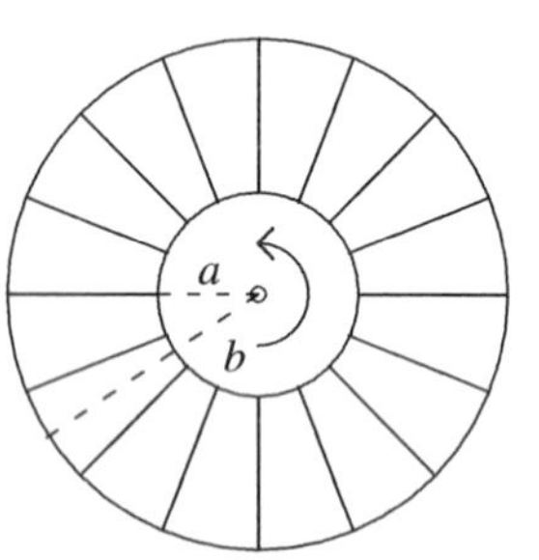

Rotor design with blades arranged like the spokes of a wheel. The positive z-axis comes out of the paper, and the indicated sense of rotation is positive.

$$Q = \int_a^b v_z(r)\, 2\pi r\, dr = UA, \tag{23.30}$$

where $A = \pi(b^2 - a^2)$ is the area of the axial rotor cross section.

The rotor channels change the azimuthal velocity field $v_\phi(r,z)$. The integral in the shaft moment (23.21) only receives contributions from the inlet at $z = z_1$ and the outlet at $z = z_2$, so the shaft power becomes

$$P = \Omega\int_a^b \rho_0 r\big(v_\phi(r,z_2) - v_\phi(r,z_1)\big)v_z(r)\, 2\pi r\, dr = \big\langle r(v_{\phi 2} - v_{\phi 1})\big\rangle \rho_0 \Omega Q. \tag{23.31}$$

In the last step we have expressed the integral in terms of the area average $\langle f\rangle = (1/A)\int_a^b f(r) 2\pi r\, dr$ over the cross section of the rotor. The shaft power is positive if the azimuthal speed generally increases through the rotor, as one would expect.

In steady operation we assume that the flow enters and leaves the spinning rotor tangentially along the blades. The rotor blades are designed such that the slope with respect to the z-axis is $\alpha_1(r)$ at the entrance and $\alpha_2(r)$ at the exit. Since the relative azimuthal rotor speed is $u_\phi = v_\phi - \Omega r$, the smoothness condition amounts to,

$$u_{\phi 1} = v_{\phi 1} - \Omega r = \alpha_1 U, \qquad u_{\phi 2} = v_{\phi 2} - \Omega r = \alpha_2 U. \tag{23.32}$$

Using these relations to eliminate the azimuthal velocities, the shaft power may be written

$$\boxed{P = \langle r(\alpha_2 - \alpha_1)\rangle \, \rho_0 U \Omega Q.} \tag{23.33}$$

The slope average $\langle r(\alpha_2 - \alpha_1)\rangle$ is a purely geometric factor which is independent of both U and Ω. Note that this expression is only valid if the smoothness condition is actually fulfilled.

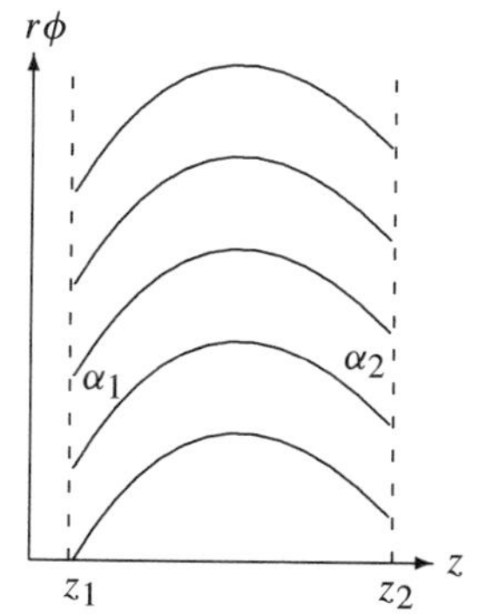

The axial flow rotor seen edgewise, folded out along a constant radius r with the r-axis going into the paper. The blades have slope α_1 relative to the z-axis at the entrance $z = z_1$ and α_2 at the exit $z = z_2$; here shown with $\alpha_1 > 0$ and $\alpha_2 < 0$.

If the inlet and exit slopes are the same $\alpha_1 = \alpha_2$, the total shaft power appears to vanish. This seems a bit puzzling, because it is well known that propellers with flat blades also require engine power to turn steadily. The explanation is that for such devices the smoothness condition is not fulfilled, and the fluid hits the propeller blades at an 'angle of attack', resulting in aerodynamic forces which oppose the motion and thus require power in the steady state (aerodynamic forces are discussed at length in chapter 29).

The actual geometric design of the blades provides us with the slopes. Here we shall for simplicity only consider the case in which the blades have slopes that grow proportionally with the radial distance r, i.e. $\alpha_1 = \beta_1 r$ and $\alpha_2 = \beta_2 r$. Under this assumption, the fluid will rotate like a solid body with $v_{\phi 1} = \Omega_1 r$ at the entrance and $v_{\phi 2} = \Omega_2 r$ at the exit, where the smoothness condition determines the angular velocities,

$$\Omega_1 = \Omega + \beta_1 U, \qquad \Omega_2 = \Omega + \beta_2 U. \tag{23.34}$$

Now the shaft power (23.33) can be calculated explicitly with the result

$$P = \tfrac{1}{2}(a^2 + b^2)(\beta_2 - \beta_1)\rho_0 U \Omega Q. \tag{23.35}$$

If the blades are not designed with uniformly growing slopes, this expression may still be used for the purpose of doing estimates.

Finally, the pressure change through the rotor is obtained from the Bernoulli function (23.23) in the rest system of the rotor. For a streamline at constant r passing through an axial channel from inlet to outlet we find

$$\frac{1}{2}\boldsymbol{u}_1^2 - \frac{1}{2}\Omega^2 r^2 + \frac{p_1}{\rho_0} = \frac{1}{2}\boldsymbol{u}_2^2 - \frac{1}{2}\Omega^2 r^2 + \frac{p_2}{\rho_0}.$$

Using $\boldsymbol{u}^2 = u_\phi^2 + u_z^2$ and $u_z = v_z = U$, we arrive at

$$p_1 - p_2 = \frac{1}{2}\rho_0(\alpha_2^2 - \alpha_1^2)U^2 = \frac{1}{2}\rho_0(\beta_2^2 - \beta_1^2)r^2 U^2. \tag{23.36}$$

The pressure difference between inlet and outlet grows in this case with the square of the distance of the channel from the axis.

Axial flow turbines are often provided with upstream *guide vanes* to add an initial azimuthal spin Ω_1 to the fluid before it enters the rotor. The guide vanes have basically the same function as the rotor blades, but since they are solidly anchored to the housing of the machine the reaction moment from the fluid on the guide vanes have no influence on the shaft power. For axial flow blowers, downstream guide vanes may similarly be used to 'straighten out' the rotating fluid before it exits from the blower.

Example 23.3.3 (Air blower): In longer road tunnels axial flow air blowers are often used to create an artificial draught along the tunnel in order to rid it of exhaust fumes, especially when traffic has stalled. Suppose an air blower with outer diameter $2b = 60$ cm and hub diameter $2a \approx 20$ cm operates at $\Omega = 600$ rpm $= 10 \times 2\pi$ s^{-1}. The blades are constructed with radially increasing slopes, and the maximal entrance angle at the tip of a blade is $-45°$, corresponding to $\alpha_1(b) = -1$ and thus $\beta_1 = \alpha_1(b)/b \approx -3.33$ m^{-1}. At the entrance the air has no spin, $\Omega_1 = 0$, and the axial flow becomes

$U = -\Omega/\beta_1 \approx 19$ m s^{-1}. At the exit, the blade slope vanishes, $\beta_2 = 0$, so that $\Omega_2 = \Omega$. The total volume flow becomes $Q \approx 4.7$ m^3 s^{-1} and the total shaft power becomes $P \approx 1$ kW. The maximal pressure increase across the blower is found at the tip of the blades, $r = b$, and is merely $\Delta p = p_2 - p_1 \approx 195$ Pa ≈ 2 mbar.

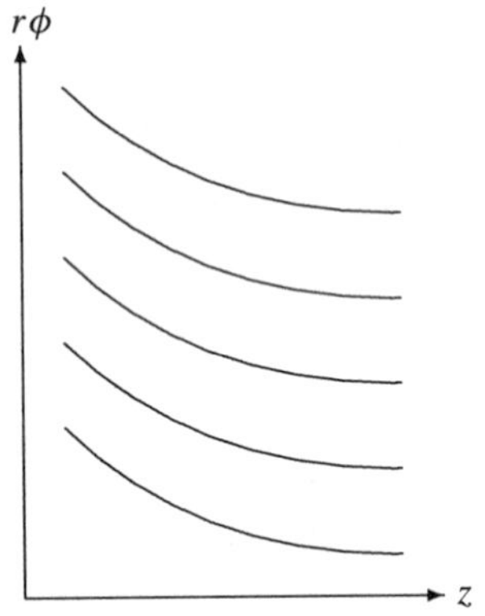

The blades in the air blower of example 23.3.3 folded out along the rotor circumference. The blades move upwards here.

Rotor scaling relations

Although one may estimate the overall behaviour of a rotor from angular momentum balance and Bernoulli's theorem, the details of the flow pattern through the actual rotor and the unavoidable viscosity of the fluid will give rise to important corrections to the shaft power and the pressure jump across the rotor, corrections that may not be easily calculable from theory. Model studies of scaled down versions of the rotor may, however, be feasible and the measured values of shaft power and pressure jump can afterwards be scaled up to yield a prediction for the actual machinery. So we need to know the scaling relations for these quantities.

A rotor of a given design is characterized by a single length scale D, the 'diameter' of the system. The fluid itself is characterized by its density ρ at some point, say at the inlet. We shall furthermore assume that the viscosity is so small that it can be disregarded. During steady operation of the rotor, the controlling parameters are the angular velocity Ω and volume flux Q. Since D controls the length scale, Ω the time scale, the only dimensionless parameter which can be constructed is,

$$q = \frac{Q}{\Omega D^3}. \tag{23.37}$$

It may be viewed as a dimensionless representation of the volume flux and is usually called the *flow coefficient*. It also represents the ratio $q \sim u/v$ between the relative flow velocity $u \sim Q/D^2$ and the absolute velocity $v \sim \Omega D$.

From (23.21) and (23.22) it follows that the shaft power scales like $P \sim \Omega D v \rho_0 Q \sim \Omega^2 D^2 \rho_0 Q$. Similarly, it follows from (23.23) that the pressure head scales like $\Delta p \sim (1/2)\rho_0 \Omega^2 D^2$. The exact equations of fluid mechanics (without viscosity) must therefore provide relationships of the form

$$\frac{P}{\Omega^2 D^2 \rho_0 Q} = f(q), \qquad \frac{\Delta p}{(1/2)\rho_0 \Omega^2 D^2} = g(q), \tag{23.38}$$

where $f(q)$ and $g(q)$ are—generally unknown—dimensionless functions of the dimensionless control variable q. For a radial rotor with nearly ideal incompressible flow, we take $D = b$ and find from (23.28) and (23.29),

$$f = 1 - \frac{a^2}{b^2} + q\frac{b}{2\pi L}(\alpha_b - \alpha_a), \qquad g = 1 - \frac{a^2}{b^2} + q^2\frac{b^4}{(2\pi L)^2}\left(\frac{1+\alpha_a^2}{a^2} - \frac{1+\alpha_b^2}{b^2}\right).$$

If viscosity is important, f and g will also depend on the Reynolds number.

Example 23.3.4 (Model turbine): For the hydraulic turbine of example 23.3.2 we take $D = b$ and find $q = 1.4$. A model turbine scaled down by a factor of 10, i.e. with $D \to 0.1D$, but operated at the same angular velocity Ω and the same value of q has a thousand times smaller volume flux $Q \to 10^{-3}\,Q$. The scaling relations now imply that $P \to 10^{-5}P$ and $\Delta p \to 10^{-2}\Delta p$, independent of the form of the generally unknown functions f and g. Measuring the values of P and Δp in the model turbine immediately allows us to infer the values for the full-scale turbine (the Reynolds numbers will, however, be different for turbine and model).

Problems

23.1 A horizontal 1 inch water pipe bends horizontally 180°. Estimate the magnitude of the force that the water exerts on the pipe, when the water speed is 1 m s^{-1}.

23.2 A firefighter bends a water hose through 90° close to the nozzle where the pressure is nearly atmospheric. The hose diameter is 5 cm and it discharges 40 litres per second. Calculate the magnitude of the force that the firefighter has to exert on the hose to bend it. Will he be able to hold the hose without using equipment?

23.3 Consider a solid body rotating with angular velocity Ω around the z-axis under the action of a moment of force $\mathcal{M}_z$. Show that the total power is $P = \Omega \mathcal{M}_z$.

24 Small-amplitude surface waves

Surface waves in the sea are created by the interaction of wind and water which somehow transforms the steady motion of the streaming air into the nearly periodic swelling and subsiding of the water. The waves appear to roll towards the coast in fairly orderly sequence of crests and troughs that is translated into the quick ebb and flow of water at the beach, so well known to all of us. On top of that there is, of course, the slow ebb and flow of the tides.

In constant 'flat-Earth' gravity, the interface between two fluids at rest is always horizontal. In moving fluids the interface can take a very complex instantaneous shape under the simultaneous influence of inertia, pressure, gravity, container shape, surface tension and viscosity. Waves controlled by pressure and gravity are naturally called *gravity waves*, whereas waves controlled by pressure and surface tension are called *capillary waves*. If the fluids have vastly different densities, as is the case for the sea and the atmosphere, one may often disregard the lighter fluid and instead consider the open surface of the heavier fluid against vacuum. For fluids of nearly equal density, for example a saline bottom layer in the sea overlayed with a brackish layer, *internal gravity waves* driven by pressure and buoyancy may arise in the interface.

This chapter is devoted to the various types of small-amplitude surface waves and the conditions under which they occur (see [42, 67, 1, 22, 36] for extended discussions of surface waves). Mathematically, small-amplitude waves are by far the easiest to deal with. More interesting and unusual wave types arise when amplitudes grow so large that the nonlinear aspects of fluid mechanics come into play. Nonlinear waves are common everyday occurrences, from the familiar run-up of waves on a beach to the less familiar sonic boom from an aircraft overhead, but the subject of truly nonlinear waves is unfortunately so mathematically challenging that we shall only treat its simpler aspects here (chapter 25).

24.1 Basic physics of surface waves

The shape of an interface between two fluids in hydrostatic equilibrium is determined by the balance between the pressure gradient and gravity throughout the interior of the fluids (chapter 7). Surface tension may also have profound influence on the shape of small fluid volumes, for example a raindrop (chapter 8). What we shall call waves in this chapter are time-dependent disturbances of a fluid interface originally in hydrostatic equilibrium.

Although surface waves may occur wherever material properties change, we shall mostly think of gravity waves. In constant gravity, the hydrostatic interface between the sea and the air is flat and horizontal, usually taken to be $z = 0$ in a flat-Earth coordinate system. A wave will disturb the surface so that its

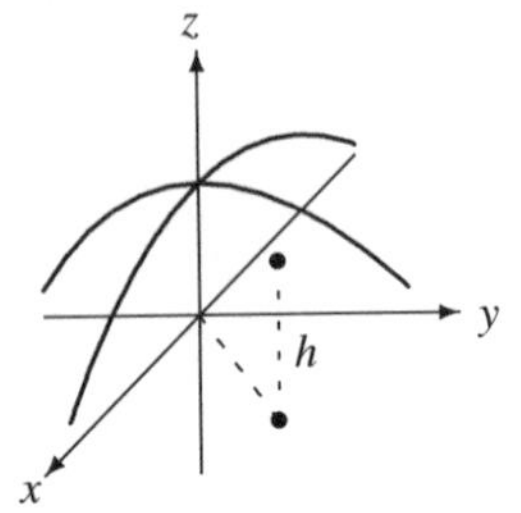

A surface wave in the flat-Earth coordinate system.

instantaneous height becomes a function of the horizontal coordinates and time,

$$z = h(x, y, t). \tag{24.1}$$

Our interest will mainly focus on trains of waves that progress periodically and in a regular pattern across the horizontal surface of the sea, although everyday experience tells us that waves can be much more complicated. In a breaking wave, the surface height is not even a single-valued function of position.

Waves can be created in many ways. The splash you make when you jump bottom-first into a swimming pool creates first a single large ring-shaped wave, perhaps followed by several smaller waves. When these waves hit the edge of the pool they are reflected and interfere with themselves to create quite chaotic patterns. In this chapter we shall, however, not be concerned much with the mechanisms by which waves are created, but rather with their internal dynamics after they have somehow been brought into existence.

Wave parameters

Any non-breaking surface wave consists locally of mounds and hollows of roughly the same size in the otherwise smooth equilibrium surface. Although a general wave can be very complex, it is convenient to describe these local features in terms of parameters that are normally reserved for harmonic waves:

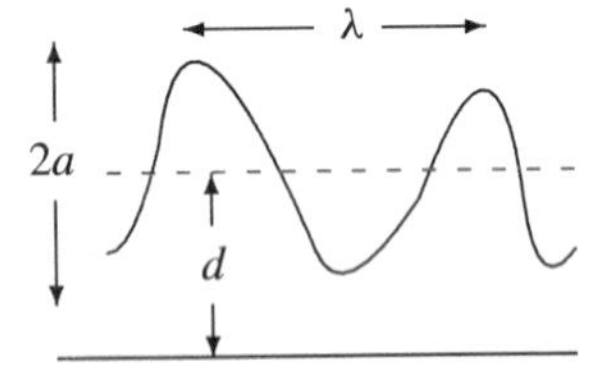

A general wave consists of mounds and hollows. Locally, the amplitude a is related to the vertical distance between maxima and minima, the wavelength λ to the horizontal size of a mound or hollow, and the period τ to the time scale for major changes in the local pattern. The depth d is the average vertical distance to the bottom.

- a—amplitude. It sets the scale for vertical variations in the height of the wave. Mostly it is taken to be the height of a mound above the equilibrium level, or equivalently the depth of a hollow below.
- λ—wavelength. This is the horizontal length scale of the wave, typically related to the width of a mound or a hollow.
- τ—period. A measure of the time scale for major changes in the wave pattern, for example the time it takes for a mound to become a hollow.
- d—depth. The vertical distance to a solid boundary, the 'bottom'.

The ratio $c = \lambda/\tau$ is called the *celerity* or *phase velocity* of the wave, and characterizes the speed with which the waveform changes shape.

If the wavelength is much greater than the depth, $\lambda \gg d$, we shall speak about *long waves* or more graphically *shallow-water waves*. Similarly, waves with wavelength much smaller than the depth, $\lambda \ll d$, are called *short waves* or *deep-water waves*. Waves with amplitude much smaller than both wavelength and depth, $a \ll \lambda, d$, are called *small-amplitude* waves.

The dispersion law

In a gravity wave the force of gravity pulls the water of a mound downwards and sets it into motion, and this motion may in turn make the water rise again. Whereas the potential energy of the wave only depends on its shape, the kinetic energy also depends on the flow velocities in the wave, and thus on the period. Therefore, if no energy is lost to friction, this continual conversion of potential energy into kinetic energy and back must provide a relation between the period of a wave and the other parameters,

$$\tau = \tau(a, \lambda, d, \ldots). \tag{24.2}$$

The precise form of this *dispersion law* for a particular type of wave is normally obtained from careful analysis of the wave dynamics, several examples of which will be given later. Here we shall only make a coarse estimate of the general form of the dispersion law using that potential and kinetic energies must be of comparable magnitudes.

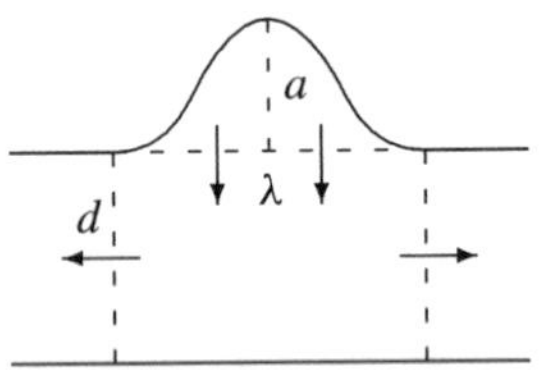

A 'waterberg' of height a and width λ rising out of a sea of depth d. When the waterberg collapses vertically, all the water in it has to disperse in the horizontal directions.

Collapse of a 'waterberg'

Suppose we have somehow created a mound of water, instantaneously at rest, for example by pulling an inverted water-filled bucket up through the flat surface of the sea. Common experience tells us that such a 'waterberg' will quickly collapse into the sea, creating instead a smaller and wider hollow which may later rise again to make an even smaller mound that in turn collapses, and so on. Eventually all traces of the initial mound will have disappeared into secondary waves running away over the surface.

Let the initial mound have height a and width λ so that its volume is of magnitude $\lambda^2 a$. During the first collapse all the water in the mound will move vertically downwards and reach the sea level in characteristic time τ. Since water is incompressible, an equal volume has to move horizontally away from the collapse region with a typical speed U. For long waves with $\lambda \gg d$, the proximity of the bottom forces all the water underneath the mound to move away horizontally. Mass conservation tells us that the volume of the mound equals the volume that streams away over a period, or $\lambda^2 a \sim \lambda d\, U\tau$. Solving for U we find,

$$U \sim \frac{a\lambda}{\tau d}. \tag{24.3}$$

Since the typical vertical speed is a/τ, the horizontal flow velocity U in a shallow sea will be much greater than the vertical by a factor $\lambda/d \gg 1$. For short waves with $\lambda \ll d$, there is no bottom to divert the water flow so that the horizontal velocity tends to be of same order of magnitude as the vertical, i.e. $U \sim a/\tau$ near the surface. From the expression for U, one may thus conclude that the deep sea may be characterized by an effective depth of the same magnitude as the wavelength, $d \sim \lambda$. These claims will be confirmed later by precise calculations, showing in fact that the effective depth of the deep sea is $d \approx \lambda/2\pi$.

The potential energy of the initial mound relative to the general level of the sea is of magnitude,

$$\mathcal{V} \sim \rho_0\lambda^2 a \cdot g_0\, a = \rho_0 g_0 a^2\lambda^2. \tag{24.4}$$

Interestingly, a hollow of depth a and width λ would have a potential energy of the same magnitude, for the simple reason that buoyancy presses the surface upward, like the hull of a ship.

The kinetic energy in the horizontal motion of the collapsing water mound becomes of magnitude

$$\mathcal{T} \sim \rho_0\lambda^2 d \cdot U^2 \sim \rho_0\frac{\lambda^4 a^2}{\tau^2 d}. \tag{24.5}$$

In the absence of dissipation, the kinetic energy must be comparable to the potential energy, $\mathcal{T} \sim \mathcal{V}$, and solving for τ we obtain the estimate of the dispersion law,

$$\boxed{\tau \sim \frac{\lambda}{\sqrt{g_0 d}}.} \tag{24.6}$$

Note that the amplitude fell out of this expression. The dispersion law is merely a coarse estimate of the overall magnitude of the collapse time. It may still be multiplied by an unknown factor of order unity which can depend on the dimensionless ratios a/λ and d/λ, and possibly on other dimensionless parameters characterizing the actual shape of the wave.

From the dispersion law we immediately get the phase velocity

$$c = \frac{\lambda}{\tau} \sim \sqrt{g_0 d}. \tag{24.7}$$

Like an echo of Toricelli's law (page 210) it is of the same order of magnitude as the free-fall velocity $\sqrt{2g_0 d}$ from height d. In shallow water where d is the true depth of the sea, the phase velocity is independent of wavelength. In deep water where the true depth is infinite, the effective depth may as pointed out above be taken to be $d \approx \lambda/2\pi$.

Example 24.1.1 (Inverted bucket): A little 'waterberg' created by lifting an inverted bucket of height $a = 30$ cm and width $\lambda = 50$ cm out of water of depth $d = 1$ m collapses in just $\tau \sim 0.15$ s.

Gravity waves are nearly ideal

The shape of a surface wave is only a manifestation of the underlying hydrodynamics, governed by the Navier–Stokes equations. From the estimate (24.3) of the horizontal flow velocity we estimate the Reynolds number in shallow water to be,

$$\mathrm{Re} = \frac{|(\boldsymbol{v}\cdot\nabla)\boldsymbol{v}|}{\nu\left|\nabla^2\boldsymbol{v}\right|} \sim \frac{U^2/\lambda}{\nu U/d^2} = \frac{ad}{\nu\tau}. \tag{24.8}$$

Here we have assumed that the advective acceleration is dominated by fast horizontal motion over a length scale λ, whereas the viscous acceleration is dominated by the vertical variation in the horizontal flow over the depth d. For deep-water waves d may as before be replaced by $\lambda/2\pi$.

The typical sea waves we encounter when swimming close to the shore at a depth of a couple of metres have amplitudes up to a metre and periods of some seconds. With $\nu \approx 10^{-6}\ \mathrm{m^2\ s^{-1}}$ for water, the Reynolds number will be in the millions, and viscosity plays essentially no role for such waves. In daily life we are otherwise quite familiar with viscous waves, for example while stirring porridge, but they are not so interesting because they quickly die out. Nearly ideal gravity waves in water simply keep rolling along. Eventually viscosity will also make these waves die away if left on their own, but that problem can be dealt with separately (see section 24.6).

Small-amplitude waves are nearly linear

The nonlinearity of the equations of fluid mechanics makes surface waves much more complex than, for example, electromagnetic waves governed by the linear Maxwell equations. The nonlinear advective acceleration of the fluid $(\boldsymbol{v} \cdot \nabla)\boldsymbol{v}$ can however often be disregarded in comparison with the local acceleration $\partial \boldsymbol{v}/\partial t$, so that the Navier–Stokes equations also become linear. For shallow-water waves we obtain the ratio of advective to local acceleration,

$$\frac{|(\boldsymbol{v} \cdot \nabla)\boldsymbol{v}|}{|\partial \boldsymbol{v}/\partial t|} \approx \frac{U^2/\lambda}{U/\tau} \approx \frac{U}{c} \approx \frac{a}{d}. \tag{24.9}$$

Quite generally we can conclude that the advective term plays no role for small-amplitude waves with $a \ll d$ (and $d \approx \lambda/2\pi$ in deep water). In short: *the Navier–Stokes equations become linear in the small-amplitude limit.*

The most general solution to a set of linear field equations with constant coefficients is a linear superposition of elementary harmonic solutions (possibly damped). We have seen this for small-amplitude vibrations in solids (chapter 14) as well as for small-amplitude pressure waves in fluids (section 17.6). In these cases the three-dimensional waves are superpositions of elementary plane waves, each of which at any given time has constant physical properties in any plane orthogonal to its direction of propagation. The same will be the case for the flow (literally) underlying small-amplitude surface waves.

24.2 Harmonic line waves

Three-dimensional plane waves have identical physical properties in every plane orthogonal to the direction of propagation. Surface waves are two-dimensional, and elementary harmonic surface waves have the same physical properties on any line orthogonal to the direction of propagation. A harmonic *line wave* is of the form

$$h = a\cos(k_x x + k_y y - \omega t + \chi) \tag{24.10}$$

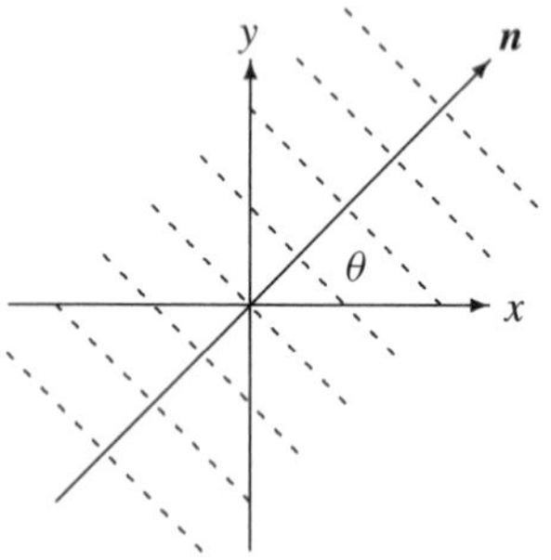

A periodic line wave on the surface. The crests are parallel lines orthogonal to the vector $\boldsymbol{n} = (\cos\theta, \sin\theta, 0)$, forming an angle θ with the x-axis.

where $\boldsymbol{k} = (k_x, k_y, 0)$ is the *wave vector*, $k = |\boldsymbol{k}| = 2\pi/\lambda$ the *wavenumber*, $\omega = 2\pi/\tau$ the *circular frequency* and χ the *phase shift*. The argument of the cosine, $\phi = k_x x + k_y y - \omega t + \chi$, is called the *phase* of the wave. The phase shift can for a single wave always be absorbed in the choice of origin of the coordinate system or of time, but differences between phase shifts may become of physical importance when waves are superposed.

The maxima or minima, or *crests* and *troughs* as they are called for surface waves, move steadily along in the direction $\boldsymbol{n} = \boldsymbol{k}/k = (\cos\theta, \sin\theta, 0)$ with the *phase velocity*,

$$c = \frac{\lambda}{\tau} = \frac{\omega}{k}. \tag{24.11}$$

For small-amplitude shallow-water waves where the dispersion law is linear, the phase velocity is independent of the wavenumber k. In general, however, the dispersion law will be nonlinear, $\tau = \tau(\lambda)$ or equivalently $\omega = \omega(k)$, as we saw for deep-water waves, and the phase velocity will depend on the wavenumber.

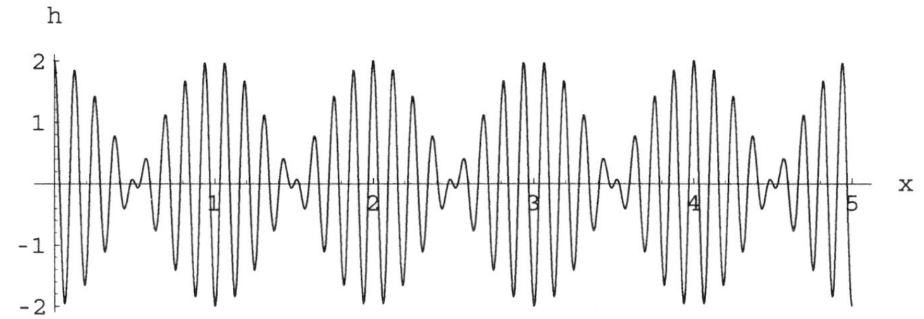

Figure 24.1. Superposition of two harmonic line waves with nearly equal wavenumbers, here at $t = 0$ for $k = 8\Delta k$, $\Delta k = 2\pi$, $a = 1$ and $\chi = 0$. The rapid oscillation of the 'carrier' wave is modulated and broken into a 'beat pattern' of wave packets of length $2\pi/\Delta k = 1$ centred at $x = n$ for all integer n.

Group velocity

Consider now two harmonic line waves which for simplicity are chosen to run along the x-axis with the same amplitudes. Their phases are $\phi_1 = k_1 x - \omega_1 t + \chi_1$ and $\phi_2 = k_2 x - \omega_2 t + \chi_2$, and using the trigonometric relation,

$$\cos\phi_1 + \cos\phi_2 = 2\cos\frac{\phi_1 + \phi_2}{2}\cos\frac{\phi_1 - \phi_2}{2}, \tag{24.12}$$

the superposition $h = h_1 + h_2$ may be written,

$$h = 2a\cos(kx - \omega t + \chi)\cos\frac{1}{2}(\Delta k\, x - \Delta\omega\, t + \Delta\chi), \tag{24.13}$$

where $k = (k_1 + k_2)/2$ etc are the average quantities for the two waves, and $\Delta k = k_1 - k_2$ etc are the differences. The first oscillating factor evidently describes a line wave moving along the x-axis with the average values of the wavenumbers, frequencies and phase shifts, but the amplitude of this wave is now *modulated* by the second factor.

If the differences are much smaller than the averages, $|\Delta k| \ll |k|$, $|\Delta\omega| \ll |\omega|$, and $|\Delta\chi| \ll |\chi|$, the second cosine factor will only slowly modulate the rapid oscillations of the first. Since the second cosine vanishes when its arguments pass through $(1/2)\pi + n\pi$ where n is an arbitrary integer, it will chop the average wave up into a string of *wave packets* of typical length $L = 2\pi/|\Delta k|$, as pictured in figure 24.1. Inside each wave packet, the crests will move with the phase velocity $c = \omega/k$, whereas the centre of each wave packet will move with the speed $\Delta\omega/\Delta k = (\omega(k_1) - \omega(k_2))/\Delta k \approx d\omega(k)/dk$. Thus, the propagation speed of a wave packet is given by the derivative of the dispersion law,

$$\boxed{c_g = \frac{d\omega}{dk},} \tag{24.14}$$

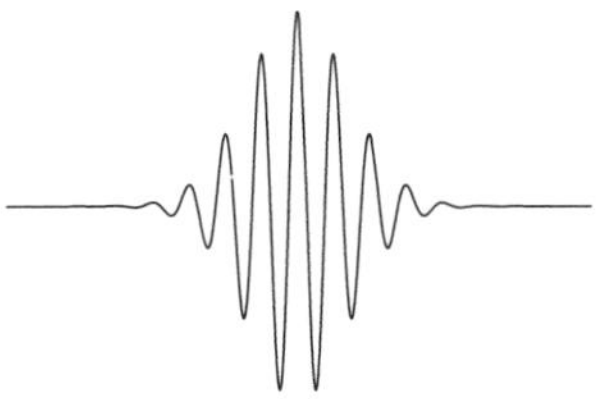

A single Gaussian wave packet.

also called the *group velocity*.

Any superposition of waves with nearly equal wave vectors will in fact form one or more wave packets moving with the group velocity (see problem 24.7). If the dispersion law is linear, the group and phase velocities are equal, but if the dispersion law is nonlinear they will be different, and the waves are said to be *dispersive*. If the group velocity is smaller than the phase velocity, $c_g < c$, the wave crests will move forward inside a wave packet as it proceeds across the surface, and conversely if it is larger.

Energy transport and group velocity

In a single wave packet, the velocity field is only non-zero in the region covered by the wave packet, so that the energy of the wave must be concentrated here and transported along the surface with the group velocity, rather than with the phase velocity. The same must be true for any superposition of single wave packets with wavenumbers taken from a narrow band of width Δk around k, such as the one shown in figure 24.1.

In the limit where the bandwidth Δk narrows down to nothing, the energy must still be transported with the group velocity, so in the end we reach the slightly strange conclusion that even in a purely monofrequent line wave the energy must be transported with the group velocity (see page 356).

24.3 Gravity waves

Small-amplitude, inviscid gravity waves in incompressible water obey the linearized Euler equation without the advective term,

$$\frac{\partial \boldsymbol{v}}{\partial t} = -\frac{1}{\rho_0}\nabla p + \boldsymbol{g}, \qquad \nabla \cdot \boldsymbol{v} = 0, \tag{24.15}$$

where $\boldsymbol{g} = (0, 0, -g_0)$. These equations do not explicitly involve the surface height, which will only come in via the boundary conditions.

Boundary conditions

At the open surface, $z = h$, there are two boundary conditions which must be fulfilled. The first is purely kinematic and expresses that a fluid particle sitting on the surface should follow the surface motion. Under the assumption of small amplitudes, $a \ll d, \lambda$, the surface is nearly horizontal everywhere, so that the vertical velocity of the fluid just below the surface should equal the vertical velocity of the surface itself,

$$\frac{\partial h}{\partial t} = v_z \qquad \text{for } z = h. \tag{24.16}$$

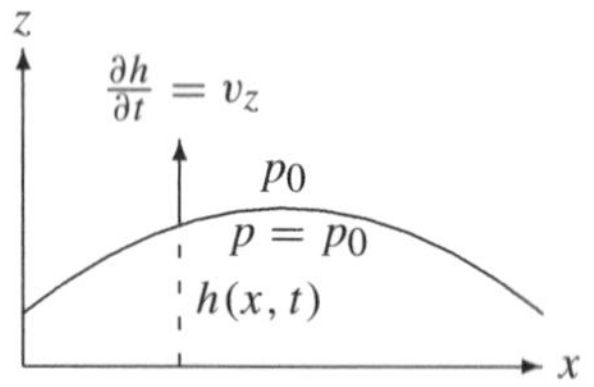

Boundary conditions for a small-amplitude wave.

The second boundary condition is dynamic and expresses the continuity of the pressure across the surface. Assuming that there is air or vacuum with constant pressure p_0 above the surface, the condition becomes

$$p = p_0 \qquad \text{for } z = h. \tag{24.17}$$

Here we have disregarded surface tension which would add a contribution to the right-hand side (see section 24.4).

Besides these, there will be further boundary conditions depending on the shape of the container.

Velocity potential

The linearized Euler equation (24.15) shows that the time derivative of the velocity field $\partial \boldsymbol{v}/\partial t$ is a gradient field. Thus, if the velocity field is initially a gradient field, it will keep on being one. In view of the linearity, the most general solution to the field equations is an irrotational (gradient) field superposed with a time-independent field, possibly containing vorticity. Such a field is of no interest for the study of wave motion, and we shall from now on focus on the time-dependent irrotational component and write it as the gradient of the *velocity potential* Ψ,

$$\boldsymbol{v} = \nabla \Psi \tag{24.18}$$

which, due to the divergence condition, has to satisfy Laplace's equation,

$$\nabla^2 \Psi = 0. \tag{24.19}$$

Inserting the gradient field into the field equation and solving for the pressure, we find the general solution,

$$p = p_0 - \rho_0 \left(g_0 z + \frac{\partial \Psi}{\partial t} \right). \tag{24.20}$$

We might, in principle, add an arbitrary function of time to the right-hand side, but such a function could, without loss of generality, be absorbed into Ψ. Taking $z = h$ we now obtain the open surface boundary conditions,

$$\frac{\partial h}{\partial t} = \nabla_z \Psi, \qquad g_0 h = -\frac{\partial \Psi}{\partial t} \qquad \text{for } z = h. \tag{24.21}$$

Applied to the solutions of the Laplace equation (24.19), these conditions determine both $\Psi(\boldsymbol{x}, t)$ and $h(x, y, t)$, and from (24.20) the pressure may be found. Since all terms of higher order in h have been left out, one may in this approximation take $z = 0$ rather than $z = h$ on the right-hand sides of these conditions.

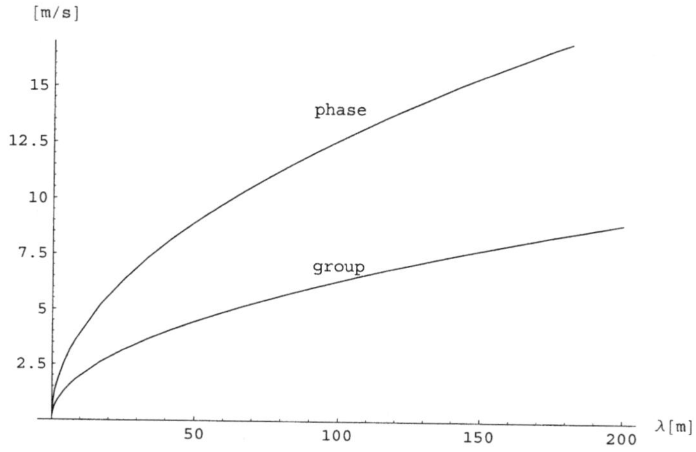

Figure 24.2. Phase and group velocities of deep-water waves as a function of wavelength. This figure corresponds to the small-wavelength part of figure 24.3. Typical ocean swells have wavelengths of $\lambda \approx 150$ m, phase velocity $c \approx 15\ \mathrm{m\,s^{-1}} \approx 55\ \mathrm{km\,h^{-1}}$ and group velocity equal to half of this. Closer to the coast the waves slow down because the water gets shallower (see figure 24.3).

Harmonic line wave solution

Suppose now that the surface wave is an elementary harmonic line wave,

$$h = a\cos(kx - \omega t) \tag{24.22}$$

and that the flow underneath is independent of y. From the first of the conditions (24.21), it follows that $\nabla_z\Psi = a\omega\sin(kx - \omega t)$ for $z = h$, and this suggests that the velocity potential at all depths will be of the form,

$$\Psi = f(z)\sin(kx - \omega t), \tag{24.23}$$

where $f(z)$ is a so far unknown function of z. The Laplace equation (24.18) takes here the form $\nabla_z^2\Psi = -\nabla_x^2\Psi$ and leads immediately to $f'' = k^2 f$. The most general solution to this equation is

$$f(z) = Ae^{kz} + Be^{-kz} \tag{24.24}$$

where A and B are constants to be determined from the boundary conditions.

Deep-water waves

In deep water the velocity field must be finite for $z \to -\infty$, implying that $B = 0$ so that $f(z) = Ae^{kz}$. The open surface boundary conditions (24.21) now lead to,

$$a\omega\sin(kx - \omega t) = k\,Ae^{kh}\sin(kx - \omega t),$$
$$g_0 a\cos(kx - \omega t) = \omega Ae^{kh}\cos(kx - \omega t).$$

In a small-amplitude wave, the wave height is small compared to the wavelength, $k\,|h| \ll 1$, so that $e^{kh} \approx 1$ on the right-hand side (a result obtained directly if one uses $z = 0$ instead of $z = h$). Solving both equations for A we get

$$A = \frac{a\omega}{k} = \frac{ag_0}{\omega}, \tag{24.25}$$

and solving the last equality for ω we obtain the *dispersion law for deep-water waves*,

$$\boxed{\omega = \sqrt{g_0 k}.} \tag{24.26}$$

In terms of period and wavelength we have $\tau = \sqrt{2\pi\lambda/g_0}$ which is of the same form as the previous estimate (24.6), except that now the numerical constant has also been determined to be $\sqrt{2\pi}$. The corresponding deep-water phase and group velocities become

$$c = \sqrt{\frac{g_0}{k}} = \sqrt{\frac{g_0\lambda}{2\pi}}, \qquad c_g = \frac{1}{2}c. \tag{24.27}$$

and are plotted in figure 24.2. Since the phase velocity is double the group velocity, the wave crests will always move forward inside a wave packet with double the speed of the wave packet.

The dispersive nature of deep-water waves have important consequences. A local surface disturbance in deep water—for example created by a storm far out at sea—usually contains more than one wavelength. The long-wave components are faster and will run ahead to arrive at the beach maybe a day or so before the slower short-wave components. The separation of wavelengths over long distances also causes the waves that arrive on the beach to be nearly monofrequent, rolling in at regular time intervals which slowly become shorter as the smaller wavelengths take over.

The complete deep-water solution for all the fields in the wave is,

$$\Psi = ace^{kz}\sin(kx - \omega t), \tag{24.28a}$$

$$v_x = a\omega e^{kz}\cos(kx - \omega t), \tag{24.28b}$$

$$v_z = a\omega e^{kz}\sin(kx - \omega t), \tag{24.28c}$$

$$p = p_0 - \rho_0 g_0 \left(z - ae^{kz}\cos(kx - \omega t)\right). \tag{24.28d}$$

The horizontal and vertical velocity components have the same scale, $a\omega = 2\pi a/\tau$, but are 90° out of phase. The fluid particles move through orbits that are approximatively circles of radius ae^{kz} at depth z.

Due to the exponential, a deep-water surface wave only influences the flow to a depth $|z| \approx 1/k = \lambda/2\pi$. What happens at the bottom has no influence on the surface waves, as long as the wavelength satisfies $\lambda \ll 2\pi d$.

Harmonic line waves at finite depth

When wavelengths become comparable to the depth of the ocean, we must take into account the shape of the bottom. For a horizontally infinite ocean with a perfectly flat impermeable bottom at constant depth $z = -d$, the only extra condition is that the vertical velocity should vanish at the bottom,

$$v_z = 0 \qquad \text{for } z = -d. \tag{24.29}$$

In the absence of viscosity, we are not at liberty to impose a no-slip condition on the horizontal velocities.

The flat-bottom boundary condition implies that $f'(-d) = 0$, or $Ae^{-kd} = Be^{kd}$, so that we have

$$f(z) = C\cosh k(z + d) \tag{24.30}$$

where C is another constant. As for deep-water waves, it is determined by the open surface boundary conditions (24.21), and we find for $|h| \ll d$,

$$C = \frac{a\omega}{k\sinh kd} = \frac{ag_0}{\omega\cosh kd}. \tag{24.31}$$

The last equality yields the dispersion law,

$$\boxed{\omega = \sqrt{g_0 k\tanh kd},} \tag{24.32}$$

with the corresponding phase and group velocities,

$$c = \sqrt{\frac{g_0}{k}\tanh kd}, \qquad c_g = \frac{1}{2}c\left(1 + \frac{2kd}{\sinh 2kd}\right). \tag{24.33}$$

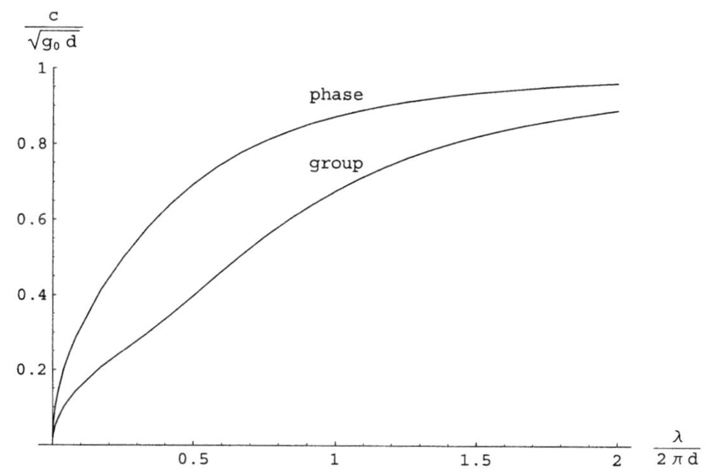

Figure 24.3. Phase and group velocities of flat-bottom gravity waves as functions of $\lambda/2\pi d$. Both velocities level out and become equal for large wavelengths, approaching the common shallow-water value, $c = c_g = \sqrt{g_0 d}$ for $\lambda \gg 2\pi d$. The influence of the finite depth is clearly notable in the group velocity for $\lambda/2\pi d \gtrsim 0.5$.

They are plotted in figure 24.3 in a dimensionless form. The phase velocity appears to curve downwards for all wavelengths, whereas the group velocity changes curvature twice. Generally, the finite depth becomes important from $\lambda \gtrsim \pi d$. For large wavelengths, $\lambda \gg 2\pi d$, both velocities approach the common shallow-water value $c = c_g = \sqrt{g_0 d}$.

The complete solution for all the fields underneath a small-amplitude harmonic line wave $h = a\cos(kx - \omega t)$ at any depth $d \gg a$ is finally,

$$\Psi = ac\frac{\cosh k(z+d)}{\sinh kd}\sin(kx - \omega t), \tag{24.34a}$$

$$v_x = a\omega\frac{\cosh k(z+d)}{\sinh kd}\cos(kx - \omega t), \tag{24.34b}$$

$$v_z = a\omega\frac{\sinh k(z+d)}{\sinh kd}\sin(kx - \omega t), \tag{24.34c}$$

$$p = p_0 - \rho_0 g_0\left(z - a\frac{\cosh k(z+d)}{\cosh kd}\cos(kx - \omega t)\right). \tag{24.34d}$$

It must be emphasized that this solution is only valid in the limit of $a \ll \lambda, d$, and that all terms of higher order in a have been left out. The solution is strictly speaking only valid in the interval $-d \le z \le 0$ (see section 24.8).

The corresponding (Lagrangian) displacement field is obtained by integrating the velocities with respect to time,

$$u_x = -a\frac{\cosh k(z+d)}{\sinh kd}\sin(kx - \omega t), \tag{24.35a}$$

$$u_z = a\frac{\sinh k(z+d)}{\sinh kd}\cos(kx - \omega t). \tag{24.35b}$$

This shows that the fluid particle orbits are ellipses of general magnitude a, which become progressively flatter as the bottom is approached for $z \to -d$. Evidently, there is no net mass transport in the direction of motion of the wave. Note that for $z = 0$ we get $u_z = h$, as we should.

Shallow-water limit

For waves with wavelength much greater than the depth, i.e. for $\lambda \gg 2\pi d$ or equivalently $kd \ll 1$, we have $\tanh kd \approx kd$ and thus obtain the *shallow-water dispersion law*,

$$\boxed{\omega = \sqrt{g_0 d}\, k,} \tag{24.36}$$

in agreement with the estimate (24.6). Shallow-water waves are *non-dispersive* with common phase and group velocities

$$c = c_g = \sqrt{g_0 d}. \tag{24.37}$$

The leading terms in the solution become in the shallow wave limit, $kd \ll 1$,

$$\Psi = \frac{a g_0}{\omega}\left(1 + \frac{k^2(z+d)^2}{2}\right)\sin(kx - \omega t), \tag{24.38a}$$

$$v_x \approx \frac{ca}{d}\cos(kx - \omega t), \tag{24.38b}$$

$$v_z \approx a\omega\left(1 + \frac{z}{d}\right)\sin(kx - \omega t), \tag{24.38c}$$

$$p \approx p_0 - \rho_0 g_0 \left(z - a\cos(kx - \omega t)\right). \tag{24.38d}$$

The horizontal velocity is the same for all z, so that all the water underneath sloshes back and forth in unison as the wave proceeds. The vertical velocity decreases and reaches zero at the bottom, as it must (this is the reason for keeping the second-order terms in Ψ). At any depth z, the pressure is just the hydrostatic pressure from the water column above, including the height of the wave.

The horizontal velocity scale, $c\,a/d = \lambda a/\tau d$, equals precisely the previously estimated shallow-water value (24.3) which is larger than the vertical velocity scale $a\omega = 2\pi a/\tau$ by a factor $\lambda/2\pi d$. This also confirms the estimates made in section 24.1.

Example 24.3.1 (Tsunami): Huge shallow-water wave trains, tsunamis (meaning harbour waves), with wavelengths up to 500 km can be generated by underwater earthquakes, landslides, volcanic eruptions or large meteorite impacts. The average depth of the oceans is about 4000 m so tsunamis move with typical speeds of a passenger jet plane, $c \approx 200$ m s^{-1} ≈ 710 km h^{-1}, at that depth, and faster at greater depths. For $\lambda = 500$ km the period becomes $\tau = \lambda/c \approx 2500$ s ≈ 42 min, but since the amplitude is small, say $a \approx 1$ m, a tsunami will be completely imperceptible to a ship at sea (today the precision of GPS positioning should make it notable). When the tsunami approaches the coastline, the water depth decreases and the wave slows down while increasing its amplitude with sometimes devastating effect on the shore.

24.4 Capillary surface waves

Surface tension is characterized by a material constant α, representing the attractive force per unit of length of the surface, or equivalently the extra energy per unit of surface area (from the missing molecular bonds). Surface tension generates a pressure jump (8.5) across any interface between two fluids, expressed through the principal radii of curvature of the surface. We saw in section 8.1 that the relative influence of surface tension and gravity in a liquid/air interface is characterized by a length (8.4), called the *capillary constant*, which is $R_c = \sqrt{\alpha/\rho_0 g_0} \approx 2.7$ mm for the water/air interface. Surface tension only plays a major role for length scales smaller than the capillary constant.

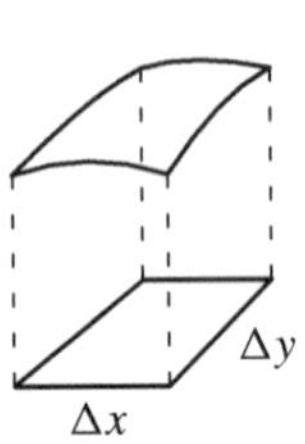

The small rectangle in the xy-plane defines a piece of the wave surface of area $A = \Delta x \Delta y$.

Pressure jump across a nearly flat surface

To calculate the pressure jump over a nearly flat surface, $z = h(x, y)$ with $|\partial h/\partial x|, |\partial h/\partial y| \ll 1$, we consider a tiny piece of the surface situated above a small rectangle between (x, y) and $(x + \Delta x, y + \Delta y)$. All four sides of this piece of surface are subject to tension from the surrounding surface, and we wish to

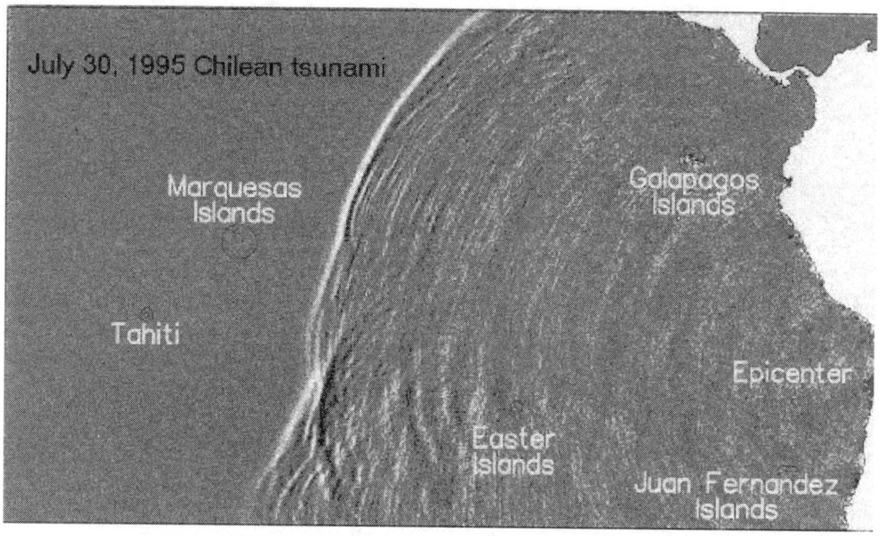

Figure 24.4. Chilean tsunami on July 30, 1995. The tsunami spread from the epicentre of an earthquake on the Chilean pacific coast.

calculate the resultant vertical force. Since the slope of the surface is small, we may disregard the slight misalignment between the vertical force and the pressure force which strictly speaking must be orthogonal to the surface. The slope of the surface at (x, y) is $\partial h/\partial x$ along x, and it follows from the geometry that the surface tension acting on the two Δy-sides of the rectangle generates a 'vertical' force,

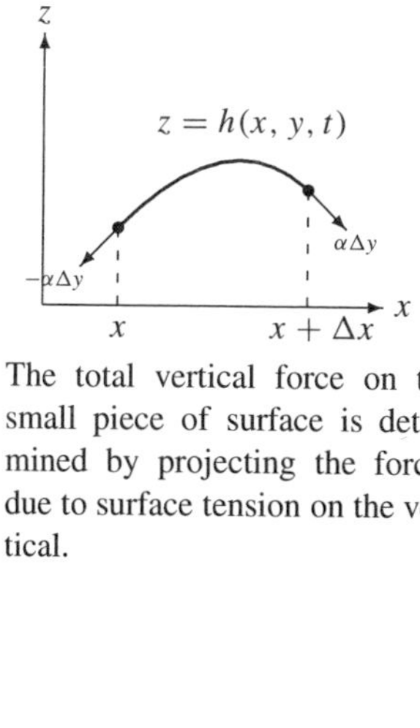

The total vertical force on the small piece of surface is determined by projecting the forces due to surface tension on the vertical.

$$\Delta\mathcal{F}_z \approx -\alpha\Delta y\frac{\partial h(x,y)}{\partial x} + \alpha\Delta y\frac{\partial h(x+\Delta x,y)}{\partial x} \approx \alpha\Delta x\Delta y\frac{\partial^2 h(x,y)}{\partial x^2}.$$

Adding the forces acting on the two Δx-sides and dividing by the area $A = \Delta x\Delta y$, we see that in order to balance the vertical force from surface tension, the pressure just below the surface must be higher by,

$$\boxed{\Delta p = -\alpha(\nabla_x^2 + \nabla_y^2)h.} \tag{24.39}$$

If the surface curves downwards in all directions at a given point, we have $(\nabla_x^2 + \nabla_y^2)h < 0$, and the extra pressure will be positive below the surface.

The curvature of a small-amplitude surface wave thus generates a pressure jump Δp at the surface. Whereas the Euler equation (24.15) is unchanged, the value of the pressure just below the surface is $p_0+\Delta p$ rather than p_0, so that the dynamic boundary condition (24.17) is replaced by,

$$p = p_0 + \Delta p \qquad \text{for } z = h. \tag{24.40}$$

Since Δp given by (24.39) is positive at a wave crest and negative at a trough, surface tension collaborates with gravity in attempting to flatten the water surface. Waves completely dominated by surface tension are called *capillary waves*.

Deep-water capillary gravity waves

Surface tension is only expected to become important for waves of very small wavelength, which except for special situations may be assumed to be deep-water waves. The velocity potential is in that case $\Psi = Ae^{kz}\sin(kx - \omega t)$, and from the kinematic boundary condition (24.16) and the modified dynamic boundary condition (24.40) we obtain in the usual way,

$$A = a\frac{\omega}{k} = a\frac{\rho_0 g_0 + \alpha k^2}{\rho_0\omega}. \tag{24.41}$$

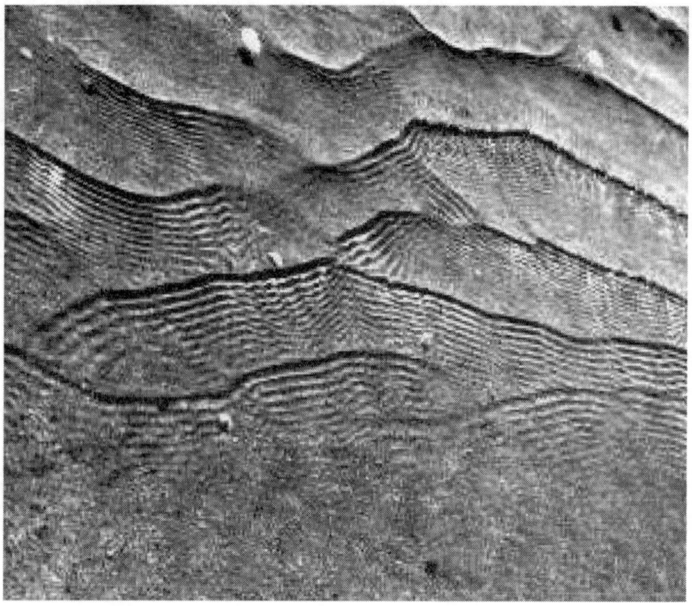

Figure 24.5. Capillary waves in front of ordinary gravity waves. Photograph by Fabrice Neyret, reproduced with permission.

This shows that the only effect of surface tension is to increase the gravitational acceleration from g_0 to,

$$g = g_0 + \frac{\alpha k^2}{\rho_0} = g_0(1 + k^2 R_c^2), \tag{24.42}$$

where $R_c = \sqrt{\alpha/\rho_0 g_0}$ is the capillary constant. As foreseen, surface tension collaborates with gravity and becomes more important than gravity for $kR_c \gtrsim 1$ or $\lambda \lesssim 2\pi R_c$ which in water is 1.7 cm.

Replacing g_0 by g in the deep-water dispersion law (24.26), we find

$$\boxed{\omega = \sqrt{g_0 k + \frac{\alpha}{\rho_0} k^3} = \sqrt{g_0 k\left(1 + k^2 R_c^2\right)}.} \tag{24.43}$$

This dispersion law agrees very well with experiments[1].

The phase and group velocities become,

$$c = \sqrt{\frac{g_0}{k}\left(1 + k^2 R_c^2\right)}, \qquad c_g = \frac{1}{2} c \frac{1 + 3k^2 R_c^2}{1 + k^2 R_c^2}, \tag{24.44}$$

and are plotted for water in figure 24.6. Note that the phase velocity has a minimum for $kR_c = 1$ (i.e. $\lambda = \lambda_c$) where the group velocity also equals the phase velocity (see problem 24.2). The minimum of the group velocity occurs at a somewhat larger wavelength, $\lambda/\lambda_c = 2.54\ldots$.

For very small wavelengths, $\lambda \ll \lambda_c$ or $kR_c \gg 1$, surface tension dominates completely, and we find the dispersion law for purely capillary waves,

$$\omega = \sqrt{\frac{\alpha k^3}{\rho_0}}, \qquad c = \sqrt{\frac{\alpha k}{\rho_0}}, \qquad c_g = \frac{3}{2} c. \tag{24.45}$$

In purely capillary waves the phase velocity is only 2/3 of the group velocity, and the wave crests appear to move backwards inside a wave packet!

[1] See for example B. Christiansen, P. Alstrøm and M. T. Levinsen, Dissipation and ordering in capillary waves at high aspect ratios, *J. Fluid Mech.* **291**, (1995) 323.

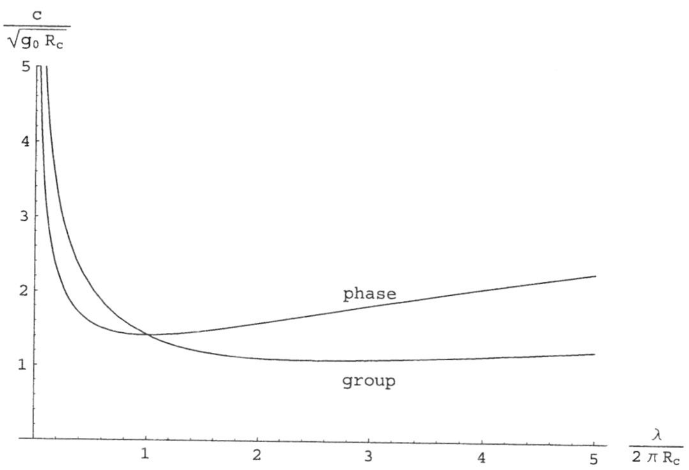

Figure 24.6. Phase and group velocities in units of $\sqrt{g_0 R_c}$ for deep-water capillary gravity waves as functions of the wavelength in units of $\lambda_c = 2\pi R_c$. In water $\sqrt{g_0 R_c} \approx 16$ cm s^{-1} and $\lambda_c \approx 1.7$ cm. Phase and group velocities cross each other at $\lambda = \lambda_c$, where the phase velocity is minimal. The group velocity has a minimum at $\lambda \approx 2.54\lambda_c$ (see problem 24.2).

24.5 Internal waves

In the ocean a heavier saline layer of water may often be found below a lighter more brackish layer, allowing so-called *internal waves* to arise at the interface. Even if the difference in density between the fluids is small, the equilibrium interface will always be horizontal with the lighter liquid situated above the heavier (as discussed previously in section 7.1). Were it somehow possible to invert the ocean so that the lighter fluid came to lie below the heavier, instability would surely arise, and the liquids would after some time find their way back to 'natural' order. As we shall see, surface tension can in fact stabilize the inverted situation in sufficiently small containers.

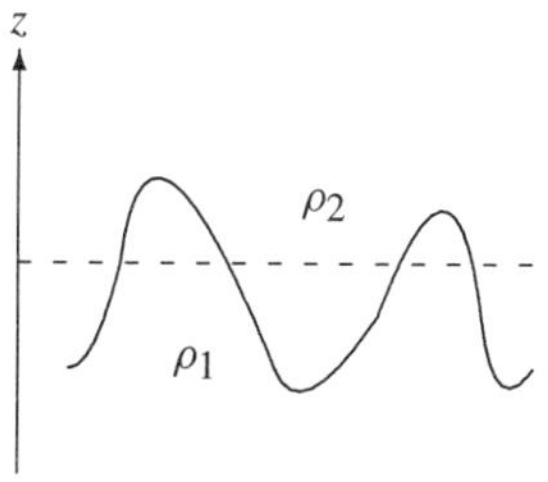

Internal waves at an interface with a heavier liquid below and a lighter above ($\rho_1 > \rho_2$).

Boundary conditions

Let the lower layer have density ρ_1 and the upper layer ρ_2 with a separating interface $z = h(x, y, t)$ between the two fluids, for which the velocity potentials are Ψ_1 and Ψ_2. The kinematic boundary conditions for small-amplitude waves express that both fluids and the separating surface must move together in the vertical direction,

$$v_{1z} = v_{2z} = \frac{\partial h}{\partial t} \qquad \text{for } z = h. \tag{24.46}$$

Including surface tension (24.39), the dynamic boundary condition becomes,

$$p_1 + \Delta p = p_2 \qquad \text{for } z = h \tag{24.47}$$

where Δp is given by (24.39), and the pressures are expressed like (24.20) in each of the fluids.

Dispersion law

Suppose again that the interface takes the form of a pure line wave, $h = a\cos(kx - \omega t)$. We shall only consider deep-water waves in which the wave flow is required to vanish far below and far above the interface. The velocity potentials are then of the form,

$$\Psi_1 = A_1 e^{+kz} \sin(kx - \omega t), \qquad \Psi_2 = A_2 e^{-kz} \sin(kx - \omega t),$$

with a notable change of sign in the exponential factors. The boundary conditions (24.46) and (24.47) imply for $k\,|h| \ll 1$ that

$$kA_1 = -kA_2 = a\,\omega, \qquad \rho_1(g_0 a - \omega A_1) + \alpha k^2 a = \rho_2(g_0 a - \omega A_2). \tag{24.48}$$

Solving these we find that

$$A_1 = -A_2 = a\frac{\omega}{k} = \frac{a}{\omega}\frac{g_0(\rho_1 - \rho_2) + \alpha k^2}{\rho_1 + \rho_2}. \tag{24.49}$$

Solving the last equality for ω we obtain the *dispersion law for deep-water internal waves,*

$$\boxed{\omega = \sqrt{\frac{g_0 k(\rho_1 - \rho_2) + \alpha k^3}{\rho_1 + \rho_2}}.} \tag{24.50}$$

If the upper density is much smaller than the lower, $\rho_2 \ll \rho_1$, these waves become ordinary deep-water capillary gravity waves, but when the densities are nearly equal, $\rho_2 \lesssim \rho_1$, the internal waves have much lower frequencies (and velocities) than waves of the same wavelength at the surface.

The capillary constant is as before defined as the length scale where the gravity contribution to the frequency is of the same magnitude as the surface tension contribution,

$$R_c = \sqrt{\frac{\alpha}{|\rho_1 - \rho_2|\,g_0}}. \tag{24.51}$$

It diverges when the densities become equal. In this limit gravity plays no role, and the internal waves have become purely capillary waves, described by (24.45) with $\rho_0 = 2\rho_1 = 2\rho_2$.

Example 24.5.1 (Brackish–saline interface in the sea): If a brackish surface layer lies above a saline layer with 4% higher density, the capillary wavelength for internal waves becomes $\lambda_c = 2\pi R_c = 8.6\ cm$. A wave with this wavelength has period $\tau = 1.2$ s and moves with phase velocity $c = 7.2\ \text{cm s}^{-1}$.

The Rayleigh–Taylor instability

When the heavier fluid lies below the lighter, $\rho_1 > \rho_2$, the frequency ω of an internal wave is always real, but if the container is quickly turned upside down, such that $\rho_1 < \rho_2$, the heavier fluid will be on top and the dispersion law may be written as,

$$\omega = \sqrt{g_0\frac{\rho_2 - \rho_1}{\rho_1 + \rho_2}k(k^2R_c^2 - 1)}, \tag{24.52}$$

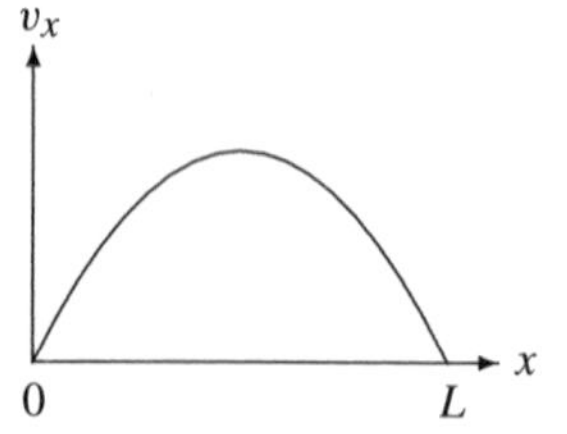

Any flow in a box of length L must obey $v_x = 0$ for $x = 0$ and $x = L$.

The argument of the square root will be negative for $kR_c < 1$, or $\lambda > \lambda_c = 2\pi R_c$. In that case, ω becomes imaginary, and the otherwise sinusoidal form, the line wave, is replaced by an exponential growth $e^{|\omega|t}$ in time. This signals an instability, called the *Rayleigh–Taylor instability.*

In an infinitely extended ocean, there is room for waves with wavelengths of any size, and the inverted situation will always be unstable. It can only be maintained for a very short while, because the smallest perturbation of the surface will lead to a run-away process that ends with the heavier liquid again arranged below the lighter.

In a finite container, there is an upper limit to the allowed wavelengths, because the boundary conditions require the horizontal velocities to vanish at the vertical walls surrounding the fluids. Any flow in a finite box-shaped container of horizontal length L must obey the boundary conditions $v_x = 0$ for both $x = 0$ and $x = L$. Linear Euler flow in a box can, like the flows we are studying here, always be resolved into a superposition of standing waves with horizontal velocity $v_x \sim \sin kx \cos \omega t$. The boundary conditions select the allowed wavenumbers to be $k = n\pi/L$ where $n = 1, 2, \ldots$ is an arbitrary integer. For $n = 1$ we obtain the largest wavelength, $\lambda = 2\pi/k = 2L$, and this shows that as long as

$$L < \frac{1}{2}\lambda_c = \pi R_c = \pi\sqrt{\frac{\alpha}{(\rho_2 - \rho_1)g_0}}, \tag{24.53}$$

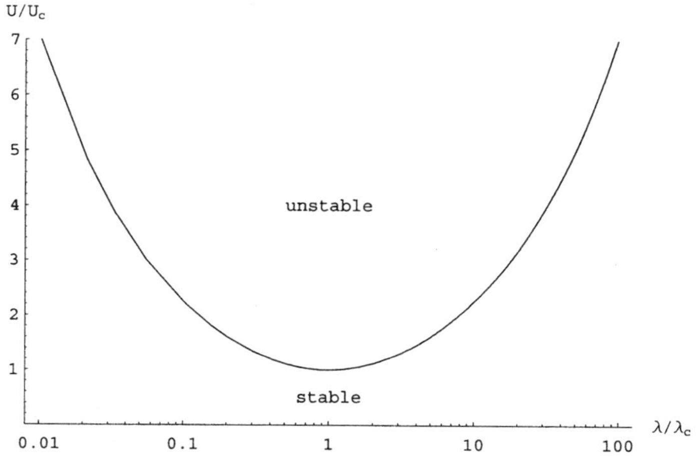

Figure 24.7. Plot of U/U_c as a function of λ/λ_c. For the water–air interface the capillary wavelength is $\lambda_c = 1.7$ cm and the critical velocity is $U_c = 7.4$ m s^{-1}. For a given velocity U, one can read off the range of unstable wavelengths from this figure.

unstable wave modes with $\lambda > \lambda_c$ cannot occur. If you invert a container with horizontal size smaller than half the capillary wavelength, the heavier liquid will remain stably on top of the lighter.

Example 24.5.2 (Home experiment): Air against water has as we have seen before a capillary wavelength of $\lambda_c = 1.7$ cm, so we require that $L < 0.85$ cm. Try it yourself with a glass tube of, for example, 5 mm diameter. It works!

The Kelvin–Helmholtz instability

Layers of inviscid fluids are capable of sliding past each other with a finite slip-velocity, if we disregard the viscous boundary layers that otherwise will soften the sharp discontinuity in velocity. When waves arise in the interface between the fluids, they will so to speak 'get in the way' of the smooth flow, leading us to expect instability at a sufficiently high slip-velocity.

Suppose the upper layer is moving with velocity U in the rest frame of the lower layer. Taking into account the slope $\nabla_x h$ of the interface, the horizontal flow in the upper layer will add $U\nabla_x h$ to the vertical velocity, such that the kinematic boundary conditions are replaced by,

$$v_{1z} = \frac{\partial h}{\partial t}, \qquad v_{2z} = \frac{\partial h}{\partial t} + U\nabla_x h \qquad \text{for } z = h. \tag{24.54}$$

The same is the case for the pressure which becomes

$$p_2 = p_0 - \rho_0 \left(g_0 h + \frac{\partial \Psi_2}{\partial t} + U\nabla_x \Psi_2 \right). \tag{24.55}$$

Putting it all together we find the boundary conditions,

$$kA_1 = a\omega, \qquad -kA_2 = a(\omega - kU),$$
$$\rho_1(g_0 a - \omega A_1) + \alpha k^2 a = \rho_2(g_0 a - (\omega - kU)A_2).$$

When combined these lead to a quadratic equation for the frequency

$$(\rho_1 + \rho_2)\omega^2 - 2\rho_2 kU\omega + \rho_2 k^2 U^2 = k((\rho_1 - \rho_2)g_0 + \alpha k^2). \tag{24.56}$$

Given the wavenumber k, the roots are real for

$$U^2 < \left(\frac{1}{\rho_1} + \frac{1}{\rho_2}\right) \frac{(\rho_1 - \rho_2)g_0 + \alpha k^2}{k}. \tag{24.57}$$

For $\rho_1 > \rho_2$, the right-hand side has an absolute minimum when $kR_c = 1$, where R_c is the capillary radius for internal waves (24.51). Selecting the minimum of the right-hand side by setting $k = 1/R_c$, the condition for absolute stability becomes

$$U < U_c = \sqrt{2g_0 R_c \left(\frac{\rho_1}{\rho_2} - \frac{\rho_2}{\rho_1}\right)}. \tag{24.58}$$

For air flowing over water the critical velocity is $U_c = 7.4\ \mathrm{m\ s^{-1}}$.

In figure 24.7 the ratio

$$\frac{U}{U_c} = \sqrt{\frac{1}{2}\left(\frac{\lambda}{\lambda_c} + \frac{\lambda_c}{\lambda}\right)} \tag{24.59}$$

is plotted as a function of λ/λ_c. For $U > U_c$ there will be a range of wavelengths around the capillary wavelength $\lambda = \lambda_c$ for which small disturbances will diverge exponentially with time. This is the *Kelvin–Helmholtz* instability which permits us at least in principle to understand how the steadily streaming wind is able to generate waves from tiny disturbances.

What actually happens to the unstable waves with their exponentially growing amplitudes, for example how they grow into the larger waves created by a storm, cannot be predicted from linear theory. It is, however, possible to say something about the statistics of wind-generated ocean waves without going into too much nonlinear theory (see section 24.7).

∗ 24.6 Energy and attenuation

Surface waves contain both potential and kinetic energy, and this energy is attenuated by several effects. First, there is viscous attenuation due to internal friction in the fluid. Secondly, there is attenuation from dissipative losses in the boundary layer that necessarily forms near the bottom, and thirdly there is dissipation due to deviations of the value of surface tension from its equilibrium value, an effect that plays a role when oil is poured on troubled waters. In this section we shall focus only on viscous attenuation.

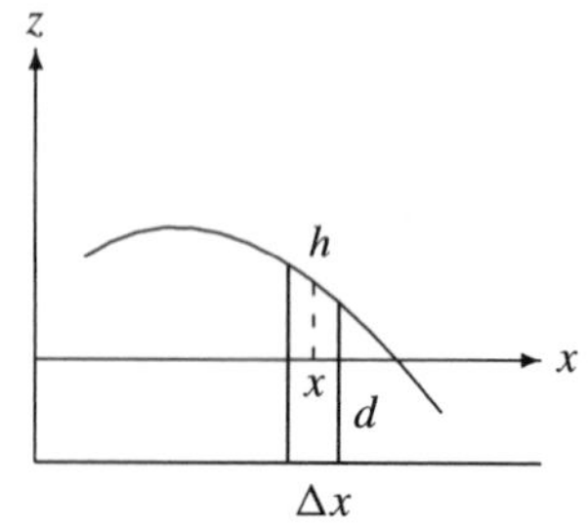

A thin water column of 'footprint' area $A = L\Delta x$ and height h in a sea of depth d.

Total energy

The first goal is to calculate the total energy in a wave. Consider a thin column of water in a gravity wave of width Δx along x and length Δy along y, so that its 'footprint' area is $A = \Delta x \Delta y$. Relative to the static water level $z = 0$, its potential energy is (apart from an additive constant),

$$\mathcal{V} = \int_V \rho_0 g_0 z \, dV = \int_{-d}^{h} \rho_0 g_0 z \, A dz = \frac{1}{2}\rho_0 g_0 (h^2 - d^2) A. \tag{24.60}$$

We shall fix the arbitrary constant by requiring the potential energy to vanish for the undisturbed surface $h = 0$, so that

$$\mathcal{V} = \tfrac{1}{2}\rho_0 g_0 h^2 A. \tag{24.61}$$

The potential energy is always positive, and rises and falls in tune with the square of the wave height $h = a\cos(kx - \omega t)$. Note that this is an exact result, also valid when the amplitude is large. In the linearized approximation where $h = a\cos(kx - \omega t)$, we find the average of the potential energy over a full period,

$$\langle \mathcal{V} \rangle = \frac{1}{\tau}\int_0^{\tau} \mathcal{V}\, dt = \frac{1}{4}\rho_0 g_0 a^2 A, \tag{24.62}$$

because the average over a squared cosine is 1/2.

The kinetic energy of the water in the column is similarly

$$\mathcal{T} = \int_V \frac{1}{2}\rho_0 v^2 \, dV = \int_{-d}^{h} \frac{1}{2}\rho_0 (v_x^2 + v_z^2)\, A dz \approx \int_{-d}^{0} \frac{1}{2}\rho_0 (v_x^2 + v_z^2)\, A dz. \tag{24.63}$$

In the last step we have assumed that the amplitude is small, $|h| \ll d$, and replaced h by 0 in the upper limit of the integral. It would in fact be wrong to keep the upper limit h because it contributes only to the higher order terms. Inserting the explicit gravity wave solution (24.34), we obtain the time average of the integrand,

$$\left\langle v_x^2 + v_z^2 \right\rangle = \frac{1}{2}\left(\frac{a\omega}{\sinh kd}\right)^2 \left(\cosh^2 k(z+d) + \sinh^2 k(z+d)\right).$$

Finally, using the relation $\cosh^2\phi + \sinh^2\phi = \cosh 2\phi$, the integral over z can be carried out, and we find after some rearrangement,

$$\langle \mathcal{T} \rangle = \frac{1}{4}\rho_0 A \left(\frac{a\omega}{\sinh kd}\right)^2 \frac{\sinh 2kd}{2k} = \frac{1}{4}\rho_0 g_0 a^2 A. \tag{24.64}$$

As expected in the general estimates of section 24.1, we have $\langle \mathcal{T} \rangle = \langle \mathcal{V} \rangle$. This is also a consequence of the much more general virial theorem (see problem 24.4).

The average of the total mechanical energy over a full period thus becomes

$$\boxed{\langle \mathcal{E} \rangle = \langle \mathcal{T} \rangle + \langle \mathcal{V} \rangle = \tfrac{1}{2}\rho_0 g_0 a^2 A.} \tag{24.65}$$

Surprisingly, the average energy per unit of surface area, $\langle \mathcal{E} \rangle / A$, only depends on the amplitude and not on the depth or wavelength. Furthermore, since the average energy is independent of both x and y, this expression is valid for a region of any size with 'footprint' area A.

Example 24.6.1 (Surface energy): A small-amplitude harmonic surface wave in water with amplitude $a = 10$ cm contains a wave energy per unit of surface area of about $\langle \mathcal{E} \rangle / A \approx 50$ J m^{-2}.

Rate of viscous dissipation

For an incompressible liquid the rate of mechanical energy loss due to viscous dissipation is given by (17.18), which for a thin vertical column of liquid with area $A = \Delta x \Delta y$, simplifies to,

$$\dot{W}_{\text{int}} = 2\eta \int_{-d}^{h} \sum_{ij} v_{ij}^2 \, A dz \approx 2\eta \int_{-d}^{0} \sum_{ij} v_{ij}^2 \, A dz, \tag{24.66}$$

where $v_{ij} = (1/2)(\nabla_i v_j + \nabla_j v_i)$ and η is the viscosity. Evaluating it for a standard line wave running along x the integrand may be recast as

$$\sum_{ij} v_{ij}^2 = (\nabla_x v_x)^2 + (\nabla_z v_z)^2 + \frac{1}{2}(\nabla_x v_z + \nabla_z v_x)^2 = 2(\nabla_x v_x)^2 + 2(\nabla_x v_z)^2.$$

In the last step have used mass conservation $\nabla_z v_z = -\nabla_x v_x$ and irrotationality $\nabla_z v_x = \nabla_x v_z$. For a harmonic wave (24.34) we may replace ∇_x by k in the time average, so that it becomes

$$\left\langle \sum_{ij} v_{ij}^2 \right\rangle = 2k^2 \left\langle v_x^2 + v_z^2 \right\rangle. \tag{24.67}$$

This is proportional to the integrand in the kinetic energy (24.63) and taking over the result (24.64) this leads to

$$\left\langle \dot{W}_{\text{int}} \right\rangle = 8\nu k^2 \langle \mathcal{T} \rangle, \tag{24.68}$$

where $\nu = \eta/\rho_0$ is the kinematic viscosity. Relative to the average of the total energy, $\langle \mathcal{E} \rangle = 2\langle \mathcal{T} \rangle$ the rate of dissipation finally becomes,

$$\boxed{\frac{\left\langle \dot{W}_{\text{int}} \right\rangle}{\langle \mathcal{E} \rangle} = 4\nu k^2.} \tag{24.69}$$

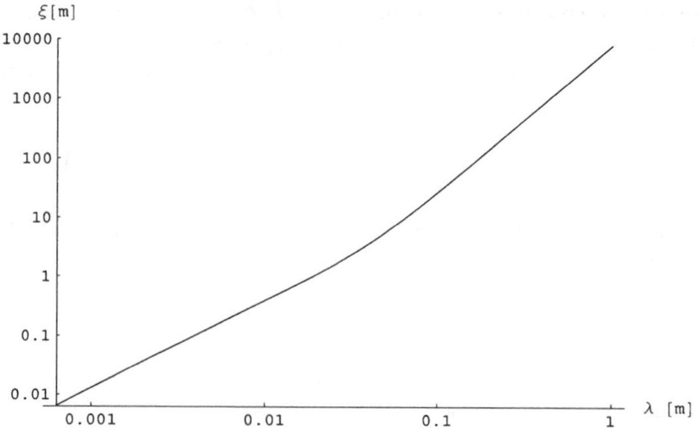

Figure 24.8. The viscous amplitude attenuation length ξ for water plotted as a function of wavelength λ. The viscosity is $\nu = 10^{-6}\ \text{m}^2\ \text{s}^{-1}$ and the surface tension $\alpha = 0.073\ \text{N m}^{-1}$. Viscous attenuation essentially only plays a role for very small wavelengths.

The dissipative energy loss grows quadratically with the wavenumber and is most important for small wavelengths, i.e. for capillary waves. These may as before be included by replacing g_0 by the effective gravity (24.42), but that does not change the above result.

The loss of energy over a wave period $\tau = 2\pi/\omega$ is $\dot{W}_{\text{int}}\,\tau$ and relative to the average energy of the wave (24.65), it becomes

$$\frac{\langle \dot{W}_{\text{int}} \rangle\, \tau}{\langle \mathcal{E} \rangle} = 4\nu k^2 \tau = 8\pi\nu \frac{k^2}{\omega} = 16\pi^2 \frac{\nu\tau}{\lambda^2}, \tag{24.70}$$

where ν is the kinematic viscosity. The condition for our calculation to be valid is that the relative attenuation should be small, or $\lambda^2/\tau \gg 16\pi^2\nu$. In deep water with $\nu \approx 10^{-6}\ \text{m}^2\ \text{s}^{-1}$, this is fulfilled for $\lambda \gg 40\ \mu\text{m}$, so that the condition should be satisfied under ordinary circumstances.

Energy and amplitude attenuation coefficients

The energy propagates, as we have argued on page 339, with the group velocity c_g rather than the phase velocity c. Dividing the energy dissipation rate (24.69) by c_g from (24.45), we obtain the spatial *energy attenuation coefficient* 2κ, defined as the fractional loss of energy per unit of length,

$$\boxed{2\kappa = \frac{4\nu k^2}{c_g}.} \tag{24.71}$$

As the energy is quadratic in the amplitude, the energy attenuation coefficient is twice the spatial *amplitude attenuation coefficient* κ. The inverse amplitude attenuation coefficient $\xi = 1/\kappa$ is called the *amplitude attenuation length* and indicates the distance over which the surface wave amplitude falls to about $e^{-1} \approx 37\%$ of its initial value. The amplitude attenuation length is plotted for water in figure 24.8 as a function of wavelength.

Example 24.6.2: In water for $\lambda = 1$ m one finds $\xi \approx 8$ km whereas for $\lambda = \lambda_c = 1.7$ cm one gets $\xi \approx 86$ cm. A raindrop hitting a lake surface thus creates a disturbance that dies out after propagating through a metre or less whereas a big object, like the human body, will make waves with much longer wavelength that continues essentially unattenuated right across the lake. In both these cases, however, the amplitude of the ring-shaped surface waves will also diminish for purely geometric reasons.

* 24.7 Statistics of wind-generated ocean waves

Waves may arise spontaneously from tiny perturbations at the wind/water interface when the wind speed surpasses the Kelvin–Helmholtz instability threshold (page 349). The continued action of the wind and nonlinear wave interactions raise the waves further, until a kind of dynamic equilibrium is reached in which the surface may be viewed as a statistical ensemble of harmonic waves with a spectrum of periods, wavelengths and amplitudes. Even if we do not understand the mechanism at play, it is nevertheless possible to draw some quite general conclusions about the wave statistics and compare with observations.

Surface height observations

A ship or buoy bobbing at a fixed position (x, y) of the ocean surface reflects the local surface height, $h(t) = h(x, y, t)$. The variations in surface height may be determined by many different techniques, for example based on accelerometers, radar or sattelites. While the wind blows steadily, a long record of $N \gg 1$ measurements $h_n = h(t_n)$ can be collected at discrete times, $t_n = n\epsilon$ $(n = 1, 2, \ldots, N)$, which for simplicity are assumed to be evenly spaced.

The underlying wave structure of the surface creates strong correlations between successive measurements of the local height. Short waves are carried on top of larger waves and so on. To get rid of such correlations, we shall for odd $N = 2M + 1$ write the record as a superposition of simple harmonics (see problem 24.8 for the precise theory of discrete Fourier transformations),

$$h_n = a_0 + \sum_{m=1}^{M} a_m \cos(\omega_m t_n - \chi_m) \tag{24.72}$$

where $\omega_m = 2\pi m/N\epsilon$ is the circular frequency, a_m the amplitude, and χ_m the phase shift of the mth harmonic. Given the $N = 2M + 1$ measured values h_n these equations may be solved for the $2M + 1$ unknowns, consisting of amplitudes and phase shifts plus the constant a_0. Note that the highest frequency that can be resolved by N observations is $\omega_M = 2\pi M/N\epsilon \approx \pi/\epsilon$.

From the data record we may calculate various averages that can be related to the parameters of the harmonic expansion. Thus, we find the average height $\langle h \rangle = \frac{1}{N}\sum_{n=1}^{N} h_n = a_0$ because all the cosines average out to zero. This shows that without any loss of generality we may always subtract the average water level a_0 from the measured heights. Assuming from now on that $\langle h \rangle = a_0 = 0$, the variance of the height becomes

$$\left\langle h^2 \right\rangle = \frac{1}{N}\sum_{n=1}^{N} h_n^2 = \frac{1}{2}\sum_{m=1}^{M} a_m^2. \tag{24.73}$$

In the last step we have used that the harmonics are uncorrelated so that the average of the product of different harmonics vanishes, whereas the average of the square of any of the cosines is 1/2 (see problem 24.8).

For large N the frequencies constitute almost a continuum and since the energy is proportional to the square of the amplitude, the *power spectrum* of the observed waves may be defined to be

$$S(\omega_m) = \frac{N\epsilon}{4\pi} a_m^2. \tag{24.74}$$

The coefficient in front has been chosen such that for large N we may write the sum over m as an integral

$$\left\langle h^2 \right\rangle = \frac{1}{2}\sum_{m=1}^{M} a_m^2 \approx \int_0^{\omega_M} S(\omega)\, d\omega, \tag{24.75}$$

where $d\omega = 2\pi/N\epsilon$ is the distance between neighbouring frequencies.

The 'canonical' form of the spectrum

The empirical spectra have a single peak at a certain frequency ω_p with a long tail towards higher frequencies and a sharp drop-off below. The position of the peak depends strongly on the wind velocity U

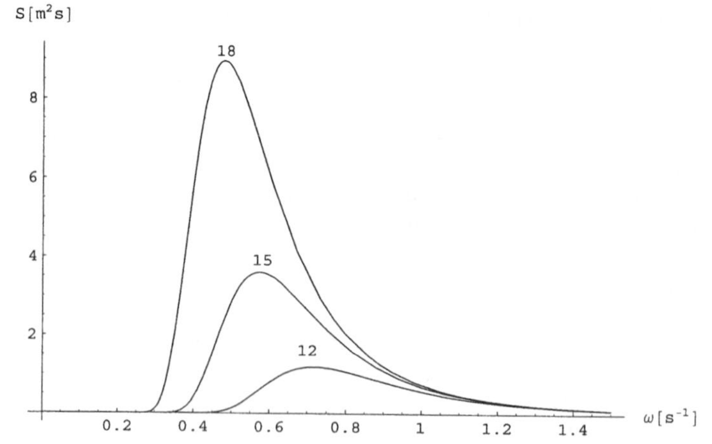

Figure 24.9. Pierson–Moskowitz wave spectrum S as a function of circular frequency ω for three different wind speeds $U = 12, 15, 20$ m s^{-1}. Note that the high-frequency tail is independent of U.

whereas the high-frequency tail appears to be the same for all U (see figure 24.9). We shall now see that it is possible to understand the general form of the spectrum using the methods of statistical mechanics.

The wind speed U sets the level of excitation of the ocean surface at large, but cannot control what happens locally so that the local wave energy E in a small neighbourhood of a fixed point in principle can take any value. But because the energy has to come from the huge reservoir of wave energy in the surrounding ocean, the probability that the local energy actually has the value E is suppressed by a canonical Boltzmann factor $e^{-\beta E}$, where the 'inverse temperature' β is a measure of the level of excitation of the ocean. Multiplying by the energy per unit of frequency $dE/d\omega$ the energy spectrum becomes

$$S \sim e^{-\beta E} \frac{dE}{d\omega}. \tag{24.76}$$

Provided the nonlinearity is not excessive, the local energy is proportional to the square of the amplitude $E \sim a^2$. Since the local energy cannot depend on U, the amplitude must be of magnitude $a \sim g_0/\omega^2$, because that is the only length scale which may be constructed from g_0 and ω. Taking $E \sim g_0^2/\omega^4$ and normalizing the frequency in the exponent by g_0/U, we get the following model for the spectrum (after redefining β),

$$\boxed{S(\omega) = \alpha \frac{g_0^2}{\omega^5} \exp\left[-\beta \left(\frac{g_0}{U\omega}\right)^4\right],} \tag{24.77}$$

where α and β are dimensionless parameters. This spectrum has indeed a sharp low-frequency cut-off, a single peak and a high-frequency tail that is independent of U (when α is independent of U).

The root-mean-square amplitude and the peak frequency are easily evaluated,

$$\sqrt{\langle h^2 \rangle} = \sqrt{\frac{\alpha}{\beta}} \frac{U^2}{2g_0}, \qquad \omega_p = \left(\frac{4\beta}{5}\right)^{1/4} \frac{g_0}{U}. \tag{24.78}$$

Note that these quantities are scaled by the only possible combinations of U and g_0 that have the right dimensions.

The Pierson–Moskowitz empirical spectrum

Assuming that the statistical equilibrium is the same everywhere on the ocean surface, the dimensionless parameters α and β can only depend on U and g_0, but since there is no dimensionless combination of U and g_0, both α and β must be constants, independent of U. Pierson and Moskowitz[2] fitted empirical spectra

[2] W. J. Pierson and L. Moskowitz, *J. Geophysical Research* **69**, (1964) 5181; L. Moskowitz, *ibid* p. 5161; W. J. Pierson, *ibid* p. 5191.

for a range of wind velocities and found the values

$$\alpha = 8.1 \times 10^{-3}, \qquad \beta = 0.74. \tag{24.79}$$

The root-mean-square amplitude and spectrum peak position are then,

$$\sqrt{\langle h^2 \rangle} = 0.11 \frac{U^2}{2g_0}, \qquad \omega_p = 0.88 \frac{g_0}{U}. \tag{24.80}$$

The actual spectrum is quite sensitive to the height at which the wind speed is determined because of air turbulence close to the surface. In the data used in the fit, the wind speed was measured about 20 m above the average surface level.

In figure 24.9 the Pierson–Moskowitz spectrum is shown for three different wind speeds. One notes how the high-frequency tails coincide, and how the low-frequency cut-off becomes sharper as the wind speed increases.

Example 24.7.1: At a wind speed of $U = 15$ m s^{-1} the period at peak is about $\tau_p = 11$ s, corresponding to a deep-water wavelength of $\lambda_p = 186$ m and a phase velocity of $c_p = 17$ m s^{-1}. The average amplitude of the waves raised by this wind is $\sqrt{\langle h^2 \rangle} = 1.2$ m. As a measure of the nonlinearity one may take $k_p\sqrt{\langle h^2 \rangle} = 0.04$ where $k_p = \omega_p^2/g_0 = 0.03$ m^{-1} is the peak wavenumber, determined from the deep-water dispersion relation (24.26). According to this estimate, the average nonlinearity at play at this wind speed is indeed quite small.

The assumed dynamic equilibrium of the ocean surface takes a long time to develop, and more so the higher the wind speed. A wind with velocity U lasting time t must have travelled over an upwind distance $L = Ut$, called the *fetch*. Even for a moderate wind at 15 m s^{-1} the sea takes about 8 h to develop, so the fetch is about 500 km. Empirically the fetch grows roughly like U^3, so that a strong wind at 30 m s^{-1} has a fetch of 4000 km. Such winds rarely manage to fully develop the equilibrium power spectrum of the sea because of the finite distance to the lee shore and the finite size of weather systems.

24.8 Global wave properties

Waves contain mass, momentum and energy, and the movement of fluid also transports these quantities around. To calculate the flux of these quantities for gravity waves at finite depth we shall use the flat-bottom fields (24.34), remembering that they represent only the lowest order of approximation. Since we do not know the nonlinear corrections to these fields, it is imperative to systematically keep only the leading order in the amplitude a in all expressions.

Mass flux

The mass flux through a vertical cut S through the wave of length L in the y-direction is,

$$\dot{M} = \int_S \rho_0 \boldsymbol{v} \cdot d\boldsymbol{S} = \int_{-d}^{h} \rho_0 v_x \, L dz \approx \int_{-d}^{0} \rho_0 v_x \, L dz. \tag{24.81}$$

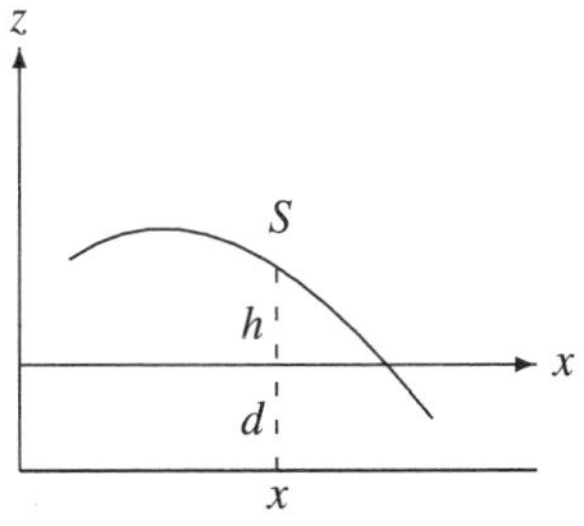

A vertical cut S through the wave (dashed). Its length is L in the y-direction.

The velocity field is of first order in a and for consistency we have in the last step replaced h by 0. Carrying out the integral by means of (24.34b) we find,

$$\dot{M} = \rho_0 acL \cos(kx - \omega t). \tag{24.82}$$

The average of the flux over a full period evidently vanishes, $\langle \dot{M} \rangle = 0$, showing that the water merely sloshes back and forth. The actual volume of water that is displaced during half a period of forward motion is $Q = aL\lambda/\pi$.

Example 24.8.1: The amount of sloshing water in a deep-water wave with amplitude 1 m and wavelength 100 m is nearly 32 m^3 per metre of transverse crest length.

Momentum flux

The total horizontal momentum transport through the vertical cut is similarly,

$$\dot{\mathcal{P}}_x = \int_S \rho_0 v_x \boldsymbol{v} \cdot d\boldsymbol{S} = \int_{-d}^{h} \rho_0 v_x^2 \, L dz \approx \int_{-d}^{0} \rho_0 v_x^2 \, L dz. \tag{24.83}$$

Averaging over a full period we get,

$$\left\langle v_x^2 \right\rangle = \frac{1}{2} \left(\frac{a\omega}{\sinh kd} \right)^2 \cosh^2 k(z+d), \tag{24.84}$$

and find upon integration,

$$\left\langle \dot{\mathcal{P}}_x \right\rangle = \frac{1}{2} \rho_0 g_0 a^2 L \frac{c_g}{c}, \tag{24.85}$$

where c and c_g are the phase and group velocities (24.33).

The average momentum flux is a measure of the average force of a wave hitting an obstacle, although the true force is strongly complicated by the shape of the obstacle and by reflected waves [47].

Example 24.8.2: Wading near the shore in water to your waist, you are hit by harmonic waves of amplitude $a = 10$ cm. If you are $L = 50$ cm wide, a deep-water wave with $c_g/c = 1/2$ will act on you with an average landwards force of $(1/4)\rho_0 g_0 L a^2 = 12$ N. That will not topple you, but a wave with five times higher amplitude could easily.

Wave power

Waves are able to do work and many ingenious schemes have been thought up for the exploitation of wave power. The power of a wave may be calculated from the rate of work performed by the pressure on a vertical cut S through the wave,

$$P = \int_S p \, \boldsymbol{v} \cdot d\boldsymbol{S} = \int_{-d}^{h} p \, v_x \, L dz \approx \int_{-d}^{0} p \, v_x \, L dz. \tag{24.86}$$

Writing the pressure (24.34d) as $p = p_0 - \rho_0 g_0 z + \rho_0 c v_x$ and using $\langle v_x \rangle = 0$, we find the average $\langle p \, v_x \rangle = \rho_0 c \left\langle v_x^2 \right\rangle$, and thus the average power

$$\boxed{\langle P \rangle = \left\langle \dot{\mathcal{P}}_x \right\rangle c = \tfrac{1}{2} \rho_0 g_0 a^2 L c_g,} \tag{24.87}$$

where the group velocity is given by (24.33). According to (24.65), the factor in front of c_g is the average wave energy $d \langle \mathcal{E} \rangle / dx$ per unit of length. Interpreting $\langle P \rangle$ as the average energy flux along x, this result is usually taken to confirm that the energy in a harmonic wave also propagates with the group velocity.

Example 24.8.3: The tsunami of example 24.3.1 with $\lambda = 500$ km, $c = c_g = 200$ m s^{-1}, and $a = 1$ m carries an average power per unit of transverse length of $\langle P \rangle / L \approx 10^6$ W m^{-1}. Such a Tsunami can really wreak havoc when it hits a coast.

Problems

24.1 Check whether the shallow-wave solution (24.38) actually satisfies the field equations (24.15). Discuss what is wrong, if not.

24.2 **(a)** Determine where the phase and group velocities (24.27) for deep-water waves cross each other (use $\alpha = 0.073$ N m^{-1}), and determine the common value at the crossing. **(b)** Determine the minimal value of the phase velocity and the corresponding wavelength. **(c)** Where is the minimum of the group velocity?

24.3 Consider a small-amplitude gravity line wave and

(a) show that

$$v_x^2 + v_z^2 = \nabla_x(\Psi v_x) + \nabla_z(\Psi v_z). \tag{24.88}$$

(b) Show that the time average satisfies

$$\left\langle v_x^2 + v_z^2 \right\rangle = \nabla_z \left\langle \Psi v_z \right\rangle. \tag{24.89}$$

(c) Use this to calculate the kinetic energy (24.64).

24.4 Prove the virial theorem

$$\langle \mathcal{T} \rangle = \frac{n}{2} \langle \mathcal{V} \rangle \tag{24.90}$$

for a single particle in periodic motion in a power potential $\mathcal{V} = kr^n$.

24.5 Justify qualitatively the common observation that waves rolling towards a beach tend to straighten out so that the wave crests become parallel to the beach.

24.6 A square jar is half filled with water of density 1 g cm^{-3} lying below oil of density 0.8 g cm^{-3}. The interface has surface tension 0.3 N m^{-1}. Determine the largest horizontal size of the jar which permits the jar to be turned around with the oil stably below the water.

$*$ **24.7** **Gaussian wave packet:** Calculate explicitly the form of a superposition of harmonic waves

$$h = \int_{-\infty}^{\infty} a(k) \cos[kx - \omega(k)t + \chi(k)]\, dk, \tag{24.91}$$

where

$$a(k) = \frac{1}{\Delta k \sqrt{\pi}} \exp\left(-\frac{(k-k_0)^2}{\Delta k^2}\right) \tag{24.92a}$$

$$\omega(k) = \omega_0 + c_g(k - k_0) \tag{24.92b}$$

$$\chi(k) = \chi_0 - x_0(k - k_0). \tag{24.92c}$$

Describe its form and determine what x_0 represents. Hint: write the wave as the real part of a complex wave and use the known Gaussian integrals.

$*$ **24.8** **Discrete Fourier transformation.**

Let h_n be a set of N generally complex numbers numbered $n = 0, 1, 2, \ldots, N-1$. Define the Fourier coefficients

$$\hat{h}_m = \frac{1}{\sqrt{N}} \sum_{n=0}^{N-1} h_n \exp\left[2\pi i \frac{nm}{N}\right]. \tag{24.93}$$

(a) Show that

$$\sum_{m=0}^{N-1} \exp\left[2\pi i \frac{nm}{N}\right] = \begin{cases} N & \text{for } n = 0 \\ 0 & \text{for } 1 < n < N. \end{cases} \tag{24.94}$$

(b) Show the reciprocity theorem

$$h_n = \frac{1}{\sqrt{N}} \sum_{m=0}^{N-1} \hat{h}_m \exp\left[-2\pi i \frac{nm}{N}\right]. \tag{24.95}$$

(c) Show Parseval's theorem

$$\sum_{n=0}^{N-1} |h_n|^2 = \sum_{m=0}^{N-1} \left|\hat{h}_m\right|^2 . \tag{24.96}$$

(d) Assume from now on that that h_n is real. Show that

$$\hat{h}_m^\times = \hat{h}_{-m} \tag{24.97}$$

where $\hat{h}_{-m}$ means $\hat{h}_{N-1-m}$.

(e) Put

$$\hat{h}_0 = \sqrt{N}\, a_0 \qquad \hat{h}_m = \tfrac{1}{2}\sqrt{N}\, a_m\, e^{i\chi_m} \tag{24.98}$$

where a_n and χ_n are real. Show that $a_{-n} = a_n$ and $\chi_{-n} = -\chi_n$. Show that for odd $N = 2M + 1$

$$h_n = a_0 + \sum_{m=1}^{M} a_m \cos(\omega_m t_n - \chi_m) \tag{24.99}$$

where $t_n = n\epsilon$, and $\omega_m = 2\pi m/N\epsilon$.

(f) Show that

$$\frac{1}{N}\sum_{n=0}^{N-1} (h_n - a_0)^2 = \frac{1}{2}\sum_{m=1}^{M} a_m^2 . \tag{24.100}$$

25

Jumps and shocks

Wading in the water near a beach and fighting to stay upright in the surf, you are evidently under the influence of nonlinear dynamics, simply because the breaking waves look so different to the smooth swells in the open sea that gave rise to them. Equally nonlinear are the dynamics behind the hydraulic jumps observed every day in the kitchen sink or on the beach, and the closely related dramatic river bores created by the rising tide at the mouth of a river and rolling far upstream.

Nonlinear dynamics also lie behind the familiar sonic boom caused by a high-speed airplane passing overhead, and the hopefully less familiar short-range shock wave created by an exploding grenade. At much larger scales one encounters the huge atmospheric shock waves released by a thermonuclear explosion, or the enormous shock waves from supernova explosions that may trigger star formation in the tenuous clouds of interstellar matter in the galaxy, a phenomenon to which we may ultimately owe our own existence.

The beauty of fluid mechanics lies in the knowledge that all these effects stem from the same nonlinear aspects of the Navier–Stokes equations. In this chapter the global laws of balance derived in chapter 22 are used to analyse the physics of fluids in the extreme limit where the nonlinearity may create discontinuities, or near discontinuities, in the properties of the fluid. There are two major classes of such phenomena. In an incompressible fluid, a *hydraulic jump* is signalled by an abrupt rise in the height of the open fluid surface of a fast stream. When compressibility is taken into account, a sufficiently violent event may create a *shock wave* in a fluid, where the flow speed abruptly drops from supersonic to subsonic. For ideal gases the analysis is, as usual, particularly simple.

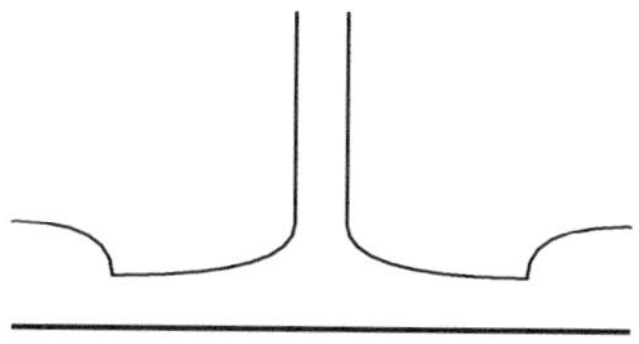

Sketch of the hydraulic jump in a kitchen sink. The water coming down from the tap splays out in a sheet which suddenly thickens.

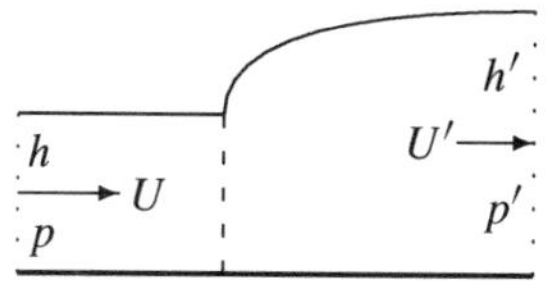

Stationary straight-line hydraulic jump (dashed). Incompressible fluid enters from the left at velocity U and height h and exits on the right at a lower velocity U' and greater height h'. The entry and exit pressures, p and p', are hydrostatic. Right at the front the flow pattern is complicated, often turbulent. The control volume encompasses the whole drawing between the dotted lines.

25.1 Hydraulic jumps

A stationary *hydraulic jump* or *step* is easily observed in a kitchen sink (see figure 25.2). The column of water coming down from the tap splays out from the impact region in a roughly circular flow pattern, and at a certain radius the thin sheet of water abruptly thickens and stays thick beyond. The transition region behind the front appears to have a narrow width and contain quite complicated flow. Strongly turbulent stationary hydraulic jumps may also arise in spillways, channelling surplus water from a dam into the river downstream.

Stationary jump in planar horizontal flow

We shall begin with an analysis of a stationary jump along a straight line orthogonal to the direction of a uniform flow over a horizontal planar surface. Although stationary jumps like the one in the kitchen sink often have curved fronts, the abruptness of the jump allows us in the leading approximation to view it as locally straight. The flow is assumed to be steady before and after the jump, whereas in the transition region there may be intermittency and turbulence. The Reynolds number is assumed to be so large that viscous friction can be ignored outside the transition region.

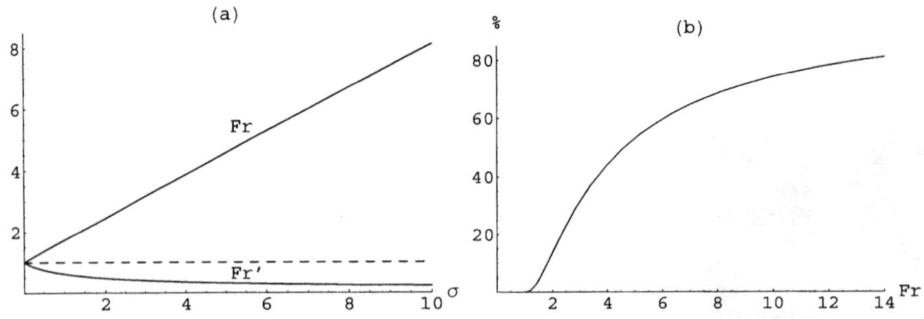

Figure 25.1. **(a)** The Froude numbers $\mathrm{Fr} = U/c$ before and $\mathrm{Fr}' = U'/c'$ after the jump, plotted as a function of the jump strength $\sigma = (h' - h)/h$. **(b)** The fraction of the incoming kinetic energy (25.15) dissipated in the stationary jump as a function of the Froude number.

Let the liquid flow towards the jump with constant uniform velocity U and constant water level h. Downstream from the jump the flow has a different velocity U' and a different water level h'. The dimensionless *strength* of the jump is defined to be the relative change in height,

$$\sigma = \frac{h' - h}{h} = \frac{h'}{h} - 1. \tag{25.1}$$

Energy balance will later show that the downstream water level must necessarily be the higher, such that the strength is always positive, $\sigma > 0$. A jump is said to be *weak* when $\sigma \ll 1$ and *strong* when $\sigma \gg 1$.

All properties of the jump may, as we shall see, be expressed in terms of the inflow parameters U, h and the strength σ. Let us choose a control volume with vertical sides and width L orthogonal to the flow. The upstream and downstream sides of the control volume are chosen orthogonal to the direction of flow and placed so far away from the transition region that both inflow and outflow are steady and uniform with velocities U and U'. Mass conservation then guarantees that the total volume flux at the outlet is the same as at the inlet,

$$Q = LhU = Lh'U'. \tag{25.2}$$

From this we get the ratio of velocities

$$\frac{U'}{U} = \frac{h}{h'} = \frac{1}{1+\sigma}. \tag{25.3}$$

Evidently, since $\sigma > 0$, the downstream velocity is always the smaller one.

Momentum balance (22.15) similarly guarantees that the net outflow of momentum from the control volume equals the total external force acting on the control volume in the direction of the flow. For an inviscid fluid the horizontal force is entirely due to the pressure acting on the two vertical sides of the control volume where the liquid enters or leaves. The pressure in the uniform planar flow that reigns in these regions is hydrostatic, given by $p = p_0 + \rho_0 g_0 (h - z)$ at the inlet and $p' = p_0 + \rho_0 g_0 (h' - z)$ at the outlet, where p_0 is the constant (atmospheric) pressure on the open liquid surface. Carrying out the integrals over the inlet and outlet as in example 4.1.2 on page 47, momentum balance becomes,

$$\rho_0 QU' - \rho_0 QU = \tfrac{1}{2}\rho_0 g_0 Lh^2 - \tfrac{1}{2}\rho_0 g_0 Lh'^2. \tag{25.4}$$

On the left-hand side one finds the difference between the momentum fluxes through the outlet and inlet of the control volume, and on the right the difference between the total pressure forces acting on the inlet and the outlet. Eliminating U' and h' using (25.1) and (25.3) the resulting equation may be solved for the entry velocity, to get

$$U = \sqrt{g_0 h}\sqrt{(1+\sigma)(1+(1/2)\sigma)}. \tag{25.5}$$

Using again (25.3) we obtain the outlet velocity

$$U' = \sqrt{g_0 h}\sqrt{\frac{1+(1/2)\sigma}{1+\sigma}}. \tag{25.6}$$

In both of these expressions the velocity scale is set by $c = \sqrt{g_0 h}$, which we recognize as the small-amplitude shallow-water wave speed (24.37) in the inflow region. Similarly, the shallow-water wave speed in the outflow region is $c' = \sqrt{g_0 h'} = c\sqrt{1+\sigma}$. Since $\sigma > 0$, it follows immediately that the various velocities obey the inequalities $U > c' > c > U'$. Note that the smallest outflow velocity is obtained in the strong jump limit, $U' \to c/\sqrt{2}$ for $\sigma \to \infty$.

The dimensionless ratio between the inflow velocity U and the shallow-water wave speed at the inlet is called the *Froude number* (plotted in figure 25.1(*a*)),

$$\mathrm{Fr} = \frac{U}{\sqrt{g_0 h}} = \sqrt{(1+\sigma)(1+(1/2)\sigma)}. \tag{25.7}$$

Solving this equation for σ we obtain the strength as a function of the Froude number, i.e. of the input values h and U,

William Froude (1810–79). *English engineer and naval architect. Discovered what are now called scaling laws, allowing predictions of ship performance to be made from studies of much smaller model ships.*

$$\sigma = \frac{\sqrt{1+8\mathrm{Fr}^2} - 3}{2} \approx \sqrt{2}\,\mathrm{Fr} - \frac{3}{2}. \tag{25.8}$$

The linear approximation is better than 3.3% for $\mathrm{Fr} > 2$ (problem 25.1). The Froude number at the outlet is similarly defined as

$$\mathrm{Fr}' = \frac{U'}{\sqrt{g_0 h'}} = \frac{\mathrm{Fr}}{(1+\sigma)^{3/2}} = \frac{\sqrt{1+(1/2)\sigma}}{1+\sigma}, \tag{25.9}$$

which is also plotted in figure 25.1(*a*). Since $\sigma > 0$, we have $\mathrm{Fr} > 1 > \mathrm{Fr}'$, and the flow is said to be *supercritical* before the jump, and *subcritical* after.

Stationary circular jump

The nearly circular stationary hydraulic jump in the kitchen sink may be viewed as being locally straight. The position of the jump is difficult to predict because it depends on the shape of the particular kitchen sink (see page 363). Here we shall simply place the jump at a certain radius $r = R$ from the centre and calculate the shape of the flow before and after.

In accordance with what we did for the planar jump, we shall also here assume that the steady, radial, inviscid flow over the kitchen sink bottom is independent of z, i.e. of the form $v_r = u(r)$. Apart from the jump region, the dynamics are governed by the constancy of the mass flux Q and of the Bernoulli function H for a streamline running along the surface,

$$Q = 2\pi r h(r) u(r), \tag{25.10a}$$

$$H = \tfrac{1}{2} u(r)^2 + g_0 h(r). \tag{25.10b}$$

In the general case this becomes a third degree equation for u (or h) with a rather messy solution.

In a typical kitchen sink experiment, as for example the one described below, the jump is usually quite strong with $\sigma \gg 1$. This implies that the local Froude number $\mathrm{Fr}(r) = u(r)/\sqrt{g_0 h(r)}$ will be large in front of the jump and from the constancy of H it follows that $u(r)$ will be nearly constant. Behind the jump the local Froude number will instead be small, which again due to the constancy of H implies that the height $h(r)$ will be nearly constant. Thus, we have approximately,

$$\begin{aligned} u(r) &= U, & h(r) &= \frac{Q}{2\pi r U}, & &\text{for } r < R. \\ u(r) &= \frac{Q}{2\pi r h'}, & h(r) &= h', & &\text{for } r > R. \end{aligned} \tag{25.11}$$

Because of mass conservation across the jump, $Uh = U'h'$, the Reynolds number $\mathrm{Re} = Uh/\nu = Q/2\pi r\nu$ is continuous and decreases everywhere inversely with the radius. To test whether the assumption of inviscid flow is justified, one may compare the water level $h(r)$ with a thickness estimate of the boundary layer, for example $\delta(r) = 3\sqrt{\nu r/u(r)}$ (see chapter 28).

In a kitchen sink experiment it is fairly easy to determine the volume discharge rate Q by collecting water in a standard household measure for a little while. From the radius a of the water jet just before it splays out at the bottom one may determine the average velocity $U = Q/\pi a^2$ which by Bernoulli's theorem

Figure 25.2. Hydraulic jump in the kitchen sink. Photograph by Philippe Belleudy, Université Joseph Fourier, reproduced with permission.

should be nearly the same as the radial velocity along the bottom of the sink. From Q, U and R all the jump parameters may then be calculated.

Example 25.1.1 (A kitchen sink experiment): In a casual home-made kitchen sink experiment (see figure 25.2 and 25.3) the discharge rate was observed to be $Q = 100\ \text{cm}^3\ \text{s}^{-1}$, the radius of the water jet $a = 0.5$ cm, and the radius of the jump $R = 7$ cm. The velocity before the jump is $U = Q/\pi a^2 = 127\ \text{cm s}^{-1}$, and using (25.11) we calculate the height just before the jump to be merely $h = h(R) = Q/2\pi RU = 0.018$ cm. The corresponding Froude number is fairly large, $\text{Fr} = U/\sqrt{g_0 h} = 30$, and using (25.8) we determine the jump strength to be $\sigma = 42$. The height after the jump $h' = (1+\sigma)h = 0.76$ cm which appears to be a bit larger than the observed height. The velocity after the jump is $U' = U/(1+\sigma) = 3\ \text{cm s}^{-1}$, corresponding to a Froude number of $\text{Fr}' = 0.11$. The Reynolds number at the jump is moderate, $\text{Re} = Uh/\nu = 264$, but the estimated boundary layer thickness at the jump, $\delta(R) = 0.065$ cm, is almost four times the water level before the jump. Viscosity must for this reason play an important role for this experiment, in particular in the region just before the jump, casting some doubt on the validity of the theory in this case.

Energy loss in the stationary jump

It would be tempting to make use of Bernoulli's theorem along a streamline going across the stationary hydraulic jump, but that is impossible because of the unruly fluid in the transition region which messes up streamlines and generates a dissipative (viscous) loss of mechanical energy.

We may nevertheless use mechanical energy balance (22.53) to calculate the rate of loss of energy from the system by keeping track of what mechanical energy goes into the control volume and what comes out. Mechanical energy balance takes the form,

$$\rho_0 Q\left(\frac{1}{2}U'^2 + \frac{1}{2}g_0 h'\right) - \rho_0 Q\left(\frac{1}{2}U^2 + \frac{1}{2}g_0 h\right) = \frac{1}{2}\rho_0 g_0 L h^2 U - \frac{1}{2}\rho_0 g_0 L h'^2 U' - P. \qquad (25.12)$$

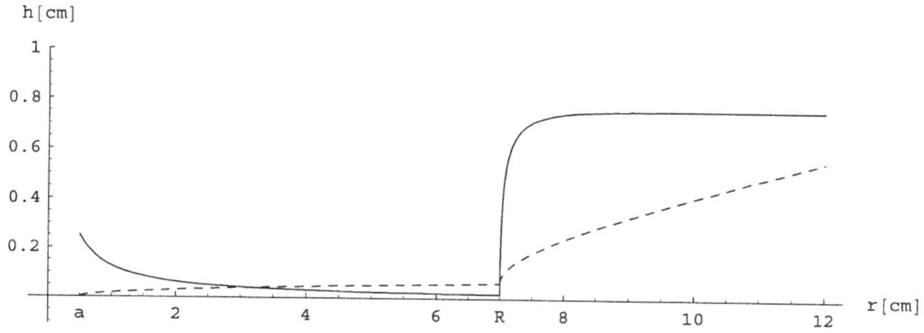

Figure 25.3. Outline of the hydraulic jump in the kitchen sink discussed in the text and in example 25.1.1. Note that the height is enlarged by a factor 4 relative to the radius. The solid line is the water level given by the model (25.11) whereas the dashed line is the estimated thickness of the boundary layer. The thickness of the jump is taken from the estimate on page 364.

On the left-hand side we have the difference between the rates of outflow and inflow of mechanical energy from the control volume, calculated from the specific mechanical energy density, $1/2\boldsymbol{v}^2 + g_0 z$, integrated over the outlet and inlet. On the right there is first the difference in rates of work of the pressure forces integrated over the inlet and outlet, and finally P, the rate of loss of energy due to the work of internal friction given by (17.18).

Solving for P we may write,

$$P = \rho_0 Q \left(\frac{1}{2}U^2 + g_0 h\right) - \rho_0 Q \left(\frac{1}{2}U'^2 + g_0 h'\right) = \rho_0 Q(H - H') \tag{25.13}$$

where H and H' are the values of the Bernoulli function (16.15) at the surface of the water before and after the jump. This clearly demonstrates that Bernoulli's theorem cannot be fulfilled when P is non-zero. Substituting the downstream quantities using (25.3) we find

$$P = \rho_0 Q g_0 h \, \frac{\sigma^3}{4(1+\sigma)}. \tag{25.14}$$

Since the rate of viscous energy loss necessarily must be positive (page 236), we conclude as promised that a stationary hydraulic jump will always have positive strength, $\sigma > 0$. Relative to the rate of kinetic energy inflow, $\dot{\mathcal{T}} = (1/2)\rho_0 Q U^2$, the *fractional dissipative loss* of kinetic energy becomes (see figure 25.1(*b*)),

$$\boxed{\frac{P}{\dot{\mathcal{T}}} = \frac{\sigma^3}{(1+\sigma)^2(2+\sigma)}.} \tag{25.15}$$

It converges as expected to unity for $\sigma \to \infty$. For $\sigma = 1$, corresponding to Fr $= 1.73$, only 8.3% of the kinetic energy is lost, whereas for $\sigma = 10$ corresponding to Fr $= 8.1$, the fractional loss is 69%. Strong hydraulic jumps are efficient dissipators of kinetic energy, and this is in fact their function in dam spillways where high speed surplus water must be slowed down before it is released into the river downstream of the dam.

Moving hydraulic jumps

Moving hydraulic jumps are seen on the beach when waves roll in, sometimes in several layers on top of each other. More dramatic *river bores* may be formed by the rising tide near the mouth of a river. When the circumstances are right such waves can roll far up the river with a nearly vertical foaming turbulent front. In the laboratory an ideal river bore can be created in a long canal with water initially at rest. When the wall in one end of the canal is set into motion with constant velocity, a bore will form and move down the canal with constant speed and constant water level.

The only difference between a river bore and a stationary hydraulic jump lies in the frame of reference. The ideal river bore is obtained in the frame where the fluid in front of the jump is at rest. Subtracting

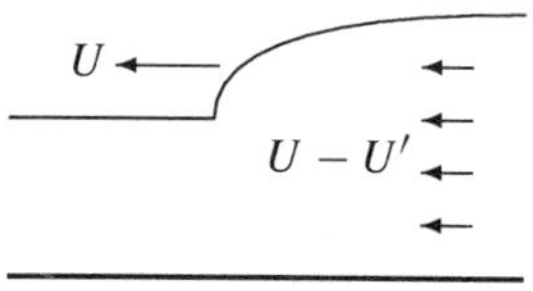

A river bore moving in from the right in water initially at rest. The water behind the front moves slower than the front itself. The velocities are obtained by subtracting U from all the velocities of the stationary jump.

Figure 25.4. The Qiantang tidal river bore is the largest in the world. Its height can be 4 m, its width 3 km, and its speed above 30 km h^{-1}. Photograph courtesy Eric Jones, Proudman Oceanographic Laboratory.

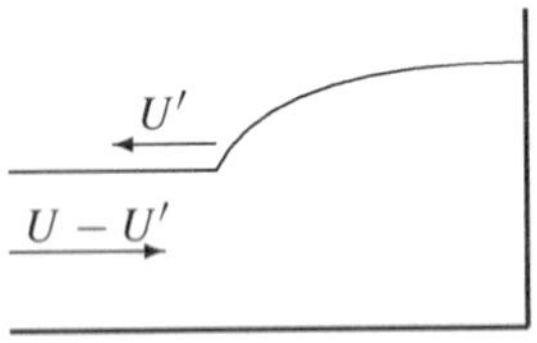

A reflection bore in a closed canal is obtained by subtracting U' from all velocities of the stationary jump.

U from all velocities (and reversing their directions), the front itself will move with velocity U, and the fluid behind the jump will move in the same direction with velocity $U - U'$, which is smaller than the shallow-water velocity $c' = \sqrt{g_0 h'}$ behind the jump for $\sigma < 2.21$.

There is also the possibility of choosing a reference frame in which the fluid behind the jump is at rest. Subtracting U' from the velocities of the stationary jump, this describes a stationary flow being reflected in a closed canal. Such a *reflection bore* moves with velocity U' out of the canal while the flow into the canal has velocity $U - U'$.

The reflection bore has in fact some bearing on the stationary jump in the kitchen sink. The layer of fluid spreads initially over the bottom of the sink with a high Froude number, and thus a roughly constant velocity and decreasing thickness. The spreading layer of fluid encounters resistance against the free flow from the sides of the sink, or from the slight slope and curvature of its bottom[1]. This creates a reflection bore with small Froude number, which moves inwards with roughly constant height and increasing velocity, until it stops when it encounters an outflow with velocity and thickness that fits a stationary jump. One may also observe that when the faucet is closed the stationary jump in the kitchen sink immediately turns into a river bore moving towards the centre.

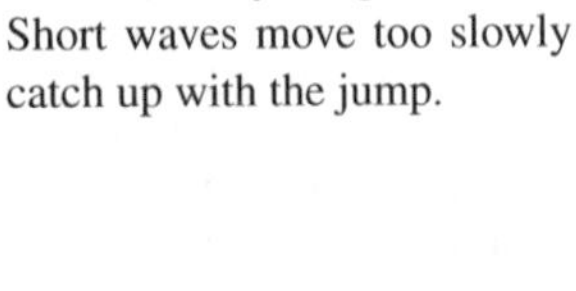

A river bore is 'pumped up' by small-amplitude surface waves of sufficiently long wavelength. Short waves move too slowly to catch up with the jump.

Thickness of a hydraulic jump

A river bore is created by the rising tide at the river mouth, and as long as the tide keeps rising, it will continue to pour more water in. The additional water supplied by the rising tide in a small time interval may be thought of as a small-amplitude surface wave riding upriver on top of the already existing bore. Although this mechanism is most obvious for the river bore, both the reflection bore and the stationary hydraulic jump must also be built up 'from behind' by small-amplitude waves.

Any small-amplitude surface wave may be resolved into a superposition of harmonic waves with a spectrum of wavelengths. Consider now a harmonic wave with wavelength λ on its way upstream towards a stationary hydraulic jump. In the rest frame of the outflow, the energy in a harmonic wave with wavenumber

[1] The viscous friction in the thinning layer of fluid may also trigger the jump; see T. Bohr, V. Putkaradze and S. Watanabe, Averaging theory for the structure of hydraulic jumps and separation in laminar free-surface flows, *Phys. Rev. Lett.* **79**, (1997) 1038.

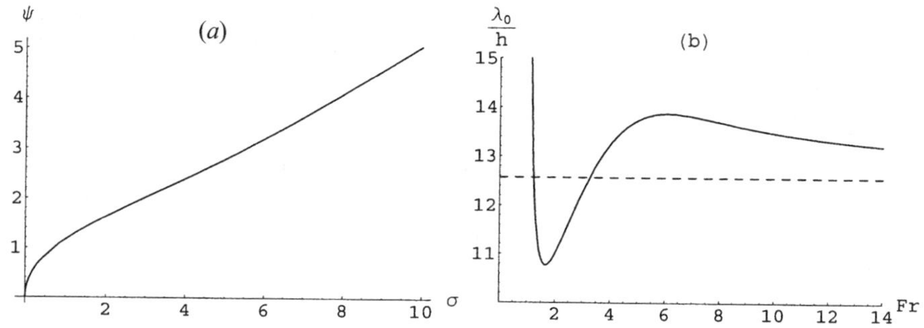

Figure 25.5. **(a)** The solution $\psi(\sigma)$ to equation (25.17). For $\sigma \to 0$ the ratio behaves as $\sqrt{\sigma}$, while for $\sigma \to \infty$ it increases as $\sigma/2$. **(b)** The ratio λ_0/h of the minimal wavelength in units of the height h before the jump as a function of Froude number Fr. For Fr $\to \infty$ the ratio becomes constant, $\lambda_0/h \to 4\pi$ (dashed line).

$k = 2\pi/\lambda$ moves with the group velocity of a gravity wave (24.33). For liquid of depth h' the group velocity becomes,

$$c'_g = \frac{1}{2}\sqrt{\frac{g_0}{k}\tanh kh'}\left(1 + \frac{2kh'}{\sinh 2kh'}\right). \tag{25.16}$$

In the rest frame of the jump, the wave propagates towards the jump with velocity $c'_g - U'$ which must be positive if the wave is ever to reach the jump. Since $c'_g \to \sqrt{g_0 h'}$ for $k \to 0$ and $U' < \sqrt{g_0 h'}$, the condition $c'_g > U'$ may always be fulfilled provided the wavenumber k is sufficiently small, i.e. the wavelength exceeds a certain minimum value, $\lambda > \lambda_0$. The minimal wavelength λ_0 is found by solving the equation $c'_g = U'$, which after division by $\sqrt{g_0 h'}$ takes the form,

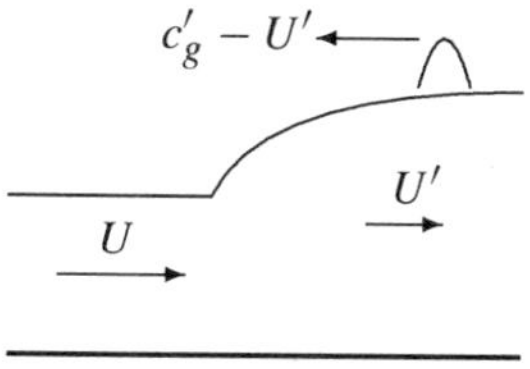

A small-amplitude wavelet moving towards the stationary jump with velocity $c' - U' > 0$.

$$\frac{1}{2}\sqrt{\frac{\tanh\psi}{\psi}}\left(1 + \frac{2\psi}{\sinh 2\psi}\right) = \mathrm{Fr}' = \frac{\sqrt{1 + 1/2\sigma}}{1 + \sigma}, \tag{25.17}$$

where $\psi = kh' = 2\pi h'/\lambda_0$. This transcendental equation may be solved for ψ and the result is shown in figure 25.5(*a*) as a function of the jump strength σ. In figure 25.5(*b*) this is converted to a plot of the ratio λ_0/h of the minimal wavelength to the *upstream* height h as a function of the Froude number Fr.

The hydraulic jump effectively acts as a high-pass filter on backflow wavelengths, only letting waves with sufficiently large wavelength, $\lambda > \lambda_0$, through to 'feed' it. The smooth part of a hydraulic jump cannot contain details much smaller than the waves that maintain it, so the minimal wavelength λ_0 may be used as a measure of the minimal thickness of the transition region.

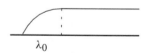

A small-amplitude gravity wave moving upstream towards the jump must have wavelength greater than λ_0, and its front cannot be sharper than that.

Example 25.1.2: A river bore with front height $h' - h = 1$ m moves up a river of depth $h = 0.5$ m. The jump strength is $\sigma = (h' - h)/h = 2$, corresponding to Fr $= 2.45$, and the front velocity becomes $U = 5.4$ m s^{-1} while the velocity of the flow behind the front is $U - U' = 3.6$ m s^{-1}. From figure 25.5(*b*) we find $\lambda_0/h = 11.6$, so that the minimal wavelength and thus the thickness becomes $\lambda_0 = 6.7$ m.

25.2 Shocks in ideal gases

An explosion in a fluid at rest creates an expanding fireball of hot gases and debris which pushes the fluid in front of it. If the velocity imparted to the fluid by the explosion is smaller than the velocity of sound in the fluid, an ordinary sound wave will run ahead of the debris and with a loud bang inform you that the explosion took place. If the initial expansion velocity of the fireball is larger than the sound velocity in the fluid, the first sign of the explosion will be the arrival of a supersonic front (here we disregard the flash of light which will arrive much earlier). The sudden jump in the properties of a fluid at the passage of a supersonic front is called a *shock*. Stationary shocks may also arise downstream from a constriction in a duct where the flow under suitable conditions discussed in section 16.4 will be supersonic. The understanding of shocks is of great importance for the design of supersonic aircraft, and of jet and rocket engines.

Stationary planar normal shocks

We shall begin by analysing stationary planar shocks, which like plane sound waves are normal to the direction of the flow. We shall later see that shocks are in fact not much thicker than the molecular length scale, allowing us to view all shocks as truly singular and locally plane. In the rest system of the shock we choose a narrow control volume just containing an area A of the shock front. Upstream from the shock the gas has velocity U, temperature T, pressure p and density ρ; downstream it has velocity U', temperature T', pressure p' and density ρ'.

There is a strong analogy between a hydraulic jump and a shock, although the *strength* of the shock is defined as the relative pressure increase,

$$\sigma = \frac{p' - p}{p} = \frac{p'}{p} - 1. \tag{25.18}$$

We shall see below that the Second Law of Thermodynamics requires that $\sigma > 0$. A shock is said to be *weak* when $\sigma \ll 1$ and *strong* when $\sigma \gg 1$.

Since the shock is so narrow there will be essentially no room for dissipation of energy, allowing us to apply Bernoulli's theorem. Using the pressure function (16.37) for an ideal gas with adiabatic index γ, we obtain three basic equations, called the *Rankine–Hugoniot* relations,

$$\rho U = \rho' U', \tag{25.19a}$$

$$\rho U^2 + p = \rho' U'^2 + p', \tag{25.19b}$$

$$\frac{1}{2} U^2 + \frac{\gamma}{\gamma - 1} \frac{p}{\rho} = \frac{1}{2} U'^2 + \frac{\gamma}{\gamma - 1} \frac{p'}{\rho'}. \tag{25.19c}$$

These relations are simple rearrangements of mass, momentum and energy balance across the shock. They may be solved explicitly for the downstream parameters in terms of the upstream ones (see problem 25.3), but it is much more convenient to express the solution in terms of the strength of the shock.

Using (25.19a) to eliminate U' in (25.19b), we obtain,

$$U^2 = \frac{\rho'}{\rho} \cdot \frac{p' - p}{\rho' - \rho}, \qquad\qquad U'^2 = \frac{\rho}{\rho'} \cdot \frac{p' - p}{\rho' - \rho}, \tag{25.20}$$

where the second equation is obtained from the first by swapping primed and unprimed variables. Inserting this into (25.19c) we find the ratio of densities,

$$\frac{\rho'}{\rho} = \frac{\gamma(p + p') + p' - p}{\gamma(p + p') - p' + p}, \tag{25.21}$$

and expressing the right-hand side in terms of the strength (25.18) we get

$$\frac{\rho'}{\rho} = \frac{U}{U'} = \frac{2\gamma + (\gamma + 1)\sigma}{2\gamma + (\gamma - 1)\sigma}. \tag{25.22}$$

The temperature ratio may now be obtained from the ideal gas law, $T'/T = (p'/\rho')/(p/\rho) = (p'/p)(\rho/\rho')$.

Since we are interested in supersonic flow, it is most convenient to express the velocities in terms of the dimensionless Mach numbers, $\mathrm{Ma} = U/c$ and $\mathrm{Ma}' = U'/c'$, where $c = \sqrt{\gamma p/\rho}$ and $c' = \sqrt{\gamma p'/\rho'}$ are the sound velocities before and after the shock. Using (25.20) we find,

$$\mathrm{Ma} = \sqrt{1 + \frac{\gamma + 1}{2\gamma} \sigma}, \qquad\qquad \mathrm{Ma}' = \sqrt{1 - \frac{\gamma + 1}{2\gamma} \frac{\sigma}{1 + \sigma}}. \tag{25.23}$$

Here the second equation is obtained from the first by swapping primed and unprimed variables, which according to the definition (25.18) amounts to replacing σ by $-\sigma/(1 + \sigma)$.

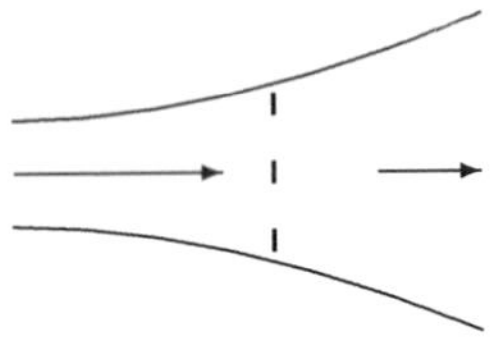

A stationary shock (dashed) in an expanding nozzle. The inflow is supersonic and the outflow subsonic.

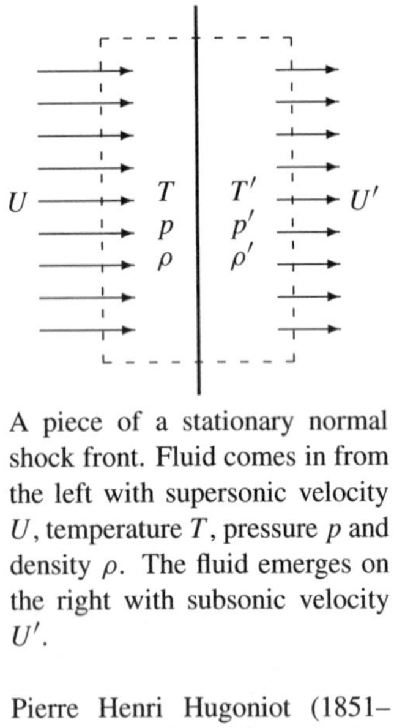

A piece of a stationary normal shock front. Fluid comes in from the left with supersonic velocity U, temperature T, pressure p and density ρ. The fluid emerges on the right with subsonic velocity U'.

Pierre Henri Hugoniot (1851–1887). *French engineer; worked mainly in the marine artillery service. Wrote only one (long) article, published in the year of his untimely death, on 'the propagation of movement in bodies, particularly in perfect gases' [33].*

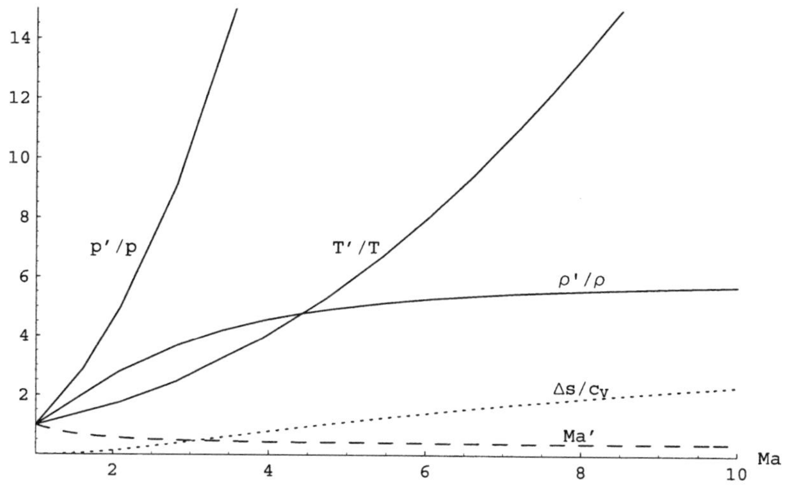

Figure 25.6. The dimensionless Rankine–Hugoniot parameters as a function of the inflow Mach number for $\gamma = 7/5$. The fully drawn curves are the ratios of pressure, density and temperature. The dashed curve is the outflow Mach number Ma′, and the dotted curve is the specific entropy jump at the shock, $\Delta s/c_V$ given in (25.24).

Up to this point, the strength may in principle range over both negative and positive values in the interval $-1 < \sigma < \infty$. The physical asymmetry between positive and negative strength becomes apparent when we calculate the change in specific entropy $\Delta s = s' - s$ across the shock. Using (C.21c) we find,

$$\frac{\Delta s}{c_V} = \log\left[\frac{p'}{p}\left(\frac{\rho'}{\rho}\right)^{-\gamma}\right] = \log(1+\sigma) - \gamma \log \frac{2\gamma + (\gamma+1)\sigma}{2\gamma + (\gamma-1)\sigma}, \tag{25.24}$$

where c_V is the specific heat constant of the gas (C.18). The right-hand side is a monotonically increasing function of σ which vanishes for $\sigma = 0$ and is thus negative for $-1 < \sigma < 0$ (see problem 25.5 and figure 25.6). However, by the Second Law of thermodynamics the specific entropy change is not permitted to be negative, and consequently we require that $\sigma \geq 0$. We have thus reached the promised conclusion that *in a stationary shock the velocity must go from supersonic* (Ma > 1) *to subsonic* (Ma′ < 1) *in the downstream direction.* The various dimensionless quantities are plotted in figure 25.6 as functions of the inflow Mach number Ma.

Oblique shock

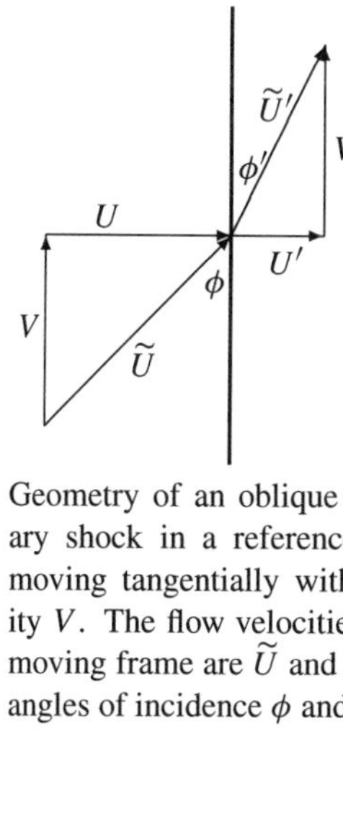

Geometry of an oblique stationary shock in a reference frame moving tangentially with velocity V. The flow velocities in the moving frame are $\widetilde{U}$ and $\widetilde{U}'$ with angles of incidence ϕ and ϕ'.

An *oblique* planar shock front may be obtained in a reference frame moving tangentially to the stationary normal shock with constant velocity V. In the moving frame the flow velocities are denoted $\widetilde{U}$ and $\widetilde{U}'$ with angle of incidence ϕ and transmission angle ϕ'. Using that the tangential velocity is the same on both sides of the shock, $V = U \cot\phi = U' \cot\phi'$, we obtain a relation between incidence and strength,

$$\frac{\tan\phi'}{\tan\phi} = \frac{U'}{U} = \frac{2\gamma + (\gamma-1)\sigma}{2\gamma + (\gamma+1)\sigma}. \tag{25.25}$$

Since $\sigma > 0$, the right-hand side is always smaller than unity, implying that the transmission angle is always smaller than the angle of incidence $\phi' < \phi$.

The geometry shows that the angle of incidence is given by $\sin\phi = U/\widetilde{U}$, and since $U \to c$ for $\sigma \to 0$, the incidence angle for a weak oblique shock obeys $\sin\phi_0 = c/\widetilde{U} = 1/\widetilde{\text{Ma}}$, where $\widetilde{\text{Ma}} = \widetilde{U}/c$ is the Mach number of the inflow. The angle ϕ_0 is called the *Mach angle*,

$$\boxed{\phi_0 = \arcsin\frac{1}{\widetilde{\text{Ma}}}.} \tag{25.26}$$

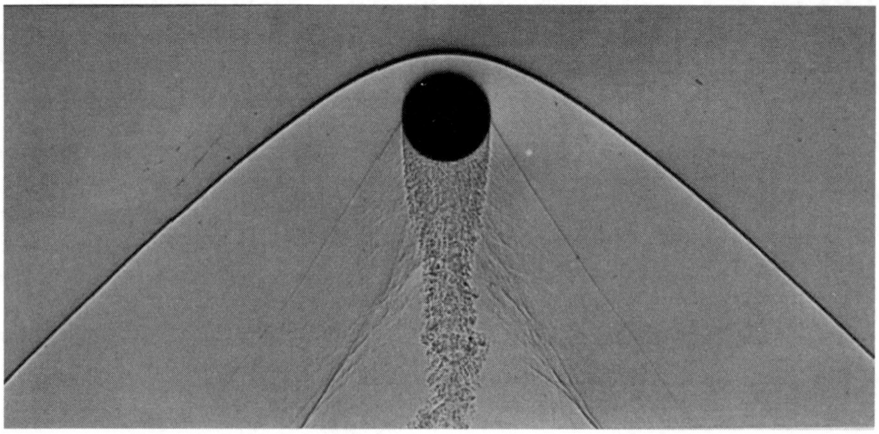

Figure 25.7. Shock waves created by a half-inch sphere moving at Ma $= 1.53$. One notes the detached bow wave which would not be there with a pointed object. The turbulence behind the sphere generates secondary weak shock fronts. Photograph by A. C. Charters.

What we perceive as a *sonic boom* is a weak shock cone with half opening angle ϕ_0 trailing a supersonic aircraft. A simple geometric construction due to Mach permits us to view the cone as the envelope of all the spherical sound waves emitted earlier from the nose of the aircraft.

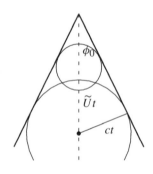

The Mach construction of the shock cone for a supersonic object. In a time t, the object moves forward a distance $\widetilde{U}t$ while the sound emitted at $t = 0$ forms a sphere of radius ct. The opening angle of the envelope of all spheres is $\sin\phi_0 = ct/\widetilde{U}t = c/\widetilde{U} = 1/\widetilde{\text{Ma}}$.

Example 25.2.1: For an aircraft at $\widetilde{\text{Ma}} = 2$, the Mach angle is $\phi_0 = 30\,°$, so when you hear the sonic boom of the aircraft passing overhead at $h = 20$ km altitude, it is already a distance $d = h\cot\phi_0 \approx 35$ km beyond your position.

Moving normal shock

A stationary planar normal shock may be converted to a normal shock moving through a gas at rest by choosing a reference frame moving with velocity U. In this frame, the previously incoming gas will now be at rest, whereas the shock front itself moves in the opposite direction with velocity U, and the gas behind the front moves in the same direction with velocity $U - U'$ which may or may not be supersonic. Locally, this describes a blast wave arising from a violent explosion, to be discussed in the following section.

Front thickness

Since the gas is completely at rest before the passage of the moving front, it is clear that a moving shock must be 'fed from behind', like a hydraulic jump, but because air is non-dispersive all small-amplitude waves move upstream at the same speed, $c' = \sqrt{\gamma p'/\rho'}$. Consequently there will be no lower limit to the wavelength of disturbances running upstream towards the front, and from a macroscopic point of view the shock front has vanishing thickness, as long as viscosity can be disregarded.

The only quantity with the dimension of length that may naturally be constructed from the front velocity U and the kinematic viscosity ν is,

$$\delta \sim \frac{\nu}{U}. \tag{25.27}$$

For a weak shock in the atmosphere under normal conditions we have $\nu \approx 1.6 \times 10^{-5}$ m^2 s^{-1} and $U \approx c \approx 340$ m s^{-1}, leading to $\delta \sim 45$ nm, so that the viscous thickness of the front is comparable to the mean free path in the gas. This conclusion actually invalidates the calculation, but for all practical purposes a shock front may be assumed to have zero thickness.

Strong shock limit

In the limit of large shock strength $\sigma \gg 1$, the density ratio ρ'/ρ, the velocity ratio U'/U, the ratio $p'/\rho U^2$ and Ma′ all approach constant values,

$$\frac{\rho'}{\rho} = \frac{U}{U'} \to \frac{\gamma+1}{\gamma-1}, \qquad \frac{p'}{\rho U^2} \to \frac{2}{\gamma+1}, \qquad \mathrm{Ma}' \to \sqrt{\frac{\gamma-1}{2\gamma}}. \tag{25.28}$$

For a diatomic gas with $\gamma = 7/5$ we find $\rho'/\rho \to 6$ and $\mathrm{Ma}' \to 1/\sqrt{7} \approx 0.38$. The actual Mach number of the flow behind a moving shock in a stationary gas is $(U - U')/c' \to 1/\sqrt{2\gamma(\gamma-1)}$ which becomes $5/2\sqrt{7} \approx 0.94$ for $\gamma = 7/5$. The flow behind the front is only marginally subsonic in a diatomic gas, whereas in a multi-atomic gas with $\gamma = 4/3$ the flow is marginally supersonic.

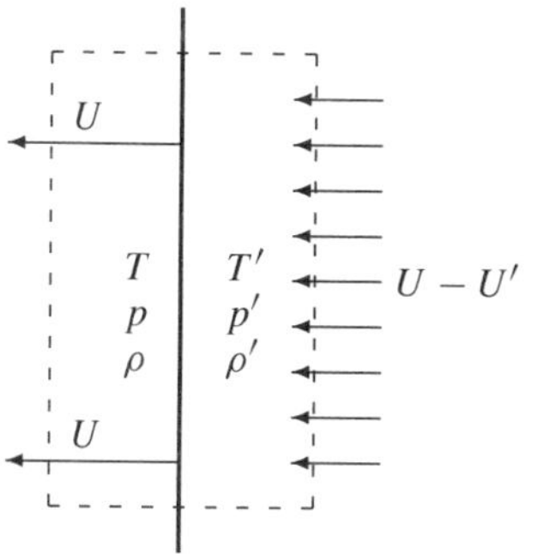

A planar shock moving towards the left with supersonic velocity U into a gas at rest. The fluid behind the front moves to the left with velocity $U - U'$ which may or may not be supersonic.

25.3 Atmospheric blast wave

A large explosion in the atmosphere generates a *blast wave* bounded by a spherical supersonic shock front. Such blast waves are mostly invisible, but films of nuclear or thermonuclear bomb explosions show that the physical conditions can become so extreme that the blast wave can be seen as a rapidly expanding spherical fireball, appearing right after the initial flash but before the mushroom cloud erupts. In this section we shall investigate the time evolution of such blast waves in the atmosphere, following the road laid out by G. I. Taylor in the 1940s[2].

Radius of the strong shock front

Let the atmosphere initially be at rest with density ρ_0 and pressure p_0. The blast deposits almost instantly a huge amount of energy E_0 within a tiny region of radius a, which initially contains the possibly ionized gases produced in the blast, as well as the solid remains of the bomb if there are any. The huge pressure in the initial fireball creates a shock front expanding at supersonic speed. After some time t, the shock front has become nearly spherical with a radius $R(t)$ that is very large compared to the initial size a. The volume inside the front contains essentially all of the initial energy E_0 in the form of 'shocked' air with only little contamination from the bomb itself. At this stage the shock has become a purely atmospheric phenomenon and all details about its origin in any particular explosion have been 'forgotten'.

Under these conditions the radius of the shock front $R(t)$ will be determined by the equations of gas dynamics, and can only depend on time t, the total energy E_0, and the atmospheric parameters ρ_0 and p_0. In a strong shock the ambient pressure p_0 is negligible compared to the pressures inside, implying that $R(t)$ should be finite in the limit of $p_0 \to 0$, and thus only depend on t, E_0 and ρ_0. Since E_0/ρ_0 is measured in units of of J kg^{-1} m^{-3} = m^5 s^{-2}, the only possible form of the relationship between size and time is

$$\boxed{R(t) = A\left(\frac{E_0 t^2}{\rho_0}\right)^{1/5}}, \tag{25.29}$$

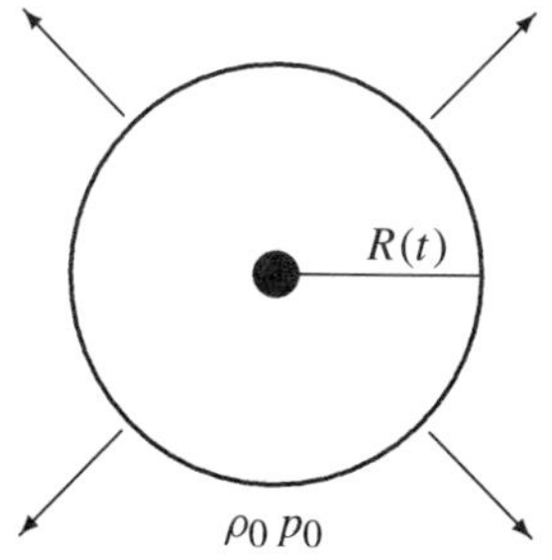

A spherical shock front in the atmosphere some time t after the detonation of a bomb (black circle). Its current radius is $R(t)$, which is much larger than the initial blast region. The atmosphere is initially at rest with density ρ_0 and pressure p_0. The volume of the sphere consists almost entirely of air.

where A is a numerical constant which as we shall see below is very close to unity.

It is remarkable that a phenomenon as violent as an atomic explosion can be captured in such a simple relationship. Taylor used it in 1947 to estimate the yield of the first nuclear bomb to be about 10^{14} J from a time-lapse sequence of recently released photographs, a feat which created some embarrassment with the security authorities and earned him a mild admonishment, although he had done nothing wrong [8].

[2]G. I. Taylor, The formation of a blast wave by a very intense explosion (I. Theoretical discussion, II. The atomic explosion of 1945), *Proc. Roy. Soc.* **A201**, (1950) 159–186. Taylor actually formulated the theory in 1941, but first published it in 1950. Photographs of the first atomic test were declassified by the US Atomic Energy Commission in 1947, although its yield remained classified.

Figure 25.8. Trinity fireball 16 milliseconds after detonation. The sharp outline of the fireball indicates that it coincides with the shock front. Later the shock front will detach from the fireball. Source: White Sands Missile Range.

Physical parameters in the strong shock limit

In the strong shock limit we may use the strong shock results (25.28) to determine the physical quantities just inside the front,

$$\frac{\rho_1}{\rho_0} = \frac{\gamma+1}{\gamma-1}, \qquad \frac{v_1}{U} = \frac{2}{\gamma+1}, \qquad \frac{p_1}{\rho_0 U^2} = \frac{2}{\gamma+1}, \tag{25.30}$$

where $U = dR/dt = (2/5)R/t$ is the shock front velocity and $v_1 = U - U' = (1 - U'/U)U$ is the radial flow velocity just behind the front. The temperature is obtained from the ideal gas law,

$$T_1 = \frac{p_1}{\rho_1}\frac{M_{\text{mol}}}{R_{\text{mol}}} = \frac{2(\gamma-1)}{(\gamma+1)^2}\frac{M_{\text{mol}}U^2}{R_{\text{mol}}} \tag{25.31}$$

where R_{mol} is the molar gas constant and M_{mol} is the molar weight of the gas. Note that none of these quantities depend on the ambient pressure p_0. In the following we shall use the value $\gamma = 7/5$ for diatomic gas, although the violent initial shock may dissociate a fraction of the molecules.

Example 25.3.1 (Trinity): The first atomic bomb test, codenamed *Trinity*, was carried out at Alamogordo, New Mexico on July 16, 1945. Its yield was $E_0 \approx 8.5 \times 10^{13}$ J which is roughly equivalent to 17 000 tons of the high explosive TNT. Taking $A = 1$ and $\gamma = 7/5$ we find that after $t = 15$ ms the fireball has expanded to a radius $R \approx 110$ m with the front moving at a speed of $U \approx 2900$ m s^{-1}. The pressure just inside the front is $p_1 \approx 85$ atm, the density $\rho_1 \approx 7.2$ kg m^{-3}, and the temperature $T_1 \approx 4200$ K.

Isentropic radial gas dynamics

A spherically invariant isentropic flow in an ideal gas is described by a purely radial velocity field $\boldsymbol{v} = v(r,t)\boldsymbol{e}_r$, a density field $\rho(r,t)$, a pressure field $p(r,t)$ and the field of specific entropy $s = c_V \log(p\rho^{-\gamma})$.

In the absence of gravity and viscosity, the fields obey the dynamic equations,

$$\frac{\partial v}{\partial t} + v\frac{\partial v}{\partial r} = -\frac{1}{\rho}\frac{\partial p}{\partial r}, \tag{25.32a}$$

$$\frac{\partial \rho}{\partial t} + \frac{1}{r^2}\frac{\partial\left(r^2 \rho v\right)}{\partial r} = 0, \tag{25.32b}$$

$$\frac{\partial s}{\partial t} + v\frac{\partial s}{\partial r} = 0. \tag{25.32c}$$

The first is the Euler equation, the second the continuity equation on radial form, and the last expresses that the flow is isentropic, meaning that the specific entropy is constant along a comoving (particle) orbit. The last equation also shows that if the initial state is *homentropic* with spatially constant specific entropy $\partial s/\partial r = 0$, it will remain so forever. In a strong blast entropy is only produced right at the front, and that does not lead to a homentropic state of the air behind the front.

Strong self-similar shock

In the strong shock limit, $\sigma \to \infty$ and $p_0 \to 0$, the only parameter with dimension of length is the radius of the shock front $R(t)$, which we for a moment pretend not to know explicitly. The proper non-dimensional radial variable for all the fields is therefore,

$$\xi = \frac{r}{R(t)}. \tag{25.33}$$

Including suitable dimensional coefficients depending only on $R(t)$ and ρ_0, the fields are assumed to be of the form,

$$v = \dot{R}(t)u(\xi), \tag{25.34a}$$

$$\rho = \rho_0 f(\xi), \tag{25.34b}$$

$$p = \rho_0 \dot{R}(t)^2 q(\xi), \tag{25.34c}$$

where a dot indicates differentiation after time. With this assumption, the spatial variation of the fields is self-similar at all times.

Inserting these fields into the equations of motion (25.32) we obtain three coupled ordinary first-order differential equations, and using a prime to denote differentiation after ξ, we find

$$\alpha u + (u - \xi)u' = -\frac{q'}{f}, \tag{25.35a}$$

$$(u - \xi) f' = -f u' - \frac{2uf}{\xi}, \tag{25.35b}$$

$$2\alpha + (u - \xi)\left(\frac{q'}{q} - \gamma\frac{f'}{f}\right) = 0, \tag{25.35c}$$

where

$$\alpha = \frac{R\ddot{R}}{\dot{R}^2}. \tag{25.36}$$

The α-terms on the left-hand side of (25.35) stem from the explicitly time-dependent prefactors of the fields. Since α only depends on t and the other functions only on ξ, it follows from these equations that α must be a constant, independent of time. The solution to (25.36) is then a power law

$$R(t) \sim t^{1/(1-\alpha)}. \tag{25.37}$$

The value of α cannot be determined from the dynamic equations alone but depends on the boundary conditions imposed on the solution, in particular the condition that the radial velocity u must vanish at $\xi = 0$. We shall determine it below by requiring that the excess energy E_0 is constant.

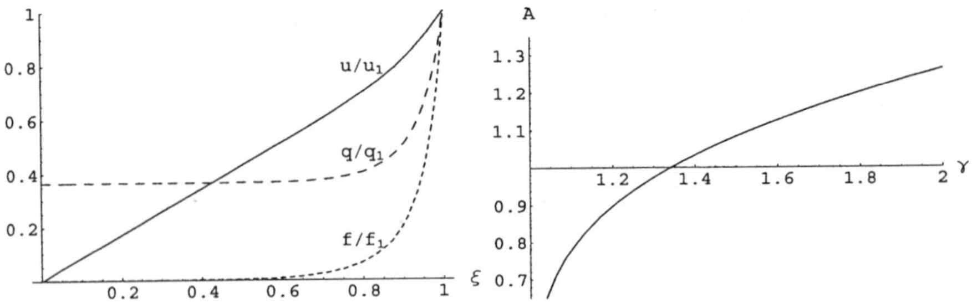

(a) (b)

Figure 25.9. **(a)** The numeric solution to the dynamic equations (25.35) for $\gamma = 7/5$. All quantities are normalized to their maximal values at $\xi = 1$. **(b)** The dimensionless coefficients A in the blast radius (25.29) as a function of γ.

Numerical solution

The three ordinary first-order differential equations (25.35) need three boundary conditions which may be determined at $\xi = 1$ from the strong shock properties (25.30),

$$f_1 = \frac{\gamma+1}{\gamma-1}, \qquad u_1 = \frac{2}{\gamma+1}, \qquad q_1 = \frac{2}{\gamma+1}. \tag{25.38}$$

Although it is possible to find an analytic solution [62], it turns out to be quite complicated, and it is much easier to integrate the differential equations numerically. The numeric solution is plotted in figure 25.9(a) for $\gamma = 7/5$ with the functions normalized by their values at $\xi = 1$. Evidently there are two distinct regions in a strong shock, a *core* for $\xi < 0.7$ where the pressure is nearly constant with $q_0/q_1 \approx 0.37$ and the velocity increases linearly, and a *shell* for $\xi > 0.7$ where the velocity and pressure rise rapidly to meet the required values at the front.

Excess energy

The total energy of the gas in the volume inside the shock front consists of the kinetic energy of the moving gas plus its internal energy. Subtracting the internal energy of the gas before the explosion we find from (22.60) on page 320 the excess of energy inside the shock front,

$$E_0 = \int_0^{R(t)} \left(\frac{1}{2}\rho(r,t)v(r,t)^2 + \frac{p(r,t) - p_0}{\gamma - 1} \right) 4\pi r^2\, dr. \tag{25.39}$$

Barring radiative losses, this energy must be constant and equal to the extra energy added to the atmosphere in the point-like explosion. It takes quite a bit of algebraic work to demonstrate explicitly from the dynamics (25.32) and the strong shock properties (25.30) that the time derivative of this expression indeed vanishes.

Inserting the self-similar fields (25.34), the excess energy (25.39) becomes in the limit of $p_0 \to 0$,

$$E_0 = \rho_0 R^3 \dot{R}^2 K(\gamma), \tag{25.40}$$

where $K(\gamma)$ is the dimensionless integral,

$$K(\gamma) = 4\pi \int_0^1 \left(\frac{1}{2} f(\xi)u(\xi)^2 + \frac{q(\xi)}{\gamma - 1} \right) \xi^2\, d\xi. \tag{25.41}$$

It is immediately clear from (25.40) that $E_0 \sim R^3\dot{R}^2 \sim t^{(3+2\alpha)/(1-\alpha)}$, so for the power law (25.37) to lead to a constant energy, the exponent must be $\alpha = -3/2$, and thus $R \sim t^{2/5}$. This confirms the dimensional analysis (25.29), and inserting $\dot{R} = (2/5)R/t$ in (25.40) we also obtain an expression for the dimensionless constant,

$$A = \left(\frac{4}{25} K(\gamma) \right)^{-1/5}. \tag{25.42}$$

In figure 25.9(b) the numeric solution is plotted for a range of γ values. For $\gamma = 7/5$ we have $A = 1.03$, which justifies taking $A \approx 1$ in our earlier estimates.

The weakening shock

As the shock front expands, it decreases in strength until it no longer satisfies the conditions for the strong shock approximation used above. The characteristic strength for the breakdown of the strong shock approximation may be chosen to be $\sigma_s = 2\gamma/(\gamma - 1)$ where the two terms in the denominator of the density ratio (25.22) become equal. For $\gamma = 7/5$ we have $\sigma_s = 7$ and thus $p_1 = (1 + \sigma_s)p_0 = 8$ atm. Using the strong shock formalism in the *Trinity* example 25.3.1 this is estimated to happen at time $t \approx 110$ ms when the front radius is $R \approx 240$ m and its velocity $U \approx 900$ m s^{-1}.

Continuing the expansion beyond this point, the core pressure will decrease until it reaches the ambient pressure p_0. At this point the core can no longer perform work on the surrounding atmosphere and will stop expanding. The core is very hot due to its large entropy with a correspondingly low density, and the buoyancy of the low-density core will make it rise like a thermal bubble, creating thereby the stalk of the mushroom cloud. We may estimate the time when the core expansion ceases by equating the core pressure $p = \rho_0 U^2 q_0$ with atmospheric pressure p_0. This is probably a rather bad approximation, but in lieu of a better one we find in the *Trinity* example 25.3.1 that it happens at time $t \approx 260$ ms when the front radius is $R \approx 340$ m and its velocity has dropped to $U \approx 525$ m s^{-1}.

After this point the shock continues as a spherical wave in the form of a thin shell. When such a front passes a given point, the pressure first rises above atmospheric pressure, followed by a characteristic suction phase where the pressure drops below atmospheric pressure.

Problems

25.1 Show that the linear approximation in (25.8) is better than 3.3% for $\mathrm{Fr} > 2$.

25.2 Discuss mass conservation for the river bore and the reflection bore.

25.3 Verify that the solution of the Rankine–Hugoniot relation (25.19) is

$$U' = \frac{\gamma - 1}{\gamma + 1} U + \frac{2\gamma}{\gamma + 1} \frac{p}{\rho U}, \tag{25.43a}$$

$$\frac{1}{\rho'} = \frac{\gamma - 1}{\gamma + 1} \frac{1}{\rho} + \frac{2\gamma}{\gamma + 1} \frac{p}{\rho^2 U^2}, \tag{25.43b}$$

$$p' = -\frac{\gamma - 1}{\gamma + 1} p + \frac{2}{\gamma + 1} \rho U^2. \tag{25.43c}$$

25.4 In a general material, the third Rankine–Hugoniot relation (25.19c) becomes

$$\frac{1}{2}U^2 + \frac{p}{\rho} + u = \frac{1}{2}U'^2 + \frac{p'}{\rho'} + u', \tag{25.44}$$

where u and u' are the specific internal energies before and after the shock front.

Show that it follows from the Rankine–Hugoniot relations that,

$$u' - u = \frac{(p + p')(\rho' - \rho)}{2\rho\rho'}. \tag{25.45}$$

This identity goes under the name of Hugoniot's equation.

25.5 Show that the entropy change (25.24) is a growing function of σ for $\sigma > -1$.

25.6 The largest Hydrogen bomb ever detonated in the atmosphere had a yield of 50 megatons of TNT (about 2.5×10^{17} J). Estimate the radius and velocity of the shock front 1 s after the explosion, and estimate the pressure, density and temperature just behind the shock front. Show that the strong shock approximation is valid at this time.

26 Whirls and vortices

Whirls and vortices are common features of real fluids. Stirring the coffee, you create a circulating motion, a whirl that dies out after some time when you stop stirring. A fast spinning vortex may form at the drain of a bathtub when it is emptied, and can remain quasi-stable as long as there is enough water in the tub. Normally, whirls and vortices are invisible far away from free boundaries, but in the bathtub the drain vortex is made visible by soap remains clouding the water and the depression it creates at the surface. It may even become audible, because water falling through the drain sucks air down from the surface.

In the atmosphere above heated ground, whirls and vortices arise all the time. Mostly they are small and invisible, but sometimes they pick up dust and debris and appear as dust devils. Larger dust devils, in some countries called sky-pumps, are known to pick up haystacks and scatter them, or to throw tables around in sidewalk cafes. When the heat-driven air vortices grow really big and nasty, they become tornadoes. The force powering heat-driven vortices is fundamentally the same as the force that maintains a bathtub vortex, namely gravity. Whereas a bathtub vortex is driven by gravity acting on the water going down the drain, a tornado is maintained by the buoyancy of hot air draining skywards.

Vortices are also found in the wakes of moving objects. From the tips of the wings of aircraft there will always trail long invisible vortices (to be discussed in chapter 29). Large passenger aircraft taking off or landing create strong and fairly stable vortices, capable of overturning smaller planes following after. 'Beware of vortex' is a common warning issued by flight controllers to small aircraft when taking off or landing behind heavy passenger planes.

The chapter sets the stage with a discussion of the structure of free cylindrical vortices in an incompressible fluid, and continues with a presentation of the basic vortex dynamics in non-viscous fluids. The remainder of the chapter is devoted to models for non-singular laminar vortices driven by secondary flow. Extensive treatment of vortex structure and dynamics may be found in [60, 45].

26.1 Free cylindrical vortices

The spindle-driven vortex discussed on page 262 was powered by a rotating cylindrical spindle, which delivered the work necessary to overcome viscous friction in the fluid. Although the flow pattern (18.74) did not depend explicitly on the viscosity of the fluid, a finite viscosity was nevertheless necessary for the spindle to be able to 'crank up' the vortex, starting from fluid at rest.

In a truly inviscid fluid, we would not be able to crank up the vortex, but once it had somehow been established there would be no viscous losses and the vortex would keep spinning forever. The solid spindle could then with impunity be replaced by a *core* made from the same fluid as the vortex and, for example, rotating as a solid body with the same angular velocity Ω as the spindle.

The Rankine vortex

William John Macquorn Rankine (1820–72). *Scottish civil engineer and physicist. Worked on thermodynamics (with Kelvin), in particular steam engine theory. Created a now defunct absolute temperature based on Fahrenheit degrees.*

The *Rankine vortex* with core radius c is constructed in precisely this way,

$$v_\phi = \begin{cases} \Omega\, r & 0 \le r \le c \\ \dfrac{\Omega\, c^2}{r} & c \le r < \infty \end{cases}. \tag{26.1}$$

The vorticity vanishes outside the core and is constant inside (see problem 18.22).

In the absence of gravity, the pressure $p = p(r)$ can only depend on r, and the radial pressure gradient must according to (18.60a) provide the centripetal force necessary to maintain the circular fluid motion,

$$\frac{dp}{dr} = \rho_0 \frac{v_\phi^2}{r}, \tag{26.2}$$

sometimes called *cyclostrophic balance*. Integrating this equation with the Rankine vortex we obtain,

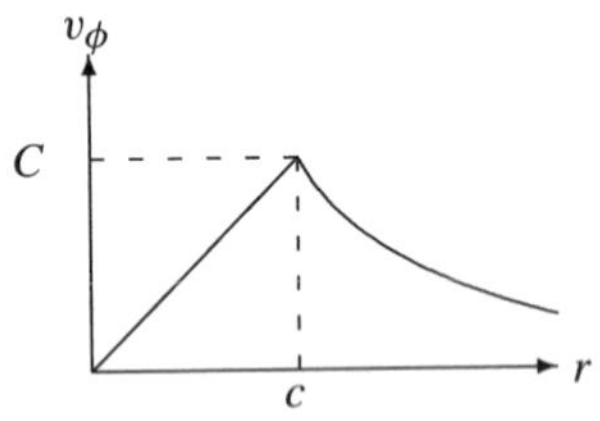

Sketch of the velocity field for the Rankine vortex. In a viscous fluid the cusp at the core boundary $r = c$ indicates a discontinuity in shear stress which is physically unacceptable.

$$p = -\rho_0 \int_r^\infty \frac{v_\phi^2}{r}\, dr = -\frac{1}{2}\rho_0 \Omega^2 c^2 \begin{cases} 2 - \dfrac{r^2}{c^2} & r < c \\ \dfrac{c^2}{r^2} & r > c \end{cases} \tag{26.3}$$

where the pressure has been normalized to vanish at infinity. The shape of the pressure is shown as the dashed curve in figure 26.1(*b*).

The shear stress calculated from (18.65) becomes

$$\sigma_{\phi r} = \begin{cases} 0 & r < c \\ -2\eta\Omega \dfrac{c^2}{r^2} & r > c \end{cases}. \tag{26.4}$$

The cusp in the Rankine vortex at the core boundary creates a finite jump in stress at $r = c$, which is physically unacceptable because it violates Newton's third law. The Rankine vortex is only meaningful in the limit of a truly inviscid fluid.

Interpolating vortex

Horace Lamb (1849–1934). *British physicist. Contributed mainly to acoustics and fluid mechanics. His textbooks remained for many years the standard in these fields.*

It is possible to find a smooth interpolating function which coincides with the Rankine vortex for $r \ll c$ and $r \gg c$, for example the *Lamb vortex*,

$$v_\phi = \frac{\Omega\, c^2}{r}\left(1 - e^{-r^2/c^2}\right), \tag{26.5}$$

which approaches the Rankine vortex for both $r \ll c$ and for $r \gg c$. The shear stress is continuous everywhere and the problem with Newton's third law has disappeared. The velocity field is shown in figure 26.1(*a*).

Integrating (26.2) with the interpolating vortex and requiring the pressure to vanish at infinity one gets,

$$p = -\frac{1}{2}\rho_0 \Omega^2 c^2 F\left(\frac{r^2}{c^2}\right), \tag{26.6}$$

where $F(\xi)$ is a purely mathematical function,

$$F(\xi) = \int_\xi^\infty \left(1 - e^{-x}\right)^2 \frac{dx}{x^2}. \tag{26.7}$$

With pressure normalized to vanish at infinity and using that $F(0) = 2\log 2$, the central pressure becomes

$$p_0 = -\frac{1}{2}\rho_0 \Omega^2 c^2 F(0) = -\rho_0 \Omega^2 c^2 \log 2. \tag{26.8}$$

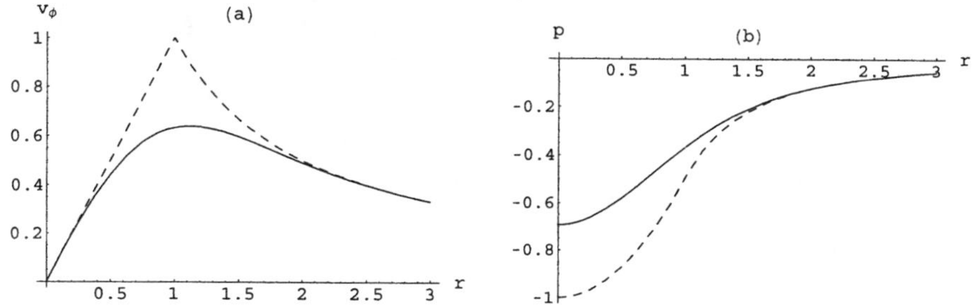

Figure 26.1. The interpolating vortex (solid line) and the Rankine vortex (dashed) for $c = \Omega = \rho_0 = 1$ **(a)** The velocity field. **(b)** The pressure field, which is somewhat shallower in the interpolating vortex than in the Rankine vortex.

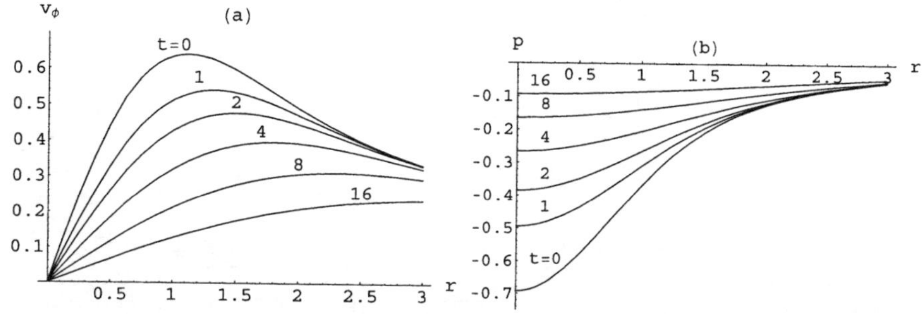

Figure 26.2. Time evolution of the Oseen–Lamb vortex (26.10) for $c = \Omega = \rho_0 = 1$ and $\nu = 0.1$ at selected times. The core grows larger with time while the central angular velocity (i.e. the slope of v_ϕ at $r = 0$) becomes smaller. **(a)** Azimuthal velocity. **(b)** Pressure.

The pressure is, of course, lower at the centre than at infinity (see figure 26.1(*b*)), and for a liquid vortex in constant gravity this would create a central depression in the open surface of the same shape (see section 26.6).

The resolution of the stress problem comes at a price, because the interpolating vortex cannot be stable on its own in a viscous fluid. The shear stress will cause dissipation of kinetic energy everywhere in the fluid and make the vortex spin down. The largest shear stress is found just outside the core and will slow down the core's rotation while expanding its radius to conserve angular momentum. We shall now see that the Lamb vortex with a time-dependent core radius is an exact solution to the Navier–Stokes equations.

Diffusive spin-down of vortex

Carl Wilhelm Oseen (1879–1944). *Swedish mathematical physicist. Worked on relativity. Influential member of the Nobel committee, he was instrumental in awarding Einstein the prize in 1921 for the photoelectric effect, rather than for relativity.*

Assuming that the azimuthal velocity field retains its purely cylindrical form $v_\phi(r, t)$ at all times, we may go through the same steps as lead to (18.60) and derive the time-dependent Navier–Stokes equation in cylindrical coordinates,

$$\boxed{\frac{\partial v_\phi}{\partial t} = \nu \frac{\partial}{\partial r}\left(\frac{1}{r}\frac{\partial (r v_\phi)}{\partial r}\right),} \tag{26.9}$$

with the pressure given by (26.2). This equation is quite analogous to the momentum diffusion equation (17.5), and given any initial field $v_\phi(r, 0)$ it will always produce a solution.

By insertion one may directly verify that the solution which for $t = 0$ starts out as the interpolating vortex (26.5) is,

$$v_\phi(r, t) = \frac{C}{r}\left(1 - \exp\left(-\frac{r^2}{c^2 + 4\nu t}\right)\right), \tag{26.10}$$

Figure 26.3. Tornado near LaPlata, Maryland, on April 28, 2002. One of the largest tornados ever to hit the eastern part of the US.

where C is a constant. It is called the *Oseen–Lamb vortex* and its time evolution is plotted in figure 26.2. Whereas the asymptotic circulation is constant, the core radius, $c(t) = \sqrt{c^2 + 4\nu t}$, grows with time, while the angular velocity of the core, $\Omega(t) = \partial v_\phi/\partial r|_{r=0} = C/(c^2 + 4\nu t)$, decreases. The time it takes for the vortex to double its original core radius is $t = 3c^2/4\nu$, and the core overtakes any point r, well outside the original core, in a time $t \gtrsim r^2/4\nu$. These times are recognized as typical momentum diffusion time scales (see page 233), but in this case the quantity that diffuses is actually angular momentum.

Finally, it should be mentioned that there are many other exact solutions to (26.9), some of which are found in problems 26.9 and 26.10.

Hermann Ludwig Ferdinand von Helmholtz (1821–94). *German physician and physicist. Invented in 1851 the ophthalmoscope (for studying the eye) and published an influential Handbook of physiological optics. Was deeply engaged in the physiology of perception and the physiological basis for the theory of music. In physics his main contribution was in vortex theory. Worked as professor of physics in Berlin from 1871.*

Example 26.1.1 (Bathtub vortex): A water vortex with core radius $c = 3$ cm doubles the radius of its viscous core in the time $3c^2/4\nu = 675$ s, a little more than 11 min. This is surprisingly large, but it should be remembered that the estimate is only valid for laminar flow. Turbulence make most real-life bathtub vortices spin down faster, unless they are fed by inflow.

26.2 Ideal vortex dynamics

Vortex lines were defined (on page 219) to be curves that always follow the instantaneous vorticity field, $\boldsymbol{\omega} = \nabla \times \boldsymbol{v}$. Like streamlines, vortex lines cannot cross each other, and since the vorticity field is a 'curl', it is rigorously divergence-free, $\nabla \cdot \boldsymbol{\omega} = 0$. This implies that vortex lines cannot emerge from anywhere, but must form closed curves, or curves coming from and going to the boundaries of the flow (which may be spatial infinity). Vortex dynamics can be quite complicated, and we shall here only present a few general results mainly due to Helmholtz.

Vortex tubes

The set of vortex lines that pass through the points of a closed curve at a given time form together a *vortex tube* which may curve and bend in space. We shall now prove that whatever shape it takes, it will nevertheless have the same total flux of vorticity through every cross section. To prove this we just use Gauss' theorem on the closed surface formed by a stretch of vortex tube between two cross sections, S_1 and S_2. Since the divergence of the vorticity field vanishes, the closed surface integral over this stretch of tube must vanish,

$$\oint_S \boldsymbol{\omega} \cdot d\boldsymbol{S} = \int_{S_2} \boldsymbol{\omega} \cdot d\boldsymbol{S} - \int_{S_1} \boldsymbol{\omega} \cdot d\boldsymbol{S} = 0. \tag{26.11}$$

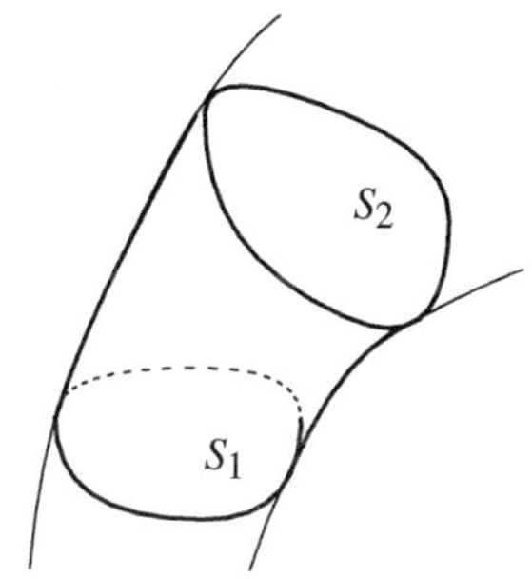

The flux of vorticity is the same in every cross section of a vortex tube.

The sides of the tube do not contribute to the integral, because the normals to the surface elements are orthogonal to the vortex lines, i.e. to the vorticity field.

The common flux through any cross section of a vortex tube is called the *strength* of the tube, and by Stokes theorem (16.57) this flux equals the circulation around any curve circumnavigating the tube along its surface. From Kelvin's circulation theorem (16.59) we know that in an ideal fluid the circulation around a comoving curve is constant in time. A comoving vortex tube thus retains its strength in the course of time, allowing us to view a vortex tube (and even a single vortex line) as an individual material 'object' that is advected, 'blown along', with the ambient flow.

The vortex core

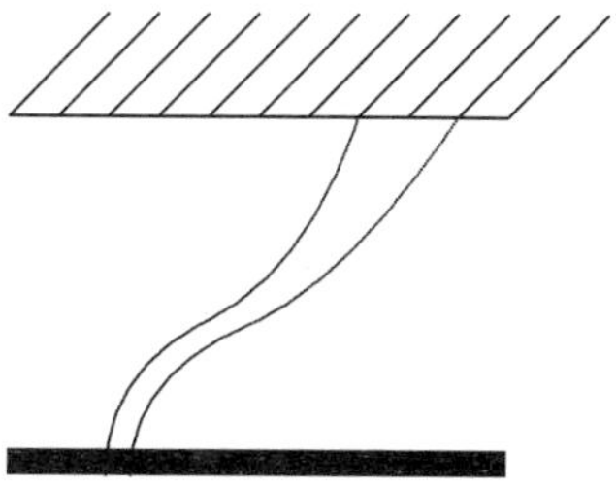

A tornado funnel is a vortex connecting the ground and the thundercloud. Wind speed differences at different altitudes usually make the tornado turn and twist.

Vortices often have a central vortex tube, called the *core* of the vortex, containing almost all of the vorticity, with the flow being nearly irrotational outside, like in the Rankine vortex (26.1). From Stokes' theorem it then follows that the circulation is the same along any closed curve that circumnavigates the core. Since the core region is a vortex tube, it cannot disappear but only move around, and a vortex in an ideal fluid thus acquires an identity of its own. It may connect to the boundaries (possibly at infinity), but often it will take the form of a closed loop, which like a smoke ring drifts in the wind created by itself.

In the extreme and singular limit, where the vortex core becomes infinitely thin and its vorticity field infinitely high while the strength remains finite, it is called a *vortex filament*. A vortex filament is entirely characterized by its strength and the curve that describes the whereabouts of the core. In the preceding section we saw that viscosity will make the core of a free vortex expand and eventually cause the whole vortex to disappear, but if the core radius is small compared to the radius of curvature of the filament and other length scales in the flow, a real vortex may in fact behave as a vortex filament over long periods of time.

> The indestructibility of vortex filaments in ideal flow inspired Lord Kelvin in 1867 to suggest that atoms could be microscopic ring vortices in the ideal ether, and that vortex theory could be the foundation for what today would be called a 'theory of everything'. This early 'string theory' of matter was quickly abandoned but inspired instead the development of mathematical knot theory.

The line vortex

The simplest vortex filament is a *line vortex* in which the core is a straight line. In either cylindrical or Cartesian coordinates the velocity field becomes,

$$\boxed{\boldsymbol{v} = \frac{C}{r}\,\boldsymbol{e}_\phi = C\frac{(-y, x, 0)}{x^2 + y^2}.} \tag{26.12}$$

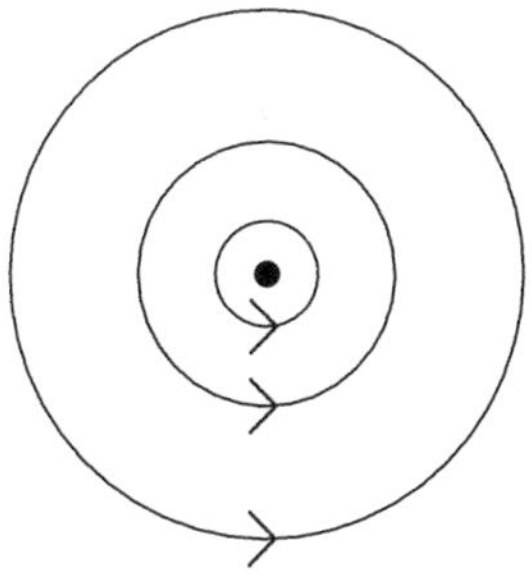

Irrotational flow outside the core of a line vortex. The circulation is the same around any streamline.

Comparing with the Rankine vortex (26.1) we see that the constant C is obtained in the limit where $\Omega \to \infty$ and $c \to 0$ such that Ωc^2 remains finite. Integrating around the vortex along a streamline of radius r, we find the circulation $\Gamma = 2\pi r v_\phi = 2\pi C$. The circulation is thus independent of r as expected from Stokes' theorem and the irrotational character of the flow.

Outside the singular core the velocity field of a line vortex has vanishing vorticity and must therefore be a gradient field, $\boldsymbol{v} = \nabla\Psi$, where Ψ is the velocity potential. From the gradient operator expressed

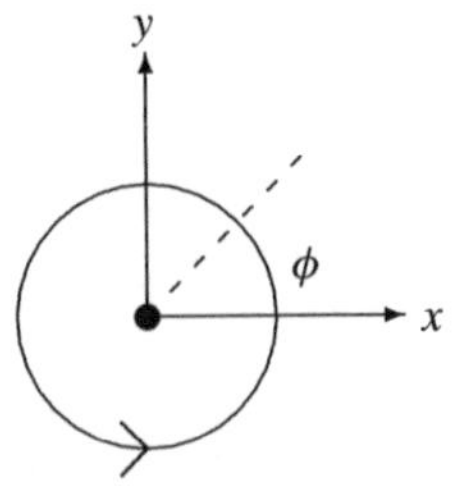

The multivalued angular dependence of the velocity potential of an elementary vortex filament reflects the non-vanishing vorticity carried in its core.

in cylindrical coordinates (B.6) it follows immediately that the gradient of the azimuthal angle ϕ is $\nabla\phi = \boldsymbol{e}_\phi/r$, such that the velocity potential of a line vortex must be,

$$\boxed{\Psi = C\phi = C \arctan\frac{y}{x}.} \tag{26.13}$$

The azimuthal angle ϕ is a multivalued function which increases by 2π each time a curve circles around the vortex core, reflecting of course the non-vanishing value of the circulation $\Gamma = 2\pi C$ arising from the infinite vorticity in the singular core.

Although the field of a line vortex is the gradient of a single-valued velocity potential in any simply connected region of space, the singular core carrying the vorticity makes the space around a line vortex multiple-connected. Every closed curve may topologically be characterized by its winding number, the number of times it circles the core. The winding number stays the same regardless of how the curve is deformed as long as it does not cross the core.

26.3 Parallel line vortices

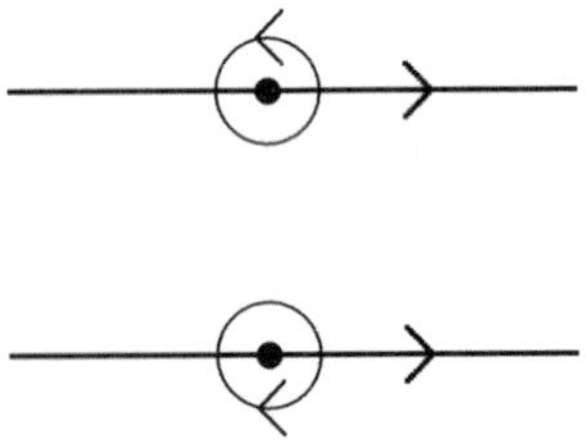

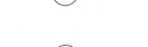

Two line vortices of equal but opposite strength blow each other in the same direction with the same speed, here towards the right. The same picture could also describe a cross section through the centre of a smoke ring.

Vortices are not always loners like the bathtub vortex or the tornado. The advection of one vortex filament in the flow created by all the others and even by different parts of itself gives rise to a number of interesting, beautiful and often quite counter-intuitive phenomena. Although the linearity of potential flow allows us to add the contributions from each filament, the end result can become quite complex. For parallel line vortices the analysis is much simpler.

Complex notation

A collection of parallel line vortices is essentially two-dimensional and two-dimensional problems are often facilitated by using complex notation. If we define the complex position and velocity fields,

$$z = x + iy, \qquad\qquad w = v_x - iv_y, \tag{26.14}$$

it follows from the Cartesian form of the line vortex (26.12) that the complex velocity field of a vortex centred at the origin is a simple pole,

$$w = C\frac{-y - ix}{x^2 + y^2} = -iC\frac{1}{x + iy} = -i\frac{C}{z}. \tag{26.15}$$

The field of a vortex centred in another point is obtained by shifting the origin to that point.

Consider now two counter rotating parallel line vortices of strength C and $-C$, a distance $2b$ apart. Each vortex will blow the other along with the common velocity $C/2b$, and thus keep a constant distance $2b$ between their centers. If they are positioned at $y = \pm b$, the instantaneous complex velocity field becomes,

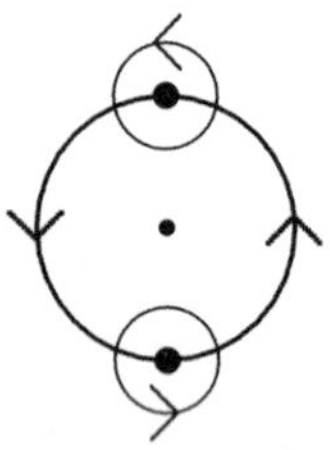

Two parallel line vortices of equal strength force each other to move with their cores on a common circle. Up to six vortex filaments may participate in the dance.

$$w = -i\frac{C}{z - ib} + i\frac{C}{z + ib} = \frac{2Cb}{z^2 + b^2}. \tag{26.16}$$

The usual velocity fields v_x and v_y may be calculated from the real and imaginary parts of this expression (problem 26.3), and the streamlines of this field are shown in figure 26.4(a). At the origin the velocity is $2C/b$ which is four times the drift velocity, whereas at long distance from the vortices, $r \gg b$, the velocity field vanishes as $1/r^2$.

The instantaneous field of a pair of corotating vortices of strength C a distance $2b$ apart will instead blow each other in opposite directions, making the two vortex cores dance around on a circle of radius b with angular velocity satisfying $\Omega b = C/2b$, or $\Omega = C/2b^2$. If they are positioned at $y = \pm b$ the instantaneous field is,

$$w = -i\frac{C}{z - ib} - i\frac{C}{z + ib} = -i\frac{2Cz}{z^2 + b^2}. \tag{26.17}$$

The streamlines of this field are shown in figure 26.4(b). At the origin the velocity field vanishes, whereas at long distances from the vortices it becomes the field of a single vortex with strength $2C$.

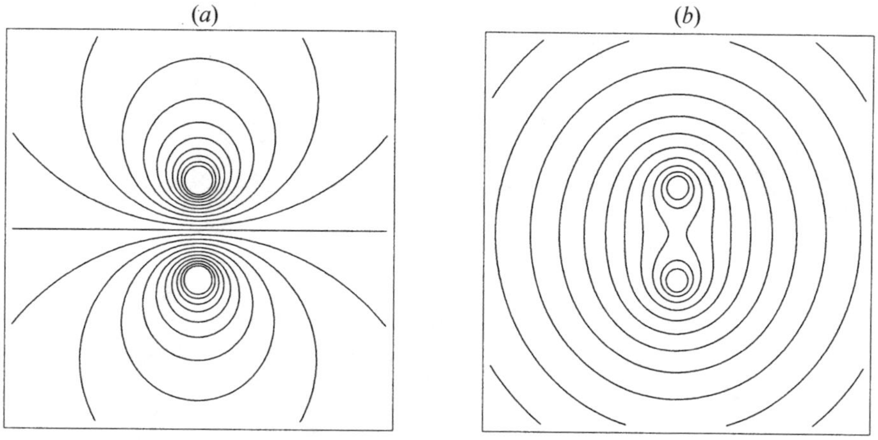

Figure 26.4. Streamlines around a pair of parallel line vortices in a fluid asymptotically at rest. **(a)** Counter-rotating vortices. The cores move horizontally at the same speed. **(b)** Corotating vortices. The cores move in a circle around the centre of the figure.

The equations of motion for an arbitrary collection of line vortices may easily be written down (problem 26.5) and solved for many symmetric initial configurations, but even if the vortices move in a regular fashion, their orbits are not necessarily stable towards small perturbations. For example, only two to six line vortices on a ring form a stable pattern.

* The von Karman vortex street

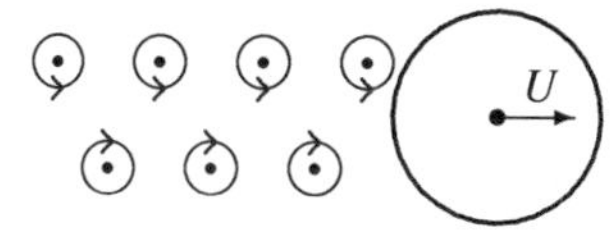

The von Karman vortex street created by a cylinder moving through a fluid at rest with velocity U. The vortex chains follow the cylinder at a somewhat lower speed.

Consider a long object, for example a cylinder, moving through a fluid at rest with constant velocity and with its axis orthogonal to the direction of motion. At fairly low Reynolds numbers (well above unity) the viscous boundary layers will detach and form two counter-rotating eddies behind the object. At higher Reynolds numbers around 100 the flow behind the object becomes unstable, causing the eddies alternately to detach at regular time intervals. Thus, in the wake of the moving object one may in a region of Reynolds numbers observe a beautiful alternating pattern of free counter-rotating vortices shed by the object. This pattern is called the *von Karman vortex street.* At still higher Reynolds numbers the instabilities become chaotic and lead to turbulence.

First we consider an infinite collection of corotating vortices of unit strength spaced regularly with interval $\Delta x = \pi$ along the x-axis. The symmetry of the chain guarantees that these vortices are not mutually advected by each other, and the complex velocity field becomes,

$$w = -i \sum_{n=-\infty}^{\infty} \frac{1}{z - n\pi} = -i \cot z. \tag{26.18}$$

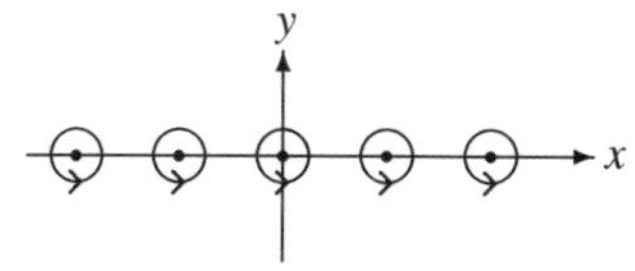

A chain of corotating vortices of equal strength placed at regular intervals along the x-axis.

The proof that the sum equals the cotangent only requires complex function theory at a fairly elementary level (problem 26.6).

The von Karman vortex street is now modelled by two vortex chains of strength $-C$ at $y = -b$ and $+C$ at $y = b$. In each chain the vortices are spaced regularly with interval $2a$, but one of the chains is shifted by a with respect to the other. The total field becomes

$$w = iC \sum_n \frac{1}{z - 2na + ib} - iC \sum_n \frac{1}{z - (2n+1)a - ib}. \tag{26.19}$$

Using $\cot(z + \pi/2) = -\tan z$ we find the total field,

$$w = i\frac{C\pi}{2a}\left(\cot\frac{\pi(z+ib)}{2a} + \tan\frac{\pi(z-ib)}{2a}\right). \tag{26.20}$$

The streamlines of this field are plotted in figure 26.6 for aspect ratio $b/a = 0.28$ (see below).

Figure 26.5. Instantaneous picture of vortex street behind circular cylinder at Re = 105. Photograph by Sadatoshi Taneda.

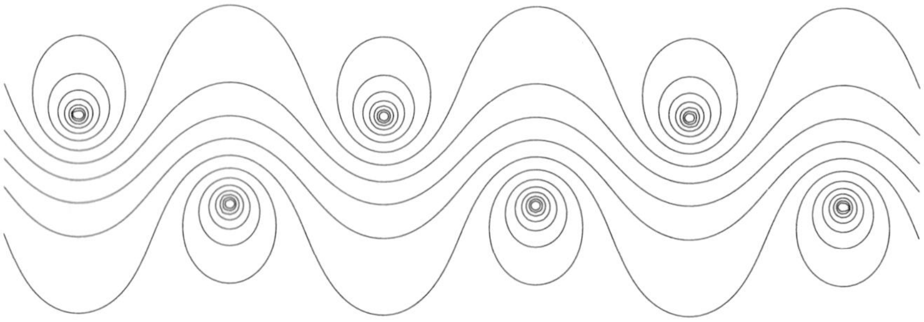

Figure 26.6. Streamlines for the von Karman vortex street in a fluid asymptotically at rest with $b/a = 0.28$. The whole pattern blows itself horizontally at constant speed.

Evaluating the velocity field of the upper chain on the positions $z = na - ib$ of the vortices of the lower chain, we obtain the velocity of the lower chain,

$$\boxed{W = \frac{C\pi}{2a} \tanh \frac{\pi b}{a}.} \tag{26.21}$$

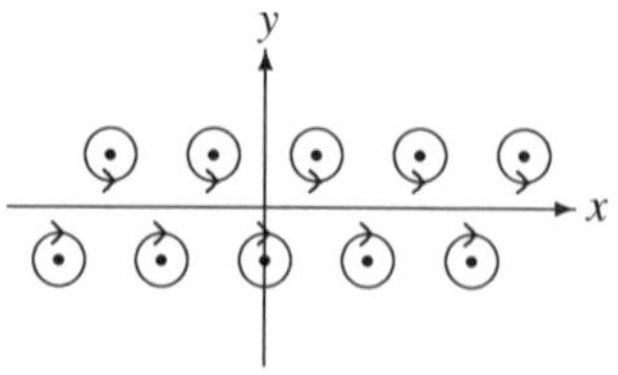

Modelling the von Karman vortex street as two vortex chains of opposite strength.

This is also the velocity of the upper chain, so that in a fluid at rest the whole vortex system propagates horizontally with this speed. If the object creating the vortex street is moving with velocity U along x through the fluid, the velocity of the vortex street becomes $U - W$ relative to the object. The frequency of vortex shedding from the object becomes,

$$f = \frac{U - W}{2a} = \frac{U}{2a}\left(1 - \frac{\pi C}{2Ua} \tanh \frac{\pi b}{a}\right). \tag{26.22}$$

The expression in parenthesis is in fact independent of the velocity U of the object because the eddies giving rise to the detached vortices have a size comparable b and thus a strength of the order of $C \sim Ub$.

A closer analysis of the equations of motion for the vortex street reveals that it is unstable to infinitesimal perturbations of the vortex positions for all values of the aspect ratio b/a, except one [60,

Figure 26.7. Arctic vortex street captured on June 6, 2001. The 300 km long north–south vortex street is formed downwind from the island Jan Mayen situated 650 km northeast of Iceland. The vortex street is created by the 2.2 km high Beerenberg volcano on the island. Image Credit: NASA/GSFC/LaRC/JPL, MISR Team.

p. 133]

$$\sinh \frac{\pi b}{a} = 1, \tag{26.23}$$

or $b/a = 0.2805\ldots$, close to the observed aspect ratios. This is in fact expected to hold for a large class of similar vortex streets[1].

Vibrations driven by periodic vortex shedding

The vortices periodically shed from a moving object, or a stationary object in a uniform flow, act back on the object with a force of alternating sign. Such a force is capable of driving sustained vibrations in the object as well as in the fluid surrounding it. When the driving frequency is close to a natural vibration frequency of the object, strong vibrations may be resonantly excited, a phenomenon well known from the 'singing' of a taut wire in cross wind.

The *Strouhal number* is a dimensionless measure of the vortex shedding frequency, defined from the velocity U of the object and its effective diameter d,

$$\text{Sr} = \frac{fd}{U}. \tag{26.24}$$

For the von Kàrmàn vortex street with aspect ratio $b/a = 0.28$, $d \approx 2b$ and $C \approx Ub$ we find $\text{Sr} \approx 0.2$ from (26.22), which is a typical value.

Theodore von Karman (1881–1963). *Influential Hungarian-American engineer. Lived from 1930 in the US, and became in 1944 cofounder of the Jet Propulsion Laboratory at the California Institute of Technology. Made major contributions to the understanding of fluid mechanics, aircraft structures, rocket propulsion, and soil erosion. A crater on the Moon bears his name today.*

Example 26.3.1 (Piano wire): A piano wire of diameter $d = 1$ mm in a cross wind of $U = 3 \text{ m s}^{-1}$ has a Reynolds number $\text{Re} \approx 200$. Taking $\text{Sr} \approx 0.2$ the vortex shredding frequency becomes $f = 600$ Hz which will become audible if it is close to the natural frequency of the piano wire.

A spectacular case of large-amplitude wind-driven vibrations caused the collapse of the Tacoma Narrows bridge (Puget Sound, Washington, USA) on November 7, 1940. Although the vibrations were originally attributed to resonant vortex shedding, it was later realized that this could not be the case. The collapse happened when the amplitude of a torsional mode of the bridge grew beyond the limit of the structural tolerance (see figure 26.8). On the day of the collapse the wind speed was $U \approx 20 \text{ m s}^{-1}$. With a span thickness of $d \approx 4$ m and a Strouhal number 0.2, this leads to a vortex shedding period of about $1/f = 1$ s, which is too far from the observed 5 s period of the torsional mode to be involved in the collapse. The mechanism actually driving the amplitude of the torsional oscillation towards collapse was later shown to be a kind of aerodynamic flutter[2].

Vincez Štrouhal (1850–1922). *Czech mathematical physicist.*

[1] J. Jimenez, On the linear stability of the inviscid Kàrmàn vortex street, *J. Fluid. Mech.* **178**, (1987) 177.

[2] A fairly recent account of the underlying physics may be found in K. Y. Billah and R. H. Scanlan, Resonance,

Figure 26.8. Tacoma Narrows bridge in torsional oscillation. Image Credit: University of Washington Libraries, Special Collections, PH Coll 290-31.

26.4 Steady vortex sustained by secondary flow

Intuitively, it seems as if the natural tendency for the free vortex core to expand under the influence of viscosity could be counteracted by a sufficiently strong steady radial inflow, v_r. Since the radial inflow of fluid cannot accumulate at the centre of the vortex, there must also be a steady axial outflow, v_z.

The basic mechanism shaping a steady vortex in the presence of radial inflow is conservation of angular momentum. A fluid particle of constant mass dM moving radially inwards at sufficiently high speed will conserve its angular momentum $d\mathcal{L}_z = rv_\phi \, dM$, such that it arrives at a radial distance r with a velocity $v_\phi \sim 1/r$. Provided the radial inflow is sufficiently strong, it will be able to maintain a constant azimuthal flow in the general shape of a line vortex, $v_\phi \sim 1/r$, at least far from the core where the influence of viscosity is small.

Vortex equations

In the presence of secondary flow, the azimuthal equation for steady flow contains non-vanishing advective terms on the left-hand side. We shall again assume that the azimuthal velocity, $v_\phi = v_\phi(r)$, only depends on r, but that $v_r = v_r(r, z)$ and $v_z = v_z(r, z)$ in principle may also depend on z. Using the by now familiar methods of appendix B it is seen that the advective part of the azimuthal equation only has two non-vanishing terms,

$$\boldsymbol{e}_\phi \cdot (\boldsymbol{v} \cdot \nabla)\boldsymbol{v} = v_r \nabla_r v_\phi + \frac{v_\phi}{r} v_r = \frac{v_r}{r} \frac{d(rv_\phi)}{dr},$$

such that the azimuthal equation may now be written,

$$\boxed{\frac{v_r}{r} \frac{d(rv_\phi)}{dr} = \nu \frac{d}{dr}\left(\frac{1}{r} \frac{d(rv_\phi)}{dr}\right).} \tag{26.25}$$

Since v_ϕ only depends on r, it follows immediately from this equation that the radial flow must also be independent of z, i.e. $v_r = v_r(r)$.

Tacoma Narrows bridge failure, and undergraduate physics textbooks, *Am. J. Phys.* **59**, (1991) 118. Photographs and a dramatic film clip of the collapse are readily available at many sites on the internet.

The relation between radial and axial flow is given by the equation of continuity, which in cylindrical coordinates takes the form

$$\frac{1}{r}\frac{\partial(r v_r)}{\partial r} + \frac{\partial v_z}{\partial z} = 0. \tag{26.26}$$

Solving for v_z we obtain,

$$\boxed{v_z(r, z) = w(r) - \frac{z}{r}\frac{d(r v_r(r))}{dr},} \tag{26.27}$$

where $w(r)$ is an arbitrary function which specifies the axial flow at $z = 0$. The second term represents the accumulated radial inflow.

The Burgers vortex

The vortex equation (26.25) allows us to find the radial inflow necessary to maintain *any* azimuthal flow $v_\phi(r)$. Inserting your favorite vortex $v_\phi(r)$ into (26.25), you immediately find the radial flow $v_r(r)$, and afterwards you can calculate the axial flow v_z from (26.27), given your favourite choice of $w(r)$.

The complete flow field which maintains the interpolating vortex (26.5) is (for $w = 0$),

$$v_\phi = \frac{\Omega c^2}{r}(1 - e^{-r^2/c^2}), \tag{26.28a}$$

$$v_r = -\frac{2\nu}{c^2} r, \tag{26.28b}$$

$$v_z = \frac{4\nu}{c^2} z. \tag{26.28c}$$

It is called the *Burgers vortex* [54] and is in fact an exact solution to the full set of steady-flow Navier–Stokes equations (see problem 26.7). The pressure is

$$p = -\frac{1}{2}\rho_0 \Omega^2 c^2 F\left(\frac{r^2}{c^2}\right) - \frac{2\nu^2}{c^4}\left(r^2 + 4z^2\right) \tag{26.29}$$

with $F(\xi)$ given in (26.7).

The scale of the secondary flow is set by the inverse of the viscous core expansion time, $4\nu/c^2$, but that is not particularly surprising, since the purpose of the secondary flow is precisely to counteract core expansion.

Johannes Martinus Burgers (1895–1981). *Dutch physicist. Worked on turbulence, vortex theory, sedimentation, gas dynamics, shock waves and plasma physics. The Burgers equation, the Burgers vortex and the Burgers vector are today standard terms in the physics vocabulary [54].*

Vortex with localized axial jet

In the Burgers vortex, the axial flow v_z is independent of r, and that makes it very different from naturally born vortices, such as the bathtub vortex, where the axial downflow must converge upon a narrow drain hole, or the tornado where the upflow primarily takes place inside a narrow funnel. Here we shall consider the extreme case of a vortex with a localized *axial jet* at $r = 0$. This vortex may be viewed as a generalization of the line vortex (26.12) to include a steady inflow.

Demanding that $v_z = 0$ outside the vortex core $r > 0$, it follows from (26.27) that $w = 0$ and $d(r v_r)/dr = 0$, such that the radial inflow must be of the form[3],

$$v_r = -\frac{q}{r}, \tag{26.30}$$

where q is a positive number. The total flux of fluid coming in through a stretch of the vortex of length L is $Q = 2\pi r L(-v_r) = 2\pi L q$.

Solving the vortex equation (26.25) with this radial inflow, we obtain

$$v_\phi = \frac{C}{r} + A\, r^{1-2\alpha}, \tag{26.31}$$

[3] A number of exact and approximative solutions of this general family are analysed in V. Shtern, A. Borissov and F. Hussain, Vortex sinks with axial flow: Solution and applications, *Phys. Fluids* **9**, (1997) 2941.

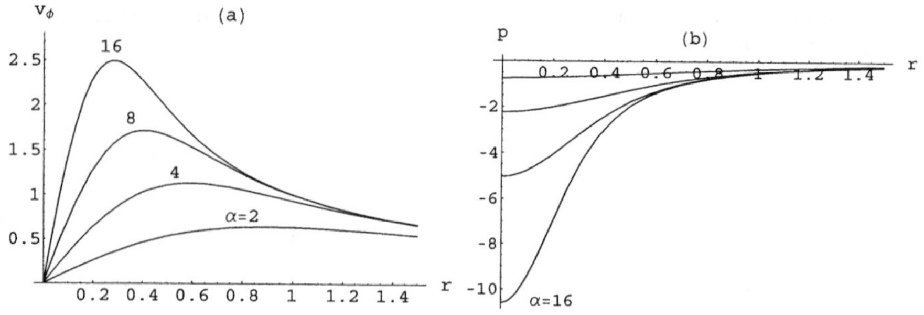

Figure 26.9. Vortex with extended axial jet (26.33) for $\alpha = 2, 4, 8, 16$. The parameters are $a = 1$, $\nu = 0.01$ and $C = 1$. **(a)** Azimuthal flow v_ϕ. **(b)** Pressure p, normalized to vanish at infinity.

where C and A are constants, and

$$\alpha = \frac{q}{2\nu}, \tag{26.32}$$

is a dimensionless measure of the inflow, related to the radial Reynolds number, $\mathrm{Re}_r = r\,|v_r|/\nu = q/\nu = 2\alpha$.

The second term in (26.31) decays faster than the first at infinity only for $\alpha > 1$, and this confirms that the radial inflow must be larger than a certain minimum to maintain a flow in the shape of a line vortex at large distances. In the following we assume this to be the case.

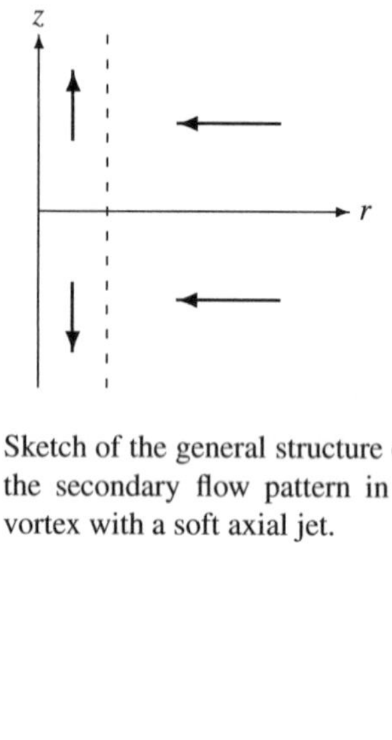

Sketch of the general structure of the secondary flow pattern in a vortex with a soft axial jet.

Extended axial jet

A vortex with an extended axial jet, a 'drain pipe', may be created by choosing a radial flow that interpolates smoothly between the singular jet and the Burgers vortex, for example,

$$v_r = -\frac{q}{r}(1 - e^{-r^2/a^2}). \tag{26.33}$$

The axial flow now follows from (26.27) with $w = 0$,

$$v_z = z\,\frac{2q}{a^2}\exp\left(-\frac{r^2}{a^2}\right), \tag{26.34}$$

which vanishes rapidly for $r > a$, as it should. The azimuthal field and the pressure can only be calculated numerically and is pictured in figure 26.9 for a selection of values of α.

There is, as usual, a price to be paid for the construction. Although the solution interpolates perfectly between two exact steady-flow solutions to the Navier–Stokes equation, it is itself not an exact solution in the transition region near the edge of the 'drain pipe' at $r = a$.

26.5 Advective spin-up of a vortex

It is a common observation that a vortex can be 'spun up' by draining fluid from its central region at a steady rate. We shall now analyse this process for a vortex with a localized axial jet and a steady radial inflow and no vertical flow for $r > 0$,

$$v_r = -\frac{q}{r}, \qquad\qquad v_z = 0. \tag{26.35}$$

Such a flow may be approximately realized in a bathtub far from the drain hole, $r \gg a$, when water is continually being resupplied to keep a constant asymptotic water level L. The total volume influx is then $Q = 2\pi L q$ which must equal the flux through the drain hole. In the following section we shall study what happens to the bathtub vortex when the surface shape near the drain hole is taken into account.

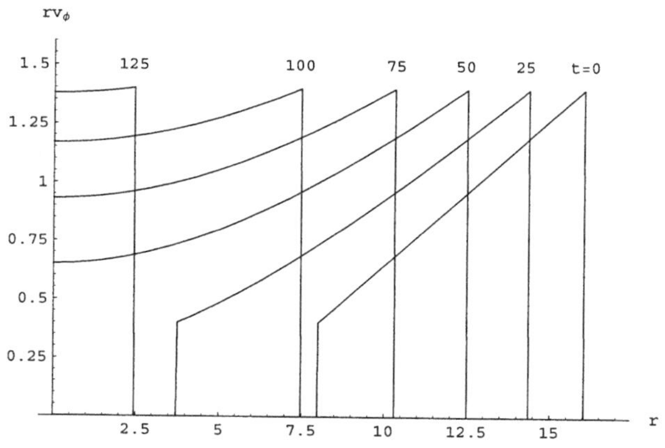

Figure 26.10. 'Movie' of advective spin-up of a vortex with a singular axial jet. The specific angular momentum rv_ϕ is plotted as a function of radial distance for selected times (shown above the curves), starting with $t = 0$. The initial shape is progressively flattened and eventually, for $t = 128$, it has completely gone down the drain at the left.

Inflow of angular momentum

Assuming that the radial Reynolds number is large $\mathrm{Re}_r = q/\nu \gg 1$, the time-dependent Euler equation for the azimuthal velocity field becomes

$$\frac{\partial v_\phi}{\partial t} + \frac{v_r}{r}\frac{\partial (r v_\phi)}{\partial r} = 0. \tag{26.36}$$

With the given radial inflow (26.35) the most general solution is,

$$v_\phi(r,t) = \frac{C\left(\sqrt{r^2 + 2qt}\right)}{r}, \tag{26.37}$$

where $C(\cdot)$ is an arbitrary function, representing the initial distribution of specific angular momentum $C(r) = rv_\phi(r, 0)$ well away from the drain, $r \gg a$.

At any sufficiently late time $t \gg a^2/2q$, the specific angular momentum is nearly independent of r in the central region $a \ll r \ll \sqrt{2q\,t}$, where the velocity field to a good approximation is that of a line vortex, $v_\phi \approx C\left(\sqrt{2qt}\right)/r$. As time goes by, the line vortex shape thus spreads out from the central region of the vortex, albeit with a time-dependent circulation 'constant' $C\left(\sqrt{2qt}\right)$, and the initial velocity profile is probed to farther and farther distances. The size of the region of line vortex shape grows as $\sqrt{2qt}$, reminiscent of a diffusive process although it has nothing to do with that.

A steady vortex is never reached unless the initial specific angular momentum distribution $C(r)$ approaches a constant for $r \to \infty$. These arguments show that it is in fact impossible to spin-up a truly steady vortex by means of steady radial inflow, for the simple reason that it requires infinite angular momentum to be present in the flow to begin with. If the angular momentum is initially finite, all of it will sooner or later go down the drain in a non-rotating container. In a rotating container angular momentum may be continually supplied from the moment of force exerted on the fluid by the container walls, resulting eventually in a steady flow. This case was discussed in some detail in section 20.5.

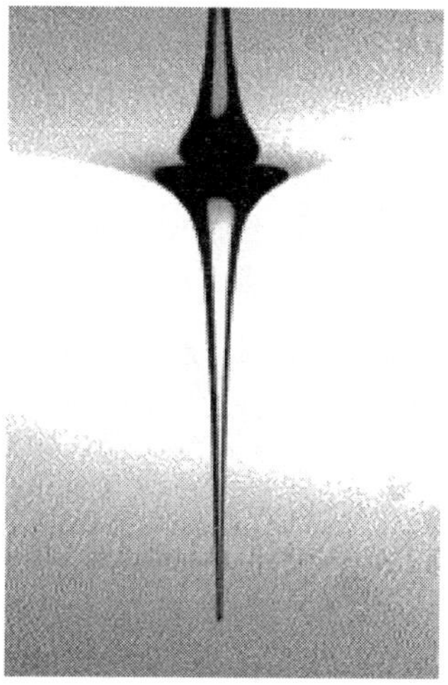

Needle-like bathtub vortex created in the laboratory (see footnote 1 on page 287).

Bathtub vortex: The process of getting out of a real bathtub cannot avoid leaving a certain amount of angular momentum in the water. It is this angular momentum that the inflow picks up and sends down the drain as a spinning vortex while the bathtub is being emptied (see problem 26.12). A real bathtub vortex may change speed and even stop-up and reverse its sense of rotation depending on the details of how you got out of the water. In a bathtub of ordinary size the incidental initial angular momentum distribution is normally much greater than that provided by the slowly rotating Earth. The question of the Earth's influence on the sense of rotation of a real bathtub vortex was discussed in section 20.6.

∗ 26.6 Bathtub-like vortices

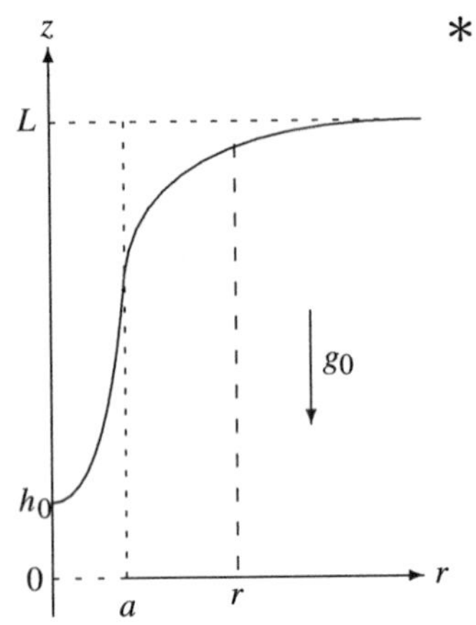

Bathtub-like liquid vortex with open surface. At $z = 0$ there is a drain-hole of radius a, and the 'tub' has essentially infinite extension in the radial direction. There is a central depression of height h_0 and the asymptotic liquid level is $z = L$.

A bathtub vortex is an isolated liquid vortex with an open surface, sustained by radial inflow and powered by gravity[4]. The most conspicuous feature of such a vortex is the central depression which may even penetrate the drain and make audible sounds. In the laboratory such vortices can be created in rotating containers (see section 20.5), but in this section we shall ignore the complication of container rotation and simply assume that the primary flow in the vortex is infinitely extended with the shape of a line vortex, $rv_\phi \to C$, at large distances. This guarantees an unlimited supply of mass and angular momentum, such that the vortex may be truly steady and not disappear down the drain as in real bathtubs.

Bathtub equations

In a flat-Earth coordinate system with gravity directed towards negative z, the vortex is drained through a circular region, a 'drain hole' of radius $r = a$ situated at $z = 0$. The open liquid surface of the vortex is assumed to be rotationally invariant, $z = h(r)$, and we shall again assume that the primary flow is cylindrical, $v_\phi = v_\phi(r)$, at least well above the bottom of the container. It follows as before that the azimuthal flow must obey the azimuthal equation (26.25), implying that the radial flow is cylindrical, $v_r = v_r(r)$, and the axial flow is given by (26.27) with $w(r)$ being the drain flow. Besides these equations there are two boundary conditions on the open surface, one kinematic and one dynamic.

The open liquid surface with a depression at the centre forces the radial inflow into a region of smaller height and thereby speeds it up in comparison with flow in the axial jet vortex. Consequently, there must exist a relation between the height of the surface, $h(r)$, and the radial inflow $v_r(r)$. The quantitative form of this relation is derived from the fact that the streamlines must follow the surface,

$$v_z(r, h(r)) = v_r(r)\frac{dh(r)}{dr}. \tag{26.38}$$

Using (26.27) with $z = h$ this c relation may be written,

$$\boxed{\frac{1}{r}\frac{d(rv_r h)}{dr} = w.} \tag{26.39}$$

This *kinematic condition* is exact in the cylindrical approximation used here.

For 'sufficiently small' secondary flow the pressure gradient will be the only term in the Navier–Stokes equation that can balance gravity and deliver the centripetal force necessary for the circulating flow,

$$\frac{\partial p}{\partial z} = -\rho_0 g_0, \qquad \frac{\partial p}{\partial r} = \rho_0\frac{v_\phi^2}{r}. \tag{26.40}$$

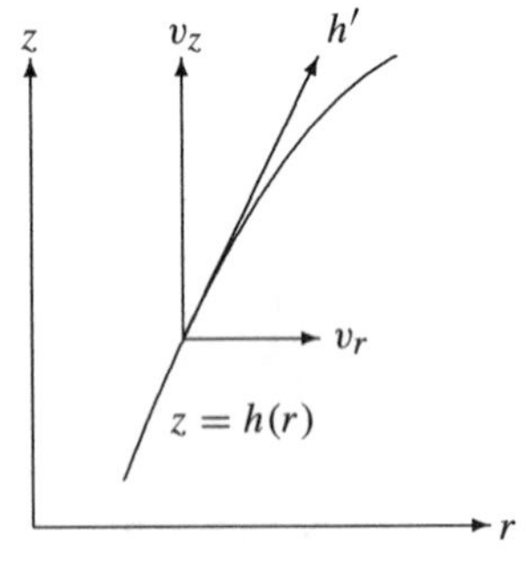

The streamlines must have the same slope $h' = dh/dr$ as the free surface.

On the right-hand sides of both of these equations we have dropped all terms that depend on the radial and axial flow. Given that the pressure is constant $p = p_0$ at the open surface $z = h(r)$, it follows from the first of these equations that the hydrostatic pressure takes the form, $p(r, z) = p_0 + \rho_0 g_0(h(r) - z)$. This expression can only be valid well above the drain hole where the secondary velocities are small. Inserting this into the second equation we find,

$$\boxed{\frac{dh}{dr} = \frac{v_\phi^2}{g_0 r}.} \tag{26.41}$$

A simple geometrical construction reveals the meaning of this *dynamic condition*: the tangential component of gravity balances the tangential component of the centrifugal force acting on a fluid particle sitting on the rotating surface.

Choice of drain flow

The three bathtub equations (26.25), (26.39) and (26.41) connect the four fields, v_ϕ, v_r, w and h. Although the Navier–Stokes equation in principle could also provide a fourth equation (see problem 26.7), it is in view

[4]The analysis in this section is inspired by T. S. Lundgren, The vortical flow above the drain-hole in a rotating vessel, *J. Fluid Mech.* **155**, (1985) 381.

of the many approximations most convenient to impose a reasonable choice for the drain flow $w(r)$, and then solve the bathtub equations for the remaining three fields. This has the advantage that the kinematic condition (26.39) can be integrated,

$$r v_r h = \int_0^r s\, w(s)\, ds. \tag{26.42}$$

Here we have used that v_r and h must both be finite for $r = 0$.

To simulate a drain hole of radius a we shall choose a 'soft plug flow'

$$\boxed{w = -We^{-r^2/a^2},} \tag{26.43}$$

where W is the outflow velocity at the centre of the drain. Since the total flux through the drain is

$$Q = -\int_0^\infty w(r)\, 2\pi r dr = \pi a^2 W, \tag{26.44}$$

the average drain velocity is also W. Integrating (26.42) we find,

$$\boxed{r v_r h = -\tfrac{1}{2} W a^2 (1 - e^{-r^2/a^2}).} \tag{26.45}$$

In a rotating frame it will also be necessary to include an upflow from the Ekman layer at the bottom (see section 20.5).

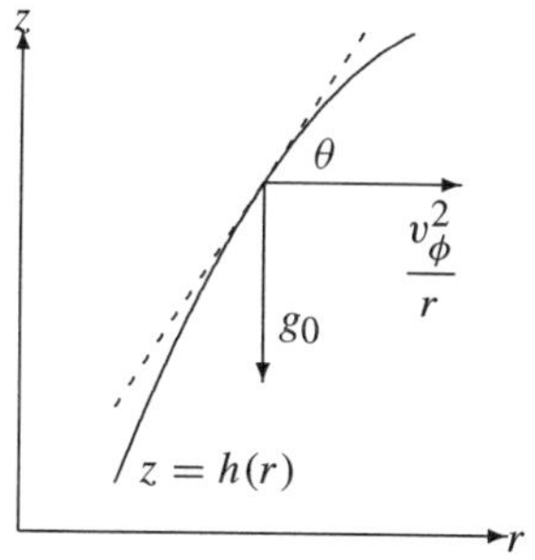

A fluid particle sitting on the rotating surface of the vortex is subject to gravity and centrifugal force. The projections of these forces on the surface tangent (dashed) cancel in the leading approximation, $g_0 \sin\theta = (v_\phi^2/r)\cos\theta$, where $\tan\theta = dh/dr$.

Estimating the depth of the central depression

The core of the bathtub vortex is assumed to rotate as a solid body with angular velocity Ω, whereas far from the core the flow is assumed to be that of a line vortex with circulation constant C, approximately a Rankine vortex,

$$v_\phi \approx \begin{cases} \Omega r & r \ll c \\ \dfrac{C}{r} & r \gg c \end{cases}. \tag{26.46}$$

The core radius c is estimated by matching these expressions,

$$c \approx \sqrt{\frac{C}{\Omega}}, \tag{26.47}$$

and is assumed to lie inside the drain radius, $c < a$.

From (26.41) it then follows that

$$h \approx \begin{cases} h_0 + \dfrac{\Omega^2}{2g_0} r^2 & r \ll c \\ L - \dfrac{C^2}{2g_0} \dfrac{1}{r^2} & r \gg c \end{cases}. \tag{26.48}$$

At the match point $r = c$ we obtain vortex height at the core radius in two ways

$$h_c \approx h_0 + \frac{\Omega^2 c^2}{2g_0} \approx L - \frac{C^2}{2g_0\, c^2},$$

and using the core radius estimate (26.47) we find the depth of the depression and the height at the core radius,

$$L - h_0 \approx \frac{C^2}{g_0 c^2}, \qquad h_c \approx \frac{L + h_0}{2}. \tag{26.49}$$

This makes the estimate of the vortex shape (26.48) as well as its derivative continuous at the match point.

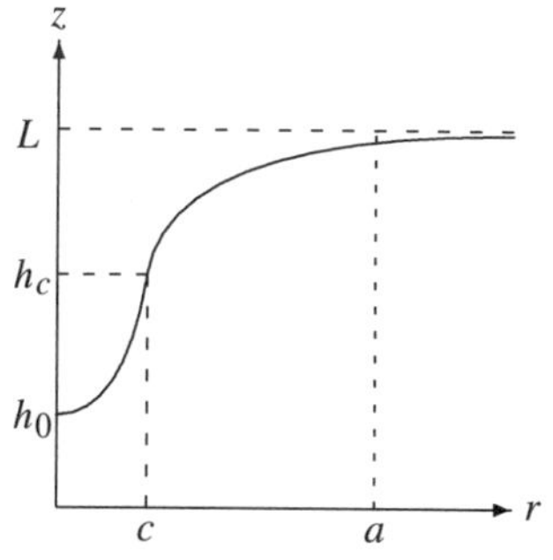

Estimated shape of the bathtub vortex.

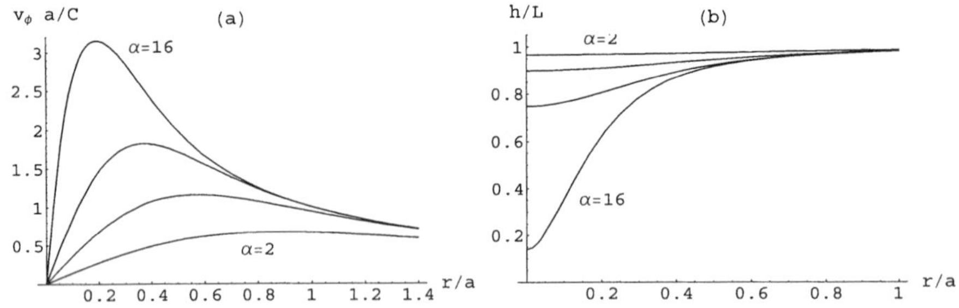

Figure 26.11. Bathtub vortex for $\beta = 0.0217$ and $\alpha = 2, 4, 8, 16$. The core radius corresponds roughly to the top in v_ϕ and shrinks with growing secondary flow (α) whereas the depression deepens. **(a)** Azimuthal velocity. **(b)** Vortex surface height.

Knowing h we may obtain v_r from (26.45) and insert it into (26.25). Quite generally the transition between the core and the outer vortex happens when $r v_r \approx -2\nu$, as was the case for the Burgers vortex (26.28). Taking $r = c$ in (26.45) and expanding the exponential to lowest order, we find

$$c^2 \approx \frac{4\nu h_c}{W} \approx \frac{2\nu(L + h_0)}{W}. \tag{26.50}$$

Inserting this result in the expression for the depth of the depression, we get

$$L - h_0 \approx \frac{C^2}{g_0} \frac{W}{2\nu(L + h_0)}, \tag{26.51}$$

and solving for h_0, we finally arrive at,

$$\boxed{h_0 \approx \sqrt{L^2 - \frac{C^2 W}{2 g_0 \nu}}.} \tag{26.52}$$

If the argument of the square root becomes negative, there is no solution because the central dip in the surface has plunged right through the drain, thereby violating the basic assumptions behind the model (see also problem 26.16).

Example 26.6.1 (Bathtub vortex): A bathtub is filled with water to a height $L = 50$ cm. On exiting, the bather accidentally imparts a small rotation to the water, corresponding to an average specific angular momentum of $C \approx 20\ \text{cm}^2\ \text{s}^{-1}$. When the plug of radius $a = 2$ cm is pulled, the water is found to drain at a rate of $Q \approx 1$ litre per second. A needle-like vortex forms which according to the above expression is $L - h_0 \approx 20$ cm deep. The core diameter becomes $c \approx 3$ mm and the core is estimated to rotate about 160 turns per second!

Numeric integration

To solve the coupled equations numerically, it is most convenient to use a non-dimensional formalism. Let us define the dimensionless spatial variable,

$$\xi = \frac{r^2}{a^2}, \tag{26.53}$$

and the dimensionless height and azimuthal variables,

$$h = L g(\xi), \qquad r v_\phi = C u(\xi), \tag{26.54}$$

where L is the asymptotic level of liquid and C the asymptotic circulation constant. Eliminating v_r by means of (26.45) we obtain from the azimuthal vortex equation (26.25) and the centrifugal condition (26.41) the coupled differential equations,

$$u'' = -\alpha\, u' \frac{1 - e^{-\xi}}{\xi\, g}, \qquad g' = \beta \frac{u^2}{\xi^2}, \tag{26.55}$$

where

$$\alpha = \frac{Wa^2}{4L\nu}, \qquad \beta = \frac{C^2}{2g_0 La^2}, \tag{26.56}$$

are dimensionless constants. Although the vortex is defined by four parameters, W, C, a and L, its dynamics are controlled by only two dimensionless parameters. Whereas α is related to the radial Reynolds number, $\mathrm{Re}_r = 2\alpha$, the constant β is related to the Froude number in the azimuthal flow at the edge of the drain.

The coupled differential equations (26.55) must be solved with the boundary conditions $u = 0$ for $\xi = 0$ and $g, u \to 1$ for $\xi \to \infty$. The results are shown in figure 26.11 for a particular choice of β and a selection of radial Reynolds numbers. In its gross features the shape is virtually indistinguishable from the vortex with extended axial jet in figure 26.9. The condition for the vortex not to plunge through the drain in (26.52) becomes $\alpha\beta \lesssim 0.25$, but the numerical calculation indicates that it is actually $\alpha\beta \lesssim 0.36$ (see problem 26.16).

Problems

26.1 **(a)** Calculate the angular momentum and kinetic energy of the core of the Rankine vortex per unit of axial length. **(b)** Calculate the same quantities outside the core for $c < r < R$.

26.2 **(a)** Find the surface shape of a liquid Rankine vortex in constant gravity. **(b)** What is the depth of the depression? **(c)** Calculate the depth for a Rankine vortex with core radius $c = 1$ cm, rotating 10 times a second.

26.3 Calculate the real velocity fields v_x and v_y of the counter- and corotating vortex pairs, (26.16) and (26.17).

* **26.4** The streamline plots in figure 26.4 are made as contour plots of the stream function ψ, defined such that

$$v_x = \nabla_y \psi, \qquad v_y = -\nabla_x \psi. \tag{26.57}$$

Find the stream function for counter- and corotating vortex pairs of equal strength (use the answer to problem 26.3).

26.5 Consider a collection of parallel vortices enumerated by n with constant strengths C_n and the cores moving along individual orbits $z = z_n(t)$. **(a)** Show that the equations of motion of the centers are,

$$\frac{dz_n^\times(t)}{dt} = -i \sum_{m \neq n} \frac{C_m}{z_n(t) - z_m(t)}, \tag{26.58}$$

where $^\times$ stands for complex conjugation. **(b)** Verify that the orbits of a pair of vortices of opposite strength satisfy these equations of motion. **(c)** The same for a pair of the same strength.

* **26.6** Define the function

$$f_N(z) = \cot z - \sum_{n=-N}^{N} \frac{1}{z - \pi n} \tag{26.59}$$

and show that $f(z) = \lim_{N\to\infty} f_N(z)$ vanishes everywhere in the complex plane. Hints: **(a)** Show that f is antisymmetric, $f(-z) = -f(z)$. **(b)** Show that $f(z)$ is periodic with period $\Delta x = \pi$. **(c)** Show that $f(z)$ is holomorphic in the strip $-\pi/2 < x < \pi/2$. **(d)** Show that $f(z)$ vanishes at the boundaries of the strip. Finally, a general theorem of complex analysis states that a function which is holomorphic in a region and vanishes on the boundaries vanishes identically. By periodicity it also vanishes in all other strips.

$*$ **26.7** Assume that the velocity fields $v_{r,\phi,z}$ in cylindrical coordinates only depend on r and z. Show that the Navier–Stokes equations for incompressible flow with $\rho_0 = 1$ and no gravity become

$$\frac{\partial v_\phi}{\partial t} + v_r \frac{\partial v_\phi}{\partial r} + \frac{v_r v_\phi}{r} + v_z \frac{\partial v_\phi}{\partial z} = \nu \left(\frac{\partial^2 v_\phi}{\partial r^2} + \frac{1}{r}\frac{\partial v_\phi}{\partial r} - \frac{v_\phi}{r^2} + \frac{\partial^2 v_\phi}{\partial z^2} \right), \tag{26.60a}$$

$$\frac{\partial v_r}{\partial t} + v_r \frac{\partial v_r}{\partial r} - \frac{v_\phi^2}{r} + v_z \frac{\partial v_r}{\partial z} = \nu \left(\frac{\partial^2 v_r}{\partial r^2} + \frac{1}{r}\frac{\partial v_r}{\partial r} - \frac{v_r}{r^2} + \frac{\partial^2 v_r}{\partial z^2} \right) - \frac{\partial p}{\partial r}, \tag{26.60b}$$

$$\frac{\partial v_z}{\partial t} + v_r \frac{\partial v_z}{\partial r} + v_z \frac{\partial v_z}{\partial z} = \nu \left(\frac{\partial^2 v_z}{\partial r^2} + \frac{1}{r}\frac{\partial v_z}{\partial r} + \frac{\partial^2 v_z}{\partial z^2} \right) - \frac{\partial p}{\partial z}, \tag{26.60c}$$

$$\frac{\partial v_r}{\partial r} + \frac{v_r}{r} + \frac{\partial v_z}{\partial z} = 0. \tag{26.60d}$$

Verify that the Burgers vortex (26.28) and the vortex with a localized axial jet (page 385) are exact solutions.

26.8 Calculate the streamlines for the Burgers vortex.

26.9 **(a)** Show that the so-called Taylor vortex

$$v_\phi(r, t) = \tau \frac{r}{t^2} e^{-r^2/4\nu t}, \tag{26.61}$$

is a solution to (26.9) where τ is a constant with dimension of time. **(b)** Calculate the total angular momentum per unit of axial length of the Taylor vortex and show that it is constant in time.

26.10 Assume that a free cylindrical vortex only depends on the dimensionless variable

$$\xi = \frac{r^2}{4\nu t} \tag{26.62}$$

and assume that

$$v_\phi(r, t) = \frac{F(t)}{r} f(\xi), \tag{26.63}$$

where $F(t)$ is a time-dependent factor.

(a) Find a differential equation for $f(\xi)$.

(b) Show that $F(t) \sim t^{-\alpha}$ where α is a dimensionless parameter.

(c) Show that the solution is

$$f_\alpha(\xi) = \sum_{n=0}^{\infty} \frac{(n+\alpha)!}{n!\,\alpha!} (-1)^n \frac{\xi^{n+1}}{(n+1)!}. \tag{26.64}$$

(d) Find a closed form for $\alpha = 0, 1, 2$ and draw the vortex shapes.

26.11 Assume that a vortex at $t = 0$ has the shape of an interpolating vortex (26.5) with radius b, and that there is a radial inflow identical to that of the Burgers vortex (26.28) corresponding to a radius a. Determine how the azimuthal flow of the original vortex converges upon the interpolating function of radius a.

26.12 Assume that the water in a bathtub originally rotates like a solid body, $v_\phi = \Omega r$, far from the drain. Determine the far field at a later time and the rate at which angular momentum flows in towards the drain.

26.13 Show that the vortex equation (26.25) may be solved explicitly by quadrature when $v_r(r)$ is given.

26.14 Match the Burgers vortex and the singular vortex directly at $r = a$ and determine the fields as a function of $\alpha = q/2\nu$ when the azimuthal field and its derivative must be continuous.

26.15 **(a)** Show that the divergence condition (26.60d) can be solved by means of a stream function $\psi(r, z)$ satisfying

$$\frac{\partial\psi}{\partial z} = rv_r, \qquad \frac{\partial\psi}{\partial r} = -rv_z. \tag{26.65}$$

(b) Show that the streamlines satisfy

$$\psi(r, z) = \text{const.} \tag{26.66}$$

(c) Show that the stream function for a cylindrically invariant vortex is

$$\psi(r, z) = rv_r(r)z - \int rw(r)dr. \tag{26.67}$$

26.16 Show that the depth of the depression is

$$L - h_0 = \int_0^\infty \frac{v_\phi^2}{g_0 r}\, dr = k\frac{C^2}{g_0 c^2} \tag{26.68}$$

and determine a value for k from the interpolating vortex. How does this change the estimate (26.52) for the central height of the vortex?

27

Lubrication

The most important technological invention of all time must be the wheel and its bearing. From the earliest times it was realized that friction in the bearing was considerably lowered by lubricating it with viscous fluid. Wooden bearings might even catch fire if not lubricated. Fat from pigs, olive oil and mineral oil turned out to work much better than water.

When you deal a pack of cards on a table with a smooth surface, it is easy to overestimate the speed the cards must be thrown with. Suddenly, several cards in a row slide easily over the surface and land on the floor. The reason is that a lubricating layer of air has formed between the card and the surface of the table, and has caused the friction you expected when throwing the card to drop nearly away. Air also lubricates the tiny gap between the magnetic pickup heads and the rapidly spinning hard-disk in your computer and prevents the heads from crashing into the surface. Water is used to lubricate children's slides in amusement parks, and sports like ice skating or curling depend crucially on a thin lubricating film of water.

In the fluid-filled gap between a moving object and a nearby solid wall, viscosity plays a dominant role because the velocity gradients normal to the surface grow large compared to the gradients parallel to the surface. When the gap widens towards the front of the moving object, as is normally the case, viscous friction will drag fluid into the gap, creating a pressure that can become surprisingly high. As the gap narrows, this pressure will in the end become sufficient to keep the object afloat in the lubricant, thereby securing a smooth ride.

In this chapter we shall analyse incompressible flow in narrow gaps in the creeping flow approximation, first making general estimates and afterwards solving the equations for creeping flow along the lines laid out by Reynolds already in 1886.

27.1 Physics of lubrication

Far from any boundaries, the flow around an object of size L moving with velocity U through an otherwise stationary fluid of density ρ and viscosity η is characterized by the Reynolds number

$$\mathrm{Re} \approx \frac{\rho U L}{\eta}. \tag{27.1}$$

In many everyday situations—running, swimming, driving, washing dishes or babies—the Reynolds number is very large, $\mathrm{Re} \gg 1$, and the flow may be considered nearly ideal (see chapter 16).

The situation is completely different when the body moves with velocity U close to a stationary solid boundary. If the gap width is of size $d \ll L$, the no-slip condition forces the flow velocity to change rapidly from 0 to U over this distance, and the Reynolds number in the gap becomes,

$$\mathrm{Re}_{\mathrm{gap}} \approx \frac{|\rho(\boldsymbol{v}\cdot\boldsymbol{\nabla})\boldsymbol{v}|}{|\eta\nabla^2\boldsymbol{v}|} \approx \frac{\rho U^2/L}{\eta U/d^2} = \left(\frac{d}{L}\right)^2 \mathrm{Re}, \tag{27.2}$$

because the advective acceleration in the numerator is dominated by the flow variation along the gap, whereas the Laplacian in the denominator is dominated by the flow variation across the gap.

Even if the free-flow Reynolds number (27.1) is large, viscous forces will dominate the flow in the gap when the distance d becomes so small that $\mathrm{Re_{gap}} \ll 1$, or

$$d \ll \delta = \frac{L}{\sqrt{\mathrm{Re}}}. \tag{27.3}$$

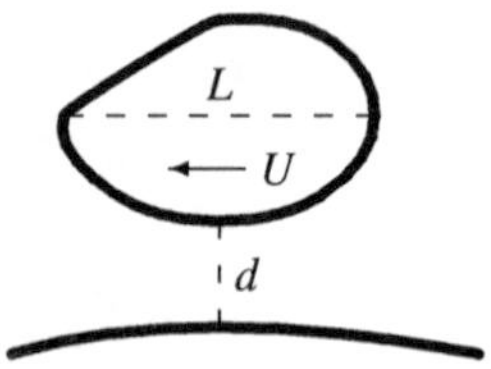

Object moving close to a nearly flat wall. The gap is assumed to be so narrow that creeping flow conditions prevail.

In chapter 28 we shall see that δ is a measure of the thickness of the boundary layer surrounding a moving body. The conditions for creeping flow will thus be fulfilled when the gap lies well inside the boundary layer. In the following we shall assume this to be the case.

Example 27.1.1 (Playing card): A typical playing card has size 7 cm $\times$ 10 cm, and since we do not know how it moves we shall take $L \approx 8$ cm. Skimming through the air above a horizontal table at $U \approx 1\ \mathrm{m\,s^{-1}}$, it has $\mathrm{Re} \approx 5 \times 10^3$ and thus $\delta \approx 1$ mm. The condition for creeping flow is fulfilled when the distance to the table is somewhat less than δ, say $d = 0.3$ mm.

When the solid boundary is (nearly) flat, the flow will be steady in the rest frame of the body where the wall moves with velocity U. We may then use the equations for steady incompressible creeping flow (19.1) without gravity

$$\nabla p = \eta \nabla^2 \boldsymbol{v}, \qquad \nabla \cdot \boldsymbol{v} = 0. \tag{27.4}$$

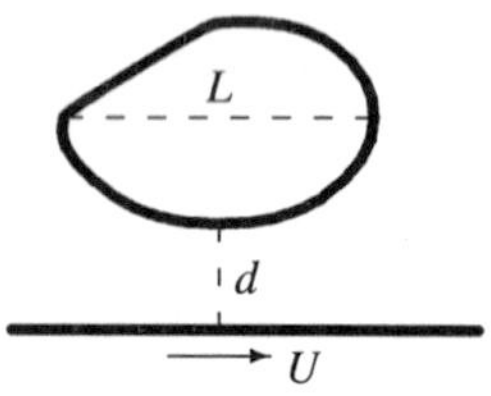

In the reference frame where the boundary moves with velocity U, the object is stationary and the flow is steady.

Contrary to creeping flow far from the boundaries, where lift and drag are of roughly the same magnitude (section 19.1), we shall now see that the flow in a narrow gap will generate a pressure that is much larger than the shear stress. This is in fact the secret behind lubrication: the lift will balance the weight of the body so that it floats in the fluid without generating a drag of comparable magnitude.

Estimate of lift

The magnitude of the pressure in the gap can be estimated from the equation for creeping flow (27.4). Since the double derivative across the gap dominates the Laplacian, we estimate the pressure gradient to be $|\nabla p| \approx \eta\, U/d^2$. Multiplying by the length of the gap L, we obtain the magnitude of the pressure variations along the gap, $|\Delta p| \approx \eta U L/d^2$. Finally multiplying by the gap area A, we get an estimate for the lift,

$$\mathcal{L} \approx f \frac{\eta U L A}{d^2}. \tag{27.5}$$

Here we have put in a dimensionless prefactor f of order unity, which depends on the geometry and orientation of the body. The flow around the upper part of the body will also contribute an aerodynamic lift (see page 446). It is, however, independent of d and will in the limit of $d \to 0$ always be dominated by the lift from the gap.

For simplicity we shall assume that the underside of the body is reasonably flat, such that it makes sense to speak about an *angle of attack* α with the boundary (in section 27.2 we shall deal with the general case of an angle of attack that varies along the body). The sign of this angle is chosen to be positive when the gap widens towards the front of the body, and intuitively one expects in this case a positive lift that drives the body away from the boundary. For vanishing angle of attack, the flow will essentially become planar velocity-driven Couette flow (see page 232), which does not generate a lift. For negative angle of attack the lift is also expected to be negative, causing the body to be sucked towards the boundary rather than pushed away from it.

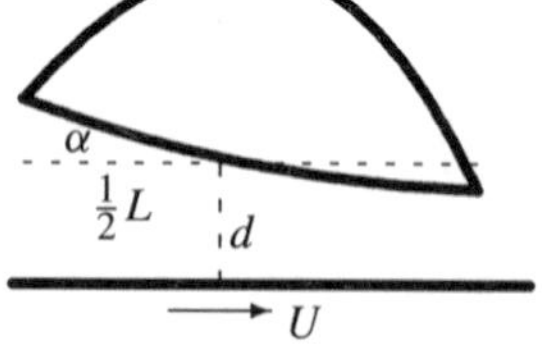

Body with relatively flat underside and a small positive angle of attack α. In this case the lift will be upwards. The rear of the body will touch the ground at the rear if $\alpha L/2 \approx d$ where d is the average width.

These arguments indicate that the prefactor in the leading approximation should be proportional to the angle of attack, $f \sim \alpha$. Taking into account that the magnitude of this angle is limited by the requirement that the ends of the body should not touch the ground, $|\alpha| \lesssim 2d/L$, and since f is at most of order unity we estimate that

$$f \approx \alpha \frac{L}{2d}. \tag{27.6}$$

The calculations in section 27.3 shall confirm that this is a reasonable estimate for small angles of attack, $|\alpha| \lesssim 2d/L$.

Example 27.1.1 (Playing card, continued): If the playing card has mass $M \approx 2$ g, and provided the table surface is smooth, the card will sink into the boundary layer. If the shape factor is taken to be $f \approx 0.4$, the lift equals the weight for $d \approx 0.4$ mm. The corresponding angle of attack becomes a miniscule $\alpha \approx 0.25\,°$. The card probably has to bend slightly upwards at both ends for steady lift to be generated. Professional card players avoid bending their cards and also cover the table with rough green felt cloth. You never see a card skimming the table surface and landing on the floor in their company.

Estimate of skin drag

In the gap the normal velocity gradient is U/d, such that the shear stress becomes of magnitude $\eta U/d$. Multiplying with the area A of the gap, we obtain an estimate for the *skin drag* on the body from fluid friction in the gap,

$$\mathcal{D}_{\text{skin}} \approx \frac{\eta U A}{d}. \tag{27.7}$$

As for lift, there will also be drag from the flow around the body outside the gap, but the drag from the fluid in the gap will always dominate in the limit of $d \to 0$.

Both lift and skin drag grow with decreasing gap size, but the lift grows faster than the drag and eventually comes to dominate it. The ratio of lift to skin drag is estimated to be,

$$\frac{\mathcal{L}}{\mathcal{D}_{\text{skin}}} \approx f\frac{L}{d} \approx \frac{1}{2}\alpha\left(\frac{L}{d}\right)^2. \tag{27.8}$$

The drag can only dominate the lift if the angle of attack is very small, $\alpha \lesssim 2(d/L)^2$, but that will require careful tuning of the angle of attack, which is usually not possible.

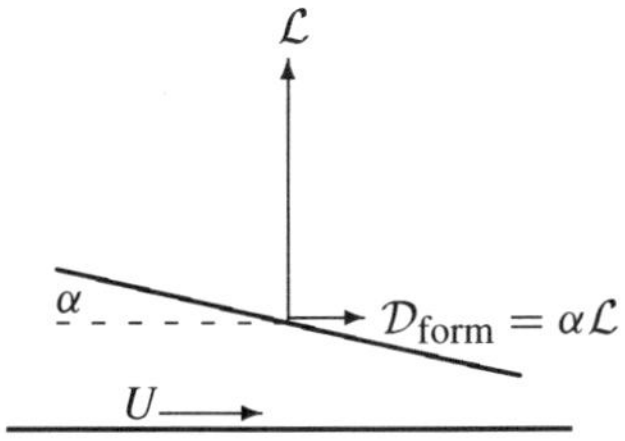

The pressure force acts orthogonally to the surface and gives rise to normal lift as well as tangential form drag.

Example 27.1.1 (Playing card, continued): For the playing card the actual ratio of lift to skin drag is $\mathcal{L}/\mathcal{D}_{\text{skin}} \approx 70$ for $f = 0.4$. Lift would equal skin drag for $f \approx d/L$, and from the balance of lift to weight we get the corresponding flying height $d \approx 6\ \mu$m. No tables are that smooth and the card will touch down long before this level is reached.

Form drag

Besides frictional skin drag there is also a *form drag* from the pressure variations in the gap. Since the pressure force is orthogonal to the body surface, the form drag may be estimated from the angle of attack and the lift,

$$\mathcal{D}_{\text{form}} \approx \alpha\mathcal{L}. \tag{27.9}$$

The form drag will always be positive, independent of the sign of α, because lift is proportional to α. Drag forces should, of course, never be able to accelerate a body.

Contrary to the skin drag, which (to leading order) is independent of the angle of attack, the form drag is quadratic in α and vanishes for $\alpha = 0$. Using the estimates above we obtain the ratio of form to skin drag

$$\frac{\mathcal{D}_{\text{form}}}{\mathcal{D}_{\text{skin}}} \approx f\alpha\frac{L}{d} \approx 2f^2. \tag{27.10}$$

Since this is at most of order unity, we conclude that although the form drag (and lift) for fixed α varies like $1/d^3$, it can never really win over skin drag because of the geometrical constraint, $|f| \lesssim 1$.

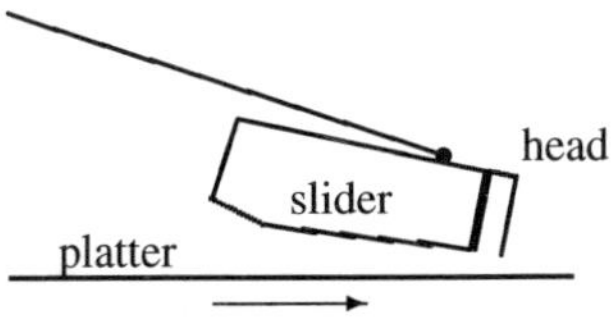

Sketch of the head-to-disk interface in a hard disk. The platter rotates towards the left and drags air into the gap between the slider and the surface, and thereby prevents the slider from touching the platter. The elastic actuator arm counteracts the lift force from the air in the gap. The head itself is here positioned at the rear end of the slider.

Example 27.1.2 (Magnetic read/write heads): The continued sophistication in the design of read/write heads and platter surfaces has been a major cause for the enormous improvement in hard disk performance over the last 30 years. A typical modern (2002) hard disk has a platter diameter of about 9 cm and runs at a speed of about 7000 rpm, leading to average platter surface speeds of $U \approx 16\ \text{m s}^{-1} \approx 60\ \text{km h}^{-1}$. The read/write heads sit on the tip of an actuator arm that can roam over the rotating platters and exchange data with the magnetic surfaces. A typical read/write head is formed as a flat wing or 'slider' with size $L \approx 1$ mm, for which the Reynolds number comes to about

Re ≈ 1000, and the maximal gap size for creeping flow comes to $\delta \approx 30\ \mu$m. The need for increased data density demands smaller and smaller flying height for the actual read/write device sitting on the rear of the slider. Today (2002) it is about $0.15\ \mu$m (and even smaller), implying that the slider flies deeply inside the boundary layer. Taking the average slider gap height to be a conservative $d = 1\ \mu$m the Reynolds number in the gap becomes $\mathrm{Re}_{\mathrm{gap}} = 10^{-3}$ which is far within the creeping flow regime. The geometry makes the angle of attack $\alpha = 0.1$ degrees and the geometric prefactor $f = 0.85$. The average excess pressure in the gap becomes surprisingly high, $\Delta p \approx \mathcal{L}/A \approx 2.5$ bar, implying a lubrication lift force on the slider of $\mathcal{L} \approx 0.25$ N. The ratio of lift to skin plus form drag becomes $\mathcal{L}/\mathcal{D} \approx 350$. The lift force corresponds to a weight of 25 g, which must be provided by the elastic actuator arm to keep the flying height constant. If the total mass of the actuator arm and slider is of the order of grams, such a large actuator force explains why modern hard disks can tolerate rather large accelerations, in this example perhaps 10–20 times the acceleration of gravity.

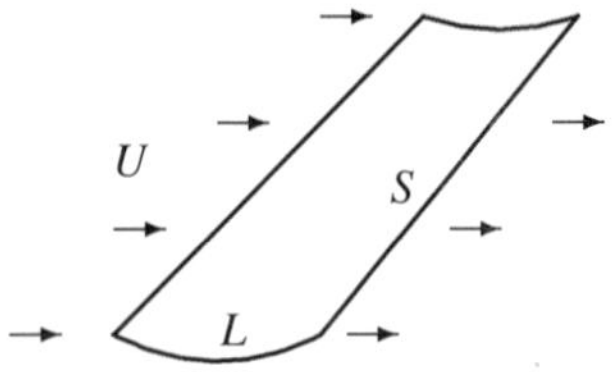

A two-dimensional wing with chord length L and span S. The wing is stationary above a flat ground which moves with speed U.

27.2 Creeping flow in a long narrow gap

Creeping flow near a solid boundary is much more amenable to analytic calculations than real aerodynamics (chapter 29). For simplicity we shall consider the essentially two-dimensional case of a stationary 'wing' of constant shape and 'chord length' L along the x-axis. The wing 'span' along z is denoted S such that the total wing area is $A = LS$. The 'ground' is chosen to be perfectly flat at $y = 0$, and moves with constant velocity U along the x-axis. The height of the wing above the ground is given by a function $y = h(x)$ for all z, which is assumed to be slowly varying $|h'(x)| \ll 1$. The free-flow Reynolds number (27.1) is furthermore assumed to be so large and the height so small that the creeping approximation is valid, $h(x) \lesssim \delta \ll L$.

Under these conditions we need only to retain the dominant derivatives after y in the Laplacian of (27.4). Dropping all second derivatives after x we get a simplified set of equations for the flow in the gap,

$$\frac{\partial p}{\partial x} = \eta \frac{\partial^2 v_x}{\partial y^2}, \qquad \frac{\partial p}{\partial y} = \eta \frac{\partial^2 v_y}{\partial y^2}, \qquad \frac{\partial v_y}{\partial y} = -\frac{\partial v_x}{\partial x}. \tag{27.11}$$

We shall now see that the right-hand side of the second equation is so small in the creeping flow approximation that the pressure to leading order is independent of y, a conclusion allowing us to solve the gap equations analytically.

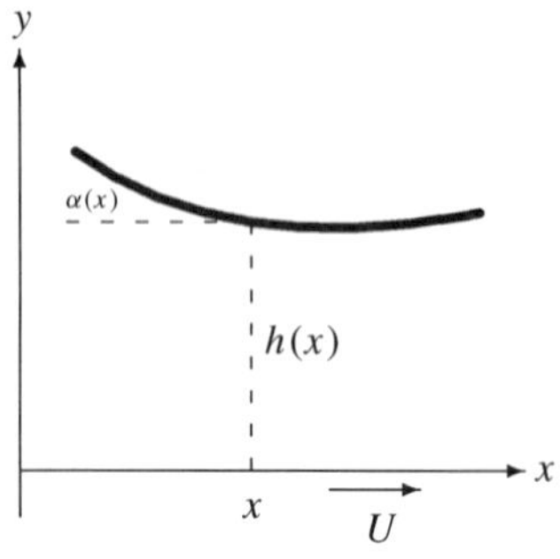

Local geometry of creeping 'flight' in the xy-plane. The 'ground' moves with velocity U relative to the 'wing' which flies at height $y = h(x)$. The local angle of attack is $\alpha(x) = -h'(x)$.

Solving the gap equations

Replacing the derivatives by ratios of suitable scales in the first gap equation, we estimate the pressure variation along the gap to be $\Delta_x p \sim \eta U L/d^2$ where d is the average value of $h(x)$. From the continuity equation we get $|v_y| \sim Ud/L$, and from the second gap equation we find the pressure variation across the gap, $\Delta_y p \sim \eta |v_y|/d \sim \eta U/L$. This shows that the ratio of the pressure variations across and along the gap is $\Delta_y p/\Delta_x p \sim (d/L)^2$, which by assumption is tiny. The conclusion is that to this order of approximation we may take the pressure in the gap to be only a function of x.

Inserting $p = p(x)$ in the first gap equation it may immediately be integrated with the boundary conditions $v_x = U$ at $y = 0$ and $v_x = 0$ at $y = h(x)$, yielding

$$v_x = U\left(1 - \frac{y}{h(x)}\right) - \frac{p'(x)}{2\eta} y(h(x) - y), \tag{27.12}$$

where $p'(x)$ is the pressure gradient. The solution is a superposition of velocity-driven planar flow (17.6) and pressure-driven planar flow (18.5), with a variable plate distance $h(x)$ and pressure gradient $p'(x)$.

The moving ground at $y = 0$ drags fluid along in the direction of positive x. The rate at which the fluid is dragged along becomes (per unit of length in the z-direction),

$$Q = \int_0^{h(x)} v_x(x, y)\, dy = \frac{1}{2} U h(x) - \frac{p'(x) h(x)^3}{12\eta}. \tag{27.13}$$

Since the fluid is incompressible, Q is independent of x, and solving for the pressure gradient we find,

$$\boxed{p'(x) = 6\eta \left(\frac{U}{h(x)^2} - \frac{2Q}{h(x)^3}\right),} \tag{27.14}$$

from which the pressure may be obtained by straightforward integration.

Eliminating the pressure gradient in the velocity (27.12), we find

$$v_x = U\frac{(h-y)(h-3y)}{h^2} + Q\frac{6y(h-y)}{h^3} \tag{27.15}$$

where we for clarity have suppressed the explicit dependence on x, now entirely due to the slowly varying gap height $h(x)$. Finally, we insert v_x into the continuity equation and integrate over y with the condition $v_y = 0$ for $y = 0$, and get

$$v_y = -2h'\frac{Uh-3Q}{h^4}y^2(h-y), \tag{27.16}$$

where $h'(x)$ is the local derivative of the height. The appearance of the tiny height derivative h' confirms that $|v_y| \ll |v_x|$.

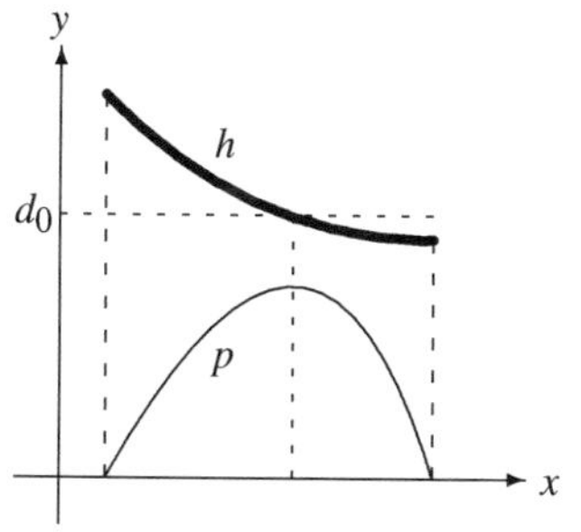

Sketch of the gap pressure $p(x)$ in relation to d for a wing with positive angle of attack.

Effective gap width

The pressure change along the gap is calculated from the pressure gradient,

$$\Delta p = \int_L p'(x)\,dx = 6\eta\left(U\int_L \frac{dx}{h(x)^2} - 2Q\int_L \frac{dx}{h(x)^3}\right). \tag{27.17}$$

The wing is supposed to move in a fluid which would have constant pressure, were it not for the disturbance created by the wing itself. Assuming that the pressure $p(x)$ is (nearly) the same at both ends of the wing, $\Delta p \approx 0$, we obtain from (27.17) a relation between the discharge rate and the velocity,

$$Q = \frac{1}{2}Ud_0, \qquad d_0 = \frac{\langle h^{-2}\rangle}{\langle h^{-3}\rangle}, \tag{27.18}$$

where $\langle F\rangle = (1/L)\int_L F(x)dx$ denotes the average of a function $F(x)$ along the gap. The relation between Q and U depends on the shape of the gap only through the parameter d_0, with dimension of length. We shall call d_0 the *effective gap width*. For flat plates with constant gap height, $h(x) = h_0$, we get $d_0 = h_0$, and this shows that d_0 represents the width of a gap between parallel plates, carrying the same discharge rate as the actual gap.

The pressure gradient (27.14) may now be written

$$p' = 6\eta U\frac{h-d_0}{h^3}. \tag{27.19}$$

Thus, in regions where $h(x) > d_0$ the pressure will rise, whereas it will fall in regions where $h(x) < d_0$. The pressure always has an extremum in the gap.

The velocity field may also be written

$$v_x = U\left(1-\frac{y}{h}\right)\left(1-\frac{3y(h-d_0)}{h^2}\right). \tag{27.20}$$

Whereas the first factor is always positive, the last may have either sign.

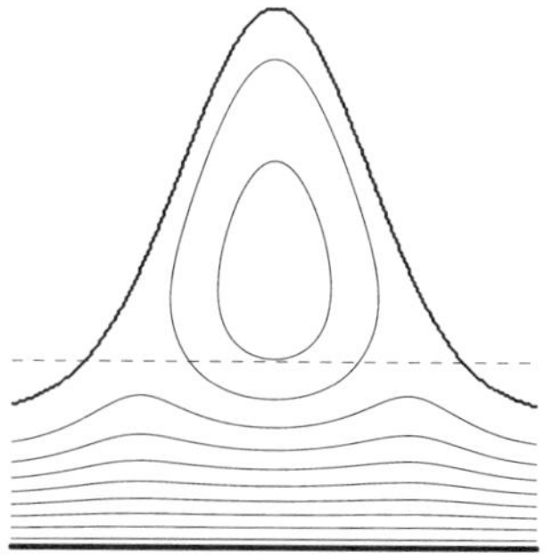

Pattern of flow reversal under a Gaussian bump $h = 1 + 3\exp(-x^2)$ in the interval $-2 < x < 2$. The effective height is $d_0 \approx 1.38$ (dashed line). The pressure first drops, then rises under the bump and finally drops again.

Flow reversal

Evidently the last factor is positive for small y, but may for a given x vanish and become negative for $y > h^2/3(h-d_0)$. When this point lies inside the gap, $0 < y < h$, the fluid close to the wing will flow *against* the direction of ground motion. Evidently this is only possible for $h < 3(h-d_0)$, or

$$h(x) > \frac{3}{2}d_0. \tag{27.21}$$

In regions where this condition is fulfilled, 'rollers' of counter-rotating fluid will appear. Typically, this happens when the gap widens locally, because the discharge rate is mainly determined by the narrow parts of the gap, making the effective gap width d_0 considerably lower than the maximal height of the bump.

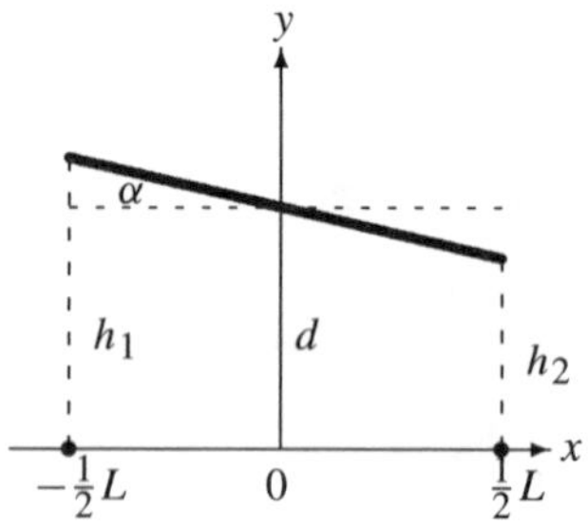

Geometry of flat wing with constant angle of attack α and average height d. Varying α rotates the wing around its midpoint.

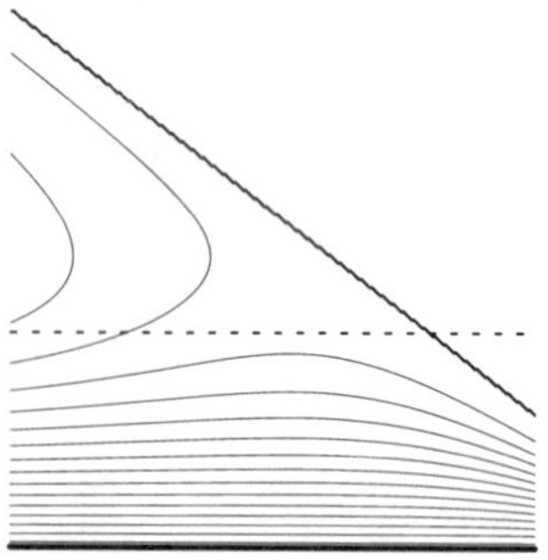

Flow reversal under flat wing for $\gamma = 0.6$ and $L/2d = 10$. The dashed line indicates the effective width d_0. The scale of the ordinate is exaggerated by a factor 10.

27.3 Flat wing

Up to this point, everything has been valid for an arbitrary wing shape. We shall now specialize to the case of a flat wing with constant angle of attack α and average height d (the prototype slider bearing of example 27.1.2)

$$h = d - \alpha x, \tag{27.22}$$

with $-L/2 \le x \le L/2$. Instead of the parameters α and h_0, it is more convenient to specify the extreme heights, h_1 and h_2, at the front and back of the wing. The average height is then $d = (h_1 + h_2)/2$ and the angle of attack $\alpha = (h_1 - h_2)/L$. It is also convenient to define the dimensionless parameter

$$\gamma = \frac{\alpha L}{2d} = \frac{h_1 - h_2}{h_1 + h_2}, \tag{27.23}$$

which ranges over the interval, $-1 < \gamma < 1$. Conversely, we have

$$h_1 = (1+\gamma)d, \qquad h_2 = (1-\gamma)d. \tag{27.24}$$

All averages over powers of h may be evaluated once and for all,

$$\langle h^n \rangle = \frac{h_1^{n+1} - h_2^{n+1}}{(n+1)\alpha L}, \qquad \langle h^{-1} \rangle = \frac{1}{\alpha L} \log \frac{h_1}{h_2}. \tag{27.25}$$

Using this the effective gap width becomes,

$$d_0 = \frac{2h_1 h_2}{h_1 + h_2} = (1 - \gamma^2)d. \tag{27.26}$$

It is never larger than d, and the discharge rate $Q = U d_0/2$ vanishes as expected, if either end of the wing touches the ground for $\gamma = \pm 1$.

For positive angle of attack $\gamma > 0$, the condition for flow reversal (27.21) is easiest to fulfill at the front of the wing, where the left-hand side is largest. Solving the inequality for $x = -L/2$ we find $\gamma > 1/3$. A sufficiently large positive angle of attack will always cause flow reversal at the front of the wing. Likewise, for sufficiently large negative angle of attack, $\gamma < -1/3$, there will instead be flow reversal at the rear of the wing. In between for $-1/3 < \gamma < 1/3$ there is no flow reversal.

Pressure and lift

In view of the linearity of the height function (27.22), the pressure gradient (27.19) may now be integrated to yield,

$$p = \frac{3\eta U(2h - d_0)}{\alpha h^2} - \frac{3\eta U}{\alpha d}, \tag{27.27}$$

where the constant secures that the pressure vanishes at both ends of the wing.

The total lift from the pressure in the gap becomes,

$$\mathcal{L} = \int_{-L/2}^{L/2} p(x)\, S dx = \frac{\eta U L A}{d^2} \cdot \frac{3}{2\gamma^2} \left(\log \frac{1+\gamma}{1-\gamma} - 2\gamma \right). \tag{27.28}$$

The γ-dependent factor on the right represents the influence of the orientation of the wing which we called f in the estimate (27.5), and for $|\gamma| \ll 1$ we have $f \approx \gamma$ in agreement with the estimate. It is plotted in figure 27.1 as a function of γ. The linear region where $f \approx \gamma$ extends over the interval $-0.7 < \gamma < 0.7$ and is a very good approximation in most cases of interest.

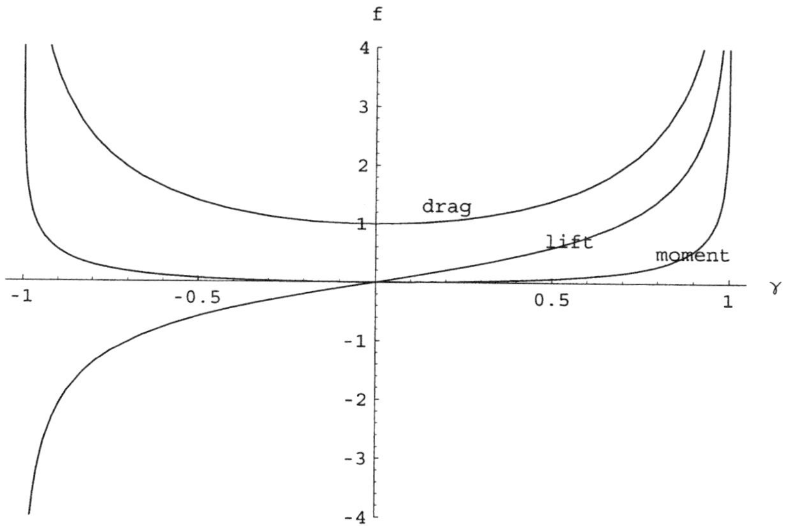

Figure 27.1. Orientation-dependent shape factors for lift , drag and moment for a flat wing as a function of $\gamma = \alpha L/2d$. All quantities diverge logarithmically for $|\gamma| \to 1$.

Drag and power

As discussed in section 27.1 the drag on the wing has two components, skin drag arising from viscous friction and form drag arising from pressure forces projected on the direction of motion. The 'ground' also experiences a skin drag, but no form drag because the ground is aligned with the direction of motion. Newton's third law tells us that the drag forces on the wing and on the ground must be equal and opposite, and this allows us to calculate the total drag on the wing, $\mathcal{D} = \mathcal{D}_{\text{skin}} + \mathcal{D}_{\text{form}}$, from the skin drag alone on the ground.

The shear stress on the flat ground is

$$\sigma_{xy}\big|_{y=0} = \eta \left[\frac{\partial v_x}{\partial y} + \frac{\partial v_y}{\partial x}\right]_{y=0} \approx \eta \frac{\partial v_x}{\partial y} = -\eta U \frac{4h - 3d_0}{h^2}. \tag{27.29}$$

Integrating the shear stress over the ground surface, and changing sign in accordance with Newton's third law, we obtain the total drag on the wing,

$$\mathcal{D} = -\int_{-L/2}^{L/2} \sigma_{xy}\big|_{y=0}\, S dx = \frac{\eta U A}{d} \cdot \frac{1}{\gamma}\left(2 \log \frac{1+\gamma}{1-\gamma} - 3\gamma\right). \tag{27.30}$$

The leading part of this expression agrees with the estimate (27.7). The drag estimate is modified by a γ-dependent factor which converges to unity for $\gamma \to 0$ and like the lift diverges logarithmically for $|\gamma| \to 1$ (see figure 27.1).

The external forces that keep the ground moving with constant velocity U must be equal to the drag. Their rate of work is

$$P = \mathcal{D} U. \tag{27.31}$$

Since the wing is at rest, no other external forces perform work on the system, so this must also equal the total rate of dissipation into heat. In figure 27.1 the shape factor for the rate of dissipation as a function of γ is the same as the shape factor for the drag.

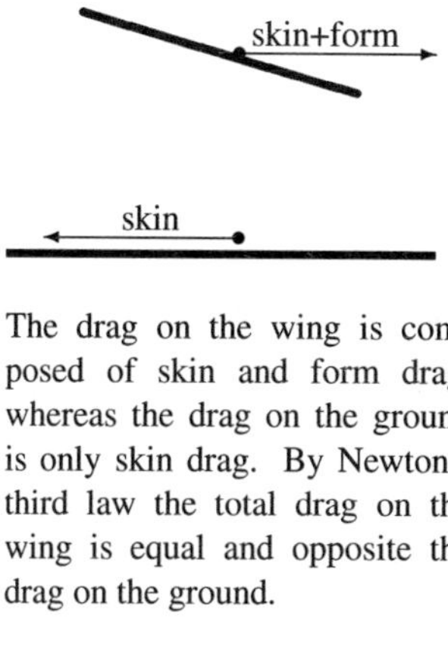

The drag on the wing is composed of skin and form drag, whereas the drag on the ground is only skin drag. By Newton's third law the total drag on the wing is equal and opposite the drag on the ground.

Moment

The pressure forces also create a turning moment around the centre of the wing,

$$\mathcal{M}_z = \int_{-L/2}^{L/2} x p(x)\, S dx = \frac{\eta U L^2 A}{d^2} \cdot \frac{3}{8\gamma^3}\left((3-\gamma^2) \log \frac{1+\gamma}{1-\gamma} - 6\gamma\right). \tag{27.32}$$

The γ-dependent factor is also plotted in figure 27.1. It is always positive, but mostly very small and vanishes like $(1/5)\gamma^2$ for $\gamma \to 0$. If the angle of attack α is positive, the positive moment tends to rotate the wing towards the horizontal, whereas for negative α the moment tends to turn the wing further into the ground, destabilizing the flight.

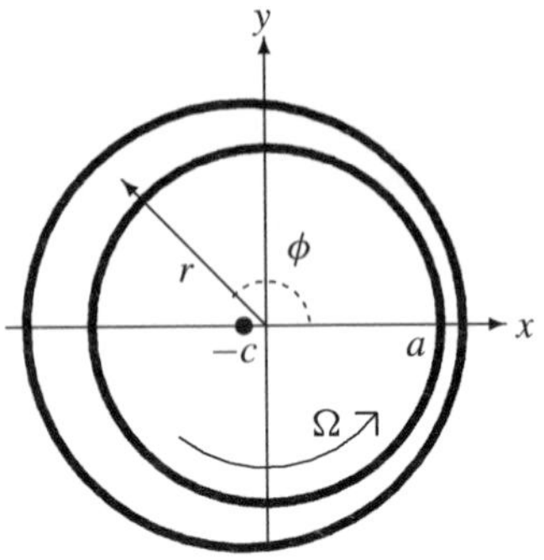

Geometry of off-centre journal bearing. The inner cylinder has radius a and the outer b. The outer cylinder is shifted to the left by an amount c. The inner cylinder rotates with constant angular velocity Ω in the counter-clockwise direction.

Example 27.3.1 (Playing card): A completely flat unbent playing card thrown with a positive angle of attack will slowly sink further and further towards the table surface while the tiny moment rotates it towards the horizontal and the lift becomes still smaller (here we ignore again any forces acting on the upper side of the card). Thrown with a negative angle of attack, the playing card will get sucked towards the table at an increasing rate because the positive moment makes the angle of attack still more negative. Eventually, the card may catch on surface irregularities and turn over, showing its value, to the dismay of the players.

27.4 Loaded journal bearing

In section 18.6 we discussed the case of laminar flow between two concentric rotating cylinders—the prototypical *journal bearing*. If the inner shaft or the outer sleeve (bushing) carries a load orthogonal to the shaft, the cylinders will no more be concentric, although we shall assume that they are still parallel with respect to each other.

In a non-rotating journal bearing the lubricating fluid will be squeezed out and the shaft will come into direct contact with the sleeve at a point opposite the direction of the load. When the shaft (or the sleeve) is brought into rotation, fluid will be dragged along due to the no-slip condition and forced into the narrow part of the gap, thereby creating a pressure that tends to lift the shaft away from the sleeve. As we shall see, the lift will not be directed orthogonally to the point of contact at rest.

In this section we only discuss laminar flow in the gap. A rotating journal bearing is also prone to the centrifugal instabilities discussed in section 18.7 on page 262 with formation of Taylor vortices and more complicated structures.

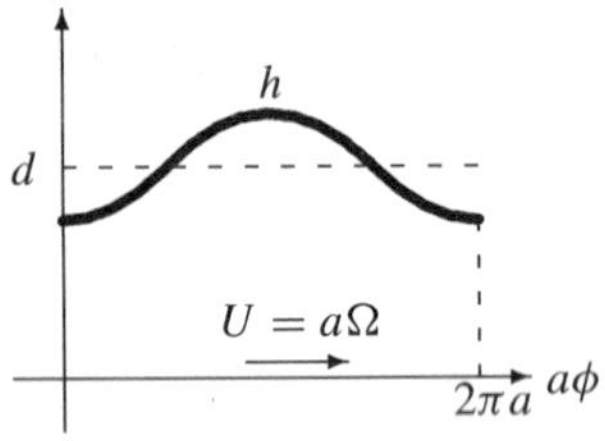

'Flattened' gap between the cylinders. The inner cylinder replaces the flat boundary along the x-axis with $x \to a\phi$, and the space between the cylinders is described by $y \to s = r - a$. The velocity of the inner cylinder is $U = a\Omega$

Narrow gap approximation

Let the inner cylinder have radius a and the outer radius $b > a$ with a difference $d = b - a$ that is assumed to be tiny, $d \ll a$. In a coordinate system with the z-axis coinciding with the axis of the inner cylinder, we may without loss of generality assume that the point of closest approach takes place on the positive x-axis. Denoting the centre of the outer cylinder $x = -c$, the points of the outer cylinder are determined by the equation $(x + c)^2 + y^2 = b^2$. In standard cylindrical coordinates this becomes $r^2 + c^2 + 2rc\cos\phi = b^2$, which to first order in the small quantity c has the solution $r = b - c\cos\phi$, and the width of the gap $h = r - a$ between the cylinders becomes

$$h = d - c\cos\phi. \tag{27.33}$$

It is convenient to introduce the dimensionless eccentricity parameter,

$$\gamma = \frac{c}{d} \tag{27.34}$$

which must lie in the interval $-1 \le \gamma \le 1$.

The shaft has length S, area $A = 2\pi a S$, and rotates at constant angular velocity Ω with surface velocity $U = a\Omega$. Disregarding the possibility that lubricant may be squeezed out along the z-axis (a non-trivial technical detail), the problem is essentially two-dimensional. We shall assume that the longitudinal field vanishes, $v_z = 0$, and the azimuthal and radial fields, v_ϕ and v_r, depend only on r and ϕ. In this approximation, the general theory of flow in a narrow gap (section 27.2) may be brought into play, replacing x by $a\phi$ and y by $s = r - a$. The narrow gap between the cylinders has essentially been converted to a flat-wall gap of length $L = 2\pi a$ with strictly periodic boundary conditions.

The effective gap width defined in (27.18) may now be evaluated by averaging over ϕ, with the result (see problem 27.6),

$$d_0 = d\frac{2(1-\gamma^2)}{2+\gamma^2}. \tag{27.35}$$

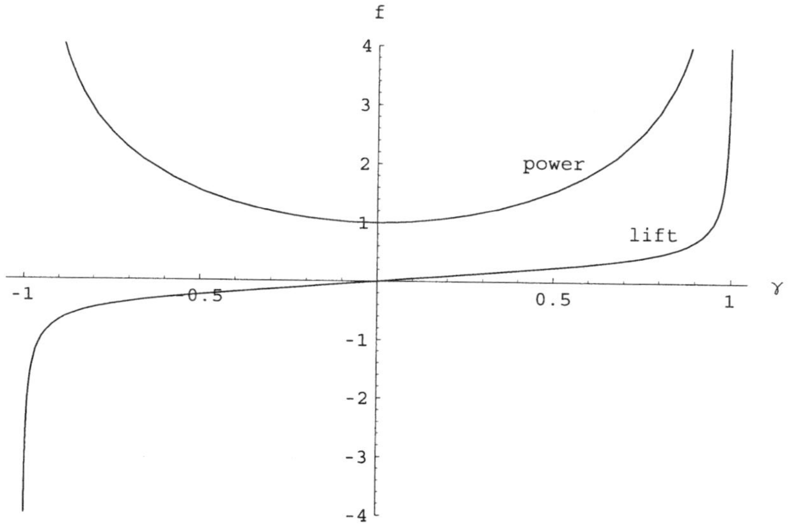

Figure 27.2. Orientation-dependent shape factors for lift and dissipated power in a journal bearing as a function of the eccentricity γ. Both quantities diverge as the inverse square root for $|\gamma| \to 1$.

As for the flat plate the volume discharge is $Q = Ud_0/2$ and it vanishes as expected for $\gamma = \pm 1$ when the two cylinders come into contact.

Pressure and lift

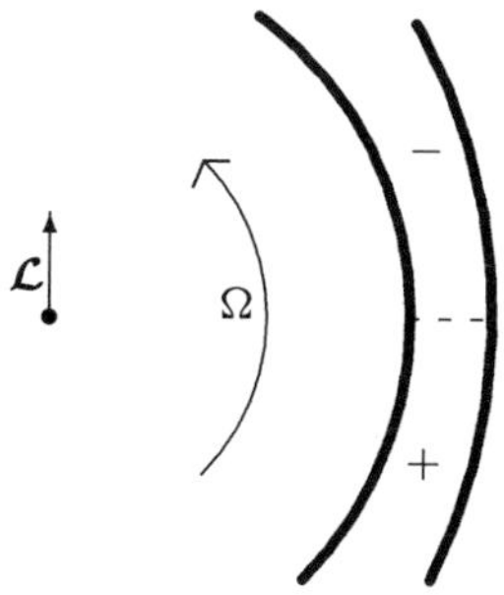

The pressure generated by the shaft's rotation is asymmetric with respect to the point of closest approach. The total lift $\mathcal{L}$ on the shaft is always parallel with the direction of motion at the point of closest approach.

The pressure derivative is as before given by (27.19), and integrating over ϕ one obtains the expression (most easily checked by differentiation with respect to ϕ),

$$p = -\frac{6\eta U a}{d^2} \frac{\gamma(2 - \gamma\cos\phi)\sin\phi}{(2+\gamma^2)(1-\gamma\cos\phi)^2}. \tag{27.36}$$

Evidently the pressure vanishes at the point of closest approach $\phi = 0$ and at the opposite point $\phi = \pi$. If $\gamma > 0$, the pressure is positive in the lower half of the gap and negative in the upper. The up-down antisymmetry of the pressure under $\phi \to -\phi$ implies that the average pressure is zero, $\langle p \rangle = 0$.

In Cartesian coordinates the surface element of the inner cylinder is $d\boldsymbol{S} = (\cos\phi, \sin\phi, 0)\, a d\phi\, dz$. The up-down antisymmetry of p shows that the lift along x vanishes,

$$\mathcal{L}_x = \oint_{r=a} (-p) dS_x = -\frac{A}{2\pi}\int_0^{2\pi} p\,\cos\phi\, d\phi = 0. \tag{27.37}$$

Replacing the cosine by a sine the lift on the shaft along y becomes (problem 27.6),

$$\mathcal{L}_y = \oint_{r=a} (-p) dS_y = \frac{\eta U L A}{d^2} \cdot \frac{3\gamma}{\pi(2+\gamma^2)\sqrt{1-\gamma^2}}, \tag{27.38}$$

where $L = 2\pi a$, $A = LS$ and $U = \Omega a$. The γ-dependent expression represents the shape factor f in our estimate (27.5). The behaviour of the shape factor is shown in figure 27.2. It is nearly linear in the interval $-0.9 < \gamma < 0.9$, outside which it diverges rapidly. The divergence permits, in principle, the bearing to carry any load by adjusting γ to be sufficiently close to unity.

It is perhaps a bit counterintuitive that the lift is always parallel with the direction of motion at the point of closest approach, rather than orthogonal to it. This is, as we have seen, a consequence of the antisymmetry of the pressure with respect to the radial direction at closest approach. When the shaft just starts to rotate from rest, the direction of lift will thus be orthogonal to the direction of load, and with no other forces at play, this lift will tend to shift the point of closest approach sideways in the direction of motion of the shaft surface. As the point of closest approach moves away the lift changes direction, until it reaches a point where radial direction is orthogonal to the direction of the load. The actual distance at closest approach $d(1-\gamma)$ is determined by the balance between load and lift, and the larger the load the closer the value of γ must be to unity.

Moment of drag and power

The shear stress $\sigma_{\phi r}$ at the surface of the shaft is given by the general expression (27.29). In this case it makes no sense to speak about a drag force, but rather the moment of the shear stress around the centre of the shaft becomes (problem 27.6)

$$\mathcal{M}_z = \frac{A}{2\pi}\int_0^{2\pi} a\sigma_{\phi r}\,d\phi = -\frac{\eta U A\, a}{d}\cdot\frac{2(1+2\gamma^2)}{\sqrt{1-\gamma^2}\,(2+\gamma^2)}. \tag{27.39}$$

It is negative, as one would expect, and the rate of work performed by the external moment that keeps the shaft turning against viscosity is obtained by multiplying with $-\Omega$,

$$P = \frac{\eta U A\Omega\, a}{d}\cdot\frac{2(1+2\gamma^2)}{\sqrt{1-\gamma^2}\,(2+\gamma^2)}. \tag{27.40}$$

For $\gamma \to 0$ the dissipated power approaches the result for the unloaded bearing (18.72), whereas for $|\gamma| \to 1$ it diverges (like lift) as an inverse square root. The large dissipation in heavily loaded journal bearings can be reduced by employing rollers or balls that keep the shaft centred in its bushing.

The ratio of dissipated power to lift,

$$\frac{P}{\mathcal{L}_y} = \frac{1+2\gamma^2}{3\gamma}\,\Omega d, \tag{27.41}$$

is finite for $|\gamma| \to 1$. Since a heavily loaded bearing has $|\gamma| \approx 1$, the ratio becomes $P/\mathcal{L}_y \approx \Omega d$ in this limit, and since the lift must equal the load, this makes it easy to calculate the dissipated power.

Flow reversal

The narrow gap between the cylinders looks like a flat gap with a bump opposite the point of closest contact. The discussion on page 399 indicates that a stationary flow-reversed 'roller' may arise at this point. The reversal condition (27.21) becomes,

$$1-\gamma\cos\phi > 3\frac{1-\gamma^2}{2+\gamma^2}. \tag{27.42}$$

Taking $\phi = \pi$ the roller will appear when $\gamma^2 + 3\gamma - 1 > 0$. Solving the quadratic inequality this is the case for $|\gamma| > (\sqrt{13}-3)/2 \approx 0.303$.

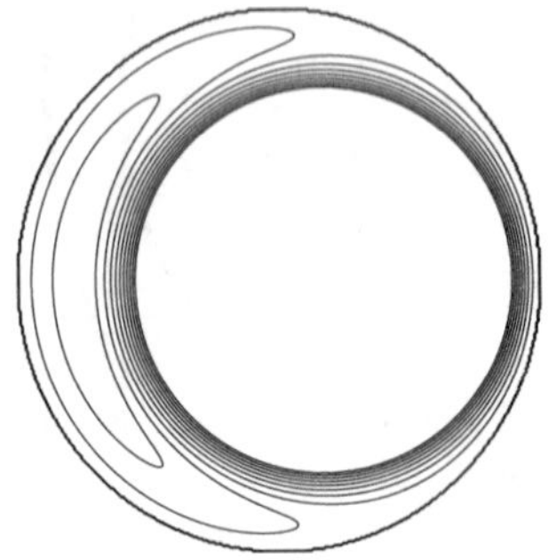

Creeping flow pattern in loaded journal bearing with $d/a = 0.4$ and $\gamma = 0.6$. The flow is reversed and forms a counter-rotating patch of fluid opposite the point of closest approach.

Problems

27.1 Estimate the gap height for creeping horizontal flight in constant gravity when the angle of attack is α and the body has mass M.

27.2 Show explicitly that the fluid discharge rate Q (equation (27.13)) is independent of x.

27.3 Find the conditions under which the velocity (27.15) has an extremum in the gap for a given x.

27.4 Define the gap average

$$\langle F\rangle = \frac{1}{L}\int_L F(x)\,dx, \tag{27.43}$$

and express all dynamic gap quantities in terms of averages.

27.5 Consider a nearly flat gap with $h(x) = d(1 + \chi(\xi))$ where $d = \langle h \rangle$ is the average height, and $\xi = x/L$. **(a)** Using the results of problem 27.4, calculate for $|\chi| \ll 1$ the leading non-trivial approximation to all the dynamic quantities. **(b)** Compare with the exact flat wing results.

∗ **27.6** Show that for $a > |b|$

$$\frac{1}{2\pi}\int_0^{2\pi} \frac{d\phi}{a - b\cos\phi} = \frac{1}{\sqrt{a^2 - b^2}}, \tag{27.44}$$

and use this to derive the integrals

$$\int_0^{2\pi} \frac{1}{1-\gamma\cos\phi}\frac{d\phi}{2\pi} = \frac{1}{\sqrt{1-\gamma^2}}, \qquad \int_0^{2\pi} \frac{1}{(1-\gamma\cos\phi)^2}\frac{d\phi}{2\pi} = \frac{1}{(1-\gamma^2)^{3/2}},$$

$$\int_0^{2\pi} \frac{1}{(1-\gamma\cos\phi)^3}\frac{d\phi}{2\pi} = \frac{2+\gamma^2}{2(1-\gamma^2)^{5/2}}, \qquad \int_0^{2\pi} \frac{\cos\phi}{(1-\gamma\cos\phi)^2}\frac{d\phi}{2\pi} = \frac{\gamma}{(1-\gamma^2)^{3/2}},$$

$$\int_0^{2\pi} \frac{\cos\phi}{(1-\gamma\cos\phi)^3}\frac{d\phi}{2\pi} = \frac{3\gamma}{2(1-\gamma^2)^{5/2}}.$$

28 Boundary layers

Between the two extremes of sluggish creeping flow at low Reynolds number and lively ideal flow at high, there is a regime in which neither is dominant. At large Reynolds number, the flow will be nearly ideal almost everywhere, except near solid boundaries where the no-slip condition requires the speed of the fluid to match the speed of the boundary wall. Here transition layers will arise in which the flow velocity changes rapidly from the velocity of the wall to the velocity of the flow in the fluid at large. Boundary layers are typically thin compared to the radii of curvature of the solid walls, and that simplifies the basic equations.

In a boundary layer the character of the flow thus changes from creeping near the boundary to ideal well outside. The most interesting and also most difficult physics characteristically takes place in such transition regions. But humans live out their lives in nearly ideal flows of air and water at Reynolds numbers in the millions with boundary layers only millimetres thick, and are normally not conscious of them. Smaller animals eking out an existence at the surface of a stone in a river may be much more aware of the vagaries of boundary layer physics which may influence their body shapes and internal layout of organs.

Boundary layers serve to insulate bodies from the ideal flow that surrounds them. They have a 'life of their own' and may separate from the solid walls and wander into regions containing only fluid. Detached layers may again split up, creating complicated unsteady patterns of whirls and eddies. Advanced understanding of fluid mechanics begins with an understanding of boundary layers. Systematic boundary layer theory was initiated by Prandtl in 1904 and has in the twentieth century become a major subtopic of fluid mechanics [61, 64, 79].

Ludwig Prandtl (1875–1953). *German physicist, often called the father of aerodynamics. Contributed to wing theory, streamlining, compressible subsonic airflow and turbulence.*

In this chapter we shall mainly focus on the theory of incompressible laminar boundary layers without heat flow. A semi-empirical discussion of turbulent boundary layers is also included.

28.1 Physics of boundary layers

The no-slip condition forces the velocity of a fluid to vanish at a static solid wall. Under many—but not all—circumstances, the transition between the rapid flow at high Reynolds number in the fluid at large and the stagnation at the wall will take place in a thin *boundary layer* hugging the wall. Close to the wall, the velocity field is so small that the flow pattern will always be laminar, in fact creeping, with the velocity field rising linearly from zero. The laminar flow may extend all the way to the edge of the boundary layer, or it may become turbulent if the Reynolds number is sufficiently large.

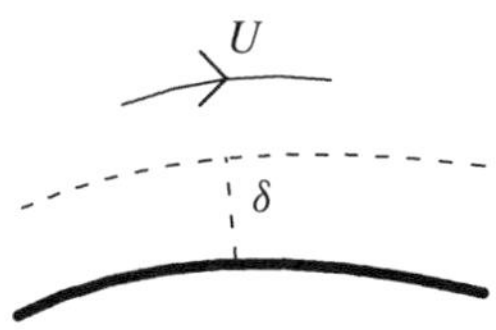

The transition from zero velocity at the wall to the mainstream velocity U mostly takes place in a layer of finite thickness δ.

Laminar boundary layer thickness

Denoting the typical velocity of the mainstream flow by U, the Reynolds number is as usual $\mathrm{Re} \approx UL/\nu$ where L is the length scale for significant changes in the flow, determined by the geometry of bodies and containers. We shall always assume that it is large, $\mathrm{Re} \gg 1$. The effective Reynolds number in a steady laminar boundary layer of thickness δ can be estimated from the ratio of advective to viscous terms in the Navier–Stokes equation,

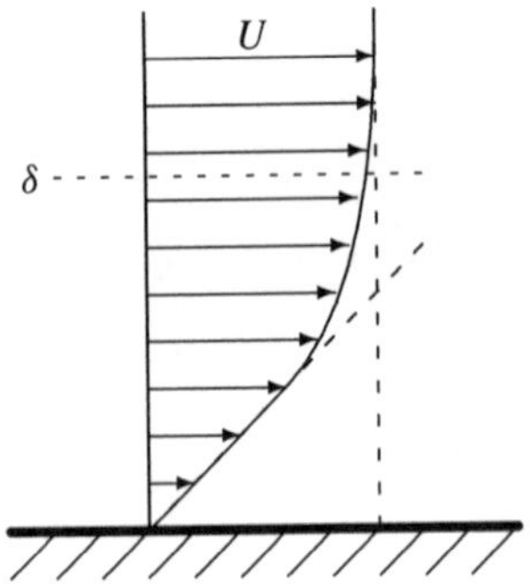

Sketch of the flow in a laminar boundary layer with constant mainstream flow. The velocity rises linearly close to the solid wall but veers off to match the mainstream flow velocity U at a characteristic distance δ from the wall. The precise layer thickness depends on what one means by 'matching' the mainstream flow.

$$\frac{|(\boldsymbol{v}\cdot\nabla)\boldsymbol{v}|}{|\nu\nabla^2\boldsymbol{v}|} \sim \frac{U^2/L}{\nu U/\delta^2} = \frac{\delta^2}{L^2}\mathrm{Re}. \tag{28.1}$$

Here the numerator is estimated from the change in mainstream velocity along the wall over a distance L, using that the flow in a laminar layer must follow the geometry of the body. The denominator is estimated from the rapid change in velocity across the thickness δ of the boundary layer.

Since the boundary layer represents the transition region from essentially non-viscous flow at large with $\mathrm{Re} \gg 1$ to creeping flow near the wall with $\mathrm{Re} \ll 1$, the *boundary layer thickness* for steady laminar flow may be estimated by requiring the effective Reynolds number (28.1) to be around unity, leading to

$$\boxed{\delta \sim \sqrt{\frac{\nu L}{U}} = \frac{L}{\sqrt{\mathrm{Re}}}.} \tag{28.2}$$

This estimate is valid up to a coefficient of order unity which will be discussed later (section 28.4). For large mainstream Reynolds number, $\mathrm{Re} \gg 1$, the thickness of the boundary layer will thus be considerably smaller than the typical length scale of the mainstream flow.

Example 28.1.1: The Reynolds numbers for flows we encounter in daily life easily reach into the millions, making the laminar boundary layer thickness smaller than a thousandth of the scale of the flow. Jogging or swimming, one hardly notes the existence of boundary layers that are only millimetres thick. The pleasant tingling skin sensation you experience from streaming air or water comes presumably from the complex and turbulent flow at larger scale generated by the irregular shape of your body.

Wall shear stress

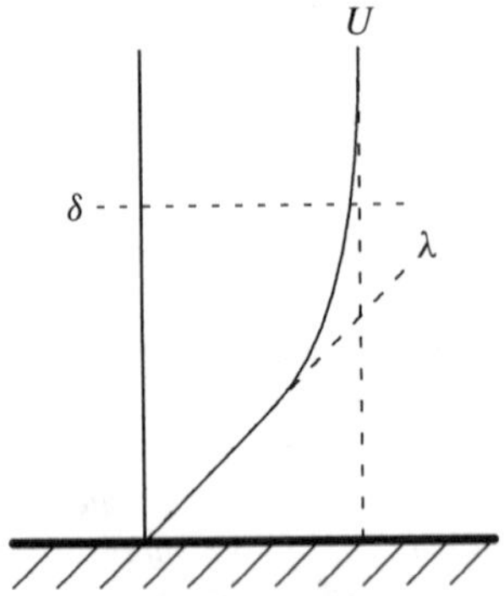

The stress on the wall of a laminar boundary layer is determined by the slope $\lambda \approx U/\delta$ of the linearly rising velocity near the wall.

In the laminar boundary layer, the velocity rises linearly with the distance from the boundary in the same way as in planar velocity-driven flow. The normal velocity gradient at the wall is approximately U/δ, and multiplying by the viscosity we obtain an estimate of the shear stress on the wall,

$$\sigma_{\mathrm{wall}} \approx \eta\frac{U}{\delta} \sim \frac{\rho_0 U^2}{\sqrt{\mathrm{Re}}}. \tag{28.3}$$

The wall stress thus increases as $U^{3/2}$ with increasing mainstream velocity U and as $\sqrt{\nu}$ with increasing viscosity ν.

Initial viscous growth

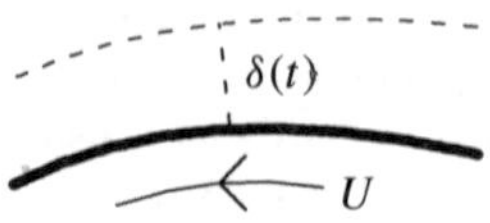

The wall is suddenly set into motion. After a time t, the velocity of the fluid at the edge of the growing boundary layer has changed from 0 to U.

When a body at rest is suddenly set into motion at $t = 0$ with velocity U, the fluid in its immediate vicinity will have to follow along to satisfy the no-slip boundary condition. Large velocity gradients and therefore large stresses will arise in the fluid next to the body, and these stresses will cause fluid layers farther out also to be dragged along. Eventually this process may come to an end and the flow will become steady. In the beginning the newly created boundary layer is extremely thin, so that the general geometry of the flow and the shape of the body cannot matter. This indicates that suddenly created boundary layers always start out their growth in the same universal way.

To estimate the universal growth of the boundary layer thickness $\delta(t)$, we use that at time t the fluid at the edge of the boundary layer will have changed its velocity from 0 to U, making the local acceleration of order U/t. The ratio between the local and advective acceleration then becomes,

$$\frac{|\partial\boldsymbol{v}/\partial t|}{|(\boldsymbol{v}\cdot\nabla)\boldsymbol{v}|} \sim \frac{U/t}{U^2/L} = \frac{L}{Ut}. \tag{28.4}$$

The time the fluid takes to pass the body is L/U. For times much shorter than this, $t \ll L/U$, the local acceleration dominates the advective acceleration term, and the boundary layer will continue to grow. In a 'young' boundary layer the advective acceleration can thus be disregarded, and the physics is controlled by the ratio of local to viscous acceleration,

$$\frac{|\partial \boldsymbol{v}/\partial t|}{\left|\nu \nabla^2 \boldsymbol{v}\right|} \approx \frac{U/t}{\nu U/\delta^2} = \frac{\delta^2}{\nu t}. \tag{28.5}$$

Requiring this to be of order unity, we find for $t \ll L/U$,

$$\boxed{\delta \sim \sqrt{\nu t}.} \tag{28.6}$$

A suddenly created boundary layer always starts out like this, growing with the square root of time. This behaviour is typical of viscous diffusion processes (page 233).

After the initial universal growth, the boundary layer comes to depend on the general geometry of the flow for $t \approx L/U$, when the estimate reaches the steady layer thickness (28.2). It takes more careful analysis to see whether the flow eventually 'goes steady', or whether instabilities arise, leading to a radical change in the character of the flow, such as boundary layer separation or turbulence.

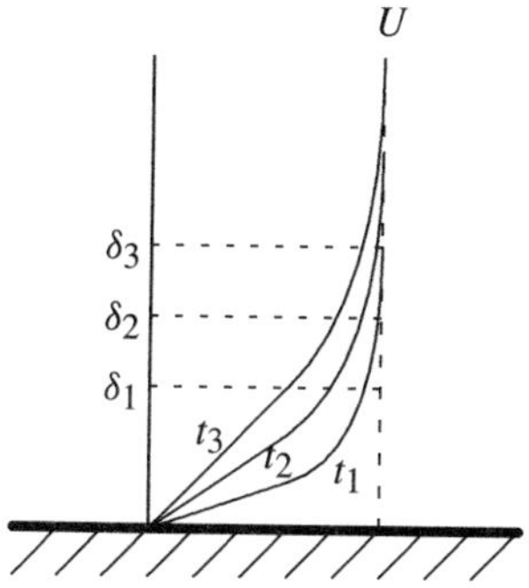

Initial growth of a laminar boundary layer. The three velocity profiles correspond to increasing times $t_1 < t_2 < t_3$ and increasing thicknesses $\delta_1 < \delta_2 < \delta_3$.

Influence of body geometry

The simplest geometry in which a steady boundary layer can be studied is a semi-infinite plate with its edge orthogonal to a uniform mainstream flow (solved analytically in section 28.4). Here the only possible length scale is the distance x from the leading edge, so we must have

$$\delta \sim \sqrt{\frac{\nu x}{U}}. \tag{28.7}$$

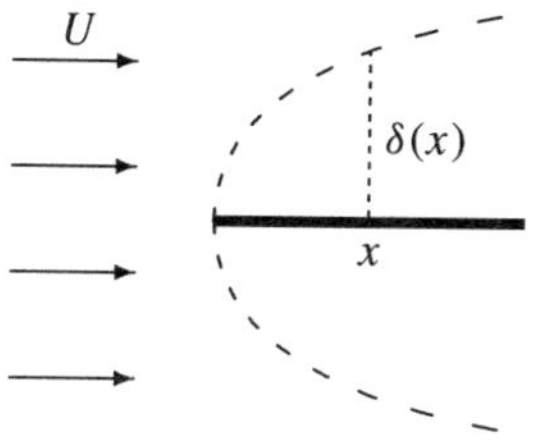

A semi-infinite plate in an otherwise uniform flow. The dashed curve is the estimated parabolic boundary layer shape.

This shows that the boundary layer grows thicker downstream, even if the mainstream flow is completely uniform and independent of x. Disregarding sound waves, the time $t = x/U$ it takes the flow to pass through the distance x is also the earliest moment that the leading edge can causally influence the flow near x. Intuitively one might say that the universal viscous growth of the boundary layer is curtailed by the encounter with the blast of undisturbed fluid coming in from afar.

Boundary layers have a natural tendency towards *downstream thickening*, because they build up along a body in a cumulative fashion. Having reached a certain thickness, a boundary layer acts as a not-quite-solid 'wall' on which another boundary layer will form. The thickness of a steady boundary layer is also strongly dependent on whether the mainstream flow is accelerating or decelerating along the body, behaviour which in turn is determined by the geometry. If the mainstream flow accelerates, i.e. grows with x, the boundary layer tends to remain thin. This happens at the front of a moving body, where the fluid must speed up to get out of the way. Conversely, towards the rear of the body, where the mainstream flow again decelerates in order to 'fill up the hole' left by the passing body, the boundary layer becomes rapidly thicker, and may even separate from the body, creating an unsteady, even turbulent, *trailing wake*. The von Kàrmàn vortex street (page 381) is an example of periodic unsteady flow in the wake of a body.

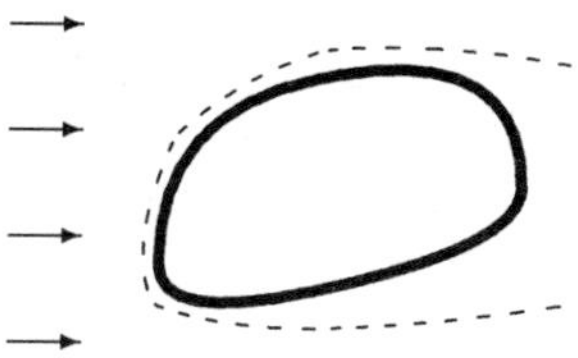

Sketch of the boundary layer around a bluff body in steady uniform flow. On the windward side the boundary layer is thin, whereas it widens and tends to separate on the lee side. In the channel formed by the separated boundary layer, unsteady flow patterns may arise.

Merging boundary layers

The increase of a boundary layer's thickness with downstream distance implies that the boundary layer around an infinite body must become infinitely thick or at least so thick that it fills out all the available space. In steady planar flow between moving plates (section 17.1), we saw that the velocity profile of the fluid interpolates linearly between the plate velocities, and there is nothing like a boundary layer with finite thickness near the plates. Likewise, in pressure-driven steady planar flow or pipe flow (sections 18.2 and 18.4), the exact shape of the velocity profile is parabolic (as long as the flow is laminar), whatever the viscosity of the fluid. Again we see no trace of a finite boundary layer in the exact solutions. Completely merged boundary layers are, however, only found in infinite systems. In a pipe of finite length the boundary layers grow out from the walls, starting at the entrance to the pipe and eventually merge at a distance, called the *entrance length*, which we estimated on page 254.

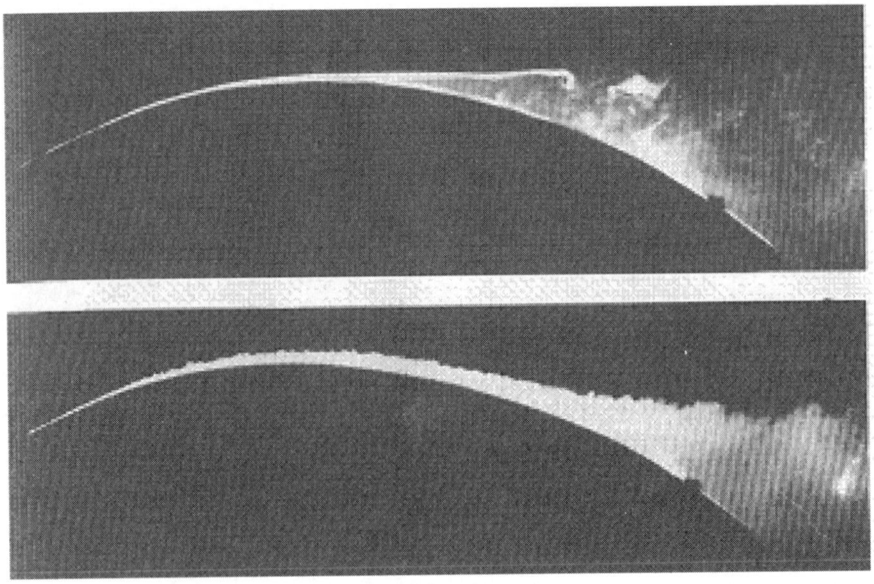

Figure 28.1. Laminar and turbulent boundary layers. The laminar layer (top) separates on the crest while the turbulent boundary layer separates further downstream. This is the phenomenon behind the drag crisis (page 274). Reproduced from M. R. Head, in *Flow Visualisation II* ed W. Merzkirch, Hemisphere, Washington 1982, pp 399–403.

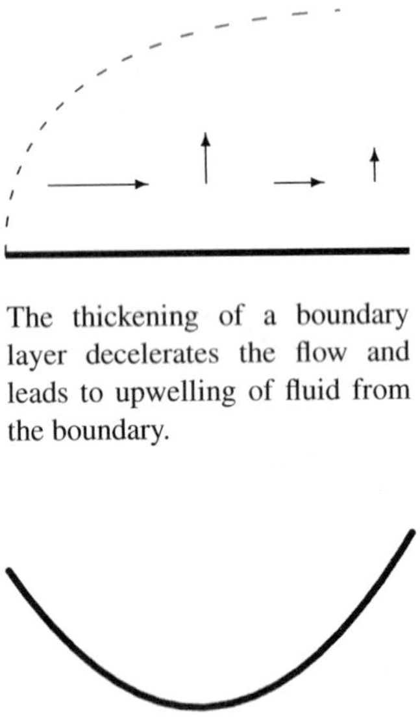

The thickening of a boundary layer decelerates the flow and leads to upwelling of fluid from the boundary.

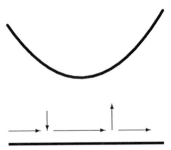

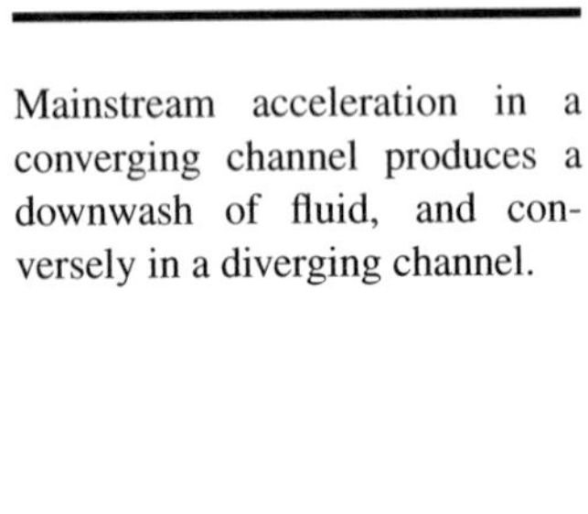

Mainstream acceleration in a converging channel produces a downwash of fluid, and conversely in a diverging channel.

Upwelling and downwash

Inside a boundary layer, at a fixed distance from a flat solid wall with a uniform mainstream velocity U, the flow decelerates downstream as the boundary layer thickens due to the action of viscosity. Mass conservation requires a compensating *upwelling* of fluid into the fluid at large. If, on the other hand, the boundary is permeable and fluid is sucked down through it at a constant rate, the upflow can be avoided, and a steady boundary layer of constant thickness may be created (problem 28.1).

The mainstream flow is determined by bodies and containers that guide the fluid and will generally not be uniform but rather accelerate or decelerate along the boundaries. An accelerating mainstream flow will counteract the natural deceleration in the boundary layer and may even overwhelm it, leading to a *downwash* towards the boundary. Mainstream acceleration thus tends to stabilize a boundary layer so that it has less tendency to thicken, and may lead to constant or even diminishing thickness. Conversely, if the mainstream flow decelerates, this will add to the natural deceleration in the boundary layer and increase its thickness as well as the upwelling.

Separation

Even at very moderate mainstream deceleration, the upwelling can become so strong that at some point the fluid flowing in the mainstream direction cannot feed it. Some of the fluid in the boundary layer will then have to flow against the mainstream flow. Between the forward and reversed flows there will be a separation line ending in a stagnation point on the wall. Such *flow reversal* was also noted in lubrication (page 399), although there are no boundary layers in creeping flow.

In the region of reversed flow, the velocity still has to vanish right at the boundary. Moving up from the boundary wall, the flow first moves backwards with respect to the mainstream, but farther from the wall it must again turn back to join up with the mainstream. The velocity gradient must accordingly be negative right at the boundary in the reversal region, corresponding to a negative wall stress σ_{wall}. At the separation point the wall stress must necessarily vanish, but flow reversal can in principle take place entirely within the boundary layer without proper separation, for example at a dent in the wall.

Turbulence

Boundary layers can also become turbulent. Turbulence efficiently mixes fluid in all directions. The orderly layers of fluid that otherwise isolate the wall from the mainstream flow all but disappear, and on average, the mainstream velocity will press much closer to the wall. Turbulence typically sets in downstream from the front of body when the laminar boundary layer has grown so thick that the *local Reynolds number* (for typical flow variations of size δ),

$$\mathrm{Re}_\delta = \frac{U\delta}{\nu}, \tag{28.8}$$

becomes large enough, say in the thousands. In a laminar boundary layer, the estimate (28.2) shows that $\mathrm{Re} \sim \mathrm{Re}_\delta^2$, so that turbulence will not arise in the boundary layer until the mainstream Reynolds number reaches a million or more, which incidentally is just about the range in which humans and many of their machines operate.

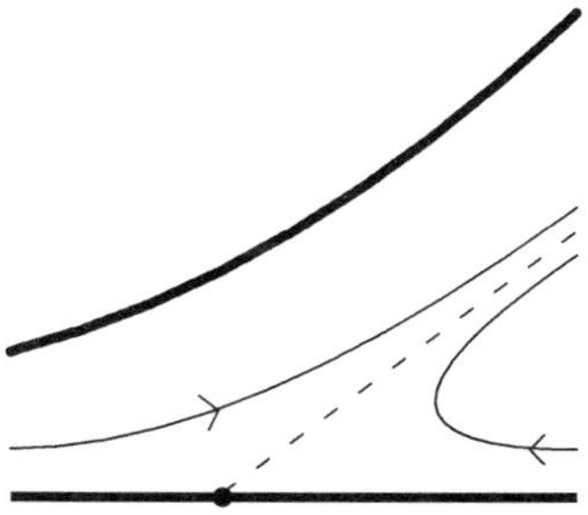

Flow reversal and boundary layer separation in a diverging channel with decelerating flow.

Although the turbulent velocity fluctuations press close to the wall, there will always remain a thin viscous, nearly laminar, sublayer close to the wall, in which the average velocity gradient normal to the wall rises linearly with distance. Since the mainstream velocity on average comes much closer to the wall, the average wall stress will be much larger than in a completely laminar boundary layer. The skin drag on a body is consequently expected to increase when the boundary layer becomes turbulent, though other changes in the flow may interfere and instead cause an even larger drop in the form drag at a particular value of the Reynolds number, as we saw in the discussion of the 'drag crisis' (page 274).

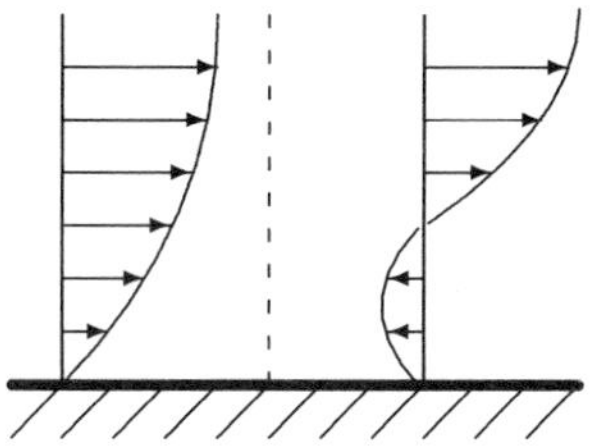

Velocity profiles before and after the separation point (dashed line).

28.2 The Stokes layer

The initial growth of the boundary layer at a flat plate suddenly set into motion (Stokes first problem) must also follow the universal law (28.6). In this case there is no intrinsic length scale for the geometry, and the transition to geometry-dependent steady flow cannot take place. The planar boundary layer, called the *Stokes layer*, can for this reason be expected to provide a clean model for the universal viscous growth.

Analytic solution

As usual it is best to view the system from the reference frame in which the plate and the fluid initially move with the same velocity U and the plate is suddenly stopped at $t = 0$. Assuming that the flow is planar with $v_x = v_x(y, t)$ and $v_y = 0$, the Navier–Stokes equation for incompressible flow reduces to the momentum-diffusion equation (17.5), which is repeated here for convenience

$$\frac{\partial v_x}{\partial t} = \nu \frac{\partial^2 v_x}{\partial y^2}. \tag{28.9}$$

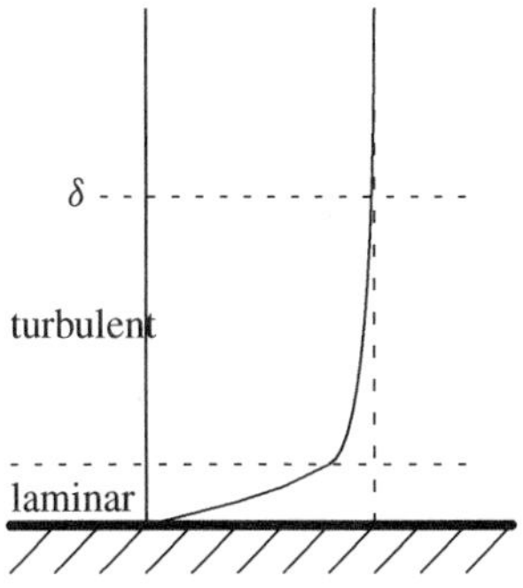

In a turbulent boundary layer there will always be a thin nearly laminar sublayer, in which the velocity profile rises linearly from the wall.

The linearity of this equation guarantees that the velocity everywhere must be proportional to U, and since there is no intrinsic length or time scale in the definition of the problem, the velocity field must be of the form,

$$v_x(y,t) = U f(s), \qquad s = \frac{y}{2\sqrt{\nu t}}. \tag{28.10}$$

The so far unknown function $f(s)$ should obey the boundary conditions $f(0) = 0$ and $f(\infty) = 1$. The factor two in the denominator is just a convenient choice.

Upon insertion of (28.10) into (28.9) we are led to an ordinary second-order differential equation for $f(s)$,

$$\boxed{f''(s) + 2sf'(s) = 0.} \tag{28.11}$$

Viewed as a first-order equation for $f'(s)$, it has the unique solution $f'(s) \sim \exp(-s^2)$. Integrating this expression once more over s and applying the boundary conditions, the final result becomes

$$f(s) = \frac{2}{\sqrt{\pi}} \int_0^s e^{-u^2}\, du = \mathrm{erf}(s), \tag{28.12}$$

where erf$(\cdot)$ is the well-known error function, shown in figure 28.2(a).

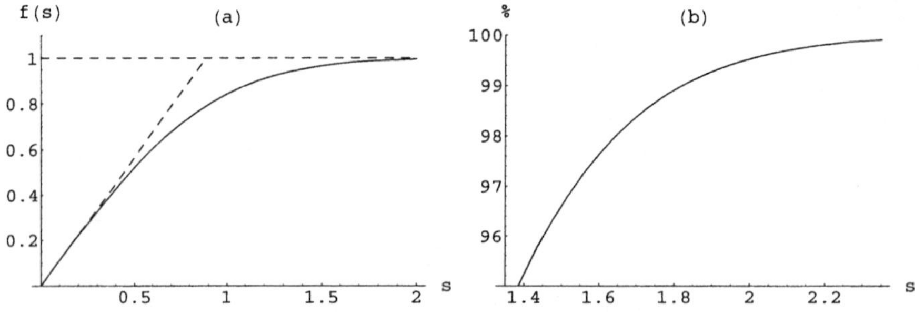

Figure 28.2. **(a)** The Stokes layer shape function $f(s)$. The sloping dashed line is tangent at $s = 0$ with inclination $f'(0) = 2/\sqrt{\pi}$. **(b)** Detail near $f(s) = 1$ in %.

Gaussian tail

For large values the error function approaches unity with a Gaussian tail, $1 - f(s) \sim \exp(-s^2) = \exp(-y^2/4\nu t)$, typical of momentum diffusion. The Gaussian tail extends all the way to spatial infinity for *any* positive time, $t > 0$, but how can that be, when the plate was only brought to stop at time $t = 0$? Will it take a finite time for this event to propagate to spatial infinity? The short answer is that we have assumed the fluid to be incompressible, and this—fundamentally untenable—assumption will in itself entail infinite signal speeds. At a deeper level, a diffusion equation like (28.9) is the statistical continuum limit of the dynamics of random molecular motion in the fluid, and although extremely high molecular speeds are strongly damped, they may in principle occur. The effective limit to diffusion speed is, as discussed before, always set by the finite speed of sound.

Vorticity

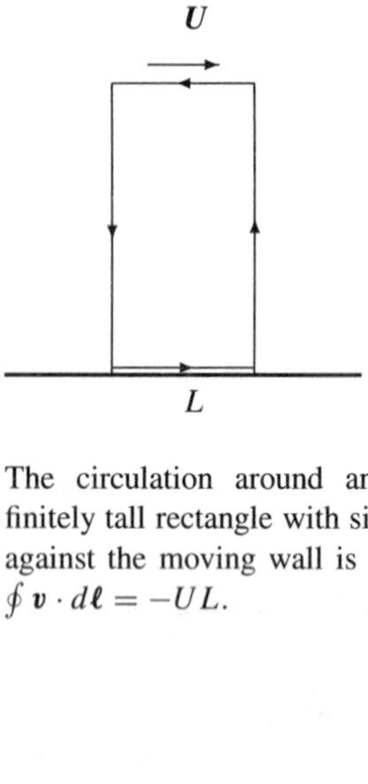

The circulation around an infinitely tall rectangle with side L against the moving wall is $\Gamma = \oint \boldsymbol{v} \cdot d\boldsymbol{\ell} = -UL$.

The vorticity field has only one component

$$\omega_z(y, t) = -\frac{\partial v_x(y, t)}{\partial y} = -\frac{U f'(s)}{2\sqrt{\nu t}} = -\frac{U}{\sqrt{\pi \nu t}} e^{-y^2/4\nu t}. \qquad (28.13)$$

When the plate was still moving for $t < 0$, stopped, the flow was everywhere irrotational. Afterwards there is evidently vorticity everywhere in the boundary layer. Where did that come from?

Consider a (nearly) infinite rectangle with support of length L on the plate. By Stokes' theorem the total flux of vorticity (or circulation) through the rectangle is $\Gamma = \int \boldsymbol{\omega} \cdot d\boldsymbol{S} = \oint \boldsymbol{v} \cdot d\boldsymbol{\ell}$. The fluid velocity always vanishes on the plate, is orthogonal to the sides and approaches the constant U at infinity, so that we obtain $\Gamma = -UL$. Since the circulation is constant in time, vorticity is not generated inside the boundary layer itself during its growth, but rather at the plate surface during the instantaneous deceleration to zero velocity. If the plate did not stop with infinite deceleration, but followed a gentler road $U(t)$ from U to 0, the circulation $\Gamma(t) = (U(t) - U)L$ would also have decreased gently from 0 to $-UL$. The conclusion is that vorticity is generated at the plate surface during the deceleration, and afterwards it diffuses away from the plate and into the fluid at large without changing the total circulation.

Thickness

The velocity field is self-similar because it only depends on the dimensionless variable $s = y/2\sqrt{\nu t}$. At different times the velocity profiles only differ by the 'vertical' length scale $2\sqrt{\nu t}$. There is no cut-off in the infinitely extended Gaussian tail and therefore no 'true' thickness δ. Conventionally, one defines the boundary layer thickness to be the distance where the velocity has reached 99% of the terminal velocity. The solution to $f(s) = 0.99$ is $s = 1.82\ldots$ (see figure 28.2(b)), such that

$$\delta_{99} = 1.82\ldots * 2\sqrt{\nu t} \approx 3.64\sqrt{\nu t}. \qquad (28.14)$$

In section 28.7 we shall meet other more physical definitions of boundary layer thickness.

28.3 Boundary layer theory

When Prandtl introduced the concept of boundary layers he pointed out that there were simplifying features, allowing for less complicated equations. The greater simplicity comes from the assumption of nearly ideal mainstream flow with $\mathrm{Re} \gg 1$, which according to the estimate (28.2) implies that boundary layers are thin, i.e. $\delta \ll L$ where L is the length scale for variations in the mainstream flow.

We shall—as Prandtl did—consider only the two-dimensional case with an infinitely extended planar boundary wall at $y = 0$ and a unidirectional mainstream flow along x. In the absence of viscosity, the incompressible fluid would slip along the boundary, $y = 0$, with a slowly varying *slip-velocity*, $v_x = U(x)$. Leaving out gravity, it follows from Bernoulli's theorem (16.15) that there must be an associated slip-flow pressure at the boundary,

$$P(x) = P_0 - \tfrac{1}{2}\rho_0 U(x)^2, \tag{28.15}$$

where P_0 is a constant. The slip-flow pressure simply reflects the variation in slip-velocity along the boundary.

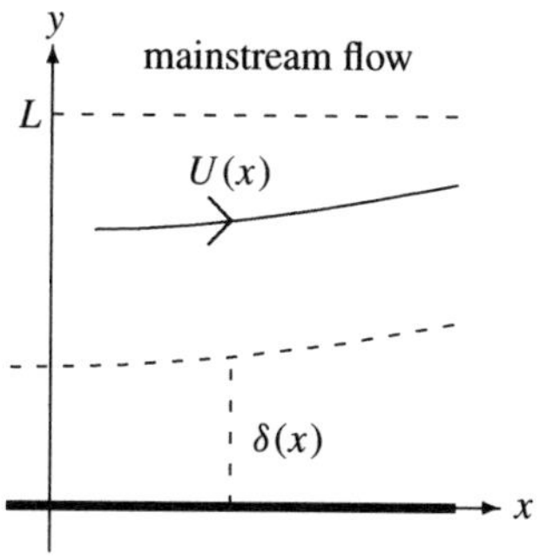

Geometry of two-dimensional planar boundary flow. In the absence of viscosity there would be a slowly varying slip-flow $U(x)$ along the boundary. Viscosity interposes a thin boundary layer of thickness $\delta(x)$ between the slip-flow and the boundary.

The Prandtl equations

In viscous flow the no-slip condition demands that the true velocity must change rapidly from $v_x = 0$ right at the boundary $y = 0$ to $v_x = U(x)$ outside the boundary layer $y \gg \delta$. Formulated more carefully, the slip-flow velocity $U(x)$ and boundary pressure $P(x)$ should now be understood as describing the flow in the region $\delta \ll y \ll L$, well outside the boundary layer but still so close to the boundary that the mainstream flow depends mainly on x. In the mainstream proper, for $y \gtrsim L$, flow and pressure come to depend on the general flow geometry with other length scales for major flow variations along both x and y.

The continuity equation,

$$\frac{\partial v_x}{\partial x} + \frac{\partial v_y}{\partial y} = 0 \tag{28.16}$$

determines the upflow v_y both inside and outside the boundary layer. Integrating over y, and using the boundary condition $v_y = 0$ for $y = 0$, we obtain the exact relation,

$$\boxed{v_y(x, y) = -\frac{\partial}{\partial x}\int_0^y v_x(x, y')\, dy'.} \tag{28.17}$$

Since major flow variations take place on the length scales L along x and δ along y, this equation permits us to estimate the upflow to be of magnitude $v_y \sim U\,\delta/L \sim U/\sqrt{\mathrm{Re}}$. The upflow inside the boundary layer will thus for $\mathrm{Re} \gg 1$ be much smaller than the slip-flow.

Mass conservation implies that the upflow inside the boundary layer continues into the slip-flow region $\delta \ll y \ll L$. Writing $v_x = U - (U - v_x)$ in (28.17) we obtain,

$$v_y(x, y) \approx -y\frac{dU(x)}{dx} + \frac{dQ(x)}{dx} \qquad \text{for } \delta \ll y \ll L, \tag{28.18}$$

where

$$Q(x) = \int_0^y (U(x) - v_x(x, y))\, dy \approx \int_0^L (U(x) - v_x(x, y))\, dy. \tag{28.19}$$

In the last step we have replaced the upper limit y by L under the assumption that $U - v_x$ vanishes rapidly outside the boundary layer for $y \gtrsim \delta$.

The quantity $Q(x)$ represents the volume of the slip-flow *displaced* by the boundary layer per unit of time (and per unit of length along z). The upflow (28.18) in the slip-flow region thus has two contributions, one from the variations in slip-flow and one from the boundary layer itself. The latter represents the natural upwelling in the boundary layer discussed on page 410.

In two dimensions the steady-flow Navier–Stokes equations (18.1) become

$$v_x \frac{\partial v_x}{\partial x} + v_y \frac{\partial v_x}{\partial y} = -\frac{1}{\rho_0}\frac{\partial p}{\partial x} + \nu \left(\frac{\partial^2 v_x}{\partial x^2} + \frac{\partial^2 v_x}{\partial y^2} \right), \tag{28.20a}$$

$$v_x \frac{\partial v_y}{\partial x} + v_y \frac{\partial v_y}{\partial y} = -\frac{1}{\rho_0}\frac{\partial p}{\partial y} + \nu \left(\frac{\partial^2 v_y}{\partial x^2} + \frac{\partial^2 v_y}{\partial y^2} \right). \tag{28.20b}$$

In either of these equations, the double derivative after y is proportional to $1/\delta^2$, whereas the double derivative after x is proportional to $1/L^2$, making it a factor $1/\text{Re}$ smaller and thus negligible for $\text{Re} \gg 1$. Setting $v_x \approx U$ and $v_y \approx U\delta/L$ we estimate from any of the remaining terms in the second equation that the normal pressure gradient is $\partial p/\partial y \approx \rho_0 U^2 \delta/L^2$. Finally, multiplying this expression with δ we obtain the pressure variation across the boundary layer $\Delta_y p \approx \delta\, \partial p/\partial y \approx \rho_0 U^2 \delta^2/L^2 \sim \rho_0 U^2/\text{Re}$. For $\text{Re} \gg 1$ this is much smaller than the typical variation in slip-flow pressure $\Delta P \approx \rho_0 U^2$ and may be disregarded. In this approximation the true pressure in the boundary layer equals the slip-flow pressure everywhere, $p(x, y) \approx P(x)$. The slip-flow pressure appears to be 'stiff', and penetrates the boundary layer to act directly on the boundary.

Inserting $p = P$ in (28.20a) and dropping the second order derivative after x, we arrive at *Prandtl's momentum equation*,

$$\boxed{v_x \frac{\partial v_x}{\partial x} + v_y \frac{\partial v_x}{\partial y} = U \frac{dU}{dx} + \nu \frac{\partial^2 v_x}{\partial y^2}.} \tag{28.21}$$

Since v_y is given in terms of v_x by (28.17), we have obtained a single integro-differential equation which for any given $U(x)$ determines v_x, subject to the boundary conditions $v_x = 0$ for $y = 0$ and $v_x \to U$ for $y \to \infty$. The preceding analysis shows that the correction terms to this equation are of order $1/\text{Re}$. The Prandtl approximation, however, breaks down near a separation point, where the upflow becomes comparable to the mainflow.

28.4 The Blasius layer

Paul Richard Heinrich Blasius (1883–1970). *German physicist, a student of Prandtl's. Worked on boundary-layer drag and on smooth pipe resistance.*

The generic example of a steady laminar boundary layer is furnished by a semi-infinite plate with its edge orthogonal to a uniform flow with constant velocity U, a problem first solved by Blasius in 1908.

Self-similarity

As the main variation in v_x happens across the boundary layer, it may be convenient to measure y in units of the thickness of the boundary layer, estimated to be of order $\delta \sim \sqrt{\nu x/U}$ in (28.7). Let us for this reason write the velocity in dimensionless self-similar form,

$$v_x(x, y) = U f(s), \qquad s = y\sqrt{\frac{U}{2\nu x}}, \tag{28.22}$$

where $f(s)$ must satisfy the boundary conditions, $f(0) = 0$ and $f(\infty) = 1$. The factor of two in the square root is conventional. In principle, the function could also depend on the dimensionless variable $\text{Re}_x = Ux/\nu$, but the correctness of the above assumption will be justified by finding a solution satisfying the boundary conditions.

Upflow

From the equation of continuity (28.17) we obtain

$$v_y(x, y) = -\frac{\partial}{\partial x}\int_0^y v_x(x, y')\, dy' = -\frac{\partial}{\partial x}\left[\sqrt{2U\nu x}\; g(s)\right].$$

Here we have for convenience defined the integral

$$g(s) = \int_0^s f(s')\, ds' \tag{28.23}$$

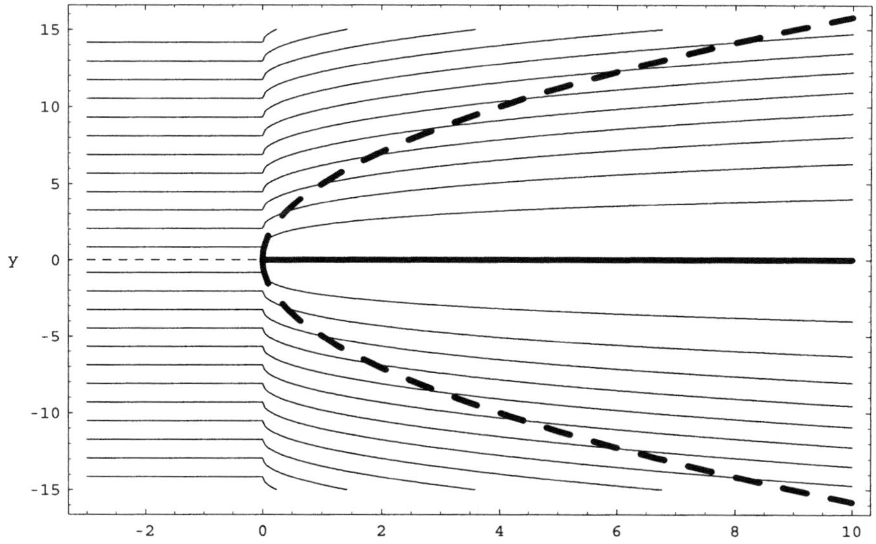

Figure 28.3. Streamlines around a semi-infinite thin plate with fluid flowing uniformly in from the left (see problem 28.4). Units are chosen so that $U = \nu = 1$. The thin dashed streamline terminates in a stagnation point. The heavy dashed curve indicates the 99% thickness, $y = \delta = 5\sqrt{x}$. The kink in the streamlines at $x = 0$ signals breakdown of the Prandtl approximation in this region.

such that $f(s) = g'(s)$. Carrying out the differentiation, we obtain the upflow from the layer,

$$v_y(x, y) = h(s)\sqrt{\frac{U\nu}{2x}}, \tag{28.24}$$

with

$$h(s) = sf(s) - g(s). \tag{28.25}$$

The asymptotic value $h(\infty) = \lim_{s\to\infty}(s - g(s))$ determines the total upflow injected into the mainstream by the boundary layer.

Blasius' equation

Finally, v_y is inserted into the Prandtl equation (28.21) and using that in this case U is constant, we obtain a single third-order ordinary differential equation, called the *Blasius equation*,

$$\boxed{g'''(s) + g(s)g''(s) = 0,} \tag{28.26}$$

which must be solved with the boundary conditions $g(0) = 0$, $g'(0) = f(0) = 0$, and $g'(\infty) = f(\infty) = 1$. Numeric integration yields the results shown in figure 28.3 and 28.4.

> **Numeric trick:** The condition $g'(\infty) = 1$ seems at first a bit troublesome to implement numerically. The trick is first to find the solution $\widetilde{g}(s)$ which satisfies the Blasius equation with with $\widetilde{g}''(0) = 1$. This solution converges at infinity to a value $a^2 \equiv \widetilde{g}'(\infty) = 1.65519\ldots$ instead of unity. The correct solution is finally obtained by the transformation,
>
> $$g(s) = \frac{1}{a}\widetilde{g}\left(\frac{s}{a}\right). \tag{28.27}$$
>
> It is a simple matter to verify that this function also satisfies the Blasius equation.

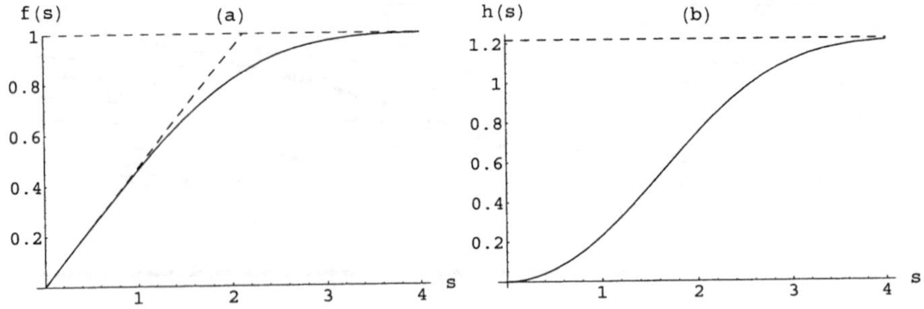

Figure 28.4. **(a)** The self-similar Blasius shape function $f(s)$ from (28.22). The dashed line in pane (a) has slope $f'(0) = 0.46960\ldots$ and is tangent at $s = 0$. It crosses unity at $s = 2.1295$. **(b)** The function $h(s)$ from (28.24). The dashed line indicates the asymptotic value $h(\infty) = 1.21678\ldots$ which determines the upwelling of fluid from the boundary layer into the mainstream flow.

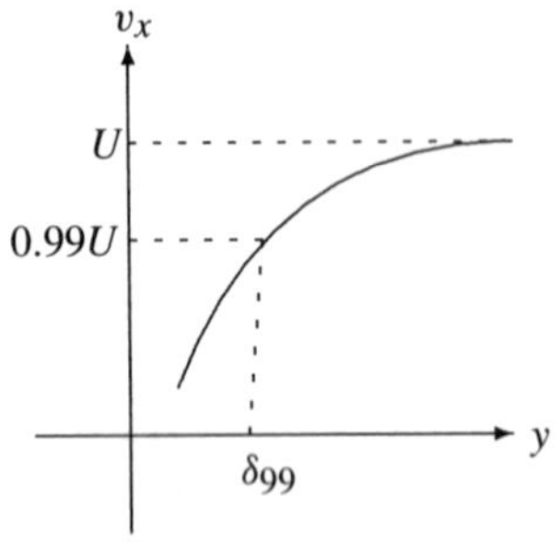

Definition of conventional thickness δ_{99} as the distance from the wall where the velocity has reached 99% of the mainstream velocity U.

Thickness and local Reynolds number

As for the Stokes layer (section 28.2) the conventional thickness of the Blasius layer is defined to be the distance $y = \delta$ where the velocity has reached 99% of the slip-flow velocity. The solution of $f(s) = 0.99$ is $s = 3.4719\ldots$, so the thickness becomes,

$$\delta_{99} = 3.4719\ldots * \sqrt{\frac{2\nu x}{U}} \approx 4.91\sqrt{\frac{\nu x}{U}}. \tag{28.28}$$

Typically, one uses $\delta \approx 5\sqrt{\nu x/U}$ for estimates. In dimensionless form, this may be expressed in terms of the local Reynolds number,

$$\mathrm{Re}_\delta = \frac{U\delta}{\nu} \approx 5\sqrt{\mathrm{Re}_x}, \tag{28.29}$$

where as before $\mathrm{Re}_x = Ux/\nu$ is the 'downstream' Reynolds number. Turbulence typically sets in around $\mathrm{Re}_x \approx 5 \times 10^5$, corresponding to $\mathrm{Re}_\delta \approx 3500$.

Wall shear stress and friction coefficient

The wall shear stress becomes

$$\sigma_{\text{wall}} = \eta \left.\frac{\partial v_x}{\partial y}\right|_{y=0} = \eta U \sqrt{\frac{U}{2\nu x}}\, f'(0). \tag{28.30}$$

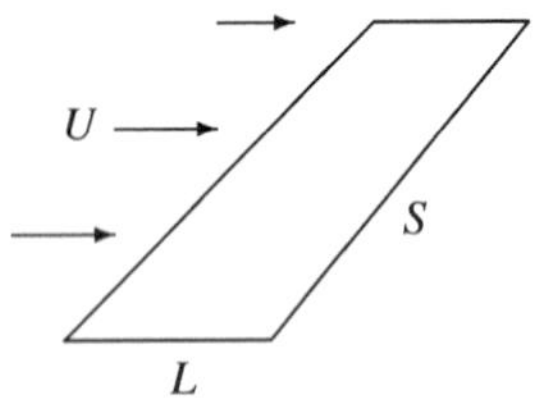

A flat, thin wing aligned with the flow only experiences skin drag.

It is customary also to make the wall stress dimensionless by dividing with $(1/2)\rho_0 U^2$ to get the so-called *local friction coefficient*,

$$C_f \equiv \frac{\sigma_{\text{wall}}}{(1/2)\rho_0 U^2} = f'(0)\sqrt{\frac{2}{\mathrm{Re}_x}} \approx \frac{0.664}{\sqrt{\mathrm{Re}_x}}. \tag{28.31}$$

In a turbulent boundary layer (section 28.5), this expression is replaced by a semi-empirical power law of the same general form but with a different power and coefficient. Note the singularity at $x = 0$, which signals breakdown of the boundary layer approximation at the leading edge of the plate.

Laminar skin drag on a flat wing

Consider now an infinitely thin rectangular 'wing' with 'chord' L in the direction of flow, and a 'span' S orthogonal to the flow (in the z-direction). Such an object generates no form drag, when it is aligned with the flow, and disregarding the influence of the rear and side edges of the wing, the total (skin) drag is obtained by integrating $\sigma_{\text{wall}} = (1/2)\rho_0 U^2 C_f(x)$ over both sides of the plate,

$$\mathcal{D} = 2\int_0^L \sigma_{\text{wall}}(x)\, S dx = 2\eta U S f'(0) \int_0^L \sqrt{\frac{U}{2\nu x}}\, dx = 4\eta U S f'(0)\sqrt{\frac{UL}{2\nu}}.$$

This becomes more transparent when expressed in terms of the dimensionless *drag coefficient* for the wing,

$$C_{\mathcal{D}} = \frac{\mathcal{D}}{(1/2)\rho_0 U^2 A} = \frac{8f'(0)}{\sqrt{2\text{Re}}} \approx \frac{2.6565}{\sqrt{\text{Re}}}, \tag{28.32}$$

where $A = LS$ is the wing's area and $\text{Re} = UL/\nu$ is its Reynolds number. The skin drag coefficient decreases with the square root of the Reynolds number, and is of little importance in most everyday situations with Reynolds number in the millions. Mostly the skin drag is overwhelmed by the form drag coefficient which does not decrease but rather stays constant for $\text{Re} \to \infty$.

Example 28.4.1 (Weather vane): A little rectangular metal weather vane with sides $L = 30$ cm and $S = 20$ cm in a $U = 10$ m s^{-1} wind has $\text{Re} = UL/\nu \approx 2 \times 10^5$, well below the onset of turbulence. The drag coefficient becomes $C_{\mathcal{D}} \approx 6 \times 10^{-3}$, and the total skin drag $\mathcal{D} \approx 0.02$ N, when the vane is aligned with the wind. This drag corresponds to a weight of merely 2 g, whereas the form drag and the other aerodynamic forces that align the vane are much stronger. One should certainly not dimension the support of the vane on the basis of the laminar skin drag!

28.5 Turbulent boundary layer in uniform flow

Sufficiently far downstream from the leading edge, the Reynolds number, $\text{Re}_x = Ux/\nu$, will eventually grow so large that the boundary layer becomes turbulent. Empirically, the transition happens for $5 \times 10^5 \lesssim \text{Re}_x \lesssim 3 \times 10^6$, depending on the circumstances, for example the uniformity of the mainstream flow and the roughness of the plate surface. We shall in the following discussion take $\text{Re}_x = 5 \times 10^5$ as the nominal transition point.

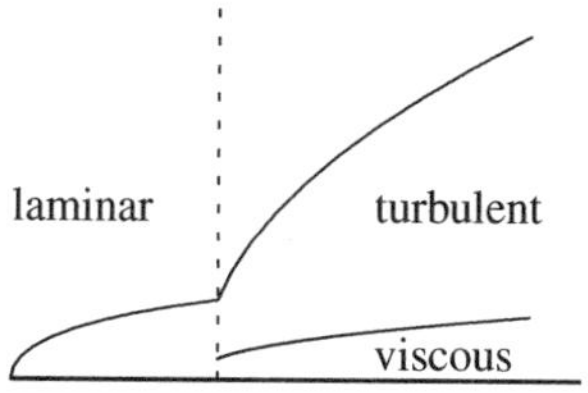

Sketch of the thickness of the boundary layer from the leading edge through the transition region. At the transition (dashed line), the turbulent layer grows rapidly whereas the viscous sublayer only grows slowly.

The line of transition across the plate is not a straight line parallel with the z-axis, but rather an irregular, time-dependent, jagged, even fractal interface between the laminar and turbulent regions. This is also the case for the extended, nearly 'horizontal' interface between the turbulent boundary layer and the fluid at large. Such intermittent and fractal behaviour is common to the onset of turbulence in all systems.

Friction coefficient

In a turbulent boundary layer, the true velocity field $\boldsymbol{v}$ fluctuates in all directions and in time around some mean value. Even very close to the wall, there will be notable fluctuations. The no-slip condition nevertheless has to be fulfilled and a thin sublayer dominated by viscous stresses must exist close to the wall. In this viscous sublayer the average velocity $\bar{v}_x$ rises linearly from the surface with a slope, $\partial \bar{v}_x/\partial y|_{y=0} = \bar{\sigma}_{\text{wall}}/\eta$ that can be determined from drag measurements.

A decent semi-empirical expression for the friction coefficient of a turbulent boundary layer was already given by Prandtl (see for example [80, 79] for details)

$$C_f \equiv \frac{\bar{\sigma}_{\text{wall}}}{1/2\rho_0 U^2} \approx \frac{0.027}{\text{Re}_x^{1/7}}. \tag{28.33}$$

The turbulent friction coefficient thus decreases much slower than the corresponding laminar friction coefficient (28.31). The two expressions cross each other at $\text{Re}_x \approx 7800$ which is far below the transition to turbulence, implying a jump from $C_f \approx 9.4 \times 10^{-4}$ to $C_f \approx 4.1 \times 10^{-3}$ at $\text{Re}_x \approx 5 \times 10^5$. Turbulent boundary layers thus cause much more skin drag than laminar boundary layers (by a factor of more than four at the nominal transition point).

In figure 28.5 the friction coefficient is plotted across the laminar and turbulent regimes. The transition from laminar to turbulent is in reality not nearly as sharp as shown here, partly because of the average over the jagged transition line. Eventually, for sufficiently large Re_x, the roughness of the plate surface makes the friction coefficient nearly independent of viscosity and thus of Re_x.

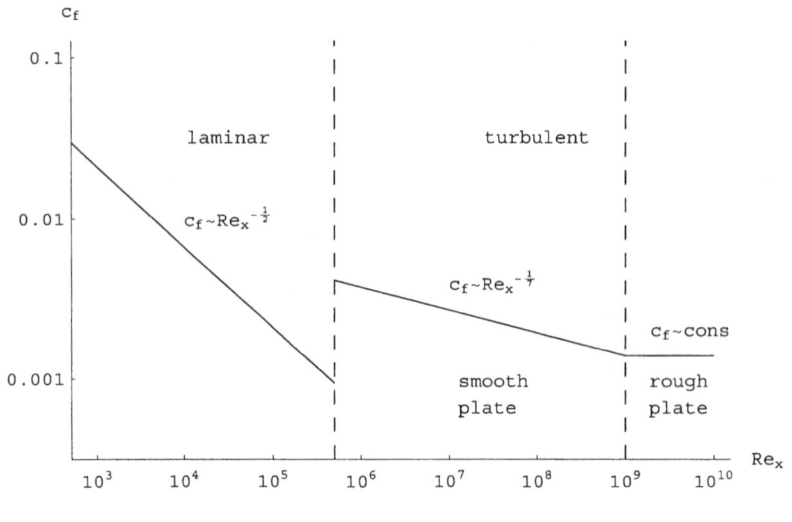

Figure 28.5. Schematic plot of the local friction coefficient $C_f = \bar{\sigma}_{\text{wall}}(x)/(1/2)\rho_0 U^2$ across the laminar and turbulent regions as a function of the downstream Reynolds number $\text{Re}_x = Ux/\nu$. The transitions at the nominal points $\text{Re}_x = 5 \times 10^5$ and at $\text{Re}_x = 10^9$ are in reality softer than shown here. The position of the second transition and the terminal value of C_f depend on the roughness of the plate surface (see [80] for details).

Drag on a flat wing

Let us again consider a finite 'wing' of size $A = L \times S$. For sufficiently large Reynolds number $\text{Re} = UL/\nu$, the boundary layer will always become turbulent some distance downstream from the the leading edge, and the drag will in general be dominated by the turbulent boundary layer's larger friction coefficient. Assuming a fully turbulent boundary layer, the dimensionless turbulent drag coefficient becomes (including both sides of the plate)

$$C_{\mathcal{D}} = \frac{2}{1/2\rho_0 U^2 A} \int_0^L \bar{\sigma}_{\text{wall}}(x)\, S dx = \frac{0.063}{\text{Re}^{1/7}}. \tag{28.34}$$

If the leading laminar boundary layer cannot be disregarded, this expression is somewhat modified.

> **Example 28.5.1 (Flag blowing in the wind):** A $A = 2 \times 2\ \text{m}^2$ flag in a $U = 10\ \text{m s}^{-1}$ wind has a Reynolds number of $\text{Re} \approx 1.3 \times 10^6$, well inside the turbulent region. The laminar skin drag coefficient is $C_{\mathcal{D}} \approx 2.3 \times 10^{-3}$ whereas the turbulent skin drag coefficient is $C_{\mathcal{D}} \approx 8.4 \times 10^{-3}$. The turbulent skin drag is only $\mathcal{D} \approx 1.68$ N which seems much too small to keep the flag straight. But flags tend to flap irregularly in the wind, thereby adding a much larger average form drag to the total drag, and giving them even in a moderate wind the nearly straight form that we admire so much.

Local drag and momentum balance

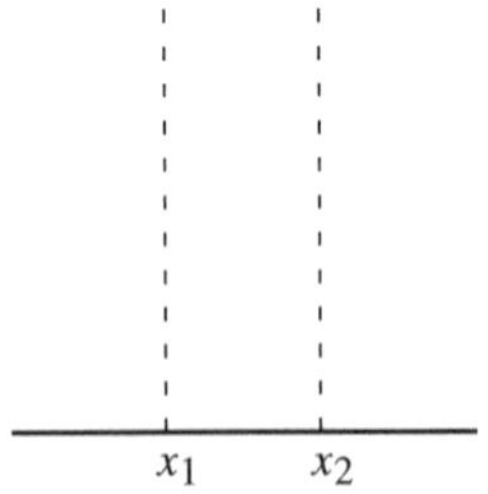

The drag on the plate between x_1 and x_2 must equal the rate of loss of momentum from the fluid between the dashed lines.

Momentum balance guarantees that the drag on any section of the plate, say for $x_1 < x < x_2$, must equal the rate of momentum loss from the slice of fluid above this interval, independent of whether the fluid is laminar or turbulent. Formally, we may use the Prandtl equations to derive such a relation for an infinitesimal slice of the boundary layer, as was first done by von Karman in 1921.

For constant slip-flow it follows trivially from (28.16) and (28.21) that

$$-\nu \frac{\partial^2 v_x}{\partial y^2} = \frac{\partial((U - v_x)v_x)}{\partial x} + \frac{\partial(v_y(U - v_x))}{\partial y}. \tag{28.35}$$

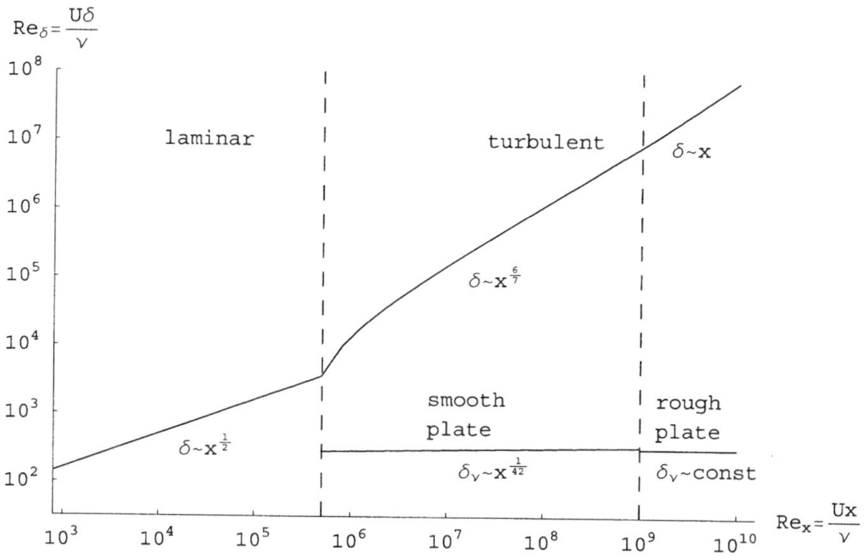

Figure 28.6. Schematic plot of the dimensionless 'true' thickness, represented by $\mathrm{Re}_\delta = U\delta/\nu$, as a function of downstream distance x, represented by $\mathrm{Re}_x = Ux/\nu$. Also shown is the thickness of the viscous sublayer δ_ν. The momentum thickness δ_{mom} is obtained from the refined expression (28.42) and is everywhere roughly a factor 10 smaller than the 'true' thickness δ. The real transition at the nominal value, $\mathrm{Re}_x = 5 \times 10^5$, is even softer than shown here. The transition from smooth to rough plate at $\mathrm{Re} = 10^9$ is barely visible.

Integrating over all y and using the boundary conditions, we see that the second term on the right-hand side does not contribute, and we obtain

$$\nu \left.\frac{\partial v_x}{\partial y}\right|_{y=0} = \frac{d}{dx}\int_0^\infty (U - v_x)v_x\, dy. \tag{28.36}$$

The quantity on the left-hand side is simply $\sigma_{\mathrm{wall}}/\rho_0$, and the integral on the right-hand side is the flux of lost momentum. In section 28.7 we shall make a more systematic study of such relations.

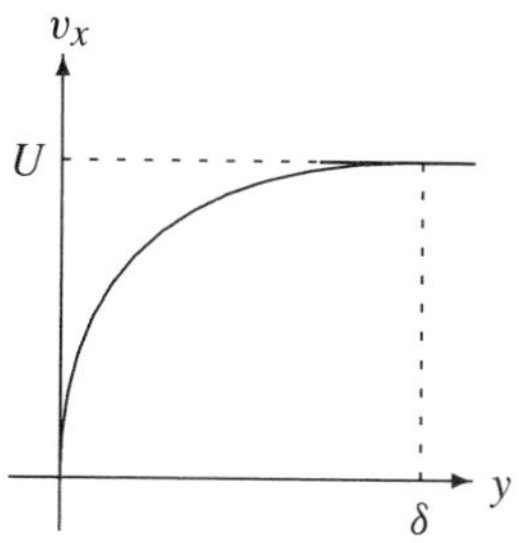

The turbulent velocity profile is approximately a power $\bar{v}_x \sim y^\gamma$ with $\gamma \approx 1/7$. The vertical tangent at $y = 0$ is unphysical, because it implies infinite wall stress. A finite wall stress is provided by a thin viscous sublayer.

Turbulent velocity profile and thickness

Empirically, the flat-plate turbulent boundary layer profile outside the thin viscous sublayer is decently described by the simple model (due to Prandtl [79]),

$$\frac{v_x}{U} = \left(\frac{y}{\delta}\right)^{1/7}, \qquad (0 \lesssim y < \delta). \tag{28.37}$$

Ignoring the sublayer which cannot contribute much to the integral, we obtain

$$\int_0^\delta (U - v_x)v_x\, dy = \frac{7}{72}U^2\delta. \tag{28.38}$$

Although the thickness δ is not known at this stage, we may use the von Karman relation (28.36) to relate it to the friction coefficient, for which we have the semi-empirical expression (28.33). In this way we obtain an ordinary differential equation for the thickness,

$$\frac{d\delta}{dx} = \frac{36}{7}C_f. \tag{28.39}$$

In a fully turbulent boundary layer, we may integrate this equation using (28.33) with the initial value $\delta = 0$ at $x = 0$, to get

$$\delta = \frac{36}{7}\int_0^x C_f\, dx \approx 0.16\frac{\nu}{U}\mathrm{Re}_x^{6/7}. \tag{28.40}$$

Expressing the thickness in dimensionless form by means of the local Reynolds number, we finally have

$$\boxed{\mathrm{Re}_\delta \equiv \frac{U\delta}{\nu} = 0.16\,\mathrm{Re}_x^{6/7}.} \tag{28.41}$$

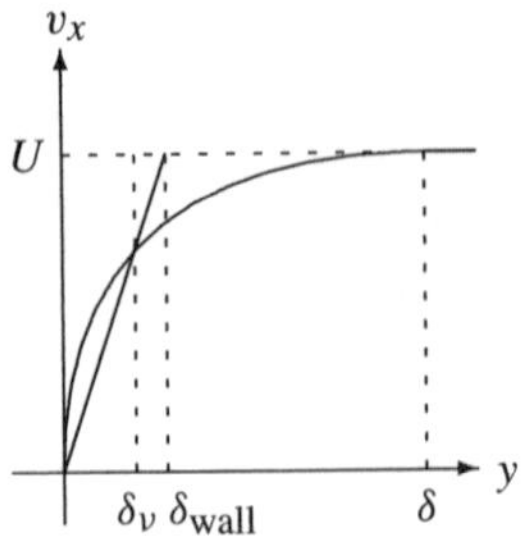

The 'true' thickness δ_ν of the viscous sublayer is obtained from the intercept between Prandtl's power profile (28.37) and the linearly rising velocity in the sublayer. The kink at $y = \delta_\nu$ is unphysical, because it gives rise to a small jump in shear stress that violates Newton's third law (slightly). The real transition from viscous sublayer to turbulent main layer is softer than shown here.

The jump in the local Reynolds number at the nominal transition point $x = x_0$ where $\mathrm{Re}_{x_0} = 5 \times 10^5$, is only apparent. A more precise expression for the local Reynolds number may be obtained by using the Blasius result for $0 \le x \le x_0$, and integrating the turbulent expression only for $x > x_0$,

$$\mathrm{Re}_\delta = \begin{cases} 5\sqrt{\mathrm{Re}_x} & x < x_0 \\ 5\sqrt{\mathrm{Re}_{x_0}} + 0.16\left(\mathrm{Re}_x^{6/7} - \mathrm{Re}_{x_0}^{6/7}\right) & x > x_0. \end{cases} \tag{28.42}$$

By construction, this expression is continuous across the nominal transition point (see figure 28.6).

The viscous sublayer

It is also possible to get an estimate of the thickness δ_ν of the viscous sublayer from the intercept between the linearly rising field, $v_x = y\bar{\sigma}_{\rm wall}/\eta$, in the sublayer and the power law (28.37). Demanding continuity at $y = \delta_\nu$, we get

$$\frac{\bar{\sigma}_{\rm wall}}{\eta}\delta_\nu = U\left(\frac{\delta_\nu}{\delta}\right)^{1/7}.$$

Solving this equation for δ_ν and inserting $\sigma_{\rm wall}$ from (28.33) and δ from (28.41) we obtain the remarkable expression,

$$\frac{U\delta_\nu}{\nu} = 206\mathrm{Re}_x^{1/42}. \tag{28.43}$$

At the nominal transition point $\mathrm{Re}_x = 5 \times 10^5$, this becomes 282, which grows to 337 at $\mathrm{Re}_x = 10^9$. The sublayer thickness is also plotted in figure 28.6 and its variation with Re_x is barely perceptible.

It is now also possible to calculate what fraction of the terminal velocity the fluid has achieved at the 'edge' of the sublayer,

$$\frac{v_x|_{y=\delta_\nu}}{U} = 2.8\mathrm{Re}_x^{-5/42}. \tag{28.44}$$

At the nominal transition point it is 0.59 and falls by roughly a factor 2 to 0.24 at $\mathrm{Re}_x = 10^9$.

* 28.6 Self-similar boundary layers

In the two preceding sections we have only discussed the case of constant slip-flow, but now we turn to the study of slip-flows that vary with x. In this section we shall focus on the generalization of the self-similar laminar Blasius solution to non-constant flat-plate slip-flow $U(x)$. The assumption of self-similarity has before allowed us to convert the partial differential equations of fluid mechanics into ordinary differential equations. We shall now see that self-similarity implies that the Prandtl equations essentially only permit power law slip-flows of the form $U \sim x^m$. The class of such slip-flows is on the other hand sufficiently general to illustrate what happens when a slip-flow accelerates ($m > 0$) or decelerates ($m < 0$), although it cannot handle the separation phenomenon.

The Falkner–Skan equation

A self-similar flow is defined to be of the form,

$$v_x = U(x)f(s), \qquad s = \frac{y}{\delta(x)}, \tag{28.45}$$

where $f(s)$ is a dimensionless function of the dimensionless variable. The velocity scale is set by $U(x)$ and the scale of the boundary layer thickness by $\delta(x)$. As for the Blasius layer we choose the boundary conditions to be $f(0) = 0$ and $f(\infty) = 1$. In the following we suppress the x-dependence wherever possible.

It follows immediately from the equation of continuity (28.17) that

$$v_y = -\frac{\partial}{\partial x}(U\delta g), \tag{28.46}$$

where $g(s)$ is again given by (28.23). Substituting v_x and v_y into the Prandl equation (28.21) we obtain after a little algebra the coupled ordinary differential equations,

$$f'' + \alpha g f' + \beta(1 - f^2) = 0, \qquad g' = f, \tag{28.47}$$

where a prime indicates the derivative with respect to s. Using a dot to denote the derivative after x, the purely numerical coefficients become,

$$\alpha = \frac{\delta\dot{\delta}U + \delta^2\dot{U}}{\nu}, \qquad \beta = \frac{\delta^2\dot{U}}{\nu}, \tag{28.48}$$

and may by construction be x-dependent. The crucial point is now that since $f(s)$ and $g(s)$ only depend on s, it follows from the differential equation (28.47) that both α and β must in fact be independent of x.

The definitions (28.48) may thus be viewed as two coupled ordinary differential equations for δ and U with constant values of α and β. To solve them we first note that they imply $d(\delta^2 U)/dx = (2\alpha - \beta)\nu$. For $\alpha > 0$ we may without loss of generality rescale δ such that we get $\alpha = 1$. Integrating we now obtain $\delta^2 U = (2 - \beta)\nu x$ where for $\beta \neq 2$ we choose the origin of x such that integration constant vanishes. Inserting this into the second equation in (28.48), it follows that $\beta = (2-\beta)x\dot{U}/U$, implying that $U \sim x^m$ with $m = \beta/(2 - \beta)$. Putting it all together, the allowed class is given by,

$$U(x) = Ax^m, \qquad \delta(x) = \sqrt{\frac{2\nu x}{(1+m)U(x)}}, \tag{28.49}$$

where $A > 0$ is a constant. For $m = 0$ this reduces to the Blasius case with $\delta = \sqrt{2\nu x/U}$.

The resulting nonlinear third-order differential equation,

$$\boxed{g''' + gg'' + \beta(1 - g'^2) = 0,} \tag{28.50}$$

with $\beta = 2m/(1 + m)$ was introduced in 1931 by Falkner and Skan[1], and the one-parameter family of solutions has since been extensively studied [61, 64]. A selection of profiles are shown in figure 28.7 for a few values of m. We leave it as an exercise to discuss the other families of solutions with $\alpha \leq 0$ or $\beta = 2$.

Upflow

A non-constant inviscid slip-flow $v_x = U(x)$ generates by itself (and mass conservation) an upflow $v_y = -y\dot{U}(x)$. The part of the upflow due to the presence of a boundary layer is consequently $v_y + y\dot{U}$, which may be written in the form

$$v_y + y\dot{U}(x) = \frac{\nu}{\delta}h(s), \tag{28.51}$$

$$h(s) = sf(s) - g(s) + \beta s(1 - f(s)). \tag{28.52}$$

This reduces to the Blasius result (28.24) for $\beta = 0$. Since for $s \to \infty$ we expect that $f(s) \to 1$ with an exponential tail, this function converges for $s \to \infty$ to a finite value, $h(\infty) = \lim_{s\to\infty}(s - g(s))$.

[1] V. M. Falkner and S.W.Skan, Some approximate solutions of the boundary layer equations, *Phil. Mag.* **12**, (1931) 865.

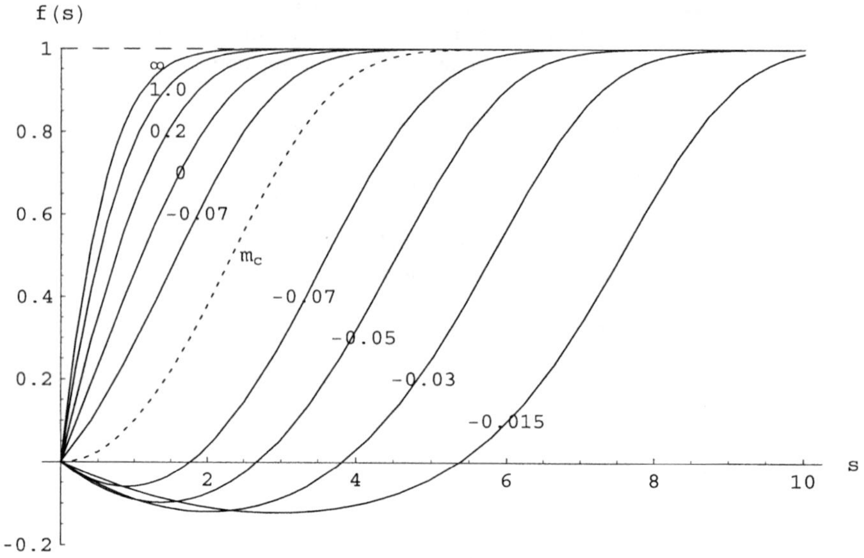

Figure 28.7. Self-similar Falkner–Skan velocity profiles $f(s)$ for select values of $m_c < m < \infty$ where $m_c = -0.0904286$. The Blasius profile is obtained for $m = 0$. Note that there are two solutions for each value of m in the interval $m_c < m < 0$, one of which has reversed flow close to the wall. The critical profile with $m = m_c$ is shown dashed.

Numeric method and results

Numeric integration of the Falkner–Skan equation is reasonably straightforward using a 'ballistic' method of integration. The integration process is initiated with $g(0) = f(0) = 0$ and $f'(0) = \mu$, and a search is made for the value of the slope μ that yields $f(\infty) = 1$. In the same way as there are two elevation angles that may be used to hit a target with a cannon ball, there may be more than one solution to the Falkner–Skan equation for certain values of m.

In figure 28.7 the velocity profile is shown for a selection of m-values and in figure 28.8 the wall slope $f'(0)$ and asymptotic upflow $h(\infty)$ are plotted as a function of m in the interval $-0.1 < m < 0.2$. The most conspicuous feature of the figures is the existence of two solutions in the interval $m_c < m < 0$ where $m_c = -0.0904286\ldots$. One solution has positive slope and the other negative, indicating backflow in the boundary layer. The slope vanishes at the critical point $m = m_c$, where the positive solution joins with the negative with vertical tangent.

Accelerating and decelerating slip-flow

In the region $0 < m < \infty$ the slip-velocity increases downstream. The solutions are unique and all resemble the Blasius solutions, except that the thickness scale

$$\delta(x) \sim x^{(1-m)/2} \tag{28.53}$$

grows slower than $\sqrt{x}$. For $m > 1$ the layer is even suppressed by the accelerating flow and gets thinner with x.

For $-1 < m < 0$ the slip-flow decelerates and the thickness grows faster than $\sqrt{x}$. There are, as mentioned, precisely two solutions in the interval $m_c < m < 0$. Just below the critical point, in the interval $-1/3 < m < m_c$, there are no solutions at all, but for any m in the interval $-1 < m < -1/3$ there appears to be from one to three distinct solutions with different values of the slope [64]. These solutions oscillate and overshoot while converging upon $f(\infty) = 1$. They are of no interest in the context of flat plate boundary layers.

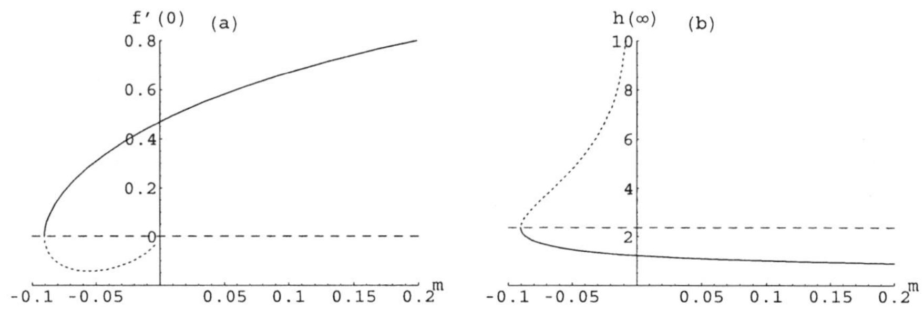

Figure 28.8. **(a)** The wall slope $f'(0)$ as a function of m in the interval $-0.1 \leq m \leq 0.2$. It is double-valued for negative m, and the negative solution (dotted) joins continuously with vertical tangent to the positive solution at the critical point $m_c = -0.0904286\ldots$. **(b)** The limiting upflow $h(\infty)$ as a function of m. It is also double-valued for negative m and takes the value $h(\infty) = 2.35885\ldots$ for $m = m_c$.

Separation?

Self-similar flows can strictly speaking not be used to model separation, because the velocity profile by construction has the same general shape for all x. If we let m slide down towards the critical value m_c, it nevertheless looks very much as if separation does take place (all over the x-axis at the same time). At the critical point, where the positive solution joins with the backflowing negative solution, there is a singularity with infinite slope derivative $f''(m_c)$. Such singularities are generic for the Prantdl equations (see section 28.8).

28.7 Exact results for varying slip-flow

A varying slip-flow $U(x)$ will strongly influence the flow in the boundary layer. Accelerating flow with $dU/dx > 0$ tends to suppress the boundary layer, so that its downstream thickness grows slower than the $\sqrt{x}$ of the Blasius layer. Sufficiently strong acceleration may even make the boundary layer become thinner downstream. Conversely, if the slip-flow decelerates, $dU/dx < 0$, the thickness will grow faster than $\sqrt{x}$, and sufficiently strong deceleration may lift the boundary layer off the plate and make it wander into the mainstream as a separated boundary layer.

In this section we shall establish some relations that are valid for any exact solution to Prandtl's equations. These relations will be useful for the discussion of the separation phenomenon to be taken up in section 28.8. Although we shall always think of a flat plate boundary layer, the following discussion is also valid for slowly curving walls, such as the much-studied flow around a cylinder.

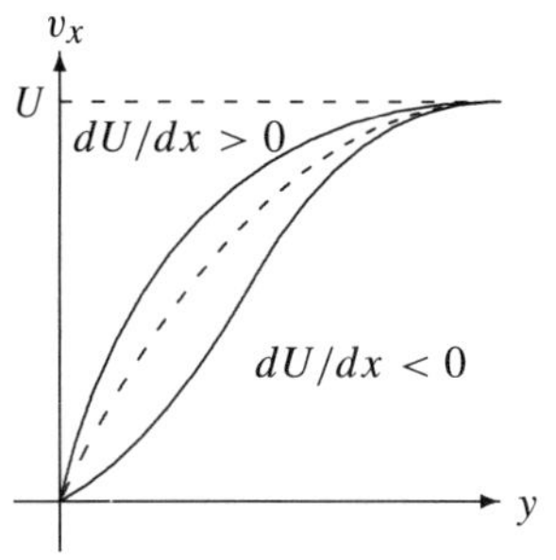

In accelerating flow, $dU/dx > 0$, the wall curvature is negative, whereas in decelerating flow the curvature is positive. The dashed curve sketches the velocity profile for vanishing wall curvature.

Exact wall derivatives

At the wall, $y = 0$, we know that both v_x and v_y must vanish, and that the derivative of the velocity at the wall,

$$\lambda = \left.\frac{\partial v_x}{\partial y}\right|_{y=0} \tag{28.54}$$

is in general non-vanishing, except at a separation point, where it has to vanish. The wall vorticity is $\omega_z = -\lambda$.

Setting $y = 0$ in the Prandtl equation (28.21) we immediately get the double derivative, also called the *wall curvature*,

$$\nu \left.\frac{\partial^2 v_x}{\partial y^2}\right|_{y=0} = -U\frac{dU}{dx}. \tag{28.55}$$

Its direct relation to the slip-flow opens up a qualitative discussion of the shape of the velocity profile. If the slip-flow accelerates ($dU/dx > 0$), the wall curvature will be negative and favor the approach of the velocity towards its terminal value, U. Conversely, if the slip-velocity decreases ($dU/dx < 0$), the wall curvature will be positive and adversely affect the approach to terminal velocity. This forces an inflection point into the velocity profile and raises the need for including higher derivatives to secure the turn-over towards the asymptotic slip-flow. For constant slip-flow, i.e. the Blasius case, we have $dU/dx = 0$, and the wall curvature vanishes.

The higher order wall derivatives may be calculated by differentiating the Prandtl equation repeatedly with respect to y. Differentiating once, we find

$$\nu \left. \frac{\partial^3 v_x}{\partial y^3} \right|_{y=0} = 0, \tag{28.56}$$

and once more

$$\nu \left. \frac{\partial^4 v_x}{\partial y^4} \right|_{y=0} = \lambda \frac{d\lambda}{dx}. \tag{28.57}$$

Clearly, this process can be continued indefinitely to obtain all wall derivatives of v_x depending only on U, λ and their derivatives.

Exact integral relations

We have already derived a relation (28.36) from momentum balance in uniform flow. For general varying slip-flow we first rewrite the Prandtl equation (28.21) in the form,

$$-\nu \frac{\partial^2 v_x}{\partial y^2} = (U - v_x)\frac{dU}{dx} + \frac{\partial[v_x(U - v_x)]}{\partial x} + \frac{\partial[v_y(U - v_x)]}{\partial y}. \tag{28.58}$$

Integrating this equation over y from 0 to ∞, and using the boundary values $v_y \to 0$ for $y \to 0$ and $U - v_x \to 0$ for $y \to \infty$, we obtain the general von Karman relation

$$\boxed{\nu \left. \frac{\partial v_x}{\partial y} \right|_{y=0} = \frac{dU}{dx} \int_0^\infty (U - v_x)\, dy + \frac{d}{dx} \int_0^\infty (U - v_x) v_x \, dy.} \tag{28.59}$$

It states that the drag on any infinitesimal interval of the plate equals the rate of momentum loss from the slice of fluid above the interval.

We may similarly derive a relation expressing kinetic energy balance by multiplying the Prandtl equation with v_x, and rewriting it in the form,

$$\nu \left(\frac{\partial v_x}{\partial y} \right)^2 = \frac{1}{2} \nu \frac{\partial^2 (v_x^2)}{\partial y^2} + \frac{1}{2} \frac{\partial ((U^2 - v_x^2) v_x)}{\partial x} + \frac{1}{2} \frac{\partial (v_y (U^2 - v_x^2))}{\partial y}. \tag{28.60}$$

Integrating this over all y and using the boundary conditions, we get

$$\boxed{\nu \int_0^\infty \left(\frac{\partial v_x}{\partial y} \right)^2 dy = \frac{1}{2} \frac{d}{dx} \int_0^\infty (U^2 - v_x^2) v_x \, dy.} \tag{28.61}$$

This relation states that the rate of heat dissipation in any infinitesimal slice equals the rate of loss of kinetic energy from the fluid. It is also possible to derive further relations for angular momentum balance and thermal energy balance [79, 61].

The integrands in the momentum and energy balance equations, (28.59) and (28.61), may be interpreted physically in terms of flow properties. The expression $U - v_x$ is the volume flux of fluid displaced by the plate, the expression $\rho_0(U - v_x)v_x$ is the flux of 'lost momentum' caused by the presence of the plate, $1/2\rho_0(U^2 - v_x^2)v_x$ is the flux of 'lost kinetic energy', and $\eta(\partial v_x/\partial y)^2$ is the density of heat dissipation.

Dynamic thicknesses

It is convenient to introduce dynamic length scales (or thicknesses) related to each of these quantities (and the wall stress),

$$\frac{1}{\delta_{\text{wall}}} = \frac{1}{U} \left. \frac{\partial v_x}{\partial y} \right|_{y=0}, \tag{28.62a}$$

$$\delta_{\text{disp}} = \frac{1}{U} \int_0^\infty (U - v_x)\, dy, \tag{28.62b}$$

$$\delta_{\text{mom}} = \frac{1}{U^2} \int_0^\infty (U - v_x) v_x \, dy, \tag{28.62c}$$

$$\delta_{\text{ener}} = \frac{1}{U^3} \int_0^\infty (U^2 - v_x^2) v_x \, dy, \tag{28.62d}$$

$$\frac{1}{\delta_{\text{heat}}} = \frac{1}{U^2} \int_0^\infty \left(\frac{\partial v_x}{\partial y} \right)^2 dy. \tag{28.62e}$$

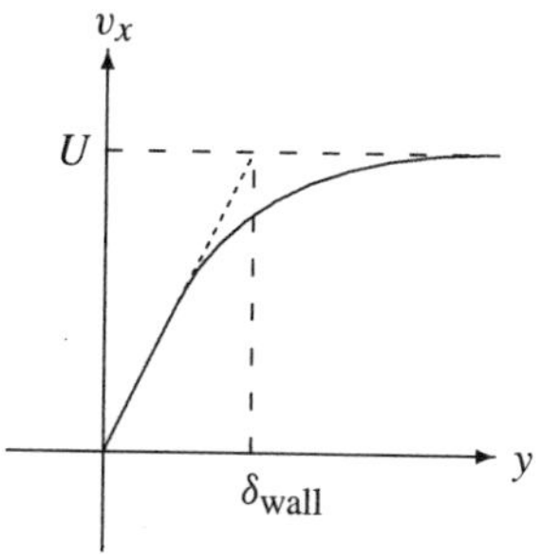

Definition of wall stress thickness from the intercept between the linearly rising velocity at small y and the mainstream velocity U.

In terms of these thicknesses, the momentum and energy balance equations now take the compact and quite useful forms,

$$\frac{\nu U}{\delta_{\text{wall}}} = U \frac{dU}{dx} \delta_{\text{disp}} + \frac{d\left(U^2 \delta_{\text{mom}}\right)}{dx}, \tag{28.63a}$$

$$\frac{\nu U^2}{\delta_{\text{heat}}} = \frac{1}{2} \frac{d(U^3 \delta_{\text{ener}})}{dx} \tag{28.63b}$$

We emphasize that these relations, like the wall derivatives (28.55)–(28.57), are fulfilled for any exact solution to the boundary layer equations.

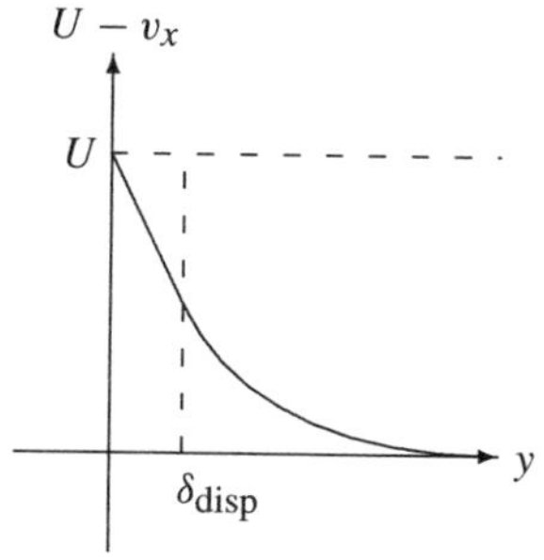

Definition of displacement thickness from the integral of the flux of volume loss, $U - v_x$.

The Blasius thicknesses

For the Blasius layer, self-similarity make all thicknesses proportional to the same basic scale, $\delta = \sqrt{2\nu x/U}$. Numeric integration yields,

$$\begin{aligned} \delta_{\text{wall}} &\approx 2.129\,\delta, & \delta_{\text{disp}} &\approx 1.217\,\delta, & \delta_{\text{mom}} &\approx 0.470\,\delta, \\ \delta_{\text{ener}} &\approx 0.738\,\delta, & \delta_{\text{heat}} &\approx 2.708\,\delta, & \delta_{99} &\approx 3.472\,\delta. \end{aligned} \tag{28.64}$$

The self-similarity thus guarantees that the ratios between any thicknesses are pure numbers independent of x. The integral relations (28.63) simplify in this case to,

$$2\delta_{\text{wall}}\delta_{\text{mom}} = \delta_{\text{heat}}\delta_{\text{ener}} = 2\delta^2. \tag{28.65}$$

These relations are of course fulfilled for the numeric values above.

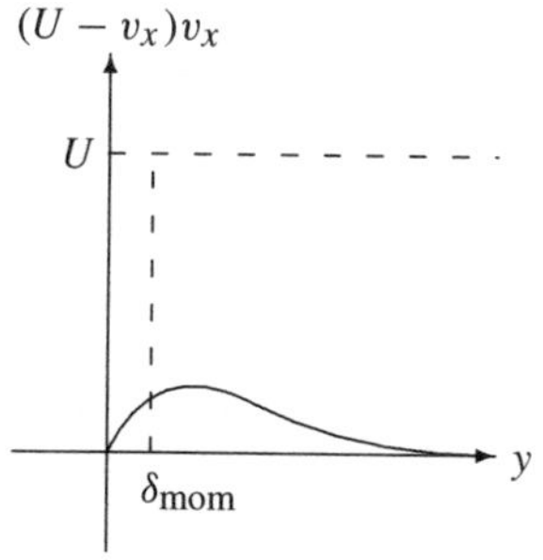

Definition of momentum thickness from the integral of the flux of momentum loss, $(U - v_x)v_x$.

28.8 Laminar boundary layer separation

When a separating boundary layer takes off into a decelerating mainstream, the character of the mainstream flow is profoundly changed, thereby actually invalidating the Prandtl approximation. Careful analysis has revealed that this is a generic problem, to which the boundary layer equations respond by developing an unphysical singularity at the point of separation. This so-called *Goldstein singularity* [64] prevents us in general from using boundary layer theory to connect the regions before and after separation.

Although Prandtl's boundary layer theory for this reason is useless for separation problems, the existence of a singularity is nevertheless believed to be an indicator of boundary layer separation in the general vicinity of the point where the singularity occurs. During the twentieth century the problem of predicting the singular separation point for boundary layers around variously shaped objects has been of great importance to fluid mechanics, for fundamental as well as technological reasons. It has proven to be a challenging problem, to say the least [61, 64]. A number of approximative schemes have been proposed

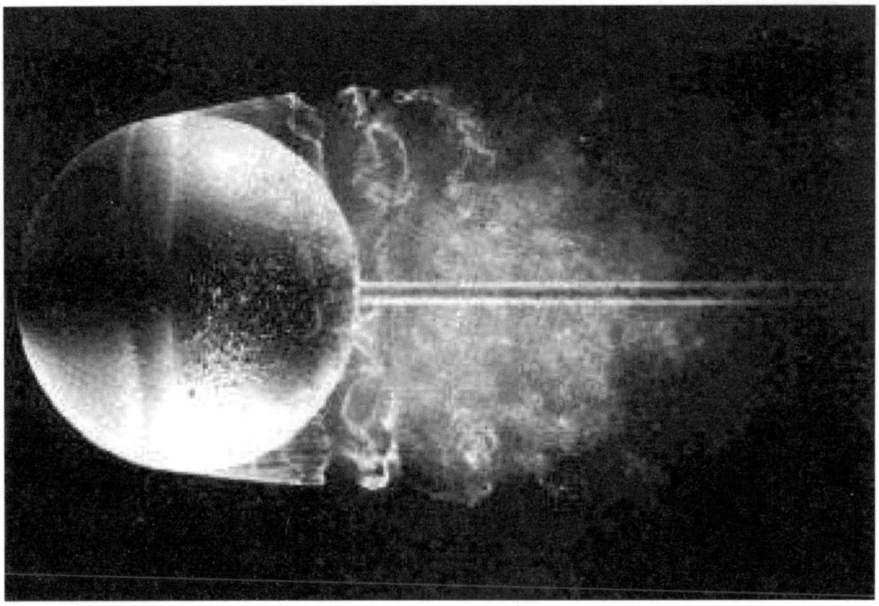

Figure 28.9. Laminar boundary layer separation from a sphere at Re = 15 000. Note how the boundary layer separates at the forward facing half of the sphere. ONERA photograph, H. Werle *Rech. Aerospace* **198-5** (1980) 35–49.

and tried out, and with suitable empirical input, they compare reasonably well with analytic or numeric calculations [79].

The Goldstein singularity is unavoidable as long as we persist in the belief that we can employ Prandtl's equations and also specify the slip-flow velocity as we wish. The price to pay for avoiding the singularity is that the Prandtl equations must be replaced by the Navier–Stokes equations and that the mainstream flow cannot be fully specified in advance, but has to be allowed to be influenced by what happens deep inside the boundary layer. Since separation originates in the innermost viscous 'deck' of the boundary layer, viscosity thus takes a decisive part in selecting the presumed inviscid flow at large, again emphasizing that inviscid flow solutions are not unique, and that inviscid flow is indeed an ideal.

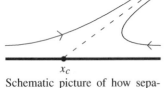

Schematic picture of how separation is thought to take place in a decelerating flow. The mainstream flow is profoundly changed by the separation both upstream and downstream from the separation point.

In the last half of the twentieth century, it has been conclusively demonstrated through theoretical analysis and numerical simulation that the Navier–Stokes equations do not lead to any boundary layer singularities and do in fact smoothly connect the regions before and after separation. The most successful method is in fact a natural extension of Prandtl's idea of dividing the flow into two 'decks', a viscous transition layer and an inviscid slip-flow layer, that in the end are 'stitched together' to form a complete boundary layer. In the 'triple deck' approach the transition layer is further subdivided into a near-wall viscous sublayer and a second viscous layer interpolating between the sublayer and the slip-flow. Unfortunately, there does not seem to be any simple way of presenting this modern 'interactive' boundary layer theory [61, 64, 68].

Sydney Goldstein (1903–89). *British mathematical physicist. Worked on numerical solutions to the steady-flow laminar boundary layer equations.*

In this section we shall first justify that the Goldstein singularity exists, and make a primitive attempt to determine its position in a number of cases where the exact position is known (table 28.1). In the remainder of the section we shall see that it is possible to predict the position of the singularity to an accuracy better than 1% from momentum and energy balance alone without any empirical input. A simplified version of this model yields an accuracy better than 3%.

The wall-anchored model

The simplest model of boundary layer flow is obtained by approximating the velocity profile with a fourth-order polynomial in y constructed from the exact wall derivatives (page 423)

$$v_x = \lambda\, y - \frac{U\dot{U}}{2\nu}\, y^2 + \frac{\lambda\dot{\lambda}}{24\nu}\, y^4, \tag{28.66}$$

where a dot is used to denote differentiation with respect to x. Evidently, this expression is exact for $y \to 0$, but for $y \to \infty$ where all three terms diverge, there is of course a problem. In decelerating slip-flow ($\dot{U} < 0$), the second-order term is always positive, and the fourth-order term is always negative just upstream from the separation point, because the slope λ is positive and decreasing towards zero at separation. After an initial rise governed by the first- and second-order terms, the fourth-order term must eventually pull down the profile to minus infinity, unless we 'catch' it at the top.

We 'catch' the profile by requiring the velocity to join smoothly at maximum with the given slip-flow $U(x)$, i.e. $v_x = U(x)$ and $\partial v_x/\partial y = 0$ at $y = \delta$, resulting in the conditions,

$$\lambda\,\delta - \frac{U\dot{U}}{2\nu}\delta^2 + \frac{\lambda\dot{\lambda}}{24\nu}\delta^4 = U, \tag{28.67a}$$

$$\lambda - \frac{U\dot{U}}{\nu}\delta + \frac{\lambda\dot{\lambda}}{6\nu}\delta^3 = 0. \tag{28.67b}$$

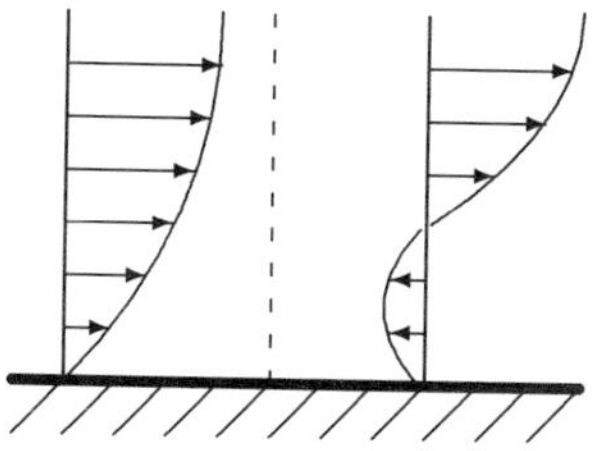

Velocity profiles before and after the separation point (dashed line).

Eliminating $\lambda\dot{\lambda}$ between these equations we obtain an algebraic relation between λ and δ,

$$\lambda = \frac{4U}{3\delta}\left(1 + \frac{\dot{U}\delta^2}{4\nu}\right). \tag{28.68}$$

This shows that there is in fact only one free parameter in the problem, say δ, satisfying a first-order differential equation (see problem 28.6). It also follows from this relation that since $\lambda = 0$ at a separation point the thickness $\delta_c = \delta(x_c)$ will be finite here,

$$\delta_c = 2\sqrt{-\frac{\nu}{\dot{U}_c}}, \tag{28.69}$$

where $\dot{U}_c = \dot{U}(x_c)$ is the 'deceleration' at the separation point ($\dot{U}_c < 0$).

From the second condition (28.67b) the behaviour of $\lambda\dot{\lambda}$ can be now determined in the vicinity of the separation point, where

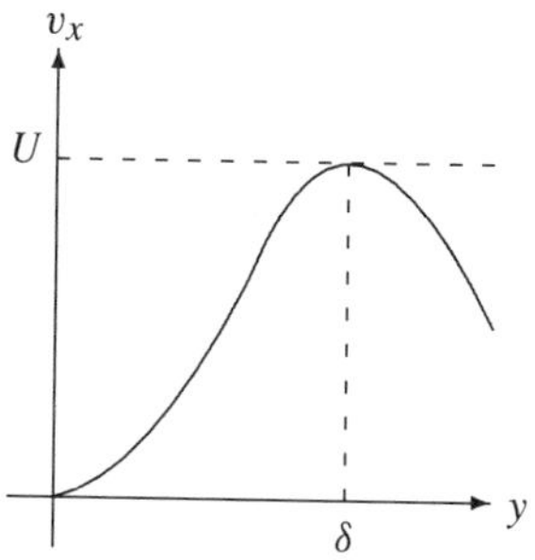

Wall-anchored fourth-order polynomial joins continuously with $v_x = U$ at its maximum $y = \delta$. The continuation beyond δ drops to $-\infty$ in a decelerating slip-flow and is unphysical.

$$2\lambda\dot{\lambda} \approx -3\frac{U_c\dot{U}_c^2}{\nu}. \tag{28.70}$$

Integrating over x using $\lambda = 0$ for $x = x_c$ we arrive at,

$$\lambda \approx \kappa\sqrt{x_c - x}, \qquad \kappa = -\dot{U}_c\sqrt{\frac{3U_c}{\nu}}, \tag{28.71}$$

which clearly demonstrates the existence of the Goldstein singularity.

Near the wall just upstream from the separation point, the velocity profile becomes,

$$\boxed{v_x \approx \lambda\,y \approx \kappa\,y\sqrt{x_c - x},} \tag{28.72}$$

and from mass conservation (28.17) we determine the corresponding upflow

$$v_y \approx -\frac{1}{2}\frac{d\lambda}{dx}y^2 \approx \frac{\kappa}{4}\frac{y^2}{\sqrt{x_c - x}}. \tag{28.73}$$

Evidently, the upflow diverges for all y at the separation point. Apart from being totally unphysical, this shows that it is not possible to solve the separation problem within the Prandtl approximation itself, because one of the conditions for this approximation, $|v_y| \ll |v_x|$, fails miserably near the separation point.

The model is analysed further in problem 28.6. The separation points obtained from numeric integration of this model are listed in the third column (marked 'wall') in table 28.1 for nine decelerating slip-flows. They agree rather poorly with the exact results (second column), overshooting by up to 50%. The poor performance of the model must be ascribed to the much too solid anchoring of the boundary layer to the wall, which tends to generate errors in the shape of the velocity profile in the bulk of the boundary layer.

Table 28.1. Table of decelerating slip-flows and the positions of their Goldstein singularities ('separation points') calculated in various models discussed in the text. The exact values x_c in the second column are taken from [79]. The separation points determined by the wall-anchored fourth-order polynomial (28.66) are listed in the third column, and have typical errors of 40%. The fourth column is obtained by Pohlhausen's method (28.99). It has errors less than 25% and is in all cases better than the wall-anchored approximation. In the fifth column the separation points are determined from both momentum and energy balance (28.63) with typical errors smaller than 1%. Finally, in the last column, the separation points are derived from the simple approximation (28.85) with the typical error being less than 3%.

	$U(x)$	x_c	wall	wall+mom	mom+ener	approximation
1.	$1-x$	0.120	0.176	0.157	0.121	0.123
2.	$\sqrt{1-x}$	0.218	0.314	0.277	0.219	0.221
3.	$(1-x)^2$	0.064	0.094	0.084	0.064	0.065
4.	$(1+x)^{-1}$	0.151	0.230	0.214	0.154	0.158
5.	$(1+x)^{-2}$	0.071	0.107	0.098	0.072	0.074
6.	$1-x^2$	0.271	0.365	0.312	0.270	0.268
7.	$1-x^4$	0.462	0.565	0.491	0.460	0.449
8.	$1-x^8$	0.640	0.729	0.652	0.643	0.621
9.	$\cos x$	0.389	0.523	0.447	0.386	0.383

The Pohlhausen family of profiles

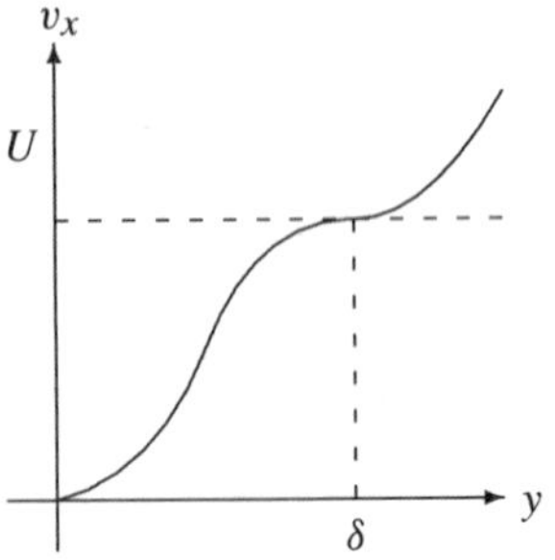

The fourth-order polynomial joins smoothly with $v_x = U$ with a horizontal inflection point at $y = \delta$. The continuation beyond $y = \delta$ which diverges to $+\infty$ for $\mu < 3$ is not used.

The wall-anchored approximation (28.66) suffers from an unnatural jump in the curvature $\partial^2 v_x/\partial y^2$ at the edge of the boundary layer at $y = \delta$. A smoother fourth-order polynomial was introduced by Pohlhausen in 1921, in which the coefficients are fixed by demanding the field to approach $v_x = U$ at $y = \delta$ with vanishing first- *and* second-order derivatives,

$$\frac{v_x}{U} = \mu\frac{y}{\delta} + 3(2-\mu)\left(\frac{y}{\delta}\right)^2 - (8-3\mu)\left(\frac{y}{\delta}\right)^3 + (3-\mu)\left(\frac{y}{\delta}\right)^4. \tag{28.74}$$

It contains two unknown functions: the dimensionless slope at the wall $\mu(x)$ and the layer thickness $\delta(x)$. One may readily verify that this is indeed $\mathcal{O}\left((y-\delta)^3\right)$ for $y \to \delta$.

Note that for $\mu < 2$, the second- and fourth-order terms are always positive. It is the negative third-order term, which secures the smooth turn-over towards the slip-flow at $y = \delta$. A selection of this family of profiles is shown in figure 28.10. The profiles are well defined for all μ, both before the separation point ($\mu > 0$) and after ($\mu < 0$). After separation they exhibit backflow, as one would expect, but the Goldstein singularity prevents us from connecting these two regions of solutions.

Pohlhausen solved the model by imposing the wall-curvature condition (28.55) together with the von Karman relation (28.63a). The details are given in problem 28.7, and the separation point predictions are tabulated in the fourth column (marked 'wall+mom') of table 28.1. The agreement is still rather poor with typical errors of 25%.

Momentum and energy balance

The nearly self-similar form of the Pohlhausen family (28.74) implies that all thicknesses are polynomials in μ scaled by δ. The wall, displacement and momentum thicknesses become

$$\frac{\delta}{\delta_{\text{wall}}} = \mu, \qquad \frac{\delta_{\text{disp}}}{\delta} = \frac{2}{5}\left(1-\frac{\mu}{8}\right), \qquad \frac{\delta_{\text{mom}}}{\delta} = \frac{4}{35}\left(1+\frac{\mu}{12}-\frac{5\mu^2}{144}\right), \tag{28.75}$$

and the energy and heat thicknesses,

$$\frac{\delta_{\text{ener}}}{\delta} = \frac{876}{5005}\left(1+\frac{\mu}{12}-\frac{253\mu^2}{10512}-\frac{7\mu^3}{3504}\right), \qquad \frac{\delta}{\delta_{\text{heat}}} = \frac{48}{35}\left(1-\frac{\mu}{12}+\frac{\mu^2}{16}\right). \tag{28.76}$$

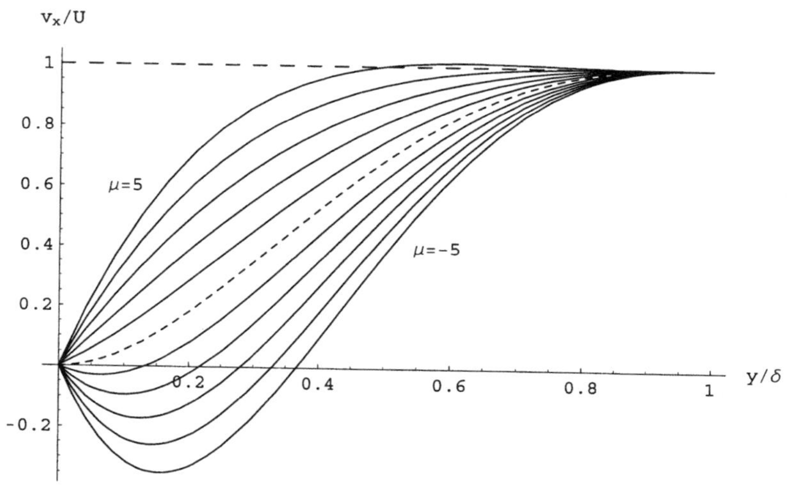

Figure 28.10. Profiles of the Pohlhausen family for integer values of μ from -5 to $+5$. The dashed curve is the profile at separation, corresponding to $\mu = 0$, and the profiles with $\mu < 0$ have backflow.

Inserting these into the integral relations (28.63), one arrives at two rather horrible coupled first-order differential equations for δ and μ. These differential equations are derived in problem 28.8 and solved numerically with the results shown in the column marked 'mom+ener' of table 28.1. They agree with the exact singularities to better than 1% in nearly all cases, demonstrating that together momentum and energy balance do predict the position of the Goldstein singularity with acceptable precision.

Approximative solution

We now turn to the question of whether it is possible to simplify the model and still retain a reasonable predictive ability. The approximation is based on the observation that the dimensionless ratios $\delta_{\rm mom}/\delta_{\rm ener}$ and $\delta_{\rm ener}/\delta_{\rm heat}$ are nearly constant because the terms linear in μ cancel in the ratios (see also figure 28.11).

Multiplying the energy relation (28.63b) with $4U^3\delta_{\rm ener}$, we may rewrite it as

$$\frac{d(U^6\delta_{\rm ener}^2)}{dx} = 4\nu U^5 \frac{\delta_{\rm ener}}{\delta_{\rm heat}}. \tag{28.77}$$

Approximating the ratio by a constant, we may integrate the equation to get,

$$U^6\delta_{\rm ener}^2 \approx 4\nu \frac{\delta_{\rm ener}}{\delta_{\rm heat}} \int_0^x U(x)^5\, dx. \tag{28.78}$$

The integral defines a slip-flow length scale

$$L(x) = \frac{1}{U(x)^5} \int_0^x U(x)^5\, dx, \tag{28.79}$$

and the energy relation may now be expressed as

$$\boxed{\delta_{\rm heat}\delta_{\rm ener} = 4\frac{\nu L(x)}{U(x)},} \tag{28.80}$$

which is of the same general form as the Blasius result (28.65), corresponding to $L(x) = x$.

The momentum equation (28.63a) is analogously multiplied with $2U^4\delta_{\rm mom}$, and rewritten in the form,

$$2\nu U^5 \frac{\delta_{\rm mom}}{\delta_{\rm wall}} = 2U^5\dot{U}\left(\frac{\delta_{\rm disp}}{\delta_{\rm mom}} - 1\right)\delta_{\rm mom}^2 + \frac{d(U^6\delta_{\rm mom}^2)}{dx}. \tag{28.81}$$

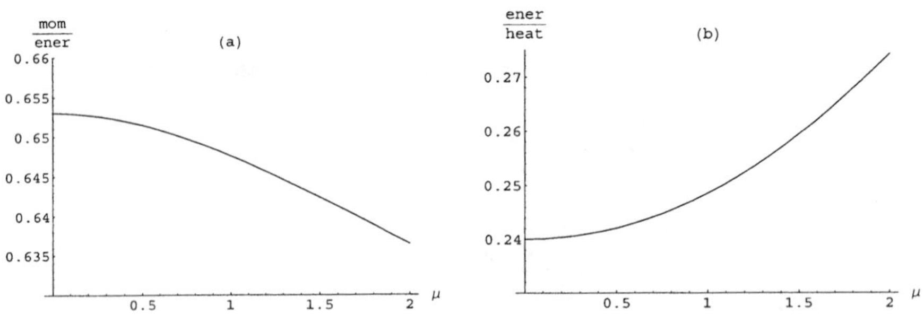

Figure 28.11. Nearly constant thickness ratios in the Pohlhausen family. **(a)** The ratio $\delta_{\rm mom}/\delta_{\rm ener}$ varies by about 1.3% around its mean value between $\mu = 0$ and $\mu = 2$. **(b)** The ratio $\delta_{\rm ener}/\delta_{\rm heat}$ varies by about 6.6% around its mean value in the same interval.

Writing $\delta^2_{\rm mom} = \delta^2_{\rm ener}(\delta^2_{\rm mom}/\delta^2_{\rm ener})$ and approximating the ratio $\delta^2_{\rm mom}/\delta^2_{\rm ener}$ by a constant, we may use (28.77) to get the following algebraic relation between the various thicknesses

$$2\nu\frac{\delta_{\rm mom}}{\delta_{\rm wall}} = 2\dot U\left(\frac{\delta_{\rm disp}}{\delta_{\rm mom}} - 1\right)\delta^2_{\rm mom} + 4\nu\frac{\delta^2_{\rm mom}}{\delta_{\rm ener}\delta_{\rm heat}}. \tag{28.82}$$

Rewriting this expression by means of (28.80) it becomes a dimensionless relation,

$$\boxed{\frac{\dot U L}{U} = \frac{\delta_{\rm ener}\delta_{\rm heat} - 2\delta_{\rm mom}\delta_{\rm wall}}{4\delta_{\rm wall}(\delta_{\rm disp} - \delta_{\rm mom})} \equiv R(\mu).} \tag{28.83}$$

Since the left-hand side depends only on the given velocity $U(x)$, and the right-hand side is a known rational function of $\mu(x)$ in the Pohlhausen model (see problem 28.9), this equation may be solved for $\mu(x)$. Having obtained $\mu(x)$ one may then search for a separation point where $\mu(x) = 0$. For the Blasius layer where there is no separation point, the left-hand side vanishes and we obtain instead from the right-hand side $\delta_{\rm heat}\delta_{\rm ener} = 2\delta_{\rm wall}\delta_{\rm mom}$, a relation which we have already derived in (28.65).

At the separation point $\mu \to 0$ the wall thickness diverges, $\delta_{\rm wall} \to \infty$, while the other thicknesses remain finite, such that

$$\frac{\dot U L}{U} \to -\frac{1}{2}\left.\frac{\delta_{\rm mom}}{\delta_{\rm disp} - \delta_{\rm mom}}\right|_{\mu=0} = -\frac{1}{5}, \tag{28.84}$$

where on the right-hand side we have used the Pohlhausen family value $\delta_{\rm disp}/\delta_{\rm mom} = 7/2$ for $\mu = 0$. Inserting (28.80) we obtain an equation that the separation point $x = x_c$ must satisfy,

$$\boxed{\frac{\dot U(x_c)}{U(x_c)^6}\int_0^{x_c} U(x)^5\,dx = -\frac{1}{5}.} \tag{28.85}$$

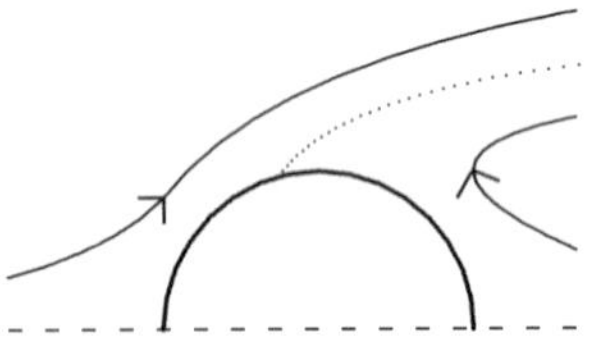

Sketch of the much-studied separation flow around a cylinder (of which only half is shown here). The separating streamline is dotted.

Solving this equation for the usual test cases we find the results shown in the last column of table 28.1. The errors are typically less than 3% which is as good as the semi-empirical methods [79]. The model also predicts $\mu(x)$ via (28.83) and thus all the thickness ratios (28.75) and (28.76). The absolute thickness scale $\delta(x)$ is finally obtained from (28.80).

Separation from cylinder

Some of the more interesting slip-flows first accelerate and then decelerate. Among them the notorious cylinder in a uniform cross-wind, which has been the favourite target for boundary layer research for nearly a hundred years. The cylinder presents two difficulties, as does in fact every realistic separation problem. The first is the question of what is the correct slip-flow. Since separation interacts with the mainstream flow, this is not so simple. The second is the technical question of calculating a precise value for the separation point in a slip-flow that may not be purely decelerating.

If the external flow around the cylinder is taken to be the potential flow, the slip-flow is given by (16.66b), i.e. $U(x) = 2U_0 \sin x$ where U_0 is the strength of the uniform cross-wind and x is the angle of observation (the upwind direction corresponds to $x = 0$). The approximative equation (28.85) immediately yields $x_c = 1.800 = 103\,^\circ$. The exact separation point for this slip-flow (obtained by numerical calculation) is known to be $x_c = 1.823 \approx 104\,^\circ$ (see [79]), so in this case the approximation works to a precision only slightly worse than 1%.

The real flow around the cylinder is, however, not potential flow because of the vorticity induced into the main flow by the boundary layer, especially in the separation region and beyond. Hiemenz (1911) determined experimentally that the slip-flow at a Reynolds number $\mathrm{Re} = U_0 a/\nu = 9500$ was well described by the odd polynomial [79, p. 272],

$$U = U_0(1.814x - 0.271x^3 - 0.0471x^5). \tag{28.86}$$

Now the approximative equation (28.85) predicts separation at $x_c = 1.372 = 78.6\,^\circ$ which is only about 2% away from the value $x_c = 80.5\,^\circ$ measured by Himenz. Even if agreement is obtained, this is not very impressive, because the slip-flow itself has not been calculated from first principles. Most of the information about the separation from the cylinder lies in fact in the empirical slip-flow.

Problems

28.1 **(a)** Show that for constant slip-flow velocity it is possible to obtain a boundary layer of constant thickness on an infinitely extended plate, if fluid is sucked through the plate at a constant rate. **(b)** Discuss what happens if fluid is pushed through instead.

* **28.2** Show that the planar flow generated by a plate at rest for $t < 0$ and moving with velocity $U(t)$ for $t > 0$, is

$$v_x(y,t) = \int_0^t \left(1 - \mathrm{erf}\left(\frac{y}{2\sqrt{\nu(t-t')}}\right)\right) \dot{U}(t')\, dt', \tag{28.87}$$

where $\dot{U}(t) = dU(t)/dt$. Hint: Show that it satisfies the equation of motion and the boundary conditions.

28.3 Show that the Blasius solution satisfies

$$\int_0^\infty (1 - f(s))\, ds = h(\infty) \tag{28.88a}$$

$$\int_0^\infty f(s)(1 - f(s))\, ds = 2f'(0). \tag{28.88b}$$

28.4 The streamlines in figure 28.3 may be obtained directly from the Blasius solution.

(a) Show that the streamlines are solutions to (with s given by (28.22))

$$\frac{dy}{dx} = \sqrt{\frac{\nu}{Ux}}\left(s - \frac{g(s)}{f(s)}\right). \tag{28.89}$$

(b) Show that this may be written

$$\frac{ds}{dx} = -\frac{1}{2x}\frac{g(s)}{f(s)}. \tag{28.90}$$

(c) Show that a streamline satisfies

$$g(s) = \frac{C}{\sqrt{x}} \tag{28.91}$$

where C is a constant.

(d) Show that the explicit solution is

$$y = \sqrt{\frac{\nu x}{U}}\, g^{-1}\left(y_0\sqrt{\frac{U}{\nu x}}\right) \tag{28.92}$$

where g^{-1} is the inverse function of g and y_0 is the intercept with the y-axis for $x = 0$.

* **28.5** Assume that a two-dimensional flow is of the form

$$v_x = U(x) \tag{28.93}$$

$$v_y = -y\frac{dU(x)}{dx} \tag{28.94}$$

all over space and not just near a boundary. Derive a third-order differential equation for U and discuss its possible solutions.

Can one choose $U(x)$ freely if one adds $V(x)$ to v_y?

* **28.6** Consider the wall-anchored model (28.66). Parametrize the model with a dimensionless slope μ,

$$\lambda = \mu\frac{U}{\delta}, \qquad \delta = \sqrt{\frac{\nu(3\mu - 4)}{\dot{U}}}. \tag{28.95}$$

(a) Show that μ must satisfy the differential equation

$$\dot{\mu} = \frac{24}{\mu}\frac{2-\mu}{8-3\mu}\frac{\dot{U}}{U} - \mu\frac{4-3\mu}{8-3\mu}\left(2\frac{\dot{U}}{U} + \frac{\ddot{U}}{\dot{U}}\right) \tag{28.96}$$

and that the initial condition is $\mu(0) = 4/3$.

(b) Show that in the interval $0 < \mu < 4/3$ the equation may be approximated by

$$\dot{\mu} = \frac{6}{\mu}\frac{\dot{U}}{U} \tag{28.97}$$

and solve it.

(c) Obtain an equation for the separation point and **(d)** compare the resulting values for the nine cases in table 28.1.

* **28.7** Consider the Pohlhausen family of velocity profiles (28.74).

(a) Show that the wall-curvature condition implies that

$$\delta = \sqrt{\frac{6\nu(\mu - 2)}{\dot{U}}}. \tag{28.98}$$

(b) Show that the von Karman relation leads to the differential equation

$$\dot{\mu} = 42\frac{48 - 20\mu + 3\mu^2}{(4-\mu)(24+25\mu)}\frac{\dot{U}}{U} - (2-\mu)\frac{144 + 12\mu - 5\mu^2}{(4-\mu)(24+25\mu)}\left(\frac{\ddot{U}}{\dot{U}} - 4\frac{\dot{U}}{U}\right), \tag{28.99}$$

and integrate this equation numerically to obtain the separation points in the fourth column of table 28.1.

(c) Is there a Goldstein singularity in this model?

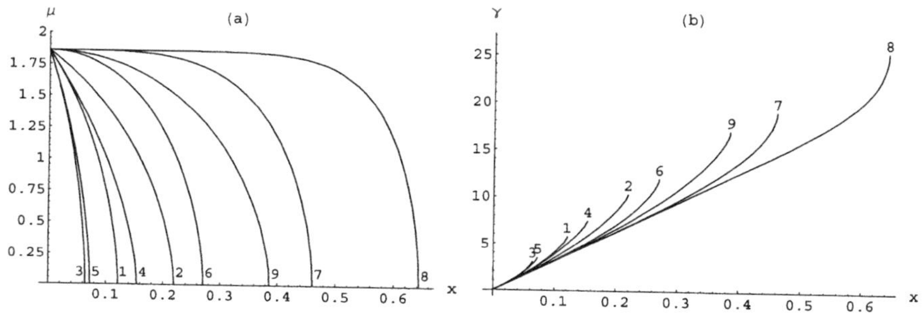

Figure 28.12. Solutions to the coupled equations (28.101) for the nine fields in table 28.1. **(a)** The slope μ as a function of x. All curves have vertical tangent at the separation point, indicating a singularity. **(b)** The parameter γ defined in (28.100) as a function of x. The linear envelope $\gamma \approx 32\,x$ corresponds to the Blasius thickness $\delta \sim \sqrt{x}$.

28.8 Consider the Pohlhausen family of velocity profiles (28.74) and introduce the auxiliary variable

$$\gamma = \frac{\delta^2 U}{\nu}, \qquad (28.100)$$

which has the dimension of length. For $x = 0$ we assume $\delta(0) = \gamma(0) = 0$.

(a) Show that the energy and momentum relations (28.63) become,

$$4(6 - 5\mu)\gamma\dot{\mu} + (144 + 12\mu - 5\mu^2)\dot{\gamma} + 15(96 - 6\mu - \mu^2)\gamma\frac{\dot{U}}{U} = 2520\mu, \qquad (28.101a)$$

$$2(876 - 506\mu - 63\mu^2)\gamma\dot{\mu} + (10512 + 876\mu - 253\mu^2 - 21\mu^3)\left(\dot{\gamma} + 5\gamma\frac{\dot{U}}{U}\right)$$
$$= 6864(48 - 4\mu + 3\mu^2) \qquad (28.101b)$$

where a dot means differentiation with respect to x.

(b) Show that the derivative $\dot{\gamma}(0)$ satisfies the two relations,

$$\dot{\gamma}(0) = \frac{2520\mu}{144 + 12\mu - 5\mu^2} = \frac{6864(48 - 4\mu + 3\mu^2)}{10512 + 876\mu - 253\mu^2 - 21\mu^3} \qquad (28.102)$$

where $\mu = \mu(0)$ is the starting value of μ.

(c) Show that this fourth-order algebraic equation for μ has four real roots, with the one closest to zero being

$$\mu(0) = 1.85685\ldots, \qquad \dot{\gamma}(0) = 31.3955\ldots. \qquad (28.103)$$

(d) Solve the coupled differential equations numerically with these initial values and produce figure 28.12 for the velocities in table 28.1 as a function of x.

28.9 Show that the right-hand side of (28.83) is the rational function,

$$R = \frac{-1976832 + 1103760\,\mu + 50796\,\mu^2 - 42581\,\mu^3 + 2085\,\mu^4}{2860\,(72 - 15\,\mu + \mu^2)\,(48 - 4\,\mu + 3\,\mu^2)}. \qquad (28.104)$$

Show that the lowest root in the numerator is $\mu = 1.85685\ldots$.

29 Subsonic flight

The take off of a large airplane never fails to impress passengers and bystanders alike. After building up speed during a brief half-minute run, gravity lets go and the plane marvellously lifts off. For this to happen, it is obvious that the engines and the airflow must together generate a vertical force which is larger than the weight of the aircraft. After becoming airborne the airplane accelerates further for a while, and then goes into a fairly steep steady climb until it levels off at its cruising altitude. Aloft in level flight at constant speed, the aerodynamic lift must nearly balance the weight, whereas the engine thrust almost entirely goes to oppose the drag. What is not obvious to most passengers is how the lift depends on the forward speed, the angle of attack and the shape of the airframe, especially the wings.

The explanation of aerodynamic lift is in fact quite simple, even if it was only in the beginning of the twentieth century—about the same time as the first generation of airplanes were built—that the details became understood. In nearly ideal flow, pressure is the dominant stress acting on any surface, so to get lift the pressure must on average be higher underneath the wing than above. Bernoulli's theorem then implies that the airspeed must be higher above the wing than below, effectively creating a circulation around the wing, a kind of bound vortex superimposed on the general airflow. Without this circulation, caused by the shape and flying attitude of the wing, there can be no lift.

In this chapter we shall only study the most basic theory for subsonic flight with an emphasis on concepts and estimates. Aerodynamics is a huge subject (see for example [4, 23]) of importance for all objects moving through the air, such as rifle bullets, rockets, airplanes, cars, birds and sailing ships, and to some extent for submarines moving through water. The potential for triumphant rise and tragic fall unavoidably associated with flying machines makes aerodynamics different from most other branches of science.

29.1 Aircraft controls

Historically aircraft design went through many phases with sometimes weird shapes emerging, especially during the nineteenth century. In the twentieth century, where sustained powered flight was finally attained, most of the design problems were solved through systematic application of theory and experiment. The history of the evolution of aerodynamics, the courageous men and their wonderful flying machines, is dramatic to say the least (see for example [3]).

Control surfaces

The symmetry plane and wing plane of normal winged aircraft.

The majority of all winged aircraft that have ever been built are symmetric under reflection in a midplane. The wings are typically placed in a plane orthogonal to the midplane, but often swept somewhat backwards and a bit upwards. On the wings, and also on the horizontal and vertical stabilizing wing-like surfaces found at the tail-end of most aircraft, there are smaller movable *control surfaces*, connected physically or

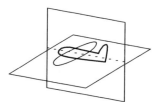

electronically to the 'stick' and the 'pedals' in the cockpit. Still smaller movable sections of the control surfaces allow the pilot to *trim* the aircraft. When an aircraft is trimmed for steady flight, the cockpit controls are relaxed and do not require constant application of force to keep the airplane steady. At cruising speed the aircraft is typically handled with quite small movements of the controls, often carried out by the autopilot, whereas at low speeds, for example during take off and landing, much larger moves are necessary.

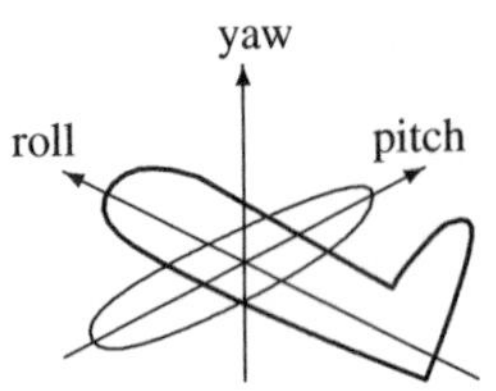

The three axes of a normal winged airplane. The aircraft rolls around the longitudinal axis, pitches around the lateral axis, and yaws around the directional axis. The nose lies at the end of the longitudinal axis and the wings in the plane of the longitudinal and lateral axes.

Main aircraft axes

The symmetry plane and the wing plane of normal aircraft define three orthogonal axes. The first is the *longitudinal* axis, running along the body of the aircraft in the intersection of the midplane and the wing plane. Rotation around this axis is called *roll*, and is controlled by the *ailerons* usually found at the trailing edge of the wings near the wing tips. When the pilot moves the stick from side to side the ailerons move oppositely to each other and create a rolling moment around the longitudinal axis. The second is the *lateral* axis which lies in the wing plane and is orthogonal to the midplane. Rotation around this axis is called *pitch* and is normally controlled by the *elevator*, usually found in the tail of the aircraft. The pilot moves the stick forward and backward to create a pitching moment around the lateral axis. The third is the *directional* axis which is orthogonal to both the wings and the body, and thus vertical in straight level flight. Rotation around this axis is called *yaw* and normally controlled by the *rudder*, usually also placed in the tail end. In conventional aircraft, the pilot presses foot pedals to move the rudder and create a yawing moment around the directional axis. The Wright brothers were the first to introduce controls for all three axes of their aircraft [3, p. 243].

Wilbur and Orwille Wright (1867–1912, 1871–1948). *American flight pioneers. From their bicycle shop in Dayton, Ohio, the inseparable brothers carried out systematic empirical investigations of the conditions for flight, beginning in 1896. Built gliders and airfoil models, wind tunnels, engines and propellers. They finally succeeded in performing the first heavier-than-air, manned, powered flight on December 17, 1903, at Kitty Hawk, North Carolina.*

Take off, cruise and landing

The take off of a normal passenger aircraft begins with a *run* typically lasting half a minute. In modern aircraft with a nose wheel, the body stays horizontal during the whole run, whereas in older aircraft with a tail wheel or slider, the tail would lift up well into the run first and then make the body nearly horizontal. Having reached sufficient speed for flight, about 250–300 km h^{-1} for large passenger jet planes, the pilot gently pulls the stick back and thereby raises the elevator, creating a pitching moment that lifts the nose wheel off the runway while the main undercarriage stays in contact with it. After a bit of acceleration in this attitude the undercarriage also leaves the runway, and the aircraft is airborne. For safety reasons the aircraft should not lift-off until the speed is somewhat above the minimal speed for flight. In older airplanes, the actual lift-off was almost imperceptible, whereas the powerful engines of modern aircraft make the lift-off much more notable through the rather steep climb angle that the aircraft is capable of assuming immediately afterwards. The climb normally lasts until the aircraft has reached cruising altitude, typically 10 000 m, at which point it levels off and accelerates further until it reaches cruising speed, around 800–900 km h^{-1} for a modern jet. At normal temperature and pressure the sound speed is about 1200 km h^{-1} but at cruising altitude the fall in temperature has reduced it by about 10%—as can be seen from equation (16.9). The airspeed is thus about 85% of sound speed, also called Mach 0.85.

Landing is by far the hardest part of flying. The aircraft has to be brought down to the ground and the speed must be reduced. At low speed the aircraft controls need to be worked harder than at high speed, and random winds and turbulence influence the aircraft much more. Keeping the air speed above stall speed is uppermost in the pilot's mind, because a stall at low altitude makes the airplane crash into the ground. Landing speeds are comparable to take off speeds, but the aircraft has to be maneuvered into the narrow space that the runway presents, and in all kinds of weather. Landing lengths can be made shorter than take off lengths by diverting jet exhaust into the forward direction or reversing propeller blades, in addition to the application of wheel brakes.

Example 29.1.1 (Boeing 747-200): Jet engines develop nearly constant thrust (force) at a given altitude such that their power (energy output per unit of time) increases proportionally with airspeed, all other factors being equal. Propeller engines yield instead roughly constant power so that the thrust decreases inversely with airspeed. Ignoring air resistance (drag), the constant thrust from jet engines translates into nearly constant acceleration during the take off run. A typical large passenger jet airliner (Boeing 747-200) has a maximal weight of 374 000 kg and four engines that together yield a maximum thrust of 973 000 N, corresponding to a runway acceleration of 2.6 m s^{-2} when fully loaded. At this acceleration the plane reaches take off speed of 290 km h^{-1} in 31 s after a run of about 1250 m. Actual take off length is somewhat larger because of drag and rolling friction. For safety reasons, runways are required to be at least twice that length, typically between 3 and 4 km.

Extreme flying

An aircraft can in principle move through the air in any attitude—and some pilots enjoy making their planes do exactly that—but there is an intended normal flying attitude with the wings nearly horizontal and orthogonal to the airflow. In this attitude, the aircraft is designed such that the flow of air over the wings and body of the aircraft is as laminar as possible, because laminar flow yields the largest lift force and smallest drag.

In other attitudes, steep climb, dive, roll, loop, tight turn, spin, tail-glide, sideways crabbing etc, the airflow over the wings may become turbulent resulting in almost complete loss of lift. When that happens, the aircraft is said to have *stalled*. Stalling an aircraft in level flight at sufficient altitude is a common–and fun–training exercise. First, the engine power is cut to make the aircraft slow down. While the airspeed is falling the pilot slowly pulls back the stick to pitch the nose upwards so that the aircraft keeps constant altitude. This can of course not continue, and at a critical point the laminar flow over the wings is lost and replaced by turbulence. The aircraft suddenly and seemingly by its own volition pitches its nose downwards and begins to pick up speed in a dive. A modern aircraft normally recovers all by itself and goes into a steady glide at a somewhat lower altitude. A stall close to the ground can be catastrophic, as the many hang-glider accidents can confirm (the first fatal one happened in 1896 and cost the life of flight pioneer Otto Lilienthal).

Otto Lilienthal (1848–1896). *German engineer. One of the great pioneers of manned flight. Over more than two decades he carried out systematic studies of lift and drag for many types of wing surfaces and demonstrated among other things the superiority of cambered airfoils. Constructed (and exported!) manned gliders, and also took out patents on such flyers in 1893. Stalled and crashed from a height of about 17 m outside Berlin on August 9, 1896. Whether he would actually have invented powered flight before the Wright brothers did in 1903 is not clear [3]*

> Most aircraft are today equipped with mechanical stall detection devices near the leading edge of the wings, and audible stall warnings are frequently heard in aircraft cockpits during landing, just before touchdown. The warnings indicate that a stall in the wing flow is imminent, although the aircraft will usually not go into a proper stall before touching down.

Other situations may arise in which only a part of the lifting surfaces stall. In a tight turn at low speed, the inner wing may stall whereas the outer wing keeps flying, and the aircraft goes into a vertical *spin*. In the early days of flight it was nearly impossible to recover from such a situation which could easily arise if the aircraft was damaged, for example in air combat. In those days pilots were not equipped with parachutes, and they often saw no other way out than jumping from the airplane, rather than burn with it. Today's passenger aircraft are not cleared for spin, but it can be fun to take a modern small aircraft that *is* cleared for aerobatics into a spin at sufficient altitude, for example by pulling hard back and sideways at the stick just before it otherwise would go into a normal stall, as described above. Most people find the experience quite unpleasant and disorienting, especially due to the weightlessness that is felt while the aircraft slowly tumbles over before it goes into a proper spin. Again, modern aircraft are so stable that they tend to slip out of a spin by themselves if the controls are left free.

29.2 Aerodynamic forces and moments

There are several stages in the process of getting to understand flight. The first of these concerns the global forces and moments that act on a moving body completely immersed in a nearly ideal fluid such as air. Initially we put no constraints on the shape of the object or on the motion of the air relative to the object, although mostly we shall think of an aircraft under normal flight conditions, and discuss only the forces acting on it. Apart from scattered comments we leave the discussion of moments to more specialized treatments [4].

Total force

The total force $\mathcal{F}$ acting on a body determines the acceleration of its centre of mass. The only way a fluid can act on an immersed body is through contact forces, described by the stress tensor $\boldsymbol{\sigma} = \{\sigma_{ij}\}$. Including the weight $M\boldsymbol{g}_0$ and engine thrust $\boldsymbol{T}$ the total force becomes,

$$\boxed{\mathcal{F} = \boldsymbol{T} + M\boldsymbol{g}_0 + \mathcal{R}} \qquad (29.1)$$

where

$$\mathcal{R} = \oint_{\text{body}} \boldsymbol{\sigma} \cdot d\boldsymbol{S}, \tag{29.2}$$

is the resultant of all contact forces acting on the body, also called the *reaction force* (chapter 23). In principle this includes hydrostatic buoyancy forces (chapter 5), which serve to diminish the effective gravitational mass of a body. For heavier-than-air flying, buoyancy can normally be disregarded.

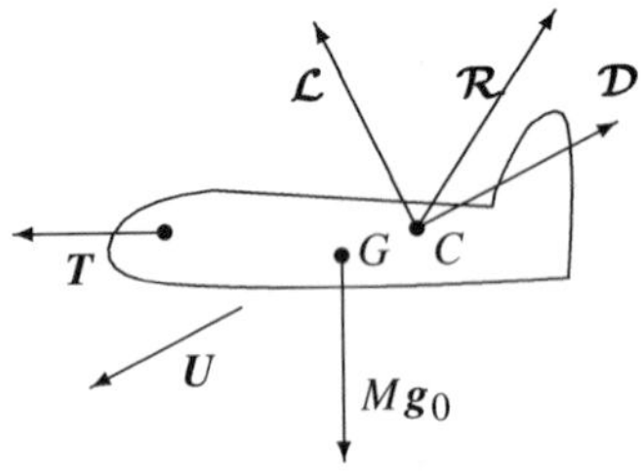

Sketch of the forces acting on a body moving with instantaneous centre-of-mass velocity $\boldsymbol{U}$. The thrust $\boldsymbol{T}$ propels the object forward, gravity $M\boldsymbol{g}_0$ pulls it down, and the aerodynamic reaction force $\mathcal{R}$ may be resolved into lift $\mathcal{L}$ and drag $\mathcal{D}$. Note that the reaction $\mathcal{R}$ and its components are plotted as acting on a single point C, called the aerodynamic centre, although this concept may not always be meaningful. For stability the centre of thrust should lie forward of the aerodynamic centre.

Lift and drag

It is convenient to resolve the reaction force into two components, the *lift* which is orthogonal to the instantaneous centre-of-mass velocity $\boldsymbol{U}$ of the aircraft, and the *drag* which is parallel with it,

$$\mathcal{R} = \mathcal{L} + \mathcal{D}. \tag{29.3}$$

Lift and drag thus satisfy

$$\mathcal{L} \cdot \boldsymbol{U} = 0, \qquad \mathcal{D} \times \boldsymbol{U} = \boldsymbol{0}. \tag{29.4}$$

The drag always acts in the opposite direction of the centre-of-mass velocity whereas the lift may take any direction orthogonal to it.

Lift may even point directly downwards. It is for this reason dangerous to fly an aircraft inverted close to the ground, because the gut reaction of pulling the stick towards you to get away from the ground will generate an extra lift which sends you directly into the ground. During banked turns an airplane also generates lift away from the vertical, creating in this way the centripetal force necessary to change its direction.

* Total moment of force

The total moment of all forces acting on the body is

$$\boxed{\mathcal{M} = \mathcal{M}_T + \boldsymbol{x}_G \times M\boldsymbol{g}_0 + \mathcal{M}_R,} \tag{29.5}$$

where $\mathcal{M}_T$ is the moment of thrust, $\boldsymbol{x}_G$ the centre of gravity of the body, and the moment of the contact forces is,

$$\mathcal{M}_R = \oint_{\text{body}} \boldsymbol{x} \times \boldsymbol{\sigma} \cdot d\boldsymbol{S}. \tag{29.6}$$

The total moment depends on the choice of origin for the coordinate system, but if the total force $\mathcal{F}$ on the body vanishes, the total moment becomes independent of the origin. In that case one may calculate the total moment around any convenient point, for example the centre of gravity. The individual contributions to the total moment will depend on the choice of origin, even if their sum is independent.

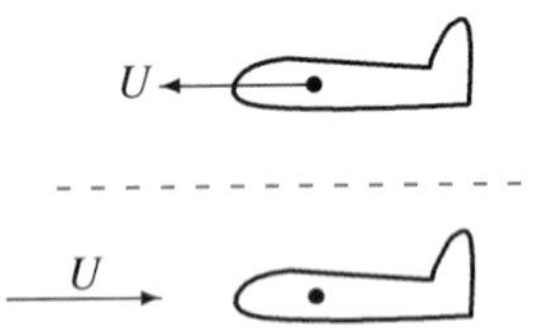

The two situations above are physically equivalent. In the upper drawing the centre-of-mass moves with velocity U to the left. In the lower, the centre-of-mass does not move, but the surrounding air moves with velocity U to the right at great distances from the object.

29.3 Steady flight

In steady flight the aircraft moves with constant centre-of-mass velocity in a non-accelerated frame of reference, so that the sum of all forces must vanish

$$\boldsymbol{T} + M\boldsymbol{g}_0 + \mathcal{L} + \mathcal{D} = \boldsymbol{0}. \tag{29.7}$$

Even if passenger comfort demands that the pilot tries to achieve nearly vanishing total force on an airplane, irregular motion of the air may buffet the plane around. In extreme cases, unannounced clear air turbulence may suddenly cause unfettered passengers to fly around inside the cabin. We shall disregard such phenomena and assume that the aircraft is capable of flying with a steady velocity through an atmosphere that would have been at rest were it not for the moving aircraft. Since forces in Newtonian mechanics are the same in all inertial reference frames, we are free to work in the rest-frame of the aircraft where the air asymptotically moves at constant speed.

The orientation of an aircraft with respect to the direction of flight (and the vertical) is called its *flying attitude*. The main attitude parameter responsible for lift is the *angle of attack* α, also called the *angle of incidence*, between the airflow and the plane of the aircraft, formed by the longitudinal and lateral axes. In normal flight at high speed the angle of attack is usually quite small, typically a couple of degrees.

The angle of attack α is the angle between airflow U and the plane of the aircraft.

In the same way as floating bodies, ships and icebergs, should be in stable hydrostatic equilibrium, aircraft should preferably also be dynamically stable in steady flight, meaning that a small perturbation of the aircraft's steady flying attitude should generate a moment counteracting the perturbation. In general this requires the centre of thrust to lie forward of the aerodynamic centre.

Most modern aircraft are dynamically stable when properly trimmed, and that is very good for amateur pilots, but in military fighter planes, dynamic stability is sometimes traded for maneuverability. Certain modern fighter planes can in fact only maintain a stable attitude through corrections continually applied to the control surfaces by a fast computer.

Steady climb

After acceleration and take off a powered aircraft normally goes into a steady climb forming a constant positive *climb angle* (or *angle of ascent*) θ with the horizon. Cockpit instruments usually indicate the vertical velocity, called the *rate of ascent* or *climb rate* $U \sin\theta$, rather than the climb angle. Having reached cruising altitude the pilot reduces power to a fuel-economic setting and the aircraft levels off with $\theta = 0$ and a tiny angle of attack consistent with steady flight. In this phase of flight lift must nearly balance weight and thrust must nearly balance drag. Approaching its destination, the power is reduced, though usually not cut completely off, and the aircraft goes into a powered descent with negative θ. Just before landing, power is lowered to near zero, the aircraft flares out almost horizontally with $\theta \approx 0$ and nose up under a fairly large angle of attack before the final touchdown.

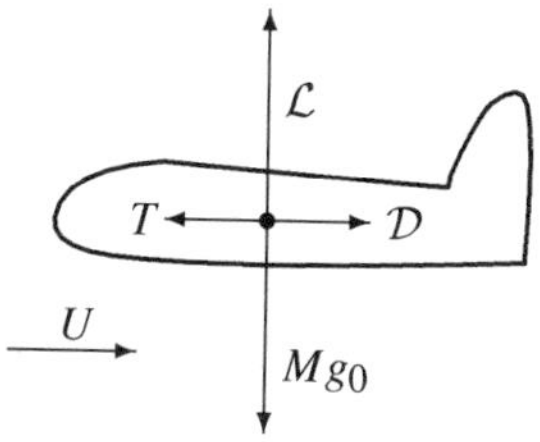

Sketch of forces in steady horizontal powered flight with a small angle of attack. Lift balances weight and drag balances thrust.

Assuming that the thrust is directed along the longitudinal axis, as is normally the case for fixed-wing aircraft[1], we obtain the following expressions for lift and drag by projecting the equation of force balance (29.7) on the direction of motion and the direction orthogonal to it,

$$\mathcal{D} = T\cos\alpha - Mg_0\sin\theta, \qquad \mathcal{L} = -T\sin\alpha + Mg_0\cos\theta. \tag{29.8}$$

Given the values of all the parameters on the right-hand sides, we may calculate the values of lift and drag that are required to keep the aircraft in steady flight.

This is, however, not really the way a pilot operates an aircraft. Typically, the pilot selects a power setting for the engine(s) and a certain rate of climb, and then waits until the aircraft steadies on a certain airspeed, angle of attack and angle of climb. This procedure, of course, presupposes that there are such solutions within the aircraft's flight envelope for the specified values of power and climb rate (if not, the aircraft will stall). We shall see later that aerodynamic theory allows us to calculate lift and drag for a given aircraft in terms of the airspeed, the angle of attack and the air density. The steady flight equations (29.8) may then be solved for the airspeed U and the angle of attack α, given the air density, the weight, the engine power, and the climb rate. There is the further complication that for a given engine setting the thrust T tends to fall inversely with airspeed for propeller engines whereas it stays more or less constant for jets. On top of that, the air temperature, pressure and density all vary with altitude.

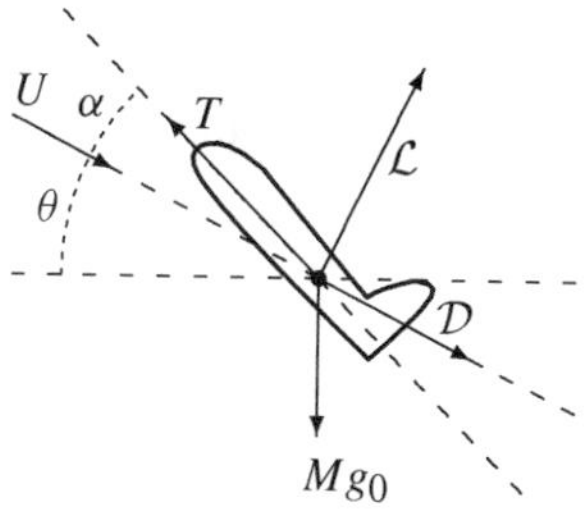

Forces acting on an aircraft in powered steady climb at an angle θ with angle of attack α. All the forces are assumed to lie in the symmetry plane of the aircraft. For convenience we have moved all forces to the centre of gravity.

During steady climb with a small angle of attack, $\alpha \ll 1$, the steady flight equations (29.8) may be written

$$\sin\theta \approx \frac{T - \mathcal{D}_0}{Mg_0}, \qquad \mathcal{L} \approx Mg_0\cos\theta, \tag{29.9}$$

where $\mathcal{D}_0$ is the residual drag at zero angle of attack. Thus, the ratio of the *excess of power* $T - \mathcal{D}_0$ to the weight of the aircraft Mg_0 determines the climb angle. To get a finite positive angle of climb, the thrust

[1] In other types of aircraft the engine thrust can also have a component orthogonal to the longitudinal axis which also contributes to lift. In the extreme case of a VTOL (vertical take-off and landing) aircraft, for example a helicopter, there is almost no other lift, and the engine thrust balances by itself both drag and weight.

must not only overcome the drag but also part of the weight of the airplane. From the climb angle one can afterwards calculate the lift that the airflow over the wings and body of the aircraft necessarily must generate to obtain a steady climb.

Example 29.3.1: During initial climb, speed is fairly low, and if the residual drag can be ignored relative to thrust it follows that $\sin\theta \lesssim T/Mg_0$. For the fully loaded Boeing 747-200 of example 29.1.1 we find $T/Mg_0 \approx 0.27$ and thus $\theta \lesssim 15\,°$.

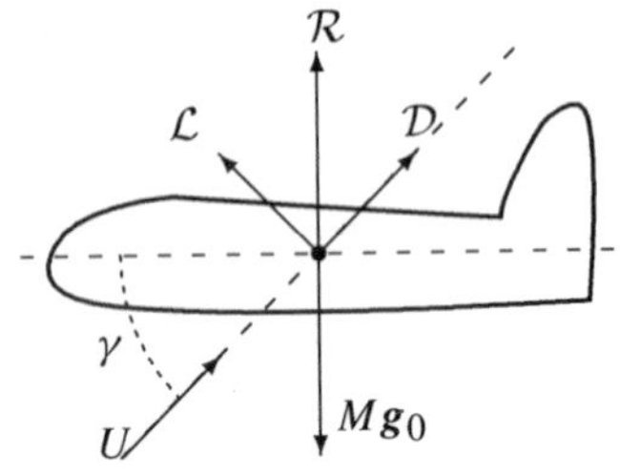

Sketch of forces in unpowered steady descent at a glide angle γ. Lift and drag collaborate to balance the weight. The airplane can glide with many different angles of attack (here shown with $\alpha = \gamma$) but there is a best angle of attack which yields the smallest glide angle, or equivalently the highest glide ratio.

Unpowered steady descent

Most freely falling objects quickly reach a constant terminal velocity. Stones fall vertically, whereas aircraft, paper gliders, paragliders and parachutists in free fall will attain sometimes large horizontal speeds. An aircraft in unpowered flight is able to glide towards the ground with constant velocity and constant rate of descent. It is part of early training for pilots to learn how to handle their craft in unpowered steady descent, and usually the aircraft is so dynamically stable that it by itself ends up in a steady glide if the engine power is cut and the controls are left free. Paper gliders on the other hand often go through a series of swooping dives broken by stalls, or spiral towards the ground in a spin.

During steady unpowered descent the air hits the aircraft from below at an angle γ with the horizontal, called the *glide angle*, corresponding to a *glide slope* $\tan\gamma$. The ratio of horizontal to vertical air speed is called the *glide ratio*, and equals the inverse of the glide slope, i.e. $\cot\gamma$. An aircraft can glide steadily with different airspeeds for a large range of angles of attack. The angle of attack that yields maximal glide ratio determines how far an aircraft at best can reach by gliding down from a given altitude h, also called its *glide range*, $x = h\cot\gamma$.

Typical commercial aircraft have best glide ratios of 10–20 with 17 for the Boeing 747 and 8 for Concorde. So, if the engines cut out at an altitude of 10 km, the pilot has to look for a place to land inside 100–200 km, depending on the aircraft. These glide ratios are comparable to those of gliding birds like the swift with glide ratio 10 and soaring birds like the albatross with glide ratio 20. Modern sailplanes may reach glide ratios around 30–55 and in extreme cases even higher. The space shuttle with its stubbed wings and large weight is a rather bad glider, and approaches the runway at a glide angle $\gamma = 19\,°$, corresponding to a glide ratio of around 3, comparable to that of a sparrow. The human body is even worse, with a best glide ratio of about unity.

In steady unpowered descent the aerodynamic reaction force must be equal and opposite to the weight of the aircraft, or in size $\mathcal{R} = Mg_0$. Resolving the reaction force into lift and drag we find,

$$\mathcal{L} = Mg_0\cos\gamma, \qquad \mathcal{D} = Mg_0\sin\gamma. \tag{29.10}$$

These equations could also have been obtained from the steady flight equations (29.8) with $T = 0$ and $\theta = -\gamma$. From the glide angle and the weight of an aircraft, we may thus determine both the lift and drag that acts on it in this flight condition. The glide ratio evidently equals the ratio of lift to drag in unpowered descent,

$$\cot\gamma = \frac{\mathcal{L}}{\mathcal{D}}. \tag{29.11}$$

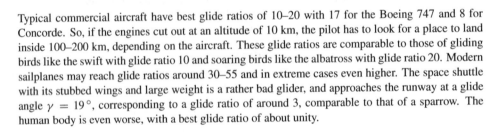

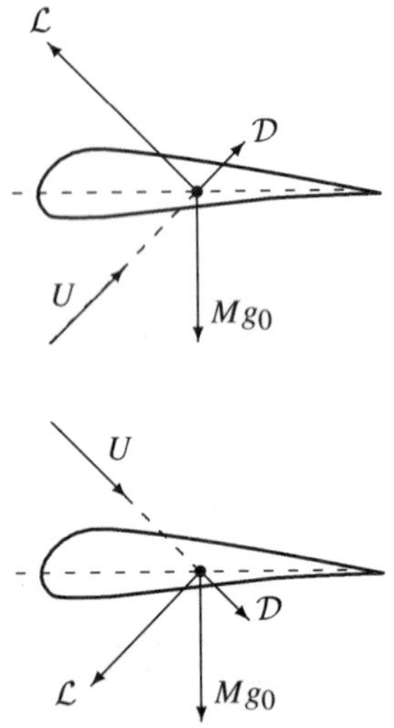

Sketch of forces during bird wing downstroke (top) and upstroke (bottom) in forward flight. The total force does not vanish but accelerates the bird. The lift provides forward thrust in both cases when the drag is not too large.

Aerodynamics tells us (see the following section) that the ratio of lift to drag essentially only depends on the angle of attack, so the best glide ratio is obtained by choosing that angle of attack which maximizes $\mathcal{L}/\mathcal{D}$.

Note that neither lift nor drag are horizontal in unpowered descent. The lift is tilted forward and the drag backwards.

The forward tilt of the lift also allows us to understand broadly how birds and insects generate thrust in level flight by flapping their wings straight up and down. In this unsteady flight mode there is no instantaneous balance of aerodynamic forces and gravity, but during the downstroke the air will hit the wing from below and generate a tilted lift, propelling the bird forward (and upwards), provided the drag is not too large. During upstroke the picture is inverted, and air hits the wing from above, but the lift is still tilted forwards and thus again propels the bird forwards, provided drag does not overwhelm it. Insects and birds that hover instead of flying horizontally get lift by interacting with vortices created at the leading edge of the wings during the downstroke and by other mechanisms[2].

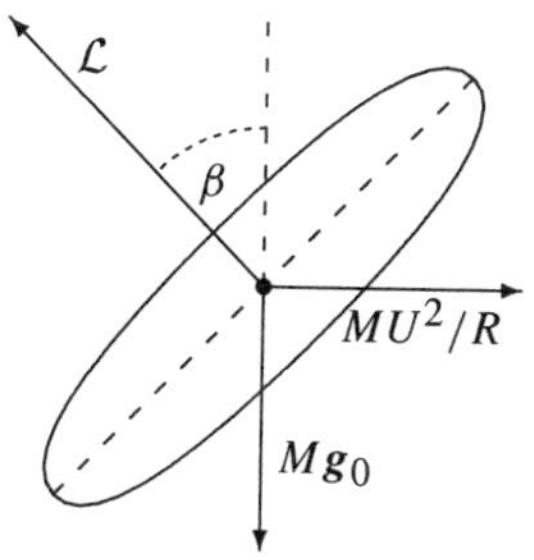

Sketch of forces in the transverse plane during a steady horizontal banked turn, here with tilt angle $\beta \approx 45\,°$.

Horizontal banked turn

Consider an aircraft flying steadily under power with velocity U in a horizontal circle of radius R. From the (rotating) rest frame of the aircraft, the air again flows steadily past with velocity U, but now there is also a centrifugal force MU^2/R directed away from the centre of the circle. The engine thrust is assumed to balance the drag, and the lift must therefore balance the vector sum of the weight and the centrifugal force.

Denoting by β the angle between the vertical and the lift vector, and projecting the lift on the horizontal as well as the vertical directions, we obtain the equations,

$$\mathcal{L}\cos\beta = Mg_0 \qquad\qquad \mathcal{L}\sin\beta = \frac{MU^2}{R}. \tag{29.12}$$

The lift divides out in the ratio of these equations, and we get

$$\tan\beta = \frac{U^2}{g_0 R}. \tag{29.13}$$

Airplanes are normally tilted (banked) through precisely this angle β during turns, such that the floor of the aircraft remains orthogonal to the lift. In such a *clean turn*, the effective gravity experienced inside the airplane is (in units of standard gravity)

$$\frac{g_{\text{eff}}}{g_0} = \frac{\mathcal{L}}{Mg_0} = \frac{1}{\cos\beta}, \tag{29.14}$$

also called the *load factor* or the *g-factor*.

Example 29.3.2: In a clean 60° banked turn, one pulls a g-factor of 2. Fighter jets may generate g-factors up towards 10, corresponding to bank angles $\beta \lesssim 84\,°$. To avoid passenger discomfort, most commercial aircraft rarely bank beyond 15° with a nearly imperceptible increase in load factor of about 4%. At a speed of about 900 km h^{-1} and $\beta \approx 15\,°$, the clean turn diameter is $2R = 2U^2 \cot\beta/g_0 \approx$ 48 km, and a full turn at this speed takes $2\pi R/U \approx 10$ min.

29.4 Estimating lift

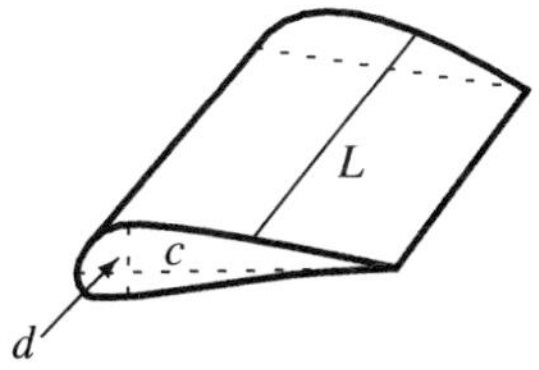

A wing is characterized by three lengths: the span L, the chord c, and the thickness d. The wing profile depicted here also carries aerodynamic twist.

Aerodynamic lift in nearly ideal flow is almost entirely caused by pressure differences between the upper and lower wing surfaces. In this section we shall describe the basic physics of lift and estimate its properties from relatively simple physical arguments, and in the following section we shall make similar estimates of the various contributions to drag. It should, however, be borne in mind that we would rather want to calculate lift and drag from fluid mechanics, in terms of the angle of attack, velocity, air density and the shape of the wing. Such theoretical knowledge makes it possible to predict which parameter intervals allow an aircraft to become airborne and sustain steady flight. In section 29.7 we explicitly calculate the lift for thin airfoils.

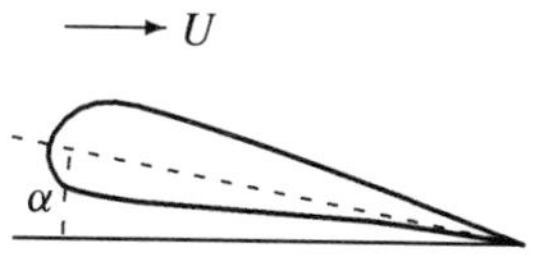

The angle of attack for an airfoil is conventionally defined to be the angle between the airflow and chord line.

Wing and airfoil geometry

An airplane wing may be characterized by three different length scales: the tip-to-tip length or *span* L, the transverse width or *chord* c and the thickness d. A wing can only in the coarsest of approximations be viewed as a rectangular box. Typical wings are both thin and long, $d \ll c \ll L$. Many wings *taper* towards the tip and are swept back towards the rear. Other wing shapes are also found, for example the delta-wing of supersonic aircraft such as Concorde. For a rectangular wing, the dimensionless number L/c is called the *aspect ratio*. For tapering and unusually shaped wings where the true cord length may be ill-defined, one may instead use the average cord length $c = A/L$, where A is the planiform area of the wing (the area

[2]See for example R. B. Srygley and A. L. R. Thomas, Unconventional lift-generating mechanisms in free-flying butterflies, *Nature* **420**, (2002) 660 and references therein. See also [20, 69].

Figure 29.1. NASA's Helios prototype electrically powered flying wing. Image Credit: NASA Dryden flight research center, photo by Carla Thomas.

of the wing's 'shadow' on the wing plane), so that the aspect ratio becomes $L/c = L^2/A$. The wing may furthermore twist slightly along the span leading to a varying angle of attack. This is, in particular, true for propellers that basically are wings mounted on a rotating shaft.

The transverse wing profile, also called the *airfoil*, is normally slightly curved, or *cambered*, along the cord, with a soft leading edge and a sharp trailing edge. The angle of attack of an airfoil is defined to be the angle between the asymptotic airflow and the *chord line* which is a straight line of length c connecting the leading and trailing edges. Depending on how the wings are attached to the aircraft there may be a small difference between the angles of attack of the wing and the aircraft.

Example 29.4.1: The Boeing 747-400 has a wing span of $L \approx 64$ m and a wing area $A \approx 520$ m^2, leading to an average cord length of $c \approx 8$ m and an aspect ratio of $L/c \approx 8$. For comparison, the albatross with its narrow long wings has an aspect ratio of about 20, at par with modern sailplanes. At the extreme end one finds NASAs solar-cell powered flyer Helios which has an aspect ratio of nearly 31. Incidentally, a man with his arms stretched out as wings has an aspect ratio of about 20, so aspect ratio is not everything.

Wing loading

Suppose an aircraft with thin and almost planar wings flies horizontally under a small angle of attack. The chordwise Reynolds number,

$$\mathrm{Re}_c = \frac{Uc}{\nu}, \tag{29.15}$$

is a dimensionless combination of the asymptotic flow speed U, the chord c and the kinematic viscosity of air, $\nu = \eta/\rho$ which characterizes the airflow around the wing. It will always be assumed to be very large, of the order of many millions, so that the slip-flow pattern around the wings just outside the omnipresent

Figure 29.2. The famous Cessna 150 from the First Production in 1959. Reproduced with permission from the Cessna 150–152 Club.

boundary layers may be taken to be very nearly ideal. In section 28.3 it was shown that the slip-flow pressure then penetrates the boundary layer and acts directly on the upper and lower wing surfaces.

In this situation the only way to obtain lift is by the slip-flow pressure p_a immediately above the wing being generally lower than the slip-flow pressure p_b immediately below the wing (see figure 29.3). The difference in average pressures above and below the wing may be estimated from the total lift per unit of wing area $\mathcal{L}/A$, also called the *wing loading*,

$$\Delta p_{ab} \equiv \langle p_a \rangle - \langle p_b \rangle \approx -\frac{\mathcal{L}}{A}, \tag{29.16}$$

where the brackets indicate averaging over the upper or lower wing surfaces. In steady level flight, lift equals weight $\mathcal{L} = M g_0$, and wing loading is easy to calculate from the aircraft parameters.

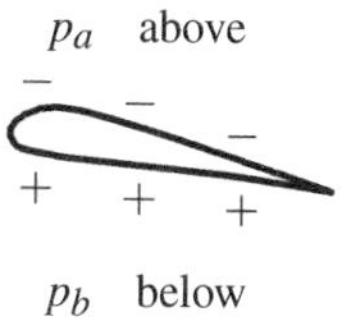

The pressure is on average lower above the wing than below. This is what carries the aircraft.

Example 29.4.2 (Cessna 150): The popular Cessna 150 two-seater has a wing span of $L = 10.2$ m, wing area $A = 14.8$ m^2, and thus an average chord of $c \approx 1.45$ m, and an aspect ratio of $L/c \approx 7.0$. With a maximum take off mass of $M = 681$ kg, the wing loading becomes $\mathcal{L}/A \approx 451$ Pa which is merely 0.45% of atmospheric pressure at sea level. Cruising with $U = 180$ km h^{-1} at sea level, the Mach number is Ma ≈ 0.15 and the Reynolds number $\mathrm{Re}_c \approx 5 \times 10^6$. The approximation of nearly ideal flow is very well fulfilled with boundary layers only millimetres thick.

Example 29.4.3 (Boeing 747-400): The cruising speed for the Boeing 747-400 is $U \approx 900$ km h^{-1} $\approx$ 250 m s^{-1} at the normal cruising altitude of $z = 10$ km. With an average chord length of $c \approx 8$ m, the chordwise Reynolds number becomes $\mathrm{Re}_c \approx 6 \times 10^7$ when we take $\nu \approx 3.4 \times 10^{-5}$ m^2 s^{-1} at the cruising altitude. The maximal take off mass is $M \approx 400\,000$ kg distributed over a wing area of about $A \approx 520$ m^2, leading to a wing loading of about $\mathcal{L}/A \approx 7500$ Pa, which is only 7.5% of atmospheric pressure at sea level, but about 25% of the actual pressure at the normal cruising altitude (see section 4.6).

Velocity differences

In nearly ideal flow Bernoulli's theorem may be used to relate the pressure and flow velocity well outside the boundary layers. As previously discussed (section 16.4 on page 213) air is effectively incompressible at speeds much lower than the speed of sound c_S, i.e. for small Mach number Ma $= U/c_S$, say Ma $\lesssim 0.3$. The following analysis of lift in incompressible flow with constant density ρ_0 thus excludes passenger jets cruising at Ma $\approx$ 0.8–0.9 but we shall return to this question towards the end of the section.

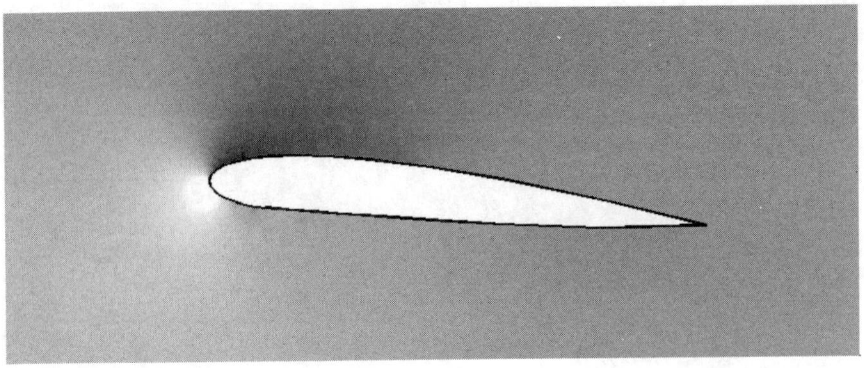

Figure 29.3. Pressure distribution around an airfoil at Reynolds number $\mathrm{Re}_c = 10\,000$ and angle of attack $\alpha \approx 6\,°$, obtained by numeric simulation. Note the higher stagnation pressure (light) at the leading edge and the lower lifting pressure (dark) above the wing. It typically acts about one quarter of the chord length downstream from the leading edge of the wing. The pressure below the wing is not much higher than the pressure at infinity.

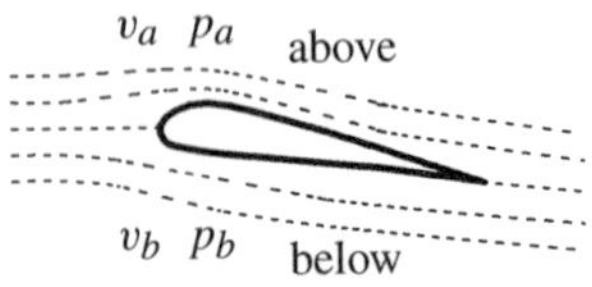

Sketch of streamlines around a wing with positive lift. The wing profile and angle of attack accelerates the flow across the top of the wing and retards it below, and thus creates a lower pressure above than below. Note the presence of front and rear stagnation points with vanishing airspeed. Note also the downwash behind the wing.

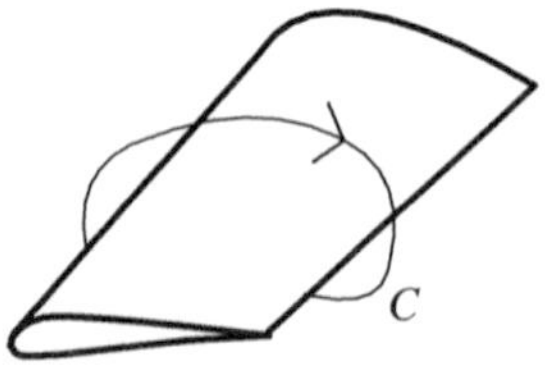

The integration contour for wing circulation.

Any streamline coming in from afar will start out with the same velocity U and pressure P, and Bernoulli's theorem (16.15) relates these values to the velocity and pressure along the streamline. For a streamline passing immediately above the wing well outside the boundary layer we find,

$$\frac{1}{2}U^2 + \frac{P}{\rho_0} = \frac{1}{2}v_a^2 + \frac{p_a}{\rho_0}, \tag{29.17}$$

where v_a is the flow velocity. A thin wing will not change the flow velocity much, and using $v_a \approx U$ we find to leading order in $v_a - U$,

$$p_a - P = \tfrac{1}{2}\rho_0(U^2 - v_a^2) = \tfrac{1}{2}\rho_0(U + v_a)(U - v_a) \approx \rho_0 U(U - v_a).$$

The assumption $|v_a - U| \ll U$ must necessarily break down near the leading and trailing ends of the wing, where there are stagnation points with vanishing flow speed, but for a sufficiently thin planar wing these end effects can be disregarded. Averaging this relation over the upper wing surface, and subtracting a similar expression for a streamline running immediately below the wing, we get

$$\langle p_a \rangle - \langle p_b \rangle \approx -\rho_0 U(\langle v_a \rangle - \langle v_b \rangle). \tag{29.18}$$

Combining this result with (29.16) we obtain an estimate of the difference in average flow velocities above and below the wing in units of the mainstream flow,

$$\Delta v_{ab} \equiv \langle v_a \rangle - \langle v_b \rangle \approx -\frac{\Delta p_{ab}}{\rho_0 U} \approx \frac{\mathcal{L}}{\rho_0 U A}. \tag{29.19}$$

For the Cessna 150 of example 29.4.2 we find $\Delta v_{ab} \approx 7\ \mathrm{m\,s^{-1}}$ which is well below the cruising speed $U \approx 54\ \mathrm{m\,s^{-1}}$.

Circulation and lift

The difference in average flow velocities above and below the wing may also be estimated from the circulation around the wing along a contour C which encircles the wing in the direction of the asymptotic airflow on top and against it below (just outside the boundary layers),

$$\Gamma = \oint_C \boldsymbol{v} \cdot d\boldsymbol{\ell} \approx c\ \langle v_a \rangle - c\ \langle v_b \rangle = c\,\Delta v_{ab}. \tag{29.20}$$

Using (29.19) with $A = cL$ we finally obtain the relation,

$$\boxed{\mathcal{L} \approx \rho_0 U L \Gamma.} \tag{29.21}$$

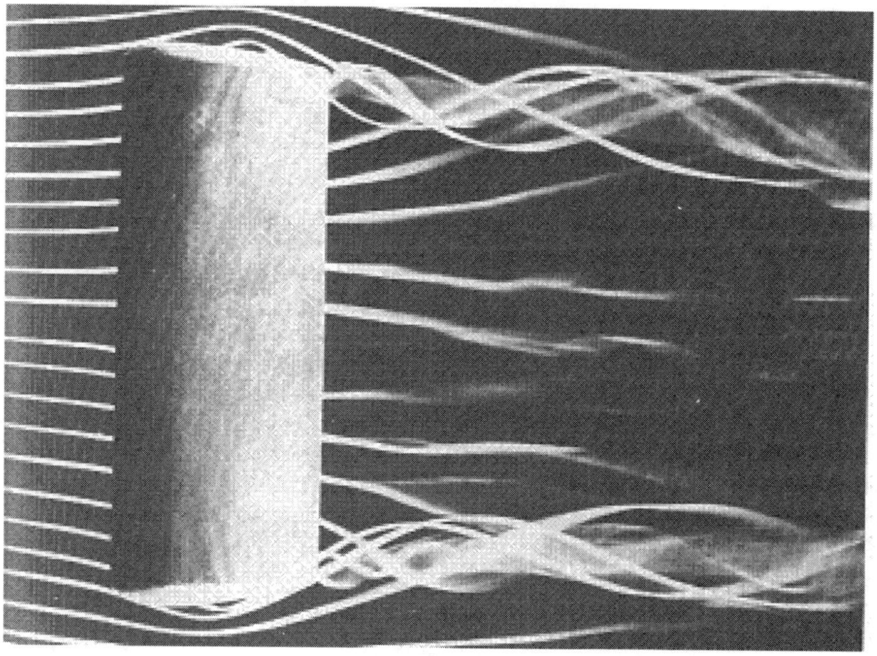

Figure 29.4. Trailing wing tip vortices at Re = 100 000. Reproduced from M. R. Head, in *Flow Visualisation II* ed W. Merzkirch, Hemisphere, Washington, 1982, pp 399–403.

This is the famous *Kutta–Joukowsky* theorem from the beginning of the twentieth century. Here derived from estimates, we shall see in section 29.6 (page 456) that in nearly ideal and nearly irrotational flow this relation is in fact exact when Γ is replaced by the average circulation $\langle\Gamma\rangle$ along the wing.

The realization that lift and circulation are two sides of the same coin, was probably *the single most important insight into the mechanics of flight*. With this in hand, the road was opened for calculating the lift produced by any specific airfoil for which the circulation could be evaluated in asymptotically uniform, irrotational ideal flow.

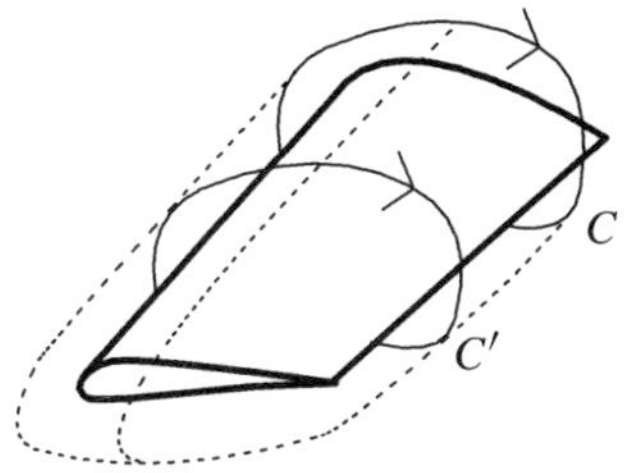

Sketch of vortex bound to the (left) wing of an aircraft. The circulation is the same for the two curves C and C', provided there is no flux of vorticity through the surface bounded by these curves. Note how C' can be slid off the tip of the wing and shrunk to a point if the flow is truly irrotational (which it therefore cannot be).

The horseshoe vortex system

In nearly ideal irrotational flow, the circulation is the same around any curve encircling the wing, because Stokes' theorem relates the difference in circulation between two such curves to the flux of vorticity (which is assumed to vanish) through the surface bounded by the two curves. The lift-generating circulation thus forms a *bound vortex* that ideally cannot leave the wing. For an infinitely long wing this creates no problem, but for a wing of finite span, the assumption of vanishing vorticity has to break down, because one of the curves may be 'slid over' the tip of the wing and shrunk to a point with no circulation. The inescapable conclusion is that since lift requires non-zero circulation, vorticity must come off somewhere along the finite span of a real wing.

The shedding of vorticity from a wing of finite span depends strongly on its shape. A wing that tapers towards the tip will shed vorticity everywhere along its trailing edge, though most near the tip. If the wing is rectangular with constant chord, the vorticity will tend to appear very close to the tip. In any case, the vorticity coming off the wing is blown backwards with respect to the direction of flight, forming a *trailing vortex* in continuation of the bound vortex. Alternatively one may see the trailing vortex as created by the flow around the tip seeking to equalize the higher pressure underneath the wing with the lower pressure above. Together with the bound vortex the two trailing vortices coming off the wing tips form a horseshoe-shaped vortex system accompanying all winged aircraft in flight.

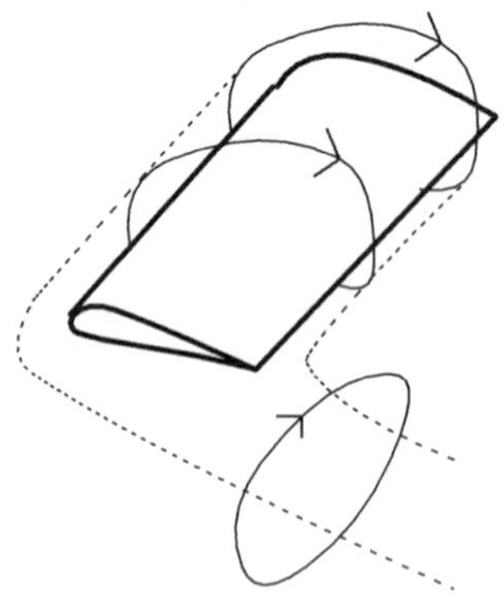

The bound vortex turning into a trailing vortex at the wing tip.

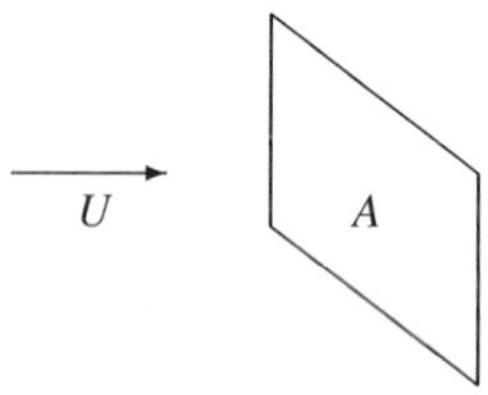

The momentum flux through the area A from the uniform flow with velocity U is the product of the momentum density $\rho_0 U$ and the volume flux UA.

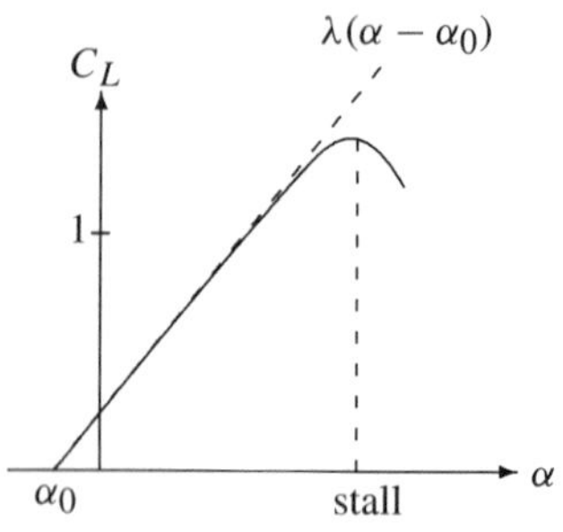

Sketch of a lift curve rising linearly until it veers off rather sharply at an angle of typically 15–20 °, signalling stall. Beyond this angle, lift drops precipitously, and so does the airplane.

Lift coefficient

A wing's ability to produce lift is characterized by the dimensionless *lift coefficient*,

$$\boxed{C_L = \frac{\mathcal{L}}{(1/2)\rho_0 U^2 A}}. \tag{29.22}$$

The denominator is proportional to the momentum flux $\rho_0 U \cdot UA$ through an area A orthogonal to the mainstream direction, and thus sets the scale for the total force that the incoming airstream can exert on this area. The factor $1/2$ in the denominator is conventional. Knowing the lift coefficient, we can immediately determine the average velocity difference (29.19) $\Delta v_{ab} = 1/2 C_L U$.

Being dimensionless, the lift coefficient can only depend on dimensionless quantities, such as the angle of attack α, the Reynolds number $\mathrm{Re}_c = Uc/\nu$, the aspect ratio L/c, and other dimensionless quantities characterizing the shape of the wing. The mainly empirical studies of wing behaviour by scientists and engineers in the last half of the nineteenth century, up to and including the Wright brothers, led to the understanding that the angle of attack was the most important parameter in the lift coefficient. The dependence on the other dimensionless parameters was found to be weaker, in fact so weak that it was mostly ignored before 1900.

The weak dependence of the lift coefficient on the Reynolds number and wing shape parameters allows us to conclude that the lift itself,

$$\mathcal{L} = \tfrac{1}{2}\rho_0 U^2 A C_L, \tag{29.23}$$

is directly proportional to the air density, the wing area and to the square of the velocity. At take off and especially during approach to landing, where speeds are low, the pilot can increase the wing area by means of *flaps*. Since lift always nearly equals the constant weight of the aircraft, it follows from this expression that the increase in area will be compensated by a decrease in required airspeed (for a fixed angle of attack). Fully extended flaps also have a considerably larger angle of attack than the wing itself, increasing thereby the lift coefficient and leading to a further reduction in the required landing speed.

Dependence on angle of attack

Empirically, the lift coefficient is surprisingly linear in the angle of attack,

$$C_L \approx \lambda(\alpha - \alpha_0), \tag{29.24}$$

where $\lambda = dC_L/d\alpha$ is called the *lift slope*, and α_0 is the angle of attack at which the lift vanishes. In section 29.7 we shall see theoretically that for thin airfoils the slope is universally $\lambda = 2\pi$ (with the angle of attack measured in radians). The zero-lift angle α_0 depends mainly on the shape of the airfoil, and is usually small and negative, for example $\alpha_0 \approx -2\,°$ for the Cessna 150 of example 29.4.2.

It follows from the above equations and the constancy of required lift (equal to the weight of the aircraft) that the relative angle of attack $\alpha - \alpha_0$ must vary inversely with the square of the velocity, $\alpha - \alpha_0 \sim 1/U^2$. With decreasing airspeed the required angle of attack rises rapidly until at some critical value boundary layer separation no longer takes place at the trailing edge of the wing but instead suddenly moves forward towards the leading edge, accompanied by turbulence over nearly all of the upper wing surface. The end result is a dramatic loss of lift and a large increase in drag, a phenomenon called *stall*, which was described on page 437.

Stall typically happens at a critical angle of attack, called the *stall angle*, of the order of $\alpha_{\text{stall}} \approx 15°$–$20\,°$ for normal aircraft. Whereas the lift slope and zero-lift angle are essentially independent of the Reynolds number in the linear regime, the stall angle increases a bit with the increasing Reynolds number. For special aircraft the stall angle can be fairly high, for example $35\,°$ for delta-winged aircraft such as the space-shuttle or Concorde. In such aircraft the higher stall angle is offset by a smaller lift slope, say $\lambda \approx 3$ (per radian) rather than 2π. In order to get sufficient lift, these aircraft are forced to take off and land under remarkably high angles of attack.

Dependence on aspect ratio

The shedding of vorticity from finite wings makes the lift slope depend on the aspect ratio L/c. An expression useful for estimating this effect for thin airfoils is [4, p. 380]

$$\lambda \approx \frac{2\pi}{1+2\frac{c}{L}}, \tag{29.25}$$

which in the limit of the infinite aspect ratio, $L/c \to \infty$, converges upon 2π.

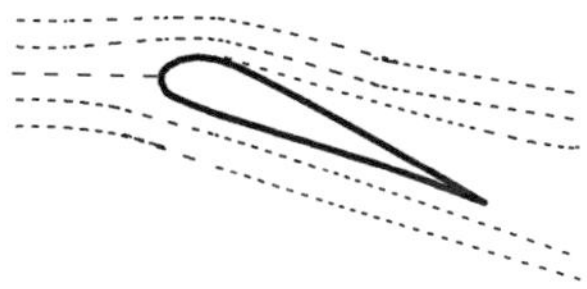

Sketch of the flow around an airfoil at large angle of attack. The boundary layer separates near the leading edge and replaces the previously laminar flow above the wing with turbulent flow yielding small lift and large drag.

Example 29.4.4 (Cessna 150): For the Cessna 150 (page 443) at cruising speed $U = 180$ km h^{-1} we have $\Delta v_{ab}/U = 0.15$, implying a lift coefficient $C_L = 0.34$. With aspect ratio $L/c \approx 7$ the lift slope is $\lambda = 4.9$, and the relative angle of attack becomes $\alpha - \alpha_0 = 4.0\,°$. For this airfoil $\alpha_0 = -2\,°$, and the true angle of attack becomes $\alpha = 2\,°$. This airfoil has a stall angle of $\alpha_{\text{stall}} = 16\,°$, corresponding to a stall velocity $U_{\text{stall}} \approx 80$ km h^{-1}, although stall can be delayed somewhat by enlarging the wing area by means of flaps.

* Dependence on Mach number

Even if modern passenger jets fly below the speed of sound, their Mach number $\text{Ma} = U/c_S$ is so close to unity at cruising speed that there will be major corrections to the lift. To find these corrections we go back to the expression for the divergence of the velocity field (16.33) on page 214 in compressible Euler flow. Inserting $\boldsymbol{v} = \boldsymbol{U} + \Delta\boldsymbol{v}$ and expanding to first order in the presumed small velocity $\Delta\boldsymbol{v}$ we find,

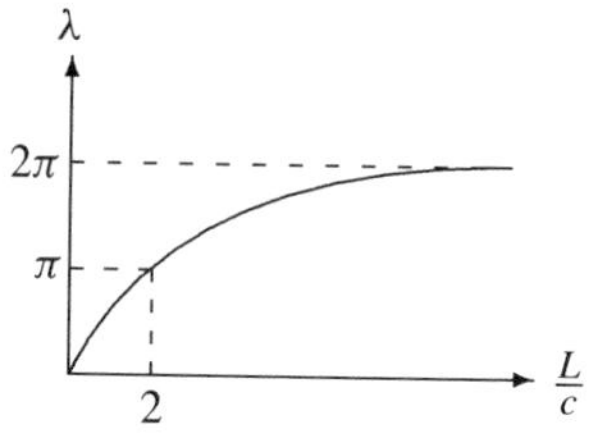

Sketch of the variation of the lift slope with aspect ratio.

$$\boldsymbol{\nabla}\cdot\Delta\boldsymbol{v} \approx \frac{\boldsymbol{U}\cdot(\boldsymbol{U}\cdot\boldsymbol{\nabla})\Delta\boldsymbol{v}}{c_S^2} = \frac{U^2}{c_S^2}\nabla_x \Delta v_x, \tag{29.26}$$

where in the last step we used $\boldsymbol{U} = U\boldsymbol{e}_x$. In ideal, irrotational flow the velocity field is the gradient of the velocity potential, $\Delta\boldsymbol{v} = \boldsymbol{\nabla}\Psi$. Inserting this in the above equation we arrive at,

$$(1-\text{Ma}^2)\nabla_x^2\Psi + \nabla_y^2\Psi + \nabla_z^2\Psi = 0. \tag{29.27}$$

This shows that Ψ is a solution to the incompressible Laplace equation (16.61), provided the x-coordinate is replaced by $x \to x' = x/\sqrt{1-\text{Ma}^2}$.

Since lift stems from an integral of the pressure over x along the chord this rule immediately yields the lift in subsonic compressible flow,

$$\mathcal{L} = \frac{\mathcal{L}_0}{\sqrt{1-\text{Ma}^2}}, \tag{29.28}$$

where $\mathcal{L}_0$ is the lift in incompressible flow for $\text{Ma} \to 0$. This *Prandtl–Glauert rule* is valid up to a critical Mach number of about 0.7 where the flow over part of the airfoil may become sonic [4, 3]. Applying it nevertheless to the Boeing 747 of example 29.4.3 we find $\mathcal{L} \approx 2.2\mathcal{L}_0$ at cruising speed, a quite sizeable increase in lift which at a fixed velocity can be offset by diminishing the angle of attack.

Hermann Glauert (1892–1934). *Leading British aerodynamicist. Worked at the Royal Aircraft Establishment on airfoil and propeller theory, and on the autogyro. Derived the Mach number correction in 1927 in the way done here, independent of Prandtl who had discussed such a rule in the early 1920s.*

29.5 Estimating drag

Whereas lift has one cause, namely the pressure difference between the upper and lower wing surfaces, drag has several. First, there is skin friction from the air flowing over the wing. Second, there is form drag due to the wing obstructing the free air flow and leaving a trail of turbulent air behind, and third there is induced drag from the vortices that always trail the wing tips. For real aircraft, the body shape and various protrusions (radio antenna, Pitot tube, rivet heads, etc) also add to drag.

As for lift, it is convenient and customary to discuss drag in terms of the dimensionless drag coefficient

$$C_D = \frac{\mathcal{D}}{(1/2)\rho_0 U^2 A}, \tag{29.29}$$

which has the same denominator as the lift coefficient (29.22).

Figure 29.5. Prandtl–Glauert condensation around a B1 bomber passing the sound barrier. Crossing the Prandtl–Glauert 'singularity' at Mach 1, the pressure drops sharply and creates this condensation cloud in humid air. Photograph by Gregg Stansbery.

The denominator can be understood in the following way. At the leading edge of a wing there is always a streamline that ends with vanishing airspeed (stagnation). The pressure increase at the stagnation point relative to infinity is $\Delta p = (1/2)\rho_0 U^2$, according to Bernoulli's theorem. If the wing were raised squarely into the oncoming airflow, it would present an area A to this pressure, and assuming that the turbulent 'dead' air behind the wing exerts essentially no extra pressure on the back of the wing, the total drag force would be $\Delta p\, A = (1/2)\rho_0 U^2 A$, which is the denominator. If this argument is right, we predict $C_D \approx 1$ for a thin flat plate with its face into the wind, and that agrees in fact quite reasonably with both theory and experiment. For a circular disc we have $C_D = 1.17$ (see section 19.4 on page 275).

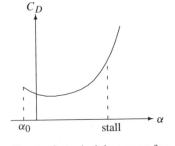

Sketch of a typical drag curve for a cambered airfoil as a function of the angle of attack. Note that the drag of such an airfoil actually decreases for small angles of attack. Beyond the stall angle, the drag coefficient rises steeply.

Skin friction

Close to the wing surfaces there are thin boundary layers (see chapter 28 on page 407), in which the flow velocity changes rapidly from zero right at the skin of the wing to the mainstream airspeed just outside. The maximal thickness of the boundary layer on a wing may be estimated from the flat plate laminar Blasius solution (28.28) or from semi-empirical turbulent expression (28.41),

$$\frac{\delta}{c} \approx \begin{cases} 5\mathrm{Re}_c^{-1/2} & \text{laminar} \\ 0.16\mathrm{Re}_c^{-1/7} & \text{turbulent} \end{cases}. \tag{29.30}$$

Boundary layers do not have the same thickness over the wing surface, but are generally thinnest at the leading edge of the wing and become thicker towards the rear. Usually, the Reynolds number is so high that the boundary layers also develop turbulence somewhere downstream from the leading edge. For aircraft with chordwise Reynolds numbers in the millions and chords of the order of metres, a fully laminar boundary layer is only millimetres thick whereas a fully turbulent layer is an order of magnitude thicker.

To estimate the skin friction we again use flat-plate Blasius' result (28.32) in the laminar regime and the semi-empirical expression (28.34) in the turbulent,

$$C_D^{\text{skin}} = \frac{\mathcal{D}_{\text{skin}}}{(1/2)\rho_0 U^2 A} \approx \begin{cases} 2.65 \ \mathrm{Re}_c^{-1/2} & \text{laminar} \\ 0.063 \ \mathrm{Re}_c^{-1/7} & \text{turbulent} \end{cases}. \tag{29.31}$$

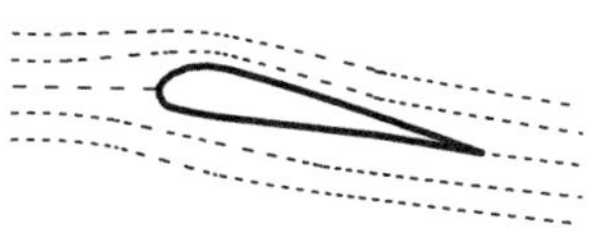

The stagnating streamline at the leading edge terminates in a point with zero flow velocity. Bernoulli's theorem implies that the pressure increase at the forward stagnation point is $\Delta p = 1/2\rho_0 U^2$.

The skin drag coefficient always decreases with increasing Reynolds number, but like the thickness it varies much more slowly in the turbulent region than in the laminar. Turbulent drag is considerably larger than laminar drag, but precise theoretical prediction of skin drag is quite hard because it is difficult to predict the line along the span where the boundary layer becomes turbulent. This is one of the reasons why wind tunnel experiments, and in more recent times numeric (CFD) simulations, have been and still are so important for aerodynamics engineering.

Example 29.5.1 (Cessna 150): For the Cessna 150 of example 29.4.2 at cruising speed with $\mathrm{Re}_c \approx 5 \times 10^6$, the estimate of the maximal laminar boundary layer thickness becomes $\delta \approx 3.4$ mm whereas the estimate of the maximal turbulent thickness becomes $\delta \approx 26$ mm. The corresponding laminar skin drag coefficient is $C_D^{\text{skin}} \approx 0.0012$ whereas the turbulent one is about six times larger, $C_D^{\text{skin}} \approx 0.0070$. The true skin drag is probably closest to this value.

Form drag

Boundary layers thicken and become turbulent towards the rear of the wing and leave a trailing turbulent wake behind. The initial thickness of the trailing wake is comparable to the boundary layer thickness (here strongly exaggerated).

The flow around a highly streamlined body, such as a thin wing narrowing down into a sharp trailing edge, will be nearly ideal everywhere, except in the boundary layers. It has been pointed out before (and we shall prove it in the following section) that a body in a truly ideal, irrotational flow does not experience any drag at all, independent of its shape. Both skin friction and form drag therefore owe their existence to viscosity, but where skin friction is due to shear stresses in the boundary layer, form drag arises from changes in the pressure distribution over the body caused by the presence of boundary layers.

Airfoil boundary layers tend to become turbulent at some point downstream from the leading edge of the wing. At the sharp trailing edge the boundary layers separate from the wing and continue as a *trailing wake* (see figures 29.6 and 29.7). The unsteady turbulent wake found immediately behind the trailing edge of a wing expands slowly and eventually calms down and becomes steady and laminar at some downstream distance from the wing. Further downstream the laminar wake continues to expand by viscous diffusion at a considerably faster rate than the turbulent wake. In section 29.8 we shall determine the general form of the field in the distant laminar wake.

Inside the trailing wake immediately behind the body, the pressure will be lower than the stagnation pressure $\Delta p = 1/2\rho_0 U^2$ at the leading edge (see figure 29.3), and this pressure difference is the cause of drag. The thickness of the turbulent wake immediately after the trailing edge of the wing may be estimated from the boundary layer thickness δ, leading to the form drag estimate $\mathcal{D}_{\text{form}} \sim \Delta p L \delta$. In terms of the drag coefficient we thus find for the turbulent case,

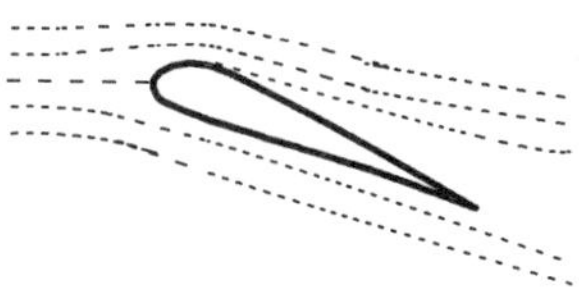

Sketch of the flow around an airfoil at large angle of attack. The boundary layer separates near the leading edge and replaces the previously laminar flow above the wing with turbulent flow yielding small lift and large drag.

$$C_D^{\text{form}} \sim \frac{\delta}{c} \sim \mathrm{Re}_c^{-1/7}. \tag{29.32}$$

Form drag usually amounts to a finite fraction of skin friction for streamlined objects where the boundary layers are thin everywhere.

With growing angle of attack, flow separation may occur on the upper side of the wing at some point before the trailing edge of the airfoil, thereby increasing form drag and diminishing lift. At a certain angle of attack, the point of separation for the turbulent boundary layer on the top side of the wing may suddenly shift forward from the trailing edge, creating a highly turbulent region above the wing. This leads to loss of almost all of the lift and at the same time an increased drag. The wing and the aircraft are then said to have *stalled*.

The efforts of aircraft designers between the world wars in the twentieth century were mainly directed towards form drag reduction by streamlining. A smaller drag generally implies higher top speed, greater payload capacity and better fuel economy. Besides streamlining of lift surfaces, drag reduction was also accomplished by internalizing the wing support structure and the undercarriage, and providing the engines with carefully designed cowlings.

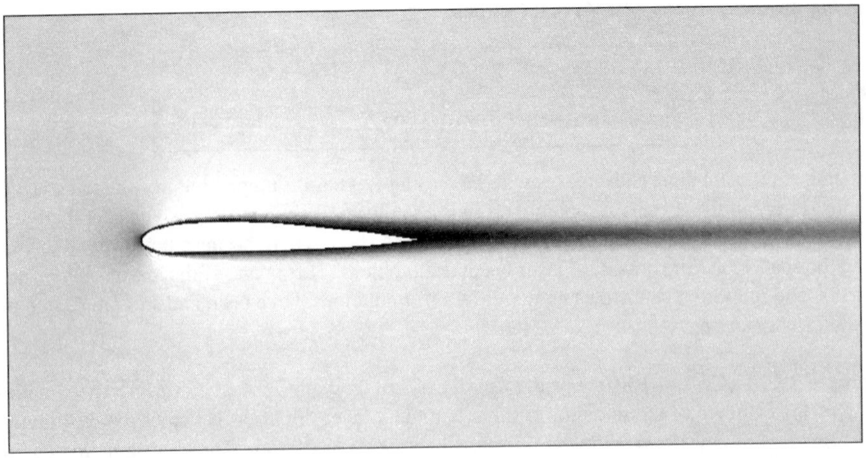

Figure 29.6. Horizontal velocity distribution (v_x) around an airfoil at Reynolds number $\text{Re}_c = 10\,000$ and $\alpha \approx 1\,°$, obtained by numeric simulation. Note the faster flow above the wing (light), the stagnating flow at the leading edge (dark), and the strong slowdown of the flow in the boundary layers and the trailing wake (very dark). The boundary layers are thin and laminar and thicken towards the rear especially on the upper surface where the initial acceleration of the air is followed by deceleration. There is no turbulence in the boundary layers at a Reynolds number as low as this. At more realistic Reynolds numbers in the millions, the boundary layers are mostly turbulent and about an order of magnitude thinner than here. Well behind the airfoil the wake has a thickness comparable to the boundary layers. The slow viscous expansion of the laminar wake is not visible at the scale of this figure.

Induced drag

The two vortices trailing from the wing tips of an aircraft rotate in opposite directions and carry roughly the same circulation Γ as the vortex bound to the wing. They are created at a rate determined by the speed U of the airplane and persist indefinitely in a truly ideal fluid. In a viscous fluid they spin down and dissolve after a certain time.

The process of 'spinning up' and 'feeding out' the trailing vortices from the wing-tips of the aircraft is accompanied by a continuous loss of energy, which causes an extra drag on the aircraft. We can estimate the order of magnitude of this drag from the kinetic energy contained in the core of a vortex with circulation Γ and core radius a. Since the maximal flow speed is of order $v_\phi \sim \Gamma/2\pi a$, the kinetic energy of two vortex segments of length Δx becomes of magnitude $\Delta\mathcal{T} \sim \rho_0 v_\phi^2 \pi a^2 \Delta x \sim \rho_0 \Gamma^2 \Delta x$ (dropping all simple numeric factors). This loss of energy must cause a drag on the aircraft of magnitude

$$\mathcal{D}_{\text{induced}} = \frac{\Delta\mathcal{T}}{\Delta x} \sim \rho_0 \Gamma^2 \sim \rho_0 U^2 c^2 C_L^2,$$

where we in the last estimate have used $\Gamma = 1/2 U c\, C_L$, obtained from the relation between lift and circulation (29.21) and the definition of the lift coefficient (29.22). The estimate of the induced drag coefficient thus becomes

$$\boxed{C_D^{\text{induced}} = \frac{\mathcal{D}_{\text{induced}}}{1/2\rho_0 U^2 A} \sim \frac{c}{L} C_L^2.} \tag{29.33}$$

Classical wing theory (for example [4, p. 369]) yields an expression of precisely this form but roughly a factor π smaller. Since induced drag is a byproduct of the lift-generating flow around a finite wing, it is also called *drag due to lift*. It is the unavoidable price we have to pay for wings of finite span.

Induced drag is normally smaller than skin drag, but grows rapidly with increased angle of attack and may win over skin drag at low speeds. This happens, for example, at take off and landing where the angle of attack is large and the skin friction small. Most importantly, induced drag decreases with increasing aspect ratio L/c, explaining why large aspect ratios are preferable, up to the point where the sheer length of the wing begins to compromise the strength of the wing structure.

Example 29.5.2 (Cessna 150): For the Cessna 150 of example 29.5.1 cruising in level flight with $C_L \approx 0.3$ and aspect ratio 7 we find $C_{\text{induced}} = 0.004$ (including the factor $1/\pi$). The induced drag is thus about half of the turbulent skin drag. At half this speed the relative angle of attack is four times bigger, so that the induced drag coefficient becomes 16 times bigger whereas the turbulent skin drag coefficient stays roughly constant.

Lift-to-drag ratio

The total drag coefficient C_D is the sum of all the contributions from various sources: skin drag, form drag, induced drag, etc. The lift-to-drag ratio, in French called the finesse,

$$\frac{\mathcal{L}}{\mathcal{D}} = \frac{C_L}{C_D}, \tag{29.34}$$

is a measure of the *aerodynamic efficiency* of an airplane. Like lift and drag, the lift-to-drag ratio is strongly dependent on the angle of attack and less on the Reynolds number and aspect ratio. The quadratic growth of induced drag as a function of angle of attack normally overcomes the linear rise in lift and creates a maximum in $\mathcal{L}/D$ for a certain angle of attack, typically at about half the stall angle. In view of the difficulty in making theoretical estimates of drag on the airframe, empirical lift-to-drag curves are usually provided for a particular aircraft to document its performance.

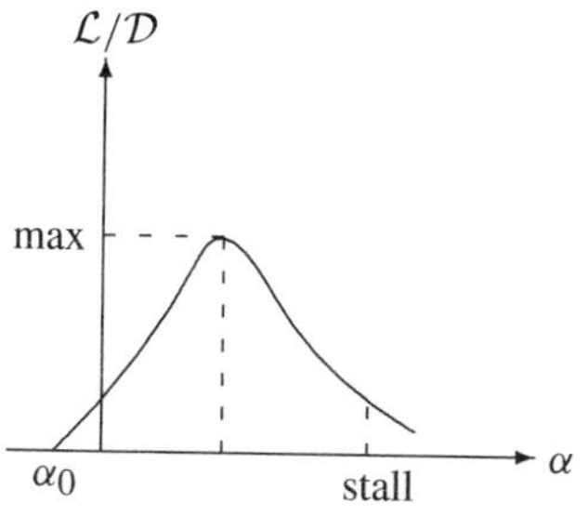

Sketch of a typical lift-to-drag curve for an airplane. The maximal value of lift to drag determines the best glide ratio and the best glide speed.

Example 29.5.3 (Cessna 150): For the Cessna 150 the quoted maximal lift-to-drag ratio is 8.4 at a speed $U_{\text{glide}} \approx 110 \text{ km h}^{-1}$. According to (29.11) the corresponding best glide angle is about $\gamma \approx 6.8°$. From the known lift $\mathcal{L} \approx M g_0 \cos\gamma$ it then follows that the lift coefficient is $C_L \approx 0.78$, and using the known lift slope $\lambda = 4.9$, the angle of attack becomes $\alpha \approx 7.2°$, making the aircraft nearly horizontal in the glide. Cruising in level flight with $U = 180 \text{ km h}^{-1}$, the lift coefficient is instead $C_L \approx 0.34$, and the lift-to-drag ratio is reduced to about 4.5.

* Breguet's range equation

Louis-Charles Breguet (1880–1955). *French engineer. First to recognize the importance of the lift-to-drag ratio for airplane performance. He built an airplane assembly factory which manufactured the famous Breguet 14 bomber for the French forces during World War I. In 1919 he founded the airline company that later became Air France.*

Given an amount of fuel, the endurance and range of an aircraft depend on the aerodynamic efficiency of the airframe and the engine efficiency. The latter is usually measured by the *specific fuel consumption*, defined as the rate of fuel consumption (by weight) per unit of produced thrust T,

$$f = -\frac{1}{T}\frac{d(Mg_0)}{dt}, \tag{29.35}$$

where M is the mass of the aircraft. We shall assume that f, which has dimension of inverse time, is a constant characterizing the engine performance, independent of the thrust it delivers.

Since $\mathcal{L} \approx Mg_0$ and $\mathcal{D} \approx T$ in level flight, the thrust can be determined from the lift-to-drag ratio $T = Mg_0/(\mathcal{L}/D)$, and inserting this expression, we find

$$f \approx -\frac{\mathcal{L}}{\mathcal{D}}\frac{1}{M}\frac{dM}{dt}. \tag{29.36}$$

As the fuel is spent, the mass of the aircraft decreases and with it the required lift. If the angle of attack is kept constant, the velocity has to decrease, but the lift-to-drag ratio will be constant. Integrating the above equation from $t = 0$, where the fuel tank is full and the mass of the aircraft is M_0, to time t, where the tank is empty and the mass is $M_0 - \Delta M$, we obtain the aircraft's *endurance*, i.e. the length of time it can fly on a given amount of fuel,

$$t = \frac{1}{f}\frac{\mathcal{L}}{\mathcal{D}}\log\frac{M_0}{M_0 - \Delta M}. \tag{29.37}$$

The gentle growth of the logarithm shows that the endurance is not increased significantly even if the fuel is a sizable fraction of the aircraft's total mass.

Similarly, writing $dM/dt = U dM/dx$ in (29.36) and integrating, we obtain an equation for the distance an aircraft can fly on a given amount of fuel, called *Breguet's range equation*,

$$\boxed{x = \frac{U}{f}\frac{\mathcal{L}}{\mathcal{D}}\log\frac{M_0}{M_0 - \Delta M}.} \tag{29.38}$$

In deriving this equation we have assumed constancy of $U\mathcal{L}/\mathcal{D}$ as the fuel is being spent, such that U and $\mathcal{L}/\mathcal{D}$ may be taken to be the initial values for the fully loaded aircraft.

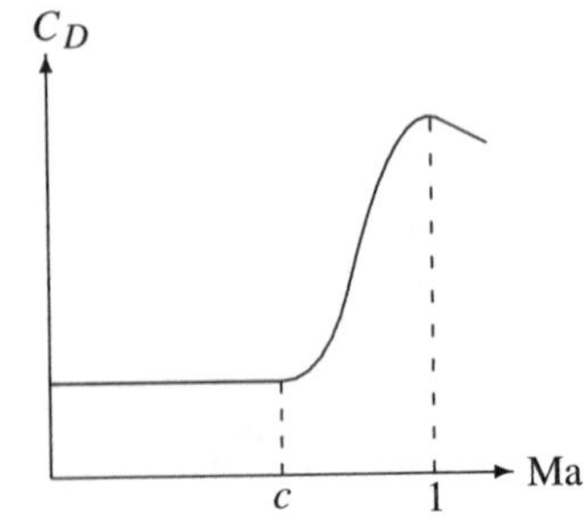

Sketch of the drag coefficient as a function of Mach number. The drag is essentially constant up to the critical point Ma_c after which it rises dramatically towards the 'sound barrier' at Ma = 1.

Example 29.5.4 (Cessna 150): The Cessna 150 has a usable fuel capacity of $\Delta M = 61$ kg and maximal mass of $M = 681$ kg. Cruising at $U = 180$ km h^{-1} the fully loaded aircraft uses fuel at a rate of about 16 kg h^{-1}. Taking $\mathcal{L}/\mathcal{D} \approx 5$, the total thrust is $T \approx Mg_0/(\mathcal{L}/\mathcal{D}) \approx 1300$ N, and the specific fuel consumption becomes $f \approx 3.2 \times 10^{-5}$ s^{-1}. The predicted endurance becomes $t \approx 4$ h and the range $x \approx 720$ km, in decent agreement with known values.

The sound barrier

We have previously derived the Prandtl–Glauert rule (29.28) for the dependence of the lift on the Mach number. That result was based on the assumption of ideal flow and does not apply to skin and form drag which as we have discussed is always caused by viscosity.

Empirically, the drag coefficient is constant up to the critical Mach number $\mathrm{Ma}_c \approx 0.7$ where part of the flow over the wing becomes sonic. After this point the drag begins to increase rapidly with increasing speed. At the time of the Second World War and just after, aircraft came close to the speed of sound and became exposed to the violent stresses that reign here, stresses that could lead to breakup in the air. Although common sense told the engineers that the drag could not actually diverge when the aircraft reached sound speed, it was not clear whether it could mount to such high values that sound velocity would become a barrier, in practice impenetrable with the engines and airframes available at that time. History of course tells us that the 'sound barrier' was passed on October 14, 1947 with a rocket-propelled experimental aircraft (see [4] for more details).

∗ 29.6 Lift, drag and the trailing wake

An often recurring question that can lead to heated discussions is whether an airplane stays aloft in steady flight because of the pressure differences between the upper and lower wing surfaces, or whether it gets lift from diverting momentum downwards. The general treatment of momentum balance in section 22.5 on page 311 indicates that the total contact force on the airplane should be balanced by an opposite momentum flux at great distances, where all stresses have died away. Either position is in fact tenable in a discussion, but as we shall see the correct answer is more subtle than might be guessed at first glance.

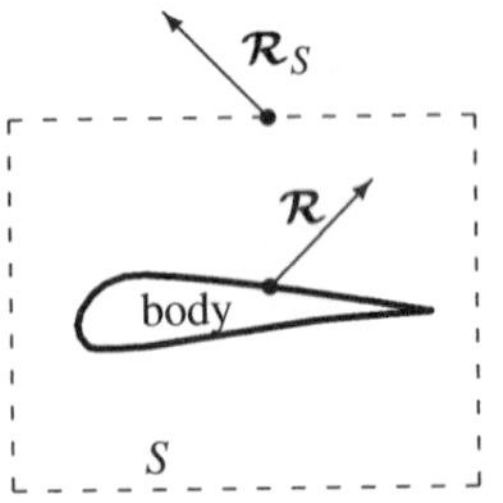

Streamlined body and enclosing box S. The box does not need to be rectangular as it is here but can be a volume of any shape. The reaction forces on body and box are for illustration purposes set to act on arbitrarily chosen points on the surfaces.

Momentum balance in a box

Let the steadily moving body—an aircraft or wing—be surrounded with a huge imagined 'box' of any shape S, and let the volume of air between the body surface and the box be our control volume. As can be seen from figure 29.7 this box will cut through the trailing wake somewhere behind the body. Disregarding gravity, the total force acting on the control volume of air consists of the contact forces on the two bounding surfaces,

$$\boldsymbol{\mathcal{F}} = -\boldsymbol{\mathcal{R}} + \oint_S \boldsymbol{\sigma} \cdot d\boldsymbol{S}, \tag{29.39}$$

where $\boldsymbol{\mathcal{R}}$ is the air's reaction force (29.2) on the body. For the moment we do not split up the reaction force into lift and drag.

In steady flow the total momentum of the air contained in the control volume remains constant, $d\boldsymbol{\mathcal{P}}/dt = \mathbf{0}$, apart from small time-dependent contributions from the fluctuating velocity field in the turbulent wake (which we shall ignore here). Since there can be no momentum flux through the impermeable surface of the body, it follows from momentum balance (23.2) that the total force on the air in the control volume must equal the flux of momentum out of the box $\boldsymbol{\mathcal{F}} = \oint_S \rho \boldsymbol{v}\boldsymbol{v} \cdot d\boldsymbol{S}$. Solving for the reaction force and assuming for simplicity that the air is effectively incompressible, $\rho = \rho_0$, we find,

$$\boxed{\boldsymbol{\mathcal{R}} = -\rho_0 \oint_S \boldsymbol{v}\boldsymbol{v} \cdot d\boldsymbol{S} + \oint_S \boldsymbol{\sigma} \cdot d\boldsymbol{S},} \tag{29.40}$$

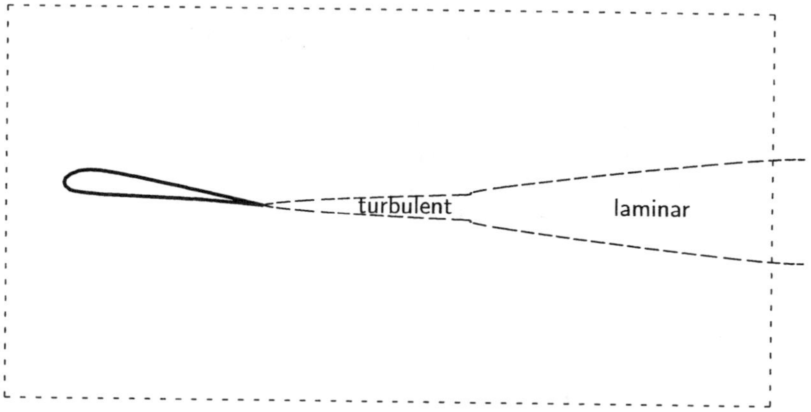

Figure 29.7. Sketch of a body (airfoil) and its trailing wake. Initially the trailing wake is turbulent, but expands slowly and becomes laminar some distance downstream. The thickness of the wake is greatly exaggerated compared to the distance from the body. The dashed box surrounding the system (but crossing through the trailing wake) is used in the text to define a control volume of air between the surfaces of the body and the box.

where $\boldsymbol{\sigma} = \{\sigma_{ij}\}$ is the incompressible stress tensor $\sigma_{ij} = -p\delta_{ij} + \eta(\nabla_i v_j + \nabla_j v_i)$. The following analysis can also be carried through for barotropic compressible air (problem 29.5).

The total reaction force on the body can thus be calculated from the pressure and velocity fields, all evaluated at the surface of any box surrounding the body. It must be emphasized that this result is exact, valid for any shape and size of body and box.

Box at spatial infinity

Now let the box expand to huge distances in all directions such that the velocity field on its surface approaches the asymptotic value, $\boldsymbol{v} \to \boldsymbol{U}$. Setting $\boldsymbol{v} = \boldsymbol{U} + \Delta\boldsymbol{v}$ in the first (momentum) term of (29.40), it becomes to first order in $\Delta\boldsymbol{v}$,

$$-\rho_0 \oint_S (\boldsymbol{U}\boldsymbol{U} + \boldsymbol{U}\Delta\boldsymbol{v} + \Delta\boldsymbol{v}\boldsymbol{U}) \cdot d\boldsymbol{S}.$$

Given that the total vector area of a closed surface always vanishes, $\oint d\boldsymbol{S} = \boldsymbol{0}$, together with global mass conservation, $\oint \Delta\boldsymbol{v} \cdot d\boldsymbol{S} = 0$, the first two terms in the integrand do not contribute to the integral.

If we think of the 'box' as a huge sphere with radius r and surface area $4\pi r^2$, the velocity field correction must decay like $|\Delta\boldsymbol{v}| \sim 1/r^2$ at great distances outside the trailing wake for any contribution to survive in the limit. Inside the wake the velocity field behaves differently (see section 29.8) but the general conclusions remain valid. The velocity derivatives in the stress tensor must consequently vanish like $|\nabla\Delta\boldsymbol{v}| \sim 1/r^3$, and cannot contribute to the second term in (29.40). Pressure is thus the only stress component that has a possibility of surviving in this limit, and the reaction force on the body may be written,

$$\boxed{\boldsymbol{\mathcal{R}} = -\rho_0 \oint_S \Delta\boldsymbol{v}\,\boldsymbol{U} \cdot d\boldsymbol{S} - \oint_S \Delta p\, d\boldsymbol{S}.} \tag{29.41}$$

Since a constant pressure yields no contribution, we have replaced the pressure by the residual pressure $\Delta p = p - p_0$ where p_0 is the constant asymptotic pressure. The residual pressure can only contribute to the integral if it decays as $\Delta p \sim 1/r^2$ outside the trailing wake, which we shall see that it does.

We are now in a position to answer the question of whether there remains a pressure contribution to lift far from the moving body. Although the derivation of the above equation shows that the *sum* of the momentum flow and pressure contributions is independent of the choice of the box shape, each term by itself may (and will) depend on it. *The limiting value of the pressure contribution may depend on how the box is taken to infinity.*

If, for example, we choose a cube or sphere and let it expand uniformly in all directions, there will usually be a residual pressure contribution to lift, even in the limit of an infinite box (see section 29.8 for an explicit calculation). Alternatively, one may choose a box in the form of a huge cylinder with radius R and length L, oriented with its axis parallel to the asymptotic flow $\boldsymbol{U}$. The pressure integral over the end caps cannot contribute to lift, because they are orthogonal to the velocity. If we now let the radius R become infinite, *before* the end caps are moved off to infinity, the pressure integral over the cylinder surface will behave like the area $2\pi LR$ times the pressure $p \sim 1/R^2$. It thus vanishes like L/R for $R \to \infty$, leaving no pressure contribution to lift in the limit (see [38]).

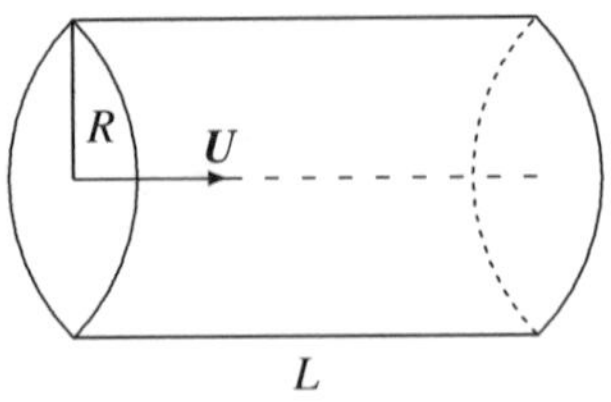

Box in the shape of a cylinder of radius R and length L with axis parallel to the asymptotic velocity $\boldsymbol{U}$.

Lift and drag

Let us rearrange (29.41) in the form

$$\boldsymbol{\mathcal{R}} = -\rho_0 \oint_S \boldsymbol{U} \times (\Delta \boldsymbol{v} \times d\boldsymbol{S}) - \oint_S (\Delta p + \rho_0 \Delta \boldsymbol{v} \cdot \boldsymbol{U})\, d\boldsymbol{S}. \tag{29.42}$$

The first integral is orthogonal to the asymptotic velocity and represents the lift,

$$\boxed{\boldsymbol{\mathcal{L}} = -\rho_0 \oint_S \boldsymbol{U} \times (\boldsymbol{v} \times d\boldsymbol{S}).} \tag{29.43}$$

Here we have also replaced $\Delta \boldsymbol{v}$ by $\boldsymbol{v}$, using $\oint d\boldsymbol{S} = \boldsymbol{0}$.

In regions where the flow is non-viscous and all streamlines connect to spatial infinity, the pressure excess is determined by Bernoulli's theorem,

$$\Delta p = \tfrac{1}{2}\rho_0 (\boldsymbol{U}^2 - \boldsymbol{v}^2) \approx -\rho_0 \boldsymbol{U} \cdot \Delta \boldsymbol{v}. \tag{29.44}$$

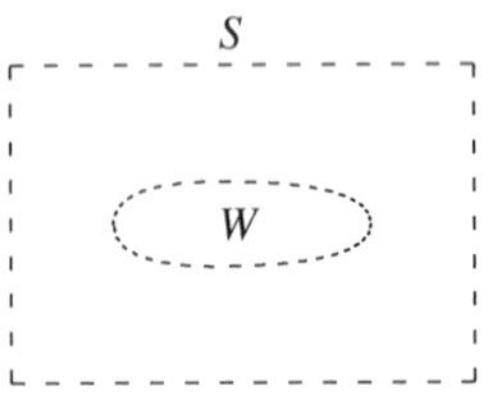

The downstream face of the box cuts the wake in a region W chosen to be orthogonal to the asymptotic flow $\boldsymbol{U}$.

The contributions to the second term in (29.42) can only come from the region W, where the box cuts through the wake. Choosing W to be planar and orthogonal to $\boldsymbol{U}$, its surface element $d\boldsymbol{S}$ will be parallel with $\boldsymbol{U}$, and the second term becomes a pure drag,

$$\boxed{\boldsymbol{\mathcal{D}} = -\int_W (\Delta p + \rho_0 \Delta \boldsymbol{v} \cdot \boldsymbol{U})\, d\boldsymbol{S},} \tag{29.45}$$

where $d\boldsymbol{S}$ is the area element of W. If we lift the restriction that W is a planar part of S orthogonal to $\boldsymbol{U}$, the formulae for lift and drag become slightly different (problem 29.6). Otherwise they are valid for all kinds of bodies moving steadily through an incompressible fluid at subsonic speed.

d'Alembert's paradox: a gift to powered flight

We have previously (page 226) shown that a cylinder or sphere in irrotational (potential) flow experiences no drag. In irrotational flow, Bernoulli's theorem is fulfilled everywhere, such that the drag (29.45) must vanish. D'Alembert's paradox must therefore be valid in full generality: *there is no drag on a body of arbitrary shape in completely irrotational flow.* But bodies moving through viscous fluid cannot help leaving a narrow trailing wake containing vorticity, and equation (29.45) immediately resolves the paradox: *the drag on a body stems entirely from the trailing wake.*

In general, the narrower the wake the smaller the drag will be. As we have seen in the estimates of the preceding section, drag is typically an order of magnitude smaller than lift for properly streamlined bodies, such as airfoils. This indicates that one should rather treat d'Alembert's 'paradox' as a *theorem* about the near vanishing of drag for streamlined bodies at high Reynolds number. This theorem is in fact what makes flying technically possible with engines producing a thrust much smaller than the weight of the aircraft. Without d'Alembert's theorem, the Wright brothers would never have had a chance of flying at Kitty Hawk in December 1903, given the puny engine power then available to them.

Lift and vorticity

The box S used in the lift integral (29.43) is assumed to be of essentially infinite size. Now let S' be another closed surface surrounding the body somewhere inside the box S. From Gauss' theorem we obtain

$$\left(\oint_S - \oint_{S'}\right) \boldsymbol{U} \times (\boldsymbol{v} \times d\boldsymbol{S}) = -\int_V \boldsymbol{U} \times (\nabla \times \boldsymbol{v})\, dV = -\int_V \boldsymbol{U} \times \boldsymbol{\omega}\, dV,$$

where V is the region between the two surfaces, and $\boldsymbol{\omega} = \nabla \times \boldsymbol{v}$ is the vorticity field. In the extreme case we may take S' to be the body surface itself, where the velocity and thus the integral over S' vanishes because of the no-slip condition. It then follows from the above equation that the lift (29.43) is also given by the integral of the vorticity field over all of the air space,

$$\boxed{\mathcal{L} = \rho_0 \boldsymbol{U} \times \int_{\text{air}} \boldsymbol{\omega}\, dV.} \tag{29.46}$$

This integral can only receive contributions from the regions of non-vanishing vorticity, i.e. from the boundary layers and the trailing wake. Again we conclude that *without vorticity created by friction, there can be no lift*!

More generally, if there is no vorticity found in V, the integral over the box S equals the integral over S'. The original box may in other words be deformed into any other closed surface as long as it crosses no region containing vorticity. The box at infinity has now served its purpose and may be forgotten. In the following the surface S in (29.43) may be taken to be simply any surface surrounding the body, as long as there is no vorticity *outside* S.

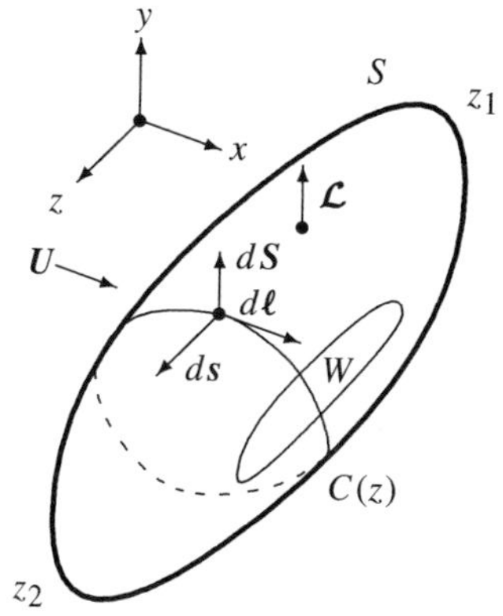

The surface S may always be sliced into a sequence of oriented planar curves $C(z)$ parallel with the xy-plane for $z_1 \le z \le z_2$.

Lift and circulation

It is useful to introduce a 'natural' coordinate system with the x-axis along the direction of the asymptotic velocity $\boldsymbol{U} = U\boldsymbol{e}_x$ and the y-axis along the lift $\mathcal{L} = \mathcal{L}\boldsymbol{e}_y$. Working out the cross products, the lift (29.43) becomes

$$\mathcal{L} = \rho_0 U \oint_S (\boldsymbol{v} \times d\boldsymbol{S})_z = \rho_0 U \oint_S v_x dS_y - v_y dS_x, \tag{29.47}$$

together with the condition

$$\oint_S v_x dS_z - v_z dS_x = 0, \tag{29.48}$$

expressing that there should be no lift along the z-direction. For a symmetric aircraft in normal horizontal flight, this condition is automatically fulfilled.

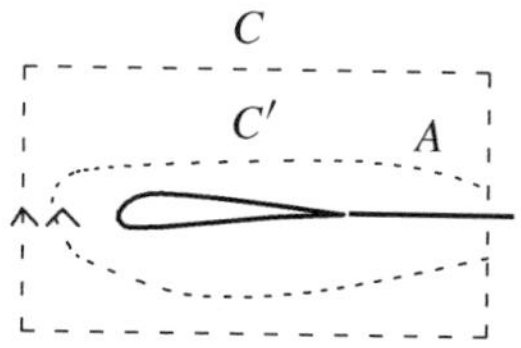

The contours C and C' cross the trailing wake along the same curve (at infinity) and have the same circulation as long as the area A between the curves carries no flux or vorticity.

The closed surface S may always be sliced into a set of planar closed contours $C(z)$ parallel with the xy-plane, and parametrized by the z-coordinate in some interval $z_1 \le z \le z_2$. The contours are given negative orientation in the xy-plane, i.e. clockwise as seen from positive z-values. Let now $d\boldsymbol{\ell} = (dx, dy, 0)$ be a line element on a point of the curve $C(z)$; it is evidently a tangent vector to the surface S. Let $d\boldsymbol{s} = (0, dy, dz)$ be another tangent vector to the surface in the yz-plane with the same y-coordinate dy. Then the outward pointing surface element becomes $d\boldsymbol{S} = d\boldsymbol{s} \times d\boldsymbol{\ell} = (-dydz, dzdx, -dxdy)$ and thus

$$v_x dS_y - v_y dS_x = (v_x dx + v_y dy)\, dz = \boldsymbol{v} \cdot d\boldsymbol{\ell}\, dz.$$

This shows that the lift (29.47) may be written as an integral over z

$$\boxed{\mathcal{L} = \rho_0 U \int_{z_1}^{z_2} \Gamma(z)\, dz,} \tag{29.49}$$

with an integrand given by the circulation around $C(z)$,

$$\Gamma(z) = \oint_{C(z)} \boldsymbol{v} \cdot d\boldsymbol{\ell}. \tag{29.50}$$

If the contour is deformed, Stokes' theorem (16.57) tells us that the circulation is unchanged provided the contour is swept through an area A devoid of vorticity. Since we have assumed that vorticity is only found in the boundary layers and the trailing wake, the contour may be freely deformed, as long as the piece of the contour that crosses the wake is kept fixed, and the new contour does not pass through the wake or into the boundary layers. This is of course the same conclusion as was reached in the preceding subsection.

Again it should be emphasized that no approximations have been made, and that this result is exactly valid, as long as the wake-crossing takes place at great distance from the body along a line parallel with the y-axis.

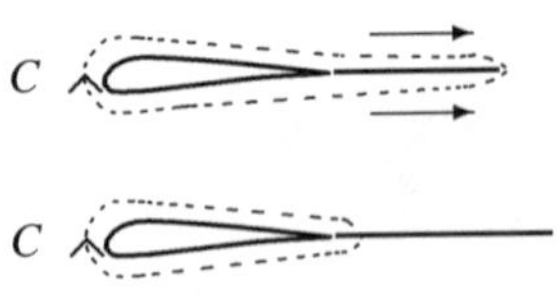

The dashed contour C hugs the wing profile closely, but is still attached to the distant part where it crosses the wake (top). For nearly infinite Reynolds number, the velocity is very nearly the same above and below the wake, allowing us to cut off the tail (bottom).

The general Kutta–Joukowsky theorem

The lift integral (29.49) may be written in the form of a generalized version of the Kutta–Joukowsky theorem (29.21),

$$\boxed{\mathcal{L} = \rho_0 U L \langle \Gamma \rangle ,} \tag{29.51}$$

where

$$\langle \Gamma \rangle = \frac{1}{L} \int_{z_1}^{z_2} \Gamma(z)\, dz, \tag{29.52}$$

is the circulation along z averaged over the wing span $L = z_2 - z_1$. The only difference is that the integration contour $C(z)$ used to calculate $\Gamma(z)$ in (29.50) has to cross the wake far away from the airfoil, whereas in the original Kutta–Joukowsky theorem it is supposed to hug the airfoil tightly all the way around. We shall now see how to get rid of the 'tail' of the contour when the chordwise Reynolds number, $\mathrm{Re}_c = Uc/\nu$ and the aspect ratio L/c are very large.

Martin Wilhelm Kutta (1867–1944). *German mathematician. Probably best known for his extension of a method developed by Runge for numeric solutions to differential equations. Obtained the first analytic result for lift, and effectively discovered the relation between lift and circulation in 1902*

Nikolai Yegorovich Joukowsky (1847–1921). *Russian mathematician and physicist (also spelled Zhukovskii). Constructed the first Russian wind tunnel in 1902 and many others early in the twentieth century. Found and used the relation between lift and circulation in 1906*

In this limit the flow around the wing becomes nearly ideal and irrotational, and the boundary layers turn into a 'skin' of vorticity covering the airfoil with nearly vanishing thickness $\delta \sim c/\sqrt{\mathrm{Re}_c}$. Downstream from the airfoil, the skin continues into the trailing wake which forms a horizontal sheet, also of nearly vanishing thickness δ. Physically, the flow velocity in the wake cannot become infinite, so that the downstream volume flux in the wake, which per unit of span is of order $v_x\delta$, must itself vanish in the limit of infinite Reynolds number. It then follows from mass conservation that the orthogonal velocity v_y must be the same above and below the sheet, for otherwise fluid would accumulate in the wake.

The pressure must also be the same above and below the trailing wake sheet because of Newton's third law. Combining these two results with Bernoulli's theorem, which states that $p + 1/2\rho_0(v_x^2 + v_y^2 + v_z^2)$ takes the same value everywhere outside the wake, we conclude that $v_x^2 + v_z^2$ must be the same just above and below the wake. The span-wise induced flow v_z is connected to the shedding of vorticity along the span, especially the wing-tip vortices. When the aspect ratio is large, this flow will be tiny compared to the downstream flow, i.e. $|v_z| \sim v_x c/L \ll v_x$, so that it may be ignored in the Bernoulli function. Consequently, v_x itself takes the same value just above and below the trailing sheet. The two oppositely directed contributions to the integral (29.50) running along the tail of C thus tend to cancel each other for large Reynolds number and aspect ratio.

Since the orthogonal velocity v_y may not be infinite inside the wake, the part of the integral from the contour passing through the sheet will be of order of magnitude $v_y\delta$ and thus vanish in the limit. The contribution from the tail of the integration contour can thus be ignored in the leading approximation and we may let it circle the wing while tightly hugging the airfoil profile. Finally, we have arrived at the (generalized) Kutta–Joukowsky theorem.

∗ 29.7 Two-dimensional airfoil theory

Most wings have fairly large aspect ratios in the vicinity of $L/c \approx 7$–20 with airfoil cross sections that taper gently towards the wing tips. For nearly infinite aspect ratio and nearly constant cross section, there is very little induced flow along z towards the wing tips, so that the flow becomes essentially two-dimensional,

$$\boldsymbol{v} = (v_x(x, y), v_y(x, y), 0). \tag{29.53}$$

For such an airfoil, the circulation will be independent of z. Although it would be possible to simplify the following calculations using complex notation as in section 26.3, we shall use the physically more transparent real notation.

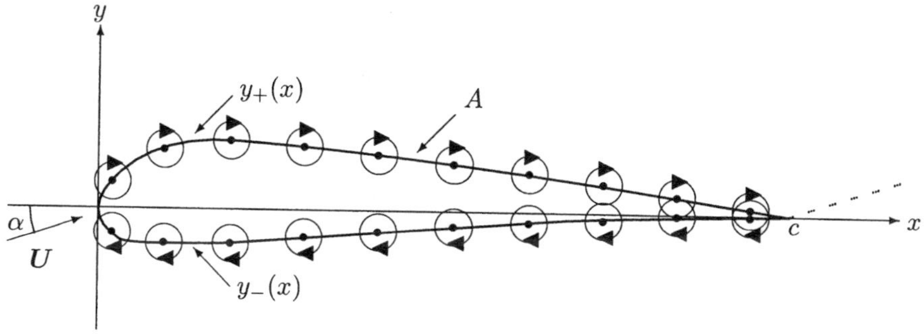

Figure 29.8. The irrotational velocity field of the wing is viewed as a superposition of elementary line vortices with singular cores arranged around the outline A of the airfoil. On the top of the airfoil, the vortices tend to increase the local velocity above the asymptotic flow $\boldsymbol{U}$, and conversely at the bottom. The airfoil is positioned with the chord-line on the x-axis and the y-axis at the leading edge. Its geometry is described by two functions $y_\pm(x)$ with $0 \leq x \leq c$ where c is the chord length. The asymptotic flow is $\boldsymbol{U} = U(\cos\alpha, \sin\alpha)$, where α is the angle of attack. The trailing wake, indicated by the dashed line, also forms an angle α with the x-axis. The z-axis comes out of the plane.

The field of the vortex sheet

In the limit of nearly infinite Reynolds number, vorticity only exists in the infinitesimally thin boundary skins of the airfoil. Outside these skins and outside the infinitesimal sheet of the trailing wake, we assume that the flow is irrotational, described by a velocity potential Φ that satisfies Laplace's equation $\nabla^2\Phi = 0$. Being a linear equation, its solutions may be superposed. All the nonlinearity of the original Euler equation has been collected in the Bernoulli pressure $p = 1/2\rho_0(U^2 - \boldsymbol{v}^2)$. The additivity of potential flows makes it possible to view the irrotational flow outside the boundary layers as arising from a superposition of the asymptotic flow $\boldsymbol{U}$ and the field generated by the sheet of vorticity covering the wing surface.

Due to the two-dimensional nature of the flow, the vortex sheet making up the skin may be understood as a collection of elementary line vortex cores running parallel with the z-axis (see figure 29.8). The contribution from the velocity field of a line vortex passing through the origin of the coordinate system with the core parallel with the z-axis is of the well-known form (26.12),

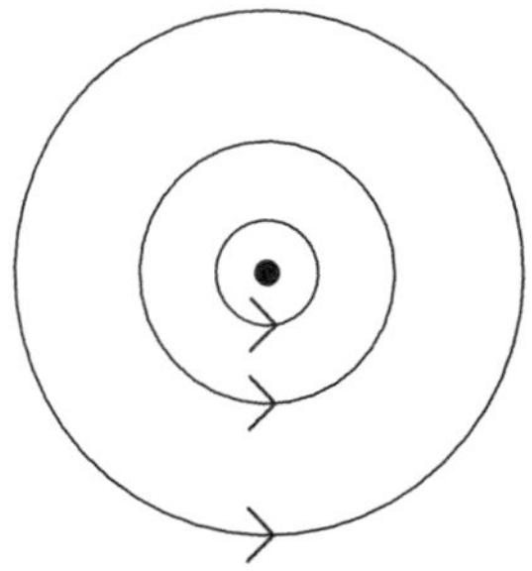

Outside the singular core of a line vortex the flow is irrotational with circular streamlines. The field around any collection of line vortices is obtained by adding their individual fields together.

$$\boldsymbol{v} = \frac{\Gamma}{2\pi}\frac{(-y, x)}{x^2 + y^2}, \tag{29.54}$$

where Γ is its circulation.

Denoting the infinitesimal strength of the vortex passing through the point (x', y') on the airfoil outline A by $d\Gamma'$, the complete velocity field at an arbitrary point (x, y) becomes a curve integral around the airfoil outline A,

$$\boxed{\boldsymbol{v}(x, y) = \boldsymbol{U} + \oint_A \frac{(-y + y', x - x')}{(x - x')^2 + (y - y')^2}\frac{d\Gamma'}{2\pi}.} \tag{29.55}$$

Mathematically, the points of the airfoil outline should be parametrized as a pair of functions, $(x, y) = (f_x(\theta), f_y(\theta))$, of a running parameter θ in some interval, say $\theta_1 \leq \theta \leq \theta_2$, beginning and ending at the cusp. The circulation element then becomes $d\Gamma = \gamma(\theta)\, d\theta$ where $\gamma(\theta)$ is the circulation density. Wherever possible we shall suppress this elaborate, but mathematically more concise, notation.

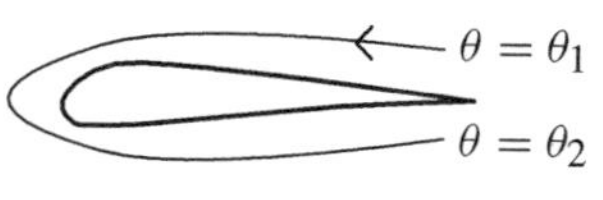

The airfoil outline may be parametrized with a parameter θ running over the interval $\theta_1 \leq \theta \leq \theta_2$.

The Kutta condition

Near the front and rear ends of the airfoil there are stagnation points where the velocity field vanishes. In ideal flow there may be more than one velocity field solution satisfying the Euler equation (16.1) and the boundary conditions. Such solutions may have different stagnation points and thus different circulation. It is even possible to find a solution with vanishing circulation (and lift). We shall see below that if the rear stagnation point is not situated right at the cusp of the trailing edge, unacceptable infinite velocity

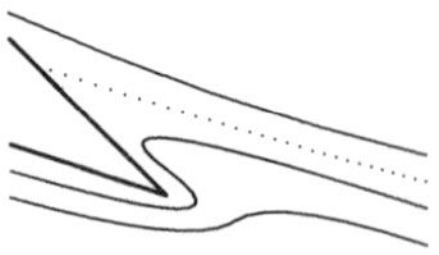

Separating flow pattern near the trailing end of an airfoil with stagnation point before the cusp. Such a flow may have vanishing lift.

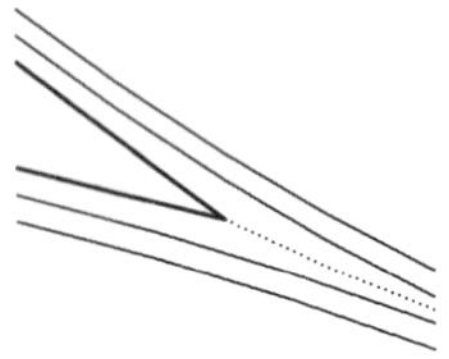

Highly laminar flow near the trailing end of an airfoil with stagnation point at the cusp.

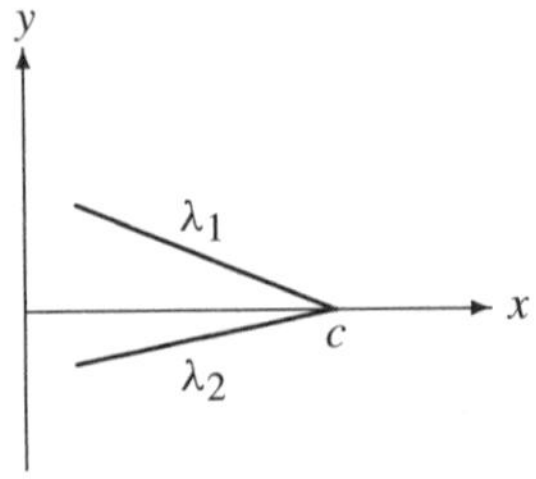

The cusp consists of two straight lines with different slopes $dy/dx = \lambda_1 = \dot{y}_1/\dot{x}_1$ and $dy/dx = \lambda_2 = \dot{y}_2/\dot{x}_2$.

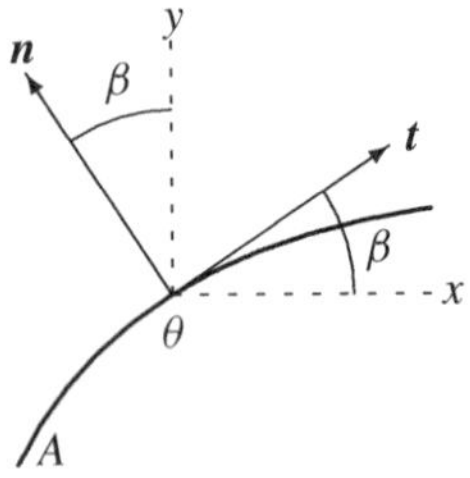

At any point $(x, y) = (f_x(\theta), f_y(\theta))$ of the airfoil outline, the tangent angle is denoted β and the slope $\lambda = \tan\beta = \dot{y}/\dot{x}$. The tangent vector may be chosen to be $\boldsymbol{t} = (\dot{x}, \dot{y})$ and the normal $\boldsymbol{n} = (-\dot{y}, \dot{x})$.

field values will arise at the cusp. The *Kutta condition* (1902) enforces that the trailing edge cusp (c in figure 29.8) is actually a stagnation point. The condition thus repairs a mathematical problem in truly ideal flow by selecting a particular solution. In the real world, a streamlined airfoil under a small angle of attack in nearly ideal flow will, in fact, fulfill the Kutta condition because of viscous friction in the boundary layers which selects a unique laminar solution.

The problem arises when we attempt to calculate the velocity field at a point (x, y) on the airfoil outline itself, because the integrand of (29.55) is formally infinite for $(x', y') = (x, y)$. At any point where the airfoil is smoothly varying, there is in fact no problem and the integral is finite. To see this we cut out a small parameter interval $\theta - \epsilon < \theta' < \theta + \epsilon$ around the singularity in the integrand. Assuming smoothness in this interval we may expand $x - x' \approx (\theta - \theta')\dot{x}$ and $y - y' = (\theta - \theta')\dot{y}$ where $\dot{x} = df_x(\theta)/d\theta$ and $\dot{y} = df_y(\theta)/d\theta$. The leading contribution to the integral from this interval then becomes

$$\Delta\boldsymbol{v} \approx \gamma(\theta)\frac{(-\dot{y}, \dot{x})}{\dot{x}^2 + \dot{y}^2}\int_{\theta-\epsilon}^{\theta+\epsilon}\frac{1}{\theta - \theta'}\frac{d\theta'}{2\pi}. \tag{29.56}$$

Here the integral vanishes because of symmetry around the singularity (mathematically it is a principal-value integral).

The argument fails, however, at the trailing edge for $\theta = \theta_{1,2}$ where the airfoil has a sharp cusp. Here we find instead the contributions,

$$\Delta\boldsymbol{v} \approx \gamma_1\frac{(-\dot{y}_1, \dot{x}_1)}{\dot{x}_1^2 + \dot{y}_1^2}\int_{\theta_1}^{\theta_1+\epsilon}\frac{1}{\theta_1 - \theta'}\frac{d\theta'}{2\pi} + \gamma_2\frac{(-\dot{y}_2, \dot{x}_2)}{\dot{x}_2^2 + \dot{y}_2^2}\int_{\theta_2-\epsilon}^{\theta_2}\frac{1}{\theta_2 - \theta'}\frac{d\theta'}{2\pi},$$

where the coefficients are evaluated at $\theta = \theta_{1,2}$. Since the slopes $(\dot{x}_1, \dot{y}_1)$ and $(\dot{x}_2, \dot{y}_2)$ are different, the divergent integrals will in general not cancel. Only if the vortex density vanishes from both sides of the cusp, $\gamma_1 = \gamma_2 = 0$, will the singular contribution disappear.

Given that in ideal flow the velocity must be tangential to the airfoil outline A, the circulation becomes

$$\Gamma = \oint_A \boldsymbol{v}(x, y)\cdot d\boldsymbol{\ell} = \oint_A |\boldsymbol{v}|\, d\ell. \tag{29.57}$$

Locally each little line element $d\ell$ of A contributes the infinitesimal amount,

$$d\Gamma = \gamma\, d\theta = |\boldsymbol{v}|\, d\ell, \tag{29.58}$$

to the circulation. Since $d\ell/d\theta$ is regular on each side of the cusp, the vanishing of the vortex density γ is equivalent to the vanishing of the velocity field $\boldsymbol{v}$ at the cusp, which is the Kutta condition.

The fundamental airfoil equation

The vortex density $\gamma(\theta)$ must be chosen such that the streamlines follow the airfoil outline. This is equivalent to requiring the normal component of the velocity field to vanish on the impermeable airfoil surface,

$$\boldsymbol{v}\cdot\boldsymbol{n} = 0, \tag{29.59}$$

where $\boldsymbol{v} = \boldsymbol{v}(x, y)$ is the slip-velocity and $\boldsymbol{n} = \boldsymbol{n}(x, y)$ is the normal at the point (x, y) on A. For every point on A we thus get one scalar condition, and together with the Kutta condition this is sufficient to determine the vortex density $\gamma(\theta)$.

We are not at liberty to impose a no-slip condition on the field, because the Euler equation (16.1) is only of first order in spatial derivatives. Although the field (29.55) exists both inside and outside the airfoil outline, the outside solution now fulfills the Euler equation and obeys the correct boundary conditions for inviscid flow around a solid body. Consequently, we may with impunity replace the region inside the airfoil outline with a solid body.

For convenience, the airfoil is positioned with its chord-line on the x-axis such that the asymptotic velocity becomes $\boldsymbol{U} = U(\cos\alpha, \sin\alpha)$ (see figure 29.8). In the θ-parametrization, the tangent vector to the

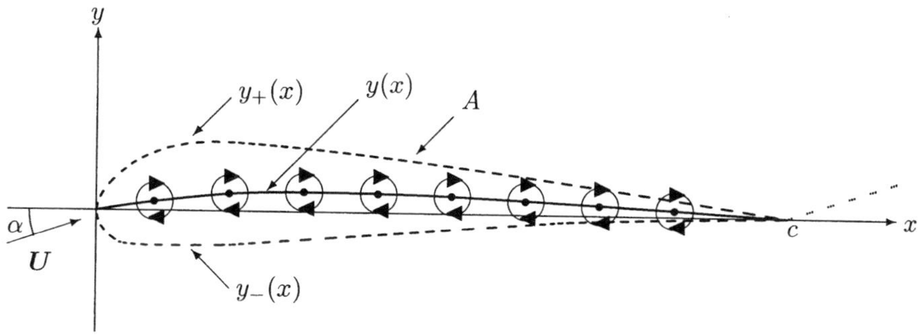

Figure 29.9. A thin airfoil is represented by a single layer of vorticity $\gamma(x) = \gamma_+(x) + \gamma_-(x)$ along the camber line $y(x) = 1/2(y_+(x) + y_-(x))$.

airfoil outline at the point θ is $\boldsymbol{t} = (\dot{x}, \dot{y}) = d(x, y)/d\theta$, and the normal may be taken to be $\boldsymbol{n} = (-\dot{y}, \dot{x})$. The boundary condition (29.59) then takes the explicit form

$$U(-\dot{x}\sin\alpha + \dot{y}\cos\alpha) = \oint_A \frac{\dot{x}(x - x') + \dot{y}(y - y')}{(x - x')^2 + (y - y')^2} \frac{d\Gamma'}{2\pi}, \tag{29.60}$$

where now both $(x, y) = (f_x(\theta), f_y(\theta))$ and $(x', y') = (f_x(\theta'), f_y(\theta'))$ are points on A and $d\Gamma' = \gamma(\theta')d\theta'$.

Marvellously this equation can be integrated over θ. Using that for $\theta = \theta_1$ we must have $x = c$ and $y = 0$, it may be verified by differentiation after θ that the following expression is the correct integral,

$$\boxed{U((c - x)\sin\alpha + y\cos\alpha) = \frac{1}{2}\oint_A \log\frac{(x - x')^2 + (y - y')^2}{(c - x')^2 + y'^2} \frac{d\Gamma'}{2\pi}.} \tag{29.61}$$

This is the *fundamental equation of two-dimensional airfoil theory*. Given the parametrized airfoil geometry through the functions $(x, y) = (f_x(\theta), f_y(\theta))$, this integral equation should be solved for the vortex density $\gamma(\theta) = d\Gamma/d\theta$.

Having done that, the total circulation may afterwards be obtained by integrating the result, $\Gamma = \int_{\theta_1}^{\theta_2} \gamma(\theta)\, d\theta$. Finally, inserting this into the Kutta–Joukowsky theorem (29.51), we obtain the lift. We have thus established a precise analytic or numeric procedure which for ideal flow will yield the lift as a function of the geometry of the airfoil.

Thin airfoil approximation

Even if modern airfoils are much thicker than the airfoils of the early airplanes, the ratio d/c of thickness to chord is rarely more than 10–15%. Using x as the parameter, a decent approximation for thin airfoils is obtained by replacing the double layer of circulation density $\gamma_\pm(x)$ on the two halves of the airfoil outline by a single layer with circulation density $\gamma(x)$ distributed along a *camber line* $y(x)$, where

$$\gamma(x) = \gamma_+(x) + \gamma_-(x), \qquad y(x) = \tfrac{1}{2}(y_+(x) + y_-(x)). \tag{29.62}$$

For a thin airfoil with $d/c \ll 1$ we always have $|y(x) - y(x')| \ll |x - x'|$, so that the fundamental airfoil equation (29.61) reduces to the much simpler equation,

$$\boxed{U((c - x)\sin\alpha + y(x)\cos\alpha) = \frac{1}{2\pi}\int_0^c \log\frac{|x - x'|}{c - x'}\, \gamma(x')\, dx'.} \tag{29.63}$$

Given the camber line $y(x)$, this linear integral equation must be solved for $\gamma(x)$.

The circulation can be obtained directly by using the relation,

$$\int_0^c \frac{c}{(c - x)\sqrt{x(c - x)}} \log\frac{|x - x'|}{c - x'}\, dx = -2\pi, \tag{29.64}$$

which is true for all $c > 0$ and all x' between 0 and c. It takes a fair bit of complex analysis to prove this (see problem 29.8), but it may easily be checked numerically. Using this result in (29.63) we get the circulation

$$\Gamma = \int_0^c \gamma(x')dx' = -Uc\int_0^c \frac{(c-x)\sin\alpha + y(x)\cos\alpha}{(c-x)\sqrt{x(c-x)}}\,dx. \tag{29.65}$$

Since $\int_0^c dx/\sqrt{x(c-x)} = \pi$, we finally obtain

$$\Gamma = -Uc\left(\pi\sin\alpha + \cos\alpha\int_0^c \frac{y(x)}{(c-x)\sqrt{x(c-x)}}\,dx\right). \tag{29.66}$$

The integral converges because $y(x) \sim c - x$ for $x \to c$.

Taking into account that the contour of integration in the Kutta–Joukowsky theorem (29.51) is clockwise and not counterclockwise as assumed in the above calculation, the lift is $\mathcal{L} = -\rho_0 U L \Gamma$. The lift coefficient may now be written,

$$\boxed{C_L = \frac{\mathcal{L}}{1/2\rho_0 U^2 c L} = 2\pi\frac{\sin(\alpha-\alpha_0)}{\cos\alpha_0},} \tag{29.67}$$

where the zero-lift angle α_0 is defined from the integral,

$$\tan\alpha_0 = -\frac{1}{\pi}\int_0^c \frac{y(x)}{(c-x)\sqrt{x(c-x)}}\,dx. \tag{29.68}$$

Normally, airfoils have positive camber, $y(x) > 0$, so that $\alpha_0 < 0$. For a flat plate we evidently have $\alpha_0 = 0$ because $y(x) = 0$. For small angles of attack, $|\alpha|, |\alpha_0| \ll 1$, the lift coefficient takes the form (29.24) with lift slope $\lambda = 2\pi$. In problem 29.9 the integral is worked out for a simple non-trivial case.

* 29.8 The distant laminar wake

At short distances the velocity field is strongly dependent on the shape and attitude of a moving body, but far from the body such details are lost. It is, as we shall now see, possible to determine the general form of the laminar velocity field at large distances from the body in terms of the lift and drag that the body produces (see also [38, p. 67]). The analysis in this section should be viewed as the natural continuation of d'Alembert's theorem to fluids that are not perfectly inviscid. Such fluids will not 'close up' behind the moving body, but instead—as we have discussed above—leave a trailing wake, a disturbance that never dies completely out even at huge distance behind the body. In the real unruly and turbulent atmosphere, the trailing wake from a passing airplane will of course only be notable for a finite distance.

Oseen's approximation

Sufficiently far from the body, the velocity field is laminar and approximatively equal to the asymptotic value $\boldsymbol{U}$ both inside and outside the trailing wake. Inserting $\boldsymbol{v} = \boldsymbol{U} + \Delta\boldsymbol{v}$ into the steady flow Navier–Stokes equation without gravity,

$$(\boldsymbol{v}\cdot\nabla)\boldsymbol{v} = -\frac{1}{\rho_0}\nabla p + \nu\nabla^2\boldsymbol{v}, \tag{29.69}$$

we obtain to first order in $\Delta\boldsymbol{v}$,

$$(\boldsymbol{U}\cdot\nabla)\Delta\boldsymbol{v} = -\frac{1}{\rho_0}\nabla\Delta p + \nu\nabla^2\Delta\boldsymbol{v}. \tag{29.70}$$

The linearity of this equation allows us to superpose its solutions. Let us write the velocity difference as a sum,

$$\Delta\boldsymbol{v} = \boldsymbol{u} + \nabla\Phi, \tag{29.71}$$

where Φ is a generalized velocity potential and $\boldsymbol{u}$ is the contribution from the vorticity in the trailing wake. Choosing Φ as a solution to

$$(\boldsymbol{U} \cdot \nabla)\Phi = -\frac{\Delta p}{\rho_0} + \nu \nabla^2 \Phi, \tag{29.72}$$

and using this equation to eliminate p in (29.70), it follows that the field $\boldsymbol{u}$ must satisfy

$$\boxed{(\boldsymbol{U} \cdot \nabla)\boldsymbol{u} = \nu \nabla^2 \boldsymbol{u}.} \tag{29.73}$$

The incompressibility condition, $\nabla \cdot \Delta \boldsymbol{v} = 0$, yields a further relation,

$$\boxed{\nabla^2 \Phi = -\nabla \cdot \boldsymbol{u}.} \tag{29.74}$$

This equation determines Φ, given a solution $\boldsymbol{u}$ to (29.73), and then the pressure is obtained from (29.72).

Flow inside the wake

The trailing wake is assumed to be narrow compared to the distance to the body. In a coordinate system with the x-axis along the asymptotic velocity, $\boldsymbol{U} = U\boldsymbol{e}_x$, we may assume that $x \gg |y|, |z|$, inside the wake. In the now familiar way, it follows that the double x-derivative in the Laplacian of (29.73) is small compared with the y, z-derivatives, and the equation for $\boldsymbol{u}$ becomes

$$U\frac{\partial \boldsymbol{u}}{\partial x} = \nu \left(\frac{\partial^2}{\partial y^2} + \frac{\partial^2}{\partial z^2} \right) \boldsymbol{u}. \tag{29.75}$$

This is a standard diffusion equation of the same form as the momentum diffusion equation (17.5) with two transverse dimensions and 'time' $t = x/U$. Note that t is also the time it takes for the asymptotic flow to reach the position x downstream from the body.

At distances much larger than the body size, $x \gg L$, the body appears as a point particle with no discernable shape, situated at the origin of the coordinate system. By insertion into the above equation one may verify that the following expression is an exact 'shapeless' solution,

$$\boldsymbol{u} = \frac{\boldsymbol{A}}{x} \exp\left(-U\frac{y^2 + z^2}{4\nu x} \right), \tag{29.76}$$

where $\boldsymbol{A} = (A_x, A_y, A_z)$ is a constant vector. It is in fact also the most general solution at large downstream distance x (see problem 29.10). Evidently, the distant wake has a Gaussian shape in the transverse directions with a narrow width $\delta = \sqrt{4\nu x/U}$. The width of the laminar wake is, however, the same in both transverse directions, confirming that there is no imprint of the original shape of the object on the Gaussian form of the distant wake. For $|y|, |z| \lesssim \delta$, the solution decays as x^{-1} along the wake, rather than the expected $r^{-2} \approx x^{-2}$. This is consistent with the area of the wake being of order $\delta^2 \sim x$ such that the volume flux, i.e. the integral of $\boldsymbol{u}$ over the cross section of the wake, remains finite for $x \to \infty$. The terms that have been left out in the above solution by dropping the x-derivatives in the Laplacian are a further factor x^{-1} smaller than the above solution and cannot contribute in the limit.

The generalized potential is determined by solving (29.74). Consistently leaving out the double derivatives with respect to the Laplacian, it becomes

$$\left(\frac{\partial^2}{\partial y^2} + \frac{\partial^2}{\partial z^2} \right) \Phi = -\nabla \cdot \boldsymbol{u}. \tag{29.77}$$

On the right-hand side one cannot leave out the $\nabla_x u_x$ contribution to the divergence because it is only a factor $x^{-1/2}$ smaller than the others. It may—with some effort—be verified by insertion that the following potential is an exact solution to this equation inside the wake,

$$\Phi = \frac{2\nu}{U} \left(-\frac{A_y y + A_z z}{y^2 + z^2} + \frac{A_x}{2x} \right) \left(1 - \exp\left(-U\frac{y^2 + z^2}{4\nu x} \right) \right). \tag{29.78}$$

Here the first term in the first parenthesis is of order $x^{-1/2}$ and the second of order x^{-1}. The leading corrections from leaving out the double x-derivatives in the Laplacian are of order $x^{-3/2}$. Only the exponential in the second parenthesis represents a true solution to the inhomogeneous equation (29.77), to which one may add an arbitrary solution to Laplace's equation $\nabla^2\Phi = 0$. Here we have added the solution which makes the potential non-singular for $y, z \to 0$.

Drag and lift

The pressure is obtained in the same approximation from (29.72),

$$\Delta p = -\rho_0 U \frac{\partial \Phi}{\partial x} + \rho_0 \nu \left(\frac{\partial^2}{\partial y^2} + \frac{\partial^2}{\partial z^2}\right)\Phi = \rho_0 \nu \frac{A_x}{x^2}. \tag{29.79}$$

Since it decays like x^{-2} and the area of the wake is $\delta^2 \sim x$, it cannot contribute to drag for $x \to \infty$, so that the leading contribution to the integrand of (29.45) becomes

$$\Delta p + \rho_0 U \Delta v_x \approx \rho_0 U u_x. \tag{29.80}$$

In the last step we have dropped the pressure and the potential derivative $\nabla_x \Phi \sim x^{-3/2}$ which are both negligible compared to $u_x \sim x^{-1}$. Integrating over all y, z, we find from (29.45)

$$\mathcal{D} = -\rho_0 U \iint u_x \, dydz = -4\pi\rho_0 \nu A_x \tag{29.81}$$

which fixes the coefficient A_x. The errors committed in extending the integral over the wake to all values of y and z are exponentially small.

The lift is obtained from the complete circulation integral (29.50),

$$\Gamma(z) = \oint_{C(z)} (\boldsymbol{u} + \nabla\Phi) \cdot d\boldsymbol{\ell} = \oint_{C(z)} \boldsymbol{u} \cdot d\boldsymbol{\ell} = -\int u_y dy.$$

In the second step we have used that Φ is single-valued so that $\oint \nabla\Phi \cdot d\boldsymbol{\ell} = 0$, and in the third that $\boldsymbol{u}$ vanishes outside the wake. The minus sign stems from the contour running through the wake against the direction of the y-axis. Inserting this result into (29.49) we find

$$\mathcal{L} = -\rho_0 U \iint u_y \, dydz = -4\pi \nu \rho_0 A_y, \tag{29.82}$$

which fixes A_y. Similarly, since there is no lift in the z-direction, we must have $A_z = 0$.

The complete three-dimensional field configuration inside the wake has now been obtained in terms of the lift and drag that the body generates,

$$\boxed{\begin{aligned} \boldsymbol{u} &= -\frac{(\mathcal{D}, \mathcal{L}, 0)}{4\pi\rho_0 \nu x} \exp\left(-U\frac{y^2+z^2}{4\nu x}\right), \\ \Phi &= \frac{1}{4\pi\rho_0 U x}\left(\frac{2xy}{y^2+z^2}\mathcal{L} - \mathcal{D}\right)\left(1 - \exp\left(-U\frac{y^2+z^2}{4\nu x}\right)\right). \end{aligned}} \tag{29.83}$$

The correction terms are all of order x^{-1} relative to the leading terms.

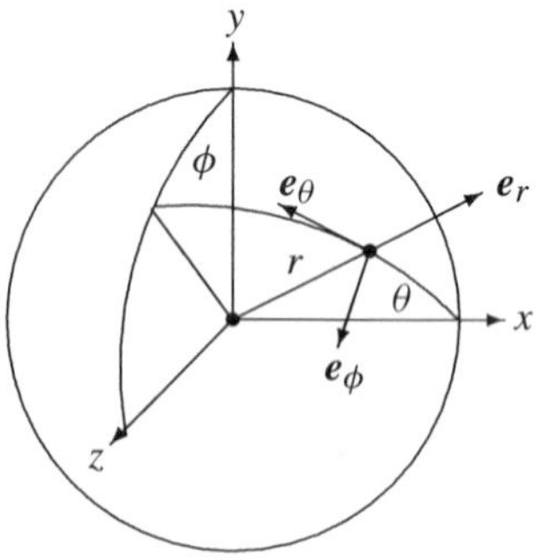

Spherical coordinates and their tangent vectors for the far field outside the wake.

Flow outside the wake

Outside the wake, the flow is assumed to be irrotational with $\Delta \boldsymbol{v} = \nabla\Phi$ and $\nabla^2\Phi = 0$. We have before argued that $\Delta \boldsymbol{v} \sim 1/r^2$ at large distances, and consequently we must have $\Phi \sim 1/r$. In spherical coordinates with the polar axis in the x-direction and the null-meridian in the xy-plane, we may thus write

$$\Phi = \frac{F(\theta, \phi)}{r}. \tag{29.84}$$

The spherical Laplacian (B.16) implies that F has to satisfy

$$\left(\sin^2\theta\frac{\partial^2}{\partial\theta^2}+\cos\theta\sin\theta\frac{\partial}{\partial\theta}+\frac{\partial^2}{\partial\phi^2}\right)F=0. \tag{29.85}$$

In view of the periodicity in ϕ, the complete solution may be written as a Fourier series

$$F=A_0(\theta)+\sum_{n=1}^{\infty}A_n(\theta)\cos n\phi+B_n(\theta)\sin n\phi, \tag{29.86}$$

where the coefficients A_n and B_n satisfy the equation

$$\left(\sin^2\theta\frac{d^2}{d\theta^2}+\cos\theta\sin\theta\frac{d}{d\theta}-n^2\right)A_n=0. \tag{29.87}$$

Surprisingly, this equation has a complete set of exact solutions $(\tan(\theta/2))^{\pm n}$ (see problem 29.11). Since F has to be regular at $\theta=\pi$, where the tangent diverges, the exponents must be non-positive, $n\le 0$. Furthermore, for $\theta\to 0$, where $A_n\sim\theta^{-n}$, this solution has to join continuously with the inside solution (29.83), which for $\delta\ll y,z\ll x$ behaves like

$$\Phi\approx\frac{1}{4\pi\rho_0 Ux}\left(\frac{2xy}{y^2+z^2}\mathcal{L}-\mathcal{D}\right)\approx\frac{\mathcal{L}}{2\pi\rho_0 U}\frac{\cos\phi}{r\,\theta}-\frac{\mathcal{D}}{4\pi\rho_0 Ur}. \tag{29.88}$$

This shows that the only possible exponents are $n=0,1$ with $A_0=-\mathcal{D}/4\pi\rho_0 U$, $A_1=\mathcal{L}/4\pi\rho_0 U$ and $B_1=0$. Thus, the potential far from the body outside the wake becomes

$$\boxed{\Phi=\frac{\mathcal{L}\cos\phi\cot(\theta/2)-\mathcal{D}}{4\pi\rho_0 Ur}.} \tag{29.89}$$

It joins continuously with the field inside the wake.

Pressure and lift

Since the leading contribution to pressure vanishes far downstream inside the wake, it plays no role in the drag which as shown by (29.81) is entirely due to a loss of fluid momentum in the trailing wake. The lift contribution from pressure stems on the other hand entirely from the outside solution. Using Bernoulli's theorem and the spherical derivatives (B.14), the outside pressure becomes, $\Delta p=-\rho_0 U\nabla_x\Phi=-\rho_0 U(\cos\theta\nabla_r-\sin\theta\nabla_\theta)\Phi$. After a bit of algebra this reduces to,

$$\boxed{\Delta p=-\frac{\mathcal{D}+\mathcal{L}\sin\theta\cos\phi}{4\pi r^2}.} \tag{29.90}$$

It is immediately clear that the spherically symmetric first term cannot contribute to the total pressure force. Also since $dS_x=\cos\theta\cdot r^2\sin\theta d\theta d\phi$, the second term, which is linear in $\cos\phi$, cannot produce a force in the x-direction, i.e. a drag.

Due to its ϕ-dependence the second term is, however, negative for $y>0$ and positive for $y<0$, and must therefore produce a lift. Using $dS_y=\sin\theta\cos\phi\cdot r^2\sin\theta d\theta d\phi$, we find

$$-\oint\Delta p\,dS_y=\frac{\mathcal{L}}{4\pi}\iint\sin^3\theta\cos^2\phi\,d\theta d\phi=\frac{1}{3}\mathcal{L}. \tag{29.91}$$

Pressure thus produces one third of the lift, even for an infinite sphere. The remaining two thirds of lift stems from momentum flux in the trailing wake. As discussed before (page 452), the partition of lift between pressure and momentum flux depends on the choice of integration surface at infinity.

Problems

29.1 The Concorde airliner has a powerplant of four engines that together develop 677 kN with afterburner. It maximal take off mass is 185 000 kg and its take off speed 360 km h^{-1}. Ignore drag and estimate the runway **(a)** acceleration, **(b)** time, and **(c)** length.

29.2 Show that the tangent to the bank angle in a horizontal banked turn is 2π times the ratio between the time it takes to fall freely from rest to velocity U (with no air resistance) divided by the time T it takes to make a complete turn.

29.3 Show that the induced drag is smaller if the single trailing vortex is divided into a number of smaller vortices coming off the wing.

29.4 The quoted take off speed for the Cessna 150 with mass 681 kg is about 120 km h^{-1}, and the take off length is 225 m in about 20 s. Assuming that the engine power is maximal and a constant 75 kW, calculate the fraction of this power that is converted into thrust. Calculate the take off time and compare with the quoted value. Any comments?

29.5 Determine how the calculations in section 29.6 are modified when air is assumed to be a barotropic compressible fluid which asymptotically has constant density ρ_0.

29.6 Calculate lift and drag when it is not assumed that the wake is cut orthogonally to the asymptotic velocity.

$*$ **29.7** Show explicitly that the sheet vortex field (29.55) has circulation

$$\Gamma = \oint_C \boldsymbol{v} \cdot d\boldsymbol{\ell} = \oint_A d\Gamma \tag{29.92}$$

where C is an arbitrary curve completely surrounding the airfoil A.

$*$ **29.8** Show that

$$\int_0^1 \frac{1}{(1-t)\sqrt{t(1-t)}} \log \frac{|t-x|}{1-x}\, dt = -2\pi, \tag{29.93}$$

for $0 < x < 1$.

$*$ **29.9** Assume that the chord line beginning in $x = 0$ has vertical tangent. **(a)** Show that if the airfoil is smooth near $x = 0$ then

$$y_{\pm}(x) = \pm a\sqrt{x} + 2\lambda x + \mathcal{O}\left(x^{3/2}\right) \tag{29.94}$$

where a and λ are constants. **(b)** The simplest example of a thin wing camber function fulfilling the various conditions is therefore,

$$y(x) = 2\frac{\lambda}{c} x(c-x). \tag{29.95}$$

Calculate the zero-lift angle α_0 for this airfoil as a function of λ.

$*$ **29.10** Consider the N-dimensional diffusion equation in the variables $x = (x_1, \ldots, x_N)$

$$\frac{\partial F}{\partial t} = \sum_{n=1}^{N} \frac{\partial^2 F}{\partial x_n^2}. \tag{29.96}$$

(a) Show that with initial data $F(x, 0) = F_0(x)$, the solution at time t is

$$F(x,t) = (4\pi t)^{-N/2} \int F_0(y) \exp\left(-\frac{(x-y)^2}{4t}\right) d^N y \tag{29.97}$$

where $(x-y)^2 = \sum_n (x_n - y_n)^2$.

(b) Show that if $F_0(y)$ is bounded or decreases at least as rapidly as a Gaussian for $|y| \to \infty$, the the solution for $t \to \infty$ is

$$F(x,y) = (4\pi t)^{-N/2} \exp\left(-\frac{x^2}{4t}\right) \int F_0(y)\, d^N y. \tag{29.98}$$

∗ **29.11** Find all solutions to

$$\left(\sin^2\theta\frac{d^2}{d\theta^2}+\cos\theta\sin\theta\frac{d}{d\theta}-n^2\right)f=0. \tag{29.99}$$

30 Heat

The conservation of energy is one of the most fundamental laws of physics. It first arose as an almost trivial consequence of Newton's equations of motion but was later expanded into the present all-embracing notion during the development of thermodynamics by Carnot, Joule, Kelvin, Clausius and others in the middle of the nineteenth century. Thermodynamics has turned out to be one of the most durable physical theories ever constructed. Although the basic laws are extremely simple, the fundamental concepts are abstract and often hard to connect with the complex reality of physics and chemistry, and thermodynamics is often perceived as very difficult at the first encounter.

The *First Law of Thermodynamics* states that there exists an abstract quantity, called *energy*, which is independent of the previous history of the system, and which does not change with time for an isolated system. Energy is abstract in the sense that we cannot afterwards tell how it was put into the system, whether by heat, by work or by other means. The *Second Law of Thermodynamics* states that there is another abstract quantity, called *entropy*, which is likewise independent of the prehistory of the system, and which cannot diminish with time for a system in isolation. Entropy is perhaps best understood as a measure of *disorder*, informing us about the partition of the stored energy between the random motion of molecules that we call heat and other forms of energy more directly convertible to work. The inexorable growth of entropy with time parallels the intuitive notion that disorder spontaneously increases for a system in isolation.

In this chapter only the First Law of Thermodynamics for continuous systems is discussed, completing thereby the analysis of laws of balance begun in chapter 22. For more complete presentations of continuum thermodynamics and the Second Law, see [7, 14, 37, 24, 72]. Convection, the dynamic interplay of flow and heat, will be treated in chapter 31.

30.1 Energy balance

The *First Law of Thermodynamics* postulates (1) that the energy $\mathcal{E}$ of a physical system is only a function of the instantaneous state, i.e. of the macroscopic fields which characterize the state of the system, and (2) that energy is conserved in a system that is isolated from its environment. For a non-isolated system, any change in energy can therefore be accounted for by the actions of the environment. In terms of the heat Q transferred to the system and the work W performed on the system, the First Law takes the form,

$$\boxed{\Delta\mathcal{E} = Q + W.} \tag{30.1}$$

Whereas heat and work may depend on the history of how the environment interacted with the system, their sum does not, because energy is only a function of the instantaneous state. If the environment can influence the system in other ways, for example through chemical reactions, further terms have to be included on the right-hand side.

First Law in continuum physics

Continuous systems are of course also physical systems, so the First Law must by default be valid in continuum physics, provided the possibility of material flow through the surface of the control volume is taken properly into account. In an arbitrary moving control volume, the First Law turns into an equation of balance for the material rate of change of the total energy,

$$\boxed{\frac{D\mathcal{E}}{Dt} = \dot{Q} + \dot{W},} \tag{30.2}$$

where $\dot{Q}$ is the rate at which the environment transfers heat to the system, and $\dot{W}$ is the rate of work performed on the system by the environment.

It is a basic assumption in continuum physics that energy is an extensive quantity, transported by material particles. The energy of a control volume is as any other extensive quantity an integral,

$$\mathcal{E} = \int_V \rho\,\epsilon\,dV, \tag{30.3}$$

where the energy density $\rho\,\epsilon = d\mathcal{E}/dV$ is the product of the mass density ρ and the *specific energy* $\epsilon = d\mathcal{E}/dM$. The material time derivative of the energy then obeys Reynolds theorem (22.10) with $q \to \epsilon$,

$$\frac{D\mathcal{E}}{Dt} = \frac{d\mathcal{E}}{dt} + \oint_S \rho\,\epsilon\,(\boldsymbol{v} - \boldsymbol{v}_S) \cdot d\boldsymbol{S}. \tag{30.4}$$

The last term represents the net outflow of energy from the control volume.

Rate of work

The environment can perform work on a system through long-range body forces such as gravity and electromagnetism, and through contact forces acting on the surface of the control volume. The rate of work of the contact forces is,

$$\dot{W} = \oint_S \boldsymbol{v} \cdot \boldsymbol{\sigma} \cdot d\boldsymbol{S}, \tag{30.5}$$

where $\boldsymbol{\sigma} = \{\sigma_{ij}\}$ is the stress field. If external body forces, for example gravity, also act on the material in the control volume, their rate of work must be added to the right-hand side. Alternatively, a static external gravitational field can be included as a potential energy term in the total energy (see section 22.8 on page 318), and the same is possible for external electrostatic forces. When all body forces are 'internalized' as potential energy, we only need to account for the environment's work on the system through contact forces, as given by the above expression. We shall assume this to be the case in the following.

Rate of heat transfer

The everyday experience of handling a hot potato tells us that heat can be carried along with the movement of matter, but that possibility is already taken care of in the equation of energy balance (30.2) through the use of the material derivative, effectively turning the arbitrary control volume into a comoving volume. When we heat water to make tea, it becomes equally clear that heat can also be conducted through the solid bottom of the kettle. Heat transfer by conduction is not an advective transport phenomenon tied to the motion of material particles, but is rather a diffusion process by which the warmer material particles of the wall 'infect' the colder particles of the water with heat.

Macroscopically we describe the flow of heat through a surface by a current density (vector) field $\boldsymbol{q}(\boldsymbol{x}, t)$, defined such that the amount of heat which flows through a surface element $d\boldsymbol{S}$ in the time interval δt is $\delta Q = \boldsymbol{q}\delta t \cdot d\boldsymbol{S}$. The total rate of heat transfer *into* the system thus becomes,

$$\dot{Q} = -\oint_S \boldsymbol{q} \cdot d\boldsymbol{S} + \int_V h\,dV, \tag{30.6}$$

where we have also allowed for the possibility that heat may be produced at a rate $h(\boldsymbol{x}, t)$ per unit of volume by chemical, nuclear or other processes.

Example 30.1.1 (Geothermal heat flow): The Earth produces heat in the interior by radioactivity, and the average geothermal heat current is $q_C \approx 0.06$ W m^{-2} for the continents and $q_S \approx 0.10$ W m^{-2} for the seas [34, p. 192]. Given that the Earth's mean radius is $a = 6371$ km, and that the continents cover a fraction $f \approx 29\%$ of the surface, the total rate of heat flow out of the Earth is $\oint_S \boldsymbol{q} \cdot d\boldsymbol{S} = 4\pi a^2(f q_C + (1-f)q_S) \approx 45 \times 10^{12}$ W. For comparison, the present world power consumption is about 13×10^{12} W.

Internal energy and heat equation

Separating out the kinetic energy $\mathcal{T}$ and the potential energy $\mathcal{V}$, we shall write the total energy as,

$$\boxed{\mathcal{E} = \mathcal{T} + \mathcal{V} + \mathcal{U},} \tag{30.7}$$

where the remainder $\mathcal{U}$ is called the *internal energy*. Since energy is an extensive quantity, the internal energy must also be an integral over a density,

$$\mathcal{U} = \int_V \rho\, u\, dV, \tag{30.8}$$

where $u = d\mathcal{U}/dM$ is the *specific internal energy*. Using the usual expressions for the kinetic energy (22.34) and the potential energy (22.50) the specific total energy becomes

$$\boxed{\epsilon = \frac{1}{2}v^2 + \Phi + u.} \tag{30.9}$$

At this point we do not know the nature of the system and cannot be more explicit about the precise form of the specific internal energy.

Using energy balance (30.2) and mechanical energy balance (22.53) we obtain

$$\frac{D\mathcal{U}}{Dt} = \frac{D\mathcal{E}}{Dt} - \frac{D\mathcal{T}}{Dt} - \frac{D\mathcal{V}}{Dt} = \dot{W} + \dot{Q} - \widetilde{P},$$

where $\widetilde{P}$ is the reduced power, i.e. the total power (22.37) without the contribution due to the rate of work of gravity. Assuming as before that there are no other volume forces at play than static gravity, this reduces to the global equation of *internal energy balance*,

$$\boxed{\frac{D\mathcal{U}}{Dt} = \dot{Q} + \dot{W}_{\text{int}},} \tag{30.10}$$

where W_{int} is the rate of work against internal stresses (22.38). It represents the rate at which kinetic energy is converted into internal energy by internal stresses, also called *dissipation*.

Transforming the surface integral in (30.6) into a volume integral by means of Gauss' theorem, and making use of (22.11) with $q \to u$, it follows that the specific internal energy must obey the local equation,

$$\boxed{\rho \frac{Du}{Dt} = -\nabla \cdot \boldsymbol{q} + h + \sum_{ij} \sigma_{ij} \nabla_j v_i,} \tag{30.11}$$

where the last term is the density of dissipation. Given explicit expressions for the specific internal energy u, the current of heat conduction $\boldsymbol{q}$, the rate of heat production per unit of volume h, and the internal stress field $\boldsymbol{\sigma}$, this equation becomes a dynamic equation for the temperature field, called the *heat equation*.

Jean Baptiste Joseph Fourier (1768–1830). *French scientist who made fundamental contributions to mathematics (Fourier series) and to the theory of heat.*

30.2 Heat equation for isotropic matter at rest

The Second Law of Thermodynamics implies that in an isolated system heat will spontaneously stream from regions of higher to regions of lower temperatures (and conversely: this general rule is equivalent to the Second Law). We expect for this reason that the heat flow $\boldsymbol{q}$ is locally related to the gradient of the temperature field ∇T. In isotropic matter, the simplest local relationship is *Fourier's law of heat conduction*,

$$\boxed{\boldsymbol{q} = -k\nabla T,} \tag{30.12}$$

where the sign has been chosen so that heat is conducted from hot to cold. The positive constant k is called the *thermal conductivity* of the material, and is measured in units of watts per kelvin per metre. For water it is $k \approx 0.6\ \mathrm{W\,K^{-1}\,m^{-1}}$ and for air $k \approx 0.025\ \mathrm{W\,K^{-1}\,m^{-1}}$.

In the following we shall always assume Fourier's law to be valid.

Example 30.2.1 (Geothermal gradient): The average continental geothermal heat flow is $q_c \approx 0.06\ \mathrm{W\,m^{-2}}$ [34]. Taking the thermal conductivity of bedrock to be $k \approx 2\ \mathrm{W\,K^{-1}\,m^{-1}}$ [41], the average geothermal temperature gradient in the upper crust becomes $|\nabla T| \approx q_c/k \approx 0.03\ \mathrm{K\,m^{-1}}$, or 30 kelvin per kilometre. It must be emphasized that the geothermal gradient varies strongly from place to place because of variations in the composition of the upper crust.

Fourier's equation

The simplest of all materials is an incompressible homogeneous isotropic fluid (or solid) at rest with constant mass density field, $\rho = \rho_0$. In such a material there is only one free thermodynamic variable, which may be taken to be the temperature field $T = T(\boldsymbol{x}, t)$, and any other local thermodynamic quantity, for example the specific energy $u = u(T)$, becomes a local function of the temperature. We shall for simplicity assume that the specific energy is linear in the temperature, $u = c_0 T$, or at least linear in a certain temperature range. The constant c_0 is the *specific heat capacity* of the material, defined as the amount of heat necessary to raise a unit of mass by one unit of temperature[1].

Inserting $u = c_0 T$ and $\boldsymbol{v} = \boldsymbol{0}$ into the equation of local internal energy balance (30.11), and using Fourier's law of heat conduction (30.12), we obtain the heat equation for isotropic matter at rest,

$$\rho_0 c_0 \frac{\partial T}{\partial t} = k\nabla^2 T + h. \tag{30.13}$$

Dividing by $\rho_0 c_0$ it takes (for $h = 0$) the form of a standard diffusion equation, called *Fourier's equation*

$$\boxed{\frac{\partial T}{\partial t} = \kappa\nabla^2 T,} \tag{30.14}$$

with *heat diffusivity*,

$$\kappa = \frac{k}{\rho_0 c_0}. \tag{30.15}$$

For water we have $\kappa \approx 1.4 \times 10^{-7}\ \mathrm{m^2\,s^{-1}}$ which is about 6 times smaller than the momentum diffusivity (kinematic viscosity) ν. Somewhat surprisingly, the heat diffusivity of air turns out to be 140 times larger than for water (see page 475).

Why is it then that we use air for insulation in thermoglass windows, bed covers and winter coats—rather than sleeping and walking in wetsuits? The explanation is that although heat diffuses about 140 times quicker in air than in water, the actual heat current (30.12) is for a given temperature gradient not determined by the diffusivity but by the thermal conductivity, k, which is about 25 times larger in water than in air. Consequently, you lose less heat in a fur coat than in a wetsuit even if the cold penetrates the fur coat much faster. The role of a fur coat or wetsuit is mainly to prevent advection of heat by air or water currents which will rapidly remove the warm fluid adjacent to your skin, increasing thereby the temperature gradient at the skin and thus the heat flow from your body.

[1] There is a small ambiguity of language when we use the verb 'heat' to mean both to 'raise the temperature' and to 'add heat'. It is perfectly possible to add heat to a system without raising its temperature, for example when boiling water. Conversely, the temperature may rise without heat transfer when you pump your bicycle tyre.

Heat diffusion time

The analogy between Fourier's equation (30.14) and the planar momentum diffusion equation (17.5) on page 232 allows us immediately to take the solution (17.12) and adapt it to a uniform planar Gaussian temperature distribution in the yz-plane. Let the planar temperature field initially at $t = 0$ be Gaussian, $T = T_0 + \Theta \exp(-x^2/a^2)$, where Θ is the temperature excess at $x = 0$. At a later time the temperature field will then be,

$$T(x,t) = T_0 + \Theta \frac{a}{\sqrt{a^2 + 4\kappa t}} \exp\left(-\frac{x^2}{a^2 + 4\kappa t}\right). \tag{30.16}$$

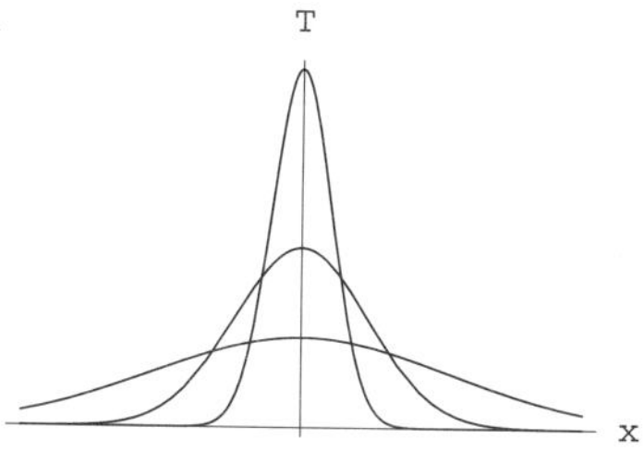

Heat diffusion makes a Gaussian temperature distribution widen and diminish in height as time goes by.

It may of course be verified by direct insertion that this solution indeed satisfies Fourier's equation. The temperature distribution thus remains Gaussian at all times with a width that grows like $\sqrt{a^2 + 4\kappa t}$ with time. Since the environment transfers no heat to the system and performs no work on it (because the material is everywhere at rest), the total internal energy in the heated material must remain constant while it spreads away from the central region (problem 30.1).

At large times, $t \gg a^2/4\kappa$, the Gaussian factor $\exp(-x^2/4\kappa t)$ becomes universal for all bounded temperature distributions. The strong fall-off of the Gaussian makes it appear as if the temperature excess $\Delta T = T - T_0$ expands with a well-defined front. At the distance $x = 2\sqrt{\kappa t}$ the Gaussian has fallen to $e^{-1} \approx 37\%$ of its central value. Depending on the application it may be more convenient to choose a more conservative front, for example $x = 3.5\sqrt{\kappa t}$ where the Gaussian is only about 5% of its central value.

Conversely, one may also characterize heat diffusion by the characteristic time it takes for the front to reach a distance $x = L \gg a$. At the 37% level, this time becomes,

$$t \approx \frac{L^2}{4\kappa}. \tag{30.17}$$

For a more conservative definition of the front, a correspondingly smaller estimate may be used, for example $t \approx L^2/12\kappa$ for the 5% level. Such estimates are in fact very general and may be used to get an idea of the heat diffusion time in any system.

Example 30.2.2 (Hot porridge): Everybody has painfully learnt early in life that the centre of a bowl of hot porridge may remain hot for a long time, even as the periphery grows cold. Taking the heat diffusivity of porridge to be equal to that of water, the time it takes for a spherical ball of hot porridge of radius $a \approx 5$ cm to reach room temperature is estimated from (30.17) to be 4500 s, or about an hour and a quarter. Parents have found a solution to this problem (see example 30.3.1 below).

Planar heat wave

Similarly, we may adapt the plane wave solution (17.14). Suppose the temperature is forced to oscillate sinusoidally in the plane $z = 0$ with period τ and amplitude Θ around the mean temperature T_0, such that $T = T_0 + \Theta \cos(2\pi t/\tau)$ at $z = 0$. Recasting the solution (17.14) in terms of these variables, with $\omega \to 2\pi/\tau$ and $k \to 1/d$, it becomes a damped heat wave in which the temperature at depth z is,

$$T = T_0 + \Theta\, e^{-z/d} \cos\left(2\pi \frac{t}{\tau} - \frac{z}{d}\right), \qquad d = \sqrt{\frac{\kappa\tau}{\pi}}, \tag{30.18}$$

where d is the penetration depth. The corresponding wavelength is $\lambda = 2\pi d$, and at a depth of one wavelength the damping factor is $e^{-2\pi} = 1.8 \times 10^{-3}$.

Example 30.2.3 (Annual soil temperature variation): The surface temperature of soil follows the annual variations in atmospheric temperature with period $\tau = 1$ year $= 3.2 \times 10^7$ s. The thermal conductivity of soil consisting of sand, stones, and clay, is probably not unlike that of water with, say, $k \approx 0.5$ W K^{-1} m^{-1}, density $\rho \approx 2$ g cm^{-3}, and specific heat capacity about $c_0 \approx 2$ J g^{-1} K^{-1}, implying a heat diffusivity of around $\kappa \approx 1.3 \times 10^{-7}$ m^2 s^{-1}, which is nearly identical to that of water. The soil penetration depth becomes $d = 1.1$ m, meaning that the amplitude of the temperature variations at this depth has fallen to $e^{-1} = 37\%$ of the surface amplitude (see figure 30.1). Since the wavelength is $2\pi \approx 6$ times the penetration depth, the surface temperatures will be delayed by about 2 months at $z = d$.

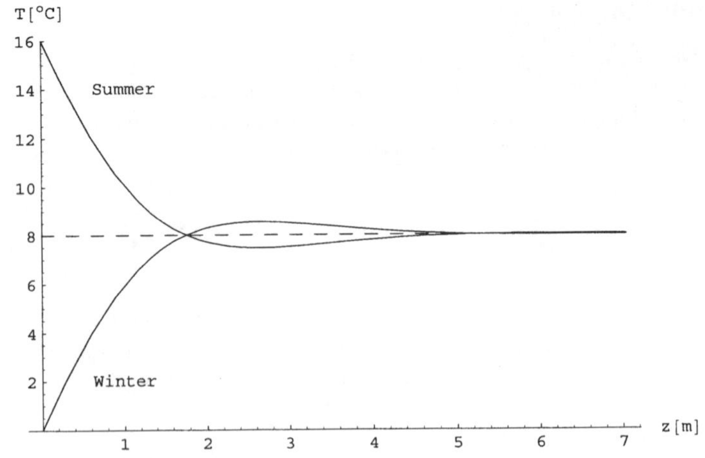

Figure 30.1. Calculated ground penetration of mean summer and winter temperatures in Denmark as a function of depth z (example 30.2.3). It takes nearly two months to penetrate one metre of depth, and beyond $z = 5$ m there is no yearly variation.

Denmark for example, has a temperate Northern climate with mean temperature $T_0 \approx 8°$ C and a yearly variation of $\Theta \approx 8°$ C. Freezing temperatures occur often in winter, even if the average barely gets below $0°$ C, but a freezing spell will have to last more than 2 months to penetrate to a depth of one metre, and that is highly improbable. Frost-free depth is accordingly defined to be 90 cm for water mains and the foundations of houses.

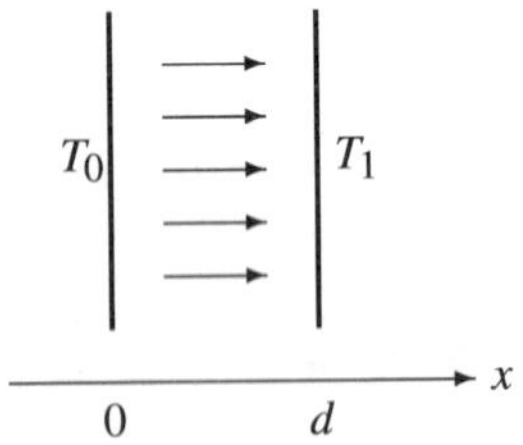

Steady heat flow between parallel plates at different temperatures. The plates continue far above and below the section shown here. The fluid is at rest and there is no gravity.

Steady planar heat flow

Suppose a slab of isotropic material at rest is enclosed between two infinitely extended flat plates held at different temperatures, T_0 and $T_1 = T_0 + \Theta$. The planar symmetry suggests that the temperature field only depends on the transverse coordinate, $T = T(x)$. Inserting this into Fourier's equation we obtain $\nabla_x^2 T = 0$ which has the solution

$$T(x) = T_0 + \Theta \frac{x}{d}, \tag{30.19}$$

where d is the distance between the plates. Although heat conduction is necessary for establishing a flow of heat, the final distribution does not depend on the value of the coefficient of thermal conductivity. As we shall now see, thermal conductivity determines instead the time scale for the appearance of a steady flow pattern. This is completely analogous to velocity-driven planar flow between moving plates which does not depend on the viscosity (section 17.2 on page 231), although viscosity does determine the time scale for the flow to settle down into a steady pattern.

Steady heat production

If heat is produced at a constant rate h per unit of volume and if the environment has constant properties, an equilibrium will eventually be attained in which the heat produced in a volume equals the amount of heat that leaves the volume. The temperature field will then become time-independent and obey the steady heat equation (30.13),

$$k\nabla^2 T + h = 0. \tag{30.20}$$

Given the heat production and suitable boundary conditions this equation determines the temperature distribution. Often h will itself vary with temperature because the heat producing processes are temperature-dependent.

Suppose nevertheless that heat is produced at a fixed rate $h(\boldsymbol{x}, t) = h_0$ inside a sphere of radius a, making the total heat rate $\dot{Q} = (4/3)\pi a^3 h_0$. The spherical solution to the steady heat equation (30.20) then becomes,

$$T = T_s + \frac{h_0}{6k}(a^2 - r^2), \tag{30.21}$$

where T_s is the (constant) surface temperature. The temperature is (naturally) highest in the centre for $h_0 > 0$. Averaging over the sphere, the mean temperature becomes (see problem 30.3),

$$\langle T \rangle = T_s + \frac{h_0}{15k}a^2. \tag{30.22}$$

The average temperature excess in the sphere is thus $2/5$ of the temperature difference between the centre and the surface.

Example 30.2.4 (Geothermal heat production): The total geothermal heat output of the Earth is $\dot{Q} \approx 45 \times 10^{12}$ W (see example 30.1.1), implying an average (radioactive) heat production rate $h_0 \approx 4 \times 10^{-8}$ W m^{-3}. If the Earth were made from uniform material with $k \approx 2$ W K^{-1} m^{-1}, the central temperature excess would be $T_c - T_s \approx 140\,000$ K. This estimate fails miserably in comparison with geophysical models which place the central temperature at 4000–6000 K, demonstrating that the Earth is a highly non-uniform object with respect to heat transfer. In the liquid mantle rapid convection rather than slow conduction is the dominant mechanism of heat transfer, although thermal conductivity also increases with depth.

Example 30.2.5 (Human skin temperature): The average heat output from a human being is $\dot{Q} \approx 100$ W (for example used for dimensioning cooling systems for concert halls). With a typical mass of $M \approx 70$ kg and density equal to that of water $\rho_0 \approx 1000$ kg m^{-3} the volume becomes $V = 0.07$ m^3, implying an average heat production density rate $h_0 \approx 1.4$ kW m^{-3}. In the 'spherical approximation' a human body of this mass would have radius $a \approx 0.25$ m, and using (30.22) with $k = 0.6$ W K^{-1} m^{-1} we obtain the difference between the average temperature and the skin temperature $\langle T \rangle - T_s \approx 10$ K. Taking $\langle T \rangle = 37°$ C, the skin temperature is predicted to be $27°$ C. The conclusion is that a naked human being should be able to survive 'indefinitely' in water of this temperature, a result which agrees decently with common experience (see also problem 30.6).

30.3 Heat equation for fluids in motion

The dominant effect of the velocity of a moving fluid is advective transport of internal energy, expressed through the material derivative on the left-hand side of the equation of local internal energy balance (30.11). The last term on the right-hand side also depends on the velocity but it only plays a role for compressible or viscous fluids. Even then, dissipative heating will only be important under extreme circumstances, for example at the leading edges of a supersonic aircraft or a spacecraft re-entering the atmosphere.

Incompressible inviscid fluid

For an incompressible inviscid fluid the heat equation takes (in the absence of heat production) almost the same form as Fourier's equation (30.14),

$$\boxed{\frac{\partial T}{\partial t} + (\boldsymbol{v} \cdot \nabla)T = \kappa \nabla^2 T.} \tag{30.23}$$

The extra advective term on the left-hand side expresses that heat moves along with the flow of matter, whereas the right-hand side expresses that it also streams down temperature gradients as for a fluid at rest. We shall see below that this heat equation can essentially always be maintained when the fluid is nearly inviscid and the flow speed is small compared to the velocity of sound.

The Péclet number

The time it takes for heat to be advected by the flow of fluid with velocity U through a distance L is $t_{\rm adv} \sim L/U$ whereas the time it takes for it to diffuse through the same distance is $t_{\rm diff} \sim L^2/\kappa$, disregarding all purely numeric factors. The ratio of the time scales for diffusion and flow is an important dimensionless quantity, called the *Péclet number*, $\mathrm{Pe} \sim t_{\rm diff}/t_{\rm adv} \sim UL/\kappa$.

More formally, and in analogy with the Reynolds number, the Péclet number is defined to be the ratio of the advective to diffusive terms in the heat equation,

Jean-Claude Eugene Péclet (1793–1857). *French physicist. One of the first scholars of Ecole Normale (Paris); known for his clarity of style, sharp-minded views and well-performed experiments.*

$$\mathrm{Pe} = \frac{|(\boldsymbol{v}\cdot\nabla)T|}{\left|\kappa\nabla^2 T\right|} \approx \frac{UL}{\kappa}. \tag{30.24}$$

The value of the Péclet number determines whether heat flow in a fluid is dominated by advection or by conduction. For large Péclet numbers the right-hand side of the heat equation (30.23) can be disregarded, and the heat equation now simply says that the temperature is constant around any comoving material particle. The smallness of the heat diffusivity in most materials usually makes the Péclet number quite large for flows of ordinary dimensions and flow speeds. For small Péclet number, the advective term can be disregarded, and the heat equation reverts to that of a fluid at rest.

Example 30.3.1 (Cooling hot porridge): The parental solution to the problem of the hot porridge (see example 30.2.2) is to stir it a few times with a spoon. Stirring sets the porridge into motion and the heat (or rather the internal energy) is carried along with the porridge by advection. Mixing the hot and cold regions increases the local temperature gradients, resulting in a large diffusive heat flow which quickly evens out the temperature differences and makes the porridge eatable. During stirring, layers of hot and cold porridge are interwoven with each other, lowering the typical distance scale to perhaps $L = 1$ cm. Taking $U = 10\ \mathrm{cm\ s^{-1}}$ and the diffusivity to be approximatively that of water, $\kappa \approx 1.4\times10^{-7}\ \mathrm{m^2\ s^{-1}}$, we find a Péclet number of $\mathrm{Pe} \approx 7000$, showing that heat conduction can safely be ignored while you stir. Diffusive cooling due to the larger temperature gradients mainly takes place when you stop stirring after a few seconds. The heat diffusion time (30.17) in the stirred porridge is now estimated to be of the order of minutes.

Compressible ideal gas

In section 22.9 on page 319 we calculated the specific internal energy (22.60) of a compressible ideal gas with adiabatic index γ (see also appendix C). As for incompressible fluid, the specific energy density was found to be linear in the absolute temperature[2],

$$u = c_v T, \qquad c_v = \frac{1}{\gamma-1}\frac{R}{M_{\rm mol}}, \tag{30.25}$$

where the constant c_v is *the specific heat at constant density (i.e. volume)*. It represents—as the first equation shows—the amount of *energy* that is necessary to raise a unit of mass of the gas by one unit of temperature. If the density (volume) is held constant, so that no work is performed on the gas, this is the same as the amount of *heat* that must be transferred to a unit mass of the gas to raise its temperature by one unit.

If on the other hand the pressure is held constant, the amount of *heat* that must be transferred to a unit mass of the gas to raise its temperature by one unit is instead,

$$c_p = c_v + \frac{R}{M_{\rm mol}} = \frac{\gamma}{\gamma-1}\frac{R}{M_{\rm mol}}, \tag{30.26}$$

called the *specific heat at constant pressure* or the *isobaric specific heat*.

Disregarding viscous friction and heat production, local internal energy balance (30.11) becomes,

$$\boxed{\rho c_v\left(\frac{\partial T}{\partial t} + (\boldsymbol{v}\cdot\nabla)T\right) = k\nabla^2 T - p\nabla\cdot\boldsymbol{v},} \tag{30.27}$$

[2]Although this expression was derived under the assumption of adiabatic (isentropic) processes, it must be correct in general because energy is only a function of the final state and not of the particular road that leads to this state, whether adiabatic or not.

where the pressure is given by the ideal gas law, $p = (R/M_{\rm mol})\rho T$. The first source term on the right-hand side represents the local inflow of heat by conduction whereas the last represents the work of local compression (because it is positive when $\nabla \cdot \boldsymbol{v} < 0$). We shall justify below that if the flow velocity is much smaller than the speed of sound, the local compression source can be 'moved over' to the left-hand side, replacing the specific heat at constant volume by the specific heat at constant pressure,

$$\boxed{\rho c_p \left(\frac{\partial T}{\partial t} + (\boldsymbol{v} \cdot \nabla) T \right) = k \nabla^2 T.} \qquad (30.28)$$

It is of the same form as for an incompressible fluid, except for the use of the specific heat at constant pressure, but it should be remembered that this is only an approximation (valid at small flow speeds) to the correct equation (30.27).

The heat diffusivity is now defined as,

$$\kappa = \frac{k}{\rho c_p}. \qquad (30.29)$$

For air at normal temperature and pressure, we have $\kappa \approx 2 \times 10^{-5}\ \mathrm{m^2\ s^{-1}}$ which as mentioned before is about 140 times larger than that of water.

Derivation of the heat equation (30.28)

Using the equation of state $p = (R/M_{\rm mol})\rho T$ we get,

$$\frac{Dp}{Dt} = \frac{R}{M_{\rm mol}} \left(\frac{D\rho}{Dt} T + \rho \frac{DT}{Dt} \right) = -p \nabla \cdot \boldsymbol{v} + (c_p - c_v) \rho \frac{DT}{Dt}.$$

In the last step we have used the equation of continuity (15.28) to eliminate the density derivative. The heat equation (30.27) may now be written

$$\rho c_p \frac{DT}{Dt} - \frac{Dp}{Dt} = k \nabla^2 T. \qquad (30.30)$$

For steady nearly ideal flow and in the absence of gravity, Bernoulli's theorem (16.15) relates a change in pressure to a change in flow velocity, $\Delta p \approx -1/2 \rho \Delta(v^2)$. A change in the absolute temperature is related to a change in the velocity of sound (16.9), from which we get $\Delta T \sim \Delta(c_s^2) M_{\rm mol}/R\gamma$. The ratio of the two terms on the left-hand side of the heat equation may thus be estimated as,

$$\frac{\left| \dfrac{Dp}{Dt} \right|}{\left| \rho c_p \dfrac{DT}{Dt} \right|} \approx \frac{\Delta p}{\rho c_p \Delta T} \sim \frac{\Delta(v^2)}{\Delta(c_S^2)} \sim \frac{v^2}{c_S^2}. \qquad (30.31)$$

In the last steps we have disregarded all factors of order unity. This argument suggests that the material pressure derivative can be disregarded in steady flow at high Reynolds number as long as the flow speed is much smaller than the speed of sound (which also makes the flow effectively incompressible; see section 16.4 on page 213). For general unsteady flow there will be further conditions on the rate of change of the pressure, $\partial p/\partial t$, and on the field of gravity (see [7] for a more careful discussion).

General isotropic fluid

In a general isotropic fluid, the specific internal energy will depend on both the temperature and the mass density, $u = u(T, \rho)$, and derivation of the heat equation becomes considerably more involved, requiring the full apparatus of thermodynamics. An account close in spirit to the presentation in this chapter is found in [7]. The end result is that for nearly inviscid and nearly incompressible flows the heat equation again takes the form (30.28) with heat diffusivity determined from the isobaric specific heat of the fluid (30.29). In liquids there is not a great difference between the specific heats at constant volume and at constant pressure.

The internal friction caused by the stresses in a viscous fluid leads to a loss of kinetic energy which reappears as source of internal energy (i.e. internal work) in the last term in the equation of internal energy balance (17.30). Including the dissipative power density (17.18) for general Newtonian fluids, the heat equation becomes,

$$\rho c_p \left(\frac{\partial T}{\partial t} + (\boldsymbol{v} \cdot \nabla) T \right) = k \nabla^2 T + h + 2\eta \sum_{ij} v_{ij}^2 + \zeta (\nabla \cdot \boldsymbol{v})^2 \tag{30.32}$$

where h is the local rate of heat production, v_{ij} the symmetric strain rate (17.29), η the shear viscosity and ζ is the bulk viscosity.

In most everyday flows heat production by dissipation will be negligible in comparison with advection and conduction. In extreme situations with strong compression and shear, temperatures can become extremely high. A hypersonic object like a 'shooting star' may even burn up completely when entering the atmosphere at more than 10 km s^{-1}.

30.4 Advective cooling or heating

When you take a walk on a cold day, your heat loss is amplified by wind which removes the warm air near your body and creates a large temperature gradient at the surface of the skin, resulting in a larger conductive heat transfer from your body to the air. In meteorology this is known as *wind chill*, and the local 'wind chill temperature' is often announced by weather forecasters during winter. A similar phenomenon must of course occur in a hot desert wind, although the local 'wind burn' is not a regular part of a summer weather forecast. In a sauna with air at 120 ° C, it is well known that one should not move around too fast.

Water chill is even more important. A cold water current moving with the same speed and temperature as a cold wind results in much stronger cooling because of the 25 times higher thermal conductivity of water. This is why you are only able to survive naked for minutes in streaming water at 0 ° C whereas you may survive for hours in a wind of that temperature. Similarly, hot water scalds you much faster than hot air of the same temperature.

In this section we shall discuss the limit where the temperature does not influence the motion of the fluid. In that case the flow of heat takes place against the background of a mass flow, completely controlled by the external forces that drive the fluid. This limit is often called the limit of *forced* convection to distinguish it from the opposite limit, *free* convection, where the motion of the fluid is entirely caused by temperature differences (to be discussed at length in chapter 31).

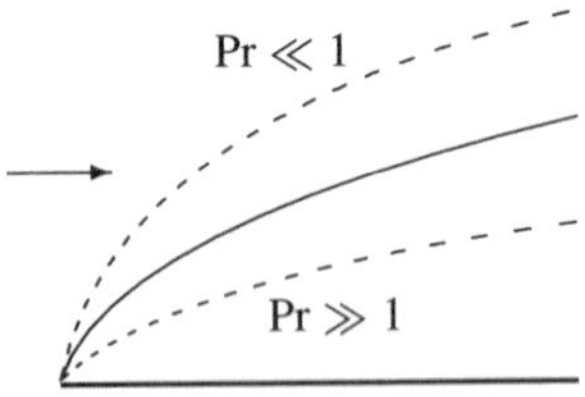

Advective cooling of a plate with wind coming in from the left. The boundary for the mass flow (solid line) and the heat fronts (dashed) for large and small Prandtl numbers.

The Prandtl number

Advective cooling (or heating) of the surface of a body involves both momentum and heat diffusion along the normal to the surface. In a time t after the start of the flow, momentum diffusion reaches a characteristic distance $\delta_{\text{mass}} \sim \sqrt{\nu t}$ from the surface whereas heat diffusion reaches $\delta_{\text{heat}} \sim \sqrt{\kappa t}$. The ratio between momentum and heat diffusivities is for this reason an important dimensionless quantity, called the *Prandtl number*,

$$\text{Pr} = \frac{\nu}{\kappa}. \tag{30.33}$$

When the Prandtl number is large, temperature variations will take place well inside the usual boundary layer, whereas if the Prandtl number is small, the temperature distribution spreads well beyond the boundary layer.

> In contrast to other dimensionless numbers, for example the Reynolds or Péclet numbers, the Prandtl number is a property of the fluid rather than of the flow. In gases it is of order unity, for example $\text{Pr} = 0.73$ for air at normal temperature and pressure. In liquids it may take a wide range of values: in water it is about 6, whereas in liquid metals it is quite small, for example 0.025 for mercury. For isolating liquids like oil it may be quite large, of the order of 1000.

Wind chill estimate

In a steady flow with velocity scale U boundary layers will stop growing after the time $t \sim L/U$ it takes for the fluid to move across the downstream length L of the body. The scale of the heat boundary layer thickness thus becomes $\delta_{\text{heat}} \sim \sqrt{\kappa L/U}$, and the temperature gradient at the surface is expected to be $|\nabla T| \approx \Theta/\delta_{\text{heat}}$, where Θ is the temperature excess of the body relative to the fluid at large. From Fourier's law (30.12) we estimate that the rate of loss of heat from a body surface of area A is,

$$\dot{Q} \sim k\frac{\Theta}{\delta_{\text{heat}}}A \sim k\Theta A\sqrt{\frac{U}{\kappa L}}. \qquad (30.34)$$

We shall show below in an exact calculation that this is indeed of the right form.

The main consequence of the above estimate is that the heat loss grows with the square root of the velocity, thus confirming the observation that a higher velocity has the same effect as a larger temperature excess. Let the actual temperature excess be $\Theta = T_1 - T_0$ where T_1 is the temperature of the exposed surface and T_0 is the wind temperature, and let the actual wind velocity be U. Then the heat loss will be the same for a temperature excess $\Theta^* = T_1 - T_0^*$ and a wind speed U^*, satisfying $\Theta^*\sqrt{U^*} = \Theta\sqrt{U}$. Solving for the (fictive) wind temperature T_0^* we find,

$$T_0^* = T_1 - (T_1 - T_0)\sqrt{\frac{U}{U^*}}. \qquad (30.35)$$

Apart from corrections for high wind speeds, this is essentially identical to the first *wind chill formula* by Siple and Passel (1945), who—after measuring cooling rates of water in plastic containers—chose the nominal values $T_1 \approx 33\,^\circ$ C and $U^* \approx 5$ m s^{-1}. The Siple–Passel formula was used by the *US National Weather Service* from 1973 but was in 2001 replaced by a more conservative expression based on modern theory and experiments.

Paul Allen Siple (1908–68). *American Antarctic explorer. Accompanied (as an Eagle Scout) the first Byrd expedition to Antarctica in 1928–30. Participated in Byrd's second expedition 1933–35 as a chief biologist. Coined the term 'wind chill' in 1939.*

Example 30.4.1: At $0\,^\circ$ C a wind speed of $U = 10$ m s^{-1} corresponds to a fictive wind temperature of $T_0^* \approx -14\,^\circ$ C according to (30.35). The modern formula yields instead $T_0^* \approx -7\,^\circ$ C.

Heat flow in the Blasius boundary layer

The simplest model of advective cooling is furnished by the steady-flow Blasius boundary layer discussed in section 28.4 on page 414, with the added condition that the plate is held at a constant temperature $T = T_0 + \Theta$ which is different from the ambient temperature T_0. Since by assumption the mass flow is not influenced by the heat flow we may take the exact Blasius solution and insert it into the steady flow heat equation,

$$(\boldsymbol{v}\cdot\nabla)T = \kappa\nabla^2 T. \qquad (30.36)$$

In the boundary layer approximation the double derivative ∇_x^2 can be disregarded in the Laplacian, so that the equation takes the same form as the Prandtl equation (28.21)

$$v_x\frac{\partial T}{\partial x} + v_y\frac{\partial T}{\partial y} = \kappa\frac{\partial^2 T}{\partial y^2}. \qquad (30.37)$$

Assuming that the temperature only depends on the scaling variable $s = y\sqrt{U/2\nu x}$, we may write (a general discussion of heat flow in boundary layers is found in [61]),

$$T(s) = T_0 + \Theta\, F(s), \qquad (30.38)$$

where $F(s)$ is a dimensionless function. In the same way as for Blasius' equation (28.26) it follows that,

$$F''(s) + \Pr g(s) F'(s) = 0, \qquad (30.39)$$

where $g(s)$ is the Blasius solution (28.23). Solving it with the boundary conditions $F(0) = 0$ and $F(\infty) = 1$, we find

$$F(s) = 1 - \frac{H(s)}{H(\infty)}, \qquad (30.40)$$

where

$$H(s) = \int_0^s e^{-\Pr G(u)}\,du, \qquad G(s) = \int_0^s g(u)\,du. \qquad (30.41)$$

The function $F(s)$ is plotted in figure 30.2*a* for three different values of the Prandtl number and illustrates clearly how the heat front depends on this number. Note that for $\Pr = 1$ we have $1 - F(s) = g'(s) = f(s)$ according to (28.26).

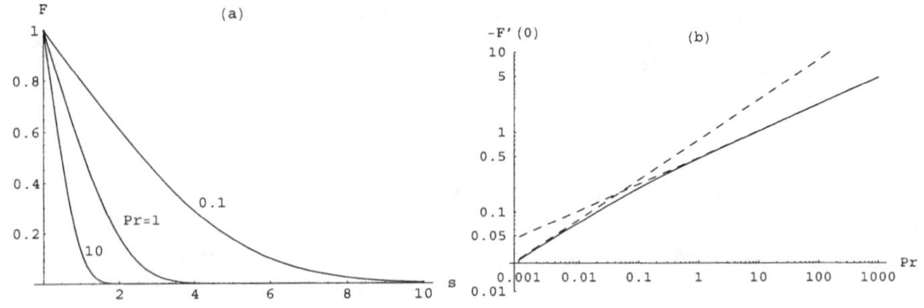

Figure 30.2. **(a)** The function $F(s)$ for three different values of the Prandtl number. **(b)** The slope $F'(0)$ as a function of Pr. The dashed lines indicate the asymptotic power law behaviours (see problem 30.7).

Total rate of heat loss

The gradient of the temperature field on the plate

$$\nabla_y T\big|_{y=0} = \Theta F'(0)\sqrt{\frac{U}{2\nu x}}, \tag{30.42}$$

determines the heat flow $q_y = -k\nabla_y T$ according to Fourier's law. Integrating q_y along an area A of the plate of length L in the x-direction and width A/L in the z-direction, we find the total rate of heat loss,

$$\dot{Q} = \frac{A}{L}\int_0^L q_y(x)dx = -F'(0)k\Theta A\sqrt{\frac{2U}{\nu L}}. \tag{30.43}$$

The slope $-F'(0) = 1/H(\infty)$ is plotted as a function of the Prandtl number in figure 30.2*b*. Note that this is not quite of the same form as the previous estimate (30.34) because it has ν instead of κ in the denominator of the square root. The explanation is that for Pr $\ll 1$ the slope varies like the square root of the Prandtl number, $-F'(0) \approx 0.8\sqrt{\text{Pr}}$, bringing the estimate into agreement with the exact calculation (see problem 30.7).

Example 30.4.2 (Human heat loss): The grown-up human body has a skin surface area of about $A \approx 2$ m^2. A naked human standing with shoulders aligned with the wind will roughly present a (two-sided) plate area with $L \approx 0.5$ m and $A/L \approx 4$ m. For air with Pr $= 0.73$ we have $-F'(0) = 0.42$, so that in a wind with $U = 1$ m s^{-1} and a temperature $\Theta = 10$ K below the skin temperature we find the heat loss rate $\dot{Q} \approx 100$ W. Since the human body produces heat at this rate, a skin temperature of 27 ° C can thus be maintained essentially indefinitely in a gentle breeze with velocity 1 m s^{-1} and temperature 17 ° C, not unlike what you find on a Scandinavian beach on a summer day. In this calculation we have ignored natural convection and evaporative heat losses.

Under the same conditions in water where $-F'(0) = 0.87$ the rate of heat loss becomes enormous, $\dot{Q} \approx 23$ kW, and will almost instantly cool the skin to the temperature of the water. Everybody is familiar with the (relatively mild) skin shock that is experienced when one jumps into water as warm as 17 ° C. At this temperature the loss of heat will eventually lead to severe hypothermia and death in the course of some hours, depending on what you wear and how you behave.

Problems

30.1 Show that the total internal energy is conserved for planar heat diffusion (30.16).

30.2 Show that the spherical temperature distribution

$$T(r) = T_0 + \Theta\left(\frac{a^2}{a^2+4\kappa t}\right)^{3/2}\exp\left(-\frac{r^2}{a^2+4\kappa t}\right) \tag{30.44}$$

is a solution to Fourier's equation (30.14).

30.3 Show that the average of r^n over a sphere of radius a is,

$$\langle r^n \rangle = \frac{3}{n+3} a^n. \tag{30.45}$$

30.4 Consider two plates at $y = y_1$ and $y = y_2$ and fixed temperatures T_1 and T_2. Show that if there is incompressible fluid at rest between the plates, the temperature in the fluid is,

$$T = T_1 + (T_2 - T_1)\frac{y - y_1}{y_2 - y_1}. \tag{30.46}$$

Show that this is also a solution if the fluid is inviscid and moves steadily along x.

30.5 Consider two coaxial cylinders with radii a_1 and a_2 and incompressible fluid at rest between. Show that the temperature distribution between the cylinders is

$$T = T_1 + (T_2 - T_1)\frac{\log(r/a_1)}{\log(a_2/a_1)}. \tag{30.47}$$

Show that this is also true if the fluid is inviscid and moves steadily along z.

30.6 The blood circulation in the human body actively attempts to maintain constant temperature everywhere in the body. Consider a 'spherical human' of radius a and mass M with constant heat production h_0 everywhere in the body (example 30.2.5). The circulation maintains constant temperature T_c in the core of the body of radius $c < a$. Calculate the temperature drop $\Theta = T_c - T_s$ in the skin and estimate its value for an average 'skin thickness' of $a - c = 5$ cm.

30.7 Show (for heat flow in the Blasius layer) that the asymptotic behaviour of $-F'(0) = 1/H(\infty)$ is

$$-F'(0) \approx \begin{cases} 0.80\mathrm{Pr}^{1/2} & \mathrm{Pr} \to 0 \\ 0.48\mathrm{Pr}^{1/3} & \mathrm{Pr} \to \infty. \end{cases} \tag{30.48}$$

31 Convection

Convection is a major driving agent behind most of the weather phenomena in the atmosphere, from ordinary cyclones to hurricanes, thunderstorms and tornadoes. Continental drift on Earth as well as transport of heat from the centre of a planet or star to the surface are also mainly driven by convection. On a smaller scale, we use heat convection in the home to create a natural circulation which transports heat around the rooms from the radiators of the central heating system. The circulation of water in the central heating system also used to be driven by convection, but is today mostly driven by pumps.

The mechanism behind convection rests on a combination of material properties and gravity. Most fluids, even those we call incompressible, tend to expand when the temperature is raised, leading to a slight decrease in density. Were it not for gravity, the minuscule changes in density caused by local temperature variations would be of very little consequence, but gravity makes the warmer and lighter fluid buoyant relative to the colder and heavier, and the buoyancy forces will attempt to set the fluid into motion. Such heat-driven motion is called *convection*, and will in fact arise naturally in the presence of gravity wherever there are sufficiently large local variations in the temperature of a fluid.

In this chapter we shall first discuss some examples of steady laminar convection flows driven by time-independent temperature differences on the container boundaries. Afterwards we shall address the thermal instabilities characterizing the onset of convection. Of particular interest are the Rayleigh–Bénard instabilities in a horizontal layer of fluid heated from below.

31.1 Convection

Convection is caused by buoyancy forces in combination with the tendency for most materials to expand when heated[1]. The coupled partial differential equations controlling the interplay of heat and motion are generally so complex that exact solutions are completely out of the question. Numeric simulations are, however, possible and used wherever practical problems have to be solved. Analytic insight into convection is mainly obtained from an approximation developed by Boussinesq in 1902.

[1] Concentration gradients in mixed fluids can also cause convective flow. In this book we reserve the word 'convection' to denote a flow that is mainly driven by temperature differences in conjunction with buoyancy, whereas 'advection' is used to denote heat transport in a flow mainly driven by other forces. In practice both mechanisms are at play, and sometimes it is useful instead to distinguish between *free* and *forced* convection.

Thermal expansion coefficient

The (isobaric) *coefficient of thermal expansion* is for all kinds of isotropic matter defined as the relative decrease in density per unit of temperature rise (at constant pressure),

$$\alpha = -\frac{1}{\rho}\left(\frac{\partial\rho}{\partial T}\right)_p. \tag{31.1}$$

It is a material 'constant' and for ideal gases where $\rho \propto p/T$ one finds $\alpha = 1/T$, i.e. about $\alpha \approx 3 \times 10^{-3}\ \mathrm{K}^{-1}$ at room temperature. For most liquids it is also of this magnitude, the exception being water which has $\alpha \approx 2.5 \times 10^{-4}\ \mathrm{K}^{-1}$ at $25\,°$ C. Water is in many respects exceptional with a negative expansion coefficient between 0 and $4\,°$ C, and a solid phase (ice) that is lighter than the liquid.

Turning the above definition around we may calculate the change in density,

$$\Delta\rho = -\alpha\Delta T\rho, \tag{31.2}$$

due to a small change in temperature satisfying $|\alpha\Delta T| \ll 1$. In a constant field of gravity $\boldsymbol{g}_0$, this density change causes an extra gravitational force density, a buoyancy term $\Delta\rho\boldsymbol{g}_0$ for $\Delta T > 0$, to appear on the right-hand side of the Navier–Stokes equation.

In a flow with velocity scale U, length scale L and temperature variation scale Θ, the dimensionless ratio of the buoyancy term to the advective term becomes,

$$\mathrm{Ri} = \frac{|\Delta\rho\boldsymbol{g}_0|}{|\rho(\boldsymbol{v}\cdot\boldsymbol{\nabla})\boldsymbol{v}|} \approx \alpha\Theta\frac{g_0 L}{U^2}. \tag{31.3}$$

Lewis Fry Richardson (1881–1953). *British physicist. The first to apply the method of finite differences to predict weather.*

It is called the *Richardson number*, and when this number is small, advection will dominate over convection. Conversely, when it becomes of order unity the flow will be driven by convection with a typical speed $U \sim \sqrt{\alpha\Theta}\sqrt{g_0 L}$, which is the product of the small quantity $\sqrt{\alpha\Theta} \ll 1$ and the free-fall velocity $\sqrt{g_0 L}$ from height $L/2$.

The Boussinesq approximation

Valentin Joseph Boussinesq (1842–1929). *French physicist and mathematician. Contributed to many aspects of hydrodynamics: whirlpools, solitary waves, drag, advective cooling and turbulence.*

Suppose an effectively incompressible fluid is initially at rest in constant gravity $\boldsymbol{g}_0 = (0, 0, -g_0)$ with constant density ρ_0, temperature T_0 and hydrostatic pressure $p = p_0 - \rho_0 g_0 z$. At a certain time the boundary temperatures are changed, and the resulting flow of heat changes the temperature in the fluid and thereby its density, eventually resulting in a convective flow with velocity field $\boldsymbol{v}$.

The main assumption behind the Boussinesq approximation is that the temperature variations are small on the scale set by the thermal expansion coefficient, i.e. $|\alpha\Delta T| \ll 1$ where $\Delta T = T - T_0$, and the change in density is to first order given by (31.2) with $\rho = \rho_0$. Adding the buoyancy term $\Delta\rho\boldsymbol{g}_0 = -\rho_0\alpha\Delta T\boldsymbol{g}_0$ to the Navier–Stokes equation, and cancelling out the normal hydrostatic pressure by writing $p = p_0 - \rho_0 g_0 z + \Delta p$, the Boussinesq equations for an effectively incompressible fluid become,

$$\frac{\partial\Delta T}{\partial t} + (\boldsymbol{v}\cdot\boldsymbol{\nabla})\Delta T = \kappa\nabla^2\Delta T, \tag{31.4a}$$

$$\frac{\partial\boldsymbol{v}}{\partial t} + (\boldsymbol{v}\cdot\boldsymbol{\nabla})\boldsymbol{v} = -\frac{\boldsymbol{\nabla}\Delta p}{\rho_0} + \nu\nabla^2\boldsymbol{v} - \alpha\Delta T\boldsymbol{g}_0, \tag{31.4b}$$

$$\boldsymbol{\nabla}\cdot\boldsymbol{v} = 0. \tag{31.4c}$$

The complete and correct derivation of the Boussinesq approximation is however not without subtlety (see for example [72, p. 188]).

Steady convection in open vertical slot heated on one side

We have previously (page 472) discussed the steady heat flow in a fluid at rest between two plates, one of which was situated at $x = 0$ with temperature T_0 and the other at $x = d$ with higher temperature $T_1 = T_0 + \Theta$. The result was that the temperature rises linearly across the slot. If the plates are vertical, buoyancy forces will act on the heated fluid and unavoidably set it into motion. For definiteness, the plates are assumed to be large but finite with the openings at the top and bottom connected to a reservoir of the same fluid at the same temperature T_0 as the cold plate. This provides the correct hydrostatic pressure at the top and bottom of the slot, a pressure which is necessary to prevent the fluid in the slot from 'falling out' under its own weight, even before it is heated. Heating will cause the fluid in the slot to rise and merge with the fluid in the reservoir, and it seems reasonable to expect that a steady flow of heat and fluid may come about, in which the fluid rises fastest near the warm plate. The reservoir is assumed to be so large that it never changes its temperature, except near the exit from the slot.

From the planar symmetry of the configuration we expect that the velocity field is vertical everywhere, and that it and the temperature field depend only on x,

$$\boldsymbol{v} = (0, 0, v_z(x)), \qquad T = T(x). \tag{31.5}$$

Under these assumptions, there will be no advective contribution to the heat equation (31.4a), which becomes $\nabla_x^2 \Delta T = 0$. The temperature thus varies linearly with x across the slot, and using the boundary conditions we get,

$$\Delta T = \Theta \frac{x}{d}, \tag{31.6}$$

just as in the static case (30.19).

From the assumed form of the velocity field it also follows that the advective term in (31.4b) vanishes, and from the x, y-components of this equation we conclude that $\nabla_x \Delta p = \nabla_y \Delta p = 0$, such that the pressure can only depend on z. From the z-component we then get

$$\frac{1}{\rho_0} \nabla_z \Delta p(z) = \nu \nabla_x^2 v_z(x) + \alpha \Theta \frac{x}{d} g_0.$$

Since the left-hand side depends only on z and the right-hand side only on x, both sides of this equation are constant, and since the pressure excess Δp must vanish at the top and bottom of the slot, it must vanish everywhere, $\Delta p = 0$. Using the no-slip boundary conditions on the plates the solution becomes,

$$\boxed{v_z = \frac{\alpha \Theta g_0 d^2}{6\nu} \frac{x}{d} \left(1 - \frac{x^2}{d^2}\right).} \tag{31.7}$$

Note that the steady flow pattern is independent of the heat diffusivity κ as it would be in forced convection.

The maximal velocity in the slot is found a bit to the right of the middle, at $x = d/\sqrt{3}$. The average velocity in the slot becomes

$$U = \frac{1}{d} \int_0^d v_z(x)\, dx = \frac{\alpha \Theta g_0 d^2}{24\nu}. \tag{31.8}$$

Due to the spurious absence of advection the velocity field scale is set by viscosity rather than by advection. It disagrees with our earlier estimate $\sqrt{\alpha \Theta g_0 L}$ and is a factor $(1/2)\alpha\Theta$ smaller than the average steady fall velocity through the slot, calculated from (18.8) by setting $G = \rho_0 g_0$. The Reynolds number corresponding to this velocity is

$$\mathrm{Re} = \frac{Ud}{\nu} = \frac{1}{24} \cdot \frac{\alpha \Theta g_0 d^3}{\nu^2}. \tag{31.9}$$

The second factor on the right is called the *Grashof number* and denoted Gr. Formally, the Richardson number (3.13) becomes $\mathrm{Ri} = 24/\mathrm{Re}$.

Example 31.1.1: For water with $d \approx 1$ cm, $\Theta \approx 10$ K we find $U \approx 12$ cm s^{-1}. The corresponding Reynolds number is $\mathrm{Re} \approx 1400$, indicating that the flow should be laminar, as was assumed implicitly in the above calculation. For air the velocity is $U \approx 8$ cm s^{-1} and the Reynolds number is $\mathrm{Re} \approx 50$.

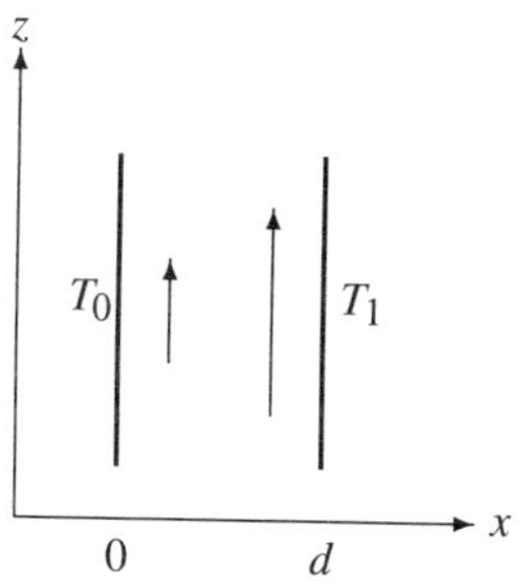

Steady convection between vertical parallel plates held at different temperatures. The plates continue far above and below the section shown here.

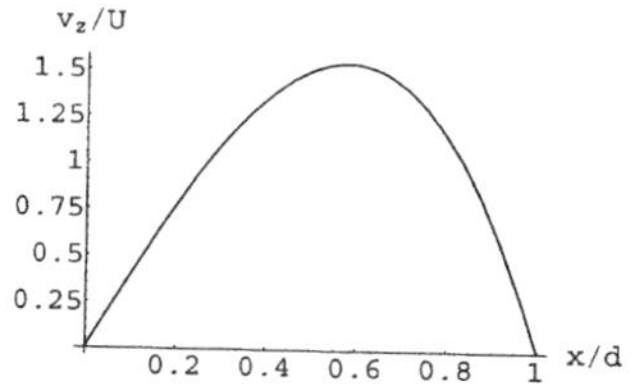

Convective velocity profile in units of the width of the slot and the average velocity. Its shape is reminiscent of the pressure-driven planar flow (18.5) although skewed a bit towards the right i.e. the hot plate.

Franz Grashof (1826–93). *German engineer who sought to transform machine building into a proper science.*

Entrance length for heat

The rate at which heat is transported by convection from the slot into the reservoir may be calculated from the extra internal energy carried by the fluid as it exits the slot,

$$\dot{Q} = \int_0^d \rho_0 c_p \Delta T\, v_z L dx = \frac{\rho_0 c_p \alpha \Theta^2 g_0 d^3 L}{45\nu}, \tag{31.10}$$

where L is the size of the slot in the y-direction. This raises a puzzle because the temperature gradient is constant, $\nabla_x \Delta T = \Theta/d$, across the slot, and Fourier's law (30.12) then implies that the same amount of heat is added to the fluid at the warm plate as is removed at the cold. Consequently, no net heat is added to the fluid from the plates, in blatant contradiction with the above calculation and common experience.

What is wrong is the assumption that the solutions (31.6) and (31.7) are valid at the bottom of the slot where the fluid enters. Here the temperature gradient cannot be constant, because mass conservation in the steady state forces the cold fluid to enter with the same average velocity U given by (31.8). Since it takes a certain amount of time, $t \approx d^2/4\kappa$, for the heat supplied by the warm plate to diffuse across the slot (see equation (30.17)), the fluid will have moved through a vertical distance

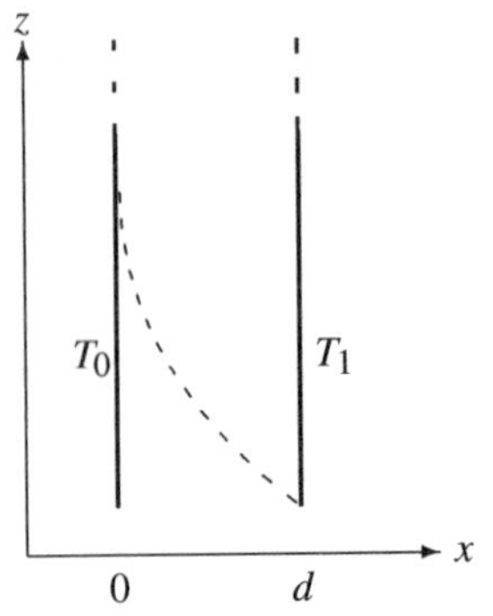

The heat front (dashed) at the entrance to the vertical slot. The heat transferred to the fluid at the warm plate must diffuse across the slot but is at the same time advected upwards with average velocity U. It makes contact with the cold plate after having moved a vertical distance ℓ, called the entrance length for heat.

$$\ell \approx Ut = \frac{\alpha \Theta g_0 d^4}{96 \kappa \nu}, \tag{31.11}$$

before the heat comes into contact with the cold plate. For consistency we should compare the heat loss at the exit (31.10) with the total rate of heat transferred into the fluid in the entrance region. It may be estimated from Fourier's law (30.12) applied to the entrance area $L\ell$. In this region the heated fluid only extends about halfway across the slot leading to a heat flow estimate $\dot{Q} \sim \ell L \cdot k\,\Theta/d$, and this is indeed of the same size as the heat loss (31.10) at the exit.

The entrance length may also be written,

$$\ell \approx \frac{\text{Ra}}{96} d, \tag{31.12}$$

where the dimensionless quantity,

$$\text{Ra} = \frac{\alpha \Theta g_0 d^3}{\nu \kappa}, \tag{31.13}$$

is the famous *Rayleigh number*. In units of the slot width the entrance length for heat ℓ/d is about 1% of the Rayleigh number whereas the viscous entrance length (21.22) in units of the slot width is estimated to be about 2% of the Reynolds number.

Example 31.1.2: In the preceding example 31.1.1 the Rayleigh number for water becomes Ra $= 2 \times 10^5$, and we get an enormous entrance length $\ell \approx 21$ m, whereas for air we find Ra ≈ 900 and a much more manageable $\ell \approx 10$ cm. For comparison the viscous entrance length is 27 cm in water and 1 cm in air.

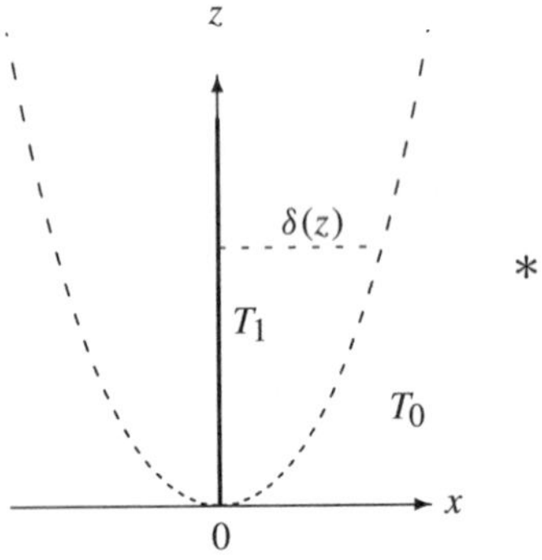

Outline of thermal boundary layers forming on both sides of a thin plate with constant temperature $T_1 = T_0 + \Theta$, placed vertically in an infinite sea of fluid originally at rest with temperature T_0. The layer has a z-dependent thickness $\delta(z)$.

* Thermal boundary layer

If the entrance length is much greater than the height of the slot, $\ell \gg h$, the heated fluid will never reach the cold plate before it exits from the slot. In this limit, the cold plate can be ignored, and the appropriate model is instead that of a warm vertical plate with constant temperature placed in a sea of cold fluid. The heated fluid rising along the plate then forms a *thermal boundary layer*, and we shall now determine the steady laminar flow pattern in such a boundary layer by combining the Boussinesq approximation with Prandtl's boundary layer approximation (section 28.3 on page 413).

The coordinate system is chosen with the positive z-axis along the plate. Replacing ℓ by z and d by $\delta(z)$ in the estimate of the heat entrance length (31.11) we obtain an estimate of the z-dependent thickness of the boundary layer (apart from a dimensionless numerical factor),

$$\delta(z) \sim \left(\frac{\kappa \nu z}{\alpha \Theta g_0} \right)^{1/4}. \tag{31.14}$$

Since for $z \to \infty$ we have $\delta/z \propto z^{-3/4}$ the boundary layer may indeed be viewed as thin, except for a region near the leading edge of the plate.

Under these circumstances we may apply the Prandtl formalism to the Boussinesq equations (31.4) and discard the double derivative after z in the Laplace operators together with the pressure excess Δp. These simplifications lead to the following (Boussinesq–Prandtl) equations for the velocity field, $\boldsymbol{v} = (v_x, 0, v_z)$,

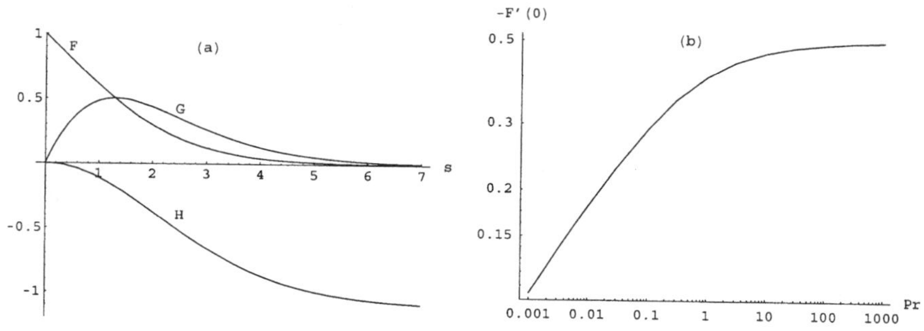

Figure 31.1. Structure of the self-similar thermal boundary layer for Pr $= 1$. **(a)** Plot of the functions $F(s)$, $G(s)$, and $H(s)$. Note that asymptotically for $s \to \infty$ there is a horizontal flow towards the plate which feeds the convective upflow. **(b)** Doubly logarithmic plot of the heat slope at the plate, $-F'(0)$, as a function of the Prandtl number.

and the temperature excess, $\Delta T = T - T_0$,

$$(v_x \nabla_x + v_z \nabla_z)\Delta T = \kappa \nabla_x^2 \Delta T, \tag{31.15a}$$

$$(v_x \nabla_x + v_z \nabla_z)v_z = \nu \nabla_x^2 v_z + \alpha \Delta T g_0, \tag{31.15b}$$

$$\nabla_x v_x + \nabla_z v_z = 0. \tag{31.15c}$$

These equations must be solved with the boundary conditions that $\Delta T = \Theta$ and $v_x = v_z = 0$ for $x = 0$, and ΔT, $v_z \to 0$ for $x \to \infty$.

Since there is no other possible length scale for x than $\delta(z)$ we shall assume that the fields only depend on x through the variable $x/\delta(z)$, or

$$s = \left(\frac{\alpha \Theta g_0}{\kappa \nu z}\right)^{1/4} x. \tag{31.16}$$

Apart from dimensional prefactors, the fields are parametrized with dimensionless functions of this dimensionless variable,

$$\Delta T = \Theta F(s), \qquad v_z = \sqrt{\alpha \Theta g_0 z}\, G(s), \qquad v_x = \left(\frac{\alpha \Theta \kappa \nu g_0}{z}\right)^{1/4} H(s), \tag{31.17}$$

and the field equations become coupled differential equations in s alone,

$$\frac{4}{\sqrt{\mathrm{Pr}}} F'' + (sG - 4H)F' = 0, \tag{31.18}$$

$$4\sqrt{\mathrm{Pr}} G'' + (sG - 4H)G' + 4F - 2G^2 = 0, \tag{31.19}$$

$$4H' + 2G - sG' = 0, \tag{31.20}$$

where Pr $= \nu/\kappa$ is the Prandtl number. These equations can be solved numerically with the boundary conditions $F(0) = 1$, $G(0) = H(0) = 0$, and $F(\infty) = G(\infty) = 0$. The result is shown in figure 31.1(a) for Pr $= 1$. Interestingly, the solution has an asymptotic horizontal flow towards the plate (represented by $H(\infty) = -1.10941\ldots$ for Pr $= 1$), rather than a vertical upflow from below, as might have been expected. The divergence of the horizontal velocity for $z \to 0$ is a spurious consequence of the Prandtl approximation.

The rate of heat flow out of one side of a plate of dimensions $L \times h$ is obtained from Fourier's Law (30.12),

$$\dot{Q} = -\int_{L\times h} k \left.\frac{\partial \Delta T}{\partial x}\right|_{x=0} dydz = -\frac{4}{3} F'(0) \left(\frac{\alpha \Theta g_0 h^3}{\kappa \nu}\right)^{1/4} k\,\Theta L. \tag{31.21}$$

The slope $-F'(0)$ is shown in figure 31.1b as a function of the Prandtl number. The quantity in parenthesis is the Rayleigh number for the height of the plate. Note that the heat loss, $\dot{Q} \sim \Theta^{5/4}$, grows a little faster than linearly.

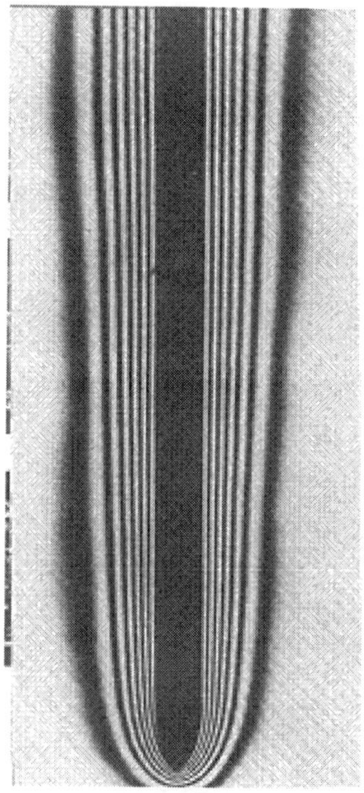

Photograph of isotherms around a thermal boundary layer in air. The Grashof number is about Gr $\approx 5 \times 10^6$ some distance from the end of the plate. Reproduced from E. R. G. Eckert and E. Soehngen *US Air Force Tech. Report* 5747.

Example 31.1.3 (Heat radiator): A heat radiator consisting of a single plate of height $h = 70$ cm and width $L = 1$ m is kept at $T = 65\,^\circ$ C and placed in a room at $T_0 = 20\,^\circ$ C. For air we have $\Pr = 0.73$ and $F'(0) = -0.388$ leading to a total heat flow $\dot{Q} = 235$ W from the radiator (including both sides). The maximal vertical velocity $U = \max v_z$ is quite naturally found at the top of the radiator, $z = h$, and inspection of figure 31.1(a) yields $U = \max_x v_z(x, h) \approx 0.5\sqrt{\alpha\Theta g_0 h} \approx 50$ cm s^{-1} at $s = 1.5$ corresponding to a boundary layer width $\delta \approx 1$ cm. The horizontal Reynolds number becomes $\mathrm{Re}_\delta = U\delta/\nu \approx 340$, so there should be no turbulence in this boundary layer. The asymptotic horizontal inflow is merely $v_x|_{x\to\infty} \approx -1$ cm s^{-1} at $z = 5$ cm above the bottom of the radiator.

Example 31.1.4 (Human heat loss): For a naked grown-up human the skin surface area is $A \approx 2$ m^2, and standing up the height $h \approx 2$ m and $L \approx 1$ m. Taking $\Theta = 10$ K we find the heat loss $\dot{Q} \approx 40$ W. Since this is less than half the heat production of 100 W, a human being should easily be able to maintain a skin temperature of 27 ° C in calm air at 17 ° C, perhaps even by sweating a little (see example 30.2.5). As soon as there is even a very gentle wind, the heat loss grows and rapidly begins to chill the body (see example 30.4.2).

31.2 Convective instability

Fluids with horizontal temperature variations, such as the vertical slot discussed above, cannot remain in hydrostatic equilibrium but must immediately start to convect (see problem 31.1). The situation is, however, quite different if the fluid is only subject to vertical temperature variations $T = T(z)$. In the following argument we assume that the fluid is incompressible, so that we do not run into problems with the natural temperature lapse of the 'atmospheric' kind, discussed in section 4.6.

If the ambient vertical temperature rises with height ($dT/dz > 0$), hydrostatic equilibrium is stable because a blob of fluid that is quickly displaced upwards will arrive with lower temperature and higher density than its new surroundings, and thus experience a downwards buoyancy force, tending to bring it down again. Hydrostatic equilibrium may, however, not be stable if the temperature falls with height ($dT/dz < 0$) because a blob that is suddenly displaced upwards into a region of lower temperature will arrive with lower density than its new surroundings and thus experience an upwards buoyancy force which tends to drive it further upwards.

Were it not for drag and heat loss, the displaced blob would rise with ever-increasing velocity. Drag from the surrounding fluid grows with the upwards blob velocity, linearly to begin with. Conductive heat loss lowers the excess temperature of the blob and thereby its buoyancy. Both of these effects are proportional to the surface area of the blob whereas buoyancy is proportional to the volume. Consequently, we expect that large blobs of fluid tend to be more unstable and rise faster than small. This indicates that there is a critical blob size below which blobs are not capable of rising at all.

In this section we shall estimate the critical blob size, and see that it may be expressed as a critical Rayleigh number. In the following sections we shall calculate the critical value of the Rayleigh number for the onset of instability in a horizontal slot.

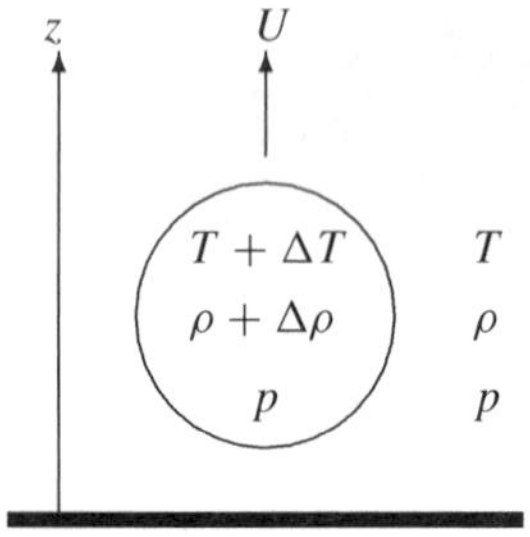

A spherical blob of fluid moving upwards with constant velocity U. If the temperature gradient is negative ($dT/dz < 0$) the temperature of the moving blob will be larger than its surroundings ($\Delta T > 0$). The pressure is assumed to be the same inside and outside the blob.

Stability estimate for spherical blob of fluid

For simplicity we begin with an incompressible fluid at rest in constant gravity g_0 with a constant negative vertical temperature gradient, $dT/dz = -G$, so that the temperature field is of the form $T(z) = T_0 - Gz$. A constant gradient could, as we have seen, be created in a horizontal slot with a fixed temperature difference between the lower and upper plates.

Imagine now that a spherical blob of fluid with radius a is set into upwards motion with a small steady velocity $U > 0$. This is, of course, a thought experiment, and we do not speculate on the technological difficulties in creating and maintaining such a blob. While it slowly rises towards lower and lower temperatures, the warmer blob will transfer its excess heat to the colder environment over a typical diffusion time $t \sim a^2/4\kappa$ (see equation (30.17)). In this time the blob rises through the height $\Delta z \approx Ut \sim Ua^2/4\kappa$, and the environment cools by $\Delta T \sim G\Delta z \sim GUa^2/4\kappa$. In the steady state the competition between the falling temperature of the environment and the loss of heat from the blob should lead to a time-independent temperature excess of size ΔT, which in turn determines the buoyancy force.

Unfortunately the estimate of ΔT is a bit weak because the Péclet number $\mathrm{Pe} = 2aU/\kappa$ vanishes in the limit of vanishing U. This implies that for sufficiently small U advection of heat will be negligible compared to diffusion, and that the heat escaping from the blob will spread far beyond the blob radius and thereby raise the temperature of the environment. The rising sphere thus finds itself surrounded by a large 'cocoon' of fluid (of its own making) that is warmer than the environment and therefore provides smaller buoyancy than would be the case if the temperature of the environment reigned all the way to the surface of the sphere. In order to calculate the buoyancy force we must know the temperature distribution inside the blob relative to the temperature at its surface.

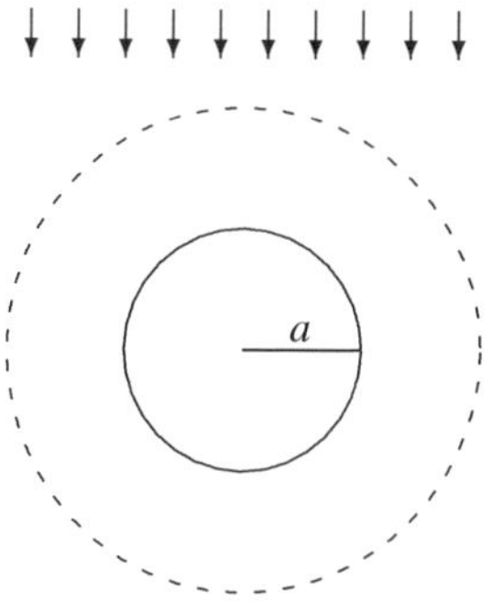

The heat front reaches far beyond the blob radius when the blob velocity is tiny. Here the blob is viewed in its rest frame, where asymptotically there is a uniform downwards wind $-U$ carrying a consistently lower temperature.

It is most convenient to go to the rest frame of the blob where the flow outside the blob is steady with the temperature of the environment dropping at a constant rate. The true temperature field inside the blob must then be of the form,

$$T' = T_0 - G\,(z + Ut) + \Delta T, \tag{31.22}$$

where by assumption the temperature excess field ΔT is time-independent. Inside the blob, T' must obey Fourier's heat equation for at fluid at rest (30.14), which under the given assumptions becomes

$$-GU = \kappa \nabla^2 \Delta T. \tag{31.23}$$

Seeking a spherical solution, we find $\Delta T = -(GU/6\kappa)r^2 + \text{const}$, and the difference between the temperature inside the blob and on its surface becomes

$$\delta T = \Delta T - \Delta T|_{r=a} = \frac{GU}{6\kappa}(a^2 - r^2). \tag{31.24}$$

This is indeed of the same order of magnitude as the previous estimate $\Delta T \sim GUa^2/\kappa$. The total upwards buoyancy force is obtained from the density change $\delta\rho = -\alpha\delta T\rho_0$ inside the blob,

$$\mathcal{F}_B = \int_V \delta\rho(-g_0)dV = \rho_0 g_0 \alpha \int_0^a \delta T(r) 4\pi r^2\, dr = \frac{4\pi\rho_0 g_0 \alpha G a^5 U}{45\kappa}. \tag{31.25}$$

Evidently, the buoyancy grows like the fifth power of the radius because its volume grows like the third power and the diffusion time like the second.

The viscous drag on a solid sphere in slow steady motion is given by Stokes Law (19.12). Disregarding the internal flow in the blob we estimate that

$$\mathcal{F}_D = 6\pi\eta a U. \tag{31.26}$$

This is valid for small Reynolds number $\text{Re} = 2aU/\nu \ll 1$, a condition which is always fulfilled in the limit of vanishing U.

If the buoyancy is smaller than the drag, $\mathcal{F}_B < \mathcal{F}_D$, the sphere cannot continue to rise on its own. In dimensionless form, the stability condition becomes

$$\boxed{\frac{\mathcal{F}_B}{\mathcal{F}_D} = \frac{2}{135}\frac{g_0\alpha G a^4}{\kappa\nu} < 1,} \tag{31.27}$$

where $\nu = \eta/\rho_0$ is the kinematic viscosity (momentum diffusivity) of the fluid. Evidently, this inequality puts an upper limit on the size of stable blobs.

The critical Rayleigh number

In terms of the blob diameter $d = 2a$, the stability condition (31.27) may be written as a condition on the Rayleigh number,

$$\text{Ra} \equiv \frac{g_0\alpha G d^4}{\kappa\nu} < 1080. \tag{31.28}$$

Tracing back over the preceding calculation, we see that the large critical value $\text{Ra}_c = 1080$ on the right-hand side is mainly due to the 'cocoon' of warm fluid carried along with the blob and diminishing its buoyancy. The critical Rayleigh number for blobs of general globular shape may presumably always be taken to be around 1000, whereas blobs with radically different shapes, for example long cylinders, will have quite different critical Rayleigh numbers, although typically they will be large.

Example 31.2.1 (Pot of water): Consider a pot of water with depth $h = 10$ cm on a warm plate held at 50° C in a room with temperature 20 ° C. The temperature difference is $\Theta \approx 30$ K and the gradient $G = \Theta/h = 300\ \text{K m}^{-1}$. The stability limit for spherical blobs is then obtained from (31.28) and becomes $d \lesssim 3.7$ mm. Convective currents are thus expected to arise spontaneously everywhere in the pot. If instead there is heavy porridge in the pot with heat properties like water but kinematic viscosity, say $\nu \approx 1\ \text{m}^2\ \text{s}^{-1}$, the critical diameter becomes $d \approx 12$ cm. Such blobs cannot find room in the container, and only a little convection is expected.

If the geometry of a fluid container cannot accommodate blobs larger than a certain diameter d, and if the Rayleigh number for this diameter is below the critical value, the fluid in the container will be stable with the given negative temperature gradient. The critical Rayleigh number depends, however, strongly on the geometry of the container, and cannot in general be calculated analytically. In the following section we shall determine it for the simplest of all geometries, the horizontal slot.

Estimate of terminal blob speed

If the Rayleigh number for a blob is larger than the critical value, $\text{Ra} > \text{Ra}_c$, the blob will accelerate upwards with larger and larger speed. For large Reynolds numbers, form drag on a sphere, $\mathcal{F}_D \approx (1/4)\rho_0 \pi a^2 U^2$ (see page 272), grows quadratically with velocity, and will eventually balance the buoyancy force $\mathcal{F}_B$. The terminal speed determined by solving $\mathcal{F}_D = \mathcal{F}_B$ becomes for $\text{Ra} \gg \text{Ra}_c$,

$$U \approx \frac{2}{45} \frac{g_0 \alpha G d^3}{\kappa}, \tag{31.29}$$

where d is the blob diameter. The shape of a rising blob is, however, strongly influenced by the high speeds, so this is only a coarse estimate.

Example 31.2.2: In example 31.2.1 a water blob with $d = 1$ cm will reach a terminal speed of $U \approx 23\ \text{cm s}^{-1}$, in reasonable agreement with daily experience.

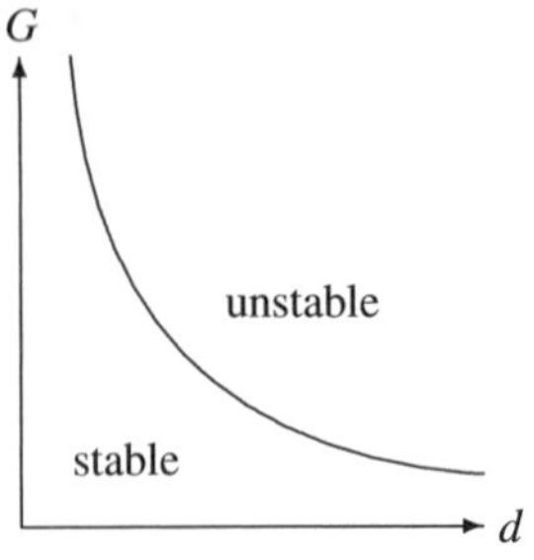

Sketch of the stability plot for heat convection in a system with negative temperature gradient G and size d. The critical surface is $G \sim d^{-4}$.

* 31.3 Linear stability analysis of convection

The onset of instability in dynamical systems is usually determined by linearizing the dynamical equations around a particular 'baseline' state that may or may not be unstable. The solutions to the linearized dynamics represent the possible fluctuations around the baseline state, and if no fluctuation can grow indefinitely with time, the baseline state is said to be stable. The existence of a single run-away fluctuation mode indicates on the other hand that the baseline state is unstable. In the space of parameters that control the system, the condition that all fluctuations are damped leads to an inequality like (31.28), which in the limit of equality defines a *critical surface*, separating the stable region in parameter space from the unstable.

Linearized dynamics of flow and heat

In the present case the baseline state is simply an incompressible fluid at rest in hydrodynamic equilibrium with a vertical temperature distribution of constant negative gradient, $T = T_0 - Gz$. The pressure must obey the equations of hydrodynamic equilibrium (4.20) on page 50 with the modified density $\rho = \rho_0(1 - \alpha(T - T_0))$ of the heated fluid. Solving the hydrostatic equilibrium equations we find $p = p_0 - \rho_0 g_0 z - 1/2 \rho_0 g_0 \alpha G z^2$ where p_0 is the pressure at $z = 0$.

A small velocity perturbation $\boldsymbol{v}$ will generate small corrections to the fields, ΔT and Δp, so that the true temperature and pressure fields become,

$$T = T_0 - Gz + \Delta T, \qquad p = p_0 - \rho_0 g_0 z - \tfrac{1}{2}\rho_0 g_0 \alpha G z^2 + \Delta p. \tag{31.30}$$

To first order in the small quantities ΔT, Δp and $\boldsymbol{v}$, the heat equation (30.23) becomes,

$$\boxed{\frac{\partial \Delta T}{\partial t} - G v_z = \kappa \nabla^2 \Delta T.} \tag{31.31}$$

Dissipation does not contribute because it is of second order in $\boldsymbol{v}$. Adding the buoyancy term $-\alpha\Delta T\boldsymbol{g}_0 = \alpha\Delta T g_0 \boldsymbol{e}_z$ to the incompressible Navier–Stokes equation (17.16) we obtain to first order in the small quantities,

$$\boxed{\frac{\partial \boldsymbol{v}}{\partial t} = -\frac{\nabla \Delta p}{\rho_0} + \nu\nabla^2\boldsymbol{v} + \alpha\Delta T g_0 \boldsymbol{e}_z.} \tag{31.32}$$

The advective acceleration is absent because it is of second order in $\boldsymbol{v}$. Finally the velocity field must satisfy the divergence condition,

$$\boxed{\nabla\cdot\boldsymbol{v} = 0.} \tag{31.33}$$

These five coupled partial linear differential equations should now be solved for the five fluctuation fields, ΔT, Δp and $\boldsymbol{v}$, with the appropriate boundary conditions for the particular geometry under study.

Fourier transformation

Fourier transformation is the method of choice for solving homogeneous linear partial differential equations with constant coefficients. All fields are assumed to be superpositions of elementary harmonic waves of the form $\exp(\lambda t + i\boldsymbol{k}\cdot\boldsymbol{x})$ where $\boldsymbol{k}$ is a real wave vector and λ may be a complex number. For a single harmonic wave, we obtain from the linearized dynamics,

$$\lambda\Delta\widetilde{T} - G\widetilde{v}_z = -\kappa\boldsymbol{k}^2\Delta\widetilde{T}, \tag{31.34a}$$

$$\lambda\widetilde{\boldsymbol{v}} = -\frac{i\boldsymbol{k}}{\rho_0}\Delta\widetilde{p} - \nu\boldsymbol{k}^2\widetilde{\boldsymbol{v}} + \alpha\Delta\widetilde{T}g_0\boldsymbol{e}_z, \tag{31.34b}$$

$$\boldsymbol{k}\cdot\widetilde{\boldsymbol{v}} = 0 \tag{31.34c}$$

where now $\Delta\widetilde{T}$, $\Delta\widetilde{p}$ and $\widetilde{\boldsymbol{v}}$ denote the amplitudes of the harmonic waves. Solving the first equation for $\widetilde{v}_z$, and dotting the second equation with $\boldsymbol{k}$ (using the third), we obtain,

$$\widetilde{v}_z = \frac{\lambda + \kappa\boldsymbol{k}^2}{G}\Delta\widetilde{T}, \qquad \frac{\Delta\widetilde{p}}{\rho_0} = -\alpha\Delta\widetilde{T}g_0\frac{ik_z}{\boldsymbol{k}^2}.$$

Inserting this into the z-component of (31.34b), we find a linear equation for $\Delta\widetilde{T}$, which only has a non-trivial solution for

$$(\lambda + \nu\boldsymbol{k}^2)(\lambda + \kappa\boldsymbol{k}^2) = \alpha G g_0\left(1 - \frac{k_z^2}{\boldsymbol{k}^2}\right). \tag{31.35}$$

This equation expresses that the determinant of the system of five linear algebraic equations (31.34) must vanish.

Being a quadratic equation in λ it always has two roots,

$$\lambda = -\frac{1}{2}(\nu + \kappa)\boldsymbol{k}^2 \pm \frac{1}{2}\sqrt{(\nu - \kappa)^2\left(\boldsymbol{k}^2\right)^2 + 4\alpha G g_0\left(1 - \frac{k_z^2}{\boldsymbol{k}^2}\right)}. \tag{31.36}$$

Both roots are real and one of the roots is evidently negative whereas the other may be positive. For stability this root should also be negative, which means that

$$(\nu + \kappa)\boldsymbol{k}^2 > \sqrt{(\nu - \kappa)^2\left(\boldsymbol{k}^2\right)^2 + 4\alpha G g_0\left(1 - \frac{k_z^2}{\boldsymbol{k}^2}\right)}.$$

Squaring this inequality and rearranging, it may be written,

$$\frac{\alpha G g_0}{\kappa\nu} < \frac{\left(\boldsymbol{k}^2\right)^3}{\boldsymbol{k}^2 - k_z^2}. \tag{31.37}$$

The right-hand side depends on the geometry of the fluid container. If d is the typical length scale for the geometry, the right-hand side scales as $|\boldsymbol{k}|^4 \sim d^{-4}$. If there is no intrinsic length scale ($d = \infty$), the minimum of the right-hand side is zero, and there can be no stability, just as we concluded for the rising blobs.

For finite d we multiply both sides by d^4. The inequality now becomes a condition on the dimensionless Rayleigh number of the same form as (31.28),

$$\boxed{\mathrm{Ra} \equiv \frac{g_0 \alpha G d^4}{\kappa \nu} < \frac{(k_x^2 + k_y^2 + k_z^2)^3}{k_x^2 + k_y^2} d^4.} \tag{31.38}$$

The finite geometry imposes constraints on the possible wave vectors $\boldsymbol{k}$ on the right-hand side. Among all the allowed fluctuations, the one that leads to the smallest value of the right-hand side is called the *critical fluctuation*, and the corresponding value of the right-hand side is called *critical Rayleigh number* Ra_c. It defines the upper limit to the convective stability of the baseline state.

Critical fluctuations

When the stability condition (31.38) is fulfilled with a non-vanishing right-hand side, all fluctuations are exponentially damped in time, and the fluid will essentially stay at rest if perturbed slightly. If a fluctuation violates the stability condition its amplitude will grow exponentially with time, resulting in more complicated flow patterns, even turbulence, about which linear stability analysis has nothing to say.

Right at the critical point where the inequality becomes an equality the largest of the two stability exponents (31.36) must vanish, $\lambda = 0$. This indicates that the critical fluctuations are time-independent (for a rigorous proof of this assertion see [37]). In this case it is better to revert to ordinary space where the critical fluctuations must obey the steady-flow versions of the linearized dynamic equations (31.31)–(31.33),

$$-G v_z = \kappa \nabla^2 \Delta T, \tag{31.39a}$$

$$\frac{\nabla \Delta p}{\rho_0} = \nu \nabla^2 \boldsymbol{v} + \alpha \Delta T g_0 \boldsymbol{e}_z, \tag{31.39b}$$

$$\nabla \cdot \boldsymbol{v} = 0. \tag{31.39c}$$

These equations may in fact be combined into a single equation for the temperature excess ΔT. Taking the divergence of the second equation we first get,

$$\frac{1}{\rho_0} \nabla^2 \Delta p = \alpha g_0 \nabla_z \Delta T, \tag{31.40}$$

and using this equation and (31.39a), v_z and Δp can be eliminated from the z-component of the second equation, and we obtain,

$$\boxed{(\nabla^2)^3 \Delta T = \frac{\alpha G g_0}{\kappa \nu} (\nabla^2 - \nabla_z^2) \Delta T.} \tag{31.41}$$

This equation is equivalent to the condition of the vanishing determinant (31.35) for $\lambda = 0$, and should be solved with the boundary conditions for the geometry of the system. Being a kind of eigenvalue equation, each solution determines a value of the coefficient $\alpha G g_0 / \kappa \nu$, and the one that yields the smallest value determines the point where convection first begins, i.e. the critical Rayleigh number.

Henri Bénard (1874–1939). *French physicist. Discovered hexagonal convection patterns in thin layers of whale oil in 1900. Such cellular convective structures have later been named Bénard cells.*

A critical fluctuation is apparently another steady solution to the combined heat and mass flow problem. It must, however, be kept in mind that an essential assumption behind linear stability analysis is that the fluctuation amplitudes are infinitesimal so that nonlinear terms can be disregarded. Nonlinear terms tend in fact to be beneficial and exert a stabilizing influence on the critical fluctuation such that it is able to persist somewhat above the critical value without diverging exponentially. This is why critical fluctuations can be observed at all.

* 31.4 Rayleigh–Bénard convection

Warming a horizontal layer of fluid from below is a common task in the kitchen as well as in industry. It was first investigated experimentally by Bénard in 1900 and later analysed theoretically by Rayleigh in 1916.

Figure 31.2. Bénard–Marangoni convection pattern. Image courtesy Manuel G. Velarde, Universidade Complutense, Madrid.

The most conspicuous feature of the heated fluid is that convection breaks the original planar symmetry, thereby creating characteristic convection patterns. That the symmetry must break is fairly clear, because it is impossible for all the fluid in the layer to start to rise simultaneously. A localized fluctuation current which begins to rise will have to veer off into the horizontal direction because of the horizontal boundaries. We shall see below that at the onset of convection the flow breaks up into an infinite set of 'rollers' with alternating sense of rotation.

Much later, in 1956, it was understood that the beautiful hexagonal surface tessellation observed by Bénard in a thin layer of heated whale oil was not caused by buoyancy alone but was driven by the interplay between buoyancy and temperature-dependent surface tension, a phenomenon now called Bénard–Marangoni convection. Here we shall only discuss clean Rayleigh–Bénard convection in layers of fluid so thick that the Marangoni effect can be disregarded.

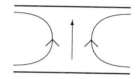

A rising flow in a horizontal layer of fluid has to veer off horizontally both at the top and the bottom.

Carlo Marangoni (1840–1925). *Italian physicist. Investigated surface tension effects using oil drops spread on water.*

General solution

Let the horizontal layer of incompressible fluid have thickness d and be subject to a constant negative temperature gradient G. The boundaries are chosen symmetrically at $z = \pm d/2$, for reasons that will become clear in the following. Since the flow has to veer off at the boundaries, the fields must depend on z, implying that $k_z \neq 0$ in the stability condition (31.38). The wavenumbers k_x and k_y can in principle take any real values because of the infinitely extended planar symmetry, but since the right-hand side of the stability condition diverges for both $k_x^2 + k_y^2 \to 0$ and $k_x^2 + k_y^2 \to \infty$, the minimum must occur at a finite value of $k_x^2 + k_y^2$. This argument demonstrates that the critical solution must have a periodic horizontal structure.

For given k_x and k_y we may without loss of generality rotate the coordinate system to obtain $k_y = 0$ and $k_x > 0$, implying that the fields only depend on x and z but not on y. Since there is no geometric constraint in the x-direction, we are still free to Fourier transform along x, such that the most general form of the temperature excess takes the form,

$$\Delta T = \Theta \cos k_x x f(z) \tag{31.42}$$

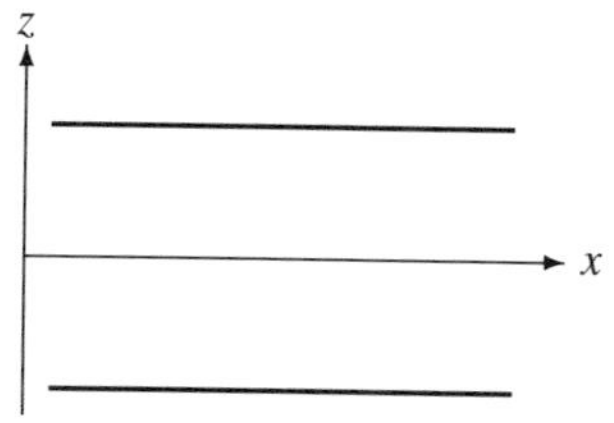

The plates are placed symmetrically at $z = \pm d/2$.

where Θ is a constant, and $f(z)$ is a (so far unknown) dimensionless function of z. From (31.39a) we find the vertical velocity,

$$v_z = -\frac{\kappa\Theta}{G}\cos k_x x(\nabla_z^2 - k_x^2)f(z), \tag{31.43}$$

and from the divergence condition (31.39c) we get the horizontal velocity,

$$v_x = \frac{\kappa\Theta}{Gk_x}\sin k_x x(\nabla_z^2 - k_x^2)f'(z), \tag{31.44}$$

where $f'(z) = df(z)/dz$. The third velocity component v_y does not participate, and it can be shown (problem 31.3) that it must vanish, $v_y = 0$.

Inserting ΔT into the determinant equation (31.41) we get a sixth-order ordinary differential equation for this function,

$$\left(\nabla_z^2 - k_x^2\right)^3 f(z) = -\frac{\alpha G g_0}{\kappa\nu}k_x^2 f(z). \tag{31.45}$$

Using this relation one may verify that the pressure excess,

$$\frac{\Delta p}{\rho_0} = -\frac{\kappa\nu\Theta}{Gk_x^2}\cos k_x x(\nabla_z^2 - k_x^2)^2 f'(z), \tag{31.46}$$

satisfies (31.40).

In a given physical context the actual solution depends on the boundary conditions imposed on the fields. We shall always assume that the boundaries are perfect conductors of heat such that $\Delta T = 0$ for $z = \pm d/2$. For the velocity fields the boundary conditions depend on whether the boundaries are solid plates or free open surfaces. We shall (as Rayleigh did in 1916) first analyse the latter case which is by far the simplest.

Two free boundaries

The simplest choice which satisfies the temperature boundary conditions $\Delta T = 0$ for $z = \pm d/2$ is,

$$f(z) = \cos k_z z, \qquad k_z = \frac{(1+2n)\pi}{d} \tag{31.47}$$

where $n = 0, 1, 2, \ldots$ is an integer. Inserting this into (31.45) and solving for the Rayleigh number we obtain

$$\mathrm{Ra} \equiv \frac{\alpha G g_0 d^4}{\kappa\nu} = \frac{(k_z^2 + k_x^2)^3}{k_x^2}d^4. \tag{31.48}$$

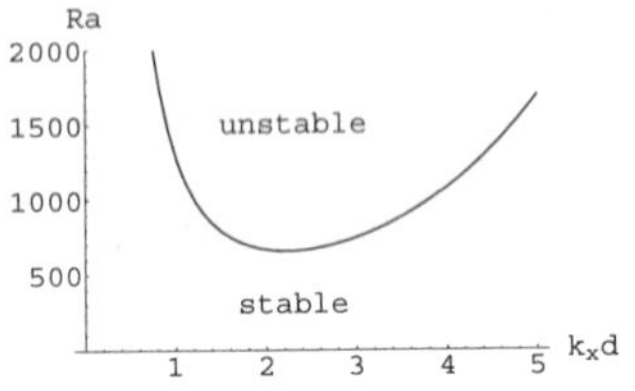

Plot of Ra versus $k_x d$ for $n = 0$. The minimum Ra $= 27\pi^4/4 \approx 658$ is found at $k_x d = \pi/\sqrt{2} \approx 2.22$.

The minimum of the right-hand side is found for $k_x = k_z/\sqrt{2}$ and $n = 0$, such that the critical Rayleigh number is,

$$\boxed{\mathrm{Ra}_c = \frac{27}{4}\pi^4 \approx 657.511\ldots.} \tag{31.49}$$

The complete critical solution becomes (with $k_x = \pi/d\sqrt{2}$ and $k_z = \pi/d$),

$$\Delta T = \Theta\cos k_x x\cos k_z z, \tag{31.50a}$$
$$v_x = \sqrt{2}U\sin k_x x\sin k_z z, \tag{31.50b}$$
$$v_y = 0, \tag{31.50c}$$
$$v_z = U\cos k_x x\cos k_z z, \tag{31.50d}$$

where $U = 3\pi^2\kappa\Theta/2Gd^2$.

The solution is pictured in the three panels of figure 31.3. The flow pattern (middle panel) consists of an infinite sequence of nearly elliptic 'rollers' with aspect ratio $\sqrt{2}$. The temperature pattern (top panel) is 90° out of phase with the flow pattern. This confirms the intuition that the central temperature should be higher when fluid transports heat from the warm lower boundary towards the cold upper boundary, and conversely.

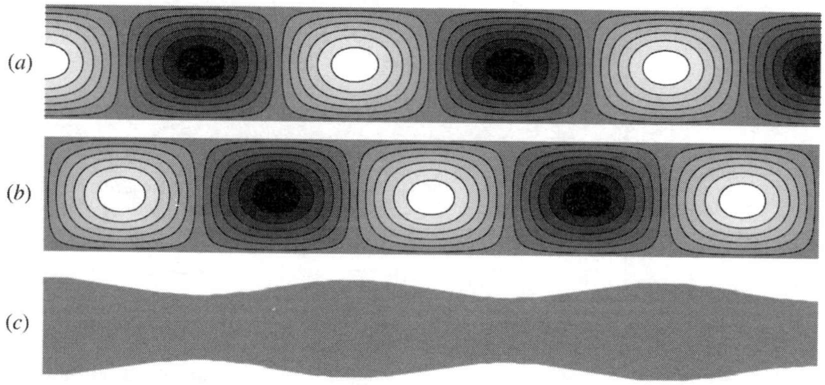

Figure 31.3. Critical fields in the horizontal layer of fluid with free boundaries (equation (31.50)). The steady flow pattern consists of an infinity of approximatively elliptical rolls of rotating fluid with aspect ratio $\sqrt{2}$ and alternating sense of rotation and temperature excess. **(a)** Contour plot of the temperature field ΔT with high temperature indicated by white. **(b)** Streamlines for the steady flow (v_x, v_z) with white indicating clockwise rotation. **(c)** Deformation of the originally parallel boundaries (strongly exaggerated).

From the above solution we immediately obtain the shear stress,

$$\sigma_{xz} = \eta(\nabla_x v_z + \nabla_z v_x) = \frac{\pi U \eta}{\sqrt{2}d} \sin k_x x \cos k_z z. \tag{31.51}$$

It evidently vanishes for $z = \pm d/2$, and since we trivially have $\sigma_{yz} = 0$, both boundaries are completely free of shear. There is no practical problem in arranging the upper boundary to be shear-free; that is in fact what we do when we cook. A shear-free lower boundary is on the contrary rather unphysical, so the main virtue of the shear-free model is that it is easy to solve.

The pressure excess in the critical solution may be calculated from (31.46),

$$\Delta p = \frac{2}{3\pi} \alpha \Theta \rho_0 g_0 d \cos k_x x \sin k_z z, \tag{31.52}$$

so the excess in the normal stress becomes

$$\Delta\sigma_{zz} = -\Delta p + 2\eta \nabla_z v_z = -\frac{10}{9\pi} \alpha \Theta \rho_0 g_0 d \cos k_x x \sin k_z z. \tag{31.53}$$

It does not vanish at the boundaries, showing the solution is not perfect. The non-vanishing normal stress can, however, be compensated by hydrostatic pressure if the layer thickness is allowed to vary a bit. Dividing by $\rho_0 g_0$ we find the required shift at the two boundaries,

$$\Delta z = -\left.\frac{\Delta\sigma_{zz}}{\rho_0 g_0}\right|_{z=\pm d/2} = \pm \frac{10}{9\pi} \alpha \Theta d \cos k_x x. \tag{31.54}$$

The shape of the deformed layer is shown in the bottom panel of figure 31.3.

Two solid boundaries

A horizontal slot bounded by two solid plates is easy to set up experimentally. Numerous experiments have been carried out in the twentieth century and agree very well with the theoretical results [13].

For simplicity we choose the plate distance $d = 1$ in the following analysis. The fundamental equation (31.45) is an ordinary sixth-order differential equation with constant coefficients, implying that the solution is a superposition of exponentials $e^{\lambda z}$ where λ is a root of the sixth-order algebraic equation,

$$(\lambda^2 - k_x^2)^3 = -\mathrm{Ra}\, k_x^2. \tag{31.55}$$

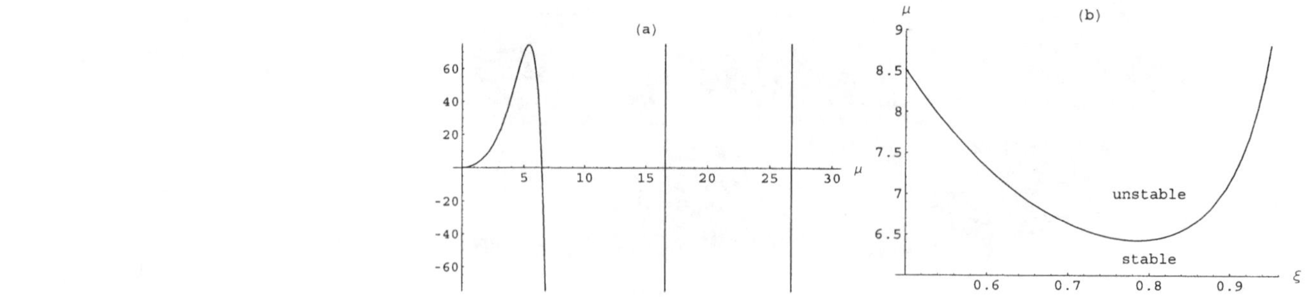

Figure 31.4. Solution for horizontal slot with solid boundaries. **(a)** The value of the determinant as a function of $\mu = \mathrm{Ra}^{1/4}$ for $\xi = 0.75$. One notes the regularly spaced solutions where the determinant crosses zero. **(b)** Stability plot for the lowest branch as a function of ξ. The minimum $\mu_c = 6.42846\ldots$ for $\xi_c = 0.785559\ldots$ determines the critical Rayleigh number.

The six roots are evidently,

$$\lambda = \pm\sqrt{k_x^2 + \mathrm{Ra}^{1/3} k_x^{2/3} \sqrt[3]{-1}}, \tag{31.56}$$

where $\sqrt[3]{-1} = -1, (1 \pm i\sqrt{3})/2$ is any one of the three third roots of -1. Parametrizing $k_x = \mu\xi^3$ with $\mu = \mathrm{Ra}^{1/4}$, the roots may be written $\lambda = \pm\mu\xi\sqrt{\xi^4 + \sqrt[3]{-1}}$. Writing the roots in the form $\lambda = \pm i\mu_0$ and $\lambda = \pm\mu_1 \pm i\mu_2$, we find,

$$\mu_0 = \mu\xi\sqrt{1 - \xi^4}, \tag{31.57a}$$

$$\mu_1 = \frac{1}{2}\mu\xi\sqrt{1 + 2\xi^4 + 2\sqrt{1 + \xi^4 + \xi^8}}, \tag{31.57b}$$

$$\mu_2 = \frac{1}{2}\mu\xi\sqrt{-1 - 2\xi^4 + 2\sqrt{1 + \xi^4 + \xi^8}}. \tag{31.57c}$$

These quantities are all real for $0 < \xi < 1$.

The boundary conditions are as before $\Delta T = 0$ together with $v_x = v_z = 0$ at $z = \pm 1/2$. Since the boundary conditions as well as the fundamental equation (31.45) are invariant under a change of sign of z, it follows that the solutions are either symmetric (even) or antisymmetric (odd) in z. In the even case we have,

$$f(z) = A\cos\mu_0 z + B\cosh\mu_1 z\cos\mu_2 z + C\sinh\mu_1 z\sin\mu_2 z, \tag{31.58}$$

where A, B and C are constants. From the general solution (31.42), (31.43) and (31.44) we see that $f(z)$, $f''(z)$, and $f'''(z) - k_x^2 f'(z)$ must vanish for $z = 1/2$.

These conditions provide three homogenous equations for the determination of A, B and C. Such equations only have a non-trivial solution if their (3×3) determinant vanishes. It takes a bit of algebra to show that the determinant is proportional to,

$$\begin{aligned}\det(\mu, \xi) \propto\ & \mu_0(\cosh\mu_1 + \cos\mu_2)\sin\frac{\mu_0}{2} \\ & + ((\mu_1 + \sqrt{3}\,\mu_2)\sinh\mu_1 + (\mu_2 - \sqrt{3}\,\mu_1)\sin\mu_2)\cos\frac{\mu_0}{2}.\end{aligned} \tag{31.59}$$

Solving the transcendental equation, $\det(\mu, \xi) = 0$, yields a family of solutions $\mu = \mu(\xi)$, as shown in figure 31.4(a). The minimum of the lowest branch determines the critical values $\mu_c = \mathrm{Ra}_c^{1/4} = 6.42846\ldots$ and $\xi_c = 0.785559\ldots$ (see figure 31.4(b)). The critical Rayleigh number becomes (see also problem 31.4 for an approximative calculation)

$$\boxed{\mathrm{Ra}_c = 1707.76\ldots} \tag{31.60}$$

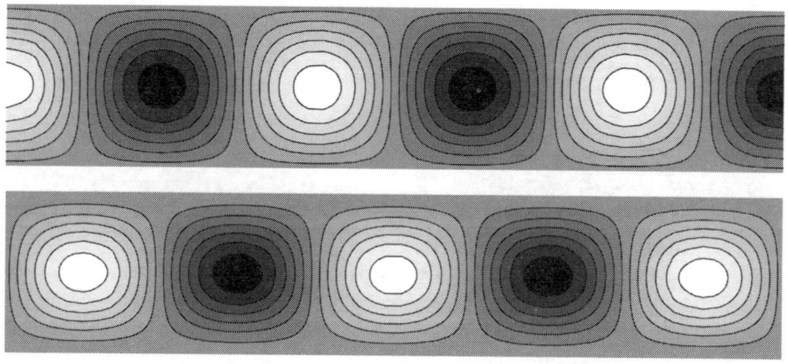

Figure 31.5. Critical fields in a horizontal layer with solid boundaries. The steady flow pattern consists of an infinity of approximately circular cylindrical rolls of fluid with alternating sense of rotation and alternating temperature excess. **(a)** Contour plot of the temperature field ΔT with high temperature indicated by white. **(b)** Streamlines for the steady flow (v_x, v_z) with white indicating clockwise rotation. Note how the streamlines 'shy away' from the solid walls because of the no-slip conditions.

and the corresponding wavenumbers,

$$\mu_0 = 3.9737\ldots \qquad \mu_1 = 5.19439\ldots \qquad \mu_2 = 2.12587\ldots. \tag{31.61}$$

Finally, solving the boundary conditions at the critical point, the coefficients become (apart from an overall factor),

$$A = 1 \qquad B = 0.120754\ldots \qquad C = 0.00132946\ldots. \tag{31.62}$$

From these values the actual fields ΔT, v_z and v_x may be determined. As can be seen from figure 31.5 the critical flow pattern consists of an infinity of roughly circular cylindrical rolls, which because of the no-slip conditions appear to 'shy away' from the boundaries.

It should be mentioned that in the antisymmetric case,

$$f(z) = D\sin\mu_0 z + E\cosh\mu_1 z\sin\mu_2 z + F\sinh\mu_1 z\cos\mu_2 z, \tag{31.63}$$

is treated in the same way and leads to a critical Rayleigh number of $\mathrm{Ra}_c \approx 17610.4\ldots$ which is uninteresting because it is (much) larger than the even solution.

Solid bottom and free top

This is the situation most often found in the household and industry. Since the boundary conditions are asymmetric, the solution is a superposition of all six possibilities,

$$\begin{aligned} f(z) = {} & A\cos\mu_0 z + B\cosh\mu_1 z\cos\mu_2 z + C\sinh\mu_1 z\sin\mu_2 z \\ & + D\sin\mu_0 z + E\cosh\mu_1 z\sin\mu_2 z + F\sinh\mu_1 z\cos\mu_2 z. \end{aligned} \tag{31.64}$$

Although more complicated, the solution is found in the same way as before, and the critical values are $\mu_c = 5.75986\ldots$ and $\xi_c = 0.775115\ldots$, and thus the Rayleigh number is,

$$\boxed{\mathrm{Ra} = 1100.65\ldots.} \tag{31.65}$$

This value is probably by accident nearly the same value as the estimate for a rising bubble (31.28). The wavenumbers for this solution are

$$\mu_0 = 3.56895\ldots, \qquad \mu_1 = 4.55531\ldots, \qquad \mu_2 = 1.8947\ldots, \tag{31.66}$$

and the coefficients,

$$\begin{aligned} & A = 1, && B = 0.086726\ldots, && C = -0.00956513\ldots, \\ & D = 0.216993\ldots, && E = 0.00778275\ldots, && F = -0.08632\ldots, \end{aligned} \tag{31.67}$$

again apart from an overall factor. The flow pattern is shown in figure 31.6.

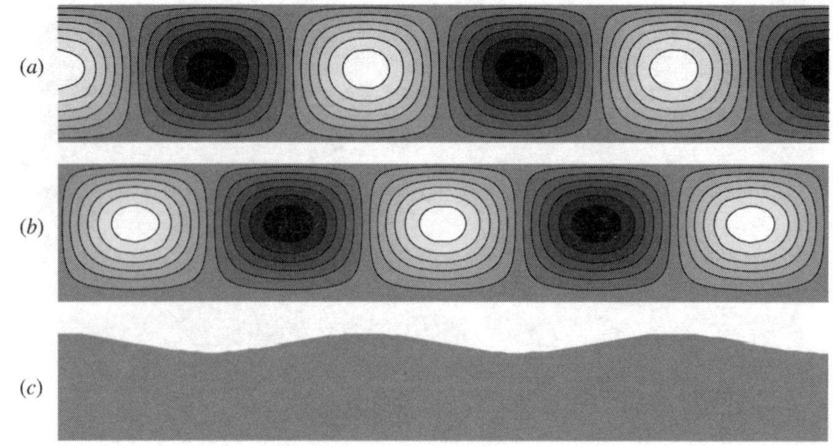

Figure 31.6. Critical fields in the horizontal layer of fluid with solid bottom and free top. **(a)** Contour plot of the temperature field ΔT with high temperature indicated by white. **(b)** Streamlines for the steady flow with white indicating clockwise rotation. Note how the no-slip condition makes the streamlines 'shy away' from the bottom while it 'hugs' the free surface at the top. **(c)** Deformation of the originally flat upper boundary (strongly exaggerated).

Energy balance?

Where does the energy to drive the rolls come from? The steadily rotating fluid could in principle be set to do useful work, and according to the First Law of Thermodynamics this work must be taken from the heat flowing between the plates. In effect the plates act as heat reservoirs and the convection as a heat engine converting heat to work by means of the buoyancy of warm fluid. In the present setup all the work done by the rotating fluid is actually dissipated back into heat by internal viscous forces, so in the steady state the energy of the fluid is constant, and no steady inflow of heat into the system is required. The local heat flow through the boundaries will, however, be uneven because of the local variations in the temperature gradient.

Convective pattern formation

The spontaneous formation of convection patterns in otherwise featureless geometries is a common occurrence. The nonlinear terms which have been left out in the linear approximation will exert a stabilizing influence on the patterns such that they are able to persist at Rayleigh numbers somewhat larger than the critical one. At still larger Rayleigh numbers, the rolls of the critical pattern will develop further instabilities and eventually turbulent convection may result (for an account of convection patterns with numerous photographs see [72, 21, 5]).

Problems

31.1 Show that there cannot be hydrostatic equilibrium in vertical gravity with horizontal temperature differences. Estimate the speed with which the fluid rises.

31.2 Calculate the ratio between the exit heat flow $\dot{Q}$ and the heat flow $\dot{Q}_0$ out of the warm plate in a vertical slot as a function of the plate dimensions. Show that this ratio is

$$\mathrm{Nu} = \frac{\dot{Q}}{\dot{Q}_0} = \frac{1}{45} \cdot \frac{d}{h} \cdot \mathrm{Ra} \tag{31.68}$$

where Ra is the Rayleigh number. This number is called the Nusselt number.

31.3 Show that $v_y = 0$ in the Rayleigh–Bénard solution.

* **31.4** Show that an approximate solution to the vanishing determinant (31.59) for Rayleigh–Bénard flow in a horizontal slot is given by the equation,

$$\mu = 2\frac{\mu}{\mu_0}\left(\pi - \arctan\frac{\mu_1 + \sqrt{3}\mu_2}{\mu_0}\right), \tag{31.69}$$

where the right-hand side is only a function of ξ. Numerical minimization of this function leads to $\mu = 6.44397\ldots$ for $\xi = 0.787942\ldots$, corresponding to a critical Rayleigh number $\text{Ra}_c = 1724$.

32 Turbulence

Turbulence is so commonplace that we hardly notice it in our daily lives. The size of humans and their speed of locomotion in air or water bring the typical Reynolds numbers into the millions, whether running or swimming, but most of us never think of the trail of turbulent air we leave in our wake. Our vehicles move at much higher speeds and generate more turbulence, although great efforts have been made to limit the extent of the turbulent wake by streamlining cars, airplanes, ships and submarines. In strong winds the lower part of the atmosphere may become strongly turbulent, but patches of turbulent air can in fact be found at any altitude, normally in connection with cloud formations, although also in clear air. Most people have experienced the quite unpleasant buffeting in an airplane exposed to atmospheric turbulence, particularly at the approach to an airport in heavy weather.

Even under the steadiest of conditions, for example the draining of a large water cistern through a long open pipe, turbulence will completely fill the pipe when the Reynolds number becomes sufficiently large. The actual road from laminar flow to turbulence as a function of a slowly increasing Reynolds number is complex and goes through many stages of sometimes intermittent flow patterns, not yet fully understood. Eventually, at sufficiently high Reynolds number, the turbulent flow seems to settle down into a homogeneous and isotropic state where the small-scale fluctuations in the flow appear to be of the same nature throughout the pipe, except for a stretch near the entrance and a thin layer very close to the wall. Such featureless *fully developed turbulence* is much more amenable to analysis than the flow stages leading up to it.

In this closing chapter of the book we shall only touch on the most basic and most successful aspects of turbulence, a subject which for a long time has been and still is an important topic of basic research. The many modern textbooks on the subject may be consulted to obtain a deeper insight than can be provided here [6, 71, 72, 40, 25, 39, 61, 58].

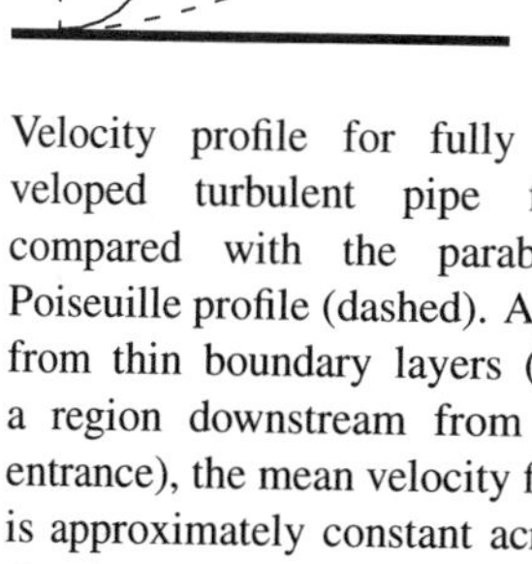

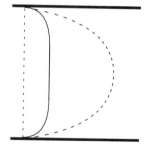

Velocity profile for fully developed turbulent pipe flow compared with the parabolic Poiseuille profile (dashed). Apart from thin boundary layers (and a region downstream from the entrance), the mean velocity field is approximately constant across the pipe.

32.1 Fully developed turbulence

Turbulence typically appears as a fluctuating velocity component on top of a mainstream flow with high Reynolds number. In the discussion of the phenomenology of stationary turbulent pipe flow (section 18.5 on page 255), it was pointed out that when turbulence has become fully developed some distance downstream from the pipe entrance, it proceeds plug-like down the pipe with a fairly uniform mainstream velocity distribution across the pipe, apart from a thin layer near the wall where the velocity drops to zero. In the reference frame of the mainstream flow and at length scales much smaller than the diameter of the pipe, the bulk of the fluid appears to be in a state of uniform agitation. In the following section we shall estimate the general statistical properties of this state.

Specific rate of dissipation

A stationary flow can only be maintained if the environment continually performs work on the fluid to balance the unavoidable viscous dissipation. In pipe flow the external work is performed by the pressure drop Δp maintained between its ends, and if the pipe carries a volume flux Q, the rate of work (power) of the pressure drop is $P = \Delta p\, Q$ (see page 253). Dividing by the total mass M of the fluid in the pipe, the *specific rate of dissipation* averaged over the whole pipe becomes,

$$\varepsilon = \frac{P}{M} = \frac{\Delta p\, Q}{M} = \frac{U^3}{a} C_f(\text{Re}). \tag{32.1}$$

In the last step we have used that $M = \pi a^2 L \rho_0$, $\Delta p = \sigma_{\text{wall}} 2L/a$ and $Q = \pi a^2 U$ where a is the pipe radius, σ_{wall} the wall stress, and U the average velocity of the flow. The Reynolds number is $\text{Re} = 2aU/\nu$ and from the definition of the dimensionless friction coefficient (18.49) the last expression follows.

In the smooth-pipe turbulent region, $4 \times 10^3 \lesssim \text{Re} \lesssim 10^5$, we have seen (page 257) that the friction coefficient decreases slowly as $\text{Re}^{-1/4}$ with increasing Reynolds number. Beyond this region, the friction coefficient becomes constant, although its actual value depends on the roughness of the pipe surface. The astonishing conclusion following from (32.1) is that *the specific dissipation rate is independent of the viscosity for sufficiently large Reynolds number*. Although ultimately caused by viscous friction, the specific rate of dissipation is finite in the limit of vanishing viscosity!

Example 32.1.1 (Concrete water pipe): A large concrete pipe with diameter $2a = 1$ m carries water at a discharge rate of $Q \approx 10\ \text{m}^3\ \text{s}^{-1}$. The average velocity is $U \approx 13\ \text{m s}^{-1}$, corresponding to a Reynolds number of $\text{Re} = 1.3 \times 10^7$ well into the rough pipe region. With a roughness scale of 1 mm, the friction coefficient calculated from (18.51) is $C_f(\infty) \approx 5 \times 10^{-3}$, leading to a specific dissipation $\varepsilon \approx 21\ \text{W kg}^{-1}$. Perhaps it does not sound much, but the dissipation rate (per unit of pipe length) is $\rho_0 \pi a^2 \varepsilon \approx 16\ \text{kW m}^{-1}$, which is about 4000 times larger than it would have been had the flow remained laminar with a friction coefficient $C_f \approx 16/\text{Re} \approx 1.3 \times 10^{-6}$.

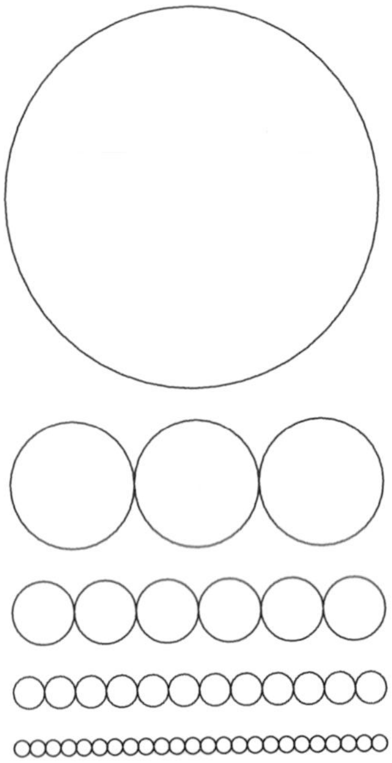

Naive sketch of the energy cascade. A large eddy is thought to break up into smaller eddies while conserving both mass and energy. The break-up stops at the dissipation scale.

32.2 The energy cascade

In the model we shall use here the fluctuating velocity component is viewed as composed of fairly localized flow structures, generically called 'eddies', covering a wide range of sizes. The nonlinearity of fluid mechanics is assumed to transform ('break up') the eddies into smaller eddies, while conserving mass and kinetic energy, until they become so small that they are wiped out by viscous friction, and their energy dissipated into heat (a picture owed to L. F. Richardson, 1922).

In an isolated patch of turbulence the kinetic energy will in this way be continually sapped 'from below' until the turbulence dies away and the flow becomes laminar and in the end will stop completely. We have all seen this happen after filling a bucket with water from a wide-open tap. Under steady external conditions the kinetic energy that is lost to heat will instead be continually re-supplied by the work of the external forces.

Scaling laws

The energy cascade plays out in an interval of eddy sizes, $\lambda_d \lesssim \lambda \lesssim L$, where the upper cutoff L is related to the flow geometry (for example the pipe diameter), and the lower cut-off λ_d represents the dissipation length scale where viscosity 'kicks in'. Far above the dissipation cut-off, $\lambda \gg \lambda_d$, the eddy dynamics should be independent of the viscosity, whereas far below the geometry cut-off, $\lambda \ll L$, the actual flow geometry should not matter for the eddy dynamics.

In the so-called *inertial range* satisfying both conditions, $\lambda_d \ll \lambda \ll L$, the only parameters that can be at play in the eddy dynamics are the size λ of an eddy and the time τ it takes for an eddy of this size to transform into smaller eddies. In the inertial range the magnitude of any other quantity must be related to these parameters, essentially by dimensional arguments. The 'break-up' of an eddy is, for example, bound to move some fluid a distance of magnitude λ in time τ, leading to an estimate of the velocity variation in an eddy, $u \sim \lambda/\tau$.

The energy-cascade model yields a relation between the time scale τ and the length scale λ by way of the specific dissipation rate ε. To see this we focus on a collection of eddies of size λ with total mass dM

and total kinetic energy $d\mathcal{T} \sim dMu^2$. In a time interval τ all these eddies break up into smaller eddies, making the rate of energy turn-over $d\dot{\mathcal{T}} \sim d\mathcal{T}/\tau \sim dMu^2/\tau$. Since both mass and energy are assumed to be conserved in the cascade that follows, the energy turn-over rate must be the same all the way down to the smallest scales, and thus in the end nearly equal to the viscous dissipation rate in the original amount of fluid, $dP = \varepsilon dM$. Using $u \sim \lambda/\tau$ and $dP \approx d\dot{\mathcal{T}}$ we find,

$$\varepsilon = \frac{dP}{dM} \approx \frac{d\dot{\mathcal{T}}}{dM} \sim \frac{u^2}{\tau} \sim \frac{\lambda^2}{\tau^3}. \tag{32.2}$$

Since ε has dimension $[\varepsilon] = \mathrm{W\ kg^{-1}} = \mathrm{m^2\ s^{-3}}$, one might think that this estimate could have been derived from a dimensional argument alone, but that would be wrong. The crucial input from the energy-cascade model is that the terminal rate of dissipation dP may be set equal to the kinetic energy turn-over rate $d\dot{\mathcal{T}}$ for any collection of eddies, independent of their size scale λ.

Solving (32.2) for the time scale, we find the λ-dependent result,

$$\boxed{\tau \sim \varepsilon^{-1/3}\lambda^{2/3},} \tag{32.3}$$

and from this we obtain the velocity scale,

$$\boxed{u \sim \frac{\lambda}{\tau} \sim (\varepsilon\lambda)^{1/3},} \tag{32.4}$$

also known as *Kolmogorov's scaling law*.

A basic assumption in the energy-cascade model is that the turbulence is homogeneous and isotropic in the rest frame of the mainstream flow. Even if ultimately maintained by the work of steady external forces, the physics in the inertial range only depends on the background velocity U through the specific dissipation rate ε, which for pipe flow is given by (32.1). At the scale L of the geometry, it follows from the scaling law that the velocity of the largest eddies is $u_L \sim (\varepsilon L)^{1/3}$. This is, however, not necessarily the same as the mainstream velocity U. Inserting the pipe flow expression (32.1) with $a \sim L$, we find $u_L/U \sim C_f^{1/3}$. In example 32.1.1 we have $C_f = 5 \times 10^{-3}$, such that u_L in that case is nearly an order of magnitude smaller than U.

Statistics

The fundamental concept necessary for establishing a statistical model of turbulence is the fraction $dF = dM/M$ of the total fluid mass (or volume) residing in eddies with sizes in a small interval $d\lambda$ around λ. The eddy distribution over sizes, $dF/d\lambda$, will in general depend on λ as well as on the system scale L and the viscosity ν, but in the inertial range, $\lambda_d \ll \lambda \ll L$, the eddy distribution should only depend on the basic scale parameters, λ and τ. Since $dF/d\lambda$ has dimension of inverse length, it must be of the form,

$$\frac{dF}{d\lambda} \sim \frac{1}{\lambda} \qquad (\lambda_d \ll \lambda \ll L). \tag{32.5}$$

The integral of this distribution diverges for both $\lambda \to 0$ and $\lambda \to \infty$. So although this eddy distribution looks universal, it cannot be properly normalized without introducing a (slowly varying) non-universal normalization factor depending logarithmically on the limits of the inertial range[1]. We shall ignore such factors here (see, however, problem 32.2).

Energy spectrum

The kinetic energy residing in the eddies with size λ may now be estimated as $d\mathcal{T} \sim u^2 dM$. Using $dM = MdF$, and dividing $d\mathcal{T}$ by $Md\lambda$ we obtain the distribution of the specific kinetic energy according

[1]Non-universal logarithmic factors have been introduced phenomenologically by Y. Gagnes and B. Castaing, A universal representation without global scaling invariance of energy spectra in developed turbulence, *C. R. Acad. Sci. Paris* **212**, (1991) 441.

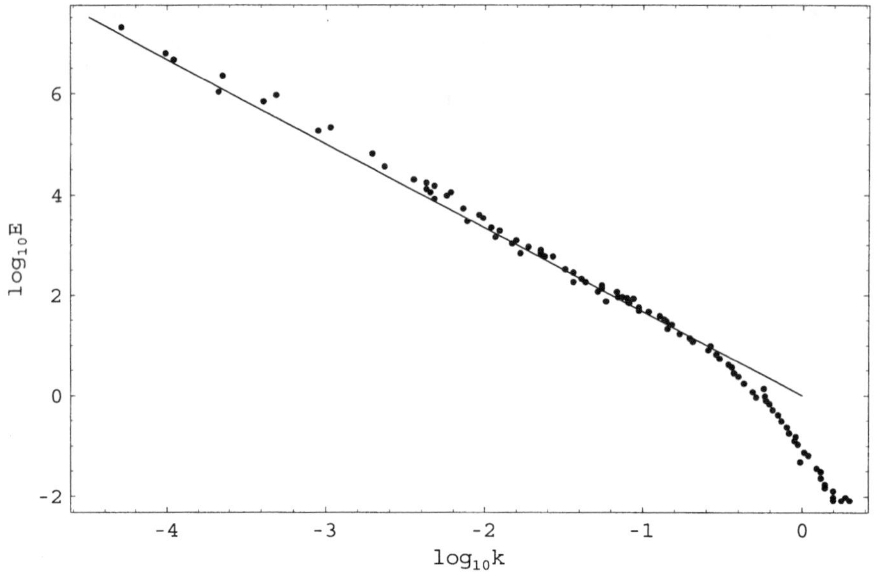

Figure 32.1. The logarithm of the turbulent energy spectrum E as a function of the logarithm of the wavenumber k (suitably normalized). Data extracted from the classical experiment carried out by towing a probe after a ship through the Discovery Passage of Western Canada under sometimes quite difficult conditions (footnote 2 on page 502). The largest Reynolds number is about 3×10^8 yielding more than three decades in the inertial range. The data agrees very well with Kolmogorov's $-5/3$ law (solid line) up to the beginning of the dissipation range at abscissa -1.

to eddy size,

$$\frac{1}{M}\frac{d\mathcal{T}}{d\lambda} \sim \frac{u^2}{\lambda} \sim \frac{\lambda}{\tau^2} \sim \varepsilon^{2/3}\lambda^{-1/3}. \tag{32.6}$$

To find the total kinetic energy, we must integrate this expression over all λ. Since the integral of this expression is convergent for $\lambda \to 0$ and diverges for $\lambda \to \infty$, we cut it off at the system scale L and find the total kinetic energy per unit of mass,

$$\frac{\mathcal{T}}{M} \sim (\varepsilon L)^{2/3} \sim u_L^2. \tag{32.7}$$

where u_L is the eddy velocity for $\lambda = L$. The dependence on the system scale shows that in spite of the fact that the distribution function (32.6) decreases slowly with wavelength most of the turbulent kinetic energy resides in the largest eddies near the system scale cut-off, rather than in the small eddies near the dissipation cut-off.

Andrei Nikolaevich Kolmogorov (1903–1987). *Russian mathematician. Made major contributions to a wide range of subjects: Markov processes, Lebesgue measure theory, axiomatic foundation of probability theory, turbulence, and dynamical systems.*

This is even better seen by expressing the kinetic energy distribution in terms of the inverse length scale, the 'wavenumber' $k = 2\pi/\lambda$. Using $d\lambda \sim dk/k^2$ we obtain the specific energy spectrum of turbulence as a function of wavenumber,

$$\boxed{E(k) \equiv \frac{1}{M}\frac{d\mathcal{T}}{dk} \sim \varepsilon^{2/3}k^{-5/3}.} \tag{32.8}$$

This is Kolmogorov's famous $-5/3$ power law from 1941. The negative exponent shows that the spectrum is indeed dominated by the smallest wavenumbers, $k \sim 1/L$, corresponding to the largest length scales, $\lambda \approx L$.

The left-out purely numeric constant in front of the spectrum is called the *Kolmogorov constant* and empirically determined to be about 1.5. Many experiments have confirmed the validity of the Kolmogorov spectrum in the inertial range, for example the one plotted in figure 32.1[2].

[2] H. L. Grant, R. W. Stewart and A. Moilliet, Turbulent spectra from a tidal channel, *J. Fluid Mech.* **12**, (1962)

The dissipation scale

The rate of viscous dissipation is calculated from the velocity gradients in the flow in equation (17.18) on page 236. Estimating the dissipation rate in a collection of eddies of mass dM and size λ we find,

$$dP \sim \eta\,|\nabla \boldsymbol{v}|^2\,dV = \nu\,|\nabla \boldsymbol{v}|^2\,dM \sim \nu\left(\frac{u}{\lambda}\right)^2 M\frac{d\lambda}{\lambda} \sim M\frac{\nu}{\lambda\tau^2}\,d\lambda.$$

Dividing by $Md\lambda$ we find the specific dissipation spectrum (per unit of size),

$$\frac{1}{M}\frac{dP}{d\lambda} \sim \frac{\nu}{\lambda\tau^2} \sim \nu\,\varepsilon^{2/3}\lambda^{-7/3}. \tag{32.9}$$

The largest dissipation evidently takes place at the smallest scales. To get the total specific dissipation ε we must integrate over λ, and since the integral diverges at the smallest scales, we cut it off at λ_d and get,

$$\varepsilon = \frac{P}{M} \approx \int_{\lambda_d}^{\infty} \frac{1}{M}\frac{dP}{d\lambda}\,d\lambda \sim \nu\varepsilon^{2/3}\lambda_d^{-4/3}. \tag{32.10}$$

Solving for λ_d we arrive at the *Kolmogorov dissipation scale*,

$$\boxed{\lambda_d \sim \varepsilon^{-1/4}\nu^{3/4}.} \tag{32.11}$$

In practice the inertial scaling laws may be assumed to be valid about an order of magnitude inside inertial range cut-offs, i.e. for $10\lambda_d \lesssim \lambda \lesssim 0.1L$.

At the dissipation scale the corresponding time and velocity scales become,

$$\tau_d \sim \varepsilon^{-1/2}\nu^{1/2}, \qquad u_d \sim \varepsilon^{1/4}\nu^{1/4}. \tag{32.12}$$

At this scale, everything can be expressed in terms of the specific rate of dissipation ε and the viscosity ν.

For the pipe flow in example 32.1.1 we find $\lambda_d \sim 15\ \mu$m, $\tau_d \sim 220\ \mu$s and $u_d \sim 7$ cm s^{-1}. The geometric scale is $L \approx 1$ m, implying that the inertial range should be 1 mm $\lesssim \lambda \lesssim$ 100 mm, covering a factor 100, or two full decades. This illustrates that it may be necessary to go to very large Reynolds numbers to get three or more decades of length scales in the inertial range. The broadest inertial ranges have in fact been reached in large natural systems, such as the ocean and the atmosphere (see figure 32.1).

Eddy viscosity

The random turbulent motion also transfers momentum between adjacent regions of fluid in a way which resembles molecular momentum diffusion, estimated in equation (17.2). This naturally leads to the idea that turbulent diffusion at a length scale λ likewise might be described by a kind of effective *eddy viscosity* ν_λ, which in the inertial range is determined by the eddy length and time scales,

$$\nu_\lambda \sim \frac{\lambda^2}{\tau} \sim u\lambda \sim \varepsilon^{1/3}\lambda^{4/3}. \tag{32.13}$$

The eddy viscosity grows monotonically with the size of the eddies and is maximal at the system scale, $\nu_L \sim \varepsilon^{1/3}L^{1/3}$, a value which must also characterize the average eddy viscosity. The smallest eddy viscosity $\nu_\lambda \sim \nu$ is obtained for $\lambda \sim \lambda_d$, showing that at the dissipation scale momentum diffusion due to eddies is of the same magnitude as molecular momentum diffusion.

Non-dimensional formulation

It is sometimes useful to convert to a dimensionless formalism by introducing the Reynolds number for eddies of size λ,

$$\mathrm{Re}_\lambda = \frac{u\lambda}{\nu} \sim \frac{\nu_\lambda}{\nu} \sim \frac{\epsilon^{1/3}\lambda^{4/3}}{\nu}. \tag{32.14}$$

241. This classical paper contains a delightful account of the difficulties encountered while attempting experimentally to verify a simple theoretical prediction in the real world of ships at sea.

At the dissipation scale $\lambda \sim \lambda_d$ we find as expected $\mathrm{Re}_\lambda \sim 1$, such that the ratios between the scale parameters and the dissipation scales may be written,

$$\frac{\lambda}{\lambda_d} \sim \mathrm{Re}_\lambda^{3/4}, \qquad \frac{\tau}{\tau_d} \sim \mathrm{Re}_\lambda^{1/2}, \qquad \frac{u}{u_d} \sim \mathrm{Re}_\lambda^{1/4}. \tag{32.15}$$

These results are valid in the inertial range $1 \ll \mathrm{Re}_\lambda \ll \mathrm{Re}_L = u_L L/\nu$.

Numerical simulation of turbulence

In a numerical simulation of a turbulent flow the grid should be so fine-grained that the dissipation scale can be resolved in all directions. Consequently, the number N of grid points must obey

$$N \gtrsim \left(\frac{L}{\lambda_d}\right)^3 \sim \mathrm{Re}_L^{9/4}. \tag{32.16}$$

In the pipe flow example 32.1.1 we have $\mathrm{Re}_L \approx 2.7 \times 10^6$, implying that to simulate the turbulent flow in the pipe one would have to use at least 3×10^{14} grid points. Even with a computer executing 10^{12} floating point operations per second, it would still take of the order of hours to do a single update of the whole grid, because each grid point requires quite a few operations. A complete simulation probing all time scales would require at least $\tau_L/\tau_d = \sqrt{\mathrm{Re}_L} \approx 1600$ updates, and might easily take a full year of computation.

Direct numeric simulation of realistic turbulent flows belongs among the hardest computational problems. Fortunately the need to find practical solutions to the technical problems presented by cars, ships and airplanes have led to a number of more or less *ad hoc* approximations that yield quite decent results.

Critique of Kolmogorov's theory

The derivation of the energy spectrum from scaling considerations depends only on the existence of a process by which the energy of turbulent motion at a given length scale is continually redistributed over smaller length scales until it is, in the end, dissipated by viscous forces. The detailed mechanism underlying this process is not known for sure, although more than 50 years of intense research has produced a wealth of ideas and results [6, 72, 40, 25, 39]. Numerical simulations have revealed that fully developed turbulence contains vortex filaments, and it is believed that stretching of these vortices plays a central role in transferring the vortex energy to smaller length scales.

Many objections, both mathematical and physical, have been raised against the energy-cascade model [25]. One physical objection is that fully developed turbulent flow is not as featureless as assumed here, but is observed experimentally to contain large-scale coherent structures, not included in the Kolmogorov model [32]. Another is that the Kolmogorov theory fails to take into account the conspicuous intermittency of turbulence, especially in the unstable regime near the transition from laminar to turbulent flow.

In view of the numerous experimental verifications of the Kolmogorov energy spectrum, one is, however, driven to conclude that the law must contain an element of truth, whatever its origin.

32.3 Mean flow and fluctuations

In this section we shall formalize the analysis of turbulence, and for simplicity we continue to assume that the turbulent flow is driven by steady boundary conditions, necessary for obtaining a statistically stationary state. In the preceding scaling analysis it was tacitly assumed that the mainstream flow did not influence the fluctuating component of the velocity field. We shall now see that the fluctuations do in fact exert a decisive influence on the mainstream field, although mathematically it is fraught with difficulties.

Mean values

The mainstream field is defined as a suitable average of the true velocity field, for example over a large time interval T starting at any time t,

$$\langle \boldsymbol{v}(\boldsymbol{x}, t)\rangle = \lim_{T\to\infty} \frac{1}{T}\int_0^T \boldsymbol{v}(\boldsymbol{x}, t+s)\, ds. \tag{32.17}$$

All other fields may similarly be averaged over time, for example the pressure,

$$\langle p(\boldsymbol{x}, t)\rangle = \lim_{T\to\infty} \frac{1}{T}\int_0^T p(\boldsymbol{x}, t+s)\, ds. \tag{32.18}$$

These limits are assumed to exist and be time-independent, although that is by no means a foregone conclusion. Defined in this way, the mean value of any field can in principle be determined experimentally with any desired precision from the average of a sufficiently large number of measurements of the instantaneous field value near the point $\boldsymbol{x}$ over a sufficiently long time interval T. The meaning of 'near' and 'sufficiently' depends on the measurement precision, as was discussed in chapter 1.

The fluctuating part of a field is defined as the difference between the field and its mean value. For the velocity and pressure fields we introduce special symbols for the fluctuations,

$$\boldsymbol{u}(\boldsymbol{x}, t) = \boldsymbol{v}(\boldsymbol{x}, t) - \langle \boldsymbol{v}(\boldsymbol{x}, t)\rangle, \tag{32.19a}$$

$$q(\boldsymbol{x}, t) = p(\boldsymbol{x}, t) - \langle p(\boldsymbol{x}, t)\rangle. \tag{32.19b}$$

By definition these fluctuations have vanishing means, $\langle \boldsymbol{u}\rangle = \boldsymbol{0}$ and $\langle q\rangle = 0$.

As in statistical mechanics, it is often convenient for more formal analysis to replace the time average of a field by the average of the field over a suitable statistical ensemble of fields. That this leads to the same mean values as time averages is a deep result which we shall not discuss here. In the following we only assume that there is a well-defined statistical procedure for calculating mean values, denoted by angular brackets $\langle \cdots \rangle$. The system is said to be in *statistical equilibrium* when all mean values are time-independent.

Mean field equations

Even if quite different from laminar flow, turbulent flow is also assumed to be governed by the usual Navier–Stokes equations. To avoid cumbersome notation we shall in the following shift the fluctuations out of the basic fields, $\boldsymbol{v} \to \boldsymbol{v} + \boldsymbol{u}$ and $p \to p + q$, such that the incompressible Navier–Stokes equations become (in the absence of gravity),

$$\frac{\partial(\boldsymbol{v}+\boldsymbol{u})}{\partial t} + (\boldsymbol{v}+\boldsymbol{u})\cdot\nabla(\boldsymbol{v}+\boldsymbol{u}) = -\frac{1}{\rho_0}\nabla(p+q) + \nu\nabla^2(\boldsymbol{v}+\boldsymbol{u}), \tag{32.20a}$$

$$\nabla\cdot(\boldsymbol{v}+\boldsymbol{u}) = 0. \tag{32.20b}$$

It must be emphasized that in these equations $\boldsymbol{v}$ and p now denote the mean fields whereas $\boldsymbol{u}$ and q denote time-dependent fluctuations with vanishing means. For generality we here also allow the mean fields to be time-dependent.

Taking the mean of the second equation and using $\langle \boldsymbol{v}\rangle = \boldsymbol{v}$ and $\langle \boldsymbol{u}\rangle = \boldsymbol{0}$ we get $\nabla\cdot\boldsymbol{v} = 0$. This shows that the mean of an incompressible flow is also incompressible, a consequence of the linearity of the divergence condition. Using this result in (32.20b) it follows in turn that the fluctuation field is itself incompressible, $\nabla\cdot\boldsymbol{u} = 0$. From the mean of the first equation, we get similarly,

$$\frac{\partial\boldsymbol{v}}{\partial t} + (\boldsymbol{v}\cdot\nabla)\boldsymbol{v} + \langle(\boldsymbol{u}\cdot\nabla)\boldsymbol{u}\rangle = -\frac{1}{\rho_0}\nabla p + \nu\nabla^2\boldsymbol{v}. \tag{32.21}$$

The last term on the left-hand side stems from the nonlinear advective acceleration, and creates a coupling between the fluctuations and the mean fields. Using $\nabla\cdot\boldsymbol{u} = 0$, it may be rewritten as a tensor divergence,

$$(\boldsymbol{u}\cdot\nabla)u_i = \sum_j u_j\nabla_j u_i = \sum_j \nabla_j(u_i u_j), \tag{32.22}$$

and moving the mean of this term to the right-hand side of (32.21), we arrive at the *mean field equations* (or *Reynolds equations*) for turbulent flow,

$$\boxed{\begin{aligned}\frac{\partial\boldsymbol{v}}{\partial t} + (\boldsymbol{v}\cdot\nabla)\boldsymbol{v} &= -\frac{1}{\rho_0}\nabla p + \nu\nabla^2\boldsymbol{v} - \nabla\cdot\langle\boldsymbol{u}\boldsymbol{u}\rangle.\\ \nabla\cdot\boldsymbol{v} &= 0.\end{aligned}} \tag{32.23}$$

The fluctuation field itself satisfies an equation which may be derived from the Navier–Stokes equation (32.20a) by subtracting the first mean field equation. Rewriting the advective term using (32.22) we obtain the fluctuation equations,

$$\frac{\partial \boldsymbol{u}}{\partial t} + (\boldsymbol{v} \cdot \nabla)\boldsymbol{u} + (\boldsymbol{u} \cdot \nabla)\boldsymbol{v} + \nabla \cdot (\boldsymbol{uu} - \langle \boldsymbol{uu} \rangle) = -\frac{1}{\rho_0}\nabla q + \nu \nabla^2 \boldsymbol{u}, \tag{32.24a}$$

$$\nabla \cdot \boldsymbol{u} = 0. \tag{32.24b}$$

The split of the fields into means and fluctuations has thus led to two sets of coupled differential equations rather than one. They are of course together equivalent to the original Navier–Stokes equations so nothing has really been gained, except increased complexity!

The closure problem

The mean field equations (32.23) would by themselves form a closed system of dynamic equations, if only we could somehow express the second-order fluctuation moment $\langle u_i u_j \rangle$ in terms of the mean field and its derivatives. Unfortunately, as we shall now see, that is in general not possible.

One might, for example, attempt to determine the second-order moment $\langle u_i u_j \rangle$ from first principles by using the above fluctuation field equations (32.24) to calculate $\partial(u_i u_j)/\partial t$, and afterwards take the mean. But such a procedure would introduce an unknown third-order moment of the form $\langle u_i u_j u_k \rangle$ through the nonlinear fluctuation term $\nabla \cdot (\boldsymbol{uu})$ (see problem 32.3 for details). Again one could use (32.24) to get an equation for the third-order moment, only to find that it involves a fourth-order moment, and so on. Even if in this way an infinite set of equations can in principle be written down for the moments, each equation will refer to a moment of higher order than the one it is meant to determine. The set of moment equations does not close.

The closure problem is inherent in the statistical treatment of nonlinear systems of equations, and a large part of the theoretical studies of turbulence have focused on modelling the moments of the fluctuations to obtain closure. Sometimes symmetries and dimensional considerations permit general conclusions to be drawn in simple flow geometries where the symmetry is high and the number of parameters small, but a fundamental theory of turbulence allowing for effective closure in any geometry is yet to be discovered.

Reynolds stresses

The first Reynolds equation takes the same form as the Navier–Stokes equation with an effective stress tensor,

$$\sigma^*_{ij} = -p\,\delta_{ij} + \eta(\nabla_i v_j + \nabla_j v_i) - \rho_0 \langle u_i u_j \rangle. \tag{32.25}$$

The last term was introduced by Reynolds in 1894 and is appropriately called the *Reynolds stress tensor*.

Apart from the fact that we cannot in general calculate the Reynolds stresses, they appear in the Reynolds equation for the mean velocity field on equal footing with the mean pressure and the mean viscous stresses. Since the Reynolds equation in analogy with the Navier–Stokes equation may be understood as expressing local momentum balance, the Reynolds stresses also act as forces by transferring momentum across internal surfaces cutting through the fluid. The only difference is that the Reynolds stresses must vanish at solid boundaries due to the no-slip condition for the fluctuations $\boldsymbol{u} = \boldsymbol{0}$, and that prevents us from measuring them directly. There are, however, many ways of measuring the velocity fluctuations over a long time interval, and from such data the mean values $\langle u_i u_j \rangle$, as well as the mean value of any other product, can be obtained.

Velocity defect

It has been mentioned before that in fully developed turbulent flow, the mean velocity does not vary much across the geometry, except very close to the walls where the no-slip condition has to be fulfilled. If $\boldsymbol{U}$ is a typical mean velocity in the bulk of the turbulent flow, the quantity

$$\Delta \boldsymbol{v} = \boldsymbol{v} - \boldsymbol{U}, \tag{32.26}$$

is called the *velocity defect*. In the bulk of the turbulent flow it is expected to be small, $\Delta v \ll U$, while it will be large near a solid wall where the no-slip condition requires $\boldsymbol{v} = \boldsymbol{0}$. The magnitude of the velocity defect therefore determines the magnitude of the velocity gradients, $|\nabla \boldsymbol{v}| \sim \Delta v/L$, in the bulk of the turbulent flow.

From the Kolmogorov scaling analysis in the preceding section we know that average eddy velocity must be of the same order of magnitude as the maximal eddy velocity u_L such that $|\langle \boldsymbol{uu}\rangle| \sim u_L^2$. The effective turbulent viscosity ν_{turb} can now be estimated from the ratio of the Reynolds stresses to the viscous stresses,

$$\nu_{\text{turb}} \sim \frac{|\langle \boldsymbol{uu}\rangle|}{|\nabla \boldsymbol{v}|} \sim \frac{u_L^2}{\Delta v/L} = \frac{u_L^2 L}{\Delta v}. \tag{32.27}$$

Since the eddy viscosity (32.13) grows rapidly with the eddy scale, the largest eddies will dominate the turbulent viscosity. Setting $\nu_{\text{turb}} \sim \nu_L = L u_L$ in the above estimate we conclude that $\Delta v \sim u_L$.

These arguments support that *the velocity defect in the bulk of the turbulent flow is of roughly the same magnitude as the turbulent velocity fluctuations*. In the discussion of simple turbulent flows in sections 32.7 and 32.8 we shall verify this claim.

Effective Reynolds number for turbulence

The condition for turbulent viscosity to dominate over the molecular viscosity is that the Reynolds number of the system-scale eddies is large, $\text{Re}_L = u_L L/\nu \sim \nu_{\text{turb}}/\nu \gg 1$. There is in fact no turbulence unless this condition is fulfilled.

We are now also in a position to estimate the effective turbulent Reynolds number, defined as the ratio between the advective mean field contribution and the dominant Reynolds stress contribution,

$$\text{Re}_{\text{turb}} = \frac{|(\boldsymbol{v}\cdot\nabla)\boldsymbol{v}|}{|\nabla\cdot\langle \boldsymbol{uu}\rangle|} \sim \frac{U\Delta v/L}{u_L^2/L} \sim \frac{U}{u_L} = \frac{\text{Re}}{\text{Re}_L}, \tag{32.28}$$

where as before $\text{Re} = UL/\nu$. In the following we shall see that the right-hand side is not usually very large, in realistic systems somewhere between 10 and 40. This is comforting, since it would be completely meaningless if Re_{turb} were to become so large that the mean field equations became unstable and themselves developed turbulence!

32.4 Turbulence near a smooth solid wall

Turbulent flow along a solid wall appears at first sight to be rather uniform and reach all the way to the wall, but that cannot be completely true. Viscosity demands that the mean velocity field as well as the fluctuation field must vanish exactly on the wall. The conflict between viscosity's insistence on a no-slip condition and turbulence's tendency to even out mean velocity differences leads to the formation of a two-decked boundary layer structure. Thus, in any turbulent boundary flow there will be an *inner layer* which secures the fulfillment of the no-slip condition, and an *outer layer* that implements the interface to the mainstream flow. These main layers may be further subdivided into sublayers.

The inner and outer layers must *overlap* in a finite region, because in a viscous flow there cannot be a sharp break anywhere in the true velocity field or in any of its derivatives. It is the requirement of a finite overlap region that connects the inner and outer layers, and thereby the main flow with the stresses that the fluid exerts on the wall. It must be emphasized that due to the finite overlap, there will be no clearly discernable break between the inner and outer layers. There is, as we shall see, instead a fairly well-defined transition between the two sublayers of the inner layer.

In the following analysis we shall make the same basic assumptions about the boundary flow as in the laminar case (chapter 28). The total thickness δ of the boundary layer will always be assumed to be much smaller than the mainstream geometry scale $\delta \ll L$. We may then take the wall to be locally flat, coinciding with the xz-plane, $y = 0$, in a Cartesian coordinate system. We shall also choose the x-axis along the mainstream mean flow direction such that $v_z = 0$ throughout the boundary layer. The non-vanishing velocity components v_x and v_y depend mainly on y, but may change slowly with x (and z) over the mainstream geometry scale L.

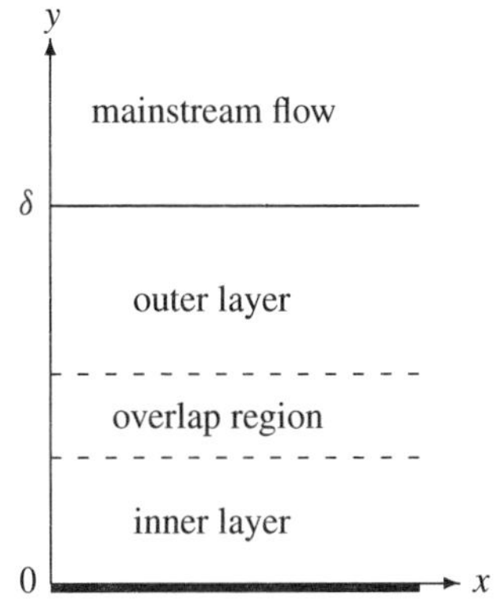

Sketch of the turbulent boundary layer structure near a solid wall. The inner layer interfaces to the wall whereas the outer layer interfaces to the mainstream flow. The inner and outer layers overlap in a finite region between the dashed lines. There is no clear break between the inner and outer layers.

32.5 Inner layer structure

The fundamental dynamic quantity that controls the flow very near the wall is the mean wall shear stress,

$$\sigma_{\text{wall}} = \langle\sigma_{xy}\rangle_{y=0} = \eta \, \nabla_y v_x \big|_{y=0} \equiv \rho_0 u_0^2. \tag{32.29}$$

Here we have for convenience introduced a parameter u_0, called the *friction velocity*, and it is also convenient to define a third length scale,

$$\delta_0 = \frac{\nu}{u_0}, \tag{32.30}$$

called the *viscous length scale*. At this point we do not know what u_0 and δ_0 mean, except that given ρ_0 and ν they contain exactly the same information as the wall stress. Like the wall stress they may also change slowly with x over distances on the mainstream scale L. We shall soon see that u_0 sets the scale for the velocity and δ_0 the scale for the thickness of an extremely thin viscous and nearly laminar inner sublayer that is always present closest to the wall.

The ratio of inner and outer length scales is an important dimensionless number, called the *wall Reynolds number*[3],

$$\text{Re}_0 = \frac{\delta}{\delta_0} = \frac{u_0\delta}{\nu}. \tag{32.31}$$

It is quite different from the conventional Reynolds number of a boundary layer (see page 411),

$$\text{Re}_\delta = \frac{U_\delta \delta}{\nu}, \tag{32.32}$$

where U_δ is the scale of the streamwise velocity at the 'edge' of the outer boundary layer $y \sim \delta$. In turbulent flows the wall Reynolds number is always large, typically $\text{Re}_0 \gtrsim 100$, whereas the outer Reynolds number Re_δ typically is between 20 and 40 times larger, i.e. in the thousands.

Viscous linear sublayer

The finiteness of the mean velocity gradient at the wall,

$$\nabla_y v_x \big|_{y=0} = \frac{u_0^2}{\nu} = \frac{u_0}{\delta_0}, \tag{32.33}$$

ensures that very near the wall the streamwise velocity will grow linearly with y. In units of u_0 this *linear law of the wall* may be written,

$$\boxed{\frac{v_x}{u_0} = \frac{y}{\delta_0}.} \tag{32.34}$$

Empirically, the linear law of the wall is found to be quite accurate for $0 \le y/\delta_0 \lesssim 5$, as may be seen in figure 32.2. The average velocity in this region is $2.5u_0$, and this demonstrates that the friction velocity u_0 sets the scale of the velocity in the viscous sublayer, and δ_0 the scale of its thickness.

Example 32.1.1 (Concrete water pipe, continued): In this example (page 500) the wall stress is $\sigma_{\text{wall}} = (1/2)\rho_0 U^2 C_f \approx 400$ Pa and thus $u_0 \approx 0.6$ m s^{-1} which is only about 5% of the mainstream flow velocity $U = 13$ m s^{-1}. The wall length scale becomes $\delta_0 = 1.6\ \mu$m, and taking $\delta = a = 0.5$ m, we get $\text{Re}_0 = 318\,000$ for this highly turbulent flow. The viscous length scale is much smaller than the 1 mm roughness scale assumed for concrete and this shows that the idea of a clean viscous sublayer cannot hold up in this case. Rough walls require a slight modification of the theory which will not be discussed here [61, p. 526].

[3] The wall Reynolds number may in the Kolmogorov picture be interpreted as the Reynolds number of the largest eddies at the edge of the boundary layer, and is often used to characterize experiments and numerical simulations. In the literature it is also frequently denoted Re_τ.

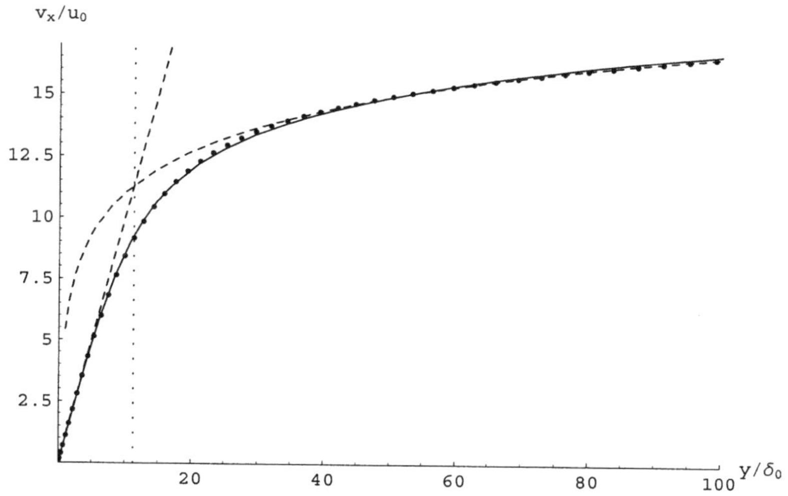

Figure 32.2. Plot of v_x/u_0 against y/δ_0 in the near-wall region $0 \le y/\delta_0 \lesssim 100$ at $\mathrm{Re}_0 = 590$. The data points are obtained from direct numerical simulation of channel flow by Moser, Kim and Mansour (see footnote 4 on page 509). The dashed straight line is the velocity in the linear sublayer (32.34), and the dashed curve is a fit to the velocity in the logarithmic sublayer (32.36) with $A = 2.42$ and $B = 5.42$. The two approximations cross at $y/\delta_0 \approx 11$, indicated by the vertical dotted line. The solid line is the interpolation (32.46) with $A = 2.42$, $B = 5.57$, $\alpha = 9.92 \times 10^{-4}$ and $\beta = 4.02 \times 10^{-4}$. One notes that the fit is not perfect. The imperfections can be seen more clearly in figure 32.4.

Turbulent logarithmic sublayer

Deeply inside the boundary layer but well beyond the linear region, i.e. for $\delta_0 \ll y \ll \delta$, the flow will be turbulent but because of the proximity of the wall, the wall friction velocity u_0 still sets the velocity scale. In keeping with the discussion in the preceding sections we expect that the mean velocity slope $\nabla_y v_x$ in this region is quite small and essentially independent of the viscosity. It can then only depend on u_0 and y, and must for dimensional reasons be of the form,

$$\nabla_y v_x = A \frac{u_0}{y}, \tag{32.35}$$

where A is a dimensionless constant. Integrating this expression we obtain the *logarithmic law of the wall,*

$$\boxed{\frac{v_x}{u_0} = A \log \frac{y}{\delta_0} + B,} \tag{32.36}$$

where B is another dimensionless constant, and where we have normalized the logarithm by the viscous length scale $\delta_0 = \nu/u_0$ to make its argument dimensionless. The logarithmic law of the wall goes back to von Karman and Prandtl in the early 1930s, and the inverse $\kappa = 1/A$ is called the *Karman constant.*

The received wisdom is that the logarithmic law of the wall is reasonably accurate for $y/\delta_0 \gtrsim 30$ with $A \approx 2.4$ and $B \approx 5$ in all wall-bounded flows. In figure 32.2 the logarithmic law is compared with numerical channel flow simulation data[4] at $\mathrm{Re}_0 = 590$, and a reasonable fit is obtained in the region $30 \lesssim y/\delta_0 \lesssim 100$ with $A = 2.42$ and $B = 5.42$. In recent years this wisdom has, however, been challenged. Highly precise compressed air pipe flow experiments[5] indicate that the strict logarithmic law first sets in for $y/\delta_0 \gtrsim 500$, and that the constants are $A = 2.29$ and $B = 6.13$. In the intermediate region

[4] R. D. Moser, J. Kim and N. N. Mansour, Direct numerical simulation of turbulent channel flow up to $\mathrm{Re}_\tau = 590$, *Phys. Fluids* **11**, (1999) 943. The authors are thanked for making their data files publicly available on the internet.

[5] M. V. Zagarola and A. J. Smits, Scaling of the mean velocity profile for turbulent pipe flow, *Phys. Rev. Lett.* **78**, (1997) 239. The authors are thanked for making their data from the Princeton SuperPipe experimental facility publicly available on the internet.

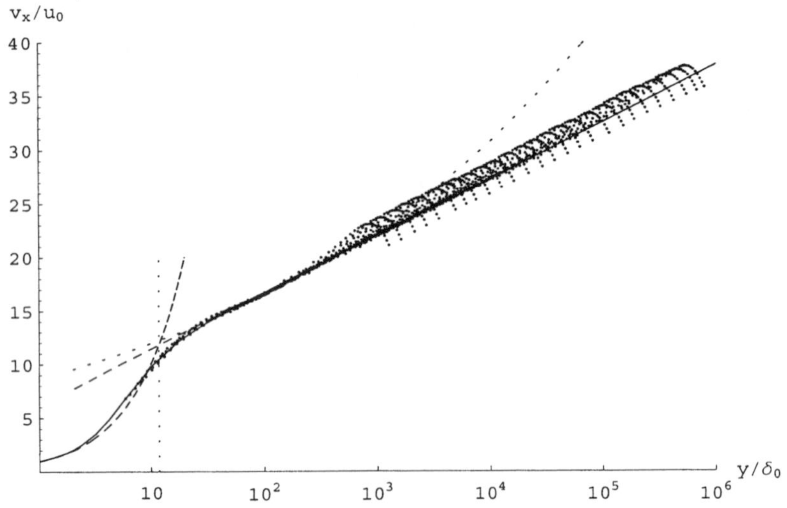

Figure 32.3. Semilogarithmic plot of v_x/u_0 against y/δ_0 at high Reynolds number. The data points are obtained from the experiments by Zagarola and Smits on the Princeton SuperPipe facility at 26 different wall Reynolds numbers ranging from $\mathrm{Re}_0 \approx 851$ to $\mathrm{Re}_0 \approx 529\,000$ (see footnote 5 on page 509). The 'bumps' represent the wake function for pipe flow (see figure 32.7 and section 32.8). The solid curve is the interpolation (32.47) with $A = 2.29$, $B = 6.13$, $\alpha = -0.0189$, $\beta = 0.0214$. The dashed curve is the velocity in the linear sublayer (32.34), and the dashed straight line the velocity in the logarithmic sublayer (32.36) with $A = 2.29$ and $B = 6.13$. The dotted curve is the power law approximation (32.37). Note that the data points (and the fitted interpolation) overshoot the linear rise in the outer part of the viscous sublayer.

$30 \lesssim y/\delta_0 \lesssim 500$ it is found that the velocity profile may be approximated by a power law,

$$\frac{v_x}{u_0} = 8.70 \left(\frac{y}{\delta_0}\right)^{0.137} . \tag{32.37}$$

Alternatively this region may be modelled with slowly sliding values of A and B. Older experiments typically took place at such low wall Reynolds numbers that the truly logarithmic region was never reached, and this explains perhaps the discrepancies between the values of A and B found in different experiments. The SuperPipe data and the various approximations to it are plotted in figure 32.3.

Complete inner law of the wall

In the whole inner layer $0 \le y \ll \delta$ the flow should not depend on δ, so that we may express the streamwise mean velocity v_x in terms of the friction velocity u_0 and the dimensionless parameter y/δ_0,

$$\boxed{\frac{v_x}{u_0} = f\left(\frac{y}{\delta_0}\right)}, \tag{32.38}$$

where $f(s)$ is a purely numeric function of its variable $s = y/\delta_0$ in the interval $0 \le s \ll \mathrm{Re}_0$. Since $\mathrm{Re}_0 > 100$ in all turbulent flows, and mostly much larger, this should leave sufficient room for the logarithmic sublayer to be realized in the interval $30 \lesssim s \ll \mathrm{Re}_0 = \delta/\delta_0$.

From the preceding analysis we already know the asymptotic behaviour of $f(s)$ for small and not too large s,

$$f(s) = \begin{cases} s & (0 \le s \lesssim 5) \\ A \log s + B & (30 \lesssim s \ll \mathrm{Re}_0) \end{cases} . \tag{32.39}$$

As discussed above, A and B may change slightly in the region $30 \lesssim s \lesssim 500$, but for $A \approx 2.5$ and $B \approx 5$ these expressions cross each other at $s \approx 11$. Although we do not know the exact shape of the function in the crossover region, $5 \lesssim s \lesssim 30$, this anchoring in the extremes does not leave much liberty for variation in between. Below we shall present the simplest possible interpolation which connects the two extreme behaviours.

The mean field in the y-direction is determined by the divergence condition $\nabla_x v_x + \nabla_y v_y = 0$, and may in the inner wall layer easily be verified to be,

$$v_y = -y\, f\left(\frac{y}{\delta_0}\right) \nabla_x u_0. \tag{32.40}$$

The ratio of the two velocities is consequently

$$\frac{v_y}{v_x} = -\frac{y \nabla_x u_0}{u_0}. \tag{32.41}$$

Since $\nabla_x u_0 \sim u_0/L$, the magnitude of the ratio is $v_y/v_x \sim y/L$ which is negligible in the whole inner wall region $0 \leq y \ll \delta$.

Reynolds shear stress in the inner layer

In the preceding analysis we have only used dimensional arguments and mass conservation, and completely ignored the Reynolds equations (32.23) which after all control the dynamics. Since all the x-dependence in the inner wall layer stems from the wall friction velocity u_0, any x-derivative must be of magnitude $\nabla_x \sim 1/L$. Similarly, all y-dependence is governed by δ_0 such that $\nabla_y \sim 1/\delta_0$. In the limit of $L \to \infty$ we may drop all x-derivatives (and therefore also v_y) in the first Reynolds equation (32.23), and get

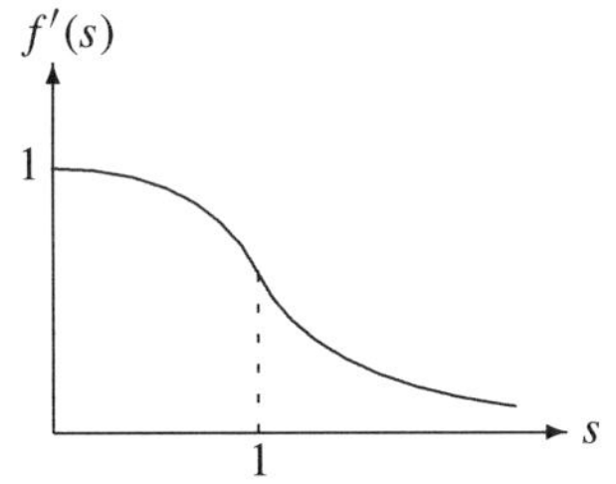

Sketch of the slope $f'(s)$.

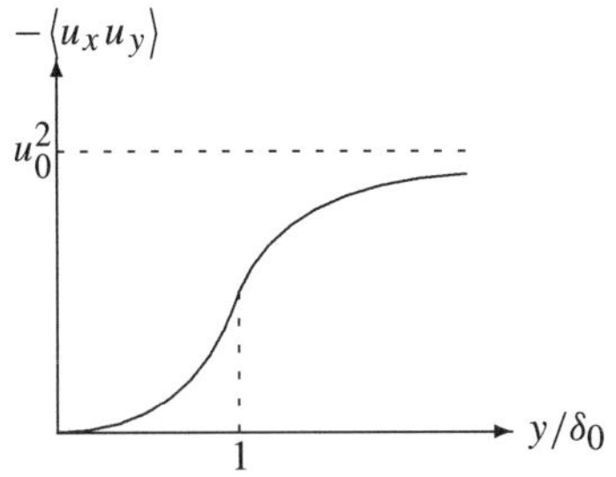

Sketch of $-\langle u_x u_y \rangle$ as a function of y/δ_0.

$$0 = \nu \nabla_y^2 v_x - \nabla_y \langle u_x u_y \rangle, \tag{32.42a}$$

$$\nabla_y p = -\rho_0 \nabla_y \left\langle u_y^2 \right\rangle. \tag{32.42b}$$

Integrating these equations over y and using the boundary condition given by the wall shear stress (32.29), they become

$$u_0^2 = \nu \nabla_y v_x - \langle u_x u_y \rangle, \tag{32.43a}$$

$$p = p_0 - \rho_0 \left\langle u_y^2 \right\rangle, \tag{32.43b}$$

where p_0 is the pressure on the wall. The last equation shows that in distinction to the laminar case, turbulence causes a (small) drop in the pressure as we move away from the wall.

Finally, inserting the mean field (32.38) into (32.43a), we get,

$$\boxed{\langle u_x u_y \rangle = -u_0^2 \left(1 - f'\left(\frac{y}{\delta_0}\right)\right).} \tag{32.44}$$

The slope $f'(s)$ decreases from $f' = 1$ at $s = 0$ to $f' = 0$ for $s = \infty$, so for $\delta_0 \ll y \ll \delta$ the fluctuation moment is constant $\langle u_x u_y \rangle \approx -u_0^2$. Due to the negative sign, the fluctuations u_x and u_y are always statistically anti-correlated, meaning that if $u_x > 0$ then $u_y < 0$, and conversely.

How fast does $\langle u_x u_y \rangle$ vanish in the viscous sublayer? For physical reasons all velocity derivatives must be finite at the wall, so that all components of the fluctuation field (u_x, u_y, u_z) must vanish at least as fast as y. Combined with the divergence condition $\nabla \cdot \boldsymbol{u} = \nabla_x u_x + \nabla_y u_y + \nabla_z u_z = 0$, it follows that $\nabla_y u_y \sim y$ and therefore $u_y \sim y^2$, whereas $u_x \sim y$ and $u_z \sim y$. From this we get $\langle u_x u_y \rangle \sim y^3$, and putting in the dimensional factors this may be written,

$$\langle u_x u_y \rangle \approx -\alpha\, u_0^2 \left(\frac{y}{\delta_0}\right)^3, \tag{32.45}$$

where α is a numerical constant. In terms of the dimensionless shape function $f(s)$ this translates into $f'(s) \approx 1 - \alpha s^3$ and $f(s) = s - (1/4)\alpha s^4$. Usually, it is assumed that α is positive, but the overshooting of the pipe flow data in figure 32.3 relative to the linear rise in the viscous sublayer indicates that α might in fact be negative (if the experimental values can be trusted this close to the wall).

Interpolation between linear and logarithmic sublayers

There are many ways of generating an interpolating function $f(s)$ connecting the extremes $s \ll 1$ and $s \gg 1$. A 'minimal' interpolation for the slope which takes into account that both $f' = 1 - \alpha s^3$ for $s \ll 1$ and $f' = A/s$ for $s \gg 1$ is, for example,

$$\boxed{f'(s) = \frac{1}{1 + (\alpha + \beta)\, s^3} + \frac{A\beta\, s^3}{A + \beta\, s^4},} \tag{32.46}$$

where β is another numeric constant. One may verify that its integral is,

$$\begin{aligned} f(s) &= \frac{1}{4} A \log\left(1 + \frac{\beta}{A} s^4\right) \\ &+ \frac{1}{18\gamma}\left(\pi\sqrt{3} - 6\sqrt{3}\arctan\frac{1 - 2\gamma s}{\sqrt{3}} + 3\log\frac{(1+\gamma s)^2}{1 - \gamma s + \gamma^2 s^2}\right), \end{aligned} \tag{32.47}$$

where $\gamma = (\alpha + \beta)^{1/3}$. For $s \to \infty$ this approaches the logarithmic law of the wall (32.36) with

$$B = \frac{2\pi}{3\sqrt{3}}(\alpha + \beta)^{-1/3} - \frac{1}{4} A \log\frac{A}{\beta}. \tag{32.48}$$

Any choice of constants A, B, α and β must satisfy this relation.

Older data indicated that the constants were $A = 2.44$, $B = 5.00$, $\alpha = 0.61 \times 10^{-3}$ and $\beta = 1.43 \times 10^{-3}$ [61, p. 523]. In figure 32.2 the precise numerical simulations yield a good fit for $A = 2.42$, $B = 5.57$, $\alpha = 9.92 \times 10^{-4}$ and $\beta = 4.02 \times 10^{-4}$. The recent high precision experiments on pipe flow plotted in figure 32.3 lead instead to $A = 2.29$, $B = 6.13$, $\alpha = -0.0189$, and $\beta = 0.0214$. The interpolation is in this case very precise for all s and as good as the power law approximation (32.37) in the region $30 \lesssim s \lesssim 500$.

Turbulent viscosity

The local turbulent shear viscosity $\nu_{\rm turb}$ is defined by writing,

$$-\left\langle u_x u_y \right\rangle = \nu_{\rm turb} \nabla_y v_x. \tag{32.49}$$

Inserting this into (32.43a) and making use of (32.44) we find

$$\frac{\nu_{\rm turb}}{\nu} = \frac{1}{f'(y/\delta_0)} - 1 \approx \begin{cases} \alpha\left(\dfrac{y}{\delta_0}\right)^3 & (y \ll \delta_0) \\[2ex] \dfrac{1}{A}\dfrac{y}{\delta_0} & (y \gg \delta_0) \end{cases}. \tag{32.50}$$

In accordance with intuition, the turbulent viscosity in the inner layer thus vanishes rapidly for $y \to 0$ and diverges linearly for $y \to \infty$.

32.6 Outer layer structure

Eventually, at a scale set by the boundary layer thickness δ, the logarithmic growth of the mean velocity field v_x will come under the influence of the mainstream flow, causing deviations from the universal inner law of the wall (32.38). It is for obvious reasons not possible to give a general description of all mainstream flows, but as long as the boundary layer is thin compared to the mainstream geometry, $\delta \ll L$, it is possible to analyse the flow in the outer layer.

Wake function

It was previously (page 506) pointed out that in a fully developed turbulent flow the mean velocity only varies slowly across the geometry, and that the velocity defect is of the same order of magnitude as the turbulent fluctuations. In the outer layer $\delta_0 \ll y \lesssim \delta$ we shall interpret this to mean that the difference $v_x - u_0 f(y/\delta_0)$ between the velocity and the universal inner law (32.38) is itself of magnitude u_0. Thus, we may in the whole boundary layer $0 \le y \lesssim \delta$ write,

$$\boxed{\frac{v_x}{u_0} = f\left(\frac{y}{\delta_0}\right) + w\left(\frac{y}{\delta}\right),} \tag{32.51}$$

where the so-called *wake function* $w(s)$ is a dimensionless function of the dimensionless variable $s = y/\delta$. Since the first term ensures that the logarithmic law $f = A\log(y/\delta_0) + B$ is fulfilled for $y \gg \delta_0$, and since there must be a finite overlap between the inner and outer layers, the wake function has to vanish for $s \to 0$.

The wake function is assumed to be independent of the viscosity, but may depend on the streamwise coordinate x, either explicitly or implicitly through the x-dependence of the thickness $\delta = \delta(x)$ (we suppress this dependence whenever it is safe to do so).

Friction law

Suppose that we somehow know the velocity $v_x = U_\delta$ for $y = \delta$. Taking $y = \delta$ in (32.51) and using $\text{Re}_0 = \delta/\delta_0 = u_0\delta/\nu \gg 1$ we obtain the so-called *friction law* (originally due to Prandtl),

$$\boxed{\frac{U_\delta}{u_0} = f(\text{Re}_0) + C \approx A\log \text{Re}_0 + B + C,} \tag{32.52}$$

where $C = w(1)$ is a dimensionless 'constant' that in general depends on the flow geometry. Turbulent flows typically have $\text{Re}_0 \gtrsim 100$ such that $U_\delta/u_0 \gtrsim 17$ for $A = 2.5$, $B = 5.0$ and $C = 0$. The ratio U_δ/u_0 grows very slowly with Re_0 and normally takes values between 20 and 40.

Velocity defect

Subtracting the friction law (32.52) from the velocity (32.51) we obtain an exact expression for the velocity defect in units of the friction velocity,

$$\frac{v_x - U_\delta}{u_0} = f\left(\frac{y}{\delta_0}\right) - f(\text{Re}_0) - C + w\left(\frac{y}{\delta}\right). \tag{32.53}$$

In the outer layer where $\delta_0 \ll y \lesssim \delta$, we may replace f by the logarithmic law of the wall and get the simple result, called the *velocity defect function*,

$$\boxed{\frac{v_x - U_\delta}{u_0} \approx A\log\frac{y}{\delta} - C + w\left(\frac{y}{\delta}\right).} \tag{32.54}$$

It should be noted that this expression only depends on y/δ and vanishes for $y = \delta$ because $C = w(1)$. The constant B has fallen out here and thus has no influence on the velocity defect in the whole outer layer. Different flow geometries only differ in their outer layers, and unless high precision data exist for the viscous inner layers, it is the velocity defect that is usually compared with experiments.

32.7 Planar turbulent flows

Symmetry and dimensional arguments play an especially important role in the analysis of stationary turbulence in simple flow geometries where the symmetry is high and the number of parameters small. In this section we shall apply the theory of turbulent wall flows to model the turbulent flow profile in planar pressure-driven (channel Poiseuille) flow and velocity-driven (channel Couette) flow. The laminar solutions for these geometries have previously been obtained in sections 18.2 and 17.2.

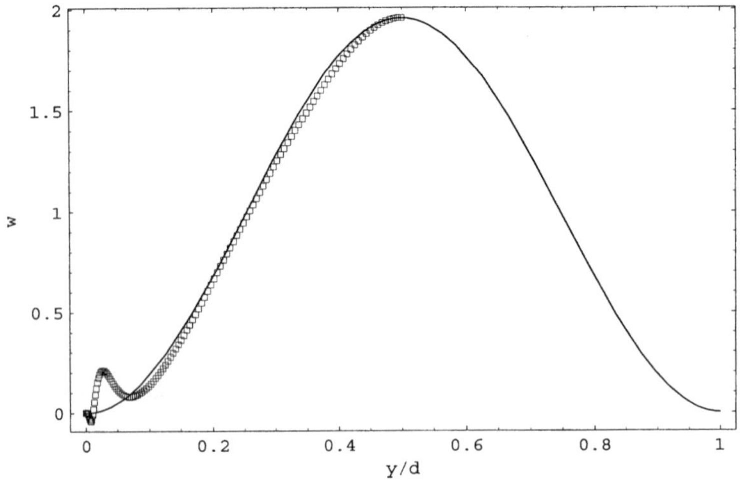

Figure 32.4. The wake function for planar pressure-driven (Poiseuille) flow as a function of y/d for $\text{Re}_0 = 587.19$. The data points are obtained from the same source as in figure 32.2, using (32.56) with parameters determined by the fit in figure 32.2, i.e. $A = 2.42$, $B = 5.57$, $\alpha = 9.92 \times 10^{-4}$ and $\beta = 4.02 \times 10^{-4}$. The solid curve is $w = C \sin^2(\pi y/d)$ with $C = 1.95$. The wiggle on the left corresponds to the nearly invisible imperfections in the fits of figures 32.2 and 32.5.

Both planar geometries are infinitely extended along the flow direction and orthogonally to it, so that formally we may take $L = \infty$. The infinite longitudinal extent allows us to assume that the outer boundary layers have expanded to completely fill the interior of the geometry with a *turbulent core*, smoothly connecting the outer layer of one wall with that of the opposite wall.

Pressure-driven planar flow

In the planar Poiseuille or channel flow set up both plates are fixed at distance d, and the fluid is driven between them by a constant pressure gradient. In this case we take $\delta = d/2$, so that the defect velocity scale equals the maximal flow velocity in the midplane between the plates, $U_\delta = U_{\max}$. We shall soon relate this velocity to the average velocity between the plates. The two plates are completely equivalent and must satisfy,

$$v_x(d - y) = v_x(y), \tag{32.55}$$

because the channel flow set up is symmetric around the midplane.

A representation of the same nature as (32.51) which respects this symmetry and has the correct near-wall behaviour at both plates is

$$\boxed{\frac{v_x}{u_0} = f\left(\frac{y(d-y)}{\delta_0 d}\right) + w\left(\frac{y}{d}\right).} \tag{32.56}$$

The wake function is symmetric, $w(1 - y/d) = w(y/d)$, and must vanish on the plates, $w(0) = w(1) = 0$. A convenient choice that fits with data is, for example, $w = C \sin^2(\pi y/d)$.

Taking $y = \delta = d/2$ and using the logarithmic law of the wall (32.36), we find the friction law,

$$\frac{U_{\max}}{u_0} = f\left(\frac{\text{Re}_0}{2}\right) + w\left(\frac{1}{2}\right) \approx A \log \frac{\text{Re}_0}{2} + B + C, \tag{32.57}$$

where $\text{Re}_0 = \delta/\delta_0 = d/2\delta_0$ and $C = w(1/2)$. Subtracting this from (32.56) we obtain the defect function in the turbulent core ($y \gg \delta_0$ and $d - y \gg \delta_0$),

$$\frac{v_x - U_{\max}}{u_0} \approx A \log \frac{4y(d-y)}{d^2} - C + w\left(\frac{y}{d}\right). \tag{32.58}$$

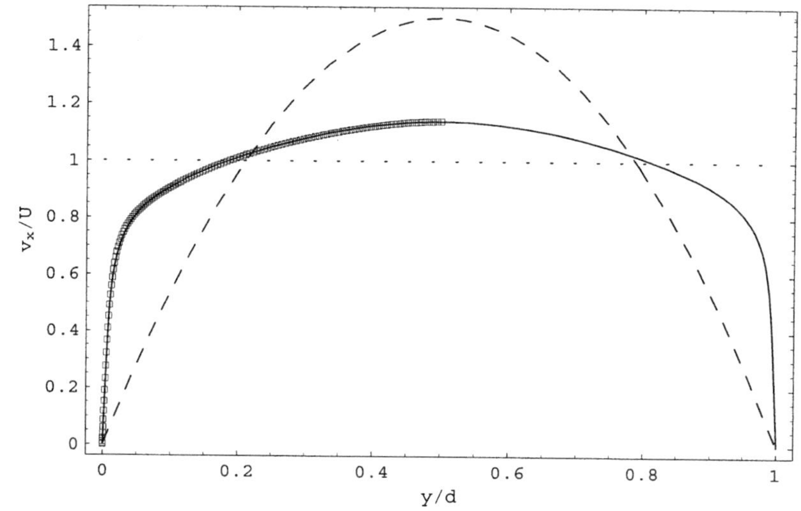

Figure 32.5. The mean velocity field for planar pressure-driven (Poiseuille) flow in units of the average velocity as a function of the y-coordinate for $\text{Re}_0 = 587.19$. The data points are selected from high precision numerical simulation results (see footnote 4 on page 509). The solid curve is obtained from the procedure described in the text, and the resulting fit is excellent. The dashed curve is the laminar solution $v_x/U = 6y(1-y)$ with the same average velocity.

It should be noted that this function is symmetric and only depends on the variable y/d, and that the constant B has fallen out.

In figure 32.4 the wake function is calculated from (32.56) and the high precision simulation data at $\text{Re}_0 = 587.19$ (footnote 4 on page 509) with the parameters determined in the fit shown in figure 32.2 (i.e. $A = 2.42$, $B = 5.57$, $\alpha = 9.92 \times 10^{-4}$ and $\beta = 4.02 \times 10^{-4}$). The result compares well with the expression $w = C\sin^2(\pi y/d)$ for $C = 1.95$. From the friction law it follows that $U_{\text{max}}/u_0 = 21.26$, and in view of the small value of $u_0/U_{\text{max}} \approx 0.05$ the wake function has only about 10% influence on the velocity profile in the centre. Small deviations such as the small wiggle close to the wall in figure 32.4 become essentially invisible in the velocity profile plotted in figures 32.2 and 32.5. The wake function exposes in by far the best way the inadequacy of the wall interpolation function (32.46).

In planar channel flow the controlling parameter is customarily chosen to be the average velocity (rather than the maximal),

$$U = \frac{1}{d}\int_0^d v_x(y)\,dy. \tag{32.59}$$

Inserting the velocity from (32.56) we find

$$\frac{U}{u_0} = \int_0^1 \big(f(2\text{Re}_0 s(1-s)) + w(s)\big)\,ds, \tag{32.60}$$

and dividing this into (32.56) we find the curve for v_x/U shown in figure 32.5. A precise value of this integral must be obtained numerically.

An approximate value can be found using the logarithmic law $f(s) = A\log s + B$, for which the integral becomes

$$\boxed{\frac{U}{u_0} \approx A\log(2\text{Re}_0) - 2A + B + \frac{1}{2}C.} \tag{32.61}$$

Inserting the values $A = 2.42$, $B = 5.57$ and $C = 1.95$ we get for $\text{Re}_0 = 587.19$ the value $U/u_0 = 18.79$ whereas numeric integration yields $U/u_0 = 18.67$, differing by merely half a per cent.

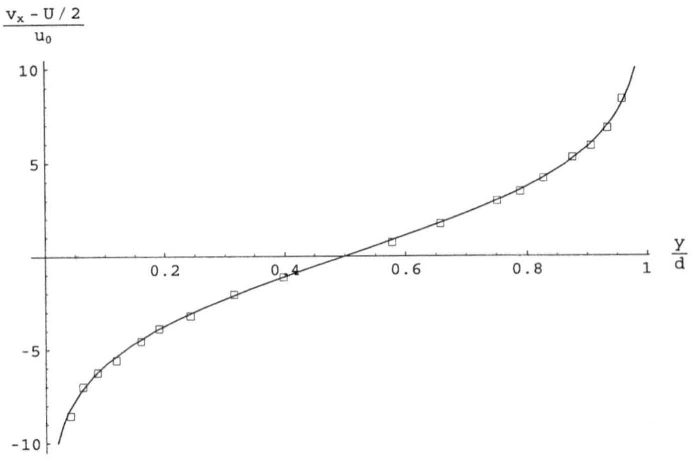

Figure 32.6. The velocity defect function in planar velocity-driven (Couette) flow as a function of y/d. With $w = 0$, $A = 2.44$ and $C = 0.56$, there is good agreement between (32.66) and older data at $\mathrm{Re}_0 = 733$ (extracted from [61, fig. 17.2]). The constant B cannot be determined because of the lack of near-wall data.

Velocity-driven planar flow

In the planar Couette flow set up one plate is fixed at $y = 0$ and the other moves along the x-axis with velocity U at a fixed distance $y = d$. The two plates are completely equivalent in their respective rest systems, implying that the mean velocity field must satisfy the symmetry relation,

$$v_x(d - y) = U - v_x(y). \tag{32.62}$$

The boundary layer thickness is for symmetry reasons chosen as $\delta = d/2$, making the defect velocity equal to the central velocity $U_\delta = U/2$.

A wake function representation which respects this symmetry and has the correct near-wall behaviour at both plates is quite complicated,

$$\boxed{\frac{v_x}{u_0} = \frac{U}{u_0}\frac{y}{d} + f\left(\frac{y}{\delta_0}\right) - f\left(\frac{d-y}{\delta_0}\right) + \left(1 - \frac{2y}{d}\right) f\left(\frac{d}{\delta_0}\right) + w\left(\frac{y}{d}\right).} \tag{32.63}$$

The first term is identical to the laminar velocity profile and the fourth secures that the no-slip condition is fulfilled at both plates. The wake function is antisymmetric, $w(1 - y/d) = -w(y/d)$, and must vanish in the centre and the limits, $w(0) = w(1/2) = w(1) = 0$. The simplest choice is $w(s) = 0$ for all s.

In this case we cannot obtain the friction law by setting $y = \delta = d/2$, because both sides of the equation automatically vanish at the centre. The friction law follows instead from the observation that the central velocity gradient $\nabla_y v_x$ should be independent of the viscous length δ_0, and that is only possible when the gradients of the first and fourth terms cancel, so that

$$\frac{U}{2u_0} = f(2\mathrm{Re}_0) + C \approx A\log(2\mathrm{Re}_0) + B + C, \tag{32.64}$$

where $\mathrm{Re}_0 = \delta/\delta_0 = d/2\delta_0$ and C is an arbitrary constant. The defect function now becomes,

$$\frac{v_x - \frac{1}{2}U}{u_0} = f\left(\frac{y}{\delta_0}\right) - f\left(\frac{d-y}{\delta_0}\right) - \left(1 - \frac{2y}{d}\right) C + w\left(\frac{y}{d}\right). \tag{32.65}$$

In the turbulent core ($y \gg \delta_0$ and $d - y \gg \delta_0$) this simplifies to,

$$\frac{v_x - U/2}{u_0} \approx A\log\frac{y}{d-y} - \left(1 - \frac{2y}{d}\right) C + w\left(\frac{y}{d}\right), \tag{32.66}$$

which is perfectly symmetric and depends only on y/d. The constant B has again fallen out.

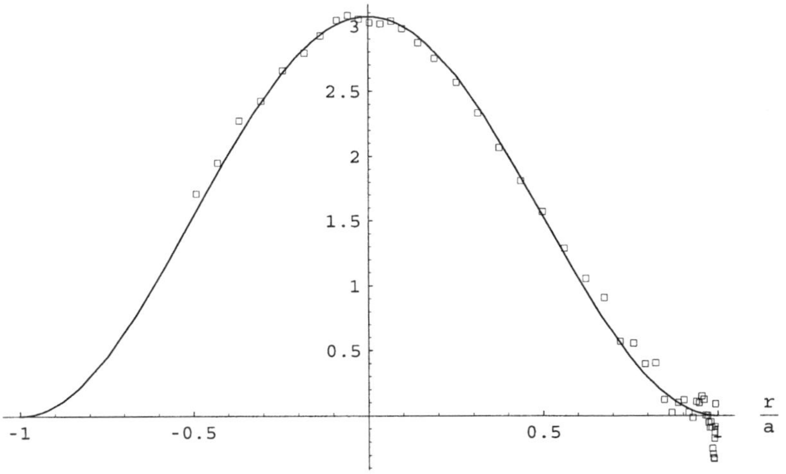

Figure 32.7. Pipe wake function for $\mathrm{Re}_0 = 851$ corresponding to $\mathrm{Re} = 31\,577$. The data is taken from Zagarola and Smits (see footnote 5 on page 509). The wake function data is calculated using (32.68) with the parameters determined in figure 32.3, i.e. $A = 2.29$, $B = 6.13$, $\alpha = -0.0189$, $\beta = 0.0214$. The solid profile is $w(s) = C\cos^2(\pi r/2a)$ with $C = 3.07$. The fit is excellent, except close to the wall at $r = a$.

In figure 32.6 a comparison has been made with older experimental data on the velocity defect at $\mathrm{Re}_0 = 733$ (extracted from [61, fig. 17.2]). With $w(s) = 0$ a good fit can be obtained for $A = 2.44$ and $C = 0.56$. Since the data does not cover the wall region, the constant B cannot be determined.

32.8 Turbulent pipe flow

In a pipe of radius a, the longitudinal extent of the flow is infinite, whereas the orthogonal extent $2\pi a$ is finite. Provided the viscous sublayer is much thinner than the radius $\delta_0 \ll a$, the flat-wall theory should also apply to this case. A pipe carrying turbulent fluid looks in the rz-plane very much like planar pressure-driven flow with $d = 2a$ when the coordinate y is replaced by $a - r$, the velocity $v_x(y)$ replaced by $v_z(r)$, and the range of the r-coordinate is extended to the full interval $-a \le r \le a$. As defect scales we choose the radius, $\delta = a$, and the central velocity $U_\delta = U_{\max}$. We shall impose reflection symmetry on the field,

$$v_z(-r) = v_z(r), \tag{32.67}$$

to make it continue smoothly across the centerline at $r = 0$.

Wake function representation

A representation that obeys this symmetry and has the right behaviour at the walls $r = \pm a$ is quite analogous to the channel flow expression (32.56),

$$\boxed{\frac{v_x}{u_0} = f\left(\frac{a^2 - r^2}{2a\delta_0}\right) + w\left(\frac{r}{a}\right)} \tag{32.68}$$

where the wake function must vanish at the wall, $w(\pm 1) = 0$. The friction law is obtained for $r = 0$,

$$\frac{U_{\max}}{u_0} = f\left(\frac{\mathrm{Re}_0}{2}\right) + w(1) \approx A \log\frac{\mathrm{Re}_0}{2} + B + C \tag{32.69}$$

with $C = w(0)$ and $\mathrm{Re}_0 = a/\delta_0$. Subtracting we obtain the defect function,

$$\frac{v_x - U_{\max}}{u_0} \approx A\log\left(1 - \frac{r^2}{a^2}\right) - C + w\left(\frac{r}{a}\right), \tag{32.70}$$

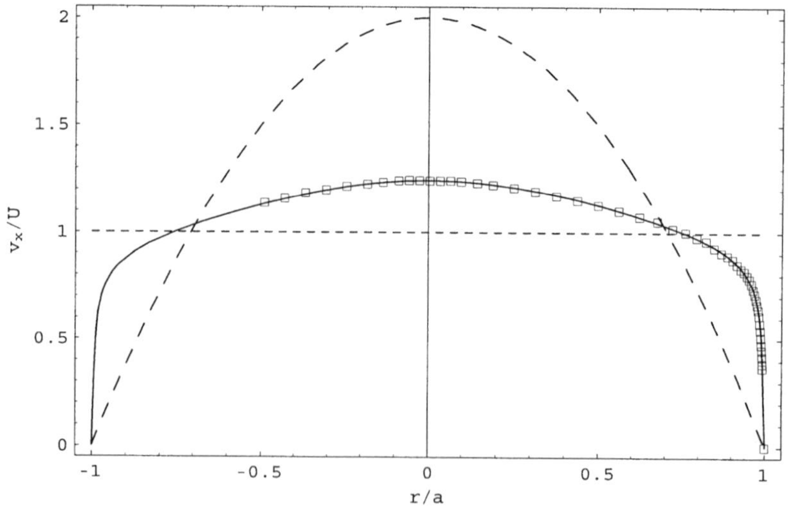

Figure 32.8. Pipe flow velocity profile for $\mathrm{Re}_0 = 851$ corresponding to $\mathrm{Re} = 31\,577$. The data is taken from Zagarola and Smits (see footnote 5 on page 509). The solid profile is calculated from (32.68) with parameters determined in figure 32.3 i.e. $A = 2.29$, $B = 6.13$, $\alpha = -0.0189$, $\beta = 0.0214$. The dashed curve is the laminar profile carrying the same volume discharge.

where the approximation is valid in the turbulent core for $a - r \gg \delta_0$ and $a + r \gg \delta_0$. Again we note that it is independent of the viscosity.

In figure 32.7 the wake function is determined from the Princeton SuperPipe data at $\mathrm{Re}_0 = 851$, corresponding to $\mathrm{Re} = 31\,577$ (see footnote 5 on page 509). The function $w = C\cos^2(\pi r/2a)$ with $C = 3.07$ fits very well. The value of C is slightly Reynolds number dependent.

Velocity profile

In pipe flow the control parameter is usually taken to be the easily measurable average velocity,

$$U = \frac{1}{\pi a^2}\int_0^a v_z(r) 2\pi r\, dr. \tag{32.71}$$

Inserting (32.68) we obtain,

$$\frac{U}{u_0} = \int_0^1 \left(f\left(\tfrac{1}{2}\mathrm{Re}_0(1-s^2)\right) + w(s)\right) 2s\, ds. \tag{32.72}$$

The integral is best done numerically, but a useful approximation is obtained by using the logarithmic law of the wall, $f(s) = A\log s + B$, and $w(s) = C\cos^2(\pi r/2a)$,

$$\frac{U}{u_0} \approx A\log\frac{\mathrm{Re}_0}{2} - A + B + \frac{\pi^2 - 4}{2\pi^2}C. \tag{32.73}$$

As for pressure-driven planar flow, this expression must be expected to differ from the true value, because the integral depends on the flow profile near the wall. In figure 32.8 the velocity profile is shown with the same parameters as in figure 32.7. The value obtained by numerical integration is $U/u_0 = 18.505$ whereas the approximation above yields $U/u_0 \approx 18.636$, with a difference of less than one per cent.

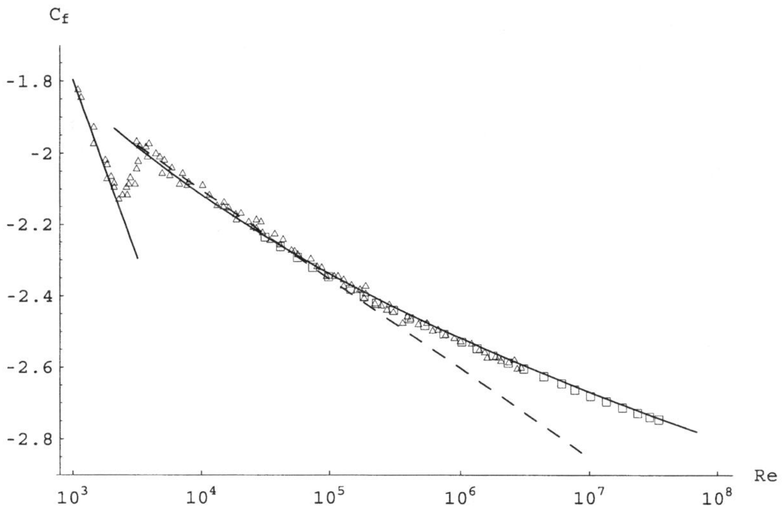

Figure 32.9. The friction coefficient for smooth-pipe flow as a function of the Reynolds number. The solid straight line is the laminar result $C_f = 16/\mathrm{Re}$, and the solid curve is the parametric solution obtained from (32.74a) with $A = 2.29$ and $E = 3.03$. The dashed line is the Blasius smooth-pipe expression $C_f = 0.079\mathrm{Re}^{-1/4}$ (page 257). The data points are extracted from [61, fig. 1.4] (triangles) and from the tables of Zagarola and Smits (footnote 5 on page 509).

Reynolds number and friction coefficient

The traditional pipe flow Reynolds number $\mathrm{Re} = 2aU/\nu = 2\mathrm{Re}_0 U/u_0$ and friction coefficient $C_f = \sigma_{\mathrm{wall}}/(1/2)\rho_0 U^2 = 2(u_0/U)^2$ may now be expressed in terms of Re_0,

$$\mathrm{Re} = 2\mathrm{Re}_0(A \log \mathrm{Re}_0 + E), \tag{32.74a}$$

$$C_f = \frac{2}{(A \log \mathrm{Re}_0 + E)^2} \tag{32.74b}$$

where the value $U/u_0 = 18.505$ corresponds to $E \approx 3.03$. Together these equations constitute a parametric representation of the friction coefficient C_f as a function of the Reynolds number Re.

In figure 32.9 the parameter values $A = 2.29$ and $E = 3.03$ determined above are used to calculate the friction coefficient. The agreement with data is best from $\mathrm{Re} = 4 \times 10^3$ to $\mathrm{Re} = 3 \times 10^7$, whereas the Blasius power law (18.50) only fits well up to $\mathrm{Re} = 10^5$. A very nice closed empirical approximation to the data is afforded by the expression,

$$C_f = \frac{0.416}{(\log \mathrm{Re} - 1.98)^2}, \tag{32.75}$$

which is virtually indistinguishable from the data in the interval $10^4 \lesssim \mathrm{Re} \lesssim 10^8$.

32.9 Turbulent boundary layer in uniform flow

Boundary layers always arise between a nearly ideal laminar mainstream flow and any solid wall that bounds it. At sufficiently high Reynolds number the flow in a boundary layer inevitably turns turbulent while the mainstream flow remains laminar and nearly ideal. The turbulent boundary layer thickness is defined by the transition in the mean velocity from turbulence back to laminar mainstream flow, although this transition is associated with strong intermittency which we ignore here (see figure 32.10). We shall also ignore the thin *viscous superlayer* in which the final transition to the true slip-flow takes place.

In section 28.4 we analysed the beautiful laminar Blasius solution, and in this section we shall extend the analysis to the turbulent case in the light of our new understanding of the near-wall flow structure.

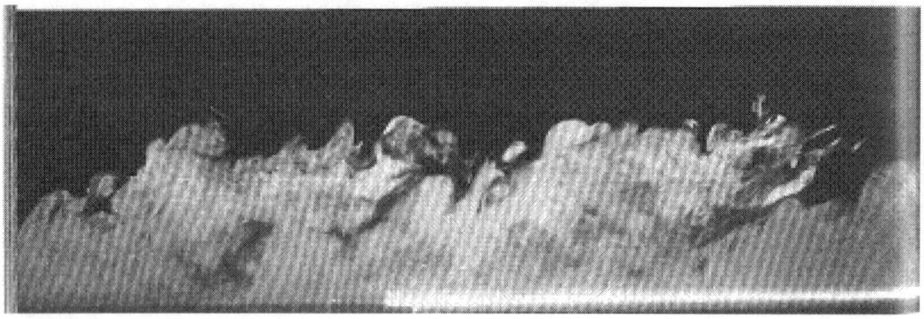

Figure 32.10. Turbulent boundary layer in a wind tunnel at $\mathrm{Re}_\delta = 4000$, showing the strong irregularity of the instantaneous interface to the laminar mainstream flow. Reproduced from R. E. Falco *Phys. Fluids* **20** (1977) 5124–32.

The analysis goes along the same lines as in the preceding section but turns out to be somewhat more complicated because of x-dependence.

The uniform flow set up

In the uniform (Blasius) set up a flow $\boldsymbol{v} = U\boldsymbol{e}_x$ impinges upon the edge of a semi-infinite plate at zero angle of incidence. The boundary layer is characterized by three different Reynolds numbers.

$$\mathrm{Re}_x = \frac{Ux}{\nu}, \qquad \mathrm{Re}_\delta = \frac{U\delta}{\nu}, \qquad \mathrm{Re}_0 = \frac{u_0\delta}{\nu}. \tag{32.76}$$

The first and largest is the traditional mainstream Reynolds number, the second the local Reynolds number, and the third and smallest the wall Reynolds number.

The friction law (32.52) may be written as a relation,

$$\boxed{\mathrm{Re}_\delta = \mathrm{Re}_0(A\log\mathrm{Re}_0 + B + C),} \tag{32.77}$$

between the local Reynolds number and the wall Reynolds number. Traditionally the local friction coefficient is defined as,

$$C_f = \frac{\sigma_{\text{wall}}}{(1/2)\rho_0 U^2} = 2\left(\frac{u_0}{U}\right)^2, \tag{32.78}$$

and inserting the friction law it also becomes a function of the wall Reynolds number,

$$\boxed{C_f = \frac{2}{(A\log\mathrm{Re}_0 + B + C)^2}.} \tag{32.79}$$

Together (32.77) and (32.79) constitute a parametric representation of the friction coefficient C_f as a function of the local Reynolds number Re_δ, basically of the same form as for pipe flow. This function depends only implicitly on x through the still unknown thickness $\delta = \delta(x)$. To connect Re_δ and C_f with the mainstream Reynolds number Re_x we need yet another relation.

Momentum balance

The theory of two-dimensional laminar boundary layers with thickness $\delta(x)$ and slip-flow $U(x)$ was developed in section 28.3. The mean flow in a turbulent boundary layer must likewise satisfy a set of simplified boundary layer equations, derived from the Reynolds equations (32.23) by leaving out all x-derivatives of the mean field and fluctuations in comparison with y-derivatives. Under these conditions we find that the turbulent version of Prandtl's momentum equation (28.21) on page 414 only differs from the laminar version by an extra term $-\nabla_y\langle u_x u_y\rangle$ on the right-hand side.

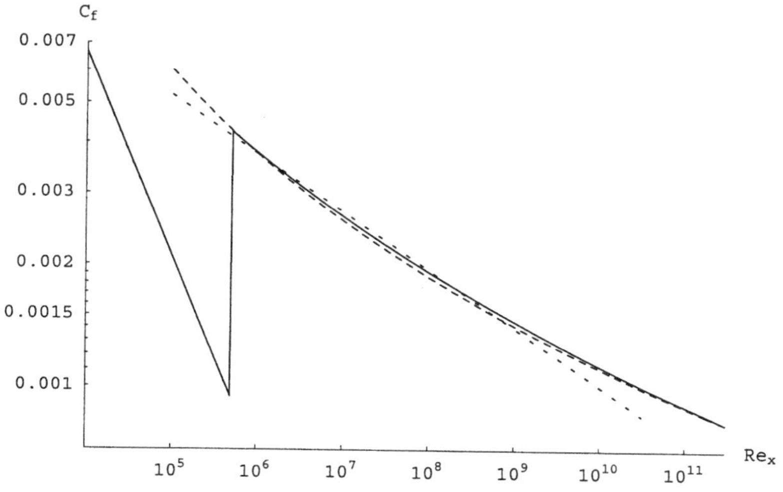

Figure 32.11. The local smooth-plate friction coefficient (solid line) in uniform flow as a function of the Reynolds number $\mathrm{Re}_x = Ux/\nu$. The straight line on the left is the Blasius laminar result (28.31). The vertical straight line at $\mathrm{Re}_x = 5 \times 10^5$ indicates the transition between the laminar and turbulent regimes, but the transition is in fact softer than shown here. The fully drawn curve on the right is the numerical solution of (32.79) and (32.82). The dashed curve that nearly coincides with the fully drawn curve is the approximation (32.83). The dotted straight line is the power law approximation (28.33) which fits well for $\mathrm{Re}_x \lesssim 10^9$.

The Reynolds stresses are only non-vanishing in the turbulent region of the boundary layer. In the viscous sublayer as well as in the viscous superlayer interfacing to the slip-flow, we have $\langle u_x u_y \rangle = 0$. In deriving the von Karman integral relation (28.35) for the mean flow by integration of the modified momentum equation over all y there will, for this reason, be no contribution from the extra term, and the relation takes in uniform flow in exactly the same form as before,

$$\nu \left.\frac{\partial v_x}{\partial y}\right|_{y=0} = \frac{d}{dx} \int_0^\infty (U - v_x) v_x \, dy. \tag{32.80}$$

Here we may cut off the integral at $y = \delta$ because $v_x = U$ for $y \gtrsim \delta$.

The left-hand side of this equation equals u_0^2, whereas the right-hand side is more complicated. To get an idea about its structure we disregard the viscous sublayer and ignore the wake function by setting $w = C = 0$, such that $v_x = U + u_0 A \log(y/\delta)$. The integral is now elementary, and the momentum equation takes the form

$$u_0^2 = \frac{d\big(A u_0 \delta (U - 2A u_0)\big)}{dx}.$$

Since $u_0 \delta = \nu \mathrm{Re}_0$, and since the friction law (32.52) for $C = 0$ yields $U/u_0 = f(\mathrm{Re}_0)$ where $f(s) = A \log \mathrm{Re}_0 + B$, the right-hand side depends only on Re_0. Differentiating through Re_0 and introducing $\mathrm{Re}_x = Ux/\nu$ in place of x, the above differential equation may be written,

$$\frac{d\mathrm{Re}_x}{d\mathrm{Re}_0} = A\big(f(\mathrm{Re}_0)^2 - 2A f(\mathrm{Re}_0) + 2A^2\big). \tag{32.81}$$

Integrating this over Re_0 we obtain the desired relation,

$$\boxed{\mathrm{Re}_x = A \mathrm{Re}_0 \big(f(\mathrm{Re}_0)^2 - 4A f(\mathrm{Re}_0) + 6A^2\big),} \tag{32.82}$$

apart from an undetermined additive constant. From this relation the local Reynolds number (32.77) and the friction coefficient (32.79) may be determined as functions of Re_x. It is, of course, possible to include

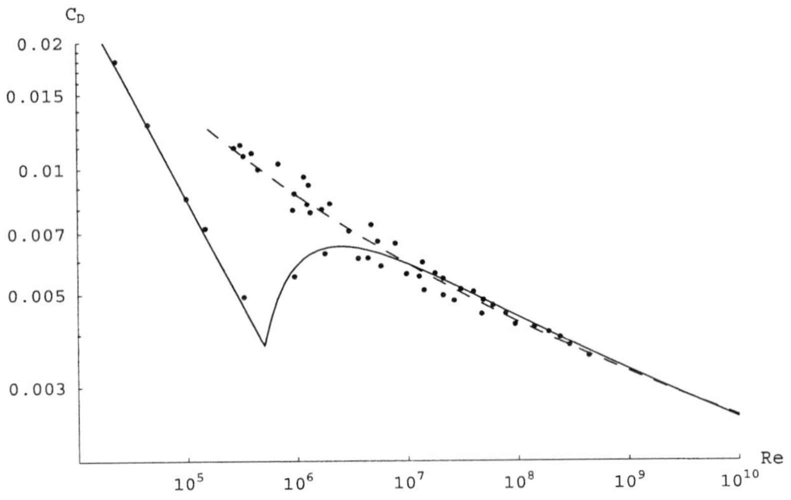

Figure 32.12. The drag coefficient for a flat wing (including both sides) as a function of the Reynolds number. The fully drawn curve is obtained by integrating the friction coefficient in figure 32.11. The straight line on the left is the laminar result (28.32). The cusp indicates the nominal transition point at $\mathrm{Re} = 5 \times 10^5$. The dashed curve is the turbulent drag approximation (32.85). The data points are extracted from [61, fig. 1.3]. There is some 'hysteresis' in the data, indicating that the transition point is not particularly well defined.

the constant C and take into account the viscous sublayer and the wake function to derive a more refined and more complicated relation.

In figure 32.11 the friction coefficient is plotted as a function of Re_x. In the region $5 \times 10^5 \lesssim \mathrm{Re}_x \lesssim 10^{12}$ it is very well approximated by the asymptotic expression [79, p. 432]

$$C_f = \frac{0.455}{(\log \mathrm{Re}_x - 2.81)^2}. \tag{32.83}$$

The power law (28.31) is marked as a dashed straight line, and only fits up to $\mathrm{Re}_x \lesssim 10^9$.

Plate drag coefficient

The drag coefficient for a flat 'wing' of downstream length L and cross-stream width W is obtained by averaging the friction coefficient over the plate (including a factor 2 for both sides),

$$C_D = \frac{\mathcal{D}}{(1/2)\rho_0 U^2 L W} = \frac{2}{L}\int_0^L C_f(x)\,dx = \frac{2}{\mathrm{Re}}\int_0^{\mathrm{Re}} C_f(\mathrm{Re}_x)d\mathrm{Re}_x. \tag{32.84}$$

In the last step we have introduced the Reynolds number of the plate $\mathrm{Re} = UL/\nu$ and converted the integral over x into an integral over Re_x.

The integral can be carried out numerically using the friction coefficient plotted in figure 32.11. The result is shown in figure 32.12 and compared with data. A nice asymptotic approximation is (see [79, p. 433]),

$$C_D = \frac{1.046}{(\log \mathrm{Re} - 2.81)^2}. \tag{32.85}$$

It is shown as the dashed curve in figure 32.12. Surprisingly, it continues to fit the data some distance below the critical point. One may interpret this phenomenon as a kind of 'hysteresis' in the transition back to laminar flow when the Reynolds number decreases.

32.10 Turbulence modelling

In the preceding sections we have analysed a few systems so simple that their behaviour in the turbulent regime is severely constrained by symmetry and dimensional arguments. It must be emphasized that this approach is not fundamental but rather semi-empirical, as witnessed by the appearance of numerical constants that have to be determined from comparison with data. Some of these constants are justifiably believed to be universal, whereas others depend on the general flow configuration.

Ideally, we would like to calculate these constants from first principles. Since we have at hand the fundamental Navier–Stokes equations governing all Newtonian fluids, it is 'just' a question of solving them for any particular flow geometry. Direct numerical simulation of turbulent flow is indeed possible and yields believable results with high precision as in the case of channel flow (page 514). Combined with symmetry and dimensional analysis the numerical constants may in this sense be determined from first principles without any recourse to real experiments. Unfortunately, the hoped-for universality is not perfect and different numerical and real experiments disagree somewhat on the precise values of the constants.

Theoretically, the closure problem is the real obstacle to solving the Navier–Stokes equations with turbulence. Barring direct numerical simulation and cheaper variants thereof such as large eddy simulation, the only way out of this conundrum seems to be to break off the otherwise infinite set of fluctuation equations, and thereby enforce closure. There are, however, many ways of doing so, and in the last 100 years numerous closure modelling schemes have appeared, sometimes introducing many new adjustable parameters. It is not the intention here to review this enormous field, which is well represented in the excellent modern textbooks on turbulence [58, 40, 25, 39]

Problems

32.1 Investigate the conditions under which the mean value of the time derivative vanishes $\langle \partial \boldsymbol{v}/\partial t\rangle = \mathbf{0}$.

32.2 Try to normalize the eddy probability distribution (32.5) in the inertial range. How does it change the Kolmogorov law and the microscopic scale?

32.3 Show that the following equation is satisfied by the second-order fluctuation moment tensor $\langle u_i u_j\rangle$,

$$\frac{\partial \langle u_i u_j\rangle}{\partial t} + (\boldsymbol{v}\cdot\nabla)\langle u_i u_j\rangle + \langle u_i \boldsymbol{u}\rangle\cdot\nabla v_j + \langle u_j \boldsymbol{u}\rangle\cdot\nabla v_i + \nabla\cdot\langle \boldsymbol{u} u_i u_j\rangle$$
$$= -\frac{1}{\rho_0}\left(\langle u_i \nabla_j q\rangle + \langle u_j \nabla_i q\rangle\right) + \nu\left(\nabla^2\langle u_i u_j\rangle - 2\langle \nabla u_i\cdot\nabla u_j\rangle\right). \qquad (32.86)$$

Discuss the closure problem.

32.4 Investigate the convergence of the following iteration scheme for the friction law (32.52)

$$\Lambda_n = A\log \mathrm{Re}_\delta + B + C - A\log\Lambda_{n-1}, \qquad \Lambda_0 = 1, \qquad (32.87)$$

where Λ_n is the nth approximant to $\Lambda = U_\delta/u_0$. Find the approximant sequence for $A = 2.44$, $B = 5$, $C = 0$ and $\mathrm{Re}_\delta = 10^5$.

A

Newtonian particle mechanics

The equations of continuum mechanics are derived by a systematic application of Newton's laws to systems that nearly behave as if they consisted of idealized point particles. It is for this reason useful here to recapitulate the basic mechanics of point particles, and to derive the global laws that so often are found to be useful.

The global laws state for any collection of point particles that

- *the rate of change of momentum equals force*
- *the rate of change of angular momentum equals the moment of force*
- *the rate of change of kinetic energy equals power*

Even if these laws are not sufficient to determine the dynamics of a physical system, they represent seven individual constraints on the motion of any system of point particles, independent of how complex it is. They are equally valid for continuous systems when it is properly taken into account that the number of particles in a body may change with time.

In the main text a certain familiarity with Newtonian mechanics is assumed throughout. This appendix only serves as a reminder and as a refresher. It can in no way substitute for a proper course in Newtonian mechanics.

A.1 Dynamic equations

In Newtonian mechanics, a physical system or *body* is understood as a collection of a certain number N of point particles numbered $n = 1, 2, \ldots, N$. Each particle obeys Newton's second law,

$$m_n \frac{d^2 \boldsymbol{x}_n}{dt^2} = \boldsymbol{f}_n, \tag{A.1}$$

where m_n denotes the (constant) mass of the nth particle, $\boldsymbol{x}_n$ its instantaneous position, and $\boldsymbol{f}_n$ the instantaneous force acting on the particle. Due to the mutual interactions between the particles, the forces may depend on the instantaneous positions and velocities of all the particles, including themselves,

$$\boldsymbol{f}_n = \boldsymbol{f}_n\left(\boldsymbol{x}_1, \ldots, \boldsymbol{x}_N, \frac{d\boldsymbol{x}_1}{dt}, \ldots, \frac{d\boldsymbol{x}_N}{dt}, t\right). \tag{A.2}$$

The forces will in general also depend on parameters describing the external influences from the system's environment, for example the Earth's gravity. The explicit dependence on t in the last argument of the force usually derives from such time-dependent external influences. It is, however, often possible to view the environment as just another collection of particles and include it in a larger *isolated* body without any influences from the environment.

The dynamics of a collection of particles thus becomes a web of coupled second-order differential equations in time. In principle, these equations may be solved numerically for all values of t, given initial positions and velocities for all particles at a definite instant of time, say $t = t_0$. Unfortunately, the large number of molecules in any macroscopic body usually presents an insurmountable obstacle to such an endeavor. Even for smaller numbers of particles, *deterministic chaos* may effectively prevent any long-term numeric integration of the equations of motion.

A.2 Force and momentum

A number of quantities describe the system as a whole. The total mass of the system is defined to be

$$M = \sum_n m_n, \tag{A.3}$$

and the total force

$$\mathcal{F} = \sum_n \boldsymbol{f}_n. \tag{A.4}$$

Note that these are true definitions. Nothing in Newton's laws tells us that it is physically meaningful to add masses of different particles, or worse, forces acting on different particles. As shown in problem A.1, there is no obstacle to making a different definition of total force.

> The choice made here is particularly convenient for particles moving in a constant field of gravity, such as we find on the surface of the Earth, because the gravitational force on a particle is directly proportional to the mass of the particle. With the above definitions, the total gravitational force, the weight, becomes proportional to the total mass. This additivity of weights, the observation that a volume of flour balances an equal volume of flour, independent of how it is subdivided into smaller volumes, goes back to the dawn of history.

Having made these definitions, the form of the equations of motion (A.1) tells us that we should also define the average of the particle positions weighted by the corresponding masses

$$\boldsymbol{x}_M = \frac{1}{M} \sum_n m_n \boldsymbol{x}_n. \tag{A.5}$$

For then the equations of motion imply that

$$M \frac{d^2 \boldsymbol{x}_M}{dt^2} = \mathcal{F}. \tag{A.6}$$

Formally, this equation is of the same form as Newton's second law for a single particle, so *the centre of mass moves like a point particle under the influence of the total force.* But before we get completely carried away, it should be remembered that the total force depends on the positions and velocities of all the particles, not just on the centre of mass position $\boldsymbol{x}_M$ and its velocity $d\boldsymbol{x}_M/dt$. The above equation is in general *not* a solvable equation of motion for the centre of mass.

> There are, however, important exceptions. The state of a stiff body is characterized by the position and velocity of its centre of mass, together with the body's orientation and its rate of change. If the total force on the body does not depend on the orientation, the above equation truly becomes an equation of motion for the centre of mass. It is fairly easy to show that for a collection of spherically symmetric stiff bodies, the gravitational forces can only depend on the positions of the centers of mass, even if the bodies rotate. It was Newton's good fortune that planets and stars to a good approximation behave like point particles.

It is convenient to reformulate the above equation by defining the total *momentum* of the body,

$$\mathcal{P} = \sum_n m_n \frac{d\boldsymbol{x}_n}{dt} = M \frac{d\boldsymbol{x}_M}{dt}. \tag{A.7}$$

Like the total force it is a purely *kinematic* quantity, depending only on the particle velocities, calculated as the sum over the individual momenta $m_n d\boldsymbol{x}_n/dt$ of each particle. The equation of motion (A.6) implies that the total momentum obeys the equation

$$\boxed{\frac{d\mathcal{P}}{dt} = \mathcal{F}.} \tag{A.8}$$

Again it should be noted that this equation cannot be taken as an equation of motion, except in very special circumstances. It should rather be viewed as a *constraint* (or rather three since it is a vector equation) that follows from the true equations of motion, independent of what form the forces take. This constraint is particularly useful if the total momentum is known to be constant, or equivalently the centre of mass has constant velocity, because then the total force must vanish.

A.3 Moment of force and angular momentum

Similarly, the total *moment of force* acting on the system is defined as

$$\mathcal{M} = \sum_n \boldsymbol{x}_n \times \boldsymbol{f}_n. \tag{A.9}$$

Like the total force, it is a *dynamic* quantity calculated from the sum of the individual moments of force acting on the particles.

The corresponding kinematic quantity is the total *angular momentum*,

$$\mathcal{L} = \sum_n \boldsymbol{x}_n \times m_n \frac{d\boldsymbol{x}_n}{dt}. \tag{A.10}$$

Differentiating with respect to time we find

$$\frac{d\mathcal{L}}{dt} = \sum_n m_n \left(\frac{d\boldsymbol{x}_n}{dt} \times \frac{d\boldsymbol{x}_n}{dt} + \boldsymbol{x}_n \times \frac{d^2\boldsymbol{x}_n}{dt^2} \right).$$

The first term in parenthesis vanishes because the cross product of a vector with itself always vanishes, and using the equations of motion in the second term, we obtain

$$\boxed{\frac{d\mathcal{L}}{dt} = \mathcal{M}.} \tag{A.11}$$

Like the equation for total momentum and total force, (A.8), this equation is also a constraint that must be fulfilled, independent of the nature of the forces acting on the particles. Angular momentum has to do with the state of rotation of the body as a whole. If the total angular momentum is known to be constant, as for a non-rotating body, the total moment of force must vanish. This is also what lies behind the lever principle.

> From the earliest times levers have been used to lift and move heavy weights, such as the stones found in stone-age monuments. A primitive lever is simply a long stick with one end wedged under a heavy stone. Applying a small 'man-sized' force orthogonal to the other end of the stick, the product of the long arm and the small force translates into a much larger force at the end of the small arm wedged under the stone. The total moment vanishes, when the stick is not moving.

The moment of force and the angular momentum both depend explicitly on the choice of origin of the coordinate system. These quantities might as well have been calculated around any other fixed point $\boldsymbol{c}$, leading to

$$\mathcal{M}(\boldsymbol{c}) = \sum_n (\boldsymbol{x}_n - \boldsymbol{c}) \times \boldsymbol{f}_n = \mathcal{M} - \boldsymbol{c} \times \mathcal{F}, \tag{A.12}$$

$$\mathcal{L}(\boldsymbol{c}) = \sum_n (\boldsymbol{x}_n - \boldsymbol{c}) \times m_n \frac{d(\boldsymbol{x}_n - \boldsymbol{c})}{dt} = \mathcal{L} - \boldsymbol{c} \times \mathcal{P}. \tag{A.13}$$

This shows that if the total force vanishes, the total moment of force becomes independent of the choice of origin, and similarly if the total momentum vanishes, the total angular momentum will be independent of the choice of origin.

If a point $\boldsymbol{c}$ exists such that $\boldsymbol{\mathcal{M}}(\boldsymbol{c}) = \boldsymbol{0}$ we get $\boldsymbol{\mathcal{M}} = \boldsymbol{c} \times \boldsymbol{\mathcal{F}}$. In this case, the point $\boldsymbol{c}$ is called the *centre of action* or *point of attack* for the total force $\boldsymbol{\mathcal{F}}$. In general, there is no guarantee that a centre of action exists, since it requires the total force $\boldsymbol{\mathcal{F}}$ to be orthogonal to the total moment $\boldsymbol{\mathcal{M}}$. Even if the centre of action exists, it is not unique because any other point $\boldsymbol{c} + k\boldsymbol{\mathcal{F}}$ with arbitrary k is as good a centre of action as $\boldsymbol{c}$. In constant gravity where $\boldsymbol{f}_n = m_n \boldsymbol{g}_0$, it follows immediately that the centre of mass is also the centre of action for gravity, or as it is usually called, the centre of gravity.

A.4 Power and kinetic energy

Forces generally perform work on the particles they act on. The total rate of work or *power* performed by the forces acting on all the particles making up a body is

$$P = \sum_n \boldsymbol{f}_n \cdot \frac{d\boldsymbol{x}_n}{dt}. \tag{A.14}$$

Note that there is a dot-product between the force and the velocity. In non-Anglo-Saxon countries this is called *effect* rather than power.

The corresponding kinematic quantity is the total *kinetic energy*,

$$\mathcal{T} = \frac{1}{2} \sum_n m_n \left(\frac{d\boldsymbol{x}_n}{dt} \right)^2, \tag{A.15}$$

which is the sum of individual kinetic energies of each particle. Differentiating with respect to time and using the equations of motion (A.1), we find

$$\boxed{\frac{d\mathcal{T}}{dt} = P.} \tag{A.16}$$

The rate of change of the kinetic energy equals the power.

A.5 Internal and external forces

The force acting on a particle may often be split into an internal part due to the other particles in the system and an external part due to the system's environment,

$$\boldsymbol{f}_n = \boldsymbol{f}_n^{\text{int}} + \boldsymbol{f}_n^{\text{ext}}. \tag{A.17}$$

The internal forces, in particular gravitational forces, are often two-particle forces with $\boldsymbol{f}_{n,n'}$ denoting the force that particle n' exerts on particle n. The total internal force on particle n thus becomes

$$\boldsymbol{f}_n^{\text{int}} = \sum_{n'} \boldsymbol{f}_{n,n'}. \tag{A.18}$$

Most two-particle forces also obey Newton's third law, which states that the force from n' on n is equal and opposite to the force from n on n',

$$\boldsymbol{f}_{n,n'} = -\boldsymbol{f}_{n',n}. \tag{A.19}$$

Although the external forces may themselves stem from two-particle forces of this kind, this is ignored as long as the nature of the environment is unknown.

Strong theorems follow from the above assumptions about the form of the internal forces. The first is that the total internal force vanishes,

$$\boldsymbol{\mathcal{F}}^{\text{int}} = \sum_n \boldsymbol{f}_n^{\text{int}} = \sum_{n,n'} \boldsymbol{f}_{n,n'} = \boldsymbol{0} \tag{A.20}$$

where we have used the antisymmetry (A.19). This expresses the simple fact that you cannot lift yourself up by your bootstraps; Baron Münchausen notwithstanding.

The momentum rate equation (A.8) thus takes the form

$$\frac{d\mathcal{P}}{dt} = \mathcal{F}^{\text{ext}}, \tag{A.21}$$

showing that it is sufficient to know the total external force acting on a system in order to calculate its rate of change of momentum. The details of the internal forces can be ignored as long as they are of the two-particle kind and obey Newton's third law.

Under the same assumptions, the internal moment of force becomes

$$\mathcal{M} = \sum_{n,n'} \boldsymbol{x}_n \times \boldsymbol{f}_{n,n'} = \frac{1}{2} \sum_{n,n'} (\boldsymbol{x}_n - \boldsymbol{x}_{n'}) \times \boldsymbol{f}_{n,n'}. \tag{A.22}$$

This does not in general vanish, except for the case of *central forces* where

$$\boldsymbol{f}_{n,n'} \sim \boldsymbol{x}_n - \boldsymbol{x}_{n'}. \tag{A.23}$$

Gravitational forces (and others) are of this type. Provided the internal forces stem from central two-particle forces, the total moment of force equals the external moment, such that

$$\frac{d\mathcal{L}}{dt} = \mathcal{M}^{\text{ext}}. \tag{A.24}$$

This rule is, however, not on nearly the same sure footing as the corresponding equation for the momentum rate (A.21).

Finally, there is not much to be said about the kinetic energy rate (A.16), which in general has non-vanishing internal and external contributions.

A.6 Hierarchies of particle interactions

Under what circumstances can a collection of point particles be viewed as a point particle? The dynamics of the solar system may to a good approximation be described by a system of interacting point particles, although the planets and the Sun are in no way pointlike in our own scale. In the scale of the whole universe, even galaxies are sometimes treated as point particles.

A point particle approximation may be in place as long as the internal cohesive forces that keep the interacting bodies together are much stronger than the external forces. In addition to mass and momentum, such a point particle may also have to be endowed with an intrinsic angular momentum (spin), and an intrinsic energy. The material world appears in this way as a hierarchy of approximately point-like interacting particles, from atoms to galaxies, at each level behaving as if they had no detailed internal structure. Corrections to the ideal point-likeness can later be applied to add more detail to this overall picture. Over the centuries this extremely reductionist method has shown itself to be very fruitful, but it is an open (scientific) question whether it can continue indefinitely.

Problems

A.1 Try to define the total force as $\mathcal{F}' = \sum_n m_n \boldsymbol{f}_n$ rather than (A.4), and investigate what this entails for the global properties of a system. Can you build consistent mechanics on this definition?

A.2 Show that the total momentum is $\mathcal{P} = M d\boldsymbol{x}_M/dt$ where $\boldsymbol{x}_M$ is the centre of mass position.

B

Curvilinear coordinates

The distance between two points in Euclidean space takes the simplest form in Cartesian coordinates. The geometry of concrete physical problems may, however, make non-Cartesian coordinates more suitable as a basis for analysis, even if the distance becomes more complicated in the new coordinates. Since the new coordinates are nonlinear functions of the Cartesian coordinates, they define three sets of intersecting curves, and are for this reason called *curvilinear coordinates*.

At a deeper level, it is often the *symmetry* of a physical problem that points to the most convenient choice of coordinates. Cartesian coordinates are well suited to problems with translational invariance, cylindrical coordinates for problems that are invariant under rotations around a fixed axis, and spherical coordinates for problems that are invariant or partially invariant under arbitrary rotations. Elliptic and hyperbolic coordinates are also of importance but will not be discussed here (see [7, p. 455]).

B.1 Cylindrical coordinates

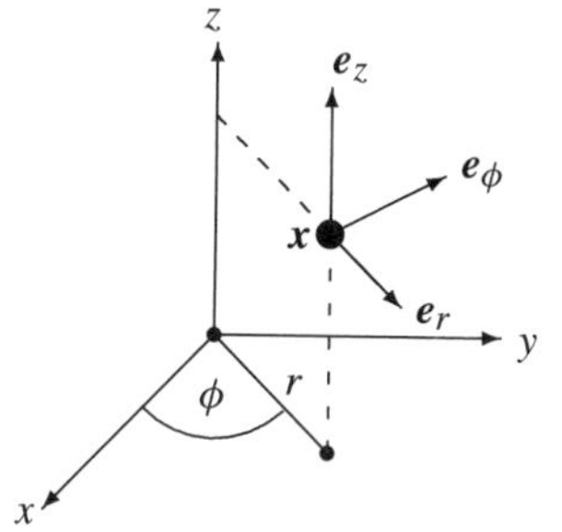

Cylindrical coordinates and basis vectors.

The relation between Cartesian coordinates x, y, z and cylindrical coordinates r, ϕ, z is given by

$$x = r\cos\phi, \tag{B.1a}$$
$$y = r\sin\phi, \tag{B.1b}$$
$$z = z \tag{B.1c}$$

with the range of variation $0 \le r < \infty$ and $0 \le \phi < 2\pi$. The two first equations simply define polar coordinates in the xy-plane[1]. The last is rather trivial but included to emphasize that this is a transformation in three-dimensional space.

Curvilinear basis

The *curvilinear basis vectors* are defined from the tangent vectors, obtained by differentiating the Cartesian position after the cylindrical coordinates,

$$\boldsymbol{e}_r = \frac{\partial \boldsymbol{x}}{\partial r} = (\cos\phi, \sin\phi, 0), \tag{B.2a}$$
$$\boldsymbol{e}_\phi = \frac{1}{r}\frac{\partial \boldsymbol{x}}{\partial \phi} = (-\sin\phi, \cos\phi, 0), \tag{B.2b}$$
$$\boldsymbol{e}_z = \frac{\partial \boldsymbol{x}}{\partial z} = (0, 0, 1). \tag{B.2c}$$

[1] Some texts use Θ instead of ϕ as the conventional name for the polar angle in the plane. Various arguments can be given one way or the other by comparing with spherical coordinates. But what's in a name? A polar angle by any name still works as sweet.

As may be directly verified, they are orthogonal and normalized everywhere, and thus define a local curvilinear basis with an orientation that changes from place to place. An arbitrary vector field may therefore be resolved in this basis

$$\boldsymbol{V} = \boldsymbol{e}_r V_r + \boldsymbol{e}_\phi V_\phi + \boldsymbol{e}_z V_z, \tag{B.3}$$

where the vector coordinates

$$V_r = \boldsymbol{V} \cdot \boldsymbol{e}_r, \qquad V_\phi = \boldsymbol{V} \cdot \boldsymbol{e}_\phi, \qquad V_z = \boldsymbol{V} \cdot \boldsymbol{e}_z, \tag{B.4}$$

are the projections of $\boldsymbol{V}$ on the local basis vectors.

Resolution of the gradient

The derivatives after cylindrical coordinates are found by differentiation through the Cartesian coordinates

$$\frac{\partial}{\partial r} = \frac{\partial x}{\partial r}\frac{\partial}{\partial x} + \frac{\partial y}{\partial r}\frac{\partial}{\partial y} = \cos\phi\frac{\partial}{\partial x} + \sin\phi\frac{\partial}{\partial y},$$

$$\frac{\partial}{\partial \phi} = \frac{\partial x}{\partial \phi}\frac{\partial}{\partial x} + \frac{\partial y}{\partial \phi}\frac{\partial}{\partial y} = -r\sin\phi\frac{\partial}{\partial x} + r\cos\phi\frac{\partial}{\partial y}.$$

From these relations we may calculate the projections of the gradient operator $\nabla = (\partial_x, \partial_y, \partial_z)$ on the cylindrical basis, and we obtain

$$\nabla_r = \boldsymbol{e}_r \cdot \nabla = \frac{\partial}{\partial r}, \tag{B.5a}$$

$$\nabla_\phi = \boldsymbol{e}_\phi \cdot \nabla = \frac{1}{r}\frac{\partial}{\partial \phi}, \tag{B.5b}$$

$$\nabla_z = \boldsymbol{e}_z \cdot \nabla = \frac{\partial}{\partial z}. \tag{B.5c}$$

Conversely, the gradient may be resolved on the basis

$$\boxed{\nabla = \boldsymbol{e}_r\nabla_r + \boldsymbol{e}_\phi\nabla_\phi + \boldsymbol{e}_z\nabla_z = \boldsymbol{e}_r\frac{\partial}{\partial r} + \boldsymbol{e}_\phi\frac{1}{r}\frac{\partial}{\partial \phi} + \boldsymbol{e}_z\frac{\partial}{\partial z}.} \tag{B.6}$$

Together with the only non-vanishing derivatives of the basis vectors

$$\frac{\partial \boldsymbol{e}_r}{\partial \phi} = \boldsymbol{e}_\phi, \tag{B.7a}$$

$$\frac{\partial \boldsymbol{e}_\phi}{\partial \phi} = -\boldsymbol{e}_r, \tag{B.7b}$$

we are now in possession of all the necessary tools for calculating in cylindric coordinates.

The Laplacian

An operator which often occurs in differential equations is the *Laplace operator* or *Laplacian*,

$$\nabla^2 = \nabla_x^2 + \nabla_y^2 + \nabla_z^2 = \frac{\partial^2}{\partial x^2} + \frac{\partial^2}{\partial y^2} + \frac{\partial^2}{\partial z^2}. \tag{B.8}$$

In cylindrical coordinates this operator takes a different form, which may be found by squaring the resolution of the gradient (B.6). Keeping track of the order of the operators and basis vectors we get

$$\begin{aligned}
\nabla^2 &= (\boldsymbol{e}_r\nabla_r + \boldsymbol{e}_\phi\nabla_\phi + \boldsymbol{e}_z\nabla_z) \cdot (\boldsymbol{e}_r\nabla_r + \boldsymbol{e}_\phi\nabla_\phi + \boldsymbol{e}_z\nabla_z) \\
&= (\boldsymbol{e}_r\nabla_r + \boldsymbol{e}_\phi\nabla_\phi + \boldsymbol{e}_z\nabla_z) \cdot \boldsymbol{e}_r\nabla_r \\
&+ (\boldsymbol{e}_r\nabla_r + \boldsymbol{e}_\phi\nabla_\phi + \boldsymbol{e}_z\nabla_z) \cdot \boldsymbol{e}_\phi\nabla_\phi \\
&+ (\boldsymbol{e}_r\nabla_r + \boldsymbol{e}_\phi\nabla_\phi + \boldsymbol{e}_z\nabla_z) \cdot \boldsymbol{e}_z\nabla_z \\
&= \nabla_r^2 + \frac{1}{r}\nabla_r + \nabla_\phi^2 + \nabla_z^2.
\end{aligned}$$

In the second line we have distributed the first factor on the terms of the second, and in going to the last line we have furthermore distributed the terms of the first factor, using the orthogonality of the basis and taking into account that differentiation with respect to ϕ may change the basis vectors according to (B.7).

Finally, using (B.5) we arrive at the cylindrical Laplacian,

$$\boxed{\nabla^2 = \frac{\partial^2}{\partial r^2} + \frac{1}{r}\frac{\partial}{\partial r} + \frac{1}{r^2}\frac{\partial^2}{\partial \phi^2} + \frac{\partial^2}{\partial z^2},} \tag{B.9}$$

expressed in terms of the usual partial derivatives.

B.2 Spherical coordinates

The treatment of spherical coordinates follows much the same pattern as cylindrical coordinates. Spherical or polar coordinates consist of the radial distance r, the polar angle θ and the azimuthal angle ϕ. If the z-axis is chosen as the polar axis and the x-axis as the origin for the azimuthal angle, the transformation from spherical to Cartesian coordinates becomes,

$$x = r\sin\theta\cos\phi, \tag{B.10a}$$
$$y = r\sin\theta\sin\phi, \tag{B.10b}$$
$$z = r\cos\theta. \tag{B.10c}$$

The domain of variation for the spherical coordinates is $0 \le r < \infty$, $0 \le \theta \le \pi$ and $0 \le \phi < 2\pi$.

Curvilinear basis

The normalized tangent vectors along the directions of the spherical coordinate are,

$$\boldsymbol{e}_r = \frac{\partial \boldsymbol{x}}{\partial r} = (\sin\theta\cos\phi, \sin\theta\sin\phi, \cos\theta), \tag{B.11a}$$
$$\boldsymbol{e}_\theta = \frac{1}{r}\frac{\partial \boldsymbol{x}}{\partial \theta} = (\cos\theta\cos\phi, \cos\theta\sin\phi, -\sin\theta), \tag{B.11b}$$
$$\boldsymbol{e}_\phi = \frac{1}{r\sin\theta}\frac{\partial \boldsymbol{x}}{\partial \phi} = (-\sin\phi, \cos\phi, 0). \tag{B.11c}$$

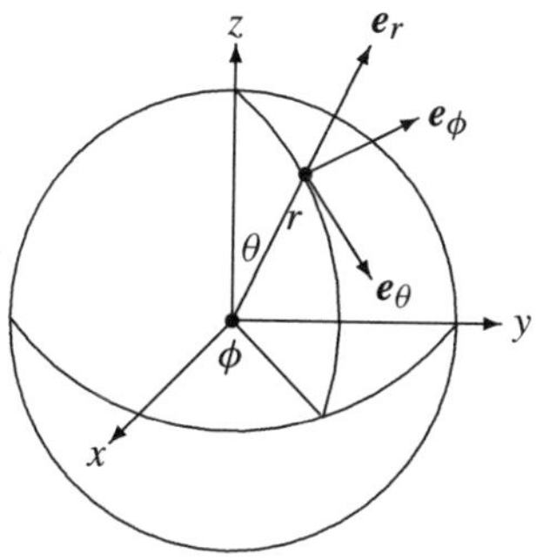

Spherical coordinates and their basis vectors.

They are orthogonal, such that an arbitrary vector field may be resolved in these directions,

$$\boldsymbol{V} = \boldsymbol{e}_r V_r + \boldsymbol{e}_\theta V_\theta + \boldsymbol{e}_\phi V_\phi \tag{B.12}$$

with $V_a = \boldsymbol{e}_a \cdot \boldsymbol{V}$ for $a = r, \theta, \phi$.

Resolution of the gradient

The gradient operator may also be resolved on the basis,

$$\boxed{\nabla = \boldsymbol{e}_r\nabla_r + \boldsymbol{e}_\theta\nabla_\theta + \boldsymbol{e}_\phi\nabla_\phi,} \tag{B.13}$$

where

$$\nabla_r = \boldsymbol{e}_r \cdot \nabla = \frac{\partial}{\partial r}, \tag{B.14a}$$
$$\nabla_\theta = \boldsymbol{e}_\theta \cdot \nabla = \frac{1}{r}\frac{\partial}{\partial \theta}, \tag{B.14b}$$
$$\nabla_\phi = \boldsymbol{e}_\phi \cdot \nabla = \frac{1}{r\sin\theta}\frac{\partial}{\partial \phi}. \tag{B.14c}$$

The non-vanishing derivatives of the basis vectors are

$$\frac{\partial \boldsymbol{e}_r}{\partial \theta} = \boldsymbol{e}_\theta, \qquad \frac{\partial \boldsymbol{e}_r}{\partial \phi} = \sin\theta\, \boldsymbol{e}_\phi, \tag{B.15a}$$

$$\frac{\partial \boldsymbol{e}_\theta}{\partial \theta} = -\boldsymbol{e}_r, \qquad \frac{\partial \boldsymbol{e}_\theta}{\partial \phi} = \cos\theta \boldsymbol{e}_\phi, \tag{B.15b}$$

$$\frac{\partial \boldsymbol{e}_\phi}{\partial \phi} = -\sin\theta\, \boldsymbol{e}_r - \cos\theta\, \boldsymbol{e}_\theta. \tag{B.15c}$$

These are all the relations necessary for calculations in spherical coordinates.

The Laplacian

The Laplacian (B.8) becomes in this case

$$\begin{aligned}
\boldsymbol{\nabla}^2 &= (\boldsymbol{e}_r \nabla_r + \boldsymbol{e}_\theta \nabla_\theta + \boldsymbol{e}_\phi \nabla_\phi) \cdot (\boldsymbol{e}_r \nabla_r + \boldsymbol{e}_\theta \nabla_\theta + \boldsymbol{e}_\phi \nabla_\phi) \\
&= (\boldsymbol{e}_r \nabla_r + \boldsymbol{e}_\theta \nabla_\theta + \boldsymbol{e}_\phi \nabla_\phi) \cdot \boldsymbol{e}_r \nabla_r \\
&+ (\boldsymbol{e}_r \nabla_r + \boldsymbol{e}_\theta \nabla_\theta + \boldsymbol{e}_\phi \nabla_\phi) \cdot \boldsymbol{e}_\theta \nabla_\theta \\
&+ (\boldsymbol{e}_r \nabla_r + \boldsymbol{e}_\theta \nabla_\theta + \boldsymbol{e}_\phi \nabla_\phi) \cdot \boldsymbol{e}_\phi \nabla_\phi \\
&= \left(\nabla_r^2 + \frac{1}{r}\nabla_r + \frac{\sin\theta}{r\sin\theta}\nabla_r\right) + \left(\nabla_\theta^2 + \frac{\cos\theta}{r\sin\theta}\nabla_\theta\right) + \nabla_\phi^2.
\end{aligned}$$

In the last step we have used the orthogonality and the derivatives (B.15). Finally, using (B.14) this becomes

$$\boxed{\boldsymbol{\nabla}^2 = \frac{\partial^2}{\partial r^2} + \frac{2}{r}\frac{\partial}{\partial r} + \frac{1}{r^2}\frac{\partial^2}{\partial \theta^2} + \frac{\cos\theta}{r^2\sin\theta}\frac{\partial}{\partial\theta} + \frac{1}{r^2\sin^2\theta}\frac{\partial^2}{\partial\phi^2}} \tag{B.16}$$

in standard notation.

C Thermodynamics of ideal gases

An ideal gas is a nice 'laboratory' for understanding the thermodynamics of a fluid with a non-trivial equation of state. In this section we shall recapitulate the conventional thermodynamics of an ideal gas with constant heat capacity. For more extensive treatments, see for example [35, 12].

C.1 Internal energy

In section 4.1 we analysed Bernoulli's model of a gas consisting of essentially non-interacting point-like molecules, and found the pressure $p = (1/3)\rho\, v^2$ where v is the root-mean-square average molecular speed. Using the ideal gas law (4.26) the total molecular kinetic energy contained in an amount $M = \rho V$ of the gas becomes,

$$\tfrac{1}{2} M v^2 = \tfrac{3}{2} p V = \tfrac{3}{2} n R T, \qquad \text{(C.1)}$$

where $n = M/M_{\text{mol}}$ is the number of moles in the gas. The derivation in section 4.1 shows that the factor 3 stems from the three independent translational degrees of freedom available to point-like molecules. The above formula thus expresses that in a mole of a gas there is an internal kinetic energy $(1/2)RT$ associated with each translational degree of freedom of the point-like molecules.

Whereas monatomic gases like argon have spherical molecules and thus only three translational degrees of freedom, diatomic gases like nitrogen and oxygen have stick-like molecules with two extra rotational degrees of freedom orthogonal to the bridge connecting the atoms, and multiatomic gases like carbon dioxide and methane have three extra rotational degrees of freedom. According to the *equipartition theorem* of statistical mechanics these degrees of freedom will also carry a kinetic energy $(1/2)RT$ per mole. Molecules also possess vibrational degrees of freedom that may become excited at high temperatures, but we shall disregard them here.

The *internal energy* of n moles of an ideal gas is defined to be,

$$\boxed{\mathcal{U} = \frac{k}{2}\, n R T,} \qquad \text{(C.2)}$$

where k is the number of molecular degrees of freedom. A general result of thermodynamics (Helmholtz' theorem [35, p. 154]) guarantees that for an ideal gas $\mathcal{U}$ cannot depend on the volume but only on the temperature. Physically a gas may dissociate or even ionize when heated, and thereby change its value of k, but we shall for simplicity assume that k is in fact constant with $k = 3$ for monatomic, $k = 5$ for diatomic and $k = 6$ for multiatomic gases. For mixtures of gases the number of degrees of freedom is the molar average of the degrees of freedom of the pure components (see problem C.1).

C.2 Heat capacity

Suppose that we raise the temperature of the gas by δT without changing its volume. Since no work is performed, and since energy is conserved, the necessary amount of heat is $\delta Q = \delta\mathcal{U} = C_V \delta T$ where the constant,

$$C_V = \frac{k}{2} nR, \tag{C.3}$$

is naturally called the *heat capacity at constant volume*.

If instead the pressure of the gas is kept constant while the temperature is raised by δT, we must also take into account that the volume expands by a certain amount δV and thereby performs work on its surroundings. The necessary amount of heat is now larger by this work, $\delta Q = \delta U + p\delta V$. Using the ideal gas law (4.26) we have for constant pressure $p\delta V = \delta(pV) = nR\delta T$. Consequently, the amount of heat which must be added per unit of increase in temperature at constant pressure is

$$C_p = C_V + nR, \tag{C.4}$$

called the *heat capacity at constant pressure*. It is always larger than C_V because it includes the work of expansion.

The adiabatic index

The dimensionless ratio of the heat capacities,

$$\gamma = \frac{C_p}{C_V} = 1 + \frac{2}{k}, \tag{C.5}$$

is for reasons that will become clear in the following called the *adiabatic index*. It is customary to express the heat capacities in terms of γ rather than k,

$$C_V = \frac{1}{\gamma - 1} nR, \qquad C_p = \frac{\gamma}{\gamma - 1} nR. \tag{C.6}$$

Given the adiabatic index, all thermodynamic quantities for n moles of an ideal gas are completely determined. The value of the adiabatic index is $\gamma = 5/3$ for monatomic gases, $\gamma = 7/5$ for diatomic gases and $\gamma = 4/3$ for multiatomic gases.

C.3 Entropy

When neither the volume nor the pressure are kept constant, the heat that must be added to the system in an infinitesimal process is,

$$\delta Q = \delta U + p\delta V = C_V \delta T + nRT\frac{\delta V}{V}. \tag{C.7}$$

It is a mathematical fact that there exists no function, $Q(T, V)$, which has this expression as differential (see problem C.2). It may on the other hand be directly verified (by insertion) that

$$\delta S = \frac{\delta Q}{T} = C_V \frac{\delta T}{T} + nR\frac{\delta V}{V}, \tag{C.8}$$

can be integrated to yield a function,

$$\boxed{S = C_V \log T + nR \log V + \text{const.}} \tag{C.9}$$

called the *entropy* of the amount of ideal gas. Being an integral the entropy is only defined up to an arbitrary constant. The entropy of the gas is, like its energy, an abstract quantity which cannot be directly measured. But since both quantities depend on the measurable thermodynamic quantities, ρ, p and T, that characterize the state of the gas, we can calculate the value of energy and entropy in any state. But why bother to do so?

The two fundamental laws of thermodynamics

The reason is that the two fundamental laws of thermodynamics are formulated in terms of the energy and the entropy. Both laws concern processes that may take place in an *isolated* system which is not allowed to exchange heat with or perform work on the environment.

The *First Law* states that the energy is unchanged under any process in an isolated system. This implies that the energy of an open system can only change by exchange of heat or work with the environment. We actually use this law implicitly in deriving heat capacities and entropy.

The *Second Law* states that the entropy cannot decrease. In the real world, the entropy of an isolated system must in fact grow. Only if all the processes taking place in the system are completely *reversible* at all times, will the entropy stay constant. Reversibility is an ideal which can only be approached by very slow *quasi-static* processes, consisting of infinitely many infinitesimal reversible steps. Essentially all real-world processes are *irreversible* to some degree.

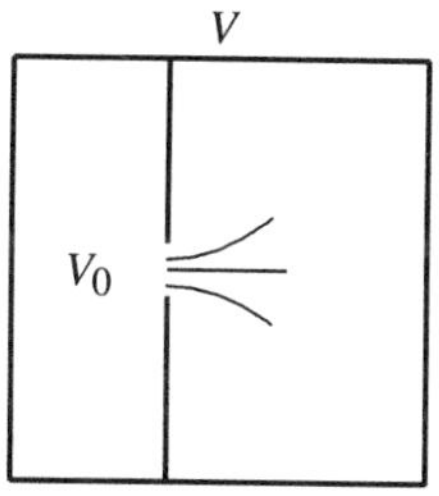

A compartment of volume V_0 inside an isolated box of volume V. Initially, the compartment contains an ideal gas with vacuum in the remainder of the box. When the wall breaks, the gas expands by itself to fill the whole box. The reverse process would entail a decrease in entropy and never happens by itself.

Example C.3.1 (Joule's expansion experiment): An isolated box of volume V contains an ideal gas in a walled-off compartment of volume V_0. When the wall is opened, the gas expands into vacuum and fills the full volume V. The box is completely isolated from the environment, and since the internal energy only depends on the temperature, it follows from the First Law that the temperature must be the same before and after the event. The change in entropy then becomes

$$\Delta \mathcal{S} = (C_V \log T + nR \log V) - (C_V \log T + nR \log V_0) = nR \log(V/V_0)$$

which is evidently positive (because $V/V_0 > 1$). This result agrees with the Second Law, which thus appears to be unnecessary.

The strength of the Second Law becomes apparent when we ask the question of whether the air in the box could ever—perhaps by an unknown process to be discovered in the far future—by itself enter the compartment of volume V_0, leaving vacuum in the box around it. Since such an event would entail a negative change in entropy which is forbidden by the Second Law, it never happens.

James Prescott Joule (1818–1889). *English physicist. Gifted experimenter who was the first to demonstrate the equivalence of mechanical work and heat, a necessary step on the road to the First Law. Demonstrated (in continuation of earlier experiments by Gay-Lussac) that the irreversible expansion of a gas into vacuum does not change its temperature.*

Isentropic processes

Any process in an open system which does not exchange heat with the environment is said to be *adiabatic*. If the process is furthermore reversible, it follows that $\delta Q = 0$ in each infinitesimal step, so that $\delta \mathcal{S} = \delta Q/T = 0$. The entropy (C.9) must in other words stay constant in any reversible, adiabatic process. Such a process is for this reason called *isentropic*.

By means of the adiabatic index (C.5) we may write the entropy (C.9) as,

$$\mathcal{S} = C_V \log\left(T V^{\gamma-1}\right) + \text{const.} \tag{C.10}$$

From this it follows that

$$T\,V^{\gamma-1} = \text{const}, \tag{C.11}$$

for any isentropic process in an ideal gas. Using the ideal gas law to eliminate $V \sim T/p$, this may be written equivalently as,

$$T^\gamma p^{1-\gamma} = \text{const.} \tag{C.12}$$

Eliminating instead $T \sim pV$, the isentropic condition takes its most common form,

$$\boxed{p\,V^\gamma = \text{const.}} \tag{C.13}$$

Note that the constants are different in these three equations.

Example C.3.2: When the air in a bicycle pump is compressed from V_0 to V_1 (while you block the valve with your finger), the adiabatic law implies that $p_1 V_1^\gamma = p_0 V_0^\gamma$. For $p_0 = 1$ atm and $V_1 = V_0/2$ we find $p_1 = 2.6$ atm. The temperature simultaneously rises about 100 degrees, but the hot air quickly becomes cold again during the backstroke. One may wonder why the fairly rapid compression stroke may be assumed to be reversible, but as long as the speed of the piston is much smaller than the velocity of sound, this is in fact a reasonable assumption. Conversely, we may conclude that the air expands with a velocity close to the speed of sound when the wall is opened in example C.3.1.

Isothermal versus isentropic bulk modulus

The bulk modulus of a strictly isothermal ideal gas with $p = \rho R T_0/M_{\rm mol}$ is equal to the pressure,

$$K_T = \rho \left(\frac{\partial p}{\partial \rho}\right)_T = p. \tag{C.14}$$

Here the index T (in the usual thermodynamic way of writing derivatives) signals that the temperature must be held constant while we differentiate.

In terms of the mass density $\rho = M/V$, the isentropic condition may be written in any of three different ways (with three different constants),

$$\boxed{p\,\rho^{-\gamma} = \text{const}, \qquad T\rho^{1-\gamma} = \text{const}, \qquad T^\gamma p^{1-\gamma} = \text{const}.} \tag{C.15}$$

Using the first we find the isentropic bulk modulus of an ideal gas,

$$K_S = \rho \left(\frac{\partial p}{\partial \rho}\right)_S = \gamma\, p, \tag{C.16}$$

where the index S now signals that the entropy must be held constant. The distinction between the isothermal and isentropic bulk modulus is necessary in all materials, but for nearly incompressible liquids there is not a great difference between K_S and K_T.

Among Isaac Newton's great achievements was the first calculation of the speed of sound in air, using the ideal gas law with constant temperature. His result did not agree with experiment, because normal sound waves oscillate so rapidly that compression and expansion are essentially isentropic processes. In section 16.2 we find that the speed of sound is $c = \sqrt{K/\rho}$, such that the ratio between the isentropic and isothermal sound velocities is $c_S/c_T = \sqrt{\gamma}$. For air with $\gamma \approx 1.4$ this amounts to an 18% error in the sound velocity. Much later, in 1799, Laplace derived the correct value for the speed of sound.

C.4 Specific quantities

In classical thermodynamics we always think of a macroscopic volume of matter with the same thermodynamic properties throughout the volume. Volume, mass, energy, entropy and the specific heats are all *extensive* quantities, meaning that the amount of any such quantity in a composite system is the sum of the amounts in the parts. Pressure, temperature and density are in contrast *intensive* quantities, that cannot be added when a system is put together from its parts.

In continuum physics, an intensive quantity becomes a field that may vary from place to place, whereas an extensive quantity becomes an integral over the density of the quantity. Since a material particle with a fixed number of molecules has a fixed mass (subject to the reservations set down in chapter 1), the natural field to introduce to describe an extensive quantity like the energy is the *specific internal energy*, $u = d\mathcal{U}/dM$, which is the amount of energy per unit of mass in the neighbourhood of a given point. The actual energy density becomes $d\mathcal{U}/dV = \rho\, u$, and the total energy in a volume

$$\mathcal{U} = \int_V \rho\, u\, dV. \tag{C.17}$$

The specific energy is an intensive quantity like temperature, pressure or density.

Similarly, we define the *specific heat* as the local heat capacity per unit of mass. Since the heat capacities (C.6) of an ideal gas are directly proportional to the mass $M = nM_{\rm mol}$, the specific heats of an ideal gas become,

$$c_V = \frac{1}{\gamma - 1}\frac{R}{M_{\rm mol}}, \qquad c_p = \frac{\gamma}{\gamma - 1}\frac{R}{M_{\rm mol}}. \tag{C.18}$$

They are constants which only depend on the properties of the gas. For air we have $c_V = 718\ \text{J K}^{-1}\,\text{kg}^{-1}$ and $c_p = 1005\ \text{J K}^{-1}\,\text{kg}^{-1}$. From (C.2) we obtain after dividing by M,

$$u = c_V T. \tag{C.19}$$

The specific energy of an ideal gas is the specific heat times the absolute temperature.

Finally, we define the *specific entropy*, $s = dS/dM$, from which the total entropy may be calculated as an integral,

$$\mathcal{S} = \int_V \rho\, s\, dV. \tag{C.20}$$

For an ideal gas, the specific entropy is obtained by dividing (C.10) by $M = nM_{\text{mol}}$. It may be written in three different forms related by the ideal gas law,

$$s = c_V \log\left(T\rho^{1-\gamma}\right) + \text{const} \tag{C.21a}$$

$$= c_V \log\left(T^\gamma p^{1-\gamma}\right) + \text{const} \tag{C.21b}$$

$$= c_V \log\left(p\,\rho^{-\gamma}\right) + \text{const}. \tag{C.21c}$$

Note, however, that the constants are different in the three cases.

Problems

C.1 An ideal gas mixture contains $n = \sum_i n_i$ moles.

(a) Show that the mixture obeys the equation of state (4.27) when the molar mass of the mixture is defined as

$$M_{\text{mol}} = \sum_i c_i M_{\text{mol},i} \tag{C.22}$$

where $c_i = n_i/n$ of the molar fraction of the ith component.

(b) Show that the number of degrees of freedom of a mixture is

$$k = \sum_i c_i k_i, \tag{C.23}$$

where k_i is the degrees of freedom of the ith component.

C.2 (a) Show that for a function $Q = Q(T, V)$ the differential takes the form $dQ = AdT + BdV$ where $\partial A/\partial V = \partial B/\partial T$. (b) Prove that this is not fulfilled for (C.7).

Answers to problems

An answer may not represent an explicit or complete solution to a problem, but only some useful hints on how to get there. Some problems have not been answered, either because it is too easy or because the result is already contained in the formulation of the problem.

1 Continuous matter

1.1

(a) This is the usual binomial distribution

$$\mathcal{P}_M(n) = \binom{M}{n} p^n (1-p)^{M-n}$$

(b) The average of n is

$$\langle n \rangle = \sum_{n=0}^{M} n\mathcal{P}_M(n) = pM \sum_{n=1}^{M} \mathcal{P}_{M-1}(n-1) = pM \sum_{n=0}^{M-1} \mathcal{P}_{M-1}(n) = pM$$

(c) Similarly we get

$$\langle n(n-1) \rangle = \sum_{n=0}^{M} n(n-1)\mathcal{P}_M(n) = pM \sum_{n=1}^{M} (n-1)\mathcal{P}_{M-1}(n-1) = p^2 M(M-1)$$

from which we get

$$\langle (n - \langle n \rangle)^2 \rangle = \langle n(n-1) \rangle + \langle n \rangle - \langle n \rangle^2 = p(1-p)M$$

1.2 The surface is the difference between the cube and an inner cube of side length $M-2$, such that $K = M^2 - (M-2)^3$. For large M we have $\Delta N \approx \sqrt{6}M = \sqrt{6}N^{1/3}$ so that $\Delta N/N \approx \sqrt{6}N^{-2/3} \approx 2.5N^{-2/3}$.

1.3 If the sphere has radius R the number of molecules in the volume is $N = \frac{4}{3}\pi R^3/L_{\text{mol}}^3$ whereas the number of molecules at the surface is $N_S = 4\pi R^2/L_{\text{mol}}^2 = 4\pi(3N/4\pi)^{2/3}$. For $N \gg 1$ we have $N_S \ll N$. If the surface molecules are randomly in or out of the volume, one gets $\Delta N \approx \sqrt{N_S} = \sqrt{4\pi}(3N/4\pi)^{1/3}$ and thus $\Delta N/N \simeq (36\pi)^{1/6}N^{-2/3} \simeq 2.2N^{-2/3}$.

1.4 The CMS-velocity is

$$\boldsymbol{v}_c = \frac{1}{N}\sum_n \boldsymbol{v}_n = \boldsymbol{v} + \frac{1}{N}\sum_n \boldsymbol{u}_n$$

The average is obviously $\boldsymbol{v}$ and its fluctuation is

$$\Delta v_c^2 = \langle(\boldsymbol{v}_c - \boldsymbol{v})^2\rangle = \frac{1}{N^2}\sum_{n,n'}\langle \boldsymbol{u}_n \cdot \boldsymbol{u}_{n'}\rangle = \frac{1}{N^2}\sum_n \langle \boldsymbol{u}_n^2\rangle = \frac{1}{N}v_0^2$$

where in the second step it has been used that the velocities are uncorrelated.

1.5 Solving the equation $L_{\text{micro}} = \lambda$ we get

$$\epsilon = (\pi\sqrt{2})^{3/2}\frac{d^3}{L_{\text{mol}}^3} \approx 10\frac{d^3}{L_{\text{mol}}^3}. \tag{1.A1}$$

Under normal conditions with $\rho \approx 1.2$ kg m^{-3} we have $L_{\text{mol}} \approx 10d$ so that $\epsilon = 10^{-2}$. A 100 times smaller density requires a 100 times better precision or $\epsilon = 10^{-4}$.

2 Space and time

2.1 a, b) Trivial. c) Follows from the definition that the distance between $\boldsymbol{a}$ and $\boldsymbol{b}$ is the smallest number of steps. Any path going through a point $\boldsymbol{c}$ cannot be shorter.

2.2

(a) $|\boldsymbol{a}| = 7$, $|\boldsymbol{b}| = 5$

(b) $\boldsymbol{a}\cdot\boldsymbol{b} = -6$

(c) $\boldsymbol{a}\times\boldsymbol{b} = (-24, -18, -17)$

(d) $\boldsymbol{ab} = \begin{pmatrix} 6 & -8 & 0 \\ 9 & -12 & 0 \\ -18 & 24 & 0 \end{pmatrix}$

2.3 Yes, $\boldsymbol{a} - \boldsymbol{b} + \boldsymbol{c} = \boldsymbol{0}$.

2.4 In an Earth-centred Cartesian coordinate system with z-axis towards the North pole at latitude $\delta = 90°$ and x-axis towards Greenwich at longitude $\alpha = 0$ we have $x = (a+h)\cos\alpha\cos\delta$, $y = (a+h)\sin\alpha\cos\delta$, and $z = (a+h)\sin\delta$ where a is the sea-level radius of the Earth. Using the invariance of the distance the square of the distance function becomes $d^2 = (x_1 - x_2)^2 + (y_1 - y_2)^2 + (z_1 - z_2)^2$ which may be recast into $d^2 = (a+h_1)^2 + (a+h_2)^2 - 2(a+h_1)(a+h_2)(\cos\delta_1\cos\delta_2\cos(\alpha_1 - \alpha_2) + \sin\delta_1\sin\delta_2)$.

2.5

(a) Use that $0 \le |\boldsymbol{a} + s\boldsymbol{b}|^2 = |\boldsymbol{a}|^2 + s^2|\boldsymbol{b}|^2 + 2s\boldsymbol{a}\cdot\boldsymbol{b}$. The minimum of the right-hand side is obtained for $s = -\boldsymbol{a}\cdot\boldsymbol{b}/\boldsymbol{b}^2$, and the inequality follows.

(b) Using the preceding result we have $(\boldsymbol{a}+\boldsymbol{b})^2 = \boldsymbol{a}^2 + \boldsymbol{b}^2 + 2\boldsymbol{a}\cdot\boldsymbol{b} \le \boldsymbol{a}^2 + \boldsymbol{b}^2 + 2|\boldsymbol{a}||\boldsymbol{b}| = (|\boldsymbol{a}| + |\boldsymbol{b}|)^2$, and the inequality follows.

2.6 Use that the determinant is unchanged under transposition and that the determinant of a product of matrices equals the product of the determinant.

2.10 Assume first that the vectors $\boldsymbol{a}$, $\boldsymbol{b}$ and $\boldsymbol{c}$ are linearly independent. In that case the cross products on the right-hand side will also be linearly independent, so that $\boldsymbol{d}$ can be written as a linear combination of these cross products. Check that coefficients are given by the dot products of $\boldsymbol{d}$ with $\boldsymbol{a}$, $\boldsymbol{b}$ and $\boldsymbol{c}$. Finally, one must discuss the cases where $\boldsymbol{a}$, $\boldsymbol{b}$ og $\boldsymbol{c}$ are linearly dependent.

2.14 Replace the six vectors in problem 2.6 by $\boldsymbol{a} \to \boldsymbol{e}_i$, $\boldsymbol{b} \to \boldsymbol{e}_j$, $\boldsymbol{c} \to \boldsymbol{e}_k$, and $\boldsymbol{d} \to \boldsymbol{e}_l$, $\boldsymbol{e} \to \boldsymbol{e}_m$, and $\boldsymbol{f} \to \boldsymbol{e}_n$.

2.16 Under a general transformation $x_i' = \sum_j a_{ij} x_j$, we use the chain rule for differentiation,

$$\nabla_j = \frac{\partial}{\partial x_j} = \sum_k \frac{\partial x_k'}{\partial x_j} \frac{\partial}{\partial x_k'} = \sum_k a_{kj} \nabla_k'. \tag{2.A1}$$

Multiplying by a_{ij} and summing we get

$$\sum_j a_{ij} \nabla_j = \sum_{jk} a_{ij} a_{kj} \nabla_k' = \nabla_i' \tag{2.A2}$$

where the orthogonality of the transformation matrix has been used in the last step.

2.17 The transformed trace is $\sum_i T_{ii}' = \sum_{ijk} a_{ij} a_{ik} T_{jk} = \sum_{jk} \delta_{jk} T_{jk} = \sum_j T_{jj}$, showing that the trace is invariant.

2.18 The transformed tensor is, $\delta_{ij}' = \sum_{kl} a_{ik} a_{jl} \delta_{kl} = \sum_k a_{ik} a_{jk} = \delta_{ij}$.

2.20 The transformed relation is $W_i' = \sum_j T_{ij}' V_j'$. Using that $\mathbf{W}$ and $\boldsymbol{V}$ are vectors we have $\sum_k a_{ik} W_k = \sum_{jl} T_{ij}' a_{jl} V_l$, or $\sum_{kl} a_{ik} T_{kl} V_l = \sum_{jl} T_{ij}' a_{jl} V_l$. Since it is valid for all $\boldsymbol{V}$, we can remove V_l and get $\sum_{kl} a_{ik} T_{kl} = \sum_{jl} T_{ij}' a_{jl}$. Finally multiplying with a_{jl} and summing over l we get from the orthogonality of the transformation matrix, $T_{ij}' = \sum_{kl} a_{ik} a_{jl} T_{kl}$, which proves that $\mathbf{T}$ is a tensor.

2.21 Differentiate $(\boldsymbol{x} - \boldsymbol{y})^2 = (\boldsymbol{f}(\boldsymbol{x}) - \boldsymbol{f}(\boldsymbol{y}))^2$ after $\boldsymbol{x}$ to obtain $\boldsymbol{x} - \boldsymbol{y} = (\boldsymbol{f}(\boldsymbol{x}) - \boldsymbol{f}(\boldsymbol{y})) \cdot \mathbf{A}(\boldsymbol{x})$ with $\mathbf{A}(\boldsymbol{x}) = \{a_{ij}(\boldsymbol{x})\}$ and $a_{ij} = \partial f_i / \partial x_j$. Differentiate again after $\boldsymbol{y}$ to obtain $-\mathbf{1} = -\mathbf{A}(\boldsymbol{y})^\top \cdot \mathbf{A}(\boldsymbol{x})$. This means that $\mathbf{A}(\boldsymbol{x})^{-1} = \mathbf{A}(\boldsymbol{y})^\top$. The left-hand side depends only on $\boldsymbol{x}$ and the right-hand side only on $\boldsymbol{y}$ which implies that both sides are independent of $\boldsymbol{x}$ and $\boldsymbol{y}$, i.e. the matrix $\mathbf{A}$ is a constant, and orthogonal. Integrating $a_{ij} = \partial f_i / \partial x_j$ one gets $\boldsymbol{f}(\boldsymbol{x}) = \mathbf{A} \cdot \boldsymbol{x} + \boldsymbol{b}$.

2.22 Let $\mathbf{A}_z(\phi)$ be the matrix of the simple rotation (2.36) through an angle ϕ around the z-axis. Then the three Euler angles ϕ, θ and ψ determine any rotation matrix as a product $\mathbf{A} = \mathbf{A}_z(\psi) \cdot \mathbf{A}_y(\theta) \cdot \mathbf{A}_z(\phi)$.

3 Gravity

3.2 The centripetal acceleration in a circular orbit must equal the force of gravity, $v^2/r = GM/r^2$ leading to $v = \sqrt{GM/r} = \sqrt{-\Phi} = \sqrt{g_0 a^2/r}$. At ground level the velocity becomes $v = v_{\text{esc}}/\sqrt{2} = 7.9\ \text{km s}^{-1}$ where $v_{\text{esc}} = 11.2\ \text{km s}^{-1}$ is the escape velocity.

3.3 Earth's true rotation period $T = T_0 * 365/366$ is a bit shorter than $T_0 = 24$ hours because the orbital motion adds one full revolution in one year. Taking $v = \Omega r$ where $\Omega = 2\pi/T$ we find from the equality of centripetal acceleration and gravity that

$$r\Omega^2 = g_0 \frac{a^2}{r^2}. \tag{3.A1}$$

which solved for r/a yields

$$\frac{r}{a} = \left(\frac{g_0}{a\Omega^2}\right)^{1/3} \approx 6.613. \tag{3.A2}$$

The orbit height is $h = r - a \approx 5.613a \approx 35,800$ km.

3.4 At the height z above the ground the force on a small piece dz of the line is

$$d\mathcal{F} = \left(-g_0 \frac{a^2}{(a+z)^2} + (a+z)\Omega^2\right) \rho A \, dz \tag{3.A3}$$

where Ω is the angular velocity in the stationary orbit and the second term represents the centrifugal force. Since this only vanishes for $z = h$, the total force is maximal at the satellite. Integrating the force from 0 to h, we find the maximal force

$$\mathcal{F} = \int_0^h d\mathcal{F}(z) = \rho A h \left(-g_0 \frac{a}{a+h} + \Omega^2 \left(a + \frac{1}{2}h\right)\right). \tag{3.A4}$$

The absolute value of the tension-to-density ratio becomes,

$$\frac{\sigma}{\rho} = h\left(g_0 \frac{a}{a+h} - \Omega^2\left(a + \frac{1}{2}h\right)\right) \approx 4.8 \times 10^7 \text{ m}^2\,\text{s}^{-2} \tag{3.A5}$$

The tensile strength of a beryllium–copper alloy of density $\rho = 8230$ kg m^{-3} can go as high as $\sigma \approx 1.4$ GPa, leading to $\sigma/\rho \approx 1.7 \times 10^5$ m^2 s^{-2}, a factor nearly 300 below the required value. Ropes based on carbon fibers are expected to approach this value.

3.5

(a) Minimal kinetic energy: $\frac{1}{2}v_{\text{esc}}^2 \approx 63$ (km s)$^{-2}$ $= 63 \times 10^6$ J kg^{-1}.

(b) Melting, heating and evaporating ice about $\approx 3.6 \times 10^6$ J kg^{-1}.

3.6 Energy conservation: $\frac{1}{2}\dot{r}^2 + \Phi(r) = \Phi(a)$. Use (3.31).

(a) $v_0 = -\dot{r}|_{r=0} = a\sqrt{2(\Phi(a) - \Phi(0))} = a\sqrt{\frac{4}{3}\pi\rho_0 G} = \sqrt{g_0 a} = 7.9$ km s^{-1}.

(b) $t_0 = \int_0^a \frac{dr}{\sqrt{2(\Phi(a) - \Phi(r))}} = \int_0^a \frac{dr}{\sqrt{\frac{4}{3}\pi\rho_0 G(a^2 - r^2)}} = \frac{\pi a}{2v_0} = 1267$ s.

3.7 **(a)** From (3.17) we get

$$g(r) = -\frac{4}{3}\pi G \begin{cases} r\rho_1 & r \le a_1 \\ \dfrac{a_1^3}{r^2}\rho_1 + \left(r - \dfrac{a_1^3}{r^2}\right)\rho_2 & a_1 < r \le a \\ \dfrac{a_1^3\rho_1 + (a^3 - a_1^3)\rho_2}{r^2} & r > a. \end{cases} \tag{3.A6}$$

and from (3.28)

$$\Phi(r) = -\frac{2}{3}\pi G \begin{cases} (3a_1^2 - r^2)\rho_1 + 3(a^2 - a_1^2)\rho_2 & r \le a_1 \\ 2\dfrac{a_1^3}{r}\rho_1 + \left(3a^2 - r^2 - 2\dfrac{a_1^3}{r}\right)\rho_2 & a_1 \le r \le a \\ 2\dfrac{a_1^3}{r}\rho_1 + 2\dfrac{a^3 - a_1^3}{r}\rho_2 & r \ge a. \end{cases} \tag{3.A7}$$

(b) It follows from $|g(a_1)| > |g(a)|$, that $a_1\rho_1 > (a_1^3\rho_1 + (a^3 - a_1^3)\rho_2)/a^2$ which may be rewritten in the form of the inequality (3.44). For the Earth the left-hand side becomes 1.42 and the right-hand side 1.18, so the inequality is fulfilled.

3.8 Cut out a small sphere $|\boldsymbol{x}' - \boldsymbol{x}| \leq a$ around the point $\boldsymbol{x}$. Let a be so small that $\rho(\boldsymbol{x}')$ does not vary appreciably within this sphere. Then we get the contribution to gravity from the small sphere

$$\Delta \boldsymbol{g}(\boldsymbol{x}) = -G \int_{|\boldsymbol{x}'-\boldsymbol{x}|\leq a} \frac{\boldsymbol{x} - \boldsymbol{x}'}{|\boldsymbol{x} - \boldsymbol{x}'|^3} \rho(\boldsymbol{x}')\, dv' \approx -G\rho(\boldsymbol{x}) \int_{|\boldsymbol{x}'-\boldsymbol{x}|\leq a} \frac{\boldsymbol{x} - \boldsymbol{x}'}{|\boldsymbol{x} - \boldsymbol{x}'|^3}\, dv' = \boldsymbol{0}.$$

The last integral vanishes because of the spherical symmetry (it is a vector with no direction to point in).

3.9 A small volume is invariant under a rotation $dV' = dV$ and so is the amount of mass contained in it, $dM' = dM$. By the definition (3.1) we have $dM' = \rho'(\boldsymbol{x}')dV'$ and $dM = \rho(\boldsymbol{x})dV$ and from that $\rho'(\boldsymbol{x}') = \rho(\boldsymbol{x})$.

3.10 By the definition (3.5) we have $d\mathcal{F}' = \boldsymbol{g}'(\boldsymbol{x}')\, dM'$ and $d\mathcal{F} = \boldsymbol{g}(\boldsymbol{x})dM$. The force on a small volume is a vector and transforms according to $d\mathcal{F}' = \mathbf{A} \cdot d\mathcal{F}$ where $\mathbf{A}$ is the rotation matrix, and the mass element is invariant $dM' = dM$. From this we get $\boldsymbol{g}'(\boldsymbol{x}') = \mathbf{A} \cdot \boldsymbol{g}(\boldsymbol{x})$.

3.11 Multiplying (3.13) by $\boldsymbol{e}_r = \boldsymbol{x}/r$ and using (3.16) one gets

$$g(r) = -G \int_{|\boldsymbol{x}'|\leq a} \frac{\boldsymbol{x} \cdot (\boldsymbol{x} - \boldsymbol{x}')}{r\,|\boldsymbol{x} - \boldsymbol{x}'|^3} \rho(\boldsymbol{x}')\, dv'.$$

Introducing $s = |\boldsymbol{x}'|$ and the angle θ between $\boldsymbol{x}$ and $\boldsymbol{x}'$, so that $dv' = 2\pi \sin\theta d\theta s^2 ds$, this becomes

$$g(r) = -2\pi G \int_0^a \rho(s) s^2 ds \int_{-1}^1 d\cos\theta \frac{r - s\cos\theta}{(r^2 + s^2 - 2rs\cos\theta)^{\frac{3}{2}}}.$$

Integrating over $u = \cos\theta$ one obtains

$$\int_{-1}^1 du \frac{r - su}{(r^2 + s^2 - 2rsu)^{\frac{3}{2}}} = -\frac{\partial}{\partial r} \int_{-1}^1 du \frac{1}{\sqrt{r^2 + s^2 - 2rsu}}$$

$$= \frac{\partial}{\partial r} \left[\frac{\sqrt{r^2 + s^2 - 2rsu}}{rs} \right]_{u=-1}^1 = \frac{\partial}{\partial r} \frac{|r - s| - (r + s)}{rs}$$

$$= -2\frac{\partial}{\partial r} \begin{cases} \dfrac{1}{r} & r > s \\ \dfrac{1}{s} & r < s \end{cases} = \begin{cases} \dfrac{2}{r^2} & r > s \\ 0 & r < s \end{cases}$$

which leads to the desired result (3.17).

3.12

(a) $g(r) = -4\pi G \dfrac{A}{3+\alpha} r^{1+\alpha}$, $\Phi(r) = 4\pi G \dfrac{A}{2+\alpha} \left(\dfrac{r^{2+\alpha}}{3+\alpha} - a^{2+\alpha} \right)$.

(b) $\alpha > -3$.

(c) $-3 < \alpha < -1$.

3.13 Use equation (3.17). Setting $u = r/a$ one gets

$$M(r) = \int_0^r \rho(s) 4\pi s^2\, ds = 4\pi\rho_0 \int_0^r e^{-s/a} s^2\, ds = 4\pi\rho_0 a^3 (2 - (2 + 2u + u^2)e^{-u}).$$

Similarly, using (3.30) one finds

$$\int_r^\infty s\rho(s)\, ds = \rho_0 \int_r^\infty s e^{-s/a}\, ds = \rho_0 a^2 (1 + u)e^{-u}$$

and from this

$$\Phi = -\frac{4\pi G\rho_0 a^3}{r} (2(1 - e^{-u}) - ue^{-u}).$$

3.14 Line distribution $\rho dv = \lambda dz$ along the z-axis. Put $r = \sqrt{x^2 + y^2}$.

(a) Substitute $z' = z - r \sinh \psi$

$$\Phi = -G \int_{-a}^{a} \frac{\lambda dz'}{\sqrt{r^2 + (z - z')^2}} = -G\lambda(\sinh^{-1} \frac{z+a}{r} - \sinh^{-1} \frac{z-a}{r})$$

where $\sinh^{-1} u = \log(u + \sqrt{u^2 + 1})$ is the inverse hyperbolic sine. Then one gets

$$g_z = -\frac{\partial \Phi}{\partial z} = G\lambda \left(\frac{1}{\sqrt{(r^2 + (z+a)^2}} - \frac{1}{\sqrt{(r^2 + (z-a)^2}} \right)$$

$$g_r = -\frac{\partial \Phi}{\partial r} = -G\lambda \frac{1}{r} \left(\frac{z+a}{\sqrt{(r^2 + (z+a)^2}} - \frac{z-a}{\sqrt{(r^2 + (z-a)^2}} \right).$$

(b) For $r \to \infty$: $\Phi \to -G\frac{2a\lambda}{r}$, $g_z \to -G2a\lambda\frac{z}{r^2}$, $g_r \to -G\frac{2a\lambda}{r^2}$.

(c) For $a \to \infty$: $\Phi \to -2G\lambda \log \frac{a}{r}$, $g_z \to -2G\lambda\frac{z}{a^2}$, $g_r \to -\frac{2G\lambda}{r}$.

3.15 Use cylindrical coordinates (r, ϕ, z).

(a)

$$\Phi = -G\sigma \int_0^a s ds \int_0^{2\pi} d\phi \frac{1}{\sqrt{z^2 + r^2 + s^2 - 2rs\cos\phi}}.$$

(b) $\Phi = -2\pi G\sigma(\sqrt{z^2 + a^2} - |z|)$.

(c) $\Phi \to -G\frac{\sigma\pi a^2}{|z|}$.

(d) $\Phi \to -2\pi G(a - |z|)$ for $a \to \infty$.

4 Fluids at rest

4.1 Put $h_1 = 9$ m, $h_2 = 6$ m and $a = 12$ m. Atmospheric pressure is p_0.

(a) On one side $p_1 = p_0 + \rho_0 g_0(h_1 - z)$, on the other $p_2 = p_0 + \rho_0 g_0(h_2 - z)$.

(b) $\mathcal{F}_1 = \int_0^{h_1}(p_1 - p_0)a\,dz = \frac{1}{2}h_1^2 a\rho_0 g_0$. $\mathcal{F} = \mathcal{F}_1 - \mathcal{F}_2 \approx 2.7 \times 10^6$ N.

(c) $\mathcal{M}_1 = \int_0^{h_1} z(p_1 - p_0)a\,dz = \frac{1}{6}h_1^3 a\rho_0 g_0$. $\mathcal{M} = \mathcal{M}_1 - \mathcal{M}_2 \approx 10^7$ Nm.

(d) $z = \mathcal{M}/\mathcal{F} = 3.8$ m.

4.2 Put $h = 3$ m and $d = 2a = 30$ cm. The horizontal pressure force on the hemisphere is equal to the pressure force on the vertical plane through the centre of the sphere. The linear rise of pressure with depth makes the pressure act with its average value at the centre. So the force becomes $\rho_0 g_0 h\pi d^2/4 \approx 2100$ N, corresponding to the weight of 210 kg.

If you do not like this argument, it is also possible with some effort to integrate the pressure force directly in spherical coordinates

$$\boldsymbol{\mathcal{F}} = -\int_{\text{half-sphere}} (p - p_0)\,d\boldsymbol{S} = -\int_{\text{half-sphere}} (p - p_0)\boldsymbol{e}_r\,dS$$

$$= -a^2 \int_0^{\pi} d\theta \int_{-\pi/2}^{\pi/2} d\phi \sin\theta\,(p - p_0)\,\boldsymbol{e}_r.$$

Using

$$p = p_0 + \rho_0 g_0 (h - a\cos\theta)$$

where h is the depth of the centre of the lamp, we obtain for the x-component

$$\begin{aligned}\mathcal{F}_x &= -a^2 \int_0^\pi d\theta \int_{-\pi/2}^{\pi/2} d\phi \, \sin\theta \, \rho_0 g_0 (h - a\cos\theta) \sin\theta \cos\phi \\ &= -2a^2 \rho_0 g_0 \int_0^\pi d\theta \sin^2\theta (h - a\cos\theta) \\ &= -\rho_0 g_0 h \pi a^2.\end{aligned}$$

4.3 Put $h_1 = 6$ m, $p_1 = 1.6$ atm and $h_2 = 3$ m, $p_2 = 2.8$ atm. From $p_1 - p_2 = -\rho_0 g_0 (h_1 - h_2)$ we get $\rho_0 \approx 4100$ kg m^{-3}, and from $p_1 - p_0 = -(h_1 - h_0)\rho_0 g_0$ we get $h_0 = 7.5$ m.

4.4 (a) The surface inside the tube will be at the same level $h_1 + h_2$ as in the jar. (b) The heavy liquid in the tube must initially rise to the same level h_1 as in the jar. When the light liquid is poured in, the surface of the heavy liquid in the tube must rise further to balance the weight of the light and rise to a height $h_1 + h_2\rho_2/\rho_1 < h_1 + h_2$.

4.5 Solving for the pressure we find

$$P = \frac{nRT}{V - nb} - \frac{n^2 a}{V^2}. \tag{4.A1}$$

(a) Differentiating we get

$$K_T = -V\left(\frac{\partial p}{\partial V}\right)_T = \frac{nRTV}{(V - nb)^2} - \frac{2an^2}{V^2}. \tag{4.A2}$$

(b) It can become negative for

$$nRT < \frac{2an^2(V - nb)^2}{V^3}. \tag{4.A3}$$

When $K < 0$ the gas must condense.

4.6

(a) $K = n(p + B)$.

(b) Put $h = \dfrac{n}{n-1}\dfrac{p_0 + B}{\rho_0 g_0} = 35$ km.
Then $\rho = \rho_0(1 - z/h)^{1/(n-1)}$ and $p + B = (p_0 + B)(1 - z/h)^{n/(n-1)}$.

(c) $p \approx 1100$ bar and $(\rho - \rho_0)/\rho_0 \approx 4.5\%$ at $z = -10.924$ km.

4.7 When K is constant it follows from (4.33) that $p = K \log\rho + \text{const}$. Inserting this into the local hydrostatic equation (4.20c), it may be solved with the result,

$$\rho = \frac{\rho_0}{1 + \dfrac{z}{h_1}}, \tag{4.A4}$$

$$p = p_0 - K \log\left(1 + \frac{z}{h_1}\right), \tag{4.A5}$$

where

$$h_1 = \frac{K}{\rho_0 g_0} = \frac{K}{p_0} h_0. \tag{4.A6}$$

The pressure and density become singular for $z = -h_1$, which for water is $h_1 \approx 237$ km.

4.8

(a) $p(z) = \rho_0 g_0 z$.

(b) $0 \leq \frac{1}{A}\int_A (z - I_1)^2\, dS = I_2 - I_1^2$.

(c) $\mathcal{F} = \int_A p(z)\, dS = \rho_0 g_0 I_1$.

(d) $\mathcal{M} = \int_A p(z) z\, dS = \rho_0 g_0 I_2$.

(e) $z_P = \frac{\mathcal{M}}{\mathcal{F}} = \frac{I_2}{I_1} \geq I_1 = z_M$.

(f) Since the width of the triangle is bz/h at depth z, the area of the triangle is $A = \int_0^h b\frac{z}{h}\, dz = \frac{1}{2}hb$ (which is of course well known). It then follows that $I_1 = \frac{2}{3}h$ and $I_2 = \frac{1}{2}h^2$, and thus $z_M = \frac{2}{3}h$ and $z_P = \frac{3}{4}h$. Clearly $z_P > z_M$.

4.9 The pressure must be continuous across the boundary and thus of the form

$$p = p_b \begin{cases} (\rho/\rho_1)^\gamma & \text{in the core} \\ (\rho/\rho_2)^\gamma & \text{in the mantle} \end{cases} \tag{4.A7}$$

where ρ_1 is the density in the core and ρ_2 the density at the mantle at the boundary. The common pressure on both sides of the boundary is p_b.

5 Buoyancy

5.1 Volume 0.04 m^3. Density 2500 kg m^{-3}.

5.2 The same mass displaces two different volumes $M = \rho_1 V_1 = \rho_2 V_2$. The difference in volumes is only due to the stem, $\pi a^2 d = V_2 - V_1 = M(1/\rho_2 - 1/\rho_1)$. The result is $d = 20$ mm.

5.3 (a) Displacement $M_1 + M_2 = \rho_0((1 - f)V_1 + V_2)$ with $M_1 = \rho_1 V_1$ and $M_2 = \rho_2 V_2$. Then

$$\frac{M_1}{M_2} = \frac{1 - \dfrac{\rho_0}{\rho_2}}{(1 - f)\dfrac{\rho_0}{\rho_1} - 1} = 2.36$$

(b) $f \leq 1 - \rho_1/\rho_0 = 0.35$.

5.4 The half upper side is $b = h \tan\alpha$. We define $a = L/2$ and let d be the draught.

(a) The origin of the coordinates is chosen at the peak with area function

$$A(z) = 2Lz\tan\alpha \qquad (0 < z < h). \tag{5.A1}$$

The total volume is $V = \int_0^h A(z)\, dz = Lhb = Lh^2\tan\alpha$, and the centre of gravity $z_G = \frac{1}{V}\int_0^h zA(z)\, dz = \frac{2}{3}h$. Putting $h = d$ the submerged volume becomes $V_0 = Ld^2\tan\alpha$ and the centre of buoyancy becomes $z_B = \frac{2}{3}d$. The second-order moment is $I_0 = \frac{2}{3}L(d\tan\alpha)^3$ according to (5.29), and the metacentric height $z_M = \frac{2}{3}d + \frac{2}{3}d\tan^2\alpha = \frac{2}{3}d/\cos^2\alpha$. The stability condition (5.33) takes the form $d/h > \cos^2\alpha$. Archimedes' Law yields $\rho_1 V = \rho_0 V'$ or $\rho_1/\rho_0 = (d/h)^2$. The condition on the density ratio is $\rho_1/\rho_0 > \cos^4\alpha$.

(b) The origin of coordinates is again chosen at the peak of the cone. The area function is now $-A(z)$ in the interval $-h < z < 0$. The parameter d is for convenience chosen to be the 'antidraught' such that the true draught is $h - d$. The total volume is again $V = Lh^2 \tan\alpha$, and the centre of gravity $z_G = -\frac{2}{3}h$. The submerged volume is $V_0 = L(h^2 - d^2)\tan\alpha$ and the centre of buoyancy $z_B = -\frac{2}{3}(h^3 - d^3)/(h^2 - d^2)$. The waterline integral is the same as before, and the metacentric height $z_M = -\frac{2}{3}(h^3 - d^3/\cos^2\alpha)/(h^2 - d^2)$. The stability condition becomes again $d/h > \cos^2\alpha$ but since $\rho_1/\rho_0 = V_0/V = 1 - (d/h)^2$ the density condition becomes $\rho_1/\rho_0 < 1 - \cos^4\alpha$.

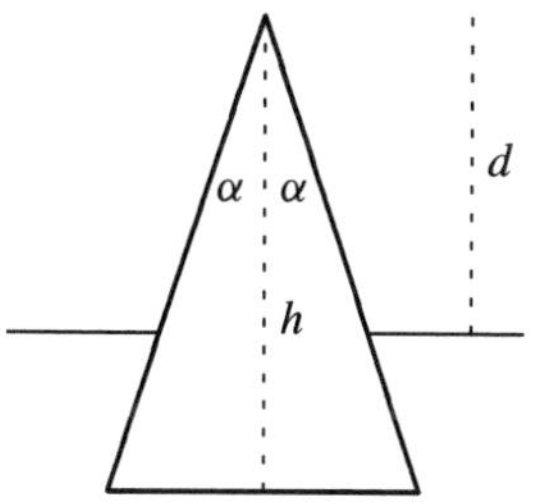

Triangle with the peak vertically upwards. Note that d is not the draft in this case.

(c) Combining the two conditions we get $\cos^4\alpha < \rho_1/\rho_0 < 1 - \cos^4\alpha$. This is only possible for $\cos\alpha < 2^{-1/4}$ or $\alpha > 33\,°$.

5.5 Using the same method as in problem 5.4 we find **(a)** $\rho_1/\rho_0 > \cos^6\alpha$, **(b)** $\rho_1/\rho_0 < 1 - \cos^6\alpha$, and **(c)** $\alpha > 27\,°$.

5.6 **(a)** The effective potential $H = g_0 z + w(p)$ is constant before the body enters. Afterwards $H + \Delta H = g_0 z + \Delta\Phi + w(p + \Delta p)$ is constant. Expanding w to lowest order, it follows that $\Delta H = \Delta\Phi + \Delta p/\rho$ is constant. At large distances $\Delta\Phi \to 0$ and $\Delta p \to 0$ so that $\Delta H = 0$. This proves the claim. **(b)** On the surface of a spherical body we have $\Delta\Phi = -g_1 a$ according to (3.30). Since the density $\rho(z)$ falls with increasing z, Δp will be smaller on the top of the sphere than on the bottom, and thus increase the buoyancy force.

5.7 **(a)** The extra pressure $\Delta p(\boldsymbol{x})$ created by the field of the spheres is determined by

$$w(p_0 + \Delta p) + \Phi_1 + \Phi_2 = w(p_0), \tag{5.A2}$$

with the potentials of the two spheres

$$\Phi_1 = -\frac{g_1 a^2}{\sqrt{x^2 + y^2 + z^2}}, \qquad \Phi_2 = -\frac{g_1 a^2}{\sqrt{x^2 + y^2 + (D - z)^2}}, \tag{5.A3}$$

where $g_1 = GM/a^2$ is the surface gravity of the spheres. Expanding to first order in Δp, we obtain

$$\Delta p = -\rho_0(\Phi_1 + \Phi_2) \tag{5.A4}$$

where $\rho_0 = \rho(p_0)$ is the density in the absence of the spheres. At the surface of sphere 1, $r = a$, we obtain the extra pressure

$$\Delta p_1 = \rho_0 a g_1 \left(1 + \frac{a}{D} + \left(\frac{a}{D}\right)^2 \cos\theta\right), \tag{5.A5}$$

where we have expanded to leading non-trivial order in a/D, and where θ is the polar angle of the point at the surface of sphere 1. Integrating over the surface of sphere 1, we obtain the extra bouyancy in the z-direction

$$\mathcal{F}_1 = \int_{r=a} (-\Delta p_1)dS_z = -2\pi a^2 \int_0^\pi \Delta p_1(\theta)\cos\theta\sin\theta\, d\theta = -\frac{4\pi}{3}\rho_0 g_1 \frac{a^5}{D^2}, \tag{5.A6}$$

which is a repulsion.

(b) The gravitational attraction from sphere 2 on sphere 1 is

$$\mathcal{G}_1 = \frac{GM^2}{D^2} = \frac{4\pi}{3}\rho_1 g_1 \frac{a^5}{D^2} \tag{5.A7}$$

and the ratio becomes $\mathcal{F}_1/\mathcal{G}_1 = -\rho_0/\rho_1$.

(c) If $\rho_1 = \rho_0$, then the total force becomes $\mathcal{F}_1 + \mathcal{G}_1 = 0$, as one would expect.

5.8 Using Gauss' theorem, the moment of buoyancy is

$$(\mathcal{M}_B)_i = -\int_S \sum_{jk} \epsilon_{ijk} x_j p \, dS_k = -\int_V \sum_{jk} \epsilon_{ijk} \nabla_k (x_j p) \, dV$$
$$= -\int_V \sum_{jk} \epsilon_{ijk} x_j \nabla_k p \, dV = -\int_V (\boldsymbol{x} \times \nabla p)_i \, dV.$$

Using local hydrostatic equilibrium (4.19) we get,

$$\mathcal{M}_B = -\int_V \boldsymbol{x} \times \rho_{\text{fluid}} \boldsymbol{g} \, dV. \tag{5.A8}$$

5.9 The total force $\mathcal{F} = (M_{\text{body}} - M_{\text{fluid}}) \boldsymbol{g}_0$. The total moment is $\mathcal{M} = \boldsymbol{x}_G \times M_{\text{body}} \boldsymbol{g}_0 - \boldsymbol{x}_B \times M_{\text{fluid}} \boldsymbol{g}_0 = \boldsymbol{x}_0 \times \mathcal{F}$ where

$$\boldsymbol{x}_0 = \frac{M_{\text{body}} \boldsymbol{x}_G - M_{\text{fluid}} \boldsymbol{x}_B}{M_{\text{body}} - M_{\text{fluid}}}. \tag{5.A9}$$

5.10 Use a Lagrange multiplier λ to handle the normalization condition, and find the extrema of the unconstrained form $I_{xx} n_x^2 + 2 I_{xy} n_x n_y + I_{yy} n_y^2 - \lambda (n_x^2 + n_y^2)$. The extremum conditions are $I_{xx} n_x + I_{xy} n_y = \lambda n_x$ and $I_{xy} n_x + I_{yy} n_y = \lambda n_y$, which is the eigenvalue equation.

5.11 Let the small angle of tilt around the principal axis be α. The angular momentum along the tilt axis is $J\dot{\alpha}$ and using (5.32) the equation of motion becomes a harmonic equation,

$$J \ddot{\alpha} = -\alpha (z_M - z_G) M g_0 \tag{5.A10}$$

from which we read off the oscillation frequency.

5.12 Under a rotation of the coordinate system by ϕ, the coordinates transform according to (2.36). The components of the area moment tensor transform as

$$I_{x'x'} = I_{xx} \cos^2 \phi + I_{yy} \sin^2 \phi + 2 I_{xy} \sin \phi \cos \phi \tag{5.A11}$$
$$I_{y'y'} = I_{xx} \sin^2 \phi + I_{yy} \cos^2 \phi - 2 I_{xy} \sin \phi \cos \phi \tag{5.A12}$$
$$I_{x'y'} = (I_{yy} - I_{xx}) \sin \phi \cos \phi + I_{xy} (\cos^2 \phi - \sin^2 \phi). \tag{5.A13}$$

If the area has a discrete symmetry such there will be at least one angle $\phi \neq 0, \pi$ such that $I_{x'x'} = I_{xx}$, $I_{y'y'} = I_{yy}$ and $I_{x'y'} = I_{xy}$. For this angle we thus have

$$(I_{yy} - I_{xx}) \sin^2 \phi + 2 I_{xy} \sin \phi \cos \phi = 0 \tag{5.A14}$$
$$(I_{xx} - I_{yy}) \sin^2 \phi - 2 I_{xy} \sin \phi \cos \phi = 0 \tag{5.A15}$$
$$(I_{yy} - I_{xx}) \sin \phi \cos \phi - 2 I_{xy} \sin^2 \phi = 0. \tag{5.A16}$$

The two first equations are the same, and the last two can only be satisfied if $I_{xx} = I_{yy}$ and $I_{xy} = 0$ when $\phi \neq 0, \pi$.

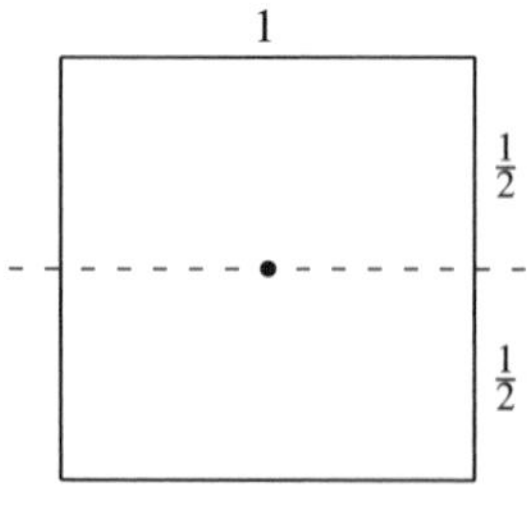

Here the cube floats with two faces horizontal and four vertical. This configuration is unstable. The x-axis goes into the paper.

5.13 The centre of gravity is at the centre of the cube $z_G = 0$ and the submerged volume is $V_0 = 1/2$ in all orientations. For symmetry reasons the centre of buoyancy is vertically below the centre of gravity in all three cases:

(a) The block is floating with two faces horizontal and the other faces vertical. Here we use (5.35) with $a = b = c = d = 1/2$ so that $z_B = -\frac{1}{4}$ and find the metacentric height $z_M = -(1/12)$ which is below the centre of gravity. This configuration is manifestly unstable.

(b) The block is floating with one horizontal edge below the water, one above the water, and two in the waterline. In this configuration the waterline area is a rectangle with sides 1 and $\sqrt{2}$. Taking the x-axis along the horizontal edges, we find the moments

$$I_{xx} = \int_{-\frac{1}{2}}^{\frac{1}{2}} dx \int_{-\frac{1}{2}\sqrt{2}}^{\frac{1}{2}\sqrt{2}} dy\, y^2 = \frac{1}{6}\sqrt{2} \tag{5.A17}$$

$$I_{yy} = \int_{-\frac{1}{2}}^{\frac{1}{2}} dx \int_{-\frac{1}{2}\sqrt{2}}^{\frac{1}{2}\sqrt{2}} dy\, x^2 = \frac{1}{12}\sqrt{2} \tag{5.A18}$$

$$I_{xy} = 0. \tag{5.A19}$$

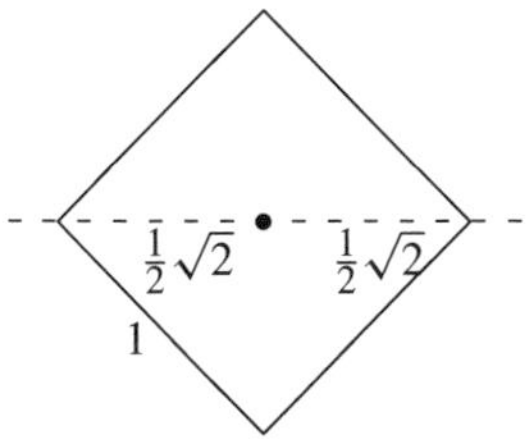

Here the cube floats with one edge horizontal and two edges in the waterline. This configuration is marginally stable (or unstable). The x-axis goes into the paper.

The centre of buoyancy is given by (5.21) with $A(z) = \sqrt{2} + 2z$ at depth z,

$$z_B = 2\int_{-\frac{1}{2}\sqrt{2}}^{0} z(\sqrt{2} + 2z)\, dz = -\frac{1}{6}\sqrt{2}. \tag{5.A20}$$

Thus, the metacentre (or the smallest moment) is at

$$z_M = z_B + \frac{I_{yy}}{V_0} = 0. \tag{5.A21}$$

This floating configuration is thus marginally stable for rotations around the y-axis which lower a corner and raise another. That brings us to the last configuration.

(c) The block is floating with one corner vertically below the centre of the cube. In this case the waterline area is a hexagon with sides of length $(1/2)\sqrt{2}$. Because of the symmetry we may calculate the area moment around any axis we choose, for example one that connects two opposite corners of the hexagon, with length $\sqrt{2}$. Integrating over the first quadrant we have we have from the geometry of the hexagon

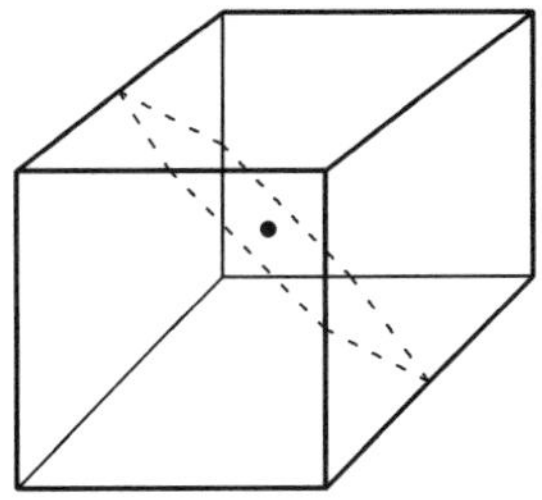

Here the cube floats with the lower left corner vertically below the centre of gravity (gravity points downwards to the left in this picture). The waterline area is a regular hexagon (dashed).

$$I_0 = 4\int_0^{\frac{1}{4}\sqrt{6}} dy \int_0^{\frac{1}{2}\sqrt{2}-y\frac{1}{3}\sqrt{3}} dx\, y^2 \tag{5.A22}$$

$$= 4\int_0^{\frac{1}{4}\sqrt{6}} y^2\left(\frac{1}{2}\sqrt{2} - y\frac{1}{3}\sqrt{3}\right) dy = \frac{5\sqrt{3}}{64}. \tag{5.A23}$$

The centre of buoyancy is of the same form as before given by (5.21), but in this case it is a bit harder to determine the area $A(z)$ at depth z because the shape of the area changes from a triangle to a hexagon at $z = -(1/6)\sqrt{3}$. For $-(1/2)\sqrt{3} < z < -(1/6)\sqrt{3}$ the shape is an equilateral triangle. Since its side length must vary linearly with z from $s = 0$ at $z = -(1/2)\sqrt{3}$ to $s = \sqrt{2}$ for $z = -(1/6)\sqrt{3}$, we have $s = (3/2)\sqrt{2}+z\sqrt{6}$ and area $A(z) = (1/4)\sqrt{3}\,s^2$. For $-(1/6)\sqrt{3} < z < 0$ the (irregular) hexagon can be obtained from this triangle by removing smaller equilateral triangles from the corners. Since their side length must vary linearly with z from $t = (1/2)\sqrt{2}$ for $z = 0$ to $t = 0$ for $z = -(1/6)\sqrt{3}$, it must be $t = (1/2)\sqrt{2}+z\sqrt{6}$. The area is thus $A(z) = (1/4)\sqrt{3}(s^2-3t^2)$ for $-(1/6)\sqrt{3} < z < 0$. One may verify that the two area functions reproduce the volume correctly. The centre of buoyancy becomes $z_B = -(13/96)\sqrt{3}$ and the metacentre

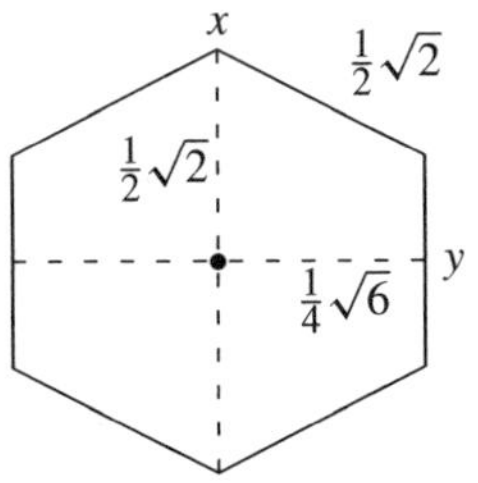

Hexagonal waterline area.

$$z_M = z_B + \frac{I_0}{V_0} = \frac{1}{48}\sqrt{3}. \tag{5.A24}$$

Thus, the metacentre is above the centre of gravity ($z_G = 0$) and the cube floats stably in this configuration.

6 Planets and stars

6.1 Put $\boldsymbol{g}(\boldsymbol{x}) = \boldsymbol{a}p(\boldsymbol{x})$ where $\boldsymbol{a}$ is an arbitrary constant vector. Then $\nabla\cdot\boldsymbol{g} = \boldsymbol{a}\cdot\nabla p$ and (4.22) follows from (6.4). Conversely from (4.22) with $p \to g_x$ we get $\oint_S g_x\, dS_x = \int_V \nabla_x g_x\, dV$ and adding the two other directions we obtain (6.4).

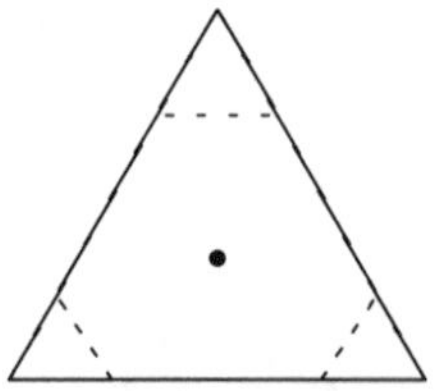

The general hexagon is an equilateral triangle with small equilateral triangles cut off at the corners.

6.2 The equation follows from $\nabla w = (1/\rho)\nabla p$. It also follows from the constancy of $H = \Phi + w$, proven on page 54.

6.3 Integrate (6.16) using the field (3.A6), and we find

$$p(r) = \begin{cases} p_c - \frac{2}{3}\pi G\rho_1^2 r^2 & \text{for } r \le a_1, \\ p_c - \frac{2}{3}\pi G\rho_1^2 a_1^2 - \frac{2}{3}\pi G\rho_2^2(r^2 - a_1^2) & \\ \qquad -G\rho_0(\rho_1 - \rho_2)a_1^3\left(\frac{1}{a_1} - \frac{1}{r}\right) & \text{for } a_1 \le r \le a. \end{cases} \tag{6.A1}$$

Apart from an additive constant and overall normalization, this is exactly the same as the minus the gravitational potential (see problem 6.3 and figure 3.4). Setting $r = a$, the central pressure becomes

$$p_c = \frac{2}{3}\pi G\rho_1^2 a_1^2 + \frac{2}{3}\pi G\rho_2^2(a^2 - a_1^2) + \frac{4}{3}\pi G\rho_2(\rho_1 - \rho_2)a_1^3\left(\frac{1}{a_1} - \frac{1}{a}\right). \tag{6.A2}$$

Again the total mass may be used to establish a relation between the densities and the radii, $M_0 = \frac{4}{3}\pi(a_1^3\rho_1 + (a^3 - a_1^3)\rho_2)$. For the Earth the central pressure becomes $p_c = 3.2$ MBar.

6.4 Eliminate ρ_0 in (6.20) by means of $M_0 = (4/3)\pi\rho_0 a^3$.

6.5 Use that $T \sim p^{1-1/\gamma}$ in an adiabatic process.

6.6

(a) The density is $\rho \sim T^{1/(\gamma-1)} \sim r^{\alpha/(\gamma-1)}$ and from (6.22) one gets $\alpha - 2 = \dfrac{\alpha}{\gamma - 1}$, which implies $\alpha = 2\dfrac{\gamma-1}{\gamma-2}$. Thus $\alpha < 0$ for $1 < \gamma < 2$.

(b) $M \sim \int r^2\rho\, dr \sim r^{3+\alpha/(\gamma-1)} \sim r^{(3\gamma-4)/(\gamma-2)}$. The exponent is negative for $1 < \gamma < 2$ so that the mass is never finite.

6.7 Use the approximation (6.23).

6.8

(a) Insert into (6.29).

(b) $p \sim (1 + \xi^2/3)^{-3}$, $\rho \sim (1 + \xi^2/3)^{-5/2}$.

(c) $M \sim \int \rho r^2\, dr \sim \int (1 + \xi^2/3)^{-5/2}\xi^2\, d\xi \approx \int \xi^{-3} d\xi$ converges for $\xi \to \infty$.

6.9 The gravitational energy is $-E_{grav}/M \approx 25\,000$ J g^{-1}. Heating and melting iron takes only about 8000 J g^{-1}. The Earth definitely melted while its bulk accumulated.

6.10 Inside the planet we find from (6.40) with $M(r) = \frac{4}{3}\pi r^2\rho_0$,

$$E_1 = -\frac{1}{2}G\int_0^a \frac{M(r)^2}{r^2}\, dr = -\frac{1}{10}\frac{GM_0^2}{a}. \tag{6.A3}$$

Outside we get

$$E_2 = -\frac{1}{2}G\int_a^\infty \frac{M_0^2}{r^2}\, dr = -\frac{1}{2}\frac{GM_0^2}{a}. \tag{6.A4}$$

The ratio is $E_1/E_2 = 1/5$ so $5/6 = 84\%$ of the gravitational energy is found outside the planet.

6.11 Put $\boldsymbol{g} = -\boldsymbol{x}/|\boldsymbol{x}|^3 = \nabla(1/|\boldsymbol{x}|)$, corresponding to $M = G = 1$ in (6.2), and use Gauss' theorem (6.4),

$$\oint_S \nabla \frac{1}{|\boldsymbol{x}|} \cdot d\boldsymbol{S} = \int_V \nabla^2 \frac{1}{|\boldsymbol{x}|} = \begin{cases} -4\pi & (\boldsymbol{x} \in V) \\ 0 & (\boldsymbol{x} \notin V). \end{cases} \tag{6.A5}$$

Show that $\nabla^2 |\boldsymbol{x}|^{-1} = 0$ for $\boldsymbol{x} = 0$ to complete the proof.

7 Hydrostatic shapes

7.1 The centrifugal acceleration is $a\Omega^2 = g_0$ so that $\Omega = \sqrt{g_0/a}$. The potential is $\Phi = -\frac{1}{2}a^2\Omega^2$ at the cylinder. Use (3.36) to get $v_{\text{esc}} = \sqrt{ag_0} \approx 313 \text{ m } s^{-1}$.

7.2 Let Δh be the change in sea level due to Δp. Use the constancy of (4.21) in the water close to the surface to get $H = \Delta p/\rho_0 + g_0 \Delta h = 0$. Then $\Delta h = -\Delta p/\rho_0 g_0 \approx -20$ cm.

7.3 Equipotential surfaces and isobars always rise and fall with the Moon's position. **(a)** No change in surface pressure, because the sea level follows along (in the static approximation). **(b)** The maximal change in pressure is $\Delta p \approx \rho_0 \Delta \Phi = \rho_0 g_0 H_0 \approx 6$ Pa where ρ_0 is the atmospheric density at sea level and H_0 is the tidal range (7.15).

7.4 Range of tidal bulge $(1/\sqrt{3}) \le \cos\theta \le 1$, or $0 \le \theta \le \theta_0 \approx 55\,°$. The volume of tidal bulge $\int_0^{\theta_0} h(\theta) 2\pi a \sin\theta \, a d\theta = 2\pi a^2 \int_{1/\sqrt{3}}^1 h(\theta) d\cos\theta = \frac{4\pi a^2}{9\sqrt{3}} H_0 \approx 17.7 \times 10^3 \text{ km}^3$.

7.5 Use (7.15) to get $m_{\text{satt}}/m_{\text{Moon}} = (D_{\text{satt}}/D_{\text{Moon}})^3 \approx 1.33 \times 10^{-3}$ or $m_{\text{satt}} \approx 9.76 \times 10^{19}$ kg.

7.6 Use $h = (\cos^2\theta - (1/3))H_0$ such that $\langle h \rangle = \left(\left\langle \cos^2\theta \right\rangle - (1/3)\right) H_0$. Averaging over a period we get $\left\langle \cos^2\theta \right\rangle = \sin^2\delta \sin^2\delta_0 + (1/2)\cos^2\delta \cos^2\delta_0$ from (7.16).

7.7 Without loss of generality we may place the Moon's orbit in the xz-plane. Then the centrifugal potential becomes $\Phi_{\text{centrifugal}} = -\frac{1}{2}\Omega^2(x^2 + (z-d)^2) = -\frac{1}{2}\Omega^2(x^2 + z^2 - 2zd + d^2)$ where $d = Dm/(m+M)$ is the distance to the centre-of-mass. Neither the constant nor the linear term in z can raise the water, so we may effectively take $\Phi_{\text{centrifugal}} = -(1/2)\Omega^2(x^2+z^2) = -(1/2)\Omega^2(a^2 - y^2)$. Including this potential, the y^2-term yields an extra range, $\Delta H_0 = (1/2)\Omega^2 a^2/g_0 \approx 14$ m, on top of the already calculated range.

7.8 Write $f(r)(3\cos^2\theta - 1) = \dfrac{f(r)}{r^2}(2z^2 - x^2 - y^2)$ and use $\nabla^2(uv) = u\nabla^2 v + v\nabla^2 u + 2\nabla u \cdot \nabla v$. Result $g = \dfrac{d^2 f}{dr^2} + \dfrac{2}{r}\dfrac{df}{dr} - 6\dfrac{f}{r^2}$.

8 Surface tension

8.1 The pressure jump across the bubble surface is $\Delta p = 4\alpha/a \approx 20$ Pa. The capillary length is as for massive spheres defined by the length scale where the hydrostatic pressure change inside the bubble matches the pressure jump, $R_c = \sqrt{2\alpha/\rho_0 g_0}$, where ρ_0 is the density of air. Numerically it becomes $R_c = 16$ cm. The bubble radius $a = 3$ cm is much smaller than this, and the bubble should be quite spherical.

8.2 In cylindrical coordinates the radial perturbation is of the form $r = a - b\cos\phi$ with $\phi = 2\pi z/\lambda$ for $b \ll a$ and $b \ll \lambda$. **(a)** The volume of a period becomes

$$V = \int_0^\lambda \pi r^2\, dz = \pi\lambda\left(a^2 + \frac{1}{2}b^2\right). \tag{8.A1}$$

The area becomes to lowest significant order in b/λ,

$$A = \int_0^\lambda 2\pi r\sqrt{1 + \left(\frac{dr}{dz}\right)^2}\, dz \approx 2\pi a\lambda\left(1 + \frac{\pi^2 b^2}{\lambda^2}\right). \tag{8.A2}$$

(b) To keep the volume constant we must vary a together with b. Defining the constant volume $V = \pi a_0^2\lambda$, we find

$$a = \sqrt{a_0^2 - \frac{1}{2}b^2} \approx a_0\left(1 - \frac{b^2}{4a_0^2}\right). \tag{8.A3}$$

Inserting this in the area we get

$$A = 2\pi a_0\lambda\left(1 + \frac{\pi^2 b^2}{\lambda^2} - \frac{b^2}{4a_0^2}\right). \tag{8.A4}$$

For $\lambda < 2\pi a_0$ the area grows with b because of the perturbation.

8.3 **(a)** Put $x = r\cos\phi$ and $y = r\sin\phi$. A circle with radius R and centre at $z = R$ has in the rz-plane the equation $R^2 = r^2 + (z-R)^2 \approx r^2 + R^2 - 2zR$ for $r \ll R$, or $z = r^2/2R$. Comparing with the polynomial one finds $1/R = \partial^2 z/\partial r^2 = 2(a\cos^2\phi + b\sin^2\phi + 2c\cos\phi\sin\phi)$. **(b)** The extrema are determined from the vanishing of $\partial(1/R)/\partial\phi = 2(-(a-b)\sin 2\phi + 2c\cos 2\phi)$, or $\tan 2\phi = (a-b)/2c$. The solutions are $\phi = \phi_0$ and $\phi = \phi_0 + \pi/2$ where $\phi_0 = \frac{1}{2}\arctan[(a-b)/2c]$.

8.4 Expanding to second order around $(x, y, z) = (x_0, 0, z_0)$ we find

$$\Delta z = \alpha\Delta x + \frac{1}{2}\beta\Delta x^2 + \frac{\alpha}{2x_0}y^2, \tag{8.A5}$$

where $\Delta z = z - z_0$, $\Delta x = x - x_0$, $\alpha = f'(x_0) = \tan\theta$, and $\beta = f''(x_0)$. Introduce a local coordinate system with coordinates ξ and η in $(x_0, 0, z_0)$

$$\Delta x = \xi\cos\theta + \eta\sin\theta \tag{8.A6}$$

$$\Delta z = -\xi\sin\theta + \eta\cos\theta. \tag{8.A7}$$

Substituting and solving for η keeping up to second-order terms,

$$\eta = \frac{1}{2}\beta\cos^3\theta\,\xi^2 + \frac{\sin\theta}{2x_0}y^2. \tag{8.A8}$$

Hence

$$\frac{1}{R_1} = \frac{\partial^2\eta}{\partial\xi^2} = \beta\cos^3\theta, \qquad \frac{1}{R_2} = \frac{\partial^2\eta}{\partial y^2} = \frac{\sin\theta}{x_0}. \tag{8.A9}$$

But

$$\beta = \frac{d^2 z}{dx^2} = \frac{d\tan\theta}{dx} = \frac{1}{\cos^2\theta}\frac{d\theta}{dx} = \frac{1}{\cos^2\theta}\frac{ds}{dx}\frac{d\theta}{ds} = \frac{1}{\cos^3\theta}\frac{d\theta}{ds} \tag{8.A10}$$

proving that $1/R_1 = d\theta/ds$.

8.5 Differentiating after s we get

$$\frac{dV}{ds} = 2\pi r \frac{dr}{ds}\left(2\frac{\sin\theta}{r} + z - \frac{2}{R_0}\right) + \pi r^2\left(2\frac{\cos\theta}{r}\frac{d\theta}{ds} - 2\frac{\sin\theta}{r^2}\frac{dr}{ds} + \frac{dz}{ds}\right) = \pi r^2 \frac{dz}{ds}.$$

Integrating from 0 to s and using that $V(0) = 0$ we get

$$V(s) = \int_0^s \pi r(s')^2 \frac{dz}{ds'}\, ds' \tag{8.A11}$$

which is indeed the volume.

9 Stress

9.1 The normal reaction is the weight N and the tangential reaction is $T = \mu N$. The angle is given by $\tan\alpha = T/N = \mu$.

9.2 The kinetic energy of the car is $\mathcal{T} = \frac{1}{2}mv^2$ and the maximal friction without skidding is $\mathcal{F} = \mu_0 m g_0$. Since the force is constant the braking distance is $d_0 = \mathcal{T}/\mathcal{F} = v^2/2\mu_0 g_0 \approx 44\ m$. Skidding we have $\mathcal{F} = \mu m g_0$, so the distance becomes $d = d_0\mu_0/\mu \approx 56\ m$.

9.3 **(a)** $\sigma = F/NA = 391$ Pa. **(b)** $\sigma = 80\,000$ Pa $= 0.8$ bar.

9.4 The pressure at the bottom in the middle of the mountain where it is highest is $p \approx \rho g_0 h$ where h is its height. Consequently, the maximal value of h is $\sigma/\rho g_0 = 10$ km. On Mars the maximal height is 27 km.

9.5 The characteristic equation is $-\lambda^3 + 3\tau\lambda^2 = 0$. Eigenvalues $\lambda = 3\tau$ and $\lambda = 0$ (doubly degenerate). Eigenvectors $\boldsymbol{e}_1 = (1, 1, 1)/\sqrt{3}$, $\boldsymbol{e}_2 = (-2, 1, 1)/\sqrt{6}$ and $\boldsymbol{e}_3 = (0, -1, 1)/\sqrt{2}$, or any linear combination of the last two.

9.6 Let the stress tensor be diagonal in a given coordinate system. Under a small rotation through an angle ϕ

$$x' = x - \phi y \qquad\qquad y' = y + \phi x \tag{9.A1}$$

we find

$$\sigma'_{yx} = \phi\sigma_{xx} - \phi\sigma_{yy} = \phi(\sigma_{xx} - \sigma_{yy}). \tag{9.A2}$$

Since that has to vanish, we must have $\sigma_{xx} = \sigma_{yy}$ and similarly for the other components.

9.7

$$\boldsymbol{\sigma}' = \mathbf{A}\cdot\boldsymbol{\sigma}\cdot\mathbf{A}^\top = \begin{pmatrix} 91/5 & -6 & -12/5 \\ -6 & 5 & -8 \\ -12/5 & -8 & 84/5 \end{pmatrix}.$$

9.8 **(a)** The average of $\langle n_i n_j\rangle$ over all directions of $\boldsymbol{n}$ does not itself depend on any direction, so that it must be proportional to Kronecker's delta, $\langle n_i n_j\rangle = k\delta_{ij}$. The constant k is determined by taking the trace of both sides, $1 = \langle \boldsymbol{n}^2\rangle = 3k$. **(b)** The average of the normal stress is $\left\langle\sum_{ij}\sigma_{ij}n_i n_j\right\rangle = \frac{1}{3}\sum_k \sigma_{ii} = -p$.

9.9

(a) The body starts to move when the elastic force equals the maximal static friction, that is for $ks = \mu_0 m g_0$ or $s = \mu_0 m g_0/k$.

(b) When the body is at the point x at time t, the actual stretch is $s + vt - x$. The equation of motion becomes

$$m\ddot{x} = k(s + vt - x) - \mu m g_0.$$

(c) Define $y = x - vt - s + \mu m g_0/k = x - vt - (1-r)s$. Then $\ddot{y} = -\omega^2 y$ which has the solution $y = A\cos\omega t + B\sin\omega t + Ct + D$. The particular solution follows from the initial conditions $x = \dot{x} = 0$ for $t = 0$.

(d) The velocity is

$$\dot{x} = v(1 - \cos\omega t) + (1-r)s\omega\sin\omega t = 2v\sin^2\frac{\omega t}{2} + 2(1-r)s\omega\sin\frac{\omega t}{2}\cos\frac{\omega t}{2},$$

which vanishes for the first time after start when

$$\tan\frac{\omega t}{2} = -\frac{(1-r)s\omega}{v},$$

so that $\omega t_0 = 2\pi - 2\alpha$ where

$$\alpha = \arctan\frac{(1-r)s\omega}{v}.$$

The other possibility is $\sin(\omega t/2) = 0$ happens later, for $\omega t = 2\pi$.

(e) The stretch is

$$\begin{aligned} s + vt - x &= rs + \frac{v}{\omega}\sin\omega t + (1-r)s\cos\omega t \\ &= rs + \frac{v}{\omega\cos\alpha}\sin(\omega t + \alpha) \end{aligned}$$

The minimum stretch is $s_1 = rs - v/\omega\cos\alpha = s(r - (1-r)/\sin\alpha)$ at $t = t_1$ where $\omega t_1 + \alpha = 3\pi/2$. For the minimum stretch to be positive, one must require that $r > (1-r)/\sin\alpha$, and since $\sin\alpha < 1$ this inequality can only be fulfilled for $r > 1 - r$ or $r > 1/2$. Notice that $\omega(t_1 - t_0) = \alpha - \pi/2$ and $\alpha < \pi/2$, so that the minimum always happens before the block stops.

(f) When the block stops at $t = t_0$ the stretch is $s_0 = s(2r - 1)$. The stretch grows to $s_0 + v\Delta t$ a time Δt after the body stops. The body starts to move again when the stretch becomes s, or $v\Delta t = s - s_0 = 2(1-r)s$, which is positive as expected. The circular frequency of the jumping motion is

$$\Omega = \frac{2\pi}{t_0 + \Delta t} = \frac{\omega}{1 + (\tan\alpha - \alpha)/\pi}.$$

When $v \to 0$ we have $\alpha \to \pi/2$ and $\Omega \to 0$, in accordance with intuition.

9.10 Writing out the six terms of the determinant, the characteristic equation becomes $\det[\boldsymbol{\sigma} - \lambda\mathbf{1}] = -\lambda^3 + I_1\lambda^2 - I_2\lambda + I_3$. An asymmetric stress tensor also has the invariant $I_4 = \sum_{ij}\sigma_{ij}(\sigma_{ij} - \sigma_{ji})$ which vanishes for a symmetric stress tensor.

10 Strain

10.1

$$\{\nabla_j u_i\} = \alpha\begin{pmatrix} 5 & -1 & 3 \\ 1 & 8 & 0 \\ -3 & 4 & 5 \end{pmatrix}, \qquad u_{ij} = \alpha\begin{pmatrix} 5 & 0 & 0 \\ 0 & 8 & 2 \\ 0 & 2 & 5 \end{pmatrix}.$$

10.2

$$\nabla \cdot \boldsymbol{u} = 3\alpha + 2\beta(x + y + z)$$
$$\nabla \times \boldsymbol{u} = (-2\alpha, -2\alpha, -2\alpha)$$

and

$$\{u_{ij}\} = \begin{pmatrix} \alpha + 2\beta x & \alpha & \alpha \\ \alpha & \alpha + 2\beta y & \alpha \\ \alpha & \alpha & \alpha + 2\beta z \end{pmatrix} = \alpha \begin{pmatrix} 1 & 1 & 1 \\ 1 & 1 & 1 \\ 1 & 1 & 1 \end{pmatrix} + 2\beta \begin{pmatrix} x & 0 & 0 \\ 0 & y & 0 \\ 0 & 0 & z \end{pmatrix}.$$

10.3

$$\{u_{ij}\} = \begin{pmatrix} A & C & 0 \\ C & -B & 0 \\ 0 & 0 & 0 \end{pmatrix}.$$

For the volume to be unchanged we must have $\nabla \cdot \boldsymbol{u} = \sum_i u_{ii} = 0$ or $A = B$.

10.4 The strain tensor becomes

$$u_{ij} = \alpha \begin{pmatrix} 0 & 1 & 0 \\ 1 & 0 & 0 \\ 0 & 0 & 0 \end{pmatrix}. \tag{10.A1}$$

The characteristic polynomial is $\det(\mathbf{u} - \lambda\mathbf{1}) = -\lambda(\lambda^2 - \alpha^2)$. The eigenvalues are $\lambda = \pm\alpha$ and $\lambda = 0$. The (unnormalized) eigenvectors are $(1, 1, 0)$, $(1, -1, 0)$ and $(0, 0, 1)$. The eigenvalues are the relative changes in length along these directions.

10.5 **(a)** The strain gradients become

$$\{\nabla_j u_i\} = \alpha \begin{pmatrix} 0 & 2y & 0 \\ y & x & 0 \\ 0 & 0 & 0 \end{pmatrix}, \tag{10.A2}$$

and are small for $|\alpha| \ll 1/L$. Cauchy's strain tensor becomes in the same approximation

$$\{u_{ij}\} = \alpha \begin{pmatrix} 0 & \frac{3}{2}y & 0 \\ \frac{3}{2}y & x & 0 \\ 0 & 0 & 0 \end{pmatrix}. \tag{10.A3}$$

(b) The eigenvalue equation becomes $\lambda^2 - x\lambda - 9/4y^2 = 0$, and has the solution $\lambda = (1/2)(x \pm \sqrt{x^2 + 9y^2})$. The corresponding (unnormalized) eigenvectors are $(3y, x \pm \sqrt{x^2 + 9y^2}, 0)$ and $(0, 0, 1)$.

10.6 Let $\boldsymbol{c} = \boldsymbol{a} + \boldsymbol{b}$. Then $2\boldsymbol{a} \cdot \boldsymbol{b} = \boldsymbol{c}^2 - \boldsymbol{a}^2 - \boldsymbol{b}^2$ and thus

$$\delta(\boldsymbol{ab}) = |\boldsymbol{c}|\,\delta\,|\boldsymbol{c}| - |\boldsymbol{a}|\,\delta\,|\boldsymbol{a}| - |\boldsymbol{b}|\,\delta\,|\boldsymbol{b}| = \boldsymbol{c}^2 u_{cc} - \boldsymbol{a}^2 u_{aa} - \boldsymbol{b}^2 u_{bb}. \tag{10.A4}$$

10.7 We have

$$\begin{aligned} a_i' &= a_i + \sum_j \nabla_j u_i a_j \\ &= a_i + \sum_j \frac{1}{2}(\nabla_j u_i - \nabla_i u_j) a_j + \sum_j \frac{1}{2}(\nabla_j u_i + \nabla_i u_j) a_j \\ &= a_i + \sum_{jk} \epsilon_{jik} \phi_k a_j + \sum_j u_{ij} a_j \\ &= (\boldsymbol{a} + \boldsymbol{\phi} \times \boldsymbol{a} + \mathbf{U} \cdot \boldsymbol{a})_i\,. \end{aligned}$$

The second term corresponds to a rotation by $|\boldsymbol{\phi}|$ around the axis $\boldsymbol{\phi}/|\boldsymbol{\phi}|$.

10.8 We must solve

$$\nabla_x u_x = \nabla_y u_y = \nabla_z u_z = 0$$
$$\nabla_y u_z + \nabla_z u_y = \nabla_z u_x + \nabla_z u_z = \nabla_x u_y + \nabla_y u_x = 0.$$

From the first we get that u_x can only depend on y and z, and for the second derivatives we get

$$\nabla_y^2 u_x = -\nabla_y \nabla_x u_y = -\nabla_x \nabla_y u_y = 0$$
$$\nabla_z^2 u_x = -\nabla_z \nabla_x u_z = -\nabla_x \nabla_z u_z = 0$$
$$\nabla_y \nabla_z u_x = -\nabla_y \nabla_x u_z = -\nabla_x \nabla_y u_z = \nabla_x \nabla_z u_y = \nabla_z \nabla_x u_y = -\nabla_z \nabla_y u_x.$$

From the last equation we get $\nabla_y \nabla_z u_x = 0$. Consequently, we must have $u_x = A + Dy + Ez$ and similar results for u_y and u_z. The vanishing of the shear strains relates some of the constants.

10.9 Successive needle transformations

$$a_i'' = \sum_j (\delta_{ij} + \nabla_j' u_i') a_j' = \sum_{jk} (\delta_{ij} + \nabla_j' u_i')(\delta_{jk} + \nabla_k u_j) a_k.$$

This shows that to lowest order the displacement gradients are simply added

$$\nabla_j u_i'' = \nabla_j u_i' + \nabla_j u_i \tag{10.A5}$$

and the same is the case for the strain tensor,

$$u_{ij}'' = u_{ij}' + u_{ij}. \tag{10.A6}$$

10.11 Use that the determinant of a product of matrices is the product of the determinant to get $\det \delta_{ij} + 2u_{ij} = \det |\delta_{ij} + \nabla_j u_i|^2$.

10.12 Define the Jacobian,

$$J_{ij} = \frac{\partial x_i'}{\partial x_j}. \tag{10.A7}$$

The condition for vanishing strain is that the metric is the identity,

$$\sum_k J_{ik} J_{jk} = \sum_k J_{ki} J_{kj} = \delta_{ij}. \tag{10.A8}$$

Thus $\mathbf{J}$ is everywhere an orthogonal matrix. Differentiating after x_l we get

$$\sum_k \frac{\partial J_{ik}}{\partial x_l} J_{jk} + \sum_k J_{ik} \frac{\partial J_{jk}}{\partial x_l} = 0 \tag{10.A9}$$

or after multiplying with J_{jm} and summing

$$\frac{\partial J_{im}}{\partial x_l} = -\sum_{jk} J_{ik} J_{jm} \frac{\partial J_{jk}}{\partial x_l}. \tag{10.A10}$$

Now we use that

$$\frac{\partial J_{jk}}{\partial x_l} = \frac{\partial^2 x_j'}{\partial x_k \partial x_l} = \frac{\partial J_{jl}}{\partial x_k} \tag{10.A11}$$

and by repeated applications of the rules we find

$$\begin{aligned}\frac{\partial J_{im}}{\partial x_l} &= -\sum_{jk} J_{ik}J_{jm}\frac{\partial J_{jl}}{\partial x_k} = \sum_{jk} J_{ik}J_{jl}\frac{\partial J_{jm}}{\partial x_k}\\ &= \sum_{jk} J_{ik}J_{jl}\frac{\partial J_{jk}}{\partial x_m} = -\sum_{jk} J_{ik}J_{jk}\frac{\partial J_{jl}}{\partial x_m}\\ &= -\frac{\partial J_{il}}{\partial x_m} = -\frac{\partial J_{im}}{\partial x_l}.\end{aligned}$$

Consequently, all the derivatives of J_{im} must vanish, so that it is a constant orthogonal matrix, i.e. a rotation (here excluding reflections). The only other possible transformation is a bodily translation which may be added to the rotation.

11 Elasticity

11.1 Expanding to first order around an equilibrium configuration $r = a$, we find for r close to a

$$\mathcal{F}(r) = \mathcal{F}(a) + (r-a)\mathcal{F}'(a).$$

In equilibrium $\mathcal{F}(a) = 0$ and thus $\mathcal{F} = (r-a)\mathcal{F}'(a)$ which is Hooke's law with $k = \mathcal{F}'(a)$.

11.2 **(a)** Trivial rewriting using (11.15). **(b)** The trace of the first term vanishes.

11.3 **(a)** Divergence and curl

$$\begin{aligned}\nabla\cdot\boldsymbol{u} &= \nabla_x u_x + \nabla_y u_y + \nabla_z u_z = 3\alpha + 2\beta(x+y+z)\\ \nabla\times\boldsymbol{u} &= (-2\alpha, -2\alpha, -2\alpha) = -2\alpha(1,1,1)\end{aligned}$$

(b) Displacement gradient

$$\{\nabla_i u_j\} = \begin{pmatrix}\alpha+2\beta x & 0 & 2\alpha\\ 2\alpha & \alpha+2\beta y & 0\\ 0 & 2\alpha & \alpha+2\beta z\end{pmatrix} \tag{11.A1}$$

Strain tensor

$$\{u_{ij}\} = \begin{pmatrix}\alpha+2\beta x & \alpha & \alpha\\ \alpha & \alpha+2\beta y & \alpha\\ \alpha & \alpha & \alpha+2\beta z\end{pmatrix} = \alpha\begin{pmatrix}1&1&1\\1&1&1\\1&1&1\end{pmatrix} + 2\beta\begin{pmatrix}x&0&0\\0&y&0\\0&0&z\end{pmatrix} \tag{11.A2}$$

(c) Stress tensor

$$\{\sigma_{ij}\} = 2\mu\alpha\begin{pmatrix}1&1&1\\1&1&1\\1&1&1\end{pmatrix} + 3\alpha\lambda\begin{pmatrix}1&0&0\\0&1&0\\0&0&1\end{pmatrix} \tag{11.A3}$$

11.4 Assuming that there is no transverse or shear deformation, and taking the x-axis along the beam axis, the only component of strain is u_{xx}. Inserting this into (11.9) one gets $\sigma_{xx} = (2\mu+\lambda)u_{xx} = P$ and $\sigma_{yy} = \sigma_{zz} = \lambda u_{xx} = \lambda P/(2\mu+\lambda)$. The displacement field is $u = xP/(2\mu+\lambda)$ and $u_y = u_z = 0$. Since it satisfies the equilibrium equation and the boundary conditions, it is the right solution.

11.5 The boundary conditions are fulfilled with the constant values $\sigma_{xx} = P$, $u_{yy} = 0$, and $\sigma_{zz} = 0$. Inserting this into (11.18b) we get $\sigma_{yy} = \nu P$, and using (11.18a,c) we get $u_{xx} = (1-\nu^2)P/E$, and $u_{zz} = -\nu(1+\nu)P/E$. The displacement field becomes $u_x = xu_{xx}$ and $u_z = zu_{zz}$. Since it satisfies the boundary conditions and the equilibrium equations, it is the right solution.

11.6

(a) Use that

$$u_{xx} = \nabla_x u_x, \qquad u_{yy} = \nabla_y u_y, \qquad u_{xy} = \tfrac{1}{2}(\nabla_x u_y + \nabla_y u_x) \tag{11.A4}$$

(b) The equilibrium equations become,

$$\nabla_x \sigma_{xx} + \nabla_y \sigma_{xy} = 0, \qquad \nabla_x \sigma_{yx} + \nabla_y \sigma_{yy} = 0. \tag{11.A5}$$

(c) Solution follows by insertion.

(d) The strain tensor becomes

$$E u_{xx} = \sigma_{xx} - \nu\sigma_{yy} = \nabla_y^2\phi - \nu\nabla_x^2\phi, \tag{11.A6}$$

$$E u_{yy} = \sigma_{yy} - \nu\sigma_{xx} = \nabla_x^2\phi - \nu\nabla_y^2\phi, \tag{11.A7}$$

$$E u_{zz} = -\nu(\sigma_{xx} + \sigma_{yy}) = -\nu(\nabla_x^2\phi + \nabla_y^2\phi), \tag{11.A8}$$

$$E u_{xy} = (1+\nu)\sigma_{xy} = -(1+\nu)\nabla_x\nabla_y\phi, \tag{11.A9}$$

The equation now follows by insertion into (11.35).

(e) The equations to be solved are,

$$u_{xx} = 2x, \qquad u_{yy} = -2\nu x, \qquad u_{zz} = -2\nu x, \tag{11.A10}$$

$$u_{xy} = -2(1+\nu)y, \qquad u_{xz} = 0, \qquad u_{yz} = 0. \tag{11.A11}$$

First the diagonal elements are solved on their own,

$$u_x = x^2, \qquad u_y = -2\nu xy, \qquad u_z = -2\nu xz. \tag{11.A12}$$

Here $u_{yz} = 0$ automatically. To fulfill $u_{xz} = 0$ an extra term is necessary in u_x, so that

$$u_x = x^2 + \nu z^2, \qquad u_y = -2\nu xy, \qquad u_z = -2\nu xz, \tag{11.A13}$$

Finally add another term to u_x to obtain the correct expression for u_{xy},

$$u_x = x^2 + \nu z^2 - (2+\nu)y^2, \qquad u_y = -2\nu xy, \qquad u_z = -2\nu xz, \tag{11.A14}$$

To this may be added an arbitrary solid translation or rotation.

11.7 The absolute minimum of the coefficient of the first term happens for $\alpha = 1/3$.

11.9 **(a)** Follows from the symmetry of σ_{ij} and u_{ij}. **(b)** In order for $\Sigma_{ij}\delta u_{ij}\sigma_{ij} = \Sigma_{ijkl}\delta u_{ij}\lambda_{ijkl}u_{kl}$ to become a total differential $\delta((1/2)\Sigma_{ijkl}\lambda_{ijkl}u_{ij}u_{kl})$, the (9×9)-matrix $\lambda_{(ij)(kl)}$ must be symmetric in its index pairs.

12 Solids at rest

12.2 Insert the displacement field into the Navier–Cauchy equation (12.2) and obtain the equations

$$2\mu A + 2(\lambda+\mu)(A+B+C) = 0 \tag{12.A1}$$

$$2\mu B + 2(\lambda+\mu)(A+B+C) = 0 \tag{12.A2}$$

$$k + 2\mu C + 2(\lambda+\mu)(A+B+C) = 0 \tag{12.A3}$$

Since the determinant is non-vanishing, this set of three equations has a unique solution for A, B, and C.

12.3 (**a**) Let the rear end of the bullet be flat and moving according to $x = x(t)$ with $x = 0$ for $t = 0$ reckoned from the back end $x = 0$ of the barrel. The volume is $V = Ax$ where $A = \pi a^2$ is the cross section of the barrel. The equation of motion for the bullet is $m\ddot{x} = pA$. The isentropic expansion obeys $pV^\gamma =$ const. or $px^\gamma = p_0 x_0^\gamma$ where p_0 is the initial pressure. The equation of motion takes the form

$$m\ddot{x} = p_0 A \left(\frac{x_0}{x}\right)^\gamma . \tag{12.A4}$$

Multiplying by $\dot{x}$ and integrating we get

$$\dot{x}^2 = \frac{2p_0 A x_0}{(\gamma - 1)m}\left(1 - \left(\frac{x_0}{x}\right)^{\gamma-1}\right), \tag{12.A5}$$

where the constant has been determined such that $\dot{x} = 0$ for $x = x_0$.

(**b**) For $x = L$ we have $\dot{x} = U$, leading to

$$p_0 = \frac{(\gamma - 1)mU^2}{2Ax_0}\left(1 - \left(\frac{x_0}{L}\right)^{\gamma-1}\right)^{-1} . \tag{12.A6}$$

Numerically this becomes $p_0 \approx 9700$ bar. The exit pressure is $p_1 = p_0(x_0/L)^\gamma \approx 15$ bar.

(**c**) The initial temperature is $T_0 = p_0 M_{\rm mol}/R\rho_0 \approx 3500$ K and the final temperature $T_1 \approx 550$ K.

(**d**) The strains are for $r = a$, $b = a + d$, $P = p_0$, $E = 205$ GPa and $\nu = 29\%$,

$$u_{rr} = \frac{a^2}{b^2 - a^2}\left(1 - \nu - (1 + \nu)\frac{b^2}{a^2}\right)\frac{P}{E} \approx -5300 \text{ bar}, \tag{12.A7a}$$

$$u_{\phi\phi} = \frac{a^2}{b^2 - a^2}\left(1 - \nu + (1 + \nu)\frac{b^2}{a^2}\right)\frac{P}{E} \approx 7000 \text{ bar}, \tag{12.A7b}$$

$$u_{zz} = -2\nu\frac{a^2}{b^2 - a^2}\frac{P}{E} \approx \; -690 \text{ bar}. \tag{12.A7c}$$

The tensile strength of steel is 5500 bar. Since $u_{\phi\phi}$ is just above this value, the barrel may blow up.

12.4 We use that u_{zz} is linear in x and y, of the form

$$u_{zz} = \nabla_z u_z = \alpha - \beta_x x - \beta_y y$$

and consequently

$$u_z = a_z - \phi_y x + \phi_x y + \alpha z - \beta_x xz - \beta_y yz$$

where the coefficients are all constants. Using this result, we find $\nabla_x u_x = \nabla_y u_y = -\nu(\alpha - \beta_x x - \beta_y y)$. Integrating and demanding that $u_{xy} = u_{xz} = u_{zy} = 0$, we obtain the most general form

$$u_x = a_x - \phi_z y + \phi_y z - \alpha\nu x + \frac{1}{2}\beta_x(z^2 - \nu(x^2 - y^2)) - \beta_y \nu xy, \tag{12.A8a}$$

$$u_y = a_y + \phi_z x - \phi_x z - \alpha\nu y + \frac{1}{2}\beta_y(z^2 - \nu(y^2 - x^2)) - \beta_x \nu xy, \tag{12.A8b}$$

$$u_z = a_z - \phi_y x + \phi_x y + \alpha z - \beta_x xz - \beta_y yz. \tag{12.A8c}$$

Here (a_x, a_y, a_z) represents simple translations of the body, and (ϕ_x, ϕ_y, ϕ_z) simple rotations around the coordinate axes. The coefficient α corresponds to a uniform stretching, and only (β_x, β_y) represents bending into the coordinate directions.

12.5 Same general solution as for the pressurized tube but with a and b interchanged

$$A = -\frac{1}{2(\lambda+\mu)}\frac{b^2}{b^2-a^2}P = -(1+\sigma)(1-2\sigma)\frac{b^2}{b^2-a^2}\frac{P}{E},$$
$$B = -\frac{1}{2\mu}\frac{a^2b^2}{b^2-a^2}P = -(1+\sigma)\frac{a^2b^2}{b^2-a^2}\frac{P}{E}.$$

The rest is straightforward.

12.6 (a) The centrifugal force density is radial and given by $f_r = \rho_0\Omega^2 r$. (b) The general solution to (12.59) is

$$u_r = Ar + \frac{B}{r} - \frac{1}{8}\frac{\rho_0\Omega^2}{\lambda+2\mu}r^3 \tag{12.A9}$$

where A and B are integration constants. Use that u_r must be finite for $r=0$ and $\sigma_{rr}=0$ for $r=a$. The final solution becomes

$$u_r = \frac{1}{8}\frac{\rho_0\Omega^2 a^2}{\lambda+2\mu}r\left(3-2\nu-\frac{r^2}{a^2}\right). \tag{12.A10}$$

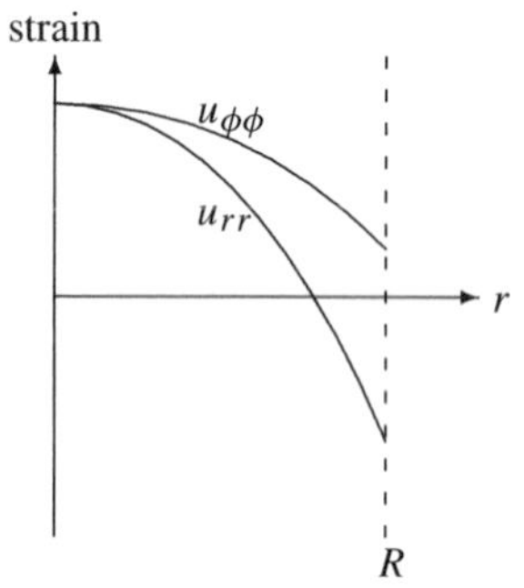

Sketch of strains in the massive rotating cylinder.

(c) The strains are

$$u_{rr} = \frac{1}{8}\frac{\rho_0\Omega^2 a^2}{\lambda+2\mu}\left(3-2\nu-3\frac{r^2}{a^2}\right), \quad u_{\phi\phi} = \frac{1}{8}\frac{\rho_0\Omega^2 a^2}{\lambda+2\mu}\left(3-2\nu-\frac{r^2}{a^2}\right). \tag{12.A11}$$

The radial strain is positive for $r=0$, vanishes for $r = a\sqrt{1-2\nu/3}$, and is negative for $r=a$. (d) Breakdown happens for $r=0$ where the extension and tension is maximal.

12.7 Letting $x \to x-\alpha$ in (12.20), we get

$$\begin{aligned}
u_x &= \frac{\nu}{2R}\alpha^2 & -\frac{\nu}{R}\alpha x & \quad +\frac{1}{2R}(z^2+\nu(x^2-y^2)),\\
u_y &= & -\frac{\nu}{R}\alpha y & \quad +\frac{\nu}{R}xy,\\
u_z &= & \frac{1}{R}\alpha z & \quad -\frac{1}{R}xz.
\end{aligned}$$

The first column represents a simple translation and the second a uniform stretching of the form (11.25).

13 Computational elastostatics

13.1 We assume a linear combination

$$\nabla_x^+ f(x) = af(x) + bf(x+\Delta x) + cf(x+2\Delta x).$$

Expand to second order and require the coefficient of $f(x)$ and $\nabla_x^2 f(x)$ to vanish and the coefficient of $\nabla_x f(x)$ to be 1, to get

$$\begin{aligned}
a+b+c &= 0\\
\frac{1}{2}b+2c &= 0\\
b+2c &= 1.
\end{aligned}$$

The solution is $a=-3/2$, $b=2$, and $c=-1/2$.

14 Vibrations

14.1 When the Lamé coefficients depend on position we get three extra terms involving their gradients

$$\rho\frac{\partial^2 u_i}{\partial t^2} = f_i + \mu\nabla^2 u_i + (\lambda+\mu)\nabla_i\nabla\cdot\boldsymbol{u} + (\nabla\cdot\boldsymbol{u})\nabla_i\lambda + (\nabla_i\boldsymbol{u})\cdot\nabla\mu + (\nabla\mu\cdot\nabla)u_i.$$

14.2 The total wave is

$$\boldsymbol{u} = (k, 0, k_L)e^{ik_L z} + A_T(k_T, 0, k)e^{-ik_T z} + A_L(k, 0, -k_L)e^{-ik_L z}, \tag{14.A1}$$

from which one derives

$$\sigma_{xz} = \frac{1}{2}i\mu(2k(1-A_L)k_L + (k^2-k_T^2)A_T, \tag{14.A2}$$

$$\sigma_{zz} = -i\mu((k^2-k_T^2)(1+A_L) + 2kk_T A_T. \tag{14.A3}$$

They vanish for

$$A_L = \frac{4k^2k_Lk_T - (k^2-k_T^2)^2}{4k^2k_Lk_T + (k^2-k_T^2)^2}, \qquad A_T = -\frac{4kk_T(k^2-k_T^2)}{4k^2k_Lk_T + (k^2-k_T^2)^2}. \tag{14.A4}$$

Next one puts $k_L = k\cot\theta$, $k_T = k\cot\theta'$ and use Snell's law $\sin\theta'/\sin\theta = c_T/c_L = q$.

14.3 The fields have the form

$$u_y = Ae^{\kappa_T z}, \tag{14.A5a}$$

$$u'_y = A'e^{-\kappa'_T z}. \tag{14.A5b}$$

The boundary conditions $u'_y = u_y$ and $\sigma'_{yz} = \sigma_{yz}$ at $z = 0$ lead to the equations,

$$A' = A, \qquad \mu'\kappa'_T A' = -\mu\kappa_T A. \tag{14.A6}$$

In view of κ_T and κ'_T being positive these equations do not have a solution. Thus, there are no interface waves of this kind. At a free surface, we would instead have $A' = 0$, but can only require $\sigma_{yz} = 0$ which leads to $A = 0$.

14.4 It is easiest to reintroduce the stress tensor and write (14.14) as

$$\rho_0\omega^2 u_i = -\sum_j \nabla_j\sigma_{ij}. \tag{14.A7}$$

Multiplying it with the complex conjugate field $u_i^\times$ and summing over i we get

$$\rho_0\omega^2|\boldsymbol{u}|^2 = -\sum_{ij} u_i^\times\nabla_j\sigma_{ij} = -\sum_{ij}\nabla_j(u_i^\times\sigma_{ij}) + \sum_{ij}\nabla_j u_i^\times\sigma_{ij}.$$

Integrating over the body this becomes

$$\rho_0\omega^2\int_V |\boldsymbol{u}|^2\,dV = -\oint_S\sum_{ij}u_i^\times\sigma_{ij}\,dS_j + \int_V u_{ij}^\times\sigma_{ij}\,dV, \tag{14.A8}$$

where we have used in the last term the symmetry of the stress tensor. The surface integral vanishes because $\boldsymbol{u} = \boldsymbol{0}$ or $\boldsymbol{\sigma}\cdot\boldsymbol{n} = \boldsymbol{0}$ at the surface. The integral on the right-hand side is essentially equal to the elastic energy which is real and positive definite. Hence ω^2 is real and positive.

14.5 **(a)** For an arbitrary vector field $\boldsymbol{u}$ we first find a solution ψ to the equation

$$\nabla^2\psi = \nabla\cdot\boldsymbol{u}. \tag{14.A9}$$

Then we put

$$\boldsymbol{u}_L = \nabla\psi \tag{14.A10}$$

$$\boldsymbol{u}_T = \boldsymbol{u} - \boldsymbol{u}_L. \tag{14.A11}$$

Clearly, $\nabla\times\boldsymbol{u}_L = \boldsymbol{0}$ and $\nabla\cdot\boldsymbol{u}_T = 0$.
(b) Inserting the mixed field (14.3) into (14.2) we obtain

$$\boldsymbol{F} = \rho\frac{\partial^2\boldsymbol{u}_L}{\partial t^2} - (\lambda+2\mu)\nabla^2\boldsymbol{u}_L = -\rho\frac{\partial^2\boldsymbol{u}_T}{\partial t^2} + \mu\nabla^2\boldsymbol{u}_T. \tag{14.A12}$$

This field has no curl and no divergence, $\nabla\times\boldsymbol{F} = \boldsymbol{0}$, $\nabla\cdot\boldsymbol{F} = \boldsymbol{0}$. From the double-cross rule (2.67) it must satisfy $\nabla^2\boldsymbol{F} = \boldsymbol{0}$. Determine a field $\boldsymbol{G}$ satisfying $\rho\partial^2\boldsymbol{G}/\partial^2 t = \boldsymbol{F}$ together with $\nabla\times\boldsymbol{G} = \boldsymbol{0}$ and $\nabla\cdot\boldsymbol{G} = 0$. Such a field can always be found by integrating $\boldsymbol{F}$ twice over time. Then define new longitudinal and transverse fields

$$\boldsymbol{u}'_L = \boldsymbol{u}_L - \boldsymbol{G}, \qquad \boldsymbol{u}'_T = \boldsymbol{u}_T + \boldsymbol{G}. \tag{14.A13}$$

These fields also have $\nabla\times\boldsymbol{u}'_L = \boldsymbol{0}$ and $\nabla\cdot\boldsymbol{u}'_T = 0$, satisfy (14.2) and yield the same solution as the others: $\boldsymbol{u} = \boldsymbol{u}_L + \boldsymbol{u}_T = \boldsymbol{u}'_L + \boldsymbol{u}'_T$.

15 Fluids in motion

15.3 Leonardo's law tells us that $Av = A_1v_1 + A_2v_2$. The ratio of the rates in the two pipes is $A_1v_1/A_2v_2 = 2$, and since $A_1/A_2 = 9/4$ we get $v_1/v_2 = 8/9$. The total rate is $Av = 3A_2v_2$ so that $v_2/v = 4/3$ and $v_1/v = 32/27$.

15.4 Differentiating through all the time-dependence, one gets

$$\frac{d\rho(\boldsymbol{x}(t),t)}{dt} = \frac{d\boldsymbol{x}(t)}{dt}\cdot\frac{\partial\rho(\boldsymbol{x},t)}{\partial\boldsymbol{x}} + \frac{\partial\rho(\boldsymbol{x},t)}{\partial t} = \boldsymbol{v}(\boldsymbol{x},t)\cdot\nabla\rho(\boldsymbol{x},t) + \frac{\partial\rho(\boldsymbol{x},t)}{\partial t} = \frac{D\rho}{Dt}.$$

15.5 **(a)** Let Q be the total volume of flow in the stream. Then the average velocity in the x-direction is $v_x(x) = Q/h(x)d$. **(b)** The advective acceleration is estimated as $w = (\boldsymbol{v}\cdot\nabla)\boldsymbol{v} \approx v_x dv_x/dx$. **(c)** Constant acceleration w implies $v_x \approx \sqrt{2wx}$ for a suitable choice of origin and orientation of the x-axis. Hence $h(x) \sim 1/\sqrt{x}$ is the shape of the curve.

15.6 **(a)** Let Q be the total volume of flow in the stream. Then the average velocity in the x-direction is $v_x(x) = Q/\pi a(x)^2$. **(b)** The advective acceleration is estimated as $w \approx v_x dv_x/dx$. **(c)** Constant acceleration w implies $v_x \approx \sqrt{2wx}$ for a suitable choice of origin and orientation of the x-axis. Hence $a(x) \sim 1/x^{1/4}$ is the shape of the tube.

15.7 Define the vector field

$$\boldsymbol{f}'(\boldsymbol{v}) = \frac{\partial f(\boldsymbol{v})}{\partial\boldsymbol{v}}. \tag{15.A1}$$

Then

$$\frac{\partial\rho}{\partial t} = -\frac{3}{t}\rho - \frac{M_0}{t^3}\frac{\boldsymbol{x}}{t^2}\cdot\boldsymbol{f}'\left(\frac{\boldsymbol{x}}{t}\right), \tag{15.A2}$$

$$\rho\nabla\cdot\boldsymbol{v} = \frac{3}{t}\rho, \tag{15.A3}$$

$$(\boldsymbol{v}\cdot\nabla)\rho = \frac{M_0}{t^3}\frac{\boldsymbol{x}}{t^2}\cdot\boldsymbol{f}'\left(\frac{\boldsymbol{x}}{t}\right). \tag{15.A4}$$

The sum of the three right-hand sides vanishes which means that the equation of continuity (15.24) is satisfied.

15.8 In a small interval of time, δt, a material particle with a small volume dV is displaced to fill out another volume dV', the size of which may be calculated from the Jacobi determinant of the infinitesimal mapping $\boldsymbol{x}' = \boldsymbol{x} + \boldsymbol{v}\delta t$,

$$\frac{dV'}{dV} = \left|\frac{\partial \boldsymbol{x}'}{\partial \boldsymbol{x}}\right| = \begin{vmatrix} \frac{\partial x'}{\partial x} & \frac{\partial y'}{\partial x} & \frac{\partial z'}{\partial x} \\ \frac{\partial x'}{\partial y} & \frac{\partial y'}{\partial y} & \frac{\partial z'}{\partial y} \\ \frac{\partial x'}{\partial z} & \frac{\partial y'}{\partial z} & \frac{\partial z'}{\partial z} \end{vmatrix} = \begin{vmatrix} 1 + \nabla_x v_x \delta t & \nabla_x v_y \delta t & \nabla_x v_z \delta t \\ \nabla_y v_x \delta t & 1 + \nabla_y v_y \delta t & \nabla_y v_z \delta t \\ \nabla_z v_x \delta t & \nabla_z v_y \delta t & 1 + \nabla_x v_z \delta t \end{vmatrix}.$$

To first order in δt, only the diagonal elements contribute to the determinant, and we find

$$\frac{dV'}{dV} \approx (1 + \nabla_x v_x \delta t)(1 + \nabla_y v_y \delta t)(1 + \nabla_z v_z \delta t) \tag{15.A5}$$

$$\approx 1 + \nabla_x v_x \delta t + \nabla_y v_y \delta t + \nabla_z v_z \delta t = 1 + \boldsymbol{\nabla} \cdot \boldsymbol{v}\, \delta t. \tag{15.A6}$$

The change in volume is $\delta(dV) = dV' - dV$, and after dividing by δt the rate of change of such a *comoving* volume becomes

$$\frac{D(dV)}{Dt} = \boldsymbol{\nabla} \cdot \boldsymbol{v}\, dV. \tag{15.A7}$$

15.9 The local 'continuity equations' for mass and momentum are

$$\frac{\partial \rho}{\partial t} + \boldsymbol{\nabla} \cdot (\rho \boldsymbol{v}) = J,$$

$$\frac{\partial(\rho v_i)}{\partial t} + \boldsymbol{\nabla} \cdot (\rho v_i \boldsymbol{v}) = \rho g_i.$$

The cosmological equations then become

$$\dot{\rho} = -3H\rho + J \tag{15.A8a}$$

$$\dot{H} + H^2 = J\frac{H}{\rho} - \frac{4\pi}{3} G\rho. \tag{15.A8b}$$

Clearly there is a steady-state solution

$$J = 3H\rho \tag{15.A9a}$$

$$\rho = \frac{9H^2}{4\pi G} = 6\rho_c. \tag{15.A9b}$$

From $H = H_0 = 75$ km s^{-1} Mpc^{-1} $= 2.4 \times 10^{-18}$ s^{-1} one gets $\rho = 6.3 \times 10^{-26}$ kg m^{-3} and $J = 4.6 \times 10^{-43}$ kg m^{-3} s^{-1}, corresponding to the creation of nine protons per cubic kilometre per year. Not much!

16 Nearly ideal flow

16.1 Use that $v_0 \approx \sqrt{3p_0/\rho_0}$ and $c_0 = \sqrt{\gamma R T_0 / M_{\text{mol}}}$, and the ideal gas law. For air $\gamma \approx 7/5$ so that $v_0/c_0 \approx \sqrt{15/7} \approx 1.46$.

16.2

(a) Leonardo's law says $\pi a^2 U_0 = \pi b^2 U_1$, so that

$$U_1 = \frac{a^2}{b^2} U_0 = 5.74 \text{ m/s}. \tag{16.A1}$$

(b) Bernoulli's theorem applied to a streamline through the center leads to,

$$\frac{1}{2}U_0^2 + \frac{p_0}{\rho_0} = \frac{1}{2}U_1^2 + \frac{p_1}{\rho_0}. \tag{16.A2}$$

Consequently the pressure drop in the center becomes,

$$\Delta p = p_0 - p_1 = \frac{1}{2}\rho_0\left(\frac{a^4}{b^4} - 1\right)U_0^2 = 3973 \text{ Pa}. \tag{16.A3}$$

(c) The pressure difference must balance the weight of the mercury minus its buoyancy, $\Delta p = (\rho_1 - \rho_0)g_0 h$, so that

$$h = \frac{\Delta p}{(\rho_1 - \rho_0)g_0} = 3.2 \text{ cm}. \tag{16.A4}$$

(d) The mercury level is $h/2$ over the initial level. The condition becomes $h < 2(c + d)$, or

$$U_0 < 2\sqrt{\frac{b^4(c+d)g_0}{b^4 - a^2}\frac{\rho_1 - \rho_0}{\rho_0}} = 9.7 \text{ m/s}. \tag{16.A5}$$

16.3 The exit speed is given by Torricelli's law $U = \sqrt{2g_0 h} = 2 \text{ m s}^{-1}$. The total volume rate is $Q = \pi a^2 U = 0.16 \text{ l s}^{-1}$ so that the time it takes to fill $V = 10$ litre is $t = V/Q = 64$ s.

16.4 Leonardo's law tells us that the average velocity at the top of the barrel is $v_0 = (A/A_0)v$, where v is the average velocity in the spout. Bernoulli's theorem now says

$$\frac{1}{2}v_0^2 + \frac{p_0}{\rho_0} + g_0 h = \frac{1}{2}v^2 + \frac{p_0}{\rho_0} \tag{16.A6}$$

so that with $\kappa = A/A_0$

$$v = \sqrt{\frac{2g_0 h}{1 - \kappa^2}}. \tag{16.A7}$$

The speed is slightly higher than the speed of free fall.

16.5 Take two streamlines that pass from the surface in the barrel through the two spouts. Then we have

$$\frac{p_0}{\rho_0} + g_0 h = \frac{1}{2}v_1^2 + \frac{p_0}{\rho_0} = \frac{1}{2}v_2^2 + \frac{p_0}{\rho_0} \tag{16.A8}$$

or $v_1 = v_2$.

16.6 The solution to the differential equation (16.22) with initial condition $z = h$ at $t = 0$ is

$$z = \left(\sqrt{h} - \frac{1}{2}t\frac{A}{A_0}\sqrt{2g_0}\right)^2 = h\left(1 - \frac{t}{T}\right)^2 \tag{16.A9}$$

where T is the total time (16.23) for emptying the barrel.

16.7 **(a)** The volume flux is

$$Q = \int_0^a v(r)2\pi r\, dr = \frac{\pi a^2 U}{1 + \kappa}. \tag{16.A10}$$

Since $U = \sqrt{2g_0 h}$ the reduction factor is $1/(1 + \kappa)$.

16.8 For an ideal gas $\rho = p/C$ where $C = RT_0/M_{\rm mol}$ and thus $w = C \log p$.
(a) Bernoulli's theorem becomes

$$\frac{1}{2}U^2 + C \log p = C \log p_0, \tag{16.A11}$$

where p is the ambient pressure and p_0 the stagnation pressure in the tube,

$$p_0 = p \exp\left(\frac{M_{\rm mol} U^2}{2RT_0}\right). \tag{16.A12}$$

(b) Taking $T_0 = 223$ K and $M_{\rm mol} = 29$ g mol^{-1}, we obtain $(p_0 - p)/p = 63\%$.

16.9 Consider the difference between two surfaces S_1 and S_2, both having the curve C as perimeter and oriented consistently,

$$\int_{S_1} \nabla \times \boldsymbol{v} \cdot d\boldsymbol{S} - \int_{S_2} \nabla \times \boldsymbol{v} \cdot d\boldsymbol{S} = \oint_S \nabla \times \boldsymbol{v} \cdot d\boldsymbol{S} = \int_V \nabla \cdot (\nabla \times \boldsymbol{v})\, dV = 0.$$

Here $S = S_1 - S_2$ is the closed surface formed by the two open surfaces with an extra minus sign because of the requirement that the closed surface should have an outwardly oriented normal. In the last step, Gauss' theorem has been used to convert the integral over S to an integral over the volume V contained in S.

16.10 Using that $\nabla \times (\boldsymbol{\Omega} \times \boldsymbol{x}) = 2\boldsymbol{\Omega}$ it follows from Stokes theorem that for all $\boldsymbol{\Omega}$

$$\oint_C \boldsymbol{\Omega} \cdot \boldsymbol{x} \times d\boldsymbol{\ell} = \oint_C \boldsymbol{\Omega} \times \boldsymbol{x} \cdot d\boldsymbol{\ell} = \int_S 2\boldsymbol{\Omega} \cdot d\boldsymbol{S}. \tag{16.A13}$$

16.11 The proof is trivial because for a barotropic fluid $\nabla p = \rho \nabla w$ where w is the pressure potential.

16.12 The proof is straightforward because from (16.54) we get $\nabla(\partial\Psi/\partial t + H) = \boldsymbol{0}$ where H is given by (16.35).

16.13 From (16.70) we get the critical velocity

$$U = \sqrt{\frac{3\pi}{5}\left(\frac{\rho_1}{\rho_0} - 1\right) a g_0}, \tag{16.A14}$$

which becomes merely $U \approx 7$ cm s^{-1} for the worm.

16.14 In spherical coordinates the integral becomes

$$\begin{aligned}\int_{r \geq a} (\boldsymbol{v} - \boldsymbol{U})^2\, dV &= \int_{r \geq a} \left((v_r - U_r)^2 + (v_\theta - U_\theta)^2\right) dV \\ &= \int_a^\infty dr \int_0^\pi r d\theta \int_0^{2\pi} r \sin\theta d\phi\, U^2 \frac{a^6}{4r^6}\left(1 + 3\cos^2\theta\right) \\ &= \frac{2}{3}\pi a^3 U^2.\end{aligned}$$

16.15 From Lenoardo's law we have

$$v_x = \frac{Q}{h(x) - b(x)} \tag{16.A15}$$

where Q is a constant. From mass conservation, $\nabla_z v_z = -\nabla_x v_x$ it follows that $v_z = f(x) - z\nabla_x v_x$. At the bottom $z = b$ we have the boundary condition $v_z = v_x b'$ and at the surface $z = h$ we have $v_z = v_x h'$. Either of these determine $f(x)$.

16.16 Writing out the double sum,

$$\boldsymbol{v} \cdot (\boldsymbol{v} \cdot \nabla)\boldsymbol{v} = \sum_{ij} v_i v_j \nabla_j v_i \tag{16.A16}$$

we find

$$|\boldsymbol{v} \cdot (\boldsymbol{v} \cdot \nabla)\boldsymbol{v}|^2 \leq \sum_{ij}(v_i v_j)^2 \sum_{kl}(\nabla_l v_k)^2 = |\boldsymbol{v}|^4 \, |\nabla \boldsymbol{v}|^2 . \tag{16.A17}$$

16.17 **(a)** Define $\rho = \rho_0 + \Delta\rho$ and $p = p_0 + \Delta p$. Using that $\nabla p_0 = \rho_0 \boldsymbol{g}$, the Euler and continuity equations become to first order in the small quantities,

$$\rho_0 \frac{\partial \boldsymbol{v}}{\partial t} = \boldsymbol{g}\Delta\rho - \nabla\Delta p, \tag{16.A18a}$$

$$\frac{\partial \Delta\rho}{\partial t} = -\nabla \cdot (\rho_0 \boldsymbol{v}). \tag{16.A18b}$$

Using $\Delta p = c_0^2 \Delta\rho$ we get

$$\frac{\partial^2 \Delta p}{\partial t^2} = c_0^2 \frac{\partial^2 \Delta\rho}{\partial t^2} = -c_0^2 \nabla \cdot \left(\rho_0 \frac{\partial \boldsymbol{v}}{\partial t}\right) = c_0^2 \nabla^2 \Delta p - c_0^2 \nabla \cdot (\boldsymbol{g}\Delta\rho) \tag{16.A19}$$

and using $\nabla \cdot \boldsymbol{g} = 0$ the wave equation follows.

(b) The ratio of the two terms on the right-hand side of the wave equation becomes for a wave of wavelength λ

$$\frac{\left|c_0^2 \nabla^2 \Delta p\right|}{\left|c_0^2 (\boldsymbol{g} \cdot \nabla)(\Delta p / c_0^2)\right|} \approx \frac{c_0^2 \, |\Delta p| \, /\lambda^2}{g_0 \, |\Delta p| \, /\lambda} \approx \frac{c_0^2}{g_0 \lambda}. \tag{16.A20}$$

The condition for ignoring gravity is that $g_0\lambda \ll c_0^2$ or $\lambda \ll c_0^2/g_0$. In the atmosphere this becomes $\lambda \ll 12$ km which is quite reasonable in view of the height of the atmosphere.

17 Viscosity

17.1 In an isentropic gas we have $p \sim \rho^\gamma$ and $p \sim \rho T$, so that $\rho \sim T^{1/(\gamma-1)}$ and $\nu \sim T^{1/2-1/(\gamma-1)}$. For monatomic gases $\gamma = 5/3$ and $\nu \sim T^{-1}$, for diatomic $\gamma = 7/5$ and $\nu \sim T^{-2}$, and for multiatomic $\gamma = 4/3$ and $\nu \sim T^{-5/2}$.

17.2 One finds from (17.10) and (17.11) $d \approx 3$ μm and $t_0 \approx 11$ s. The layer seems a bit thin compared to, for example curling (example 17.2.1). The tire pattern probably influences the 'ice grip' considerably.

17.3 **(a)** The total flux along x per unit of length along z is

$$Q(t) = \int_{-\infty}^{\infty} v_x(y, t)\, dy. \tag{17.A1}$$

From (17.5) we get by integrating over y

$$\frac{dQ}{dt} = \nu \int_{-\infty}^{\infty} \frac{\partial^2 v_x}{\partial y^2}\, dy = 0, \tag{17.A2}$$

because $\partial v_x / \partial y$ vanishes at infinity.
(b) The total momentum per unit of length is

$$\mathcal{P} = \int_{-\infty}^{\infty} \rho_0 v_x(y, t)\, dy = \rho_0 Q \tag{17.A3}$$

and is constant because Q is.
(c) The kinetic energy per unit of length in both x and z is

$$\mathcal{T} = \frac{1}{2}\rho_0 \int_{-\infty}^{\infty} v_x(y,t)^2\, dy. \tag{17.A4}$$

In the Gaussian case this becomes

$$\mathcal{T} = \frac{1}{2}\rho_0 U^2 \frac{a^2}{a^2+2\nu t}\int_{-\infty}^{\infty} \exp\left(-2\frac{y^2}{a^2+2\nu t}\right) dy \tag{17.A5}$$

$$= \frac{1}{2}\rho_0 U^2 \sqrt{\frac{\pi}{2}}\frac{a^2}{\sqrt{a^2+2\nu t}}. \tag{17.A6}$$

It vanishes like $t^{-1/2}$ for $t \to \infty$.

17.4 **(a)** $L \approx 10$ km, $U \approx 1$ m/s, $\mathrm{Re} \approx 10^{10}$.
(b) $L \approx 30$ m, $U \approx 30$ m/s, $\mathrm{Re} \approx \times 10^9$.
(c) $L \approx 1000$ km, $U \approx 10$ m/s, $\mathrm{Re} \approx 10^{12}$.
(d) $L \approx 500$ km, $U \approx 50$ m/s, $\mathrm{Re} \approx 3 \times 10^{12}$.
(e) $L \approx 1$ km, $U \approx 100$ m/s, $\mathrm{Re} \approx 10^{10}$.
(f) $L \approx 50$ m, $U \approx 10$ m/s, $\nu \approx 1$ m^2/s, $\mathrm{Re} \approx 500$.
(g) $L \approx 10^3$ km, $U \approx 10$ cm/y, $\nu \approx 10^6$ m^2/s, $\mathrm{Re} \approx 3 \times 10^{-6}$.

17.5 Write the pressure correction (17.31) in the form

$$\Delta p = \left(c_0^2 + \frac{\zeta}{\rho_0}\frac{\partial}{\partial t}\right)\Delta\rho \tag{17.A7}$$

and apply the operator in parenthesis to (17.34), using that this operator commutes with both spatial and time derivatives.

17.6 One verifies explicitly that the given expression satisfies the equation of motion (17.5). The constant in front is determined by requiring

$$\int_{-\infty}^{\infty} v_x(y,t)\, dy = \int_{-\infty}^{\infty} v_x(y,0)\, dy.$$

For $t \to 0$ the Gaussian becomes infinitely narrow (a δ-function) and thus $v_x(y,t) \to v_x(y,0)$. Finally, assuming that $v_x(y,0) = 0$ for $|y| \le a$ one gets for $|y| \to \infty$ and $4\nu t \gg a^2$

$$v_x(y,t) \approx \frac{1}{2\sqrt{\pi\nu t}}\exp\left(-\frac{y^2}{4\nu t}\right)\int_{-\infty}^{\infty} v_x(y',0)\, dy'. \tag{17.A8}$$

18 Plates and pipes

18.1 The Navier–Stokes equation becomes

$$(\boldsymbol{v}\cdot\boldsymbol{\nabla})\boldsymbol{v} = -\frac{1}{\rho_0}\boldsymbol{\nabla} p^* + \nu\boldsymbol{\nabla}^2\boldsymbol{v}. \tag{18.A1}$$

18.2 Use the general solution (18.4) and the no-slip conditions to get

$$v_x = \frac{G}{2\eta}y(d-y) + U\frac{y}{d}$$

The maximum happens at $y = \dfrac{d}{2} + \dfrac{U\eta}{Gd}$ and lies between the plates for $2U\eta < Gd^2$.

18.3 Let the pressure gradient be G along the x-direction and the relative plate velocity U along the z-direction. Assume that the field is of the form $\boldsymbol{v} = (v_x(y), 0, v_z(y))$. Then the Navier–Stokes equations imply that p and v_x are identical to the planar pressure driven flow, whereas v_z is identical to the velocity driven flow.

18.4 If pressure were used to drive the planar sheet, there would have to be a linearly falling pressure along the open surface. But that is impossible because the open surface requires constant pressure.

18.5 Using the same method as in section 18.3 we find in the two layers the equations,

$$\nabla_y p = -\rho_1 g_0 \cos\theta, \qquad \nabla_z p = 0, \tag{18.A2}$$

$$\nabla_y p = -\rho_2 g_0 \cos\theta, \qquad \nabla_z p = 0. \tag{18.A3}$$

from which we get the pressure (using continuity)

$$p = \begin{cases} p_0 - \rho_1 g_0 y \cos\theta & (0 < y < d_1) \\ p_0 - (\rho_1 d_1 + \rho_2(y - d_1)) g_0 \cos\theta & (d_1 < y < d_1 + d_2) \end{cases}. \tag{18.A4}$$

Similarly, for the velocity field one gets in general

$$v_x(y) = \begin{cases} -K_1 y^2 + A_1 y + B_1 & (0 < y < d_1) \\ -K_2 y^2 + A_2 y + B_2 & (d_1 < y < d_1 + d_2) \end{cases} \tag{18.A5}$$

where $K_1 = g_0 \sin\theta / 2\nu_1$ and $K_2 = g_0 \sin\theta / 2\nu_2$. Demanding no-slip at the solid boundaries and continuity of v_x and $\sigma_{yx} = \eta \nabla_y v_x$ at the interface one finds

$$v_x(y) = \begin{cases} (K_1(d_1 - y) + A d_2) y & (0 < y < d_1) \\ (K_2(d_1 - y) + A d_1)(d_1 + d_2 - y) & (d_1 < y < d_1 + d_2) \end{cases} \tag{18.A6}$$

where $A = (K_1 d_1 - K_2 d_2)/(d_1 + d_2)$.

18.6 Drag and dissipation are

$$\mathcal{D} = \Delta p L d = G L W d = 12 \eta U \frac{LW}{d} \tag{18.A7}$$

$$P = U\mathcal{D} = 12 \eta U^2 \frac{LW}{d}. \tag{18.A8}$$

18.7 On the plate the shear stress is

$$\sigma_{xy} = \eta \left. \nabla_y v_x \right|_{y=0} = \rho_0 g_0 \sin\theta h. \tag{18.A9}$$

The drag is obtained by multiplying with the plate area

$$\mathcal{D} = \sigma_{xy} L W = \rho_0 g_0 \sin\theta L W d = M g_0 \sin\theta \tag{18.A10}$$

where $M = LWh$ is the mass of the fluid. The dissipation is $P = M g_0 U \sin\theta = U\mathcal{D}$.

18.8 The effective pressure gradient is $G = \rho_0 g_0$ and the Reynolds number (18.30) becomes

$$\mathrm{Re} = \frac{g_0 a^3}{4\nu^2}. \tag{18.A11}$$

Solving for a we find

$$a = \left(\frac{4\nu^2}{g_0} \mathrm{Re} \right)^{1/3} \approx 74\ \mu\mathrm{m} \times \mathrm{Re}^{1/3}. \tag{18.A12}$$

For Re = 2300 one finds $a = 1$ mm.

18.9 **(a)** The simplest way is to recognize that the mass dimension (kg) contained in ρ and η can only be removed by forming the ratio $\nu = \eta/\rho$ of dimension $\mathrm{m^2\,s^{-1}}$. Since Q has dimension of $\mathrm{m^3\,s^{-1}}$, the time unit can only be removed by forming the ratio Q/ν which has dimension of m. Finally, dividing by a, we get the dimensionless number $Q/\nu a$ which is proportional to the Reynolds number. **(b)** Since the dimension of the pressure gradient is $[G] = \mathrm{Pa\,m^{-1}} = \mathrm{kg\,s^{-2}\,m^{-2}}$ we get rid of the mass dimension by dividing by the mass density to get $[G/\rho] = \mathrm{m\,s^{-2}}$. Then $[G/\rho\nu^2] = \mathrm{m^{-3}}$ and $Ga^3/\rho\nu^2$ is dimensionless. For laminar pipe flow this is also proportional with the Reynolds number (18.30).

18.10 Since pressure differences $\Delta p = \rho_0 Q R$ are additive when put in series whereas Q is the same, R is additive in series. Since the mass flux $Q = \Delta p/\rho_0 R$ is additive over the branches of a parallel connection, whereas Δp is the same, reciprocal resistance is additive in parallel connections.

18.12 **(a)** Use the no-slip boundary conditions on (18.25). **(b)** The shear stress is

$$\sigma_{zr}(r) = \eta \frac{dv_z}{dr} = -\frac{1}{2} G r + \frac{\eta A}{r}.$$

The total drag per unit of length on the two inner surfaces becomes $\sigma_{zr}(a) 2\pi a - \sigma_{zr}(b) 2\pi b = \pi G (b^2 - a^2)$.

18.13 **(a)** Insert the fields into the steady-flow equations and verify that they are fulfilled and that the boundary conditions are fulfilled. **(b)** For $a = b$ we get the circular pipe field and for $b \to \infty$ we get the planar field. **(c)** Using the area element in elliptic coordinates the flux becomes

$$Q = \int_0^1 dr \int_0^{2\pi} d\theta\, abr\, v_z(r) = \frac{\pi G a^3 b^3}{4\eta(a^2 + b^2)}. \tag{18.A13}$$

(d) $U = Q/\pi ab$. **(e)** The drag is $\mathcal{D} = GL\pi ab$.

18.14 **(a)** Use the general solution (18.25) with $G = 0$ and boundary conditions $v_z = U$ for $r = a$ and $v_z = 0$ for $r = b$, to get

$$v_z = U \frac{\log \dfrac{b}{r}}{\log \dfrac{b}{a}},$$

(b) The discharge rate is

$$\begin{aligned} Q &= \int_a^b v_z 2\pi r\, dr = 2\pi \frac{U}{\log(b/a)} \int_a^b (\log b - \log r) r\, dr \\ &= \frac{\pi(b^2 - a^2)}{2 \log \frac{b}{a}} U - \pi a^2 U. \end{aligned}$$

Numerically this is $Q = 20.3\ \mathrm{cm^3/s}$.
(c) Since $Q = \pi(c^2 - a^2)U$ we get

$$c = \sqrt{\frac{b^2 - a^2}{2 \log \frac{b}{a}}} = 2.7\ \mathrm{mm} \tag{18.A14}$$

(d) The longitudinal stress from the liquid on the surface of the wire becomes

$$\sigma_{zr} = \eta\, \nabla_r v_z|_{r=a} = -\frac{\eta U}{a \log \frac{b}{a}} \approx -62\ \mathrm{Pa}. \tag{18.A15}$$

The total friction force

$$\mathcal{F}_z = 2\pi a L \sigma_{zr} = -\frac{\eta U L}{\log \frac{b}{a}} \approx -0.08\ \mathrm{N}. \tag{18.A16}$$

(e) There are many ways to answer this question, for example by calculating the ratio between the friction force and the weight of a cylinder of the liquid of radius a

$$\frac{|\mathcal{F}_z|}{\rho_0 g_0 \pi a^2 L} \approx \frac{\eta U}{\rho_0 g_0 \pi a^2 \log \frac{b}{a}} \approx 2 \tag{18.A17}$$

It was only marginally correct to disregard gravity.

18.15 **(a)** Under the same assumptions used to derive (18.43) with h replaced by $z+L$ and $U_1 = 4\nu L/a^2$, one has $U(z) = \sqrt{g_0(z+L)+c^2} - c$ with $c = 4\nu L/a^2$. Using Leonardo's law the differential equation becomes

$$\frac{dz}{dt} = -\frac{a^2}{b^2} U(z). \tag{18.A18}$$

(b) It has the solution

$$T = \frac{b^2}{a^2}\int_0^h \frac{dz}{U(z)} = \frac{2b^2}{a^2 g_0}\left(U(h) - U(0) + c\log\frac{U(h)}{U(0)}\right) \tag{18.A19}$$

as may be verified by differentiation after h.

18.19 In cylindrical coordinates assume that the flow field is radial, $\boldsymbol{v} = v_r(r)\,\boldsymbol{e}_r$ outside the pipe. Volume conservation implies that $v_r 2\pi r L$ is the same for all r. Hence $v_r(r) = Q/2\pi r$ where Q is the volume flow through the pipe wall per unit of pipe length.

18.20 **(a)** The moment of force on the cylinder is the sum of the torque exerted by the wire and by the fluid, i.e. for $d \ll a$,

$$M_z = -\tau\phi - 2\pi\eta\dot{\phi}a^3\frac{L}{d}. \tag{18.A20}$$

The moment of inertia is Ma^2 and the angular momentum of the inner cylinder $\mathcal{L}_z = Ma^2\dot{\phi}$. The equation of motion then becomes $d\mathcal{L}_z/dt = \mathcal{M}_z$. **(b)** Insert $\phi \sim e^{\lambda t}$ to get

$$Ma^2\lambda^2 = -\tau - 2\pi\eta a^3\frac{L}{d}\lambda, \tag{18.A21}$$

which has the solutions

$$\lambda = -\gamma \pm \sqrt{\gamma^2 - \frac{\tau}{Ma^2}}, \qquad \gamma = \frac{\pi\eta a L}{Md}. \tag{18.A22}$$

For $\gamma < \tau/Ma^2$ the solutions are oscillating but decay with the rate γ. For $\gamma > \tau/Ma^2$ the solutions are exponentially damped with the leading exponent $\lambda = -\gamma + \sqrt{\gamma^2 - \tau/Ma^2}$. Critical damping occurs for $\gamma = \tau/Ma^2$.

18.21 $P = 2\pi\eta\Omega^2 a^3 L/d \approx 10$ W.

18.22 The vorticity field is

$$\boldsymbol{\omega} = \nabla \times \boldsymbol{v} = (\boldsymbol{e}_r\nabla_r + \boldsymbol{e}_\phi\nabla_\phi) \times v_\phi(r)\boldsymbol{e}_\phi = \left(\frac{dv_\phi}{dr} + \frac{v_\phi}{r}\right)\boldsymbol{e}_z = \boldsymbol{0}.$$

18.23 **(a)** Using (18.26) and (17.18) we obtain the density of dissipation,

$$2\eta\sum_{ij} v_{ij}^2 = 2\eta(v_{rz}^2 + v_{zr}^2) = \eta(\nabla_r v_z)^2 = \frac{G^2}{4\eta}r^2 \tag{18.A23}$$

and dividing by ρ_0 the local specific rate of dissipation.
(b) Integrating we get the total rate of dissipation

$$P = 2\eta\int_V \sum_{ij} v_{ij}^2 2\pi r L\, dr = \frac{\pi G^2 a^4 L}{8\eta} = QGL = Q\Delta p.$$

19 Creeping flow

19.1 Velocity $U = 11\ \mu\text{m s}^{-1}$, Reynolds number $\text{Re} = 10^{-4}$, settling time $t = 12$ years.

19.2 From (19.14) we get for $\rho_1 \gg \rho_0$

$$-\frac{dz}{dt} = \frac{2}{9}\rho_1 \frac{a^2 g_0}{\eta(z)}, \tag{19.A1}$$

where the height-dependent viscosity $\eta(z)$ is found from (17.3) and (4.43),

$$\eta(z) = \eta_0 \sqrt{1 - \frac{z}{h_2}}. \tag{19.A2}$$

From the differential equation we get

$$t_0 = \int_0^z \frac{9}{2} \frac{\eta_0}{\rho_1 a^2 g_0} \sqrt{1 - \frac{z'}{h_2}}\, dz' = \frac{3\eta_0 h_2}{\rho_1 a^2 g_0}\left(1 - \left(1 - \frac{z}{h_2}\right)^{3/2}\right). \tag{19.A3}$$

Taking $2a = 1\ \mu\text{m}$, $z = 10$ km and $\rho_1 = 2.5\ \text{g cm}^{-3}$ we get $t_0 \approx 4$ years.

19.3 The equation of motion for a particle of mass m_1 and radius a is

$$m_1 \ddot{z} = -(m_1 - m_0) g_0 - 6\pi\eta a \dot{z} \tag{19.A4}$$

where z is a vertical coordinate, $m_1 = 4/3\pi a^3 \rho_1$ and $m_0 = 4/3\pi a^3 \rho_0$. The solution is

$$\dot{z} = -U(1 - e^{-t/\tau}) \tag{19.A5}$$

where U is the terminal velocity (19.14) and the time constant,

$$\tau = \frac{2}{9}\frac{a^2}{\nu}\frac{\rho_1}{\rho_0} = \frac{U}{g_0}\frac{\rho_1}{\rho_1 - \rho_0}. \tag{19.A6}$$

19.4 The equation of motion for a particle of mass m_1 and radius a is

$$m_1 \ddot{z} = -m_1 g_0 + \tfrac{1}{2} C_D \rho_0 A \dot{z}^2 \tag{19.A7}$$

where z is a vertical coordinate, $m_1 = 4/3\pi a^3 \rho_1$. For the downwards velocity $v = -\dot{z}$ we get

$$\dot{v} = g_0\left(1 - \frac{v^2}{U^2}\right), \tag{19.A8}$$

where U is given by (19.23). The solution is

$$v = U \tanh \frac{g_0 t}{U}, \tag{19.A9}$$

with an exponential time constant of $\tau = U/2g_0$.

19.5 The pressure contribution to Stokes drag is

$$\mathcal{D}_p = -\int_{r=a} p\, dS_z = -\int_0^\pi p \cos\theta\, 2\pi a^2 \sin\theta\, d\theta = 2\pi\eta U a \tag{19.A10}$$

which is 1/3 of the total drag.

19.6 **(a)** The discharge is at $\theta = \pi/2$

$$Q = \int_a^b (-v_\theta) 2\pi r \, dr = \pi (b-a)^2 \left(1 + \frac{a}{2b}\right) U.$$

(b) The ratio is

$$\frac{Q}{\pi (b^2 - a^2) U} = \frac{b-a}{a+b} \left(1 + \frac{a}{2b}\right).$$

(c) The ratio vanishes because of the no-slip condition which requires the velocity to vanish at the surface of the sphere.

19.7

(a) Write $\boldsymbol{x} = r\boldsymbol{e}_r$ and use (B.15) to obtain

$$\frac{d\boldsymbol{x}}{dt} = \frac{dr}{dt}\boldsymbol{e}_r + r\frac{d\theta}{dt}\boldsymbol{e}_\theta.$$

(b) Combine the differential equations to obtain

$$\frac{d\theta}{dr} = -\frac{B}{A} \tan\theta$$

which is a solvable first-order equation. The integral over r is carried out by means of partial fractions

$$\frac{B}{A} = \frac{r^2 + \frac{1}{4}ar + \frac{1}{4}a^2}{r(r-a)(r+\frac{1}{2}a)} = -\frac{1}{2r} + \frac{1}{r-a} + \frac{1}{2r+a}.$$

(c) For $r \to \infty$ we get $d \to r\sin\theta = \sqrt{x^2 + y^2}$.

(d) Put $\theta = \pi/2$ to get $d = (r-a)\sqrt{1 + a/2r}$ where r is the point of closest approach.

19.8 Let $\boldsymbol{v}$ be the velocity field in the rest frame of the body. The total work of the body on the fluid is in the rest frame of the asymptotic fluid

$$\oint_S \sum_{ij} (v_i - U_i)(-\sigma_{ij}) \, dS_j = \sum_i U_i \oint_S \sum_j \sigma_{ij} \, dS_j = \boldsymbol{U} \cdot \boldsymbol{\mathcal{R}} = U\mathcal{D} \tag{19.A11}$$

because $v_i = 0$ at the surface of the body.

19.9 **(a)** Follows from linearity of the field equations and the pressure independence of the boundary conditions.
(b) The field equations are

$$\nabla^2 A_{ij} = \nabla_i Q_j \tag{19.A12}$$

$$\sum_i \nabla_i A_{ij} = 0. \tag{19.A13}$$

The boundary conditions are for $|\boldsymbol{x}| \to \infty$

$$A_{ij}(\boldsymbol{x}) \to \delta_{ij} \tag{19.A14}$$

$$Q_i(\boldsymbol{x}) \to 0. \tag{19.A15}$$

At the surface of the body the velocity field must vanish, $\boldsymbol{n} \cdot \mathbf{A}(\boldsymbol{x}) = 0$.
(c) The stress tensor is

$$\sigma_{ij} = -p\delta_{ij} + \eta(\nabla_i v_j + \nabla_j v_i) = \eta \sum_k \tau_{ijk} U_k \tag{19.A16}$$

where

$$\tau_{ijk} = -\delta_{ij} Q_k + \nabla_i A_{jk} + \nabla_j A_{ik}. \tag{19.A17}$$

The total reaction force on the body with surface $\mathcal{S}$ is

$$\mathcal{R}_i = \oint_{\mathcal{S}} \sum_j \sigma_{ij} dS_j = \eta \sum_k U_k \oint_{\mathcal{S}} \tau_{ijk} dS_j. \tag{19.A18}$$

This shows that the tensor,

$$S_{ik} = \oint_{\mathcal{S}} \tau_{ijk} dS_j \tag{19.A19}$$

may be understood as a form factor, such that

$$\mathcal{R}_i = \eta \sum_k S_{ik} U_k. \tag{19.A20}$$

20 Rotating fluids

20.1 **(a)** The centrifugal acceleration is $g_0 = a\Omega^2$ so $\Omega = \sqrt{g_0/a} \approx 0.44\ \mathrm{s}^{-1}$ and $\tau = 2\pi/\Omega \approx 14$ s. **(b)** For a horizontally moving ball the Coriolis force is vertical, if the table is aligned with the ring and vanishes if the table is orthogonal to the ring. At typical velocity of $v \approx 4\ \mathrm{m\ s}^{-1}$ the Coriolis acceleration is $g_c = 2\Omega v \approx 3.6\ \mathrm{m\ s}^{-2}$ which is 1/3 of standard gravity. It does not make the ball deviate to the side, but changes the vertical acceleration considerably, depending on whether the ball flies along the direction of motion of the ring or against it. **(c)** With the basket plate mounted orthogonally to the ring, the game is influenced in the same way as for Ping Pong. With the basket plate parallel with the ring, the Coriolis force makes the ball deviate sideways in its upwards motion.

20.2 The pressure becomes according to (20.16)

$$p = \begin{cases} p_0 - 2\Omega U_0 \rho_0 y - \rho_0 g_0 z & \text{(upper)} \\ p_1 - 2\Omega U_1 \rho_1 y - \rho_1 g_1 z & \text{(lower)} \end{cases}, \tag{20.A1}$$

where p_0 and p_1 are integration constants determined by the thickness of the layers (demanding continuity at, say, $y = 0$). The pressures must be equal at both sides of the interface for all y, so that the interface obeys the equation,

$$z = \frac{p_1 - p_0 - 2\Omega(U_1\rho_1 - U_0\rho_0)y}{(\rho_1 - \rho_0)g_0}. \tag{20.A2}$$

The difference between the shore levels at $y = \pm d/2$ is

$$\Delta z = -\frac{2\Omega(U_1\rho_1 - U_0\rho_0)d}{(\rho_1 - \rho_0)g_0} \approx +107\ \mathrm{cm} \tag{20.A3}$$

which is opposite and five times bigger than the difference (20 cm) at the open surface.

20.3 **(a)** The transformation matrix $a_{ij} = (\boldsymbol{a}_i)_j$ is orthogonal $\sum_k a_{ik}a_{jk} = \sum_k a_{ki}a_{kj} = \delta_{ij}$. Differentiating the last after time we get $\sum_k \dot{a}_{ki}a_{kj} + \sum_k a_{ki}\dot{a}_{kj} = 0$. This shows that $\Omega_{ij} = \sum_k \dot{a}_{ki}a_{kj}$ is antisymmetric $\Omega_{ij} = -\Omega_{ji}$ so that we may put $\Omega_{ij} = \sum_k \epsilon_{ijk}\Omega_k$. Using orthogonality we have trivially $(\dot{\boldsymbol{a}}_i)_j = \dot{a}_{ij} = \sum_{mk} \dot{a}_{mj}a_{mk}a_{ik} = \sum_l \epsilon_{jkl}\Omega_l a_{ik} = (\boldsymbol{a}_i \times \boldsymbol{\Omega})_j = -(\boldsymbol{\Omega} \times \boldsymbol{a}_i)_j$.

(b) Differentiating after time the velocity becomes

$$\dot{\boldsymbol{x}} = -\dot{\boldsymbol{c}} + \sum_i \dot{\boldsymbol{a}}_i x'_i + \sum_i \boldsymbol{a}_i \dot{x}'_i = -\dot{\boldsymbol{c}} - \boldsymbol{\Omega} \times \boldsymbol{x} + \sum_i \boldsymbol{a}_i \dot{x}'_i. \tag{20.A4}$$

Differentiating once more the acceleration becomes

$$\begin{aligned}\ddot{x} &= -\ddot{c} - \dot{\Omega} \times x - \Omega \times \dot{x} + \sum_i \dot{a}_i \dot{x}'_i + \sum_i a_i \ddot{x}'_i \\ &= -\ddot{c} - \dot{\Omega} \times x - \Omega \times \dot{x} - \Omega \times \sum_i a_i \dot{x}'_i + \sum_i a_i \ddot{x}'_i \\ &= -\ddot{c} - \dot{\Omega} \times x - \Omega \times \dot{x} - \Omega \times (\dot{x} + \dot{c} + \Omega \times x) + \sum_i a_i \ddot{x}'_i\end{aligned}$$

or

$$\ddot{x} = -\ddot{c} - \Omega \times \dot{c} - \dot{\Omega} \times x - 2\Omega \times \dot{x} - \Omega \times (\Omega \times x) + \sum_i a_i \ddot{x}'_i. \tag{20.A5}$$

(c) Using $m\ddot{x}' = f'$ and $f = \sum_i a_i f'_i$ we get (20.41).

21 Computational fluid dynamics

21.1 The last term is easily integrated, because

$$\delta \int v \cdot \nabla \tilde{p} \, dV = \int dV [\delta v \cdot \nabla \tilde{p} + v \cdot \nabla \delta \tilde{p}] = \int dV \delta v \cdot \nabla \tilde{p}$$

where we have used Gauss' theorem in the last step, and dropped the surface terms.

The middle term is also easily integrated, because

$$\delta \int dV \frac{1}{2} \sum_{ij} (\nabla_i v_j)^2 = \int dV \sum_{ij} \nabla_i v_j \nabla_i \delta v_j = \int dV [-\nabla^2 v] \cdot \delta v$$

where we again have used Gauss' theorem and discarded boundary terms.

The problem arises from the inertia term $\delta v \cdot (v \cdot \nabla) v = \sum_{ij} \delta v_i v_j \nabla_j v_i$. Assume that the integral is an expression of the form $\sum_{ijkl} a_{ijkl} v_i v_j \nabla_k v_l$ with suitable coefficients a_{ijkl} satisfying $a_{ijkl} = a_{jikl}$. Varying the velocity and again dropping boundary terms we get (suppressing the integral as well as the sums over repeated indices)

$$\begin{aligned}\delta(a_{ijkl} v_i v_j \nabla_k v_l) &= 2a_{ijkl} \delta v_i v_j \nabla_k v_l + a_{ijkl} v_i v_j \nabla_k \delta v_l \\ &= 2a_{ijkl} \delta v_i v_j \nabla_k v_l - 2a_{ijkl} \delta v_l v_j \nabla_k v_i \\ &= 2(a_{ijkl} - a_{ljki}) \delta v_i v_j \nabla_k v_l.\end{aligned}$$

In order for this to reproduce the desired result $\delta v_i v_j \nabla_j v_i$ we must have

$$a_{ijkl} - a_{ljki} = \tfrac{1}{2} \delta_{il} \delta_{kj} \tag{21.A1}$$

but that is impossible because the left-hand side is antisymmetric under interchange of i and l whereas the right-hand side is symmetric.

21.2 Under a small variation $\delta p(x)$ we find

$$\delta \mathcal{E} = \int_V (\nabla \tilde{p} \cdot \nabla \delta \tilde{p} + s \delta \tilde{p}) \, dV = \int_V \left(-\nabla^2 p + s\right) \delta \tilde{p} \, dV \tag{21.A2}$$

where the surface terms in the integral have been dropped (assuming either $\tilde{p} = 0$ or $n \cdot \nabla \tilde{p} = 0$ on the surface). This vanishes only for arbitrary variations when the Poisson equation is fulfilled. Choosing

$$\delta \tilde{p} = \epsilon (\nabla^2 \tilde{p} - s) \tag{21.A3}$$

will make $\delta \mathcal{E}$ negative and thus make the field converge towards the desired solution.

21.3 In the pressure stress $\sigma_{xx} = -p + 2\eta \nabla_x v_x$ both terms belong to the 00-grid, and similarly for σ_{yy}. The shear stress $\sigma_{xy} = \eta(\nabla_x v_y + \nabla_y v_x)$ belongs to the 11-grid.

22 Global laws of balance

22.1 **(a)** Use the result of section 22.1 to get $\omega = 0.03\ \mathrm{s}^{-1}$ and thus $T = 2\pi/\omega \approx 200$ s. **(b)** Using the dissipation rate from Poiseuille flow (18.36) the energy balance becomes,

$$\frac{d\mathcal{E}}{dt} = -8\pi\eta v^2 L, \tag{22.A1}$$

leading to the equation of motion

$$\frac{d^2z}{dt^2} = -\frac{2g_0}{L}z - \frac{8\nu}{a^2}\frac{dz}{dt}. \tag{22.A2}$$

Inserting $z \sim e^{\lambda t}$, the solution to the characteristic equation is

$$\lambda = -\gamma \pm \sqrt{\gamma^2 - \omega^2} \tag{22.A3}$$

where $\gamma = 4\nu/a^2 \approx 0.04\ \mathrm{s}^{-1}$. The slowest decay rate $\lambda = -0.015\ \mathrm{s}^{-1}$, so the water will not oscillate at all but come to rest exponentially fast in a characteristic time $1/\gamma \approx 66$ s. **(c)** Yes.

22.2 With initial conditions $v = 0$ for $t = 0$, the solution to (22.24) becomes

$$u = -g_0 t - U \log\left(1 - \frac{Qt}{M_0}\right). \tag{22.A4}$$

Integrating $u = dz/dt$ with initial condition $z = 0$ at $t = 0$, the attained height becomes

$$z = -\frac{1}{2}g_0 t^2 + Ut + \frac{UM_0}{Q}\left(1 - \frac{Qt}{M_0}\right)\log\left(1 - \frac{Qt}{M_0}\right). \tag{22.A5}$$

At the end of the burn at $t = t_1 = (M_0 - M_1)/Q$, the velocity and height are

$$u_1 = -g_0 t_1 - U \log\frac{M_1}{M_0}, \qquad z_1 = -\frac{1}{2}g_0 t_1^2 + Ut_1 + \frac{UM_1}{Q}\log\frac{M_1}{M_0}. \tag{22.A6}$$

After this moment, the ballistic orbit becomes

$$z = z_1 + u_1(t - t_1) - \frac{1}{2}g_0(t - t_1)^2. \tag{22.A7}$$

The height is maximal at

$$t_2 = t_1 + \frac{u_1}{g_0}, \qquad z_2 = z_1 + \frac{u_1^2}{2g_0}. \tag{22.A8}$$

After t_2 the orbit is

$$z = z_2 - \tfrac{1}{2}g_0(t - t_2)^2 \tag{22.A9}$$

so that the rocket reaches the ground again at

$$t_3 = t_2 + \sqrt{\frac{2z_2}{g_0}}. \tag{22.A10}$$

22.3 Integrating (22.46) one gets the total volume,

$$V(t) = \int_0^t A\, v(t)\, dt = \sqrt{2g_0 h}\, A\tau \log\cosh\frac{t}{\tau}. \tag{22.A11}$$

22.4 The total mechanical energy of the water in the cistern is

$$\mathcal{E} \approx \frac{1}{2}\rho_0 A_0 h\, v_0^2 + \frac{1}{2}\rho_0 A L v^2 + \frac{1}{2}\rho_0 A_0 g_0 h^2 \approx \frac{1}{2}\rho_0 A L_0 v^2 + \frac{1}{2}\rho_0 A_0 g_0 h^2. \tag{22.A12}$$

Evidently the potential energy is constant. Choosing the gravitational potential to vanish at the exit, the flux of energy out of the control volume becomes

$$\oint_S \rho_0 \epsilon_{\text{mech}}\, \boldsymbol{v} \cdot d\boldsymbol{S} \approx \rho_0 \left(\frac{1}{2}v^3 A - \frac{1}{2}v_0^3 A_0 - g_0 A_0 h v_0\right) \approx \rho_0 A v \left(\frac{1}{2}v^2 - g_0 h\right). \tag{22.A13}$$

Since the atmospheric pressure performs no net work on the fluid, $A_0 p_0 v_0 = A v p_0$ the mechanical energy must be conserved, and we find

$$\rho_0 A L_0 v \frac{dv}{dt} + \rho_0 A v \left(\frac{1}{2}v^2 - g_0 h\right) = 0 \tag{22.A14}$$

which is the correct equation of motion (22.45).

22.5 Integrating over the half space, the kinetic energy of the transition region is

$$\mathcal{T}' = \int_a^\infty \frac{1}{2}\rho_0 v(r)^2\, 2\pi r^2\, dr = \pi \rho_0 a^3 v^2. \tag{22.A15}$$

The ratio of this kinetic energy to the kinetic energy in the cistern, $\mathcal{T}_0 = \frac{1}{2}\rho_0 A_0 h v_0^2 = \rho_0 A^2 h v^2 / 2A_0$, is

$$\frac{\mathcal{T}'}{\mathcal{T}_0} = 2\frac{b^2}{ah} \tag{22.A16}$$

which can be large or small depending on h. The ratio to the kinetic energy in the pipe $\mathcal{T}_1 = (1/2)\pi a^2 L \rho_0 v^2$ is

$$\frac{\mathcal{T}'}{\mathcal{T}_1} = \frac{2a}{L} \tag{22.A17}$$

which is generally small.

22.6 Choose a control volume consisting of all the water in the system between the two open surfaces. The open surface in the cistern is fixed whereas the open surface in the pipe is moving. The total kinetic energy in the system is (under the usual assumptions),

$$\mathcal{T} = \frac{1}{2}\rho_0 A_0 h v_0^2 + \frac{1}{2}\rho_0 A (L + x) v^2. \tag{22.A18}$$

Using Reynolds' transport theorem the material derivative of the kinetic energy is

$$\frac{D\mathcal{T}}{Dt} = \frac{d\mathcal{T}}{dt} - \frac{1}{2}\rho_0 A_0 v_0^3 \tag{22.A19}$$

because the system only gains kinetic energy through the open water surface of the cistern. The total power of gravity is,

$$P = \rho_0 g_0 A_0 h v_0 + \rho_1 g_1 A (L + x) v, \tag{22.A20}$$

where $g_1 = g_0 \sin\alpha$ is the projection of gravity on the pipe slope. The total power of the pressure on the open surfaces vanishes because of Leonardo's law, $A_0 v_0 = A v$.

Kinetic energy balance now leads to the differential equation

$$\frac{d^2x}{dt^2} = \frac{2g_0(h + (L + x)\sin\alpha) - (dx/dt)^2}{2(L_0 + x)}, \tag{22.A21}$$

which can be solved numerically.

22.7 (**a**) Putting $\boldsymbol{x}_n = \boldsymbol{x} + \boldsymbol{x}'_n$ and $\boldsymbol{v}_n = \boldsymbol{v} + \boldsymbol{v}'_n$ we get

$$\boldsymbol{\mathcal{L}} = \sum_n m_n \boldsymbol{x}_n \times \boldsymbol{v}_n = M\boldsymbol{x} \times \boldsymbol{v} + \sum_n m_n \boldsymbol{x}'_n \times \boldsymbol{v}'_n. \tag{22.A22}$$

The average of the second term vanishes because of the lack of correlation between $\boldsymbol{x}'_n$ and $\boldsymbol{v}'_n$, so that

$$\langle \boldsymbol{\mathcal{L}} \rangle = M\boldsymbol{x} \times \boldsymbol{v}. \tag{22.A23}$$

(**b**) The total kinetic energy is

$$\mathcal{T} = \sum_n \frac{1}{2} m_n \boldsymbol{v}_n^2 = \frac{1}{2} M\boldsymbol{v}^2 + \sum_n \frac{1}{2} m_n \boldsymbol{v}_n'^2. \tag{22.A24}$$

Its average becomes

$$\langle \mathcal{T} \rangle = \frac{1}{2} M\boldsymbol{v}^2 + \sum_n \frac{1}{2} m_n \left\langle \boldsymbol{v}_n'^2 \right\rangle = \frac{1}{2} M\boldsymbol{v}^2 + \frac{3}{2} MU^2. \tag{22.A25}$$

The last term represents the internal energy.

23 Reaction forces and moments

23.1 The force is found from (23.6)

$$\mathcal{F} = 2\rho_0 v^2 A. \tag{23.A1}$$

The cross section of the pipe is $A \approx 5\ \text{cm}^2$, so we get $\mathcal{F} \approx 1$ N.

23.2 The cross section is $A = \pi d^2/4 = 2 \times 10^{-3}\ \text{m}^2$ and the average velocity $U \approx Q/A = 20\text{m s}^{-1}$. The reaction force becomes $|\boldsymbol{\mathcal{R}}| = \sqrt{2}\rho_0 A U^2 \approx 1150$ N which is equivalent to a weight of 115 kg. Unless he is very strong, the firefighter certainly needs equipment.

23.3 Write the total moment as

$$\boldsymbol{\mathcal{M}} = \int_V \boldsymbol{x} \times \boldsymbol{f}^* \, dV \tag{23.A2}$$

where $\boldsymbol{f}^*$ is the effective density of force, including all contact forces. Then

$$\Omega \mathcal{M}_z = \boldsymbol{\Omega} \cdot \boldsymbol{\mathcal{M}} = \int_V \boldsymbol{\Omega} \cdot (\boldsymbol{x} \times \boldsymbol{f}^*) \, dV = \int_V (\boldsymbol{\Omega} \times \boldsymbol{x}) \cdot \boldsymbol{f}^* \, dV = \int_V \boldsymbol{v} \cdot \boldsymbol{f}^* \, dV = P$$

where we have used $\boldsymbol{v} = \boldsymbol{\Omega} \times \boldsymbol{x}$ for a solidly rotating body.

24 Surface waves

24.1 From the solution (24.38) we find

$$\frac{\partial v_z}{\partial t} + \frac{1}{\rho_0} \nabla_z p + g_0 = -a\omega^2 \left(1 + \frac{z}{d}\right) \cos(kx - \omega t) \tag{24.A1}$$

which ought to vanish. Since the finite-depth solution does satisfy the field equations, the problem must lie in the higher-order terms in kz we have dropped in the shallow-water limit.

24.2 **(a)** The waves cross for $c = c_g$ or $kR_c = 1$, i.e. for $\lambda = \lambda_c = 1.7$ cm in water. The common velocity is $c = c_g = \sqrt{2g_0 R_c} = 23$ cm s^{-1}. **(b)** The minimum of the phase velocity is obtained for

$$\frac{dc}{dk} \sim (kR_c)^2 - 1 = 0 \tag{24.A2}$$

which also occurs for $kR_c = 1$. **(c)** The minimum of the group velocity is obtained from

$$\frac{dc_g}{dk} \sim 3(kR_c)^4 + 6(kR_c)^2 - 1 = 0, \tag{24.A3}$$

or $kR_c = \sqrt{-1 + 2/\sqrt{3}} = 0.39\ldots$ or $\lambda/2\pi R_c \approx 2.54\ldots$.

24.3

(a) Use mass conservation $\nabla_x v_x + \nabla_z v_z = 0$ to get

$$\nabla_x(\Psi v_x) + \nabla_z(\Psi v_z) = v_x \nabla_x \Psi + \psi \nabla_x v_x + v_z \nabla_z \Psi + \Psi \nabla_z v_z = v_x^2 + v_z^2.$$

(b) Since Ψv_x is a periodic function of $x - ct$ we have

$$\langle \nabla_x(\Psi v_x) \rangle = \frac{1}{\tau}\int_0^\tau \nabla_x(\Psi v_x)\,dt = \frac{1}{\lambda}\int_0^\lambda \nabla_x(\Psi v_x)\,dx = \left[\Psi v_x\right]_0^\lambda = 0.$$

(c) Integrate over z to get

$$\int_{-d}^0 \left\langle v_x^2 + v_z^2 \right\rangle dz = \int_{-d}^0 \nabla_z \langle \Psi v_z \rangle\,dz = \langle \Psi v_z \rangle_{z=0} = \frac{1}{2} ac \coth kd\, a\omega = \frac{1}{2}a^2 g_0.$$

Multiplying with $(1/2)\rho_0 A$ we obtain (24.64).

24.4 The kinetic energy averaged over a period is

$$\langle \mathcal{T} \rangle = \frac{1}{\tau}\int_0^\tau \frac{1}{2} m\dot{\boldsymbol{x}}^2\,dt = -\frac{1}{\tau}\int_0^\tau \frac{1}{2} m\boldsymbol{x}\cdot\ddot{\boldsymbol{x}}\,dt \tag{24.A4}$$

$$= \frac{1}{2\tau}\int_0^\tau \boldsymbol{x}\cdot\frac{\partial \mathcal{V}}{\partial \boldsymbol{x}}\,dt = \frac{n}{2}\langle \mathcal{V} \rangle \tag{24.A5}$$

where we have integrated partially, used the periodicity of the orbit, and Newton's second law.

24.5 Consider a wave rolling in at an angle towards the beach. Since for shallow-water waves we have $c \sim \sqrt{d}$, the phase velocity of the part of a wave farther from the beach is greatest, causing the part of the crest that is farther out to approach the coastline faster than the crest closer to the beach.

24.6 Using (24.53) we get $L < 3.9$ cm.

24.7 The wave becomes

$$h = \mathcal{R}e\int_{-\infty}^{\infty} a(k)\exp[i(kx - \omega(k)t + \chi(k))]\,dk$$

$$= \frac{1}{\Delta k\sqrt{\pi}}\mathcal{R}e\int_{-\infty}^{\infty}\exp\left(i(k_0x - \omega_0 t + \chi_0) + i(k - k_0)(x - c_g t - x_0) - \frac{(k-k_0)^2}{\Delta k^2}\right)dk$$

$$= \frac{1}{\sqrt{\pi}}\mathcal{R}e\int_{-\infty}^{\infty}\exp\left(i(k_0x - \omega_0 t + \chi_0) + iu\Delta k(x - c_g t - x_0) - u^2\right)du$$

$$= \frac{1}{\sqrt{\pi}}\mathcal{R}e\int_{-\infty}^{\infty}\exp\left[i(k_0x - \omega_0 t + \chi_0) - \left(u - \frac{i}{2}\Delta k(x - c_g t - x_0)\right)^2 - \frac{1}{4}\Delta k^2(x - c_g t - x_0)^2\right]du$$

$$= \frac{1}{\sqrt{\pi}}\mathcal{R}e\int_{-\infty}^{\infty}\exp\left[i(k_0x - \omega_0 t + \chi_0) - u^2 - \frac{1}{4}\Delta k^2(x - c_g t - x_0)^2\right]du$$

$$= \cos(k_0x - \omega_0 t + \chi_0)\exp\left[-\frac{1}{4}\Delta k^2(x - c_g t - x_0)^2\right].$$

In the second line we have substituted $k = k_0 + u\Delta k$ and in the third we have rearranged the resulting quadratic form. In the fourth we shift $u \to u + (i/2)\Delta k(x - c_g t - x_0)$ and in the fifth we use that $\int_{-\infty}^{\infty} \exp(-u^2)\, du = \sqrt{\pi}$.

The wave contains a single wave packet with a Gaussian envelope of width $\sim 1/\Delta k$ with the centre moving along $x = x_0 + c_g t$. The phase shift derivative $x_0 = -d\chi/dk$ determines the position of the centre at $t = 0$.

24.8

(a) For $n = 0$ it is trivial. For $n \neq 0$, the sum is geometric with progression factor $F = \exp(2\pi i n/N)$

$$\sum_{m=0}^{N-1} \exp\left[2\pi i \frac{nm}{N}\right] = \sum_{m=0}^{N-1} F^m = \frac{1 - F^N}{1 - F} = 0 \tag{24.A6}$$

because $F \neq 1$ but $F^N = 1$.

(b) Write the last expression as a double sum

$$h_n = \frac{1}{N} \sum_{m=0}^{N-1} \sum_{k=0}^{N-1} h_k \exp\left[2\pi i \frac{(k-n)m}{N}\right] \tag{24.A7}$$

and do the m-sum first.

(c) Do the triple sum

$$\sum_n |h_n|^2 = \frac{1}{N} \sum_{n,m,k} \hat{h}_m \hat{h}_k^\times \exp\left[2\pi i \frac{(k-m)n}{N}\right]. \tag{24.A8}$$

Do the sum over n first.

25 Jumps and shocks

25.1 Expand the equation to first order in $1/\text{Fr}^2$

$$\sigma = \frac{\sqrt{1 + 8\text{Fr}^2} - 3}{2} \approx \sqrt{2}\text{Fr} - \frac{3}{2} + \frac{\sqrt{2}}{16\text{Fr}} + \cdots, \tag{25.A1}$$

the last term is 3.3% of the leading terms for $\text{Fr} = 2$.

25.2 For the river bore mass conservation takes the form

$$U(h' - h) = (U - U')h'. \tag{25.A2}$$

This demonstrates that the amount of water flowing in the front itself equals the amount of water flowing in the bore.

Similarly for a reflection bore we have

$$(U - U')h = U'(h' - h). \tag{25.A3}$$

This demonstrates that the amount of water flowing towards the wall equals the amount of water flowing away from it in the front.

25.5 The derivative

$$\frac{d(\Delta s/c_V)}{d\sigma} = \frac{(\gamma^2 - 1)\sigma^2}{(1+\sigma)(\gamma^2(\sigma+2)^2 - \sigma^2)} \tag{25.A4}$$

has singularities for $\sigma = -1, -2\gamma/(\gamma+1), -2\gamma/(\gamma-1)$. The last two are always smaller than -1, so the entropy change is a growing function for $\sigma > -1$.

25.6 $R = 2900$ m, $U = 1160$ m s^{-1}, $p_1 = 13$ atm, $\rho_1 = 7.2$ kg m^{-3}, $T_1 = 650$ K. Since $p_1 > 8p_0$, the strong shock approximation is still valid.

26 Whirls and vortices

26.1 **(a)** Inside the core the angular momentum becomes per unit of axial length

$$\frac{d\mathcal{L}_z}{dz} = \int_0^c \rho_0 r v_\phi \, 2\pi r dr = \frac{1}{2}\pi\rho_0\Omega c^4. \qquad (26.A1)$$

The kinetic energy becomes per unit of length

$$\frac{d\mathcal{T}}{dz} = \int_0^c \frac{1}{2}\rho_0 v_\phi^2 \, 2\pi r dr = \frac{1}{4}\pi\rho_0\Omega^2 c^4. \qquad (26.A2)$$

(b) Outside the core the angular momentum becomes

$$\frac{d\mathcal{L}_z}{dz} = \int_c^R \rho_0 r v_\phi \, 2\pi r dr = \pi\rho_0\Omega c^2(R^2 - c^2). \qquad (26.A3)$$

The kinetic energy becomes,

$$\frac{d\mathcal{T}}{dz} = \int_c^R \frac{1}{2}\rho_0 v_\phi^2 \, 2\pi r dr = \pi\rho_0\Omega^2 c^4 \log\frac{R}{c}. \qquad (26.A4)$$

26.2 **(a)** In the presence of gravity the pressure is $p = p_R - \rho_0 g_0 z$ where p_R is the Rankine pressure (26.3). Requiring it to be constant for $z = h(r)$, we get $h(r) = L + p_R(r)/\rho_0 g_0$ where L is the asymptotic height. **(b)** The depth of the depression is $d = L - h(0) = \Omega^2 c^2/g_0$. **(c)** For $\Omega = 63$ s^{-1} we get $d = 4$ cm.

26.3 The fields are

$$\boldsymbol{v} = C\frac{(-y+d, x, 0)}{x^2+(y-d)^2} - C\frac{(-y-d, x, 0)}{x^2+(y+d)^2}. \qquad (26.A5)$$

$$\boldsymbol{v} = C\frac{(-y+d, x, 0)}{x^2+(y-d)^2} + C\frac{(-y-d, x, 0)}{x^2+(y+d)^2}. \qquad (26.A6)$$

26.4 For counter-rotating and corotating vortices we have respectively,

$$\psi = -\tfrac{1}{2}C\left(\log(x^2+(y-d)^2) - \log(x^2+(y+d)^2)\right) \qquad (26.A7)$$

$$\psi = -\tfrac{1}{2}C\left(\log(x^2+(y-d)^2) + \log(x^2+(y+d)^2)\right) \qquad (26.A8)$$

26.5 **(a)** The instantaneous complex velocity field of all the vortices becomes,

$$w = -i\sum_n \frac{C_n}{z - z_n(t)}. \qquad (26.A9)$$

Each vortex is advected with the velocity field created by all the others, and using that $dz/dt = v_x + iv_y = w^\times$, we find the desired result. **(b)** For the counter-rotating vortices the orbits must satisfy

$$\frac{dz_+^\times}{dt} = i\frac{C}{z_+ - z_-}, \qquad \frac{dz_-^\times}{dt} = -i\frac{C}{z_- - z_+}. \qquad (26.A10)$$

This is fulfilled for $z_\pm = Ut \pm ib$ with $U = C/2b$. **(c)** The corotating vortices satisfy the equations

$$\frac{dz_+^\times}{dt} = -i\frac{C}{z_+ - z_-}, \qquad \frac{dz_-^\times}{dt} = -i\frac{C}{z_- - z_+}. \qquad (26.A11)$$

This is fulfilled for $z_\pm = \pm ibe^{i\Omega t}$ with $\Omega = C/2b^2$.

26.6 **(a)** Changing $n \to -n$ in the sum we get

$$f_N(-z) + f_N(z) = \sum_{n=-N}^{N} \frac{1}{-z - \pi n} - \sum_{n=-N}^{N} \frac{1}{-z + \pi n} = 0 \,. \tag{26.A12}$$

(b) Shifting the sum one gets

$$f_N(z+\pi) - f_N(z) = \sum_{n=-N}^{N} \left(\frac{1}{z + \pi - \pi n} - \frac{1}{z - \pi n} \right) = \frac{1}{z + \pi(N+1)} - \frac{1}{z - \pi N} \,.$$

For fixed z, the right-hand side vanishes for $N \to \infty$. **(c)** In the strip the only singularities are the pole $1/z$ of the cotangent and the pole $-1/z$ from the sum, and they cancel. **(d)** We know from periodicity and antisymmetry that $f(\pi/2) = f(-\pi/2) = -f(\pi/2)$ and thus $f(\pi/2) = 0$. For arbitrary fixed x in the strip, the leading behaviour for fixed $|y| \gg 1$ is

$$f_N(x+iy) \approx \cot(iy) + \sum_{n=1}^{N} \frac{2iy}{y^2 + \pi^2 n^2} \approx \mp i + 2iy \int_0^N \frac{dn}{y^2 + \pi^2 n^2}$$

$$\approx \mp i + \frac{2iy}{\pi\,|y|} \arctan \frac{N\pi}{|y|} \approx \mp i \pm i \frac{2}{\pi} \arctan \frac{N\pi}{|y|}$$

which converges to 0 for $N \to \infty$.

26.8 The streamlines are obtained by solving (15.2) in cylindrical coordinates,

$$r\dot{\phi} = v_\phi, \qquad \dot{r} = v_r, \qquad \dot{z} = v_z. \tag{26.A13}$$

The last two are elementary to integrate with the result

$$r = r_0 e^{-t/2t_a}, \qquad z = z_0 e^{t/t_a}, \tag{26.A14}$$

which after elimination of t becomes

$$z = z_0 \left(\frac{r_0}{r} \right)^2 . \tag{26.A15}$$

26.9 **(a)** Insert and verify. **(b)**

$$\frac{d\mathcal{L}_z}{dz} = \int_0^\infty r \rho_0 v_\phi(r,t) 2\pi r \, dr = 16\pi \tau v^2 \rho_0 . \tag{26.A16}$$

26.10

(a) Insert v_ϕ into (26.9) to obtain

$$\frac{d^2 f(\xi)}{d\xi^2} + \frac{df(\xi)}{d\xi} = \frac{t}{F(t)} \frac{dF(t)}{dt} \frac{1}{\xi} f(\xi). \tag{26.A17}$$

(b) The t-dependent factor must be a constant, $-\alpha$, so $F(t) \sim t^{-\alpha}$.

(c) Insert and verify that the series expansion satisfies

$$\frac{d^2 f(\xi)}{d\xi^2} + \frac{df(\xi)}{d\xi} + \frac{\alpha}{\xi} f(\xi) = 0. \tag{26.A18}$$

The expansion is a confluent hypergeometric function.

(d) For integer α the functions are Laguerre polynomials multiplied with $e^{-\xi}$. The first few such solutions are

$$f_0(\xi) = 1 - e^{-\xi}, \tag{26.A19a}$$

$$f_1(\xi) = \xi e^{-\xi}, \tag{26.A19b}$$

$$f_2(\xi) = \xi(1 - \xi/2) e^{-\xi}. \tag{26.A19c}$$

The Oseen–Lamb vortex corresponds to $f_0(\xi)$ and the Taylor vortex (problem 26.9) to $f_1(\xi)$.

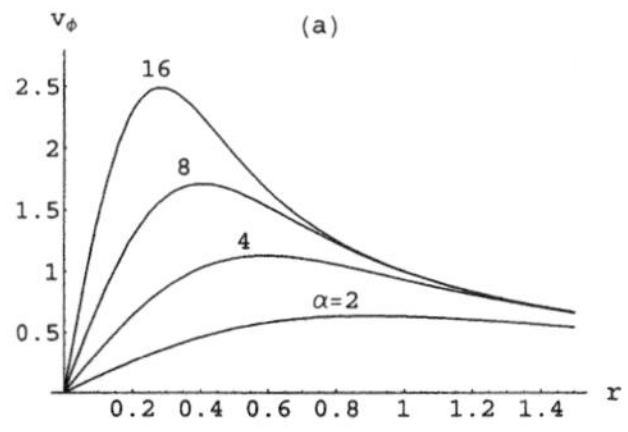

Family of self-similar vortex shapes $f_\alpha(\xi)$ with α in steps of 0.5 from 0 (top) to 3 (bottom).

26.11 The equation of motion is

$$\frac{\partial v_\phi}{\partial t} + \frac{v_r}{r}\frac{\partial (r v_\phi)}{\partial r} = \nu \frac{\partial}{\partial r}\left(\frac{1}{r}\frac{\partial (r v_\phi)}{\partial r}\right). \tag{26.A20}$$

Assume that the azimuthal flow is of the form

$$r v_\phi = C(1 - e^{-r^2/\lambda(t)}). \tag{26.A21}$$

Then λ must satisfy

$$\dot{\lambda} + \frac{4\nu}{a^2}\lambda = 4\nu. \tag{26.A22}$$

Solving this with the boundary condition $\lambda(0) = b^2$ we find

$$\lambda(t) = a^2 + (b^2 - a^2)e^{-4\nu t/a^2}. \tag{26.A23}$$

26.12 Using (26.37) we obtain

$$v_\phi = \Omega\left(r + \frac{2qt}{r}\right). \tag{26.A24}$$

The rate of angular momentum flowing through a cylinder of radius r and length L is

$$r v_\phi \, 2\pi r v_r L = -Q(2qt + r^2) \tag{26.A25}$$

where $Q = 2\pi q L$ is the volume discharge of water through the drain.

26.13 Integrating the vortex equation (26.25) once one gets,

$$\frac{1}{r}\frac{d(r v_\phi)}{dr} = 2\Omega \exp\left[\frac{1}{\nu}\int_0^r v_r(r')\,dr'\right] \tag{26.A26}$$

where the normalization has been fixed by $v_\phi/r \to \Omega$ for $r \to 0$. Integrating once more, using $v_\phi = 0$ for $r = 0$, one gets

$$v_\phi = \frac{\Omega}{r}\int_0^r \exp\left[\frac{1}{\nu}\int_0^{r_2} v_r(r_1)\,dr_1\right] 2r_2 dr_2. \tag{26.A27}$$

26.14 The continuity of the radial field requires

$$c = \frac{a}{\sqrt{\alpha}}. \tag{26.A28}$$

The azimuthal field becomes

$$v_\phi = \begin{cases} \dfrac{C}{r}\left(1 - e^{-r^2/c^2}\right), & (r < a) \\[2ex] \dfrac{C}{r}\left(1 + \frac{e^{-\alpha}}{\alpha - 1}\left(1 - \alpha\left(\frac{a^2}{r^2}\right)^{\alpha-1}\right)\right), & (r > a) \end{cases} \tag{26.A29}$$

where $C = \Omega c^2$.

26.15 **(a)** Trivial. **(b)** The streamlines obey $d\boldsymbol{x}/dt = \boldsymbol{v}$. Write $\boldsymbol{x} = r\boldsymbol{e}_r + z\boldsymbol{e}_z$ and differentiate to get,

$$\frac{dr}{dt} = v_r, \qquad r\frac{d\phi}{dt} = v_\phi, \qquad \frac{dz}{dt} = v_z. \tag{26.A30}$$

Then

$$\frac{d\psi}{dt} = \frac{\partial\psi}{\partial r}\frac{dr}{dt} + \frac{\partial\psi}{\partial z}\frac{dz}{dt} = 0. \tag{26.A31}$$

(c) Calculate the derivatives of ψ.

26.16 Inserting (26.5) we find

$$k = \int_0^\infty \frac{\left(1 - e^{-s^2}\right)^2}{s^3} ds = \log 2. \tag{26.A32}$$

This changes the estimate of the central depression to

$$h_0 \approx \sqrt{L^2 - k\frac{C^2 W}{2g_0 \nu}}, \tag{26.A33}$$

or $\alpha\beta < 0.25/k = 0.36$ in dimensionless variables.

27 Lubrication

27.1 Equate lift and weight

$$\alpha \frac{L}{2d} \frac{\eta U L A}{d^2} = M g_0. \tag{27.A1}$$

Solve for d

$$d = \left(\frac{\alpha \eta U L^2 A}{2 M g_0}\right)^{1/3}. \tag{27.A2}$$

27.2 Differentiate the integral in (27.13) after x

$$\frac{dQ}{dx} = h' v_x|_{y=h} + \int_0^h \frac{\partial v_x}{\partial x} dy = -\int_0^h \frac{\partial v_y}{\partial y} dy = 0.$$

27.3 The extremum of v_x as a function of y is found at

$$y = \frac{h(4h - 3d_0)}{6(h - d_0)}$$

which must be in the interval $0 < y < h$. This is only the case when $h < 3d_0/4$ or $h > 3d_0/2$.

27.4 One finds

$$d_0 = \frac{\left\langle h^{-2}\right\rangle}{\left\langle h^{-3}\right\rangle}, \tag{27.A3}$$

$$\mathcal{L} = A\langle p\rangle = -A\left\langle x p'\right\rangle = -6\eta U A\left\langle\frac{x(h - d_0)}{h^3}\right\rangle, \tag{27.A4}$$

$$\mathcal{D} = -A\left\langle \sigma_{xy}\big|_{y=0}\right\rangle = \eta U A\left\langle\frac{4h - 3d_0}{h^2}\right\rangle, \tag{27.A5}$$

$$\mathcal{M}_z = A\langle x p\rangle = -\frac{1}{2}A\left\langle x^2 p'\right\rangle = -3\eta U A\left\langle\frac{x^2(h - d_0)}{h^3}\right\rangle. \tag{27.A6}$$

27.5 **(a)** They become

$$d_0 \approx d\left(1 - 3\left\langle\chi^2\right\rangle\right), \tag{27.A7}$$

$$\mathcal{L} \approx -6\frac{\eta U A L}{d^2}\langle\xi\chi\rangle, \tag{27.A8}$$

$$\mathcal{D} \approx \frac{\eta U A}{d}\left(1 + 4\left\langle\chi^2\right\rangle\right), \tag{27.A9}$$

$$\mathcal{M}_z \approx -3\frac{\eta U A L^2}{d^2}\left\langle\xi^2\left(\chi - 3\left(\chi^2 - \left\langle\chi^2\right\rangle\right)\right)\right\rangle. \tag{27.A10}$$

(b) For the flat wing we have $\chi = -2\gamma\xi$ with $-1/2 < \xi < 1/2$,

$$\left\langle \xi^2 \right\rangle = \frac{1}{12}, \qquad \left\langle \xi^3 \right\rangle = 0, \qquad \left\langle \xi^4 \right\rangle = \frac{1}{80}. \tag{27.A11}$$

Then it follows that

$$\left\langle \chi^2 \right\rangle = \frac{1}{3}\gamma^2, \qquad \langle \xi\chi \rangle = -\frac{1}{6}\gamma, \qquad \left\langle \xi^2\chi \right\rangle = 0, \qquad \left\langle \xi^2\chi^2 \right\rangle = \frac{1}{20}\gamma^2, \tag{27.A12}$$

and from these the leading approximations are obtained.

27.6 Use the complex variable $z = e^{i\phi}$ and write for arbitrary real $a > |b|$

$$\int_0^{2\pi} \frac{1}{a - b\cos\phi}\frac{d\phi}{2\pi} = \oint \frac{dz}{2\pi i}\frac{2}{2az - b(1+z^2)}$$

with the integration contour along the unit circle in the complex plane. The roots of the denominator are $z = (a \pm \sqrt{a^2 - b^2})/b$. The root with the negative sign lies inside the contour and we obtain the result

$$\int_0^{2\pi} \frac{1}{a - b\cos\phi}\frac{d\phi}{2\pi} = \frac{1}{\sqrt{a^2 - b^2}} \tag{27.A13}$$

from the residue at the pole.The other integrals are obtained by differentiation of both sides of this expression after a or b, and afterwards taking $a = 1$ and $b = \gamma$.

28 Boundary layers

28.1 **(a)** Assume that $v_x(y)$ and $v_y(y)$ only depend on y, not on x. The boundary conditions are $v_x(0) = 0$, $v_x(\infty) = U$ and $v_y(0) = -V$. The continuity equation shows that $dv_y(y)/dy = 0$ so $v_y(y) = -V$ everywhere. The Prandtl equation for v_x becomes

$$-V\frac{dv_x}{dy} = \nu\frac{d^2v_x}{dy^2}$$

with the solution

$$v_x(y) = U(1 - e^{-Vy/\nu}).$$

(b) For $V < 0$ the v_x grows exponentially $y \to \infty$ which is impossible, since it has to converge to U.

28.2 Since $\mathrm{erf}(0) = 0$ we have $v_x(0,t) = \int_0^t \dot{U}(t')\,dt' = U(t)$. Use $\mathrm{erf}(\infty) = 1$ and the Stokes solution to derive

$$\left(\frac{\partial}{\partial t} - \nu\frac{\partial^2}{\partial y^2}\right)v_x = -\int_0^t \left(\frac{\partial}{\partial t} - \nu\frac{\partial^2}{\partial y^2}\right)\mathrm{erf}\left(\frac{y}{2\sqrt{\nu(t-t')}}\right)\dot{U}(t')\,dt' = 0.$$

28.3 **(a)** Just use the definition of $h(s)$

$$\int_0^\infty (1 - f(s))\,ds = \lim_{s\to\infty}(s - g(s)) = \lim_{s\to\infty}(sf(s) - g(s)) = h(\infty). \tag{28.A1}$$

(b) Use $f(s) = g'(s)$, carry out the partial integration and use Blasius' equation (28.26)

$$\int_0^\infty f(s)(1 - f(s))\,ds = \int_0^\infty (1 - f(s))\,dg(s) = \int_0^\infty g(s)g''(s)\,ds$$
$$= -2\int_0^\infty g'''(s)\,ds = 2g''(0) = 2f'(0).$$

28.5 Insert the field in the Navier–Stokes equations (28.20) to get (take $\rho_0 = 1$)

$$\frac{\partial p}{\partial x} = -U\frac{dU}{dx} + \nu\frac{d^2U}{dx^2} \tag{28.A2}$$

$$\frac{\partial p}{\partial y} = y\left(U\frac{d^2U}{dx^2} - \left(\frac{dU}{dx}\right)^2 - \nu\frac{d^3U}{dx^3}\right). \tag{28.A3}$$

Use the cross derivative $\partial^2 p/\partial x\partial y = \partial^2 p/\partial y\partial x$ to obtain

$$\frac{d}{dx}\left(U\frac{d^2U}{dx^2} - \left(\frac{dU}{dx}\right)^2 - \nu\frac{d^3U}{dx^3}\right) = 0 \tag{28.A4}$$

leading to

$$\nu\frac{d^3U}{dx^3} = U\frac{d^2U}{dx^2} - \left(\frac{dU}{dx}\right)^2 + K \tag{28.A5}$$

where K is a constant. This is not generally solvable, except for special cases, such as $U = a - bx$ or $U \sim 1/x$.

If you put $v_y = V(x) - yU(x)$ then $\partial p/\partial y$ is still linear in y and you just get an extra equation for V.

28.6 **(a)** Insert the parametrizations into (28.67). **(b)** Retain the leading terms for $\mu \to 0$. The solution is (with $U(0) = 1$ and $\mu(0) = 4/3$)

$$\mu^2 = \frac{16}{9} + 12\log U. \tag{28.A6}$$

(c) At separation $x = x_c$ we have $\mu = 0$, leading to $U_c = \exp(-4/27) = 0.862303$. **(d)** Solving this equation for the nine cases one obtains $x_c = 0.138, 0.256, 0.071, 0.160, 0.077, 0.371, 0.609, 0.780, 0.531$. The average deviation from the exact values is only about 20% but that must be fortuitous.

29 Subsonic flight

29.1 **(a)** 3.7 m s^{-2}, **(b)** 27 s, **(c)** 1400 m.

29.2 Write (29.13) as

$$\tan\beta = \frac{U}{g_0}\frac{U}{R} \tag{29.A1}$$

and use that $U/R = 2\pi/T$.

29.3 Imagine that there are N small vortices, each carrying Γ/N of the total bound vorticity. The total drag becomes $\mathcal{D}_N \approx N\rho_0(\Gamma/N)^2/4\pi \approx \mathcal{D}_1/N$.

29.4 From Newton's second law we get $MdU/dt = T = P/U$, where P is the thrust power. Dividing with $U = dx/dt$ we find $MdU/dx = P/U^2$. Integrating the solution becomes $x = MU^3/3P$. Inserting the given values one gets $P = 37$ kW which is 50% of the engine power. The computed take off time is similarly $t = MU^2/2P \approx 10$ s. This is only half the quoted value, so the engine power is not converted into thrust at a constant rate during the run.

29.5 Instead of (29.40) we have

$$\mathcal{R} = -\oint_S \rho \boldsymbol{v}\, \boldsymbol{v} \cdot d\boldsymbol{S} + \oint_S \boldsymbol{\sigma} \cdot d\boldsymbol{S}. \tag{29.A2}$$

At large distances where $\rho = \rho_0 + \Delta\rho$ and $\boldsymbol{v} = \boldsymbol{U} + \Delta\boldsymbol{v}$ we get to first order,

$$\mathcal{R} = -\oint_S \Delta\rho \boldsymbol{U}\boldsymbol{U} \cdot d\boldsymbol{S} - \rho_0 \oint_S \boldsymbol{U}\Delta v \cdot d\boldsymbol{S} - \rho_0 \int_S \Delta\boldsymbol{v}\boldsymbol{U} \cdot d\boldsymbol{S} - \oint_S \Delta p\, d\boldsymbol{S}.$$

Mass conservation becomes in the same approximation

$$\nabla \cdot (\rho\boldsymbol{v}) = \rho_0 \nabla \cdot \Delta\boldsymbol{v} + (\boldsymbol{U} \cdot \nabla)\Delta\rho = 0. \tag{29.A3}$$

This makes the two first terms in $\mathcal{R}$ cancel and we arrive at the incompressible result (29.41). At great distances from the body where the velocity corrections are tiny, barotropic fluids are effectively incompressible and Bernoulli's theorem takes the usual form.

29.6 Let $\boldsymbol{e}_U = \boldsymbol{U}/U$ be a unit vector in the direction of the asymptotic flow. Projecting on (29.42) we find

$$\mathcal{D} = \oint_S (p + \rho_0 \Delta\boldsymbol{v} \cdot \boldsymbol{U})\boldsymbol{e}_U \cdot d\boldsymbol{S}. \tag{29.A4}$$

Using that $\mathcal{L} = \mathcal{R} - \boldsymbol{e}_U(\boldsymbol{e}_U \cdot \mathcal{R}) = -\boldsymbol{e}_U \times (\boldsymbol{e}_U \times \mathcal{R})$ the lift takes the form

$$\mathcal{L} = -\rho_0 \oint_S \boldsymbol{U} \times (\Delta\boldsymbol{v} \times d\boldsymbol{S}) + \oint_S (\Delta p + \rho_0 \Delta\boldsymbol{v} \cdot \boldsymbol{u})\, \boldsymbol{e}_U \times (\boldsymbol{e}_U \times d\boldsymbol{S}). \tag{29.A5}$$

The last term evidently vanishes if S cuts the wake in a planar region orthogonal to the asymptotic velocity, so that $d\boldsymbol{S} \sim \boldsymbol{e}_U$.

29.7 The circulation becomes

$$\begin{aligned}\Gamma &= \frac{1}{2\pi}\oint_C (dx, dy) \cdot \oint_A \frac{(-y + y', x - x')}{(x - x')^2 + (y - y')^2}\, d\Gamma(x', y') \\ &= \oint_A d\Gamma(x', y') \frac{1}{2\pi} \oint_C \frac{(x - x')dy - (y - y')dx}{(x - x')^2 + (y - y')^2}.\end{aligned}$$

Using complex notation, $z = x + iy$, the integrand becomes

$$\frac{1}{2\pi}\oint_C \frac{(x - x')dy - (y - y')dx}{(x - x')^2 + (y - y')^2} = \oint_C \frac{x - x' - i(y - y')}{(x - x')^2 + (y - y')^2} \frac{dx + idy}{2\pi i} = \mathcal{R}e \oint_C \frac{1}{z - z'} \frac{dz}{2\pi i} = 1$$

where we in the last step have used Cauchy's residue theorem for the pole z' inside C.

29.8 Replace x by complex $z = x + iy$, so that the integral is the real part of

$$I(z) = \int_0^1 \frac{1}{(1 - t)\sqrt{t(1 - t)}} \log \frac{t - z}{1 - z}\, dt \tag{29.A6}$$

for $y \to 0$ and $0 < x < 1$. Perform a partial integration to get

$$\begin{aligned}I &= -2\int_0^1 \sqrt{\frac{t}{1 - t}} \frac{dt}{t - z} = -2\int_0^1 \left(1 + \frac{z}{t - z}\right) \frac{dt}{\sqrt{t(1 - t)}} \\ &= -2\pi - 2z \int_0^1 \frac{1}{t - z} \frac{dt}{\sqrt{t(1 - t)}}.\end{aligned}$$

We must now show that the real part of the last integral vanishes for $y \to 0$ and $0 < x < 1$.

From Cauchy's theorem we have

$$\frac{1}{\sqrt{z(z-1)}} = \oint_z \frac{1}{\sqrt{t(t-1)}} \frac{1}{t-z} \frac{dt}{2\pi i}$$
$$= \left(\int_{0+i\epsilon}^{1+i\epsilon} - \int_{0-i\epsilon}^{1-i\epsilon} \right) \frac{1}{\sqrt{t(t-1)}} \frac{1}{t-z} \frac{dt}{2\pi i}$$

because there is a cut along $0 < t < 1$ on the real axis. Using that $\sqrt{t(t-1+i\epsilon)} = \pm i\sqrt{t(1-t)}$, we get the desired integral,

$$\int_0^1 \frac{1}{t-z} \frac{dt}{\sqrt{t(1-t)}} = -\frac{\pi}{\sqrt{z(z-1)}}. \tag{29.A7}$$

Now letting $y \to 0 \pm i\epsilon$ for $0 < x < 1$ the right-hand side becomes purely imaginary.

29.9 **(a)** The smoothness guarantees that x is a polynomial of y near $x = 0$,

$$x = \frac{y^2}{a^2} + by^3 + \mathcal{O}\left(y^4\right). \tag{29.A8}$$

Solving for y to third order in $\pm\sqrt{x}$ one gets the desired result with $\lambda = -(1/4)a^4 b$. **(b)** The integral becomes

$$\tan\alpha_0 = -\frac{2\lambda}{\pi c} \int_0^c \sqrt{\frac{x}{c-x}}\, dx = -\lambda. \tag{29.A9}$$

29.10

(a) Direct insertion confirms that the diffusion equation is satisfied. For $t \to 0$, the Gaussian becomes very sharply peaked at $y = x$, so that

$$F(x,t) \approx (4\pi t)^{-N/2} F_0(x) \int \exp\left(-\frac{(x-y)^2}{4t}\right) d^N y. \tag{29.A10}$$

The integral is now standard and leads to $F(x,t) = F_0(x)$ in the limit of $t \to 0$.

(b) First write the solution as

$$F(x,t) = (4\pi t)^{-N/2} e^{-x^2/4t} \int F_0(y) e^{-x\cdot y/2t - y^2/4t}\, d^N y. \tag{29.A11}$$

The first exponential shows that the solution is only non-vanishing for $|x| < \sqrt{t}$. Assume that $F_0(y)$ is only non-vanishing for $|y| < a$. Then for $|x| < \sqrt{t}$ and $t \gg a^2$ the exponential inside the integral can be disregarded.

29.11 First write the equation as

$$\left(\sin\theta \frac{d}{d\theta}\left(\sin\theta \frac{d}{d\theta}\right) - n^2\right) f = 0. \tag{29.A12}$$

Define $\xi = \log\tan(\theta/2)$ which has $d\xi = d\theta/\sin\theta$. Then the equation becomes

$$\left(\frac{d^2}{d\xi^2} - n^2\right) f = 0 \tag{29.A13}$$

which has the solutions

$$f = e^{\pm n\xi} = (\tan(\theta/2))^{\pm n}. \tag{29.A14}$$

30 Heat

30.1 The total internal energy contained in an infinite column of material along x with area A in the yz-plane is,

$$\mathcal{U} = \int_V \rho_0 u \, dV = \rho_0 c_0 A \int_{-\infty}^{\infty} T(x,t)\, dx = \rho_0 c_0 A a T_0 \sqrt{\pi}.$$

As expected, its value is independent of time.

30.4 The temperature does not depend on x (or z) because of symmetry so that Fourier's equation becomes

$$\frac{d^2T}{dy^2} = 0 \tag{30.A1}$$

which has the solution $T = A + By$. The constants A and B are determined by the boundary conditions. If the fluid moves steadily with velocity $v_x(y)$, the advective term vanishes, $(\boldsymbol{v} \cdot \nabla)T = 0$.

30.5 The temperature does not depend on x (or z) because of symmetry so that Fourier's equation becomes

$$\frac{1}{r}\frac{d}{dr}\left(r\frac{dT}{dr}\right) = 0 \tag{30.A2}$$

which has the solution $T = A + B\log r$. The constants A and B are determined by the boundary conditions. If the fluid moves steadily with velocity $v_z(r)$, the advective term vanishes, $(\boldsymbol{v} \cdot \nabla)T = 0$.

30.6 Using the notation of example 30.2.5, the temperature in the sphere is given by (30.21) and thus,

$$\Theta = T_c - T_s = \frac{h_0}{6k}(a^2 - c^2) \tag{30.A3}$$

where k is the thermal conductivity of water. Numerically this comes to $\Theta \approx 9$ K.

30.7 We have

$$H(\infty) = \int_0^\infty \exp(-\mathrm{Pr}\, G(s))\, ds, \qquad G(s) = \int_0^s g(u)\, du. \tag{30.A4}$$

For Pr $\to 0$ the integrand is dominated by large values of $G(s)$ which in turn is dominated by large values of $g(s)$. Since $g(s) \approx s$ for $s \to \infty$ we have $G(s) \approx 1/2s^2$ and thus

$$H(\infty) \approx \int_0^\infty e^{-\frac{1}{2}\mathrm{Pr}\, s^2} = \sqrt{\frac{\pi}{2\mathrm{Pr}}}. \tag{30.A5}$$

For Pr $\to \infty$ the small values of $G(s)$ dominate. Since for $s \to 0$ we have $g(s) \approx 1/2f'(0)s^2$ and $G(s) \approx 1/6f'(0)s^3$ thus

$$H(\infty) \approx \int_0^\infty e^{-\frac{1}{6}f'(0)\mathrm{Pr}\, s^3}\, ds = \left(\frac{6}{f'(0)\mathrm{Pr}}\right)^{1/3} \Gamma\left(\frac{4}{3}\right) \tag{30.A6}$$

where Γ is the gamma-function.

31 Convection

31.1 Use the Boussinesq equations for zero velocity. The buoyancy term varies in the horizontal direction, implying that the pressure excess also has a horizontal variation. Thus, the isobars are not horizontal, which they have to be for any hydrostatic solution in constant gravity. The balance between buoyancy and friction forces for the heated fluid between the plates may be estimated as,

$$\alpha\Theta\rho_0 g_0 A d \sim \eta\frac{U}{d}A, \tag{31.A1}$$

where A is the plate area. Apart from numerical factors this is of the same form as (31.8).

31.2 The flow out of the warm plate is

$$\dot{Q}_0 = Lh \cdot k\frac{\Theta}{d}, \tag{31.A2}$$

where h is the height of the plate. Dividing (31.10) by this we find

$$\frac{\dot{Q}}{\dot{Q}_0} = \frac{\alpha\Theta g_0 d^4}{45\kappa\nu h} = \frac{1}{45}\cdot\frac{d}{h}\cdot \mathrm{Ra}. \tag{31.A3}$$

31.3 From the y-component of (31.39b) it follows that $(\nabla_z^2 - k^2)v_y = 0$ such that $v_y \propto \exp(\pm kz)$. Since k is fixed by the solution of the other fields, the boundary conditions on v_y cannot be fulfilled except for $v_y = 0$.

31.4 From the exact solution we have $\cosh\mu_1 \approx \sinh\mu_1 \approx 90$ so that to a precision of about 1% the determinant (31.59) becomes,

$$det(\mu,\xi) \propto \mu_0 \sin\frac{\mu_0}{2} + (\mu_1 + \sqrt{3}\,\mu_2)\cos\frac{\mu_0}{2}. \tag{31.A4}$$

The determinant vanishes for

$$\tan\frac{\mu_0}{2} = -\frac{\mu_1 + \sqrt{3}\mu_2}{\mu_0} \tag{31.A5}$$

which has the solution

$$\mu_0 = 2\left(\pi - \arctan\frac{\mu_1 + \sqrt{3}\mu_2}{\mu_0}\right). \tag{31.A6}$$

Numerical minimization leads to the quoted results.

32 Turbulence

32.1 From definition (32.17) we find

$$\left\langle\frac{\partial \boldsymbol{v}}{\partial t}\right\rangle = \lim_{T\to\infty}\frac{1}{T}\int_0^T \frac{\partial \boldsymbol{v}}{\partial t}(\boldsymbol{x}, t+s)\,ds = \lim_{T\to\infty}\frac{1}{T}\int_0^T \frac{\partial \boldsymbol{v}(\boldsymbol{x}, t+s)}{\partial s}\,ds$$
$$= \lim_{T\to\infty}\frac{\boldsymbol{v}(\boldsymbol{x}, t+T) - \boldsymbol{v}(\boldsymbol{x}, t)}{T} = \boldsymbol{0}$$

when $|\boldsymbol{v}(\boldsymbol{x}, t)|$ is bounded for all $\boldsymbol{x}$ and t. This condition is normally fulfilled for physical systems of finite size.

32.2 If the distribution is maintained to the limits of the inertial range, the normalized distribution becomes

$$\frac{dF}{d\lambda} = \frac{N}{\lambda}, \qquad \frac{1}{N} = \log\frac{L}{\lambda_d}. \tag{32.A1}$$

This clearly destroys the universality of the Kolmogorov law because the Kolmogorov 'constant' becomes dependent on the ratio of the macroscopic and microscopic scales.

32.3 From (32.24) one finds

$$\left\langle \frac{\partial(u_i u_j)}{\partial t} \right\rangle + \left\langle u_i(\boldsymbol{v}\cdot\nabla)u_j + u_j(\boldsymbol{v}\cdot\nabla)u_i \right\rangle$$
$$+ \left\langle u_i(\boldsymbol{u}\cdot\nabla)v_j + u_j(\boldsymbol{u}\cdot\nabla)v_i \right\rangle + \left\langle u_i(\boldsymbol{u}\cdot\nabla)u_j + u_j(\boldsymbol{u}\cdot\nabla)u_i \right\rangle$$
$$= -\frac{1}{\rho_0}\left\langle u_i \nabla_j q + u_j \nabla_i q \right\rangle + \nu \left\langle u_i \nabla^2 u_j + u_j \nabla^2 u_i \right\rangle.$$

Rearranging this equation it becomes the desired equation (32.86), which does not close because of the third-order fluctuation moment $\langle u_i u_j u_k \rangle$ occurring in the last term on the left-hand side. There is also a problem in the terms $\langle u_i \nabla_j q\rangle + \langle u_j \nabla_i q\rangle$ and $\langle \nabla_k u_i \nabla_k u_j \rangle$ which is not caused by lack of closure in the moments. Such terms can, in principle, be handled by calculating the more general moments $\langle u_i(\boldsymbol{x}_1, t) q(\boldsymbol{x}_2, t)\rangle$ and $\langle u_i(\boldsymbol{x}_1, t) u_j(\boldsymbol{x}_2, t)\rangle$, and afterwards taking the limit of $\boldsymbol{x}_2 \to \boldsymbol{x}_1$.

32.4 Write $\Lambda_n = \Lambda(1+\delta_n)$ and expand to first order in the precision δ_n to get

$$\delta_n = -\frac{A}{\Lambda}\delta_{n-1}. \tag{32.A2}$$

This shows that the approximate sequence converges rapidly for $\Lambda \gg A$. With the given values one finds

$$\Lambda_n = 1,\ 33.0915,\ 24.5533,\ 25.2815,\ 25.2102,\ 25.2171,\ 25.2164,\ 25.2165,\ 25.2164\ldots$$

after which point it stays constant to within this precision.

A Newtonian particle mechanics

A.1 Defining the alternative total mass to be $M' = \sum_n m_n^2$ and the alternative 'centre of mass' to be $\boldsymbol{X}' = \sum_n m_n^2 \boldsymbol{x}_n / \sum_n m_n^2$, the global equation becomes of the same form as before, $M'(d^2\boldsymbol{X}'/dt^2) = \boldsymbol{\mathcal{F}}'$. Since the fundamental equations (A.1) are unchanged, all the physical consequences must be unchanged and cannot depend on definitions.

B Curvilinear coordinates

C Thermodynamics of ideal gases

C.1 **(a)** The total mass is $M = nM_{\rm mol} = \sum_i M_i = \sum_i n_i M_{{\rm mol},i} = n\sum_i c_i M_{{\rm mol},i}$. Using $V = M/\rho$ in (4.26) we get the result. **(b)** The total energy is $\mathcal{U} = \sum_i U_i = \sum_i (1/2) k_i n_i RT = (1/2)knRT$.

C.2 **(a)** The differential of a function is

$$dQ = \frac{\partial Q}{\partial T}dT + \frac{\partial Q}{\partial V}dV \tag{C.A1}$$

so that $A = \partial Q/\partial T$ and $B = \partial Q/\partial B$, then $\partial A/\partial V = \partial B/\partial T = \partial^2 Q/\partial V\partial T$.

(b) We have $A = C_V$ and $B = nRT/V$, and thus $\partial A/\partial V = 0$ and $\partial B/\partial T = nR/V \neq 0$.

List of literature

The following list is a collection of mostly recent textbooks and monographs that can be found in almost all university libraries. The list is certainly not comprehensive but contains those texts that have proven particularly useful in writing the present book. Some of the entries have personal comments. Not all of the entries are actually referred to in the main text.

[1] D. J. Acheson: *Elementary Fluid Dynamics*, Oxford University Press, 1990.

An excellent physics-oriented exposition of most of the interesting phenomena of fluid mechanics with lots of historical notes.

[2] J. D. Anderson, Jr: *Computational Fluid Dynamics*, McGraw-Hill, 1995.

[3] J. D. Anderson, Jr: *A History of Aerodynamics*, Cambridge University Press, 1997.

A wonderfully warm and comprehensive history of aerodynamics at a level suitable for university students.

[4] J. D. Anderson, Jr: *Aerodynamics*, McGrawHill, 2001.

A fine and readable book on aerodynamics.

[5] P. Ball: *The Self-made Tapestry*, Oxford University Press, 1997.

[6] G. Batchelor: *The Theory of Homogeneous Turbulence*, Cambridge University Press (1953).

[7] G. K. Batchelor: *An Introduction to Fluid Dynamics*, Cambridge University Press, 1967.

A classic text on fluid mechanics with emphasis on both qualitative, semiquantitative and analytic arguments. .

[8] G. Batchelor: *The Life and Legacy of G. I. Taylor*, Cambridge University Press (1996).

A loving biography by a friend and former PhD student.

[9] G. Boxer: *Fluid Mechanics*, The MacMillan Press Ltd, 1988.

[10] D. Braess: *Finite Elements*, Cambridge University Press, 2001.

[11] H. B. Callen: *Thermodynamics*, John Wiley, 1960.

[12] Y. A. Cengel and M. A. Boles: *Thermodynamics, an engineering approach*, McGraw-Hill (2002).

[13] S. Chandrasekhar: *Hydrodynamic and hydromagnetic stability*, Dover (1981).

[14] D. S. Chandrasekharaiah and L. Debnath: *Continuum Mechanics*, Academic Press, 1994.

A thorough presentation of continuum mechanics with numerous examples and problems.

[15] I. B. Cohen: *The Birth of a New Physics*, Penguin, 1985.

[16] Jørgen Christensen-Dalsgaard: *The 'standard' Sun*, Space Science Reviews **85**,19 (1998) and private communication.

[17] I. Doghri: *Mechanics of Deformable Solids*, Springer, 2000.

[18] J. F. Douglas, J. M. Gasiorek, and J. A. Swaffield: *Fluid Mechanics, 4.th edition*, Prentice Hall, 2001.

A rather technical book with a rich collection of examples.

[19] P. G. Drazin and W. H. Reid: *Hydrodynamic stability*, Cambridge University Press (1981).

[20] R. Dudley: *The Biomechanics of Insect Flight*, Princeton University Press, 200.

[21] M. Van Dyke: *An Album of Fluid Motion*, The Parabolic Press, Stanford California, 1982.

[22] T. E. Faber: *Fluid Dynamics for Physicists*, Cambridge University Press, 1995.

[23] A. Filippone: *Advanced Topics in Aerodynamics*, Electronic book at URL 'http://aerodyn.org'.

A truly modern book on aerodynamics, CFD, propulsion systems and related technology.

[24] R. W. Fox and A. T. McDonald: *Introduction to Fluid Mechanics (third edition)*, Wiley, 1985.

An excellent textbook with numerous practical examples and excercizes.

[25] U. Frisch: *Turbulence*, Cambridge University Press (1995).

[26] A. E. Green and J. E. Adkins: *Large Elastic Deformations and Non-Linear Continuum Mechanics*, Oxford University Press, 1960.

[27] A. E. Green and W. Zerna: *Theoretical Elasticity*, Dover Publications, 1992.

[28] H. P. Greenspan: *The Theory of Rotating Fluids*, Cambridge University Press, 1968.

The standard treatise on rotating fluids.

[29] M. Griebel, T. Dornseifer, and T. Neunhoeffer: *Numerical Simulation in Fluid Dynamics*, SIAM, 1998.

A good 'do-it-yourself' book for implementing a fluid dynamics simulator capable of handling a number of problems in two and three dimensions.

[30] E. Guyon, J-P. Hulin, L. Petit, and C. D. Mitescu: *Physical hydrodynamics*, Oxford University Press (2001).

[31] W. F. Hughes and J. A. Brighton: *Fluid Dynamics*, McGraw-Hill, 1991.

An excellent account of fluid dynamics with emphasis on engineering.

[32] J. C. R. Hunt and J. C. Vassilicos: *Turbulence Structure and Vortex Dynamics*, Cambridge University Press (2000).

[33] J. N. Johnson and R. Chéret: *Classic Papers in Shock Compression Science*, Springer (1998).

Contains papers and biographies of the major players in shock wave theory.

[34] G. W. C. Kaye and T. H. Laby: *Tables of Physical and Chemical Constants (16th edition)*, Longman Group Ltd, 1995.

[35] D. Kondepudi and I. Prigogine: *Modern Thermodynamics*, Wiley (1998).

[36] H. Lamb: *Hydrodynamics*, Cambridge University Press (1993). First published 1879.

[37] L. D. Landau and E. M. Lifshitz: *Theory of Elasticity*, Pergamon Press, 1986.

[38] L. D. Landau and E. M. Lifshitz: *Fluid Mechanics, 2nd edition*, Butterworth-Heinemann, 1987.

[39] M. Lesieur: *Turbulence in Fluids. Third Edition*, Kluwer Academic Publishers (1997).

[40] P. Libby: *Introduction to Turbulence*, Taylor and Francis (1996).

[41] D. R. Lide (ed): *Handbook of Chemistry and Physics (77th Edition)*, CRC Press, 1996.

[42] J. Lighthill: *Waves in fluids*, Cambridge University Press (1978).

The classic text on waves.

[43] C. C. Lin: *Hydrodynamic stability*, Cambridge University Press (1955).

[44] A. E. H. Love: *A Treatise on the Mathematical Theory of Elasticity*, Cambridge University Press, 1906.

[45] A. Maurel and P. Petitjeans: *Vortex Structure and Dynamics*, Springer, 2000.

Lecture notes from a workshop held in Rouen 1999.

[46] B. Massey (revised by J. Ward-Smith): *Mechanics of Fluids*, Stanley Thornes (Publishers) Ltd, 1998.

A fine book intended for engineering students.

[47] C. C. Mei: *The applied dynamics of ocean surface waves*, World Scientific, 1989.

[48] P. Melchior: *The Tides of the Planet Earth*, Pergamon Press, 1978.

[49] C. W. Misner, K. S. Thorne, and J. A. Wheeler: *Gravitation*, Freeman, 1973.

[50] N. I. Mushkhelishvili: *Some Basic Problems of the Mathematical Theory of Elasticity*, P. Noordhoof Ltd, Holland (1953).

A comprehensive mathematical presentation of linear elasticity theory in particular for planar systems.

[51] M. N. L. Narasimhan: *Principles of Continuum Mechanics*, Wiley, 1993.

A rather formal but comprehensive treatment of the fundamental aspects of continuum mechanics.

[52] National Institute of Standards and Technology (NIST): `http://physics.nist.gov`.

[53] I. Newton: *The Principia, translated by I. B. Cohen and A. Whitman*, University of California Press (1999).

[54] T. M. Nieuwstadt and J. A. Steketee: *Selected papers of J. M. Burgers*, Kluver Academic Publishers (1995).

[55] J. Pedlosky: *Geophysical Fluid Dynamics*, Springer, 1987.

[56] A. Peters, K. Y. Chung, and S. Chu: *Measurement of gravitational acceleration by dropping atoms*, Nature 400, 849 (1999).

[57] D. Pnueli and C. Gutfinger: *Fluid Mechanics*, Cambridge University Press, 1992.

[58] S. B. Pope: *Turbulent Flows*, Cambridge University Press (2003).

An excellent and very accessible account of all aspects of turbulent flows.

[59] W. H. Press, S. A. Teukolsky, W. T. Vetterling, and B. P. Flannery: *Numerical recipes in C*, Cambridge University Press, 1992.

The standard text for numeric methods.

[60] P. G. Saffman: *Vortex Dynamics*, Cambridge University Press, 1992.

[61] H. Schlichting, K. Gersten: *Boundary-Layer Theory*, 8th Edition, Springer, 2000.

[62] L. Sedov: *Similarity and Dimensional Methods in Mechanics*, Academic Press, New York (1959).

[63] S. L. Shapiro and S. A. Teukolsky: *Black Holes, White Dwarfs, and Neutron Stars*, John Wiley & Sons, 1983.

[64] I. J. Sobey: *Introduction to Interactive Boundary Layer Theory*, Oxford University Press, 2000.

[65] I. S. Sokolnikoff: *Mathematical Theory of Elasticity*, McGraw-Hill Book Company, Inc., 1956.

A very clear and comprehensive treatment of theoretical elasticity, in particular variational principles and numerical methods.

[66] R. W. Soutas-Little: *Elasticity*, Dover Publications, 1999.

[67] J. J. Stoker: *Water waves*, Wiley, 1992.

Mathematically oriented text on mainly shallow surface waves in water.

[68] V. Sychev, A. Ruban, V. Sychev, and G. Korolev: *Asymptotic theory of separated flows*, Cambridge University Press (1998).

[69] H. Tennekes: *The Simple Science of Flight*, The MIT Press, 1997.

A remarkable book covering most aspects of flight from miniscule gnats to huge jumbo jets.

[70] G. A. Tokaty: *A History and Philosopy of Fluid Mechanics*, Dover Publications, 1994.

[71] A. A. Townsend: *The Structure of Turbulent Shear Flow*, Cambridge University Press, 1956.

[72] D. J. Tritton: *Physical Fluid Dynamics*, Oxford University Press (1988).

A phenomenologically oriented approach to fluid mechanics with a wealth of details, photographs, pictures, and drawings. A good book to get an easily understandable account of nearly any aspect of fluid behavior.

[73] H. K. Versteeg and W. Malalasekera: *An introduction to Computational Fluid Dynamics*, Prentice Hall, 1995.

[74] S. Vogel: *Life's Devices*, Princeton University Press (1988).

[75] S. Vogel: *Life in Moving Fluids*, Princeton University Press,1994.

A truly wonderful book on the life styles of the numerous organisms that swim and fly.

[76] S. Vogel: *Cats' Paws and Catapults*, Penguin Books, 1998.

[77] A. Walz: *Boundary Layers of Flow and Temperature*, The MIT Press, 1969.

[78] S. Weinberg: *Gravitation and Cosmology*, John Wiley, 1972.

[79] F. M. White: *Viscous Fluid Flow*, McGraw-Hill, 1991.

[80] F. M. White: *Fluid Mechanics*, McGraw-Hill, 1999.

An excellent and detailed book with emphasis on engineering aspects of fluid mechanics.

Index